Class Field Theory and *L* Functions

Class Field Theory and *L* Functions

Foundations and Main Results

Franz Halter-Koch

CRC Press
Taylor & Francis Group
Boca Raton London New York

CRC Press is an imprint of the
Taylor & Francis Group, an **informa** business
A CHAPMAN & HALL BOOK

First edition published 2022

by CRC Press
6000 Broken Sound Parkway NW, Suite 300, Boca Raton, FL 33487-2742

and by CRC Press
4 Park Square, Milton Park, Abingdon, Oxon, OX14 4RN

CRC Press is an imprint of Taylor & Francis Group, LLC

ISBN: 978-1-138-58358-0 (hbk)
ISBN: 978-1-032-20265-5 (pbk)
ISBN: 978-0-429-50657-4 (ebk)

DOI: 10.1201/9780429506574

Typeset in CMR10
by KnowledgeWorks Global Ltd.

Dedicated to my family

Contents

Preface

The theory of algebraic numbers and algebraic functions is one of the most impressive creations of 19th century mathematics and culminated in the development of class theory in the first half of the 20th century. A comprehensive description of this history is in W. Narkiewicz's book [45]. Since that time class field theory was enhanced by the introduction of structural facilities from topology, harmonic analysis and homological algebra. A contemporary presentation of the theory can be found among others in the articles of J.-P. Serre [55] and J. Tate [60], and in the books of J. Neukirch [48], [46], A. Weil [64], F. Lorenz [40], S. Lang [38] and N. Childress [6].

The shortage of all these excellent books lies in the fact that they do not present the topological and cohomological requirements (which usually get short in standard algebraic courses) in great detail. Therefore several years ago I planned to publish a self-contained volume on algebraic number theory including class field theory together with a detailed presentation of all necessary requirements. However, the material turned out to be too comprehensive to be published in one volume. Thus I decided to split my presentation. The first part, containing the classical theory of algebraic numbers and algebraic functions, just appeared (see [18]) and comprises the basic methods and results from ideal and divisor theory, valuation theory and elementary analytic theory. Based on this, the present volume contains local and global class field theory in Chapters 4 to 6, and the theory of Artin L functions (both for number fields and function fields) in Chapter 7. As to the foundations beyond classical algebraic number theory, the present volume contains elaborate presentations of the essentials from the theory of topological groups (Chapter 1), group cohomology (Chapter 2), and the theory of simple algebras (Chapter 3).

I decided to develop class field theory on the basis of the algebra-theoretic concept of H. Hasse, following in essential parts the presentation of F. Lorenz (see [39] for the local and [40] for the global theory). As to the local theory, important sources are the books of J.-P. Serre [57] and I.B. Fesenko and S.V. Vostokov [12]. Moreover, I focussed on the theory of Lubin-Tate, for which K. Iwasawa's book [26] was an essential guide. In global theory, I carefully discussed the various types of ray class groups and holomorphy domains and their influence on class fields.

The theory of Artin L functions is based on old seminar notes [19]. Here the proof of the functional equation of the Dirichlet L series for number fields follows the classical lines of E. Landau [35] and T. Tatuzawa [61] and the function field case is based on M. Deuring [9] and C. Moreno [42]. Corresponding to the style of this book, I refrained from a presentation of Tate's proof of the functional equation, since otherwise I would have been forced to present the foundations from harmonic analysis, which would be beyond the scope of this book. I provide a self-contained presentation of the tools from representation theory, and I close this volume with applications to prime decomposition and density results.

Preliminaries

Throughout the reader is assumed to have a strong background in classical algebra and is familiar with the concepts and results of the basic theory of algebraic numbers and algebraic functions. For this I shall refer on every occasion to my volume [18]. Alternatively, every textbook containing the fundamentals of algebraic numbers and algebraic functions will serve the same goal. In particular, I have in mind the first 4 chapters in S. Lang's book [38] for the theory of number fields, the book of Serre [57] or the article of Fröhlich [14] for the local theory, and H. Stichtenoth's book [58] together with Chapter 1 in [36] for the theory of function fields.

There is comprehensive literature on algebraic number fields and function fields which presents additional and alternative material on the subject and also might partially serve as a basis for the present volume. My favorites are the books of J. Neukirch [48], A. Weil [64], W. Narkiewicz [44], M. Fried and M. Jarden [13] (in particular Chapter 3), H. Koch [33], G. J. Janusz [29], J.W.S. Cassel [5], J.-P. Serre [57] and M. Deuring [9], just to present a subjective selection.

Acknowledgements

I am much obliged to Florian Kainrath who carefully read the whole manuscript and not only corrected numerous misspellings, but several more or less serious mathematical errors. In addition, I give my thanks to Alfred Geroldinger for many encouraging conversations over the years supporting this work.

Although I retired from my university duties several years ago, I appreciate still having the facilities of the Mathematical Institute of the University of Graz at my disposal.

Author

Franz Halter-Koch is professor emeritus at the University of Graz, Austria. He studied Mathematics at the Universities of Hamburg and Graz, where in 1968 he got his PhD. He held professorships at the Universities of Cologne, Essen and Graz, where he retired in 2008. He is the author of *Ideal Systems* (Marcel Dekker, 1998), *Quadratic Irrationals* (CRC Press, 2013), *An Invitation to Algebraic Numbers and Algebraic Functions* (CRC Press, 2020), and the co-author of *Non-Unique Factorizations* (CRC Press, 2006). His current research interests focus on algebraic number theory with emphases on class field theory, quadratic irrationals and non-unique factorizations.

Notation

In this introductory paragraph we review the notations used throughout the volume. In particular, we review the relevant concepts from [18] concerning Dedekind domains, valuation theory, algebraic number fields and algebraic function fields.

A. General

We denote by

- $\mathbb{N}_0$ the set of non-negative integers;
- $\mathbb{N} = \mathbb{N}_0 \setminus \{0\}$ the set of positive integers;
- $\mathbb{P}$ the set of prime numbers;
- $\mathbb{Z}$ the ring of integers, and $\mathbb{Q}$ the field of rational numbers;
- $\mathbb{R}$ the field of real numbers, and $\mathbb{C}$ the field of complex numbers;
- $\mathbb{F}_p = \mathbb{Z}/p\mathbb{Z}$ the field of p elements for a prime p;
- $|A| \in \mathbb{N}_0 \cup \{\infty\}$ the number of elements of a set A;
- $A^\bullet$ the set of non-zero elements of a set A which potentially contains a zero element.

Let A and B be sets. The notions $A \subset B$ and $A \supset B$ comprise equality; we write $A \subsetneq B$ or $A \supsetneq B$ to exclude equality. For a map $f\colon A \to B$ and a subset S of A we denote by $f \restriction S\colon S \to B$ the restriction of f to S. For a set A, we denote by $\mathbb{P}(A)$ the set of all subsets of A and by $\mathbb{P}_{\rm fin}(A)$ the set of all finite subsets of A.

We use the self-explaining notations $\mathbb{R}_{>0}$, $\mathbb{R}_{\ge 0}$, $\mathbb{Q}_{>0}$, $\mathbb{N}_{\ge 2}$ etc. For $a,\, b \in \mathbb{Z}$, we set $[a,b] = \{x \in \mathbb{Z} \mid a \le x \le b\}$ if there is no danger of confusion with the real interval, and we set $[a,b] = \emptyset$ if $a > b$.

For $x \in \mathbb{R}$, we set $\lfloor x \rfloor = \max\{g \in \mathbb{Z} \mid g \le x\}$, $\lceil x \rceil = \min\{g \in \mathbb{Z} \mid g \ge x\}$, and we denote by $\operatorname{sgn}(x) \in \{0, \pm 1\}$ its sign. For a complex number z, we denote by $\Re(z)$ its real part, by $\Im(z)$ its imaginary part, and we normalize $\sqrt{z}$ so that $\sqrt{z} \ge 0$ if $z \in \mathbb{R}_{\ge 0}$, and $\Im(\sqrt{z}) > 0$ if $z \in \mathbb{C} \setminus \mathbb{R}_{\ge 0}$. We denote by e Euler's constant and by $\mathsf{i} = \sqrt{-1}$ the imaginary unit.

In asymptotic results, we use $\ll$-notation of Vinogradov. More generally, we say that an assertion holds for $x \gg 1$ if there is some $x_0 \ge 1$ such that the assertion holds for all $x \ge x_0$.

We use the symbol $\mathbf{0}$ to denote the set $\{0\}$, and we also write $\mathbf{0} = (0, \ldots, 0)$ for the zero row and $\mathbf{0} = (0, \ldots, 0)^{\mathrm{t}}$ for the zero column provided that the meaning is clear from the context. In a similar way, we use the symbol $\mathbf{1}$.

For a multiplicative group G we usually denote by $1 = 1_G$ the unit element, and in Galois theory we shall occasionally denote by 1 the identical automorphism.

We are careless in the notation of direct sums. Usually we will not explicitly distinguish between outer and inner direct sums when the meaning is clear from the context. In contrast, we handle direct products more carefully. For subgroups A, B of a (multiplicative abelian) group G, we denote by $A{\cdot}B$ their direct inner product (so that $A \times B \to A{\cdot}B$, $(a, b) \mapsto ab$ is an isomorphism).

By a ring R we always mean a ring possessing a unit element $1 = 1_R$, and (unless otherwise stated) by an R-module we mean a left R-module. For a ring R we denote by $\mathsf{M}_{m,n}(R)$ and $\mathsf{M}_n(R)$ the matrix rings. For an extension $R \subset S$ of commutative rings we denote by $\operatorname{cl}_S(R)$ the integral closure of R in S. For a field K, we denote by $\mu(K)$ the group of all roots of unity in K, for $n \in \mathbb{N}$, we set $\mu_n(K) = \{x \in K \mid x^n = 1\}$, and for a prime p, we denote by μ_{p^∞} the group of all units of p-power order.

For a field K, we denote by $\overline{K}$ an algebraic closure, by K_{sep} a separable closure, and by $G_K = \operatorname{Gal}(\overline{K}/K) = \operatorname{Gal}(K_{\mathrm{sep}}/K)$ its absolute Galois group (see [18, Definition and Theorem 1.4.8]). Unless stated otherwise, all algebraic extensions of K are assumed to lie inside $\overline{K}$. We denote by K_{ab} the maximal abelian extension of K (inside K_{sep}). For a finite field extension L/K we denote by $\mathsf{N}_{L/K}$ the norm and by $\mathsf{Tr}_{L/K}$ the trace for L/K.

B. Dedekind domains [18, Ch. 2]

1. Let $\mathfrak{o}$ be a domain and $K = \mathsf{q}(\mathfrak{o})$ its quotient field. We denote by

- $\mathcal{P}(\mathfrak{o})$ the set of all non-zero prime ideals of $\mathfrak{o}$;
- $\mathcal{J}(\mathfrak{o})$ the (multiplicative) group of all invertible fractional ideals of $\mathfrak{o}$;
- $\mathcal{J}'(\mathfrak{o})$ the submonoid of all invertible (integral) ideals of $\mathfrak{o}$;
- $\mathcal{H}(\mathfrak{o})$ the subgroup of all fractional principal ideals of $\mathfrak{o}$;

- $\mathcal{H}'(\mathfrak{o}) = \mathcal{H}(\mathfrak{o}) \cap \mathcal{I}(\mathfrak{o})$ the submonoid of all non-zero principal ideals of $\mathfrak{o}$;

- $\mathcal{C}(\mathfrak{o}) = \mathcal{I}(\mathfrak{o})/\mathcal{H}(\mathfrak{o})$ the ideal class group (or Picard group) of $\mathfrak{o}$.

These concepts are connected by the exact sequence

$$\mathbf{1} \to \mathfrak{o}^\times \xrightarrow{\iota} K^\times \xrightarrow{\partial} \mathcal{I}(\mathfrak{o}) \xrightarrow{\rho} \mathcal{C}(\mathfrak{o}) \to \mathbf{1},$$

where ι denotes the inclusion, $\partial(a) = a\mathfrak{o}$ for all $a \in K^\times$, $\operatorname{Im}(\partial) = \mathcal{H}(\mathfrak{o})$, and $\rho(\mathfrak{a}) = \mathfrak{a}\mathcal{H}(\mathfrak{o}) = [\mathfrak{a}] \in \mathcal{C}(\mathfrak{o})$ is the ideal class containing $\mathfrak{a}$.

2. Let $\mathfrak{o}$ be a Dedekind domain. Then $\mathfrak{o} \ne K$, $\mathcal{P}(\mathfrak{o}) = \max(\mathfrak{o})$ is the set of all maximal ideals of $\mathfrak{o}$, $\mathcal{I}(\mathfrak{o})$ is the free abelian group and $\mathcal{I}'(\mathfrak{o})$ is the free abelian multiplicative monoid with basis $\mathcal{P}(\mathfrak{o})$ (hence $\mathcal{I}(\mathfrak{o}) \cong \mathbb{Z}^{(\mathcal{P}(\mathfrak{o}))}$ and $\mathcal{I}'(\mathfrak{o}) \cong \mathbb{N}_0^{(\mathcal{P}(\mathfrak{o}))}$).

For $\mathfrak{p} \in \mathcal{P}(\mathfrak{o})$ let $\mathsf{k}_\mathfrak{p} = \mathfrak{o}/\mathfrak{p}$ be the residue class field, $v_\mathfrak{p} \colon K \to \mathbb{Z} \cup \{\infty\}$ the $\mathfrak{p}$-adic valuation of K and alike by $v_\mathfrak{p} \colon \mathcal{I}(\mathfrak{o}) \to \mathbb{Z}$ the $\mathfrak{p}$-adic valuation of $\mathcal{I}(\mathfrak{o})$.

For $\mathfrak{a} \in \mathcal{I}(\mathfrak{o})$ we denote by $\operatorname{supp}(\mathfrak{a}) = \{\mathfrak{p} \in \mathcal{P}(\mathfrak{o}) \mid v_\mathfrak{p}(\mathfrak{a}) = 0\}$ the support of $\mathfrak{a}$.

For $\mathfrak{m} \in \mathcal{I}'(\mathfrak{o})$, we denote by $\mathcal{I}^\mathfrak{m}(\mathfrak{o}) = \{\mathfrak{a} \in \mathcal{I}(\mathfrak{o}) \mid \operatorname{supp}(\mathfrak{a}) \cap \operatorname{supp}(\mathfrak{m}) = \emptyset\}$ the group of all fractional ideals coprime to $\mathfrak{m}$. It is the free abelian group with basis $\mathcal{P}^\mathfrak{m}(\mathfrak{o}) = \mathcal{P}(\mathfrak{o}) \cap \mathcal{I}^\mathfrak{m}(\mathfrak{o})$.

Suppose that $\mathsf{k}_\mathfrak{p}$ is finite for all $\mathfrak{p} \in \mathcal{P}(\mathfrak{o})$. Then $\mathfrak{N}(\mathfrak{a}) = (\mathfrak{o} : \mathfrak{a}) < \infty$ for all $\mathfrak{a} \in \mathcal{I}'(\mathfrak{o})$, and we extend $\mathfrak{N}$ to a homomorphism $\mathfrak{N} \colon \mathcal{I}(\mathfrak{O}) \to \mathbb{Q}_{>0}$ (the absolute norm).

3. Let $\mathfrak{o}$ be a Dedekind domain, $K = \mathsf{q}(\mathfrak{o})$, L/K a finite separable field extension and $\mathfrak{O} = \mathrm{cl}_L(\mathfrak{o})$. For $\mathfrak{P} \in \mathcal{P}(\mathfrak{O})$ and $\mathfrak{p} = \mathfrak{P} \cap \mathfrak{o}$ we write $\mathfrak{P} \,|\, \mathfrak{p}$, we denote by $e(\mathfrak{P}/\mathfrak{p})$ the ramification index and by $f(\mathfrak{P}/\mathfrak{p} = [\mathsf{k}_\mathfrak{P} : \mathsf{k}_\mathfrak{p}]$ the residue class degree of $\mathfrak{P}$ over $\mathfrak{p}$. If L/K is Galois and $G = \operatorname{Gal}(L/K)$, then $\mathsf{k}_\mathfrak{P}/\mathsf{k}_\mathfrak{p}$ is normal, and for $\sigma \in G$, we denote by $\sigma_\mathfrak{P} \in \operatorname{Gal}(\mathsf{k}_\mathfrak{P}/\mathsf{k}_\mathfrak{p})$ the residue class automorphism.

The ideal norm $\mathcal{N}_{\mathfrak{O}/\mathfrak{o}} \colon \mathcal{I}(\mathfrak{O}) \to \mathcal{I}(\mathfrak{o})$ is the unique homomorphism which satisfies $\mathcal{N}_{\mathfrak{O}/\mathfrak{o}}(\mathfrak{P}) = \mathfrak{p}^{f(\mathfrak{P}/\mathfrak{p})}$ for all $\mathfrak{P} \in \mathcal{P}(\mathfrak{O})$ and $\mathfrak{p} = \mathfrak{P} \cap \mathfrak{o}$. If $\mathfrak{o} = \mathbb{Z}$ and $\mathfrak{A} \in \mathcal{I}(\mathfrak{O})$, then $\mathcal{N}_{\mathfrak{o}/\mathbb{Z}}(\mathfrak{A}) = \mathfrak{N}(\mathfrak{A})\mathbb{Z}$.

C. Valuation theory [18, Ch. 5]

1. Let K be a field. An absolute value of K is a map $|\cdot| \colon K \to \mathbb{R}_{\ge 0}$ such that

- $|a| > 0$ for all $a \in K^\times$, and $|a| > 1$ for some $a \in K$;

- $|a + b| \le C \max\{|a|, |b|\}$ for some $C \in \mathbb{R}_{\ge 1}$ and $|ab| = |a|\,|b|$ for all $a, b \in K$.

An absolute value $|\cdot|$ is called non-Archimedian if $C = 1$, Archimedian otherwise, and it is called discrete if $|K^\times|$ is a discrete subgroup of $\mathbb{R}_{>0}$. A valued field $K = (K, |\cdot|)$ is field together with an absolute value (then $(x, y) \mapsto |x-y|$ is a metric on K which induces the valuation topology). Two absolute values $|\cdot|, |\cdot|'$ of a field K are called equivalent if they induce the same topology (equivalently, $|\cdot|' = |\cdot|^\rho$ for some $\rho \in \mathbb{R}_{>0}$). Every Archimedian absolute value is equivalent to an absolute value satisfying the triangle inequality $|a + b| \le |a| + |b|$ for all $a, b \in K$.

Every valued field $(K, |\cdot|)$ has an (up to topological isomorphisms) unique completion $(\widehat{K}, |\cdot|)$ (we denote the extension of $|\cdot|$ to $\widehat{K}$ again by $|\cdot|$). If $(K, |\cdot|)$ is complete and $\overline{K}$ is an algebraic closure of K, then $|\cdot|$ has a unique extension to an (equally denoted) absolute value $|\cdot|\colon \overline{K} \to \mathbb{R}_{\ge 0}$.

2. Let $(K, |\cdot|)$ be a non-Archimedian valued field. We denote by

- $\mathcal{O}_K = \{x \in K \mid |x| \le 1\}$ its valuation domain;
- $\mathfrak{p}_K = \{x \in K \mid |x| < 1\}$ its valuation ideal;
- $\mathsf{k}_K = \mathcal{O}_K/\mathfrak{p}_K$ its residue class field;
- $U_K = \mathcal{O}_K^\times = \{x \in K \mid |x| = 1\}$ its unit group.

If $|\cdot|$ is discrete and $\rho \in (0,1)$ is such that $|K^\times| = \langle\rho\rangle$, then we call

$$v_K\colon K \to \mathbb{Z} \cup \{\infty\}, \quad \text{defined by} \quad v_K(x) = \frac{\log|x|}{\log\rho}$$

the associated discrete valuation. In this case,

- $\mathcal{O}_K = \{x \in K \mid v_K(x) \ge 0\}$ is a discrete valuation domain (i. e., a local Dedekind domain);
- $\mathfrak{p}_K = \{x \in K \mid v_K(x) > 0\} = \pi\mathcal{O}_K$ for every prime element π of $\mathcal{O}_K$ (satisfying $|\pi| = \rho$ and $v_K(\pi) = 1$);
- $U_K = U_K^{(0)} = \{x \in K \mid v_K(x) = 0\}$. For $n \in \mathbb{N}$ we denote by $U_K^{(n)} = 1 + \mathfrak{p}_K^n$ the group of principal units of level n.

Note that up to equivalence $|\cdot|$ is uniquely determined by v_K, and we also call (K, v_K) a discrete valued field.

3. Let $(K, |\cdot|)$ be a discrete valued field. Then its completion $\widehat{K}$ is a discrete valed field as well, $v_{\widehat{K}} \restriction K = v_K$, and we shall always write v_K instead of $v_{\widehat{K}}$. Clearly, $\mathcal{O}_{\widehat{K}} \cap K = \mathcal{O}_K$, every prime element of $\mathcal{O}_K$ is also a prime element of $\mathcal{O}_{\widehat{K}}$, and $\mathsf{k}_K = \mathsf{k}_{\widehat{K}}$.

4. Let $(K, |\cdot|)$ be a complete discrete valued field and L/K a finite extension. Then we denote by $e(L/K) = e(\mathfrak{p}_L/\mathfrak{p}_K) = (|L^\times| : |K^\times|)$ the ramification index and by $f(L/K) = f(\mathfrak{p}_L/\mathfrak{p}_K) = [\mathsf{k}_L : \mathsf{k}_K]$ the residue class degree of L/K. If L/K is separable, we denote by $\mathcal{N}_{L/K} = \mathcal{N}_{\mathcal{O}_L/\mathcal{O}_K}$ the ideal norm, by $\mathfrak{D}_{L/K}$ the different, by $\mathfrak{C}_{L/K} = \mathfrak{D}_{L/K}^{-1}$ the Dedekind complementary module and by $\mathfrak{d}_{L/K} = \mathcal{N}_{L/K}(\mathfrak{D}_{L/K})$ the discriminant of L/K.

5. Let $\mathfrak{o}$ be a Dedekind domain and $K = \mathsf{q}(\mathfrak{o})$. For $\mathfrak{p} \in \mathcal{P}(\mathfrak{o})$ and $\rho \in (0,1)$ let $|\cdot|_{\mathfrak{p},\rho} \colon K \to \mathbb{R}_{\geq 0}$ be defined by $|x|_{v,\rho} = \rho^{v_\mathfrak{p}(x)}$. Then $(K, |\cdot|_{\mathfrak{p},\rho})$ is a discrete valued field, $v_\mathfrak{p}$ is the associated discrete valuation, $\mathfrak{o}_\mathfrak{p}$ its valuation domain, $\mathfrak{p}\mathfrak{o}_\mathfrak{p}$ its valuation ideal and $\mathsf{k}_\mathfrak{p} = \mathfrak{o}/\mathfrak{p} = \mathfrak{o}_\mathfrak{p}/\mathfrak{p}\mathfrak{o}_\mathfrak{p}$ its residue class field. We denote by $K_\mathfrak{p}$ a completion of $(K, |\cdot|_{\mathfrak{p},v})$; it is a complete discrete valued field with valuation domain $\widehat{\mathfrak{o}}_\mathfrak{p}$, valuation ideal $\widehat{\mathfrak{p}} = \mathfrak{p}\widehat{\mathfrak{o}}_\mathfrak{p}$, and residue class field $\mathsf{k}_\mathfrak{p}$. Its discrete valuation is an extension of $v_\mathfrak{p}$ and is again denoted by $v_\mathfrak{p}$.

6. We denote by $|\cdot|_\infty$ the ordinary absolute value of $\mathbb{C}$. For a prime p we denote by $|\cdot|_p \colon \mathbb{Q} \to \infty$ the p-adic absolue value, defined by

$$|a|_p = p^{-v_p(a)} \quad \text{for } a \in \mathbb{Q}^\times, \quad \text{and} \quad |0|_p = 0,$$

where $v_p(a)$ is the p-adic exponent of a. $(\mathbb{Q}, |\cdot|_p)$ is a discrete valued field, the localization $\mathbb{Z}_{(p)} = (\mathbb{Z}\setminus p\mathbb{Z})^{-1}\mathbb{Z}$ is its valuation domain, $p\mathbb{Z}_{(p)}$ its valuation ideal, $\mathbb{F}_p = \mathbb{Z}/p\mathbb{Z}$ its residue class field and $v_p \colon \mathbb{Q} \to \mathbb{Z} \cup \{\infty\}$ the associated discrete valuation. The completion $\mathbb{Q}_p$ of $(\mathbb{Q}, |\cdot|_p)$ is the field of p-adic numbers; its valuation domain $\mathbb{Z}_p$ is the domain of p-adic inters, $p\mathbb{Z}_p$ is its valuation ideal, $\mathbb{F}_p = \mathbb{Z}_p/p\mathbb{Z}_p$ is its residue class field, and v_p has a unique extension to an (equally denoted) discrete valuation $v_p \colon \mathbb{Q}_q \to \mathbb{Z} \cup \{\infty\}$. We set $U_p = U_{\mathbb{Q}_p}$ and $U_p^{(n)} = 1 + p^n\mathbb{Z}_p = U_{\mathbb{Q}_p}^{(n)}$ for all $n \in \mathbb{N}$.

D. Algebraic number fields [18, Ch. 3 and Sec. 5.9]

The set $\overline{\mathbb{Q}}$ of all algebraic numbers in $\mathbb{C}$ is an algebraic closure of $\mathbb{Q}$, and by an algebraic number field we always mean an over $\mathbb{Q}$ finite subfield of $\overline{\mathbb{Q}}$.

1. Let K be an algebraic number field, and $\mathrm{Hom}_\mathbb{Q}(K, \mathbb{C}) = \{\sigma_1, \ldots, \sigma_n\}$, where $[K : \mathbb{Q}] = n = r_1 + 2r_2$ such that $\sigma_i(K) \subset \mathbb{R}$ for $i \in [1, r_1]$, and $\sigma_{r_1+r_2+i}(x) = \overline{\sigma_{r_1+i}(x)}$ for all $x \in K$ and $i \in [1, r_2]$. The normalized Archimedian absolute values of K are for $i \in [1, r_1 + r_2]$ defined by

$$|x|_{\infty,i} = |\sigma_i(x)|_\infty^{l_i} \qquad \text{where } l_i = \begin{cases} 1 & \text{if } i \in [1, r_1], \\ 2 & \text{if } i \in [r_1 + 1, r_1 + r_2]. \end{cases}$$

2. Let K be an algebraic number field. Its domain of integers $\mathfrak{o}_K = \mathrm{cl}_K(\mathbb{Z})$ is a Dedekind domain, and we set $\mathcal{P}_K = \mathcal{P}(\mathfrak{o}_K)$, $\mathcal{J}_K = \mathcal{J}(\mathfrak{o}_K)$, $\mathcal{J}'_K = \mathcal{J}'(\mathfrak{o}_K)$, $\mathcal{H}'_K = \mathcal{H}(\mathfrak{o}_K)$, $\mathcal{C}_K = \mathcal{C}(\mathfrak{o}_K)$ and $U_K = \mathfrak{o}_K^\times$. For $\mathfrak{m} \in \mathcal{J}'_K$ we set $\mathcal{J}_K^\mathfrak{m} = \mathcal{J}^\mathfrak{m}(\mathfrak{o}_K)$, $\mathcal{J}_K'^\mathfrak{m} = \mathcal{J}'_K \cap \mathcal{J}_K^\mathfrak{m}$ and $\mathcal{P}_K^\mathfrak{m} = \mathcal{P}_K \cap \mathcal{J}_K^\mathfrak{m}$. Recall that $\mathcal{J}_K'^\mathfrak{m}$ is the free (multiplicative) abelian monoid and $\mathcal{J}_K^\mathfrak{m}$ is the free (multiplicative) abelian group with basis $\mathcal{P}_K^\mathfrak{m}$. We denote by

- d_K the discriminant,
- $h_K = |\mathcal{C}_K|$ the class number,
- R_K the regulator, and by
- $w_K = |\mu(K)|$ the number of roots of unity in K.

3. Let L/K be an extension of algebraic number fields. Then $\mathfrak{o}_L = \mathrm{cl}_L(\mathfrak{o}_K)$, and we denote by

- $\mathcal{N}_{L/K} = \mathcal{N}_{\mathfrak{o}_L/\mathfrak{o}_K} : \mathcal{I}_L \to \mathcal{I}_K$ the ideal norm,
- $\mathfrak{D}_{L/K} = \mathfrak{D}_{\mathfrak{o}_L/\mathfrak{o}_K} \in \mathcal{I}'_L$ the (relative) different,
- $\mathfrak{C}_{L/(K} = \mathfrak{D}_{L/K}^{-1} \in \mathcal{I}_L$ the Dedekind complementary module, and by
- $\mathfrak{d}_{L/K} = \mathfrak{d}_{\mathfrak{o}_L/\mathfrak{o}_K} \in \mathcal{I}'_K$ the (relative) discriminant

of L/K. In particular, $\mathfrak{d}_{K/\mathbb{Q}} = \mathsf{d}_K\mathbb{Z}$.

E. Algebraic function fields [18, Ch. 6]

1. Let q be a prime power and $\mathbb{F}_q$ the finite field of q elements. By a **function field** $K/\mathbb{F}_q$ we mean a finitely generated field extension of transcendence degree 1 such that $\mathbb{F}_q$ is its field of constants (i. e., $\mathbb{F}_q$ is relatively algebraically closed in K).

Let $K/\mathbb{F}_q$ be a function field. A prime divisor P of K is the valuation ideal of a discrete absolute value $|\cdot|\colon K \to \mathbb{R}_{\ge 0}$ (called the P-adic absolute value) such that $|\cdot| \restriction \mathbb{F}_q^\times = 1$. Then $(K, |\cdot|)$ is a discrete valued field, and we denote by v_P its discrete valuation. Then $P = \{x \in K \mid v_P(x) > 0\}$, $\mathcal{O}_P = \{x \in K \mid v_P(x) \ge 0\}$ is its valuation domain and $\mathsf{k}_P = \mathcal{O}_P/P$ is the residue class field. It is a finite extension of $\mathbb{F}_p$, and $\deg(P) = [\mathsf{k}_P : \mathbb{F}_q]$ is the degree of P.

Let $\mathbb{P}_K = \mathbb{P}_{K/F}$ be the set of all prime divisors of K. Then the divisor group $\mathbb{D}_K$ is the free additive abelian group with basis $\mathbb{P}_K$. For $P \in \mathbb{P}_K$ let $v_P\colon \mathbb{D}_K \to \mathbb{Z} \cup \{\infty\}$ also denote the P-adic valuation of $\mathbb{D}_K$. If $D \in \mathbb{D}_K$, then $v_P(D) = 0$ for almost all $P \in \mathbb{P}_K$,

$$D = \sum_{P \in \mathbb{P}_K} v_P(D)P, \quad \text{and} \quad \deg(D) = \sum_{P \in \mathbb{P}_K} v_P(D)\deg(P) \ \text{ is the degree of } D.$$

We denote by $\mathbb{D}'_K = \{D \in \mathbb{D}_K \mid v_P(D) \ge 0 \text{ for all } v \in \mathbb{P}_K\}$ the submonoid of effective divisors. For $x \in K^\times$ we denote by $(x) = (x)^K$ the principal divisor of x, by $(K^\times)$ the group of all principal divisors and by $\mathcal{C}_K = \mathbb{D}_K/(K^\times)$ the divisor class group. The map $\deg\colon \mathbb{D}_K \to \mathbb{Z}$ is an epimorphism, and $(K^\times) \subset \mathrm{Ker}(\deg)$. Consequently $\deg$ induces an (equally denoted) epimorphism $\deg\colon \mathcal{C}_K \to \mathbb{Z}$. For $n \in \mathbb{Z}$ we denote by $\mathbb{D}_K^n$ resp. $\mathcal{C}_k^n$ the set of all divisors resp. divisor classes of degree n. Then $h_K = |\mathcal{C}_K^0| < \infty$.

For a divisor $D \in \mathbb{D}_K$, the set

$$\mathcal{L}(D) = \{x \in K \mid v_P(x) \geq -v_P(D) \ \text{ for all } \ P \in \mathbb{P}_K\}$$

is a vector space over $\mathbb{F}_q$, and $\dim(D) = \dim_{\mathbb{F}_q}(\mathcal{L}(D)$ is the dimension of D; it only depends on the divisor class $[D] = D + (K^\times) \in \mathcal{C}_K$.

Let $g = g_K$ be the genus of K. There exists a unique divisor class $\boldsymbol{w} \in \mathcal{C}_P$ such that, for every $D \in \mathbb{D}_K$ and $W \in \boldsymbol{w}$ we have

$$\dim(D) = \deg(D) - g + 1 + \dim(W - D) \quad \text{(Riemann-Roch theorem)}.$$

We call $\boldsymbol{w}$ the canonical class and every divisor $W \in \boldsymbol{w}$ a canonical divisor (see [18, Section 6.4]).

2. Let $K/\mathbb{F}_q$ be a function field and $P \in \mathbb{P}_K$. Then $P = t\mathcal{O}_P$ for every $t \in K$ such that $v_P(t) = 1$. The formal Laurent series field $K_P = \mathsf{k}_P\,(\!(t)\!)$ is the completion of the discrete valued field (K, v_P) and $v_P = \mathsf{o}_t$ is just the order valuation on K_P .

1

Topological groups and infinite Galois theory

A basic knowledge of topological groups and the Galois theory of infinite field extensions is indispensable for a deeper understanding of higher arithmetic. Since these topics usually get short in introductory courses of algebra and number theory, we present the necessary material in this chapter, only assuming the fundamentals of general topology and classical Galois theory.

Beyond the basic concepts we offer a detailed study of injective and projective limits, of profinite groups, of infinite Galois theory and of the duality theory of locally compact abelian groups. We do not prove Pontrjagin's duality theorem in its full generality but we do this in all arithmetically relevant cases. As to convergence theory, we could not resolve upon the use of filters since almost all arithmetically relevant applications deal with sequences. Thus, we decided to present an (almost) satisfactory theory of completeness using sequences. In several places I referred to my previous book [18] to show how topological concepts shed light on facts which remained unsatifactory there.

Almost the whole material of this chapter can be found scattered in the literature. The volumes which primarily influenced my presentation are [2], [30], [54], [24], [51] and [50].

By a **topology** $\mathcal{T}$ on a set X we mean the set of all open subsets of X such that $X = (X, \mathcal{T})$ is a topological space.

Unless otherwise specified, all groups are assumed to be multiplicative.

1.1 Basics of topological groups

For subsets A, B of a (multiplicative) semigroup G, let

$$AB = \{ab \mid a \in A,\ b \in B\}$$

and, for all $n \in \mathbb{N}_0$,

$$A^n = \{a^n \mid a \in G\} \text{ and } A^{[n]} = \{a_1 \cdot \ldots \cdot a_n \mid a_1, \ldots, a_n \in A\}.$$

DOI: 10.1201/9780429506574-1

If G is a group, then $A^{-1} = \{a^{-1} \mid a \in A\}$. If $n \in \mathbb{N}$, then A^n also denotes the cartesian product $A \times \ldots \times A$, and for an ideal $\mathfrak{a}$ of a ring, $\mathfrak{a}^n$ denotes the ideal generated by $\mathfrak{a}^{[n]} = \mathfrak{a} \cdot \ldots \cdot \mathfrak{a}$. In any case, the individual meaning should be clear from the context.

Definition 1.1.1. A **topological group** $G = (G, \mathcal{T})$ is a group G, together with a topology $\mathcal{T}$ on G such that the map $G \times G \to G$, $(x, y) \mapsto xy^{-1}$, is continuous [equivalently, the maps $(x, y) \mapsto xy$ and $x \mapsto x^{-1}$ are both continuous]. Note that the map $G \to G$, $x \mapsto x^{-1}$ is even a homeomorphism. We say that G has **subgroup topology** if the open subgroups are a fundamental system of neighbourhoods of 1 in G. Every group endowed with the discrete topology is a topological group.

Let G and G' are topological groups. We denote by $\mathrm{Hom}^{\mathsf{c}}(G, G')$ the set of all continuous homomorphisms $f \colon G \to G'$. A map $f \colon G \to G'$ is called a **topological isomorphism** if it is a homeomorphism and a group isomorphism.

Every subgroup of a topological group [having subgroup topology], endowed with the subspace topology, is again a topological group [having subgroup topology]. The direct product of a family of topological groups [having subgroup topology], endowed with the product topology, is again a topological group [having subgroup topology].

If G is a topological group, $n \in \mathbb{N}$ and $G_1, \ldots, G_n$ are subgroups of G, then G is called the **topological (inner) direct product** of $G_1, \ldots, G_n$ if it is the (inner) direct product, $G = G_1 \cdot \ldots \cdot G_n$, and if it carries the product topology [equivalently, the map $\phi \colon G_1 \times \ldots \times G_n \to G$, defined by $\phi(g_1, \ldots, g_n) = g_1 \cdot \ldots \cdot g_n$, is a topological isomorphism].

In the following (lengthy) Theorem 1.1.2 we gather some elementary properties of topological groups.

Theorem 1.1.2. *Let G be a topological group and $\mathfrak{U}$ a fundamental system of neighbourhoods of 1 in G.*

1. *Let $a \in G$. Then the left multiplication $l_a \colon G \to G$ and the right multiplication $r_a \colon G \to G$, defined by*

$$l_a(x) = ax \quad \text{and} \quad r_a(x) = xa \quad \text{for all} \quad x \in G,$$

are homeomorphisms, and the sets $\{aU \mid U \in \mathfrak{U}\}$ and $\{Ua \mid U \in \mathfrak{U}\}$ are fundamental systems of neighbourhoods of a in G. In particular, the topology of G is uniquely determined by $\mathfrak{U}$.

2. *Let $U \in \mathfrak{U}$ and $a \in G$. Then there exists some $V \in \mathfrak{U}$ such that $VV \subset U$, $VV^{-1} \subset U$ and $aVa^{-1} \subset U$. If $V \in \mathfrak{U}$ and $VV^{-1} \subset U$, then $\overline{V} \subset U$. In particular:*

(a) *If $U \in \mathfrak{U}$, then U° and U^{-1} are also neighbourhoods of* 1, *and there exists some $V \in \mathfrak{U}$ such that $V^{-1} \subset U$.*

(b) *For every $a \in G$, the closed neighbourhoods of a are a fundamental system of neighbourhoods of a.*

3. If A is a subset of G, then

$$\overline{A} = \bigcap_{U \in \mathfrak{U}} AU. \quad \textit{In particular,} \quad \overline{\{1\}} = \bigcap_{U \in \mathfrak{U}} U.$$

4. Let U and V be subsets of G.

(a) *If U is open, then UV is open, too.*

(b) *If U and V are both quasicompact, then UV is quasicompact.*

(c) *If U is quasicompact and $n \in \mathbb{N}$, then U^{-1}, U^n and $U^{[n]}$ are quasicompact.*

(d) *If U is quasicompact and V is open, then the set*

$$B = \{b \in G \mid Ub \subset V\}$$

is open.

Proof. 1. The map $l_a = (x \mapsto ax) = [x \mapsto (a, x) \mapsto ax]$ is continuous, bijective, and $l_a^{-1} = l_{a^{-1}}$. Hence l_a is a homeomorphism, and thus $\{aU \mid U \in \mathfrak{U}\} = l_a(\mathfrak{U})$ is a fundamental system of neighbourhoods of a. The same holds for the right multiplication with a.

2. As the maps $(x, y) \mapsto xy$, $(x, y) \mapsto xy^{-1}$ and $x \mapsto axa^{-1}$ are continuous, there exists some $V \in \mathfrak{U}$ such that $VV \subset U$, $VV^{-1} \subset U$ and $aVa^{-1} \subset U$. If $U \in \mathfrak{U}$, then U° is a neighbourhood of 1 by the very definition, and U^{-1} is a neighbourhood of 1, since $x \mapsto x^{-1}$ is a homeomorphism. If $V \in \mathfrak{U}$ is such that $VV^{-1} \subset U$, then $V^{-1} \subset VV^{-1} \subset U$.

Now let $V \in \mathfrak{U}$, $VV^{-1} \subset U$ and $a \in \overline{V}$. Then $aV \cap V \neq \emptyset$, hence $av_1 = v_2$ for some $v_1, v_2 \in V$, and consequently $a = v_2v_1^{-1} \in VV^{-1} \subset U$. Hence $\overline{V} \subset U$, and therefore the closed neighbourhoods of 1 are a fundamental system of neighbourhoods of 1. By 1., the same is true for every $a \in G$.

3. Let $A \subset G$, $z \in \overline{A}$ and $U \in \mathfrak{U}$. If $V \in \mathfrak{U}$ is such that $V^{-1} \subset U$, then $zV \cap A \neq \emptyset$ and therefore $z \in AV^{-1} \subset AU$.

Conversely, suppose that $z \in AU$ and thus $zU^{-1} \cap A \neq \emptyset$ for all $U \in \mathfrak{U}$. If $V \in \mathfrak{U}$ is arbitrary and $U \in \mathfrak{U}$ satisfies $U^{-1} \subset V$, then $zV \cap A \supset zU^{-1} \cap A \neq \emptyset$ and thus $z \in \overline{A}$.

4. (a) Let U be open. Clearly, Uv is open for all $v \in V$, and thus

$$UV = \bigcup_{v \in V} Uv \quad \text{is open.}$$

(b) Let U and V be quasicompact. Then $U \times V$ is quasicompact, and the map $\mu \colon G \times G \to G$, defined by $\mu(x, y) = xy$, is continuous. Consequently $UV = \mu(U \times V)$ is quasicompact.

(c) Let U be quasicompact. As the maps $x \mapsto x^{-1}$ and $x \mapsto x^n$ are continuous, the sets U^{-1} and U^n are quasicompact. By a simple induction using (b) it follows that $U^{[n]}$ is quasicompact.

(d) Let U be quasicompact and V be open. It suffices to prove that for every $b \in B$ there exists some $W \in \mathfrak{U}$ such that $bW \subset B$. Thus let $b \in B$, hence $Ub \subset V$. For every $u \in U$ there exists an open neighbourhood $W_u' \in \mathfrak{U}$ such that $ubW_u' \subset V$, and there exists an open neighbourhood $W_u \in \mathfrak{U}$ such that $W_uW_u \subset W_u'$. Then obviously

$$Ub \subset \bigcup_{u \in U} ubW_u,$$

and since Ub is quasicompact, there exist some $n \in \mathbb{N}$ and $u_1, \ldots, u_n \in U$ such that $Ub \subset u_1bW_{u_1} \cup \ldots \cup u_nbW_{u_n}$. Then $W = W_{u_1} \cap \ldots \cap W_{u_n} \in \mathfrak{U}$, and

$$UbW \subset \bigcup_{i=1}^{n} u_ibW_{u_i}W_{u_i} \subset \bigcup_{i=1}^{n} u_ibW_{u_i}' \subset V, \quad \text{hence } bW \subset B. \qquad \square$$

We continue with an investigation of the structure of subgroups of a topological group.

Theorem 1.1.3. *Let G be a topological group and H a subgroup of G.*

1. *The following assertions are equivalent:*

(a) H is open.

(b) H is a neighbourhood of 1.

(c) $H^\circ \neq \emptyset$.

2. *$\overline{H}$ is a subgroup of G, and if H is a normal subgroup of G, then so is $\overline{H}$.*

3. *If H is open, then H is closed. If H is closed and $(G{:}H) < \infty$, then H is open.*

If H is open and G is quasicompact, then $(G{:}H) < \infty$.

4. *If G is Hausdorff and H is discrete, then H is closed.*

Proof. 1. (a) $\Rightarrow$ (b) $\Rightarrow$ (c) Obvious.

(c) $\Rightarrow$ (a) If $c \in H^\circ$, then $1 \in c^{-1}H^\circ$, and therefore

$$H = \bigcup_{a \in H} ac^{-1}H^\circ \quad \text{is open.}$$

2. The map $\psi\colon G \times G \to G$, defined by $\psi(x,y) = xy^{-1}$, is continuous, and $\psi(H \times H) = H$. Hence $\psi(\overline{H} \times \overline{H}) = \psi(\overline{H \times H}) \subset \overline{H}$, and therefore $xy^{-1} \in \overline{H}$ for all $x,\, y \in \overline{H}$. Thus $\overline{H}$ is a subgroup of G.

Now let H be a normal subgroup of G and $a \in G$. Then the map $\theta\colon G \to G$, defined by $\theta(x) = a^{-1}xa$, is continuous, and $\theta(H) = H$. Hence $\theta(\overline{H}) = a^{-1}\overline{H}a \subset \overline{H}$, and thus $\overline{H}$ is also a normal subgroup of G.

3. Let M be the union of the left cosets aH of H which are distinct from H. Then $H = G \setminus M$, and $(G\!:\!H) = |\{aH \mid a \in G\}|$. If H is open, then M is open and thus H is closed. If $(G\!:\!H) < \infty$ and H is closed, then M is closed, and thus H is open. If H is open and G is quasicompact, then $\{aH \mid a \in G\}$ is an open covering of G consisting of pairwise disjoint sets, and thus it is finite.

4. Let G be Hausdorff, H be discrete and $a \in G \setminus H$. Let $\mathfrak{U}$ be the set of all neighbourhoods of 1 in G, and assume to the contrary that $aU \cap H \neq \emptyset$ for all $U \in \mathfrak{U}$. As H is discrete, there exists some $W \in \mathfrak{U}$ such that $W^{-1}W \cap H = \{1\}$. As $aW \cap H \neq \emptyset$, there exists some $w \in W$ such that $aw \in H$. In particular, $w \neq 1$, and thus there exists some $V \in \mathfrak{U}$ such that $V \subset W$ and $w \notin V$. Again $aV \cap H \neq \emptyset$, and if $v \in V$ is such that $av \in H$, then $v \neq w$ and $(av)^{-1}aw = v^{-1}w \in W^{-1}W \cap H$ and $v^{-1}w \neq 1$, a contradiction. □

In the following Theorem 1.1.4 we investigate the distinctive properties of the fundamental systems of neighbourhoods of the unity.

Theorem 1.1.4. *Let G be a group.*

1. Let $\mathfrak{U} \neq \emptyset$ be a set of non-empty subsets of G with the following properties:

(a) For every $U_1, U_2 \in \mathfrak{U}$, there exists some $U \in \mathfrak{U}$ such that $U \subset U_1 \cap U_2$.

(b) For every $U \in \mathfrak{U}$ and $a \in G$ there exists some $V \in \mathfrak{U}$ such that $VV \subset U$, $VV^{-1} \subset U$ and $aVa^{-1} \subset U$.

Then there exists a unique topology $\mathcal{T}$ on G such that $(G, \mathcal{T})$ is a topological group and $\mathfrak{U}$ is a fundamental system of neighbourhoods of 1.

2. Let $\Sigma \neq \emptyset$ be a set of normal subgroups of G such that 1.(a) *holds. Then there exists a unique topology $\mathcal{T}$ on G such that $(G, \mathcal{T})$ is a topological group with subgroup topology and Σ is a fundamental system of open neighbourhoods of* 1.

Proof. 1. Uniqueness follows from 1.1.2.1. To prove existence, note first that (b) implies $1 \in U$ for all $U \in \mathfrak{U}$, and therefore $V^{-1} \subset VV^{-1}$ for all $V \in \mathfrak{U}$. Let $\mathcal{T}$ be the set of all subsets O of G with the following property:

For every $x \in O$ there exists some $V \in \mathfrak{U}$ such that $xV \subset O$.

Then $\mathcal{T}$ is closed under arbitrary unions and finite intersections. Hence it is a topology, and it suffices to prove:

a. For all $a \in G$ the set $\{aU \mid U \in \mathfrak{U}\}$ is a fundamental system of neighbourhoods of a.

b. The map $\psi\colon G\times G \to G$, defined by $\psi(x,y) = xy^{-1}$, is continuous.

Proof of **a.** Let $a \in G,\ U \in \mathfrak{U}$ and

$$O = \{x \in G \mid xW \subset aU \text{ for some } W \in \mathfrak{U}\}.$$

Then $a \in O,\ O \subset aU$, and we assert that $O \in \mathcal{T}$ (then aU is a neighbourhood of a). We must prove that for every $x \in O$ there exists some $V \in \mathfrak{U}$ such that $xV \subset O$.

Thus let $x \in O,\ W \in \mathfrak{U}$ such that $xW \subset aU$ and $V \in \mathfrak{U}$ such that $VV \subset W$. If $y \in xV$, then $yV \subset xVV \subset xW \subset aU$, hence $y \in O$ and consequently $xV \subset O$.

If Y is any neighbourhood of a, then $a \in O \subset Y$ for some $O \in \mathcal{T}$, and (as above) there exists some $U \in \mathfrak{U}$ such that $aU \subset O \subset Y$. Hence $\{aU \mid U \in \mathfrak{U}\}$ is a fundamental system of neighbourhoods of a.

Proof of **b.** Let $(a,b) \in G\times G$ and $U \in \mathfrak{U}$. Then there exists some $W \in \mathfrak{U}$ such that $bWb^{-1} \subset U$ and some $V \in \mathfrak{U}$ such that $VV^{-1} \subset W$. Now it follows that

$$(aV)(bV)^{-1} = aVV^{-1}b^{-1} \subset aWb^{-1} = ab^{-1}bWb^{-1} \subset ab^{-1}U.$$

2. By 1. Σ is a fundamental system of neighbourhoods of 1, and by 1.1.3.1 the sets $U \in \Sigma$ are open. □

Let G be a topological group, H a subgroup of G, $G/H = \{aH \mid a \in G\}$ the space of left cosets endowed with the quotient topology and $\pi\colon G \to G/H$ the residue class map, defined by $\pi(a) = aH$ for all $a \in G$. A subset Ω of G/H is open [closed] if and only if $\pi^{-1}(\Omega)$ is open [closed] in G. In particular, if U is a subset of G, then $\pi(U)$ is open [closed] in G/H if and only if UH is open [closed] in G. Obviously, the natural isomorphism $G \stackrel{\sim}{\to} G/\{1\}$ is a topological isomorphism.

Theorem 1.1.5. *Let G be a topological group, H a subgroup of G and denote by $\pi\colon G \to G/H$ the residue class map.*

1. π is continuous and open.

2. Let K be a subgroup of G such that $H \subset K$ (then K/H is a subset of G/H).

(a) The quotient topology on K/H coincides with the subspace topology induced by G/H.

(b) $\overline{K/H} = \overline{K}/H \subset G/H$.

3. If $a \in G$, then $\psi_a\colon G/H \to G/H$, defined by $\psi_a(xH) = axH$ for all $x \in G$, is a homeomorphism.

4. Let H be a normal subgroup of G. Then G/H is a topological group, and if G has subgroup topology, then G/H has subgroup topology, too.

5. G/H is discrete if and only if H is open, and G/H is Hausdorff if and only if H is closed. In particular: G is discrete if and only if $\{1\}$ is open, and G is Hausdorff if and only if $\{1\}$ is closed.

Proof. 1. By definition, π is continuous. If U is an open subset of G, then $\pi^{-1}(\pi(U)) = UH$ is open by 1.1.2.4(a), and therefore $\pi(U)$ is open. Hence π is open.

2. (a) If K/H has the subspace topology induced by G/H, then the map $\pi \restriction K \colon K \to K/H$ is continuous and surjective, and we show that it is open (then K/H has the quotient topology induced by $\pi \restriction K$). Let U be a (relatively) open subset of K, say $U = V \cap K$ for some open subset V of G. Then $\pi(V)$ is open in G/H, and we shall prove that $\pi(V) \cap \pi(K) = \pi(V \cap K)$ (then $(\pi \restriction K)(U) = \pi(V) \cap \pi(K)$ is open in K/H and we are done). The inclusion $\pi(V \cap K) \subset \pi(V) \cap \pi(K)$ is obvious. For the reverse inclusion, let $v \in V$, $k \in K$ and $\pi(v) = \pi(k) \in \pi(V) \cap \pi(K)$. Then $vH = kH$, hence $k' = k^{-1}v \in H \subset K$, $v = kk' \in V \cap K$ and $\pi(v) \in \pi(V \cap K)$.

(b) As $\pi^{-1}(\overline{K}/H) = \overline{K}$, it follows that $\overline{K}/H$ is closed, and $K/H \subset \overline{K}/H$ implies $\overline{K/H} \subset \overline{K}/H$. Since π is continuous, we obtain

$$\overline{K}/H = \pi(\overline{K}) \subset \overline{\pi(K)} = \overline{K/H}.$$

3. Let $a \in G$, and define $\tau_a \colon G \to G$ and $\tau_a^* \colon G/H \to G/H$ by $\tau_a(x) = ax$ and $\tau_a^*(xH) = axH$. Then we obtain the commutative diagram

$$\begin{array}{ccc} G & \xrightarrow{\ \tau_a\ } & G \\ {\scriptstyle\pi}\downarrow & & \downarrow{\scriptstyle\pi} \\ G/H & \xrightarrow{\ \tau_a^*\ } & G/H. \end{array}$$

If W is an open subset of G/H, then $\pi^{-1}(\tau_a^{*-1}(W)) = \tau_a^{-1}(\pi^{-1}(W)$ is open in G, hence $\tau_a^{*-1}(W)$ is open in G/H, and therefore τ_a^* is continuous. Since τ_a^* is bijective and $\tau_a^{*-1} = \tau_{a^{-1}}^*$, it follows that τ_a^* is a homeomorphism.

4. Let the maps $\mu \colon G{\times}G \to G$ and $\mu^* \colon G/H{\times}G/H \to G/H$ be defined by $\mu(x,y) = xy^{-1}$ and $\mu^*(xH, yH) = xy^{-1}H$ for all $x,\ y \in G$. Then μ is continuous and we must prove that μ^* is continuous. We consider the commutative diagram

$$\begin{array}{ccc} G{\times}G & \xrightarrow{\ \mu\ } & G \\ {\scriptstyle\pi\times\pi}\downarrow & & \downarrow{\scriptstyle\pi} \\ G/H{\times}G/H & \xrightarrow{\ \mu^*\ } & G/H. \end{array}$$

As π is continuous, open and surjective, the same is true for $\pi\times\pi$, and consequently $G/H\times G/H$ has the quotient topology induced by $\pi\times\pi$. If W is an open subset of G/H, then $(\pi\times\pi)^{-1}(\mu^{*-1}(W)) = \mu^{-1}(\pi^{-1}(W))$ is open in $G\times G$, and thus $\mu^{*-1}(W)$ is open in $G/H\times G/H$. Hence μ^* is continuous.

Assume now that G has subgroup topology, and let $\mathfrak{U}$ be a fundamental system of neighbourhoods of 1 in G consisting of open subgroups. Then $\{UH/H \mid U \in \mathfrak{U}\}$ is a fundamental system of neighbourhoods of 1 in G/H consisting of open subgroups of G/H.

5. G/H is discrete if and only if $\{\pi(a)\} \subset G/H$ is open for all $a \in G$. If $a \in G$, then $\pi^{-1}(\{\pi(a)\}) = aH$ if open [closed] if and only if H is open [closed]. Hence H is open if and only if G/H is discrete, and H is closed if and only if every singleton in G/H is closed. In particular, if G/H is Hausdorff, then H is closed.

Assume now that H is closed. We prove first that for every point of G/H the closed neighbouhoods are a fundamental system of neighbourhoods. By 3., it suffices to do this for $\pi(1)$. Thus let Y be any neighbourhood of $\pi(1)$ in G/H. Then $U = \pi^{-1}(Y)$ is a neighbourhood of 1 in G, and there exists some neighbourhood V of 1 in G such that $VV^{-1} \subset U$. Then $\pi(V)$ is a neighbourhood of $\pi(1)$, and we assert that $\overline{\pi(V)} \subset Y$. Indeed, let $x \in G$ such that $\pi(x) \in \overline{\pi(V)}$. Since $\pi(xV)$ is a neighbourhood of $\pi(x)$, it follows that $\pi(xV) \cap \pi(V) \neq \emptyset$, and there exist v_1, $v_2 \in V$ such that $\pi(xv_1) = \pi(v_2)$. It follows that $\pi(x) = \pi(v_2v_1^{-1}) \in \pi(VV^{-1}) \subset \pi(U) = Y$.

Now let x, $y \in G$ such that $\pi(x) \neq \pi(y)$. Then $G/H \setminus \{\pi(x)\}$ is a neighbourhood of $\pi(y)$. Let Y be a closed neighbourhood of $\pi(y)$ such that $Y \subset G/H\setminus\{\pi(x)\}$. Then $G/H\setminus Y$ is a neighbourhood of $\pi(x)$ which is disjoint from Y.

The final assertions follow with $H = \mathbf{1}$. □

Theorem 1.1.6. *Let $f\colon G \to G'$ be a homomorphism of topological groups and $1' = 1_{G'}$. Let H be an open subgroup of G, $\mathfrak{U}$ a fundamental system of neighbourhoods of 1 in G and $\mathfrak{U}'$ a fundamental system of neighbourhoods of $1'$ in G'.*

1. *The following assertions are equivalent:*

   (a) *f is continuous.*

   (b) *f is continuous in 1.*

   (c) *For all $U' \in \mathfrak{U}'$, the set $f^{-1}(U')$ is a neighbourhood of 1 in G.*

   (d) *$f\restriction H\colon H \to G'$ is continuous.*

   In particular, if $\mathrm{Ker}(f)$ is open in G, then f is continuous.

2. *The following assertions are equivalent:*

   (a) *f is open.*

   (b) *For every $U \in \mathfrak{U}$ the set $f(U)$ is a neighbourhood of $1'$ in G'.*

(c) $f\restriction H\colon H \to G'$ is open.

3. Let N be a normal subgroup of G, $N \subset \mathrm{Ker}(f)$, and let $f^*\colon G/N \to G'$ be defined by $f^*(xN) = f(x)$ for all $x \in G$. Then f^* is continuous [open] if and only if f is continuous [open]. If f is closed, then f^* is closed, too.

4. Let $f^*\colon G/\mathrm{Ker}(f) \to f(G)$ be the isomorphsm induced by f. Then f^* is a topological isomorphism if and only if $f\colon G \to f(G)$ is continuous and open.

Proof. 1. (a) $\Rightarrow$ (b) $\Leftrightarrow$ (c) and (a) $\Rightarrow$ (d) are obvious.

(b) $\Rightarrow$ (a) Let $a \in G$, and define the maps $\tau\colon G \to G$ and $\tau'\colon G' \to G'$ by $\tau(x) = a^{-1}x$ for all $x \in G$ and $\tau'(x') = f(a)x'$ for all $x' \in G'$. Then τ and τ' are homeomorphisms, $\tau(a) = 1$, $\tau'(1') = f(a)$, and since f is continuous in 1, it follows that $f = \tau' \circ f \circ \tau$ is continuous in a.

(d) $\Rightarrow$ (c) If $U' \in \mathfrak{U}'$, then $(f\restriction H)^{-1}(U') = f^{-1}(U') \cap H$ is a neighbourhood of 1 in H and thus in G, since H is open.

2. (a)$\Rightarrow$ (c) is obvious.

(c) $\Rightarrow$ (b) Let $U \in \mathfrak{U}$. Then there exists an open neighbourhood V of 1 such that $V \subset U \cap H$. As $f(V)$ is open and $1' \in f(V) \subset f(U)$, it follows that $f(U)$ is a neighbourhood of $1'$.

(b) $\Rightarrow$ (a) Let O be an open subset of G. If $a \in O$, then there exists some $U \in \mathfrak{U}$ such that $aU \subset O$, and then $f(a)f(U) = f(aU) \subset f(O)$. Hence $f(a)f(U)$ and thus also $f(O)$ is a neighbourhood of $f(a)$. Since $f(O)$ is a neighbourhood of $f(a)$ for all $a \in O$, it is open.

3. Let $\pi\colon G \to G/N$ be the residue class epimorphism. Then $f = f^* \circ \pi$, and since G/N carries the quotient topology induced by π, it follows that f^* is continuous if and only if f is continuous. If f^* is open, then f is open by 1.1.5.1. Now let f be open [closed] and W an open [closed] subset of G/N. Then $\pi^{-1}(W)$ is open [closed] in G, and therefore $f^*(W) = f(\pi^{-1}(W))$ is open [closed] in G'. Hence f^* is open [closed].

4. By 3., applied with $N = \mathrm{Ker}(f)$ and $f(G)$ instead of G'. □

Definition and Theorem 1.1.7. A homomorphism $f\colon G \to G'$ of topological groups is called **strict** if it is continuous and the (equally denoted) induced surjective homomorphism $f\colon G \to f(G)$ is open.

Let $f\colon G \to G'$ be a continuous homomorphism of topological groups, and let $f^*\colon G/\mathrm{Ker}(f) \to f(G)$ be the induced homomorphism, defined by $f^*(x\mathrm{Ker}(f)) = f(x)$ for all $x \in G$.

1. If f is sujective, then f is strict if and only if f is open.

2. f is strict if and only if $f^*\colon G/\mathrm{Ker}(f) \to f(G)$ is a topological isomorphism.

3. If G is quasicompact and $f(G)$ is Hausdorff, then f is strict.

4. Let $f\colon G \to G'$ be strict and surjective, and let $g\colon G' \to G''$ be another strict homomorphisms into a topological group G''. Then $g{\circ}f$ is strict.

Proof. 1. is obvious, and 2. follows by 1.1.6.4.

3. $f^*\colon G/\mathrm{Ker}(g) \to f(G)$ is a continuous isomorphism. As $G/\mathrm{Ker}(f)$ is compact and G' is Hausdorff, f^* a topological isomorphism, and thus f is strict by 1.

4. $g{\circ}f$ is continuous, and $g{\circ}f(G) = g(G')$. As f is open by 1. and the map $g\colon G' \to g(G')$ is open, it follows that $g{\circ}f\colon G \to g(G') = g{\circ}f(G)$ is open, and thus $g{\circ}f$ is strict. □

Theorem 1.1.8. *Let G be a topological group.*

1. Let N be a normal subgroup of G, $\pi\colon G \to G/N$ the residue class epimorphism, and let H be any subgroup of G.

(a) $\pi(H) = NH/N \subset G/N$, and on $\pi(H)$ the subgroup topology induced by G/N coincides with the quotient topology induced from NH.

(b) Let $\phi\colon H/N \cap H \to NH/N$ be defined by $\phi(x(N \cap H)) = xN$ for all $x \in H$. Then ϕ is a continuous isomorphism which need not be open.

2. Let N and H be normal subgroups of G such that $H \subset N \subset G$. Then $\phi\colon G/N \to (G/H)/(N/H)$, defined by $\phi(xN) = (xH)(N/H)$ for all $x \in G$, is a topological isomorphism.

We identify: $G/N = (G/H)/(N/H)$.

3. Let Γ be a topological group, $\pi\colon G \to \Gamma$ a continuous epimorphism, and let $j\colon \Gamma \to G$ be a continuous homomorphism such that $\pi{\circ}j = \mathrm{id}_\Gamma$. Then the map $j\colon \Gamma \to j(\Gamma)$ is a topological isomorphism, and if $N = \mathrm{Ker}(\pi)$, then $G = N{\cdot}j(\Gamma)$ is the topological inner direct product of N and $j(\Gamma)$.

Proof. 1.(a) By 1.1.5.2, applied with $N \subset NH \subset G$ instead of $K \subset H \subset G$.

(b) $\pi \restriction H$ is continuous, $\mathrm{Ker}(\pi \restriction H) = N \cap H$, $\mathrm{Im}(\pi \restriction H) = NH/N$, and ϕ is the group isomorphism induced by $\pi \restriction H$. Hence ϕ is continuous by 1.1.6.3.

To see that ϕ need not be open, consider $G = \mathbb{R}$, $N = \mathbb{Z}$ and $H = \theta\mathbb{Z}$ for some $\theta \in \mathbb{R} \setminus \mathbb{Q}$. Then $\phi\colon \theta\mathbb{Z} = \theta\mathbb{Z}/\mathbb{Z} \cap \theta\mathbb{Z} \overset{\sim}{\to} \mathbb{Z} + \theta\mathbb{Z}/\mathbb{Z}$ is a continuous isomorphism but not a homeomorphism (and thus not open). Indeed, $\theta\mathbb{Z}$ is discrete, but $\mathbb{Z}+\theta\mathbb{Z}/\mathbb{Z}$ is dense in $\mathbb{Q}/\mathbb{Z}$ and thus not discrete.

2. The map

$$\pi^*\colon G \xrightarrow{\pi'} G/H \xrightarrow{\pi''} (G/H)/(N/H), \text{ defined by } \pi^*(x) = (xH)(N/H),$$

is the composition of two continuous and open residue class epimorphisms and thus it is itself a continuous and open epimorphism. Since $\operatorname{Ker}(\pi^*) = N$ it follows that ϕ is the isomorphism induced by π^*, and 1.1.6.4 implies that ϕ is a topological isomorphism.

3. Since $\pi\circ j = \mathrm{id}_\Gamma$, the map $j\colon G \to j(\Gamma)$ is a continuous monomorphism, and as $\pi \restriction j(\Gamma) = j^{-1}\colon j(\Gamma) \to \Gamma$ is continuous, the map $j\colon \Gamma \to j(\Gamma)$ is a topological isomorphism. Thus it suffices to prove that the homomorphism

$$\Psi\colon N\times\Gamma \to G, \quad \text{defined by} \quad \Psi(n,c) = nj(c) \ \text{ for all } \ (n,c) \in N\times\Gamma,$$

is a topological isomorphism.

If $a \in G$, then $\pi(a\,(j\circ\pi)(a^{-1})) = \pi(a)\pi(a^{-1}) = 1 \in \Gamma$. Hence it follows that $a\,(j\circ\pi)(a^{-1}) \in N$, and we define $\rho\colon G \to N$ by $\rho(a) = a\,(j\circ\pi)(a^{-1})$ for all $a \in G$. Then ρ is continuous, $\rho\restriction N = \mathrm{id}_N$, and if $c \in \Gamma$, then

$$\rho\circ j(c) = j(c)(j\circ\pi\circ j(c^{-1})) = 1 \in N.$$

If $\Phi\colon G \to N\times\Gamma$ is defined by $\Phi(a) = (\rho(a), \pi(a))$ for all $a \in G$, then it follows that $\Psi\circ\Phi = \mathrm{id}_G$, $\Phi\circ\Psi = \mathrm{id}_{N\times\Gamma}$, and as Φ and Ψ are continuous, Ψ is a topological isomorphism. □

Theorem 1.1.9.

1. A topological group G is locally compact if and only if it is Hausdorff and possesses a compact neighbourhood of 1.

2. Let G be a locally compact topological group, H a closed subgroup of G and $\pi\colon G \to G/H$ the residue class map.

(a) H and G/H are locally compact.

(b) If C' is a compact subset of G/H, then there exists a compact subset C of G such that $C' = \pi(C)$.

3. Let $f\colon G \to G'$ be a continuous homomorphism of locally compact topological groups, and let C be a compact neighbourhood of 1 *in G such that $f(C)$ is a neighbourhood of the unit element $1'$ of G'. Then f is open.*

Proof. 1. Obvious by 1.1.2.1.

2. (a) Let K be a compact neighbourhood of 1 in G. Then $K\cap H$ is a compact neighbourhood of 1 in H, and for every $a \in G$ the set $\pi(aK)$ is a compact neighbourhood of $\pi(a)$ in G/H. Since both H and G/H are Hausdorff (see 1.1.5.5), they are locally compact.

(b) Let C' be a compact subset of G/H), and for $y \in C'$ let V_y be a compact neighbourhood of some $x \in \pi^{-1}(y)$. If $y \in C'$, then $\pi(V_y)$ is a neighbourhood of y, hence $y \in \pi(V_y)^\circ$, and thus $\{\pi(V_y)^\circ \mid y \in C'\}$ is an open covering of C'. Therefore there exists a finite subset Y of C' such that

$$C' \subset \bigcup_{y \in Y} \pi(V_y)^\circ \subset \pi(V), \quad \text{where} \quad V = \bigcup_{y \in Y} V_y.$$

Since V is compact and $\pi^{-1}(C')$ is closed, it follows that $C = V \cap \pi^{-1}(C')$ is compact, and $C' = \pi(C)$.

3. By 1.1.3.1, the group $H = \langle C \rangle$ is an open subgroup of G, and the group $f(H) = \langle f(C) \rangle$ is an open subgroup of G'. By 1.1.6.2, it suffices to prove:

$(*)$ For every neighbourhood U of 1 in H the set $f(U)$ is a neighbourhood of $1'$.

For the proof of $(*)$ we use:

Theorem of Baire (see [2, Ch. IX, §5.3]) Let X be a locally compact topological space, $(A_n)_{n \ge 1}$ a sequence of closed subsets of X and A their union. If $A_n^\circ = \emptyset$ for all $n \ge 1$, then $A^\circ = \emptyset$.

Proof of $(*)$ Let U be a neighbourhood of 1 in H and V a compact neighbourhood of 1 in H such that $VV^{-1} \subset U$. We assert that there is a sequence $(x_n)_{n \ge 1}$ in H such that

$$H = \langle C \rangle = \bigcup_{n \ge 1} (C \cup C^{-1})^{[n]} = \bigcup_{n \ge 1} x_n V.$$

Indeed, for each $n \ge 1$ the set $(C \cup C^{-1})^{[n]}$ is compact by 1.1.2.4(c), and thus it is the union of finitely many left cosets xV, where $x \in C \cup C^{-1}$. Consequently, H is the union of countably many such cosets, and thus there exists a sequence $(x_n)_{n \ge 1}$ in H such that

$$H = \bigcup_{n \ge 1} x_n V, \quad \text{and consequently} \quad f(H) = \bigcup_{n \ge 1} f(x_n V).$$

Together with V the sets $f(x_n V)$ are compact and thus closed in G'. Since $f(H)^\circ \ne \emptyset$, the theorem of Baire shows that $f(x_n V)^\circ = f(x_n) f(V)^\circ \ne \emptyset$ for some $n \in \mathbb{N}$ and thus $f(V)^\circ \ne \emptyset$. Hence

$$1' \in f(V)^\circ [f(V)^\circ]^{-1} \subset f(VV^{-1}) \subset f(U),$$

and as $f(V)^\circ [f(V)^\circ]^{-1}$ is open, $f(U)$ is a neighbourhood of $1'$. □

Recall that the (connected) component of 1 in a topological group G is the largest connected subset of G containing 1. A subset of a topological space which is both closed and open is called **clopen**.

Theorem 1.1.10. *Let G be a topological group and C the component of* 1 *in* G.

1. If $a \in G$, then Ca is the component of a. If $C = \mathbf{1}$, then G is totally disconnected. If U is a clopen neighbourhood of 1 in G, then $C \subset U$.

2. For a subset K of G the following assertions are equivalent:

(a) $K = C$.

(b) K is a connected closed normal subgroup of G, and G/K is totally disconnected.

3. If G is Hausdorff and has a fundamental system of neighbourhoods of 1 consisting of clopen sets, then G is totally disconnected. In particular, if G is Hausdorff and has subgroup topology, then G is totally disconnected.

Proof. 1. The map $\tau_a\colon G \to G$, $x \mapsto xa$ is a homeomorphism satisfying $Ca = \tau_a(C)$. Hence Ca is the component of a, and if $C = \mathbf{1}$, then G is totally disconnected. If U is a clopen neighbourhood of 1, then $U \cap C$ is a clopen neighbourhood of 1 relative C, and therefore $C = U \cap C \subset U$.

2. (a) $\Rightarrow$ (b) Being a component, C is closed and connected. Hence $C{\times}C$ is connected, and as $\psi\colon G{\times}G \to G$, defined by $\psi(x, y) = xy^{-1}$, is continuous, $\psi(C{\times}C)$ is also connected. Now $1 \in \psi(C{\times}C)$ implies $\psi(C{\times}C) \subset C$, and thus C is a subgroup of G.

If $a \in G$, then the map $\theta\colon G \to G$, defined by $\theta(x) = axa^{-1}$, is continuous, and therefore $\theta(C) = aCa^{-1}$ is connected. Since $1 \in aCa^{-1}$ it follows that $aCa^{-1} \subset C$, and therefore C is a normal subgroup of G.

We assert that there is no connected subset Γ of G/C such that $\pi(1) \in \Gamma$ and $|\Gamma| \geq 2$ (then G/C is totally disconnected by 1.). Thus let $\pi\colon G \to G/C$ be the residue class epimorphism, $\Gamma \subset G/C$, $\pi(1) \in \Gamma$ and $|\Gamma| \geq 2$. If $X = \pi^{-1}(\Gamma)$, then $C \subsetneq XC$, hence XC is not connected, and there exist open subsets U, V of G such that $XC \subset U \cup V$, $U \cap XC \neq \emptyset$, $V \cap XC \neq \emptyset$ and $U \cap V \cap XC = \emptyset$. Hence $\Gamma \subset \pi(U) \cup \pi(V)$, $\Gamma \cap \pi(U) \neq \emptyset$, $\Gamma \cap \pi(V) \neq \emptyset$, and we must prove that $\pi(U) \cap \pi(V) \cap \Gamma = \emptyset$. If $x \in X$, then $xC \subset (U \cap xC) \cup (V \cap xC)$. As xC is connected, it follows that $U \cap xC = \emptyset$ or $V \cap xC = \emptyset$, and consequently $\pi(x) \notin \pi(U)$ or $\pi(x) \notin \pi(V)$. Hence $\pi(U) \cap \pi(V) \cap \Gamma = \emptyset$, and Γ is not connected.

(b) $\Rightarrow$ (a) By the proof of (a) $\Rightarrow$ (b), C is a normal subgroup of G, and by definition $K \subset C$. Since C is connected, C/K is connected, too. But as $C/K \subset G/K$, it follows that C/K is totally disconnected. Hence $|C/K| = 1$ and $C = K$.

3. Let $\mathfrak{U}$ be a fundamental system of neighbourhoods of 1 consisting clopen sets. Then $C \subset U$ for all $U \in \mathfrak{U}$ by 1. and thus $C = \mathbf{1}$ by 1.1.2.3.

If G has subgroup topology, then the open subgroups are a fundamental system of neighbourhoods of 1 in G, and by 1.1.3.3 open subgroups are closed. $\square$

1.2 Topological rings, topological fields and real vector spaces

Definition and Remarks 1.2.1. A **topological ring** $R = (R, \mathcal{T})$ is a ring R, together with a topology $\mathcal{T}$ on R such that the maps

$$\begin{cases} R\times R & \to \ R \\ (x,y) & \mapsto \ x-y \end{cases} \quad \text{and} \quad \begin{cases} R\times R & \to \ R \\ (x,y) & \mapsto \ xy \end{cases}$$

are continuous (then the additive group $(R, +)$ is a topological group).

A **topological division ring** $K = (K, \mathcal{T})$ is a division ring K, together with a topology $\mathcal{T}$ such that $(K, \mathcal{T})$ is a topological ring, and $K^\times$, endowed with the subspace topology, is a topological group.

If R is a commutative topological ring, then every polynomial $g \in R[X_1, \ldots, X_n]$ induces an (equally denoted) continuous function $g\colon R^n \to R$, $\boldsymbol{x} \mapsto g(\boldsymbol{x})$.

If R is a topological field and $h = g^{-1}f \in R(X_1, \ldots, X_n)$ is a rational function, where $g,\ f \in R[X_1, \ldots, X_n]$, $g \neq 0$ and $P = \{\boldsymbol{x} \in R^n \mid g(\boldsymbol{x}) \neq 0\}$, then h induces an (equally denoted) continuous function $h\colon R^n \setminus P \to R$, $\boldsymbol{x} \mapsto h(\boldsymbol{x}) = g(\boldsymbol{x})^{-1}f(\boldsymbol{x})$.

Every subring [subfield] of a topological ring [of a topological field] is itself a topological ring [a topological field]. The direct product of a family of topological rings is a topological ring. The unit group of a topological ring, endowed with the subspace topology, need not be a topological group (an example will be given in 5.2.5.**2**). However, if K is a topological division ring and D is a subring of K, then $D^\times$ is a subgroup of $K^\times$, and thus $D^\times$ endowed with the subspace topology is a topological group.

Every subfield of $\mathbb{C}$ endowed with the subspace topology is a topological field. In [18, Remarks 5.2.1] we noted informally that a valued field $(K, |\cdot|)$, endowed with its valuation topology, is a topological field (for a precise argument see 1.4.1.**4.B** below).

Theorem and Definition 1.2.2. *Let R be a ring and $\mathfrak{a}$ an ideal of R. Then there exists a (uniquely determined) topology $\mathcal{T}$ on R such that $(R, \mathcal{T})$ is a topological ring and $\{\mathfrak{a}^n \mid n \in \mathbb{N}\}$ is a fundamental system of neighbourhoods of* 0.

This topology is called the **$\mathfrak{a}$-adic topology**; by definition, it is a subgroup topology.

Proof. By 1.1.4.2 it follows that $(R, +)$ is a topological group and the set $\{\mathfrak{a}^n \mid n \in \mathbb{N}\}$ is a fundamental system of neighbourhoods of 0.

If $a,\ b \in R$ and $n \in \mathbb{N}$, then $(a+\mathfrak{a}^n)(b+\mathfrak{a}^n) \subset ab+\mathfrak{a}^n$. Hence multiplication is continuous and R is a topological ring. □

The following example refers to [18, Definition 5.3.1 and Theorem 5.3.2] and connects the notions introduced there with the concepts of topological groups and rings.

Example 1.2.3. Let $(K, |\cdot|)$ be a discrete valued field endowed with the valuation topology. Let $\mathcal{O}_K$ be its valuation domain, $U_K = \mathcal{O}_K^\times$ its unit group, $\mathfrak{p}_K$ its valuation ideal and $U_K^{(n)} = 1 + \mathfrak{p}_K^n$ for all $n \in \mathbb{N}$. Then $\{\mathfrak{p}_K^n \mid n \in \mathbb{N}\}$ is a fundamental system of neighbourhoods of 0 and $\{U_K^{(n)} \mid n \in \mathbb{N}\}$ is a fundamental system of neighbourhoods of 1. Hence the valuation topology induces on $\mathcal{O}_K$ the $\mathfrak{p}_K$-adic topology and on $K^\times$ a subgroup topology.

Definition and Theorem 1.2.4. *Let R be a topological ring.*
Let $\varepsilon \colon R^\times \to R \times R$ be defined by $\varepsilon(u) = (u, u^{-1})$. The initial topology on $R^\times$ induced by ε is called the **unit topology**. Explicitly, a subset U of $R^\times$ is open in the unit topology if and only if $U = \varepsilon^{-1}(W)$ for some open subset W of $R \times R$.

Let $R^\times$ be equipped with the unit topology.

1. *$R^\times$ is a topological group, and the inclusion $\iota \colon R^\times \hookrightarrow R$ is continuous. In particular, the unit topology on $R^\times$ is finer than the subspace topology, and equality holds if and only if $R^\times$ equipped with the subspace topology is a topological group.*

2. *Let R' be a subring of R. Then $R'^\times \subset R^\times$, and the unit topology of $R'^\times$ is the subspace topology induced by the unit topology of $R^\times$. If R' is open in R, then $R'^\times$ is open in $R^\times$.*

3. *Let $f \colon R \to R'$ be a continuous homomorphism of topological rings. Then the induced homomorphism $f \restriction R^\times \colon R^\times \to R'^\times$ is continuous with respect to the unit topologies.*

4. *Let $(R_i)_{i \in I}$ be a family of topological rings. If*

$$R = \prod_{i \in I} R_i, \quad \text{then} \quad R^\times = \prod_{i \in I} R_i^\times,$$

and if every $R_i^\times$ carries the unit topology, then the product topology on $R^\times$ is the unit topology of $R^\times$

Proof. 1. We prove that $m \colon R^\times \times R^\times \to R^\times$, defined by $m(x, y) = xy^{-1}$, is continuous. We consider the commutative diagram

$$\begin{array}{ccc} R^\times \times R^\times & \xrightarrow{\quad m \quad} & R^\times \\ {\scriptstyle \varepsilon \times \varepsilon}\big\downarrow & & \big\downarrow{\scriptstyle \varepsilon} \\ (R \times R) \times (R \times R) & \xrightarrow{\quad m_2 \quad} & R \times R, \end{array}$$

where $m_2\colon (R\times R)\times(R\times R) \to R\times R$ is defined by $m_2(u,u_1,v,v_1) = (uv_1, vu_1)$. Then m_2 is continuous. If U is an open subset of $R^\times$, then $U = \varepsilon^{-1}(V)$ for some open subset V of $R\times R$, and then $m^{-1}(U) = (m_2\circ(\varepsilon\times\varepsilon))^{-1}(V)$ is open. Hence m is continuous.

If $p_1\colon R\times R \to R$ is the projection onto the first component, then the map $\iota = p_1\circ\varepsilon$ is continuous, and therefore the unit topology on $R^\times$ is finer than the subspace topology. If $R^\times$ is a topological group with respect to the subspace topology, then ε is continuous, hence the subspace topology is finer than the unit topology, and thus the two topologies are equal. The converse is obvious.

2. We consider the commutative diagram

$$\begin{array}{ccc} R'^\times & \xrightarrow{\;i\;} & R^\times \\ {\scriptstyle\varepsilon_0}\big\downarrow & & \big\downarrow{\scriptstyle\varepsilon} \\ R'\times R' & \xrightarrow{\;j\;} & R\times R, \end{array}$$

where i and j are the inclusions and $\varepsilon_0 = \varepsilon\restriction R'^\times$. A subset U of $R'^\times$ is open in the unit topology if and only if $U = (j\circ\varepsilon_0)^{-1}(W)$ for some open subset W of $R\times R$, and U is open in the subspace topology induced by $R^\times$ if and only if $U = (\varepsilon\circ i)^{-1}(W)$ for some open subset W of $R\times R$. Since $j\circ\varepsilon_0 = \varepsilon\circ i$, the two topologies are equal.

If R' is open in R, then $R'\times R'$ is open in $R\times R$, and thus $R'^\times = \varepsilon^{-1}(R'\times R')$ is open in $R^\times$.

3. We consider the commutative diagram

$$\begin{array}{ccc} R^\times & \xrightarrow{\;f\restriction R^\times\;} & R'^\times \\ {\scriptstyle\varepsilon_0}\big\downarrow & & \big\downarrow{\scriptstyle\varepsilon'} \\ R\times R & \xrightarrow{\;f\times f\;} & R'\times R', \end{array}$$

where $\varepsilon(u) = (u, u^{-1})$ for $u \in R^\times$ and $\varepsilon'(u') = (u', u'^{-1})$ for $u' \in R'^\times$. If U' is open in $R'^\times$ in the unit topology, then $U' = \varepsilon'^{-1}(W')$ for some open subset W' of $R'\times R'$, and therefore $(f\restriction R^\times)^{-1}(U') = ((f\times f)\circ\varepsilon_0)^{-1}(W'))$ is open in $R^\times$ in the unit topology. Hence f is continuous with respect to the unit topologies.

4. Let $\varepsilon\colon R^\times \to R\times R$ be defined by $\varepsilon(u) = (u, u^{-1})$. For $i \in I$, we define $\varepsilon_i\colon R_i^\times \to R_i\times R_i$ by $\varepsilon_i(u_i) = (u_i, u_i^{-1})$, and let $p_i\colon R \to R_i$ be the projection. Then there are commutative diagrams

$$\begin{array}{ccc} R^\times & \xrightarrow{\;\varepsilon\;} & R\times R \\ {\scriptstyle p_i\restriction R^\times}\big\downarrow & & \big\downarrow{\scriptstyle p_i\times p_i} \\ R_i^\times & \xrightarrow{\;\varepsilon_i\;} & R_i\times R_i \end{array}$$

for all $i \in I$. As $R\times R$ carries the topology induced by the family $(p_i\times p_i)_{i\in I}$, the unit topology on $R^\times$ is induced by the family $((p_i\times p_i)\circ\varepsilon)_{i\in I}$. On the other hand, for each $i \in I$, the unit topology on $R_i^\times$ is induced by ε_i, and thus the product topology on $R^\times$ is induced by the family $(\varepsilon_i\circ p_i\restriction R^\times)_{i\in I}$. Since $\varepsilon_i\circ(p_i\restriction R^\times) = (p_i\times p_i)\circ\varepsilon$ for all $i \in I$, the two topologies are equal. □

We close this section with some simple remarks concerning the topology of real vector spaces.

Let V be a finite-dimensional vector space over $\mathbb{R}$. Then any two norms on V are equivalent, and V carries the topology induced by any of its norms which we call the **natural topology** of V. With it, V is a locally compact topological group, and the map $\mathbb{R}\times V \to V$, $(\lambda, v) \mapsto \lambda v$ is continuous. In the natural topology every $\mathbb{R}$-homomorphism $\varphi\colon V \to V'$ between finite-dimensional vector spaces over $\mathbb{R}$ is continuous. The following simple lemmma provides the connection between the natural topology and the abstract topological concepts.

Lemma 1.2.5. *Let $\varphi\colon V \to V'$ be an $\mathbb{R}$-epimorphism between finite-dimensional vector spaces and $W = \mathrm{Ker}(\varphi)$. Then φ is open, and the quotient topology on V/W is the natural topology. In particular, every $\mathbb{R}$-isomorphism of finite-dimensional vector spaces over $\mathbb{R}$ is a topological isomorphism.*

Proof. Let W' be a subspace of V such that $V = W\oplus W'$, and let $p\colon V \to W'$ be the projection. Then $\varphi = \phi\circ p$, where $\phi\colon W' \to V'$ is an isomorphism. As p is open and ϕ is a homeomorphism, it follows that φ open. If we endow V/W with the quotient topology, then φ induces a topological isomorphism $\varphi^*\colon V/W \overset{\sim}{\to} V'$ by 1.1.6.4. On the other hand, φ^* is an $\mathbb{R}$-isomorphism and thus it is a homeomorphism for the natural topology. Consequently, on V/W the quotient topology and the natural topology coincide. □

Theorem 1.2.6. *Let D be a closed (additive) subgroup of $\mathbb{R}$. If $D \ne \mathbf{0}$, then either $D = \mathbb{R}$, or $D = a\mathbb{Z}$, where $a = \min(D\cap\mathbb{R}_{>0}$.*

Proof. Let $D \ne \mathbf{0}$ and $a = \inf(D\cap\mathbb{R}_{>0}$.

CASE 1: $a \notin D$. We assert that $D = \mathbb{R}$. Thus let $x \in \mathbb{R}$, let $(a_n)_{n\ge0}$ be a strictly monotonically decreasing sequence in D such that $(a_n)_{n\ge0} \to a$, and for $n \ge 0$ let $\lambda_n = a_n - a_{n+1}$ and $p_n = \lfloor\lambda_n^{-1}x\rfloor$. Then it follows that $0 \le x - p_n\lambda_n = \lambda_n(\lambda_n^{-1}x - p_n) \le \lambda_n$, hence $(\lambda_n)_{n\ge0} \to x$, and therefore $x \in D$, since D is closed.

CASE 2: $a \in D$. Then $a\mathbb{Z} \subset D$, and we assert that equality holds. Indeed, if $x \in D$, let $k \in \mathbb{Z}$ be such that $ka \le x < (k+1)a$. Then $0 \le x - ka < a$ and $x - ka \in D$, which implies $x - ka = 0$ and $x \in a\mathbb{Z}$. □

1.3 Inductive and projective limits

In order to present the following definitions and results in a sufficiently general scope, we deal with *objects* and associated *morphisms* to denote one of the following categories (however avoiding the general language of category theory). We denote by R an arbitrary and by k a commutative ring.

objects	associated morphisms
sets	maps
monoids	monoid homomorphisms
groups	group homomorphisms
rings	ring homomorphisms
R-modules	R-module homomorphisms
k-algebras	k-algebra homomorphisms
topological spaces	continuous maps
topological groups	continuous group homomorphisms

A morphism $\phi\colon X \to X'$ is called an **isomorphism** if there exists a morphism $\phi'\colon X' \to X$ such that $\phi' \circ \phi = \mathrm{id}_X$ and $\phi \circ \phi' = \mathrm{id}_{X'}$. For algebraic objects a morphism is an isomorphism if and only if it is bsijective. For topological objects a morphism is an isomorphism if and only if it is a homeomorphism (i. e., it is bijective and either open or closed). However, we shall frequently use the following fact: If X is a quasicompact and X' is a Hausdorff topologial space, then every continuous bijective map $f\colon X \to X'$ is closed and thus a homeomorphism.

A **directed set** $(I, \leq)$ is a non-empty set I together with a binary relation $\leq$ such that

- for all $i,\, j \in I$ we have $i \leq j$ and $j \leq i$ if and only if $i = j$;
- for all $i,\, j,\, k \in I$, if $i \leq j$ and $j \leq k$, then $i \leq k$;
- for all $i,\, j \in I$ there exists some $k \in I$ such that $i \leq k$ and $j \leq k$.

A subset I' of a directed set $(I, \leq)$ is called **cofinal** if for every $i \in I$ there is some $i' \in I'$ such that $i' \geq i$. If I' is a cofinal subset of a directed set I, then I' is directed with respect to the induced partial ordering.

Let X be any set. A subset Σ of $\mathbb{P}(X)$ is called **upwards** resp. **downwards directed** if $(\Sigma, \subset)$ resp. $(\Sigma, \supset)$ is directed. In particular, $\mathbb{P}(X)$ and $\mathbb{P}_{\mathrm{fin}}(X)$ are both upwards and downwards directed.

Definition and Remarks 1.3.1.

1. An **inductive** or **direct system** (of objects and associated morphisms) over a directed set $(I, \le)$ is a pair

$$(X_i, \psi_{j,i})_I = \big((X_i)_{i\in I},\, (\psi_{j,i}\colon X_j \to X_i)_{j\le i}\big),$$

consisting of a family of objects $(X_i)_{i\in I}$, together with a family of associated morphisms $(\psi_{j,i}\colon X_j \to X_i)_{j\le i}$ with the following properties:

- $\psi_{i,i} = \mathrm{id}_{X_i}$ for all $i \in I$;
- $\psi_{k,i} = \psi_{j,i}\circ\psi_{k,j}$ for all $i,\ j,\ k \in I$ such that $k \le j \le i$.

If $(X_i, \psi_{j,i})_I$ is an inductive system, we set

$$X = \biguplus_{i\in I} X_i\times\{i\},$$

and for two pairs $(x_i, i),\ (x_j, j) \in X$ we define $(x_i, i) \sim (x_j, j)$ if there exists some $k \in I$ such that $k \ge i,\ \ k \ge j$, and $\psi_{i,k}(x_i) = \psi_{j,k}(x_j)$ [then the same is true for all $k' \in I$ such that $k' \ge k$, since

$$\psi_{i,k'}(x_i) = \psi_{k,k'}\circ\psi_{i,k}(x_i) = \psi_{k,k'}\circ\psi_{j,k}(x_j) = \psi_{j,k'}(x_j)\,].$$

$\sim$ is an equivalence relation on X. Indeed, it is of course reflexive and symmetric. To prove that it is transitive, suppose that $(x_i, i) \sim (x_j, j)$ and $(x_j, j) \sim (x_k, k)$. Then there is some $l \in I$ is such that $l \ge i,\ \ l \ge j\ \ l \ge k$, and $\psi_{i,l}(x_i) = \psi_{j,l}(x_j) = \psi_{k,l}(x_k)$, which implies $(x_i, i) \sim (x_k, k)$. For a pair $(x_i, i) \in X$ we denote by $[x_i, i] \in X/{\sim}$ its equivalence class, and we call

$$X/{\sim} \;=\; \varinjlim\,(X_i, \psi_{j,i})_I \;=\; \varinjlim_{i\in I} X_i \;=\; \varinjlim X_i$$

the **inductive** or **direct limit** of the inductive system $(X_i, \psi_{j,i})_I$. For every $j \in I$ we define the map

$$\psi_j\colon X_j \to \varinjlim X_i \quad\text{by}\quad \psi_j(x_j) = [x_j, j] \quad\text{for all}\quad x_j \in X_j.$$

Then $\psi_j = \psi_i\circ\psi_{j,i}$ for all $i,\ j \in I$ such that $j \le i$, and

$$\varinjlim X_i \;=\; \bigcup_{j\in I} \psi_j(X_j).$$

The maps $\psi_j\colon X_j \to \varinjlim X_i$ are called the **injections** into the inductive limit (though they need not be injective).

An inductive system $(X_i, \psi_{j,i})_I$ is called **injective** if all $\psi_{j,i}$ are injective. Then the injections $\psi_j\colon X_j \to \varinjlim X_i$ are also injective. Indeed, suppose that $j \in J,\ \ x_j,\ x_j' \in X_j$ and $\psi_j(x_j) = \psi_j(x_j')$. Then we get $\psi_{j,i}(x_j) = \psi_{j,i}(x_j')$ for some $i \ge j$, and therefore $x_j = x_j'$.

If $(X_i, \psi_{j,i})_I$ is an inductive system over the directed set $(I, \leq)$, and if $(Y_i \subset X_i)_{i\in I}$ is a family of subobjects such that $\psi_{j,i}(Y_j) \subset Y_i$ for all $j \leq i$, then $(Y_i, \psi_{j,i} \restriction Y_j)_I$ is itself an inductive system (called a **subsystem** of $(X_i, \psi_{j,i})$), and $\varinjlim Y_i \subset \varinjlim X_i$.

2. Next we show how the various structures from the objects and morphisms of an inductive system carry over to the inductive limit.

First, let $(X_i, \psi_{j,i})_I$ be an inductive system of topological spaces. Then we endow $\varinjlim X_i$ with the final topology induced by the family of injections $(\psi_j \colon X_j \to \varinjlim X_i)_{j\in I}$. Explicitly, a subset X of $\varinjlim X_i$ is open if and only if $\psi_j^{-1}(X)$ is open in X_j for all $j \in I$. In particular, the sets $\psi_j(X_j)$ are open in $\varinjlim X_i$. Consequently, an inductive limit of an inductive system of topological spaces (and continuous maps) is a topological space, and the injections into the inductive limit are continuous.

Now let for all $i \in I$ a law of composition $*_i \colon X_i{\times}X_i \to X_i$ be given, and let $\psi_{j,i} \colon X_j \to X_i$ be a $(*_j, *_i)$-homomorphism for all $i \geq j$ (i. e., $\psi_{j,i}(x_j *_j x'_j) = \psi_{j,i}(x_j) *_i \psi_{j,i}(x'_j)$ for all $x_j,\ x'_j \in X_j$). Then we define a law of composition $*$ on $\varinjlim X_i$ as follows: For $i,\ j \in I,\ \ x_i \in X_i$ and $x_j \in X_j$, we choose some $k \in I$ such that $k \geq i,\ \ k \geq j$, and we define

$$[x_i, i] * [y_j, j] = [\psi_{i,k}(x_i) *_k \psi_{j,k}(y_j), k].$$

It is easily checked that this definition does not depend on the various choices. With this definition of composition laws, the inductive limit of an inductive system of (abelian) monoids [(abelian) groups, (commutative) rings] is again an (abelian) monoid [an (abelian) group, a (commutative) ring]. Note, however, that the inductive limit of an arbitrary inductive system of topological groups need not be a topological group (the composition need not be continuous).

If $(X_i, \varphi_{j,i})_I$ is an inductive system of R-modules, then it is easily checked that there is a unique R-module structure on $\varinjlim X_i$, such that we have $\lambda\,[x_i, i] = [\lambda x_i, i]$ for all $\lambda \in R,\ \ i \in I$ and $x_i \in X_i$. Hence the inductive limit of an inductive system of R-modules [k-algebras] is again an R-module [a k-algebra].

3. Let $(X_i, \psi_{j,i})_I,\ (X'_i, \psi'_{j,i})_I$ be inductive systems over the directed set $(I, \leq)$, and let $\psi_j \colon X_j \to \varinjlim X_i,\ \psi'_j \colon X'_j \to \varinjlim X'_i$ for $j \in I$ be the injections into the inductive limits. Let $(\varphi_i \colon X_i \to X'_i)_{i\in I}$ be a family of morphisms such that $\varphi_i \circ \psi_{j,i} = \psi'_{j,i} \circ \varphi_j$ for all $i,\ j \in I$ such that $j \leq i$. Then there exists a unique morphism

$$\varphi = \varinjlim \varphi_i \colon \varinjlim X_i \to \varinjlim X'_i,$$

such that $\varphi \circ \psi_j = \psi'_j \circ \varphi$ for all $j \in I$. It is given by

$$\varphi([x_i, i]) = [\varphi_i(x_i), i] \quad \text{for all} \quad i \in I \quad \text{and} \quad x_i \in X_i, \tag{\#}$$

and it induces for all $k \le j$ the commutative diagram

$$\begin{array}{ccccc}
\psi_k\colon X_k & \xrightarrow{\psi_{k,j}} & X_j & \xrightarrow{\psi_j} & \varinjlim X_i \\
\varphi_k \big\downarrow & & \varphi_j \big\downarrow & & \big\downarrow \varinjlim \varphi_i . \\
\psi'_k\colon X'_k & \xrightarrow{\psi'_{k,j}} & X'_j & \xrightarrow{\psi'_j} & \varinjlim X'_i
\end{array}$$

For the proof, we define φ by means of (#) and observe that this is independent of the representatives. Indeed, if $[x_i, i] = [x_j, j]$, then $\psi_{i,k}(x_i) = \psi_{j,k}(x_j)$ for some $k \ge i,\ \ k \ge j$, it follows that

$$\psi'_{i,k} \circ \varphi_i(x_i) = \varphi_k \circ \psi_{i,k}(x_i) = \varphi_k \circ \psi_{j,k}(x_j) = \psi'_{j,k} \circ \varphi_j(x_j),$$

and therefore $[\varphi(x_i), i] = [\varphi(x_j), j]$. By definition, $\varphi \circ \psi_j = \psi'_j \circ \varphi$ for all $j \in I$, and it is easily checked that φ is a morphism of the objects in question. The uniqueness of φ follows since $\varinjlim X_i$ is the union of all sets $\varphi_j(X_j)$.

Assume that $(X''_i, \psi''_{j,i})_I$ is another inductive system over $(I, \le)$ and $(\varphi'_i\colon X'_i \to X''_i)_{i \in I}$ is a family of morphisms such that $\varphi'_i \circ \psi'_{j,i} = \psi''_{j,i} \circ \varphi'_j$ for all $i,\ j \in I$ such that $j \le i$. Then

$$\varinjlim \varphi'_i \circ \varinjlim \varphi_i = \varinjlim (\varphi'_i \circ \varphi_i).$$

In particular, if $(\varphi_i)_{i \in I}$ is a family of isomorphisms, then $\varinjlim \varphi_i$ is also an isomorphism, and $(\varinjlim \varphi_i)^{-1} = \varinjlim \varphi_i^{-1}$.

We emphasise the special case where $X' = X'_i$ and $\psi'_{j,i} = \mathrm{id}_{X'}$ for all $i,\ j \in I$ such that $j \le i$. In this case we obtain the following *universal property*:

Let $(X_i, \psi_{j,i})_I$ be an inductive system over the directed set $(I, \le)$, and for $j \in I$ let $\psi_j\colon X_j \to \varinjlim X_i$ be the injection into the inductive limit. Let X' be another object and $(\varphi_i\colon X_i \to X')_{i \in I}$ a family of morphisms such that $\varphi_j = \varphi_i \circ \psi_{j,i}$ for all $i,\ j \in I$ such that $j \le i$. Then there exists a unique morphism

$$\varphi = \varinjlim \varphi_i\colon \varinjlim X_i \to X',$$

such that $\varphi \circ \psi_j = \varphi_j$ for all $j \in I$. It is given by $\varphi([x_i, i]) = \varphi_i(x_i)$ for $i \in I$ and $x_i \in X_i$. For all $j \le k$, it induces the commutative diagram

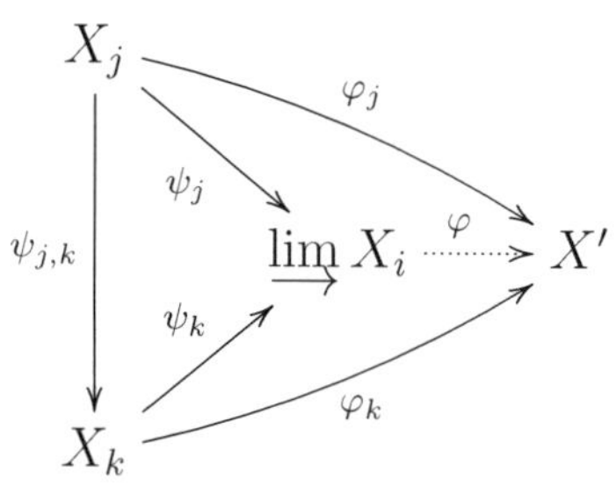

4. Let $(X_i, \psi_{j,i})_I$ be an inductive system over the directed set $(I, \le)$, and let I' be a cofinal subset of I. We set

$$X = \varinjlim_{i\in I} X_i = \{[x_i, i] \mid i \in I,\ x_i \in X_i\},$$
$$X' = \varinjlim_{i\in I'} X_i = \{[x_i, i]' \mid i \in I',\ x_i \in X_i\},$$

and we assert that there is an isomorphism $\Theta\colon X \to X'$ such that $\Theta([x_i, i]') = [x_i, i]$ for all $i \in I'$ and $x_i \in X_i$.

Obviously, $[x_i, i]' = [x_j, j]'$ implies $[x_i, i] = [x_j, j]$ for all $i,\ j \in I',\ x_i \in X_i$ and $x_j \in X_j$. Hence there exists a unique map Θ as asserted, and for all algebraic structures Θ is obviously a homomorphism.

Θ is injective: Suppose that $[x_i, i] = [x_j, j]$ for some $i,\ j \in I',\ x_i \in X_i$ and $x_j \in X_j$. Let $k \in I$ be such that $k \ge i,\ k \ge j$ and $\psi_{i,k}(x_i) = \psi_{j,k}(x_j)$, and let $k' \in I'$ be such that $k' \ge k$. Then we get $\psi_{i,k'}(x_i) = \psi_{k,k'} \circ \psi_{i,k}(x_i) = \psi_{k,k'} \circ \psi_{j,k}(x_j) = \psi_{j,k'}(x_j)$ and therefore $[x_i, i]' = [x_j, j]'$.

Θ is surjective: Let $[x_i, i] \in X$, where $i \in I$ and $x_i \in X_i$. Let $k \in I'$ be such that $k \ge i$. Then $[\psi_{i,k}(x_i), k]' \in X'$, and

$$\Theta([\psi_{i,k}(x_i), k]') = [\psi_{i,k}(x_i), k] = [x_i, i].$$

Hence Θ is an isomorphism for all algebraic structure, and we identify: $X = X'$. In particular, if I possesses a greatest element m, then $\varinjlim X_i = X_m$.

It remains to prove that in the case of topological spaces the topology induced on X by the family of injections $(\psi_i\colon X_i \to X)_{i\in I}$ coincides with the topology induced by the subfamily $(\psi_i\colon X_i \to X)_{i\in I'}$. For a subset U of X we must prove that $\psi_i^{-1}(U)$ is open in X_i for all $i \in I$ if and only if this holds for all $i \in I'$. Of course, if $\psi_i^{-1}(U)$ is open in X_i for all $i \in I$, then this is true for all $i \in I'$. Thus suppose that $\psi_j^{-1}(U)$ is open in X_j for all $j \in I'$, and let $i \in I$. If $j \in I'$ is such that $j \ge i$, then $\psi_j^{-1}(U)$ is open in X_j, and therefore $\psi_i^{-1}(U_i) = \psi_{i,j}^{-1}(\psi_j^{-1}(U))$ is open in X_i.

Examples 1.3.2. 1. Let X be an object and $\Sigma = \{X_i \mid i \in I\}$ an upwards directed set of subobjects such that $i \ne j$ implies $X_i \ne X_j$ for all $i,\ j \in I$. For $i,\ j \in I$ we define $j \le i$ if $X_j \subset X_i$, and $\psi_{j,i} = (X_j \hookrightarrow X_i)$. Then $(I, \le)$ is a directed set, $(X_i, \psi_{j,i})_I$ is an injective inductive system, and

$$\bigcup_{i\in I} X_i = \varinjlim_{i\in I} X_i.$$

Without using the (artificial) indexing, we simply write

$$(\Sigma, \le) = (X_i, \psi_{j,i})_I \quad \text{and} \quad \varinjlim_{E\in\Sigma} E = \bigcup_{E\in\Sigma} E.$$

Particular special cases:

- X is an abelian group [an R-module], and Σ is the (upwards directed) set of all finitely generaed subroups [finitely generated R-submodules].

- X is an abelian torsion group, and Σ is the (upwards directed) set of all finite subgroups.
  Conversely, if $(A_i, \psi_{j,i})_I$ is any inductive system of abelian torsion groups, then $A = \varinjlim A_i$ is an abelian torsion group (indeed, if $(\psi_i\colon A_i \to A)_{i\in I}$ is the family of injections, then $A = \{\psi_i(a_i) \mid i \in I,\ a_i \in A_i\}$ consists of elements of finite order).

- $X = k[\boldsymbol{X}]$ is a polynoial ring in an arbitrary family $\boldsymbol{X} = (X_i)_{i\in I}$ of indeterminates over a commutative ring k, and Σ is the (upwards directed) set of all polynomial rings $k[\boldsymbol{X}']$ in a finite subfamily $\boldsymbol{X}'$ of $\boldsymbol{X}$.

2. Let $(A_i)_{i\in I}$ be a family of R-modules,

$$A = \bigoplus_{i\in I} A_i, \quad \text{and for a subset } E \text{ of } I \text{ we set } A_E = \bigoplus_{i\in E} A_i$$

(so that in particular $A = A_I$). For $E,\ E' \in \mathbb{P}(I)$ with $E' \subset E$, we define

$$\psi_{E',E}\colon A_{E'} \to A_E \quad \text{by} \quad \psi_{E',E}((a_i)_{i\in E'}) = (a_i')_{i\in E}, \quad \text{where}$$
$$a_i' = \begin{cases} a_i & \text{if } i \in E', \\ 0 & \text{otherwise.} \end{cases}$$

Then $(\mathbb{P}_{\text{fin}}(I), \subset)$ is a directed set, and $(A_E, \psi_{E',E})_{\mathbb{P}_{\text{fin}}(I)}$ is an injective inductive system. For $E \in \mathbb{P}_{\text{fin}}(I)$ we set $\psi_E = \psi_{E,I}\colon A_E \to A$, and we assert that

$$\varinjlim \psi_E\colon \varinjlim_{E\in\mathbb{P}_{\text{fin}}(I)} A_E \to A$$

is an isomorphism. Indeed, if $E \in \mathbb{P}_{\text{fin}}(I)$, then $A_E' = \psi_E(A_E) = A_E \oplus \mathbf{0} \subset A$, and $(\psi_E\colon A_E \to A_E')_{E\in\mathbb{P}_{\text{fin}}}$ is a family of isomorphisms which induces an isomorphism

$$\varinjlim \psi_E\colon \varinjlim_{E\in\mathbb{P}_{\text{fin}}(I)} A_E \xrightarrow{\sim} \varinjlim_{E\in\mathbb{P}_{\text{fin}}(I)} A_E' = \bigcup_{E\in\mathbb{P}_{\text{fin}}(I)} A_E' = A.$$

We tacitly identify:

$$A = \varinjlim_{E\in\mathbb{P}_{\text{fin}}(I)} A_E.$$

3. $(\mathbb{N}, \mid)$, the positive integers equipped with the divisor relation (i. e., $m \le n$ if $m \mid n$), is a directed set. If $m,\ n \in \mathbb{N}$ are such that $m \mid n$, we define

$$\psi_{m,n}\colon \mathbb{Z}/m\mathbb{Z} \to \mathbb{Z}/n\mathbb{Z} \quad \text{by} \quad \psi_{m,n}(a + m\mathbb{Z}) = \frac{n}{m}a + n\mathbb{Z} \ \text{ for all } \ a \in \mathbb{Z}.$$

Then $(\mathbb{Z}/m\mathbb{Z}, \psi_{m,n})_{\mathbb{N}}$ is an inductive system. For $n \in \mathbb{N}$ we define

$$\varphi_n \colon \mathbb{Z}/n\mathbb{Z} \to \frac{1}{n}\mathbb{Z}/\mathbb{Z} \quad \text{by} \quad \varphi_n(a+n\mathbb{Z}) = \frac{a}{n} + \mathbb{Z} \ \text{ for all } \ a \in \mathbb{Z}.$$

Then φ_n is an isomorphism, and if $m, n \in \mathbb{N}$ are such that $m \,|\, n$, then

$$\varphi_n \circ \psi_{m,n} = \varphi_m \circ \iota_{m,n}, \quad \text{where} \quad \iota_{m,n} = \Bigl(\frac{1}{m}\mathbb{Z}/\mathbb{Z} \hookrightarrow \frac{1}{n}\mathbb{Z}/\mathbb{Z}\Bigr).$$

Hence 1.3.1.**3** yields an isomorphism

$$\varinjlim \varphi_n \colon \varinjlim \mathbb{Z}/n\mathbb{Z} \stackrel{\sim}{\to} \varinjlim \frac{1}{n}\mathbb{Z}/\mathbb{Z} = \bigcup_{n \in \mathbb{N}} \frac{1}{n}\mathbb{Z}/\mathbb{Z} = \mathbb{Q}/\mathbb{Z}.$$

4. Let p be a prime. We restrict the above considerations to p-powers. The group

$$\mathbb{Z}(p^\infty) = \bigcup_{n \in \mathbb{N}} \frac{1}{p^n}\mathbb{Z}/\mathbb{Z} \quad \text{is called } \textbf{\textit{p}-quasicyclic} \text{ or the } \textbf{Prüfer \textit{p}-group}.$$

Apparently, $\mathbb{Z}(p^\infty) = (\mathbb{Q}/\mathbb{Z})_p$ is the p-component of $\mathbb{Q}/\mathbb{Z}$, and the decomposition

$$\mathbb{Q}/\mathbb{Z} = \bigoplus_{p \in \mathbb{P}} (\mathbb{Q}/\mathbb{Z})_p$$

is the abstract form of the partial fraction expansion (see [18, Theorem 2.3.1]). Since (up to identifications by means of canonical isomorphism)

$$\frac{1}{p^n}\mathbb{Z}/\mathbb{Z} = \frac{1}{p^n}\mathbb{Z}_{(p)}/\mathbb{Z}_{(p)} = \frac{1}{p^n}\mathbb{Z}_p/\mathbb{Z}_p, \quad \mathbb{Q} = \bigcup_{n \in \mathbb{N}} \frac{1}{p^n}\mathbb{Z}_{(p)} \ \text{ and } \ \mathbb{Q}_p = \bigcup_{n \in \mathbb{N}} \frac{1}{p^n}\mathbb{Z}_p,$$

we obtain

$$\mathbb{Z}(p^\infty) = \mathbb{Q}/\mathbb{Z}_{(p)} = \mathbb{Q}_p/\mathbb{Z}_p, \quad \text{if we identify } \ \frac{a}{p^n} + \mathbb{Z} = \frac{a}{p^n} + \mathbb{Z}_{(p)} = \frac{a}{p^n} + \mathbb{Z}_p.$$

If K is an algebraically closed field and $\operatorname{char}(K) \ne p$, then

$$\mu_{p^\infty}(K) = \bigcup_{n \in \mathbb{N}} \mu_{p^n}(K) \ \text{ and } \ \mu_{p^n}(K) \cong \frac{1}{p^n}\mathbb{Z}/Z \ \text{ implies } \ \mu_{p^\infty}(K) \cong (\mathbb{Q}/\mathbb{Z})_p.$$

$\mu_{p^\infty}(K)$ is the p-component of $\mu(K)$, and if $\operatorname{char}(K) = 0$, then $\mu(K) \cong \mathbb{Q}/Z$.

Theorem 1.3.3. *Let $(X_i, \psi_{j,i})_I$, $(X_i', \psi_{j,i}')_I$, $(X_i'', \psi_{j,i}'')_I$ be inductive systems of R-modules over the directed set $(I, \le)$.*

1. Suppose that for every $i \in I$ there is an exact sequence

$$X_i' \overset{f_i}{\to} X_i \overset{g_i}{\to} X_i''$$

such that for all $i, j \in I$ with $i \geq j$ the following diagram is commutative:

$$\begin{array}{ccccc} X_j' & \xrightarrow{f_j} & X_j & \xrightarrow{g_j} & X_j'' \\ {\scriptstyle \psi_{j,i}'}\downarrow & & {\scriptstyle \psi_{j,i}}\downarrow & & \downarrow{\scriptstyle \psi_{j,i}''} \\ X_i' & \xrightarrow{f_i} & X_i & \xrightarrow{g_i} & X_i'' \end{array}$$

Then the sequence of inductive limits

$$\varinjlim_{i \in I} X_i' \overset{\varinjlim f_i}{\longrightarrow} \varinjlim_{i \in I} X_i \overset{\varinjlim g_i}{\longrightarrow} \varinjlim_{i \in I} X_i''$$

is exact.

2. Let $(f_i \colon X_i' \to X_i)_{i \in I}$ be a family of R-homomorphisms such that $\psi_{j,i} \circ f_j = f_i \circ \psi_{j,i}'$ for all $i \geq j$.

If $i \geq j$, then $\psi_{j,i}'(\mathrm{Ker}(f_j)) \subset \mathrm{Ker}(f_i)$, $\psi_{j,i}(\mathrm{Im}(f_j)) \subset \mathrm{Im}(f_i)$, and we define $\psi_{j,i}^* \colon \mathrm{Coker}(f_j) \to \mathrm{Coker}(f_i)$ by

$$\psi_{j,i}^*(x_j + \mathrm{Im}(f_j)) = \psi_{j,i}(x_j) + \mathrm{Im}(f_i).$$

Then

$$(\mathrm{Ker}(f_i), \psi_{j,i}' \restriction \mathrm{Ker}(f_j))_I, \quad (\mathrm{Im}(f_i), \psi_{j,i} \restriction \mathrm{Im}(f_j))_I \ \text{ and } \ (\mathrm{Coker}(f_i), \psi_{j,i}^*)_I$$

are inductive systems, and for $i \geq j$ we obtain the commutative diagram

$$\begin{array}{ccccccccc} \mathbf{0} & \longrightarrow & \mathrm{Ker}(f_j) & \xrightarrow{\nu_j} & X_j' & \xrightarrow{\varphi_j} & X_j & \xrightarrow{\pi_j} & \mathrm{Coker}(f_j) & \longrightarrow & \mathbf{0} \\ & & {\scriptstyle \psi_{j,i} \restriction \mathrm{Ker}(f_j)}\downarrow & & {\scriptstyle \psi_{j,i}'}\downarrow & & \downarrow{\scriptstyle \psi_{j,i}} & & \downarrow{\scriptstyle \psi_{j,i}^*} & & \\ \mathbf{0} & \longrightarrow & \mathrm{Ker}(f_i) & \xrightarrow{\nu_i} & X_i' & \xrightarrow{\varphi_i} & X_i & \xrightarrow{\pi_i} & \mathrm{Coker}(f_i) & \longrightarrow & \mathbf{0} \end{array}$$

where ν_i, ν_j are the inclusions, and π_i, π_j are the residue class homomorphisms. It induces the exact sequence

$$\mathbf{0} \to \varinjlim_{i \in I} \mathrm{Ker}(f_i) \overset{\varinjlim \nu_i}{\longrightarrow} \varinjlim_{i \in I} X_i' \overset{\varinjlim f_i}{\longrightarrow} \varinjlim_{i \in I} X_i \overset{\varinjlim \pi_i}{\longrightarrow} \varinjlim_{i \in I} \mathrm{Coker}(f_i) \to \mathbf{0}.$$

In particular, we obtain isomorphisms

$$\varinjlim_{i \in I} \nu_i \colon \varinjlim_{i \in I} \mathrm{Ker}(f_i) \overset{\sim}{\to} \mathrm{Ker}(\varinjlim f_i) \quad \text{and} \quad \mathrm{Coker}(\varinjlim f_i) \overset{\sim}{\to} \varinjlim_{i \in I} \mathrm{Coker}(f_i).$$

If all f_i are isomorphisms, then $\varinjlim f_i$ is also an isomorphism.

We identify:

$\varinjlim \mathrm{Ker}(f_i) = \mathrm{Ker}(\varinjlim f_i)$ *and* $\mathrm{Coker}(\varinjlim f_i) = \varinjlim \mathrm{Coker}(f_i)$.

Proof. 1. Let $f = \varinjlim f_i$ and $g = \varinjlim g_i$. If $[x'_j, j] \in \varinjlim X'_i$, where $j \in I$ and $x'_j \in X'_j$, then $g \circ f([x'_j, j]) = [g_j \circ f_j(x_j), j] = [0, j] = 0 \in \varinjlim X''_i$, and thus $\mathrm{Im}(f) \subset \mathrm{Ker}(g)$. For the reverse inclusion, let $[x_j, j] \in \mathrm{Ker}(g) \subset \varinjlim X_i$, where $j \in I$ and $x_j \in X_j$. Then $0 = [0, j] = g([x_j, j]) = [g_j(x_j), j]$, and there exists some $k \ge j$ such that $0 = \psi''_{j,k} \circ g_j(x_j) = g_k \circ \psi_{j,k}(x_j)$. Hence $\psi_{j,k}(x_j) \in \mathrm{Ker}(g_k) = \mathrm{Im}(f_k)$, and if $x'_k \in X'_k$ is such that $\psi_{j,k}(x_j) = f_k(x'_k)$, then $[x_j, j] = [\psi_{j,k}(x_j), k] = f([x'_k, k])$, and thus $[x_j, j] \in \mathrm{Im}(f)$.

2. The assertions are easily checked using 1. □

We proceed with the concept of projective limits. From a categorial point of view, it is dual to that of inductive limits. Its concrete description for our categories is however quite different.

Definition and Remarks 1.3.4.

1. A **projective** or **inverse system** (of objects and associated morphisms) over a directed set $(I, \le)$ is a pair

$$(X_i, \phi_{i,j})_I = \big((X_i)_{i \in I},\, (\phi_{i,j} \colon X_i \to X_j)_{i \ge j}\big),$$

consisting of a family of objects $(X_i)_{i \in I}$, together with a family of associated morphisms $(\phi_{i,j} \colon X_i \to X_j)_{i \ge j}$ with the following properties:

- $\phi_{i,i} = \mathrm{id}_{X_i}$ for all $i \in I$;
- $\phi_{i,k} = \phi_{j,k} \circ \phi_{i,j}$ for all $i,\ j,\ k \in I$ such that $i \ge j \ge k$.

A projective system $(X_i, \phi_{i,j})_I$ is called **surjective** if all $\phi_{i,j}$ are surjective.

Let $(X_i, \phi_{i,j})_I$ be a projective system over the directed set $(I, \le)$. We set

$$P = \prod_{i \in I} X_i \quad \text{and} \quad X = \Big\{ (x_i)_{i \in I} \in P \;\Big|\; \phi_{i,j}(x_i) = x_j \text{ for all } i,\ j \in I \text{ with } i \ge j \Big\}.$$

We call X the **projective** or **inverse limit** of the family $(X_i)_{i \in I}$ (with respect to the morphisms $\phi_{i,j}$), and we set

$$X = \varprojlim\,(X_i, \phi_{i,j})_I = \varprojlim_{i \in I} X_i, \qquad \text{or simply} \quad X = \varprojlim X_i.$$

The families $(x_i)_{i \in I} \in X$ are called the **threads** of the projective system.

For $j \in I$ let $p_j \colon P \to X_j$ be the projection, given by $p_j((x_i)_{i \in I}) = x_j$, and let $\phi_j = p_j \restriction \varprojlim X_i \colon \varprojlim X_i \to X_j$. The maps $\phi_j \colon X \to X_j$ are called the **projections** of the projective limit. Obviously, $\phi_{i,j} \circ \phi_i = \phi_j$ for all $i,\ j \in I$ such that $i \ge j$.

If $(X_i, \phi_{i,j})_I$ is a family of objects and morphisms (in our sense), then P is again an object; it is easily checked that $X = \varprojlim X_i$ is a subobject of P, and the projections ϕ_j are morphisms. In this way, the projective limit of a procective system of topological spaces [(abelian) monoids, (abelian) groups, topological groups, (commutative) rings, R-module, k-algebras] is again a topological space [an (abelian) monoid, an (abelian) group, a topological group, a (commutative) ring, an R-module, a k-algebra].

2. Let $(X_i, \phi_{i,j})_I$ be a projective system over the directed set $(I, \leq)$, $X = \varprojlim X_i$ and $(\phi_i \colon X \to X_i)_{i \in I}$ the family of projections. For $j \in I$ let $X'_j = \phi_j(X) \subset X_j$. If $i,\ j \in I$ and $i \geq j$, then

$$\phi_{i,j}(X'_i) = \phi_{i,j} \circ \phi_i(X) = \phi_j(X) = X'_j,$$

and we define $\phi'_{i,j} = \phi_{i,j} \restriction X'_i \colon X'_i \to X'_j$. By definition, $(X'_i, \phi'_{i,j})_I$ is a surjective projective system, $X = \varprojlim X'_i$, and the projections $\phi_j \colon X \to X'_j$ are surjective.

3. Let $(X_i, \phi_{i,j})_I$ and $(X'_i, \phi'_{i,j})_I$ be projective systems over the directed set $(I, \leq)$ with projective limits $X = \varprojlim X_i$ and $X' = \varprojlim X'_i$. For $j \in J$ let $\phi_j \colon X \to X_j$ and $\phi'_j \colon X' \to X'_j$ be the projections of the projective limits. Let $(\varphi_i \colon X'_i \to X_i)_{i \in I}$ be a family of morphisms such that $\varphi_j \circ \phi'_{i,j} = \phi_{i,j} \circ \varphi_i$ for all $i \geq j$. Then there exists a unique morphism

$$\varphi = \varprojlim \varphi_i \colon X' \to X \quad \text{such that} \quad \phi_j \circ \varphi = \varphi_j \circ \phi'_j \quad \text{for all} \quad j \in I,$$

and for all $i,\ j \in I$ such that $i \geq j$ we obtain the commutative diagram

$$\begin{array}{rcccc} \phi'_j \colon X' & \xrightarrow{\ \phi'_i\ } & X'_i & \xrightarrow{\ \phi'_{i,j}\ } & X'_j \\ {\scriptstyle \varphi = \varprojlim \varphi_i} \downarrow & & {\scriptstyle \varphi_i} \downarrow & & \downarrow {\scriptstyle \varphi_j} \\ \phi_j \colon X & \xrightarrow{\ \phi_i\ } & X'_j & \xrightarrow{\ \phi_{i,j}\ } & X_j \,. \end{array}$$

Indeed, for a thread $(x_i)_{i \in I} \in X'$ it is easily checked that $(\varphi_i(x_i))_{i \in I} \in X$, and we define $\varphi \colon X' \to X$ by $\varphi((x_i)_{i \in I}) = (\varphi_i(x_i))_{i \in I}$ for all threads $(x_i)_{i \in I} \in X'$. Then φ is a morphism with the required properties.

If $(X''_i, \phi''_{i,j})_I$ is another projective system over $(I, \leq)$, and if $(\varphi'_i \colon X''_i \to X'_i)_{i \in I}$ is a family of morphisms such that $\varphi'_j \circ \phi''_{i,j} = \phi'_{i,j} \circ \varphi'_i$ for all $i,\ j \in I$ such that $i \geq j$, then

$$\varprojlim \varphi_i \circ \varprojlim \varphi'_i = \varprojlim (\varphi_i \circ \varphi'_i).$$

If $(\varphi_i)_{i \in I}$ is a family of isomorphisms, then $\varprojlim \varphi_i$ is also an isomorphism, and obviously $(\varprojlim \varphi_i)^{-1} = \varprojlim \varphi_i^{-1}$. In particular, if $X'_i \subset X_i$ and $\varphi_i = (X'_i \hookrightarrow X_i)$ for all $i \in I$, then $X' \subset X$ and $\varphi = \varprojlim \varphi_i = (X' \hookrightarrow X)$.

We emphasise the special case where $X' = X'_i$ and $\phi'_{i,j} = \mathrm{id}_{X'}$ for all $i,\ j \in I$ such that $i \geq j$. In this case we obtain the following *universal property* of the projective limit:

Let $(X_i, \phi_{i,j})_I$ be a projective system over the directed set $(I, \leq)$, and for $j \in I$ let $\phi_j \colon \varprojlim X_i \to X_j$ be the projection of the projective limit. Let X' be any object and $(\varphi_j \colon X' \to X_j)_{j \in I}$ a family of morphisms such that $\varphi_j = \phi_{i,j} \circ \varphi_i$ for all $i \geq j$. Then there exists a unique morphism

$$\varphi = \varprojlim \varphi_i \colon X' \to \varprojlim X_i$$

such that $\phi_j \circ \varphi = \varphi_j$ for all $j \in I$. It is given by $\varphi(x) = (\varphi_i(x_i))_{i \in I}$ for all $x \in X'$, and for all $i \geq j$, it induces the commutative diagram

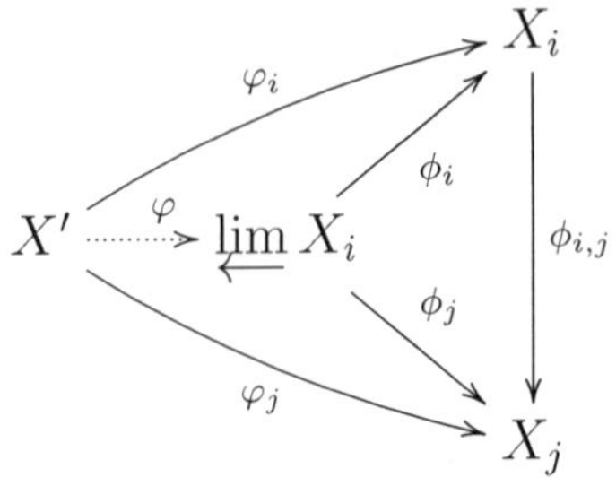

4. Let $(X_i, \phi_{i,j})_I$ be a projective system over the directed set $(I, \leq)$ such that

$$X_i = \prod_{\lambda \in \Lambda} X_{i,\lambda} \text{ for all } i \in I \text{ and } \phi_{i,j} = (\phi_{i,j,\lambda} \colon X_{i,\lambda} \to X_{j,\lambda})_{\lambda \in \Lambda} \text{ for all } i \geq j.$$

Then every thread $x \in X$ is of the form $x = ((x_{i,\lambda})_{\lambda \in \Lambda})_{i \in I} = ((x_{i,\lambda})_{i \in I})_{\lambda \in \Lambda}$, and consequently

$$\varprojlim_{i \in I} X_i = \varprojlim_{i \in I} \prod_{\lambda \in \Lambda} X_{i,\lambda} = \prod_{\lambda \in \Lambda} \varprojlim_{i \in I} X_{i,\lambda}.$$

5. Let $(X_i)_{i \in I}$ be a family of objects,

$$P = \prod_{i \in I} X_i, \quad \text{and for } F \in \mathbb{P}_{\text{fin}}(I) \text{ set } P_F = \prod_{i \in F} X_i.$$

We consider the upwards directed set $(\mathbb{P}_{\text{fin}}(I), \subset)$. If F, $F' \in \mathbb{P}_{\text{fin}}(I)$ are such that $F \subset F'$, then $P_{F'} = P_F \times P_{F' \setminus F}$, and we denote by $\phi_{F',F} \colon P_{F'} \to P_F$ the projection. Then $(P_F, \phi_{F',F})_{\mathbb{P}_{\text{fin}}(I)}$ is a projective system, and (obviously) the map

$$\Theta \colon P \to \varprojlim_{F \in \mathbb{P}_{\text{fin}}(I)} P_F, \quad \text{given by} \quad \Theta((x_i)_{i \in I}) = ((x_i)_{i \in F})_{F \in \mathbb{P}_{\text{fin}}(I)}$$

is an isomorphism (this is a counterpart to 1.3.2.**3**). We identify: $P = \varprojlim P_F$.

Theorem 1.3.5. *Let $(X_i, \phi_{i,j})_I$ be a projective system of topological spaces over the directed set $(I, \leq)$, $X = \varprojlim X_i$, $(\phi_i \colon X \to X_i)_{i \in I}$ the family of projections, and*

$$P = \prod_{i \in I} X_i.$$

1. *Let I' be a cofinal subset of I, and for every $i \in I'$ let $\mathfrak{B}_i$ be a basis of the topology of X_i. Then $\mathfrak{B} = \{\phi_j^{-1}(B_j) \mid j \in I',\ B_j \in \mathfrak{B}_j\}$ is a basis of the topology of X.*

2. If all X_i are totally disconnected, then X is totally disconnected.

3. If all X_i are Hausdorff, then X is Hausdorff, and X is closed in P.

4. If all X_i are compact and not empty, then X is compact, not empty, and

$$\phi_k(X) = \bigcap_{i \geq k} \phi_{i,k}(X_i) \quad \textit{for all} \ \ k \in I.$$

Proof. We may assume that all X_i are not empty.

1. The set

$$\mathfrak{B}^* = \Big\{ \prod_{j\in J} B_j \times \prod_{j \in I\setminus J} X_j \ \Big| \ J \in \mathbb{P}_{\rm fin}(I), \ B_j \in \mathfrak{B}_j \ \text{ for all } \ j \in J \Big\}$$

is a basis of the topology of P, and therefore $\mathfrak{B}_0^* = \{B^* \cap X \mid B^* \in \mathfrak{B}^*\}$ is a basis of the topology of X. We must prove that every set of $\mathfrak{B}_0^*$ is the union of sets in $\mathfrak{B}$. Thus let

$$B^* = \prod_{j\in J} B_j \times \prod_{j\in I\setminus J} X_j \in \mathfrak{B}^*, \ \text{ where } \ J \in \mathbb{P}_{\rm fin}(I) \ \text{ and } \ B_j \in \mathfrak{B}_j \ \text{ for all } \ j \in J,$$

and observe that

$$B^* \cap X = \bigcap_{j\in J} \phi_j^{-1}(B_j).$$

Let $i \in I'$ be such that $i \geq j$ for all $j \in J$, and set

$$U_i = \bigcap_{j\in J} \phi_{i,j}^{-1}(B_j) \subset X_i.$$

Then U_i is open in X_i,

$$\phi_i^{-1}(U_i) = \bigcap_{j\in J} (\phi_{i,j}\circ\phi_i)^{-1}(B_j) = \bigcap_{j\in J} \phi_j^{-1}(B_j) = B^* \cap X.$$

and there exists a family $(B_{i,\lambda})_{\lambda\in\Lambda}$ in $\mathfrak{B}_i$ such that

$$U_i = \bigcup_{\lambda\in\Lambda} B_{i,\lambda}, \quad \text{and then} \quad B^* \cap X = \phi_i^{-1}(U_i) = \bigcup_{\lambda\in\Lambda} \phi_i^{-1}(B_{i,\lambda}).$$

2. Suppose that all X_i are totally disconnected. Let $x = (x_i)_{i\in I} \in X$ and C the component of x. If $i \in I$, then $\phi_i(C)$ is connected, hence $\phi_i(C) = \{x_i\}$, and consequently $C = \{x\}$.

3. and 4. Assume that all X_i are Hausdorff. Then P is Hausdorff, and thus X is Hausdorff. If $i,\ j \in I$ and $i \ge j$, then

$$G_{i,j} = \{(x_i,\ x_j) \in X_i \times X_j \mid \phi_{i,j}(x_i) = x_j\} \text{ (the graph of } \phi_{i,j})$$

is closed in $X_i \times X_j$, and $G_{i,j} \ne \emptyset$. Hence

$$Y_{i,j} = G_{i,j} \times \prod_{\lambda \in I \setminus \{i,j\}} X_\lambda \quad \text{is closed in } P, \quad \text{and as } X = \bigcap_{\substack{i,\ j \in I \\ i \ge j}} Y_{i,j},$$

it follows that X is closed in P.

Assume now that all X_i are compact. Then P is compact, and therefore X is compact. We assert that every finite subfamily of $(Y_{i,j})_{i \ge j}$ has non-empty finite intersection (then it follows that $X \ne \emptyset$). Let $n \in \mathbb{N}$, $i_1, \ldots, i_n,\ j_1, \ldots, j_n \in I$ and $i_\nu \ge j_\nu$ for all $\nu \in [1, n]$. Let $i \in I$ be such that $i \ge i_\nu$ for all $\nu \in [1, n]$, and let $x_i \in X_i$. For $\nu \in [1, n]$ we set $x_{i_\nu} = \phi_{i,i_\nu}(x_i)$, $x_{j_\nu} = \phi_{i,j_\nu}(x_i) = \phi_{i_\nu,j_\nu}(x_{i_\nu}$, and for every $\lambda \in I \setminus \{i_1, \ldots, i_n,\ j_1, \ldots, j_n\}$ let $x_\lambda \in X_\lambda$ be arbitrary. Then it follows that

$$(x_k)_{k \in I} \in \bigcap_{\nu=1}^{n} Y_{i_\nu, j_\nu}.$$

It remains to prove that

$$\phi_k(X) = \bigcap_{i \ge k} \phi_{i,k}(X_i) \quad \text{for all} \quad k \in I.$$

Let $k \in I$. Then $\phi_k(X) = \phi_{i,k} \circ \phi_i(X) \subset \phi_{i,k}(X_i$ for all $i \ge k$. Thus suppose that $x \in \phi_{i,k}(X_i)$ for all $i \ge k$. For $i \in I$, let $B_i = \varphi_{i,k}^{-1}(x)$ if $i \ge k$, and $B_i = X_i$ otherwise. If $i,\ j \in I,\ \ i \ge j \ge k$, then $\varphi_{j,k} \circ \varphi_{i,j}(B_i) = \varphi_{i,k}(B_i) = B_k = \{x\}$, and therefore $\varphi_{i,j}(B_i) \subset B_j$. Hence $(B_i, \varphi_{i,j} \restriction B_i)_I$ is an projective system of non-empty compact sets, and thus $\emptyset \ne \varprojlim B_i \subset X$. If $b = (b_i)_{i \in I} \in \varprojlim B_i$ and $i \ge k$, then $\phi_k(b) = b_k = x \in \phi_k(X)$. $\square$

Theorem 1.3.6. *Let $(X_i, \phi_{i,j})_I$ be a projective system over the directed set $(I, \le)$, I' a cofinal subset of I,*

$$X = \varprojlim_{i \in I} X_i \quad \textit{and} \quad X' = \varprojlim_{i \in I'} X_i.$$

Then $\Theta\colon X \to X'$, defined by $\Theta((x_i)_{i \in I}) = (x_i)_{i \in I'}$, is an isomorphism.
We identify: $X = X'$.

Proof. Θ is injective: Let $x = (x_i)_{i \in I}$, $x' = (x'_i)_{i \in I} \in X$ and $\Theta(x) = \Theta(x')$, hence $x_i = x'_i$ for all $i \in I'$. For every $i \in I$ there exists some $i' \in I'$ such that $i' \ge i$, and then $x_i = \phi_{i',i}(x_{i'}) = \phi_{i',i}(x'_{i'}) = x'_i$. Hence $x = x'$.

Θ is surjective: Let $x' = (x_{i'})_{i' \in I'} \in X'$. For $i \in I$ let $i' \in I'$ such that $i' \ge i$, and set $x_i = \phi_{i',i}(x_{i'})$. Then x_i does not depend on the choice of i'.

Indeed, if $i'' \in I'$ is another index such that $i'' \geq i$, then there exists some $i^* \in I'$ such that $i^* \geq i'$ and $i^* \geq i''$, and

$$x_i = \phi_{i',i}(x_{i'}) = \phi_{i',i} \circ \phi_{i^*,i'}(x_{i^*}) = \phi_{i^*,i}(x_{i^*}) = \phi_{i'',i} \circ \phi_{i^*,i''}(x_{i^*}) = \phi_{i'',i}(x_{i''}).$$

If $i_1,\ i_2 \in I$ and $i_1 \geq i_2$, let $i' \in I'$ such that $i' \geq i_1$. Then

$$\phi_{i_1,i_2}(x_{i_1}) = \phi_{i_1,i_2} \circ \phi_{i',i_1}(x_{i'}) = \phi_{i',i_2}(x_{i'}) = x_{i_2},$$

hence $x = (x_i)_{i \in I} \in X$ and $x' = \Theta(x)$.

Obviously, Θ is a homomorphism (and thus an isomorphism) for all relevant algebraic structures; we identify. As to topology, let $(\phi_i \colon X \to X_i)_{i \in I}$ be the family of projections. By 1.3.5.1, the families $(\phi_i \colon X \to X_i)_{i \in I}$ and $(\phi_i \colon X \to X_i)_{i \in I'}$ induce the same topology on X. Hence Θ is a homeomorphism. □

Theorem 1.3.7. *Let $(X_i, \phi_{i,j})_I$ be a surjective projective system of topological spaces over the directed set $(I, \leq)$, $X = \varprojlim X_i$, and $(\phi_i \colon X \to X_i)_{i \in I}$ the family of projections. Assume that $\phi_{i,j}^{-1}(x_j)$ is compact for all $i,\ j \in I$ such that $i \geq j$ and all $x_j \in X_j$. Let $k \in I$.*

1. *ϕ_k is surjective, and $\phi_k^{-1}(x_k)$ is compact for all $x_k \in X_k$.*
2. *If $\phi_{j,k}$ is open for all $j \geq k$, then ϕ_k is open.*

Proof. 1. Let $x_k \in X_k$ and identify $X = \varprojlim_{i \geq k} X_i$ according to 1.3.6 by virtue of $(x_i)_{i \in I} = (x_i)_{i \geq k}$. If $i \geq j \geq k$, then $\phi_{i,j}(\phi_{j,k}^{-1}(x_k)) \subset \phi_{i,k}^{-1}(x_k)$, and therefore $(\phi_{i,k}^{-1}(x_i), \phi_{i,j} \restriction \phi_{i,k}^{-1}(x_k))_{i \geq j \geq k}$ is a projective system. We assert that

$$\phi^{-1}(x_k) = \varprojlim_{i \geq k} \phi_{i,k}^{-1}(x_k).$$

Indeed, if $x' = (x'_i)_{i \geq k} \in X$, then

$$x' \in \phi_k^{-1}(x_k) \iff x'_k = x_k \iff x'_i \in \phi_{i,k}^{-1}(x_k) \text{ for all } i \geq k \iff x' \in \varprojlim_{i \geq k} \phi_{i,k}^{-1}(x_i).$$

Since the sets $\phi_{i,k}^{-1}(x_k)$ are compact and not empty, the same is true for $\phi_k^{-1}(x_k)$, and in particular $x_k \in \phi_k(X)$.

2. For $j \geq k$ let $\mathfrak{B}_j$ be a basis of the topology of X_j. Then it suffices to prove that $\phi_k(\phi_j^{-1}(B_j))$ is open for all $j \geq k$ and $B_j \in \mathfrak{B}_j$ (since these sets are a basis for the topology of X). Thus let $j \geq k$, $B_j \in \mathfrak{B}_j$ and $B = \phi_j^{-1}(B_j) \subset X$. By 1., ϕ_j is surjective, and therefore $\phi_k(\phi_j^{-1}(B_j)) = \phi_k(B) = \phi_{j,k} \circ \phi_j(B) = \phi_{j,k}(B_j)$ is open. □

Theorem 1.3.8. *Let $(X_i,\phi_{i,j})_I$, $(X_i',\phi_{i,j}')_I$ be projective systems of compact topological spaces over the directed set $(I,\le)$, $X=\varprojlim X_i$ and $X'=\varprojlim X_i'$.*

1. *Let $(\varphi_i\colon X_i'\to X_i)_{i\in I}$ be a family of continuous surjective maps such that*

$$\phi_{i,j}\circ\varphi_i'=\varphi_j'\circ\phi_{i,j}' \quad \textit{for all}\ \ i,\ j\in I \quad \textit{with}\ \ i\ge j.$$

Then $\varprojlim\varphi_i\colon X'\to X$ is surjective.

2. *Let Y be a compact topological space, and let $(\varphi_i\colon Y\to X_i)_{i\in I}$ be a family of continuous surjective maps such that $\phi_{i,j}\circ\varphi_i=\varphi_j$ for all $i,\ j\in I$ with $i\ge j$. Then $\varprojlim\varphi_i\colon Y\to X$ is surjective.*

3. *Let Y be a closed subset of X. Then $Y=\varprojlim\ \phi_i(Y)$.*

Proof. 1. Let $x=(x_i)_{i\in I}\in X$. Then $\phi_{i,j}'(\varphi_i^{-1}(x_i))\subset\varphi_j^{-1}(x_j)$ for all $i\ge j$, and therefore $(\varphi_i^{-1}(x_i),\phi_{i,j}'\restriction\varphi_i^{-1}(x_i))_I$ is a projective system of non-empty compact topological spaces. Hence $\varprojlim\varphi_i^{-1}(x_i)\ne\emptyset$ by 1.3.5.4, and if $x'=(x_i')_{i\in I}\in\varprojlim\varphi_i^{-1}(x_i)$, then $x'\in X'$, and

$$(\varprojlim\varphi_i)(x')=(\varphi_i(x_i'))_{i\in I}=(x_i)_{i\in I}=x.$$

2. By 1. with $X_i'=Y$ and $\phi_{i,j}'=\mathrm{id}_Y$ for all $i,\ j\in I$ with $i\ge j$.

3. By 2., applied with the projective system $(\phi_i(Y),\phi_{i,j}\restriction\phi_i(Y))_I$ instead of $(X_i,\phi_{i,j})_I$, and with $\varphi_i=\phi_i\restriction Y\colon Y\to\phi_i(Y)$ for all $i\in I$, it follows that the map $\varprojlim\varphi_i=(Y\hookrightarrow\varprojlim\phi_i(Y))$ is surjective, and therefore $Y=\varprojlim\phi_i(Y)$. □

Theorem 1.3.9. *Let $(R_i,\phi_{i,j})_I$ be a projective system of topological rings over the directed set $(I,\le)$ and $R=\varprojlim R_i$. Then $(R_i^\times,\phi_{i,j}\restriction R_i^\times)_I$ is a projective system of topological groups, and $R^\times=\varprojlim R_i^\times$ (with respect to the unit topologies).*

Proof. If $i,\ j\in I$ and $i\ge j$, then $\phi_{i,j}\colon R_i\to R_j$ is a continuous ring homomorphism, and $\phi_{i,j}\restriction R_i^\times\colon R_i^\times\to R_j^\times$ is a continuous group homomorphism with respect to the unit topologies by 1.2.4.3. Hence $(R_i^\times,\phi_{i,j}\restriction R_i^\times)_I$ is a projective system of topological groups, and $\varprojlim R_i^\times\subset R$. We must prove that $\varprojlim R_i^\times=R^\times$, and that the topology of the projective limit is the unit topology.

If $(a_i)_{i\in I}\in R^\times$, then there exists some $(a_i')_{i\in I}\in R$ such that $(a_i)_{i\in I}(a_i')_{i\in I}=(1)_{i\in I}$, hence $a_ia_i'=1$ and therefore $a_i\in R_i^\times$ for all $i\in I$, and $(a_i)_{i\in I}\in\varprojlim R_i^\times$. Conversely, assume that $(a_i)_{i\in I}\in\varprojlim R_i^\times$. If $i,\ j\in I$ and $i\ge j$, then $a_i\in R_i^\times$, and it follows that $\phi_{i,j}(a_i^{-1})=\phi_{i,j}(a_i)^{-1}$. Hence $(a_i^{-1})_{i\in I}\in R$, and since $(a_i)_{i\in I}(a_i^{-1})_{i\in I}=(1)_{i\in I}$, we obtain $(a_i)_{i\in I}\in R^\times$.

Now let $\overline{R}$ be the direct product of the family $(R_i)_{i\in I}$. Then $\overline{R}^\times$ is the product of the family $(R_i^\times)_{i\in I}$, and the product topology on $\overline{R}^\times$ is the unit topology by 1.2.4.4. Since $R\subset\overline{R}$ and $R^\times\subset\overline{R}^\times$, the subspace topology and the unit topology on $R^\times$ coincide by 1.2.4.2. □

1.4 Cauchy sequences and sequentially completeness

In a topological group there is (as in any topological space) the concept of convergent sequences and (as in any uniform space) the concept of Cauchy sequences. For an utmost general theory of convergence we need the notion of filters as in [2] or of Moore-Smith sequences as in [30]. For our purposes, however, the convergence theory with ordinary sequences is sufficient.

Definition and Remarks 1.4.1. We gather some simple facts concerning the basics of the convergence theory of sequences.

1. Let $X = (X, \mathcal{T})$ be a topological space. We say that $\mathcal{T}$ resp. X satisfies the **first axiom of countablility** if every point of X has a denumerable fundamental system of neighbourbhoods. Every metric space satisfies the first axiom of countability. A topological group satisfies the first axiom of countability if 1 has a denumerable fundamental system of neighbourhoods.

2. Let X be a topological space, and for $x \in X$ let $\mathfrak{U}_x$ the set of all neighbourhoods of x.

Let $(x_n)_{n\geq 0}$ a sequence in X. For every infinite subset T of $\mathbb{N}_0$ we denote by $(x_n)_{n\in T}$ the subsequence with index set T. A point $x \in X$ is called

- a **limit poimt** of $(x_n)_{n\geq 0}$ if for every $U \in \mathfrak{U}_x$ there exists some $n_0 \in \mathbb{N}_0$ such that $x_n \in U$ for all $n \geq n_0$. In this case, we say that $(x_n)_{n\geq 0}$ **converges** to x and write $(x_n)_{n\geq 0} \to x$ or $x = \lim_{n\to\infty} x_n$;

- a **cluster point** of $(x_n)_{n\geq 0}$ if for every $U \in \mathfrak{U}_x$ the set $\{n \in \mathbb{N}_0 \mid x_n \in U\}$ is infinite.

If a subsequence of $(x_n)_{n\geq 0}$ converges to x, then x is a cluster point of $(x_n)_{n\geq 0}$. Conversely, if x is a cluser point of $(x_n)_{n\geq 0}$ and has a denumerable fundamental system of neighbourhoods, then x is the limit point of some subsequence of $(x_n)_{n\geq 0}$.

If X is Hausdorff and $(x_n)_{n\geq 0} \to x$, then x is the only cluster point of $(x_n)_{n\geq 0}$. If X is compact, then every sequence in X has a cluster point.

Let A be a subset of X. If x is a limit point of a sequence in A, then $x \in \overline{A}$. Conversely, if $x \in \overline{A}$ and x has a denumerable fundamental systemof neighbourhoods, then x is a limit point of a sequence in A.

3. Let $(X_i)_{i\in I}$ be a family of topological spaces,

$$X = \prod_{i\in I} X_i, \quad (x^{(n)} = (x_i^{(n)})_{i\in I})_{n\geq 0} \text{ a sequence in } X \text{ and } x = (x_i)_{i\in I} \in X.$$

Then $(x^{(n)})_{n\geq 0} \to x$ if and only if $(x_i^{(n)})_{n\geq 0} \to x_i$ for all $i \in I$. If I is finite, then X satisfies the first axiom of countability if and only if all X_i do so.

4. Let X and Y be topological spaces which satisfy the first axiom of countability. A map $f\colon X \to Y$ is continuous if and only if for every $x \in X$ and every sequence $(x_n)_{n\geq 0}$,

$$(x_n)_{n\geq 0} \to x \quad \text{implies} \quad (f(x_n))_{n\geq 0} \to f(x).$$

Using this together with **3.** we obtain the following criteria for topolgical groups, rings and fields.

> **A.** Let G be a group and $\mathcal{T}$ a topology on G satisfying the first axiom of countability. Then $(G,\mathcal{T})$ is a topological group if and only if $(x_n)_{n\geq 0} \to x$ and $(y_n)_{n\geq 0} \to y$ implies $(x_n y_n^{-1})_{n\geq 0} \to xy^{-1}$ for all $x,\ y \in G$ and all sequences $(x_n)_{n\geq 0}$ and $(y_n)_{n\geq 0}$ in G.
>
> **B.** Let R be a ring and $\mathcal{T}$ a topology on R satisfying the first axiom of countability. Then $(R,\mathcal{T})$ is a topological ring if and only if $(x_n)_{n\geq 0} \to x$ and $(y_n)_{n\geq 0} \to y$ implies $(x_n + y_n)_{n\geq 0} \to x + y$ and $(x_n y_n)_{n\geq 0} \to xy$ for all $x,\ y \in R$ and all sequences $(x_n)_{n\geq 0}$ and $(y_n)_{n\geq 0}$ in R.
>
> If R is even a division ring, then $(R,\mathcal{T})$ is a topological division ring if and only if $(R,\mathcal{T})$ is a topological ring and $(x_n)_{n\geq 0} \to x$ implies $(x_n^{-1})_{n\geq 0} \to x^{-1}$ for all $x \in R^\times$ and all sequences $(x_n)_{n\geq 0} \in R^\times$.

Implicity, we used these criteria already in [18, Remarks 5.2.1].

5. Let G be a topological group, $\mathfrak{U}$ a fundamental system of neighbourhoods of 1 in G, $(x_n)_{n\geq 0}$ a sequence in G and $x \in G$. Then $\{xU \mid U \in \mathfrak{U}\}$ and $\{Ux \mid U \in \mathfrak{U}\}$ are fundamental systems of neighbourhoods of x. Hence the following assertions are equivalent by the very definition of convergence:

- $(x_n)_{n\geq 0} \to x$.
- For every $U \in \mathfrak{U}$ there exists some $n_0 \geq 0$ such that $x_n \in xU$ for all $n \geq n_0$ [equivalently, $x_n \in xU$ for $n \gg 1$].
- For every $U \in \mathfrak{U}$ there exists some $n_0 \geq 0$ such that $x_n \in Ux$ for all $n \geq n_0$ [equivalently, $x_n \in Ux$ for $n \gg 1$].

Definition 1.4.2.

1. Let G be a topological group and $\mathfrak{U}$ a fundamental system of neighbourhoods of 1 in G.

   (a) A sequence $(x_n)_{n\geq 0}$ in G is called a **Cauchy sequence** if for every $U \in \mathfrak{U}$ there exists some $n_0 \geq 0$ such that $x_n x_m^{-1} \in U$ and $x_n^{-1} x_m \in U$ for all $m,\ n \geq n_0$.

(b) A subset A of G is called **sequentially complete** if every Cauchy sequence in A has a limit point in A.

2. A topological ring is called **sequentially complete** if its additive group is sequentially complete.

Theorem 1.4.3 (Properties of Cauchy sequences). *Let G be a topological group.*

1. Every convergent sequence in G is a Cauchy sequence.

2. Let $(x_n)_{n\geq 0}$ be a Cauchy sequence in G, and let $x \in G$ be a cluster point of $(x_n)_{n\geq 0}$. Then $(x_n)_{n\geq 0} \to x$.

3. Let $f\colon G \to G'$ be a continuous homomorphism into a topological group G' and $(x_n)_{n\geq 0}$ a Cauchy sequence in G. Then $(f(x_n))_{n\geq 0}$ is a Cauchy sequence in G'.

4. Let $(G_i)_{i\in I}$ be a family of topological groups and

$$G = \prod_{i\in I} G_i.$$

Let $(x^{(n)})_{n\geq 0}$ be a sequence in G and $x^{(n)} = (x_i^{(n)})_{i\in I}$ for all $n \geq 0$. Then $(x^{(n)})_{n\geq 0}$ is a Cauchy sequence if and only if $(x_i^{(n)})_{n\geq 0}$ is a Cauchy sequence for all $i \in I$.

Proof. Let $\mathfrak{U}$ be a fundamental system of neighbourhoods of 1 in G.

1. Let $(x_n)_{n\geq 0}$ be a convergent sequence in G, $x \in G$, $(x_n)_{n\geq 0} \to x$ and $U \in \mathfrak{U}$. Let $V \in \mathfrak{U}$ be such that $VV^{-1} \cup V^{-1}V \subset U$ and $n_0 \geq 0$ such that $x_n \in xV \cap Vx$ for all $n \geq n_0$. If $m,\, n \geq n_0$, then

$$x_n x_m^{-1} = (x_n x^{-1})(x_m x^{-1})^{-1} \in VV^{-1} \subset U$$

and

$$x_n^{-1} x_m = (x^{-1} x_n)^{-1}(x^{-1} x_m) \in V^{-1}V \subset U.$$

Hence $(x_n)_{n\geq 0}$ is a Cauchy sequence.

2. Let $U \in \mathfrak{U}$ and $V \in \mathfrak{U}$ such that $VV \subset U$. Let $n_0 \geq 0$ such that $x_n x_m^{-1} \in V$ and $x_n^{-1} x_m \in V$ for all $m,\, n \geq n_0$, and observe that the set $T = \{n \in \mathbb{N}_0 \mid x_n \in Vx\}$ is infinite. If $n \geq n_0$, then there exists some $m \geq n$ such that $x_m \in Vx$, and we obtain $x_n = (x_n x_m^{-1}) x_m \in VVx \subset Ux$.

3. Let U' be a neighbourhood of the unit element $1'$ of G', $U \in \mathfrak{U}$ such that $f(U) \subset U'$ and $n_0 \geq 0$ such that $x_n x_m^{-1} \in U$ and $x_n^{-1} x_m \in U$ for all $m,\, n \geq n_0$. Then we obtain $f(x_n)f(x_m)^{-1} \in f(U) \subset U'$ and $f(x_n)^{-1} f(x_m) \in f(U) \subset U'$ for all $m,\, n \geq n_0$. Hence $(f(x_n))_{n\geq 0}$ is a Cauchy sequence in G'.

4. Let first $(x^{(n)})_{n\geq 0}$ be a Cauchy sequence. If $i \in I$ and $p_i\colon G \to G_i$ denotes the projection, then p_i is a continuous homomorphism, and the sequence $(x_n^{(i)})_{n\geq 0} = (p_i(x^{(n)}))_{n\geq 0}$ is a Cauchy sequence by 3.

Let now $(x_i^{(n)})_{n\geq 0}$ be a Cauchy sequence for all $i \in I$, and $U \in \mathfrak{U}$. Then there exist a finite subset J of I and for every $j \in J$ a neighbourhood U_j of the unit element 1_j in G_j such that

$$U' = \prod_{j\in J} U_j \times \prod_{i\in I\setminus J} G_i \subset U.$$

For every $j \in J$ let $n_j \geq 0$ such that $x_j^{(n)}(x_j^{(m)})^{-1} \in U_j$ and $(x_j^{(n)})^{-1}x_j^{(m)} \in U_j$ for all $m, n \geq n_j$. If $n_0 = \max\{n_j \mid j \in J\}$, then it follows that $x^{(n)}(x^{(m)})^{-1} \in U' \subset U$ and $(x^{(n)})^{-1}x^{(m)} \in U' \subset U$ for all $m, n \geq n_0$. Hence $(x^{(n)})_{n\geq 0}$ is a Cauchy sequence. □

Theorem 1.4.4 (Sequential completeness).

1. *Let G be a topological group.*

(a) *If G is sequentially complete, then every closed subset of G is sequentially complete.*

(b) *If G is Hausdorff and satisfies the first axiom of countability, then every sequentially complete subset of G is closed.*

2. *Every locally compact topological group is sequentially complete.*

3. *The direct product of a family of sequentially complete topological groups is sequentially complete.*

4. *The projective limit of a projective system of sequentially complete Hausdorff topological groups is sequentially complete.*

Proof. 1. (a) Let G be sequentially complete and A a closed subset of G. If $(x_n)_{n\geq 0}$ is a Cauchy sequence in A, then there is some $x \in G$ such that $(x_n)_{n\geq 0} \to x$, and then $x \in \overline{A} = A$.

(b) Assume that G is Hausdorff and satisfies the first axiom of countability. Let A be a sequentially complete subset of G. If $x \in \overline{A}$, then there exists a sequence in A such that $(x_n)_{n\geq 0} \to x$. Hence $(x_n)_{n\geq 0}$ is a Cauchy sequence in A, and therefore it has a limit point $x' \in A$. Since G is Hausdorff, it follows that $x = x' \in A$.

2. Let G be a locally compact topological group, C a compact neighbourhood of 1 in G and $(x_n)_{n\geq 0}$ a Cauchy sequence in G. Then there exists some $n_0 \geq 0$ such that $y_n = x_n x_{n_0}^{-1} \in C$ for all $n \geq n_0$, and we assert that $(y_n)_{n\geq n_0}$ is a Cauchy sequence in G. Let U be a neighbourhood of 1 in G. Then there exists some neighbourhood V of 1 in G such that $V \subset U$ and $x_{n_0} V x_{n_0}^{-1} \subset U$, and there exists some $n_1 \geq n_0$ such that $x_n x_m^{-1} \in V$ and $x_n^{-1} x_m \in V$ for all $m, n \geq n_1$. Hence $y_n y_m^{-1} = x_n x_{n_0}^{-1} x_{n_0} x_m^{-1} \in V \subset U$ and $y_n^{-1} y_m = x_{n_0} x_n^{-1} x_m x_{n_0}^{-1} \in x_{n_0} V x_{n_0}^{-1} \subset U$ for all $m, n \geq n_0$, and thus $(y_n)_{n\geq n_0}$

is a Cauchy sequence in C. As C is compact, $(y_n)_{n \ge n_0}$ has a cluster point, and thus $(y_n)_{n \ge n_0}$ converges by 1.4.3.2. If $(y_n)_{n \ge n_0} \to y$, then $(x_n)_{n \ge 0} \to y x_{n_0}$.

3. Let $(G_i)_{i \in I}$ be a family of sequentially complete topological groups, G its direct product, $(x^{(n)})_{n \ge 0}$ a Cauchy sequence in G and $x^{(n)} = (x_i^{(n)})_{i \in I}$ for all $n \ge 0$. By 1.4.3.4 the sequences $(x_i^{(n)})_{n \ge 0}$ are Cauchy sequences in G_i and thus convergent for all $i \in I$. Hence $(x^{(n)})_{n \ge 0}$ is convergent, too.

4. Let $(G_i, \phi_{i,j})_I$ be an projective system of sequentially complete Hausdorff topological groups, G its projective limit and P the direct product of $(G_i)_{i \in I}$. Then G is closed in P by 1.3.5.3, and 1. implies that P and thus also G is sequentially complete. □

Let K be a topological field. Then $K^\times$, equipped with the subspace topology, is a topological group. In particular, if $(x_n)_{n \ge 0}$ is a sequence in $K^\times$ and $x \in K^\times$, then $(x_n)_{n \ge 0} \to x$ in $K^\times$ if and only if $(x_n)_{n \ge 0} \to x$ in K. For Cauchy sequences however, the situation is more subtle as the following Theorem 1.4.5 shows.

Theorem 1.4.5. *Let K be a Hausdorff topological field.*

> *1. Let $(x_n)_{n \ge 0}$ be a Cauchy sequence in $K^\times$. Then $(x_n)_{n \ge 0}$ is a Cauchy sequence in K, and 0 is not a cluster point of $(x_n)_{n \ge 0}$.*
>
> *2. If K is sequentially complete, then $K^\times$ is a sequentially complete topological group.*

Proof. Let $\mathfrak{U}$ be a fundamental system of neighbourhoods of 0 in K.

1. Let $U \in \mathfrak{U}$, and suppose that V is a closed neighbourhood of 0 such that $V + VV \subset U$ and $-1 \notin V$. Since $(x_n)_{n \ge 0}$ is a Cauchy sequence in $K^\times$ there exists some $n_0 \ge 0$ such that $x_m^{-1} x_n \in 1 + V$ for all $m, n \ge n_0$. If $n \ge n_0$, then $x_n - x_{n_0} = x_{n_0}(x_{n_0}^{-1} x_n - 1) \in x_{n_0} V$ and therefore $x_n \in x_{n_0} + x_{n_0} V$. Since $x_{n_0} + x_{n_0} V$ is closed and $0 \notin x_{n_0} + x_{n_0} V$ it follows that 0 is not a cluster point of $(x_n)_{n \ge 0}$. Now let $W \in \mathfrak{U}$ be such that $x_0 W \subset V$ and $n_1 \ge n_0$ such that $x_n^{-1} x_m \in 1 + W$ for all $m, n \ge n_1$. Then we obtain

$$\begin{aligned} x_m - x_n &= x_n(x_n^{-1} x_m - 1) \in x_n W = x_{n_0}(x_{n_0}^{-1} x_n) W \subset x_{n_0}(1 + V) W \\ &= x_{n_0} W + x_{n_0} V W \subset V + VV \subset U. \end{aligned}$$

2. Let K be sequentially complete and $(x_n)_{n \ge 0}$ a Cauchy sequence in $K^\times$. By 1., $(x_n)_{n \ge 0}$ is a Cauchy sequence in K, hence convergent, and 0 is not a cluster point of $(x_n)_{n \ge 0}$. Let $x \in K$ be such that $(x_n)_{n \ge 0} \to x$. Then $x \in K^\times$, and as $K^\times$ has the subspace topology, $(x_n)_{n \ge 0}$ converges in $K^\times$. □

Definition and Theorem 1.4.6 (Projective Completion). *Let G be a topological group and Σ a fundamental system of neighbourhoods of* 1 *consisting of open normal subgroups such that*

$$\bigcap_{U\in\Sigma} U = \mathbf{1}.$$

Then G is Hausdorff. For $U \in \Sigma$ let $\varphi_U \colon G \to G/U$ be the residue class epimorphism. If $U, V \in \Sigma$ and $U \subset V$, we define

$$\phi_{U,V} \colon G/U \to G/V \quad \textit{by} \quad \phi_{U,V}(xU) = xV \quad \textit{for all} \quad x \in G.$$

Then $(\Sigma, \supset)$ is a directed set, and if $U, V, W \in \Sigma$ are such that $U \subset V \subset W$, then $\phi_{U,W} = \phi_{V,W} \circ \phi_{U,V}$ and $\phi_{U,V} \circ \varphi_U = \varphi_V$. Hence $(G/U, \phi_{U,V})_\Sigma$ is a projective system of discrete groups. We set

$$\widetilde{G} = \varprojlim_{U\in\Sigma} G/U, \qquad \varphi = \varprojlim_{U\in\Sigma} \varphi_U \colon G \to \widetilde{G},$$

and for $V \in \Sigma$ we denote by $\phi_V \colon \widetilde{G} \to G/V$ be the projection of the projective limit. Then $\varphi(x) = (xU)_{U\in\Sigma}$ for all $x \in G$, and $\phi_V \circ \varphi = \varphi_V$ for all $V \in \Sigma$.

1. *$\widetilde{G}$ is a sequentially complete Hausdorff topological group and does not depend on the individual fundamental system Σ of neighbouhoods of* 1.

2. *$\varphi(G)$ is dense in $\widetilde{G}$, and the map $\varphi \colon G \to \varphi(G)$ is a topological isomorphism.*

3. *Suppose that either all $U \in \Sigma$ are compact, or that Σ is countable, totally ordered by $\supset$, and G is sequentially complete. Then $\varphi(G) = \widetilde{G}$.*

4. *Let H be a subgroup of G, and let $\widetilde{H} = \varprojlim_{U\in\Sigma} HU/U$. Then $\widetilde{H} = \overline{\varphi(H)}$ is a closed subgroup of $\widetilde{G}$, and $\overline{H} = \varphi^{-1}(\widetilde{H})$ is the closure of H in G.*

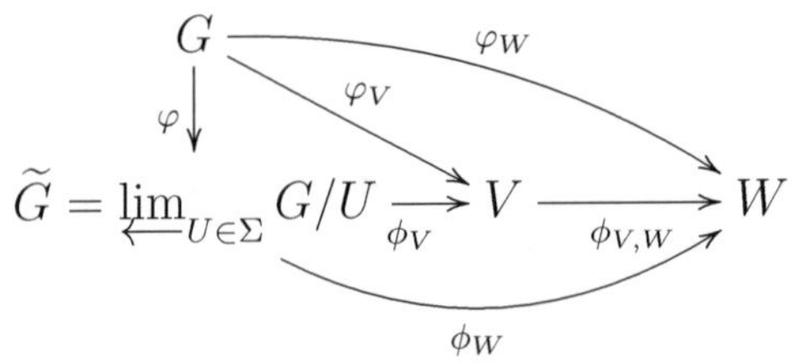

$(V, W \in \Sigma,\ V \subset W)$

The group $\widetilde{G} = \varprojlim_{U\in\Sigma} G/U$ is called the **projective completion** of G with respect to the system Σ. We identify G with $\varphi(G)$ by means of φ.

After this identification, the assertions read as follows:

2.' G is a dense subgroup of $\widetilde{G}$.

3.' If either all $U \in \Sigma$ are compact, or if Σ is countable, totally ordered by $\supset$, and G is sequentially complete, then

$$G = \widetilde{G} = \varprojlim_{U \in \Sigma} G/U.$$

4.' If H is a subgroup of G and $\widetilde{H}$ is the closure of H in $\widetilde{G}$, then $\overline{H} = \widetilde{H} \cap G$ is the closure of H in G.

Proof. $\{1\}$ is closed in G by 1.1.2.3, and thus G is Hausdorff by 1.1.5.5. For any $U, V \in \Sigma$ there exists some $W \in \Sigma$ such that $W \subset U \cap V$, and therefore $(\Sigma, \supset)$ is directed. Apparently, $(G/U, \phi_{U,V})_\Sigma$ is a projective system of discrete groups (see 1.1.5.5), $\phi_{U,V} \circ \varphi_U = \varphi_V$ for all $U, V \in \Sigma$ such that $U \subset V$, $\varphi(x) = (xU)_{U \in \Sigma} \in \widetilde{G}$ for all $x \in G$, and $\phi_V \circ \varphi = \varphi_V \colon G \to G/V$ for all $V \in \Sigma$ by 1.3.4.**3.**

1. Let $\overline{\Sigma}$ be the set of all open normal subgroups of G, ordered as above. Then Σ is cofinal in $\overline{\Sigma}$, and by 1.3.6 we get

$$\widetilde{G} = \varprojlim_{U \in \Sigma} G/U = \varprojlim_{U \in \overline{\Sigma}} G/U.$$

Hence $\widetilde{G}$ does not depend on the individual fundamental system of neighbouhoods Σ. Being discrete, the groups G/U are Hausdorff and sequentially complete, and therefore $\widetilde{G}$ is Hausdorff and sequentially complete by 1.3.5.3 and 1.4.4.4.

2. By 1.3.5.1, the set $\{\phi_U^{-1}(z) \mid U \in \Sigma,\ z \in G/U\}$ is a basis of the topology of $\widetilde{G}$. Since $\varphi^{-1}(\phi_U^{-1}(z)) = \varphi_U^{-1}(z) \neq \emptyset$ for all $U \in \Sigma$ and $z \in G/U$, it follows that $\varphi(G)$ is dense in $\widetilde{G}$. If $U \in \Sigma$, then

$$\varphi(U) = \varphi(\varphi^{-1}(\mathrm{Ker}(\phi_U)) = \mathrm{Ker}(\varphi_U) \cap \varphi(G)$$

is open relative $\varphi(G)$, and therefore $\varphi \colon G \to \varphi(G)$ is open by 1.1.6.2. But φ is continuous and

$$\mathrm{Ker}(\varphi) = \bigcap_{U \in \Sigma} U = \mathbf{1},$$

hence $\varphi \colon G \to \varphi(G)$ is a topological isomorphism.

3. Suppose first that all $U \in \Sigma$ are compact, and let $(x_U U)_{U \in \Sigma} \in \widetilde{G}$. If $U, V \in \Sigma$ and $U \subset V$, then $x_U U \subset x_V V$, and we set

$$T = \bigcap_{U \in \Sigma} x_U U.$$

If $x \in T$, then $\varphi(x) = (x_U U)_{U \in \Sigma}$, and thus we must prove that $T \neq \emptyset$. If $W \in \Sigma$, then

$$T = \bigcap_{\substack{U \in \Sigma \\ U \subset W}} x_U U \subset x_W W.$$

Since $\{x_U U \mid U \in \Sigma,\ U \subset W\}$ is a downwards directed family of closed subsets of the compact set $x_W W$, it has the finite intersection property, and consequently $T \neq \emptyset$.

Now let G be sequentially complete and $\Sigma = \{U_i \mid i \in \mathbb{N}_0\}$ such that $U_i \supsetneq U_{i+1}$ for all $i \geq 0$. Let $(x_j U_j)_{j \geq 0} \in \widetilde{G}$. We assert that $(x_i)_{i \geq 0}$ is a Cauchy sequence in G.

Indeed, if $k \geq 0$, then $U_i \subset U_j \subset U_k$ for all $i \geq j \geq k$, and $\phi_{U_i, U_j}(x_i U_i) = x_i U_j = x_j U_j$, hence $x_i^{-1} x_j \in U_j \subset U_k$ and alike $x_i x_j^{-1} = x_i (x_i^{-1} x_j)^{-1} x_i^{-1} \in U_j \subset U_k$. Since G is sequentially complete, there exists some $x \in G$ such that $(x_i)_{i \geq 0} \to x$. If $j \geq 0$, then $(x_i^{-1} x_j)_{i \geq j} \to x^{-1} x_j \in U_j$ (since U_j is closed), and thus $x_j U_j = x U_j$. Hence $\varphi(x) = (x U_j)_{j \geq 0} = (x_j U_j)_{j \geq 0}$.

4. If $U \in \Sigma$, then the natural isomorphism $H/H \cap U \to HU/U$ is continuous by 1.1.8.1 and thus it is a topological isomorphism since HU/U is discrete. We identify: $H/H \cap U = HU/U$. Since $(H \cap U)_{U \in \Sigma}$ is a fundamental system of neighbourhoods of 1 in H, it follows that

$$\widetilde{H} = \varprojlim_{U \in \Sigma} H/H \cap U = \varprojlim_{U \in \Sigma} HU/U \subset \varprojlim_{U \in \Sigma} G/U = \widetilde{G}.$$

In particular, $\widetilde{H}$ is the projective completion of H, and by definition,

$$\widetilde{H} = \widetilde{G} \cap \prod_{U \in \Sigma} HU/U \subset \prod_{U \in \Sigma} G/U.$$

Hence $\widetilde{H}$ is closed in $\widetilde{G}$, and as $\varphi(H)$ is dense in $\widetilde{H}$, it follows that $\widetilde{H} = \overline{\varphi(H)}$ is the closure of $\varphi(H)$ in $\widetilde{G}$. By 1.1.2.3 we obtain

$$\varphi^{-1}(\widetilde{H}) = \{x \in G \mid xU = xUH \text{ for all } U \in \Sigma\} = \bigcap_{U \in \Sigma} UH = \overline{H}. \qquad \square$$

1.5 Profinite groups

Definition and Theorem 1.5.1 (Profinite groups).

1. *For a topological group G the following assertions are equivalent:*

(a) *G is the projective limit of a (surjective) projective system of finite discrete groups.*

(b) *G is compact and totally disconnected.*

(c) *G is compact, and the open normal subgroups of G are a fundamental system of neighbourhoods of 1 in G.*

A topological group with these properties is called a **profinite group**.

2. Let G be a profinite group and $G^n = \{g^n \mid g \in G\}$. Then

$$\bigcap_{n\in\mathbb{N}} G^n = \mathbf{1}.$$

3. Let G be a profinite group and Σ a fundamental system of neighbourhoods of 1 in G consisting of open normal subgroups. Then $(\Sigma, \supset)$ is directed. For $U, U' \in \Sigma$ with $U \subset U'$, we denote by $\phi_{U,U'} \colon G/U \to G/U'$ and $\phi_U \colon G \to G/U$ the residue class epimorphisms.

(a) *$(G/U, \phi_{U,U'})_\Sigma$ is a projective system of finite groups, and the map*

$$\Phi\colon G \to \varprojlim_{U\in\Sigma} G/U, \quad \textit{defined by} \;\; \Phi(\sigma) = (\sigma U)_{U\in\Sigma}$$

is a topological isomorphism. We identify: $G = \varprojlim G/U$. *For $U \in \Sigma$, the residue class epimorphisms $G \to G/U$ are the projections of the projective limit.*

(b) *If H is a subgroup of G, then $\phi_U(H) = HU/U$ for all $U \in \Sigma$, and*

$$\overline{H} = \varprojlim_{U\in\Sigma} HU/U.$$

4. Closed subgroups, factor groups, direct products and projective limits of (abelian) profinite groups are again (abelian) profinite groups.

Proof. 1. (a) $\Rightarrow$ (b) If G is the projective limit of a projective system of finite discrete groups, then it is also the projective limit of a surjective such projective system by 1.3.4.**3.**, and then G is compact and totally disconnected by 1.3.5.(2. and 4.).

(b) $\Rightarrow$ (c) Let $\mathfrak{U}$ be the set of all clopen neighbourhoods of 1. We prove first:

A. $\mathfrak{U}$ is a fundamental system of neighbourhoods of 1.

Proof of **A.** The set

$$A = \bigcap_{U \in \mathfrak{U}} U$$

is closed, and we assert that $A = \mathbf{1}$. As G is totally disconneted it suffices to prove that A is connected. Thus suppose that $A = S \cup T$, where S and T are closed subsets of G, and $S \cap T = \emptyset$. We must prove that either S or T is empty. Being compact, G is normal, and thus there exist open subsets S' and T' of G such that $S' \supset S$, $T' \supset T$ and $S' \cap T' = \emptyset$. As $[G \setminus (S' \cup T')] \cap A = \emptyset$, there exist some $n \in \mathbb{N}$ and $U_1, \ldots, U_n \in \mathfrak{U}$ such that $[G \setminus (S' \cup T')] \cap U_1 \cap \ldots \cap U_n = \emptyset$. Then $V = U_1 \cap \ldots \cap U_n \in \mathfrak{U}$, and since $V = (V \cap S') \cup (V \cap T')$, we may assume that $1 \in V \cap S'$. Since $V \cap S'$ and $V \cap T'$ are open, it follows that $V \cap S' = V \setminus (V \cap T') \in \mathfrak{U}$, $A \subset V \cap S'$, and consequently we obtain $T = A \cap T \subset A \cap T' \subset V \cap S' \cap T' = \emptyset$.

Now let W be any open neighbourhood of 1. Then

$$(G \setminus W) \cap \bigcap_{U \in \mathfrak{U}} U = \emptyset,$$

and thus there exist $V_1, \ldots, V_n \in \mathfrak{U}$ such that $(X \setminus W) \cap V_1 \cap \ldots \cap V_n = \emptyset$. It follows that $V = V_1 \cap \ldots \cap V_n \in \mathfrak{U}$, and $V \subset W$. $\square$[**A.**]

It remains to prove that every $U \in \mathfrak{U}$ contains an open normal subgroup. Thus let $U \in \mathfrak{U}$. Then the set $V = \{v \in U \mid Uv \subset U\} = U \cap \{v \in G \mid Uv \subset U\}$ is open by 1.1.2.4(d), we set $H = V \cap V^{-1}$, and we assert that H is a subgroup of G. Clearly, $1 \in H$, and if $x, y \in H$, then $Uxy^{-1} \subset Uy^{-1} \subset U$, hence $xy^{-1} \in V$ and alike $yx^{-1} = (xy^{-1})^{-1} \in V$. Therefore $xy^{-1} \in V \cap V^{-1} = H$. Now H is an open subgroup of G, and therefore $(G : H) < \infty$ by 1.1.3.3. Hence H has only finitely many conjugates, and their intersection is an open normal subgroup of G contained in U.

(c) $\Rightarrow$ (a) Let Σ be the set of all open normal subgroups of G. Since all $U \in \Sigma$ are compact, 1.4.6.3′ implies

$$G = \varprojlim_{U \in \Sigma} G/U.$$

The groups G/U are discrete and compact by 1.1.5.5 (and thus finite).

2. Let Σ be the set of all open normal subgroups of G. If $U \in \Sigma$, then $(G{:}U) < \infty$ and therefore $G^n \subset U$ for some $n \in \mathbb{N}$. Since G is Hausdorff, it follows that

$$\bigcap_{n \in \mathbb{N}} G^n \subset \bigcap_{U \in \Sigma} U = \mathbf{1}.$$

3. (a) Obvious by 1.4.6.3′.

(b) By 1.3.8.3 we obtain $\overline{H} = \varprojlim_{U \in \Sigma} \phi_U(\overline{H})$. If $U \in \Sigma$, then G/U is discrete, hence $\phi_U(H) \subset \phi_U(\overline{H}) \subset \overline{\phi_U(H)} = \phi_U(H)$. Consequently we obtain $\phi_U(\overline{H}) = \phi_U(H) = HU/U$, and $\overline{H} = \varprojlim HU/U$.

4. Obviously, closed subsets and direct products of compact totally disconnected groups are again compact and totally disconnected, and by 1.3.5.(2. and 4.) the same holds for projective limits. Hence the assertions follow by 1.(b). □

Theorem 1.5.2.

1. Let $(G_i, \phi_{i,j})_I$ be an projective system of profinite groups over the directed set $(I, \leq)$, $G = \varprojlim G_i$, and let $(\phi_i \colon G \to G_i)_{i \in I}$ be the family of projections. A map $f \colon G \to Y$ into a discrete space Y is continuous if and only if there exists some $k \in I$ and a continuous map $f_k \colon G_k \to Y$ such that $f = f_k \circ \phi_k$.

2. Let G be a profinite group. For $q \in \mathbb{N}$ and a subset Z of G let $Z^q = Z \times \ldots \times Z$. A map $f \colon G^q \to Y$ into a discrete space Y is continuous if and only there exists an open normal subgroup U of G such that $f(g_1\sigma_1, \ldots, g_q\sigma_q) = f(\sigma_1, \ldots, \sigma_q)$ for all $(g_1, \ldots, g_q) \in U^q$ and $(\sigma_1, \ldots, \sigma_q) \in G^q$.

Proof. 1. By 1.5.1.4, G is a profinite group. Let $f \colon G \to Y$ be a continuous map into a discrete space Y. Then $f(G)$ is compact and discrete, hence finite, say $f(G) = \{y_1, \ldots, y_r\} \subset Y$. If $\rho \in [1, r]$, then $f^{-1}(y_\rho)$ is clopen in G, and by 1.3.5.1 there exists a subset J_ρ of I and for every $j \in J_\rho$ an open subset $B_{\rho,j}$ of G_j such that

$$f^{-1}(y_\rho) = \bigcup_{j \in J_\rho} \phi_j^{-1}(B_{\rho,j}).$$

As $f^{-1}(y_\rho)$ is compact, the open covering $\{\phi_j^{-1}(B_{\rho,j}) \mid j \in J_\rho\}$ of $f^{-1}(y_\rho)$ has a finite subcovering, and thus we may assume that the sets J_ρ are finite. Then there exists some $k \in I$ such that $k \geq j$ for all $j \in J_1 \cup \ldots \cup J_r$, and $\phi_j^{-1}(B_{\rho,j}) = \phi_k^{-1}(\phi_{k,j}^{-1}(B_{\rho,j}))$ for all $\rho \in [1, r]$ and $j \in J_\rho$. Now it follows that

$$f^{-1}(y_\rho) = \phi_k^{-1}(U_{\rho,k}), \quad \text{where} \quad U_{\rho,k} = \bigcup_{j \in J_\rho} \phi_{k,j}^{-1}(B_{\rho,j}) \subset G_k.$$

and

$$\phi_k(G) = \bigcup_{\rho=1}^{r} \phi_k(f^{-1}(y_\rho)) = \bigcup_{\rho=1}^{r} U_{\rho,k} \cap \phi_k(G).$$

If $\rho, \rho' \in [1, r]$ and $\rho \neq \rho'$, then $f^{-1}(y_\rho) \cap f^{-1}(y_{\rho'}) = \emptyset$ and therefore $U_{\rho,k} \cap U_{\rho',k} \cap \phi_k(X) = \emptyset$. Let $f_k' \colon \phi_k(G) \to Y$ be defined by $f_k'(x) = y_\rho$ for all $x \in U_{\rho,k} \cap \phi_k(G)$. Then f_k' is locally constant, hence continuous, and $f = f_k' \circ \phi_k$.

Let $V_1 = U_{1,k} \cup (G_k \setminus \phi_k(G)$ and $V_\rho = U_{\rho,k} \setminus (V_1 \cup \ldots \cup V_{\rho-1})$ for $\rho \in [2, r]$. Then $G_k = V_1 \cup \ldots \cup V_r$, V_ρ is open in G_k, $V_\rho \cap \phi_k(G) = U_{\rho,k}$ for all $\rho \in [1.r]$, and $V_\rho \cap V_{\rho'} = \emptyset$ for all $\rho, \rho' \in [1, r]$ with $\rho \neq \rho'$. Let $f_k \colon G_k \to Y$ be defined by $f_k(x) = y_\rho$ for all $x \in V_\rho$. Then f_k is locally constant, hence continuous, and since $f_k \restriction U_{\rho,k} \cap \phi_k(G) = f'_k \restriction U_{\rho,k}$ for all $\rho \in [1,r]$, we finally obtain $f = f_k \circ \phi_k$.

The converse is obvious.

2. Let Σ be the set of all open normal subgroups of G. By 1.5.1.4, it follows that $G^q = G \times \ldots \times G$ is a profinite group, and apparently $\{U^q \mid U \in \Sigma\}$ is a fundamental system of neighbourhoods of 1 in G^q consisting of open normal subgroups. Hence it follows by 1.5.1.3 that

$$G^q = \varprojlim_{U \in \Sigma} G^q/U^q.$$

Let $f\colon G^q \to Y$ be a map into a discrete space Y. If $U \in \Sigma$, then f factorizes in the form $f = (G^q \to G^q/U^q \to Y)$ if and only if

$$f(g_1\sigma_1, \ldots, g_q\sigma_q) = f(\sigma_1, \ldots, \sigma_q)$$

for all $(g_1, \ldots, g_q) \in U^q$ and $(\sigma_1, \ldots, \sigma_q) \in G^q$. Hence the assertion follows by 1. □

Theorem 1.5.3 (Existence of continuous sections). *Let G be a profinite group.*

1. *Let H and K be closed subgroups of G such that $K \subset H$, and let $\pi\colon G/K \to G/H$ be the natural epimorphism defined by $\pi(\sigma K) = \sigma H$ for all $\sigma \in G$. Then there exists a continuous map $s\colon G/H \to G/K$ satisfying $\pi \circ s = \mathrm{id}_{G/H}$.*

2. *Let $f\colon G \to G'$ be a continuous epimorphism onto a profinite group G'. Then there exists a continuous map $s\colon G' \to G$ such that $f \circ s = \mathrm{id}_{G'}$.*

Proof. 1. **a.** We assume first that $(H\!:\!K) < \infty$. Since K is relatively open in H, there exists an open normal subgroup U of G such that $U \cap H \subset K$. Let $x_1, \ldots, x_k \in G$ be such that $x_1UH, \ldots, x_kUH$ are the distinct cosets of UH in G. Then

$$G/H = \biguplus_{i=1}^{k} x_iUH/H$$

is an open covering of G/H, and it suffices to prove that, for all $i \in [1,k]$,

$$\pi \restriction x_iUK/K\colon x_iUK/K \to x_iUH/H \ \text{ is a homeomorphism} \qquad (*)$$

Indeed, if $(*)$ holds, then there exists a continuous map $s\colon G/H \to G/K$ such that $s \restriction x_iUH/H = (\pi \restriction x_iUK/K)^{-1}\colon x_iUH/H \to x_iUK/K \hookrightarrow G/K$ for all $i \in [1,k]$, and therefore $\pi \circ s(x_iuK) = \pi(x_iuH) = x_iuK$ for all $u \in U$. Hence $\pi \circ s = \mathrm{id}_{G/H}$.

Proof of $(*)$: Let $i \in [1,k]$. Obviously,

$$\pi_i = \pi \restriction x_iUK/K\colon x_iUK/K \to x_iUH/H$$

is continuous, surjective and $\pi_i(x_iuK) = x_iuH$ for all $u \in U$. If $u,\ u' \in U$ are such that $x_iuH = x_iu'H$, then $u^{-1}u' \in U \cap H \subset K$ and thus $x_iuK = x_iu'K$.

Hence π_i is bijective and thus a homeomorphism, since x_iUK/K is compact. $\square[(*)]$

b. Now we can do the general case. For closed subgroups S, S' of G such that $S' \subset S$ let $\pi_{S',S}\colon G/S' \to G/S$ and $\pi_S\colon G \to G/S$ be the natural maps defined by $\pi_{S',S}(\sigma S') = \sigma S$ and $\pi_S(\sigma) = \sigma S$ for all $\sigma \in G$, and note that $\pi_S = \pi_{S',S} \circ \pi_{S'}$. Let $\mathfrak{M}$ be the set of all pairs (S, s) consisting of a closed subgroup S of G such that $K \subset S \subset H$ and a continuous map $s\colon G/H \to G/S$ satisfying $\pi_{S,H} \circ s = \mathrm{id}_{G/H}$. Then $(H, \mathrm{id}_{G/H}) \in \mathfrak{M}$, hence $\mathfrak{M} \neq \emptyset$, and we define a partial ordering $\le$ on $\mathfrak{M}$ by setting $(S, s) \le (S', s')$ if $S' \subset S$ and $s = \pi_{S',S} \circ s'$.

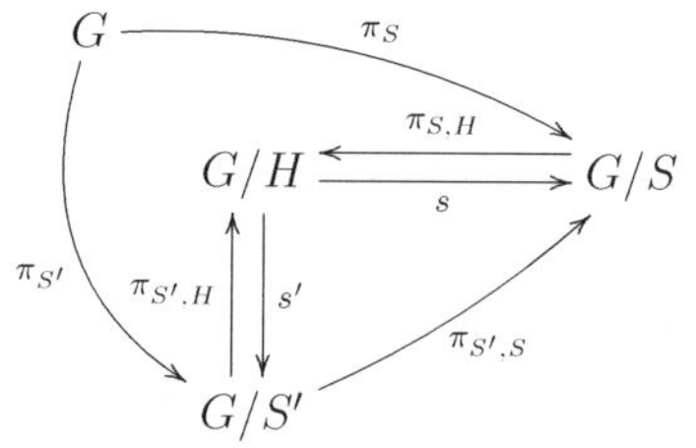

We shall apply Zorn's Lemma, and to do so we prove:

A. Every chain in $\mathfrak{M}$ has an upper bound.

Proof of **A.** Let $\mathfrak{S}$ be a chain in $\mathfrak{M}$. Then $(G/S, \pi_{S',S})_{\mathfrak{S}}$ is a projective system, and $\pi_S = \pi_{S',S} \circ \pi_{S'}$ for all (S, s), $(S', s') \in \mathfrak{S}$ such that $(S', s') \ge (S, s)$. Therefore

$$\pi' = \varprojlim_{(S,s)\in\mathfrak{S}} \pi_S : G \to \varprojlim_{(S,s)\in\mathfrak{S}} G/S$$

is a continuous epimorphism by 1.3.8.2,

$$S_0 = \mathrm{Ker}(\pi') = \bigcap_{(S,s)\in\mathfrak{S}} S$$

is a closed subgroup of G, and $S_0 \subset S$ for all $(S, s) \in \mathfrak{S}$. As G/S_0 is compact, π' induces a homeomoprhism

$$\pi\colon G/S_0 \overset{\sim}{\to} \varprojlim_{(S,s)\in\mathfrak{S}} G/S \quad \text{satisfying} \quad \pi(xS_0) = (xS)_{(S,s)\in\mathfrak{S}} \quad \text{for all} \quad x \in G.$$

As $(s\colon G/H \to G/S)_{(S,s)\in\mathfrak{S}}$ is a family of continuous maps which satisfies $\pi_{S',S} \circ s' = s$ for all (S, s), $(S', s') \in \mathfrak{S}$ such that $(S', s') \ge (S, s)$, there exists a continuous homomorphism

$$s_1 = \varprojlim_{(S,s)\in\mathfrak{S}} s\colon G/H \to \varprojlim_{(S,s)\in\mathfrak{S}} G/S$$

such that $s_1(xH) = (xS)_{(S,s)\in\mathfrak{S}}$ for all $x \in G$, and we set

$$s_0 = \pi^{-1} \circ s_1\colon G/H \to G/S_0.$$

Then s_0 is continuous. If $(S,s) \in \mathfrak{S}$ and $x \in G$, then

$$\begin{aligned}\pi_{S_0,S} \circ s_0(xH) &= \pi_{S_0,S} \circ \pi^{-1} \circ s_1(xH) = \pi_{S_0,S} \circ \pi^{-1}((xS')_{(S',s') \in \mathfrak{S}}) \\ &= \pi_{S_0,S}(xS_0) \\ &= xS = s(xH).\end{aligned}$$

In particular, $\pi_{S_0,H} \circ s_0 = \mathrm{id}_{G/H}$, and $\pi_{S_0,S} \circ s_0 = s$ for all $(S,s) \in \mathfrak{S}$. Therefore $(S_0, s_0) \in \mathfrak{M}$ is an upper bound for $\mathfrak{S}$. □[**A.**]

By **A** and Zorn's Lemma there exists a maximal element (S,s) in $(\mathfrak{M}, \le)$, and it suffices to prove that $S = K$. Assume to the contrary that $S \supsetneq K$. Let $x \in S \setminus K$ and V a relatively open normal subgroup of S such that $Vx \cap K = \emptyset$. Then $S_0 = KV$ is a relatively open normal subgroup of S, hence $(S : S_0) < \infty$, and $Vx \cap K = \emptyset$ implies $x \notin S_0$ and thus $S_0 \subsetneq S$. By **a**, there exists a continuous map $s_0 \colon G/S \to G/S_0$ satisfying $\pi_{S_0,S} \circ s_0 = \mathrm{id}_{G/S}$. Then $\overline{s} = s_0 \circ s \colon G/H \to G/S_0$ is continuous, and

$$\pi_{S_0,H} \circ \overline{s} = \pi_{S,H} \circ \pi_{S_0,S} \circ s_0 \circ s = \pi_{S,H} \circ s = \mathrm{id}_{S/H}.$$

Hence $(S_0, \overline{s}) \in \mathfrak{M}$, and since $\pi_{S_0,S} \circ \overline{s} = \pi_{S_0,S} \circ s_0 \circ s = s$, it follows that $(S_0, \overline{s}) \ge (S,s)$, though $(S_0, \overline{s}) \ne (S,s)$, a contradiction.

2. Apparently $K = \mathrm{Ker}(f)$ is a closed normal subgroup of G, and we denote by $\pi \colon G \to G/K$ the residue class epimorphism. As G/K is compact, f induces a topological isomorphism $f^* \colon G/K \to G'$ such that $f = f^* \circ \pi$. By 1, there exists a continuous map $s' \colon G/K \to G$ such that $\pi \circ s' = \mathrm{id}_{G/K}$. Then $s = s' \circ f^{*-1} \colon G' \to G$ is continuous, and $f \circ s = f^* \circ \pi \circ s' \circ f^{*-1} = \mathrm{id}_{G'}$. □

The following remarks refer to [18, Theorems 5.3.2, 5.3.3 and 5.3.6]. We connect the notions introduced there with the actual concepts of topological groups and rings.

Remarks 1.5.4. Recall that (according to [18, Definition 5.2.3]) a valued field $(K, |\cdot|)$ is complete if and only if is sequentially complete according to Definition 1.4.2.

1. Let $(K, |\cdot|)$ be a discrete valued field and $\widehat{K}$ its completion. Then $\widehat{K}$ is an (additive) sequentially complete topological group, $\mathcal{O}_{\widehat{K}}$ is a sequentially complete topological ring, and $\{\mathfrak{p}_{\widehat{K}}^n \mid n \in \mathbb{N}\}$ is a fundamental system of neighbourhoods of 0 (both in $\widehat{K}$ and $\mathcal{O}_{\widehat{K}}$).

The embedding $K \hookrightarrow \widehat{K}$ induces group isomorphisms $K/\mathfrak{p}_K^n \overset{\sim}{\to} \widehat{K}/\mathfrak{p}_{\widehat{K}}^n$ and ring isomorphisms $\mathcal{O}_K/\mathfrak{p}_K^n \overset{\sim}{\to} \mathcal{O}_{\widehat{K}}/\mathfrak{p}_{\widehat{K}}^n$ for all $n \in \mathbb{N}$. Thus the identifications made in 1.4.6 imply

$$\widehat{K} = \varprojlim_{n \in \mathbb{N}} \widehat{K}/\mathfrak{p}_{\widehat{K}}^n = \varprojlim_{n \in \mathbb{N}} K/\mathfrak{p}_K^n \quad \text{and} \quad \mathcal{O}_{\widehat{K}} = \varprojlim_{n \in \mathbb{N}} \mathcal{O}_{\widehat{K}}/\mathfrak{p}_{\widehat{K}}^n = \varprojlim_{n \in \mathbb{N}} \mathcal{O}_K/\mathfrak{p}_K^n,$$

where the embeddings $K \hookrightarrow \widehat{K}$ and $\mathcal{O}_K \hookrightarrow \mathcal{O}_{\widehat{K}}$ are defined by the identifications $x = (x + \mathfrak{p}_K^n)_{n \in \mathbb{N}}$ for all $x \in K$. In particular, the completion $\widehat{K}$ of K is just the projective completion with respect to $(\mathfrak{p}_K^n)_{n \in \mathbb{N}}$.

Moreover, $\{U_{\widehat{K}}^{(n)} \mid n \in \mathbb{N}\}$ is a fundamental system of neighbourhoods of 1 in $\widehat{K}^\times$, and $U_{\widehat{K}}/U_{\widehat{K}}^{(n)} \cong (\mathcal{O}_{\widehat{K}}/\mathfrak{p}_{\widehat{K}}^n)^\times = (\mathcal{O}_K/\mathfrak{p}_K^n)^\times \cong U_K/U_K^{(n)}$ for all $n \in \mathbb{N}$ (see [18, Theorem 2.10.5]). As $K^\times$ is a sequentially complete topological group by 1.4.5.2, we can apply 1.4.6.3′ and obtain

$$\widehat{K}^\times = \varprojlim_{n\in\mathbb{N}} \widehat{K}^\times/U_{\widehat{K}}^{(n)},$$

$$U_{\widehat{K}} = \varprojlim U_{\widehat{K}}/U_{\widehat{K}}^{(n)} \cong \varprojlim_{n\in\mathbb{N}} U_K/U_K^{(n)} \cong \varprojlim_{n\in\mathbb{N}} (\mathcal{O}_K/\mathfrak{p}_K^n)^\times,$$

and (even more general),

$$U_{\widehat{K}}^{(m)} = \varprojlim_{n\geq m} U_{\widehat{K}}^{(m)}/U_{\widehat{K}}^{(n)} \quad \text{for all } \ m \geq 0.$$

In particular, $\widehat{K}$ and $\widehat{K}^\times$ are totally disconnected topological groups.

2. Let K be a (non-Archimedian) local field (i. e., K is a complete discrete valued field with finite residue class field k_K). Then the groups $\mathcal{O}_K/\mathfrak{p}_K^n$ and $U_K/U_K^{(n)}$ are finite. Hence $\mathcal{O}_K$ and U_K are profinite groups. If $\pi \in \mathcal{O}_K$ is a prime element and $n \in \mathbb{Z}$, then $\mathfrak{p}_K^n = \pi^n\mathcal{O}_K \cong \mathcal{O}_K$, hence $\mathfrak{p}_K^n$ is a profinite group. If $n \in \mathbb{N}$, then $U_K^{(n)}$ is a closed subgroup of U_K and thus it is a profinite group, too.

3. Let p be a prime. Then $a\mathbb{Z}_p = p^{v_p(a)}\mathbb{Z}_p$ for all $a \in \mathbb{Z}^\bullet$,

$$\mathbb{Q}_p = \varprojlim_{n\in\mathbb{N}} \mathbb{Q}_p/p^n\mathbb{Z}_p = \varprojlim_{n\in\mathbb{N}} \mathbb{Q}/p^n\mathbb{Z}_{(p)} \quad \text{as additive groups,}$$

$$\mathbb{Z}_p = \varprojlim_{n\in\mathbb{N}} \mathbb{Z}_p/p^n\mathbb{Z}_p = \varprojlim_{n\in\mathbb{N}} \mathbb{Z}/p^n\mathbb{Z} \quad \text{and} \quad \mathbb{Z}_p^\times \cong \varprojlim_{n\in\mathbb{N}} (\mathbb{Z}/p^n\mathbb{Z})^\times.$$

In particular, $\mathbb{Z}_p$ is a compact totally disconnected topological ring and $\mathbb{Z}$ is a dense subring of $\mathbb{Z}_p$. Hence $\mathbb{Z}_p = \overline{\langle 1\rangle}$, and the additive group $\mathbb{Z}_p$ and the multiplicative group $\mathbb{Z}_p^\times$ are profinite groups. As above, the embedding

$$\mathbb{Z} \hookrightarrow \mathbb{Z}_p = \varprojlim_{n\in\mathbb{N}} \mathbb{Z}/p^n\mathbb{Z}$$

is given by the identification $x = (x + p^n\mathbb{Z})_{n\in\mathbb{N}}$.

Definition and Theorem 1.5.5. For $m,\, n \in \mathbb{N}$ such that $m \mid n$ we define

$$\phi_{n,m}\colon \mathbb{Z}/n\mathbb{Z} \to \mathbb{Z}/m\mathbb{Z} \quad \text{by} \quad \phi_{n,m}(a + n\mathbb{Z}) = a + m\mathbb{Z} \ \text{ for all } \ a \in \mathbb{Z}.$$

Then $(\mathbb{N}, \mid)$ is a directed set, and $(\mathbb{Z}/n\mathbb{Z}, \phi_{n,m})_{(\mathbb{N},\,\mid)}$ is a projective system of commutative rings. Therefore

$$\widehat{\mathbb{Z}} = \varprojlim \mathbb{Z}/n\mathbb{Z}$$

is a compact totally disconnected topological ring. Its additive group is called the **Prüfer group**. By 1.4.6.2 the map

$$\varphi\colon \mathbb{Z} \to \widehat{\mathbb{Z}}, \quad \text{defined by } \varphi(a) = (a + n\mathbb{Z})_{n\in\mathbb{N}}$$

induces a topological isomorphism $\varphi\colon \mathbb{Z} \to \varphi(\mathbb{Z})$, and as there we identify so that $a = (a + n\mathbb{Z})_{n\in\mathbb{N}} \in \widehat{\mathbb{Z}}$ for all $a \in \mathbb{Z}$. Then $\mathbb{Z}$ is a dense subring of $\widehat{\mathbb{Z}}$, and $\widehat{\mathbb{Z}} = \overline{\langle 1\rangle}$.

1. *There exists a unique topological isomorphism*

$$\psi\colon \widehat{\mathbb{Z}} \overset{\sim}{\to} \prod_{p\in\mathbb{P}} \mathbb{Z}_p \quad \textit{such that} \quad \psi(a) = (a)_{p\in\mathbb{P}} \quad \textit{for all} \quad a \in \mathbb{Z}.$$

We identify:
$$\widehat{\mathbb{Z}} = \prod_{p\in\mathbb{P}} \mathbb{Z}_p.$$

2. *Let $n \in \mathbb{N}$. Then $n\widehat{\mathbb{Z}}$ is an open subgroup of $\widehat{\mathbb{Z}}$, the inclusion $\mathbb{Z} \hookrightarrow \widehat{\mathbb{Z}}$ induces an isomorphism $\mathbb{Z}/n\mathbb{Z} \overset{\sim}{\to} \widehat{\mathbb{Z}}/n\widehat{\mathbb{Z}}$, and $n\widehat{\mathbb{Z}}$ is the only subgroup of $\widehat{\mathbb{Z}}$ of index n.*
We identify: $\widehat{\mathbb{Z}}/n\widehat{\mathbb{Z}} = \mathbb{Z}/n\mathbb{Z}$.

Proof. Let

$$P = \prod_{p\in\mathbb{P}} \mathbb{Z}_p, \quad \text{and for } n \in \mathbb{N} \text{ let } \quad V_n = nP = \prod_{p\in\mathbb{P}} n\mathbb{Z}_p = \prod_{p\in\mathbb{P}} p^{v_p(n)}\mathbb{Z}_p.$$

1. $\{V_n \mid n \in \mathbb{N}\}$ is a fundamental system of neighbourhood of 0 in P consisting of open subgroups, and thus $P = \varprojlim P/V_n$ by 1.5.1.3. For $n \in \mathbb{N}$, we obtain (by means of the Chinese Remainder Theorem) an isomorphism

$$\varphi_n\colon P/V_n \overset{\sim}{\to} \prod_{p\in\mathbb{P}} \mathbb{Z}_p/p^{v_p(n)}\mathbb{Z}_p = \prod_{p\in\mathbb{P}} \mathbb{Z}/p^{v_p(n)}\mathbb{Z} \overset{\sim}{\to} \mathbb{Z}/n\mathbb{Z}$$

as follows: If $\boldsymbol{a} = (a_p)_{p\in\mathbb{P}} \in P$ and $a \in \mathbb{Z}$ is such that $a \equiv a_p \bmod p^{v_p(n)}$ for all $p \,|\, n$, then $\varphi_n(\boldsymbol{a} + V_n) = a + n\mathbb{Z}$. If $m, n \in \mathbb{N}$ such that $m \,|\, n$ and $\boldsymbol{a} \in P$, then $\phi_{n,m}\circ\varphi_n(\boldsymbol{a} + V_n) = \varphi_m(\boldsymbol{a} + V_m)$. Hence (by 1.3.4.**2**) the family $(\varphi_n)_{n\in\mathbb{N}}$ induces a topological isomorphism

$$\varphi = \varprojlim \varphi_n\colon P = \varprojlim P/V_n \overset{\sim}{\to} \varprojlim \mathbb{Z}/n\mathbb{Z} = \widehat{\mathbb{Z}}$$

satisfying $\varphi((a_p)_{p\in\mathbb{P}}) = (a_n + n\mathbb{Z})_{n\in\mathbb{N}}$ for all $(a_p)_{p\in\mathbb{P}} \in P$, where $(a_n)_{n\in\mathbb{N}} \in \mathbb{Z}^{\mathbb{N}}$ is such that $a_n \equiv a_p \bmod p^{v_p(n)}$ for all $n \in \mathbb{N}$ and all primes p dividing n. In particular, if $a \in \mathbb{Z}$, then $\varphi((a)_{p\in\mathbb{P}}) = (a + n\mathbb{Z})_{n\in\mathbb{N}} = a$. Hence the topological isomorphism $\psi = \varphi^{-1}\colon \widehat{\mathbb{Z}} \to P$ satisfies $\psi(a) = (a)_{p\in\mathbb{P}}$ for all $a \in \mathbb{Z}$.

The uniqueness of ψ follows since $\mathbb{Z}$ is dense in $\widehat{\mathbb{Z}}$.

2. Let $n \in \mathbb{N}$. The inclusion $\mathbb{Z} \hookrightarrow \widehat{\mathbb{Z}}$ induces a monomorphism

$$\mathbb{Z}/n\widehat{\mathbb{Z}} \cap \mathbb{Z} \to \widehat{\mathbb{Z}}/n\widehat{\mathbb{Z}}.$$

If $a = (a)_{p\in\mathbb{P}} \in n\widehat{\mathbb{Z}} \cap \mathbb{Z} \subset nP$, then $v_p(a) \geq v_p(n)$ for all $p \in \mathbb{P}$, hence $n \,|\, a$ and $a \in n\mathbb{Z}$. Therefore $\mathbb{Z} \hookrightarrow \widehat{\mathbb{Z}}$ induces monomorphism $j\colon \mathbb{Z}/n\mathbb{Z} \to \widehat{\mathbb{Z}}/n\widehat{Z}$, and

$$\widehat{\mathbb{Z}}/n\widehat{\mathbb{Z}} = P/nP \cong \prod_{p\in\mathbb{P}} \mathbb{Z}_p/n\mathbb{Z}_p = \prod_{p\in\mathbb{P}} \mathbb{Z}_p/p^{v_p(n)}\mathbb{Z}_p = \prod_{p\in\mathbb{P}} \mathbb{Z}/p^{v_p(n)}\mathbb{Z} \cong \mathbb{Z}/n\mathbb{Z}$$

implies that $|\widehat{\mathbb{Z}}/n\widehat{\mathbb{Z}}| = |\mathbb{Z}/n\mathbb{Z}| = n$. Hence j is an isomorphism. If U is any subgroup of $\widehat{\mathbb{Z}}$ of index n, then $n\widehat{\mathbb{Z}} \subset U$, and $(\widehat{\mathbb{Z}}\colon n\widehat{\mathbb{Z}}) = n$ implies $U = n\widehat{\mathbb{Z}}$. □

Theorem and Definition 1.5.6. *Let G be a profinite group.*

1. The following assertions are equivalent:

(a) $G = \overline{\langle\gamma\rangle}$ for some $\gamma \in G$.

(b) G/U is cyclic for every open normal subgroup U of G.

(c) G is abelian and possesses for every $n \in \mathbb{N}$ at most one open subgroup of index n.

If these conditions are fulfilled, then G is called **procyclic.**
The group $\widehat{\mathbb{Z}} = \varprojlim \mathbb{Z}/n\mathbb{Z}$ and all groups $\mathbb{Z}_p = \varprojlim \mathbb{Z}/p^n\mathbb{Z}$ (for some prime p) are procyclic.

2. Let G be procyclic. Then any continuous epimorphism $g\colon G \to \widehat{\mathbb{Z}}$ is an isomorphism.

Proof. Let Σ be the set of all open normal subgroups of G. If $U,\, V \in \Sigma$ and $U \subset V$, let $\pi_{U,V}\colon G/U \to G/V$ and $\pi_U\colon G \to G/U$ be the residue class epimorphisms, so that $G = \varprojlim_{U\in\Sigma} G/U$ as in 1.5.1.3.

1. (a) $\Rightarrow$ (b) If $U \in \Sigma$, then $\langle\pi_U(\gamma)\rangle$ is dense in G/U, and as G/U is discrete, it follows that $G/U = \langle\pi(\gamma)\rangle$ is cyclic.

(b) $\Rightarrow$ (a) For $U \in \Sigma$ let $X_U = \{\gamma \in G/U \mid G/U = \langle\gamma\rangle\}$. If $U,\, V \in \Sigma$ and $U \subset V$, then $\pi_{U,V}(X_U) \subset X_V$, hence $(X_U, \pi_{U,V}{\restriction}X_U)_\Sigma$ is a projective system of non-empty finite sets, and therefore $\varprojlim_{U\in\Sigma} X_U \neq \emptyset$.

If $\gamma = (\gamma_U)_{U\in\Sigma} \in \varprojlim_{U\in\Sigma} X_U$, then $\pi_U(\langle\gamma\rangle) = \langle\pi_U(\gamma)\rangle = \langle\gamma_U\rangle = G/U$. By 1.5.1.3(b) it follows that $\overline{\langle\gamma\rangle} = \varprojlim_{U\in\Sigma} G/U = G$.

(b) $\Rightarrow$ (c) $G = \varprojlim_{U\in\Sigma} G/U$ is abelian. Assume to the contrary that there exist some $n \in \mathbb{N}$ and distinct open subgroups U_1, U_2 of index n in G. Let $U \in \Sigma$ be such that $U \subset U_1 \cap U_2$. Then G/U is cyclic, and U_1/U, U_2/U are distinct subgroups of index n in G/U, a contradiction.

(c) $\Rightarrow$ (b) If $U \in \Sigma$, then G/U is a finite abelian group which has for every $n \in \mathbb{N}$ at most one subgroup of index n; hence it is cyclic.

2. Let $g\colon G \to \widehat{\mathbb{Z}}$ be a continuous epimorphism. For every $n \in \mathbb{N}$, let $g_n'\colon G \to \mathbb{Z}/n\mathbb{Z}$ be defined by $g_n'(\tau) = g(\tau) + n\widehat{\mathbb{Z}} \in \widehat{\mathbb{Z}}/n\widehat{\mathbb{Z}} = \mathbb{Z}/n\mathbb{Z}$. Then g_n' is an epimorphism, $U_n = \mathrm{Ker}(g_n')$ is the only subgroup of index n of G, and g_n' induces an isomorphism $g_n\colon G/U_n \xrightarrow{\sim} \mathbb{Z}/n\mathbb{Z}$. If $m,\, n \in \mathbb{N}$ and $m \,|\, n$, then $U_n \subset U_m$, and there is a (natural) commutative diagram

$$\begin{array}{ccc} G/U_n & \xrightarrow{g_n} & \mathbb{Z}/n\mathbb{Z} \\ \downarrow & & \downarrow \\ G/U_m & \xrightarrow{g_m} & \mathbb{Z}/m\mathbb{Z}\,. \end{array}$$

Therefore

$$g = \varprojlim_{n\in\mathbb{N}} g_n\colon G = \varprojlim_{n\in\mathbb{N}} G/U_n \xrightarrow{\sim} \varprojlim_{n\in\mathbb{N}} \mathbb{Z}/n\mathbb{Z} = \widehat{\mathbb{Z}} \quad \text{is an isomorphism.} \qquad \square$$

1.6 Duality of abelian locally compact topological groups 1

We start with an elementary analysis of the circle group $\mathbb{T} = \{z \in \mathbb{C} \mid |z| = 1\}$ and its subgroups.

Theorem 1.6.1.

1. There is a topological isomorphism

$$c^*\colon \mathbb{R}/\mathbb{Z} \to \mathbb{T}, \text{ given by } c^*(t+\mathbb{Z}) = \mathrm{e}^{2\pi \mathrm{i} t} \text{ for all } t \in \mathbb{R}.$$

The group $\mathbb{T}_\mathbb{Q} = \{\mathrm{e}^{2\pi\mathrm{i}t} \mid t \in \mathbb{Q}\} = c^*(\mathbb{Q}/\mathbb{Z}) \cong \mathbb{Q}/\mathbb{Z}$ *is the torsion subgroup of* $\mathbb{T}$*. It is a divisible abelian group.*

2. Let T *be a subgroup of* $\mathbb{T}$*. Then either* $\overline{T} = \mathbb{T}$*, or* T *is a finite cyclic group. If* $|T| = n \in \mathbb{N}$*, then* $T = \mu_n(\mathbb{C})$ *is the group of* n*-th roots of unity.*

3. Let A *be a non-empty closed connected subset of* $\mathbb{T}$*. Then* A *is an arc (that is,* $A = c(I)$ *for some compact interval* I *of* $\mathbb{R}$*).*

4. $\{1\}$ *is the only subgroup of* $\mathbb{T}$ *contained in*

$$N_1 = \{\mathrm{e}^{2\pi\mathrm{i}t/3} \mid t \in (-1,1)\}.$$

Proof. The map $c\colon \mathbb{R} \to \mathbb{T}$, defined by $c(t) = \mathsf{e}^{2\pi \mathsf{i} t}$, is a continuous epimorphism, and $\operatorname{Ker}(c) = \mathbb{Z}$.

1. Ovbiously, c induces a continuous isomorphism $c^*\colon \mathbb{R}/\mathbb{Z} \xrightarrow{\sim} \mathbb{T}$ such that $c^*(t+\mathbb{Z}) = \mathsf{e}^{2\pi \mathsf{i} t}$ for all $t \in \mathbb{R}$. If $\pi\colon \mathbb{R} \to \mathbb{R}/\mathbb{Z}$ denotes the residue class epimorphism, then $\mathbb{R}/\mathbb{Z} = \pi([0,1])$ is compact, and therefore c^* is a topological isomorphism. As $c^*(\mathbb{Q}/\mathbb{Z}) = \mathbb{T}_{\mathbb{Q}}$, the group-theoretical properties of $\mathbb{T}_{\mathbb{Q}}$ follow from those of

$$\mathbb{Q}/\mathbb{Z}.$$

2. $T' = c^{-1}(\overline{T})$ is a closed subgroup of $\mathbb{R}$, $\mathbb{Z} \subset T'$, and $\overline{T} = c(T')$. If $T' = \mathbb{R}$, then $\overline{T} = \mathbb{T}$. If $T' \subsetneq \mathbb{R}$, then $T' = a\mathbb{Z}$ for some $a \in \mathbb{R}_{>0}$ by 1.2.6, and as $1 \in T'$, there exists some $n \in \mathbb{N}$ such that $1 = na$. Hence $\overline{T} = c(T') = \{\mathsf{e}^{2\pi \mathsf{i} g/n} \mid g \in \mathbb{Z}\} = \mu_n(\mathbb{C})$, $T = \overline{T}$ is a finite cyclic group and $|T| = n$.

3. We may assume that $A \neq \mathbb{T}$. Let $a \in \mathbb{R}$ be such that $c(a) \notin A$. As A is closed, there exists some $\varepsilon \in (0, 1/2)$ such that $c((a-\varepsilon, a+\varepsilon)) \subset \mathbb{T} \setminus A$, and then it follows that $A \subset A' = c([a+\varepsilon, a+1-\varepsilon])$. Since

$$c \restriction [a+\varepsilon, a+1-\varepsilon]\colon [a+\varepsilon, a+1-\varepsilon] \to A'$$

is a homeomorphism, it follows that $A = c(I)$ for some closed connected subset I of $[a+\varepsilon, a+1-\varepsilon]$, but then I is a compact interval of $\mathbb{R}$.

4. Let T be a subgroup of $\mathbb{T}$ contained in N_1. Then obviously T is not dense in $\mathbb{T}$, and thus $T = \mu_n(\mathbb{C})$ for some $n \in \mathbb{N}$ by 2. If $n \geq 2$, then $n = 2m + e$, where $m \in \mathbb{N}$ and $e \in \{0, 1\}$, and then

$$\frac{m}{n} = \frac{1}{2 + \frac{e}{m}} \in \Big[\frac{1}{3}, \frac{1}{2}\Big], \quad \text{and consequently} \quad \mathsf{e}^{2\pi \mathsf{i} m/n} \in \mu_n(\mathbb{C}) \setminus T. \qquad \square$$

Definition and Remarks 1.6.2. Let $G = (G, \mathcal{T})$ be a topological group.

1. A **character** of G is a continuous homomorphism $\chi\colon G \to \mathbb{T}$. We denote by $\mathsf{X}(G) = \operatorname{Hom}^{\mathsf{c}}(G, \mathbb{T})$ the group of all characters of G, equipped with pointwise multiplication, and we call $\mathsf{X}(G)$ the **character group** or the **dual group** of G. By definition, $\mathsf{X}(G)$ is a (multiplicative) abelian group, the unit character $1 = 1_G$, defined by $1_G(x) = 1$ for all $x \in G$, is the unit element of $\mathsf{X}(G)$, and for $\chi \in \mathsf{X}(G)$ the character $\overline{\chi} = \chi^{-1} \in \mathsf{X}(G)$, definied by $\overline{\chi}(x) = \overline{\chi(x)} = \chi(x)^{-1}$ for all $x \in G$, is its inverse in $\mathsf{X}(G)$.

2. Let $[G, G] = \langle\{\sigma\tau\sigma^{-1}\tau^{-1} \mid \sigma, \tau \in G\}\rangle$ be the commutator subgroup of G. Its closure $G^{(1)} = \overline{[G, G]}$ is called the **topological commutator group**, and the factor group $G^{\mathsf{a}} = G/G^{(1)}$ is called the **topological factor commutator group** of G. As $[G, G]$ is a normal subgroup of G, it follows by 1.1.3.2 that $G^{(1)}$ is a closed normal subgroup of G. If $\varphi\colon G \to A$ is a continuous homomorphism into an abelian topological group A, then $G^{(1)} \subset \operatorname{Ker}(\varphi)$, and φ induces a continuous homomorphism $\varphi^{\mathsf{a}}\colon G^{\mathsf{a}} \to A$ satisfying $\varphi^{\mathsf{a}}(\sigma G^{(1)}) = \varphi(\sigma)$ for all $\sigma \in G$.

In particular, if $\chi \in \mathsf{X}(G)$, then $\chi^{\mathsf{a}} \in \mathsf{X}(G^{\mathsf{a}})$, and the assignment $\chi \mapsto \chi^{\mathsf{a}}$ defines an isomorphism $\mathsf{X}(G) \stackrel{\sim}{\to} \mathsf{X}(G^{\mathsf{a}})$. We identify: $\mathsf{X}(G) = \mathsf{X}(G^{\mathsf{a}})$. Thus it suffices to consider character groups of abelian topological groups.

3. Let G be an abelian topological group. For a subset X of G we define its **orthogonal complement** by $X^\perp = \{\chi \in \mathsf{X}(G) \mid \chi \restriction X = 1\}$. Alike, for a subset Y of $\mathsf{X}(G)$ we define its **orthogonal complement** by

$$Y^\perp = \{x \in G \mid \chi(x) = 1 \text{ for all } \chi \in Y\} = \bigcap_{\chi \in Y} \mathrm{Ker}\chi).$$

4. Let G be a finite abelian group. Then $\mathsf{X}(G) = \mathrm{Hom}(G, \mathbb{T})$, and it is well known (see e. g. [18, Theorems 3.3.1 and 3.3.2]) that there is an isomorphism

$$u_G\colon \begin{cases} G \stackrel{\sim}{\to} \mathsf{X}(\mathsf{X}(G)) \\ x \mapsto \widehat{x} \end{cases}$$

where $\widehat{x}\colon \mathsf{X}(G) \to \mathbb{T}$ is defined by $\widehat{x}(\chi) = \chi(x)$ for all $x \in G$ and $\chi \in \mathsf{X}(G)$.

Our aim is to derive an analoguous result to 1.6.2.**4** for all locally compact abelian topological groups, and abbreviatory we call a locally compact abelian topogical group an **LCA group**. To do so, we must first endow the character group of an LCA group with a topology such that it is itself an LCA group.

Definition and Theorem 1.6.3. *Let $G = (G, \mathcal{T})$ be an* LCA *group.* For a subset C of G and a neighbourhood V of 1 in $\mathbb{T}$ we set

$$\mathsf{B}(C, V) = \mathsf{B}_G(C, V) = \{\chi \in \mathsf{X}(G) \mid \chi(C) \subset V\}.$$

If C, C' are subsets of G and V, V' are neighbourhoods of 1 in $\mathbb{T}$ such that $C \supset C'$ and $V \subset V'$, then $\mathsf{B}(C, V) \subset \mathsf{B}(C', V')$. For $\varepsilon \in (0, 1]$, we set

$$N_\varepsilon = \{\mathrm{e}^{2\pi \mathrm{i}t/3} \mid t \in (-\varepsilon, \varepsilon)\}.$$

Obviously, $\{N_\varepsilon \mid \varepsilon \in \mathbb{R}_{>0}\}$ is a fundamental system of neighbourhoods of 1 in $\mathbb{T}$.

1. *There exists a unique topology $\widehat{\mathcal{T}}$ on $\mathsf{X}(G)$ such that $(\mathsf{X}(G), \widehat{\mathcal{T}})$ is a topological group and*

$$\{\mathsf{B}(C, V) \mid C \subset G \text{ is a compact, and } V \text{ is a neighbourhood of } 1 \text{ in } \mathbb{T}\}$$

is a fundamental system of neighbourhoods of the unit character 1_G in $\mathsf{X}(G)$. $(\mathsf{X}(G), \widehat{\mathcal{T}})$ is Hausdorff, and

$$\bigcap_{\substack{C \subset G \\ C \text{ compact}}} \mathsf{B}(C, N_1) = \mathbf{1}.$$

The topological group $\mathsf{X}(G) = (\mathsf{X}(G), \widehat{\mathcal{T}})$ is called the **(topological) character group** or the **(topological) dual group** of G. Its topology $\widehat{\mathcal{T}}$ is called the **compact-open topology**.

2. The set $\{\mathsf{B}(C, N_1) \mid C \subset G$ *is compact*$\}$ *is a fundamental system of neighbourhoods of* 1_G *in* $\mathsf{X}(G)$. *In particular, if* G *is* σ*-compact (that is,* G *is the union of a denumerable family of compact sets), then* $\mathsf{X}(G)$ *satisfies the first axiom of countability.*

3. If H *is a compact subgroup of* G*, then* $H^\perp$ *is open in* $\mathsf{X}(G)$.

4. If $N = \{\mathsf{e}^{\pi \mathsf{i} t/6} \mid t \in [-1, 1]\}$ *and* C *is any compact neighbourhood of* 1 *in* G*, then* $\mathsf{B}(C, N)$ *is compact. In particular,* $\mathsf{X}(G)$ *is locally compact and therefore an* LCA *group. Moreover:*

(a) If G *is compact, then* $\mathsf{X}(G)$ *is discrete.*

(b) Let G *be discrete. Then* $\mathsf{X}(G) = \mathrm{Hom}(G, \mathbb{T})$ *is a closed subset of* $\mathbb{T}^G$*, and the compact-open topology on* $\mathsf{X}(G)$ *is induced by the product topology of* $\mathbb{T}^G$*. In particular,* $\mathsf{X}(G)$ *is compact.*

Proof. Before we start with the very proof of our assertions we establish some crucial properties of the sets N_ε.

a. If $x \in \mathbb{C}$ and $m \in \mathbb{N}$ is such that $\{x, x^2, \ldots, x^m\} \subset N_1$, then $x \in N_{1/m}$.

b. Let $\chi \in \mathrm{Hom}(G, \mathbb{T})$, $m \in \mathbb{N}$, and let U be a subset of G such that $1 \in U$ and $\chi(U^{[m]}) = \{\chi(u_1 \cdot \ldots \cdot u_m) \mid u_1, \ldots, u_m \in U\} \subset N_1$. Then $\chi(U) \subset N_{1/m}$.

c. If $\chi \in \mathrm{Hom}(G, \mathbb{T})$, then χ is continuous (and thus $\chi \in \mathsf{X}(G)$) if and only if $\chi^{-1}(N_1)$ is a neighbourhood of 1 in G.

Proof of **a.** For $u \in \mathbb{R}$ we set $\vartheta(u) = \mathsf{e}^{2\pi \mathsf{i} u/3}$ and proceed by induction on m. For $m = 1$ there is nothing to do.

$m \ge 1$, $m \to m+1$: Let $x \in \mathbb{C}$ be such that $\{x, x^2, \ldots, x^{m+1}\} \subset N_1$. Then $x \in N_{1/m}$ by the induction hypothesis, and therefore

$$x = \vartheta(t) \quad \text{for some} \quad t \in \Big(-\frac{1}{m}, \frac{1}{m}\Big).$$

It follows that

$$x^{m+1} = \vartheta((m+1)t) = \vartheta(s), \quad \text{where} \quad s \in (-1, 1),$$

and therefore $(m+1)t - s = 3u$ for some $u \in \mathbb{Z}$. Since

$$-3 \le -\frac{m+1}{m} - 1 < (m+1)t - s < \frac{m+1}{m} + 1 \le 3,$$

we obtain $u = 0$,

$$t = \frac{s}{m+1} \in \Big(-\frac{1}{m+1}, \frac{1}{m+1}\Big), \quad \text{and} \quad x \in N_{1/(m+1)}. \qquad \square[\mathbf{a.}]$$

Proof of **b.** If $x \in U$, then

$$\{x, x^2, \ldots, x^m\} \subset U^{[m]}, \quad \text{hence} \quad \{\chi(x), \chi(x)^2, \ldots, \chi(x)^m\} \subset \chi(U^{[m]}) \subset N_1,$$

and therefore $\chi(x) \in N_{1/m}$. □[**b.**]

Proof of **c.** If χ is continuous, then clearly $\chi^{-1}(N_1)$ is a neighbourhood of 1 in G. Thus let $U = \chi^{-1}(N_1)$ be a neighbourhood of 1 in G . As $\{N_{1/m} \mid m \in \mathbb{N}\}$ is a fundamental system of neighbourhoods of 1 in $\mathbb{T}$, it suffices to prove that for every $m \in \mathbb{N}$ there exists some neighbourhood V of 1 in G satisfying $\chi(V) \subset N_{1/m}$.

Let $m \in \mathbb{N}$. By 1.1.2.1, there exists an open neighbourhood V of 1 in G such that $V^{[m]} \subset U$, hence $\chi(V^{[m]}) \subset \chi(U) \subset N_1$ and therefore $\chi(V) \subset N_{1/m}$ by **b.** □[**c.**]

Now we can start with the actual proof of the theorem.

1. We check the conditions in 1.1.4.1.

(a) If C_1, C_2 are compact subsets of G and V_1, V_2 are neighbourhoods of 1 in $\mathbb{T}$, then $\mathsf{B}(C_1, V_1) \cap \mathsf{B}(C_2, V_2) \supset \mathsf{B}(C_1 \cup C_2, V_1 \cap V_2)$.

(b) Let C be a compact subset of G and V a neighbourhood of 1 in $\mathbb{T}$. Then there exists some neighbourhood V_1 of 1 in $\mathbb{T}$ such that $V_1V_1 \cup V_1V_1^{-1} \subset V$, and it follows that $\mathsf{B}(C, V_1)B(C, V_1) \cup \mathsf{B}(C, V_1)B(C, V_1)^{-1} \subset \mathsf{B}(C, V)$.

Clearly, $\psi\mathsf{B}(C, V)\psi^{-1} = \mathsf{B}(C, V)$ for all $\psi \in \mathsf{X}(G)$, and

$$\bigcap_{\substack{C \subset G \\ C \text{ compact}}} \mathsf{B}(C, N_1) = \{\chi \in \mathsf{X}(G) \mid \chi(G) \subset N_1\} = \mathbf{1} \quad \text{by 1.6.1.4.}$$

Hence $\mathsf{X}(G)$ is Hausdorff.

2. Let C_1 be a compact subset of G, V a neighbourhood of 1 in V, and $m \in \mathbb{N}$ such that $N_{1/m} \subset V$. As $\mathsf{X}(G)$ is Hausdorff, the set $C = (C_1 \cup \{1\})^{[m]}$ is compact by 1.1.2.4(b), and we assert that $\mathsf{B}(C, N_1) \subset \mathsf{B}(C_1, V)$.

Thus let $\chi \in \mathsf{B}(C, N_1)$ and $x \in C_1$. Then $x^j \in C$, hence $\chi(x)^j \in N_1$ for all $j \in [1, m]$, and therefore $\chi(x) \in N_{1/m} \subset V$ by **a.** Since $\chi(C_1) \subset V$, it follows that $\chi \in \mathsf{B}(C_1, V)$.

3. Let H be a compact subgroup of G. Then 1.6.1.4 implies that

$$H^\perp = \{\chi \in \mathsf{X}(G) \mid \chi \restriction H = 1\} = \{\chi \in \mathsf{X}(G) \mid \chi(H) \subset N_1\} = \mathsf{B}(H, N_1) \quad \text{is open.}$$

4. We first prove (a) and (b).

(a) If G is compact, then $G^\perp = \{1_G\}$ is open in $\mathsf{X}(G)$ by 3., and thus $\mathsf{X}(G)$ is discrete by 1.1.5.5. □[(a)]

(b) Let G be discrete. Then $\mathsf{X}(G) = \mathrm{Hom}(G, \mathbb{T})$ is a subgroup of $\mathbb{T}^G$. By Tychonoff's theorem $\mathbb{T}^G$ is compact, and

$$\{N_\varepsilon^C \times \mathbb{T}^{G \setminus C} \mid C \in \mathbb{P}_{\mathrm{fin}}(G),\ \varepsilon \in (0,1]\,\}$$

is a fundamental system of neighbourhoods of 1_G in $\mathbb{T}^G$. If $C \in \mathbb{P}_{\mathrm{fin}}(G)$ and $\varepsilon \in (0,1]$, then $(N_\varepsilon^C \times \mathbb{T}^{G\setminus C}) \cap \mathsf{X}(G) = \mathsf{B}(C, N_\varepsilon)$, and therefore the topology of $\mathbb{T}^G$ induces the compact-open topology on $\mathsf{X}(G)$. It remains to prove that $\mathsf{X}(G)$ is closed in $\mathbb{T}^G$, for then $\mathsf{X}(G)$ is compact.

Let $(f\colon G \to \mathbb{T}) \in \mathbb{T}^G \setminus \mathsf{X}(G)$ and $\sigma, \tau \in G$ such that $f(\sigma\tau) \ne f(\sigma)f(\tau)$. As $\mathbb{T}^G$ is Hausdorff, it follows that $f(\sigma\tau)N_\varepsilon \cap (f(\sigma)N_\varepsilon)(f(\tau)N_\varepsilon) = \emptyset$ for some $\varepsilon \in (0,1]$. We set $V = N_\varepsilon^{\{\sigma,\tau,\sigma\tau\}} \times \mathbb{T}^{G\setminus\{\sigma,\tau,\sigma\tau\}}$. Then V is an open neighbourhood of $\mathbf{1}$ in $\mathbb{T}^G$ and $fV \cap \mathsf{X}(G) = \emptyset$. Hence $\mathbb{T}^G \setminus \mathsf{X}(G)$ is open and thus $\mathsf{X}(G)$ is closed in $\mathbb{T}^G$. $\square$[(b)]

Now we can prove the main assertion. Let C be a compact neighbourhood of 1 in G, $N = \overline{N_{1/4}} = \{\mathrm{e}^{\pi i t/6} \mid t \in [-1,1]\,\} \subset \mathbb{T}$ and

$$W = \{\chi \in \mathrm{Hom}(G, \mathbb{T}) \mid \chi(C) \subset N\}.$$

If $\chi \in W$, then $\chi^{-1}(N_1) \supset \chi^{-1}(N) \supset C$, and thus χ is continuous by **c**. It follows that $W = \mathsf{B}(C, N)$, and we must prove that W is compact (then $\mathsf{X}(G)$ is locally compact, since we have already seen that it is Hausdorff).

Let G_0 be the group G with the discrete topology. Then $W \subset \mathsf{X}(G_0) \subset \mathbb{T}^G$, and we denote by $\mathcal{T}_0$ the topology on W induced by the product topology of $\mathbb{T}^G$. We shall prove:

(i) $(W, \mathcal{T}_0)$ is compact.

(ii) $\mathcal{T}_0$ is finer than the subspace topology $\mathcal{T}$ on W which is induced by $\widehat{\mathcal{T}}$.

By (i) and (ii) it follows that W is compact in $(\mathsf{X}(G), \widehat{\mathcal{T}})$.

Proof of (i) By (b), $\mathsf{X}(G_0)$ is compact, and its topology is induced by the product topology of $\mathbb{T}^G$. Hence $\mathcal{T}_0$ has the subspace topology induced by the topology of $\mathsf{X}(G_0)$. Since $W = (N^C \times \mathbb{T}^{G\setminus C}) \cap \mathsf{X}(G_0)$ is closed in $\mathsf{X}(G_0)$, it is compact with repect to $\mathcal{T}_0$. $\square$[(i)]

Proof of (ii) We must prove that for every $\chi \in W$ and every $\mathcal{T}$-neighbourhood U of χ there exists some $\mathcal{T}_0$-neighbourhood of χ contained in U. Thus let $\chi \in W$ and U a $\mathcal{T}$-neighbourhood of χ. Then there exist a compact subset B of G and $m \in \mathbb{N}$ such that $\chi\mathsf{B}(B, N_{1/m}) \cap W \subset U$. By 1.1.2.1 there exists an open neighbourhood V of 1 in G such that $V^{[2m]} \subset C$, and since $\{uV \mid u \in C\}$ is an open covering of C, there exists a finite subset F of C such that $C \subset VF$. We set $U_0 = \chi\mathsf{B}_0(F, N_{1/2m}) \cap W$, where $\mathsf{B}_0(F, N_{1/2m}) = \{\theta \in \mathrm{Hom}(G, \mathbb{T}) \mid \theta(F) \subset N_{1/2m}\}$. Then U_0 is a $\mathcal{T}_0$-neighbourhood of χ and we assert that $U_0 \subset \chi\mathsf{B}(C, N_{1/m}) \cap W$ (then

$U_0 \subset U$, and we are done). Let $\varphi = \chi\psi \in U_0$, where $\psi \in \mathrm{Hom}(G, \mathbb{T})$ and $\psi(F) \subset N_{1/2m}$. Since $\chi^{-1} = \overline{\chi} \in W$ and $\varphi \in W$, we obtain

$$\psi(C) = (\chi^{-1}\varphi)(C) \subset \chi^{-1}(C)\varphi(C) \subset NN \subset N_1.$$

In particular, $\psi^{-1}(N_1) \supset C$, hence $\psi^{-1}(N_1)$ is a neighbourhood of 1 in G, and thus $\psi \in \mathsf{X}(G)$ by **c**. Moreover, $V^{[2m]} \subset C$ implies $\psi(V^{[2m]}) \subset \psi(C) \subset N_1$, hence $\psi(V) \subset N_{1/2m}$ by **b**, and

$$\psi(C) \subset \psi(F)\psi(V) \subset N_{1/2m}N_{1/2m} = N_{1/m}.$$

If follows that $\psi \in \mathsf{B}(C, N_{1/m})$ and $\varphi = \chi\psi \in \chi\mathsf{B}(C, N_{1/m}) \cap W$. □

Definition and Theorem 1.6.4. *Let G be an* LCA *group.*

1. The group $\mathsf{X}^2(G) = \mathsf{X}(\mathsf{X}(G))$ is called the **(topological) bidual group** of G. For $x \in G$ let $\widehat{x}\colon \mathsf{X}(G) \to \mathbb{T}$ be defined by $\widehat{x}(\chi) = \chi(x)$ for all $\chi \in \mathsf{X}(G)$.
If $x \in G$, then $\widehat{x} \in \mathsf{X}^2(G)$, and the map

$$u_G\colon G \to \mathsf{X}^2(G), \quad \textit{defined by } u_G(x) = \widehat{x} \textit{ for all } x \in G,$$

is continuous.

2. *Let $f\colon G \to G'$ be a continuous homomorphism of* LCA *groups.* Then the **dual homomorphism** $\mathsf{X}(f)\colon \mathsf{X}(G') \to \mathsf{X}(G)$ is defined by $\mathsf{X}(f)(\chi) = \chi\circ f$ for all $\chi \in \mathsf{X}(G')$.
$\mathsf{X}(f)$ *is continuous, and there is a commutative diagram*

$$\begin{array}{ccc} G & \xrightarrow{u_G} & \mathsf{X}^2(G)) \\ {\scriptstyle f}\downarrow & & \downarrow{\scriptstyle \mathsf{X}^2(f)} \\ G' & \xrightarrow{u_{G'}} & \mathsf{X}^2(G'). \end{array}$$

$\mathsf{X}(\mathrm{id}_G) = \mathrm{id}_{\mathsf{X}^2(G)}$, *and if $g\colon G' \to G''$ is another continuous homomorphism of* LCA *groups, then $\mathsf{X}(g\circ f) = \mathsf{X}(f)\circ\mathsf{X}(g)$.*

Proof. 1. Let $x \in G$. Then obvioulsy $\widehat{x}(\chi_1\chi_2) = \widehat{x}(\chi_1)\widehat{x}(\chi_2)$ and if V is a neighbourhood of 1 in $\mathbb{T}$, then $\widehat{x}^{-1}(V) = \{\chi \in \mathsf{X}(G) \mid \chi(x) \in V\} = \mathsf{B}(\{1\}, V)$ is a neighbourhood of 1_G in $\mathsf{X}(G)$. Hence $\widehat{x}$ is a continuous homomorphism and thus $\widehat{x} \in \mathsf{X}^2(G)$.

Now we prove that u_G is continuous. The set

$$\{\mathsf{B}(\Gamma, N_1) \mid \Gamma \subset \mathsf{X}(G) \text{ is compact}\}$$

is a fundamental system of neighbourhoods of 1 in $\mathsf{X}^2(G)$. Hence it suffices by 1.1.6.1 to show that $u_G^{-1}(\mathsf{B}(\Gamma, N_1))$ is a neighbourhood of 1 in G for every compact subset Γ of $\mathsf{X}(G)$.

Thus let Γ be a compact subset of $\mathsf{X}(G)$ and C any compact neighbourhood of 1 in G. Then $\{\psi\mathsf{B}(C, N_{1/2}) \mid \psi \in \Gamma\}$ is an open covering of Γ, and thus there exists a finite subset F of Γ such that $\Gamma \subset F\mathsf{B}(C, N_{1/2})$. Let U be a neighbourhood of 1 in G such that $U \subset \Gamma$ and $\psi(U) \subset N_{1/2}$ for all $\psi \in F$. We assert that $U \subset u_G^{-1}(\mathsf{B}(\Gamma, N_1))$.

Indeed, if $\sigma \in U$ and $\chi \in \Gamma$, then $\chi = \psi\varphi$ for some $\psi \in F$ and $\varphi \in \mathsf{B}(\Gamma, N_{1/2})$, and $u_G(\sigma)(\chi) = \psi(\sigma)\varphi(\sigma) \in N_{1/2}N_{1/2} = N_1$, and therefore $u_G(\sigma) \in \mathsf{B}(\Gamma, N_1)$.

2. If C is a compact subset of G, then

$$\mathsf{X}(f)^{-1}(\mathsf{B}(C, N_1)) = \{\chi' \in \mathsf{X}(G') \mid \chi' \circ f \in \mathsf{B}(C, N_1)\} = \mathsf{B}(f(C), N_1).$$

As the set $\{\mathsf{B}(C, N_1) \mid C \subset G' \text{ compact}\}$ is a fundamental system of neighbourhoods of 1 in $\mathsf{X}(G)$, it follows by 1.1.6.1 that $\mathsf{X}(f)$ is continuous.

If $x \in G$ and $\chi' \in \mathsf{X}(G')$, then

$$\mathsf{X}^2(f) \circ u_G(x)(\chi') = u_G(x) \circ \mathsf{X}(f)(\chi') = \mathsf{X}(f)(\chi')(x) = \chi' \circ f(x) = u_{G'} \circ f(x).$$

Hence the diagram commutes, and the remaining assertions are obvious. □

Theorem 1.6.5. *Let G be an* LCA *group, H a closed subgroup, $\pi\colon G \to G/H$ the residue class epimorphism and $\nu = H \hookrightarrow G$.*

1. There exists a topological isomorphism

$$\widehat{\pi}\colon \mathsf{X}(G/H) \xrightarrow{\sim} H^\perp \subset \mathsf{X}(G)$$

satisfying $\widehat{\pi}(\chi) = \mathsf{X}(\pi)(\chi) = \chi \circ \pi$ for all $\chi \in \mathsf{X}(G/H)$.

2. Let $\phi\colon G \to G'$ be a surjective, continuous and open homomorphism into an LCA *group G' such that $H = \mathrm{Ker}(\phi)$. Then there eists a topological isomorphism*

$$\widehat{\phi}\colon \mathsf{X}(G') \xrightarrow{\sim} H^\perp \subset \mathsf{X}(G)$$

satisfying $\widehat{\phi}(\chi) = \mathsf{X}(\phi)(\chi) = \chi \circ \phi$ for all $\chi \in \mathsf{X}(G')$

3. Let H be open. Then there exists a topologicl isomorphism

$$\widehat{\nu}\colon \mathsf{X}(G)/H^\perp \xrightarrow{\sim} \mathsf{X}(H)$$

satisfying $\widehat{\nu}(\chi H^\perp) = \chi \restriction H$ for all $\chi \in \mathsf{X}(G)$.

For an arbitrary closed subgroup H of G we shall prove 3. using Pontrjagin's duality theorem in 1.6.7.3.

Proof. 1. By 1.6.4.2, the map $\mathsf{X}(\pi)\colon \mathsf{X}(G/H) \to \mathsf{X}(G)$ is a continuous homomorphism satisfying $\mathsf{X}(\pi)(\chi) = \chi\circ\pi \in \mathsf{X}(G)$, hence $\chi\circ\pi \restriction H = 1$ for all $\chi \in \mathsf{X}(G/H)$, and therefore $\mathrm{Im}(\mathsf{X}(\pi)) \subset H^\perp$. Conversely, if $\chi_1 \in H^\perp$, then χ_1 is constant on the cosets of H, and we define $\chi \in \mathsf{X}(G/H)$ by $\chi(xH) = \chi_1(x)$. Then $\chi \in \mathsf{X}(G)$ (since G/H carries the quotient topology), and $\chi_1 = \mathsf{X}(\pi)(\chi) \in \mathrm{Im}(\mathsf{X}(\pi))$. If $\chi \in \mathrm{Ker}(\mathsf{X}(\pi))$, then $\chi = 1_{G/H}$, since π is surjective. Therefore the map $\widehat{\pi}\colon \mathsf{X}(G/H) \to H^\perp$, defined by $\widehat{\pi}(\chi) = \mathsf{X}(\pi)(\chi) = \chi\circ\pi$ for all $\chi \in \mathsf{X}(G/H)$, is a continuous isomorphism, and we must prove that it is open. By 1.1.6.2 and 1.6.3.2, it suffices to prove that $\mathsf{X}(\pi)(\mathsf{B}(C',N_1)$ is open in $H^\perp$ for all compact subsets C' of G/H. If C' is a compact subset of G/H, then $C' = \pi(C)$ for some compact subset C of G by 1.1.9.2, and then $\mathsf{X}(\pi)(\mathsf{B}(C',N_1)) = \mathsf{X}(\pi)(\mathsf{B}(\pi(C),N_1)) = \mathsf{B}(C,N_1) \cap H^\perp$ is open in $H^\perp$.

2. As ϕ is surjective, continuous and open, there exists a topological isomorphism $\phi'\colon G/H \xrightarrow{\sim} G'$ such that $\phi'\circ\pi = \phi$ (see 1.1.6.4). If $\widehat{\pi}\colon \mathsf{X}(G/H) \xrightarrow{\sim} H^\perp$ is as in 1., then $\widehat{\phi} = \widehat{\pi}\circ\mathsf{X}(\phi')\colon \mathsf{X}(G') \to H^\perp$ is a topological isomorphism satisfying $\widehat{\phi}(\chi) = \chi\circ\phi$ for all $\chi \in \mathsf{X}(G')$.

3. By 1.6.4.2, the homomorphism $\mathsf{X}(\nu)\colon \mathsf{X}(G) \to \mathsf{X}(H)$ satisfying $\mathsf{X}(\nu)(\chi) = \chi \restriction H$ is a continuous homomorphism, and $\mathrm{Ker}(\mathsf{X}(\nu)) = H^\perp$. We prove first:

(∗) There exists some $\chi^* \in \mathrm{Hom}(G,\mathbb{T})$ such that $\chi^* \restriction H = \chi$.

Proof of (∗). Let Ω be the set of all pairs (H',χ'), where H' is a subgroup of G containing H and $\chi' \in \mathrm{Hom}(H',\mathbb{T})$ such that $\chi' \restriction H = \chi$. For $(H',\chi'), (H'',\chi'') \in \Omega$ we define $(H',\chi') \le (H'',\chi'')$ if $H' \subset H''$ and $\chi'' \restriction H' = \chi'$. Then $(\Omega,\le)$ is a partially ordered set in which every chain has an upper bound (its union). By Zorn's lemma, Ω has a maximal element (H^*,χ^*), and we assert that $H^* = G$. Assume the contrary, and let $x \in G\setminus H^*$. If $H^* \cap \langle x\rangle = \mathbf{1}$, let $t \in \mathbb{T}$ be arbitrary. If $H^* \cap \langle x\rangle \ne \mathbf{1}$, let $n \in \mathbb{N}$ be minimal such that $x^n \in H^*$, and let $t \in \mathbb{T}$ such that $\chi^*(x^n) = t^n$. Then there is a homomorphism $\widetilde{\chi}\colon H^*\langle x\rangle \to \mathbb{T}$ satisfying $\widetilde{\chi} \restriction H^* = \chi^*$ and $\widetilde{\chi}(x) = t$. This contradicts the maximality of $(H^*,\chi*)$. □[(∗)]

If $\chi^* \in \mathrm{Hom}(G,\mathbb{T})$ is such that $\chi^* \restriction H = \chi$, then $\chi^* \in \mathsf{X}(G)$ by 1.1.6.1. Hence $\mathsf{X}(\nu)$ is surjective and induces a continuous isomorphism $\widehat{\nu}\colon : \mathsf{X}(G)/H^\perp \to \mathsf{X}(H)$ satisfying $\widehat{\nu}(\chi H^\perp) = \chi \restriction H$ for all $\chi \in \mathsf{X}(G)$. If $N = \{\mathrm{e}^{\pi \mathrm{i} t/6} \mid t \in [-1,1]\}$ and C is a compact neighbourhood of $1 \in H$, then $\mathsf{B}_G(C,N)$ is a compact neighbourhood of 1_G in $\mathsf{X}(G)$ by 1.6.3.4, and $\mathsf{X}(\pi)(\mathsf{B}_G(C,N)) = \mathsf{B}_H(C,N)$ is a neighbourhood of 1_H in $\mathsf{X}(H)$. Hence $\mathsf{X}(\pi)$ is open by 1.1.9.3, and $\widehat{\nu}$ is a topological isomorphism by 1.1.6.4. □

Theorem 1.6.6 (Pontrjagin's duality theorem)**.** *Let G be an* LCA *group. Then the map $u_G\colon G \to \mathsf{X}^2(G)$ is a topological isomorphism.*
We identify: $\mathsf{X}^2(G) = G,\ x = \widehat{x}$. Consequently, $x(\chi) = \chi(x)$ forall $x \in G$ and $\chi \in \mathsf{X}(G)$.

It is well known that Pontrjagin's duality theorem 1.6.6 holds for all finite abelian groups (see [18, Section 3.3]). A proof of Pontrjagin's dualty theorem 1.6.6 in its full generality uses abstract harmonic analysis and goes beyond the scope of this book. We refer to [10, VI.(22.10.8)], [51, Theorem 3-20] or [24, Theorem (24.8)].

In the following Section 1.7 we shall investigate the dual group and prove Pontrjagin's duality for the following LCA groups: $\mathbb{R}^n$, $\mathbb{T}$, all discrete and all profinite groups, and all additive groups of non-Archimedian local fields. In particular, this list comprises all LCA groups treated in this volume.

We proceed with some consequences of Pontrjagin's duality theorem 1.6.6.

Theorem 1.6.7. *Let G be an* LCA *group and H a closed subgroup of G, such that Pontrjagin's duality theorem* 1.6.6 *holds for H and G/H.*

1. *If $x \in G \setminus H$, then there exists some $\chi \in H^\perp$ such that $\chi(x) \neq 1$. In particular:*
   - *(a) For every $x \in G \setminus \{1\}$ there exists some $\chi \in \mathsf{X}(G)$ such that $\chi(x) \neq 1$.*
   - *(b) If $x, y \in G$ and $\chi(x) = \chi(y)$ for all $\chi \in \mathsf{X}(G)$, then $x = y$.*
2. *$H^{\perp\perp} = H$.*
3. *There exists a topological isomorphism*

$$\phi\colon \mathsf{X}(G)/H^\perp \xrightarrow{\sim} \mathsf{X}(H) \quad \textit{such that} \quad \phi(\chi H^\perp) = \chi \restriction H \quad \textit{for all} \quad \chi \in \mathsf{X}(G).$$

Proof. Let $p\colon G \to G/H$ be the residue class epimorphism.

1. Let $x \in G \setminus H$. Since $1_{G/H} \neq xH \in G/H = \mathsf{X}^2(G/H)$, there exists some $\varphi \in \mathsf{X}(G/H)$ such that $(xH)(\varphi) = \varphi(xH) \neq 1$. If $\chi = \varphi \circ p \in \mathsf{X}(H)$, then $\chi \in H^\perp$, and $\chi(x) = \varphi(xH) \neq 1$.

2. If $x \in H$, then $\chi(x) = 1$ for all $\chi \in H^\perp$, and thus $x \in H^{\perp\perp}$. Conversely if $x \in G \setminus H$, then $\chi(x) \neq 1$ for some $\chi \in H^\perp$ by 1., and consequently $x \notin H^{\perp\perp}$.

3. If $i = (H \hookrightarrow G)$, then $\mathsf{X}(i)\colon \mathsf{X}(G) \to \mathsf{X}(H)$, given by $\mathsf{X}(i)(\chi) = \chi \restriction H$ for all $\chi \in \mathsf{X}(G)$, is a continuous homomorphism. Apparently, $H^\perp = \operatorname{Ker}(\mathsf{X}(i))$, and thus we obtain a continuous monomorphism

$$\phi\colon \mathsf{X}(G)/H^\perp \to \mathsf{X}(H) \ \text{ such that } \ \phi(\chi H^\perp)) = \chi \restriction H \ \text{ for all } \ \chi \in \mathsf{X}(G).$$

By 1.6.5, we get a topological isomorphism $\theta\colon \mathsf{X}(\mathsf{X}(G)/H^\perp) \xrightarrow{\sim} H^{\perp\perp} = H$, and since $(\theta \circ \mathsf{X}(\phi)\colon \mathsf{X}^2(H) = H \to \mathsf{X}(\mathsf{X}(G)/H^\perp) \xrightarrow{\sim} H) = \mathrm{id}_H$, it follows that $\mathsf{X}(\phi)$ and thus also ϕ itself is a topological isomorphism. □

Corollary 1.6.8. *Let G be an* LCA *group such that Pontrjagin's duality theorem* 1.6.6 *holds for G. Then the map*

$$\langle\,\cdot\,,\,\cdot\,\rangle\colon G\times\mathsf{X}(G)\to\mathbb{T},\quad \textit{defined by}\ \langle g,\chi\rangle=\chi(g)\ \textit{ for all }\ (g,\chi)\in G\times\mathsf{X}(G),$$

is a nondegenerate continuous pairing, i. e.,

- $\langle\,\cdot\,,\,\cdot\,\rangle$ *is continuous;*
- $\langle gg',\chi\rangle=\langle g,\chi\rangle\langle g',\chi\rangle$ *and* $\langle g,\chi\chi'\rangle=\langle g,\chi\rangle\langle g,\chi'\rangle$ *for all* $g,\ g'\in G$ *and all* $\chi,\ \chi'\in\mathsf{X}(G)$*;*
- $G^\perp=\{\chi\in\mathsf{X}(G)\mid\langle G,\chi\rangle=1\}=\mathbf{1}$*;*
- $\mathsf{X}(G)^\perp=\{g\in G\mid\langle g,\mathsf{X}(G)\rangle=1\}=\mathbf{1}$.

Proof. If $g,\ g'\in G$ and $\chi,\ \chi'\in\mathsf{X}(G)$, then

$$\langle gg',\chi\rangle=\chi(gg')=\chi(g)\chi(g')=\langle g,\chi\rangle\langle g',\chi\rangle,$$

and

$$\langle g,\chi\chi'\rangle=(\chi\chi'))g)=\chi(g)\chi'(g)=\langle g,\chi\rangle\langle g,\chi'\rangle.$$

By 1.6.5.1 and 1.6.7.1, $G^\perp\cong\mathsf{X}(G/G)=\mathsf{X}(\mathbf{1})=\mathbf{1}$, and alike $\mathsf{X}(G)^\perp=\mathbf{1}$. It remains to prove that $\langle\,\cdot\,,\,\cdot\,\rangle$ is continuous. Let $(g_0,\chi_0)\in G\times\mathsf{X}(G)$, and let V be a neighbourhood of 1 in $\mathbb{T}$. Let V_0 be a neighbourhood of 1 in $\mathbb{T}$ such that $V_0^{[3]}\subset V$, let C be a compact neighbourhood of 1 in G such that $\chi_0(C)\subset V_0$, and let W be a neighbourhood of 1_G in $\mathsf{X}(G)$ such that $\chi(g_0)\in V_0$ for all $\chi\in W$. Then $Y=W\cap\mathsf{B}(C,V_0)$ is a neighbourhood of 1_G in $\mathsf{X}(G)$, $g_0C\times\chi_0Y$ is a neighbouhood of (g_0,χ_0) in $G\times\mathsf{X}(G)$, and we assert that $\langle g_0C,\chi_0(Y)\rangle\subset V$. Indeed, if $c\in C$ and $\psi\in Y$, then

$$\langle g_0c,\chi_0\psi\rangle=\chi_0(g_0)\chi_0(c)\psi(g_0)\psi(c)\in\langle g_0,\chi_0\rangle V_0^{[3]}\subset\langle g_0,\chi_0\rangle V.\quad\square$$

Theorem 1.6.9. *Assume that Pontrjagin's duality theorem* 1.6.6 *holds for all* LCA *groups involved in the following assertions.*

1. *Let G be an* LCA *group, H a closed subgroup, $i=(H\hookrightarrow G)$ the inclusion and $p\colon G\to G/H$ the residue class epimorphism.*

   (a) *The dual homomorphism* $\mathsf{X}(i)\colon\mathsf{X}(G)\to\mathsf{X}(H)$, *given by* $\mathsf{X}(i)(\chi)=\chi\restriction H$ *for all* $\chi\in\mathsf{X}(G)$, *is a strict epimorphism, and* $\mathrm{Ker}(\mathsf{X}(i))=H^\perp$.

   (b) *The dual homomorhism* $\mathsf{X}(p)\colon\mathsf{X}(G/H)\to\mathsf{X}(G)$, *given by* $\mathsf{X}(p)(\varphi)=\varphi\circ p$ *for all* $\varphi\in\mathsf{X}(G/H)$, *is a strict monomorphism, and* $\mathrm{Im}(\mathsf{X}(p))=H^\perp$.

2. Let $f\colon G \to G'$ *be a strict homomorphism of* LCA *groups. Then the dual homomorphism* $\mathsf{X}(f)\colon \mathsf{X}(G') \to \mathsf{X}(G)$ *is strict,*

$$\mathrm{Ker}(\mathsf{X}(f)) = \mathrm{Im}(f)^{\perp} \quad \textit{and} \quad \mathrm{Im}(\mathsf{X}(f)) = \mathrm{Ker}(f)^{\perp}.$$

3. Let $G \overset{f}{\to} G' \overset{g}{\to} G''$ *be an exact sequence of* LCA *groups with strict homomorphisms. Then the dual sequence*

$$\mathsf{X}(G'') \overset{\mathsf{X}(g)}{\to} \mathsf{X}(G') \overset{\mathsf{X}(f)}{\to} \mathsf{X}(G)$$

is again an exact sequence of strict homomorphisms.

Proof. 1.(a) By 1.6.7, we get a topological isomorphism $\phi\colon \mathsf{X}(G)/H^{\perp} \overset{\sim}{\to} \mathsf{X}(H)$ such that $\phi(\chi H^{\perp}) = \chi \restriction H = \mathsf{X}(i)(\chi)$. Hence $\mathsf{X}(\chi)\colon \mathsf{X}(G) \to \mathsf{X}(H)$ is strict by 1.1.7.2, and $\mathrm{Ker}(\mathsf{X}(\chi)) = H^{\perp}$.

(b) By 1.6.5, we obtain a topological isomorphism $\widehat{p}\colon \mathsf{X}(G/H) \overset{\sim}{\to} H^{\perp}$ such that $\widehat{p}(\psi) = \psi \circ p = \mathsf{X}(p)(\chi)$. Hence $\mathsf{X}(p)$ is strict, and $\mathrm{Im}(\mathsf{X}(p)) = H^{\perp}$.

2. We split f in the form

$$f\colon G \overset{p}{\to} G/\mathrm{Ker}(f) \overset{f^*}{\to} \mathrm{Im}(f) \overset{i}{\to} G',$$

where p is the residue class homomorphism, f^* is a topological isomorphism, and i is the inclusion. Then the dual homomorphism has the form

$$\mathsf{X}(f)\colon \mathsf{X}(G') \overset{\mathsf{X}(i)}{\to} \mathsf{X}(\mathrm{Im}(f)) \overset{\mathsf{X}(f^*)}{\to} \mathsf{X}(G/\mathrm{Ker}(f)) \overset{\mathsf{X}(p)}{\to} \mathsf{X}(G),$$

where $\mathsf{X}(f^*)$ is a topological isomorphism, $\mathsf{X}(i)$ is a strict epimorphism with kernel $\mathrm{Ker}(\mathsf{X}(i)) = \mathrm{Im}(f)^{\perp}$ by 1.(a), $\mathsf{X}(p)$ is a strict monomorpism with image $\mathrm{Im}(\mathsf{X}(p)) = H^{\perp}$ by 1.(b), and therefore f is a strict homomorphism by 1.1.7.4. Now it follows that

$$\mathrm{Ker}(X(f)) = \mathrm{Ker}(\mathsf{X}(i)) = \mathrm{Im}(f)^{\perp}, \text{ and } \mathrm{Im}(\mathsf{X}(f)) = \mathrm{Im}(\mathsf{X}(p)) = \mathrm{Ker}(f)^{\perp}.$$

3. By 2, we obtain $\mathrm{Im}(\mathsf{X}(g)) = \mathrm{Ker}(g)^{\perp} = \mathrm{Im}(f)^{\perp} = \mathrm{Ker}(\mathsf{X}(g))$, and thus the dual sequence is exact. □

1.7 Duality of abelian locally compact topological groups 2

In this section, we prove Pontrjagin's duality theorem 1.6.6 for several types of LCA groups, and for them we also determine the structure of the dual group. We start with a simple lemma.

Lemma 1.7.1. *Let G be an* LCA *group. If Pontrjagin's duality theorem* 1.6.6 *holds for G, then $u_{\mathsf{X}(G)} = \mathsf{X}(u_G)^{-1}$, and Pontrjagin's duality theorem holds for $\mathsf{X}(G)$.*

Proof. Together with $u_G \colon G \to \mathsf{X}^2(G)$ the map $\mathsf{X}(u_G) \colon \mathsf{X}^3(G) \to \mathsf{X}(G)$ is a topological isomorphism. If $x \in G$, then $\widehat{x} = u_G(x) \in \mathsf{X}^2(G)$, and it satisfies $\widehat{x}(\chi) = \chi(x)$ for all $\chi \in \mathsf{X}(G)$. If $\chi \in \mathsf{X}(G)$, then $\widehat{\chi} = u_{\mathsf{X}(G)}(\chi) \in \mathsf{X}^3(G)$, and it satisfies $\widehat{\chi}(\widehat{x}) = \widehat{x}(\chi) = \chi(x)$ for all $x \in G$. Hence, for all $x \in G$ and $\chi \in \mathsf{X}(G)$,

$$\chi(x) = \widehat{\chi}(\widehat{x}) = u_{\mathsf{X}(G)}(\chi) \circ u_G(x) = \mathsf{X}(u_G) \circ u_{\mathsf{X}(G)}(\chi)(x),$$

and therefore $\mathsf{X}(u_G) \circ u_{\mathsf{X}(G)} = \mathrm{id}_{\mathsf{X}(G)}$. Hence $u_{\mathsf{X}(G)} = \mathsf{X}(u_G)^{-1}$ is a topological isomorphism, and Pontrjagin's duality theorem holds for $\mathsf{X}(G)$. □

We proceed with a conditional result concerning the additive group of locally compact valued fields. After that we prove unconditional results for the additive groups $\mathbb{R}$ and $\mathbb{Z}$ and for the compact group $\mathbb{T}$. Only in 1.7.10.3 we shall obtain an unconditional result for all locally compact valued fields.

Theorem and Definition 1.7.2. *Let $(K, |\cdot|)$ be a locally compact valued field.*

1. *If K is Archimedian, then (up to isomorphisms) either $K = \mathbb{R}$ or $K = \mathbb{C}$. If K is non-Archimedian, then (up to isomorphisms) either*

   - *$K/\mathbb{Q}_p$ is a finite extension for some prime p, or*
   - *$K = F((X))$ is the field of formal Laurent series over a finite field F.*

2. We define $\lambda_K \colon K \to \mathbb{T}$ as follows:

   (a) If $K = \mathbb{R}$, then $\lambda_{\mathbb{R}}(x) = \mathsf{e}^{2\pi \mathsf{i} x}$ for all $x \in \mathbb{R}$.

   (b) If $K = \mathbb{C}$, then $\lambda_{\mathbb{C}}(x) = \mathsf{e}^{2\pi \mathsf{i} \Re(x)}$ for all $x \in \mathbb{C}$.

   (c) If p is a prime and $K/\mathbb{Q}_p$ a finite extension, then we define λ_K by

$$\lambda_K \colon K \overset{\mathsf{Tr}_{K/\mathbb{Q}_p}}{\longrightarrow} \mathbb{Q}_p \longrightarrow \mathbb{Q}_p/\mathbb{Z}_p = (\mathbb{Q}/\mathbb{Z})_p \hookrightarrow \mathbb{Q}/\mathbb{Z} \overset{\sim}{\to} \mathbb{T}_{\mathbb{Q}} \subset \mathbb{T}.$$

   Explicitly:
   If $\alpha \in K$ and $\mathsf{Tr}_{K/\mathbb{Q}}(\alpha) = p^{-m}a + b$, where $a \in [0, p^m - 1]$, $m \in \mathbb{N}$ and $b \in \mathbb{Z}_p$, then

$$\lambda_K(\alpha) = \mathsf{e}^{2\pi \mathsf{i} a/p^m};$$

(d) Let $K = F((X))$, where F is a finite field, $\mathrm{char}(F) = p$ and $\mathbb{F}_p \subset F$. Let $\varphi\colon \mathbb{F}_p = \mathbb{Z}/p\mathbb{Z} \to \mathbb{T}$ be the character given by $\varphi(j + p\mathbb{Z}) = \mathsf{e}^{2\pi \mathsf{i} j/p}$, and consider the map (see [18, Definition 6.7.8])

$$\chi_K\colon K \overset{\mathrm{Res}_X}{\longrightarrow} F \overset{\mathsf{Tr}_{F/\mathbb{F}_p}}{\longrightarrow} \mathbb{F}_p \overset{\varphi}{\longrightarrow} \mathbb{T}.$$

Explicitly: If $\alpha \in K$,

$$\alpha = \sum_{n\in\mathbb{Z}} a_n X^n, \text{ where } a_n \in F, \text{ and } a_{-n} = 0 \text{ for } n \gg 1,$$

then $\lambda_K(\alpha) = \varphi(\mathsf{Tr}_{F/\mathbb{F}_p}(a_{-1}))$.

$\lambda_K\colon K \to \mathbb{T}$ is called the **standard character** of K.

3. Let $\lambda \in \mathsf{X}(K) \setminus \{1_K\}$, and for $y \in K$ we define $\lambda_y\colon K \to \mathbb{T}$ by $\lambda_y(x) = \lambda(yx)$ for all $x \in K$. Then $\lambda_y \in \mathsf{X}(K)$, and the map

$$\phi\colon K \to \mathsf{X}(K) \quad \textit{defined by} \quad \phi(y) = \lambda_y,$$

is a continuous and strict monomorphism. If Pontrjagin's duality theorem holds for K, then ϕ is a topological isomorphism.

Below in 1.7.10 we shall prove that Pontrjagin's duality theorem holds for every locally compact valued field. Consequently every locally compact valued field is **self-dual**, i. e., it is (up to a canonical isomorphism) its own dual group.

Proof. 1. Being locally compact, K is sequentially complete by 1.4.4.2 and thus complete. If K is Archimedian, then (up to isomorphisms) $K = \mathbb{R}$ or $K = \mathbb{C}$ by the theorem of Ostrowski (see [18, Theorem 5.2.19]). If K is non-Archimedian, then K is a local field by [18, Theorem 5.3.9], and by [18, Theorem 5.8.3], K is (up to isomorphisms) either a finite extension of some $\mathbb{Q}_p$ or the field of formal Laurent series over a finite field.

2. Apparently, λ_K is a homomorphism, and we must check that it is continuous. This is obvious in (a) and (b). In (c) it follows since $\mathsf{Tr}_{K/\mathbb{Q}_p}$ is continuous (see [18, Theorem 5.5.3]), and in (d), since Res_X is continuous (see [18, Theorem 6.7.8]),

3. Apparently, λ_y is continuous,

$$\lambda_y(x) = \lambda_x(y) = \lambda(xy), \;\; \lambda_y(x + x') = \lambda_y(x)\lambda_y(x') \text{ and } \lambda_{y+y'} = \lambda_y\lambda_{y'}$$

for all x, x', y, $y' \in K$. Moreover, $\lambda_y = 1$ if and only if $\lambda(yK) = \mathbf{1}$, and this holds if and only if $y = 0$. Consequently, the map $\phi\colon K \to \mathsf{X}(K)$ is a group monomorphism. We shall prove:

a. ϕ is continuous.

b. $\phi\colon K \to \phi(K)$ is open.

c. If Pontrjagin's duality theorem holds for K, then ϕ is surjective.

Proof of **a.** It suffices to prove that $\phi^{-1}(\mathsf{B}(C,V))$ is open for every compact subset C of K and every open neighbourhood V of 1 in $\mathbb{T}$. Let V be an open neighbourhood of 1 in $\mathbb{T}$ and C a compact subset of K. Then

$$\phi^{-1}(\mathsf{B}(C,V)) = \{y \in K \mid \lambda_y(C) \subset V\} = \{y \in K \mid yC \subset \lambda^{-1}(V)\}$$

is open by 1.1.2.4(b). □[**a.**]

Proof of **b.** For $\varepsilon \in \mathbb{R}_{>0}$ let $B_\varepsilon = \{x \in K \mid |x| < \varepsilon\}$. Then $\overline{B}_\varepsilon$ is compact, and since $\{B_\varepsilon \mid \varepsilon \in \mathbb{R}_{>0}\}$ is a fundamental system of neighbourhoods of 0 in K, it suffices to prove that $\phi(B_\varepsilon)$ is a neighbourhood of 1_G in $\phi(K)$ for all $\varepsilon \in \mathbb{R}_{>0}$. Thus let $\varepsilon \in \mathbb{R}_{>0}$, $x_0 \in K$ such that $\lambda(x_0) \neq 1$, V a neighbourhood of 1 in $\mathbb{T}$ such that $\lambda(x_0) \notin V$and $\eta = |x_0|/\varepsilon$. We assert that $\phi(B_\varepsilon) \supset \phi(K) \cap \mathsf{B}(\overline{B}_\eta, V)$ (then $\phi(B_\varepsilon)$ is a neighbourhood of 1_G in $\phi(K)$). Thus let $y \in K$ and $\phi(y) = \lambda_y \in \mathsf{B}(\overline{B}_\eta, V)$. Then $\lambda_y(\overline{B}_\eta) = \lambda(\overline{B}_{\eta|y|}) \subset V$, hence $x_0 \notin \overline{B}_{\eta|y|}$ and therefore $|x_0| > \eta|y|$. It follows that $|y| < |x_0|/\eta = \varepsilon$ and thus $y \in B_\varepsilon$.

Proof of **c.** Assume that Pontrjagin's duality theorem holds for K and thus by 1.7.1 also for $\mathsf{X}(K)$. By **a** and **b** it follows that $\phi\colon K \overset{\sim}{\to} \phi(K)$ is a topological isomorphism. Hence $\phi(K)$ is sequentially complete, and since K is σ-compact, $\mathsf{X}(K)$ satisfies the first axiom of countability by 1.6.3.2. Hence $\phi(K)$ is a closed subgroup of $\mathsf{X}(K)$ by 1.4.4.1(b), and

$$\phi(K)^\perp = \{y \in K \mid \lambda(yK) = 1\} = \mathbf{0} \subset K.$$

Now 1.6.7.2 implies $\phi(K) = \phi(K)^{\perp\perp} = \mathbf{0}^\perp = \mathsf{X}(K)$. □

Theorem 1.7.3. *For $y \in \mathbb{R}$, we define $\lambda_y\colon \mathbb{R} \to \mathbb{T}$ by $\lambda_y(x) = \mathrm{e}^{2\pi \mathrm{i} yx}$ for all $x \in \mathbb{R}$. Then $\lambda_y \in \mathsf{X}(\mathbb{R})$, the map $\phi\colon \mathbb{R} \to \mathsf{X}(\mathbb{R})$, defined by $\phi(y) = \lambda_y$ for all $y \in \mathbb{R}$, is a topological isomorphism, and $u_\mathbb{R} = \mathsf{X}(\phi)^{-1}\circ\phi\colon \mathbb{R} \overset{\sim}{\to} \mathsf{X}^2(\mathbb{R})$.*
We identify: $\phi = \mathsf{X}(\phi) = \mathrm{id}_\mathbb{R}$, and $\mathbb{R} = \mathsf{X}(\mathbb{R}) = \mathsf{X}^2(\mathbb{R})$.
Then $u_\mathbb{R} = \mathrm{id}_\mathbb{R}$, $\mathbb{Z}^\perp = \mathbb{Z}$, and Pontrjagin's duality theorem holds for $\mathbb{R}$.

Proof. By 1.7.2, it suffices to prove:
a. ϕ is surjective; **b.** $u_\mathbb{R} = \mathsf{X}(\phi)^{-1}\circ\phi$; **c.** $\mathbb{Z}^\perp = \mathbb{Z}$.

Proof of **a.** Let $\chi \in \mathsf{X}(\mathbb{R})$ and $B = \mathrm{Ker}(\chi)$. Then B is a closed subgroup of $\mathbb{R}$, and we assert that $B \neq \mathbf{0}$. Indeed, assume to the contrary that $B = \mathbf{0}$. Then χ is a continuous monomorphism, and $\chi([0,1])$ is an infinite compact connecting subset of $\mathbb{T}$ and thus an arc by 1.6.1.3. Hence there exist some $r \in \mathbb{Q}$ and $t \in (0,1)$ such that $\mathrm{e}^{2\pi \mathrm{i} r} = \chi(t)$. If $q \in \mathbb{N}$ is such that $rq \in \mathbb{Z}$, then $\chi(tq) = \chi(t)^q = \mathrm{e}^{2\pi \mathrm{i} rq} = 1$ but $tq \neq 0$, a contradiction.

If $B = \mathbb{R}$, then $\chi = 1_\mathbb{R} = \lambda_0$, and we are done. Thus suppose that $B \neq \mathbb{R}$, and consequently $B = b\mathbb{Z}$, where $b \in \min(B \cap \mathbb{R}_{>0})$ by 1.2.6. Then

$$1 = \chi(b) = \chi\Bigl(\frac{b}{2}\Bigr)^2 \ \text{ implies } \ \chi\Bigl(\frac{b}{2}\Bigr) = -1 = \chi\Bigl(\frac{b}{4}\Bigr)^2,$$

and consequently

$$\chi\Big(\frac{b}{4}\Big) = \varepsilon \mathsf{i}, \quad \text{where} \quad \varepsilon \in \{\pm 1\}.$$

CASE 1: $\varepsilon = 1$. We prove by induction on k that

$$\chi\Big(\frac{b}{2^k}\Big) = \mathsf{e}^{\pi \mathsf{i}/2^{k-1}} \quad \text{for all} \quad k \geq 2. \tag{†}$$

For $k = 2$, there is nothing to do, since $\mathsf{i} = \mathsf{e}^{\pi\mathsf{i}/2}$. Thus let $k \geq 2$ and

$$\chi\Big(\frac{b}{2^k}\Big) = \chi\Big(\frac{b}{2^{k+1}}\Big)^2 = \mathsf{e}^{\pi\mathsf{i}/2^{k-1}}. \quad \text{Then} \quad \chi\Big(\frac{b}{2^{k+1}}\Big)$$
$$= \sigma \mathsf{e}^{\pi\mathsf{i}/2^k}, \quad \text{where} \quad \sigma \in \{\pm 1\},$$

and we assume to the contrary that $\sigma = -1$. By 1.6.1.3,

$$Y = \chi\Big(\Big[\frac{b}{2^{k+1}}, \frac{b}{2^k}\Big]\Big) \quad \text{is an arc in } \mathbb{T} \text{ connecting} \quad -\mathsf{e}^{\pi\mathsf{i}/2^k} \quad \text{and} \quad \mathsf{e}^{\pi\mathsf{i}/2^{k+1}}.$$

Hence $1 \in Y$ or $-1 \in Y$, and therefore

$$\pm 1 = \chi(t) \quad \text{and} \quad \chi(2t) = \chi(t)^2 = 1 \quad \text{for some} \quad t \in \Big[\frac{b}{2^{k+1}}, \frac{b}{2^k}\Big], \quad \text{but}$$
$$2t \leq \frac{b}{2^{k-1}} < b,$$

a contradiction. Thus (†) is proved.

If $r \in \mathbb{Q}$ is a 2-adic rational number, say $r = 2^{-k}s$, where $s \in \mathbb{Z}$ and $k \geq 2$, then

$$\chi(rb) = \chi\Big(\frac{sb}{2^k}\Big) = \mathsf{e}^{\pi\mathsf{i}s/2^{k-1}} = \mathsf{e}^{2\pi\mathsf{i}r}.$$

Since the 2-adic rational numbers are dense in $\mathbb{R}$ and χ is continuous, it follows that $\chi(zb) = \mathsf{e}^{2\pi\mathsf{i}z}$ for all $z \in \mathbb{R}$, and consequently $\chi(x) = \mathsf{e}^{2\pi\mathsf{i}x/b} = \lambda_{1/b}(x)$ for all $x \in \mathbb{R}$.

CASE 2: $\varepsilon = -1$. Then

$$\chi\Big(\frac{b}{4}\Big) = -\mathsf{i} \quad \text{implies} \quad \overline{\chi}\Big(\frac{b}{4}\Big) = \mathsf{i},$$

hence $\overline{\chi} = \lambda_{1/b}$ by CASE 1, and $\chi = \lambda_{-1/b}$.

Proof of **b.** If x, $y \in \mathbb{R}$, then

$$\mathsf{X}(\phi)\circ u_{\mathbb{R}}(x)(y) = \mathsf{X}(\phi)(\widehat{x})(y) = \widehat{x}\circ\phi(y) = \phi(y)(x) = \phi(x)(y),$$

hence $\mathsf{X}(\phi)\circ u_{\mathbb{R}} = \phi$, and as $\mathsf{X}(\phi)$ is an isomorphism, we get $u_{\mathbb{R}} = \mathsf{X}(\phi)^{-1}\circ\phi$.

Proof of **c.** According to the identification $\mathbb{R} = \mathsf{X}(\mathbb{R})$ we obtain

$$\mathbb{Z}^{\perp} = \{y \in \mathbb{R} \mid \lambda_y(\mathbb{Z}) = \mathbf{1}\} = \{y \in \mathbb{R} \mid \mathsf{e}^{2\pi\mathsf{i}y\mathbb{Z}} = \mathbf{1}\} = \mathbb{Z}. \qquad \square$$

Theorem 1.7.4.

1. *For $z \in \mathbb{T}$ we define $\iota_z\colon \mathbb{Z} \to \mathbb{T}$ by $\iota_z(n) = z^n$ for all $n \in \mathbb{N}$. Then $\iota_z \in \mathsf{X}(\mathbb{Z})$, and the map $\iota\colon \mathbb{T} \to \mathsf{X}(\mathbb{Z})$, defined by $\iota(z) = \iota_z$ for all $z \in \mathbb{T}$, is a topological isomorphism.*

2. *For $n \in \mathbb{Z}$ we definie $\theta_n\colon \mathbb{T} \to \mathbb{T}$ by $\theta_n(z) = z^n$ for all $z \in \mathbb{T}$. Then $\theta_n \in \mathsf{X}(\mathbb{T})$, and the map $\theta\colon \mathbb{Z} \to \mathsf{X}(\mathbb{T})$, defined by $\theta(n) = \theta_n$ for all $n \in \mathbb{Z}$, is an isomorphism.*

3. *Pontrjagin's duality theorem holds for $\mathbb{Z}$ and $\mathbb{T}$,*

$$u_{\mathbb{T}} = \mathsf{X}(\theta)^{-1} \circ \iota\colon \mathbb{T} \overset{\sim}{\to} \mathsf{X}^2(\mathbb{T}) \quad \textit{and} \quad u_{\mathbb{Z}} = \mathsf{X}(\iota)^{-1} \circ \theta\colon \mathbb{Z} \overset{\sim}{\to} \mathsf{X}^2(\mathbb{Z}).$$

We identify: $\mathsf{X}(\mathbb{Z}) = \mathbb{T}$, $\mathsf{X}(\mathbb{T}) = \mathbb{Z}$, $\mathsf{X}^2(\mathbb{Z}) = \mathbb{Z}$, $\mathsf{X}^2(\mathbb{T}) = \mathbb{T}$, $u_{\mathbb{Z}} = \mathrm{id}_{\mathbb{Z}}$ and $u_{\mathbb{T}} = \mathrm{id}_{\mathbb{T}}$.

Proof. 1. Obviously, $\iota\colon \mathbb{T} \to \mathsf{X}(\mathbb{Z}) = \mathrm{Hom}(\mathbb{Z}, \mathbb{T})$ as defined above, is an isomorphism, and as $\mathbb{T}$ is compact, it suffices to prove that ι is coninuous. Thus let C be a compact (hence finite) subset of $\mathbb{Z}$ and V a neighbourhood of 1 in $\mathbb{T}$. Then

$$\iota^{-1}(\mathsf{B}(C, V)) = \{z \in \mathbb{T} \mid \iota_z(C) \subset V\} = \bigcap_{n \in C} \{z \in \mathbb{T} \mid z^n \in V\} \quad \text{is open in } \mathbb{T}.$$

2. We consider the epimorphism $p\colon \mathbb{R} \to \mathbb{T}$, defined by $p(x) = \mathsf{e}^{2\pi \mathrm{i} x}$. By 1.6.9 and 1.7.3, $\mathsf{X}(p)$ induces an isomorphism $\widehat{p}\colon \mathsf{X}(\mathbb{T}) \overset{\sim}{\to} \mathbb{Z}^{\perp} = \mathbb{Z}$ such that $\widehat{p}^{-1}(n)(\mathsf{e}^{2\pi \mathrm{i} x}) = \mathsf{e}^{2\pi \mathrm{i} n x}$ for all $x \in \mathbb{R}$, and therefore $\widehat{p}^{-1}(n)(t) = t^n = \theta(n)(t)$ for all $n \in \mathbb{Z}$ and $t \in \mathbb{T}$, which implies $\widehat{p}^{-1} = \theta\colon \mathsf{X}(\mathbb{T}) \to \mathbb{Z}$.

3. If $z \in \mathbb{T}$ and $n \in \mathbb{Z}$, then

$$\mathsf{X}(\theta) \circ u_{\mathbb{T}}(z)(n) = \mathsf{X}(\theta)(\widehat{z})(n) = \widehat{z} \circ \theta(n) = \theta(n)(z) = z^n = \iota(z)(n),$$

and

$$\mathsf{X}(\iota) \circ u_{\mathbb{Z}}(n)(z) = \mathsf{X}(\iota)(\widehat{n})(z) = \widehat{n} \circ \iota(z) = \iota(z)(n) = z^n = \theta(n)(z).$$

Hence $\mathsf{X}(\theta) \circ u_{\mathbb{T}} = \iota$, $\mathsf{X}(\iota) \circ u_{\mathbb{Z}} = \theta$, and consequently $u_{\mathbb{T}} = \mathsf{X}(\theta)^{-1} \circ \iota$ and $u_{\mathbb{Z}} = \mathsf{X}(\iota)^{-1} \circ \theta$ are (topological) isomorphisms. □

In the next three theorems we investigate the dual groups of direct products, inductive limits and projective limits.

Theorem 1.7.5. *For $i \in \{1, 2\}$, let G_i be an* LCA *group with unit element 1_i. For $\chi \in \mathsf{X}(G_1 \times G_2)$, let $\chi_1 \in \mathsf{X}(G_1)$ and $\chi_2 \in \mathsf{X}(G_2)$ be defined by*

$$\chi_1(x_1) = \chi(x_1, 1_2) \quad \textit{and} \quad \chi_2(x_2) = (1_1, x_2) \quad \textit{for all} \quad (x_1, x_2) \in G_1 \times G_2.$$

Then $\chi(x_1, x_2) = \chi_1(x_1)\chi_2(x_2)$ for all $\chi \in \mathsf{X}(G_1 \times G_2)$ and $(x_1, x_2) \in G_1 \times G_2$, and the map

$$h\colon \mathsf{X}(G_1 \times G_2) \to \mathsf{X}(G_1) \times \mathsf{X}(G_2), \quad \textit{defined by} \quad h(\chi) = (\chi_1, \chi_2),$$

is a topological isomorphism. If Pontrjagin's duality theorem holds for G_1 and G_2, then it also holds for $G_1 \times G_2$.

Proof. If $\chi,\ \psi \in \mathsf{X}(G_1 \times G_2)$, then

$$h(\chi\psi) = ((\chi\psi)_1, (\chi\psi)_2) = (\chi_1\psi_1, \chi_2\psi_2) = (\chi_1, \chi_2)(\psi_1, \psi_2) = h(\chi)h(\psi).$$

Hence h is a homomorphism. If $(\chi_1, \chi_2) \in \mathsf{X}(G_1) \times \mathsf{X}(G_2)$ and $\chi \in \mathsf{X}(G_1 \times G_2)$, then we get $(\chi_1, \chi_2) = h(\chi)$ if and only if $\chi(x_1, x_2) = \chi_1(x_1)\chi_2(x_2)$ for all $(x_1, x_2) \in G_1 \times G_2$. Hence h is an isomorphism.

If $C_1 \subset G_1$ and $C_2 \subset G_2$ are compact and U, V are neighbourhoods of 1 in $\mathbb{T}$ such that $VV \subset U$, then

$$h(\mathsf{B}(C_1 \times C_2, V)) \subset \mathsf{B}(C_1, V) \times \mathsf{B}(C_2, V) \subset h(\mathsf{B}(C_1 \times C_2, VV) \subset h(\mathsf{B}(C_1 \times C_2, U).$$

Hence h and h^{-1} are continuous, and thus h is a topological isomorphism.

We consider the isomorphism

$$h^2 \colon \mathsf{X}^2(G_1 \times G_2) \overset{\mathsf{X}(h)}{\to} \mathsf{X}(\mathsf{X}(G_1) \times \mathsf{X}(G_2)) \overset{h^*}{\to} \mathsf{X}^2(G_1) \times \mathsf{X}^2(G_2),$$

where h^* is the counterpart of h if we replace (G_1, G_2) with $(\mathsf{X}(G_1), \mathsf{X}(G_2))$. It satisfies $h^2(\widehat{(x_1, x_2)}) = (\widehat{x}_1, \widehat{x}_2)$ and induces the commutative diagram

$$\begin{array}{ccccc} & & G_1 \times G_2 & & \\ & \swarrow^{u_{G_1 \times G_2}} & & \searrow^{(u_{G_1}, u_{G_2})} & \\ h^2 : \mathsf{X}^2(G_1 \times G_2) & \xrightarrow[\mathsf{X}(h)]{} & \mathsf{X}(\mathsf{X}(G_1) \times \mathsf{X}(G_2)) & \xrightarrow[h^*]{} & \mathsf{X}^2(G_1) \times \mathsf{X}^2(G_2)\,. \end{array}$$

Hence, if u_{G_1} and u_{G_2} are isomorphisms, then $u_{G_1 \times G_2}$ is an isomorphism, too. □

Theorem 1.7.6. *Let G be an* LCA *group and $(I, \le)$ a directed set.*

A. *Let $(G_i, \phi_{i,j})_I$ be a surjective projective system of* LCA *groups such that*

$$G = \varprojlim_{i \in I} G_i.$$

Suppose that $\phi_{i,j}$ is open and $\mathrm{Ker}(\phi_{i,j})$ *is compact for all $i,\ j \in I$ such that $i \ge j$, and let $(\phi_i \colon G \to G_i)_{i \in I}$ be the family of projections of the projective limit.*

1. *If $i \in I$, then ϕ_i is surjective and open, $H_i = \mathrm{Ker}(\phi_i)$ is compact, and $H_i^{\perp}$ is open in $\mathsf{X}(G)$. The family $(H_i^{\perp})_{i \in I}$ in $\mathsf{X}(G)$ is upwards directed, and*

$$\mathsf{X}(G) = \bigcup_{i \in I} H_i^{\perp} = \varinjlim_{i \in I} H_i^{\perp}.$$

2. $(\mathsf{X}(G_i), \mathsf{X}(\phi_{i,j}))_I$ *is an injective inductive system, and there is a topological isomorphism*

$$\Phi = \varinjlim_{i\in I} \mathsf{X}(\phi_i) \colon \varinjlim_{i\in I} \mathsf{X}(G_i) \overset{\sim}{\to} \mathsf{X}(G)$$

satisfying $\Phi([\chi_i, i]) = \chi_i \circ \phi_i$ *for all* $i \in I$ *and* $\chi_i \in \mathsf{X}(G_i)$.
We identify: $\mathsf{X}(\varprojlim_{i\in I} G_i) = \varinjlim_{i\in I} \mathsf{X}(G_i)$.

B. *Let* $(G_i, \phi_{j,i})_I$ *be an injective inductive system of* LCA *groups such that*

$$G = \varinjlim G_i,$$

and let $(\phi_i \colon G_i \to G)_{i\in I}$ *be the family of injections. Then* $(\mathsf{X}(G_i), \mathsf{X}(\phi_{j,i}))_I$ *is a projective system of compact abelian groups, and the map*

$$\Psi \colon \mathsf{X}(G) \to \varprojlim \mathsf{X}(G_i), \quad \textit{defined by} \quad \Psi(\chi) = (\chi \circ \phi_i)_{i\in I} \quad \textit{for all} \quad \chi \in \mathsf{X}(A),$$

is a topological isomorphism.
We identify: $\mathsf{X}(\varinjlim_{i\in I} G_i) = \varprojlim_{i\in I} \mathsf{X}(G_i)$.

C. *Suppose that*

$$\textit{either} \quad G = \varprojlim_{i\in I} G_i \quad \textit{as in } 1. \quad \textit{or} \quad G = \bigcup_{i\in I} G_i = \varinjlim_{i\in I} G_i \quad \textit{as in } 2.$$

If all G_i *satisfy the duality theorem, then* G *also satisfies the duality theorem.*

Proof. **A.** 1. If $i \in I$, then ϕ_i is surjective and open and H_i is compact by 1.3.7 (indeed, if $i \ge j$, then $\phi_{i,j}$ is surjective by assumption, and if $x_j \in G_j$ and $x_i \in \phi_{i,j}^{-1}(x_j)$, then $\phi_{i,j}^{-1}(x_j) = x_i \operatorname{Ker}(\phi_{i,j})$ is compact).

By 1.6.5.2, we obtain a topological isomorphism $\widehat{\phi}_i \colon \mathsf{X}(G_i) \overset{\sim}{\to} H_i^{\perp}$ such that $\widehat{\phi}_i(\chi) = \mathsf{X}(\phi_i)(\chi) = \chi \circ \phi_i$, and $H_i^{\perp}$ is open in $\mathsf{X}(G)$ by 1.6.3.3. If i, $j \in I$ and $i \ge j$, then $\phi_j = \phi_{i,j} \circ \phi_i$, hence $H_i \subset H_j$, and therefore $H_j^{\perp} \subset H_i^{\perp}$. Consequently $(H_i^{\perp})_{i\in I}$ is a directed family of open subgroups of $\mathsf{X}(G)$, and we assert that

$$\mathsf{X}(G) = \bigcup_{i\in I} H_i^{\perp}. \tag{†}$$

For the proof of (†) let $\chi \in \mathsf{X}(G)$, $N_1 = \{\mathrm{e}^{2\pi \mathrm{i}t/3} \mid t \in (-1, 1)\}$, and set $U = \chi^{-1}(N_1)$. Then U is an open neighbourhood of 1 in G, and by 1.3.5.1 there exist some $i \in I$ and an open neighbourhood U_i of the unit element 1_i in G_i such that $H_i \subset \phi_i^{-1}(U_i) \subset U$. Then $\chi(H_i)$ is a subgroup of N_1, hence $\chi(H_i) = 1$ by 1.6.1.4, and thus $\chi \in H_i^{\perp}$. □[(†)]

2. If i, j, $k \in I$ and $i \ge j \ge k$, then $\phi_{i,k} = \phi_{j,k} \circ \phi_{i,j}$, hence we obtain $\mathsf{X}(\phi_{i,k}) = \mathsf{X}(\phi_{i,j}) \circ \mathsf{X}(\phi_{j,k})$, and there is a commutative diagram

$$\begin{array}{ccccc}
\mathsf{X}(G_i) & \xrightarrow{\ \widehat{\phi}_i\ } & H_i^{\perp} & \xrightarrow{\ \nu\ } & \mathsf{X}(G) \\
{\scriptstyle \mathsf{X}(\phi_{i,j})}\uparrow & & \uparrow{\scriptstyle \nu_{i,j}} & & \\
\mathsf{X}(G_j) & \xrightarrow{\ \widehat{\phi}_j\ } & H_j^{\perp} & &
\end{array},$$

where $\nu_{i,j}$ denotes the inclusion. Hence $(\mathsf{X}(G_i), \mathsf{X}(\phi_{i,j}))_I$ is an injective inductive system of topological groups, and the family $(\widehat{\phi}_i \colon \mathsf{X}(G_i) \to H_i^\perp)_{i\in I}$ of topological isomorphisms induces a topological isomorphism

$$\Phi = \varinjlim_{i\in I} \widehat{\phi}_i \colon \varinjlim_{i\in I} \mathsf{X}(G_i) \to \varinjlim_{i\in I} H_i^\perp = \bigcup_{i\in I} H_i^\perp = \mathsf{X}(G)$$

satisfying $\Phi([i,\chi_i]) = \chi_i \circ \phi_i$ for all $i \in I$ and $\chi_i \in \mathsf{X}(G_i)$.

B. Replacing G_i with $\phi_i(G_i)$, we may assume that $(G_i)_{i\in I}$ is an upwards directed family of open subgroups of G such that

$$G = \bigcup_{i\in I} G_i = \varinjlim_{i\in I} G_i, \quad \text{and then} \quad \phi_i = (G_i \hookrightarrow G) \ \text{ for all } \ i \in I.$$

For $i \in I$, let $\pi_i \colon G \to G/G_i$ be the residue class epimorphism, suppose that $\nu_i = (G_i \hookrightarrow G)$, and let $\widehat{\pi}_i \colon \mathsf{X}(G/G_i) \overset{\sim}{\to} G_i^\perp$ and $\widehat{\nu}_i \colon \mathsf{X}(G)/G_i^\perp \overset{\sim}{\to} \mathsf{X}(G_i)$ be the topological isomorphisms given in 1.6.5 such that $\widehat{\pi}_i(\chi) = \chi \circ \pi_i$ for all $\chi \in \mathsf{X}(G/G_i)$ and $\widehat{\nu}_i(\chi G_i^\perp) = \chi \restriction G_i$ for all $\chi \in \mathsf{X}(G)$. As G/G_i is discrete by 1.1.5.5, it follows that $G_i^\perp$ is compact by 1.6.3.4(b), and as G_i is compact, it follows that $G_i^\perp$ is open in $\mathsf{X}(G)$ by 1.6.3.3. We assert that

$$\{G_i^\perp \mid i \in I\} \quad \text{is a fundamental system of neighbourhoods of } 1_G \text{ in } \mathsf{X}(G). \tag{†}$$

For the proof of (†) we set $N_1 = \{\mathrm{e}^{2\pi \mathrm{i} t/3} \mid t \in (-1,1)\}$, and by 1.6.3.2 we must show that for every compact subset C of G there is some $i \in I$ such that $G_i^\perp \subset \mathsf{B}(C, N_1)$. If C is a compact subset of G, then there is some $i \in I$ such that $C \subset G_i$, and then $G_i^\perp = \{\chi \in \mathsf{X}(G) \mid \chi(G_i) \subset N_1\} \subset \mathsf{B}(C, N_1)$. $\square$[(†)]

For $i, j \in I$, $i \ge j$, let $\nu_{i,j} = (G_j \hookrightarrow G_i)$ and $\mathsf{X}(\nu_{i,j}) \colon \mathsf{X}(G_i) \to \mathsf{X}(G_j)$. Then $G_i^\perp \subset G_j^\perp$, and we denote by $\pi_{i,j} \colon \mathsf{X}(G)/G_i^\perp \to \mathsf{X}(G)/G_j^\perp$ the canonical epimorphism. If $i \ge j \ge k$, then $\mathsf{X}(\nu_{i,k}) = \mathsf{X}(\nu_{j,k}) \circ \mathsf{X}(\nu_{i,j})$ and $\pi_{i,k} = \pi_{j,k} \circ \pi_{i,j}$. Hence $(\mathsf{X}(G_i), \mathsf{X}(\nu_{i,j}))_I$ and $(\mathsf{X}(G)/G_i^\perp, \pi_{i,j})_I$ are projective systems, and from the commuative diagram

$$\begin{array}{ccc} \mathsf{X}(G)/G_i^\perp & \xrightarrow{\pi_{i,j}} & \mathsf{X}(G)/G_j^\perp \\ {\scriptstyle \widehat{\nu}_i}\big\downarrow & & \big\downarrow{\scriptstyle \widehat{\nu}_j} \\ \mathsf{X}(G_i) & \xrightarrow{\mathsf{X}(\nu_{i,j})} & \mathsf{X}(G_j) \end{array},$$

we derive an isomorphism

$$\widehat{\nu} = \varprojlim_{i\in I} \widehat{\nu}_i \colon \varprojlim_{i\in I} \mathsf{X}(G)/G_i^\perp \overset{\sim}{\to} \varprojlim_{i\in I} \mathsf{X}(G_i) \quad \text{satisfying}$$

$$\widehat{\nu}((\chi G_i^\perp)_{i\in I}) = (\chi \restriction G_i)_{i\in I}.$$

By 1.4.6 the map

$$\Phi \colon \mathsf{X}(G) \overset{\sim}{\to} \varprojlim_{i\in I} \mathsf{X}(G)/G_i^\perp,$$

defined by $\Phi(\chi) = (\chi G_i^\perp)_{i\in I}$ for all $\chi \in \mathsf{X}(G)$, is a topological isomorphism, and therefore $\Psi = \widehat{\nu}\circ\Phi \colon \mathsf{X}(G) \stackrel{\sim}{\to} \varprojlim_{i\in I} \mathsf{X}(G_i)$ is a topological isomorphism satisfying $\Psi(\chi) = (\chi\restriction G_i)_{i\in I} = (\chi\restriction\phi_i)_{i\in I}$ for all $\chi \in \mathsf{X}(G)$.

C. We assume first that $G = \varprojlim_{i\in I} G_i$, and we consider the isomorphism

$$h\colon \mathsf{X}^2(G) \stackrel{\mathsf{X}(\Phi)}{\to} \mathsf{X}(\varinjlim_{i\in I} \mathsf{X}(G_i)) \stackrel{\Psi^*}{\to} \varprojlim_{i\in I} \mathsf{X}^2(G_i)$$

where Ψ^* is the counterpart of Ψ if we replace G_i by $\mathsf{X}(G_i)$ for all $i \in I$. It satisfies $h\big(\widehat{(x_i)_{i\in I}}\big) = (\widehat{x}_i)_{i\in I}$ for all $(x_i)_{i\in I}$ and induces the commutative diagram

$$\begin{array}{ccccc} & & G = \varprojlim_{i\in I} G_i & & \\ & \swarrow^{u_G} & & \searrow^{(u_{G_i})_{i\in I}} & \\ h : \mathsf{X}^2(G) & \xrightarrow[\mathsf{X}(\Phi)]{} & \mathsf{X}(\varinjlim_{i\in I} \mathsf{X}(G_i)) & \xrightarrow[\Psi^*]{} & \varprojlim_{i\in I} \mathsf{X}^2(G_i)\,. \end{array}$$

Hence if all u_{G_i} are isomorphisms, then u_G is an isomorphism.

Now let $G = \varinjlim_{i\in I} G_i = \bigcup_{i\in I} G_i$ and consider the isomorphism

$$h_1\colon \mathsf{X}^2(G) \stackrel{\mathsf{X}(\Psi)^{-1}}{\to} \mathsf{X}(\varprojlim_{i\in I} \mathsf{X}(G_i)) \stackrel{\Phi^{*-1}}{\to} \varinjlim_{i\in I} \mathsf{X}^2(G_i)$$

where Φ^* is the counterpart of Φ if we replace G_i by $\mathsf{X}(G_i)$ for all $i \in I$. It satisfies $h_1\big(\widehat{[x_k,k]}\big) = [\widehat{x}_k, k]$ for all $[x_k,k] \in G$ and induces the commutative diagram

$$\begin{array}{ccccc} & & G = \varinjlim_{i\in I} G_i & & \\ & \swarrow^{u_G} & & \searrow^{[u_{G_k}]} & \\ h : \mathsf{X}^2(G) & \xrightarrow[\mathsf{X}(\Psi)^{-1}]{} & \mathsf{X}(\varprojlim_{i\in I} \mathsf{X}(G_i)) & \xrightarrow[\Phi^{*-1}]{} & \varinjlim_{i\in I} \mathsf{X}^2(G_i)\,. \end{array}$$

Hence if all u_{G_i} are isomorphisms, then u_G is an isomorphism. $\square$

Corollary 1.7.7. *$\mathsf{X}(\mathbb{Q}/\mathbb{Z}) = \widehat{\mathbb{Z}}$ and $\mathsf{X}(\widehat{\mathbb{Z}}) = \mathbb{Q}/\mathbb{Z}$, and if p is a prime, then $\mathsf{X}(\mathbb{Z}(p^\infty)) = \mathbb{Z}_p$ and $\mathsf{X}(\mathbb{Z}_p) = \mathbb{Z}(p^\infty)$.*

Proof. For $n \in \mathbb{N}$ and $a \in \mathbb{Z}$, we consider the character

$$\chi_a \in \mathsf{X}\Big(\frac{1}{n}\mathbb{Z}/\mathbb{Z}\Big), \quad \text{defined by} \quad \chi_a\Big(\frac{x}{n} + \mathbb{Z}\Big) = \mathsf{e}^{2\pi \mathsf{i} a x/n}.$$

Then the assignment $a + n\mathbb{Z} \mapsto \chi_a$ defines an isomorphism

$$\mathbb{Z}/n\mathbb{Z} \stackrel{\sim}{\to} \mathsf{X}\Big(\frac{1}{n}\mathbb{Z}/\mathbb{Z}\Big).$$

Hence we obtain (by means of the usual identifications)

$$\mathsf{X}(\mathbb{Q}/\mathbb{Z}) = \mathsf{X}\Big(\varinjlim_{n\in\mathbb{N}} \frac{1}{n}\mathbb{Z}/\mathbb{Z}\Big) = \varprojlim_{n\in\mathbb{N}} \mathsf{X}\Big(\frac{1}{n}\mathbb{Z}/\mathbb{Z}\Big) = \varprojlim_{n\in\mathbb{N}} \mathbb{Z}/n\mathbb{Z} = \widehat{\mathbb{Z}},$$

and thus also $\mathsf{X}(\widehat{\mathbb{Z}}) = \mathbb{Q}/\mathbb{Z}$. In the same way, if p is a prime, then

$$\mathsf{X}(\mathbb{Z}(p^\infty)) = \mathsf{X}\Big(\varinjlim_{n\in\mathbb{N}} \frac{1}{p^n}\mathbb{Z}/\mathbb{Z}\Big) = \varprojlim_{n\in\mathbb{N}} \mathsf{X}\Big(\frac{1}{p^n}\mathbb{Z}/\mathbb{Z}\Big) = \varprojlim_{n\in\mathbb{N}} \mathbb{Z}/p^n\mathbb{Z} = \widehat{\mathbb{Z}}_p,$$

and thus also $\mathsf{X}(\mathbb{Z}_p) = \mathbb{Z}(p^\infty)$. □

Theorem 1.7.8. *Let $(G_i)_{i\in I}$ be a family of* LCA *groups and suppose that almost all of them are compact. Let*

$$G = \prod_{i\in I} G_i,$$

and for $k \in I$ let $j_k\colon G_k \to G$ be defined by $j_k(x) = (j_k(x)_i)_{i\in I} \in G$, where

$$j_k(x)_i = \begin{cases} x & \text{if } i = k, \\ 1 & \text{if } i \ne k. \end{cases} \quad \text{for all } x \in G_k \text{ and } i \in I.$$

For $\chi \in \mathsf{X}(G)$ and $i \in I$ let $\chi_i = \chi\circ j_i \in \mathsf{X}(G_i)$. Then there exists a topological isomorphism

$$\Phi\colon \mathsf{X}(G) \overset{\sim}{\to} \bigoplus_{i\in I} \mathsf{X}(G_i) \quad \text{such that} \quad \Phi(\chi) = (\chi_i)_{i\in I} \quad \text{for all} \quad \chi \in \mathsf{X}(G).$$

If $x = (x_i)_{i\in I} \in G$ and $\chi \in \mathsf{X}(G)$, then $\chi_i(x_i) = 1$ for almost all $i \in I$, and

$$\chi(x) = \prod_{i\in I} \chi_i(x_i).$$

If all G_i satisfy Pontrjagin's duality theorem, then G also satisfies Pontrjagin's duality theorem.

Proof. If I is finite, the assertion follows by a simple induction from 1.7.5. Thus let I be infinite. Then 1.3.4.**5** shows that

$$G = \varprojlim_{F\in\mathbb{P}_{\text{fin}}(I)} G_F, \quad \text{where} \quad G_F = \prod_{i\in F} G_i \quad \text{for all} \quad F \in \mathbb{P}_{\text{fin}}(I).$$

For $F \in \mathbb{P}_{\text{fin}}(I)$ let $\Phi_F\colon G \to G_F$ be the projection of the projective limit, and let $H_F = \operatorname{Ker}(\Phi_F)$. Then 1.7.6.**A**.1 implies

$$\mathsf{X}(G) = \varinjlim_{F\in\mathbb{P}_{\text{fin}}(I)} H_F^\perp, \quad \text{and} \quad H_F^\perp = \{\chi \in \mathsf{X}(G) \mid \chi_i = 1 \text{ for all } i \in I\setminus F\}.$$

Apparently, the map

$$H_F^\perp \to \mathsf{X}(G_F) = \prod_{i\in F} \mathsf{X}(G_i), \quad \chi \mapsto (\chi_i)_{i\in F}$$

is a topological isomorphism. We identify, and by 1.7.6.**A** we obtain a topological isomorphism

$$\Phi\colon \mathsf{X}(G) = \varinjlim_{F\in\mathbb{P}_{\mathrm{fin}}(I)} H_F^\perp \xrightarrow{\sim} \varinjlim_{F\in\mathbb{P}_{\mathrm{fin}}(I)} \prod_{i\in F} \mathsf{X}(G_i) = \bigoplus_{i\in I} \mathsf{X}(G_i)$$

such that $\Phi(\chi) = (\chi_i)_{i\in I}$. Due to our identifications and the validity of the theorem for finite products, it is now clear that

$$\chi(x) = \prod_{i\in I} \chi_i(x_i) \quad \text{for all } \; x = (x_i)_{i\in I} \in G, \quad \chi = (\chi_i)_{i\in I} \in \bigoplus \mathsf{X}(G_i),$$

and that G satisfies Pontrjagin's duality theorem provided that all G_i do so. □

The following main result of this section considers important special cases of Pontrjagin's dualty theorem.

Theorem 1.7.9.

1. *Pontrjagin's duality theorm* 1.6.6 *holds*

- *for the additive groups* $\mathbb{R}^n$ *and* $\mathbb{C}^n$ *for all* $n \in \mathbb{N}$*;*
- *for the multiplicative groups* $\mathbb{T}$, $\mathbb{R}^\times$ *and* $\mathbb{C}^\times$*;*
- *for all abelian profinite groups;*
- *for all discrete abelian groups.*

2. *Let* G *be a (discrete) abelian group. Then* G *is a torsion group if and only if* $\mathsf{X}(G)$ *is a profinite group.*

3. *Let* G *be a (discrete) abelian torsion group or a profinite group. If* $\chi \in \mathsf{X}(G)$*, then* $\chi(G) \subset \mathbb{T}_\mathbb{Q}$*, and if* G *is profinite, then* $\mathsf{X}(G)$ *is a finite cyclic group. In any case,* $\mathsf{X}(G) = \mathrm{Hom}^{\mathrm{c}}(G, \mathbb{T}_\mathbb{Q})$ *is the group of all continuous homomorphisms* $G \to \mathbb{T}_\mathbb{Q}$*.*

Proof. 1. Let $n \in \mathbb{N}$. By 1.7.3 and 1.7.8, Pontrjagin's duality theorem holds for $\mathbb{R}^n$, and as $\mathbb{C}^n \cong \mathbb{R}^{2n}$, it also holds for $\mathbb{C}^n$.

By 1.7.4, Pontrjagin's duality theorem holds for $\mathbb{T}$ and $\mathbb{Z}$. Moreover, since $\mathbb{R}^\times \cong \{\pm 1\} \times \mathbb{R}$ and $\mathbb{C}^\times \cong \mathbb{T} \times \mathbb{R}$ by means of the isomorphism

$$a \mapsto \Big(\frac{a}{|a|}, \log|a|\Big),$$

Pontrjagin's duality theorem holds for $\mathbb{C}^\times$ and $\mathbb{R}^\times$ by 1.7.5.

An abelian profinite group G is the projective limit of a surjective projective system of finite abelian groups, and as Pontrjagin's duality theorem holds for all finite abelian groups, it holds for G by 1.7.6.

If G is a finitely generated abelian group, then $G \cong \mathbb{Z}^r \times F$ for some $r \in \mathbb{N}_0$ and a finite abelian group F (see [18, Theorem 2.3.6]). Hence the dualtiy theorem holds for G by 1.7.4, and 1.7.8. If G is any abelian group, then the family of its finitely generated subgroups is an upwards directed family with union G, and thus Pontrjagin's duality theorem holds for G by 1.7.6.

2. Let Σ be the upwards directed set of all finitely generated subgroups of G. Then

$$G = \bigcup_{U \in \Sigma} U = \varinjlim_{U \in \Sigma} U \quad \text{and} \quad \mathsf{X}(G) = \varprojlim_{U \in \Sigma} \mathsf{X}(U).$$

Now G is a torsion group if and only if all $U \in \Sigma$ are finite, and then $\mathsf{X}(G)$ is a profinite group. Conversely, if $\mathsf{X}(G)$ is profinite and $U \in \Sigma$, then $\mathsf{X}(U)$ is a factor group of $\mathsf{X}(G)$ by 1.6.7.1(a), hence profinite, and as it is discrete, it is finite.

3. Let $\chi \in \mathsf{X}(G)$. If G is an abelian torsion group, then $\chi(G)$ is a torsion subgroup of $\mathbb{T}$ and thus $\chi(G) \subset \mathbb{T}_{\mathbb{Q}}$. If G is profinite, then $\operatorname{Ker}(\chi) = \chi^{-1}(N_1)$ by 1.6.1.4, hence $\chi(G) \cong G/\operatorname{Ker}(\chi)$ is finite, and thus $\chi(G) = \mu_m$ for some $m \in \mathbb{N}$.

In particular, in any case we obtain $\mathsf{X}(G) = \operatorname{Hom}^{\mathsf{c}}(G, \mathbb{T}_{\mathbb{Q}})$. □

Theorem 1.7.10. *Let* $(K, |\cdot|)$ *be a locally compact valued field* (see 1.7.2).

1. If K is non-Archimedean, then Pontrjagin's duality theorem holds for the additive groups K and $\mathcal{O}_K$, and for the multiplicative groups $K^\times$ and U_K.

2. Let $\lambda \in \mathsf{X}(K) \setminus \{1_K\}$. Then the map

$$\phi\colon K \to \mathsf{X}(K), \quad \text{defined by} \quad \phi(y)(x) = \lambda(yx) \quad \text{for all} \quad x, y \in K,$$

is a topological isomorphism.

Proof. 1. By 1.5.4, we obtain

$$K = \varprojlim_{n \in \mathbb{N}} K/\mathfrak{p}_K^n, \quad \mathcal{O}_K = \varprojlim_{n \in \mathbb{N}} \mathcal{O}_K/\mathfrak{p}_K^n, \quad K^\times = \varprojlim_{n \in \mathbb{N}} K^\times/U_K^{(n)}, \quad \text{and}$$
$$U_K = \varprojlim_{n \in \mathbb{N}} U_K/U_K^{(n)}.$$

Since all these groups are projective limits of discrete groups, they satisfy Pontrjagin's duality theorem by 1.7.9 and 1.7.6.

2. By 1., 1.7.3 and 1.7.2. □

Theorem 1.7.11 (Character groups of $\mathbb{R}^\times$ and $\mathbb{C}^\times$). *For $p \in \mathbb{Z}$ and $y \in \mathbb{R}$ let*

$$\chi_{p,y}\colon \mathbb{C}^\times \to \mathbb{T} \quad \text{be defined by} \quad \chi_{p,y}(a) = \Big(\frac{a}{|a|}\Big)^p |a|^{2\pi i y} \quad \text{for all } a \in \mathbb{C}^\times.$$

Then

$$\mathsf{X}(\mathbb{C}^\times) = \{\chi_{p,y} \mid p \in \mathbb{Z},\ y \in \mathbb{R}\} \text{ and } \mathsf{X}(\mathbb{R}^\times) = \{\chi_{p,y}\restriction\mathbb{R}^\times \mid p \in \{0,1\},\ y \in \mathbb{R}\}.$$

In particular: $\mathsf{X}(\mathbb{C}^\times) \cong \mathbb{Z}\times\mathbb{R}$ *and* $\mathsf{X}(\mathbb{R}^\times) \cong \{\pm1\}\times\mathbb{R} \cong \mathbb{R}^\times$. $\chi_{0,0} = 1$ *is the only continuos character of finite order of* $\mathbb{C}^\times$, *and* $\chi_{1,0}\restriction\mathbb{R}$ *and* 1 *are the only continuous characters of finite order of* $\mathbb{R}^\times$

Proof. We consider the isomorphism

$$\phi\colon \mathbb{C}^\times \to \mathbb{T}\times\mathbb{R}, \quad \text{given by} \quad \phi(a) = \Big(\frac{a}{|a|}, \log|a|\Big) \quad \text{for all } a \in \mathbb{C}^\times.$$

By 1.7.4.2 and 1.7.3, we obtain

$$\mathsf{X}(\mathbb{T}) = \{\varepsilon_p \mid p \in \mathbb{Z}\} \cong \mathbb{Z}, \quad \text{where } \varepsilon_p(u) = u^p \text{ for all } u \in \mathbb{T},$$

and

$$\mathsf{X}(\mathbb{R}) = \{\chi_y \mid y \in \mathbb{R}\} \cong \mathbb{R}, \quad \text{where } \chi_y(x) = \mathsf{e}^{2\pi i x y} \text{ for all } x \in \mathbb{R}.$$

By means of 1.7.5 and allowing for the above identifications, it follows that $\mathsf{X}(\mathbb{C}^\times) = \{\varepsilon_p\chi_y \mid p \in \mathbb{Z},\ y \in \mathbb{R}\}$, where

$$\begin{aligned}\varepsilon_p\chi_y(a) &= \varepsilon_p\chi_y\Big(\frac{a}{|a|}\mathsf{e}^{\log|a|}\Big) = \varepsilon_p\Big(\frac{a}{|a|}\Big)\chi_y(\log|a|) = \Big(\frac{a}{|a|}\Big)^p \mathsf{e}^{2\pi i y\log|a|}\\ &= \Big(\frac{a}{|a|}\Big)^p |a|^{2\pi i y}\\ &= \chi_{p,y}(a).\end{aligned}$$

This proves the assertions for $\mathbb{C}^\times$.

By 1.6.9.1 (regarding 1.7.9), the assignment $\chi \mapsto \chi\restriction\mathbb{R}^\times$ defines an epimorphism $\mathsf{X}(\mathbb{C}^\times) \to \mathsf{X}(\mathbb{R}^\times)$. Since

$$\phi(\mathbb{R}^\times) = \{\pm1\}\times\mathbb{R} \text{ and } \{\varepsilon_p\restriction\{\pm1\} \mid p \in \mathbb{Z}\} = \{\varepsilon_0, \varepsilon_1\},$$

the assertions for $\mathbb{R}^\times$ follow. □

1.8 Infinite Galois theory

The following example 1.8.1 shows that the main theorem of Galois theory as achieved in [18, Theorem 1.5.4] for finite field extensions fails for infinite extensions and must be modified. In the sequel we tacitly make use of the Galois theory of finite field extensions (see [18, Section 1.5]).

Example 1.8.1. Let E be a finite field, $|E| = q$, $\operatorname{char}(E) = p$, $\overline{E}$ an algebraic closure of E, $G = \operatorname{Gal}(\overline{E}/E)$, $\varphi_E \in G$ the Frobenius automorphismus over E, defined by $\varphi_E(x) = x^q$ for all $x \in \overline{E}$ (see [18, Theorem 1.7.1]), and $G_0 = \langle \varphi_E \rangle$. Then $E = \overline{E}^{G_0}$, but we assert that $G_0 \subsetneq G$, contradicting the corresponding facts for finite field extensions.

Proof. For $n \in \mathbb{N}$ we set $n = p^{v_p(n)} n_0$ (so that $n_0 \in \mathbb{N}$ and $p \nmid n_0$), we let $x_n, y_n \in \mathbb{Z}$ be such that $n_0 x_n + p^{v_p(n)} y_n = 1$, and we set $a_n = n_0 x_n$. Then $a_n \equiv 0 \mod n_0$ and $a_n \equiv 1 \mod p^{v_p(n)}$ for all $n \in \mathbb{N}$. With these notations we prove:

> **A.** If $m, n \in \mathbb{N}$ and $m \,|\, n$, then $a_m \equiv a_n \mod m$, but there is no integer $a \in \mathbb{Z}$ such that $a \equiv a_n \mod n$ for all $n \in \mathbb{N}$.

Proof of **A.** Let $m, n \in \mathbb{N}$ and $m \,|\, n$. Since $v_p(m) \le v_p(n)$ and $m_0 \,|\, n_0$, we obtain $a_m - a_n \equiv 0 \mod p^{v_p(m)}$ and $a_m - a_n \equiv 0 \mod m_0$, and consequently $a_m - a_n \equiv 0 \mod m$.

Assume that there is some $a \in \mathbb{Z}$ such that $a \equiv a_n \mod n$ for all $n \in \mathbb{N}$. If $p \nmid n$, then $n_0 = n$, hence $a \equiv a_n \equiv 0 \mod n$ and thus $a = 0$. If $n = p^{v_p(n)}$ for some $e \in \mathbb{N}$, then $n_0 = 1$ and $a \equiv a_n \equiv 1 \mod p^{v_p(n)}$ and thus $a = 1$, a contradiction. $\square$[**A.**]

For $n \in \mathbb{N}$ let E_n be the unique intermediate field of $\overline{E}/E$ such that $[E_n : E] = n$ and consider the automorphism $\psi_n = \varphi_E^{a_n} \restriction E_n \in \operatorname{Gal}(E_n/E)$. If $m \in \mathbb{N}$ and $m \,|\, n$, then

$$\psi_n \restriction E_m = \varphi_E^{a_n} \restriction E_m = \varphi_E^{a_m} \restriction E_m = \psi_m \restriction E_m,$$

and therefore there exists a unique $\psi \in G$ such that $\psi \restriction E_n = \psi_n$ for all $n \in \mathbb{N}$. We assert that $\psi \notin G_0$. Assume to the contrary that $\psi = \varphi_E^a$ for some $a \in \mathbb{Z}$. If $n \in \mathbb{N}$, then $\varphi_E^a \restriction E_n = \psi \restriction E_n = \psi_n \restriction E_n = \varphi_E^{a_n} \restriction E_n$, hence $\varphi_E^{a-a_n} \restriction E_n = \operatorname{id}_{E_n}$, and therefore $a - a_n \equiv 0 \mod n$, since $\operatorname{ord}(\varphi_E \restriction E_n) = n$. Hence $a \equiv a_n \mod n$ for all $n \in \mathbb{N}$, which contradicts **A**.

In order to recover the main theorem of Galois theory for infinite extensions, we endow the Galois group with a suitable topology, the **Krull topology**, and then we obtain a bijection between intermediate fields and closed subgroups.

Remark and Definition 1.8.2. Let L/K be a (not necessarily finite) Galois extension, $G = \operatorname{Gal}(L/K)$ and $\mathcal{E}(L/K)$ the set of all over K finite Galois intermediate fields of L/K. Then

$$L = \bigcup_{E \in \mathcal{E}(L/K)} E,$$

and for $E \in \mathcal{E}(L/K)$ we define $\phi_E \colon G \to \operatorname{Gal}(E/K)$ by $\phi_E(\sigma) = \sigma \restriction E$. Then ϕ_E is an epimorphism, and $\operatorname{Ker}(\phi_E) = \operatorname{Gal}(L/E)$. If $E, E' \in \mathcal{E}(L/K)$, then

$EE' \in \mathcal{E}(L/K)$, and therefore $(\mathcal{E}(L/K), \subset)$ is a directed set. For intermediate fields E, $E' \in \mathcal{E}(L/K)$ with $E \supset E'$ we define

$$\phi_{E,E'}\colon \mathrm{Gal}(E/K) \to \mathrm{Gal}(E'/K) \text{ by } \phi_{E,E'}(\sigma) = \sigma \upharpoonright E' \text{ for all } \sigma \in \mathrm{Gal}(E/K).$$

If E, E', $E'' \in \mathcal{E}(L/K)$ and $E \supset E' \supset E''$, then $\phi_{E,E''} = \phi_{E',E''} \circ \phi_{E,E'}$, and obviously $\phi_{E'} = \phi_{E,E'} \circ \phi_E$. Therefore $(\mathrm{Gal}(E/K), \phi_{E,E'})_{\mathcal{E}(L/K)}$ is a projective system, and by 1.3.4.**3.** we obtain a homomorphism

$$\phi = \varprojlim \phi_E\colon G \to \varprojlim_{E \in \mathcal{E}(L/K)} \mathrm{Gal}(E/K) \quad \text{satisfying} \quad \phi(\sigma) = (\sigma \upharpoonright E)_{E \in \mathcal{E}(L/K)}.$$

Apparently ϕ is an isomorphism, and we identify:

$$G = \mathrm{Gal}(L/K) = \varprojlim_{E \in \mathcal{E}(L/K)} \mathrm{Gal}(E/K).$$

In this way G is a profinite group, hence compact and totally disconnected, and the homomorphisms ϕ_E are the (continuous) projections of the projective limit.

A directed subset $\mathcal{E}'$ of $\mathcal{E}(L/K$ is cofinal if and only if L is the compositum of the fields in $\mathcal{E}'$. If this is the case, then 1.3.6 shows that

$$G = \varprojlim_{E \in \mathcal{E}'} \mathrm{Gal}(E/K),$$

and $\{\mathrm{Gal}(L/E) \mid E \in \mathcal{E}'\}$ is a fundamental system of neighbourhoods of 1 in G consisting of open normal subgroups. In particular, if L/K is a finite extensions, our definition is consistent with the classical one.

Theorem 1.8.3 (Main theorem of Krull's Galois theory). *Let L/K be a Galois extension and $G = \mathrm{Gal}(L/K)$.*

1. *Let $\mathcal{Z}(L/K)$ be the set of all intermediate fields of L/K and $\mathcal{U}(G)$ the set of all closed subgroups of G. Then the maps*

$$\begin{cases} \mathcal{Z}(L/K) & \to & \mathcal{U}(G) \\ M & \mapsto & \mathrm{Gal}(L/M) \end{cases} \quad \text{and} \quad \begin{cases} \mathcal{U}(G) & \to & \mathcal{Z}(L/K) \\ H & \mapsto & L^H \end{cases}$$

are mutually inverse inclusion-reversing bijections. If Y is a subset of G, then

$$L^Y = L^{\overline{\langle Y \rangle}} \quad \text{and} \quad \mathrm{Gal}(L/L^Y) = \overline{\langle Y \rangle}.$$

2. *Let M, M' be intermediate fields of L/K, $H = \mathrm{Gal}(L/M)$ and $H' = \mathrm{Gal}(L/M')$.*

- *$M \subset M'$ if and only if $H \supset H'$.*
- *$MM' = L^{H \cap H'}$ and $H \cap H' = \mathrm{Gal}(L/MM')$.*
- *$M \cap M' = L^{\overline{\langle H \cup H' \rangle}}$ and $\overline{\langle H \cup H' \rangle} = \mathrm{Gal}(L/M \cap M')$.*

3. Let M be an intermediate field of L/K and $H = \mathrm{Gal}(L/M)$. Then M/K is Galois if and only if H is a normal subgroup of G. If this is the case, then the map

$$\rho\colon G \to \mathrm{Gal}(M/K), \quad \textit{defined by} \quad \rho(\sigma) = \sigma \restriction M,$$

is a continuous epimorphism, $\mathrm{Ker}(\rho) = H$, and ρ induces a topological isomorphism $\rho^* : G/H \stackrel{\sim}{\to} \mathrm{Gal}(M/K)$.
We identify: $G/H = \mathrm{Gal}(M/K)$, $\sigma H = \sigma \restriction M$ for all $\sigma \in G$.

Proof. 1. **a.** Let $M \in \mathcal{Z}(L/K)$ and $H = \mathrm{Gal}(L/M)$. We shall prove that H is closed in G and $M = L^H$.

Assume first that $[M:K] < \infty$, and let N be the normal closure of M/K in L. Then $[N:K] < \infty$, and thus $\mathrm{Gal}(L/N)$ is an open normal subgroup of G. Let $\mathrm{Gal}(N/M) = \{\tau_1, \ldots, \tau_m\}$, and for $j \in [1, m]$ let $\overline{\tau}_j \in G$ be such that $\overline{\tau}_j \restriction N = \tau_j$. If $\sigma \in H$, then $\sigma \restriction N = \tau_j = \overline{\tau}_j \restriction N$ for some $j \in [1, m]$, hence $\sigma \in \overline{\tau}_j \mathrm{Gal}(L/N)$, and consequently

$$H = \mathrm{Gal}(L/M) = \bigcup_{j=1}^{m} \overline{\tau}_j \mathrm{Gal}(L/N)$$

is an open and thus closed subgroup of G.

Now let $M \in \mathcal{Z}(L/K)$ be arbitrary and $\mathcal{E}_0(M/K)$ the set of all over K finite intermediate fields of M/K. Then $\mathcal{E}_0(L/K)$ is directed,

$$M = \bigcup_{E \in \mathcal{E}_0(M/K)} E, \quad \text{and therefore}$$

$$H = \mathrm{Gal}(L/M) = \bigcap_{E \in \mathcal{E}_0(M/K)} \mathrm{Gal}(L/E) \quad \text{is closed.}$$

Apparently $M \subset L^H$, and we prove equality. Let $z \in L^H$, and let N be the normal closure of $M(z)/M$ in L. Then $[N:M] < \infty$,

$$\rho\colon H = \mathrm{Gal}(L/M) \to \mathrm{Gal}(N/M), \text{ defined by } \rho(\tau) = \tau \restriction N,$$

is an epimorphism, and thus $z \in N^{\rho(H)} = N^{\mathrm{Gal}(N/M)} = M$.

b. Let $H \in \mathcal{U}(G)$ and $M = L^H$. Then $H \subset \mathrm{Gal}(L/M)$, and equality holds once we have proved that H is dense in $\mathrm{Gal}(L/M)$. Let $\sigma \in \mathrm{Gal}(L/M)$, and let U be a neighbourhood of σ in G. Then there exists an over K finite Galois intermediate field M_0 of L/K such that $\sigma\mathrm{Gal}(L/M_0) \subset U$, and we shall prove that $\sigma\mathrm{Gal}(L/M_0) \cap H \neq \emptyset$.

Let $\rho\colon G \to \mathrm{Gal}(M_0/K)$ be defined by $\rho(\tau) = \tau \upharpoonright M_0$. Then we obtain $M_0^{\rho(H)} = M \cap M_0$, and $\rho(H) = \mathrm{Gal}(M_0/M \cap M_0)$. Since moreover $\sigma \upharpoonright M_0 \in \mathrm{Gal}(M_0/M \cap M_0) = \rho(H)$, it follows that $\sigma \upharpoonright M_0 = \tau \upharpoonright M_0$ for some $\tau \in H$, and consequently $\tau \in \sigma\mathrm{Gal}(L/M_0) \cap H$.

By **a** and **b** the maps in 1. are mutually inverse bijections, and they are obviously inclusion-reversing.

c. Let Y be any subset of G. Then $H = \mathrm{Gal}(L/L^Y)$ is a closed subgroup of G, and $Y \subset H$ implies $\overline{\langle Y\rangle} \subset H$. Since $L^{\overline{\langle Y\rangle}} \subset L^Y = L^H \subset L^{\overline{\langle Y\rangle}}$, it follows that $L^Y = L^{\overline{\langle Y\rangle}}$ and $\mathrm{Gal}(L/L^Y) = \overline{\langle Y\rangle}$.

2. MM' is the smallest $M \cup M'$ containing intermediate field of L/K, and $M \cap M'$ is the largest intermediate field of L/K contained in both M and M'. On the other hand, $H \cap H'$ is the largest closed subgroup of G contained in both H and H', and $\overline{\langle H \cup H'\rangle}$ is the smallest $H \cup H'$ containing closed subgroup of G. With these observations in mind, the assertions follow by 1.

3. The extension M/K is Galois if and only if $\sigma M = M$ for all $\sigma \in G$, and by 1. this holds if and only if $\mathrm{Gal}(L/\sigma M) = \sigma H \sigma^{-1} = H$ for all $\sigma \in G$. Hence M/K is Galois if and only if H is a normal subgroup of G.

Now let H be a normal subgroup of G. Then $\rho\colon G \to \mathrm{Gal}(M/K) = \mathrm{Hom}_K(M, L)$, defined by $\rho(\sigma) = \sigma \upharpoonright M$, is an epimorphism, $\mathrm{Ker}(\rho) = H$, and if M_0 is any over K finite Galois intermediate field of M/K, then $\rho^{-1}(\mathrm{Gal}(M/M_0)) = \mathrm{Gal}(L/M_0)$ is open in G. Hence ρ is continuous and as G/H is compact, it follows that ρ induces a topological isomorphism $\rho^*\colon G/H \overset{\sim}{\to} \mathrm{Gal}(M/K)$. □

Theorem 1.8.4 (Shifting theorem of Galois theory)**.** *Let L/K be a Galois extension, N an extension field of L, K' another intermediate field of N/K and $L' = LK'$.*

1. L'/K' is a Galois extension, the map

$$\rho\colon \mathrm{Gal}(L'/K') \overset{\sim}{\to} \mathrm{Gal}(L/L \cap K') \subset \mathrm{Gal}(L/K), \text{ defined by } \rho(\sigma) = \sigma \upharpoonright L,$$

is a topological isomorphism. If moreover H is a subgroup of $\mathrm{Gal}(L/K)$, then $L'^{\rho^{-1}(H)} = L^H K'$. .

2. If K'/K is Galois, then L'/K is Galois, too. If in addition $L \cap K' = K$, then the maps

$$\begin{cases} \mathrm{Gal}(L'/K') & \overset{\sim}{\to} & \mathrm{Gal}(L/K) \\ \tau & \mapsto & \tau \upharpoonright L \end{cases}, \qquad \begin{cases} \mathrm{Gal}(L'/L) & \overset{\sim}{\to} & \mathrm{Gal}(K'/K) \\ \tau & \mapsto & \tau \upharpoonright K' \end{cases}$$

and

$$\rho\colon \begin{cases} \mathrm{Gal}(L'/K) & \overset{\sim}{\to} & \mathrm{Gal}(L/K) \times \mathrm{Gal}(K'/K) \\ \sigma & \mapsto & (\sigma \upharpoonright L, \sigma \upharpoonright K'), \end{cases}$$

are topological isomorphisms. In particular,

$$\mathrm{Gal}(L'/K) = \mathrm{Gal}(L'/K')\cdot\mathrm{Gal}(L'/L).$$

Proof. 1. The extension L'/K' is Galois (see [18, Corollary 1.4.5]), and if $\sigma \in \mathrm{Gal}(L'/K')$, then $\sigma\restriction L \in \mathrm{Hom}^K(L, N) = \mathrm{Gal}(L/K)$ and $\sigma\restriction L\cap K' = \mathrm{id}_{L\cap K'}$, which implies $\sigma\restriction L \in \mathrm{Gal}(L/L\cap K')$. Hence we obtain a homomorphism

$$\rho\colon \mathrm{Gal}(L'/K') \to \mathrm{Gal}(L/L\cap K'), \quad \text{defined by } \rho(\sigma) = \sigma\restriction L,$$

and we prove first that ρ is continuous. Thus let K_1 be any over $L\cap K'$ finite Galois intermediate field of $L/L\cap K'$. Then $K'K_1/K'$ is a finite Galois extension, and $\rho^{-1}(\mathrm{Gal}(L/K_1)) = \mathrm{Gal}(LK'/K_1K')$ is open in $\mathrm{Gal}(L'/K')$. Hence ρ is continuous.

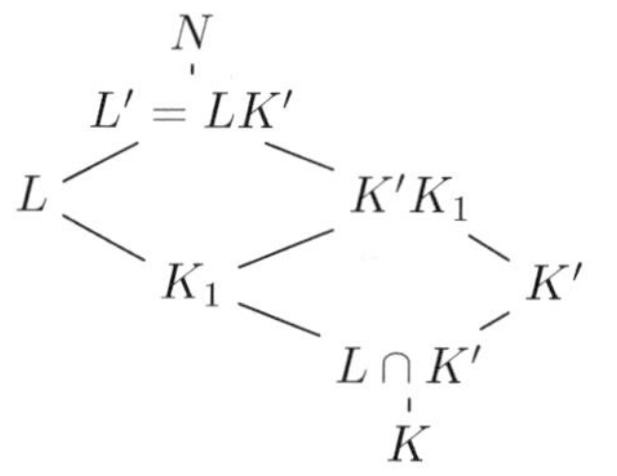

If $\sigma \in \ker(\rho)$, then $\sigma\restriction L = \mathrm{id}_L$, and $\sigma\restriction M = \mathrm{id}_M$ implies $\sigma = \mathrm{id}_{LM}$. Hence ρ is injective. If $H = \rho(\mathrm{Gal}(LM/M))$, then H is a closed subgroup of $\mathrm{Gal}(L/L\cap M)$. Apparently, $L\cap M \subset L^H$, and if $z \in L^H$, then $\sigma(z) = z$ for all $\sigma \in \mathrm{Gal}(LM/M)$, hence $z \in M$. Now $L^H = L\cap M$, $H = \mathrm{Gal}(L/L\cap M)$, and ρ is surjective. Altogether ρ is a continuous isomorphism, and as $\mathrm{Gal}(LM/M)$ is compact, ρ is a topological isomorphism.

Now let H be a subgroup of $\mathrm{Gal}(L/K)$ and $\sigma \in \mathrm{Gal}(L'/K')$. Then

$$\begin{aligned}\sigma\restriction L^HK' = \mathrm{id}_{L^HK'} &\iff \sigma\restriction L^H = \mathrm{id}_{L^H} \iff \rho(\sigma)\restriction L^H = \mathrm{id}_{L^H}\\ &\iff \rho(\sigma) \in H.\end{aligned}$$

Hence $\mathrm{Gal}(L'/L^HK') = \rho^{-1}(H)$, and therefore $L^HK' = L'^{\rho^{-1}(H)}$.

2. Let K'/K be Galois. Then L'/K' is Galois, and if $L\cap K' = K$, then the map

$$\rho\colon \mathrm{Gal}(L'/K) \stackrel{\sim}{\to} \mathrm{Gal}(L/K)\times\mathrm{Gal}(K'/K), \text{ defined by } \rho(\sigma) = (\sigma\restriction L, \sigma\restriction K'),$$

is a continuous monomorphism. As $\mathrm{Gal}(L'/K)$ is compact, it suffices to prove that ρ is surjective.

The restrictions $\mathrm{Gal}(L'/K') \to \mathrm{Gal}(L/K)$ and $\mathrm{Gal}(L'/L) \to \mathrm{Gal}(K'/K)$ are topological isomorphisms by 1. If $(\tau_1, \tau_2) \in \mathrm{Gal}(L/K)\times\mathrm{Gal}(K'/K)$, then there exists some

$$(\sigma_1, \sigma_2) \in \mathrm{Gal}(L'/K')\times\mathrm{Gal}(L'/L) \subset \mathrm{Gal}(L'/K)\times\mathrm{Gal}(L'/K)$$

such that $\sigma_1\restriction L = \tau_1$, $\sigma_2\restriction K' = \tau_2$, and consequently $\rho(\sigma_1\circ\sigma_2) = (\tau_1, \tau_2)$. □

Let $n \in \mathbb{N}$. An abelian profinite group is called **of exponent** n if $G^n = \mathbf{1}$ (that is, $\sigma^n = 1$ for all $\sigma \in G$). A field extension L/K is called **abelian of exponent** n if it is Galois and $\mathrm{Gal}(L/K)$ is abelian of exponent n.

Theorem 1.8.5. *Let L/K be a field extension and $n \in \mathbb{N}$.*

1. Let $(L_i)_{i\in I}$ be a family of over K Galois intermediate fields of L/K, and let

$$L = \prod_{i\in I} L_i$$

be its compositum. Then L/K is Galois, and the map

$$\Phi\colon \mathrm{Gal}(L/K) \to \overline{G} = \prod_{i\in I} \mathrm{Gal}(L_i/K), \quad \textit{defined by } \Phi(\sigma) = (\sigma \restriction L_i)_{i\in I},$$

is a continuous monomorphism. If all L_i/K are abelian [of exponent n], then L/K is abelian [of exponent n], too.

2. Let L/K be Galois, $G = \mathrm{Gal}(L/K)$, and let L^{a} [$L^{[n]}$] be the compositum of all over K finite abelian intermediate fields [of exponent n] of L/K. Then L^{a} [$L^{[n]}$] is the largest over K abelian intermediate field [of exponent n] of L/K,

$$\mathrm{Gal}(L/L^{\mathsf{a}}) = G^{(1)} \textit{ and } \mathrm{Gal}(L^{\mathsf{a}}/K) = G^{\mathsf{a}}$$
$$[\,\mathrm{Gal}(L/L^{[n]}) = G^{(1)}G^n \textit{ and } \mathrm{Gal}(L^{[n]}/K) = G^{\mathsf{a}}/(G^{\mathsf{a}})^n\,].$$

If K' is an intermediate field of L/K and L'^{a} [$L'^{[n]}$] is the largest over K' abelian intermediate field [of exponent n] of L/K', then $L^{\mathsf{a}} \subset L'^{\mathsf{a}}$ [$L^{[n]} \subset L'^{[n]}$].

$$\begin{array}{ccccccc}
L^{\mathsf{a}} & \text{——} & L^{\mathsf{a}}K' & \text{—} & L'^{\mathsf{a}} & \text{—} & L \\
| & & | & & & & \\
K' \cap L^{\mathsf{a}} & \text{——} & K' & & & & \\
| & & & & & & \\
K & & & & & &
\end{array}$$

Proof. 1. Apparently L/K is Galois (see [18, Theorem 1.4.8]), and Φ is a monomorphism. It remains to prove that Φ is continuous. A fundamental system of neighbourhoods of 1 in $\overline{G}$ is given by the sets of the form

$$U = \prod_{i\in J} \mathrm{Gal}(L_i/E_i) \times \prod_{i\in I\setminus J} \mathrm{Gal}(L_i/K) \subset \prod_{i\in I} \mathrm{Gal}(L_i/K),$$

where $J \in \mathbb{P}_{\mathrm{fin}}(I)$, and E_i is an over K finite Galois intermediate field of L/K for every $i \in J$. For every such set U its pre-image

$$\Phi^{-1}(U) = \mathrm{Gal}\Big(L\Big/\prod_{i\in J} E_i\Big) \text{ is open in } G, \text{ since } \prod_{i\in J} E_i\Big/K \text{ is finite Galois.}$$

Hence Φ is continuous. If all L_i/K and thus all groups $\mathrm{Gal}(L_i/K)$ are abelian [of exponent n], then $\overline{G}$ (hence $\mathrm{Gal}(L/K)$ and thus L/K) is abelian [of exponent n].

2. Let $(L_i)_{i\in I}$ be the family of all over K finite abelian intermediate fields [of exponent n] of L/K and L^{a} [$L^{[n]}$] its compositum. Then L^{a}/K [$L^{[n]}(K$] is abelian [of exponent n] by 1., and we prove that it is the largest intermediate field of L/K with this property. Thus let L_1 be any over K abelian intermediate field [of exponent n] of L/K and $a \in L_1$. Then there exists an over K finite intermediate field L_1' of L_1/K such that $a \in L_1'$. Hence we get $L_1' \in \{L_i \mid i \in I\}$ and therefore $a \in L_1' \subset L^{\mathsf{a}}$ [$a \in L_1' \subset L^{[n]}$], which implies $L_1 \subset L^{\mathsf{a}}$ [$L_1 \subset L^{[n]}$].

To calculate the Galois groups, we apply 1.8.3.3. The map

$$\rho\colon G \to \mathrm{Gal}(L^{\mathsf{a}}/K),\ \ \sigma \mapsto \sigma \restriction L^{\mathsf{a}} \ \ [\rho\colon G \to \mathrm{Gal}(L^{[n]}/K),\ \ \sigma \mapsto \sigma \restriction L^{[n]}]$$

is a continuous epimorphism, and since $\mathrm{Gal}(L^{\mathsf{a}}/K)$ [$\mathrm{Gal}(L^{[n]}/K$] is abelian [of exponent n], it follows that

$$G^{(1)} \subset \mathrm{Ker}(\rho) = \mathrm{Gal}(L/L^{\mathsf{a}}) \qquad [G^{(1)}G^n \subset \mathrm{Ker}(\rho) = \mathrm{Gal}(L/L^{[n]})]$$

and thus $L^{\mathsf{a}} \subset L^{G^{(1)}}$ [$L^{[n]} \subset L^{G^{(1)}G^n}$] (observe that $G^{(1)}G^n$ is compact by 1.1.2.4(c) and thus closed in G).

Since $L^{G^{(1)}}/K$ is Galois and $\mathrm{Gal}(L^{G^{(1)}}/K) = G/G^{(1)} = G^{\mathsf{a}}$ is abelian [$L^{G^{(1)}G^n}/K$ is Galois and $\mathrm{Gal}(L^{G^{(1)}G^n}/K) = G/G^{(1)}G^n = G^{\mathsf{a}}/(G^{\mathsf{a}})^n$ is abelian of exponent n], it follows that $L^{G^{(1)}} \subset L^{\mathsf{a}}$ [$L^{G^{(1)}G^n} \subset L^{[n]}$]. Hence $L^{G^{(1)}} = L^{\mathsf{a}}$ [$L^{G^{(1)}G^n} = L^{[n]}$], $\mathrm{Gal}(L(L^{\mathsf{a}}) = G^{(1)}$ [$\mathrm{Gal}(L/L^{[n]} = G^{(1)}G^n$] and $\mathrm{Gal}(L^{\mathsf{a}}/K) = G^{\mathsf{a}}$ [$\mathrm{Gal}(K^{[n]}/K = G^{\mathsf{a}}(G^{\mathsf{a}})^n$].

By 1.8.4.1 it follows that $L^{\mathsf{a}}K'/K'$ [$L^{[n]}K'/K'$] is a Galois extension, and that $\mathrm{Gal}(L^{\mathsf{a}}K'/K')$ [$\mathrm{Gal}(L^{[n]}K'/K')$] is isomorphic to a closed subgroup of $\mathrm{Gal}(L^{\mathsf{a}}/K)$ [$\mathrm{Gal}(L^{[n]}/K)$]. Hence $L^{\mathsf{a}}K'/K'$ [$L^{[n]}K'/K'$] is abelian [of exponent n], and $L^{\mathsf{a}} \subset L^{\mathsf{a}}K' \subset L'^{\mathsf{a}}$ [$L^{[n]} \subset L^{[n]}K' \subset L'^{[n]}$]. □

As a basic example we consider the absolute Galois group of a field K and the Galois group of its maximal abelian extension.

Remark and Example 1.8.6. Let K be a field, $\mathcal{E}(K)$ the set of all over K finite Galois extension fields E and $\mathcal{E}_{\mathrm{ab}}(K)$ the set of all over K abelian fields $E \in \mathcal{E}(K)$. As

$$K_{\mathrm{sep}} = \prod_{E\in\mathcal{E}(K)} E \quad \text{and} \quad K_{\mathrm{ab}} = \prod_{E\in\mathcal{E}_{\mathrm{ab}}(K)} E,$$

it follows that

$$G_K = \mathrm{Gal}(\overline{K}/K) = \mathrm{Gal}(K_{\mathrm{sep}}/K) = \varprojlim_{E\in\mathcal{E}(K)} \mathrm{Gal}(E/K),$$

and

$$\operatorname{Gal}(K_{\rm ab}/K) = \varprojlim_{E\in\mathcal{E}_{\rm ab}(K)} \operatorname{Gal}(E/K).$$

Moreover, by 1.8.5 we obtain

$$\operatorname{Gal}(\overline{K}/K_{\rm ab}) = G_K^{(1)} \quad\text{and}\quad \operatorname{Gal}(K_{\rm ab}/K) = G_K^{\rm a} = G_K/G_K^{(1)}.$$

Next we determine the absolute Galois group $G_E = \operatorname{Gal}(\overline{E}/E)$ of a finite field E. We use [18, Theorem 1.7.1].

Theorem 1.8.7. *Let E be a finite field, $|E| = q$, $\overline{E}$ an algebraic closure of E and $G = \operatorname{Gal}(\overline{E}/E)$. Let ϕ_E be the Frobenius automorphism over E, defined by $\phi_E(x) = x^q$ for all $x \in \overline{E}$. Then $G = \overline{\langle\phi_E\rangle}$, and there exists a unique topological isomorphism $\Theta\colon \widehat{\mathbb{Z}} \to G$ satisfying $\Theta(1) = \phi_E$.*

Proof. For $n \in \mathbb{N}$ let E_n be the unique intermediate field of $\overline{E}/E$ such that $[E_n : E] = n$, and let $\theta_n\colon \mathbb{Z}/n\mathbb{Z} \stackrel{\sim}{\to} \operatorname{Gal}(E_n/E)$ be the (unique) isomorphism satisfying $\theta_n(1+n\mathbb{Z}) = \varphi_E \restriction E_n$. If $m,\, n \in \mathbb{N}$ and $m \,|\, n$, then $E_m \subset E_n$, and we define $\rho_{n,m}\colon \operatorname{Gal}(E_n/E) \to \operatorname{Gal}(E_m/E)$ by $\rho_{n,m}(\sigma) = \sigma \restriction E_m$.

If $\phi_{n,m}\colon \mathbb{Z}/n\mathbb{Z} \to \mathbb{Z}/m\mathbb{Z}$ is the natural epimorphism as in 1.5.5, then we obtain the commutative diagram

$$\begin{array}{ccc} \mathbb{Z}/n\mathbb{Z} & \xrightarrow{\theta_n} & \operatorname{Gal}(E_n/E) \\ {\scriptstyle\phi_{n,m}}\big\downarrow & & \big\downarrow{\scriptstyle\rho_{n,m}} \\ \mathbb{Z}/m\mathbb{Z} & \xrightarrow{\theta_m} & \operatorname{Gal}(E_m/E)\,. \end{array}$$

Consequently $(\theta_n)_{n\in\mathbb{N}}$ induces a topological isomorphism

$$\Theta = \varprojlim \theta_n \colon \widehat{\mathbb{Z}} = \varprojlim \mathbb{Z}/n\mathbb{Z} \stackrel{\sim}{\to} \varprojlim \operatorname{Gal}(E_n/E) = G$$

satisfying $\Theta(1) \restriction E_m = \Theta((1+n\mathbb{Z})_{n\in\mathbb{N}}) \restriction E_m = \phi_E \restriction E_m$ for all $m \in \mathbb{N}$ and therefore $\Theta(1) = \phi_E$. From $\widehat{\mathbb{Z}} = \overline{\langle 1\rangle}$ we deduce $G = \overline{\langle\phi_E\rangle}$ and the uniqueness of Φ. $\square$

Next we determine the Galois group of the field of all roots of unity over $\mathbb{Q}$. We use [18, Theorem 1.7.5].

Theorem 1.8.8. *Let μ be the group of all roots of unity in $\mathbb{C}$. Then $\mathbb{Q}(\mu)/\mathbb{Q}$ is Galois, and there exists a topological isomorphism*

$$\Phi\colon \widehat{\mathbb{Z}}^\times \to \operatorname{Gal}(\mathbb{Q}(\mu)/\mathbb{Q})$$

satisfying $\Phi((a_n+n\mathbb{Z})_{n\in\mathbb{N}})(\zeta) = \zeta^{a_m}$ for all $(a_n+n\mathbb{Z})_{n\in\mathbb{N}} \in \widehat{\mathbb{Z}}^\times$, $m \in \mathbb{N}$ and $\zeta \in \mu_m$.

Proof. For $n \in \mathbb{N}$ let $\mathbb{Q}^{(n)} = \mathbb{Q}(\mu_n)$ be the n-th cyclotomic field. The extension $\mathbb{Q}^{(n)}/\mathbb{Q}$ is Galois, and there is an isomorphism $\Phi_n\colon (\mathbb{Z}/n\mathbb{Z})^\times \to \mathrm{Gal}(\mathbb{Q}^{(n)}/\mathbb{Q})$ satisfying $\Phi_n(a+n\mathbb{Z})(\zeta) = \zeta^a$ for all $\zeta \in \mu_n$ and $a \in \mathbb{Z}$ such that $(a,n)=1$.

$\mathbb{Q}(\mu)$ is the compositum of the family $(\mathbb{Q}^{(n)})_{n\in\mathbb{N}}$. Hence $\mathbb{Q}(\mu)/\mathbb{Q}$ is Galois, and since $\mathbb{Q}^{(m)}\mathbb{Q}^{(n)} \subset \mathbb{Q}^{(mn)}$ for all $m,\, n \in \mathbb{N}$, the family $(\mathbb{Q}^{(n)})_{n\in\mathbb{N}}$ is cofinal in the family of all over $\mathbb{Q}$ finite subfields of $\mathbb{Q}(\mu)$. Consequently,

$$\mathrm{Gal}(\mathbb{Q}(\mu)/\mathbb{Q}) = \varprojlim \mathrm{Gal}(\mathbb{Q}^{(n)}/\mathbb{Q}).$$

If $m,\, n \in \mathbb{N}$ and $m \,|\, n$, then $\mathbb{Q}^{(m)} \subset \mathbb{Q}^{(n)}$, and we obtain a commutative diagram

$$\begin{array}{ccc} (\mathbb{Z}/n\mathbb{Z})^\times & \xrightarrow{\ \Phi_n\ } & \mathrm{Gal}(\mathbb{Q}^{(n)}/\mathbb{Q}) \\ {\scriptstyle \phi'_{n,m}}\big\downarrow & & \big\downarrow{\scriptstyle \rho_{n,m}} \\ (\mathbb{Z}/m\mathbb{Z})^\times & \xrightarrow{\ \Phi_m\ } & \mathrm{Gal}(\mathbb{Q}^{(m)}/\mathbb{Q})\,, \end{array}$$

where $\rho_{n,m}(\sigma) = \sigma \restriction \mathbb{Q}^{(m)}$ for all $\sigma \in \mathrm{Gal}(\mathbb{Q}^{(n)}/\mathbb{Q})$, $\phi_{n,m}\colon \mathbb{Z}/n\mathbb{Z} \to \mathbb{Z}/m\mathbb{Z}$ is as in 1.5.5 and $\phi'_{n,m} = \phi_{n,m} \restriction (\mathbb{Z}/n\mathbb{Z})^\times \colon (\mathbb{Z}/n\mathbb{Z})^\times \to (\mathbb{Z}/m\mathbb{Z})^\times$.

By 1.3.9, we obtain $\widehat{\mathbb{Z}}^\times = \varprojlim_{n\in\mathbb{N}} (\mathbb{Z}/n\mathbb{Z})^\times$, and therefore $(\Phi_n)_{n\in\mathbb{N}}$ induces a topological isomorphism

$$\Phi = \varprojlim \Phi_n\colon \widehat{\mathbb{Z}}^\times = \varprojlim_{n\in\mathbb{N}} (\mathbb{Z}/n\mathbb{Z})^\times \overset{\sim}{\to} \varprojlim_{n\in\mathbb{N}} \mathrm{Gal}(\mathbb{Q}^{(n)}/\mathbb{Q}) = \mathrm{Gal}(\mathbb{Q}(\mu)/\mathbb{Q})$$

satisfying

$$\Phi((a_n{+}n\mathbb{Z})_{n\in\mathbb{N}}) \restriction \mathbb{Q}^{(m)} = \Phi_m(a_m{+}m\mathbb{Z}) \text{ for all } m \in \mathbb{N} \text{ and } (a_n{+}n\mathbb{Z})_{n\in\mathbb{N}} \in \widehat{\mathbb{Z}}^\times.$$

It follows that $\Phi((a_n+n\mathbb{Z})_{n\in\mathbb{N}})(\zeta) = \zeta^{a_m}$ for all threads $(a_n+n\mathbb{Z})_{n\in\mathbb{N}} \in \widehat{\mathbb{Z}}^\times$, $m \in \mathbb{N}$ and $\zeta \in \mu_m$. □

We close this section with a Galois-theoretic investigation of infinite algebraic extensions of complete non-Archimedian fields. The following two Theorems 1.8.9 and 1.8.10 supplement [18, Theorems 5.6.5 and 5.6.7] and fill a gap left there.

Theorem 1.8.9. *Let $K = (K, |\cdot|)$ be a complete non-Archimedian valued field, L/K a Galois extension and $G = \mathrm{Gal}(L/K)$. Then the residue class extension $\mathsf{k}_L/\mathsf{k}_K$ is normal, and for $\sigma \in G$ we denote by $\overline{\sigma} \in \mathrm{Gal}(\mathsf{k}_L/\mathsf{k}_K)$ the residue class automorphism induced by σ. Let T be the inertia field of L/K (i. e., T is the largest over K unramified intermediate field of L/K).*

$\mathsf{k}_L/\mathsf{k}_K$ is normal, k_T is the separable closure of k_K in k_L, the extensions T/K and $\mathsf{k}_T/\mathsf{k}_K$ are Galois, and we define

$$\rho_{L/K}\colon G \to \mathrm{Gal}(\mathsf{k}_T/\mathsf{k}_K) \ \textit{by}\ \rho(\sigma) = \overline{\sigma} \restriction \mathsf{k}_T.$$

Then $\rho_{L/K}$ is a continuous epimorphism, $\mathrm{Ker}(\rho_{L/K}) = \mathrm{Gal}(L/T)$, and $\rho_{L/K}$ induces a topological isomorphism $\rho_{T/K}\colon \mathrm{Gal}(T/K) \overset{\sim}{\to} \mathrm{Gal}(\mathsf{k}_T/\mathsf{k}_K)$ satisfying $\rho_{T/K}(\sigma) = \overline{\sigma}$ for all $\sigma \in \mathrm{Gal}(T/K)$.

Proof. We use [18, Theorem 5.6.5]. The extensions T/K and $\mathsf{k}_T/\mathsf{k}_K$ are Galois, $\mathsf{k}_L/\mathsf{k}_K$ is normal, and that k_T is the separable closure of k_K in k_L. Hence the map $\theta\colon \mathrm{Gal}(\mathsf{k}_L/\mathsf{k}_K) \to \mathrm{Gal}(\mathsf{k}_T/\mathsf{k}_K)$, defined by $\theta(\tau) = \tau \restriction \mathsf{k}_T$, is an isomorphism.

We proceed in 3 steps.

a. $\mathrm{Ker}(\rho_{L/K}) = \mathrm{Gal}(L/T)$: If $\sigma \in G$, then $\rho_{L/K}(\sigma) = \overline{\sigma \restriction T}$, and consequently $\mathrm{Gal}(L/T) \subset \mathrm{Ker}(\rho_{L/K)})$. Now let $\sigma \in \mathrm{Ker}(\rho_{L/K})$. We must prove that $\sigma \restriction T = \mathrm{id}_T$, and therefore we show that $\sigma \restriction M = \mathrm{id}_M$ for all over K finite Galois intermediate fields M of T/K. Thus let M be an over K finite Galois intermediate field of T/K. Then M/K is unramified, and the map $\rho_{M/K}\colon \mathrm{Gal}(M/K) \stackrel{\sim}{\to} \mathrm{Gal}(\mathsf{k}_M/\mathsf{k}_K)$, defined by $\rho_{M/K}(\sigma) = \overline{\sigma}$ for all $\sigma \in \mathrm{Gal}(M/K)$, is an isomorphism. Hence $\mathrm{id}_{\mathsf{k}_M} = \rho_{L/K}(\sigma) \restriction \mathsf{k}_M = \rho_{M/K}(\sigma \restriction M)$ and consequently $\sigma \restriction M = \mathrm{id}_M$. □[**a.**]

b. $\rho_{L/K}$ is continuous: Let m be an over k_K finite Galois intermediate field of $\mathsf{k}_T/\mathsf{k}_K$. Then there exists an over K finite Galois intermediate field M of T/K such that $\mathsf{k}_M = \mathsf{m}$, and since $\mathrm{Gal}(L/M) \subset \rho_{L/K}^{-1}(\mathrm{Gal}(\mathsf{k}_L/\mathsf{m}))$, it follows that $\rho_{L/K}$ is continuous in id_T. Hence $\rho_{L/K}$ is continuous by 1.1.6.1. □[**b.**]

c. $\rho_{L/K}(G) = \mathrm{Gal}(\mathsf{k}_T/\mathsf{k}_K)$. Beeing compact, $\rho_{L/K}(G)$ is closed in $\mathrm{Gal}(\mathsf{k}_T/\mathsf{k}_K)$, and it suffices to prove that $\rho_{L/K}(G)$ is dense in $\mathrm{Gal}(\mathsf{k}_T/\mathsf{k}_K)$. Let $\tau \in \mathrm{Gal}(\mathsf{k}_T/\mathsf{k}_K)$, U a neighbourhood of τ in $\mathrm{Gal}(\mathsf{k}_T/\mathsf{k}_K)$ and m an over k_K finite Galois intermediate field such that $\tau\,\mathrm{Gal}(\mathsf{k}_T/\mathsf{m}) \subset U$. Let M be an over K finite Galois intermediate field M of T/K such that $\mathsf{k}_M = \mathsf{m}$ and the assignment $\sigma \mapsto \overline{\sigma}$ defines an isomorphism $\mathrm{Gal}(M/K) \stackrel{\sim}{\to} \mathrm{Gal}(\mathsf{k}_M/\mathsf{k}_K)$. Hence there exists some $\sigma \in \mathrm{Gal}(T/K)$ such that $\rho_{L/K}(\sigma) \restriction \mathsf{k}_M = \overline{\sigma \restriction M} = \tau \restriction \mathsf{k}_M$, and thus $\rho_{L/K}(\sigma) \in \tau\,\mathrm{Gal}(\mathsf{k}_T/\mathsf{m}) \subset U$. □]**c.**]

$\rho_{L/K}$ induces a topological isomorphism

$$\rho^*_{L/K}\colon G/\mathrm{Gal}(L/T) \stackrel{\sim}{\to} \mathrm{Gal}(\mathsf{k}_T/\mathsf{k}_K)$$

such that $\rho^*_{L/K}(\sigma\mathrm{Gal}(L/T)) = \overline{\sigma} \restriction T$ for all $\sigma \in G$.

We identify $G/\mathrm{Gal}(L/T) = \mathrm{Gal}(T/K)$ by means of $\sigma\mathrm{Gal}(L/T) = \sigma \restriction T$ and observe that $\overline{\sigma \restriction T} = \overline{\sigma} \restriction \mathsf{k}_T$ for all $\sigma \in G$. Then $\rho_{L/K}$ induces a topological isomorphism $\rho_{T/K}\colon \mathrm{Gal}(T/K) \stackrel{\sim}{\to} \mathrm{Gal}(\mathsf{k}_T/\mathsf{k}_K)$ as asserted. □

Theorem 1.8.10. *Let $K = (K, |\cdot|)$ be a complete non-Archimedean valued field such that $|\mathsf{k}_K| = q < \infty$. Let $\overline{K}$ be an algebraic closure of K, K_∞ the maximal unramified extension of K in $\overline{K}$, $G = \mathrm{Gal}(K_\infty/K)$, and let $\varphi_K = \varphi_{K_\infty/K} \in G$ be the Frobenius automorphism over K* (recall from [18, Theorem 5.6.7] that φ_K is the unique automorphism of K_∞/K satisfying $\varphi_K(x) \equiv x^q \bmod \mathfrak{p}_{K_\infty}$ for all $x \in \mathcal{O}_{K_\infty}$).

Then $G = \overline{\langle\varphi_K\rangle}$, and there exists a unique topological isomorphism

$$\Theta\colon G \stackrel{\sim}{\to} \widehat{\mathbb{Z}} \quad \textit{such that} \quad \Theta(\varphi_K) = 1 \in \widehat{\mathbb{Z}}.$$

The residue class field k_{K_∞} *is the algebraic closure of* k_K*, and for* $\sigma \in G$ *we denote by* $\overline{\sigma} \in \mathrm{Gal}(\mathsf{k}_{K_\infty}/\mathsf{k}_K)$ *the residue class automorphism induced by* σ*. Then the assignment* $\sigma \mapsto \overline{\sigma}$ *defines a topological isomorphism*

$$\rho\colon G \overset{\sim}{\to} \mathrm{Gal}(\mathsf{k}_{K_\infty}/\mathsf{k}_K) \quad \textit{satisfying} \quad \rho(\varphi_K) = \phi_{\mathsf{k}_K},$$

where ϕ_{k_K} *is the Frobenius automorphism over* k_K *satisfying*

$$\mathrm{Gal}(\mathsf{k}_{K_\infty}) = \overline{\langle \phi_{\mathsf{k}_K} \rangle} \cong \widehat{\mathbb{Z}}$$

Proof. We use [18, Theorem 5.6.7]. For $n \in \mathbb{N}$ let K_n be the unique intermediate field of K_∞/K such that $[K_n : K] = n$. Then K_n/K is cyclic, and there exists a unique isomorphism $\theta_n\colon \mathbb{Z}/n\mathbb{Z} \overset{\sim}{\to} \mathrm{Gal}(K_n/K)$ satisfying $\theta_n(1+n\mathbb{Z}) = \varphi_K \restriction K_n$. If $m, n \in \mathbb{N}$ are such that $m \,|\, n$, then $K_m \subset K_n$, and we have a commutative diagram

$$\begin{array}{ccc} \mathbb{Z}/n\mathbb{Z} & \xrightarrow{\theta_n} & \mathrm{Gal}(K_n/K) \\ {\scriptstyle \phi_{n,m}}\downarrow & & \downarrow{\scriptstyle \rho_{n,m}} \\ \mathbb{Z}/m\mathbb{Z} & \xrightarrow{\theta_m} & \mathrm{Gal}(K_m/K)\,, \end{array}$$

where $\phi_{n,m}$ is the natural epimorphism. and where $\rho_{n,m}(\sigma) = \sigma \restriction K_m$ for all $\sigma \in \mathrm{Gal}(K_n/K)$. Hence $(\theta_n)_{n\in\mathbb{N}}$ induces a topological isomorphism

$$\Theta\colon G = \varprojlim \mathrm{Gal}(K_n/K) \overset{\sim}{\to} \varprojlim \mathbb{Z}/n\mathbb{Z} = \widehat{\mathbb{Z}}$$

satisfying $\Theta(\varphi_K) = \Theta((\varphi_K \restriction K_n)_{n\in\mathbb{N}}) = (1+n\mathbb{Z})_{n\in\mathbb{N}} = 1 \in \widehat{\mathbb{Z}}$. From $\widehat{\mathbb{Z}} = \overline{\langle 1 \rangle}$ we deduce $G = \overline{\langle \varphi_K \rangle}$ and the uniqueness of Θ.

By definition $\rho(\varphi_K) = \phi_{\mathsf{k}_K}$, and by 1.8.9 it follows that ρ is a topological isomorphism (note that K_∞ is the inertia field of $\overline{K}/K$). □

2

Cohomology of groups

We give an introduction to the cohomology of profinite groups without making use of the general formalism of homological algebra. We present self-contained proofs for all results used in later chapters. However we also go beyond the minimal requirements of class field theory in order to show a well-rounded picture of the subject. In particular, we give detailed proofs for results which in our opinion come short in the general literature. We carefully highlight the applications to field theory.

Main sources which influenced our presentation are [50], [54], [53], [46], [32].

2.1 Discrete G-modules

Let G be a (multiplicative) group. The elements λ of the group ring $\mathbb{Z}[G]$ have a unique representation

$$\lambda = \sum_{\sigma \in G} \lambda_\sigma \sigma, \quad \text{where } \lambda_\sigma \in \mathbb{Z}, \ \ \lambda_\sigma = 0 \ \text{ for almost all } \ \sigma \in G,$$

and the multiplication is given by

$$\Big(\sum_{\sigma \in G} \lambda_\sigma \sigma\Big)\Big(\sum_{\sigma \in G} \lambda'_\sigma \sigma\Big) = \sum_{\sigma \in G}\Big(\sum_{\substack{(\tau,\tau' \in G \\ \tau\tau' = \sigma}} \lambda_\tau \lambda'_{\tau'}\Big)\sigma.$$

We tacitly identify the unit element $1 \in G$ with $1 \in \mathbb{Z}[G]$. Then it follows that $G \subset \mathbb{Z}[G]$, $\mathbb{Z} = \mathbb{Z}1 \subset \mathbb{Z}[G]$, and $\mathbb{Z}[G]$ is a free abelian group with basis G. Let $\varepsilon_G \colon \mathbb{Z}[G] \to \mathbb{Z}$ be the unique group homomorphism satisfying $\varepsilon_G \restriction G = 1$,

$$\varepsilon_G\Big(\sum_{\sigma \in G} \lambda_\sigma \sigma\Big) = \sum_{\sigma \in G} \lambda_\sigma, \quad \text{where } \lambda_\sigma \in \mathbb{Z} \ \text{ and } \ \lambda_\sigma = 0 \ \text{ for almost all } \ \sigma \in G.$$

DOI: 10.1201/9780429506574-2

ε_G is called the **augmentation**. Its kernel $I_G = \mathrm{Ker}(\varepsilon_G)$ is called the **augmentation ideal**. If G is finite, then we call

$$N_G = \sum_{\sigma \in G} \sigma \quad \text{the \textbf{norm element} of } \mathbb{Z}[G].$$

We start with some elementary facts about integral group rings.

Theorem 2.1.1. *Let G be a group.*

1. *I_G is a free abelian group with basis $\{\sigma - 1 \mid \sigma \in G \setminus \{1\}\}$.*

2. *Let $n \in \mathbb{N}$, $\sigma_1, \ldots, \sigma_n \in G$, $m_1, \ldots, m_n \in \mathbb{Z}$ and $\sigma = \sigma_1^{m_1} \cdot \ldots \cdot \sigma_n^{m_n}$. Then*

$$\sigma - 1 = \sum_{i=1}^{n} (\sigma_i - 1)\mu_i,$$

where $\mu_i \in \mathbb{Z}[G]$ satisfies $\mu_i \equiv m_i \bmod I_G$ for all $i \in [1, n]$. In particular, if $G = \langle \sigma_1, \ldots, \sigma_n \rangle$, then

$$I_G = (\sigma_1 - 1)\mathbb{Z}[G] + \ldots + (\sigma_n - 1)\mathbb{Z}[G],$$

and if $G = \langle \sigma \rangle$ is cyclic, then $I_G = (\sigma - 1)\mathbb{Z}[G]$.

3. *Let H be a normal subgroup of G.*

   (a) *$I_H\mathbb{Z}[G]$ is an ideal of $\mathbb{Z}[G]$, and the residue class epimorphism $G \to G/H$ induces an isomorphism $\mathbb{Z}[G]/I_H\mathbb{Z}[G] \xrightarrow{\sim} \mathbb{Z}[G/H]$.*

   (b) *If H is finite, then*

$$(\sigma - 1)N_H = N_H(\sigma - 1) \ \text{ for all } \sigma \in G, \ \text{ and } N_H I_G = I_G N_H.$$

4. *Let G be finite. Then it follows that $\mathbb{Z}N_G = \mathrm{Ann}_{\mathbb{Z}[G]}(I_G)$ and $I_G = \mathrm{Ann}_{\mathbb{Z}[G]}(N_G)$.*

Proof. 1. If $\lambda \in \mathbb{Z}[G]$, then

$$\lambda = \sum_{\sigma \in G} \lambda_\sigma \sigma = \sum_{\sigma \in G} \lambda_\sigma(\sigma - 1) + \sum_{\sigma \in G} \lambda_\sigma \equiv \sum_{\sigma \in G} \lambda_\sigma \bmod I_G,$$

where $\lambda_\sigma \in \mathbb{Z}$ and $\lambda_\sigma = 0$ for almost all $\sigma \in G$. Hence it follows that

$$\lambda \in I_G \ \text{ if and only if } \ \lambda = \sum_{\sigma \in G} \lambda_\sigma(\sigma - 1) \in \langle \{\sigma - 1 \mid \sigma \in G \setminus \{1\}\} \rangle.$$

If $(\lambda_\sigma)_{\sigma \in G \setminus \{1\}}$ is a family in $\mathbb{Z}$ such that $\lambda_\sigma = 0$ for almost all $\sigma \in G \setminus \{1\}$ and

$$0 = \sum_{\sigma \in G \setminus \{1\}} \lambda_\sigma(\sigma - 1) = \sum_{\sigma \in G \setminus \{1\}} \lambda_\sigma \sigma - \Big(\sum_{\sigma \in G \setminus 1} \lambda_\sigma\Big) 1 \in \mathbb{Z}[G],$$

then $\lambda_\sigma = 0$ for all $\sigma \in G \setminus \{1\}$. Hence $\{\sigma - 1 \mid \sigma \in G \setminus \{1\}\}$ is linearly independent, and I_G is free abelian with basis $\{\sigma - 1 \mid \sigma \in G \setminus \{1\}\}$.

2. We proceed by induction on $m = |m_1| + \ldots + |m_n|$. If $m = 0$, then $\sigma = 1$ and there is nothing to do. Thus let $m \geq 0$, $\sigma = \sigma_1^{m_1} \cdot \ldots \cdot \sigma_n^{m_n}$, where $m_1, \ldots, m_n \in \mathbb{Z}$, $|m_1| + \ldots + |m_n| = m$, and suppose that

$$\sigma - 1 = \sum_{i=1}^n (\sigma_i - 1)\mu_i, \text{ where } \mu_i \in \mathbb{Z}[G] \text{ and } \mu_i \equiv m_i \bmod I_G \text{ for all } i \in [1, n].$$

If $j \in [1, n]$, then

$$\sigma_j\sigma - 1 = \sigma - 1 + (\sigma_j - 1)\sigma = \sum_{\substack{i=1 \\ i \neq j}}^n (\sigma_i - 1)\mu_i + (\sigma_j - 1)(\mu_j + \sigma),$$

$$\sigma_j^{-1}\sigma - 1 = \sigma - 1 + (\sigma_j - 1)(-\sigma_j^{-1}\sigma) = \sum_{\substack{i=1 \\ i \neq j}}^n (\sigma_i - 1)\mu_i + (\sigma_j - 1)(\mu_j - \sigma_j^{-1}\sigma),$$

and since $\mu_j + \sigma \equiv \mu_j + 1 \bmod I_G$ and $\mu_j - \sigma_j^{-1}\sigma \equiv \mu_j - 1 \bmod I_G$, the assertion follows.

3.(a) If $\sigma \in G$ and $\tau \in H$, then $(\tau - 1)\sigma = \sigma(\sigma^{-1}\tau\sigma - 1)$. Hence $I_H\mathbb{Z}[G] = \mathbb{Z}[G]I_H$ is an ideal of $\mathbb{Z}[G]$. Let $\pi\colon \mathbb{Z}[G] \to \mathbb{Z}[G/H]$ be the unique ring epimorphism satifying $\pi(\sigma) = \sigma H$ for all $\sigma \in G$, and let R be a set of representatives for $H\backslash G$ in G so that $G = \{\tau\xi \mid \tau \in H,\ \xi \in R\}$. If $\lambda \in \mathbb{Z}[G]$, then

$$\begin{aligned} \lambda = \sum_{\sigma \in G} \lambda_\sigma \sigma &= \sum_{\xi \in R} \Big(\sum_{\tau \in H} \lambda_{\tau\xi}(\tau - 1) \Big) \xi + \sum_{\xi \in R} \Big(\sum_{\tau \in H} \lambda_{\tau\xi} \Big) \xi \\ &\equiv \sum_{\xi \in R} \Big(\sum_{\tau \in H} \lambda_{\tau\xi} \Big) \xi \bmod I_H\mathbb{Z}[G], \end{aligned}$$

where $\lambda_\sigma \in \mathbb{Z}$ and $\lambda_\sigma = 0$ for almost all $\sigma \in G$. Hence $\pi(\lambda) = 0$ if and only if

$$\lambda = \sum_{\sigma \in G} \lambda_\sigma \sigma = \sum_{\xi \in R} \Big(\sum_{\tau \in H} \lambda_{\tau\xi}(\tau - 1) \Big) \xi \in I_H\mathbb{Z}[G].$$

(b) It suffices to prove that $(\sigma - 1)N_H = N_H(\sigma - 1)$. If $\sigma \in G$, then

$$\begin{aligned} (\sigma - 1)N_H &= \sum_{\tau \in H} \sigma\tau - \sum_{\tau \in H} \tau = \sum_{\tau \in H} (\sigma\tau\sigma^{-1})\sigma - \sum_{\tau \in H} \tau \\ &= \sum_{\tau \in H} \tau\sigma - \sum_{\tau \in H} \tau = N_H(\sigma - 1). \end{aligned}$$

4. Let

$$\lambda = \sum_{\sigma \in G} \lambda_\sigma \sigma \in \mathbb{Z}[G].$$

Then

$$\lambda N_G = \sum_{\sigma\in G} \lambda_\sigma \sigma N_G = \sum_{\sigma\in G} \lambda_\sigma N_G = \varepsilon_G(\lambda) N_G = \sum_{\sigma\in G} \varepsilon_G(\lambda)\sigma,$$

and

$$\lambda(\tau-1) = \sum_{\sigma\in G} \lambda_\sigma \sigma\tau - \sum_{\sigma\in G} \lambda_\sigma \sigma = \sum_{\sigma\in G} (\lambda_{\sigma\tau^{-1}} - \lambda_\sigma)\sigma \quad \text{for all } \tau \in G.$$

Therefore we obtain:

$$\lambda \in \mathrm{Ann}_{\mathbb{Z}[G]}(N_G) \iff \lambda N_G = 0 \iff \varepsilon_G(\lambda) = 0 \iff \lambda \in I_G,$$

and

$$\lambda \in \mathrm{Ann}_{\mathbb{Z}[G]}(I_G) \iff \lambda I_G = 0 \iff \lambda(\tau-1) = 0 \text{ for all } \tau \in G$$
$$\iff \lambda_\sigma = \lambda_1 \text{ for all } \sigma \in G \iff \lambda \in \mathbb{Z}N_G. \qquad \square$$

Let G be a group and $G\times A \to A$, $(\sigma, a) \mapsto \sigma a$ an operation of G on a non-empty set A [that is, $1a = a$ and $\sigma(\tau a) = (\sigma\tau)a$ for all $\sigma, \tau \in G$ and $a \in A$]. Then we set $A^G = \{a \in A \mid \sigma a = a \text{ for all } \sigma \in G\}$, and for $a \in A$ we denote by $G_a = \{\sigma \in G \mid \sigma a = a\}$ the isotropy group of a.

Theorem 2.1.2. *Let G be a profinite group and Σ a fundamental system of neighbourhoods of 1 in G consisting of open normal subgroups. Let A be a discrete topological space and $t\colon G\times A \to A$, $(\sigma,a) \mapsto t(\sigma,a) = \sigma a$ an operation of G on A. Then the following assertions are equivalent:*

(a) t is continuous.

(b) For every $a \in A$ the isotropy group G_a is an open subgroup of G.

(c) For every $a \in A$ there exists some $U \in \Sigma$ such that $a \in A^U$. Consequently,

$$A = \bigcup_{U\in\Sigma} A^U.$$

Proof. (a) ⇒ (b) If $a \in A$, then the map $t_a\colon G \to A$, defined by $t_a(\sigma) = \sigma a$, is continuous, and thus $G_a = t_a^{-1}(a)$ is open.

(b) ⇒ (c) If $a \in A$, then there exists some $U \in \Sigma$ such that $U \subset G_a$ and consequently $a \in A^U$.

(c) ⇒ (a) Let $(\sigma, a) \in G\times A$, $t(\sigma,a) = \sigma a = b$, and let $U \in \Sigma$ such that $a \in G^U$. Then $\sigma U\times\{a\}$ is a neighbourhood of (σ,a) in $G\times A$, and it satisfies $t(\sigma U\times\{a\}) = \{b\}$. $\square$

Definition and Remarks 2.1.3. Let G be a (multiplicative) group.

1. A **G-module structure** on an abelian group A is a map $G \times A \to A$ such that

$$\sigma(a+b) = \sigma a + \sigma b \quad \text{for all} \quad \sigma \in G \quad \text{and} \quad a,\, b \in A.$$

A **G-module** is an abelian group A together with a G-module structure on A.

Let A be a G-module. Then $\mu_\sigma \colon A \to A$, defined by $\mu_\sigma(a) = \sigma a$, is an automorphism of A, and the asssignment $\sigma \mapsto \mu_\sigma$ defines a group homomorphism $G \to \mathrm{Aut}(A)$ (indeed, if $\sigma,\, \tau \in G$, then $\mu_{\sigma\tau} = \mu_\sigma \circ \mu_\tau$ and $\mu_\sigma^{-1} = \mu_{\sigma^{-1}}$). A subgroup B of A is called a **G-submodule** if $\sigma B \subset B$ (then $\sigma B = B$) for all $\sigma \in G$. Then B is again a G-module, and A/B is a G-module by means of $\sigma(a+B) = \sigma a + B$ for all $\sigma \in G$ and $a \in A$. If $(A_i)_{i \in I}$ is a family of G-modules, then its (external) direct sum is a G-module with respect to the component-wise action of G.

If A is a G-module, then A becomes a $\mathbb{Z}[G]$-module by means of

$$\Big(\sum_{\sigma \in G} n_\sigma \sigma\Big) a = \sum_{\sigma \in G} n_\sigma \sigma a \quad \text{for all} \quad \sum_{\sigma \in G} n_\sigma \sigma \in \mathbb{Z}[G] \quad \text{and} \quad a \in A.$$

Conversely, if A is a $\mathbb{Z}[G]$-Modul, then the restriction of scalars makes A into a G-module.

2. A G-module A is called **trivial** if $A^G = A$ [equivalently, $\sigma a = a$ for all $a \in A$ and $\sigma \in G$]. If A is any G-module, then A^G is a trivial G-submodule of A. Every abelian group is a trivial G-module in the natural way. In particular, we always view $\mathbb{Z}$, $\mathbb{Q}$ and $\mathbb{Q}/\mathbb{Z}$ as trivial G-modules. If $G = \mathbf{1}$, then a G-module is simply an abelian group.

3. Let A and B be G-modules. A **G-homomorphism** $f \colon A \to B$ is a group homomorphism satisfying $f(\sigma a) = \sigma f(a)$ for all $\sigma \in G$ and $a \in A$. We denote by $\mathrm{Hom}_G(A, B)$ the set of all G-homomorphisms and in contrast by $\mathrm{Hom}(A, B) = \mathrm{Hom}_{\mathbb{Z}}(A, B)$ the set of all group homomorphisms (= $\mathbb{Z}$-homomorphisms) $f \colon A \to B$.

If A and B are trivial G-modules, then $\mathrm{Hom}(A, B) = \mathrm{Hom}_G(A, B)$. Endowed with pointwise addition, $\mathrm{Hom}(A, B)$ and $\mathrm{Hom}_G(A, B)$ are abelian groups.

4. For the G-module structure on a multiplicative abelian group A we use exponential notation: For $\sigma \in G$ and $a \in A$ we set $a^\sigma = \sigma a$. Then it follows that $(a^\sigma)^\tau = a^{\tau\sigma}$ for all $a \in A$ and $\sigma,\, \tau \in G$. Although most G-modules occurring in number theory are multiplicative, we shall derive the general theory with regard to a better readableness in an additive setting.

5. If $f \colon A \to B$ is a G-homomorphism, then $\mathrm{Ker}(f) \subset A$ and $\mathrm{Im}(f) \subset B$ are G-submodules, and

$$\mathbf{0} \to \mathrm{Ker}(f) \hookrightarrow A \xrightarrow{f} B \to \mathrm{Coker}(f) \to \mathbf{0}$$

is a exact sequence of G-modules.

6. Let G be a profinite group. A **discrete G-module** is a G-module A equipped with discrete topology such that the operation $t\colon G\times A \to A$, $(\sigma, a) \mapsto \sigma a$, is continuous.

Every trivial G-module is a discrete G-module, and if G is finite, then every G-module is a discrete G-module.

If A is a discrete G-module and B is a G-submodule of A, then B and A/B are again discrete G-modules. Indeed, for B this is obvious. If $a \in A$ and U is an open normal subgroup of G, then $a \in A^U$ implies $a + B \in (A/B)^U$.

7. If $(A_i)_{i\in I}$ is a family of discrete G-modules, then their direct sum A is again a discrete G-module. Indeed, if

$$a = (a_i)_{i\in I} \in A = \bigoplus_{i\in I} A_i, \quad \text{then} \quad G_a = \bigcap_{\substack{i\in I\\ a_i\neq 0}} G_{a_i} \quad \text{is an open subgroup of } G.$$

8. Let G be profinite group and A a discrete G-module. If H is a closed subgroup of G, then A is also a discrete H-Modul, and if H is a closed normal subgroup of G, then A^H is a discrete G/H-module by means of $(\sigma H)a = \sigma a$ for all $\sigma \in G$ and $a \in A^H$. Indeed, apparently A^H is a G/H-module, and if $a \in A^H$, then $(G/H)_a = G_aH/H$ is an open subgroup of G/H.

Definitions and Remarks 2.1.4. Let G be a profinite group.

1. Let H be a closed subgroup of G and B a discrete H-module. Let $\mathcal{M}_G^H(B)$ be the set of all continuous maps $f\colon G \to B$ satisfying $f(\tau g) = \tau f(g)$ for all $g \in G$ and $\tau \in H$, endowed with pointwise addition. For $\sigma \in G$ and $f \in \mathcal{M}_G^H(B)$ we define

$$\sigma f\colon G \to B \ \text{ by } \ (\sigma f)(g) = f(g\sigma),$$

and then $\sigma f \in \mathcal{M}_G^H(B)$. Indeed, if $\mu_\sigma\colon G \to G$ is defined by $\mu_\sigma(g) = g\sigma$, then μ_σ is continuous and thus $\sigma f = f\circ\mu_\sigma$ is continuous. If $g \in G$ and $\tau \in H$, then $(\sigma f)(\tau g) = f(\tau g\sigma) = \tau f(g\sigma) = \tau(\sigma f)(g)$.

If $f, f' \in \mathcal{M}_G^H(B)$ and $\sigma, \sigma' \in G$, then clearly $\sigma(f+f') = \sigma f + \sigma f'$, and if $g \in G$, then $((\sigma\sigma')f)(g) = f(g\sigma\sigma') = (\sigma' f)(g\sigma) = (\sigma(\sigma' f))(g)$, hence $(\sigma\sigma')f = \sigma(\sigma' f)$. Consequently $\mathcal{M}_G^H(B)$ is a G-module.

If $f \in \mathcal{M}_G^H(B)$, then by 1.5.2.2 there exists an open normal subgroup U of G such that f is constant on the cosets of U, and thus $G_f \supset U$. Therefore G_f is an open subgroup of G, and $\mathcal{M}_G^H(B)$ is a discrete G-module.

If B' is a H-submodule of B, then $\mathcal{M}_G^H(B')$ is a G-submodule of $\mathcal{M}_G^H(B)$.

The G-module $\mathcal{M}_G^H(B)$ is called **induced by B from H**. If $H = \mathbf{1}$ and B is an abelian group, then we set $\mathcal{M}_G(B) = \mathcal{M}_G^{\mathbf{1}}(B)$. A discrete G-module is called an **induced G-module** if it is isomorphic to $\mathcal{M}_G(B)$ for some abelian group B.

2. Let H be a closed subgroup of G. Let B and B' be discrete H-modules and $\varphi \in \mathrm{Hom}_H(B, B')$. If $f \in \mathcal{M}_G^H(B)$, then $\varphi\circ f \in \mathcal{M}_G^H(B')$. Indeed, $\varphi\circ f$ is continuous, and $(\varphi\circ f)(\tau g)=\varphi(\tau f(g))=\tau(\varphi\circ f)(g)$ for all $\tau\in H$ and $g\in G$. We define

$$\mathcal{M}_G^H(\varphi)\colon \mathcal{M}_G^H(B)\to\mathcal{M}_G^H(B') \quad\text{by}\quad \mathcal{M}_G^H(\varphi)(f)=\varphi\circ f \quad\text{for all}\quad f\in\mathcal{M}_G^H(B),$$

and we assert that $\mathcal{M}_G^H(\varphi)$ is a G-homomorphism. Clearly, $\mathcal{M}_G^H(\varphi)$ is a group homomorphism. If $f\in\mathcal{M}_G^H(B)$ and $x,\ \sigma\in G$, then

$$\mathcal{M}_G^H(\varphi)(\sigma f)(x)=(\varphi\circ\sigma f)(x)=\varphi\circ f(x\sigma)=\mathcal{M}_G^H(\varphi)(f)(x\sigma)=[\sigma\mathcal{M}_G^H(\varphi)(f)](x),$$

hence $\mathcal{M}_G^H(\varphi)(\sigma f)=\sigma\mathcal{M}_G^H(\varphi)(f)$, and therefore $\mathcal{M}_G^H(\varphi)$ is a G-homomorphism.

Obviously, $\mathcal{M}_G^H(\mathrm{id}_B)=\mathrm{id}_{\mathcal{M}_G^H(B)}$, and if B'' is another discrete H-module and $\varphi'\in\mathrm{Hom}_H(B',B'')$, then $\mathcal{M}_G^H(\varphi'\circ\varphi)=\mathcal{M}_G^H(\varphi')\circ\mathcal{M}_G^H(\varphi)$.

3. Let A be a discrete G-module. If $a\in A$, then the map $i_a\colon G\to A$, defined by $i_a(g)=ga$, is continuous, hence $i_a\in\mathcal{M}_G(A)$, and we define

$$i_A\colon A\to\mathcal{M}_G(A) \quad\text{by}\quad i_A(a)=i_a \quad\text{for all}\quad a\in A.$$

If $a,\ a'\in A$, then obviously $i_A(a+a')=i_A(a)+i_A(a')$, and if $\sigma,\ g\in G$, then

$$(\sigma i_A(a))(g)=i_A(a)(g\sigma)=g\sigma a=i_A(\sigma a)(g), \quad\text{hence}\quad \sigma i_A(a)=i_A(\sigma a).$$

Therefore i_A is a G-homomorphism. If $a\in\mathrm{Ker}(i_A)$, then $i_A(a)(1)=a=0$, and thus i_A is a monomorphism. We call $i_A\colon A\to\mathcal{M}_G(A)$ the **natural embedding** of A into the **associated induced G-module** $\mathcal{M}_G(A)$, we call and $A_1=\mathrm{Coker}(i_A)$ the **derived module** of A and denote by $j_A\colon A\to A_1$ the natural epimorphism.

If $\varphi\colon A\to B$ is a G-homomorphism of discrete G-modules, then

$$\mathcal{M}_G(\varphi)\circ i_A(a)(g)=\varphi\circ i_A(a)(g)=\varphi(ga)=g\varphi(a)=i_B\circ\varphi(a)(g)$$

for all $a\in A$ and $g\in G$. Hence

$$\begin{array}{ccccccccc}
\mathbf{0} & \longrightarrow & A & \xrightarrow{i_A} & \mathcal{M}_G(A) & \xrightarrow{j_A} & A_1 & \longrightarrow & \mathbf{0}\\
 & & \downarrow{\scriptstyle\varphi} & & \downarrow{\scriptstyle\mathcal{M}_G(\varphi)} & & & & \\
\mathbf{0} & \longrightarrow & B & \xrightarrow{i_B} & \mathcal{M}_G(B) & \xrightarrow{j_B} & B_1 & \longrightarrow & \mathbf{0}
\end{array}$$

is a commutative diagram with exact rows, and there exists a unique G-homomorphism $\varphi_1\colon A_1\to B_1$ which makes the diagram commutative. We call φ_1 the **derived homomorphism** of φ. If $\psi\colon B\to C$ is another G-homomorphism of discrete G-modules, then $\mathcal{M}_G(\psi\circ\varphi)=\mathcal{M}_G(\psi)\circ\mathcal{M}_G(\varphi)$ and $(\psi\circ\varphi)_1=\psi_1\circ\varphi_1$.

4. Let A be a discrete G-module, $n\in\mathbb{Z}$, and let $\varphi\colon A\to A$ be given by $\varphi(a)=na$ for all $a\in A$. Then $\varphi_1(a_1)=na_1$ for all $a_1\in A_1$. Indeed, let $a_1=j_A(f)$, where $f\in\mathcal{M}_G(A)$. Then

$$\varphi_1(a_1)=j_A(\mathcal{M}_G(\varphi)(f))=j_A(\varphi\circ f)=j_A(nf)=nj_A(f)=na_1.$$

Theorem 2.1.5. *Let G be a profinite group.*

1. Let H be a closed subgroup of G and $B' \overset{\varphi}{\to} B \overset{\psi}{\to} B''$ an exact sequence of discrete H-modules. Then the sequence

$$\mathcal{M}_G^H(B') \overset{\mathcal{M}_G^H(\varphi)}{\longrightarrow} \mathcal{M}_G^H(B) \overset{\mathcal{M}_G^H(\psi)}{\longrightarrow} \mathcal{M}_G^H(B'')$$

is exact, too.

2. Let $B' \overset{\varphi}{\to} B \overset{\psi}{\to} B''$ be an exact sequence of discrete G-modules. Then the derived sequence

$$B_1' \overset{\varphi_1}{\to} B_1 \overset{\psi_1}{\to} B_1''$$

is exact, too.

Proof. 1. Obviously, $\mathcal{M}_G^H(\psi)\circ\mathcal{M}_G^H(\varphi) = \mathcal{M}_G^H(\psi\circ\varphi) = 0$, and consequently $\mathrm{Im}(\mathcal{M}_G^H(\varphi) \subset \mathrm{Ker}(\mathcal{M}_G^H(\psi))$. Thus let $f \in \mathrm{Ker}(\mathcal{M}_G^H(\psi))$. Then $\psi\circ f = 0$, hence $f(G) \subset \mathrm{Ker}(\psi) = \mathrm{Im}(\varphi)$, and as $f(G)$ is compact and discrete, it is finite, say $f(G) = \{\varphi(b_1), \ldots, \varphi(b_n)\} \subset B$, where $b_i \in B'$, and let $V_i = H_{b_i}$ be the isotropy group of b_i for all $i \in [1, n]$. Let U be an open normal subgroup of G such that $U \cap H \subset V_1 \cap \ldots \cap V_n$ and f is constant on the cosets of U (see 1.5.2.2). Let $\{\sigma_1, \ldots, \sigma_t\} \subset G$ be a set of representatives for $HU\backslash G$, and for $j \in [1, t]$ let $f(\sigma_j) = \varphi(b_{\nu(j)})$, where $\nu(j) \in [1, n]$. If $\sigma \in G$, then $\sigma = \tau u \sigma_j$ for some $\tau \in H,\ u \in U$ and $j \in [1, t]$, and we assert that $\tau b_{\nu(j)}$ only depends on σ. Indeed, if $\tau u \sigma_j = \tau' u' \sigma_k$ for some $\tau,\ \tau' \in H,\ u,\ u' \in U$ and $j,\ k \in [1, t]$, then $j = k$ and $\tau^{-1}\tau' \in U \cap H$, hence $\tau b_{\nu(j)} = \tau' b_{\nu(k)}$.

We define $h\colon G \to B'$ by $h(\sigma) = \tau b_{\nu(j)}$ if $\sigma = \tau u \sigma_j \in G$ for some $\tau \in H,\ u \in U$ and $j \in [1, t]$. We claim that $h \in \mathcal{M}_G^H(B')$ and $\mathcal{M}_G^H(\varphi)(h) = f \in \mathrm{Im}(\mathcal{M}_G^H(\varphi))$.

Let $\sigma = \tau u \sigma_j \in G$, where $\tau \in H,\ u \in U$ and $j \in [1, t]$. If $v \in U$, then $v\sigma = \tau u' \sigma_j$, where $u' = \tau^{-1} v \tau u \in U$, and $h(v\sigma) = \tau b_{\nu(j)} = h(\sigma)$. Being constant on cosets of U, h is continuous, and if $y \in H$, then $h(y\sigma) = h(y\tau u\sigma_j) = y\tau b_{\nu(j)} = yh(\sigma)$. Hence $h \in \mathcal{M}_G^H(B')$, and

$$\varphi\circ h(\tau u\sigma_j) = \varphi(\tau b_{\nu(j)}) = \tau\varphi(b_{\nu(j)}) = \tau f(\sigma_j) = \tau f(u\sigma_j) = f(\tau u \sigma_j),$$

which implies $\mathcal{M}_G^H(\varphi)(h) = \varphi\circ h = f$.

2. Since $\psi_1\circ\varphi_1 = (\psi\circ\varphi)_1 = 0$, it remains to prove that $\mathrm{Ker}(\psi_1) \subset \mathrm{Im}(\varphi_1)$. For this, we consider the following commutative diagram with exact columns.

$$\begin{array}{ccccc}
B' & \overset{\varphi}{\longrightarrow} & B & \overset{\psi}{\longrightarrow} & B'' \\
{\scriptstyle i_{B'}}\big\downarrow & & {\scriptstyle i_B}\big\downarrow & & {\scriptstyle i_{B''}}\big\downarrow \\
\mathcal{M}_G(B') & \overset{\mathcal{M}_G(\varphi)}{\longrightarrow} & \mathcal{M}_G(B) & \overset{\mathcal{M}_G(\psi)}{\longrightarrow} & \mathcal{M}_G(B'') \\
{\scriptstyle j_{B'}}\big\downarrow & & {\scriptstyle j_B}\big\downarrow & & {\scriptstyle j_{B''}}\big\downarrow \\
B_1' & \overset{\varphi_1}{\longrightarrow} & B_1 & \overset{\psi_1}{\longrightarrow} & B_1'' \\
 & & \big\downarrow & & \\
 & & \mathbf{0} & &
\end{array}$$

The first row is exact by assumption and the second row is exact by 1. (where we view $B' \to B \to B''$ as an exact sequence of the underlying abelian groups).

Let $b_1 \in \operatorname{Ker}(\psi_1)$. Then $b_1 = j_B(g)$ for some $g \in \mathcal{M}_G(B)$, and we obtain $0 = \psi_1 \circ j_B(g) = j_{B''} \circ \mathcal{M}_G(\psi)(g)$. Hence $\mathcal{M}_G(\psi)(g) \in \operatorname{Ker}(j_{B''}) = \operatorname{Im}(i_{B''})$, and we assert that even $\mathcal{M}_G(\psi)(g) \in \operatorname{Im}(i_{B''} \circ \psi)$. Indeed, let $b'' \in B''$ such that $\mathcal{M}_G(\psi)(g) = i_{B''}(b'')$. If $\sigma \in G$, then

$$\mathcal{M}_G(\psi)(g)(\sigma) = \psi(g(\sigma)) = i_{B''}(b'')(\sigma) = \sigma b'',$$

and therefore $b'' = \sigma^{-1}\psi((g(\sigma)) = \psi(\sigma^{-1}g(\sigma))$. Hence there exists some $b \in B$ such that $\mathcal{M}_G(\psi)(g) = i_{B''} \circ \psi(b) = \mathcal{M}_G(\varphi) \circ i_B(b)$, and consequently $g - i_B(b) \in \operatorname{Ker}(\mathcal{M}_G(\psi)) = \operatorname{Im}(\mathcal{M}_G(\varphi))$, say $g - i_B(b) = \mathcal{M}_G(\varphi)(g')$ for some $g' \in \mathcal{M}_G(B')$. Eventually we obtain

$$b_1 = j_B(g) = j_B(g - i_B(b)) = j_B \circ \mathcal{M}_G(\varphi)(g') = \varphi_1 \circ j_{B'}(g') \in \operatorname{Im}(\varphi_1). \quad \square$$

Theorem 2.1.6. *Let G be a profinite group and H a closed subgroup of G.*

1. *Let B be a discrete H-module. Then $\mathcal{M}_G^H(B)^G$ consists of all constant maps $f\colon G \to B^H$, and the map $\varepsilon\colon \mathcal{M}_G^H(B)^G \to B^H$, defined by $\varepsilon(f) = f(1)$, is an isomorphism.*
   We identify: $\mathcal{M}_G^H(B)^G = B^H$.

2. *Let K be a closed subgroup of H and B a discrete K-module. Then the map*

$$\Phi\colon \mathcal{M}_G^K(B) \to \mathcal{M}_G^H(\mathcal{M}_H^K(B)),$$

defined by $\Phi(f)(g)(y) = f(yg)$ for all $f \in \mathcal{M}_G^K(B)$, $g \in G$ and $y \in H$, is a G-isomorphism. In particular (if $K = \mathbf{1}$ and B is an abelian group), then

$$\mathcal{M}_G(B) \cong \mathcal{M}_G^H(\mathcal{M}_H(B)).$$

Proof. 1. Let $f\colon G \to B^H$ be constant and $b = f(1) \in B^H$. If $g \in G$ and $\tau \in H$, then $f(\tau g) = b = \tau f(g)$ and therefore $f \in \mathcal{M}_G^H(B)$. If $\sigma \in G$, then $(\sigma f)(g) = f(g\sigma) = b = f(g)$ for all $g \in G$, hence $\sigma f = f$ and thus $f \in \mathcal{M}_G^H(B)^G$.

Conversely, let $f \in \mathcal{M}_G^H(B)^G$. If $\sigma \in G$, then $f(\sigma) = (\sigma f)(1) = f(1)$, and thus f is constant. If $\tau \in H$, then $\tau f(1) = f(\tau) = f(1)$, and consequently $f(1) \in B^H$. In particular, $\varepsilon\colon \mathcal{M}_G^H(B) \to B^H$ is an isomorphism.

2. Let $f \in \mathcal{M}_G^K(B)$, $g \in G$, and define $\Phi_f(g)\colon H \to B$ by $\Phi_f(g)(y) = f(yg)$. We assert that $\Phi_f(g) \in \mathcal{M}_H^K(B)$. Indeed, $\Phi_f(g)$ is continuous, and if $\tau \in K$, then $\Phi_f(g)(\tau y) = f(\tau y g) = \tau f(yg) = \tau\Phi_f(g)(y)$.

For $f \in \mathcal{M}_G^K(B)$ we define now $\Phi_f\colon G \to \mathcal{M}_H^K(B)$ by $\Phi_f(g)(y) = f(yg)$ for all $g \in G$ and $y \in H$, and we assert that $\Phi_f \in \mathcal{M}_G^H(\mathcal{M}_H^K(B))$. If $f \in \mathcal{M}_G^K(B)$, $g \in G$ and $\tau \in H$, then $\Phi_f(\tau g)(y) = f(y\tau g) = \Phi_f(g)(y\tau) = [\tau\Phi_f(g)](y)$ for all $y \in H$, and consequently $\Phi_f(\tau g) = \tau\Phi_f(g)$.

To prove that Φ_f is continuous, let $h_0 \in \mathcal{M}_H^K(B)$. Then h_0 is continuous, and as H is compact, $h_0(H)$ is finite, say $h_0(H) = \{z_1, \ldots, z_m\} \subset B$. Then

$$\Phi_f^{-1}(h_0) = \{g \in G \mid \Phi_f(g) = h_0\} = \{g \in G \mid f(yg) = h_0(y) \text{ for all } y \in H\}$$
$$= \bigcap_{j=1}^{m} V_j, \quad \text{where } V_j = \{g \in G \mid h_0^{-1}(z_j)g \subset f^{-1}(z_j)\} \text{ for all } j \in [1, m].$$

If $j \in [1, m]$, then $h_0^{-1}(z_j)$ is closed in H, hence compact, and $f^{-1}(z_j)$ is open in G. Hence V_j is open by 1.1.2.4(d), and thus $\Phi_f^{-1}(h_0)$ is open.

Consequently $\Phi\colon \mathcal{M}_G^K(B) \to \mathcal{M}_G^H(\mathcal{M}_H^K(B))$, defined by $\Phi(f) = \Phi_f$, is a well-defined map, and we prove that it is a G-isomorphism. If $f,\ f' \in \mathcal{M}_G^K(B)$, then $\Phi_{f+f'} = \Phi_f + \Phi_{f'}$, and thus Φ is a group homomorphism. If $\sigma \in G$, then

$$\Phi(\sigma f)(g)(y) = (\sigma f)(yg) = f(yg\sigma) = \Phi(f)(g\sigma)(y) = [\sigma\Phi(f)](g)(y)$$

for all $g \in G$ and $y \in H$. Hence $\Phi(\sigma f) = \sigma\Phi(f)$, and Φ is a G-homomorphism.

If $f \in \operatorname{Ker}(\Phi)$, then $\Phi_f = 0$ implies $f = 0$. Hence Φ is a monomorphism, and we must prove that it is surjective. Let $F \in \mathcal{M}_G^H(\mathcal{M}_H^K(B))$, and define $f_F\colon G \to B$ by $f_F(g) = F(g)(1)$. By 1.5.2.2, there exists an open normal subgroup U of G such that F is constant on cosets of U, and the same is true for f_F. Hence f_F is continuos. If $g \in G$ and $\tau \in K$, then

$$f_F(\tau g) = F(\tau g)(1) = \tau F(g)(1) = \tau f_F(g),$$

which implies $f_F \in \mathcal{M}_G^K(B)$. If $g \in G$ and $y \in H$, then

$$\Phi_{f_F}(g)(y) = f_F(yg) = F(yg)(1) = [yF(g)](1) = F(g)(y).$$

Hence $\Phi_{f_F} = F$, and Φ is surjective. □

Theorem 2.1.7. *Let G be a profinite group, H a closed subgroup of G and A a discrete H-module.*

1. *If U is an open normal subgroup of G, then*

   - *$A^{U\cap H}$ is a discrete HU/U-module by means of $(\tau U)a = \tau a$ for all $\tau \in H$ and $u \in U$,*
   - *$\mathcal{M}_{G/U}^{HU/U}(A^{U\cap H})$ is a discrete G-module by means of $\sigma f = (\sigma U)f$ for all $\sigma \in G$ and $f \in \mathcal{M}_{G/U}^{HU/U}(A^{U\cap H})$,*

   and there is a G-isomorphism

   $$\Phi_U\colon \mathcal{M}_{G/U}^{HU/U}(A^{U\cap H}) \to \mathcal{M}_G^H(A)^U$$

   such that $\Phi_U(f)(g) = f(gU)$ for all $f \in \mathcal{M}_{G/U}^{HU/U}(A^{U\cap H})$ and all $g \in G$.

2. Let U and V be open normal subgroups of G such that $U \subset V$. Then there exists a unique G-homomorphism

$$\theta_{V,U} \colon \mathcal{M}_{G/V}^{HV/V}(A^{V\cap H}) \to \mathcal{M}_{G/U}^{HU/U}(A^{U\cap H})$$

such that

$$\theta_{V,U}(f)(gU) = f(gV) \in A^{V\cap H} \text{ for all } f \in \mathcal{M}_{G/V}^{HV/V}(A^{V\cap H}) \text{ and } g \in G,$$

and then there is a commutative diagram

$$\begin{array}{ccc} \mathcal{M}_{G/V}^{HV/V}(A^{V\cap H}) & \xrightarrow{\Phi_V} & \mathcal{M}_G^H(A)^V \\ {\scriptstyle\theta_{V,U}}\big\downarrow & & \big\downarrow{\scriptstyle\iota} \\ \mathcal{M}_{G/U}^{HU/U}(A^{U\cap H}) & \xrightarrow{\Phi_U} & \mathcal{M}_G^H(A)^U, \end{array}$$

where $\iota = (\mathcal{M}_G^H(A)^V \hookrightarrow \mathcal{M}_G^H(A)^U)$.

3. Let Σ be a fundamental system of neighbourhoods of 1 in G consisting of open normal subgroups of G. Then $(\Sigma, \supset)$ is directed,

$$(\mathcal{M}_{G/U}^{HU/U}(A^{U\cap H}), \theta_{V,U})_\Sigma$$

is an inductive system, and there is an isomorphism

$$\varinjlim \Phi_U \colon \varinjlim_{U\in\Sigma} \mathcal{M}_{G/U}^{HU/U}(A^{U\cap H}) \overset{\sim}{\to} \varinjlim_{U\in\Sigma} \mathcal{M}_G^H(A)^U = \bigcup_{U\in\Sigma} \mathcal{M}_G^H(A)^U$$
$$= \mathcal{M}_G^H(A).$$

Proof. It suffices to prove 1. then 2. and 3. follow by the definitions.

Clearly, $A^{U\cap H}$ is a discrete HU/U-module and $\mathcal{M}_{G/U}^{HU/U}(A^{U\cap H})$ is a discrete G-module in the asserted natural way.

Let $f \in \mathcal{M}_{G/U}^{HU/U}(A^{U\cap H})$, and define $\Phi_U(f) \colon G \to A$ by $\Phi_U(f)(g) = f(gU)$ for all $g \in G$. Then $\Phi_U(f)$ is continuous,

$$\Phi_U(f)(\tau g) = f(\tau gU) = (\tau U)f(gU) = \tau\Phi_U(f)(g) \quad \text{for all } g \in G \text{ and } \tau \in H,$$

and consequently $\Phi_U(f) \in \mathcal{M}_G^H(A)$. If $x \in U$, then

$$[x\Phi_U(f)](g) = \Phi_U(f)(gx) = f(gxU) = f(gU) = \Phi_U(f)(g) \quad \text{for all } g \in G,$$

hence $x\Phi_U(f) = \Phi_U(f)$ and thus $\Phi_U(f) \in \mathcal{M}_G^H(A)^U$.

By definition, Φ_U is a group monomorphism. If $\sigma \in G$, then

$$\Phi_U(\sigma f)(g) = (\sigma f)(gU) = f(gU\sigma) = f(g\sigma U) = \Phi_U(f)(g\sigma) = [\sigma\Phi_U(f)](g)$$

for all $g \in G$, hence $\Phi_U(\sigma f) = \sigma\Phi_U(f)$, and thus Φ_U is a G-homomorphism.

It remains to prove that Φ_U is surjective. Let $h \in \mathcal{M}_G^H(A)^U$. If $\sigma \in U$ and $g \in G$, then $h(g) = (\sigma h)(g) = h(g\sigma)$, and if $\sigma \in U \cap H$, then it follows that $\sigma h(g) = h(g\sigma) = h(g)$. Hence h is constant on cosets of U, and $h(g) \in A^{U\cap H}$ for all $g \in G$. Thus there exists a function $f\colon G/U \to A^{U\cap H}$ such that $f(gU) = h(g)$ for all $g \in G$. Now it follows by the very definition that $f \in \mathcal{M}_{G/U}^{HU/U}(A^{U\cap H})$ and $\Phi_U(f) = h$. □

Theorem 2.1.8. *Let G be a profinite group, H a closed subgroup of G, G/H the (compact) coset space, $\pi\colon G \to G/H$ the residue class map, and X a (discrete) abelian group.*

1. *Let $s\colon G/H \to G$ be continuous such that $s \circ \pi = \mathrm{id}_{G/H}$ (by 1.5.3.1), and let $\mathcal{C}(G/H, X)$ be the group of all continuous maps $\varphi\colon G/H \to X$ under pointwise addition. Then there exists an H-isomorphism*

$$\Phi\colon \mathcal{M}_G(X) \xrightarrow{\sim} \mathcal{M}_H(\mathcal{C}(G/H, X))$$

satisfying $\Phi(f)(\tau)(\rho) = f(s(\rho)\tau)$ for all $f \in \mathcal{M}_G(X)$, $\tau \in H$ and $\rho \in G/H$. In particular, every induced G-module is (by restriction of scalars) an induced H-module.

2. *Let H be a normal subgroup of G. Then $\mathcal{M}_{G/H}(X)$ is a G-module by means of $\sigma f = (\sigma H)f$ for all $\sigma \in G$ and $f \in \mathcal{M}_{G/H}(X)$, and there exists a G-isomorphism*

$$\Phi\colon \mathcal{M}_{G/H}(X) \xrightarrow{\sim} \mathcal{M}_G(X)^H$$

such that $\Phi(f) = f \circ \pi$ for all $f \in \mathcal{M}_{G/H}(X)$. In particular, if $A = \mathcal{M}_G(X)$ is an induced G-module, then A^H is an induced G/H-module.

Proof. 1. For $f \in \mathcal{M}_G(X)$ and $\tau \in H$ we define $\Phi(f)(\tau)\colon G/H \to X$ by $\Phi(f)(\tau)(\rho) = f(s(\rho)\tau)$. Then $\Phi(f)(\tau) \in \mathcal{C}(G/H, X)$, and we assert that the map $\Phi(f)\colon H \to \mathcal{C}(G/H, X)$ is continuous (and thus we have $\Phi(f) \in \mathcal{M}_H(\mathcal{C}(G/H, X))$).

Let $h \in \mathcal{C}(G/H, X)$. Then $h(G/H)$ is finite, say $h(G/H) = \{x_1, \ldots, x_m\}$, and

$$\Phi(f)^{-1}(h) = \{\tau \in H \mid f_\tau = h\} = \{\tau \in H \mid f(s(\rho)\tau) = h(\rho) \text{ for all } \rho \in G/H\}$$
$$= \bigcap_{j=1}^{m} V_j, \quad \text{where } V_j = \{\tau \in H \mid s(h^{-1}(x_j))\tau \subset f^{-1}(x_j)\}.$$

If $j \in [1, m]$, then the sets $h^{-1}(x_j) \subset G/H$ and $f^{-1}(x_j) \subset G$ are clopen and therefore compact. Hence $s(h^{-1}(x_j))$ is compact, and V_j is open by 1.1.2.4(d). It follows that $\Phi(f)^{-1}(h)$ is open, and therefore $\Phi(f)$ is continuous.

If $f,\, f' \in \mathcal{M}_G(X)$, then $\Phi(f+f') = \Phi(f)+\Phi(f')$ and $\Phi(f)=0$ implies $f=0$. Hene $\Phi\colon \mathcal{M}_G(X) \to \mathcal{M}_H(\mathcal{C}(G/H),X))$ is a group monomorphism. If $f \in \mathcal{M}_G(X)$ and $y \in H$, then

$$\Phi(yf)(\tau)(\rho) = (yf)(s(\rho)\tau) = f(s(\rho)\tau y) = \Phi(f)(\tau y)(\rho) = [y\Phi(f)(\tau)](\rho)$$

for all $\tau \in H$ and $\rho \in G/H$, hence $\Phi(yf) = \, y\Phi(f)$, and Φ is an H-homomorphism.

It remains to prove that Φ is surjective. Let $h \in \mathcal{M}_H(\mathcal{C}(G/H,X)$ and define $f_0\colon G/H\times H \to X$ by $f_0(\rho,\tau) = h(\tau)(\rho)$. If $(\rho_0,\tau_0) \in G/H\times H$ and $f_0(\rho_0,\tau_0) = x_0$. Then $h(\tau_0)^{-1}(x_0)$ is an open subset of G/H, and $h^{-1}(h(\tau_0))$ is an open subset of H. Hence $h(\tau_0)^{-1}(x_0)\times h^{-1}(h(\tau_0))$ is an open subset of $G/H\times H$, and we assert that $f_0^{-1}(x_0) \supset h(\tau_0)^{-1}\times h^{-1}(h(\tau_0))$. Indeed, if $(\rho,\tau) \in h(\tau_0)^{-1}(x_0)\times h^{-1}(h(\tau_0))$, then $f_0(\rho,\tau) = h(\tau)(\rho) = h(\tau_0)(\rho) = x_0$. Since $(\rho_0,\tau_0) \in h(\tau_0)^{-1}(x_0)\times h^{-1}(h(\tau_0))$, it follows that $f_0^{-1}(x_0)$ is a neighbourhood of (ρ_0,τ_0), and therefore f_0 is continuous.

The map $s_1\colon G/H\times H \to G$, defined by $s_1(\rho,\tau) = s(\rho)\tau$ is continuous and bijective, and as $G/H\times H$ is compact, it is a homeomorphism. As the map $f = f_0\circ s_1^{-1}\colon G \to X$ is continuous, we get $f \in \mathcal{M}_G(X)$ and $\Phi(f) = h$.

2. Apparently $\mathcal{M}_{G/H}(X)$ is a G-module as asserted. If $f \in \mathcal{M}_{G/H}(X)$, then the map $f\circ\pi\colon G \to X$ is continuous. Hence $\Phi(f) \in \mathcal{M}_G(X)$, and if $\tau \in H$, then $\tau(f\circ\pi)(g) = (f\circ\pi)(g\tau) = (f\circ\pi)(g)$ for all $g \in G$ and thus $f\circ\pi \in \mathcal{M}_G(X)^H$. Obviously, Φ is a group monomorphism. If $f \in \mathcal{M}_{G/H}(X)$ and $\sigma \in G$, then

$$\Phi(\sigma f)(g) = (\sigma f)(\pi(g)) = f(\pi(g)\sigma) = f(\pi(g\sigma)) = \Phi(f)(g\sigma) = [\sigma\Phi(f)](g)$$

for all $g \in G$, hence $\Phi(\sigma f) = \sigma\Phi(f)$, and Φ is a G-homomorphism.

It remains to prove that Φ is surjective. If $h \in \mathcal{M}_G(X)^H$, $g \in G$ and $\tau \in H$, then $h(g) = (\tau h)(g) = h(g\tau)$. Hence there exists some $f \in \mathcal{M}_{G/H}(X)$ such that $\Phi(f) = f\circ\pi = h$. □

Theorem 2.1.9. *Let G be a profinite group and A a discrete G-module.*

1. Let H be an open subgroup of G, $r = (G{:}H)$, and suppose that $G/H = \{\sigma_1 H,\ldots,\sigma_r H\}$, where $\sigma_1 = 1$. Let B an H-submodule of A, suppose that

$$A = \bigoplus_{i=1}^{r} \sigma_i B = B \oplus \bigoplus_{i=2}^{r} \sigma_i B,$$

and let $\pi\colon A \to B$ the projection onto the first direct summand. Then the map

$$\Phi\colon \mathcal{M}_G^H(B) \to A, \quad \text{defined by} \quad \Phi(f) = \sum_{i=1}^{r} \sigma_i f(\sigma_i^{-1}) \quad \text{for all}$$

$$f \in \mathcal{M}_G^H(B),$$

is a G-isomorphism, and $\pi\circ\Phi(f) = f(1)$ for all $f \in \mathcal{M}_G^H(B)$.

2. Let G be finite and B a subgroup of A such that

$$A = \bigoplus_{\sigma \in G} \sigma B.$$

Then $A \cong \mathcal{M}_G(B)$, and thus A is an induced G-module.

Proof. 1. Let $B' = \{f \in \mathcal{M}_G^H(B) \mid f \restriction G \backslash H = 0\}$. Then B' is an H-submodule of $\mathcal{M}_G^H(B)$, and we define $p\colon B' \to B$ by $p(f) = f(1)$. Now we proceed in 3 steps.

I. p is an H-isomorphism.

Proof of **I.** Clearly, p is a group homomorphism, and if $f \in \mathcal{M}_G^H(B)$ and $\tau \in H$, then $p(\tau f) = (\tau f)(1) = f(\tau) = \tau f(1) = \tau p(f)$. Hence p is an H-homomorphism, and $\text{Ker}(p) = \{f \in B' \mid f \restriction H = 0\} = \mathbf{0}$. It remains to prove that p is surjective. If $b \in B$, then we define $f_b\colon G \to B$ by

$$f_b(\tau) = \begin{cases} \tau b & \text{if } \tau \in H, \\ 0 & \text{if } \tau \notin H. \end{cases}$$

Then f_b is continuous, hence $f_b \in B'$ and $p(f_b) = b$. $\square$[**I.**

II. Every $f \in \mathcal{M}_G^H(B)$ has a unique representation

$$f = \sum_{i=1}^{r} \sigma_i f_i, \quad \text{where } f_i \in B' \text{ for all } i \in [1, r].$$

Explicitly, f_i is given by

$$f_i(\tau) = \begin{cases} f(\tau\sigma_i^{-1}) & \text{if } \tau \in H, \\ 0 & \text{if } \tau \in G \setminus H. \end{cases}$$

Proof of **II.** If $f_i\colon H \to B$ is defined as above, then f_i is continuous and thus $f_i \in B'$. If $\sigma \in G$, then $H\backslash G = \{H\sigma_1^{-1}, \ldots, H\sigma_r^{-1}\}$ implies $\sigma = \tau\sigma_k^{-1}$ for some $k \in [1, r]$ and $\tau \in H$, and

$$\sum_{i=1}^{r} (\sigma_i f_i)(\sigma) = \sum_{i=1}^{r} f_i(\tau\sigma_k^{-1}\sigma_i) = f_k(\tau) = f(\tau\sigma_k^{-1}) = f(\sigma), \quad \text{hence}$$

$$f = \sum_{i \in I} \sigma_i f_i.$$

To prove uniqueness, suppose that

$$f = \sum_{i=1}^{r} \sigma_i f_i', \quad \text{where } f_i' \in B' \text{ for all } i \in [1, r].$$

Then we obtain, for all $\tau \in H$ and $k \in [1, r]$,

$$f_k(\tau) = f(\tau\sigma_k^{-1}) = \sum_{i=1}^{r} \sigma_i f_i'(\tau\sigma_k^{-1}) = f_k'(\tau) \quad \text{and thus} \quad f_k' = f_k. \qquad \square[\mathbf{II.}]$$

III. The map $\Phi\colon \mathcal{M}_G^H(B) \to A$, defined by

$$\Phi(f) = \sum_{i=1}^{r} \sigma_i f(\sigma_i^{-1}) \ \text{ for all } \ f \in \mathcal{M}_G^H(B),$$

is a G-isomorphism, and $\pi \circ \Phi(f) = f(1)$ for all $f \in \mathcal{M}_G^H(B)$.

Proof of **III.** By **II** we obtain

$$\mathcal{M}_G^H(B) = \bigoplus_{i=1}^{r} \sigma_i B',$$

and for every $i \in [1, r]$ the H-isomorphism $p \restriction B' \colon B' \to B$ induces a group isomorphism $\Phi_i \colon \sigma_i B' \to \sigma_i B$ satisfying

$$\Phi_i(\sigma_i f_i) = \sigma_i p(f_i) = \sigma_i f_i(1) = \sigma_i f(\sigma_i^{-1}) \text{ for all } f_i \in B'.$$

Hence the map $\Phi\colon \mathcal{M}_G^H(B) \to A$, defined by

$$\Phi(f) = \Phi\Big(\sum_{i=1}^{r} \sigma_i f_i\Big) = \sum_{i=1}^{r} \sigma_i f(\sigma_i^{-1}) \ \text{ for all } \ f \in \mathcal{M}_G^H(B),$$

is a group isomorphism, and $\pi \circ \Phi(f) = f(1)$ for all $f \in \mathcal{M}_G^H(B)$.

It remains to prove that Φ is a G-homomorphism. Let $\sigma \in G$ and $\pi \in \mathfrak{S}_r$ such that for all $i \in [1, r]$ we have $\sigma^{-1}\sigma_i = \sigma_{\pi(i)}\tau_i$ for some $\tau_i \in H$. If $f \in \mathcal{M}_G^H(B)$,then

$$\begin{aligned}\Phi(\sigma f) &= \sum_{i=1}^{r} \sigma_i (\sigma f)(\sigma_i^{-1}) = \sum_{i=1}^{r} \sigma_i f(\sigma_i^{-1}\sigma) = \sum_{i=1}^{r} \sigma_i f(\tau_i^{-1}\sigma_{\pi(i)}^{-1}) \\ &= \sum_{i=1}^{r} \sigma_i \tau_i^{-1} f(\sigma_{\pi(i)}^{-1}) = \sum_{i=1}^{r} \sigma\sigma_{\pi(i)} f(\sigma_{\pi(i)}^{-1}) = \sum_{i=1}^{r} \sigma\sigma_i f(\sigma_i^{-1}) = \sigma\Phi(f).\end{aligned}$$

2. By 1. for $H = \mathbf{1}$. $\square$

2.2 Cohomology groups

Definitions and Remarks 2.2.1. Let G be a profinite group, A a discrete G-module, and $q \geq 0$. Let $C^q(G, A)$ be the group of all continuous maps $x\colon G^q \to A$ equipped with pointwise addition, where in particular $G^0 = \mathbf{1}$

and $C^0(G,A) = A$. For $q \geq 0$, $x \in C^q(G,A)$ and $(\sigma_1,\ldots,\sigma_{q+1}) \in G^{q+1}$ we define $d^q x \colon G^{q+1} \to A$ by

$$\begin{aligned} d^q x(\sigma_1,\ldots,\sigma_{q+1}) = \sigma_1 x(\sigma_2,\ldots,\sigma_{q+1}) \\ + \sum_{i=1}^{q} (-1)^i x(\sigma_1,\ldots,\sigma_{i-1},\sigma_i\sigma_{i+1},\sigma_{i+2},\ldots,\sigma_{q+1}) \\ + (-1)^{q+1} x(\sigma_1,\ldots,\sigma_q) \end{aligned}$$

for all $(\sigma_1,\ldots,\sigma_{q+1}) \in G^{q+1}$. In particular, $d^0 a(\sigma) = \sigma a - a$ for all $a \in A$ and $\sigma \in G$, $\operatorname{Ker}(d^0) = A^G$, and $d^0 a \colon G \to A$ is continuous.

If $q \geq 1$ and $x \in C^q(G,A)$, then (by 1.5.2.2) there exists an open normal subgroup U of G such that x is constant on the cosets of U^q. Then $d^q x$ is constant on the cosets of U^{q+1} and thus continuous as well. Hence for every $q \geq 0$ we obtain a map

$$d^q = d^q_A \colon C^q(G,A) \to C^{q+1}(G,A),$$

which is obviously a group homomorphism. The maps d^q are called the **differentials** of the pair (G,A).

Let $\varphi \colon A \to B$ be a G-homomorphism of discrete G-modules. We set $C^0(\varphi) = \varphi$, and for $q \geq 1$ and $x \in C^q(G,A)$ we define

$$C^q(\varphi) \colon C^q(G,A) \to C^q(G,B) \quad \text{by} \quad C^q(\varphi)(x) = \varphi \circ x.$$

Then $C^q(\varphi)$ is a group homomorphism satisfying

$$d^{q-1}_B \circ C^{q-1}(\varphi) = C^q(\varphi) \circ d^{q-1}_A \colon C^{q-1}(G,A) \to C^q(G,B).$$

If φ is a monomorphism or an epimorphism, then so is $C^q(\varphi)$. If $\varphi' \colon A \to B$ and $\psi \colon B \to C$ are other G-homomorphisms of discrete G-modules, then it follows that $C^q(\varphi + \varphi') = C^q(\varphi) + C^q(\varphi')$, and $C^q(\psi \circ \varphi) = C^q(\psi) \circ C^q(\varphi)$.

Theorem 2.2.2. *Let G be a profinite group, A a discrete G-module and $q \geq 1$. Then $\operatorname{Im}(d^{q-1}) \subset \operatorname{Ker}(d^q)$, and equality holds if A is an induced G-module.*

Proof. We prove first:

I. Let X be an abelian group and $A = \mathcal{M}_G(X)$ an induced G-module. For $q \geq 1$, let $h^q \colon C^q(G,A) \to C^{q-1}(G,A)$ be defined by

$$h^q(x)(\sigma_1,\ldots,\sigma_{q-1})(g) = x(g,\sigma_1,\ldots,\sigma_{q-1})(1)$$

for all $x \in C^q(G,A)$, $(\sigma_1,\ldots,\sigma_{q-1}) \in G^{q-1}$ and $g \in G$ [for $q = 1$, this means $h^1(x)(g) = x(g)$ for all $x \in C^1(G,A)$ and $g \in G$]. Then h^q is a group monomorphism, and

$$d^{q-1} \circ h^q + h^{q+1} \circ d^q = \mathrm{id}_{C^q(G,A)} \quad \text{for all } q \geq 1.$$

Proof of **I.** Obviously h^q is a group homomorphism, and if $x \in \mathrm{Ker}(h^q)$, then

$$x(g,\sigma_1,\ldots,\sigma_{q-1})(\tau) = \tau x(g,\sigma_1,\ldots,\sigma_{q-1})(1) = \tau h^q(x)(\sigma_1,\ldots,\sigma_{q-1})(g) = 0$$

for all $g,\sigma_1,\ldots,\sigma_{q-1},\tau \in G$, hence $x = 0$, and thus h^q is a monomorphism.

If $(\sigma_1,\ldots,\sigma_q) \in G^q$, $g \in G$ and $x \in C^q(G,A)$, then

$$\begin{aligned}
d^{q-1}\circ h^q(x)(\sigma_1,\ldots,\sigma_q)(g) &= [\sigma_1 h^q(x)(\sigma_2,\ldots,\sigma_q)](g) \\
&\quad + \sum_{i=1}^{q-1}(-1)^i h^q(x)(\sigma_1,\ldots,\sigma_i\sigma_{i+1},\ldots,\sigma_q)(g) \\
&\quad + (-1)^q h^q(x)(\sigma_1,\ldots,\sigma_{q-1})(g) \\
&= h^q(x)(\sigma_2,\ldots,\sigma_q)(g\sigma_1) + \ldots \\
&= x(g\sigma_1,\sigma_2,\ldots,\sigma_q)(1) + \sum_{i=1}^{q-1}(-1)^i x(g,\sigma_1,\ldots,\sigma_i\sigma_{i+1},\ldots,\sigma_q)(1) \\
&\qquad + (-1)^q x(g,\sigma_1,\ldots,\sigma_{q-1})(1)
\end{aligned}$$

and

$$\begin{aligned}
h^{q+1}\circ d^q(x)(\sigma_1,\ldots,\sigma_q) &= d^q(x)(g,\sigma_1,\ldots,\sigma_q)(1) \\
&= gx(\sigma_1,\ldots,\sigma_q)(1) - x(g\sigma_1,\sigma_2,\ldots,\sigma_q)(1) \\
&\quad + \sum_{i=2}^{q}(-1)^i x(g,\sigma_1,\ldots,\sigma_{i-1}\sigma_i,\ldots,\sigma_q)(1) \\
&\quad + (-1)^{q+1} x(g,\sigma_1,\ldots,\sigma_{q-1})(1) \\
&= x(\sigma_1,\ldots,\sigma_q)(g) - x(g\sigma_1,\sigma_2,\ldots,\sigma_q)(1) \\
&\quad + \sum_{i=1}^{q-1}(-1)^{i+1} x(g,\sigma_1,\ldots,\sigma_i\sigma_{i+1},\ldots,\sigma_q)(1) \\
&\quad + (-1)^{q+1} x(g,\sigma_1,\ldots,\sigma_{q-1})(1),
\end{aligned}$$

and therefore

$$d^{q-1}\circ h^q(x)(\sigma_1,\ldots,\sigma_q)(g) + h^{q+1}\circ d^q(x)(\sigma_1,\ldots,\sigma_q)(g) = x(\sigma_1,\ldots,\sigma_q)(g). \quad \square[\mathbf{I.}]$$

Next we prove:

II. Let X be an abelian group and $A = \mathcal{M}_G(X)$ an induced discrete G-module. Then $\mathrm{Ker}(d^q) = \mathrm{Im}(d^{q-1})$ for all $q \ge 1$.

Proof of **II.** Let $q \ge 1$ and $x \in \mathrm{Ker}(d^q) \subset C^q(G,A)$. Then we obtain

$$x = (d^{q-1}\circ h^q + h^{q+1}\circ d^q)(x) = d^{q-1}(h^q(x)) \in \mathrm{Im}(d^{q-1})$$

and therefore $\mathrm{Ker}(d^q) \subset \mathrm{Im}(d^{q-1})$. For the reverse inclusion we prove that $d^q\circ d^{q-1} = 0$ for all $q \ge 1$ by induction on q.

$q=1$: If $(\sigma_1,\sigma_2)\in G^2$ and $a\in A$, then

$$\begin{aligned} d^1\circ d^0(a)(\sigma_1,\sigma_2) &= \sigma_1 d^0(a)(\sigma_2) - d^0(a)(\sigma_1\sigma_2) + d^0(a)(\sigma_1) \\ &= \sigma_1(\sigma_2 a - a) - (\sigma_1\sigma_2 a - a) + \sigma_1 a - a = 0. \end{aligned}$$

$q\ge 2$, $q-1\to q$: Suppose that $d^{q-1}\circ d^{q-2}=0$. Then **I** implies

$$\begin{aligned} d^{q-1} &= \mathrm{id}_{C^q(G,A)}\circ d^{q-1} = (d^{q-1}\circ h^q + h^{q+1}\circ d^q)\circ d^{q-1} = d^{q-1}\circ h^q\circ d^{q-1} \\ &\quad + h^{q+1}\circ d^q\circ d^{q-1} \\ &= d^{q-1}\circ(d^{q-2}\circ h^{q-1} + h^q\circ d^{q-1}) = d^{q-1}\circ h^q\circ d^{q-1}, \end{aligned}$$

hence $h^{q+1}\circ d^q\circ d^{q-1}=0$ and therefore $d^q\circ d^{q-1}=0$, since h^{q+1} is injective. □[**II.**]

By **II** the theorem holds for induced G-modules. Thus let A be an arbitrary discrete G-module and $i\colon A\to\mathcal{M}_G(A)$ the natural embedding of A into an induced G-module. For $q\ge 1$ we obtain the commutative diagram

$$\begin{array}{ccccc} C^{q-1}(G,A) & \xrightarrow{d^{q-1}} & C^q(G,A) & \xrightarrow{d^q} & C^{q+1}(G,A) \\ {\scriptstyle C^{q-1}(i)}\downarrow & & \downarrow{\scriptstyle C^q(i)} & & \downarrow{\scriptstyle C^{q+1}(i)} \\ C^{q-1}(G,\mathcal{M}_G(A)) & \xrightarrow{d_*^{q-1}} & C^q(G,\mathcal{M}_G(A)) & \xrightarrow{d_*^q} & C^{q+1}(G,\mathcal{M}_G(A)), \end{array}$$

where $d^\bullet=d_A^\bullet$ and $d_*^\bullet=d_{\mathcal{M}_G(A)}^\bullet$. As the bottom row is exact by **II**, it follow that $C^{q+1}(i)\circ d^q\circ d^{q-1}=d_*^q\circ d_*^{q-1}\circ C^{q-1}(i)=0$, and thus $d^q\circ d^{q-1}=0$, since $C^{q+1}(i)$ is injective. □

Definition 2.2.3. Let G be a profinite group, A a discrete G-module and $(d^q)_{q\ge1}$ the sequence of differentials of (G,A). For $q\ge 1$ we define

$$\begin{aligned} &Z^q(G,A)=\mathrm{Ker}(d^q), \quad B^q(G,A)=\mathrm{Im}(d^{q-1}) \quad \text{and} \\ &H^q(G,A)=Z^q(G,A)/B^q(G,A). \end{aligned}$$

The elements of $Z^q(G,A)$ are called **q-cocycles**, those of $B^q(G,A)$ are called **q-coboundaries**, and $H^q(G,A)$ is called the **q-th cohomology group** of G with values in A. If $\xi=x+B^q(G,A)\in H^q(G,A)$, where $x\in Z^q(G,A)$, then we call x a **defining q-cocycle** of ξ and write $\xi=[x]$.

Although we are only interested in the cases $q=1$ and $q=2$, we formulate and prove the properties of the cohomology groups for general q as far as this does not cause additional effort.

The mapping properties of $C^q(G,\cdot\,)$ induce in a natural way mapping properties of $H^q(G,\cdot\,)$ for all $q\ge 1$ as follows. Let $\varphi\colon A\to B$ be a G-homomorphism of discrete G-modules. Then $d_B^q\circ C^q(\varphi)=C^{q+1}(\varphi)\circ d_A^q$ implies $C^q(\varphi)(Z^q(G,A))\subset Z^q(G,B)$, and $d_B^{q-1}\circ C^{q-1}(\varphi)=C^q(\varphi)\circ d_A^{q-1}$ implies $C^q(\varphi)(B^q(G,A))\subset B^q(G,B)$. Hence $C^q(\varphi)$ induces a homomorphism

$$H^q(\varphi)=H^q(G,\varphi)\colon H^q(G,A)\to H^q(G,B)$$

satisfying $H^q(\varphi)(x+B^q(G,A))=C^q(\varphi)(x)+B^q(G,B)$ for all $x\in C^q(G,A)$.

If $\varphi'\colon A \to B$ and $\psi\colon B \to C$ are other G-homomorphisms of discrete G-modules, then $H^q(\varphi+\varphi') = H^q(\varphi)+H^q(\varphi')$, and $H^q(\psi\circ\varphi) = H^q(\psi)\circ H^q(\varphi)$. If $n \in \mathbb{Z}$ and $\tau_n\colon A \to A$ is defined by $\tau_n(a) = na$ for all $a \in A$, then $H^q(\tau_n)(\xi) = n\xi$ for all $\xi \in H^q(G, A)$.

Theorem 2.2.4. *Let G be a profinite group. If A is an induced G-module, then $H^q(G, A) = \mathbf{0}$ for all $q \geq 1$.*

Proof. If A is an induced G-module, then $\mathrm{Im}(d_A^{q-1}) = \mathrm{Ker}(d_A^q)$ for all $q \geq 1$ by 2.2.2. □

Remarks 2.2.5 (Cohomology groups of small dimension).
Let G be a profinite group.

$\boldsymbol{q = 1}$: $Z^1(G, A) = \mathrm{Ker}(d^1)$ is the set of all 1-cocycles of G with values in A. By definition, an 1-cocycle of G with values in A is a continuous map $x\colon G \to A$ satisfying $d^1(x) = 0$, that is,

$$x(\sigma\tau) = \sigma x(\tau) + x(\sigma) \quad \text{for all } \sigma,\, \tau \in G.$$

For $\sigma = \tau = 1$ we obtain $x(1) = 0$, and for all $\sigma \in G$ it follows that

$$0 = x(\sigma^{-1}\sigma) = \sigma^{-1}x(\sigma) + x(\sigma^{-1}), \quad \text{hence } x(\sigma^{-1}) = -\sigma^{-1}x(\sigma).$$

A 1-cocycle is called a **crossed homomorphism**.

$B^1(G, A) = \mathrm{Im}(d^0)$ is the set of 1-coboundaries of G with values in A. By definition, a 1-coboundary of G with values in A is a map $x\colon G \to A$ of the form $x(\sigma) = \sigma a - a$ for some $a \in A$. Such a map is automatically continuous and is called a **splitting 1-cocycle**.

If A is a trivial G-module, then $H^1(G, A) = Z^1(G, A) = \mathrm{Hom}^{\mathrm{c}}(G, A)$ is the group of continuous homomorphisms $G \to A$. In particular, it follows that $H^1(G, \mathbb{Q}/\mathbb{Z}) = \mathrm{Hom}^{\mathrm{c}}(G, \mathbb{Q}/\mathbb{Z})$, and the topological isomorphism $\eta\colon \mathbb{Q}/\mathbb{Z} \xrightarrow{\sim} \mathbb{T}_{\mathbb{Q}}$, defined by $\eta(t + \mathbb{Z}) = \mathsf{e}^{2\pi\mathsf{i}t}$, induces an isomorphism

$$H^1(G, \mathbb{Q}/\mathbb{Z}) \xrightarrow{\sim} \mathsf{X}(G).$$

We identify: $\mathsf{X}(G) = H^1(G, \mathbb{Q}/\mathbb{Z}) = \mathrm{Hom}^{\mathrm{c}}(G, \mathbb{Q}/\mathbb{Z})$. Accordingly we shall mostly use additive notation for characters of profinite groups.

$\boldsymbol{q = 2}$: $Z^2(G, A) = \mathrm{Ker}(d^2)$ is the set of all 2-cocycles of G with values in A. By definition, a 2-cocycle of G with values in A is a continuous maps $x\colon G \times G \to A$ satisfying the **cocycle equation** $d^2x = 0$, that is,

$$x(\nu, \sigma) + x(\nu\sigma, \tau) = \nu x(\sigma, \tau) + x(\nu, \sigma\tau) \ \text{ for all } \nu,\, \sigma,\, \tau \in G.$$

An element of $Z^2(G, A)$ is also called a **factor system** of G in A.

If $x \in Z^2(G, A)$, then the cocycle equation for $\nu = \sigma = 1$ implies that

$$x(1, \tau) = x(1, 1) \quad \text{for all} \quad \tau \in G,$$

and the cocycle equation for $\sigma = \tau = 1$ implies that

$$x(\nu, 1) = \nu x(1, 1) \quad \text{for all} \quad \nu \in G.$$

A factor system $x \in Z^2(G, A)$ is called **normalized** if $x(1, 1) = 0$ (and then $x(1, \sigma) = x(\sigma, 1) = 0$ for all $\sigma \in G$).

$B^2(G, A) = \mathrm{Im}(d^1)$ is the set of all 2-coboundaries of G with values in A. By definition, a 2-coboundary of G with values in A is a map $x\colon G \times G \to A$ for which there exists a continuous map $a\colon G \to A$ such that

$$x(\sigma, \tau) = d^1 a(\sigma, \tau) = \sigma a(\tau) - a(\sigma\tau) + a(\sigma) \quad \text{for all} \quad \sigma, \tau \in G$$

(note that the continuity of a entails the continuity of x). An element of $B^2(G, A)$ is called a **splitting factor system**.

If $a \in A$ and $a\colon G \to A$ is constant with value a, then $d^1 a(\sigma, \tau) = \sigma a$ and therefore $d^1 a(1, 1) = a$. Consequently, if $x \in Z^2(G, A)$, then $x + a \in Z^2(G, A)$, $[x] = [x + a] \in H^2(G, A)$ and $(x + a)(1, 1) = x(1, 1) + a$.

In particular, $x - x(1, 1) \in Z^2(G, A)$ is a normalized factor system, and thus every cohomology class in $H^2(G, A)$ has a normalized defining factor system.

We proceed with the investigation of some special features of finite groups.

Definition and Remarks 2.2.6. Let G be a finite group, $N_G \in G$ the norm element and $I_G \subset \mathbb{Z}[G]$ the augmentation ideal (see 2.1.1). If A is a G-module, then

$$I_G A = \sum_{\sigma \in G} (\sigma - 1)A$$

is a G-submodule of A, and the map $N_G\colon A \to A, \quad a \mapsto N_G a$ is a G-homomorphism satisfying, $N_G A \subset A^G$, and

$${}_{N_G}A = \mathrm{Ker}(N_G\colon A \to A) = \{a \in A \mid N_G a = 0\}.$$

Since $N_G(\sigma - 1) = 0$ for all $\sigma \in G$, it follows that $I_G A \subset {}_{N_G}A \subset A^G$. We define

$$H^0(G, A) = A^G / N_G A \quad \text{and} \quad H^{-1}(G, A) = {}_{N_G}A / I_G A.$$

If $\varphi\colon A \to B$ is a G-homomorphism of G-modules, then

$$\varphi(A^G) \subset B^G, \quad \varphi({}_{N_G}A) \subset {}_{N_G}B \quad \text{and} \quad \varphi(I_G A) = I_G \varphi(A) \subset I_G B.$$

Hence φ induces homomorphisms

$$\varphi^G = \varphi \restriction A^G\colon A^G \to B^G \quad \text{and} \quad H^q(\varphi)\colon H^q(G, A) \to H^q(G, B) \quad \text{for} \quad q \in \{0, -1\}.$$

If $\varphi'\colon A \to B$ and $\psi\colon B \to C$ are other G-homomorphisms of G-modules, then clearly $H^q(\varphi + \varphi') = H^q(\varphi) + H^q(\varphi')$ and $H^q(\psi \circ \varphi) = H^q(\psi) \circ H^q(\varphi)$ for $q \in \{0, -1\}$.

If $G = \langle \sigma \rangle$ is cyclic, then $I_G = (\sigma - 1)\mathbb{Z}[G]$ by 2.1.1.1, and if A is a G-module, then $I_G A = (\sigma - 1)A$.

Theorem 2.2.7. *Let G be a finite group and A a G-module.*

1. *Let H be a subgroup of G and $G/H = \{\sigma_1 H, \ldots, \sigma_r H\}$, where $r = (G{:}H)$, and $\sigma_1 = 1$. Let B be an H-submodule of A, and*

$$A = \bigoplus_{i=1}^{r} \sigma_i B = B \oplus \bigoplus_{i=2}^{r} \sigma_i B \quad \text{(hence } A \cong \mathcal{M}_G^H(B) \text{ according to 2.1.9.1).}$$

If $a = b_1 + \sigma_2 b_2 + \ldots + \sigma_r b_r \in A$, where $b_1, \ldots, b_r \in B$, we set

$$\pi(a) = b_1 \quad \text{and} \quad \nu(a) = b_1 + \ldots + b_r.$$

(a) *$\pi \restriction A^G \colon A^G \to B^H$ is an isomorphism, $\pi(N_G A) = N_H B$, and π induces an isomorphism*

$$H^0(G, A) = A^G/N_G A \overset{\sim}{\to} B^H/N_H B = H^0(H, B).$$

(b) *$\nu \colon A \to B$ is an epimorphism satisfying $\operatorname{Ker}(\nu) \subset I_G A$, $\nu({}_{N_G}A) = {}_{N_H}B$, $\nu(I_G A) = I_H B$, and ν induces an isomorphism*

$$H^{-1}(G, A) = {}_{N_G}A/I_G A \overset{\sim}{\to} {}_{N_H}B/I_H B = H^{-1}(G, B).$$

2. *If A is an induced G-module, then $H^{-1}(G, A) = H^0(G, A) = \mathbf{0}$.*

Proof. Apparently ν and π are group epimorphisms. If $B' = \sigma_2 B + \ldots + \sigma_r B$, then B' is an H-submodule of A, $A = B \oplus B'$, and $\pi \colon A \to B$ is an H-homomorphism. We prove first:

A. $A^G = (\sigma_1 + \ldots + \sigma_r) B^H$, and if $a \in A$, then

$$N_G a = (\sigma_1 + \ldots + \sigma_r) N_H \nu(a), \text{ and } \pi(N_G a) = N_H \nu(a).$$

Proof of **A.** Let $a = \sigma_1 b_1 + \ldots + \sigma_r b_r$, where $b_1, \ldots, b_r \in B$.

If $a \in A^G$ and $i \in [1, r]$, then $a = \sigma_i^{-1} a = b_i + b_i'$, where $b_i' \in B'$, and therefore $b_i = b_1$. If $\tau \in H$, then $\tau(b_1) = \tau(\pi(a)) = \pi(\tau a) = \pi(a) = b_1$. Hence we obtain $a = (\sigma_1 + \ldots + \sigma_r) b_1 \in (\sigma_1 + \ldots + \sigma_r) B^H$.

Conversely, let $a = (\sigma_1 + \ldots + \sigma_r) b$, where $b \in B^H$, and let $\sigma \in G$. Then there exists some permutation $\pi \in \mathfrak{S}_r$ such that for all $i \in [1, r]$ we have $\sigma\sigma_i = \sigma_{\pi(i)} \tau_i$ for some $\tau_i \in H$, and then

$$\sigma a = \sum_{i=1}^{r} \sigma\sigma_i b_1 = \sum_{i=1}^{r} \sigma_{\pi(i)} \tau_i b_1 = \sum_{i=1}^{r} \sigma_{\pi(i)} b_1 = \sum_{i=1}^{r} \sigma_i b_1 = a, \quad \text{hence } a \in A^G.$$

Next we calculate

$$\begin{aligned}(\sigma_1 + \ldots + \sigma_r) N_H \nu(a) &= \sum_{i=1}^{r} \sigma_i \sum_{\tau \in H} \tau \sum_{j=1}^{r} b_j \\ &= \sum_{i=1}^{r} \sum_{\tau \in H} \sigma_i \tau \nu(a) = N_G \nu(a) = N_G(a),\end{aligned}$$

since

$$N_G a - N_G \nu(a) = \sum_{j=1}^{r} N_G(\sigma_j - 1)b_j = 0.$$

Observing $G = H\sigma_1^{-1} \uplus \ldots \uplus H\sigma_r^{-1}$, it follows that

$$\pi(N_G a) = \pi\Big(\sum_{\tau\in H}\sum_{i=1}^{r} \tau\sigma_i^{-1}\sum_{j=1}^{r}\sigma_j b_j\Big) = \sum_{\tau\in H}\tau\pi\Big(\sum_{i,j=1}^{r}\sigma_i^{-1}\sigma_j b_j\Big)$$
$$= \sum_{\tau\in H}\tau\Big(\nu(a) + \sum_{\substack{i,j=1\\ i\neq j}}^{r}\sigma_i^{-1}\sigma_j b_j\Big) = N_H\nu(a), \quad \text{since}$$

$$\sum_{\substack{i,j=1\\ i\neq j}}^{r}\sigma_i^{-1}\sigma_j b_j \in B'.$$

As $N_H\nu(a) \in B$, it follows that $\pi(N_G a) = N_H\nu(a)$. □[**A.**]

Now we can do the very proof using **A** again and again.

1. Let $s\colon B \to A$ be defined by $s(b) = (\sigma_1 + \ldots + \sigma_r)b$. Then $\pi\circ s = \mathrm{id}_B$, $\pi(A^G) \subset B^H$, $s(B^H) \subset A^G$ and $s\circ\pi \restriction A^G = \mathrm{id}_{A^G}$. Therefore $\pi \restriction A^G\colon A^G \overset{\sim}{\to} B^H$ is an isomorphism. Moreover, $\pi(N_G A) = N_H\nu(A) = N_H B$, and consequently π induces an isomorphism $A^G/N_G A \overset{\sim}{\to} B^H/N_H B$.

2. $\nu({}_{N_G}A) = {}_{N_H}B$: If $a \in {}_{N_G}A$, then $0 = \pi(N_G a) = N_H\nu(a)$, and therefore $\nu(a) \in {}_{N_H}B$. If $\nu(a) \in {}_{N_H}B$, then $N_G a = (\sigma_1 + \ldots + \sigma_r)N_H\nu(a) = 0$, and therefore $a \in {}_{N_G}A$.

$\nu(I_G A) = I_H B$: As $I_H B \subset I_G A$ and $\nu \restriction B = \mathrm{id}_B$, it follows that $\nu(I_G A) \supset \nu(I_H B) = I_H B$. To prove the converse, let $a = \sigma_1 b_1 + \ldots + \sigma_r b_r \in A$, where $b_1, \ldots, b_r \in B$. If $\sigma \in G$, then there exists a permutation $\pi \in \mathfrak{S}_r$ such that for all $i \in [1, r]$ we have $\sigma\sigma_i = \sigma_{\pi(i)}\gamma_i$ for some $\gamma_i \in H$, and we obtain

$$\nu(\sigma a - a) = \nu(\sigma a) - \nu(a) = \nu\Big(\sum_{i=1}^{r}\sigma_{\pi(i)}\gamma_i b_i\Big) - \nu\Big(\sum_{i=1}^{r}\sigma_i b_i\Big)$$
$$= \sum_{i=1}^{r}\gamma_i b_i - \sum_{i=1}^{r} b_i = \sum_{i=1}^{r}(\gamma_i - 1)b_i \in I_H B.$$

Since $I_G A = \langle\{\sigma a - a \mid \sigma \in G,\ a \in A\}\rangle$, it follows that $\nu(I_G A) \subset I_H B$, and thus equality holds.

$\mathrm{Ker}(\nu) \subset I_G A$: If $a = \sigma_1 b_1 + \ldots + \sigma_r b_r \in \mathrm{Ker}(\nu)$, where $b_1, \ldots, b_r \in B$, then $0 = \nu(a) = b_1 + \ldots + b_r$, and consequently $a = (\sigma_1 - 1)b_1 + \ldots + (\sigma_r - 1)b_r \in I_G A$.

Since $\nu(I_G A) = I_H B$ and $\mathrm{Ker}(\nu) \subset I_G A$, it follows that

$$\nu^{-1}(I_H B) = I_G(A) + \mathrm{Ker}(\nu) = I_G A,$$

and ν induces an isomorphism ${}_{N_G}A/I_G A \overset{\sim}{\to} {}_{N_H}B/I_H B$.

3. If $A = \mathcal{M}_G(B) = \mathcal{M}_G^{\mathbf{1}}(B)$ for some abelian group B and $i \in \{-1, 0\}$, then 1. shows that $H^i(G, A) \cong H^i(\mathbf{1}, B) = \mathbf{0}$. $\square$

2.3 Direct sums, products and limits of cohomology groups

Theorem 2.3.1. *Let G be a profinite group and $(A_i)_{i\in I}$ a family of discrete G-modules.*

1. *Suppose that*

$$A = \bigoplus_{i\in I} A_i \quad \text{and let } (\pi_i \colon A \to A_i)_{i\in I} \text{ be the family of projections.}$$

Then A is a discrete G-module, and for all $q \geq 1$ (even for all $q \geq -1$ if G is finite) the map

$$H^q(G, A) \to \bigoplus_{i\in I} H^q(G, A_i), \quad \text{defined by} \quad \xi \mapsto (H^q(\pi_i)(\xi))_{i\in I},$$

is an isomorphism.

2. *Let G be finite, suppose that*

$$A = \prod_{i\in I} A_i \quad \text{and let } (\pi_i \colon A \to A_i)_{i\in I} \text{ be the family of projections.}$$

Then for all $q \geq -1$ the map

$$H^q(G, A) \to \prod_{i\in I} H^q(G, A_i), \quad \text{defined by} \quad \xi \mapsto (H^q(\pi_i)(\xi))_{i\in I}$$

is an isomorphism.

Proof. 1. A is a discrete G-module by 2.1.3.**7**.

Let $q \geq 0$. If $x \in C^q(G, A)$ and $i \in I$, then $\pi_i \circ x \colon G^q \to A_i$ is continuous, and we define

$$\eta^q \colon C^q(G, A) \to \bigoplus_{i\in I} C^q(G, A_i) \text{ by } \eta^q(x) = (\pi_i \circ x)_{i\in I}.$$

Apparently η^q is a monomorphism, and we assert that it is even an isomorphism. Thus let

$$(x_i)_{i\in I} \in \bigoplus_{i\in I} C^q(G, A_i) \quad \text{and} \quad x = (x_i)_{i\in I} \colon G^q \to A.$$

The set $I_0 = \{i \in I \mid x_i \neq 0\}$ is finite, and by 1.5.2.2 there exists an open normal subgroup U of G such that for all $i \in I_0$ and $\boldsymbol{\sigma} \in G^q$ the element $x_i(\boldsymbol{\sigma}) \in A_i$ only depends on the coset $\boldsymbol{\sigma}U^q$. Hence the same is true for $x(\boldsymbol{\sigma})$, and therefore x is continuous, $x \in C^q(G,A)$, and $(x_i)_{i\in I} = \eta^q(x)$. For $q \geq 1$ we obtain the commutative diagram

$$\begin{array}{ccc} C^{q-1}(G,A) & \xrightarrow[\sim]{\eta^{q-1}} & \bigoplus_{i\in I} C^{q-1}(G,A_i) \\ {\scriptstyle d_A^{q-1}}\big\downarrow & & \big\downarrow{\scriptstyle (d_{A_i}^{q-1})_{i\in I}} \\ C^q(G,A) & \xrightarrow[\sim]{\eta^q} & \bigoplus_{i\in I} C^q(G,A_i), \end{array}$$

which implies

$$\eta^q(Z^q(G,A)) = \bigoplus_{i\in I} Z^q(G,A_i) \quad \text{and} \quad \eta^q(B^q(G,A)) = \bigoplus_{i\in I} B^q(G,A_i).$$

Hence η^q induces an isomorphism

$$(H^q(\pi_i))_{i\in I} \colon H^q(G,A) \stackrel{\sim}{\to} \bigoplus_{i\in I} H^q(G,A_i)$$

such that $(H^q(\pi_i))_{i\in I}([x]) = ([\pi_i{\circ}x])_{i\in I}$ for all $x \in Z^q(G,A)$.

Now let G be finite. It follows from the very definitions that

$$A^G = \bigoplus_{i\in I} A_i^G, \quad I_GA = \bigoplus_{i\in I} I_GA_i, \quad N_GA = \bigoplus_{i\in I} N_GA_i, \quad {}_{N_G}A = \bigoplus_{i\in I} {}_{N_G}A_i,$$

and consequently

$$H^{-1}(G,A) = \bigoplus_{i\in I} H^{-1}(G,A_i) \quad \text{and} \quad H^0(G,A) = \bigoplus_{i\in I} H^0(G,A_i).$$

2. If $q \geq 0$, then obviously the map

$$\eta^q \colon C^q(G,A) \to \prod_{i\in I} C^q(G,A_i), \quad \text{defined by} \quad \eta^q(x) = (\pi_i{\circ}x)_{i\in I},$$

is an isomorphism. If $q \geq 1$, then the diagram

$$\begin{array}{ccc} C^{q-1}(G,A) & \xrightarrow[\sim]{\eta^{q-1}} & \prod_{i\in I} C^{q-1}(G,A_i) \\ {\scriptstyle d_A^{q-1}}\big\downarrow & & \big\downarrow{\scriptstyle (d_{A_i}^{q-1})_{i\in I}} \\ C^q(G,A) & \xrightarrow[\sim]{\eta^q} & \prod_{i\in I} C^q(G,A_i) \end{array}$$

commutes, hence

$$\eta^q(Z^q(G,A)) = \prod_{i\in I} Z^q(G,A_i), \quad \eta^q(B^q(G,A)) = \prod_{i\in I} B^q(G,A_i),$$

and η^q induces an isomorphism

$$(H^q(\pi_i))_{i\in I}\colon H^q(G,A) \stackrel{\sim}{\to} \prod_{i\in I} H^q(G,A_i)$$

such that $(H^q(\pi_i))_{i\in I}([x]) = ([\pi_i\circ x])_{i\in I}$ for all $x \in Z^q(G,A)$. Since

$$A^G = \prod_{i\in I} A_i^G, \quad I_GA = \prod_{i\in I} I_GA_i, \quad N_GA = \prod_{i\in I} N_GA_i \quad \text{and} \quad {}_{N_G}A = \prod_{i\in I} {}_{N_G}A_i,$$

it follows that

$$H^{-1}(G,A) = \prod_{i\in I} H^{-1}(G,A_i) \quad \text{and} \quad H^0(G,A) = \prod_{i\in I} H^0(G,A_i). \qquad \square$$

Theorem 2.3.2. *Let $(I,\le)$ be a directed set, $(G_i,\phi_{i,j})_I$ a projective system of profinite groups, $G = \varprojlim G_i$ and $(\phi_i\colon G \to G_i)_{i\in I}$ the family of projections. Let $(A_i,\psi_{j,i})_I$ be an inductive system of abelian groups, $A = \varinjlim A_i$ and $(\psi_i\colon A_i \to A)_{i\in I}$ the family of injections. For every $i \in I$ let A_i be a discrete G_i-module, and suppose that*

$$\psi_{j,i}(\phi_{i,j}(\sigma_i)a_j) = \sigma_i\psi_{j,i}(a_j) \text{ for all } i,\, j \in I \text{ with } i \ge j, \sigma_i \in G_i \text{ and } a_j \in A_j.$$

Then there exists a unique continuous map $G\times A \to A$ which makes A into a discrete G-module such that:

If $a = \psi_i(a_i)$ for some $i \in I$ and $a_i \in A_i$, then $\sigma a = \psi_i(\phi_i(\sigma)a_i)$ for all $\sigma \in G$. (#)

If $q \ge 1$ and $j \le i$, then there exists a homomorphism

$$\Theta^q_{j,i}\colon H^q(G_j,A_j) \to H^q(G_i,A_i)$$

such that $\Theta^q_{j,i}([x_j]) = [\psi_{j,i}\circ x_j \circ \phi^q_{i,j}]$ for all $x_j \in Z^q(G_j,A_j)$. Then $(H^q(G_j,A_j),\Theta^q_{j,i})_I$ is an inductive system, and there exists an isomorphism

$$\beta^q\colon \varinjlim_{i\in I} H^q(G_i,A_i) \stackrel{\sim}{\to} H^q(G,A)$$

with the following property:

If $\xi = [\xi_i,i] \in \varinjlim H^q(G_i,A_i)$, where $i \in I$ and $\xi_i = [x_i] \in H^q(G_i,A_i)$ for some $x_i \in Z^q(G_i,A_i)$, then $\beta^q(\xi) = [\psi_q\circ x_i\circ\phi_i^q] \in H^q(G,A)$.

We identify: $\varinjlim H^q(G_i,A_i) = H^q(G,A)$.

Proof. **I.** We define the action of G on A according to (#), and we prove first that this definition does not depend on i. Thus let $a \in A$, $\sigma \in G$ and $i, j \in I$

such that $a = \psi_i(a_i) = \psi_j(a_j)$ for some $a_i \in A_i$ and $a_j \in A_j$. Let $k \in I$ such that $k \geq i$, $k \geq j$ and $\psi_{i,k}(a_i) = \psi_{j,k}(a_j)$. Then

$$\begin{aligned}\psi_j(\phi_j(\sigma)a_j) &= \psi_k(\psi_{j,k}(\phi_{k,j}\circ\phi_k(\sigma)a_j)) = \psi_k(\phi_k(\sigma)\psi_{j,k}(a_j)) \\ &= \psi_k(\phi_k(\sigma)\psi_{i,k}(a_i)) = \psi_k(\psi_{i,k}(\phi_{k,i}\circ\phi_k(\sigma)a_i)) = \psi_i(\phi_i(\sigma)a_i).\end{aligned}$$

By means of this, A is a G-Modul. Indeed, if $\sigma,\, \sigma' \in G$ and $a,\, a' \in A$, then there exist some $i \in I$ and $a_i,\, a_i' \in A_i$ such that $a = \psi_i(a_i)$ and $a' = \psi_i(a_i')$. It follows that $a + a' = \psi_i(a_i + a_i')$, and now it is easily checked that $\sigma(a + a') = \sigma a + \sigma a'$ and $(\sigma\sigma')a = \sigma(\sigma' a)$. Next we prove that A is a discrete G-module. Let $a = \psi_k(a_k) \in A$ for some $k \in I$ and $a_k \in A_k$. The isotropy group $(G_k)_{a_k}$ is an open subgroup of G_k, and if $\sigma \in \phi_k^{-1}((G_k)_{a_k})$, then $\sigma a = \psi_k(\phi_k(\sigma)a_k) = \psi_k(a_k) = a$. Hence $G_a \supset \phi_k^{-1}((G_k)_{a_k})$, G_a is an open subgroup of G, and A is a discrete G-module.

II. For $q \geq 0$ and $j \leq i$ we construct $\theta^q_{j,i} \colon C^q(G_j, A_j) \to C^q(G_i, A_i)$ such that $(C^q(G_j, A_j), \theta^q_{j,i})_I$ is an inductive system, and we establish an isomorphism $\gamma^q \colon \varinjlim C^q(G_i, A_i) \to C^q(G, A)$.

If $i,\, j \in I$ and $i \geq j$, we define

$$\theta^q_{j,i} \colon C^q(G_j, A_j) \to C^q(G_i, A_i) \quad \text{and} \quad \gamma^q_j \colon C^q(G_j, A_j) \to C^q(G, A)$$

by $\theta^q_{j,i}(x_j) = \psi_{j,i}\circ x_j\circ\phi^q_{i,j}$ and $\gamma^q_j(x_j) = \psi_j\circ x_j\circ\phi^q_j$ for all $x_j \in C^q(G_j, A_j)$ (note that with $x_j \colon G^q_j \to A_j$ also the functions $\psi_{j,i}\circ x_j\circ\varphi^q_{i,j} \colon G^q_i \to A_i$ and $\psi_j\circ x_j\circ\phi_j \colon G^q \to A$ are continuous). If $i,\, j,\, k \in I$, $i \geq j \geq k$ and $x_k \in C^q(G_k, A_k)$, then

$$\theta^q_{j,i}\circ\theta^q_{k,j}(x_k) = \theta^q_{j,i}(\psi_{k,j}\circ x_k\circ\phi^q_{j,k}) = \psi_{j,i}\circ\psi_{k,j}\circ x_k\circ\phi^q_{j,k}\circ\psi^q_{i,j} = \psi_{k,i}\circ x_k\circ\phi^q_{i,k} = \theta^q_{k,i}(x_k)$$

and therefore $\theta^q_{j,i}\circ\theta^q_{k,j} = \theta^q_{k,i}$. Hence $(C^q(G_i, A_i), \theta_{j,i})_I$ is an inductive system, and we denote by $\theta^q_j \colon C^q(G_j, A_j) \to \varinjlim C^q(G_i, A_i)$ the injections into the inductive limit. If $i,\, j \in I$, $i \geq j$ and $x_j \in C^q(G_j, A_j)$, then

$$\gamma^q_i\circ\theta^q_{j,i}(x_j) = \gamma^q_i\circ\psi_{j,i}\circ x_j\circ\phi^q_{i,j} = \psi_i\circ\psi_{j,i}\circ x_j\circ\phi^q_{i,j}\circ\phi^q_i = \psi_j\circ x_j\circ\phi^q_j = \gamma^q_j(x_j),$$

and therefore $\gamma^q_i\circ\theta^q_{j,i} = \gamma^q_j$. Thus the family $(\gamma^q_j \colon C^q(G_j, A_j) \to C^q(G, A))_{j\in I}$ induces a homomorphism

$$\gamma^q = \varinjlim \gamma^q_i \colon \varinjlim C^q(G_i, A_i) \to C^q(G, A)$$

satisfying $\gamma^q\circ\theta^q_j(x_j) = \gamma^q_j(x_j)$ for all $x_j \in C^q(G_j, A_j)$. We show that γ^q is an isomorphism.

a. γ^q is injective: Let $x = \theta^q_j(x_j) \in \varinjlim C^q(G_i, A_i)$, where $j \in I$ and $x_j \in C^q(G_j, A_j)$, and suppose that $\gamma^q(x) = \psi_j\circ x_j\circ\phi^q_j = 0$. Since $x_j\circ\phi^q_j$ is continuous, it follows that $x_j\circ\phi^q_j(G^q)$ is finite, and thus there exists some $i \geq j$ such that $\psi_{j,i}\circ x_j\circ\phi^q_j(\boldsymbol{\sigma}) = 0$ for all $\boldsymbol{\sigma} \in G^q$, hence $\psi_{j,i}\circ x_j\circ\phi^q_j = 0$, and therefore

$$\theta^q_{j,i}(x_j)\circ\phi^q_i = \psi_{j,i}\circ x_j\circ\phi^q_{i,j}\circ\phi^q_i = \psi_{j,i}\circ x_j\circ\phi^q_j = 0.$$

As $\theta^q_{j,i}(x_j)\colon G^q_i \to A_i$ is continuous, the set $K_i = \{\boldsymbol{\sigma} \in G^q_i \mid \theta_{j,i}(x_j)(\boldsymbol{\sigma}) = 0\}$ is clopen, and since $\phi_i(G)^q \subset K_i$, 1.3.5.4 implies

$$\bigcap_{l \ge i} \phi_{l,i}(G_l)^q \cap (G^q_i \setminus K_i) = \phi_i(G)^q \setminus K_i = \emptyset.$$

Hence there exist indices $l_1, \ldots, l_n \ge i$ such that

$$\bigcap_{\nu=1}^{n} \phi_{l_\nu,i}(G_{l_\nu})^q \cap (G^q \setminus K_i) = \emptyset, \quad \text{and consequently} \quad \bigcap_{\nu=1}^{n} \phi_{l_\nu,i}(G_{l_\nu})^q \subset K_i.$$

If $l = \max\{l_1, \ldots, l_n\}$, then $\phi_{l,i}(G_l)^q = \phi_{l_\nu,i} \circ \phi_{l,l_\nu}(G_l)^q \subset \phi_{l_\nu,i}(G_{l_\nu})^q \subset K_i$, and consequently $\psi_{j,i} \circ x_j \circ \phi^q_{l,j} = \psi_{j,i} \circ x_j \circ \phi^q_{i,j} \circ \phi^q_{l,i} = \theta^q_{j,i}(x_j) \circ \phi^q_{l,i} = 0$. Thus we obtain

$$x = \theta^q_j(x_j) = \theta^q_l \circ \theta^q_{j,l}(x_j) = \theta^q_l \circ \psi_{j,l} \circ x_j \circ \phi^q_{l,j} = \theta^q_l \circ \psi_{i,l} \circ \psi_{j,i} \circ x_j \circ \phi^q_{l,j} = 0.$$

b. γ^q is surjective. Let $x \in C^q(G, A)$. It suffices to prove that there exist some $j \in I$ and $x_j \in C^q(G_j, A_j)$ such that $x = \psi_j \circ x_j \circ \phi^q_j$. Indeed, then it follows that $\gamma^q(\theta_j(x_j)) = \gamma^q_j(x_j) = \psi_j \circ x_j \circ \phi^q_j = x$.

Let $x(G^q) = \{a_1, \ldots, a_m\} \subset A$ and $k \in I$ such that $a_\mu = \psi_k(a_{\mu,k})$ for all $\mu \in [1, m]$ with $a_{\mu,k} \in A_k$. The sets $x^{-1}(a_1), \ldots, x^{-1}(a_m)$ are open, mutually disjoint, and they cover G^q. Let $x'\colon G^q \to A_k$ be defined by $x'(\boldsymbol{\sigma}) = a_{\mu,k}$ if $\boldsymbol{\sigma} \in x^{-1}(a_k)$. Then x' is continuous, $x = \psi_k \circ x'$, and by 1.5.2.1 there exist some $i \in I$ and a continuous map $x'_i\colon G^q_i \to A_k$ such that $x' = x'_i \circ \phi^q_i$. It follows that $x_j = \psi_{k,j} \circ x'_i \circ \phi^q_{j,i} \in C^q(G_j, A_j)$, and

$$x = \psi_k \circ x' = \psi_k \circ x'_i \circ \phi^q_i = \psi_j \circ \psi_{k,j} \circ x'_i \circ \phi^q_{j,i} \circ \phi^q_j = \psi_j \circ x_j \circ \phi^q_j.$$

III. Now we can finish the proof. For $i \in I$ let $(d^q_i)_{q \ge 0}$ be the sequence of differentials of (G_i, A_i). For $i, j \in I$ with $j \le i$ and $q \ge 1$ we obtain the obviously commutative diagram

$$\begin{array}{ccc} C^{q-1}(G_j, A_j) & \xrightarrow{\theta^{q-1}_{j,i}} & C^{q-1}(G_i, A_i) \\ {\scriptstyle d^{q-1}_j}\downarrow & & \downarrow{\scriptstyle d^{q-1}_i} \\ C^q(G_j, A_j) & \xrightarrow{\theta^q_{j,i}} & C^q(G_i, A_i)\,. \end{array} \qquad (\#)$$

Hence $\theta^q_{j,i}(Z^q(G_j, A_j)) \subset Z^q(G_i, A_i)$, $\theta^q_{j,i}(B^q(G_j, A_j)) \subset B^q(G_i, A_i)$, and consequently $\theta^q_{j,i}$ induces a homomorphism $\Theta^q_{j,i}\colon H^q(G_j, A_j) \to H^q(G_i, A_i)$ such that the following diagram with exact rows commutes.

$$\begin{array}{ccccccccc} \mathbf{0} & \longrightarrow & \mathrm{Im}(d^{q-1}_j) & \longrightarrow & \mathrm{Ker}(d^q_j) & \longrightarrow & H^q(G_j, A_j) & \longrightarrow & \mathbf{0} \\ & & {\scriptstyle \theta^q_{j,i}}\downarrow & & {\scriptstyle \theta^q_{j,i}}\downarrow & & \downarrow{\scriptstyle \Theta^q_{j,i}} & & \\ \mathbf{0} & \longrightarrow & \mathrm{Im}(d^{q-1}_i) & \longrightarrow & \mathrm{Ker}(d^q_i) & \longrightarrow & H^q(G_i, A_i) & \longrightarrow & \mathbf{0}\,. \end{array}$$

Now we apply 1.3.3 and obtain the exact sequence

$$\mathbf{0} \to \varinjlim \operatorname{Im}(d_i^{q-1}) \to \varinjlim \operatorname{Ker}(d_i^q) \to \varinjlim H^q(G_i, A_i) \to \mathbf{0},$$

and therefore

$$\mathbf{0} \to \operatorname{Im}(\varinjlim d_i^{q-1}) \to \operatorname{Ker}(\varinjlim d_i^q) \to \varinjlim H^q(G_i, A_i) \to \mathbf{0},$$

By (#) the family $(d_j^q \colon C^q(G_j, A_j) \to C^{q+1}(G_j, A_j))_{j \in I}$ yields for $q \geq 0$ a homomorphism

$$\varinjlim d_j^q \colon \varinjlim C^q(G_j, A_j) \to \varinjlim C^{q+1}(G_j, A_j),$$

which fits into the following commutative diagram.

$$\begin{array}{ccc}
\varinjlim C^q(G_j, A_j) & \xrightarrow{\varinjlim d_j^q} & \varinjlim C^{q+1}(G_j, A_j) \\
\gamma^q \downarrow & & \downarrow \gamma^{q+1} \\
C^q(G, A) & \xrightarrow{d^q} & C^{q+1}(G, A).
\end{array}$$

It follows that $\gamma^q(\operatorname{Ker}(\varinjlim d_j^q)) = \operatorname{Ker}(d^q)$, $\gamma^q(\operatorname{Im}(\varinjlim d_j^{q-1})) = \operatorname{Im}(d^{q-1})$ for all $q \geq 1$, and we obtain the following commutative diagram with exact rows in which the vertical arrows are isomorphisms.

$$\begin{array}{ccccccccc}
\mathbf{0} & \longrightarrow & \operatorname{Im}(\varinjlim d_i^{q-1}) & \longrightarrow & \operatorname{Ker}(\varinjlim d_i^q) & \longrightarrow & \varinjlim H^q(G_j, A_j) & \longrightarrow & \mathbf{0} \\
 & & \gamma^q \downarrow & & \gamma^q \downarrow & & & & \\
\mathbf{0} & \longrightarrow & \operatorname{Im}(d^{q-1}) & \longrightarrow & \operatorname{Ker}(d^q) & \longrightarrow & H^q(G, A) & \longrightarrow & \mathbf{0}.
\end{array}$$

Hence γ^q induces an isomorphism $\beta^q \colon \varinjlim H^q(G_i, A_i) \to H^q(G, A)$ as asserted. $\square$

Corollary 2.3.3. *Let G be a profinite group, $(I, \leq)$ a directed set , $(A_i, \psi_{j,i})_I$ an inductive system of discrete G-modules, $A = \varinjlim A_i$ and $(\psi_i \colon A_i \to A)_{i \in I}$ the family of injections. For every $q \geq 1$ (and even for every $q \geq -1$ if G is finite) the family $(H^q(\psi_i) \colon H^q(G, A_i) \to H^q(G, A))_{i \in I}$ induces an isomorphism*

$$\varinjlim H^q(G, \psi_i) \colon \varinjlim H^q(G, A_i) \xrightarrow{\sim} H^q(G, A). \qquad \text{We identify.}$$

Proof. For $q \geq 1$ the assertion follows by 2.3.2 (with $G_i = G$ for all $i \in I$). Thus let G be finite. Then $\psi_{j,i}(A_j^G) \subset A_i^G$, $\psi_{j,i}(N_G A_j) \subset N_G A_i$, $\psi_{j,i}({}_{N_G}A_j) \subset {}_{N_G}A_i$ and $\psi_{j,i}(I_G A_j) \subset I_G A_i$ for all $j \leq i$. Hence $\varinjlim A_i^G$, $\varinjlim N_G A_i$, $\varinjlim {}_{N_G}A_i$ and $\varinjlim I_G A_i$ are G-submodules of A by 1.3.1.**1**, and we shall prove:

a. $\varinjlim A_i^G = A^G$; **b.** $\varinjlim N_G A_i = N_G A$;
c. $\varinjlim {}_{N_G}A_i = {}_{N_G}A$; **d.** $\varinjlim I_G A_i = I_G A$.

In all cases, $\subset$ is obvious, and we show the reverse inclusions.

a. Let $x \in A^G$ and $i \in I$, $x_i \in A_i$ such that $\psi_i(x_i) = x$. Then we obtain $\psi_i(\sigma x_i) = \sigma\psi_i(x_i) = \sigma x = x = \psi_i(x_i)$ for all $\sigma \in G$, and since G is finite, there exists some $k \geq i$ such that $\sigma\psi_{i,k}(x_i) = \psi_{i,k}(\sigma x_i) = \psi_{i,k}(x_i)$ for all $\sigma \in G$, hence $x_k = \psi_{i,k}(x_i) \in A_k^G$ and $x = \psi_k(x_k) \in \varinjlim A_i^G$.

b. Let $x = N_G y \in N_G A$, where $y \in A$, and $i \in I$, $y_i \in A_i$ such that $\psi_i(y_i) = y$. Then $x = N_G\psi_i(y_i) = \psi_i(N_G y_i) \in \varinjlim N_G A_i$.

c. Let $x \in {}_{N_G}A$, and let $i \in I$ and $x_i \in A_i$ be such that $x = \psi_i(x_i)$. Then we obtain $N_G x = \psi_i(N_G x_i) = 0$, and therefore there exists some $k \geq i$ such that $\psi_{i,k}(N_G x_i) = N_G\psi_{i,k}(x_i) = 0$. Consequently $\psi_{i,k}(x_i) \in {}_{N_G}A_k$ and $x = \psi_k \circ \psi_{i,k}(x_i) \in \varinjlim {}_{N_G}A_i$.

d. Let $x \in I_G A$, and suppose that $x = (\sigma_1 - 1)y_1 + \ldots + (\sigma_m - 1)y_m$, where $m \in \mathbb{N}$, $\sigma_1, \ldots, \sigma_m \in G$ and $y_1, \ldots, y_m \in A$. Then there exist some $k \in I$ and $y_{k,1}, \ldots, y_{k,m} \in A_k$ such that $y_j = \psi_k(y_{k,j})$ for all $j \in [1, m]$. Hence it follows that $x_k = (\sigma_1 - 1)y_{k,1} + \ldots + (\sigma_m - 1)y_{k,m} \in I_G A_k$ and $x = \psi_k(x_k) \in \varinjlim I_G A_i$.

By **a**, **b**, **c**, **d** and 1.3.3 we obtain the commutative diagrams

$$\begin{array}{ccccccccc}
\mathbf{0} & \longrightarrow & \varinjlim N_G A_i & \longrightarrow & \varinjlim A_i^G & \longrightarrow & \varinjlim H^0(G, A_i) & \longrightarrow & \mathbf{0} \\
 & & \| & & \| & & \downarrow {\scriptstyle \varinjlim H^0(G,\psi_i)} & & \\
\mathbf{0} & \longrightarrow & N_G A) & \longrightarrow & A^G & \longrightarrow & H^0(G, A) & \longrightarrow & \mathbf{0}
\end{array}$$

and

$$\begin{array}{ccccccccc}
\mathbf{0} & \longrightarrow & \varinjlim I_G A_i & \longrightarrow & \varinjlim {}_{N_G}A_i & \longrightarrow & \varinjlim H^{-1}(G, A_i) & \longrightarrow & \mathbf{0} \\
 & & \| & & \| & & \downarrow {\scriptstyle \varinjlim H^{-1}(G,\psi_i)} & & \\
\mathbf{0} & \longrightarrow & I_G A) & \longrightarrow & {}_{N_G}A & \longrightarrow & H^{-1}(G, A) & \longrightarrow & \mathbf{0}
\end{array}$$

with exact rows. Thus $\varinjlim H^\nu(G, \psi_i)$ is an isomorphisms for $\nu \in \{-1, 0\}$. $\square$

Corollary 2.3.4. *Let G be a profinite group, A a discrete G-module and Σ a fundamental system of neighbourhoods of 1 in G consisting of open normal subgroups. If $U \in \Sigma$, then A^U is a G/U-module by means of $(\sigma U)a = \sigma a$ for all $\sigma \in G$ and $a \in A^U$. If $U, V \in \Sigma$, $V \subset U$, and $q \geq 1$, we define $\phi_{U,V}^q \colon H^q(G/U, A^U) \to H^q(G/V, A^V)$ as follows:*

If $x \in C^q(G/U, A^U)$ and

$$x' = [(G/V)^q \to (G/U)^q \stackrel{x}{\to} A^U \hookrightarrow A^V],$$

then $x' \in C^q(G/V, A^V)$, and $\phi_{U,V}^q([x]) = [x']$.

Then $(\Sigma, \supset)$ is directed, $(H^q(G/U, A^U), \phi^q_{U,V})_\Sigma$ is an inductive system, and there exists an isomorphism

$$\Phi\colon \varinjlim H^q(G/U, A^U) \xrightarrow{\sim} H^q(G, A)$$

with the following property:

If $\xi = [\xi_U, U] \in \varinjlim H^q(G/U, A^U)$, where $U \in \Sigma$ and $\xi_U \in H^q(G/U, A^U)$, and if $x\colon (G/U)^q \to A^U$ is defining cocycle of ξ_U, then

$$x'\colon G^q \to (G/U)^q \xrightarrow{x} A^U \hookrightarrow A$$

is a defining cocycle of $\Phi(\xi)$.

Proof. If $U \in \Sigma$, then clearly A^U is a G/U-module. Für U, $V \in \Sigma$ and $V \subset U$, let $\pi_{V,U}\colon G/V \to G/U$ be the natural epimorphism and $\psi_{U,V} = (A^U \hookrightarrow A^V)$. Since

$$A = \varinjlim_{U \in \Sigma} A^U = \bigcup_{U \in \Sigma_G} A^U \quad \text{and} \quad G = \varprojlim_{U \in \Sigma_G} G/U,$$

the assertion follows from 2.3.2. □

In 2.5.3 we shall present a structural interpretation of the isomorphism in 2.3.4, and then we shall identify: $\varinjlim H^q(G/U, A^U) = H^q(G, A)$.

2.4 The long cohomology sequence

Definition and Theorem 2.4.1. *Let G be a profinite group and*

$$\mathbf{0} \to A \xrightarrow{i} B \xrightarrow{j} C \to \mathbf{0} \qquad (*)$$

an exact sequence of discrete G-modules. We define maps

$$\delta\colon C^G \to H^1(G, A) \quad \text{and} \quad \delta^q\colon H^q(G, C) \to H^{q+1}(G, A) \quad \text{for } q \ge 1.$$

We may assume that $A \subset B$ and $i = (A \hookrightarrow B)$, and then we proceed as follows:

A. Let $c \in C^G$, $b \in B$ such that $j(b) = c$ and $y = d^0 b \in C^1(G, B)$. Then $y(\sigma) = \sigma b - b$ and $j \circ y(\sigma) = \sigma c - c = 0$, hence $y(\sigma) \in A$ for all $\sigma \in G$, and $y \in C^1(G, A)$, since y is continuous. Since $d^1 y = d^1 \circ d^0 b = 0$, we get $y \in Z^1(G, A)$, and we set that $\delta(c) = y + B^1(G, A) \in H^1(G, A)$.

B. Let $q \ge 1$ and $\xi = x + B^q(G, C) \in H^q(G, C)$, where $x \in Z^q(G, C)$. Let $x' \in C^q(G, B)$ such that $C^q(j)(x') = j \circ x' = x$, and set $y = d^q x'$. Then $y \in C^{q+1}(G, B)$, and the formula for $d^q x'$ shows that

$$j \circ y = j \circ d^q x' = d^q (j \circ x') = d^q x = 0.$$

It follows that $y(\boldsymbol{\sigma}) \in A$ for all $\boldsymbol{\sigma} \in G^{q+1}$ and all $y \in C^{q+1}(G, A)$, since y is continuous. Since $d^{q+1}y = d^{q+1}d^q x' = 0$, we get $y \in Z^{q+1}(G, A)$, and we set

$$\delta^q(\xi) = y + B^{q+1}(G, A) \in H^{q+1}(G, A).$$

In the course of the subsequent proof we shall see that the definitions of δ and δ^q above do not depend on the various choices.
The homomorphisms δ and δ^q are called the **connecting homomorphisms** for the exact sequence of discrete G-modules $(*)$.

1. $\delta\colon C^G \to H^1(G, A)$ and $\delta^q\colon H^q(G, C) \to H^{q+1}(G, A)$ are well-defined group homomorphisms and fit into the long exact sequence

$$\begin{array}{l} \mathbf{0} \to A^G \overset{i^G}{\to} B^G \overset{j^G}{\to} C^G \overset{\delta}{\to} H^1(G,A) \overset{H^1(i)}{\to} H^1(G,B) \overset{H^1(j)}{\to} \\ \overset{H^1(j)}{\to} H^1(G,C) \overset{\delta^1}{\to} H^2(G,A) \overset{H^2(i)}{\to} H^2(G,B) \overset{H^2(j)}{\to} \\ \overset{H^2(j)}{\to} H^2(G,C) \overset{\delta^2}{\to} H^3(G,A) \overset{H^3(i)}{\to} \ldots \overset{\delta^{q-1}}{\to} H^q(G,A) \overset{H^q(i)}{\to} \\ \overset{H^q(i)}{\to} H^q(G,B) \overset{H^q(j)}{\to} H^q(G,C) \overset{\delta^q}{\to} H^{q+1}(G,A) \overset{H^{q+1}(i)}{\to} \ldots \end{array}$$

2. Let

$$\begin{array}{ccccccccc} \mathbf{0} & \longrightarrow & A & \xrightarrow{i} & B & \xrightarrow{j} & C & \longrightarrow & \mathbf{0} \\ & & \downarrow{\scriptstyle f} & & \downarrow{\scriptstyle g} & & \downarrow{\scriptstyle h} & & \\ \mathbf{0} & \longrightarrow & A' & \xrightarrow{i'} & B' & \xrightarrow{j'} & C' & \longrightarrow & \mathbf{0} \end{array}$$

be a commutative diagram of G-modules with exact rows. Then there are commutative diagrams

$$\begin{array}{ccc} C^G & \xrightarrow{\delta} & H^1(G,A) \\ {\scriptstyle h^G}\downarrow & & \downarrow{\scriptstyle H^1(f)} \\ C'^G & \xrightarrow{\delta'} & H^1(G,A') \end{array} \quad \textit{and} \quad \begin{array}{ccc} H^q(G,C) & \xrightarrow{\delta^q} & H^{q+1}(G,A) \\ {\scriptstyle H^q(h)}\downarrow & & \downarrow{\scriptstyle H^{q+1}(f)} \\ H^q(G,C') & \xrightarrow{\delta'^q} & H^{q+1}(G,A') \end{array}$$

for all $q \ge 1$, where δ' and δ'^q are the connecting homomorphisms for the second exact sequence.

Proof. We assume that $A \subset B$, $i = (A \hookrightarrow B)$ and $i^G = (A^G \hookrightarrow B^G)$.

1. **a.** δ is a well-defined group homomorphism: Let $c \in C^G$, $b \in B$ such that $j(b) = c$ and $y = d^0 b$. Then $y \in Z^1(G, A)$, and $\delta(c) = y + B^1(G, A)$ does not depend on the choice of b. Indeed, let likewise $b' \in B$ such that $j(b') = c$ and $y' = d^0 b'$. Then $b' - b \in A$, $y' - y = d^0(b' - b) \in B^1(G, A)$ and $y + B^1(G, A) = y' + B^1(G, A)$.

If $c,\ c_1 \in C^G$ and $b,\ b_1 \in B$ are such that $j(b) = c$ and $j(b_1) = c_1$, then we get $j(b+b_1) = c+c_1$. If $y = d^0 b$ and $y_1 = d^0 b_1$, then $y + y_1 = d^0(b+b_1)$ and therefore $\delta(c+c_1) = y + y_1 + B^1(G,A) = \delta(c) + \delta(c_1)$.

b. If $q \ge 1$, then δ^q is a well-defined group homomorphism: For that, let $\xi \in H^q(G,C)$, say $\xi = x + B^q(G,C)$, where $x \in Z^q(G,C)$, and $x' \in C^q(G,B)$ such that $j{\circ}x' = x$. Then $y = d^q x' \in Z^{q+1}(G,A)$, and $\delta^q(\xi) = y + B^{q+1}(G,A)$ does not depend on the choice of x and x'. Indeed, let likewise $x_1 \in Z^q(G,C)$ be such that $\xi = x_1 + B^q(G,A)$, let $x_1' \in C^q(G,B)$ such that $j{\circ}x_1' = x_1$, and set $y_1 = d^q x_1'$. Since $x - x_1 \in B^q(G,C)$, there exists some $u \in C^{q-1}(G,C)$ such that $x - x_1 = d^{q-1}u$. If $u' \in C^{q-1}(G,B)$ is such that $u = j \circ u'$, then $j{\circ}(x' - x_1' - d^{q-1}u') = x - x_1 - d^{q-1}u = 0$, and $x' - x_1' - d^{q-1}u'$ is continuous. Hence $x' - x_1' - d^{q-1}u' \in C^q(G,A)$, $y - y_1 = d^q(x' - x_1' - d^{q-1}u') \in B^{q+1}(G,A)$ and $y + B^{q+1}(G,A) = y_1 + B^{q+1}(G,A)$.

If $x,\ x_1 \in Z^q(G,C)$ and $x',\ x_1' \in C^q(G,B)$ are such that $j \circ x' = x$ and $j \circ x_1' = x_1$, then $j \circ (x' + x_1') = x + x_1$. If $y = d^q x'$ and $y_1 = d^q x_1'$, then $y + y_1 = d^q(x' + x_1')$ and $\delta^q(x + x_1 + B^q(G,C)) = y + y_1 + B^{q+1}(G,C) = \delta^q(x + B^q(G,C)) + \delta^q(x_1 + B^q(G,C))$.

c. Exactness at B^G: $\operatorname{Ker}(j^G) = \operatorname{Ker}(j) \cap B^G = A \cap B^G = A^G$.

d. Exactness at C^G: If $c \in \operatorname{Im}(j^G)$, $c = j(b)$ for some $b \in B^G$, then $d^0 b = 0$ and thus $\delta(c) = 0$. Hence $\operatorname{Im}(j^G) \subset \operatorname{Ker}(\delta)$.

Now let $c \in \operatorname{Ker}(\delta) \subset C^G$, $b \in B$ such that $j(b) = c$, and $y = d^0 b$. Then we obtain $\delta(c) = y + B^1(G,A) = 0$, hence $y \in B^1(G,A)$, there exists some $a \in A$ such that $y = d^0 a$, and $d^0(b - a) = 0$. Consequently $b - a \in B^G$ and $c = j(b-a) \in \operatorname{Im}(j^G)$.

e. Exactness at $H^1(G,A)$: Let $c \in C^G$, $b \in B$ such that $c = j(b)$, and set $y = d^0 b$. Then $y \in B^1(G,B) \cap Z^1(G,A)$, $\delta(c) = y + B^1(G,A)$, and consequently $H^1(i){\circ}\delta(c) = y + B^1(G,B) = 0$. Therefore $\operatorname{Im}(\delta) \subset \operatorname{Ker}(H^1(i))$.

If $x \in Z^1(G,A)$ is such that $H^1(i)(x + B^1(G,A)) = x + B^1(G,B) = 0 \in H^1(G,B)$, then $x \in B^1(G,B)$, and there exists some $b \in B$ such that $x = d^0 b$. If $c = j(b) \in C$, then $d^0 c = j \circ d^0 b = j \circ x = 0$, hence $c \in C^G$ and $x + B^1(G,A) = \delta(c)$. Consequenty $\operatorname{Ker}(H^1(i)) \subset \operatorname{Im}(\delta)$.

f. Exactness at $H^q(G,B)$ for $q \ge 1$: Since $H^q(j){\circ}H^q(i) = H^q(j{\circ}i) = 0$, it follows that $\operatorname{Im}(H^q(i)) \subset \operatorname{Ker}(H^q(j))$.

Now let $x \in Z^q(G,B)$ be such that $x + B^q(G,B) \in \operatorname{Ker}(H^q(j))$. Then we get $0 = H^q(j)(x + B^q(G,B)) = j \circ x + B^q(G,C)$, hence $j \circ x \in B^q(G,C)$ and therefore $j \circ x = d^{q-1}x'$ for some $x' \in C^{q-1}(G,C)$. Let $y \in C^{q-1}(G,B)$ such that $x' = j \circ y$, and set $y' = x - d^{q-1}y$. Then $y' \in Z^q(G,B)$, and $j \circ y' = j \circ x - d^{q-1}(j \circ y) = d^{q-1}(x' - j \circ y) = 0$. Hence $y' \in Z^q(G,A)$, and it follows that $H^q(i)(y' + B^q(G,A)) = y' + B^q(G,B) = x + B^q(G,B)$ and $x + B^q(G,B) \in \operatorname{Im}(H^q(i))$.

g. Exactness at $H^q(G,C)$ for $q \ge 1$: If $\xi \in \operatorname{Im}(H^q(j))$, then we obtain $\xi = j{\circ}x' + B^q(G,C)$ for some $x' \in C^q(G,B)$, and $\delta^q\xi = d^q x' + B(G,A) = 0$. Hence $\operatorname{Im}(H^1(j)) \subset \operatorname{Ker}(\delta^1)$.

Now let $x \in Z^q(G,C)$ be such that $x + B^q(G,C) \in \operatorname{Ker}(\delta^q)$, and let $x' \in C^q(G,B)$ with $j\circ x' = x$. Then $0 = \delta^q(x+B^q(G,C)) = d^q x' + B^{q+1}(G,A)$, and therefore $d^q x' = d^q u$ for some $u \in C^q(G,A) \subset C^q(G,B)$. Hence $j\circ u = 0$, $x' - u \in Z^q(G,B)$ and $H^q(j)(x' - u + B^q(G,B)) = j\circ(x' - u) + B^q(G,C) = x + B^q(G,C) \in \operatorname{Im}(H^q(j))$.

h. Exactness at $H^{q+1}(G,A)$ for $q \geq 1$: Let $x \in Z^q(G,C)$, and let $x' \in C^q(G,B)$ such that $j\circ x' = x$. Then $\delta^q(x+B^q(G,C)) = d^q x' + B^{q+1}(G,A))$ and $H^{q+1}(i) \circ \delta^q(x + B^q(G,C)) = d^q x' + B^{q+1}(G,B) = 0$. Hence $\operatorname{Im}(\delta^q) \subset \operatorname{Ker}(H^{q+1}(i))$.

Now let $y \in Z^{q+1}(G,A)$ be such that $y + B^{q+1}(G,A) \in \operatorname{Ker}(H^{q+1}(i))$. Then $H^{q+1}(i)(y+B^{q+1}(G,A)) = y+B^{q+1}(G,B) = 0$, hence $y \in B^{q+1}(G,B)$, and $y = d^q x'$ for some $x' \in C^q(G,B)$. If $x = j\circ x'$, then $d^q x = j\circ d^q x' = j\circ y = 0$. Hence $x \in Z^q(G,A)$ and $y + B^{q+1}(G,A) = \delta^q(x + B^q(G,C)) \in \operatorname{Im}(\delta^q)$.

2. We may additionally assume that $A' \subset B'$, $i' = (A' \hookrightarrow B')$ and $f = g\restriction A$.

A. Let $c \in C^G$ and $b \in B$ such that $c = j(b)$. Then $d^0 b \in Z^1(G,A)$, and $\delta(c) = d^0 b + B^1(G,A)$. Hence $H^1(f)\circ\delta(c) = f\circ d^0 b + B^1(G,A') \in H^1(G,A')$. On the other hand,

$$h^G(c) = h(c) = h\circ j(b) = j'\circ g(b),\ \ d^0 g(b) = g\circ d^0 b = f\circ d^0 b \in Z^1(G,A')$$

and $\delta'\circ h^G(c) = d^0 g(b) + B^1(G,A') = f\circ d^0 b + B^1(G,A')$.

B. Let $q \geq 1$, $x \in Z^q(G,C)$ and $x' \in C^q(G,B)$ such that $j\circ x' = x$. Then we get $d^q x' \in Z^q(G,A)$, $\delta^q(x + Z^q(G,C)) = d^q x' + B^{q+1}(G,A)$, and

$$H^{q+1}(f)\circ\delta^q(x + Z^q(G,C)) = f\circ d^q x' + B^{q+1}(G,A').$$

On the other hand, $H^q(h)(x+Z^q(G,C)) = h\circ x+Z^q(G,C')$, and observing that $h\circ x = h\circ j\circ x' = j'\circ g\circ x'$, we get $\delta^q\circ H^q(h)(x+Z^q(G,C)) = d^q(g\circ x')+B^{q+1}(G,A')$, but $d^q(g\circ x') = g\circ d^q x' = f\circ d^q x'$. □

For a finite group we extend the long cohomology sequence on the left-hand side by the cohomology groups H^0 and H^{-1} as defined in 2.2.6.

Definition and Theorem 2.4.2. *Let G be a finite group and*

$$\mathbf{0} \to A \xrightarrow{i} B \xrightarrow{j} C \to \mathbf{0} \qquad (*)$$

an exact sequence of G-modules. We define maps

$$\delta^0\colon H^0(G,C) \to H^1(G,A) \quad \text{and} \quad \delta^{-1}\colon H^{-1}(G,C) \to H^0(G,A)$$

as follows.

A. We define

$$\delta^0\colon H^0(G,C) = C^G/N_G C \to H^1(G,A) \text{ by } \delta^0(c + N_G C) = \delta(c),$$

where δ is defined in 2.4.1.

B. We define $\delta^{-1}\colon H^{-1}(G,C) = {}_{N_G}C/I_GC \to H^0(G,A) = A^G/N_GA$ as follows: Let $c \in {}_{N_G}C$ and $b \in B$ such that $c = j(b)$. Then $j(N_Gb) = N_Gc = 0$, hence $N_Gb = i(a)$ for some $a \in A$, and as $i(\sigma a) = \sigma N_Gb = N_Gb = i(a)$, we get $\sigma a = a$, and $a \in A^G$. We define $\delta^{-1}(c + I_GC) = a + N_GA$.

In the course of the subsequent proof we shall see that the definitions of δ^0 and δ^{-1} do not depend on the various choices.
The homomorphisms δ^0 and δ^{-1} are called the **connecting homomorphisms** for the exact sequence of discrete G-modules $(*)$.

1. $\delta^{-1}\colon H^{-1}(G,C) \to H^0(G,A)$ and $\delta^0\colon H^0(G,C) \to H^1(G,A)$ are well-defined homomorphisms and fit into the long exact sequence

$$H^{-1}(G,A) \overset{H^{-1}(i)}{\to} H^{-1}(G,B) \overset{H^{-1}(j)}{\to} H^{-1}(G,C) \overset{\delta^{-1}}{\to} H^0(G,A) \overset{H^0(i)}{\to}$$
$$\overset{H^0(i)}{\to} H^0(G,B) \overset{H^0(j)}{\to} H^0(G,C) \overset{\delta^0}{\to} H^1(G,A) \overset{H^1(i)}{\to}$$
$$\overset{H^1(i)}{\to} H^1(G,B) \to \ldots$$

2. Let

$$\begin{array}{ccccccccc} \mathbf{0} & \longrightarrow & A & \xrightarrow{i} & B & \xrightarrow{j} & C & \longrightarrow & \mathbf{0} \\ & & {\scriptstyle f}\downarrow & & {\scriptstyle g}\downarrow & & {\scriptstyle h}\downarrow & & \\ \mathbf{0} & \longrightarrow & A' & \xrightarrow{i'} & B' & \xrightarrow{j'} & C' & \longrightarrow & \mathbf{0} \end{array}$$

be a commutative diagram of G-modules with exact rows. Then the diagrams

$$\begin{array}{ccc} H^{-1}(G,C) & \xrightarrow{\delta^{-1}} & H^0(G,A) \\ {\scriptstyle H^{-1}(h)}\downarrow & & \downarrow{\scriptstyle H^0(f)} \\ H^{-1}(G,C') & \xrightarrow{\delta'^{-1}} & H^0(G,A') \end{array} \quad \textit{and} \quad \begin{array}{ccc} H^0(G,C) & \xrightarrow{\delta^0} & H^1(G,A) \\ {\scriptstyle H^0(h)}\downarrow & & \downarrow{\scriptstyle H^1(f)} \\ H^0(G,C') & \xrightarrow{\delta'^0} & H^1(G,A') \end{array}$$

are commutative (where δ'^0 and δ'^{-1} denote the connecting homomorphisms of the second exact sequence).

Proof. We may assume that $A \subset B$ and $i = (A \hookrightarrow B)$.

1. **a.** δ^0 is a well-defined group homomorphism. For this, it suffices to prove that $N_GC \subset \mathrm{Ker}(\delta)$. Thus let $c = N_Gc_1 \in N_GC$, where $c_1 = j(b_1)$ with $b_1 \in B$. Then $c = j(N_Gb_1)$, and since $\sigma N_Gb_1 - N_Gb_1 = 0$ for all $\sigma \in G$, it follows that $\delta(c) = 0$.

b. δ^{-1} is a well-defined group homomorphism. Let $c \in {}_{N_G}C$, $b \in B$ such that $c = j(b)$ and $a = N_Gb \in A^G$. Then $\delta^{-1}(c + I_GC) = a + N_GA$ does not depend on the choice of c and b. Indeed, let likewise $c_1 \in {}_{N_G}C$ and $b_1 \in B$ such that $j(b_1) = c_1$, and $a_1 = N_Gb_1$. Then $j(b_1 - b) = c_1 - c \in I_GC = j(I_GB)$, say $j(b_1 - b) = j(b^*)$ for some $b^* \in I_G B$. Hence we get $b_1 - b - b^* \in A$ and $a_1 - a = N_G(b_1 - b) = N_G(b_1 - b + b^*) \in N_G A$.

Let $c,\, c_1 \in {}_{N_G}C,\ b,\, b_1 \in B$ such that $j(b) = c,\ j(b_1) = c_1$ and $a = N_G b$, $a_1 = N_G b_1$. Then $a,\, a_1 \in A^G,\ \delta^{-1}(c + {}_{N_G}C) = a + I_G A$, and $\delta^{-1}(c_1 + {}_{N_G}C) = a_1 + I_G A$. Since $j(b + b_1) = c + c_1$ and $N_G(b + b_1) = a + a_1$, it follows that $\delta^{-1}(c + c_1 + {}_{N_G}C) = a + a_1 + I_G(A) = \delta^{-1}(c + {}_{N_G}C) + \delta^{-1}(c_1 + {}_{N_G}C)$. Hence δ^{-1} is a group homomorphism.

c. Exactness at $H^{-1}(G, B)$: Since $H^{-1}(j) \circ H^{-1}(i) = H^{-1}(j \circ i) = 0$, it follows that $\mathrm{Im}(H^{-1}(i)) \subset \mathrm{Ker}(H^{-1}(j))$.

Now assume that $b + I_G B \in \mathrm{Ker}(H^{-1}(j))$, where $b \in {}_{N_G}B$. Then it follows that $H^{-1}(j)(b + I_G B) = j(b) + I_G C = 0$, hence $j(b) \in I_G C = j(I_G B)$, and there exists some $b_1 \in I_G B$ such that $j(b - b_1) = 0$. Consequently $a = b - b_1 \in A$ and $N_G a = 0$, hence $a \in {}_{N_G}A$ and

$$H^{-1}(i)(a + I_G A) = a + I_G B = b + I_G B \in \mathrm{Im}(H^{-1}(i)).$$

d. Exactness at $H^{-1}(G, C)$. Let $b + {}_{N_B}B \in H^{-1}(G, B)$, where $b \in {}_{N_G}B$. Then we obtain $\delta^{-1} \circ H^{-1}(j)(b + {}_{N_G}B) = \delta^{-1}(j(b) + {}_{N_G}B) = 0$, since $N_G b = 0$. Hence $\mathrm{Im}(H^{-1}(j)) \subset \mathrm{Ker}(\delta^{-1})$.

Now let $c \in {}_{N_G}C$ and $c + I_G C \in \mathrm{Ker}(\delta^{-1})$. Let $b \in B$ be such that $c = j(b)$, and set $a = N_G b \in A^G$. Then $0 = \delta^{-1}(c + I_G C) = a + N_G A$, hence $a = N_G b = N_G a_1$ for some $a_1 \in A$. It follows that $b - a_1 \in \mathrm{Ker}(N_G) = {}_{N_G}B$, $c = j(b - a_1)$, and consequently $c + I_G C = H^{-1}(j)(b - a_1 + I_G B) \in \mathrm{Im}(H^{-1}(j))$. Hence $\mathrm{Ker}(\delta^{-1}) \subset \mathrm{Im}(H^{-1}(j))$.

e. Exactness at $H^0(G, A)$. Let $c \in {}_{N_G}C,\ b \in B$ such that $j(b) = c$, and set $a = N_G b \in A^G$. Then $H^0(i) \circ \delta^{-1}(c + I_G C) = a + N_G B = 0$, and consequently $\mathrm{Im}(\delta^{-1}) \subset \mathrm{Ker}(H^0(i))$.

Now let $a \in A^G$ and $a + N_G A \in \mathrm{Ker}(H^0(i))$. Then it follows that $0 = H^0(i)(a + N_G A) = a + N_G B$ and therefore $a = N_G b$ for some $b \in B$. If $c = j(b)$, then $N_G c = j(N_G b) = j(a) = 0$, hence $c \in {}_{N_G}C$ and

$$a + N_G A = \delta^{-1}(c + I_G C) \in \mathrm{Im}(\delta^{-1}).$$

f. Exactness at $H^0(G, B)$. Since $H^0(j) \circ H^0(i) = H^0(j \circ i) = 0$, it follows that $\mathrm{Im}(H^0(i)) \subset \mathrm{Ker}(H^0(j))$.

Now let $b \in B^G$ be such that $b + N_G B \in \mathrm{Ker}(H^0(j))$. Then it follows that $0 = H^0(j)(b + N_G B) = j(b) + N_G C$, and thus $j(b) = N_G c$ for some $c \in C$. If $b_1 \in B$ is such that $c = j(b_1)$, then $j(b - N_G b_1) = 0$, and consequently we obtain $a = b - N_G b_1 \in A \cap B^G = A^G$ and

$$b + N_G B = a + N_G B = H^0(i)(a + N_G A) \in \mathrm{Im}(H^0(i)).$$

g. Exactness at $H^0(G, C)$. If $b \in B^G$, then $\delta^0 \circ H^0(j)(b + N_G B) = \delta \circ j(b) = 0$ by 2.4.1.1. Hence $\mathrm{Im}(H^0(j)) \subset \mathrm{Ker}(\delta^0)$.

Now let $c \in C^G$ and $c + N_G C \in \mathrm{Ker}(\delta^0)$. Then $0 = \delta^0(c + N_G C) = \delta(c)$, by 2.4.1.1 it follows that $c = j(b)$ for some $b \in B^G$ and thus it follows that $c + N_G C = H^0(j)(b + N_G B))$. Hence $\mathrm{Ker}(\delta^0) \subset \mathrm{Im}(H^0(j))$.

h. Exactness at $H^1(G, A)$. By 2.4.1.1, $\mathrm{Ker}(H^1(i)) = \mathrm{Im}(\delta) = \mathrm{Im}(\delta_0)$.

2. We may additionally assume that $A' \subset B'$, $i' = (A' \hookrightarrow B')$ and $f = g \restriction A$.

A. Let $c \in {}_{N_G}C$, $b \in B$ with $c = j(b)$ and $a = N_G b \in A^G$. Then

$$H^0(f) \circ \delta^{-1}(c + N_G C) = H^0(f)(a + N_G A) = f(a) + N_G A' = g(a) + N_G A'.$$

Since $h(c) = h \circ j(b) = j' \circ g(b)$ and $N_G g(b) = g(N_G b) = g(a)$, we obtain

$$\begin{aligned}\delta'^{-1} \circ H^{-1}(h)(c + I_G C) &= \delta'^{-1}(h(c) + I_G C') = g(a) + N_G A' \\ &= H^0(f) \circ \delta^{-1}(c + N_G C).\end{aligned}$$

B. The second diagram commutes because of 2.4.1.2. Indeed, if $c \in C^G$, then

$$\begin{aligned}H^1(f) \circ \delta^0(c + N_G C) &= H^1(f) \circ \delta(c) = \delta' \circ h^G(c) = \delta'^0(h(c) + N_G C') \\ &= \delta'^0 \circ H^0(h)(c + N_G C).\end{aligned}$$

□

Theorem 2.4.3 (Shapiro's lemma). *Let G be a profinite group, H a closed subgroup of G, B a discrete H-module and $q \geq 1$. Then there exists an isomorphism*

$$\Psi^q \colon H^q(G, \mathcal{M}_G^H(B)) \xrightarrow{\sim} H^q(H, B),$$

which acts as follows: Let $\xi \in H^q(G, \mathcal{M}_G^H(B))$, let $x \colon G^q \to \mathcal{M}_G^H(B)$ be a defining q-cocycle of ξ, and define $x_1 \colon H^q \to B$ by $x_1(\boldsymbol{\tau}) = x(\boldsymbol{\tau})(1)$ for all $\boldsymbol{\tau} \in H^q$. Then x_1 is a defining q-cocycle of $\Psi^q(\xi)$.

Proof. By 2.1.4.**3**, there exists an exact sequence of discrete H-modules

$$\mathbf{0} \to B \to \mathcal{M}_H(B) \xrightarrow{\pi} B_1 \to \mathbf{0}. \qquad (*)$$

By 2.2.4 and 2.4.1, it induces the exact sequences

$$\mathbf{0} \to B^H \to \mathcal{M}_H(B)^H \to B_1^H \xrightarrow{\delta} H^1(H, B) \to H^1(H, \mathcal{M}_H(B) = \mathbf{0}$$

and for all $q \geq 1$

$$\mathbf{0} = H^q(H, \mathcal{M}_H(B)) \to H^q(H, B_1) \xrightarrow{\delta^q} H^{q+1}(H, B) \to H^{q+1}(H, \mathcal{M}_H(B)) = \mathbf{0}.$$

By 2.1.5, $(*)$ induces the exact sequence of discrete G-modules

$$\mathbf{0} \to \mathcal{M}_G^H(B) \to \mathcal{M}_G^H(\mathcal{M}_H(B)) \xrightarrow{\overline{\pi}} \mathcal{M}_G^H(B_1) \to \mathbf{0}.$$

2.1.6.2 yields $\mathcal{M}_G^H(\mathcal{M}_H(B)) \cong \mathcal{M}_G(B)$, hence $H^q(G, \mathcal{M}_G^H(\mathcal{M}_H(B))) = \mathbf{0}$ for all $q \geq 1$. By 2.1.6.1, there exist isomorphisms

$$\mathcal{M}_G^H(B)^G \xrightarrow{\sim} B^H, \quad \mathcal{M}_G^H(\mathcal{M}_H(B))^G \xrightarrow{\sim} \mathcal{M}_H(B)^H \text{ and } \mathcal{M}_G^H(B_1)^G \xrightarrow{\sim} B_1^H,$$

which all are induced by the assignment $f \mapsto f(1)$. From this we obtain the following commutative diagram with exact rows in which the vertical arrows are isomorphisms induced by the assignment $f \mapsto f(1)$.

$$\begin{array}{ccccccc}
\mathcal{M}_G^H(\mathcal{M}_H(B))^G & \xrightarrow{\overline{\pi}^G} & \mathcal{M}_G^H(B_1)^G & \xrightarrow{\delta} & H^1(G, \mathcal{M}_G^H(B)) & \longrightarrow & \mathbf{0} \\
\downarrow & & \downarrow & & & & \\
\mathcal{M}_H(B)^H & \xrightarrow{\pi^H} & B_1^H & \xrightarrow{\delta} & H^1(H, B) & \longrightarrow & \mathbf{0}.
\end{array}$$

Let $\Psi_1 \colon H^1(G, \mathcal{M}_G^H(B)) \to H^1(H, B)$ be the unique isomorphism which makes the above diagram commutative. Let $\xi = \delta(f) \in H^1(G, \mathcal{M}_G^H(B))$, where $f \in \mathcal{M}_G^H(B_1)^G$, and let $f_1 \in \mathcal{M}_G^H(\mathcal{M}_H(B))$ be such that $f = \overline{\pi}(f_1)$. Then $x = d^0 f_1 \in Z^1(G, \mathcal{M}_G^H(B))$ is a defining 1-cocycle of ξ. Moreover, as $f(1) = \pi f_1(1)$, it follows that $x_1 \in Z^1(H, B)$, defined by $x_1(\sigma) = x(\sigma)(1)$, is a defining 1-cocycle of $\Psi^1(\xi)$.

Now we proceed by induction on q. Assume that the assertions hold for some $q \geq 1$, and let $\Psi^{q+1} \colon H^{q+1}(G, \mathcal{M}_G^H(B)) \to H^{q+1}(H, B)$ be the unique isomorphism which makes the following diagram commutative.

$$\begin{array}{ccccccc}
\mathbf{0} & \longrightarrow & H^q(G, \mathcal{M}_G^H(B_1)) & \xrightarrow{\delta^q} & H^{q+1}(G, \mathcal{M}_G^H(B)) & \longrightarrow & \mathbf{0} \\
& & {\scriptstyle \Psi^q}\downarrow & & \downarrow{\scriptstyle \psi^{q+1}} & & \\
\mathbf{0} & \longrightarrow & H^q(H, B_1) & \xrightarrow{\delta^q} & H^{q+1}(H, B) & \longrightarrow & \mathbf{0}.
\end{array}$$

We show that Ψ^{q+1} has the desired form. If $\xi = \delta^q(\eta) \in H^{q+1}(G, \mathcal{M}_G^H(B))$, where $\eta \in H^q(G, \mathcal{M}_G^H(B_1))$, then $\Psi^{q+1}(\xi) = \delta^q(\Psi^q(\eta))$. Let $y \in Z^q(G, \mathcal{M}_G^H(B_1))$ be a defining q-cocycle of η, and let $y_1 \colon H^q \to B_1$ be defined by $y_1(\boldsymbol{\tau}) = y(\boldsymbol{\tau})(1)$ for all $\boldsymbol{\tau} \in H^q$. Then y_1 is a defining q-cocycle of $\Psi^q(\eta)$.

If $y' \in C^q(G, \mathcal{M}_G^H(\mathcal{M}_H(B))$ is such that $y = \overline{\pi} \circ y'$, then $x = d^q y' \in Z^q(G, \mathcal{M}_G^H(B))$ is a defining $(q+1)$-cocycle of ξ. If $y_1' \colon H^q \to \mathcal{M}_H(B)$ and $x_1 \colon H^{q+1} \to \mathcal{M}_H(B)$ are defined by $y_1'(\boldsymbol{\tau}) = y'(\boldsymbol{\tau})(1)$ and $x_1 = d^q y_1'$, then $y_1 = \pi \circ y_1'$, x_1 is a defining $(q+1)$-cocycle of $\Psi^{q+1}(\xi) = \delta^q(\Psi^q(\eta))$, and $x_1(\boldsymbol{\tau}) = x(\boldsymbol{\tau})(1)$. □

Corollary 2.4.4. *Let G be a profinite group and H an open subgroup of G. Let $(G:H) = r$, and $G/H = \{\sigma_1 H, \ldots, \sigma_r H\}$, where $\sigma_1 = 1$. Let A be a discrete G-module and B an H-submodule of A such that*

$$A = \bigoplus_{i=1}^{r} \sigma_i B = B \oplus \bigoplus_{i=2}^{r} \sigma_i B.$$

Let $\pi\colon A \to B$ *be the projection onto the first summand and* $q \geq 1$*. Then there exists an isomorphism*

$$\Psi_0^q\colon H^q(G,A) \overset{\sim}{\to} H^q(H,B),$$

which acts as follows: If $\xi \in H^q(G,A)$, $x\colon G^q \to A$ *is a defining* q*-cocycle of* ξ *and* $x_1 = \pi \circ x \restriction H^q\colon H^q \to B$*, then* x_1 *is a defining* q*-cocycle of* $\Psi_0^q(\xi)$*.*

Proof. Let $\Phi\colon \mathcal{M}_G^H(B) \to A$ be the isomorphism given in 2.1.9.1, and recall that $\pi\circ\Phi(f) = f(1)$ for all $f \in \mathcal{M}_G^H(B)$. Let $\Psi^q\colon H^q(G,\mathcal{M}_G^H(B)) \overset{\sim}{\to} H^q(H,B)$ be the isomorphism given in 2.4.3 and $\Psi_0^q = \Psi^q \circ H^q(\Phi)^{-1}$.

$$\begin{array}{ccc} H^q(G,\mathcal{M}_G^H(B) & \xrightarrow{H^q(\Phi)} & H^q(G,A) \\ {\scriptstyle\Psi^q}\big\downarrow & & \big\downarrow{\scriptstyle\Psi_0^q} \\ H^q(H,B) & = & H^q(H,B) \end{array}$$

Let $\xi = H^q(\Phi)(\eta) \in H^q(G,A)$, where $\eta \in H^q(G,\mathcal{M}_G^H(B))$, and suppose that $y \in Z^q(G,\mathcal{M}_G^H(B))$ is a defining q-cocycle of η. Then $x = \Phi\circ y \in Z^q(G,A)$ is a defining q-cocycle of ξ, and $x_1 \in Z^q(H,B)$, defined by $x_1(\boldsymbol{\sigma}) = y(\boldsymbol{\sigma})(1)$ for all $\boldsymbol{\sigma} \in H^q$, is a defining q-cocycle of $\Psi_0^q(\xi)$ by 2.4.3. Since $x_1(\boldsymbol{\sigma}) = y(\boldsymbol{\sigma})(1) = \pi\circ\Phi(y(\boldsymbol{\sigma})) = \pi\circ x(\boldsymbol{\sigma})$ for all $\boldsymbol{\sigma} \in H^q$, it follows that $x_1 = \pi\circ x \restriction H^q$. $\square$

We conclude this section with the special but important case of finite cyclic groups.

Definition and Theorem 2.4.5. *Let* $G = \langle\sigma\rangle$ *be a finite cyclic group.*

1. *For every* G*-module* A *and every* $q \geq 1$*, there exists an isomorphism*

$$\Phi = \Phi_A^q\colon H^q(G,A) \overset{\sim}{\to} H^{q-2}(G,A)$$

such that for every G*-homomorphism* $\varphi\colon A \to B$ *of* G*-modules the diagram*

$$\begin{array}{ccc} H^q(G,A) & \xrightarrow{\Phi_A^q} & H^{q-2}(G,A) \\ {\scriptstyle H^q(\varphi)}\big\downarrow & & \big\downarrow{\scriptstyle H^{q-2}(\varphi)} \\ H^q(G,B) & \xrightarrow{\Phi_B^q} & H^{q-2}(G,B). \end{array}$$

commutes. If $q = 1$ *and* $x \in Z^1(G,A)$*, then*

$$\Phi_A^1(x+B^1(G,A)) = x(\sigma)+I_GA \in H^{-1}(G,A) = {}_{N_G}A/I_G(A).$$

2. If the groups $H^0(G,A)$ and $H^{-1}(G,A)$ are both finite, then we call

$$\mathsf{h}(G,A) = \frac{|H^0(G,A)|}{|H^{-1}(G,A)|}$$

the **Herbrand quotient** of A. Otherwise we say that $\mathsf{h}(G,A)$ is not defined.

Let $\mathbf{0} \to A \overset{\varphi}{\to} B \overset{\psi}{\to} C \to \mathbf{0}$ *be an exact sequence of* G*-modules.*

(a) There is an exact hexagon

$$\begin{array}{ccccc} H^0(G,A) & \xrightarrow{H^0(\varphi)} & H^0(G,B) & \xrightarrow{H^0(\psi)} & H^0(G,C) \\ {\scriptstyle\delta^{-1}}\uparrow & & & & \downarrow{\scriptstyle\Phi_A\circ\delta^0} \\ H^{-1}(G,C) & \xleftarrow[H^{-1}(\psi)]{} & H^{-1}(G,B) & \xleftarrow[H^{-1}(\varphi)]{} & H^{-1}(G,A). \end{array}$$

(b) If two of the Herbrand quotients $\mathsf{h}(G,A)$, $\mathsf{h}(G,B)$, $\mathsf{h}(G,C)$ *are defined, then so is the third, and then*

$$\mathsf{h}(G,B) = \mathsf{h}(G,A)\mathsf{h}(G,C).$$

3. For the trivial G*-module* $\mathbb{Z}$ *we have* $H^0(G,\mathbb{Z}) = \mathbb{Z}/|G|\mathbb{Z}$, $H^{-1}(G,\mathbb{Z}) = \mathbf{0}$ *and* $\mathsf{h}(G,\mathbb{Z}) = |G|$*. If* A *is a finite* G*-module, then* $\mathsf{h}(G,A) = 1$.

Proof. Let $|G| = n$ and observe that

$$A^G = \{a \in A \mid \sigma a = a\} \quad \text{and} \quad I_G A = (\sigma - 1)A.$$

1. We start with the case $q = 1$ and prove the following assertions:

A. If $x \in Z^1(G,A)$, then $x(\sigma) \in {}_{N_G}A$,

$$x(\sigma^k) = \sum_{i=0}^{k-1} \sigma^i x(\sigma) \quad \text{for all} \quad k \ge 0,$$

and $x \in B^1(G,A)$ if and only if $x(\sigma) \in I_G A$.

B. Let $a \in {}_{N_G}A$, and define

$$x\colon G \to A \quad \text{by} \quad x(\sigma^k) = \sum_{i=0}^{k-1} \sigma^i(a) \quad \text{for all } k \in [0, n-1].$$

Then $x \in Z^1(G,A)$ and $x(\sigma) = a$.

Proof of **A.** By 2.2.5, we obtain $x(1) = 0$, and if $k \ge 0$, then

$$x(\sigma^k) = \sum_{i=0}^{k-1} \sigma^i x(\sigma) \ \text{ implies } \ x(\sigma^{k+1}) = \sigma x(\sigma^k) + x(\sigma) = \sum_{i=0}^{k} \sigma^i x(\sigma).$$

Hence it follows by induction on k that the formula holds for all $k \ge 0$. In particular, $N_G x(\sigma) = x(\sigma^n) = x(1) = 0$, and therefore $x(\sigma) \in {}_{N_G}A$.

If $x \in B^1(G, A)$, then $x(\sigma) = \sigma a - a \in I_G A$ for some $a \in A$. Conversely, if $x(\sigma) \in I_G A = (\sigma - 1)A$, then $x(\sigma) = \sigma a - a$ for some $a \in A$, and for all $k \ge 1$ it follows that

$$x(\sigma^k) = \sum_{i=0}^{k-1} \sigma^i x(\sigma) = \sum_{i=0}^{k-1} (\sigma^{i+1} a - \sigma^i a) = \sigma^k a - a, \quad \text{hence } \ x \in B^1(G, A).$$

□[**A.**]

Proof of **B.** We assert that

$$x(\sigma^k) = \sum_{i=0}^{k-1} \sigma^i(a) \quad \text{for all } \ k \ge 0.$$

Indeed, if $k \ge 0$ and $k = ln + r$, where $l \in \mathbb{N}_0$ and $r \in [0, n-1]$, then **A** implies

$$\begin{aligned}\sum_{i=0}^{k-1} \sigma^i a &= \sum_{i=0}^{r-1} \sigma^i a + \sum_{j=0}^{l-1} \sum_{\nu=0}^{n-1} \sigma^{r+jn+\nu} a = \sum_{i=0}^{r-1} \sigma^i a + \sum_{j=0}^{l-1} \sum_{\nu=0}^{n-1} \sigma^{r+\nu} a \\ &= \sum_{i=0}^{r-1} \sigma^i a + l\sigma^r N_G a = \sum_{i=0}^{r-1} \sigma^i a = x(\sigma^r) = x(\sigma^k).\end{aligned}$$

For all $k,\, l \in \mathbb{N}_0$ it follows that

$$x(\sigma^k \sigma^l) = \sum_{i=0}^{k+l-1} \sigma^i a = \sigma^k \sum_{i=0}^{l-1} \sigma^i a + \sum_{i=0}^{k-1} \sigma^i a = \sigma^k x(\sigma^l) + x(\sigma^k).$$

Hence $x \in Z^1(G, A)$, and $x(\sigma) = a$ holds by the very definition. □[**B.**]

By **A** and **B** the map $\Phi_0 \colon Z^1(G, A) \to {}_{N_G}A$, defined by $\Phi_0(x) = x(\sigma)$ for all $x \in Z^1(G, A)$, is an isomorphism which satisfies $\Phi_0(B^1(G, A)) = I_G A$ and induces an isomorphism $\Phi_A^1 \colon H^1(G, A) \xrightarrow{\sim} {}_{N_G}A/I_G A = H^{-1}(G, A)$ as asserted.

Let $\varphi \colon A \to B$ be a G-homomorphism of G-modules and $x \in Z^1(G, A)$. Then

$$\begin{aligned}H^{-1}(\varphi) \circ \Phi_A^1(x + B^1(G, A)) &= H^{-1}(i)(x(\sigma) + I_G A) = \varphi \circ x(\sigma) + I_G B \\ &= \Phi_B^1(\varphi \circ x + B^1(G, B)) \\ &= \Phi_B^1 \circ H^1(\varphi)(x + B^1(G, A)).\end{aligned}$$

and therefore $H^{-1}(i) \circ \Phi_A = \Phi_B \circ H^1(i)$.

We do the general case by induction on $q \geq 1$. Let $\varphi\colon A \to B$ be a G-homomorpism of G-modules. We use the commutative diagram with exact rows given in 2.1.4.**3**:

$$\begin{array}{ccccccccc}
\mathbf{0} & \longrightarrow & A & \xrightarrow{\ i_A\ } & \mathcal{M}_G(A) & \xrightarrow{\ j_A\ } & A_1 & \longrightarrow & \mathbf{0} \\
 & & {\scriptstyle\varphi}\big\downarrow & & {\scriptstyle\mathcal{M}_G(\varphi)}\big\downarrow & & \big\downarrow{\scriptstyle\varphi_1} & & \\
\mathbf{0} & \longrightarrow & B & \xrightarrow{\ i_B\ } & \mathcal{M}_G(B) & \xrightarrow{\ j_B\ } & B_1 & \longrightarrow & \mathbf{0}\ .
\end{array}$$

It induces the incomplete cube (with the self-expanatory morhisms)

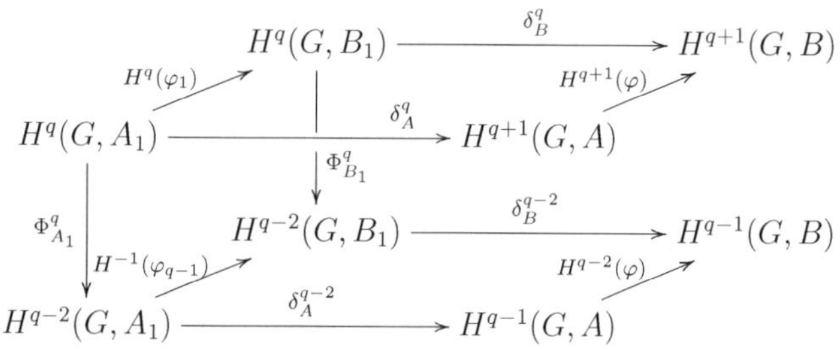

where the horizontal arrows are isomorphisms, the top and the bottom faces commute, the left vertical arrows $\Phi_{A_1}^q$ and $\Phi_{B_1}^q$ are isomorphisms and the left-hand face commutes by the induction hypothesis. Hence there exist isomorphisms

$$\Phi_A^{q+1}\colon H^{q+1}(G,A) \to H^{q-1}(G,A) \quad \text{and } \Phi_B^{q+1}\colon H^{q+1}(G,B) \to H^{q-1}(G,B)$$

which make the right-hand face commutative. This not only implies the existence of Φ_A^{q+1} but also its behavior under homomorphisms.

2. (a) We may asssume that $A \subset B$ and $\varphi = (A \hookrightarrow B)$. By 2.4.2, it suffices to prove the exactness at $H^0(G,C)$ and $H^{-1}(G,A)$. The former follows by 2.4.2.2, since $\mathrm{Ker}(\Phi_A{\circ}\delta^0) = \mathrm{Ker}(\delta^0) = \mathrm{Im}(H^0(\psi))$.

• $\mathrm{Im}(\Phi_A{\circ}\delta^0) \subset \mathrm{Ker}(H^{-1}(\varphi))$: Let $c + N_GC \in C^G/N_GC = H^0(G,C)$, where $c \in C^G$ and $c = \psi(b)$ for some $b \in B$. Then $\sigma b - b \in A$,

$$\delta^0(c + N_GC)(\sigma) = \delta(c)(\sigma) = \sigma b - b$$

by 2.4.2, and $\Phi_A{\circ}\delta^0(c + N_GC) = \sigma b - b + I_GA$. Hence we obtain

$$H^{-1}(\varphi){\circ}\Phi_A{\circ}\delta^0(c + N_GC) = \sigma b - b + I_GB = 0 \in {}_{N_G}B/I_GB = H^{-1}(G,B).$$

• $\mathrm{Ker}(H^{-1}(\varphi) \subset \mathrm{Im}(\Phi_A{\circ}\delta^0)$: Let $a \in {}_{N_G}A$ such that $a + I_GA \in \mathrm{Ker}(H^{-1}(\varphi))$. Then $0 = H^{-1}(\varphi)(a + I_GA) = a + I_GB$, hence $a = \sigma b - b$, and as above it follows that

$$a + I_GA = \sigma b - b + I_GA = \Phi_A{\circ}\delta^0(\psi(b) + N_GC) \in \mathrm{Im}(\Phi_A{\circ}\delta^0).$$

(b) If $M' \xrightarrow{f} M \xrightarrow{g} M''$ is an exact sequence of abelian groups, then it follows that $\mathrm{Im}(g) \cong M/\mathrm{Ker}(g)$, and therefore

$$|M| = |\mathrm{Ker}(g)|\,|\mathrm{Im}(g)| = |\mathrm{Im}(f)|\,|\mathrm{Im}(g)|.$$

In particular, if M' and M'' are finite, then M is finite, too.

If two of the Herbrand quotients $\mathsf{h}(G,A)$, $\mathsf{h}(G,B)$, $\mathsf{h}(G,C)$ are defined, then it follows from the above exact hexagon that the third is also defined, and

$$\begin{aligned}&|\mathrm{Im}(\delta^{-1})|\,|\mathrm{Im}(H^0(\varphi)|\,|\mathrm{Im}(H^0(\psi))|\,|\mathrm{Im}(\Phi_A\circ\delta^0)|\,|\mathrm{Im}(H^{-1}(\varphi))|\,|\mathrm{Im}(H^{-1}(\psi)|\\ &\quad = |H^0(G,A)|\,|H^0(G,C)|\,|H^{-1}(G,B)|\\ &\quad = |H^0(G,B)|\,|H^{-1}(G,A)|\,|H^{-1}(G,C)|.\end{aligned}$$

Hence we obtain

$$\mathsf{h}(G,B) = \frac{|H^0(G,B)|}{|H^{-1}(G,B)|} = \frac{|H^0(G,A)|}{|H^{-1}(G,A)|}\,\frac{|H^0(G,C)|}{|H^{-1}(G,C)|} = \mathsf{h}(G,A)\mathsf{h}(G,C).$$

3. By definition, $\mathbb{Z}^G = \mathbb{Z}$, $N_G\mathbb{Z} = n\mathbb{Z}$ and ${}_{N_G}\mathbb{Z} = I_G\mathbb{Z} = \mathbf{0}$. Hence it follows that $H^0(G,\mathbb{Z}) = \mathbb{Z}/|G|\mathbb{Z}$, $H^{-1}(G,\mathbb{Z}) = \mathbf{0}$ and $\mathsf{h}(G,\mathbb{Z}) = n$.

If A is a finite G-module, then the exact sequences

$$\mathbf{0} \to A^G \to A \xrightarrow{\sigma-1} I_GA \to \mathbf{0} \quad \text{and} \quad \mathbf{0} \to {}_{N_G}A \to A \xrightarrow{N_G} N_GA \to \mathbf{0}$$

show that $|A| = |A^G|\,|I_GA| = |{}_{N_G}A|\,|N_GA|$, hence

$$|H^0(G,A)| = \frac{|A^G|}{|N_GA|} = \frac{|{}_{N_G}A|}{|I_GA|} = |H^{-1}(G,A)|, \quad \text{and} \quad \mathsf{h}(G,A) = 1. \qquad \square$$

2.5 Restriction, inflation, corestriction and transfer

Definitions and Remarks 2.5.1. Let G be a profinite group, H a closed subgroup of G and A a discrete G-module. Then A is also a discrete H-module. For $q \ge 0$ we define

$$\mathrm{Res}^q = \mathrm{Res}^q_{G\to H,A}\colon C^q(G,A) \to C^q(H,A) \quad \text{by} \quad \mathrm{Res}^q(x) = x \restriction H^q.$$

In particular, $\mathrm{Res}^0 = \mathrm{id}_A$ is the identity map of the discrete G-module A onto the discrete H-module A. It follows from the very definition that

$$\mathrm{Res}^q \circ d^{q-1} = d^{q-1} \circ \mathrm{Res}^{q-1}\colon C^{q-1}(G,A) \to C^q(H,A) \quad \text{for all } q \ge 1.$$

Hence $\mathrm{Res}^q(Z^q(G,A)) \subset Z^q(H,A)$, $\mathrm{Res}^q(B^q(G,A)) \subset B^q(H,A)$, and Res^q induces an (equally denoted) homomorphism

$$\mathrm{Res}^q = \mathrm{Res}^q_{G\to H,A}\colon H^q(G,A) \to H^q(H,A) \quad \text{for all } q \ge 1.$$

If G is finite, then $A^G \subset A^H$, and if R is a set of representatives for $H\backslash G$, then

$$N_G a = \sum_{\tau\in H}\sum_{\rho\in R} \tau\rho a = N_H\Big(\sum_{\rho\in R}\rho a\Big) \in N_H A \quad \text{for all}$$
$$a \in G, \quad \text{hence} \quad N_G A \subset N_H A,$$

and therefore Res^0 induces an (equally denoted) homomorphism

$$\mathrm{Res}^0\colon H^0(G,A) = A^G/N_GA \to A^H/N_HA = H^0(H,A)$$

satisfying $H^0(a+N_GA) = a+N_HA$ for all $a \in A^G$. In any case we call

$$\mathrm{Res}^q = \mathrm{Res}^q_{G\to H}H^q(G,A) \quad \text{the } \textbf{restriction} \text{ to } H.$$

If $K \subset H \subset G$ are closed subgroups, then $\mathrm{Res}^q_{G\to K} = \mathrm{Res}^q_{H\to K}\circ\mathrm{Res}^q_{G\to H}$.

Theorem 2.5.2. *Let G be a profinite group, H a closed subgroup of G and $q \ge 1$ (actually $q \ge 0$ if G is finite).*

1. Let $\varphi\colon A \to B$ be a G-homomorphism of discrete G-modules. Then the diagram

$$\begin{array}{ccc} H^q(G,A) & \xrightarrow{H^q(G,\varphi)} & H^q(G,B) \\ {\scriptstyle \mathrm{Res}^q}\downarrow & & \downarrow{\scriptstyle \mathrm{Res}^q} \\ H^q(H,A) & \xrightarrow{H^q(H,\varphi)} & H^q(H,B) \end{array}$$

is commutative.

2. Let $\mathbf{0} \to A \to B \to C \to \mathbf{0}$ be an exact sequence of discrete G-modules. Then the diagrams

$$\begin{array}{ccc} C^G & \xrightarrow{\delta} & H^1(G,A) \\ {\scriptstyle \mathrm{Res}^0}\downarrow & & \downarrow{\scriptstyle \mathrm{Res}^1} \\ C^H & \xrightarrow{\delta} & H^1(H,A) \end{array} \quad \text{and} \quad \begin{array}{ccc} H^q(G,C) & \xrightarrow{\delta^q} & H^{q+1}(G,A) \\ {\scriptstyle \mathrm{Res}^q}\downarrow & & \downarrow{\scriptstyle \mathrm{Res}^{q+1}} \\ H^q(H,C) & \xrightarrow{\delta^q} & H^{q+1}(H,A) \end{array}$$

are commutative.

Proof. Simple verification using defining cocycles. □

Definition and Remarks 2.5.3. Let G be a profinite group, H a closed normal subgroup of G, $\pi\colon G \to G/H$ the residue class epimorphism and A a discrete G-module. Then A^H is a discrete G/H-module through $\pi(g)a = ga$ for all $g \in G$ and $a \in A^H$, $(A^H)^{G/H)} = A^G$, and we set $\nu_A = (A^H \hookrightarrow A)$. For $q \geq 0$ we define

$$\mathrm{Inf}^q = \mathrm{Inf}^q_{G/H \to G}\colon C^q(G/H, A^H) \to C^q(G,A) \text{ by } \mathrm{Inf}^q(x) = \nu_A \circ x \circ \pi^q\colon G^q \to A.$$

In particular, $\mathrm{Inf}^0 = (A^G \hookrightarrow A)$ is the inclusion of the discrete G/H-module A^H into the discrete G-module A. It follows from the very definition that

$$d^{q-1} \circ \mathrm{Inf}^{q-1} = \mathrm{Inf}^q \circ d^{q-1}\colon C^{q-1}(G/H, A^H) \to C^q(G,A) \quad \text{for all } q \geq 1.$$

Hence $\mathrm{Inf}^q(Z^q(G/H, A^H)) \subset Z^q(G,A)$, $\mathrm{Inf}^q(B^q(G/H, A^H)) \subset B^q(G,A)$, and Inf^q induces an (equally denoted) homomorphism

$$\mathrm{Inf}^q = \mathrm{Inf}^q_{G/H \to G}\colon H^q(G/H, A^H) \;\to\; H^q(G,A) \quad \text{for all } q \geq 1,$$

called that **inflation** from G/H to G.
If $K \subset H \subset G$ are closed normal subgroups, then $G/H = (G/K)/(H/K)$ by 1.1.8.2, and $\mathrm{Inf}^q_{G/H \to G} = \mathrm{Inf}^q_{G/H \to G/K} \circ \mathrm{Inf}^q_{G/K \to G}$.

Now we can redeem the promise made after 2.3.4 to put the isomorphism

$$\varinjlim H^q(G/U, A^U) \stackrel{\sim}{\to} H^q(G,A)$$

into a structural context.

Let Σ be a fundamental system of neighbourhoods of 1 in G consisting of open normal subgroups. Then $(\Sigma, \supset)$ is a directed set, and if U, V, $W \in \Sigma$ are such that $U \subset V \subset W$, then

$$\begin{aligned} \mathrm{Inf}^q_{G/W \to G/U} &= \mathrm{Inf}^q_{G/V \to G/U} \circ \mathrm{Inf}^q_{G/V \to G/W}, \quad \mathrm{Inf}^q_{G/V \to G} \\ &= \mathrm{Inf}^q_{G/V \to G/U} \circ \mathrm{Inf}^q_{G/U \to G}, \end{aligned}$$

hence $(H^q(G/U, A^U),\ \mathrm{Inf}^q_{G/V \to G/U})_\Sigma$ is an inductive system, and by 2.3.4

$$\Phi = \varinjlim_{U \in \Sigma} \mathrm{Inf}^q_{G/U \to G}\colon \varinjlim_{U \in \Sigma} H^q(G/U, A^U) \stackrel{\sim}{\to} H^q(G,A)$$

is an isomorphism satisfying $\Phi([\xi_U, U]) = \mathrm{Inf}_{G/U \to G}(\xi_U)$ for all $U \in \Sigma$ and all $[\xi_U, U] \in \varinjlim H^q(G/U, A^U)$. We identify:

$$H^q(G,A) = \varinjlim_{\Sigma \in U} H^q(G/U, A^U).$$

With this identification, the maps $\mathrm{Inf}^q_{G/U \to G}\colon H^q(G/U, A^U) \to H^q(G,A)$ are the injections into the inductive limit.

Theorem 2.5.4. *Let G be a profinite group, H a closed normal subgroup of G and $q \geq 1$.*

1. Let $\varphi\colon A \to B$ be a G-homomorphism of discrete G-modules. Then the diagram

$$\begin{array}{ccc} H^q(G/H, A^H) & \xrightarrow{H^q(\varphi^H)} & H^q(G/H, B^H) \\ {\scriptstyle \mathrm{Inf}^q}\downarrow & & \downarrow{\scriptstyle \mathrm{Inf}^q} \\ H^q(G, A) & \xrightarrow{H^q(\varphi)} & H^q(G, B) \end{array}$$

is commutative.

2. Let $\mathbf{0} \to A \to B \xrightarrow{j} C \to \mathbf{0}$ be an exact sequence of discrete G-modules, and let $j^H : B^H \to C^H$ be surjective. Then the diagrams

$$\begin{array}{ccc} C^G & \xrightarrow{\overline{\delta}} & H^1(G/H, A^H) \\ \| & & \downarrow{\scriptstyle \mathrm{Inf}^1} \\ C^G & \xrightarrow{\delta} & H^1(G, A) \end{array} \quad \text{and} \quad \begin{array}{ccc} H^q(G/H, C^H) & \xrightarrow{\overline{\delta}^q} & H^{q+1}(G/H, A^H) \\ {\scriptstyle \mathrm{Inf}^q}\downarrow & & \downarrow{\scriptstyle \mathrm{Inf}^{q+1}} \\ H^q(G, C) & \xrightarrow{\delta^q} & H^{q+1}(G, A) \end{array}$$

for $q \geq 1$ are commutative (where $\overline{\delta}$ and $\overline{\delta}^q$ denote the connecting homomorphisms for the exact sequence $\mathbf{0} \to A^H \to B^H \to C^H \to \mathbf{0}$ of G/H-modules).

3. Let A be a discrete G-module. Then the sequence

$$\mathbf{0} \to H^1(G/H, A^H) \overset{\mathrm{Inf}^1}{\to} H^1(G, A) \overset{\mathrm{Res}^1}{\to} H^1(H, A)$$

is exact. If $q \geq 2$ and $H^i(H, A) = \mathbf{0}$ for all $i \in [1, q-1]$, then the sequence

$$\mathbf{0} \to H^q(G/H, A^H) \overset{\mathrm{Inf}^q}{\to} H^q(G, A) \overset{\mathrm{Res}^q}{\to} H^q(H, A)$$

is also exact.

Proof. Let $\pi\colon G \to G/H$ be the residue class epimorphism.

1. Observe that $\varphi \circ \nu_A = \nu_B \circ \varphi^H \colon A^H \to B$. If $x \in Z^q(G/H, A^H)$, then

$$\begin{aligned} \mathrm{Inf}^q \circ C^q(\varphi^H)(x) &= \mathrm{Inf}^q(\varphi^H \circ x) = \nu_B \circ \varphi^H \circ x \circ \pi^q = \varphi \circ \nu_A \circ x \circ \pi^q \\ &= C^q(\varphi)(\nu_A \circ x \circ \pi^q) = C^q(\varphi) \circ \mathrm{Inf}^q(x) \in Z^q(G, A), \end{aligned}$$

and therefore $\mathrm{Inf}^q \circ H^q(\varphi^H) = H^q(\varphi) \circ \mathrm{Inf}^q$.

2. We assume that $A \subset B$, which implies $A^H \subset B^H$, $\nu_B \restriction A^H = \nu_A$ and $j \circ \nu_B = \nu_C \circ j^H$.

Let first $c \in C^G$ and $b \in B^H$ be such that $j^H(b) = c$. Define $y\colon G \to A$ by $y(\sigma) = \sigma b - b$ and $\overline{y}\colon G/H \to A$ by $\overline{y}(\pi(\sigma)) = \pi(\sigma)b - b = \sigma b - b$. Then we get $\delta(c) = [y] \in H^1(G, A)$ and $\overline{\delta}(c) = [\overline{y}] \in H^1(G/H, A^H)$.

Since $\mathrm{Inf}^1 \circ \overline{\delta}(c) = [\nu_A \circ \overline{y} \circ \pi] \in H^1(G, A)$ and $\nu_A \circ \overline{y} \circ \pi(\sigma) = y(\sigma)$, it follows that $\mathrm{Inf}^1 \circ \overline{\delta}(c) = \delta(c) \in H^1(G, A)$.

Now let $q \geq 1$, $x \in Z^q(G/H, C^H)$ and let $x' \in C^q(G/H, B^H)$ be such that $x = j^H \circ x'$. Then $d^q x' \in Z^{q+1}(G/H, A^H)$, $\overline{\delta}^q([x]) = [d^q x'] \in H^{q+1}(G/H, A^H)$, and therefore $\mathrm{Inf}^{q+1} \circ \overline{\delta}^q([x]) = [\nu_A \circ d^q x' \circ \pi^{q+1}] \in H^{q+1}(G, A)$.

On the other hand, $\mathrm{Inf}^q(x) = \nu_C \circ x \circ \pi^q = \nu_C \circ j^H \circ x' \circ \pi^q = j \circ \nu_B \circ x' \circ \pi^q$, $d^q(\nu_B \circ x' \circ \pi^q) \in Z^{q+1}(G, A)$ and $\nu_B \restriction A = \nu_A$. Hence we obtain

$$\delta^q \circ \mathrm{Inf}^q([x]) = \delta^q([\mathrm{Inf}^q(x)]) = [d^q(\nu_B \circ x' \circ \pi^q)] = [\nu_A \circ d^q x' \circ \pi^{q+1}] \in H^{q+1}(G, A),$$

and therefore $\mathrm{Inf}^{q+1} \circ \overline{\delta}^q = \delta^q \circ \mathrm{Inf}^q \colon H^q(G/H, A^H) \to H^q(G, B)$.

3. **a.** Inf^1 is injective: Let $x \in Z^1(G/H, A^H)$be such that $[x] \in \mathrm{Ker}(\mathrm{Inf}^1)$. Then $\mathrm{Inf}^1(x) = \nu_A \circ x \circ \pi \in B^1(G, A)$, and consequently there exists some $a \in A$ such that $\nu_A \circ x \circ \pi(\sigma) = \sigma a - a$ for all $\sigma \in G$. If $\tau \in H$, then

$$\tau a - a = \nu_A \circ x \circ \pi(\tau) = x(1) = 0,$$

hence $a \in A^H$ and $x(\pi(\sigma)) = \sigma a - a = \pi(\sigma)a - a$ for all $\sigma \in G$. It follows that $x \in B^1(G/H, A^H)$, and therefore $[x] = 0$.

b. If $x \in Z^1(G/H, A^H)$, then

$$\mathrm{Res}^1 \circ \mathrm{Inf}^1(x) = \mathrm{Res}^1(\nu_A \circ x \circ \pi) = \nu_A \circ x \circ \pi \restriction H = x(1) = 0,$$

and therefore $\mathrm{Im}(\mathrm{Inf}^1) \subset \mathrm{Ker}(\mathrm{Res}^1)$.

As to the converse, let $x \in Z^1(G, A)$ be such that $[x] \in \mathrm{Ker}(\mathrm{Res}^1)$, that is, $x \restriction H \in B^1(H, A)$. Then there exists some $a \in A$ such that $x(\tau) = \tau a - a$ for all $\tau \in H$. Let $y\colon G \to A$ be defined by $y(\sigma) = x(\sigma) - (\sigma a - a)$. Then $y \in x + B^1(G, A) \subset Z^1(G, A)$, and $y(\tau) = 0$ for all $\tau \in H$. If $\sigma \in G$ and $\tau \in H$, then $y(\tau\sigma) = \tau y(\sigma) + y(\tau) = \tau y(\sigma)$ and $y(\sigma\tau) = \sigma y(\tau) + y(\sigma) = y(\sigma)$. Hence there exists a continuous map $\overline{y}\colon G/H \to A$ such that $\overline{y} \circ \pi = y$. If $\tau \in H$ and $\sigma \in G$, then

$$\tau \overline{y}(\pi(\sigma)) = \tau y(\sigma) = y(\tau\sigma) = \overline{y}(\pi(\tau\sigma)) = \overline{y}(\pi(\sigma)),$$

and therefore $\overline{y}(\pi(\sigma)) \in A^H$. For all $\sigma, \tau \in G$ we obtain

$$\overline{y}(\pi(\sigma)\pi(\tau)) = \overline{y}(\pi(\sigma\tau)) = y(\sigma\tau) = \sigma y(\tau) + y(\sigma) = \pi(\sigma)\overline{y}(\pi(\tau)) + \overline{y}(\pi(\sigma)),$$

hence $\overline{y} \in Z^1(G/H, A^H)$ and $[\overline{y}] \in H^1(G/H, A^H)$.

Since $\mathrm{Inf}^1(\overline{y}) = \nu_A \circ \overline{y} \circ \pi = y$, it follows that $\mathrm{Inf}^1([\overline{y}]) = [y] = [x] \in H^1(G, A)$ and thus $\mathrm{Ker}(\mathrm{Res}^1) \subset \mathrm{Im}(\mathrm{Inf}^1)$.

c. Now assume that $q \geq 2$, $H^i(H, A) = \mathbf{0}$ for all $i \in [1, q-1]$ and proceed by induction on q. The exact sequence $\mathbf{0} \to A \to \mathcal{M}_G(A) \to A_1 \to \mathbf{0}$ (see 2.1.4.**3**) yields by 2.4.1 the exact sequence

$$\mathbf{0} \to A^H \to \mathcal{M}_G(A)^H \to A_1^H \to H^1(H, A) = \mathbf{0}.$$

By 2.1.8, it follows that $\mathcal{M}_G(A)$ is an induced H-module, and $\mathcal{M}_G(A)^H$ is an induced G/H-module. By 2.4.1 and 2.2.4, we obtain the exact sequences

$$\begin{aligned} \mathbf{0} = H^{q-1}(H, \mathcal{M}_G(A)) &\to H^{q-1}(H, A_1) \overset{\delta'^{q-1}}{\to} H^q(H, A) \\ &\to H^q(H, \mathcal{M}_G(A)) = \mathbf{0}, \\ \mathbf{0} = H^{q-1}(G, \mathcal{M}_G(A)) &\to H^{q-1}(G, A_1) \overset{\delta^{q-1}}{\to} H^q(G, A) \\ &\to H^q(G, \mathcal{M}_G(A)) = \mathbf{0} \end{aligned}$$

and

$$\begin{aligned} \mathbf{0} = H^{q-1}(G/H, M^H) \to H^{q-1}(G/H, C^H) \overset{\overline{\delta}^{q-1}}{\to} & H^q(G/H, A^H) \\ & \to H^q(G/H, \mathcal{M}_G(A)^H) = \mathbf{0}, \end{aligned}$$

which show that the connecting homomorphisms δ^{q-1}, δ'^{q-1} and $\overline{\delta}^{q-1}$ are isomorphisms. Together with 1. and 2. we obtain the following commutative diagram in which the top row is exact by the induction hypothesis, and the vertical arrows are isomorphisms:

$$\begin{array}{ccccccc} \mathbf{0} \longrightarrow & H^{q-1}(G/H, C^H) & \xrightarrow{\operatorname{Inf}^{q-1}} & H^{q-1}(G, C) & \xrightarrow{\operatorname{Res}^{q-1}} & H^{q-1}(H, C) & \\ & \overline{\delta}^{q-1}\big\downarrow & & \delta^{q-1}\big\downarrow & & \big\downarrow \delta'^{q-1} & . \\ \mathbf{0} \longrightarrow & H^q(G/H, A^H) & \xrightarrow{\operatorname{Inf}^q} & H^q(G, A) & \xrightarrow{\operatorname{Res}^q} & H^q(H, A) & \end{array}$$

Hence the bottom row is exact, too. □

Definitions and Theorem 2.5.5. *Let G be a profinite group, A a discrete G-module, H an open subgroup of G and R a set of representatives for $H\backslash G$ in G,*

$$G = \biguplus_{\rho \in R} H\rho.$$

For $\sigma \in G$ and $\rho \in R$ we set $\rho\sigma = \sigma_\rho \rho_\sigma$ (and consequently $\rho^{-1}\sigma_\rho = \sigma\rho_\sigma^{-1}$), where $\rho_\sigma \in R$ and $\sigma_\rho \in H$.

1. *(a) If $\sigma \in G$, then $\{\rho_\sigma \mid \rho \in R\} = R$.*
   (b) If $\sigma, \tau \in G$ and $\rho \in R$, then $\rho_{\sigma\tau} = (\rho_\sigma)_\tau$ and $(\sigma\tau)_\rho = \sigma_\rho \tau_{\rho_\sigma}$.
   (c) For every $\rho \in R$ the maps

$$\begin{cases} G \to R \\ \sigma \mapsto \rho_\sigma \end{cases} \quad \textit{and} \quad \begin{cases} G \to H \\ \sigma \mapsto \sigma_\rho \end{cases}$$

are continuous.

2. For $a \in A^H$ we define

$$\operatorname{Cor}^0(a) = \operatorname{Cor}^0_A(a) = \operatorname{Cor}^0_{H\to G}(a) = \sum_{\rho \in R} \rho^{-1} a.$$

(a) $\operatorname{Cor}^0 \colon A^H \to A^G$ *is a group homomorphism which does not depend on* R.

(b) Let $\varphi \colon A \to B$ *be a* G*-homomorphism of discrete* G*-modules. Then*

$$\operatorname{Cor}^0_B \circ \varphi^H = \varphi^G \circ \operatorname{Cor}^0_A \colon A^H \to B^G.$$

(c) If $a \in A^G$*, then* $\operatorname{Cor}^0 \circ \operatorname{Res}^0(a) = (G\!:\!H)a$.

(d) Let K *be an open subgroup of* G *such that* $K \subset H$*. Then*

$$\operatorname{Cor}^0_{K\to G} = \operatorname{Cor}^0_{H\to G} \circ \operatorname{Cor}^0_{K\to H} \colon A^K \to A^G.$$

3. For $q \in \{1,2\}$ we define

$$\operatorname{Cor}^q = \operatorname{Cor}^q_{H\to G,A} \colon C^q(H,A) \to C^q(G,A)$$

as follows.

- For $x \in C^1(H,A)$ and $\sigma \in G$ we set

$$\operatorname{Cor}^1(x)(\sigma) = \sum_{\rho \in R} \rho^{-1} x(\sigma_\rho).$$

- For $x \in C^2(H,A)$ and $\sigma, \tau \in G$ we set

$$\operatorname{Cor}^2(x)(\sigma,\tau) = \sum_{\rho \in R} \rho^{-1} x(\sigma_\rho, \tau_{\rho\sigma}).$$

The following assertions hold for $q \in \{1,2\}$:

(a) $\operatorname{Cor}^q \colon C^q(H,A) \to C^q(G,A)$ *is a group homomorphism satisfying*

$$\operatorname{Cor}^q(Z^q(H,A)) \subset Z^q(G,A) \quad \textit{and} \quad \operatorname{Cor}^q(B^q(H,A)) \subset B^q(G,A).$$

Hence Cor^q induces an (equally denoted) homomorphism

$$\operatorname{Cor}^q = \operatorname{Cor}^q_A = \operatorname{Cor}^q_{H\to G} \colon H^q(H,A) \to H^q(G,A),$$

called the **corestriction** from H to G.

(b) *Let $\varphi\colon A \to B$ be a G-homomorphism of discrete G-modules. Then we have the commutative diagram*

$$\begin{array}{ccc} H^q(H,A) & \xrightarrow{H^q(H,\varphi)} & H^q(H,B) \\ {\scriptstyle \mathrm{Cor}^q_A}\downarrow & & \downarrow{\scriptstyle \mathrm{Cor}^q_B} \\ H^q(G,A) & \xrightarrow{H^q(G,\varphi)} & H^q(G,B) \end{array} .$$

(c) *Let $\mathbf{0} \to A \stackrel{\varphi}{\to} B \stackrel{\psi}{\to} C \to \mathbf{0}$ be an exact sequence of discrete G-modules. Then the diagrams (with self-expanatory horizontal rows)*

$$\begin{array}{ccc} C^H & \xrightarrow{\delta_H} & H^1(H,A) \\ {\scriptstyle \mathrm{Cor}^0_C}\downarrow & & \downarrow{\scriptstyle \mathrm{Cor}^1_A} \\ C^G & \xrightarrow{\delta_G} & H^1(G,A) \end{array} \quad \text{and} \quad \begin{array}{ccc} H^1(H,C) & \xrightarrow{\delta^1_H} & H^2(H,A) \\ {\scriptstyle \mathrm{Cor}^1_C}\downarrow & & \downarrow{\scriptstyle \mathrm{Cor}^2_A} \\ H^1(G,C) & \xrightarrow{\delta^1_G} & H^2(G,A) \end{array}$$

are commutative.

(d) *The homomorphism $\mathrm{Cor}^q\colon H^q(H,A) \to H^q(G,A)$ does not depend on the set of representatives R.*

(e) *If $\xi \in H^q(G,A)$, then $\mathrm{Cor}^q_{H\to G}\circ\mathrm{Res}^q_{G\to H}(\xi) = (G\colon H)\xi$.*

(f) *Let K be another open subgroup of G such that $K \subset H$. Then*

$$\mathrm{Cor}^q_{K\to G} = \mathrm{Cor}^q_{H\to G}\circ\mathrm{Cor}^q_{K\to H}\colon H^q(K,A) \to H^q(G,A).$$

Proof. 1. (a) If $\sigma \in G$, then

$$G = \biguplus_{\rho\in R} H\rho = \biguplus_{\rho\in R} H\rho\sigma = \biguplus_{\rho\in R} H\sigma_\rho\rho_\sigma = \biguplus_{\rho\in R} H\rho_\sigma,$$

and therefore $R = \{\rho_\sigma \mid \rho \in R\}$.

(b) Let $\sigma,\,\tau \in G$. Then on the one hand $\rho\sigma\tau = \sigma_\rho\rho_\sigma\tau = \sigma_\rho\tau_{\rho_\sigma}(\rho_\sigma)_\tau$, and on the other hand $\rho\sigma\tau = (\sigma\tau)_\rho\rho_{\sigma\tau}$. Hence $\rho_{\sigma\tau} = (\rho_\sigma)_\tau$ and $(\sigma\tau)_\rho = \sigma_\rho\tau_{\rho_\sigma}$.

(c) Let $\rho \in R$. If $\sigma,\,\sigma' \in G$, then $\rho_\sigma = \rho_{\sigma'}$ if and only if $H\rho\sigma = H\rho\sigma'$, and this holds if and only if $\sigma' \in \rho^{-1}H\rho\sigma$. Since $\rho^{-1}H\rho$ is an open subgroup of G and ρ_σ depends only on the coset $\rho^{-1}H\rho\sigma$, the map $\sigma \mapsto \rho_\sigma$ is continuous by 1.5.2.2. Since $\sigma_\rho = \rho\sigma\rho_\sigma^{-1}$, the map $\sigma \mapsto \sigma_\rho$ is continuous, too.

2. (a) Let R' be another set of representatives for $H\backslash G$ in G. We may assume that $R' = \{c_\rho\rho \mid \rho \in R\}$, where $c_\rho \in H$, and then

$$\sum_{\rho'\in R'} \rho'^{-1}a = \sum_{\rho\in R} \rho^{-1}c_\rho^{-1}a = \sum_{\rho\in R} \rho^{-1}a \quad \text{for all } a \in A^H.$$

If $\sigma \in G$ and $a \in A^H$, then, using 1.(a),

$$\sigma\,\mathrm{Cor}^0(a) = \sum_{\rho\in R} \sigma\rho^{-1}a = \sum_{\rho\in R} \sigma\rho_\sigma^{-1}a = \sum_{\rho\in R} \rho^{-1}\sigma_\rho a = \sum_{\rho\in R} \rho^{-1}a = \mathrm{Cor}^0(a).$$

Hence $\mathrm{Cor}^0(A^H) \subset A^G$, and obviously $\mathrm{Cor}^0\colon A^H \to A^G$ is a homomorphism.

(b) If $a \in A^H$, then

$$\mathrm{Cor}^0_B \circ \varphi^H(a) = \sum_{\rho\in R} \rho^{-1}\varphi(a) = \varphi\Big(\sum_{\rho\in R} \rho^{-1}a\Big) = \varphi^G \circ \mathrm{Cor}^0_A(a).$$

(c) If $a \in A^G$, then

$$\mathrm{Cor}^0 \circ \mathrm{Res}^0(a) = \sum_{\rho\in R} \rho^{-1}a = |R|a = (G\!:\!H)a.$$

(d) Let S be a set of representatives for $K\backslash H$ in H. Then

$$G = \biguplus_{\rho\in R} H\rho = \biguplus_{\rho\in R}\biguplus_{\sigma\in S} K\sigma\rho,$$

and therefore SR is a set of representatives for $K\backslash G$ in G. If $a \in A^K$, then

$$\mathrm{Cor}^0_{K\to G}(a) = \sum_{\theta\in S}\sum_{\rho\in R} (\theta\rho)^{-1}a = \sum_{\rho\in R} \rho^{-1}\sum_{\theta\in S} \theta^{-1}a = \mathrm{Cor}^0_{H\to G}\circ\mathrm{Cor}^0_{K\to H}(a).$$

3. (a) If $x \in C^q(H,A)$, then $\mathrm{Cor}^q(x) \in C^q(G,A)$, since $\mathrm{Cor}^q(x)$ is continuous by 1.(c), and apparently $\mathrm{Cor}^q\colon C^q(H,A) \to C^q(G,A)$ is a group homomorphism. In the sequel we tacitly use 1. again and again.

If $x \in Z^1(H,A)$ and $\sigma,\,\tau \in G$, then

$$\begin{aligned}
\mathrm{Cor}^1(x)(\sigma\tau) &= \sum_{\rho\in R} \rho^{-1}x((\sigma\tau)_\rho) = \sum_{\rho\in R} \rho^{-1}x(\sigma_\rho\tau_{\rho_\sigma}) \\
&= \sum_{\rho\in R} \rho^{-1}(\sigma_\rho x(\tau_{\rho_\sigma}) + x(\sigma_\rho)) = \sum_{\rho\in R} \rho^{-1}\sigma_\rho x(\tau_{\rho_\sigma}) + \sum_{\rho\in R} \rho^{-1}x(\sigma_\rho) \\
&= \sum_{\rho\in R} \sigma\rho_\sigma^{-1}x(\tau_{\rho_\sigma}) + \mathrm{Cor}^1(x)(\sigma) = \sum_{\rho\in R} \sigma\rho^{-1}x(\tau_\rho) + \mathrm{Cor}^1(x)(\sigma) \\
&= \sigma\,\mathrm{Cor}^1(\tau) + \mathrm{Cor}^1(x)(\sigma),
\end{aligned}$$

which shows that $\mathrm{Cor}^1(x) \in Z^1(G,A)$.

Now let $x \in B^1(H,A)$ and $a \in A$ such that $x(\tau) = (\tau-1)a$ for all $\tau \in H$. If $\sigma \in G$, then

$$\begin{aligned}
\mathrm{Cor}^1(x)(\sigma) &= \sum_{\rho\in R} \rho^{-1}x(\sigma_\rho) = \sum_{\rho\in R} \rho^{-1}(\sigma_\rho - 1)a = \sum_{\rho\in R} \sigma\rho_\sigma^{-1}a - \sum_{\rho\in R} \rho^{-1}a \\
&= (\sigma-1)\sum_{\rho\in R} \rho^{-1}a,
\end{aligned}$$

and therefore $\mathrm{Cor}^1(x) \in B^1(G, A)$.

If $x \in Z^2(H, A)$ and ν, σ, $\tau \in G$, then

$$\begin{aligned}
\mathrm{Cor}^2(x)(\nu,\sigma) + \mathrm{Cor}^2(x)(\nu\sigma,\tau) &= \sum_{\rho\in R} \rho^{-1}[x(\nu_\rho, \sigma_{\rho_\nu}) + x((\nu\sigma)_\rho, \tau_{\rho_{\nu\sigma}})] \\
&= \sum_{\rho\in R} \rho^{-1}[x(\nu_\rho, \sigma_{\rho_\nu}) + x(\nu_\rho\sigma_{\rho_\nu}, \tau_{\rho_{\nu\sigma}})], \\
\nu\,\mathrm{Cor}^2(x)(\sigma,\tau) &= \sum_{\rho\in R} \nu\rho^{-1}x(\sigma_\rho, \tau_{\rho_\sigma}) = \sum_{\rho\in R} \nu\rho_\nu^{-1}x(\sigma_{\rho_\nu}, \tau_{\rho_{\nu\sigma}}) \\
&= \sum_{\rho\in R} \rho^{-1}\nu_\rho x(\sigma_{\rho_\nu}, \tau_{\rho_{\nu\sigma}})
\end{aligned}$$

and

$$\begin{aligned}
\mathrm{Cor}^2(\nu,\sigma\tau) &= \sum_{\rho\in R} \rho^{-1}x(\nu_\rho, (\sigma\tau)_{\rho_\nu}) = \sum_{\rho\in R} \rho^{-1}x(\nu_\rho, \sigma_{\rho_\nu}\tau_{\rho_{\nu\sigma}}) \\
&= \sum_{\rho\in R} \rho^{-1}[-\nu_\rho x(\rho_{\sigma_\nu}\tau_{\rho_{\nu\sigma}}) + x(\nu_\rho, \sigma_{\rho_\nu}) + x(\nu_\rho\sigma_{\rho_\nu}, \tau_{\rho_{\nu\sigma}})].
\end{aligned}$$

Hence $\mathrm{Cor}^2(x)(\nu,\sigma) + \mathrm{Cor}^2(x)(\nu\sigma,\tau) = \nu\,\mathrm{Cor}^2(x)(\sigma,\tau) + \mathrm{Cor}^2(x)(\nu,\sigma\tau)$, and therefore $\mathrm{Cor}^2(x) \in Z^2(G, A)$.

Now let $x \in B^2(H, A)$, and let $y \in C^1(H, A)$ such that $x = d^1y$. If σ, $\tau \in G$, then $x(\sigma,\tau) = y(\sigma\tau) - \sigma y(\tau) - y(\sigma)$, and

$$\begin{aligned}
\mathrm{Cor}^2(x)(\sigma,\tau) &= \sum_{\rho\in R} \rho^{-1}[y(\sigma_\rho\tau_{\rho_\sigma}) - \sigma_\rho y(\tau_{\rho_\sigma}) - y(\sigma_\rho)] \\
&= \sum_{\rho\in R} \rho^{-1}y((\sigma\tau)_\rho) - \sum_{\rho\in R} \sigma\rho_\sigma^{-1}y(\tau_{\rho_\sigma}) - \sum_{\rho\in R} \rho^{-1}y(\sigma_\rho) \\
&= A(\sigma\tau) - \sigma A(\tau) - A(\sigma) = d^1A(\sigma,\tau), \\
&\quad \text{where} \quad A(\sigma) = \sum_{\rho\in R} \rho^{-1}y(\sigma_\rho).
\end{aligned}$$

Hence it follows that $\mathrm{Cor}^2(x) \in B^2(G, A)$.

(b) $q = 1$: If $x \in C^1(H, A)$ and $\sigma \in G$, then

$$\begin{aligned}
\mathrm{Cor}_B^1 \circ C^1(H,\varphi)(x)(\sigma) &= \mathrm{Cor}_B^1(\varphi\circ x)(\sigma) = \sum_{\rho\in R} \rho^{-1}(\varphi\circ x)(\sigma_\rho) \\
&= \varphi\Big(\sum_{\rho\in R} \rho^{-1}x(\sigma_\rho)\Big) = C^1(G,\varphi)\circ\mathrm{Cor}_A^1(x)(\sigma),
\end{aligned}$$

and consequently $\mathrm{Cor}^1 \circ H^1(H,\varphi) = H^1(G,\varphi)\circ\mathrm{Cor}^1 \colon H^1(H, A) \to H^1(G, B)$.

$q = 2$: If $x \in C^2(H, A)$ and $\sigma,\ \tau \in G$, then

$$\mathrm{Cor}^2_B \circ C^2(H, \varphi)(x)(\sigma, \tau) = \mathrm{Cor}^2(\varphi \circ x)(\sigma, \tau) = \sum_{\rho \in R} \rho^{-1}(\varphi \circ x)(\sigma_\rho, \tau_{\rho_\sigma})$$
$$= \varphi\Big(\sum_{\rho \in R} \rho^{-1} x(\sigma_\rho, \tau_{\rho_\sigma})\Big) = C^2(G, \varphi) \circ \mathrm{Cor}^2_A(x)(\sigma, \tau),$$

and consequently $\mathrm{Cor}^2 \circ H^2(H, \varphi) = H^2(G, \varphi) \circ \mathrm{Cor}^2 \colon H^2(H, A) \to H^2(G, B)$.

(c) We may assume that $A \subset B$ and $\varphi = (A \hookrightarrow B)$.

$q = 1$: Let $c \in C^H,\ \ b \in B$ be such that $c = \psi(b)$ and $y = d^0 b$. Then $y \in Z^1(H, A),\ \ \delta_H(c) = [y] \in H^1(H, A)$ and

$$\mathrm{Cor}^1_A \circ \delta_H(c) = [\mathrm{Cor}^1_A(y)] \in H^1(G, A).$$

If $\sigma \in G$, then

$$\mathrm{Cor}^1_A(y)(\sigma) = \sum_{\rho \in R} \rho^{-1} y(\sigma_\rho) = \sum_{\rho \in R} \rho^{-1}(\sigma_\rho - 1)b = \sum_{\rho \in R} \sigma \rho_\sigma^{-1} b - \sum_{\rho \in R} \rho^{-1} b$$
$$= (\sigma - 1) \sum_{\rho \in R} \rho^{-1} b.$$

On the other hand,

$$\mathrm{Cor}^0_C(c) = \sum_{\rho \in R} \rho^{-1} c = \psi(b^*), \quad \text{where} \quad b^* = \sum_{\rho \in R} \rho^{-1} b.$$

If $y^* = d^0 b^*$, then $\delta_G \circ \mathrm{Cor}^0_C(c) = [y^*] \in H^1(G, A)$, and

$$y^*(\sigma) = (\sigma - 1) \sum_{\rho \in R} \rho^{-1} b = \mathrm{Cor}^1(y)(\sigma) \quad \text{for all} \quad \sigma \in G.$$

Consequently, $\mathrm{Cor}^1_A \circ \delta = \delta \circ \mathrm{Cor}^0_C \colon C^H \to H^1(G, A)$.

$q = 2$: Let $x \in Z^1(H, C),\ \ x' \in C^1(H, B)$ such that $\psi \circ x' = x$ and $y = d^1 x'$. Then $y \in Z^2(H, A)$,

$$\delta^1_H([x]) = [y] \in H^2(H, A) \text{ and } \mathrm{Cor}^2_A \circ \delta^1_H([x]) = [\mathrm{Cor}^2_A(y)] \in H^2(G, A).$$

If $\sigma,\ \tau \in G$, then

$$\mathrm{Cor}^2_A(y)(\sigma, \tau) = \sum_{\rho \in R} \rho^{-1} y(\sigma_\rho, \tau_{\rho_\sigma}) = \sum_{\rho \in R} \rho^{-1}[x'(\sigma_\rho \tau_{\rho_\sigma}) - \sigma_\rho x'(\tau_{\rho_\sigma}) - x'(\sigma_\rho)]$$
$$= x^*(\sigma\tau) - \sigma x^*(\tau) - x^*(\sigma) = d^1 x^*(\sigma, \tau),$$

where $x^*(\sigma) = \sum_{\rho \in R} \rho^{-1} x'(\sigma)$.

On the other hand, if $\sigma \in G$, then

$$\mathrm{Cor}^1_C(x)(\sigma) = \sum_{\rho \in R} \rho^{-1} x(\sigma_\rho) = \psi\Big(\sum_{\rho \in R} \rho^{-1} x'(\sigma_\rho)\Big) = \psi \circ x^*(\sigma),$$

and thus we obtain $\delta^1_G \circ \mathrm{Cor}^1_C([x]) = [d^1 x^*] = \mathrm{Cor}^2_A \circ \delta^1_H([x]) \in H^2(G, A)$.

(d) We consider the exact sequence $\mathbf{0} \to A \to \mathcal{M}_G(A) \to A_1 \to \mathbf{0}$ (see 2.1.4.**3**). Using (c), 2.4.1, 2.2.4 and 2.1.8.1, we obtain the commutative diagrams with exact rows

$$\begin{array}{ccccccc}
\mathcal{M}_G(A)^H & \longrightarrow & A_1^H & \xrightarrow{\delta^0} & H^1(H, A) & \longrightarrow & \mathbf{0} \\
\Big\downarrow {\scriptstyle \mathrm{Cor}^0_{\mathcal{M}_G(A)}} & & \Big\downarrow {\scriptstyle \mathrm{Cor}^0_{A_1}} & & \Big\downarrow {\scriptstyle \mathrm{Cor}^1_A} & & \\
\mathcal{M}_G(A)^G & \longrightarrow & A_1^G & \xrightarrow{\delta^0_G} & H^1(G, A) & \longrightarrow & \mathbf{0}
\end{array}$$

and

$$\begin{array}{ccccccc}
\mathbf{0} & \longrightarrow & H^1(H, A_1) & \xrightarrow{\delta^1} & H^2(H, A) & \longrightarrow & \mathbf{0} \\
 & & \Big\downarrow {\scriptstyle \mathrm{Cor}^1_{A_1}} & & \Big\downarrow {\scriptstyle \mathrm{Cor}^2_A} & & \\
\mathbf{0} & \longrightarrow & H^1(G, A_1) & \xrightarrow{\delta^1} & H^2(G, A) & \longrightarrow & \mathbf{0}
\end{array} .$$

By 2.(a), $\mathrm{Cor}^0_\bullet$ does not depend on R. Now the first diagram shows that $\mathrm{Cor}^1_\bullet$, and the second diagram shows that also $\mathrm{Cor}^2_\bullet$ does not depend on R.

(e) For a discrete G-module B and $q \in \{0, 1, 2\}$ we set

$$\theta^q_B = \mathrm{Cor}^q_{H \to G} \circ \mathrm{Res}^q_{G \to H} \colon H^q(G, B) \to H^q(G, B).$$

Again we use the exact sequence $\mathbf{0} \to A \to \mathcal{M}_G(A) \to A_1 \to \mathbf{0}$, and for $q \in \{0, 1, 2\}$ and a discrete G-module B we set

$$\theta^q_B = \mathrm{Cor}_{H \to G} \circ \mathrm{Res}_{G \to H} \colon H^q(G, B) \to H^q(G, B).$$

As above, and using 2.5.2.2, we obtain the commutative diagrams with exact rows

$$\begin{array}{ccccccc}
\mathcal{M}_G(A)^G & \longrightarrow & A_1^G & \xrightarrow{\delta^0} & H^1(G, A) & \longrightarrow & \mathbf{0} \\
\Big\downarrow {\scriptstyle \theta^0_{\mathcal{M}_G(A)}} & & \Big\downarrow {\scriptstyle \theta^0_{A_1}} & & \Big\downarrow {\scriptstyle \theta^1_A} & & \\
\mathcal{M}_G(A)^G & \longrightarrow & A_1^G & \xrightarrow{\delta^0_G} & H^1(G, A) & \longrightarrow & \mathbf{0}
\end{array}$$

and

$$\begin{array}{ccccccc}
\mathbf{0} & \longrightarrow & H^1(G, A_1) & \xrightarrow{\delta^1_H} & H^2(G, A) & \longrightarrow & \mathbf{0} \\
 & & \Big\downarrow {\scriptstyle \theta^1_{A_1}} & & \Big\downarrow {\scriptstyle \theta^2_A} & & \\
\mathbf{0} & \longrightarrow & H^1(G, A_1) & \xrightarrow{\delta^1_G} & H^2(G, A) & \longrightarrow & \mathbf{0}
\end{array} .$$

By 2.(c), $\theta^0_\bullet$ is multiplication with $(G\!:\!H)$. Now the first diagram shows that $\mathrm{Cor}^1_\bullet$ is multiplication with $(G:H)$, and the second diagram shows that the same is true for $\mathrm{Cor}^2_\bullet$.

(f) For a discrete G-module B and $q \in \{0,1,2\}$ we set

$$\theta^q_B = \mathrm{Cor}^q_{K\to G} - \mathrm{Cor}^q_{H\to G}\circ\mathrm{Cor}^q_{G\to H}\colon H^q(K,B) \to H^q(G,B).$$

Then $\theta^0_\bullet = 0$ by 2.(d), and we proceed as in (d) and (e) above to prove that $\theta^1_\bullet = \theta^2_\bullet = 0$. □

Using the method of dimension shifting (which we did not develop in this volume) it is possible to define maps $\mathrm{Cor}^q_\bullet$ for all $q \in \mathbb{N}$.

Theorem 2.5.6. *Let G be a finite group. $n = |G|$ and A a G-module. Then*

$$nH^q(G,A) = 0 \quad \textit{for all} \;\; q \in [-1,2].$$

Proof. If $q = -1$, then $H^{-1}(G,A) = {}_{N_G}A/I_GA$. If $a \in {}_{N_G}A$, then

$$\sum_{\sigma\in G} \sigma a = 0, \quad \text{and therefore} \quad na = -\sum_{\sigma\in G}(\sigma-1)a \in I_GA.$$

Hence $n(a + I_GA) = 0 \in H^{-1}(G,A)$.

If $q \in \{0,1,2\}$, then $\mathrm{Res}^q_{G\to\mathbf{1}} = 0$, and therefore it follows that

$$n\xi = \mathrm{Cor}^q_{\mathbf{1}\to G}\circ\mathrm{Res}^q_{G\to\mathbf{1}}(\xi) = 0 \;\text{ for all } \xi \in H^q(G,A)$$

by 2.5.5.[2(c) and 3(e)]. □

Definition and Theorem 2.5.7. *Let G be a profinite group and H an open subgroup of G.*

We consider the homomorphism $\mathrm{Cor}^1_{H\to G}\colon H^1(H,\mathbb{Q}/\mathbb{Z}) \to H^1(G,\mathbb{Q}/\mathbb{Z})$, built with the trivial G-module $\mathbb{Q}/\mathbb{Z}$. Recall from 2.2.5 and 1.6 that

$$H^1(G,\mathbb{Q}/\mathbb{Z}) = \mathrm{Hom}^{\mathsf{c}}(G,\mathbb{Q}/\mathbb{Z}) = \mathsf{X}(G) = \mathsf{X}(G^{\mathsf{a}}),$$

where $G^{(1)} = \overline{[G,G]}$ and $G^{\mathsf{a}} = G/G^{(1)}$. By Pontrjagin's duality theorem 1.6.6 we identify $\mathsf{X}^2(G^{\mathsf{a}}) = G^{\mathsf{a}}$ by means of $a(\chi) = \chi(a)$ for all $a \in G^{\mathsf{a}}$ and $\chi \in \mathsf{X}(G)$. Alike, $\mathsf{X}^2(H^{\mathsf{a}}) = H^{\mathsf{a}}$, and allowing all these identifications, the homomorphism

$$\mathrm{Cor}^1_{H\to G}\colon \mathsf{X}(H) = H^1(H,\mathbb{Q}/\mathbb{Z}) \to H^1(G,\mathbb{Q}/\mathbb{Z}) = \mathsf{X}(G).$$

induces the homomorphism

$$\mathsf{V}_{G\to H} = \mathsf{X}(\mathrm{Cor}^1_{H\to G})\colon G^{\mathsf{a}} \to H^{\mathsf{a}},$$

which is called the **transfer** or **Verlagerung** from G to H. Recall from 1.7.9 that we have proved Pontrjagin's duality theorem for all abelian profinite groups.

1. Let R be a set of representatives for $H\backslash G$ in G. For $\sigma \in G$ and $\rho \in R$ we set $\rho\sigma = \sigma_\rho \rho_\sigma$, where $\rho_\sigma \in R$ and $\sigma_\rho \in H$. Then

$$\mathsf{V}_{G\to H}(\sigma G^{(1)}) = \prod_{\rho\in R} \sigma_\rho H^{(1)} \in H^{\mathsf{a}} \quad \textit{for all} \ \ \sigma \in G.$$

2. Let K be an open subgroup of G such that $K \subset H$. Then

$$\mathsf{V}_{G\to K} = \mathsf{V}_{H\to K} \circ \mathsf{V}_{G\to H}.$$

3. If H is a normal subgroup of G, then

$$\mathsf{V}_{G\to H}(\sigma G^{(1)}) = \prod_{\rho\in R} \rho\sigma\rho^{-1} H^{(1)} \quad \textit{for all} \ \ \sigma \in H.$$

Proof. 1. If $\sigma \in G$, then

$$\begin{aligned}\chi\circ\mathsf{V}_{G\to H}(\sigma G^{(1)})) &= \mathsf{X}(\mathsf{V}_{G\to H})(\chi)(\sigma G^{(1)}) = \mathrm{Cor}^1_{H\to G}(\chi)(\sigma) = \prod_{\rho\in R}\chi(\sigma_\rho) \\ &= \chi\Big(\prod_{\rho\in R}\sigma_\rho H^{(1)}\Big) \quad \text{for all} \ \ \chi \in \mathsf{X}(H^{\mathsf{a}}) = \mathsf{X}(H),\end{aligned}$$

and consequently (by Pontrjagin's duality theorem)

$$\mathsf{V}_{G\to H}(\sigma G^{(1)}) = \prod_{\rho\in R} \sigma_\rho H^{(1)}.$$

2. By 2.5.5.2(d), we obtain

$$\begin{aligned}\mathsf{X}(\mathsf{V}_{H\to K}\circ\mathsf{V}_{G\to H}) &= \mathsf{X}(\mathsf{V}_{G\to H})\circ\mathsf{X}(\mathsf{V}_{H\to K}) = \mathrm{Cor}^1_{H\to G}\circ\mathrm{Cor}^1_{K\to H} \\ &= \mathrm{Cor}^1_{K\to G} = \mathsf{X}(\mathsf{V}_{G\to K}),\end{aligned}$$

and consequently $\mathsf{V}_{G\to K} = \mathsf{V}_{H\to K}\circ\mathsf{V}_{G\to H}$, again by Pontrjagin's duality theorem.

3. Let H be a normal subgroup of G. If $\sigma \in H$ and $\rho \in R$, then we have $\rho\sigma = (\rho\sigma\rho^{-1})\rho$, hence $\sigma_\rho = \rho\sigma\rho^{-1} \in H$, and thus the assertion follows by 1. □

The following principal ideal theorem of P. Furtwängler played a prominent role in the historial development of class field theory (see 6.6.9).

Theorem 2.5.8 (Principal ideal theorem)**.** *Let G be a finite group, H a subgroup of G and R a set of representatives for $H\backslash G$ in G such that $1 \in R$.*

1. $I_H + I_H I_G$ *is a free abelian group with basis*

$$T = \{(\tau - 1)\rho \mid \tau \in H \setminus \{1\},\ \rho \in R\},$$

and there exists a unique isomorphism

$$\vartheta_H \colon H^{\mathsf{a}} \to (I_H + I_H I_G)/I_H I_G$$

satisfying $\vartheta_H(\tau H^{(1)}) = \tau - 1 + I_H I_G$ *for all* $\tau \in H$.
In particular, there exists a unique isomorpism $\vartheta_G \colon G^{\mathsf{a}} \to I_G/I_G^2$ *satisfying* $\vartheta_G(\sigma G^{\mathsf{a}}) = \sigma - 1 + I_G^2$ *for all* $\sigma \in G$.

2. There exists a commutative diagram

$$\begin{array}{ccc} G^{\mathsf{a}} & \xrightarrow{\mathsf{V}_{G\to H}} & H^{\mathsf{a}} \\ {\scriptstyle\vartheta_G}\big\downarrow & & \big\downarrow{\scriptstyle\vartheta_H} \\ I_G/I_G^2 & \xrightarrow{\ s\ } & (I_H + I_H I_G)/I_H I_G\,, \end{array}$$

where

$$s(x + I_G^2) = \sum_{\rho \in R} \rho x + I_H I_G \quad \text{for all} \quad x \in I_G.$$

3. (Principal ideal theorem) *Let* $G^{(1)}$ *be abelian. Then* $\mathsf{V}_{G\to G^{(1)}}(\sigma G^{(1)}) = \mathbf{1}$ *for all* $\sigma \in G$.

Proof. By definition,

$$G = \biguplus_{\rho \in R} H\rho,$$

and for $\sigma \in G$ and $\rho \in R$ we set $\rho\sigma = \sigma_\rho \rho_\sigma$, where $\sigma_\rho \in H$ and $\rho_\sigma \in R$.

1. **a.** $I_H + I_H I_G = \langle T \rangle$.

Proof of **a.** If $\tau \in H \setminus \{1\}$ and $\rho \in R$, then $(\tau - 1)\rho = (\tau - 1) + (\tau - 1)(\rho - 1)$, hence $(\tau - 1)\rho \in I_H + I_H I_G$, and therefore $\langle T \rangle \subset I_H + I_H I_G$.

As $1 \in R$, we get $\tau - 1 \in T$ for all $\tau \in H \setminus \{1\}$ and thus $I_H \subset \langle T \rangle$. By definition, $I_H I_G = \langle \{(\tau - 1)(\tau'\rho - 1) \mid \tau,\ \tau' \in H,\ \rho \in R\} \rangle$. However, if $\tau,\ \tau' \in H$ and $\rho \in R$, then $(\tau - 1)(\tau'\rho - 1) = -(\tau - 1) - (\tau' - 1)\rho + (\tau\tau' - 1)\rho \in \langle T \rangle$, hence $I_H I_G \subset \langle T \rangle$, and consequently $I_H + I_H I_G \subset \langle T \rangle$. □[**a.**]

b. T is linearly independent.

Proof of **b.** Suppose that

$$0 = \sum_{\tau \in H\setminus\{1\}} \sum_{\rho \in R} c_{\tau,\rho}(\tau - 1)\rho = \sum_{\tau \in H\setminus\{1\}} \sum_{\rho \in R} c_{\tau,\rho}\tau\rho - \sum_{\rho \in R} \Big(\sum_{\tau \in H\setminus\{1\}} c_{\tau,\rho} \Big) \rho$$

for some $c_{\tau,\rho} \in \mathbb{Z}$. Then $c_{\tau,\rho} = 0$ for all $\tau \in H \setminus \{1\}$ and $\rho \in R$, since the elements $\tau\rho$ and ρ (for $\tau \in H \setminus \{1\}$ and $\rho \in R$) are pairwise distinct. □[**b.**]

c. Let $\vartheta'\colon H \to I_H/I_H^2$ be defined by $\vartheta'(\tau) = \tau - 1 + I_H^2$. Then ϑ' is an epimorphism and induces an isomorphism ϑ_H as asserted.

Proof of **c.** For $\tau,\, \tau' \in H$ we obtain

$$\tau\tau' - 1 = (\tau-1)(\tau'-1) + (\tau-1) + (\tau'-1) \equiv (\tau-1) + (\tau'-1) \mod I_H^2$$

and

$$\begin{aligned} \tau\tau'\tau^{-1}\tau'^{-1} - 1 &\equiv (\tau-1) + (\tau'-1) + (\tau^{-1}-1) + (\tau'^{-1}-1) \\ &\equiv (\tau-1) + (\tau'-1) - \tau^{-1}(\tau-1) - \tau'^{-1}(\tau'-1) \\ &\equiv -(\tau-1)(\tau^{-1}-1) - (\tau'-1)(\tau'^{-1}-1) \equiv 0 \mod I_H^2. \end{aligned}$$

Hence $\vartheta'\colon H \to I_H/I_H^2$ is an epimorphism such that $H^{(1)} \subset \mathrm{Ker}(\vartheta')$, and as

$$I_H^2 \subset I_H \cap I_H I_G,$$

it follows that ϑ' induces an epimorphism

$$\vartheta_H\colon H^{\mathsf{a}} \to I_H/I_H \cap I_H I_G = (I_H + I_H I_G)/I_H I_G$$

satisfying $\vartheta_H(\tau H^{(1)}) = \tau - 1 + I_H I_G$ for all $\tau \in H$.

As $I_H + I_H I_G$ is a free abelian group with basis T, there exists a unique homomorphism $\varphi'\colon I_H + I_H I_G \to H^{\mathsf{a}}$ satisfying $\varphi'((\tau-1)\rho) = \tau H^{(1)}$ for all $\tau \in H$ and all $\rho \in R$. If $\tau,\, \tau' \in H$ and $\rho \in R$, then

$$\varphi'((\tau-1)(\tau'\rho-1)) = \varphi'\big(-(\tau-1)-(\tau'-1)\rho+(\tau\tau'-1)\rho\big) = \tau^{-1}\tau'^{-1}\tau\tau' H^{(1)} = 1.$$

Hence $I_H I_G \subset \mathrm{Ker}(\varphi')$, and since $\varphi'(\tau-1) = \tau H^{(1)}$ for all $\tau \in H$, φ' induces a epimorphism $\varphi\colon I_H + I_H I_G/I_H I_G \to H^{\mathsf{a}}$ satisfying $\varphi(\tau - 1 + I_H I_G) = \tau H^{(1)}$ for all $\tau \in H$. Hence $\varphi \circ \vartheta_H = \mathrm{id}_{H^{\mathsf{a}}}$, and ϑ_H is an isomorphism.

2. If $\sigma \in G$, then

$$\vartheta_H \circ \mathsf{V}_{G\to H}(\sigma G^{(1)}) = \vartheta_H\Big(\prod_{\rho\in R} \sigma_\rho H^{(1)}\Big) = \sum_{\rho \in R} (\sigma_\rho - 1) + I_H I_G$$

and

$$s \circ \vartheta_G(\sigma G^{(1)}) = s(\sigma - 1 + I_G^2) = \sum_{\rho\in R} \rho(\sigma-1) + I_H I_G.$$

If $\rho \in R$, then

$$\begin{aligned} \rho(\sigma-1) &= (\sigma_\rho\rho_\sigma - 1) - (\rho - 1) \\ &= (\sigma_\rho - 1)(\rho_\sigma - 1) + (\sigma_\rho - 1) + (\rho_\sigma - 1) - (\rho - 1) \\ &\equiv (\sigma_\rho - 1) + (\rho_\sigma - 1) - (\rho - 1) \mod I_H I_G \end{aligned}$$

and therefore

$$\sum_{\rho\in R} \rho(\sigma-1) \equiv \sum_{\rho\in R} (\sigma_\rho - 1) + \sum_{\rho\in R} (\rho_\sigma - 1) - \sum_{\rho\in R} (\rho - 1) \equiv \sum_{\rho\in R} (\sigma_\rho - 1) \mod I_H I_G.$$

3. Let $G = \langle \sigma_1, \ldots, \sigma_n \rangle$, let $(\boldsymbol{e}_1, \ldots, \boldsymbol{e}_n)$ be the canonical basis of $\mathbb{Z}^n$, and let $\varphi\colon \mathbb{Z}^n \to G^{\mathsf{a}}$ be the unique epimorphism satisfying $\varphi(\boldsymbol{e}_i) = \sigma_i G^{(1)}$ for all $i \in [1, n]$. Let $M = (m_{i,k})_{i,k \in [1,n]} \in \mathsf{M}_n(\mathbb{Z})$ be such that $(\boldsymbol{e}_1, \ldots, \boldsymbol{e}_n)M$ is a basis of $\mathrm{Ker}(\varphi)$ and $m = \det(M)$. Then $|m| = (\mathbb{Z}^n : \mathrm{Ker}(\varphi)) = |G^{\mathsf{a}}|$, and if $k \in [1, n]$, then

$$1 = \varphi\Big(\sum_{i=1}^n \boldsymbol{e}_i m_{i,k}\Big) = \prod_{i=1}^n \sigma_i^{m_{i,k}} G^{(1)} \in G^{\mathsf{a}}, \quad \text{hence}$$

$$1 = \prod_{i=1}^n \sigma_i^{m_{i,k}} \prod_{j=1}^{t_k} \tau_{j,k} \tau'_{j,k} \tau_{j,k}^{-1} \tau_{j,k}'^{-1} \in G,$$

where $t_k \in \mathbb{N}_0$ and $\tau_{j,k}, \tau'_{j,k} \in \{\sigma_i, \sigma_i^{-1} \mid i \in [1, n]\}$. By 2.1.1.2, it follows that

$$\sum_{i=1}^n (\sigma_i - 1)\mu_{i,k} + \sum_{j=1}^{t_k} [(\tau_{j,k} - 1)\nu_{j,k} + (\tau'_{j,k} - 1)\nu'_{j,k} + (\tau_{j,k} - 1)\overline{\nu}_{j,k} + (\tau'_{j,k} - 1)\overline{\nu}_{j,k}{}' = 0$$

for all $k \in [1, n]$, where $\mu_{i,k}, \nu_{j,k}, \nu'_{j,k}, \overline{\nu}_{j,k}, \overline{\nu}'_{j,k} \in \mathbb{Z}[G]$, $\mu_{i,k} \equiv m_{i,k} \bmod I_G$, and $\nu_{j,k} \equiv \nu'_{j,k} \equiv 1 \bmod I_G$, $\overline{\nu}_{j,k} \equiv \overline{\nu}'_{j,k} \equiv -1 \bmod I_G$ for all $i, k \in [1, n]$ and $j \in [1, t_k]$. If $\tau_{j,k}^{(\prime)} \in \{\sigma_i, \sigma_i^{-1}\}$, then $\mu_{i,k} + \nu_{j,k}^{(\prime)} + \overline{\nu}_{j,k}^{(\prime)} \equiv \mu_{i,k} \bmod I_G$, and therefore

$$\sum_{i=1}^n (\sigma_i - 1)\mu_{i,k} = 0 \quad \text{for all} \quad k \in [1, n]. \tag{$*$}$$

Let $\pi\colon \mathbb{Z}[G] \to \mathbb{Z}[G^{\mathsf{a}}] = \mathbb{Z}[G]/I_{G^{(1)}}\mathbb{Z}[G]$ be the unique epimorphism satisfying $\pi \restriction \mathbb{Z} = \mathrm{id}_{\mathbb{Z}}$ and $\pi(g) = gG^{(1)}$ for all $g \in G$ (see 2.1.1.4), and set $\overline{\lambda} = \pi(\lambda)$ for all $\lambda \in \mathbb{Z}[G]$. Since $\pi(I_G) = I_{G^{\mathsf{a}}}$, we get $\overline{\mu}_{i,k} \equiv m_{i,k} \bmod I_{G^{\mathsf{a}}}$ for all $i, k \in [1, n]$, we denote by $\overline{M}^{\#}$ the adjoint matrix of $\overline{M} = (\overline{\mu}_{i,k})_{i,k \in [1,n]} \in \mathsf{M}_n(\mathbb{Z}[G^{\mathsf{a}}])$ and set $\overline{\mu} = \det(\overline{M}) \in \mathbb{Z}[G^{\mathsf{a}}]$, so that $\overline{\mu} \equiv m \bmod I_{G^{\mathsf{a}}}$.

From $(*)$ we obtain

$$\sum_{i=1}^n (\overline{\sigma}_i - 1)\overline{\mu}_{i,k} = 0 \in \mathbb{Z}[G^{\mathsf{a}}] \quad \text{for all} \quad k \in [1, n],$$

and thus $\mathbf{0} = (\overline{\sigma}_1 - 1, \ldots, \overline{\sigma}_n - 1)\overline{M}\,\overline{M}^{\#} = (\overline{\sigma}_1 - 1, \ldots, \overline{\sigma}_n - 1)\overline{\mu}$ by the very definition of $\overline{M}$. Since $G^{\mathsf{a}} = \langle \overline{\sigma}_1, \ldots, \overline{\sigma}_n \rangle$, 2.1.1.2 implies

$$I_{G^{\mathsf{a}}} = {}_{\mathbb{Z}[G^{\mathsf{a}}]}\langle \overline{\sigma}_1 - 1, \ldots, \overline{\sigma}_n - 1 \rangle,$$

hence $\overline{\mu} \in \mathrm{Ann}_{\mathbb{Z}[G^{\mathsf{a}}]}(I_{G^{\mathsf{a}}}) = \mathbb{Z}N_{G^{\mathsf{a}}}$, say $\mu = bN_{G^{\mathsf{a}}}$ for some $b \in \mathbb{Z}$. Let R be a set of representatives for G^{a} in G such that $1 \in R$. Then $|R| = |m|$, and

$$\overline{\mu} = bN_{G^{\mathsf{a}}} = b \sum_{\rho \in R} \overline{\rho} \equiv b|m| \bmod I_{G^{\mathsf{a}}}.$$

Since $I_{G^{\mathsf{a}}} \cap \mathbb{Z} = \mathbf{0}$, it follows that $b = \mathrm{sgn}(m)$, and

$$\overline{\sum_{\rho \in R} \rho(\sigma - 1)} \equiv \sum_{\rho \in R} (\overline{\sigma} - 1) \equiv 0 \bmod I_{G^{\mathsf{a}}}.$$

Since $I_{G^a} = I_G/I_{G^{(1)}}\mathbb{Z}[G]$ it follows that

$$\sum_{\rho\in R} \rho(\sigma-1) \in I_{G^{(1)}} I_G.$$

If $\sigma \in G$, then 2. (with $H = G^{(1)}$) implies

$$\vartheta_{G^{(1)}} \circ \mathsf{V}_{G\to G^{(1)}}(\sigma G^{(1)}) = \sum_{\rho\in R} \rho(\sigma-1) + I_{G^{(1)}} I_G = 0,$$

and as $\vartheta_{G^{(1)}}$ is an isomorphism by 1., we obtain $\mathsf{V}_{G\to G^{(1)}}(\sigma G^{(1)}) = 1$. □

2.6 Galois cohomology

Remarks 2.6.1. Let L/K be a Galois field extension and $G = \mathrm{Gal}(L/K)$. Then G operates on the additive group $L = (L,+)$ by means of $(\sigma,x) \mapsto \sigma x$ and on the multiplicative group $L^\times$ by means of $(\sigma,x) \mapsto x^\sigma$. With these operations, L and $L^\times$ are G-modules, and if $x \in L$, then

$$G_x = \{\sigma \in G \mid \sigma x = x\} = \mathrm{Gal}(L/K(x))$$

is an open subgroup of G. Hence L and $L^\times$ are discrete G-modules by 2.1.2, and we write $L^{(\times)}$ if we mean either L or $L^\times$.

Let $\mathcal{E}$ be an (upwards by inclusion) directed set of over K finite Galois intermediate fields of L/K such that L is its compositum. For $E \in \mathcal{E}$ we set $G_E = \mathrm{Gal}(E/K)$. Then by 1.8.2 we get

$$G = \varprojlim_{E\in\mathcal{E}} G_E,$$

$\{\mathrm{Gal}(L/E) \mid E \in \mathcal{E}\}$ is a fundamental system of neighbourhoods of 1 in G consisting of open normal subgroups, and if $E \in \mathcal{E}$, then $G_E = G/\mathrm{Gal}(L/E)$. By 2.5.3 we obtain, for all $q \ge 1$,

$$H^q(G, L^{(\times)}) = \varinjlim_{E\in\mathcal{E}} H^q(G_E, E^{(\times)}),$$

and if $E \in \mathcal{E}$, then

$$\mathrm{Inf}^q_{G_E\to G}\colon H^q(G_E, E^{(\times)}) \to H^q(G, L^{(\times)})$$

is the injection into the inductive limit.

If $[L:K] < \infty$, then

$$N_G a = \sum_{\sigma \in G} \sigma a = \mathsf{Tr}_{L/K}(a) \quad \text{for} \quad a \in L = (L,+),$$

$$N_G a = \prod_{\sigma \in G} a^\sigma = \mathsf{N}_{L/K}(a) \quad \text{for} \quad a \in L^\times,$$

and $H^0(G, L^\times) = K^\times/\mathsf{N}_{L/K}L^\times$ is the field-theoretic norm residue group.

Theorem 2.6.2 (Normal basis theorem). *Let L/K be a finite Galois extension and $G = \mathrm{Gal}(L/K)$. Then there exists some $x \in L$ such that $\{\sigma x \mid \sigma \in G\}$ is a K-basis of L.*

Proof. Let $n = [L:K]$.

CASE 1: L/K is cyclic, $G = \langle \sigma \rangle$. Then L is an (additive) $K[X]$-module by means of $fa = f(\sigma)a$ for all $f \in K[X]$ and $a \in L$. If $a \in L$, then it follows that $(X^n - 1)a = \sigma^n a - a = 0$, hence $X^n - 1 \in \mathrm{Ann}_{K[X]}L$, and we assert that $\mathrm{Ann}_{K[X]}L = (X^n - 1)K[X]$.

Let $f \in \mathrm{Ann}_{K[X]}L$, and set $f = c_0 + c_1X + \ldots + c_{n-1}X^{n-1} + (X^n - 1)h$, where $c_0, \ldots, c_{n-1} \in K$ and $h \in K[X]$. Then

$$0 = fa = (c_0 + c_1\sigma + \ldots + c_{n-1}\sigma^{n-1})a \quad \text{for all} \quad a \in L,$$

and consequently $c_0 + c_1\sigma + \ldots + c_{n-1}\sigma^{n-1} = 0 \in \mathrm{Hom}_K(L, L)$. By Dedekind's independence theorem (see [18, Theorem 1.5.2]) we obtain $c_0 = \ldots = c_{n-1} = 0$ and $f = (X^n - 1)h \in (X^n - 1)K[X]$. The structure theorem for finitely generated modules over principal ideal domains (see [18, Theorem 2.3.5]) yields an isomorphism

$$L \cong \bigoplus_{i=1}^{r} K[X]/(f_i),$$

where $f_1, \ldots, f_r \in K[X]$ are monic polynomials and $f_1 \mid f_2 \mid \ldots \mid f_r$. Since $\mathrm{Ann}_{K[X]}L = (X^n - 1)K[X] = f_rK[X]$, it follows that $f_r = X^n - 1$, and since $\dim_K K[X]/(X^n - 1) = n$, we obtain $r = 1$ and $L \cong K[X]/(X^n - 1)$. Explicitly, let $\Phi\colon K[X]/(X^n - 1) \xrightarrow{\sim} L$ be a $K[X]$-isomorphism,

$$u = 1 + (X^n - 1) \in K[X]/(X^n - 1) \quad \text{and} \quad x = \Phi(u).$$

Then $(u, Xu, \ldots, X^{n-1}u)$ is a K-basis of $K[X]/(X^n - 1)$, and consequently $(x, \sigma x, \ldots, \sigma^{n-1}x)$ is a K-basis of L.

CASE 2: L/K is not cyclic (and thus $|K| = \infty$). Let $(u_1, \ldots, u_n)$ be a K-basis of L and $G = \{1 = \sigma_1, \sigma_2, \ldots, \sigma_n\}$. Then

$$d_{L/K}(u_1, \ldots, u_n) = \det(\sigma_j(u_k)_{j,k \in [1,n]})^2 \neq 0 \text{ (see [18, Theorem 1.6.9])},$$

and therefore there exist $c_1, \ldots, c_n \in L$ such that

$$\sum_{k=1}^{n} \sigma_j(u_k)c_k = \delta_{j,1} \quad \text{for all} \quad j \in [1, n].$$

It follows that

$$\sum_{k=1}^{n} \sigma_i^{-1}\sigma_j(u_k)c_k = \delta_{i,j} \quad \text{for all} \;\; i,\, j \in [1,n],$$

and we set

$$U(X_1,\ldots,X_n) = \det\Big(\sum_{k=1}^{n} \sigma_i^{-1}\sigma_j(u_k)X_k\Big)_{i,j\in[1,n]} \in L[X_1,\ldots,X_n].$$

Then $U(c_1,\ldots,c_k) = 1$, and since K is infinite, there exists some $(a_1,\ldots,a_n) \in K^n$ such that $U(a_1,\ldots,a_n) \neq 0$. If

$$x = \sum_{k=1}^{n} a_k u_k, \quad \text{then} \quad \det(\sigma_i^{-1}\sigma_j x)_{i,j\in[1,n]} = \det\Big(\sum_{k=1}^{n} \sigma_i^{-1}\sigma_j(u_k)a_k\Big)_{i,j\in[1,n]} \neq 0,$$

and we assert that $(\sigma_1 x,\ldots,\sigma_n x)$ is a K-Basis of L. Considering that, it suffices to prove that $(\sigma_1 x,\ldots,\sigma_n x)$ is linearly independent over K. But if $b_1,\ldots,b_n \in K$, then

$$\sum_{j=1}^{n} b_j\sigma_j x = 0 \quad \text{implies} \quad \sum_{j=1}^{n} b_j\sigma_i^{-1}\sigma_j x = 0 \quad \text{for all} \;\; i \in [1,n]$$

and thus $b_1 = \ldots = b_n = 0$. □

Theorem 2.6.3 (Theorem of Hilbert–Noether). *Let L/K be a Galois extension and $G = \mathrm{Gal}(L/K)$. Then $H^1(G,L^\times) = \mathbf{1}$, and $H^q(G,L) = \mathbf{0}$ for all $q \geq 1$ (actually for all $q \geq -1$ if $[L\!:\!K] < \infty$).*

Proof. Let $\mathcal{E}$ be the set of all over K finite Galois intermediate fields of L/K. Then

$$H^q(G,L^{(\times)}) = \varinjlim_{E\in\mathcal{E}} H^q(G_E,E^{(\times)}) \quad \text{for all} \;\; q \geq 1,$$

and consequently it suffices to prove the assertions if L/K is finite. Thus suppose that $[L\!:\!K] < \infty$.

By 2.6.2, there exists some $x \in L$ such that

$$L = \bigoplus_{\sigma\in G} K\sigma x = \bigoplus_{\sigma\in G} \sigma(Kx).$$

Hence L is an induced G-module by 2.1.9.2 and thus $H^q(G,L) = \mathbf{0}$ for all $q \geq -1$ by 2.2.4 and 2.2.7.2.

If $x \in Z^1(G,L^\times)$, then $x(\sigma\tau) = x(\tau)^\sigma x(\sigma)$ for all $\sigma,\, \tau \in G$. By Dedekind's independence theorem we obtain

$$\sum_{\tau\in G} x(\tau)\tau \neq 0, \quad \text{and thus} \quad a = \sum_{\tau\in G} x(\tau)c^\tau \in L^\times \;\; \text{for some} \;\; c \in L.$$

If $\sigma \in G$, then

$$a^\sigma = \sum_{\tau \in G} x(\tau)^\sigma c^{\sigma\tau} = \sum_{\tau \in G} x(\sigma)^{-1} x(\sigma\tau) c^{\sigma\tau} = x(\sigma)^{-1} \sum_{\tau \in G} x(\tau) c^\tau = x(\sigma)^{-1} a,$$

hence $x(\sigma) = (a^{-1})^{\sigma - 1}$, $x \in B^1(G, L^\times)$, and thus $H^1(G, L^\times) = \mathbf{1}$. □

Corollary 2.6.4 (Hilbert's Theorem 90). *Let L/K be a finite cyclic field extension and* $\mathrm{Gal}(L/K) = \langle\sigma\rangle$. *Then* $\mathrm{Ker}(\mathsf{N}_{L/K} \colon L^\times \to K^\times) = L^{\times\sigma-1}$.

Proof. By 2.4.5.1 and 2.6.3, we obtain $H^{-1}(G, L^\times) = H^1(G, L^\times) = \mathbf{1}$, and therefore

$$\mathrm{Ker}(\mathsf{N}_{L/K} \colon L^\times \to K^\times)/L^{\times\sigma-1} = {}_{N_G}L^\times / I_G L^\times = H^{-1}(G, L^\times) = \mathbf{1}. \qquad \square$$

Corollary 2.6.5. *Let $K \subset L \subset \overline{L}$ be fields such that $\overline{L}/K$ and L/K are Galois. If $\overline{G} = \mathrm{Gal}(\overline{L}/K)$, $H = \mathrm{Gal}(\overline{L}/L)$ and $G = \mathrm{Gal}(L/K) = \overline{G}/H$, then there exists the exact sequence*

$$\mathbf{1} \longrightarrow H^2(G, L^\times) \overset{\mathrm{Inf}^2_{G \to \overline{G}}}{\longrightarrow} H^2(\overline{G}, \overline{L}^\times) \overset{\mathrm{Res}^2_{\overline{G} \to G}}{\longrightarrow} H^2(H, \overline{L}^\times).$$

Proof. Since $H^1(H, \overline{L}^\times) = \mathbf{1}$ by 2.6.3, the assertion follows from 2.5.4.3. □

If L/K is a Galois extension and $G = \mathrm{Gal}(L/K)$, then the group $H^2(G, L^\times)$ is called the **cohomological Brauer group** of L/K. This terminology will be justified in 3.5.4.

Next we use Galois cohomology to investigate not necessarily finite Kummer and Artin-Schreier extensions generalizing [18, Theorems 1.7.8 and 1.7.10]. We need some preparations.

Let $n \in \mathbb{N}$, G an abelian profinite group of exponent n, and let A by any cyclic group of order n, viewed as a trivial G-module. Then there is an isomorphism

$$A \overset{\sim}{\to} \frac{1}{n}\mathbb{Z}/\mathbb{Z} \hookrightarrow \mathbb{Q}/\mathbb{Z},$$

and if $\chi \in \mathsf{X}(G) = H^1(G, \mathbb{Q}/\mathbb{Z}) = \mathrm{Hom}^{\mathsf{c}}(G, \mathbb{Q}/\mathbb{Z})$, then $\chi(G) \subset \frac{1}{n}\mathbb{Z}/\mathbb{Z}$ and therefore

$$\mathsf{X}(G) = \mathrm{Hom}^{\mathsf{c}}\Big(G, \frac{1}{n}\mathbb{Z}/\mathbb{Z}\Big) \cong \mathrm{Hom}^{\mathsf{c}}(G, A) = H^1(G, A).$$

We identify: $\mathsf{X}(G) = H^1(G, A)$.

Recall that for a field K and $n \in \mathbb{N}$ denote by $\mu_n(K)$ the group of n-th roots of unity in K.

Theorem 2.6.6 (Kummer extensions). *Let $n \in \mathbb{N}$, and let K be a field which contains a primitive n-th root of unity (then $|\mu_n(K)| = n$ and $\mathrm{char}(K) \nmid n$). Let $\overline{K}$ be an algebraic closure of K, and for a subset Z of K we denote by $\sqrt[n]{Z} = \{\xi \in \overline{K} \mid \xi^n \in Z\}$.*

1. Let $Z \subset K^\times$ and $\Delta = \langle Z \cup K^{\times n}\rangle \subset K^\times$. Then it follows that $K(\sqrt[n]{Z}) = K(\sqrt[n]{\Delta}) \subset \overline{K}$, and $K(\sqrt[n]{Z})/K$ is an abelian extension of exponent n.

2. Let L/K be an abelian extension of exponent n, and suppose that $L \subset \overline{K}$, $G = \mathrm{Gal}(L/K)$ and $\Delta = L^{\times n} \cap K^\times$.

(a) $L = K(\sqrt[n]{\Delta})$, and there exists an isomorphism

$$\delta\colon \Delta/K^{\times n} \stackrel{\sim}{\to} \mathsf{X}(G) = \mathrm{Hom}^{\mathsf{c}}(G, \mu_n(K))$$

with the following property:
If $c \in \Delta$ and $\gamma \in L$ is such that $\gamma^n = c$, then

$$\delta(cK^{\times n})(\sigma) = \gamma^{\sigma - 1}$$

for all $\sigma \in G$ (independent of the choice of γ; we write $\delta(cK^{\times n})(\sigma) = \sqrt[n]{c}^{\,\sigma-1}$ or $\sigma\sqrt[n]{c} = \delta(cK^{\times n})(\sigma)\sqrt[n]{c}$ for all $c \in \Delta$).

(b) The map $\mathsf{X}(\delta)\colon G \stackrel{\sim}{\to} \mathsf{X}(\Delta/K^{\times n})$ is a topological isomorphism. In particular, if L/K is finite, then $[L{:}K] = (\Delta{:}K^{\times n})$.

(c) If Γ is a subgroup of $K^\times$ satisfying $K^{\times n} \subset \Gamma$ and $L = K(\sqrt[n]{\Gamma})$, then $\Gamma = \Delta$.

3. $K(\sqrt[n]{K^\times})$ is the largest intermediate field of $\overline{K}/K$ which is abelian of exponent n over K. If $G_K = \mathrm{Gal}(\overline{K}/K)$, then $\mathrm{Gal}(K(\sqrt[n]{K^\times})/K) = G_K^{\mathsf{a}}/(G_K^{\mathsf{a}})^{[n]}$.

Proof. We tacitly use [18, Theorem 1.7.8].

1. If $d \in Z$, then $K(\sqrt[n]{d})/K$ is a cyclic extension of degree n, and $K(\sqrt[n]{Z})$ is the compositum of all fields $K(\sqrt[n]{d})$ with $d \in Z$. Hence $K(\sqrt[n]{Z})/K$ is an abelian extension of exponent n by 1.8.5.1.

Obviously, $Z \subset \Delta$ implies $K(\sqrt[n]{Z}) \subset K(\sqrt[n]{\Delta})$. If $d \in \Delta$, then we obtain $d = d_1 \cdot \ldots \cdot d_m a^n$, where $m \in \mathbb{N}_0$, $d_1, \ldots, d_m \in Z \cup Z^{-1}$ and $a \in K^\times$. It follows that

$$K(\sqrt[n]{d}) \subset K(\sqrt[n]{d_1}, \ldots, \sqrt[n]{d_m}) \subset K(\sqrt[n]{Z}),$$

hence

$$K(\sqrt[n]{\Delta}) = \prod_{d \in \Delta} K(\sqrt[n]{d}) \subset K(\sqrt[n]{Z}), \quad \text{and therefore} \quad K(\sqrt[n]{Z}) = K(\sqrt[n]{\Delta}).$$

2. (a) By definition, $K(\sqrt[n]{\Delta}) \subset L$. For the reverse inclusion it suffices to prove that $E \subset K(\sqrt[n]{\Delta})$ for every over K finite intermediate field E of L/K. Thus let E be an over K finite intermediate field of L/K. Then $\mathrm{Gal}(E/K)$ is a factor group of G, hence a finite abelian group of exponent n, and therefore $\mathrm{Gal}(E/K) = G_1 \cdot \ldots \cdot G_k$ is the (inner) direct product of cyclic groups of exponent n. For $i \in [1,k]$, let E_i be the fixed field of $G_1 \cdot \ldots \cdot G_{i-1} \cdot G_{i+1} \cdot \ldots \cdot G_k$. Then $\mathrm{Gal}(E_i/K) \cong G_i$, and therefore we obtain $E_i = K(\delta_i)$ for some $\delta_i \in L$ satisfying $\delta_i^n \in K$ and thus $\delta_i^n \in \Delta$ for all $i \in [1,k]$. It follows that

$$E = E_1 \cdot \ldots \cdot E_k = K(\delta_1, \ldots, \delta_k) \subset K(\sqrt[n]{\Delta}),$$

and consequently $L = K(\sqrt[n]{\Delta})$.

We consider the exact sequence of discrete G-modules

$$\mathbf{1} \to \mu_n(L) \to L^\times \overset{p_n}{\to} L^{\times n} \to \mathbf{1}, \quad \text{where} \quad p_n(x) = x^n.$$

Since $\mu_n(L) = \mu_n(K) = \mu_n(L)^G$ and $(L^{\times n})^G = L^{\times n} \cap K^\times = \Delta$, the application of 2.4.1 and 2.6.3 yields the exact sequence

$$\mathbf{1} \to \mu_n(K) \to K^\times \overset{p_n \restriction K^\times}{\to} \Delta \to H^1(G, \mu_n(K)) \to H^1(G, L^\times) = \mathbf{1}$$

and as a consequnce an isomorphism

$$\delta\colon \Delta/K^{\times n} \overset{\sim}{\to} H^1(G, \mu_n(K)) = \mathrm{Hom}^{\mathsf{c}}(G, \mu_n(K)) = \mathsf{X}(G)$$

as follows:

If $c \in \Delta$ and $\gamma \in L^\times$ such that $\gamma^n = c$, then $\delta(cK^\times)(\sigma) = \gamma^{\sigma-1} \in \mu_n(K)$.

(b) By Pontrjagin's duality theorem $\mathsf{X}(\delta)\colon G \overset{\sim}{\to} \mathsf{X}(\Delta/K^{\times n})$ is a topological isomorphism, and if L/K is finite, then

$$[L\colon K] = |G| = |\mathsf{X}(\Delta/K^{\times n})| = (\Delta\colon K^{\times n}).$$

(c) Let Γ be a subgroup of $K^\times$ such that $K^{\times n} \subset \Gamma$ and $L = K(\sqrt[n]{\Gamma})$. Then $\Gamma \subset L^{\times n} \cap K^\times = \Delta$, hence $\Gamma/K^{\times n} \subset \Delta/K^{\times n}$, and we set $Y = \delta(\Gamma/K^{\times n})$. Since $Y^\perp = \{\sigma \in G \mid \gamma^{\sigma-1} = 1 \text{ for all } \gamma \in \sqrt[n]{\Gamma}\} = \mathrm{Gal}(L/K(\sqrt[n]{\Gamma})) = \mathbf{1}$, it follows by 1.6.7.3 that $Y = Y^{\perp\perp} = \mathbf{1}^\perp = \mathsf{X}(G)$, and consequently $\Gamma = \Delta$, since δ is an isomorphism.

3. Obvious by 1., 2. and 1.8.5.2. □

Theorem 2.6.7 (Artin-Schreier extensions)**.** *Let K be a field,* $\mathrm{char}(K) = p0$, *$\mathbb{F}_p \subset K$ and $\overline{K}$ an algebraic closure of K. Let $\wp(X) = X^p - X \in \mathbb{F}_p[X]$, and for a subset Z of K we set $\wp^{-1}Z = \{x \in \overline{K} \mid \wp(x) \in Z\}$.*

1. Let $Z \subset K$, and let $\Delta = \langle Z \cup \wp K \rangle \subset K$ be the (additive) subgroup of K generated by $Z \cup \wp K$. Then $K(\wp^{-1}Z) = K(\wp^{-1}\Delta)$, and $K(\wp^{-1}Z)/K$ is an abelian extension of exponent p.

2. Let L/K be an abelian extension of exponent p, $G = \mathrm{Gal}(L/K)$, $L \subset \overline{K}$ and $\Delta = \wp L \cap K$.

(a) $L = K(\wp^{-1}\Delta)$, and there exists an isomorphism

$$\delta\colon \Delta/\wp K \overset{\sim}{\to} \mathsf{X}(G) = \mathrm{Hom}^{\mathsf{c}}(G, \mathbb{F}_p)$$

with the following property:
If $c \in \Delta$ and $x \in L$ is such that $\wp(x) = c$, then

$$\delta(c + \wp K)(\sigma) = \sigma x - x$$

for all $\sigma \in G$ (independent of the choice of x).

(b) The map $\mathsf{X}(\delta)\colon G \overset{\sim}{\to} \mathsf{X}(\Delta/\wp K)$ is a topological isomorphism. In particular, if L/K is finite, then $[L{:}K] = (\Delta{:}\wp K)$.

(c) If Γ is an (additive) subgroup of K such that $\wp K \subset \Gamma$ and $L = K(\wp^{-1}\Gamma)$, then $\Gamma = \Delta$.

3. $K(\wp^{-1}K)$ is the largest intermediate field of $\overline{K}/K$ which is abelian of exponent p over K. If $G_K = \mathrm{Gal}(\overline{K}/K)$, then $\mathrm{Gal}(K(\wp^{-1}K)/K) = G_K^{\mathsf{a}}/(G_K^{\mathsf{a}})^{[p]}$.

Proof. We tacitly use [18, Theorem 1.7.10].1.

If $d \in Z$, then $K(\wp^{-1}d)/K$ is a cyclic extension of exponent p, and $K(\wp^{-1}Z)$ is the compositum of all fields $K(\wp^{-1}d)$ with $d \in Z$. Hence $K(\wp^{-1}Z)/K$ is an abelian extension of exponent p by 1.8.5.1.

Obviously, $K(\wp^{-1}Z) \subset K(\wp^{-1}\Delta)$. If $d \in \Delta$, then $d = d_1 + \ldots + d_m + \wp(a)$, where $m \in \mathbb{N}_0$, $d_1, \ldots, d_m \in Z$ and $a \in K$. It follows that

$$K(\wp^{-1}d) \subset K(\wp^{-1}d_1, \ldots, \wp^{-1}d_m) \subset K(\wp^{-1}Z),$$

hence

$$K(\wp^{-1}\Delta) = \prod_{d\in\Delta} K(\wp^{-1}d) \subset K(\wp^{-1}Z),$$

and therefore $K(\wp^{-1}Z) = K(\wp^{-1}\Delta)$.

2. (a) By definition, $K(\wp^{-1}\Delta) \subset L$. For the reverse inclusion it suffices to prove that $E \subset K(\wp^{-1}\Delta)$ for every over K finite intermediate field E of L/K. Thus let E be an over K finite intermediate field of L/K. Then $\mathrm{Gal}(E/K)$ is a factor group of G, hence a finite abelian group of exponent p, and thus $\mathrm{Gal}(E/K) = G_1 \cdot \ldots \cdot G_k$ is the (inner) direct product of cyclic groups of order p. For $i \in [1,k]$, let E_i be the fixed field of $G_1 \cdot \ldots \cdot G_{i-1} \cdot G_{i+1} \cdot \ldots \cdot G_k$. Then E_i/K is cyclic of degree p, and therefore we obtain $E_i = K(y_i)$ for some $y_i \in L$ such that $\wp(y_i) \in K$ and thus $\wp(y_i) \in \Delta$ for all $i \in [1,k]$. It follows that $E = E_1 \cdot \ldots \cdot E_k = K(y_1, \ldots, y_k) \subset K(\wp^{-1}\Delta)$, and consequently $L = K(\wp^{-1}\Delta)$.

We consider the exact sequence of discrete G-modules

$$\mathbf{0} \to \mathbb{F}_p \to L \stackrel{\wp}{\to} \wp L \to \mathbf{0}.$$

Since $(\wp L)^G = \wp L \cap K = \Delta$, we obtain (using 2.4.1 and 2.6.3) the exact sequence

$$\mathbf{0} \to \mathbb{F}_p \to K \stackrel{\wp \restriction K^\times}{\to} \Delta \to H^1(G, \mathbb{F}_p) \to H^1(G, L) = \mathbf{0}$$

and as a consequence an isomorphism

$$\delta\colon \Delta/\wp K \stackrel{\sim}{\to} H^1(G, \mathbb{F}_p) = \mathrm{Hom}^{\mathrm{c}}(G, \mathbb{F}_p) = \mathsf{X}(G)$$

with the following property:

If $c \in \Delta$ and $x \in L$ is such that $\wp(x) = c$, then $\delta(c+\wp K)(\sigma) = \sigma x - x \in \mathbb{F}_p$.

(b) By definition, $\Gamma \subset \wp L \cap K = \Delta$, hence $\Gamma + \wp K \subset \Delta + \wp K$, and we set $Y = \delta(\Gamma + \wp K) \subset \mathsf{X}(G)$. Since

$$Y^\perp = \{\sigma \in G \mid \sigma x - x = 0 \ \text{ for all } x \in \wp^{-1}\Gamma\} = \mathrm{Gal}(L/K(\wp^{-1}\Gamma)) = \mathbf{1},$$

it follows by 1.6.7.3 that $Y = \mathsf{X}(G)$, and consequently $\Gamma = \Delta$, since δ is an isomorphism.

3. By 1., 2. and 1.8.5.2. □

Definition and Remarks 2.6.8. Let L/K be a Galois extension, $\overline{L}$ an extension field of L, K' an intermediate field of $\overline{L}/K$, $L' = LK'$, and set $G = \mathrm{Gal}(L/K)$. By 1.8.4.1, L'/K' is Galois and the restriction $\sigma' \mapsto \sigma' \restriction L$ defines a monomorphism $\rho\colon G' = \mathrm{Gal}(L'/K') \to G$ such that

$$\rho(G') = \mathrm{Gal}(L/L \cap K') \subset G.$$

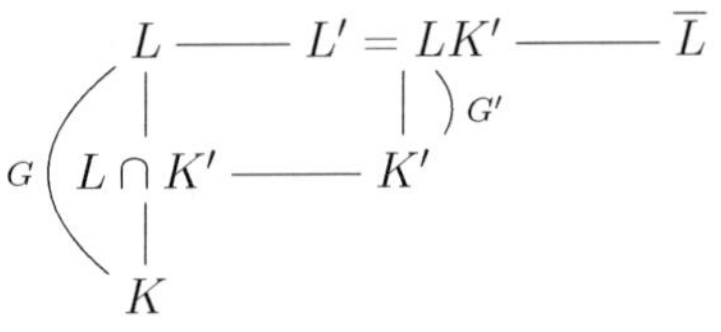

For $q \geq 1$ and $x \in C^q(G, L^\times)$ let

$$\mathrm{Res}_*^q = \mathrm{Res}^q_{L/K \to L'/K'}\colon C^q(G, L^\times) \to C^q(G', L'^\times)$$

be defined by

$$\mathrm{Res}_*^q(x)\colon G'^q \stackrel{\rho^q}{\to} G^q \stackrel{x}{\to} L^\times \hookrightarrow L'^\times \quad \text{for all } \ x \in C^q(G, L^\times).$$

If in addition $\mathrm{Res}_*^0 = (K^\times \hookrightarrow K'^\times)$, then $d^{q-1} \circ \mathrm{Res}_*^q = \mathrm{Res}_*^q \circ d^{q-1}$ for all $q \geq 1$.

Hence $\mathrm{Res}_*^q(Z^q(G,L^\times)) \subset Z^q(G',L'^\times)$, $\mathrm{Res}_*^q(B^q(G,L^\times)) \subset B^q(G',L'^\times)$, and Res_*^q induces an (equally denoted) homomorphism

$$\mathrm{Res}_*^q = \mathrm{Res}^q_{L/K\to L'/K'} \colon H^q(G,L^\times) \to H^q(G',L'^\times).$$

We call $\mathrm{Res}_*^q = \mathrm{Res}^q_{L/K\to L'/K'}$ the **restriction** from L/K to L'/K'.

If $K' \subset L$, then $L = L'$, $G' \subset G$ and

$$\mathrm{Res}^q_{L/K\to L/K'} = \mathrm{Res}^q_{G\to G'} \colon H^q(G,L^\times) \to H^q(G',L^\times).$$

Definition and Theorem 2.6.9. *Let K be a field, K_{sep} a separable closure of K, L/K a Galois extension such that $L \subset K_{\mathrm{sep}}$, K' an over K finite intermediate field of L/K, $[K'\colon K] = n$, $G = \mathrm{Gal}(L/K)$ and $G' = \mathrm{Gal}(L/K') \subset G$. Let L^{a} be the largest over K abelian intermediate field of L/K and L'^{a} the largest over K' abelian intermediate field of L/K'.*

By 1.8.5.2, we obtain $L^{\mathsf{a}} \subset L'^{\mathsf{a}}$, $G^{\mathsf{a}} = \mathrm{Gal}(L^{\mathsf{a}}/K)$ and $G'^{\mathsf{a}} = \mathrm{Gal}(L'^{\mathsf{a}}/K')$. We call

$$\big[\mathsf{V}^L_{K'/K} \colon \mathrm{Gal}(L^{\mathsf{a}}/K) \to \mathrm{Gal}(L'^{\mathsf{a}}/K')\big] = \big[\mathsf{V}_{G\to G'} \colon G^{\mathsf{a}} \to G'^{\mathsf{a}}\big]$$

the **transfer** from K to K' inside L, and we define the **(absolute) transfer** from K to K' by

$$\mathsf{V}_{K'/K} = \mathsf{V}^{K_{\mathrm{sep}}}_{K'/K} \colon G_K^{\mathsf{a}} \to G_{K'}^{\mathsf{a}},$$

where $G_K = \mathrm{Gal}(K_{\mathrm{sep}}/K)$ and $G_{K'} = \mathrm{Gal}(K'_{\mathrm{sep}}/K')$ denote the absolute Galois groups (then $G_K^{\mathsf{a}} = \mathrm{Gal}(K_{\mathrm{ab}}/K)$ and $G_{K'}^{\mathsf{a}} = \mathrm{Gal}(K'_{\mathrm{ab}}/K')$ are the Galois groups of the maximal abelian extensions).

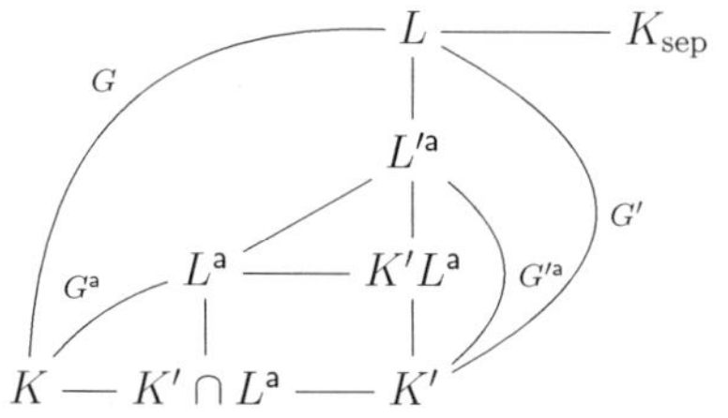

Let R be a set of representatives for $G'\backslash G$ in G. For $\sigma \in G$ and $\rho \in R$ we set $\rho\sigma = \sigma_\rho\rho_\sigma$ with $\sigma_\rho \in G'$ and $\rho_\sigma \in R$. If $\sigma \in G$, then $\sigma G^{(1)} = \sigma \restriction L^{\mathsf{a}} \in G^{\mathsf{a}}$, and

$$\mathsf{V}^L_{K'/K}(\sigma \restriction L^{\mathsf{a}}) = \prod_{\rho\in R} \sigma_\rho \restriction L'^{\mathsf{a}} \in G'^{\mathsf{a}}$$

1. *If K'' is an over K' finite intermediate field of L/K', then*

$$\mathsf{V}^L_{K''/K} = \mathsf{V}^L_{K'/K} \circ \mathsf{V}^L_{K''/K'}.$$

2. If K'/K is Galois and $\sigma \in G'$, then

$$\mathsf{V}^L_{K'/K}(\sigma \restriction L^{\mathsf{a}}) = \prod_{\rho \in R} \rho\sigma\rho^{-1} \restriction L'^{\mathsf{a}} \in G'^{\mathsf{a}}.$$

3. If $\sigma \in G$, then $\mathsf{V}^L_{K'/K}(\sigma \restriction L^{\mathsf{a}}) \restriction L^{\mathsf{a}} = \sigma^n \restriction L^{\mathsf{a}}$.

4. Let K'/K be cyclic and $\sigma \in G$ such that $\mathrm{Gal}(K'/K) = \langle \sigma \restriction K' \rangle$. Then

$$\mathsf{V}^L_{K'/K}(\sigma \restriction L^{\mathsf{a}}) = \sigma^n \restriction L'^{\mathsf{a}} \in G'^{\mathsf{a}}.$$

5. If $\sigma \in G_K$, then

$$\mathsf{V}^L_{K'/K}(\sigma \restriction K_{\mathrm{ab}}) = \mathsf{V}_{K'/K}(\sigma \restriction K_{\mathrm{ab}}) \restriction L'^{\mathsf{a}}.$$

Proof. If $\sigma \in G$, then 2.5.7.1 implies

$$\mathsf{V}^L_{K'/K}(\sigma \restriction L^{\mathsf{a}}) = \mathsf{V}_{G \to G'}(\sigma G^{(1)}) = \prod_{\rho \in R} \sigma_\rho G'^{(1)} = \prod_{\rho \in R} \sigma_\rho \restriction L'^{\mathsf{a}}.$$

1. and 2. are immediate consequences of 2.5.7.(2 and 3).

3. If $\sigma \in G$ and $\rho \in R$, then $\rho\sigma \restriction L^{\mathsf{a}} = \sigma\rho \restriction L^{\mathsf{a}}$, hence $\sigma_\rho \restriction L^{\mathsf{a}} = \sigma \restriction L^{\mathsf{a}}$, and therefore

$$\mathsf{V}^L_{K'/K}(\sigma \restriction L^{\mathsf{a}}) \restriction L^{\mathsf{a}} = \prod_{\rho \in R} (\sigma_\rho \restriction L'^{\mathsf{a}}) \restriction L^{\mathsf{a}} = \prod_{\rho \in R} \sigma \restriction L^{\mathsf{a}} = \sigma^n \restriction L^{\mathsf{a}}.$$

4. By assumption, $G/G' = \langle \sigma G' \rangle$, we set $R = \{\sigma^k \mid k \in [0, n-1]\}$, and then the assertion follows from 2.

5. For $\sigma \in G_K$ we set $\overline{\sigma} = \sigma \restriction L \in G$. If $\sigma \in G_K$, then $\sigma \in G_{K'}$ if and only if $\overline{\sigma} \in G'$. If R is a set of representatives for $G_K \backslash G_{K'}$ in G_K, then the set $\overline{R} = \{\overline{\rho} \mid \rho \in R\}$ is a set of representatives for $G \backslash G'$ in G.

As $(G_K : G_{K'}) = n = (G : G')$, the assignment $\rho \mapsto \overline{\rho}$ defines a bijective map $R \to \overline{R}$. Apparently, if $\rho \in G_K$, $\sigma \in G_{K'}$ and $\rho\sigma = \sigma_\rho\rho_\sigma$, where $\sigma_\rho \in G_{K'}$ and $\rho_\sigma \in R$, then $\overline{\rho}\,\overline{\sigma} = \overline{\sigma_\rho}\,\overline{\rho_\sigma}$, and therefore $\overline{\sigma_\rho} = \overline{\sigma}_{\overline{\rho}}$. Hence

$$\mathsf{V}_{K'/K}(\sigma \restriction K_{\mathrm{ab}}) \restriction L'^{\mathsf{a}} = \prod_{\rho \in R} (\sigma_\rho \restriction K_{\mathrm{ab}}) \restriction L'^{\mathsf{a}} = \prod_{\rho \in R} \overline{\sigma_\rho} \restriction L'^{\mathsf{a}} = \prod_{\overline{\rho} \in \overline{R}} \overline{\sigma}_{\overline{\rho}} \restriction L'^{\mathsf{a}}$$
$$= \mathsf{V}^L_{K'/K}(\sigma \restriction L^{\mathsf{a}}).$$

□

3

Simple algebras

We give a self-contained introduction to the theory of simple algebras insofar it is relevant for our foundation of class field theory. For the convenience of the reader we start with a survey of some linear algebra including tensor products which usually comes up short in basic courses on algebra. The main results of this chapter is devoted to the theory of cyclic algebras and its connection with the theory of norm residues.

Main sources which influenced the presentation of this chapter are [49], [39], [28], and [64].

3.1 Preliminaries on modules and algebras

Let K be a commutative ring.

We start by gathering some perhaps less known facts about modules over (not necessarily commutative) algebras.

Definition 3.1.1. For a ring A and a subset B of A, we call

$$\mathsf{c}_A(B) = \{a \in A \mid ax = xa \text{ for all } x \in B\}$$

the **centralizer** of B in A and $\mathsf{c}(A) = \mathsf{c}_A(A)$ the **center** of A. Apparently $\mathsf{c}_A(B)$ is a subring of A, and if B is commutative, then $\mathsf{c}_A(B) \supset B$.

A **K-algebra** is a ring A, together with a K-module structure $K \times A \to A$, given by $(\lambda, a) \mapsto \lambda a$ such that $\lambda(aa') = (\lambda a)a' = a(\lambda a')$ for all $\lambda \in K$ and $a, a' \in A$. Then the map $\varepsilon\colon K \to A$, defined by $\varepsilon(\lambda) = \lambda 1_A$ for all $\lambda \in K$, is a ring homomorphism satisfying $\varepsilon(K) \subset \mathsf{c}(A)$. The homomorphism ε is called the **structural homomorphism** of the K-algebra A. Conversely, if $\varepsilon\colon K \to A$ is a ring homomorphism such that $\varepsilon(K) \subset \mathsf{c}(A)$, then A is a K-module by means of $\lambda a = \varepsilon(\lambda)a$ for all $\lambda \in K$ and $a \in A$, and with regard to this K-module structure the ring A is a K-algebra with structural homomorphism ε. Therefore sometimes the homomorphism $\varepsilon\colon K \to A$ itself is called an K-algebra.

DOI: 10.1201/9780429506574-3

A ring A is called **simple** if $A \neq \mathbf{0}$, and $\mathbf{0}$ and A are the only ideals of A. Every division algebra is simple, and every ring homomorphism $A \to R$, where A is a simple ring and $R \neq \mathbf{0}$, is a monomorphism.

Every ring A is a $\mathbb{Z}$-algebra by means of the structural homomorphism $\varepsilon\colon \mathbb{Z} \to A$, given by $\varepsilon(g) = g1_A$ for all $g \in \mathbb{Z}$. Any homomorphism $f\colon K \to A$ of commutative rings is a K-algebra. If A is a ring and K is a subring of $\mathsf{c}(A)$, then A is a K-algebra. In particular, K itself is a K-algebra, and id_K is its structural homomorphism.

If K is a field and $A \neq \mathbf{0}$ is a K-algebra, then the structural homomorphism $\varepsilon\colon K \to A$ is a monomorphism, hence $K \cong K1_A \subset A$, and whenever it is convenient, we shall identify and assume that $K \subset A$ and $1_A = 1 \in K$.

If A is a K-algebra, then a K-**subalgebra** of A is a subring of A (containing 1_A) which is also a K-submodule of A. If B is a K-subalgebra of A, then $\mathsf{c}_A(B)$ is a K-subalgebra of A, too. If A and A' are K-algebras, then a map $f\colon A \to A'$ is called a K-**algebra homomorphism** if it is K-linear and a ring homomorphism. We denote by $\mathrm{Hom}^K(A, A')$ the set of all K-algebra homomorphisms $A \to A'$ (contrary to $\mathrm{Hom}_K(A, A')$, which denotes the set of all K-linear maps or K-module homomorphisms $A \to A'$).

For a ring A, its **opposite ring** A° is the additive group A, endowed with the multiplication given by $a \circ b = ba$ for all a, $b \in A$. If A is a K-algebra, the A° is also a K-algebra, and a subset $\mathfrak{a}$ of A is an ideal of A if and only if it is an ideal of A°. Hence A° is simple if and only if A is simple.

An A°-module is nothing but a right A-module. Thus, if M is an A°-module, then for $a \in A$ and $m \in M$ we write ma instead of am (indeed, then we have $m(aa') = (ma)a'$ for all a, $a' \in A$ and $m \in M$). If A is commutative (that is, $A = A^\circ$) and M is an A-module, then for $a \in A$ and $m \in M$ we set $ma = am$ whenever this will be convenient from the context.

Let A be a K-algebra and $n \in \mathbb{N}$. We write the elements of A^n as columns so that $A^n = \mathsf{M}_{n,1}(A)$. Let $\boldsymbol{e}_1, \ldots, \boldsymbol{e}_n \in A^n$ be the unit vectors. Then A^n is both a free A-module and a free A°-module with basis $(\boldsymbol{e}_1, \ldots, \boldsymbol{e}_n)$. The matrix algebra $\mathsf{M}_n(A)$ is again a K-algebra, and A^n is an $\mathsf{M}_n(A)$-module by means of $(a, \boldsymbol{x}) \mapsto a\boldsymbol{x}$ for all $a \in \mathsf{M}_n(A)$ and $\boldsymbol{x} \in A^n$. We denote by $1_n \in \mathsf{M}_n(A)$ the unit matrix, and, for p, $q \in [1, n]$, we define the **matrix units** $\boldsymbol{e}_{p,q} \in \mathsf{M}_n(A)$ by

$$(\boldsymbol{e}_{p,q})_{i,j} = \begin{cases} 1 & \text{if } (i,j) = (p,q), \\ 0 & \text{if } (i,j) \neq (p,q). \end{cases}$$

Then $\mathsf{M}_n(A)$ is a free A-module with basis $\{\boldsymbol{e}_{p,q} \mid p, q \in [1, n]\}$ and the basic multiplicative relations $\boldsymbol{e}_{p,q}\boldsymbol{e}_{k,l} = \delta_{q,k}\boldsymbol{e}_{p,l}$ and $\boldsymbol{e}_{p,q}\boldsymbol{e}_k = \delta_{q,k}\boldsymbol{e}_p$ for all p, q, k, $l \in [1, n]$.

The map $\mathsf{t}\colon \mathsf{M}_n(A^\circ) \to \mathsf{M}_n(A)^\circ$, defined by $\mathsf{t}(a) = a^{\mathsf{t}}$ for all $a \in \mathsf{M}_n(A^\circ)$, is a K-algebra isomorphism.

An A-module M is called **simple** if $M \neq \mathbf{0}$, and $\mathbf{0}$ and M are the only A-submodules of M.

Definition and Remarks 3.1.2. Let A and B be K-algebras.

1. Let $f\colon A \to B$ be a K-algebra homomorphism. Then every B-module M is an A-module by means of $am = f(a)m$ for all $a \in A$ and $m \in M$. Every B-submodule of M is also an A-submodule, and every B-linear map is A-linear. If f is surjective and M and both M' are B-modules, then we have $\mathrm{Hom}_B(M, M') = \mathrm{Hom}_A(M, M')$, every A-submodule of M is also a B-submodule, and M is a simple B-module if and only if it is a simple A-module.

In particular, every A-module M is a K-module. If M and M' are A-modules, then $\mathrm{Hom}_K(M, N)$, equipped with pointwise addition and scalar multiplication, is a K-module, and $\mathrm{Hom}_A(M, N)$ is a K-submodule of $\mathrm{Hom}_K(M, N)$. The endomorphism ring $\mathrm{End}_K(M) = \mathrm{Hom}_K(M, M)$ is a K-algebra, and $\mathrm{End}_A(M)$ is a K-subalgebra of $\mathrm{End}_K(M)$.

In general, $\mathrm{Hom}_A(M, N)$ and $\mathrm{End}_A(M)$ do not have the structure of an A-module. But M is an $\mathrm{End}_A(M)$-module by means of $(\varphi, m) \mapsto \varphi(m)$ for all $\varphi \in \mathrm{End}_A(M)$ and $m \in M$, and

$$\mathrm{End}_{\mathrm{End}_A(M)}(M) = \{f \in \mathrm{End}_K(M) \mid f \circ g = g \circ f \text{ for all } g \in \mathrm{End}_A(M)\}$$

is a K-subalgebra of $\mathrm{End}_K(M)$.

If M is an A-module, then the map $\mathrm{Hom}_A(A, M) \to M,\ \varphi \mapsto \varphi(1)$ is an A-module isomorphism. We identify: $\mathrm{Hom}_A(A, M) = M$.

2. An (A, B)-**bimodule** is an abelian group M together with an A-module structure $A \times M \to M,\ (a, m) \mapsto am$, and a B-module structure $B \times M \to M$, $(b, m) \mapsto bm$ such that $abm = bam$ for all $a \in A,\ b \in B$ and $m \in M$. An (A, B)-**linear** map between (A, B)-bimodules is a map which is both A-linear and B-linear.

For $n \in \mathbb{N},\ A^n$ is an (A, A°)-bimodule by means of

$$A \times A^n \to A^n,\ (a, \boldsymbol{x}) \to a\boldsymbol{x} \text{ and } A^\circ \times A^n \to A^n,\ (a, \boldsymbol{x}) \mapsto \boldsymbol{x}a.$$

Every A-module is an (K, A)-bimodule. If A is commutative then every A-module is an (A, A)-bimodule. If $f\colon A \to B$ is a K-algebra homomorphism, then B is an (A, A°)-bimodule by means of the A-action $(a, b) \mapsto f(a)b$ and the A°-action $(a, b) \mapsto bf(a)$ for all $a \in A$ and $b \in B$. In particular, A itself, $\mathsf{M}_n(A)$ and all polynomial rings $A[\boldsymbol{X}]$ are (A, A°)-bimodules.

3. Let M and N be A-modules.

a. If M is an (A, B°)-bimodule, then $\mathrm{Hom}_A(M, N)$ is a B-module by means of $(b\varphi)(m) = \varphi(mb)$ for all $b \in B$ and $m \in M$. In particular, if M is an (A, A°)-bimodule, then $\mathrm{End}_A(M)$ is an A-module, and $(a\varphi)(x) = \varphi(xa)$ for all $\varphi \in \mathrm{End}_A(M)$ and $a,\ x \in A$.

b. If N is an (A, B)-bimodule, then $\mathrm{Hom}_A(M, N)$ is a B-module by means of $(b\varphi)(m) = b\varphi(m)$ for all $b \in B$ and $m \in M$. In particular, $\mathrm{End}_K(M)$ is an A-module, and $\mathrm{End}_{\mathrm{End}_K(M)}(M)$ is an A-submodule of $\mathrm{End}_K(M)$.

4. Let M be an A-module. For $a \in A$, we define the **left multiplication** $\ell_a\colon M \to M$ by $\ell_a(m) = am$. Then $\ell_a \in \mathrm{End}_K(M) \cap \mathrm{End}_{\mathrm{End}_A(M)}(M)$, and the assignment $a \mapsto \ell_a$ defines a K-algebra homomorphism $\ell\colon A \to \mathrm{End}_{\mathrm{End}_A(M)}(M)$ with kernel $\mathrm{Ker}(\ell) = \mathrm{Ann}_A(M)$. If M is an (A, A°)-bimodule, then actually $\ell_a \in \mathrm{End}_{A^\circ}(M)$.

Alike, if M is an A°-module and $a \in A$, the **right multiplication** $r_a\colon M \to M$ is defined by $r_a(m) = ma$ for all $m \in M$. Then we have $r_a \in \mathrm{End}_K(M) \cap \mathrm{End}_{\mathrm{End}_{A^\circ}(M)}(M)$, and the assignment $a \mapsto r_a$ defines a K-algebra homomorphism $r\colon A^\circ \to \mathrm{End}_{\mathrm{End}_{A^\circ}(M)}(M)$, and $\mathrm{Ker}(r) = \mathrm{Ann}_{A^\circ}(M)$. If M is an (A, A°)-bimodule, then we even obtain $r_a \in \mathrm{End}_A(M)$, and $r_a \circ \ell_b = \ell_b \circ r_a \in \mathrm{End}_K(M)$ for all $a, b \in A$.

The following special case is used again and again.

Theorem 3.1.3. *Let A be a K-algebra. The maps*

$$\ell\colon A \to \mathrm{End}_{A^\circ}(A),\ a \mapsto \ell_a \quad \textit{and} \quad r\colon A^\circ \to \mathrm{End}_A(A),\ a \mapsto r_a$$

are K-algebra isomorphisms satisfying $\ell^{-1}\varphi = \varphi(1)$ and $r^{-1}(\varphi) = \varphi(1)$.

Proof. Observe that A is an (A, A°)-bimodule. It suffices to consider ℓ. Obviously ℓ is a K-algebra homomorphism, and if $g\colon \mathrm{End}_{A^\circ}(A) \to A$ is defined by $g(\varphi) = \varphi(1_A)$, then $g \circ \ell = \mathrm{id}_A$ and $\ell \circ g = \mathrm{id}_{\mathrm{End}_{A^\circ}(A)}$. □

Theorem 3.1.4. *Let A be a K-algebra, N an A-module and $n \in \mathbb{N}$. For $i \in [1, n]$, let $\pi_i\colon N^n \to N$ be the projection onto the i-th component and $\iota_i\colon N \to N^n$ the canonical injection into the i-th component.*

1. *The maps $\alpha_N^n\colon \mathrm{End}_A(N^n) \to \mathsf{M}_n(\mathrm{End}_A(N))$, defined by*

$$\alpha_N^n(f) = (\pi_i \circ f \circ \iota_j)_{i,j \in [1,n]}$$

and $\beta_N^n\colon \mathsf{M}_n(\mathrm{End}_A(N)) \to \mathrm{End}_A(N^n)$, defined by

$$\beta_N^n\big((f_{i,j})_{i,j\in[1,n]}\big) = \sum_{\nu,\mu=1}^{n} \iota_\nu \circ f_{\nu,\mu} \circ \pi_\mu$$

are mutually inverse K-algebra isomorphisms.
In particular, if $f \in \mathrm{End}_A(N^n)$ and $f_{i,j} = \pi_i \circ f \circ \iota_j$ for all $i, j \in [1, n]$, then

$$f = \sum_{i,j=1}^{n} \iota_i \circ f_{i,j} \circ \pi_j .$$

2. *For $f \in \mathrm{End}_K(N)$, let $f^n \in \mathrm{End}_K(N^n)$ be defined by*

$$f^n = \sum_{i=1}^{n} \iota_i \circ f \circ \pi_i, \quad \textit{i.e.,} \quad f\begin{pmatrix} x_1 \\ \vdots \\ x_n \end{pmatrix} = \begin{pmatrix} f(x_1) \\ \vdots \\ f(x_n) \end{pmatrix}.$$

Then $f \in \mathrm{End}_{\mathrm{End}_A(N)}(N)$ *if and only if* $f^n \in \mathrm{End}_{\mathrm{End}_A(N^n)}(N^n)$, *and there is a* K*-algebra isomorphism*

$$\beta\colon \mathrm{End}_{\mathrm{End}_A(N)}(N) \overset{\sim}{\to} \mathrm{End}_{\mathrm{End}_A(N^n)}(N^n), \quad \textit{given by } \beta(f) = f^n.$$

Proof. 1. Obviously, α_N^n and β_N^n are K-algebra homomorphisms and, using the identities

$$\pi_i \circ \iota_j = \delta_{i,j}\mathrm{id}_N \ \text{ for all } \ i,\, j \in [1,n] \quad \text{and} \quad \sum_{i=1}^n \iota_i \circ \pi_i = \mathrm{id}_{N^n},$$

it is easily checked that $\alpha_N^n \circ \beta_N^n = \mathrm{id}_{\mathsf{M}_n(\mathrm{End}_A(N))}$ and $\beta_N^n \circ \alpha_N^n = \mathrm{id}_{\mathrm{End}_A(N^n)}$.

2. Clearly, the assignment $f \mapsto f^n$ defines a K-algebra monomorphism $\mathrm{End}_K(N) \to \mathrm{End}_K(N^n)$. Thus it suffices to prove that for all $f \in \mathrm{End}_K(N)$:

$$f{\circ}g = g{\circ}f \ \text{ for all } \ g \in \mathrm{End}_A(N) \iff f^n{\circ}h = h{\circ}f^n \ \text{ for all } \ h \in \mathrm{End}_A(N^n).$$

Let $f \in \mathrm{End}_A(N)$.

$\Rightarrow$: Let $h \in \mathrm{End}_A(N^n)$, $h_{i,j} = \pi_i \circ h \circ \iota_j$ and $f \circ h_{i,j} = h_{i,j} \circ f$ for all $i,\, j \in [1,n]$. Then

$$f^n{\circ}h = \sum_{\nu=1}^n \iota_\nu{\circ}f{\circ}\pi_\nu{\circ}\sum_{i,j=1}^n \iota_i{\circ}h_{i,j}{\circ}\pi_j = \sum_{\nu=1}^n \iota_\nu{\circ}f{\circ}\sum_{j=1}^n h_{\nu,j}{\circ}\pi_j = \sum_{\nu,j=1}^n \iota_\nu{\circ}h_{\nu,j}{\circ}f{\circ}\pi_j$$

and

$$h{\circ}f^n = \sum_{i,j=1}^n \iota_i{\circ}h_{i,j}{\circ}\pi_j{\circ}\sum_{\nu=1}^n \iota_\nu{\circ}f{\circ}\pi_\nu = \sum_{i,j=1}^n \iota_i{\circ}h_{i,j}{\circ}f{\circ}\pi_j = f^n{\circ}h.$$

$\Leftarrow$: Let $g \in \mathrm{End}_A(N)$. Then $g^n \in \mathrm{End}_A(N^n)$,

$$(f{\circ}g)^n = f^n{\circ}g^n = g^n{\circ}f^n = (g{\circ}f)^n,$$

and therefore $f{\circ}g = g{\circ}f$. □

Definition and Remarks 3.1.5. Let A be a K-algebra.

1. A **left ideal** of A is an A-submodule of A. By a **minimal left ideal** of A we mean a minimal non-zero left ideal. Every minimal left ideal of A is a simple A-module. If $\dim_K(A) < \infty$, then A possesses minimal left ideals (indeed, every non-zero left ideal of A of minimal K-dimension is a minimal left ideal of A).

2. If M is an A-module and $X \subset M$, then the **annihilator**

$$\mathrm{Ann}_A(X) = \{a \in A \mid aX = \mathbf{0}\}$$

is a left ideal of A, and if $X^\bullet \neq \emptyset$, then $\mathrm{Ann}_A(X) \neq A$. If X is an A-submodule of M, then $\mathrm{Ann}_A(X)$ is an ideal of A. An A-module M is called **faithful** if $\mathrm{Ann}_A(M) = \mathbf{0}$.

Let A be simple. Then every minimal left ideal is a faithful simple A-module. Indeed, if L is a minimal left ideal of A, then $\mathrm{Ann}_A(L)$ is a proper ideal of A and therefore $\mathrm{Ann}_A(L) = \mathbf{0}$.

Theorem 3.1.6. *Let D be a K-division algebra, $n \in \mathbb{N}$, and for $k \in [1,n]$ let $Y_k = D\boldsymbol{e}_{1,k} \oplus \ldots \oplus D\boldsymbol{e}_{n,k}$ be the set of all matrices in $\mathsf{M}_n(D)$ with zeros outside the k-th column.*

1. *$\mathsf{M}_n(D)$ is a simple K-algebra. If $k \in [1,n]$, then the coset $Y_k = \mathsf{M}_n(D)\boldsymbol{e}_{1,k}$ is a minimal left ideal of $\mathsf{M}_n(D)$, and the map*

$$\phi_k \colon Y_k \to D^n, \quad \textit{defined by} \quad \phi_k((y_{i,j})_{i,j\in[1,n]}) = (y_{1,k}, \ldots, y_{n,k})^{\mathrm{t}},$$

is an $\mathsf{M}_n(D)$-module isomorphism. In particular, $\dim_D(Y_k) = n$, and D^n is a simple $\mathsf{M}_n(D)$-module.

2. *$\mathsf{M}_n(D) = Y_1 \oplus \ldots \oplus Y_n$, and $\mathsf{c}(\mathsf{M}_n(D)) = \mathsf{c}(D)1_n$.*

Proof. 1. Let $\mathfrak{a}$ be a non-zero ideal of $\mathsf{M}_n(D)$, $x = (x_{i,j})_{i,j\in[1,n]} \in \mathfrak{a}^\bullet$ and let p, $q \in [1,n]$ be such that $x_{p,q} \neq 0$. Then $\boldsymbol{e}_{i,j} = x_{p,q}^{-1}\boldsymbol{e}_{i,p}x\boldsymbol{e}_{q,j} \in \mathfrak{a}$ for all i, $j \in [1,n]$, and therefore $\mathfrak{a} = \mathsf{M}_n(D)$.

If $k \in [1,n]$, then it is obvious that ϕ_k is an $\mathsf{M}_n(D)$-module isomorphism. If $x = (x_{\nu,\mu})_{\nu,\mu\in[1,n]} \in Y_k^\bullet$, $q \in [1,n]$ and $x_{q,k} \neq 0$, then $x_{q,k}^{-1}\boldsymbol{e}_{p,q}x = \boldsymbol{e}_{p,k}$ for all $p \in [1,n]$. Hence $\mathsf{M}_n(D)x = Y_k$, and therefore Y_k is a minimal left ideal of $\mathsf{M}_n(D)$. In particular, $Y_k = \mathsf{M}_n(D)\boldsymbol{e}_{1,k}$, and Y_k and D^n are simple $\mathsf{M}_n(D)$-modules.

2. By definition, $Y_k = D\boldsymbol{e}_{1,k} \oplus \ldots \oplus D\boldsymbol{e}_{n,k}$, and therefore

$$\mathsf{M}_n(D) = \bigoplus_{i,j=1}^{n} D\boldsymbol{e}_{i,j} = \bigoplus_{k=1}^{n} Y_k.$$

Obviously, $\mathsf{c}(D)1_n \subset \mathsf{c}(\mathsf{M}_n(D))$. Conversely, if $x = (x_{i,j})_{i,j\in[1,n]} \in \mathsf{c}(\mathsf{M}_n(D))$ and p, q, $j \in [1,n]$, then

$$(\boldsymbol{e}_{p,q}x)_{p,j} = x_{q,j} \quad \text{and} \quad (x\boldsymbol{e}_{p,q})_{p,j} = \begin{cases} x_{p,p} & \text{if } q = j, \\ 0 & \text{if } q \neq j. \end{cases}$$

Hence $x = a1_n$ for some $a \in D$, and since

$$ab1_n = (a1_n)(b1_n) = (b1_n)(a1_n) = ba1_n \text{ for all } b \in D,$$

we obtain $a \in \mathsf{c}(D)$. □

Theorem 3.1.7 (Jacobson's density theorem). *Let A be a K-algebra, N a simple A-module and $D = \mathrm{End}_A(N)$.*

1. *D is a K-division algebra.*

2. *Let $f \in \mathrm{End}_D(N)$, $n \in \mathbb{N}$ and $z = (z_1, \ldots, z_n)^{\mathrm{t}} \in N^n$. Then there exists some $a \in A$ such that $f(z_i) = az_i$ for all $i \in [1, n]$.*

3. *Let $x_1, \ldots, x_n \in N$ be linearly independent over D and $y_1, \ldots, y_n \in N$. Then there exists some $a \in A$ such that $ax_i = y_i$ for all $i \in [1, n]$.*

Proof. 1. If $\varphi \in D^\bullet$, then $\mathrm{Ker}(\varphi) \neq N$ and $\mathrm{Im}(\varphi) \neq \mathbf{0}$. Since N is simple, we obtain $\mathrm{Ker}(\varphi) = \mathbf{0}$ and $\mathrm{Im}(\varphi) = N$. Hence φ is an isomorphism, $\varphi \in D^\times$, and D is a K-division algebra.

2. We write $N^n = N_1 \oplus \ldots \oplus N_n$ as an internal direct sum of A-submodules $N_1, \ldots, N_n$ which are A-isomorphic to N, let $J \subset [1, n]$ be maximal such that

$$Az \cap \bigoplus_{j \in J} N_j = \mathbf{0}, \quad \text{and set} \quad U = Az \oplus \bigoplus_{j \in J} N_j.$$

Then $J \subset [1, n]$, and if $j \in [1, n] \setminus J$, then $U \cap N_j \neq \mathbf{0}$ (since J is maximal), and therefore $U \cap N_j = N_j \subset U$ since N_j is simple. It follows that $N_j \subset U$ for all $j \in [1, n]$, hence $U = N^n$, and Az is a direct summand of N^n. Let $p \in \mathrm{End}_A(N^n)$ be such that $\mathrm{Im}(p) = Az$ and $p(z) = z$.

Let $f^n \colon N^n \to N^n$ be defined by $f^n((x_1, \ldots, x_n)^{\mathrm{t}}) = (f(x_1), \ldots, f(x_n))^{\mathrm{t}}$. Then, $f \in \mathrm{End}_D(N) = \mathrm{End}_{\mathrm{End}_A(N)}(N)$, hence $f^n \in \mathrm{End}_{\mathrm{End}_A(N^n)}(N^n)$ by 3.1.4.2, and therefore $f^n \circ p = p \circ f^n$. It follows that

$$f^n(z) = f^n \circ p(z) = p \circ f^n(z) \in Az, \quad \text{say} \quad f^n(z) = az \quad \text{for some} \quad a \in A,$$

and consequenly $f(z_i) = az_i$ for all $i \in [1, n]$.

3. Let $f \in \mathrm{End}_D(N)$ be such that $f(x_i) = y_i$ for all $i \in [1, n]$. Then by 2. there exists some $a \in A$ satisfying $y_i = f(x_i) = ax_i$ for all $i \in [1, n]$. □

3.2 Tensor products

Let K be a commutative ring.

In this section, we present (mostly without proofs) the basic facts about tensor products of modules and algebras over a commutative ring which we will use again and again without further reference throughout this chapter.

Definition and Remarks 3.2.1.

1. Let M and N be K-modules, F the free K-module with basis $M \times N$ and T the K-submodule of F generated by all elements of the form

$$(m+m',n)-(m,n)-(m',n), \quad (m,n+n')-(m,n)-(m,n'),$$

$$(\lambda m,n)-\lambda(m,n) \quad \text{and} \quad (m,\lambda n)-\lambda(m,n),$$

where $m,\ m' \in M,\ \ n,\ n' \in N$ and $\lambda \in K$. The quotient module

$$M \otimes_K N = F/T$$

is called the **tensor product** of M and N over K.

For $(m,n) \in M{\times}N$, we call $m{\otimes}n = (m,n){+}T \in M{\otimes}_K N$ the **elementary tensor** of m and n. By definition, $M \otimes_K N$ is a K-module, and the elementary tensors satisfy the following calculation rules for all $m,\ m' \in M,\ \ n,\ n' \in N$ and $\lambda \in K$:

$$(m+m') \otimes n = m \otimes n + m' \otimes n, \quad m \otimes (n+n') = m \otimes n + m \otimes n'$$

and

$$\lambda(m \otimes n) = \lambda m \otimes n = m \otimes \lambda n.$$

Every $z \in M \otimes_K N$ has a (generally not unique) representation in the form

$$z = \sum_{i=1}^{n} m_i \otimes n_i, \quad \text{where } n \in \mathbb{N},\ m_1,\ldots,m_n \in M \ \text{ and } \ n_1,\ldots,n_n \in N,$$

and if $a \in K$, then

$$az = \sum_{i=1}^{n} am_i \otimes n_i = \sum_{i=1}^{n} m_i \otimes an_i.$$

2. Let $M,\ N,\ P$ be K-modules. A map $\varphi\colon M{\times}N \to P$ is called K-**bilinear** if, for all $m,\ m' \in M,\ \ n,\ n' \in N$ and $\lambda \in K$ we have

$$\varphi(m+m',n) = \varphi(m,n)+\varphi(m',n), \quad \varphi(m,n+n') = \varphi(m,n)+\varphi(m,n')$$

and

$$\varphi(\lambda m,n) = \varphi(m,\lambda n) = \lambda\varphi(m,n)$$

Let $\mathcal{L}_K(M,N;P)$ be the additive group consisting of all K-bilinear maps $M \times N \to K$, endowed with pointwise addition. For $g \in \mathcal{L}_K(M,N;P)$ and $a \in K$, let $ag\colon M{\times}N \to P$ be defined by $(ag)(m,n) = g(am,n)$ $[= g(m,an)]$. Then $ag \in \mathcal{L}_K(M,N;P)$ (note that K is commutative), and the assignment $(a,g) \mapsto ag$ makes $\mathcal{L}_K(M,N;P)$ into a K-module. The tensor product is characterized by the following universal property.

For every K-bilinear map $\varphi\colon M\times N\to P$ there exists a unique K-homomorphism $g_\varphi\colon M\otimes_K N\to P$ such that

$$g_\varphi(m\otimes n)=\varphi(m,n) \text{ for all } (m,n)\in M\times N,$$

and the assignment $\varphi\mapsto g_\varphi$ defines a K-module isomorphism

$$\mathcal{L}_K(M,N;P)\stackrel{\sim}{\to}\operatorname{Hom}_K(M\otimes_K N,P).$$

3. Let $\varphi\colon M\to M'$ and $\psi\colon N\to N'$ be K-linear maps of K-modules. Then there exists a unique K-linear map

$$\varphi\otimes\psi\colon M\otimes_K N\to M'\otimes_K N'$$

such that $(\varphi\otimes\psi)(m\otimes n)=\varphi(m)\otimes\psi(n)$ for all $(m,n)\in M\times N$; it is called the **(mapping) tensor product** of φ and ψ.

4. Let M, N be K-modules. Then there are K-isomorphisms

$$\Phi\colon K\otimes_K M\stackrel{\sim}{\to}M \quad\text{satisfying}\quad \Phi(\lambda\otimes m)=\lambda m \;\text{ for all }\; (\lambda,m)\in A\times M,$$

and

$$\Psi\colon M\otimes_K N\stackrel{\sim}{\to}N\otimes_K M \quad\text{satisfying}$$
$$\Psi(m\otimes n)=n\otimes m \;\text{ für alle }\; (m,n)\in M\times N.$$

We identify: $K\otimes_K M=M$ and $M\otimes_K N=N\otimes_K M$.

5. Let R be a (not necessarily commutative) ring and N a (K,R)-bimodule. Then there is a unique R-module structure on $M\otimes_K N$ such that

$$\rho(m\otimes n)=m\otimes\rho n \text{ for all } (m,n)\in M\times N \;\text{ and }\; \rho\in R,$$

and then $M\otimes_K N$ is a (K,R)-bimodule. If R is commutative and P an R-module, then $(M\otimes_K N)\otimes_R P$ and $M\otimes_K(N\otimes_R P)$ are (K,R)-bimodules, and there exists a unique (K,R)–module isomorphism

$$\phi\colon (M\otimes_K N)\otimes_R P\to M\otimes_K(N\otimes_R P)$$

such that $\phi((m\otimes n)\otimes p)=m\otimes(n\otimes p)$ for all $(m,n,p)\in M\times N\times P$. We identify and omit the parenthesis.

Theorem 3.2.2. *Let M, N be K-modules, and let N be free with basis $(v_j)_{j\in J}$.*

1. *Every $z\in M\otimes_K N$ has a unique representation*

$$z=\sum_{j\in J}m_j\otimes v_j\,,\quad \textit{where } m_j\in M \textit{ and } m_j=0 \textit{ for almost all } j\in J.$$

2. Let A be a K-algebra. Then $A \otimes_K N$ is a free A-module, and $(1_A \otimes v_j)_{j\in J}$ is an A-basis of $A \otimes_K N$.

3. Let $f\colon M \to M'$ be an injective homomorphism of K-modules. Then the homomorphism $f \otimes \mathrm{id}_N\colon M \otimes_K N \to M' \otimes_K N$ is injective, too.

Proof. 1. It suffices to prove that the map

$$\phi\colon M^{(J)} \to M \otimes_K N, \quad \text{defined by} \quad \phi((m_j)_{j\in J}) = \sum_{j\in J} m_j \otimes v_j,$$

is bijective. The map

$$\psi'\colon M \times N \to M^{(J)}, \quad \text{defined by} \quad \psi'\Big(m, \sum_{j\in J} \lambda_j v_j\Big) = (\lambda_j m)_{j\in J}$$

is K-bilinear, and thus there exists a K-linear map

$$\psi\colon M \otimes_K N \to M^{(J)} \quad \text{such that} \quad \psi\Big(m \otimes_K \sum_{j\in J} \lambda_j v_j\Big) = (\lambda_j m)_{j\in J}.$$

It follows that

$$\begin{aligned} \phi\circ\psi\Big(m \otimes \sum_{j\in J} \lambda_j v_j\Big) &= \phi((\lambda_j m)_{j\in J}) = \sum_{j\in J} \lambda_j m \otimes v_j = \sum_{j\in J} m \otimes \lambda_j v_j \\ &= m \otimes \sum_{j\in J} \lambda_j v_j, \end{aligned}$$

and if $\iota_j\colon M \to M^{(J)}$ denotes the embedding into the j-th component, then

$$\psi\circ\phi((m_j)_{j\in J}) = \psi\Big(\sum_{j\in J} m_j \otimes v_j\Big) = \sum_{j\in J} \psi(m_j \otimes v_j) = \sum_{j\in J} \iota_j(m_j) = (m_j)_{j\in J}.$$

Hence ϕ and ψ are mutually inverse K-linear isomorphisms.

2. If $z \in A \otimes_K N$, then by 1. there exists a unique family $(a_j)_{j\in J} \in A^{(J)}$ such that

$$z = \sum_{j\in J} a_j \otimes v_j = \sum_{j\in J} a_j(1_A \otimes v_j).$$

Hence $(1_A \otimes v_j)_{j\in J}$ is an A-basis of $A \otimes_K N$.

3. Let $z \in \mathrm{Ker}(f \otimes \mathrm{id}_N)$. By 1., z has a unique representation

$$z = \sum_{j\in J} m_j \otimes v_j, \quad \text{where } m_j \in M \text{ and } m_j = 0 \text{ for almost all } j \in J.$$

Then

$$0 = (f \otimes \mathrm{id}_N)(z) = \sum_{j\in J} f(m_j) \otimes v_j,$$

hence $f(m_j) = 0$ (again by 1.) and thus $m_j = 0$ for all $j \in J$, which implies $z = 0$. □

Definitions and Remarks 3.2.3. Let A and B be K-algebras with structural homomorphisms $\varepsilon\colon K \to A$ and $\eta\colon K \to B$.

1. Let M be a K-module. Since B is a (K,B)-bimodule, it follows by 3.2.1.**5** that $M_B = M \otimes_K B$ is a B-module, and the map

$$\delta_B = \mathrm{id}_M \otimes \eta\colon M = M \otimes_K K \to M \otimes_K B = M_B$$

is K-linear. We call M_B the **base extension** of M with B. If M is a free K-module with basis $(u_i)_{i\in I}$, then M_B is a free B-module with basis $(u_i \otimes 1_B)_{i\in I}$ by 3.2.2.2, and if additionally η is injective, then δ_B is injective by 3.2.2.3. In this case, we usually identify: $u \otimes 1_B = u$ for all $u \in M$ and M is a K-submodule of M_B.

2. If $\varphi\colon M \to M'$ is K-linear, then $\varphi \otimes \mathrm{id}_B\colon M_B \to M'_B$ is B-linear. If M, M' are K-modules, then by the identifications made in 3.2.1(**4** and **5**) we obtain

$$(M \otimes_K N)_B = M_B \otimes_B N_B.$$

3. On $A \otimes_K B$, there exists a unique multiplication

$$\mu\colon (A \otimes_K B) \times (A \otimes_K B) \to A \otimes_K B$$

such that $\mu(a \otimes b, a' \otimes b') = (a \otimes b)(a' \otimes b') = aa' \otimes bb'$ for all a, $a' \in A$ and b, $b' \in B$. With this multiplication, $A \otimes_K B$ is a K-algebra with unit element $1_A \otimes 1_B$ and structural homomorhphism $\varepsilon \otimes \eta\colon K = K \otimes_K K \to A \otimes_K B$, given by

$$(\varepsilon \otimes \eta)(\lambda) = \lambda 1_A \otimes 1_B = 1_A \otimes \lambda 1_B = \lambda(1_A \otimes 1_B) \quad \text{for all} \quad \lambda \in K.$$

The K-algebra $A \otimes_K B$ is called the **(algebra) tensor product** of A and B. The maps

$$\delta_B\colon A = A \otimes_K K \overset{\mathrm{id}_A \otimes \eta}{\to} A \otimes_K B = A_B \quad \text{and}$$

$$\delta_A\colon B = K \otimes_K B \overset{\eta \otimes \mathrm{id}_B}{\to} A \otimes_K B = B_A$$

are K-algebra homomorphisms satisfying $\delta_B(a) = a \otimes 1_B$ and $\delta_A(b) = 1_A \otimes b$ for all $a \in A$ and $b \in B$.

If A is a free K-module and η is injective, then (as in **1**) δ_B is injective, and occasionally we shall identify: $A \subset A \otimes_K B$. Alike, if B is a free K-module and ε is injective, then δ_A is injective, and occasionally we shall identify: $B \subset A \otimes_K B$.

If $\varphi\colon A \to A'$ and $\psi\colon B \to B'$ are K-algebra homomorphisms, then it is easily checked that $\varphi \otimes \psi\colon A \otimes_K B \to A' \otimes_K B'$ is a K-algebra homomorpism, too.

4. Let L be a commutative K-algebra. Then $A \otimes_K L = A_L$ is an L-algebra with structural homomorphism $\delta_A\colon L \to A_L$. We call A_L the **base extension** of the K-algebra A with L.

The most interesting case is when K is a field, $A \neq \mathbf{0}$ is a finite-dimensional K-algebra and L is an extension field of K (see Section 3.4 and the examples in 3.1.6 below). In this case, $\delta_L \colon A \to A_L$ is injective, we view A as a K-subalgebra of the L-algebra A_L, and $\dim_L(A_L) = \dim_K(A)$.

We close this section with some remarks concerning matrix algebras and polynomial algebras.

Remarks 3.2.4. Let A be a K-algebra and $n \in \mathbb{N}$.

1. Recall that the K-algebra $\mathsf{M}_n(K)$ is a free K-module with basis $(\mathsf{e}_{i,j})_{i,j\in[1,n]}$, multiplication given by $\mathsf{e}_{i,j}\mathsf{e}_{\nu,\mu} = \delta_{j,\nu}\mathsf{e}_{i,\mu}$ for all $i,\ j, \nu,\ \mu \in [1,n]$, and structural homomorphism $\iota \colon K \to \mathsf{M}_n(K)$, given by $\iota(\lambda) = \lambda 1_n$ for all $\lambda \in K$. Let $p \colon \mathsf{M}_n(K) \to K$ be the K-linear map defined by $p(a) = a_{1,1}$ for all $a = (a_{i,j})_{i,j\in[1,n]} \in \mathsf{M}_n(K)$. Then $p\circ\iota = \mathrm{id}_K$, hence

$$(p \otimes \mathrm{id}_A)\circ(\iota \otimes \mathrm{id}_A) = (p\circ\iota) \otimes \mathrm{id}_A = \mathrm{id}_K \otimes \mathrm{id}_A = \mathrm{id}_A,$$

and therefore the K-algebta homomorphism

$$\iota_A = \iota \otimes \mathrm{id}_A \colon K \otimes_K A = A \to \mathsf{M}_n(K) \otimes_K A = \mathsf{M}_n(K)_A$$

is injective. As $\mathsf{M}_n(K)_A$ is a free A-module with basis $\mathsf{e}^*_{i,j} = (\mathsf{e}_{i,j} \otimes 1_A)_{i,j\in[1,n]}$, and the multiplication is given by $\mathsf{e}^*_{i,j}\mathsf{e}^*_{\nu,\mu} = \delta_{j,\nu}\mathsf{e}^*_{i,\mu}$, we identify $\mathsf{e}^*_{i,j} = \mathsf{e}_{i,j}$ and obtain

$$\mathsf{M}_n(K)_A = \mathsf{M}_n(A).$$

2. Let A and B be K-algebras and $n \in \mathbb{N}$. Then

$$\mathsf{M}_n(A) \otimes_K B \cong \mathsf{M}_n(K) \otimes_K A \otimes_K B \cong \mathsf{M}_n(A \otimes_K B).$$

We identify: $\mathsf{M}_n(A) \otimes_K B = \mathsf{M}_n(A \otimes_K B)$.

3. For $m,\ n \in \mathbb{N}$, let $(\mathsf{e}^{(n)}_{i,j})_{i,j\in[1,n]}$ be the matrix units in $\mathsf{M}_n(K)$ and $(\mathsf{e}^{(m)}_{l,k})_{l,k\in[1,m]}$ the matrix units in $\mathsf{M}_m(K)$. Then we obtain

$$\begin{aligned}(\mathsf{e}^{(n)}_{i,j} \otimes \mathsf{e}^{(m)}_{l,k})(\mathsf{e}^{(n)}_{i',j'} \otimes \mathsf{e}^{(m)}_{l',k'}) &= (\mathsf{e}^{(n)}_{i,j}\mathsf{e}^{(n)}_{i',j'} \otimes \mathsf{e}^{(m)}_{l,k}\mathsf{e}^{(m)}_{l'k'} = \delta_{j,i'}\mathsf{e}^{(n)}_{i,j'} \otimes \delta_{k,l'}\mathsf{e}^{(m)}_{l,k'} \\ &= \begin{cases} \mathsf{e}^{(n)}_{i,j'} \otimes \mathsf{e}^{(m)}_{l,k'}, & \text{if } (j,k) = (i',l'), \\ 0 & \text{otherwise.}\end{cases}\end{aligned}$$

Hence there exists a K-algebra isomorphism $\mathsf{M}_m(K) \otimes_K \mathsf{M}_n(K) \overset{\sim}{\to} \mathsf{M}_{mn}(K)$ which arises from an appropriate ordering of the rows and columns. We identify:

$$\mathsf{M}_m(K) \otimes_K \mathsf{M}_n(K) = \mathsf{M}_{mn}(K).$$

4. We finally outline the (much simpler) case of polynomial algebras. Let $n \in \mathbb{N}$ and $\boldsymbol{X} = (X_1, \ldots, X_n)$ a vector of (algebraically independent) indeterminates. Then the polynomial ring $K[\boldsymbol{X}]$ is commutative K-algebra, and if

A is any K-algebra, then we identify $K[\boldsymbol{X}]_A = A[\boldsymbol{X}]$ by means of a natural A-linear K-isomorphism. Also, if $\boldsymbol{Y}$ is another vector of indeterminates, then we identify $K[\boldsymbol{X}, \boldsymbol{Y}] = K[\boldsymbol{X}] \otimes_K K[\boldsymbol{Y}]$.

3.3 Structure of central simple algebras

Let K be a field; K-algebras are assumed to be non-zero.

If A is a K-algebra, then we usually identify $K = K1_A \subset \mathsf{c}(A) \subset A$, and we call A **central** if $\mathsf{c}(A) = K$. If L/K is a field extension, then $A_L = A \otimes_K L$ is an L-algebra, and $\dim_L(A_L) = \dim_K(A)$ by 3.2.2.2.

Theorem 3.3.1. *Let A and B be K-algebras, and let $A' \subset A$ and $B' \subset B$ be K-subalgebras. Then $\mathsf{c}_{A \otimes_K B}(A' \otimes_K B') = \mathsf{c}_A(A') \otimes_K \mathsf{c}_B(B')$. In particular,*

$$\mathsf{c}(A) \otimes_K \mathsf{c}(B) = \mathsf{c}(A \otimes_K B).$$

If A and B are both central, then $A \otimes_K B$ is central, too.

Proof. Obviously, $\mathsf{c}_A(A') \otimes_K \mathsf{c}_B(B') \subset \mathsf{c}_{A \otimes_K B}(A' \otimes_K B')$. As to the converse, let first $(u_i)_{i \in I}$ be a K-Basis of A and $c \in \mathsf{c}_{A \otimes_K B}(A' \otimes_K B')$. Then 3.2.2.1 implies

$$c = \sum_{i \in I} u_i \otimes b_i \quad \text{with a uniquely determined family} \quad (b_i)_{i \in I} \in B^{(I)}.$$

For all $b \in B'$ we obtain

$$c(1 \otimes b) = (1 \otimes b)c, \quad \text{hence} \quad \sum_{i \in I} u_i \otimes b_i b = \sum_{i \in I} u_i \otimes b b_i,$$

and therefore $b_i b = b b_i$, i. e., $b_i \in \mathsf{c}_B(B')$ for all $i \in I$. Hence $c \in A \otimes_K \mathsf{c}_B(B')$.

Let next $(v_j)_{j \in J}$ be a K-basis of $\mathsf{c}_B(B')$. Then

$$c = \sum_{j \in J} a_j \otimes v_j \text{ with a uniquely determined family } (a_j)_{j \in J} \in A^{(J)}.$$

For all $a \in A'$ we obtain

$$c(a \otimes 1) = (a \otimes 1)c \quad \text{and therefore} \quad \sum_{j \in J} a_j a \otimes v_j = \sum_{j \in J} a a_j \otimes v_j.$$

As above, we get $a_j \in \mathsf{c}_A(A')$ for all $j \in J$, and therefore $c \in \mathsf{c}_A(A') \otimes_K \mathsf{c}_B(B')$. In particular, $\mathsf{c}(A \otimes_K B) = \mathsf{c}_{A \otimes_K B}(A \otimes_K B) = \mathsf{c}_A(A) \otimes_K \mathsf{c}_B(B) = \mathsf{c}(A) \otimes_K \mathsf{c}(B)$. If A and B are both central, then

$$\mathsf{c}(A \otimes_K B) = \mathsf{c}(A) \otimes_K \mathsf{c}(B) = K \otimes_K K = K,$$

and thus $A \otimes_K B$ is central, too. □

Theorem 3.3.2 (Wedderburn). *Let A be a finite-dimensional K-algebra, M a faithful simple A-module and $D = \mathrm{End}_A(M)$.*

1. *For every finitely generated A-module N, there exists an A-module isomorphism $N \xrightarrow{\sim} M^r$ for some $r \in \mathbb{N}$. In particular:*

- *Every faithful simple A-module is isomorphic to M.*
- *Two finitely generated A-modules N and N' are A-isomorphic if and only if $\dim_K(N) = \dim_K(N')$.*

2. *A is a simple algebra, D is a K-division algebra, and there exists a K-algebra isomorphism $\phi\colon A \xrightarrow{\sim} \mathsf{M}_n(D^\circ)$, where $n = \dim_D(M)$.*

Proof. Being simple, $M = Am$ for some $m \in M$. Hence, if $(u_1, \ldots, u_n)$ is a K-basis of A, then $M = Ku_1m + \ldots + Ku_nm$, and therefore $\dim_K(M) < \infty$.

1. SPECIAL CASE: $N = A$. Let U be a K-basis of M. Then U is finite, and $\mathrm{Ann}_A(U) = \mathbf{0}$ since M is faithful. Let $B = \{m_1, \ldots, m_n\}$ be a minimal subset of U such that $\mathrm{Ann}_A(B) = \mathbf{0}$, and für $i \in [0, n]$, we set $\mathfrak{a}_i = \mathrm{Ann}_A(m_{i+1}, \ldots, m_n)$. Then $\mathfrak{a}_0 = \mathbf{0} \subset \mathfrak{a}_1 \subset \ldots \subset \mathfrak{a}_n = A$ is a sequence of left ideals of A, and we assert tat $\mathfrak{a}_{i-1} \neq \mathfrak{a}_i$ for all $i \in [1, n]$. Indeed, suppose to the contrary that $\mathfrak{a}_{i-1} = \mathfrak{a}_i$ for some $i \in [1, n]$. Then $\mathbf{0} \neq \mathrm{Ann}_A(B \setminus \{m_i\}) \subset \mathfrak{a}_i = \mathfrak{a}_{i-1} \subset \mathrm{Ann}_A(m_i)$, and thus $\mathrm{Ann}_A(B) \neq \mathbf{0}$, a contradiction.

For $i \in [1, n]$, let $M_i = \mathfrak{a}_i m_i$ and define $\varphi_i\colon \mathfrak{a}_i \to M_i$ by $\varphi_i(x) = xm_i$ for all $x \in \mathfrak{a}_i$. Then M_i is an A-submodule of M, and φ_i is an A-epimorphism with kernel $\mathrm{Ker}(\varphi_i) = \mathfrak{a}_i \cap \mathrm{Ann}_A(m_i) = \mathfrak{a}_{i-1}$. Hence φ_i induces an isomorphism $\mathfrak{a}_i/\mathfrak{a}_{i-1} \xrightarrow{\sim} M_i$. It follows that $M_i \neq \mathbf{0}$ and thus $M_i = M$. In particular, $\varphi_1\colon \mathfrak{a}_1 \to M$ is an isomorphism, and we define $\phi_i\colon \mathfrak{a}_i \to M^i$ by means of $\phi_i(x) = (xm_1, \ldots, xm_i)^{\mathrm{t}}$ for all $x \in \mathfrak{a}_i$. Then ϕ_i is A-linear,

$$\mathrm{Ker}(\phi_i) = \mathfrak{a}_i \cap \mathrm{Ann}_A(m_1, \ldots, m_i) = \mathrm{Ann}_A(B) = \mathbf{0},$$

and we prove by induction on i that ϕ_i is bijective. For $i = 1$, this is already done since $\phi_1 = \varphi_1$. Thus suppose that $i \in [2, n]$ and that $\phi_{i-1}\colon \mathfrak{a}_{i-1} \to M^{i-1}$ is already bijective. Then

$$\phi_i(\mathfrak{a}_{i-1}) = \phi_{i-1}(\mathfrak{a}_{i-1}) \times \{\mathfrak{a}_{i-1}m_i\} = M^{i-1} \times \mathbf{0} \subsetneq \phi_i(\mathfrak{a}_i) \subset M^i,$$

and since $M^i/(M^{i-1}\times\mathbf{0}) = M$ is simple, we get $\phi_i(\mathfrak{a}_i) = M^i$. For $i = n$, it follows that $\phi_n\colon \mathfrak{a}_n = A \to M^n$ is an A-module isomorphism.

GENERAL CASE: Let N be any finitely generated A-module and $\psi\colon A^r \to N$ an A-module epimorphism for some $r \in \mathbb{N}$. Then $\phi_n\colon A \overset{\sim}{\to} M^n$ induces an A-module isomorphism $\phi_n^r\colon A^r \overset{\sim}{\to} M^{nr}$, and therefore A^r is the internal direct sum of A-submodules $M_1,\ldots,M_{nr}$ which are isomorphic to M (and thus simple). Let $J \subset [1,nr]$ be maximal such that

$$\operatorname{Ker}(\psi) \cap \bigoplus_{j\in J} M_j = \mathbf{0}, \quad \text{and set} \quad P = \operatorname{Ker}(\psi) \oplus \bigoplus_{j\in J} M_j \subset A^r.$$

If $j \in [1,nr]\setminus J$, then $P \cap M_j \neq \mathbf{0}$ (by the maximality of J) and therefore $P \cap M_j = M_j \subset P$. Its follows that $M_j \subset P$ for all $j \in [1,nr]$, hence $P = A^r$, $\operatorname{Ker}(\psi) = \mathbf{0}$,

$$\psi\colon A^r = \bigoplus_{j\in J} M_j \overset{\sim}{\to} N$$

is an A-module isomorphism, and thus there exists an A-module isomorphism $N \overset{\sim}{\to} M^{|J|}$.

In particular: If N is a simple A-module and $N \cong M^r$ for some $r \in \mathbb{N}$, then $r = 1$. If N and N' are finitely generated A-modules and $r, r' \in \mathbb{N}$ are such that $N \cong M^r$ and $N' \cong M^{r'}$, then $\dim_K(N) = r\dim_K(M)$ and $\dim_K(N') = r'\dim_K(M)$. Hence $\dim_K(N) = \dim_K(N')$ if and only if $r = r'$, that is, if and only if N and N' are isomorphic A-modules.

2. By 3.1.7.1, D is a K-division algebra, and by 1. there exists an A-module isomorphism $A \overset{\sim}{\to} M^n$ for some $n \in \mathbb{N}$. Using 3.1.3 and 3.1.4.1, we obtain a K-algebra isomorphism

$$\phi\colon A \overset{\sim}{\to} \operatorname{End}_A(A)^\circ \overset{\sim}{\to} \operatorname{End}_A(M^n)^\circ \overset{\sim}{\to} \mathsf{M}_n(\operatorname{End}_A(M))^\circ = \mathsf{M}_n(D)^\circ \overset{\sim}{\to} \mathsf{M}_n(D^\circ).$$

By 3.1.6.1, $\mathsf{M}_n(D^\circ)$ is simple, $Y_1 = \mathsf{M}_n(D^\circ)\boldsymbol{e}_{1,1}$ is a minimal left ideal of $\mathsf{M}_n(D^\circ)$, and $\dim_D(Y_1) = n$. Hence A is simple, and $\phi^{-1}(Y_1)$ is a minimal left ideal of A. By 3.1.5.**1**, $\phi^{-1}(Y_1)$ is a simple A-module, and by 1. there is an A-module isomorphism $M \overset{\sim}{\to} Y_1$. Hence

$$\begin{aligned}\dim_K(D)\dim_D(M) = \dim_K(M) = \dim_K(Y_1) &= \dim_K(D)\dim_D(Y_1)\\ &= \dim_K(D)n,\end{aligned}$$

and therefore $n = \dim_D(M)$. □

Corollary 3.3.3. *Let A be a finite-dimensional K-algebra.*

1. *A is simple if and only if there exist a K-division algebra D and a K-algebra isomorphism $A \overset{\sim}{\to} \mathsf{M}_n(D)$ for some $n \in N$. If this is the case, then n and D are (up to K-algebra isomorphisms) uniquely determined by A. Moreover, D is central if and only if A is central.*

2. Let K be algebraically closed. Then A is simple if and only if $A \cong \mathsf{M}_n(K)$ for some $n \in \mathbb{N}$, and A is a K-divison algebra if and only if $A = K$.

Proof. 1. If $A \cong \mathsf{M}_n(D)$ for some $n \in \mathbb{N}$ and a K-division algebra D, then A is simple and $\mathsf{c}(A) \cong \mathsf{c}(\mathsf{M}_n(D)) = \mathsf{c}(D)1_n$ by 3.1.6. In particular, D is central if and only if A is central.

Let A be a simple K-algebra. By 3.1.5, A contains a minimal left ideal L, and (up to isomorphisms) this is the unique faithful simple A-module. By 3.3.2.2, $D = \mathrm{End}_A(L)$ is a K-division algebra, and if $n = \dim_D(L)$, then there is a K-algebra isomorphism $A \overset{\sim}{\to} \mathsf{M}_n(D^\circ)$. By a slight change of terminology (replace D° with D), there exist $n \in \mathbb{N}$, a K-division algebra D and a K-algebra isomorphism $A \overset{\sim}{\to} \mathsf{M}_n(D)$. Moreover, D (up to isomorphisms) and n are uniquely determined by A.

2. By 1., it suffices to prove that K is the only K-division algebra. Thus let A be a K-division algebra. If $x \in A$, then $[K(x):K] \le \dim_K(A)$, hence $K(x)/K$ is algebraic and therefore $K(x) = K$. □

Let A be a K-algebra and B a K-subalgebra of A. Then $A \otimes_K B^\circ$ is a K-algebra. For $a \in A$ and $b \in B$, we define

$$F_{a,b}\colon A \to A \quad \text{by} \quad F_{a,b}(x) = axb \ \text{ for all } \ a,\, x \in A \ \text{ and } \ b \in B.$$

Then $F_{a,b} \in \mathrm{End}_K(A)$, and the map $A \times B^\circ \to \mathrm{End}_K(A)$, $(a,b) \mapsto F_{a,b}$ is K-bilinear. Hence there exists a K-algebra homomorphism

$$F\colon A \otimes_K B^\circ \to \mathrm{End}_K(A)$$

such that $F(a \otimes b)(x) = axb$ for all $a,\, x \in A$ and $b \in B$. This makes A into an $A \otimes_K B^\circ$-module by means of $(a \otimes b)x = axb$ for all $a,\, x \in A$ and $b \in B$. In particular, A is an $A \otimes_K A^\circ$-module, and the $A \otimes_K A^\circ$-submodules of A are precisely the ideals of A. Consequently, A is a simple $A \otimes_K A^\circ$-module if and only if A is a simple algebra.

Theorem 3.3.4. *Let A be a K-algebra and B a K-subalgebra of A. Then the map*

$$\varepsilon\colon \mathrm{End}_{A \otimes_K B^\circ}(A) \to \mathsf{c}_A(B)^\circ, \quad \textit{defined by } \ \varepsilon(\varphi) = \varphi(1),$$

is a K-algebra isomorphism.
We identify: $\mathrm{End}_{A \otimes_K B^\circ}(A) = \mathsf{c}_A(B)^\circ$, in particular $\mathrm{End}_{A \otimes_K A^\circ}(A) = \mathsf{c}(A)$.

Proof. If $\varphi \in \mathrm{End}_{A \otimes_K B^\circ}(A)$, then $\varphi(axb) = a\varphi(x)b$ for all $a,\, x \in A$ and $b \in B$. In particular, for $a = x = 1$ it follows that $\varphi(b) = \varphi(1)b$ for all $b \in B$, and for $b = x = 1$ it follows that $\varphi(a) = a\varphi(1)$ for all $a \in A$. Hence $b\varphi(1) = \varphi(1)b$ for all $b \in B$, and consequently $\varphi(1) \in \mathsf{c}_A(B)$. By definition, ε is a K-module monomorphism. If $\varphi,\, \psi \in \mathrm{End}_{A \otimes_K B^\circ}(A)$, then $\varphi \circ \psi(1) = \varphi(\psi(1)) = \psi(1)\varphi(1)$, and therefore ε is a K-algebra homomorphism. For $c \in \mathsf{c}_A(B)$, we define

$\varphi_c \in \operatorname{End}_K(A)$ by $\varphi_c(x) = xc$. Then $\varphi_c(axb) = axbc = axcb = a\varphi_c(x)b$ for all $a \in A$ and $b \in B$, hence $\varphi_c \in \operatorname{End}_{A\otimes_K B^\circ}(A)$ and $\varepsilon(\varphi_c) = c$. Therefore ε surjective and thus a K-algebra isomorphism. $\square$

Theorem 3.3.5. *Let A and B be K-algebras.*

1. *Let A be central simple and $B = K \otimes_K B \subset A \otimes_K B$ (see 3.2.3.**2**). Let I be an ideal of $A \otimes_K B$, and let $J = I \cap B$. Then $I = A \otimes_K J$.*

2. *If $A \otimes_K B$ is simple, then both A and B are simple.*

3. *If A and B are simple and either A or B is central, then $A \otimes_K B$ is simple.*

4. *If both A and B are central simple, then $A \otimes_K B$ is central simple, too.*

Proof. 1. Clearly, $J \subset I$ implies $A \otimes_K J \subset I$. Thus let $t \in I$. Then there exsist some $n \in \mathbb{N}$, $b_1, \ldots, b_n \in B$ and $u_1, \ldots, u_n \in A$ such that $(u_1, \ldots, u_n)$ is linearly independent over K and

$$t = \sum_{i=1}^{n} u_i \otimes b_i.$$

Being a simple algebra, A is a simple $A \otimes_K A^\circ$-module, and therefore we have $\operatorname{End}_{A\otimes_K A^\circ}(A) = \mathsf{c}(A) = K$ by 3.3.4. By 3.1.7.3, there exist elements $c_1, \ldots, c_n \in A \otimes_K A^\circ$ such that $c_j u_i = \delta_{i,j} 1_A$ for all $i, j \in [1, n]$. The $A \otimes_K A^\circ$-module structure of A induces an $A \otimes_K A^\circ$-module structure on $A \otimes_K B$ by means of $c(a \otimes b) = ca \otimes b$ for all $c \in A \otimes_K A^\circ$, $a \in A$ and $b \in B$. If $a_1, a_2, a \in A$ and $b \in B$, then

$$\begin{aligned}(a_1 \otimes a_2)(a \otimes b) &= ((a_1 \otimes a_2)a) \otimes b = a_1 a a_2 \otimes b = (a_1 a \otimes b)(a_2 \otimes 1)\\ &= (a_1 \otimes 1)(a \otimes b)(a_2 \otimes 1),\end{aligned}$$

and therefore $(a_1 \otimes a_2)x = (a_1 \otimes 1)x(a_2 \otimes 1)$ for all $x \in A \otimes_K B$. In particular, $(a_1 \otimes a_2)I = (a_1 \otimes 1)I(a_2 \otimes 1) \subset I$ and thus $cI \subset I$ for all $c \in A \otimes_K A^\circ$. For all $j \in [1, n]$, we obtain

$$c_j t = \sum_{i=1}^{n} c_j u_i \otimes b_i = 1 \otimes b_j = b_j \in I \cap B = J, \quad \text{hence} \quad u_j \otimes b_j \in A \otimes_K J,$$

and therefore $t \in A \otimes_K J$.

2. Let $A \otimes_K B$ be simple, and let J be an ideal of B. Then $A \otimes_K J$ is an ideal of $A \otimes_K B$, and we assert that $(A \otimes_K J) \cap B = J$. Indeed, $A = K \oplus U$ for some K-subspace U of A, hence $A \otimes_K B = B \oplus (U \otimes_K B)$, $A \otimes_K J = J \oplus (U \otimes_K J)$,

and thererore $(A \otimes_K J) \cap B = J$. Since $A \otimes_K J = \mathbf{0}$ or $A \otimes_K J = A \otimes_K B$, it follows that $J = \mathbf{0}$ or $J = B$. Hence B is simple, and for the same reason A is simple.

3. Suppose that A is central simple and B is simple. If I is a non-zero ideal of $A \otimes_K B$, then $I = A \otimes_K J$ for some non-zero ideal J of B by 1. Hence $J = B$, $I = A \otimes_K B$, and $A \otimes_K B$ is simple.

4. The K-algebra $A \otimes_K B$ is central by 3.3.1.1 and simple by 2. □

Theorem 3.3.6 (Skolem–Noether). *Let A and B be finite-dimensional central simple K-algebras.*

1. *Let f_1, $f_2 \colon A \to B$ be K-algebra homomorphisms. Then there exists some $u \in B^\times$ such that $f_2(x) = uf_1(x)u^{-1}$ for all $x \in A$.*

2. *Let A_1 and A_2 be simple K-subalgebras of B and $f \colon A_1 \to A_2$ a K-algebra isomorphism. Then there exists some $u \in B^\times$ such that $f(x) = uxu^{-1}$ for all $x \in A_1$. In particular, $u\,\mathsf{c}_B(A_1)u^{-1} = \mathsf{c}_B(A_2)$.*

3. *Let f be a K-algebra automorphism of B. Then there exists an up to factors from $K^\times$ uniquely determined $u \in B^\times$ such that $f(x) = uxu^{-1}$ for all $x \in B$.*

Proof. 1. For $i \in \{1, 2\}$, the map $f_i \otimes \mathrm{id}_{B^\circ} \colon A \otimes_K B^\circ \to B \otimes_K B^\circ$ is a K-algebra homomorphism which makes the $(B \otimes_K B^\circ)$-module B to an $(A \otimes_K B^\circ)$-module satisfying $(a \otimes b)y = f_i(a)yb$ for all $a \in A$ and y, $b \in B$. Let B_i be the $(A \otimes_K B^\circ)$-module B equipped with the module structure induced by f_i. By 3.3.5.4, $A \otimes_K B^\circ$ is central simple, and since $\dim_K(B_1) = \dim_K(B_2)$, there exists an $A \otimes_K B^\circ$-module isomorphism $\varphi \colon B_1 \xrightarrow{\sim} B_2$ by 3.3.2.1. By definition, $\varphi(f_1(x)yb) = f_2(x)\varphi(y)b$ for all $x \in A$ and b, $y \in B$. In particular, for $y = x = 1$ we get $\varphi(b) = \varphi(1)b$ for all $b \in B$, and for $y = b = 1$ we get $\varphi(f_1(x)) = f_2(x)\varphi(1)$ for all $x \in A$. As $\varphi = \ell_{\varphi(1)} \in \mathrm{End}_{B^\circ}(B)$ is an isomorphism, we get $u = \varphi(1) \in B^\times$. Now $f_2(x)u = \varphi(f_1(x)) = uf_1(x)$, and therefore $f_2(x) = uf_1(x)u^{-1}$ for all $x \in A$.

2. We apply 1. with $f_1 = (A_1 \hookrightarrow B)$ and $f_2 = f \colon A_1 \to A_2 \hookrightarrow B$. Accordingly, there exists some $u \in B^\times$ such that $f(x) = uxu^{-1}$ for all $x \in A_1$.

3. By 2., applied with $A_1 = A_2 = B$, there exists some $u \in B^\times$ such that $f(x) = uxu^{-1}$ for all $x \in B$. If u, $v \in B^\times$ are such that $uxu^{-1} = vxv^{-1}$ (and thus $v^{-1}ux = xv^{-1}u$) for all $x \in B$, then $v^{-1}u \in \mathsf{c}(B)^\times = K^\times$. □

Theorem 3.3.7 (Centralizer theorem). *Let A be a finite-dimensional central simple K-algebra and B a simple K-subalgebra of A.*

1. *There exists a K-algebra isomorphism $A \otimes_K B^\circ \xrightarrow{\sim} \mathsf{c}_A(B) \otimes_K \mathrm{End}_K(B)$, and $\dim_K(A) = \dim_K(B)\dim_K(\mathsf{c}_A(B))$. In particular, $A \otimes_K A^\circ \cong \mathrm{End}_K(A)$.*

2. *$\mathsf{c}_A(B)$ is simple, $\mathsf{c}_A(\mathsf{c}_A(B)) = B$, and $\mathsf{c}(\mathsf{c}_A(B)) = \mathsf{c}(B)$.*

Proof. 1. For $b \in B$, we define $\ell_b, r_b \in \mathrm{End}_K(B)$ by $\ell_b(y) = by$ and $r_b(y) = yb$ for all $y \in B$. Then the maps

$$\ell\colon \begin{cases} B & \to \mathrm{End}_K(B) \\ b & \mapsto \quad \ell_b \end{cases} \quad \text{and} \quad r\colon \begin{cases} B^\circ & \to \mathrm{End}_K(B) \\ b & \mapsto \quad r_b \end{cases}$$

are non-zero K-algebra homomorphisms, and since B is simple, they are monomorphisms. We show first:

A. $\mathsf{c}_{\mathrm{End}_K(B)}(l(B)) = r(B^\circ)$.

Proof of **A.** If $\varphi \in \mathsf{c}_{\mathrm{End}_K(B)}(\ell(B))$, then $\varphi \circ \ell_b = \ell_b \circ \varphi$ for all $b \in B$, and therefore $\varphi(b) = \varphi \circ \ell_b(1) = \ell_b \circ \varphi(1) = b\varphi(1)$. Hence $\varphi = r_{\varphi(1)} \in r(B)$. Conversely, if $b \in B$, then $r_b \circ \ell_{b'} = \ell_{b'} \circ r_b$ for all $b' \in B$ and therefore $r_b \in \mathsf{c}_{\mathrm{End}_K(B)}(\ell(B))$. $\square$[**A.**]

Now let $\alpha, \beta\colon B \to A \otimes_K \mathrm{End}_K(B)$ be defined by $\alpha(b) = b \otimes \mathrm{id}_B$ and $\beta(b) = 1 \otimes \ell_b$ for all $b \in B$. Then α and β are K-algebra monomorphisms, and therefore $\alpha(B)$ and $\beta(B)$ are isomorphic K-subalgebras of $A \otimes_K \mathrm{End}_K(B)$. If $n = \dim_K(B)$, then $\mathrm{End}_K(B) \cong \mathsf{M}_n(K)$ is central simple, hence $A \otimes_K \mathrm{End}_K(B)$ is central simple, and by 3.3.6.2 we obtain

$$\mathsf{c}_{A\otimes_K \mathrm{End}_K(B)}(\beta(B)) = u\,\mathsf{c}_{A\otimes_K \mathrm{End}_K(B)}(\alpha(B))u^{-1}$$

for some $u \in (A \otimes_K \mathrm{End}_K(B))^\times$. In particular, there exists a K-algebra isomorphism

$$\mathsf{c}_{A\otimes_K \mathrm{End}_K(B)}(\beta(B)) \overset{\sim}{\to} \mathsf{c}_{A\otimes_K \mathrm{End}_K(B)}(\alpha(B)).$$

Since $\beta(B) = 1 \otimes \ell(B)$ and $\alpha(B) = B \otimes \mathrm{id}_B$, we obtain, using 3.3.1.1 and **A**,

$$\mathsf{c}_{A\otimes_K \mathrm{End}_K(B)}(\beta(B))) = \mathsf{c}_A(1) \otimes_K \mathsf{c}_{\mathrm{End}_K(B)}(\ell(B)) = A \otimes_K r(B^\circ) \cong A \otimes_K B^\circ.$$

and

$$\mathsf{c}_{A\otimes_K \mathrm{End}_K(B)}(\alpha(B)) = \mathsf{c}_A(B) \otimes_K \mathsf{c}_{\mathrm{End}_K(B)}(\mathrm{id}_B) = \mathsf{c}_A(B) \otimes_K \mathrm{End}_K(B).$$

Hence there is a K-algebra isomorphism $A \otimes_K B^\circ \overset{\sim}{\to} \mathsf{c}_A(B) \otimes_K \mathrm{End}_K(B)$. We obtain

$$\dim_K(A)\dim_K(B) = \dim_K(\mathsf{c}_A(B))\dim_K(\mathrm{End}_K(B) = \dim_K(\mathsf{c}_A(B))\dim_K(B)^2$$

and consequently $\dim_K(A) = \dim_K(B)\dim_K(\mathsf{c}_A(B))$.

If $B = A$, then $\mathsf{c}_A(A) = \mathsf{c}(A) = K$ and thus $A \otimes_K A^\circ \cong \mathrm{End}_K(A)$.

2. Since $A \otimes_K B^\circ$ is simple, it follows from 1. that $\mathsf{c}_A(B) \otimes_K \mathrm{End}_K(B)$ is simple, and, by 3.3.5.2, $\mathsf{c}_A(B)$ is simple, too. Hence we may apply 1. with $\mathsf{c}_A(B)$ instead of B and obtain

$$\dim_K(A) = \dim_K(\mathsf{c}_A(B))\dim_K(\mathsf{c}_A(\mathsf{c}_A(B))) = \dim_K(B)\dim_K(\mathsf{c}_A(B)),$$

and therefore $\dim_K(B) = \dim_K(\mathsf{c}_A(\mathsf{c}_A(B)))$. Since obviously $B \subset \mathsf{c}_A(\mathsf{c}_A(B))$, it follows that $B = \mathsf{c}_A(\mathsf{c}_A(B))$. Eventually we obtain

$$\mathsf{c}(\mathsf{c}_A(B)) = \mathsf{c}_A(\mathsf{c}_A(B)) \cap \mathsf{c}_A(B) = B \cap \mathsf{c}_A(B) = \mathsf{c}(B). \qquad \square$$

3.4 Splitting fields and the Brauer group

Let K be a field;
K-algebras are assumed to be non-zero and finite-dimensional over K.

Theorem 3.4.1. *Let A be a central K-algebra and L/K a field extension.*

1. *A_L is a central L-algebra, and A_L is simple if and only if A is simple.*

2. *If L is algebraically closed, then A is simple if and only if there exists some $n \in \mathbb{N}$ such that $A_L \cong \mathsf{M}_n(L)$.*

3. *If A is simple, then $\dim_K(A) = n^2$ for some $n \in \mathbb{N}$.*

Proof. 1. By 3.3.1, $\mathsf{c}(A_L) = \mathsf{c}(A) \otimes_K \mathsf{c}(L) = K \otimes_K L = L$. Since L is simple, 3.3.5 implies that A_L is simple if and only if A is simple.

2. By 1. and 3.3.3.2.

3. Let L be algebraically closed. If A is simple, then $A_L \cong \mathsf{M}_n(L)$ for some $n \in \mathbb{N}$ by 2., and consequently $\dim_K(A) = \dim_L(A_L) = n^2$. □

Definition 3.4.2. Let A be a central simple K-algebra, $\dim_K(A) = n^2$, and let L be an extension field of K (then A_L is a simple L-algebra by 3.4.1.1, and as in 3.2.3.**3** we view A as a K-subalgebra of A_L).

1. L is called a **splitting field** of A if there exists an L-algebra isomorphism $A_L \overset{\sim}{\to} \mathsf{M}_n(L)$ for some $n \in \mathbb{N}$, and then we say that A **splits** over L. Note that A splits over L if and only if there exists an L-algebra isomorphism $A_L \overset{\sim}{\to} \mathrm{End}_L(V)$ for some finite-dimensional vector space V over L.

2. By an **L-representation** of A we mean a K-algebra homomorphism $F\colon A \to \mathsf{M}_n(L)$.

Theorem 3.4.3. *Let A be a central simple K-algebra, $\dim_K(A) = n^2$, and let L be an extension field of K.*

1. *Let $F\colon A \to \mathsf{M}_n(L)$ be an L-representation of A. Then there exists a unique L-algebra isomorphism $F_L\colon A_L = A \otimes_K L \overset{\sim}{\to} \mathsf{M}_n(L)$ such that $F_L(a \otimes \lambda) = \lambda F(a)$ for all $a \in A$ and $\lambda \in L$.*

2. *A admits an L-representation if and only if A splits over L.*

3. *Let F and F' be L-representations of A. Then there exists an up to factors from $L^\times$ uniquely determined matrix $Y \in \mathsf{GL}_n(L)$ such that $F'(a) = YF(a)Y^{-1}$ for all $a \in A$.*

Proof. 1. The map $A \times L \to \mathsf{M}_n(L)$, $(a, \lambda) \mapsto \lambda F(a)$ is K-bilinear, and Therefore there exists a K-linear map $F_L \colon A_L = A \otimes_K L \to \mathsf{M}_n(L)$ such that $F_L(a \otimes \lambda) = \lambda F(a)$ for all $a \in A$ and $\lambda \in L$. Uniqueness is obvious.

If λ, $\mu \in L$ and a, $b \in A$, then

$$F_L(\mu(a \otimes \lambda)) = F_L(a \otimes \mu\lambda) = \mu\lambda F(a) = \mu F_L(a \otimes \lambda)$$

and

$$\begin{aligned} F_L((a \otimes \lambda)(b \otimes \mu)) &= F_L(ab \otimes \lambda\mu) = \lambda\mu F(ab) = \lambda F(a)\mu F(b) \\ &= F_L(a \otimes \lambda) F_L(b \otimes \mu). \end{aligned}$$

Hence F_L is an L-algebra homomorphism. Since A_L is simple (see 3.4.1.1), F_L is a monomorphism, and since $\dim_L(A_L) = n^2 = \dim_L(\mathsf{M}_n(L))$, it is an isomorphism.

2. If A admits an L-representation, then L is a splitting field of A by 1. Conversely, if L is a splitting field of A, then there exists an L-algebra isomorphism $h \colon A_L \xrightarrow{\sim} \mathsf{M}_n(L)$, and $F = h \upharpoonright A$ is an L-representation of A.

3. $F_L' \circ F_L^{-1} \colon \mathsf{M}_n(L) \to \mathsf{M}_n(L)$ is an L-algebra automorphism. By 3.3.6.3, there exists an up to factors from $L^\times$ uniquely determined matrix $Y \in \mathsf{GL}_n(L)$ such that $F_L' \circ F_L^{-1}(X) = YXY^{-1}$ for all $X \in \mathsf{M}_n(L) = F_L(A_L)$, and therefore

$$F_L'(a) = F_L' \circ F_L^{-1}(F_L(a)) = Y F_L(a) Y^{-1} \quad \text{for all } a \in A. \qquad \square$$

Definition and Remarks 3.4.4.

1. Let A, A' be central simple K-algebras, $A \cong \mathsf{M}_n(D)$ and $A' \cong \mathsf{M}_{n'}(D')$ with uniquely determined n, $n' \in \mathbb{N}$ and up to isomorphisms uniquely determined central K-division algebras D and D' (see 3.3.3.1). Then A and A' are called **similar**, $A \sim A'$, if D and D' are isomorphic K-algebras. The following properties follow immediately from the definition:

- $A \sim D$, and $D \sim D'$ if and only if $D \cong D'$.
- $A \cong A'$ if and only if $A \sim A'$ and $\dim_K(A) = \dim_K(A')$.
- An extension field L of K is a splitting field of A if and only if $A_L \sim L$.

Obviously, $\sim$ is an equivalence relation, and for a central simple K-algebra A we denote its similarity class by $[A]$. The set $\mathrm{Br}(K)$ of all similarity classes of K-central simple K-algebras is called that **Brauer group** of K. It is a simple exercise in set theory to check that $\mathrm{Br}(K)$ is indeed a set, and in the subsequent theorem we shall establish a group structure on $\mathrm{Br}(K)$.

Theorem 3.4.5. *Let A and B be central simple K-algebras.*

1. *For every finite-dimensional vector space V over K and every $n \in \mathbb{N}$ we have $A \sim A \otimes_K \mathrm{End}_K(V) \sim A \otimes_K \mathsf{M}_n(K)$.*

2. *The similarity class $[A \otimes_K B]$ depends only on $[A]$ and $[B]$.*

3. *Equipped with the composition*

$$([A], [B]) \mapsto [A] \cdot [B] = [A \otimes B]),$$

$\mathrm{Br}(K)$ *is a (multiplicative) abelian group with unit element $[K]$. For every central simple K-algebra A, the similarity class $[A^\circ]$ is the inverse of $[A]$ in* $\mathrm{Br}(K)$ *(and thus it also only depends on $[A]$).*

Proof. We tacitly use the properties of matrix algebras established in 3.2.4.

1. We may assume that $A = \mathsf{M}_r(D) = D \otimes_K \mathsf{M}_r(K)$ for some central K-division algebra D and $r \in \mathbb{N}$, and then $A \otimes_K \mathsf{M}_n(K) = D \otimes_K \mathsf{M}_{rn}(K) \sim A$. If V is a vector space over K and $\dim_K(V) = n$, then $\mathrm{End}_K(V) \cong \mathsf{M}_n(K)$, and therefore $A \otimes_K \mathrm{End}_K(V) \cong A \otimes_K \mathsf{M}_n(K) \sim A$.

2. Recall that $A \otimes_K B$ is central simple by 3.3.5.4. Let

$$A \cong \mathsf{M}_r(D) = D \otimes_K \mathsf{M}_r(K), B \cong \mathsf{M}_s(E) = E \otimes_K \mathsf{M}_s(K)$$

and

$$D \otimes_K E \cong \mathsf{M}_t(F) = F \otimes_K \mathsf{M}_t(K),$$

where D, E and F are (up to isomorphisms) uniquely determined K-division algebras and $r, s, t \in \mathbb{N}$. Then D and E and thus also F depend (up to isomorphisms) only on the similarity classes $[A]$ and $[B]$. Since

$$A \otimes_K B \cong D \otimes_K E \otimes_K \mathsf{M}_{rs}(K) \cong F \otimes_K \mathsf{M}_{rst}(K),$$

it follows that $[A \otimes_K B] = [F]$ also only depends on $[A]$ and $[B]$.

3. By 3.2.1(**4** and **5**), the multiplication in $\mathrm{Br}(K)$ is commutative and associative, and $[K]$ is the unit element of $\mathrm{Br}(K)$. If A is any central simple K-algebra, then $A \otimes_K A^\circ \cong \mathrm{End}_K(A) \sim K$ by 3.3.7.1, and consequently $[A^\circ] = [A]^{-1}$. □

Theorem and Definition 3.4.6. *Let L/K be a field extension and A a central simple K-algebra. Then the class $[A_L] \in \mathrm{Br}(L)$ only depends on the class $[A] \in \mathrm{Br}(K)$, and the map*

$$\mathrm{res}_{L/K} \colon \mathrm{Br}(K) \to \mathrm{Br}(L), \quad \textit{defined by} \quad \mathrm{res}_{L/K}([A]) = [A_L],$$

is a group homomorphism. In particular, L is a splitting field of A if and only if $[A] \in \mathrm{Ker}(\mathrm{res}_{L/K})$.

The group

$$\mathrm{Br}(L/K) = \mathrm{Ker}\big(\mathrm{res}_{L/K} \colon \mathrm{Br}(K) \to \mathrm{Br}(L)\big)$$

is called the **relative Brauer group** of the field extension L/K.

Proof. Let $A \cong \mathsf{M}_r(D)$ for some $r \in \mathbb{N}$ and a central K-division algebra D. Then $A_L \cong \mathsf{M}_r(D)_L = \mathsf{M}_r(D_L) \sim D_L$ by 3.2.4. Hence $[A_L] \in \mathrm{Br}(L)$ is uniquely determined by D and thus by $[A] \in \mathrm{Br}(K)$. Hence the assignment $[A] \mapsto [A_L]$ defines a map $\mathrm{res}_{L/K} \colon \mathrm{Br}(K) \to \mathrm{Br}(L)$. As $(A \otimes_K B)_L = A_L \otimes_L B_L$ for all central simple K-algebras A and B by 3.2.3.**2**, it follows that $\mathrm{res}_{L/K}$ is a group homomorphism. The remaining assertions are obvious by the very definitions. □

In the following two theorems, we show that the splitting fields of a central simple K-algebra A are closely connected with the K-subfields of A (a K-subfield of A is a K-subalgebra which is a field).

Theorem 3.4.7. *Let A be a central simple K-algebra and $\dim_K(A) = n^2$.*

1. Let L be a K-subfield of A. Then $[L\colon K] \le n$, and the following assertions are equivalent:

(a) L is a maximal commutative K-subalgebra of A.

(b) $L = \mathsf{c}_A(L)$.

(c) $[L\colon K] = n$.

If these conditions are fulfilled, then L is a splitting field of A.

2. If A is a central K-division algebra, then every commutative K-subalgebra of A is a field.

3. Let L/K be a finite field extension. Then the following assertions are equivalent:

(a) L is a splitting field of A.

(b) There exists a central simple K-algebra B such that $A \sim B$, and L is a K-subfield of B satisfying $L = \mathsf{c}_B(L)$.

Proof. 1. (a) ⇒ (b) Clearly, $\mathsf{c}_A(L) \supset L$, and if $z \in \mathsf{c}_A(L)$, then $L[z]$ is a commutative K-subalgebra of A. Hence $L[z] = L$, $z \in L$ and thus $L = \mathsf{c}_A(L)$.

(b) ⇒ (a) If L' is a commutative K-subalgebra of A and $L' \supset L = \mathsf{c}_A(L)$, then $L' \subset \mathsf{c}_A(L) = L$ and therefore $L' = L$.

(b) ⇔ (c) Note that $L \subset \mathsf{c}_A(L)$. Using 3.3.7.1, we obtain

$$n^2 = \dim_K(A) = [L\colon K]\dim_K(\mathsf{c}_A(L)) = [L\colon K]^2 \dim_L(\mathsf{c}_A(L)).$$

Hence $[L\colon K] \le n$, and equality holds if and only if $L = \mathsf{c}_A(L)$.

If $\mathsf{c}_A(L) = L$, then

$$A_L = A \otimes_K L \cong \mathsf{c}_A(L) \otimes_K \mathrm{End}_K(L) = L \otimes_K \mathrm{End}_K(L) \sim L,$$

again by 3.3.7.1, and therefore L is a splitting field of A .

2. Let A be a central K-division ring and R a commutative K-subalgebra of A. Then R is a domain, and $\dim_K(R) < \infty$. If $a \in R^\bullet$, the the map $(x \mapsto ax)$ is a K-monomorphism, hence an isomorphism and thus $a \in R^\times$. Hence R is a field.

3. (a) $\Rightarrow$ (b) Suppose that $A_L \cong \mathrm{End}_L(V)$ for some finite-dimensional vector space V over L. As $A \subset A_L \cong \mathrm{End}_L(V) \subset \mathrm{End}_K(V)$, we may assume that A is a K-subalgebra of $E = \mathrm{End}_K(V)$. Then $L \subset A_L \subset E$, and therefore $L \subset \mathsf{c}(A_L) \subset \mathsf{c}_E(A)$.

We apply 3.3.7.2 with $(A \subset E)$ instead of $(B \subset A)$. It follows that $\mathsf{c}_E(A)$ is simple, and $\mathsf{c}(\mathsf{c}_E(A) = \mathsf{c}(A) = K$. Hence $B = \mathsf{c}_E(A)^\circ$ is a central simple K-algebra, and L is a K-subfield of B. Moreover, there exists a K-algebra isomorphism

$$\mathrm{End}_K(V) \otimes_K A^\circ = E \otimes_K A^\circ \overset{\sim}{\to} B^\circ \otimes_K \mathrm{End}_K(A),$$

hence $B^\circ \sim A^\circ$ by 3.4.5.1, and thus $B \sim A$. From the above isomorphism we obtain

$$\begin{aligned} \dim_K(B) &= \frac{\dim_K(\mathrm{End}_K(V))\dim_K(A)}{\dim_K(\mathrm{End}_K(A))} = \frac{\dim_K(V)^2}{\dim_K(A)} = \frac{\dim_L(V)^2[L\!:\!K]^2}{\dim_L(A_L)} \\ &= [L\!:\!K]^2, \end{aligned}$$

and by 1. it follows that $L = \mathsf{c}_B(L)$.

(b) $\Rightarrow$ (a) By 1., L is a splitting field of B and thus of A. □

Theorem 3.4.8. *Let D be a central K-division algebra and $\dim_K(D) = n^2$.*

1. *If L is a maximal K-subfield of D, then $[L\!:\!K] = n$, $\mathsf{c}_D(L) = L$, and L is a splitting field of D.*
2. *Let L/K be a finite field extension such that L is a splitting field of D. Then $n \mid [L\!:\!K]$, and equality holds if and only if there exist a maximal K-subfield of D and a K-algebra isomorphism $K \overset{\sim}{\to} L$.*
3. *D possesses a maximal K-subfield which is separable over K.*

Proof. 1. Let L be a maximal K-subfield of D. The L is a maximal commutative K-subalgebra of D by 3.4.7.2, and all assertions hold by 3.4.7.1.

2. By 3.4.7.3, there exists a simple K-algebra B such that $B \sim D$, L is a K-subfield of B and $\mathsf{c}_B(L) = L$. By 3.4.7.1 we get $\dim_K(B) = [L\!:\!K]^2$. If $r \in \mathbb{N}$ is such that $B \cong \mathsf{M}_r(D)$, then $\dim_K(B) = r^2n^2$, hence $[L\!:\!K] = rn$, and $[L\!:\!K] = n$ holds if and only if $r = 1$, that is, if and only if $B \cong D$ and there is a K-algebra isomorphism of L onto a K-maximal subfield of D.

3. Assume to the contrary that $\mathrm{char}(K) = p > 0$ and there is no maximal K-subfield of D which is separable over K. Let L be an over K separable

K-subfield of D of maximal degree. Then L is not maximal, and consequently $E = \mathsf{c}_D(L) \supsetneq L$ by 1. If $x \in E^\bullet$, then x is purely inseparable over L, $L[x]$ is a commutative subring of D, hence a domain, and $\dim_L(L[x]) < \infty$. Therefore $L[x]$ is a field, $x \in E^\times$ and thus E is a division ring. Since $\mathsf{c}(E) = \mathsf{c}(L) = L$ (by 3.3.7.2), E is a central L-division algebra.

Let $x \in E \setminus L$ and $e \in \mathbb{N}$ minimal such that $x^{p^e} \in L$. Then $u = x^{p^{e-1}} \notin L$, and we define $\tau\colon E \to E$ by $\tau(z) = u^{-1}zu$ for all $z \in E$. Then $\tau \neq \mathrm{id}_E$, $\tau \restriction L = \mathrm{id}_L$, and $\tau^p = \mathrm{id}_E$ since $u^p \in L$. Hence $\tau - \mathrm{id}_E \neq 0$ and $(\tau - \mathrm{id}_E)^p = 0$. Let $r \in \mathbb{N}$ be maximal such that $(\tau - \mathrm{id}_E)^r \neq 0$, let $z \in E$ be such that $w = (\tau - \mathrm{id}_E)^r(z) \neq 0$, and set $v = (\tau - \mathrm{id}_E)^{r-1}(z)$. Then $(\tau - \mathrm{id}_E)(w) = 0$ and $(\tau - \mathrm{id}_E)(v) = w$, hence $\tau(w) = w$ and $\tau(v) = v + w$. Set now $y = w^{-1}v$, and let $m \in \mathbb{N}$ be such that $y^{p^m} \in L$. Then we obtain

$$\tau(y) = \frac{v+w}{w} = 1 + y \quad \text{and} \quad y^{p^m} = \tau(y^{p^m}) = \tau(y)^{p^m} = (1+y)^{p^m} = 1 + y^{p^m},$$

a contradiction. □

3.5 Factor systems and crossed products

Let K be a field;
K-algebras are assumed to be are non-zero and finite-dimensional over K.

Let L/K be a Galois extension, $G = \mathrm{Gal}(L/K)$ and $\mathrm{id}_L = 1 \in G$. Then $L^\times$ is a multiplicative discrete G-module and we use exponential notation for the G-action (see 2.1.3). Then $x^1 = x$ and $x^{\sigma\tau} = (x^\tau)^\sigma$ for all $x \in L^\times$ and $\sigma, \tau \in G$. We briefly recall the notions and results from 2.2.5 using multiplicative notation.

A factor system of G with values in $L^\times$ is a continuous map $c\colon G \times G \to L^\times$ satisfying the cocycle equation

$$c(\nu,\sigma)c(\nu\sigma,\tau) = c(\sigma,\tau)^\nu c(\nu,\sigma\tau) \quad \text{for all} \ \ \nu, \sigma, \tau \in G.$$

For a continuous map $b\colon G \to L^\times$ we define $d^1b\colon G \times G \to L^\times$ by

$$d^1b(\sigma,\tau) = b(\tau)^\sigma b(\sigma\tau)^{-1} b(\sigma) \quad \text{for all} \ \ \sigma, \tau \in G.$$

A splitting factor system of G with values in $L^\times$ is a map $c\colon G \times G \to L^\times$ of the form $c = d^1b$ for some continuous map $b\colon G \to L^\times$.

Under pointwise multiplication, the set $Z^2(G, L^\times)$ of all factor systems of G with values in $L^\times$ is an abelian group with unit element $c = 1$, the set $B^2(G,L^\times)$ of all splitting factor systems is a subgroup, and the (again muliplicatively written) factor group $H^2(G,L^\times) = Z^2(G,L^\times)/B^2(G,L^\times)$ is the 2nd cohomology group of G with values in $L^\times$. For $c \in Z^2(G,L^\times)$ we denote by $[c] = cB^2(G,L^\times) \in H^2(G,L^\times)$ the cohomology class of c.

A factor system $c \in Z^2(G,L^\times)$ is called normalized if $c(1,1) = 1$. Every cohomology class in $H^2(G,L^\times)$ contains a normalized factor system, and if c is a normalized factor system, then $c(\sigma,1) = c(1,\sigma) = 1$ for all $\sigma \in G$.

Now suppose that $[L:K] = n < \infty$, let A be an n-dimensinal vector space over L, and let $\boldsymbol{u} = (u_\sigma)_{\sigma\in G}$ be an L-basis of A such that $u_1 = 1$. Then $L \subset A$, A is a vector space over K, and $\dim_K(A) = n^2$.

Let $c \in Z^2(G,L^\times)$ be a normalized factor system. Then there exists a unique bilinear multiplication $A\times A \to A,\ \ (a,b) \mapsto ab$ such that

$$u_\sigma u_\tau = c(\sigma,\tau)u_{\sigma\tau} \quad \text{and} \quad u_\sigma\lambda = \lambda^\sigma u_\sigma \quad \text{for all } \ \sigma,\,\tau\in G \ \text{ and } \ \lambda \in L.$$

Explicitly, if $(x_\sigma)_{\sigma\in G}$, $(y_\tau)_{\tau\in G}$ are sequences in L, then

$$\Bigl(\sum_{\sigma\in G} x_\sigma u_\sigma\Bigr)\Bigl(\sum_{\tau\in G} y_\tau u_\tau\Bigr) = \sum_{\sigma,\tau\in G} x_\sigma y_\tau^\sigma c(\sigma,\tau)u_{\sigma\tau}.$$

Since $u_1 = c(1,1) = 1$, this multiplication restricted to L is the usual one. Now we prove:

A. Equipped with this multiplication, A is a K-algebra, and we set

$$A = L(c,\boldsymbol{u}).$$

If $\sigma \in G$, then $u_\sigma \in L(c,\boldsymbol{u})^\times$ and $u_\sigma^{-1} = c(\sigma^{-1},\sigma)^{-1}u_{\sigma^{-1}}$. If $\sigma,\,\tau \in G$, then $c(\sigma,\tau) = u_\sigma u_\tau u_{\sigma\tau}^{-1}$.

Proof of **A.** Let $(x_\nu)_{\nu\in G}$, $(y_\sigma)_{\sigma\in G}$, $(z_\tau)_{\tau\in G}$ be sequences in L. Then we obtain

$$\begin{aligned}\Bigl(\sum_{\nu\in G} x_\nu u_\nu \sum_{\sigma\in G} y_\sigma u_\sigma\Bigr)\sum_{\tau\in G} z_\tau u_\tau &= \sum_{\nu,\sigma\in G} x_\nu y_\sigma^\nu c(\nu,\sigma)u_{\nu\sigma}\sum_{\tau\in G} z_\tau u_\tau \\ &= \sum_{\nu,\sigma,\tau\in G} x_\nu y_\sigma^\nu z_\tau^{\nu\sigma} c(\nu,\sigma)c(\nu\sigma,\tau)u_{\nu\sigma\tau},\end{aligned}$$

$$\begin{aligned}\sum_{\nu\in G} x_\nu u_\nu\Bigl(\sum_{\sigma\in G} y_\sigma u_\sigma \sum_{\tau\in G} z_\tau u_\tau\Bigr) &= \sum_{\nu\in G} x_\nu u_\nu \sum_{\sigma,\tau\in G} y_\sigma z_\tau^\sigma c(\sigma,\tau)u_{\sigma\tau} \\ &= \sum_{\nu,\sigma,\tau\in G} x_\nu y_\sigma^\nu z_\tau^{\nu\sigma} c(\sigma,\tau)^\nu c(\nu,\sigma\tau)u_{\nu\sigma\tau},\end{aligned}$$

and the associative law follows from the cocycle equation. The distributive laws hold since the multiplication is bilinear. If $\lambda \in K$, then

$$\sum_{\nu\in G} \lambda x_\nu u_\nu \sum_{\sigma\in G} y_\sigma u_\sigma = \sum_{\nu,\sigma\in G} \lambda x_\nu y_\sigma^\nu c(\nu,\sigma)u_{\nu\sigma} = \sum_{\nu\in G} x_\nu u_\nu \sum_{\sigma\in G} \lambda y_\sigma u_\sigma,$$

and therefore $(\lambda x)y = x(\lambda y) = \lambda(xy)$ for all $x,\ y \in A$. If $\sigma \in G$, then

$$c(\sigma^{-1},\sigma)^{-1}u_{\sigma^{-1}}u_\sigma = c(\sigma^{-1},\sigma)^{-1})c(\sigma^{-1},\sigma)u_1 = 1,$$

hence $u_\sigma \in L(c,\boldsymbol{u})^\times$, and $u_\sigma^{-1} = c(\sigma^{-1},\sigma)^{-1}u_{\sigma^{-1}}$. By definition, we have $c(\sigma,\tau) = u_\sigma u_\tau u_{\sigma\tau}^{-1}$ for all $\sigma,\ \tau \in G$. □[**A.**]

B. Let $\boldsymbol{u} = (u_\sigma)_{\sigma\in G}$ and $\boldsymbol{u}' = (u'_\sigma)_{\sigma\in G}$ be L-bases of A satisfying $u_1 = u'_1 = 1$, and let $c,\ c' \in Z^2(G,L^\times)$ be normalized factor systems such that $[c] = [c']$, say $c = c'\,d^1b$ for some map $b\colon G \to L^\times$. Let $\phi \in \mathrm{Aut}_L(A)$ be the unique L-automorphism of A satisfying $\phi(u_\sigma) = b(\sigma)u'_\sigma$ for all $\sigma \in G$. Then $\phi\colon L(c,\boldsymbol{u}) \to L(c',\boldsymbol{u}')$ is a K-algebra isomorphism.

In particular, $L(c,\boldsymbol{u})$ is (up to K-algebra isomorphisms) uniquely determined by L and the cohomology class $[c] \in H^2(G,L^\times)$.

Proof of **B.** By definition,

$$c(\sigma,\tau) = c'(\sigma,\tau)b(\tau)^\sigma b(\sigma\tau)^{-1}b(\sigma) \text{ for all } \sigma,\ \tau \in G.$$

For all sequences $(x_\sigma)_{\sigma\in G}$, $(y_\tau)_{\tau\in G}$ in L we obtain

$$\begin{aligned}\phi\Big(\sum_{\sigma\in G} x_\sigma u_\sigma \sum_{\tau\in G} y_\tau u_\tau\Big) &= \phi\Big(\sum_{\sigma,\tau\in G} x_\sigma y_\tau^\sigma c(\sigma,\tau)u_{\sigma\tau}\Big)\\ &= \sum_{\sigma,\tau\in G} x_\sigma y_\tau^\sigma c(\sigma,\tau)b(\sigma\tau)u'_{\sigma\tau}\end{aligned}$$

and

$$\begin{aligned}\phi\Big(\sum_{\sigma\in G} x_\sigma u_\sigma\Big)\phi\Big(\sum_{\tau\in G} y_\tau u_\tau\Big) &= \sum_{\sigma\in G} x_\sigma b(\sigma)u'_\sigma \sum_{\tau\in G} y_\tau b(\tau)u'_\tau\\ &= \sum_{\sigma,\tau\in G} x_\sigma b(\sigma)y_\tau^\sigma b(\tau)^\sigma c'(\sigma,\tau)u'_{\sigma\tau} = \sum_{\sigma,\tau\in G} x_\sigma y_\tau^\sigma c(\sigma,\tau)b(\sigma\tau)u'_{\sigma\tau}.\end{aligned}$$ □[**B.**]

Definition 3.5.1. Let L/K be a finite Galois extension, $G = \mathrm{Gal}(L/K)$, and let $[L\colon K] = n$. For $\gamma \in H^2(G,L^\times)$, we denote by (L,G,γ) an up to K-algebra isomorphisms uniquely determined K-algebra $L(c,\boldsymbol{u})$, built with a normalized factor system $c \in \gamma$ and an L-basis $\boldsymbol{u} = (u_\sigma)_{\sigma\in G}$ of A such that $u_1 = 1$. We call (L,G,γ) a **crossed product** of L and G with cohomology class γ. If G is cyclic, we call (L,G,γ) a **cyclic K-algebra**.

Theorem 3.5.2. *Let L/K be a finite Galois extension and $[L : K] = n$. Suppose that $G = \mathrm{Gal}(L/K)$, $\gamma \in H^2(G,L^\times)$ and $A = (L,G,\gamma)$. Then A is a central simple K-algebra, $L = \mathsf{c}_A(L)$, $\dim_K(A) = n^2$ and L is a splitting field of A. Moreover, if $\gamma = 1 \in H^2(G,L^\times)$, then $A \cong \mathrm{End}_K(L) \sim K$.*

Proof. Let $c \in \gamma$ be a normalized factor system and $A = L(c, \boldsymbol{u})$ built with an L-basis $\boldsymbol{u} = (u_\sigma)_{\sigma \in G}$ of A such that $u_1 = 1$. Then $L \subset \mathsf{c}_A(L) \subset A$, and if $\mathsf{c}_A(L) = L$, then $\dim_K(A) = n^2$ and L is a splitting field of A by 3.4.7.1. Hence it suffices to prove:

a. $\mathsf{c}_A(L) = L$; **b.** $\mathsf{c}(A) = K$; **c.** A is simple; **d.** $L(1, \boldsymbol{u}) \cong \operatorname{End}_K(L)$.

Proof of **a.** It suffices to show that $\mathsf{c}_A(L) \subset L$. Let $(x_\sigma)_{\sigma \in G}$ be a sequence in L and

$$x = \sum_{\sigma \in G} x_\sigma u_\sigma \in \mathsf{c}_A(L), \quad \text{hence } 0 = ax - xa = \sum_{\sigma \in G} (a - a^\sigma) x_\sigma u_\sigma \quad \text{for all } a \in L.$$

If $\tau \in G \setminus \{1\}$, then there is some $a \in L$ such that $a^\tau \neq a$, and consequently $x_\tau = 0$. Hence $x = x_1 u_1 = x_1 \in L$. $\square$[**a.**]

Proof of **b.** Note that $\mathsf{c}(A) \subset \mathsf{c}_A(L) = L$. If $x \in \mathsf{c}(A)$ and $\sigma \in G$, then $x u_\sigma = u_\sigma x = x^\sigma u_\sigma$, hence $x^\sigma = x$, and consequently $x \in L^G = K$. $\square$[**b.**]

Proof of **c.** Let I be an non-zero ideal of A, and let

$$x = \sum_{\sigma \in G} x_\sigma u_\sigma \in I^\bullet \quad \text{(for some sequence } (x_\sigma)_{\sigma \in G} \text{ in } L)$$

be an element for which $\operatorname{supp}(x) = \{\sigma \in G \mid x_\sigma \neq 0\}$ is minimal. If $|\operatorname{supp}(x)| = 1$ and $\sigma \in G$ is such that $x_\sigma \neq 0$, then $x = x_\sigma u_\sigma \in A^\times$ and therefore $I = A$. Thus assume that $|\operatorname{supp}(x)| \geq 2$, let $\tau,\, \nu \in G$ be such that $x_\tau x_\nu \neq 0$ and $b \in L$ such that $b^\tau \neq b^\nu$. Then

$$b^\tau x - xb = \sum_{\sigma \in G} b^\tau x_\sigma u_\sigma - \sum_{\sigma \in G} x_\sigma u_\sigma b = \sum_{\sigma \in G} (b^\tau - b^\sigma) x_\sigma u_\sigma \in I^\bullet,$$

and $\operatorname{supp}(b^\tau x - xb) \subset \operatorname{supp}(x) \setminus \{\tau\}$, a contradiction. $\square$[**c.**]

Proof of **d.** Assume $c = 1$, and let $\phi \colon A \to \operatorname{End}_K(L)$ be defined by

$$\phi\Big(\sum_{\sigma \in G} x_\sigma u_\sigma\Big) = \sum_{\sigma \in G} x_\sigma \sigma \quad \text{for all sequences } (x_\sigma)_{\sigma \in G} \text{ in } L.$$

Then ϕ is L-linear, $\operatorname{Ker}(\phi) = \mathbf{0}$ by Dedekind's independence theorem, and therefore ϕ is an isomorphism since $\dim_K(\operatorname{End}_K(L)) = n^2 = \dim_K(A)$. If $(x_\sigma)_{\sigma \in G}$, $(y_\tau)_{\tau \in G}$ are any sequences in L, then

$$\begin{aligned}\phi\Big(\sum_{\sigma \in G} x_\sigma u_\sigma \sum_{\tau \in G} y_\tau u_\tau\Big) &= \phi\Big(\sum_{\sigma,\tau \in G} x_\sigma y_\tau^\sigma u_{\sigma\tau}\Big) = \sum_{\sigma,\tau \in G} x_\sigma y_\tau^\sigma \sigma\tau \\ &= \Big(\sum_{\sigma \in G} x_\sigma \sigma\Big)\Big(\sum_{\tau \in G} y_\tau \tau\Big) = \phi\Big(\sum_{\sigma \in G} x_\sigma u_\sigma\Big)\phi\Big(\sum_{\tau \in G} y_\tau u_\tau\Big).\end{aligned}$$

Hence ϕ is a K-algebra isomorphism. $\square$[**d.**]. $\square$

Definition and Theorem 3.5.3. *Let L/K be a (not necessarily finite) Galois extension and $G = \mathrm{Gal}(L/K)$. Let A be a central simple K-algebra which splits over L, and $\dim_K(A) = n^2$.*
For a matrix $Y = (y_{\nu,\mu})_{\nu,\mu\in[1,n]} \in \mathsf{M}_n(L)$ and $\sigma \in G$ we set

$$Y^\sigma = (y^\sigma_{\nu,\mu})_{\nu,\mu\in[1,n]}.$$

Then $\mathsf{M}_n(L)$ is a G-module, for all $\sigma \in G$ the map $Y \mapsto Y^\sigma$ is a K-algebra automorphism of $\mathsf{M}_n(L)$, and $\mathsf{M}_n(L)^G = \mathsf{M}_n(K)$. If $Y = (y_{\nu,\mu})_{\nu,\mu\in[1,n]} \in \mathsf{M}_n(L)$, then the isotropy group

$$G_Y = \{\sigma \in G \mid Y^\sigma = Y\} = \bigcap_{\nu,\mu=1}^{n} G_{y_{\nu,\mu}}$$

is an open subgroup of G, and thus $\mathsf{M}_n(L)$ is a discrete G-module by 2.1.2. In the sequel we associate with A a cohomology class $\gamma_A \in H^2(G, L^\times)$.

Let $F\colon A \to \mathsf{M}_n(L)$ be an L-representation of A. For $\sigma \in G$, be define $F^\sigma\colon A \to \mathsf{M}_n(L)$ by $F^\sigma(a) = F(a)^\sigma$. Then F^σ is again an L-representation of A, and below (in **A**) we shall prove that

$$U = \{\sigma \in G \mid F^\sigma = F\} \text{ is an open subgroup of } G.$$

Consequently, if $\sigma \in G$, then F^σ only depends on the coset σU, and by 3.4.3.3 there exists a matrix $Y(\sigma) \in \mathsf{GL}_n(L)$ (which again only depends on the coset σU) such that $F(a)^\sigma = Y(\sigma)F(a)Y(\sigma)^{-1}$ for all $a \in A$.

For all σ, $\tau \in G$ and $a \in A$ it follows that

$$\begin{aligned} F(a)^{\sigma\tau} = [F(a)^\tau]^\sigma = [Y(\tau)F(a)Y(\tau)^{-1}]^\sigma &= Y(\tau)^\sigma Y(\sigma)F(a)Y(\sigma)^{-1}Y(\tau)^{-\sigma} \\ &= Y(\sigma\tau)F(a)Y(\sigma\tau)^{-1}. \end{aligned}$$

Hence there exists a uniquely determined element $c(\sigma, \tau) \in L^\times$ such that

$$Y(\tau)^\sigma Y(\sigma) = c(\sigma, \tau)Y(\sigma\tau).$$

Now we can state the assertions of the theorem.

1. *$c\colon G^2 \to L^\times$ is a factor system, and the cohomology class $\gamma_L(A) = [c] \in H^2(G, L^\times)$ only depends on A and L.*
We call $\gamma_L(A)$ the **cohomology class** of A with respect to L.

2. *Let $\overline{L}$ be an extension field of L which is Galois over K, let $\overline{G} = \mathrm{Gal}(\overline{L}/K)$ and $H = \mathrm{Gal}(\overline{L}/L)$ (we identify $G = \overline{G}/H$ by means of $\overline{\sigma} \restriction L = \overline{\sigma}H$ for all $\overline{\sigma} \in \overline{G}$). Then it follows that $\gamma_{\overline{L}}(A) = \mathrm{Inf}^2_{G\to\overline{G}}(\gamma_L(A))$.*

3. *If $[L\!:\!K] < \infty$, $\gamma \in H^2(G, L^\times)$ and $A = (L, G, \gamma)$ is a crossed product of L and G with γ, then $\gamma = \gamma_L(A)$.*

Proof. As announced we show initially:

> **A.** Let $F\colon A \to \mathsf{M}_n(L)$ be an L-representation of A. Then the isotropy group $U = \{\sigma \in G \mid F^\sigma = F\}$ is an open subgroup of G.

Proof of **A.** Let $N = n^2$ and $(e_1, \ldots, e_N)$ be a K-basis of A. If $a \in A$, say

$$a = \sum_{i=1}^{n} c_i a_i, \quad \text{where} \quad c_i \in K,$$

then

$$F(a) = \sum_{i=1}^{N} c_i F(e_i), \quad \text{and} \quad F^\sigma(a) = \sum_{i=1}^{N} c_i F(e_i)^\sigma \quad \text{for all} \quad \sigma \in G.$$

Hence $F^\sigma(a) = F(a)$ for all $a \in A$ if and only if $F(e_i)^\sigma = F(e_i)$ for all $i \in [1, n]$, and therefore

$$\{\sigma \in G \mid F^\sigma = F\} = \bigcap_{i=1}^{N} G_{F(e_i)}$$

is an open subgroup of G. □[**A.**]

1. We prove first that $c\colon G\times G \to L^\times$ is continuous. Let $(\sigma, \tau) \in G\times G$. By construction $Y(\sigma)$ is constant on σU, $Y(\tau)$ is constant on τU and $Y(\sigma\tau)$ is constant on $\sigma\tau U$. The set $V = \{(x, y) \in \sigma U \times \tau U \mid xy \in \sigma\tau U\}$ is an open neighbourhood of (σ, τ) in $G\times G$, and c is constant on V. Hence c is continuous in (σ, τ).

If ν, σ, $\tau \in G$, then

$$\begin{aligned} Y(\nu\sigma\tau) &= Y(\tau)^{\nu\sigma} Y(\nu\sigma) c(\nu\sigma, \tau)^{-1} = Y(\tau)^{\nu\sigma} Y(\sigma)^\nu Y(\nu) c(\nu, \sigma)^{-1} c(\nu\sigma, \tau)^{-1} \\ &= Y(\sigma\tau)^\nu Y(\nu) c(\nu, \sigma\tau)^{-1} = [Y(\tau)^\sigma Y(\sigma) c(\sigma, \tau)^{-1}]^\nu\, Y(\nu) c(\nu, \sigma\tau)^{-1} \\ &= Y(\tau)^{\nu\sigma} Y(\sigma)^\nu Y(\nu) c(\sigma, \tau)^{-\nu} c(\nu, \sigma\tau)^{-1}, \end{aligned}$$

hence $c(\nu, \sigma) c(\nu\sigma, \tau) = c(\sigma, \tau)^\nu c(\nu, \sigma\tau)$, and c is a factor system.

Next we show that the cohomology class $[c] \in H^2(G, L^\times)$ does not depend on F. Thus let F, $F'\colon A \to \mathsf{M}_n(L)$ be L-representations. By 3.4.3.3, there exists an up to factors from $L^\times$ uniquely determined matrix $T \in \mathsf{GL}_n(L)$ such that $F'(a) = TF(a)T^{-1}$ for all $a \in A$.

For $\sigma \in G$, let $Y(\sigma)$, $Y'(\sigma) \in \mathsf{GL}_n(L)$ and c, $c' \in Z^2(G, L^\times)$ such that

$$F(a)^\sigma = Y(\sigma)F(a)Y(\sigma)^{-1}, F'(a)^\sigma = Y'(\sigma)F'(a)Y'(\sigma)^{-1},$$
$$Y(\tau)^\sigma Y(\sigma) = c(\sigma, \tau) Y(\sigma\tau)$$

and $Y'(\tau)^\sigma Y'(\sigma) = c'(\sigma, \tau) Y'(\sigma\tau)$ for all $a \in A$ and σ, $\tau \in G$.

For $\sigma \in G$ and $a \in A$ we obtain

$$\begin{aligned} F'(a)^\sigma &= Y'(\sigma)F'(a)Y'(\sigma)^{-1} = Y'(\sigma)TF(a)T^{-1}Y'(\sigma)^{-1} \\ &= T^\sigma F(a)^\sigma T^{-\sigma} = T^\sigma Y(\sigma)F(a)Y(\sigma)^{-1}T^{-\sigma}, \end{aligned}$$

and by the uniqueness of T we obtain $T^\sigma Y(\sigma) = Y'(\sigma)Tb(\sigma)$, where $b(\sigma) \in L^\times$. If σ, $\tau \in G$, then

$$\begin{aligned} Y(\sigma\tau) &= Y(\tau)^\sigma Y(\sigma)c(\sigma,\tau)^{-1} = [T^{-\tau}Y'(\tau)Tb(\tau)]^\sigma\,[T^{-\sigma}Y'(\sigma)Tb(\sigma]\,c(\sigma,\tau)^{-1} \\ &= T^{-\sigma\tau}Y'(\tau)^\sigma b(\tau)^\sigma Y'(\sigma)Tb(\sigma)c(\sigma,\tau)^{-1} \\ &= T^{-\sigma\tau}Y'(\sigma\tau)c'(\sigma,\tau)b(\tau)^\sigma Tb(\sigma)c(\sigma,\tau)^{-1} \\ &= Y(\sigma\tau)b(\sigma\tau)^{-1}c'(\sigma,\tau)b(\tau)^\sigma b(\sigma)b(\sigma\tau)^{-1}c(\sigma,\tau)^{-1}, \end{aligned}$$

and consequently $c(\sigma,\tau) = c'(\sigma,\tau)b(\tau)^\sigma b(\sigma\tau)^{-1}b(\sigma) = c'(\sigma,\tau)d^1b(\sigma,\tau)$. Hence it follows that $[c'] = [c] \in H^2(G,L^\times)$.

2. Let $F\colon A \to \mathsf{M}_n(L)$ be an L-representation and $i = (\mathsf{M}_n(L) \hookrightarrow \mathsf{M}_n(\overline{L}))$ the inclusion. Then $\overline{F} = i\circ F\colon A \to \mathsf{M}_n(\overline{L})$ is an $\overline{L}$-representation of A. For $\sigma \in G$, let $Y(\sigma) \in \mathrm{GL}_n(L)$ be such that $F(a)^\sigma = Y(\sigma)F(a)Y(\sigma)^{-1}$ for all $a \in A$, and for σ, $\tau \in G$ let $c(\sigma,\tau) \in Z^2(G,L^\times)$ be such that

$$Y(\tau)^\sigma Y(\sigma) = c(\sigma,\tau)Y(\sigma\tau). \text{ Then } [c] = \gamma_L(A) \in H^2(G,L^\times).$$

If $\overline{\sigma}$, $\overline{\tau} \in \overline{G}$ are such that $\sigma = \overline{\sigma}\restriction L$, $\tau = \overline{\tau}\restriction L \in G$, we set $\overline{Y}(\overline{\sigma}) = Y(\sigma)$ and $\overline{c}(\overline{\sigma},\overline{\tau}) = c(\sigma,\tau)$. Then we obtain $\overline{c} = \mathrm{Inf}^2_{\overline{G}\to G}(c) \in Z^2(\overline{G},\overline{L}^\times)$ by 2.5.3. Since $\overline{F}(a)^{\overline{\sigma}} = \overline{Y}(\overline{\sigma})\overline{F}(a)\overline{Y}(\overline{\sigma})^{-1}$ and $\overline{Y}(\overline{\tau})^{\overline{\sigma}}\overline{Y}(\overline{\sigma}) = \overline{c}(\overline{\sigma},\overline{\tau})\overline{Y}(\overline{\sigma}\,\overline{\tau})$ for all $a \in A$ and $\overline{\sigma}$, $\overline{\tau} \in \overline{G}$, it follows that $[\overline{c}] = \gamma_{\overline{L}}(A) = \mathrm{Inf}^2_{G\to\overline{G}}(\gamma_L(A))$.

3. Let $[L\colon K] = n < \infty$, $\gamma \in H^2(G,L^\times)$ and $A = (L,G,\gamma) = L(c,\boldsymbol{u})$, built with a normalized factor system $c \in \gamma$ and an L-basis $\boldsymbol{u} = (u_\sigma)_{\sigma\in G}$ of A such that $u_1 = 1$. For $a \in A$ we define $F(a) = (F(a)_{\sigma,\kappa})_{\sigma,\kappa\in G} \in \mathsf{M}_n(L)$ by

$$u_\sigma a = \sum_{\mu\in G} F(a)_{\sigma,\mu}u_\mu \quad \text{for all } \ \sigma \in G.$$

By definition, $F\colon A \to \mathsf{M}_n(L)$ is injective and K-linear, and if a, $b \in A$, then

$$\begin{aligned} (u_\sigma a)b &= \sum_{\mu\in G} F(a)_{\sigma,\mu} \sum_{\kappa\in G} F(b)_{\mu,\kappa}u_\kappa = \sum_{\kappa\in G}[F(a)F(b)]_{\sigma,\kappa}u_\kappa \\ &= u_\sigma(ab) = \sum_{\kappa\in G} F(ab)_{\sigma,\kappa}u_\kappa. \end{aligned}$$

Hence $F(ab) = F(a)F(b)$, and F is an L-representation of A. If $\alpha \in L^\times$ and σ, $\tau \in G$, then

$$u_\sigma(\alpha u_\tau) = \sum_{\mu\in G} F(\alpha u_\tau)_{\sigma,\mu}u_\mu = \alpha^\sigma c(\sigma,\tau)u_{\sigma\tau},$$

and therefore

$$F(\alpha u_\tau)_{\sigma,\mu} = \begin{cases} \alpha^\sigma c(\sigma,\tau) & \text{if } \mu = \sigma\tau, \\ 0 & \text{otherwise.} \end{cases}$$

For $\nu \in G$ we define $Y(\nu) = (Y(\nu)_{\mu,\kappa})_{\mu,\kappa\in G} \in \mathsf{GL}_n(L)$ by

$$Y(\nu)_{\mu,\kappa} = \begin{cases} c(\nu,\mu) & \text{if } \nu\mu = \kappa, \\ 0 & \text{otherwise,} \end{cases}$$

and we shall prove:

B. If $\nu \in G$, then $F(a)^\nu = Y(\nu)F(a)Y(\nu)^{-1}$ for all $a \in A$.

Proof of **B.** By linearity, it suffices to prove that

$$[F(\alpha u_\tau^\nu)Y(\nu)]_{\sigma,\mu} = [Y(\nu)F(\alpha u_\tau)]_{\sigma,\mu} \quad \text{for all} \quad \alpha \in L^\times \quad \text{and } \nu,\ \tau,\ \sigma,\ \mu \in G.$$

Let $\alpha \in L^\times$ and ν, τ, σ, $\mu \in G$. Then

$$[F(\alpha u_\tau^\nu)Y(\nu)]_{\sigma,\mu} = \sum_{\kappa\in G} F(\alpha u_\tau^\nu)_{\sigma,\kappa} Y(\nu)_{\kappa,\mu},$$

and for all $\kappa \in G$ we obtain

$$F(\alpha u_\tau^\nu)_{\sigma,\kappa} Y(\nu)_{\kappa,\mu} = \begin{cases} \alpha^{\nu\sigma} c(\sigma,\tau)^\nu c(\nu,\sigma\tau) & \text{if } \kappa = \sigma\tau \text{ and } \mu = \nu\kappa, \\ 0 & \text{otherwise,} \end{cases}$$

and therefore

$$[F(\alpha u_\tau^\nu)Y(\nu)]_{\sigma,\mu} = \begin{cases} \alpha^{\nu\sigma} c(\sigma,\tau)^\nu c(\nu,\sigma\tau) & \text{if } \mu = \nu\sigma\tau, \\ 0 & \text{otherwise.} \end{cases}$$

On the other hand,

$$[Y(\nu)F(\alpha u_\tau)]_{\sigma,\mu} = \sum_{\kappa\in G} Y(\nu)_{\sigma,\kappa} F(\alpha u_\tau)_{\kappa,\mu},$$

for all $\kappa \in G$ we obtain

$$Y(\nu)_{\sigma,\kappa} F(\alpha u_\tau)_{\kappa,\mu} = \begin{cases} c(\nu,\sigma)\alpha^\kappa c(\kappa,\tau) & \text{if } \nu\sigma = \kappa \text{ and } \mu = \tau\kappa, \\ 0 & \text{otherwise,} \end{cases}$$

and therefore

$$[Y(\nu)F(\alpha u_\tau)]_{\sigma,\mu} = \begin{cases} \alpha^{\nu\sigma} c(\nu,\sigma)c(\nu\sigma,\tau) & \text{if } \mu = \nu\sigma\tau, \\ 0 & \text{otherwise.} \end{cases}$$

Since $c \in Z^2(G, L^\times)$, we get $c(\nu,\sigma)c(\nu\sigma,\tau) = c(\sigma,\tau)^\nu c(\nu,\sigma\tau)$ for all ν, σ, $\tau \in G$ by the cocycle equation, and consequently

$$[F(\alpha u_\tau^\nu)Y(\nu)]_{\sigma,\mu} = [Y(\nu)F(\alpha u_\tau)]_{\sigma,\mu} \text{ for all } \nu,\ \tau,\ \sigma,\ \mu \in G.$$

$\square$[**B.**]

Now let $c^* \in Z^2(G, L^\times)$ be such that $Y(\tau)^\sigma Y(\sigma) = Y(\sigma\tau)c^*(\sigma,\tau)$ for all $\sigma, \tau \in G$, hence $\gamma_L(A) = [c^*] \in H^2(G, L^\times)$. For all σ, τ, ν, $\mu \in G$ we obtain

$$[Y(\tau)^\sigma Y(\sigma)]_{\nu,\mu} = \sum_{\kappa\in G} Y(\tau)^\sigma_{\nu,\kappa} Y(\sigma)_{\kappa,\mu},$$

where

$$Y(\tau)^\sigma_{\nu,\kappa} Y(\sigma)_{\kappa,\mu} = \begin{cases} c(\tau,\nu)^\sigma c(\sigma,\kappa) & \text{if } \tau\nu = \kappa \text{ and } \sigma\kappa = \mu, \\ 0 & \text{otherwise,} \end{cases}$$

and therefore

$$[Y(\tau)^\sigma Y(\sigma)]_{\nu,\mu} = \begin{cases} c(\tau,\nu)^\sigma c(\sigma,\tau\nu) & \text{if } \sigma\tau\nu = \mu, \\ 0 & \text{otherwise.} \end{cases}$$

If σ, τ, ν, $\mu \in G$ are such that $\sigma\tau\nu = \mu$, then $(Y(\sigma\tau)_{\nu,\mu} = c(\sigma\tau,\nu)$, and consequently

$$c^*(\sigma,\tau) = c(\tau,\nu)^\sigma c(\sigma,\tau\nu)c(\sigma\tau,\nu)^{-1} = c(\sigma,\tau)c(\sigma\tau,\nu)c(\sigma\tau,\nu)^{-1} = c(\sigma,\tau)$$

for all σ, $\tau \in G$. Hence $c^* = c$ and $\gamma_L(A) = \gamma$. □

Theorem 3.5.4. *Let L/K be a (not necessarily finite) Galois extension and $G = \mathrm{Gal}(L/K)$. Then there exists a unique group isomorphism*

$$\Theta_{L/K}\colon \mathrm{Br}(L/K) \overset{\sim}{\to} H^2(G, L^\times)$$

such that $\Theta_{L/K}([A]) = \gamma_L(A)$ for every central simple K-algebra A which splits over L. In particular, if A and B are central simple K-algebras which split over L, then the following assertions hold:

1. $\gamma_L(A \otimes_K B) = \gamma_L(A)\gamma_L(B)$.

2. $\gamma_L(A^\circ) = \gamma_L(A)^{-1}$, and $\gamma_L(A) = [1] \in H^2(G, L^\times)$ if and only if $A \sim K$.

3. $\gamma_L(A) = \gamma_L(B)$ if and only if $A \sim B$.

Proof. We first prove the assertions 1., 2., and 3. Let A and B be central simple K-algebras which split over L

1. Let $F\colon A \to \mathsf{M}_n(L)$ and $G\colon B \to \mathsf{M}_m(L)$ be L-representations. As the map $F{\times}G\colon A{\times}B \to \mathsf{M}_n(L)\otimes_K \mathsf{M}_m(L)$, defined by $(F{\times}G)(a,b) = F(a)\otimes G(b)$, is K-bilinear, there exists a unique K-linear map

$$F \otimes G\colon A \otimes_K B \to \mathsf{M}_n(L) \otimes_L \mathsf{M}_m(L) = \mathsf{M}_{mn}(L)$$

such that $(F \otimes G)(a \otimes b) = F(a) \otimes G(b)$ for all $a \in A$ and $b \in B$. If $a,\ a' \in A$ and $b,\ b' \in B$, then

$$\begin{aligned}(F \otimes G)((a \otimes b)(a' \otimes b')) &= (F \otimes G)(aa' \otimes bb') = F(aa') \otimes G(bb') \\ &= F(a)F(a') \otimes G(b)G(b') \\ &= (F(a) \otimes G(b))(F(a') \otimes G(b')) \\ &= (F \otimes G)(a \otimes b)(F \otimes G)(a' \otimes b'),\end{aligned}$$

and therefore $F \otimes G$ is an L-representation of $A \otimes_K B$.

For $\sigma \in G$ let $Y(\sigma) \in \mathsf{GL}_n(L)$, $W(\sigma) \in \mathsf{GL}_m(L)$ and $c_A,\ c_B \in Z^2(G, L^\times)$ such that

$$F(a)^\sigma = Y(\sigma)F(a)Y(\sigma)^{-1}, \qquad G(b)^\sigma = W(\sigma)G(b)W(\sigma)^{-1},$$

$$Y(\tau)^\sigma Y(\sigma) = c_A(\sigma,\tau)Y(\sigma\tau) \quad \text{and} \quad W(\tau)^\sigma W(\sigma) = c_B(\sigma,\tau)W(\sigma\tau)$$

for all $a \in A,\ \ b \in B$ and $\sigma,\ \tau \in G$. Then $\gamma_L(A) = [c_A]$ and $\gamma_L(B) = [c_B]$. For all $a \in A,\ \ b \in B$ and $\sigma,\ \tau \in G$ it follows that

$$\begin{aligned}(F \otimes G)(a \otimes b)^\sigma &= F(a)^\sigma \otimes G(b)^\sigma = Y(\sigma)F(a)Y(\sigma)^{-1} \otimes W(\sigma)G(b)W(\sigma)^{-1} \\ &= [Y(\sigma) \otimes W(\sigma)]\,[F(a) \otimes G(b)]\,[Y(\sigma \otimes W(\sigma)]^{-1}\end{aligned}$$

and

$$[Y(\tau) \otimes W(\tau)]^\sigma [Y(\sigma) \otimes W(\sigma)] = c^*(\sigma,\tau)Y(\sigma\tau) \otimes W(\sigma\tau)$$

in $\mathsf{GL}_{mn}(L)$, where $c^* \in Z^2(G, L^\times)$ and $\gamma_L(A \otimes_K B) = [c^*] \in H^2(G, L^\times)$. On the other hand,

$$\begin{aligned}&[Y(\tau) \otimes W(\tau)]^\sigma [Y(\sigma) \otimes W(\sigma)] = Y(\tau)^\sigma Y(\sigma) \otimes W(\tau)^\sigma W(\sigma) \\ &\quad = c_A(\sigma,\tau)Y(\sigma\tau) \otimes c_B(\sigma,\tau)W(\sigma\tau) = c_A(\sigma,\tau)c_B(\sigma,\tau)Y(\sigma\tau) \otimes W(\sigma\tau).\end{aligned}$$

Hence $c^* = c_A c_B$ and $\gamma_L(A \otimes_K B) = \gamma_L(A)\gamma_L(B) \in H^2(G, L^\times)$.

2. It suffices to prove that $\gamma_L(A) = [1]$ if and only if $A \sim K$. Indeed, by 3.4.5.3 it follows that $A \otimes_K A^\circ \sim K$, hence $[1] = \gamma_L(A \otimes_K A^\circ) = \gamma_L(A)\gamma_L(A^\circ)$, and therefore $\gamma_L(A^\circ) = \gamma_L(A)^{-1}$.

Assume first that $A \sim K$, let $\phi\colon A \xrightarrow{\sim} \mathsf{M}_n(K)$ be a K-algebra isomorphism and $i = (\mathsf{M}_n(K) \hookrightarrow \mathsf{M}_n(L))$. Then $F = i \circ \phi\colon A \to \mathsf{M}_n(L))$ is an L-representation satisfying $F^\sigma = F$ for all $\sigma \in G$. Hence $c = 1$ and $\gamma_L(A) = [1]$.

As to the converse, suppose that $\gamma_L(A) = [1]$, and let $F\colon A \to \mathsf{M}_n(L)$ be an L-representation. We prove first:

I. For every $\sigma \in G$ there exists some $Y(\sigma) \in \mathsf{GL}_n(L)$ such that

$$F(a)^\sigma = Y(\sigma)F(a)Y(\sigma)^{-1} \quad \text{and} \quad Y(\tau)^\sigma Y(\sigma) = Y(\sigma\tau)$$

for all $a \in A$ and $\sigma,\ \tau \in G$.

Proof of **I.** For $\sigma \in G$ let $Y_1(\sigma) \in \mathsf{GL}_n(L)$ and $c \in Z^2(G, L^\times$ such that $F(a)^\sigma = Y_1(\sigma)F(a)Y_1(\sigma)^{-1}$ for all $a \in A$, and $Y_1(\tau)^\sigma Y_1(\sigma) = c(\sigma,\tau)Y_1(\sigma\tau)$ for all σ, $\tau \in G$. Then $\gamma_L(A) = [c] = [1] \in H^2(G, L^\times)$, and therefore there exists a continuous map $b\colon G \to L^\times$ such that $c(\sigma,\tau) = d^1b(\sigma,\tau) = b(\tau)^\sigma b(\sigma\tau)^{-1}b(\sigma)$ for all σ, $\tau \in G$, and for $\sigma \in G$, we set $Y(\sigma) = b(\sigma)^{-1}Y_1(\sigma) \in \mathsf{GL}_n(L)$. Then $F(a)^\sigma = Y(\sigma)F(a)Y(\sigma)^{-1}$ for all $a \in A$, and

$$\begin{aligned} Y(\tau)^\sigma Y(\sigma) &= b(\tau)^{-\sigma}Y_1(\tau)^\sigma b(\sigma)^{-1}Y_1(\sigma) = b(\sigma\tau)^{-1}c(\sigma,\tau)^{-1}Y_1(\tau)^\sigma Y_1(\sigma) \\ &= b(\sigma\tau)^{-1}Y_1(\sigma\tau) = Y(\sigma\tau) \quad \text{for all} \quad \sigma,\, \tau \in G. \end{aligned} \qquad \square[\mathbf{I.}]$$

For $\boldsymbol{u} = (u_1, \ldots, u_n)^{\mathsf{t}} \in L^n$ and $\tau \in G$ we set $\tau(\boldsymbol{u}) = (\tau(u_1), \ldots, \tau(u_n))^{\mathsf{t}}$, and we define $\varphi\colon L^n \to L^n$ by

$$\varphi(\boldsymbol{u}) = \sum_{\tau \in G} Y(\tau)\tau(\boldsymbol{u}) \quad \text{for all} \quad \boldsymbol{u} \in L^n.$$

Then φ is K-linear (in general not L-linear), and we prove:

II. $L\varphi(L^n) = L^n$.

Proof of **II.** Assume to the contrary that $L\varphi(L^n) \subsetneq L^n$. Then there exists some $\boldsymbol{v} \in L^n \setminus \{\boldsymbol{0}\}$ such that $\boldsymbol{v}^{\mathsf{t}}\varphi(\boldsymbol{u}) = 0$ for all $\boldsymbol{u} \in L^n$. For $i \in [1, n]$ and all $x \in L$ we obtain

$$0 = \boldsymbol{v}^{\mathsf{t}}\varphi(x\boldsymbol{e}_i) = \sum_{\tau \in G} \boldsymbol{v}^{\mathsf{t}}Y(\tau)\tau(x)\boldsymbol{e}_i = \Big[\sum_{\tau \in G}(\boldsymbol{v}^{\mathsf{t}}Y(\tau))_i\tau\Big]x, \quad \text{hence}$$

$$\sum_{\tau \in G}(\boldsymbol{v}^{\mathsf{t}}Y(\tau))_i\tau = 0,$$

and consequently $\boldsymbol{v}^{\mathsf{t}}Y(\tau)_i = 0$ by Dedekind's independence theorem (see [18, Theorem 1.5.2]). It follows that $\boldsymbol{v}^{\mathsf{t}}Y(\tau) = 0$ and finally $\boldsymbol{v} = \boldsymbol{0}$, a contradiction. $\square$[**II.**]

Now let $\boldsymbol{u}_1, \ldots, \boldsymbol{u}_n \in L^n$ be such that $(\varphi(\boldsymbol{u}_1), \ldots, \varphi(\boldsymbol{u}_n))$ are linearly independent over L, and set $U = (\boldsymbol{u}_1, \ldots, \boldsymbol{u}_n)^{\mathsf{t}} \in \mathsf{GL}_n(L)$. Then

$$Z = (\varphi(\boldsymbol{u}_1), \ldots, \varphi(\boldsymbol{u}_n))^{\mathsf{t}} = \sum_{\tau \in G} U^\tau Y(\tau) \in \mathsf{GL}_n(L),$$

$$Z^\sigma Y(\sigma) = \sum_{\tau \in G} U^{\sigma\tau}Y(\tau)^\sigma Y(\sigma) = \sum_{\tau \in G} U^{\sigma\tau}Y(\sigma\tau) = Z \quad \text{for all} \quad \sigma \in G,$$

and the map $F^*\colon A \to \mathsf{M}_n(L)$, defined by $F^*(a) = ZF(a)Z^{-1}$ for all $a \in A$, is an L-representation. For all $a \in A$ and $\sigma \in G$ we obtain

$$F^*(a)^\sigma = Z^\sigma F(a)^\sigma Z^{-\sigma} = Z^\sigma Y(\sigma)F(a)Y(\sigma)^{-1}Z^{-\sigma} = ZF(a)Z^{-1} = F^*(a),$$

hence $F^*(a) \in \mathsf{M}_n(K)$, and $F^*\colon A \to \mathsf{M}_n(K)$ is a K-algebra isomorphism.

3. By 1. and 2. we infer

$$\begin{aligned} A \sim B &\iff A \otimes_K B^\circ \sim B \otimes_K B^\circ \sim K \\ &\iff \gamma_L(A \otimes_K B^\circ) = \gamma_L(A)\gamma_L(B)^{-1} = [1] \\ &\iff \gamma_L(A) = \gamma_L(B). \end{aligned}$$

Now we can prove the main assertion. Uniqueness is obvious by the very definition. By 1., 2., and 3., there exists a monomorphism $\Theta_{L/K} \colon \mathrm{Br}(L/K) \to H^2(G, L^\times)$ such that $\Theta_{L/K}([A]) = \gamma_L(A)$ for every central simple K-algebra A, and it remains to prove that it is surjective.

Let $\mathcal{E}$ be the set of all intermediate fields of L/K which are finite Galois over K, and for $E \in \mathcal{E}$ set $G_E = \mathrm{Gal}(E/K)$. Then 2.6.1 implies

$$H^2(G, L^\times) = \varinjlim_{E \in \mathcal{E}} H^2(G_E, E^\times),$$

and for each $E \in \mathcal{E}$ the map $\mathrm{Inf}^2_{G_E \to G} \colon H^2(G_E, E^\times) \to H^2(G, L^\times)$ is the injection into the inductive limit.

Assume now that $\xi \in H^2(G, L^\times)$, and let $E \in \mathcal{E}$ and $\xi_E \in H^2(G_E, E^\times)$ be such that $\xi = \mathrm{Inf}^2_{G_E \to G}(\xi_E)$. By 3.5.2, it follows that the central simple K-algebra $A = (E, G_E, \xi_E)$ splits over E, hence $\gamma_E(A) = \xi_E$ by 3.5.3.3, and by 3.5.3.2 we obtain

$$\gamma_L(A) = \mathrm{Inf}^2_{G_E \to G}(\gamma_E(A)) = \mathrm{Inf}^2_{G_E \to G}(\xi_E) = \xi. \qquad \square$$

Corollary 3.5.5. *Let L/K be a finite Galois extension and $G = \mathrm{Gal}(L/K)$.*

1. *Let A be a central simple K-algebra which splits over L. Then there exists a unique $\gamma \in H^2(G, L^\times)$ such that $A \sim (L, G, \gamma)$.*

2. *For all γ, $\delta \in H^2(G, L^\times)$, the following assertions hold:*

(a) $(L, G, \gamma) \otimes_K (L, G, \delta) \sim (L, G, \gamma\delta)$.

(b) $(L, G, \gamma)^\circ \sim (L, G, \gamma^{-1})$.

(c) $(L, G, \gamma) \sim (L, G, \delta)$ *if and only if* $\gamma = \delta$.

Proof. 1. If $\gamma = \gamma_L(A)$ and $A' = (L, G, \gamma)$, then $\gamma_L(A') = \gamma = \gamma_L(A)$ by 3.5.3.3, and therefore $A \sim A'$ by 3.5.4.3. Uniqueness follows by 2.(c).

2. Assume that γ, $\delta \in H^2(G, L^\times)$, $A = (L, G, \gamma)$, and $B = (L, G, \delta)$, hence $\gamma = \gamma_L(A)$ and $\delta = \gamma_L(B)$. Then the following assertions hold by 3.5.4:

(a) $\gamma_L(A \otimes_K B) = \gamma_L(A)\gamma_L(B) = \gamma\delta$, and consequently $A \otimes B \sim (L, G, \gamma\delta)$.

(b) $\gamma_L(A^\circ) = \gamma_L(A)^{-1} = \gamma^{-1}$, and consequently $A^\circ \sim (L, G, \gamma^{-1})$.

(c) $A \sim B$ if and only if $\gamma = \gamma_L(A) = \gamma_L(B) = \delta$. $\square$

Theorem 3.5.6. *Let L/K be a finite Galois extension and $G = \mathrm{Gal}(L/K)$. Let K'/K be any field extension such that both L and K' are contained in some field $\overline{L}$, and set $L' = LK'$.*

$$\begin{array}{ccccc} & L & \text{——} & L' = LK' & \text{—} \ \overline{L} \\ & | & & | \ )\, G' & \\ G\,(& L \cap K' & \text{——} & K' & \\ & | & & & \\ & K & & & \end{array}$$

Then L'/K' is a finite Galois extension, and the assignment $\sigma' \mapsto \sigma' \restriction L$ defines an isomorphism $G' = \mathrm{Gal}(L'/K') \xrightarrow{\sim} \mathrm{Gal}(L/L \cap K') \subset G$.

1. *If A is a central simple K-algebra which splits over L, then $A_{K'}$ splits over L', and there is a homomorphism*

$$\mathrm{res}_{L/K}^{L'/K'} : \mathrm{Br}(L/K) \to \mathrm{Br}(L'/K'), \quad \textit{given by} \quad \mathrm{res}_{L/K}^{L'/K'}([A]) = [A_{K'}].$$

It induces the following commutative diagram in which i denotes the inclusion, $\mathrm{res}_{K'/K}$ is defined in 3.4.6, and $\mathrm{Res}^2_{L/K \to L'/K'}$ is defined in 2.6.8:

$$\begin{array}{ccccc} \mathrm{Br}(K) & \xleftarrow{\ i\ } & \mathrm{Br}(L/K) & \xrightarrow{\Theta_{L/K}} & H^2(G, L^\times) \\ {\scriptstyle \mathrm{res}_{K'/K}} \downarrow & & \downarrow {\scriptstyle \mathrm{res}_{L/K}^{L'/K'}} & & \downarrow {\scriptstyle \mathrm{Res}^2_{L/K \to L'/K'}} \\ \mathrm{Br}(K') & \xleftarrow{\ i\ } & \mathrm{Br}(L'/K') & \xrightarrow{\Theta_{L'/K'}} & H^2(G', L'^\times)\,. \end{array}$$

In particular, $\mathrm{Res}^2_{L/K \to L'/K'}(\gamma_L(A)) = \gamma_{L'}(A_{K'})$ for every central simple K-algebra which splits over L.

2. *Assume that $K' \subset L$ (hence $L' = L$), let K'/K be Galois, $G' = \mathrm{Gal}(L/K')$ and $G/G' = \mathrm{Gal}(K'/K)$. Then*

$$\mathrm{res}_{L/K}^{L/K'} = \mathrm{res}_{K'/K} \restriction \mathrm{Br}(L/K) \colon \mathrm{Br}(L/K) \to \mathrm{Br}(L/K'),$$

and there is the following commutative diagram with exact rows in which i denotes the inclusion:

$$\begin{array}{ccccccc} \mathbf{1} & \longrightarrow & \mathrm{Br}(K'/K) & \xrightarrow{\ i\ } & \mathrm{Br}(L/K) & \xrightarrow{\mathrm{res}_{L/K}^{L/K'}} & \mathrm{Br}(L/K') \\ & & {\scriptstyle \Theta_{K'/K}} \downarrow & & {\scriptstyle \Theta_{L/K}} \downarrow & & \downarrow {\scriptstyle \Theta_{L/K'}} \\ \mathbf{1} & \longrightarrow & H^2(G/G', K'^\times) & \xrightarrow{\mathrm{Inf}^2_{G/G' \to G}} & H^2(G, L^\times) & \xrightarrow{\mathrm{Res}^2_{G \to G'}} & H^2(G', L^\times) \end{array}.$$

In particular, $\mathrm{Res}^2_{G \to G'}(\gamma_L(A)) = \gamma_L(A_{K'})$ for every central simple K-algebra which splits over L.

Proof. 1. Let A be a central simple K-algebra which splits over L. Then $A_{K'}$ splits over L', since $(A_{K'})_{L'} = A_{L'} = (A_L)_{L'} \sim L'$. Hence $A \mapsto A_{K'}$ induces a homomorphism $\mathrm{res}_{L/K}^{L'/K'} = \mathrm{res}_{K'/K} \restriction \mathrm{Br}(L/K) \colon \mathrm{Br}(L/K) \to \mathrm{Br}(L'/K')$, and by this definition the left-hand side of the diagram commutes.

To prove that also the right-hand side of the diagram commutes, let A be a central simple K-algebra which splits over L, and let $F \colon A \to \mathsf{M}_n(L)$ be an L-representation. For $\sigma \in G$ let $Y(\sigma) \in \mathsf{GL}_n(L)$ and $c \in Z^2(G, L^\times)$ be such that $F(a)^\sigma = Y(\sigma)F(a)Y(\sigma)^{-1}$ and $Y(\tau)^\sigma Y(\sigma) = c(\sigma,\tau)Y(\sigma\tau)$ for all $a \in A$ and $\sigma, \tau \in G$. Then $\Theta_{L/K}([A]) = \gamma_L(A) = [c] \in H^2(G, L^\times)$. Let $c' \colon G' \times G' \to L'^\times$ be defined by $c'(\sigma',\tau') = c(\sigma' \restriction L, \tau' \restriction L)$ for all σ', $\tau' \in G'$. Then $c' = \mathrm{Res}^2_{L/K \to L'/K'}(c) \in Z^2(G', L'^\times)$ and therefore

$$\mathrm{Res}^2_{L/K \to L'/K'} \circ \Theta_{L/K}([A]) = [c'] \in H^2(G', L'^\times).$$

The map $A \times K' \to \mathsf{M}_n(L')$, $(a,\lambda) \mapsto F(a)\lambda$, is K-bilinear, and therefore there exists a K-linear map $F_{K'} \colon A \otimes_K K' = A_{K'} \to \mathsf{M}_n(L')$ such that $F_{K'}(a \otimes z) = F(a)z$ for all $a \in A$ and $z \in K'$. Obviously, $F_{K'}$ is K'-linear, and it is easily checked that $F_{K'}$ is an L'-representation of $A_{K'}$. For $\sigma' \in G'$ let

$$Y'(\sigma') = Y(\sigma' \restriction L) \in \mathsf{GL}_n(L) \subset \mathsf{GL}_n(L').$$

If $a \in A$, $\lambda \in K'$ and σ', $\tau' \in G'$, then $F_{K'}(a \otimes \lambda)^{\sigma'} = Y'(\sigma')F_{K'}(a \otimes \lambda)Y'(\sigma')^{-1}$ and $Y'(\tau')^{\sigma'} Y'(\sigma') = c'(\sigma',\tau')Y'(\sigma'\tau')$. Hence it follows that

$$\Theta_{L'/K'} \circ \mathrm{res}_{L/K}^{L'/K'}([A]) = \gamma_{L'}(A_{K'}) = [c'] \in H^2(G', L'^\times).$$

2. By 1., applied with $L = L'$, the right-hand side of the diagram commutes since $\mathrm{Res}^2_{G \to G'} = \mathrm{Res}^2_{L/K \to L/K'}$. The left-hand side commutes by 3.5.3.2 and 3.5.4 since

$$\mathrm{Inf}^2_{G/G' \to G} \circ \Theta_{K'/K}([A]) = \mathrm{Inf}^2_{G/G' \to G}(\gamma_{K'}(A)) = \gamma_L(A) = \Theta_{L/K}([A]).$$

The vertical arrows are isomorphisms by 3.5.4, and the bottom row is exact by 2.5.4.3. Hence the top row is exact, too. □

Theorem 3.5.7. *Let A be a central simple K-algebra. Then A has a splitting field which is a finite Galois extension of K. If L/K is any finite Galois extension such that A splits over L and $[L:K] = n$, then $[A]^n = [K] \in \mathrm{Br}(K)$.*

In particular, if K_{sep} is a separable closure of K, then

$$\mathrm{Br}(K) = \mathrm{Br}(K_{\mathrm{sep}}/K) = \bigcup_{\substack{K \subset E \subset K_{\mathrm{sep}} \\ E/K \text{ finite Galois}}} \mathrm{Br}(E/K),$$

and $\mathrm{Br}(K)$ is a torsion group.

Proof. Let $A \cong \mathsf{M}_r(D)$, where $r \in \mathbb{N}$ and D is a central K-division algebra. By 3.4.8, there exists a maximal subfield L_0 of D which is separable over K, and then L_0 is a splitting field of D and thus of A. Hence every Galois extension of K containing L_0 is a splitting field of A. In particular, there exists an over K finite Galois splitting field of A, $K_{\rm sep}$ is a splitting field of A, and

$$\mathrm{Br}(K) = \mathrm{Br}(K_{\rm sep}/K) = \bigcup_{\substack{K\subset E\subset K_{\rm sep} \\ E/K \text{ finite Galois}}} \mathrm{Br}(E/K).$$

Let L/K be a finite Galois extension such that A splits over L, let $G = \mathrm{Gal}(L/K)$ and $[L:K] = n$. By 3.5.4, there exists an isomorphism $\Theta_{L/K}\colon \mathrm{Br}(L/K) \overset{\sim}{\to} H^2(G, L^\times)$, by 2.5.6 it follows that

$$\Theta_{L/K}([A]^n) = \Theta_{L/K}([A])^n = [1] \in H^2(G, L^\times),$$

and therefore $[A]^n = [K]$, sincee $\Theta_{L/K}$ is an isomorphism. □

3.6 Cyclic algebras

Let K be a field;
K-algebras are assumed to be are non-zero and finite-dimensional over K.

Definition and Theorem 3.6.1. *Let L/K be a cyclic field extension such that $[L{:}K] = n$ and $G = \mathrm{Gal}(L/K) = \langle\tau\rangle$.*

For $a \in K^\times$ let $c_{a,\tau}\colon G\times G \to L^\times$ be defined by

$$c_{a,\tau}(\tau^i, \tau^j) = \begin{cases} 1 & \text{if } i+j < n, \\ a & \text{if } i+j \ge n. \end{cases}$$

1. *Let $a \in K^\times$.*

(a) *$c_{a,\tau}$ is a factor system.*

The cohomology class $\gamma_{a,\tau} = [c_{a,\tau}] \in H^2(G, L^\times)$ is called the **cyclic cohomology class** for L/K with parameter a. A cyclic K-algebra $(L, G, \gamma_{a,\tau})$ is uniquely determined by L, τ and a, and we set $(L, \tau, a) = (L, G, \gamma_{a,\tau})$.
If $A = (L, \tau, a)$, then $\gamma_L(A) = \Theta_{L/K}([A]) = \gamma_{a,\tau} = [c_{a,\tau}]$ by 3.5.3 and 3.5.4.

(b) Let $A = L(c_{a,\tau}, \boldsymbol{u})$, *where* $\boldsymbol{u} = (u_{\tau^i})_{i\in[0,n-1]}$ *is an* L*-basis of* A *satisfying* $u_1 = 1$. *Then* $u_{\tau^i} = u_\tau^i$ *for all* $i \in [0, n-1]$, *and* $u_\tau^n = a$.

2. Let A *be a* K*-algebra such that* L *is a subfield of* A, *and let* $u \in A$ *such that* $u\lambda = \lambda^\tau u$ *for all* $\lambda \in L$, $(u^i)_{i\in[0,n-1]}$ *is an* L*-basis of* A, *and* $u^n = a \in K^\times$. *Then* $A = (L, \tau, a)$.

3. Let $\gamma \in H^2(G, L^\times)$, *and let* $A = (L, G, \gamma) = L(c, \boldsymbol{u})$, *built with an* L*-basis* $\boldsymbol{u} = (u_{\tau^i})_{i\in[0,n-1]}$ *satisfying* $u_1 = 1$ *and a normalized factor system* $c \in \gamma$. *Then* $(u_\tau^i)_{i\in[0,n-1]}$ *is also an* L*-basis of* A,

$$u_\tau^n = \prod_{\nu=0}^{n-1} c(\tau^\nu, \tau) = a \in K^\times, \quad \textit{and} \quad A = (L, \tau, a).$$

Proof. 1. (a) Let $\phi\colon G \to \mathbb{Q}$ be defined by

$$\phi(\tau^i) = \left\lfloor \frac{i}{n} \right\rfloor \quad \text{for all } \ i \in [0, n-1].$$

Then $c(\sigma, \rho) = a^{\phi(\sigma)+\phi(\rho)-\phi(\sigma\rho)}$ for all σ, $\rho \in G$, and we must prove that

$$c(\nu, \sigma)c(\nu\sigma, \rho) = c(\sigma, \rho)c(\nu, \sigma\rho) \quad \text{for all } \ \nu,\ \sigma,\ \rho \in G.$$

But this is equivalent to the obvious identity

$$[\phi(\nu) + \phi(\sigma) - \phi(\nu\sigma)] + [\phi(\nu\sigma) + \phi(\tau) - \phi(\nu\sigma\tau)]$$
$$= [\phi(\sigma) + \phi(\tau) - \phi(\sigma\tau)] + [\phi(\nu) + \phi(\sigma\tau) - \phi(\nu\sigma\tau)].$$

(b) By definition,

$$u_{\tau^i}\lambda = \lambda^{\tau^i} u_{\tau^i} \quad \text{for all } \ \lambda \in L \ \text{ and } \ i \in [0, n-1],$$

and

$$u_{\tau^i}u_{\tau^j} = c_{a,\tau}(\tau^i, \tau^j)u_{\tau^{i+j}} \quad \text{for all } \ i,\ j \in [0, n-1].$$

In particular, $u_{\tau^{i+1}} = u_{\tau^i}u$ for all $i \in [0, n-2]$ and $u_{\tau^{n-1}}u = au_{\tau^n} = a$. By a simple induction we obtain $u_{\tau^i} = u^i$ for all $i \in [0, n-1]$, and consequently $u^n = a$.

2. By a simple induction we obtain $u^i\lambda = \lambda^{\tau^i} u^i$ for all $\lambda \in K$ and $i \ge 0$. For $i \in [0, n-1]$ we set $u_{\tau^i} = u^i$. If i, $j \in [0, n-1]$, then this definition implies

$$u_{\tau^i}u_{\tau^j} = u^{i+j} = \begin{cases} u_{\tau^{i+j}} & \text{if } \ i+j \le n-1 \\ au^{i+j-n} = au_{\tau^{i+j}} & \text{if } \ i+j \ge n. \end{cases}$$

Hence $u_{\tau^i}u_{\tau^j} = c_{a,\tau}(\tau^i, \tau^j)u_{\tau^{i+j}}$ for all i, $j \in [0, n-1]$, and therefore we obtain $A = (L, \tau, a)$.

3. Again by a simple induction we infer $u_\tau^i\lambda = \lambda^{\tau^i} u_\tau^i$ for all $\lambda \in L$ and $i \ge 0$. In particular, we obtain $u_\tau^n\lambda = \lambda u_\tau^n$ for all $\lambda \in L$, hence $u_\tau^n \in \mathsf{c}_A(L) = L$ by 3.5.2, and $(u_\tau^n)^\tau u_\tau = u_\tau u_\tau^n$. It follows that $(u_\tau^n)^\tau = u_\tau^n$ and consequently $a = u_\tau^n \in K^\times$.

For all $\lambda \in L$ and $i \in [0, n-1]$ we obtain $\lambda^{\tau^i} = u_\tau^i \lambda u_\tau^{-i} = u_{\tau^i}\lambda u_{\tau^i}^{-1}$ and therefore $u_{\tau^i}^{-1}u_\tau^i\lambda = \lambda u_{\tau^i}^{-1}u_\tau^i$. Hence $u_{\tau^i}^{-1}u_\tau^i \in \mathsf{c}_A(L) = L$, and therefore $u_\tau^i = b_i u_{\tau^i}$ for some $b_i \in L^\times$. As $(u_{\tau^i})_{i\in[0,n-1]}$ is an L-basis of A, it follows that $(u^i)_{i\in[0,n-1]}$ is an L-basis of A, too, and $A = (L, \tau, a)$ by 2.

To prove the product formula, we show by induction that

$$u_\tau^i = \prod_{\nu=1}^{i-1} c(\tau^\nu, \tau) u_{\tau^i} \quad \text{for all } \ i \ge 1.$$

For $i = n$ the assertion follows since $u_{\tau^n} = u_1 = 1$. For $i = 1$ there is nothing to do. If $i \ge 1$, then

$$u_\tau^i = \prod_{\nu=1}^{i-1} c(\tau^\nu, \tau) u_{\tau^i} \quad \text{implies}$$

$$u_\tau^{i+1} = u_\tau^i u_\tau = \prod_{\nu=1}^{i-1} c(\tau^\nu, \tau) u_{\tau^i} u_\tau = \prod_{\nu=1}^{i} c(\tau^\nu, \tau) u_{\tau^{i+1}}. \qquad \square$$

Theorem 3.6.2. *Let L/K be a cyclic extension and $G = \mathrm{Gal}(L/K) = \langle\tau\rangle$. Then the maps $\beta\colon K^\times \to \mathrm{Br}(L/K)$ and $\overline{\beta}\colon K^\times \to H^2(G, L^\times)$, defined by*

$$\beta(a) = [(L, \tau, a)] \quad \textit{and} \quad \overline{\beta}(a) = \gamma_{a,\tau},$$

are epimorphisms such that $\mathrm{Ker}(\beta) = \mathrm{Ker}(\overline{\beta}) = \mathsf{N}_{L/K}L^\times$ and $\overline{\beta} = \Theta_{L/K}\circ\beta$. They induce isomorphisms

$$\beta_{L/K}\colon K^\times/\mathsf{N}_{L/K}L^\times \xrightarrow{\sim} \mathrm{Br}(L/K) \quad \textit{and} \quad \overline{\beta}_{L/K}\colon K^\times/\mathsf{N}_{L/K}L^\times \xrightarrow{\sim} H^2(G, L^\times)$$

satisfying $\beta_{L/K}(a\mathsf{N}_{L/K}L^\times) = [(L, \tau, a)]$ and $\overline{\beta}_{L/K}(a\mathsf{N}_{L/K}L^\times) = \gamma_{a,\tau}$ for all $a \in K^\times$.

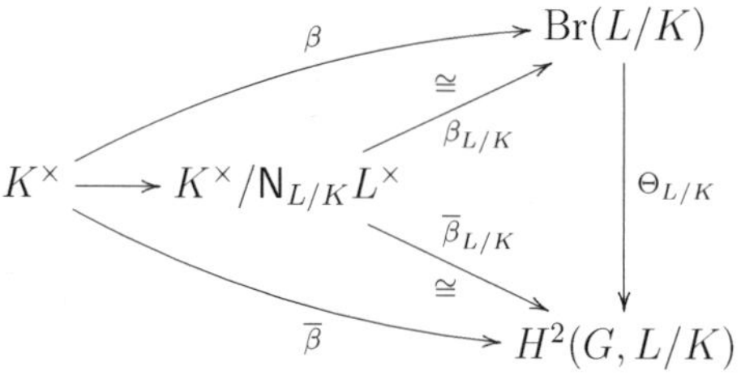

In particular, the following assertions hold:

1. *For every central simple K-algebra A which splits over L there exists some $a \in K^\times$ such that $A \sim (L, \tau, a)$.*

2. *Let $a,\, b \in K^\times$.*

(a) $(L, \tau, a) \otimes_K (L, \tau, b) \sim (L, \tau, ab)$.

(b) $(L, \tau, a) \sim (L, \tau, b)$ *holds if and only if* $b \in a\mathsf{N}_{L/K}L^\times$, *and* $(L, \tau, a) \sim K$ *holds if and only if* $a \in \mathsf{N}_{L/K}L^\times$.

Proof. It suffices to prove the main assertion (then 1. and 2. are obvious). Let $n = [L\!:\!K]$.

If $a,\, b \in K^\times$, then $c_{a,\tau}c_{b,\tau} = c_{ab,\tau}$, hence $\gamma_{a,\tau}\gamma_{b,\tau} = \gamma_{ab,\tau} \in H^2(G, L^\times)$, and $\overline{\beta}$ is a homomorphism. If $a \in K^\times$ and $A = (L, \tau, a) = (L, G, \gamma_{\tau,a})$, then

$$\Theta_{L/K}\circ\beta(a) = \Theta_{L/K}([A]) = \gamma_L(A) = \gamma_{\tau,a} = \overline{\beta}(a) \quad \text{by 3.5.3.3 and 3.5.4.}$$

Since $\Theta_{L/K}$ is an isomorphism, it follows that also β is a homomorphism and $\mathrm{Ker}(\beta) = \mathrm{Ker}(\overline{\beta})$. Hence it suffices to prove the following assertions:

a. β is surjective.

b. $\mathrm{Ker}(\overline{\beta}) \subset \mathsf{N}_{L/K}L^\times$.

c. $\mathsf{N}_{L/K}L^\times \subset \mathrm{Ker}(\beta)$.

Proof of **a.** Let A be a central simple K-algebra which splits over L, and let $\gamma = \gamma_L(A) \in H^2(G, L^\times)$. Then $A \sim (L, G, \gamma)$ by 3.5.3.3, hence $A = (L, \tau, a)$ for some $a \in K^\times$ by 3.6.1.3, and $[A] = \beta(a)$. □[**a.**]

Proof of **b.** Let $a \in \mathrm{Ker}(\overline{\beta})$. Then $c_{a,\tau} \sim 1$, and there exists some map $b\colon G \to L^\times$ such that $d^1b = c_{a,\tau}$. For all $i,\, j \in [0, n-1]$ we obtain

$$d^1b(\tau^i, \tau^j) = b(\tau^j)^{\tau^i} b(\tau^{i+j})^{-1} b(\tau^i) = c_{\tau,a}(\tau^i, \tau^j) = \begin{cases} 1 & \text{if } i+j \le n-1, \\ a & \text{if } i+j \ge n. \end{cases}$$

For $i = j = 0$ we get $b(1)^\tau b(1)^{-1} b(1) = b(1)^\tau = 1$, and therefore $b(1) = 1$. If $i \in [0, n-2]$ and $j = 1$, then $1 = d^1b(\tau^i, \tau) = b(\tau)^{\tau^i} b(\tau^{i+1})^{-1} b(\tau^i)$, hence $b(\tau^{i+1}) = b(\tau^i)b(\tau)^{\tau^i}$, and by a simple induction it follows that

$$b(\tau^{i+1}) = \prod_{\nu=0}^{i} b(\tau)^{\tau^\nu} \quad \text{for all } i \in [0, n-2].$$

Since $b(\tau^n) = b(1) = 1$ we get

$$a = d^1b(\tau^{n-1}, \tau) = b(\tau)^{\tau^{n-1}} b(\tau^n)^{-1} b(\tau^{n-1}) = b(\tau)^{\tau^{n-1}} \prod_{\nu=0}^{n-2} b(\tau)^{\tau^\nu}$$
$$= \mathsf{N}_{L/K}(b(\tau)),$$

and therefore $a \in \mathsf{N}_{L/K}L^\times$.

Proof of **c.** Let $y \in L^\times$, $a = \mathsf{N}_{L/K}(y)$ and $A = (L, \tau, a) = L(c_{a,\tau}, \boldsymbol{u})$, where $\boldsymbol{u} = (u_{\tau^i})_{i \in [0,n-1]}$ and $u_1 = 1$. If $u = u_\tau$, then we get $u_{\tau^i} = u^i$ for all $i \in [0, n-1]$ and $u^n = a$ by 3.6.1.1(b). Hence $u^i\lambda = \lambda^{\tau^i} u^i$ for all $i \geq 0$ and $\lambda \in L$, and it follows by induction on i that

$$(y^{-1}u)^i = \prod_{\nu=0}^{i-1} y^{-\tau^\nu} u^i \quad \text{for all} \quad i \geq 0.$$

Indeed, for $i \in \{0,1\}$ there is nothing to do, and if $i \geq 1$, then

$$(y^{-1}u)^i = \prod_{\nu=0}^{i-1} y^{-\tau^\nu} u^i$$

implies

$$(y^{-1}u)^{i+1} = \prod_{\nu=0}^{i-1} y^{-\tau^\nu} u^i y^{-1} u = \prod_{\nu=0}^{i-1} y^{-\tau^\nu} y^{-\tau^i} u^{i+1} = \prod_{\nu=0}^{i} y^{-\tau^\nu} u^{i+1}.$$

Hence the sequence $((y^{-1}u)^i)_{i \in [0,n-1]}$ is also an L-basis of A, we have $(y^{-1}u)\lambda = y^{-1}\lambda^\tau u = \lambda^\tau (y^{-1}u)$ for all $\lambda \in K$, and

$$(y^{-1}u)^n = \prod_{\nu=0}^{n-1} y^{-\tau^\nu} u^n = \mathsf{N}_{L/K}(y^{-1})a = 1.$$

By 3.6.1.2 it follows that $A = (L, \tau, 1) \sim K$. $\square$

Corollary and Definition 3.6.3 (Theorem of Frobenius)**.** Let $\iota = (z \mapsto \overline{z})$ be the complex conjugation (so that $\mathrm{Gal}(\mathbb{C}/\mathbb{R}) = \langle \iota \rangle$). Then the cyclic algebra

$$\mathbb{H} = (\mathbb{C}, \iota, -1)$$

is called the **Hamilton quaternion algebra**. By definition (with $\mathsf{j} = u_\iota$) it follows that $(1, \mathsf{j})$ is a $\mathbb{C}$-basis of $\mathbb{H}$, $\mathsf{j}^2 = -1$ and $z\mathsf{j} = \mathsf{j}\,\overline{z}$ for all $z \in \mathbb{C}$. In particular, $\mathsf{k} = \mathsf{ij} = -\mathsf{ji}$, and $(1, \mathsf{i}, \mathsf{j}, \mathsf{k})$ is an $\mathbb{R}$-basis of $\mathbb{H}$.

$\mathbb{H}$ *is a central K-division algebra, and* $\mathrm{Br}(\mathbb{R}) = \{[\mathbb{R}], [\mathbb{H}]\}$ (Theorem of Frobenius).

Proof. By 3.6.2 there is an isomorphism

$$\beta_{\mathbb{C}/\mathbb{R}} \colon \mathbb{R}^\times/\mathsf{N}_{\mathbb{C}/\mathbb{R}}\mathbb{C}^\times = \mathbb{R}^\times/\mathbb{R}_{>0} \overset{\sim}{\to} \mathrm{Br}(\mathbb{C}/\mathbb{R}) = \mathrm{Br}(\mathbb{R}),$$

given by $\beta_{\mathbb{C}/\mathbb{R}}(a\mathbb{R}_{>0}) = [(\mathbb{C}, \iota a)]$ for all $a \in \mathbb{R}^\times$. Since $\{1, -1\}$ is a set of representatives for $\mathbb{R}^\times/\mathbb{R}_{>0}$, $\mathbb{R} \sim (\mathbb{C}, \iota, 1)$ and $\mathbb{H} = (\mathbb{C}, \iota, -1\}$, it follows that $\mathrm{Br}(\mathbb{R}) = \{[\mathbb{R}], [\mathbb{H}]\}$. $\square$

We close this section with an explicit description of Res and Inf for cyclic algebras.

Theorem 3.6.4. *Let L/K be a cyclic field extension, and suppose that $G = \mathrm{Gal}(L/K) = \langle\tau\rangle$. Let K'/K be an arbitrary field extension such that L and K' are contained in some field $\overline{L}$, and set $L' = LK' \subset \overline{L}$.*

$$\begin{array}{ccccc} L & \text{---} & L' = LK' & \text{---} & \overline{L} \\ | & & |\ \langle\tau^d\rangle & & \\ \langle\tau\rangle\quad L \cap K' & \text{---} & K' & & \\ d\,| & & & & \\ K & & & & \end{array}$$

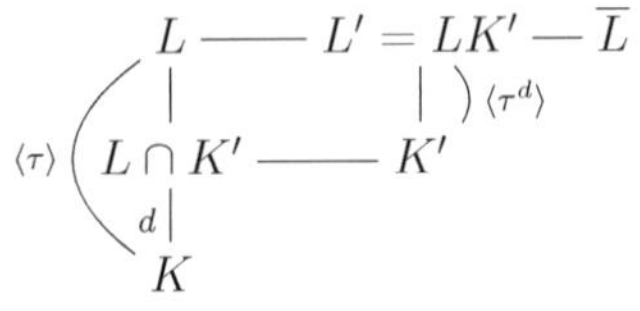

Then L'/K' is a finite Galois extension, and the assignment $\sigma' \mapsto \sigma' \restriction L$ induces an isomorphism $G' = \mathrm{Gal}(L'/K') \overset{\sim}{\to} \mathrm{Gal}(L/L \cap K') \subset G$. Hence L'/K' is cyclic, and if $d = [L \cap K' : K]$, then $\mathrm{Gal}(L/L \cap K') = \langle\tau^d\rangle$. We identify: $\mathrm{Gal}(L'/K') = \langle\tau^d\rangle$.

If $a \in K^\times$, then $\mathrm{Res}^2_{L/K\to L'/K'}(c_{a,\tau}) = c_{a,\tau^d}$, and if $A = (L,\tau,a)$ is a cyclic K-algebra, then $A_{K'} \sim (L',\tau^d,a)$.

Proof. If $m = [L' : K']$, then $[L : K] = dm$, and for $a \in K^\times$ and $i, j \in [0, m-1]$ we obtain

$$\begin{aligned}\mathrm{Res}^2_{L/K\to L'/K'}(c_{a,\tau})(\tau^{di},\tau^{dj}) &= c_{a,\tau}(\tau^{di}\restriction L, \tau^{dj}\restriction L) = c_{a,\tau}(\tau^{di},\tau^{dj}) \\ &= \begin{cases} 1 & \text{if } i+j < m, \\ a & \text{if } i+j \ge m \end{cases} \\ &= c_{a,\tau^d}(\tau^{di},\tau^{dj}),\end{aligned}$$

and therefore $\mathrm{Res}^2_{L/K\to L'/K'}(c_{a,\tau}) = c_{a,\tau^d}$ (since $c_{a,\tau^d} = c_{a,\tau} \restriction \langle\tau^d\rangle \times \langle\tau^d\rangle$ is a factor system of $\langle\tau^d\rangle$).

If $A = (L,\tau,a)$, then

$$\gamma_L(A) = [c_{\tau,a}],\ \gamma_{L'}(A_{K'}) = [\mathrm{Res}_{L/K\to L'/K'}(c_{\tau,a})] = [c_{\tau^d,a}]$$

by 3.5.6, and conseuently $A_{K'} = (L,\tau,a) \otimes_K K' \sim (L',\tau^d,a)$. □

Theorem 3.6.5. *Let $K \subset M \subset L$ be fields such that L/K is Galois, M/K is cyclic, $[M : K] = m$, $[L : M] = d$ and $[L : K] = n = md$. Let $G = \mathrm{Gal}(L/K)$ and $\mathrm{Gal}(M/K) = \langle\tau\rangle$. Let $a \in K^\times$, and let $A = (M,\tau,a)$ be a cyclic K-algebra.*

1. *For $\nu \in G$, let $i_\nu \in [0, m-1]$ be such that $\nu \restriction M = \tau^{i_\nu}$, and let $\overline{c} \in Z^2(G, L^\times)$ be defined by*

$$\overline{c}(\nu,\mu) = \begin{cases} 1 & \text{if } i_\nu + i_\mu < m, \\ a & \text{if } i_\nu + i_\mu \ge m, \end{cases}.$$

Then $A \sim (L, G, [\overline{c}])$.

2. Let L/K *be cyclic,* $\mathrm{Gal}(L/K) = \langle\theta\rangle$ *and* $\theta \restriction M = \tau$. *Then* $A \sim (L, \theta, a^d)$.

Proof. 1. We apply 3.5.3. By definition, $A = (M, \tau, a) = (M, \langle\tau\rangle, [c_{a,\tau}])$, and therefore $\gamma_M(A) = [c_{a,\tau}]$. If $\overline{c} = \mathrm{Inf}^2_{\langle\tau\rangle\to G}(c_{a,\tau}) \in Z^2(G, L^\times)$, then $\gamma_L(a) = [\overline{c}]$, and $A \sim (L, G, [\overline{c}])$. For all ν, $\mu \in G$ it follows that

$$\overline{c}(\nu, \mu) = c_{a,\tau}(\nu \restriction M, \mu \restriction M) = c_{a,\tau}(\tau^{i_\nu}, \tau^{i_\mu}) = \begin{cases} 1 & \text{if } i_\nu + i_\mu < m, \\ a & \text{if } i_\nu + i_\mu \geq m. \end{cases}$$

2. By 1., we may assume that $A = L(\overline{c}, \boldsymbol{u})$, where $\boldsymbol{u} = (u_{\theta^i})_{i\in[0,n-1]}$ is an L-basis of A such that $u_1 = 1$, $u_{\theta^i}\lambda = \lambda^{\theta^i} u_{\theta^i}$ for all $\lambda \in L$ and $i \in [0, n-1]$, and $u_{\theta^i} u_{\theta^j} = \overline{c}(\theta^i, \theta^j) u_{\theta^{i+j}}$ for all i, $j \in [0, n-1]$. We set $u = u_\theta$ and prove that $u^n = a^d$. Then it follows that $A \sim (L, \theta, a^d)$ by 3.6.1.2.

For $i \geq 0$ let $[i]_m \in [0, m-1]$ be such that $i \equiv [i]_m \bmod m$. For all i, $j \in [0, n-1]$, we obtain

$$\overline{c}(\theta^i, \theta^j) = c_{a,\tau}(\tau^i, \tau^j) = \begin{cases} 1 & \text{if } [i]_m + [j]_m < m, \\ a & \text{if } [i]_m + [j]_m \geq m. \end{cases}$$

We prove by induction: If $i = km + j$ for some $k \geq 0$ and $j \in [0, m-1]$, then $u^i = a^k u_{\theta^i}$. For $i = 0$ there is nothing to do. Thus suppose that $i = km + j$ for some $k \geq 0$ and $j \in [0, m-1]$, and assume that $u^i = a^k u_{\theta^i}$. Then

$$u^{i+1} = a^k u_{\theta^i} u_\theta = a^k \overline{c}(\theta^i, \theta) u_{\theta^{i+1}} = \begin{cases} a^k u_{\theta^{i+1}} & \text{if } i < m-1, \\ a^{k+1} u_{\theta^{i+1}} & \text{if } i = m-1, \end{cases}$$

and therefore $u^n = a^d u_{\theta^n} = a^d u_1 = a^d$. □

4

Local class field theory

We present the main results of local class field theory based on the theory of simple algebras and the structure of the Brauer group of a local field. This approach, going back to H. Hass and E. Noether, is maybe neither the shortest nor the simplest one, but it has the appeal showing connections between various mathematical theories which at a first glance seem to be disjointed. In the sequel we explore in detail the explicit description of abelian extensions by means of formal groups (Lubin-Tate theory). We briefly sketch the descriptions of the reciprocity law by B. Dwork and J. Neukirch, but we do not go into the cohomological theory using cup products.

Our main references for this chapter are [12], [55], [57], [39, §32], [48], [26], [6].

We tacitly use the basic theory of complete discrete valued fields including the theory of higher ramification. We explicitly refer to [18, Ch. 5], alternatively one may consult [57] or [14]. Throughout we observe that Pontrjagin's duality theorem 1.6.6 holds for all profinite groups (see 1.7.9.1).

We briefly recall the basic definitions for discrete valued fields.

Let K be a discrete valued field, $\overline{K}$ an algebraic closure, $K_{\rm sep}$ a separable closure, $K_{\rm ab}$ the largest abelian extension of K, and suppose that $K_{\rm ab} \subset K_{\rm sep} \subset \overline{K}$. We call $K_{\rm ab}$ the **maximal abelian extension** of K and $G_K = \mathrm{Gal}(K_{\rm sep}/K) = \mathrm{Gal}(\overline{K}/K)$ the **absolute Galois group** of K. Then 1.8.5.2 implies $\mathrm{Gal}(K_{\rm sep}/K_{\rm ab}) = G_K^{(1)} = \overline{[G,G]}$ and $\mathrm{Gal}(K_{\rm ab}/K) = G_K/G_K^{(1)} = G_K^{\mathsf{a}}$. We tacitly assume that all algebraic extensions of K lie inside $\overline{K}$.

Let $v_K\colon K \to \mathbb{Z} \cup \{\infty\}$ be the discrete valuation of K, $\mathcal{O}_K$ the valuation domain, $\mathfrak{p}_K$ the valuation ideal, $U_K = U_K^{(0)} = \mathcal{O}_K^\times$ the unit group, and let $\mathsf{k}_K = \mathcal{O}_K/\mathfrak{p}_K$ be the residue class field of K. For $n \ge 1$, let $U_K^{(n)} = 1 + \mathfrak{p}_K^n$ be the group of principal units of level n.

Let $|\cdot|\colon K \to \mathbb{R}_{\ge 0}$ be an absolute of K belonging to v_K. Then there is some $\rho \in (0,1)$ such that $|a| = \rho^{v_K(a)}$ for all $a \in K$ (where $\rho^\infty = 0$). In particular, $\pi \in K$ is a prime element of $\mathcal{O}_K$ (i. e., $\mathfrak{p}_K = \pi\mathcal{O}_K$) if and only if $|\pi| = \rho$. If $|\mathsf{k}_K| = q < \infty$, then we call $|\cdot|$ **normalized** if $\rho = q^{-1}$.

Let K be complete and $|\cdot|\colon \overline{K} \to \mathbb{R}_{\ge 0}$ the (unique and equally denoted) extension of K to an absolute value of $\overline{K}$. If L/K is a finite field extension and $[L\colon K] = n$, then $|a| = \sqrt[n]{|\mathsf{N}_{L/K}(a)|}$ for all $a \in L$, and $(L, |\cdot|)$ is complete.

DOI: 10.1201/9780429506574-4

If $e = e(L/K)$ is the ramification index and $f = f(L/K)$ is the residue class degree, then $n = ef$, $v_L \restriction K = ev_K$, and $v_K \circ \mathsf{N}_{L/K} = fv_L$.

By a **local field** we always mean a non-Archimedean local field, i. e., a complete discrete valued field K with finite residue class field k_K.

Let K be a local field, $p = \mathrm{char}(\mathsf{k}_K)$ and $|\mathsf{k}_K| = q$. Then K is (up to isomorphisms) either a finite extension of $\mathbb{Q}_p$, or $K = \mathbb{F}_q((X))$, and every finite extension of K is again a local field.

Special case $K = \mathbb{Q}_p$: $\mathcal{O}_{\mathbb{Q}_p} = \mathbb{Z}_p$, $\mathfrak{p}_{\mathbb{Q}_p} = p\mathbb{Z}_p$, $U_{\mathbb{Q}_p}^{(n)} = U_p^{(n)} = 1 + p^n\mathbb{Z}_p$ for all $n \in \mathbb{N}$, $U_{\mathbb{Q}_p} = U_p = \mathbb{Z}_p^\times$, $\mathsf{k}_{\mathbb{Q}_p} = \mathbb{F}_p$, and $v_{\mathbb{Q}_p} = v_p$ the p-adic exponent. The p-adic absolute value $|\cdot|_p \colon \mathbb{Q}_p \to \mathbb{R}_{\ge 0}$ is given by $|x|_p = p^{-v_p(x)}$ if $x \in \mathbb{Q}_p^\times$, and $|0| = 0$. We extend it to $|\cdot|_p \colon \overline{\mathbb{Q}}_p \to \mathbb{R}_{\ge 0}$.

Special case $K = \mathbb{F}_q((X))$: $\mathcal{O}_{\mathbb{F}_q((X))} = \mathbb{F}_q[\![X]\!]$, $\mathfrak{p}_{\mathbb{F}_q((X))} = X\mathbb{F}_q[\![X]\!]$, $U^{(n)}_{\mathbb{F}_q((X))} = 1 + X^n\mathbb{F}_q[\![X]\!]$ for $n \in \mathbb{N}$, $\mathsf{k}_{\mathbb{F}_q((X))} = \mathbb{F}_q$, and $v_{\mathbb{F}_q((X))} = \mathsf{o}_X$ is the order valuation.

Let again K be a local field. We denote by K_∞ the maximal unramified extension of K (inside $\overline{K}$), and by $\varphi_K \in \mathrm{Gal}(K_\infty/K)$ the Frobenius automorphism over K, defined by $\varphi_K(x) \equiv x^q \bmod \mathfrak{p}_{K_\infty}$ for all $x \in \mathcal{O}_{K_\infty}$ (see 1.8.10, and [18, Theorems 5.6.5 and 5.6.7]). Then $\mathrm{Gal}(K_\infty/K) = \overline{\langle \varphi_K \rangle} \cong \widehat{\mathbb{Z}}$, for every $n \in \mathbb{N}$ the fixed field $K_n = K_\infty^{\langle \varphi_K^n \rangle}$ of φ_K^n is the unique unramified extension of K of degree n, and $\mathrm{Gal}(K_n/K) = \langle \varphi_K \restriction K_n \rangle$. In particular:

Let L be any unramified extension of K and $[L:K] = m < \infty$. Then $L = K_m$, $\varphi_L = \varphi_K^m$, and if $\varphi_{L/K} = \varphi_K \restriction L$, then $\mathrm{Gal}(L/K) = \langle \varphi_{L/K} \rangle$ and $\varphi_{L/K}(a) \equiv a^q \bmod \mathfrak{p}_L$ for all $a \in \mathcal{O}_L$. If $K \subset M \subset L$ is an intermediate field, then $\varphi_{L/K} \restriction M = \varphi_{M/K}$ and $\varphi_{L/M} = \varphi_{L/K}^{[M:K]}$.

4.1 The Brauer group of a local field

Let K be a local field;
K-algebras are assumed to be non-zero and finite-dimensional over K.

Definition and Remark 4.1.1. Let A be a K-algebra. For $a \in A$ we define $\mu_a^A \in \mathrm{End}_K(A)$ by $\mu_a^A(x) = ax$ for all $x \in A$, and we call

$$\mathsf{N}_{A/K}(a) = \det(\mu_a^A) \in K$$

the **norm** of a.

If $(u_1, \ldots, u_n)$ is a K-basis of A and $(au_1, \ldots, au_n) = (u_1, \ldots, u_n)M$ for some matrix $M \in \mathsf{M}_n(K)$, then $\mathsf{N}_{A/K}(a) = \det(M)$.

If L is a K-subfield of A and $\dim_L(A) = m$, then $\mathsf{N}_{A/K}(a) = \mathsf{N}_{L/K}(a)^m$ for all $a \in L$ (indeed, $A = L^{\oplus n}$ as a K-space; if $a \in L$, then $\mu_a^A = (\mu_a^L)^{\oplus n}$, and therefore $\det(\mu_a^A) = \det(\mu_a^L)^n$). In particular, if A/K is a field extension, then $\mathsf{N}_{A/K}$ has the usual meaning in field theory.

Theorem 4.1.2. *Let D be a central K-division algebra and $d = \dim_K(D)$.* For $a \in D$ we define

$$|a|_D = \sqrt[d]{|\mathsf{N}_{D/K}(a)|} \in \mathbb{R}_{\geq 0}.$$

Then $|D^\times| = \langle \rho_D \rangle$ for some $\rho_D \in (0,1)$, $|\cdot|_D \restriction L = |\cdot|$ for every K-subfield L of D, and for all a, $b \in D$, the following assertions hold:

1. *$|a|_D = 0$ if and only if $a = 0$.*
2. *$|ab|_D = |a|_D\,|b|_D$.*
3. *$|a+b|_D \leq |a|_D + |b|_D$.*

In particular, $|\cdot|_D \colon D \to \mathbb{R}_{\geq 0}$ is a $|\cdot|$-compatible norm of D, and $(D, |\cdot|_D)$ is complete (see [18, Theorem 5.2.8]).

Proof. If $\rho \in (0,1)$ is such that $|K^\times| = \langle \rho \rangle$, then $|D^\times| \subset \langle \rho^{1/d} \rangle$, and therefore $|D^\times| = \langle \rho_D \rangle$ for some $\rho_D \in (0,1)$.

Let L be a K-subfield of D, $[L\colon K] = n$ and $\dim_L(D) = m$. Then $d = nm$, and if $a \in L$, then

$$|a|_D = \sqrt[d]{|\mathsf{N}_{D/K}(a)|} = \sqrt[d]{|\mathsf{N}_{L/K}(a)^m|} = \sqrt[n]{|\mathsf{N}_{L/D}(a)|} = |a|.$$

Now let a, $b \in D$. Then 2. is obvious by the definition, and clearly $|0|_D = 0$. If $a \in D^\times$, then $aa^{-1} = 1$ implies $|a|_D|a^{-1}|_D = |1|_D = 1$ and therefore $|a| > 0$. For the proof of 3. we may assume that $a \neq 0$. Then $K(a^{-1}b)$ is a subfield of D, and

$$\begin{aligned} |a+b|_D &= |a|_D|1+a^{-1}b|_D = |a|_D|1+a^{-1}b| \leq |a|_D(1+|a^{-1}b|) \\ &= |a|_D(1+|a^{-1}b|_D) = |a|_D + |b|_D. \end{aligned}$$

□

Theorem 4.1.3. *A central K-division algebra has a maximal over K unramified K-subfield.*

Proof. Assume to the contrary that D is a central K-division algebra such that every maximal K-subfield of D is ramified over K. Let L be an over K unramified K-subfield of D of maximal degree. Then L is not a maximal K-subfield of D, and therefore $E = \mathsf{c}_D(L) \supsetneq L$ by 3.4.7.1. If $a \in E^\bullet$, then $a^{-1} \in E$, hence E is a division algebra, and by 3.3.7.2 we get $\mathsf{c}(E) = \mathsf{c}(L) = L$.

Hence E is a central L-division algebra, and every L-subfield F of E is fully ramified over L.

Let $d = \dim_L(E)$, define $|\cdot|_E\colon E \to \mathbb{R}_{\ge 0}$ by $|a|_E = \sqrt[d]{|\mathsf{N}_{E/L}(a)|}$, let $|E^\times|_E = \langle \rho_E \rangle$, where $\rho_E \in (0,1)$, and let $\pi \in E$ be such that $|\pi| = \rho_E$. We shall prove that $E = L(\pi)$. Then E is a field and $L = \mathsf{c}(E) = E$, a contradiction.

Let $x \in E$, $n \in \mathbb{N}_0$ such that $|\pi^n x|_E \le 1$, and set $z = \pi^n x$. We construct recursively a sequence $(a_n)_{n\ge 0}$ in $\mathcal{O}_L$ such that

$$\Big| z - \sum_{i=0}^{n-1} a_i \pi^i \Big|_E \le \rho^n \quad \text{for all } n \ge 0.$$

For $n = 0$ there is nothing to do. Thus let $n \ge 0$, $a_0, \ldots, a_{n-1} \in \mathcal{O}_L$ such that

$$\Big| z - \sum_{i=0}^{n-1} a_i \pi^i \Big|_E \le \rho_E^n, \quad \text{and set} \quad z' = \pi^{-n}\Big(z - \sum_{i=0}^{n-1} a_i \pi^i \Big) \in E.$$

Then $|z'| \le 1$, $F = L(z')$ is an L-subfield of E, and thus F/L is fully ramified. Hence $\mathsf{k}_F = \mathsf{k}_L$, and since $z' \in \mathcal{O}_F$ there exists some $a_n \in \mathcal{O}_L$ such that $z' - a_n \in \mathfrak{p}_F$. Then $|z' - a_n|_E = |z' - a_n| < 1$, hence $|z' - a_n|_E \le \rho$ and

$$\Big| z - \sum_{i=0}^{n} a_i \pi^i \Big|_E = |\pi^n (z' - a_n)|_E \le \rho^{n+1}.$$

Now it follows that

$$z = \sum_{n=0}^{\infty} a_n \pi^n \in L(\pi), \quad \text{and consequently} \quad x = \pi^{-n} z \in L(\pi). \qquad \square$$

Corollary 4.1.4. *Every central simple K-algebra has a splitting field which is a finite unramified extension of K.*

Proof. Let A be a central simple K-algebra, say $A \cong \mathsf{M}_r(D)$ for some central K-division algebra D and $r \in \mathbb{N}$. By 4.1.3, D has an over K unramified maximal subfield L. Hence L is a splitting field of D and thus of A by 3.4.8.1. $\square$

Definition and Theorem 4.1.5. *Let A be a central simple K-algebra, L an over K finite unramified splitting field of A and $[L:K] = n$. Then we have $A \sim (L, \varphi_{L/K}, a)$ for some $a \in K^\times$, and the residue class*

$$\mathrm{inv}_K(A) = \frac{v_K(a)}{n} + \mathbb{Z} \in \mathbb{Q}/\mathbb{Z}$$

depends only on the similarity class $[A] \in \mathsf{Br}(K)$.

$\mathrm{inv}_K(A) = \mathrm{inv}_K([A]) \in \mathbb{Q}/\mathbb{Z}$ is called the **Hasse invariant** of A (resp. of $[A]$).

Proof. By 3.6.2.1, it follows that $A \sim (L, \varphi_{L/K}, a)$ for some $a \in K^\times$, and we must prove that $\mathrm{inv}_K(A)$ only depends on $[A]$. Thus let $a_1,\, a_2 \in K^\times$ and $L_1,\, L_2$ be over K finite unramified extension fields such that $[L_1:K] = n_1$, $[L_2\!:\!K] = n_2$ and $(L_1, \varphi_{L_1/K}, a_1) \sim (L_2, \varphi_{L_2/K}, a_2)$. We must prove that

$$\frac{v_K(a_1)}{n_1} - \frac{v_K(a_2)}{n_2} \in \mathbb{Z}.$$

Let $L = L_1L_2$, $[L : L_1] = d_1$ and $[L : L_2] = d_2$. Then L/K is unramified, hence cyclic, and $n = [L : K] = n_1d_1 = n_2d_2$. For $i \in \{1,2\}$ we get $\varphi_{L_i/K} = \varphi_{L/K} \restriction L_i$, and therefore $(L_i, \varphi_{L_i/K}, a_i) \sim (L, \varphi_{L/K}, a_i^{d_i})$ by 3.6.5.2. Hence

$$(L, \varphi_{L/K}, a_1^{d_1}) \sim (L, \varphi_{L/K}, a_2^{d_2}) \quad \text{and} \quad a_1^{d_1}\mathsf{N}_{L/K}L^\times = a_2^{d_2}\mathsf{N}_{L/K}L^\times.$$

Since $v_K(\mathsf{N}_{L/K}L^\times) = n\mathbb{Z}$, it follows that $v_K(a_1^{d_1}) \equiv v_K(a_2^{d_2}) \bmod n$, and

$$\frac{v_K(a_1)}{n_1} - \frac{v_K(a_2)}{n_2} = \frac{d_1v_K(a_1) - d_2v_K(a_2)}{n} = \frac{v_K(a_1^{d_1}) - v_K(a_2^{d_2})}{n} \in \mathbb{Z}. \quad \square$$

Theorem 4.1.6. *The map* $\mathrm{inv}_K \colon \mathsf{Br}(K) \to \mathbb{Q}/\mathbb{Z}$ *is an isomorphism. In particular:*

1. *Every finite subgroup of* $\mathsf{Br}(K)$ *is cyclic, and for every* $n \in \mathbb{N}$ *there is precisely one subgroup of* $\mathsf{Br}(K)$ *of order* n.
2. *If* A *and* B *are central simple* K*-algebras, then*

- $\mathrm{inv}_K(A \otimes_K B) = \mathrm{inv}_K(A) + \mathrm{inv}_K(B)$.
- $\mathrm{inv}_K(A) = 0$ *if and only if* $A \sim K$.

Proof. For $i \in \{1,2\}$, let A_i be a central simple K-algebra and L_i an over K finite unramified splitting field of A_i. Then $L = L_1L_2$ is an over K finite unramified splitting field of both A_1 and A_2, and we set $n = [L : K]$. If $A_i \sim (L, \varphi_{L/K}, a_i)$, then $A_1 \otimes_K A_2 \sim (L, \varphi_{L/K}, a_1a_2)$ by 3.6.2.2(a), and

$$\begin{aligned}\mathrm{inv}_K(A_1 \otimes_K A_2) &= \frac{v_K(a_1a_2)}{n} + \mathbb{Z} = \Big(\frac{v_K(a_1)}{n} + \mathbb{Z}\Big) + \Big(\frac{v_K(a_2)}{n} + \mathbb{Z}\Big) \\ &= \mathrm{inv}_K(A_1) + \mathrm{inv}_K(A_2).\end{aligned}$$

Hence $\mathrm{inv}_K \colon \mathsf{Br}(K) \to \mathbb{Q}/\mathbb{Z}$ is a group homomorphism.

Next let A be a central simple K-algebra such that $\mathrm{inv}_K(A) = 0 \in \mathbb{Q}/\mathbb{Z}$. Let L be an over K finite unramified splitting field of A, $[L : K] = n$ and $a \in K^\times$ such that $A \sim (L, \varphi_{L/K}, a)$. It follows that $v_K(a) \equiv 0 \bmod n$, hence $a \in \mathsf{N}_{L/K}L^\times$ (see [18, Theorem 5.6.9]), and $A \sim K$ by 3.6.2.2(b). Hence inv_K is a monomorphism.

For $n \in \mathbb{N}$ let L/K be an unramified extension such that $[L\colon K] = n$, π a prime element of $\mathcal{O}_K$ and $A = (L, \varphi_{L/K}, \pi)$. Then

$$\mathrm{inv}_K(A) = \frac{1}{n} + \mathbb{Z}, \quad \text{and since} \quad \mathbb{Q}/\mathbb{Z} = \Big\langle \Big\{ \frac{1}{n} + \mathbb{Z} \;\Big|\; n \in \mathbb{N} \Big\} \Big\rangle$$

it follows that inv_K is surjective.

1. and 2. are now obvious. □

Definition 4.1.7. For later use (in 6.2.5) we define the Hasse invariant inv_K also for $K = \mathbb{R}$ and $K = \mathbb{C}$, and to this end we regard $\mathbb{C}/\mathbb{R}$ as an unramified extension. As $\mathsf{Br}(\mathbb{C}) = \mathbf{1}$, we set $\mathrm{inv}_{\mathbb{C}} = 0$, and for a central simple $\mathbb{R}$-algebra A we set

$$\mathrm{inv}_{\mathbb{R}}(A) = 0 \text{ if } A \sim \mathbb{R}, \quad \text{and} \quad \mathrm{inv}_{\mathbb{R}}(A) = \frac{1}{2} + \mathbb{Z} \text{ if } A \sim \mathbb{H} \quad \text{(see 3.6.3)}.$$

Theorem 4.1.8. *Let A be a central simple K-algebra, L/K a finite field extension and $n = [L\colon K]$.*

1. $\mathrm{inv}_L(A_L) = n\,\mathrm{inv}_K(A)$.
2. *L is a splitting field of A if and only if*

$$A^{\otimes n} = A \otimes_K \ldots \otimes_K A \sim K.$$

Proof. 1. By 4.1.4, A has an over K finite unramified splitting field L'. We set $\widetilde{L} = LL'$ and $d = [\widetilde{L}\colon L]$. Then $\widetilde{L}/L$ is unramified, and $\widetilde{L}$ is a splitting field of A_L.

CASE **a**: L/K is unramified. Then also $\widetilde{L}/K$ is unramified, therefore $A \sim (\widetilde{L}, \varphi_{\widetilde{L}/K}, a)$ for some $a \in K^\times$, and $A_L \sim (\widetilde{L}, \varphi^n_{\widetilde{L}/K}, a) = (\widetilde{L}, \varphi_{\widetilde{L}/L}, a)$ by 3.6.4. Hence

$$\mathrm{inv}_L(A_L) = \frac{v_L(a)}{d} + \mathbb{Z} = n\Big(\frac{v_K(a)}{dn} + \mathbb{Z}\Big) = n\,\mathrm{inv}_K(A).$$

CASE **b**: L/K is fully ramified. Then $d = [\widetilde{L}\colon L] = [L'\colon K]$, $[\widetilde{L}\colon L'] = n$, $\widetilde{L}/L'$ is fully ramified and therefore $\mathsf{k}_{\widetilde{L}} = \mathsf{k}_{L'}$. Hence $\varphi_{\widetilde{L}/L} \restriction L' = \varphi_{L'/K}$, and if $A \sim (L', \varphi_{L'/K}, a)$, where $a \in K^\times$, then $A_L \sim (\widetilde{L}, \varphi_{\widetilde{L}/L}, a)$ by 3.6.4. Consequently,

$$\mathrm{inv}_L(A_L) = \frac{v_L(a)}{d} + \mathbb{Z} = \frac{n v_K(a)}{d} + \mathbb{Z} = n\Big(\frac{v_K(a)}{d} + \mathbb{Z}\Big) = n\,\mathrm{inv}_K(A).$$

CASE **c**: L/K is arbitrary. Let T be the inertia field of L/K. Then T/K is unramified and L/T is fully ramified (see [18, Corollary 5.6.6]). Using CASE **a** and CASE **b**, we obtain

$$\begin{aligned}\mathrm{inv}_L(A_L) &= \mathrm{inv}_L((A_T)_L) = [L\!:\!T]\,\mathrm{inv}_T(A_T) = [L\!:\!T]\,[T\!:\!K]\,\mathrm{inv}_K(A)\\ &= [L\!:\!K]\,\mathrm{inv}_K(A).\end{aligned}$$

2. On the one hand $A_L \sim L$ if and only if $\mathrm{inv}_L(A_L) = 0$, and on the other hand $\mathrm{inv}_L(A_L) = n\,\mathrm{inv}_K(A) = \mathrm{inv}_K(A^{\otimes n}) = 0$ if and only if $A^{\otimes n} \sim K$. □

Theorem 4.1.9. *Let L/K be a finite field extension and $n = [L:K]$. Then the homomorphism $\mathrm{res}_{L/K}\colon \mathsf{Br}(K) \to \mathsf{Br}(L)$ is surjective,*

$$\mathrm{inv}_L \circ \mathrm{res}_{L/K} = n\,\mathrm{inv}_K \quad \textit{and} \quad \mathrm{inv}_K(\mathsf{Br}(L/K)) = \frac{1}{n}\mathbb{Z}/\mathbb{Z}.$$

In particular, $\mathsf{Br}(L/K) = \mathrm{Ker}(\mathrm{res}_{L/K})$ is the cyclic subgroup of order n of $\mathsf{Br}(K)$.

Proof. From 4.1.8.1 we obtain the following commutative diagram with exact rows (where i denotes the inclusion).

$$\begin{array}{ccccccccc}
\mathbf{1} & \longrightarrow & \mathrm{Br}(L/K) & \xrightarrow{\ i\ } & \mathrm{Br}(K) & \xrightarrow{\mathrm{res}_{L/K}} & \mathrm{Br}(L) & & \\
 & & & & \Big\downarrow{\scriptstyle \mathrm{inv}_K} & & \Big\downarrow{\scriptstyle \mathrm{inv}_L} & & \\
\mathbf{0} & \longrightarrow & \frac{1}{n}\mathbb{Z}/\mathbb{Z} & \xrightarrow{\ i\ } & \mathbb{Q}/\mathbb{Z} & \xrightarrow{\cdot n} & \mathbb{Q}/\mathbb{Z} & \longrightarrow & \mathbf{0}
\end{array}$$

By 4.1.6, inv_K and inv_L are isomorphisms. Hence $\mathrm{res}_{L/K}$ is surjective,

$$\mathrm{inv}_K(\mathrm{Br}(L/K)) = \frac{1}{n}\mathbb{Z}/\mathbb{Z},$$

and $\mathsf{Br}(L/K) = \mathrm{Ker}(\mathrm{res}_{L/K})$ is the cyclic subgroup of order n of $\mathsf{Br}(K)$. □

Theorem 4.1.10. *Let L/K be a finite separable extension.*

1. Let L/K be cyclic and $G = \mathrm{Gal}(L/K)$. Let $[L:K] = n$ and $e = e(L/K)$. Then

$$H^0(G, L^\times) = K^\times/\mathsf{N}_{L/K}L^\times \cong \mathbb{Z}/n\mathbb{Z}, \quad \mathsf{h}(G, L^\times) = n$$

and

$$H^0(G, U_L) = U_K/\mathsf{N}_{L/K}U_L \cong H^1(G, U_L) \cong \mathbb{Z}/e\mathbb{Z}.$$

2. Let $\widetilde{L}$ be a Galois closure of L/K. Then

$$(K^\times : \mathsf{N}_{L/K}L^\times) \le [\widetilde{L}:K] \quad \textit{and} \quad (U_K : \mathsf{N}_{L/K}U_L) \le e(\widetilde{L}/K).$$

3. $\mathsf{N}_{L/K}L^\times$ is an open subgroup of finite index of $K^\times$, and there exists some $m \in \mathbb{N}$ such that $U_K^{(m)} \subset \mathsf{N}_{L/K}U_L$.

Proof. 1. Recall that $H^{-1}(G, L^\times) = H^1(G, L^\times) = \mathbf{1}$ by 2.6.3 and 2.4.5.1, and

$$K^\times/\mathsf{N}_{L/K}L^\times = H^0(G, L^\times) \cong H^2(G, L^\times) \cong \mathrm{Br}(L/K) \cong \frac{1}{n}\mathbb{Z}/\mathbb{Z}$$

by 2.4.5.1, 3.5.4 and 4.1.9.1. Hence $\mathsf{h}(G, L^\times) = n$. From the exact sequence

$$\mathbf{1} \to U_L \to L^\times \stackrel{v_L}{\to} \mathbb{Z} \to \mathbf{0}$$

we obtain (using 2.4.2 and 2.4.5.3) the exact sequence

$$\mathbf{0} = H^{-1}(G, \mathbb{Z}) \stackrel{\delta^{-1}}{\to} H^0(G, U_L) \to H^0(G, L^\times) \stackrel{H^0(v_L)}{\to} H^0(G, \mathbb{Z}) \stackrel{\delta^0}{\to}$$
$$\stackrel{\delta^0}{\to} H^1(G, U_L) \to H^1(G, L^\times) = \mathbf{1}.$$

It shows that $H^0(G, U_L) \cong \mathrm{Ker}(H^0(v_L))$ and $H^1(G, U_L) \cong \mathrm{Coker}(H^0(v_L))$. Hence it suffices to prove that $|\mathrm{Ker}(H^0(v_L))| = |\mathrm{Coker}(H^0(v_L))| = e$.

The homomorphism $H^0(v_L) \colon K^\times/\mathsf{N}_{L/K}L^\times \to \mathbb{Z}/n\mathbb{Z}$ is given by

$$H^0(v_L)(a\mathsf{N}_{L/K}L^\times) = v_L(a) + n\mathbb{Z} = ev_K(a) + n\mathbb{Z} \quad \text{for all } \ a \in K^\times.$$

Hence $\mathrm{Im}(H^0(v_L)) = e\mathbb{Z}/n\mathbb{Z}$, $\mathrm{Coker}(H^0(v_L)) = \mathbb{Z}/e\mathbb{Z}$ and

$$|\mathrm{Ker}(H^0(v_L))| = \frac{|K^\times/\mathsf{N}_{L/K}L^\times|}{\mathrm{Im}(H^0(v_L))|} = \frac{n}{n/e} = e.$$

2. Assume first that L/K is Galois (hence $L = \widetilde{L}$), $n = [L:K] > 1$, and the assertion holds for all Galois extensions of smaller degree. By [18, Corollary 5.5.4] the extension L/K is solvable, and therefore there exists an intermediate field K' of L/K such that K'/K is cyclic and $[L:K'] < n$. Hence $(K'^\times : \mathsf{N}_{L/K'}L^\times) \le [L:K']$ and $(U_{K'} : \mathsf{N}_{L/K'}U_L) \le e(L/K')$ by the induction hypothesis. Using 1., we get

$$\begin{aligned}(K^\times : \mathsf{N}_{L/K}L^\times) &= (K^\times : \mathsf{N}_{K'/K}K'^\times)(\mathsf{N}_{K'/K}K'^\times : \mathsf{N}_{K'/K}\mathsf{N}_{L/K'}L^\times) \\ &\le [K':K]\,[K'^\times : \mathsf{N}_{L/K'}L^\times] \le [K':K]\,[L:K'] = [L:K]\end{aligned}$$

and

$$\begin{aligned}(U_K : \mathsf{N}_{L/K}U_L) &= (U_K : \mathsf{N}_{K'/K}U_{K'})(\mathsf{N}_{K'/K}U_{K'} : \mathsf{N}_{K'/K}\mathsf{N}_{L/K'}U_L) \\ &\le e(K'/K)\,(U_{K'} : \mathsf{N}_{L/K'}U_L) \le e(K'/K)e(L/K') = e(L/K).\end{aligned}$$

If L/K is arbitrary, then

$$(K^\times : \mathsf{N}_{L/K}L^\times) \le (K^\times : \mathsf{N}_{L/K}\mathsf{N}_{\widetilde{L}/L}\widetilde{L}^\times) = (K^\times : \mathsf{N}_{\widetilde{L}/K}\widetilde{L}^\times) \le [\widetilde{L}:K]$$

and

$$(U_K : \mathsf{N}_{L/K}U_L) \le (U_K : \mathsf{N}_{L/K}\mathsf{N}_{\widetilde{L}/L}U_{\widetilde{L}}) = (U_K : \mathsf{N}_{\widetilde{L}/K}U_{\widetilde{L}}) \le e(\widetilde{L}/K).$$

3. Since $\mathsf{N}_{L/K} \colon L^\times \to K^\times$ is continuous and U_L is compact, it follows that $\mathsf{N}_{L/K}U_L$ is compact and thus closed in U_K. But $(U_K : \mathsf{N}_{L/K}U_L) < \infty$, hence $\mathsf{N}_{L/K}U_L$ is an open neighbourhood of 1 in U_K (see 1.1.3). As $\{U_K^{(m)} \mid m \in \mathbb{N}\}$ is a fundamental system of neighbourhoods of 1, we get $U_K^{(m)} \subset \mathsf{N}_{L/K}U_L$ for some $m \in \mathbb{N}$.

Since $(K^\times : \mathsf{N}_{L/K}L^\times) < \infty$ it suffices to prove that $\mathsf{N}_{L/K}L^\times$ is closed in $K^\times$. Let $(x_n)_{n\ge 0}$ be a sequence in $\mathsf{N}_{L/K}L^\times$ such that $(x_n)_{n\ge 0} \to x \in K$. Then $|x_n| = |x|$ for all $n \gg 1$, and we may assume that $|x_n| = |x|$ for all $n \ge 0$. Then we obtain $x_0^{-1}x_n \in U_K \cap \mathsf{N}_{L/K}L^\times = \mathsf{N}_{L/K}U_L$, $(x_0^{-1}x_n)_{n\ge 0} \to x_0^{-1}x$, hence $x_0^{-1}x \in \mathsf{N}_{L/K}U_L$ and $x \in \mathsf{N}_{L/K}L^\times$. □

4.2 The local reciprocity law

Let K be a local field.

Let G be a profinite group. By the identifications made in 2.2.5, its (topological) character group

$$\mathsf{X}(G) = \mathrm{Hom}^{\mathsf{c}}(G, \mathbb{Q}/\mathbb{Z}) = H^1(G, \mathbb{Q}/\mathbb{Z})$$

is an additive abelian group. If $\chi \in \mathsf{X}(G)$, then $\chi(G)$ is a finite cyclic subgroup of $\mathbb{Q}/\mathbb{Z}$ by 1.7.9.3, and $\mathrm{ord}(\chi) = |\chi(G)|$ is the order of χ. If $G^{(1)} = \overline{[G,G]}$ is the topological commutator group and $G^{\mathsf{a}} = G/G^{(1)}$, then $\mathsf{X}(G) = \mathsf{X}(G^{\mathsf{a}})$, and we idenify $\mathsf{X}^2(G) = \mathsf{X}^2(G^{\mathsf{a}}) = G^{\mathsf{a}}$ by means of $g(\chi) = \chi(g)$ for all $g \in G$ and $\chi \in \mathsf{X}(G)$ (see 1.6.6 and 1.7.9.1).

Let L/K be a Galois extension and $G = \mathrm{Gal}(L/K)$. Let $L^{\mathsf{a}} = K_{\mathrm{ab}} \cap L$ be the largest over K abelian intermediate field of L/K. Then it follows that $G^{(1)} = \mathrm{Gal}(L/L^{\mathsf{a}})$, $\mathrm{Gal}(L^{\mathsf{a}}/K) = G^{\mathsf{a}}$, and $K_{\mathrm{ab}} = (K_{\mathrm{sep}})^{\mathsf{a}}$ is the maximal abelian extension of K.

Definition and Theorem 4.2.1. *Let L/K be a Galois extension and* $G = \mathrm{Gal}(L/K)$, $\chi \in \mathsf{X}(G)$ *and* $n_\chi = \mathrm{ord}(\chi)$. *Let* $L_\chi = L^{\mathrm{Ker}(\chi)}$ *be the fixed field of* $\mathrm{Ker}(\chi)$, *and set* $G_\chi = \mathrm{Gal}(L_\chi/K) = G/\ker(\chi)$. *Then χ induces an (equally denoted) isomorphism*

$$\chi\colon G_\chi \overset{\sim}{\to} \mathrm{Im}(\chi) = \frac{1}{n_\chi}\mathbb{Z}/\mathbb{Z} \subset \mathbb{Q}/\mathbb{Z},$$

and we denote by $\sigma_\chi \in G_\chi$ *the unique automorphism satisfying*

$$\chi(\sigma_\chi) = \frac{1}{n_\chi} + \mathbb{Z} \in \mathbb{Q}/\mathbb{Z}.$$

1. The map

$$\langle\,\cdot\,,\,\cdot\,\rangle = \langle\,\cdot\,,\,\cdot\,\rangle_K\colon K^\times \times \mathsf{X}(G) \to \mathbb{Q}/\mathbb{Z}, \quad \text{defined by} \quad \langle a, \chi\rangle = \mathrm{inv}_K(L_\chi, \sigma_\chi, a),$$

is a continuous pairing, i. e.,

- $\langle\,\cdot\,,\,\cdot\,\rangle$ *is continuous (if $\mathbb{Q}/\mathbb{Z}$ carries the discrete topology);*
- *if* $a, a' \in K^\times$ *and* $\chi, \chi_1, \chi_2 \in \mathsf{X}(G)$, *then*

$$\langle aa', \chi\rangle = \langle a, \chi\rangle + \langle a', \chi\rangle \quad \text{and} \quad \langle a, \chi_1 + \chi_2\rangle = \langle a, \chi_1\rangle + \langle a, \chi_2\rangle.$$

2. If $\chi \in \mathsf{X}(G)$ *and* $a \in K^\times$, *then* $\langle a, \chi\rangle = 0 \in \mathbb{Q}/\mathbb{Z}$ *if and only if* $a \in \mathsf{N}_{L_\chi/K} L_\chi^\times$.

3. There exists a unique continuous homomorphism

$$(\,\cdot\,, L/K)\colon K^\times \to G^{\mathsf{a}}$$

satisfying

$$\chi(a, L/K) = \langle a, \chi\rangle = \mathrm{inv}_K(L_\chi, \sigma_\chi, a) \quad \text{for all } a \in K^\times \text{ and } \chi \in \mathsf{X}(G).$$

The map $(\,\cdot\,, L/K)\colon K^\times \to G^{\mathsf{a}} = \mathrm{Gal}(L^{\mathsf{a}}/K)$ is called the (local) **norm residue symbol** for L/K.
For $L = K_{\mathrm{sep}}$, the map

$$(\,\cdot\,, K) = (\,\cdot\,, K_{\mathrm{sep}}/K)\colon K^\times \to G_K^{\mathsf{a}} = \mathrm{Gal}(K_{\mathrm{ab}}/K)$$

is called the (local) **universal norm residue symbol** for K.

4. Let L_0 be an over K Galois intermediate field of L/K. Then $L_0^{\mathsf{a}} = L_0 \cap L^{\mathsf{a}}$, *and*

$$(a, L_0/K) = (a, L/K) \restriction L_0^{\mathsf{a}} \quad \text{for all } a \in K^\times.$$

5. If $[L{:}K] < \infty$, *then* $(\mathsf{N}_{L/K}(z), L/K) = 1 \in G^{\mathsf{a}}$ *for all* $z \in L^\times$.

Proof. $\chi(G)$ is cyclic of order n_χ. Hence χ induces an (equally denoted) isomorphism

$$\chi\colon G_\chi = G/\ker(\chi) \overset{\sim}{\to} \operatorname{Im}(\chi) = \frac{1}{n_\chi}\mathbb{Z}/\mathbb{Z} \subset \mathbb{Q}/\mathbb{Z}$$

as asserted, and $G_\chi = \operatorname{Gal}(K_\chi/K) = \langle \sigma_\chi \rangle$.

1. We defer the proof of the continuity to 2.

If $a,\ a' \in K^\times$ and $\chi \in \mathsf{X}(G)$, then $(L_\chi, \sigma_\chi, aa') \sim (L_\chi, \sigma_\chi, a) \otimes_K (L_\chi, \sigma_\chi, a')$ by 3.6.2.2(a), and consequently $\langle aa', \chi \rangle = \langle a, \chi \rangle + \langle a', \chi \rangle$.

Now let $a \in K^\times$ and $\chi_1,\ \chi_2,\ \chi_3 \in \mathsf{X}(G)$ satisfy $\chi_3 = \chi_1 + \chi_2$. We shall prove that

$$(L_{\chi_1}, \sigma_{\chi_1}, a) \otimes_K (L_{\chi_2}, \sigma_{\chi_2}, a) \sim (L_{\chi_3}, \sigma_{\chi_3}, a). \qquad (*)$$

Then it follows by 4.1.6.1 that

$$\begin{aligned}\langle a, \chi_3 \rangle &= \operatorname{inv}_K(L_{\chi_3}, \sigma_{\chi_3}, a) = \operatorname{inv}_K(L_{\chi_1}, \sigma_{\chi_1}, a) + \operatorname{inv}_K(L_{\chi_2}, \sigma_{\chi_2}, a) \\ &= \langle a, \chi_1 \rangle + \langle a, \chi_2 \rangle.\end{aligned}$$

Proof of $(*)$. For $k \in \{1,2,3\}$ and $\nu \in G$ we set

$$\nu \restriction L_{\chi_k} = \sigma_{\chi_k}^{i_k(\nu)}, \quad \text{where } i_k(\nu) \in [0, n_{\chi_k} - 1], \quad \text{and } c_k(\nu) = \frac{i_k(\nu)}{n_{\chi_k}} \in \mathbb{Q}.$$

Then $\chi_k(\nu) = c_k(\nu) + \mathbb{Z} \in \mathbb{Q}/\mathbb{Z}$, and $(L_{\chi_k}, \sigma_{\chi_k}, a) \sim (L, G, [\overline{c}_k])$ for some factor system $\overline{c}_k \in Z^2(G, L^\times)$ such that, for all $\nu,\ \mu \in G$ by 3.6.5.1.

$$\overline{c}_k(\nu,\mu) = a^{c_k(\nu)+c_k(\mu)-c_k(\nu\mu)} = \begin{cases} 1 & \text{if } i_k(\nu) + i_k(\mu) < n_{\chi_k}, \\ a & \text{if } i_k(\nu) + i_k(\mu) \ge n_{\chi_k}. \end{cases}$$

Then $(L_{\chi_1}, \sigma_{\chi_1}, a) \otimes_K (L_{\chi_2}, \sigma_{\chi_2}, a) \sim (L, G, [\overline{c}_1\overline{c}_2])$, and if $\nu,\ \mu \in G$, then

$$(\overline{c}_1\overline{c}_2)(\nu,\mu) = a^{[c_1(\nu)+c_1(\mu)-c_1(\nu\mu)]+[c_2(\nu)+c_2(\mu)-c_2(\nu\mu)]}.$$

If $\tau \in \{\nu, \mu, \nu\mu\}$, then $c_1(\tau) + c_2(\tau) + \mathbb{Z} = \chi_1(\tau) + \chi_2(\tau) = \chi_3(\tau) = c_3(\tau) + \mathbb{Z}$, hence $c_1(\tau) + c_2(\tau) - c_3(\tau) \in \mathbb{Z}$ and $b(\tau) = a^{c_1(\tau)+c_2(\tau)-c_3(\tau)} \in K^\times$. For all $\nu, \mu \in \mathbb{Z}$ we get $(\overline{c}_1\overline{c}_2\overline{c}_3{}^{-1})(\nu,\mu) = b(\nu)b(\mu)b(\nu\mu)^{-1}$, hence $\overline{c}_1\overline{c}_2\overline{c}_3^{\,-1} \in B^2(G, L^\times)$, and therefore $[\overline{c}_1\overline{c}_2] = [\overline{c}_3] \in H^2(G, L^\times)$. Eventually we obtain

$$\begin{aligned}(L_{\chi_3}, \sigma_{\chi_3}, a) &\sim (L, G, [\overline{c}_3]) \sim (L, G, [\overline{c}_1]) \otimes_K (L, G, [\overline{c}_2]) \\ &\sim (L_{\chi_1}, \sigma_{\chi_1}, a) \otimes_K (L_{\chi_2}, \sigma_{\chi_2}, a). \qquad \square[(*)]\end{aligned}$$

2. Let $\chi \in \mathsf{X}(G)$ and $a \in K^\times$. By 4.1.6.2 and 3.6.2.2(b),

$$\langle a, \chi \rangle = \operatorname{inv}_K(L_\chi, \sigma_\chi, a) = 0 \iff (L_\chi, \sigma_\chi, a) \sim K \iff a \in \mathsf{N}_{K_\chi/K} K_\chi^\times.$$

Now we catch up on the continuity of the pairing $\langle\,\cdot\,,\,\cdot\,\rangle\colon K^\times \times \mathsf{X}(G) \to \mathbb{Q}/\mathbb{Z}$. If $(a,\chi) \in K^\times \times \mathsf{X}(G)$, then $V = a\mathsf{N}_{K_\chi/K}K_\chi^\times \times \{\chi\}$ is an open neighbourhood of (a,χ) (since $\mathsf{N}_{K_\chi/K}K_\chi^\times$ is open in $K^\times$ and $\mathsf{X}(G)$ is discrete), and $\langle\,\cdot\,,\,\cdot\,\rangle$ is constant on V. Being locally constant, $\langle\,\cdot\,,\,\cdot\,\rangle$ is continuous.

3. Apparently, the pairing $K^\times \times \mathsf{X}(G) \to \mathbb{Q}/\mathbb{Z}$ induces a homomorphism

$$(\,\cdot\,, L/K)\colon K^\times \to \mathsf{X}^2(G) = G^{\mathsf{a}}$$

such that $(a, L/K)(\chi) = \chi(a, L/K) = \langle a, \chi\rangle$ for all $a \in K^\times$ and $\chi \in \mathsf{X}(G)$. We defer the proof of its continuity to 5.

4. Clearly, $L_0^{\mathsf{a}} = L_0 \cap L^{\mathsf{a}}$. Let $U = \mathrm{Gal}(L/L_0)$, $G_0 = G/U = \mathrm{Gal}(L_0/K)$, and let $\rho\colon G \to G_0$ be the natural epimorphism, given by $\rho(\sigma) = \sigma U = \sigma \restriction L_0$ for all $\sigma \in G$. Let $\chi_0 \in \mathsf{X}(G_0)$ and $\chi = \chi_0 \circ \rho \in \mathsf{X}(G)$. Then $U \subset \mathrm{Ker}(\chi)$, and $\mathrm{Ker}(\chi_0) = \mathrm{Ker}(\chi)/U$. Hence

$$(L_0)_{\chi_0} = L_\chi,\;\; (G_0)\chi_0 = G_0/\mathrm{Ker}(\chi_0) = G/\mathrm{Ker}(\chi) = G_\chi \text{ and } \sigma_{\chi_0} = \sigma_\chi.$$

If $a \in K^\times$, then $\langle a, \chi_0\rangle = \mathrm{inv}_K(L_{\chi_0}, \sigma_{\chi_0}, a) = \mathrm{inv}_K(L_\chi, \sigma_\chi, a) = \langle a, \chi\rangle$. Thus we obtain

$$\chi_0((a, L/K) \restriction L_0^{\mathsf{a}}) = \chi(a, L/K) = \langle a, \chi\rangle = \langle a, \chi_0\rangle = \chi_0(a, L_0/K)$$

for all $\chi_0 \in \mathsf{X}(G_0)$, and consequently $(a, L/K) \restriction L_0^{\mathsf{a}} = (a, L_0/K)$ by 1.6.7.1(b).

5. Let $[L\colon K] < \infty$ and $z \in L^\times$. Then

$$\chi(\mathsf{N}_{L/K}(z), L/K) = \langle \mathsf{N}_{L/K}(z), \chi\rangle = \langle \mathsf{N}_{L_\chi/K}(\mathsf{N}_{L/L_\chi}(z), \chi\rangle) = 0 \in \mathbb{Q}/\mathbb{Z}$$

for all $\chi \in \mathsf{X}(G)$ by 2., and therefore $(\mathsf{N}_{L/K}(z), L/K) = 1 \in G$ by 1.6.7.1(a).

Finally we prove the continuity of $(\,\cdot\,, L/K)\colon K^\times \to G^{\mathsf{a}}$ for an arbitrary Galois extension L/K. By 1.1.6.1 it suffices to prove that $(\,\cdot\,, L/K)^{-1}(U)$ is open for every open normal subgroup U of G^{a}. Thus let U be an open normal subgroup of G^{a}, let $L_0 = (L^{\mathsf{a}})^U$ be its fixed field, and $G_0 = \mathrm{Gal}(L_0/K)$. Then L_0/K is a finite Galois extension, and if $a \in \mathsf{N}_{L_0/K}L_0^\times$, then $1 = (a, L_0/K) = (a, L/K) \restriction L_0$, and therefore $(a, L/K) \in U$. Hence $\mathsf{N}_{L_0/K}L_0^\times \subset (\,\cdot\, L/K)^{-1}(U)$, and as $\mathsf{N}_{L_0/K}L_0^\times$ is open by 4.1.10.3, it follows that $(\,\cdot\,,\, L/K)^{-1}(U)$ is open, too. □

Corollary 4.2.2. *Let L/K be a Galois extension and $a \in K^\times$. Then*

$$(a, L^{\mathsf{a}}/K) = (a, L/K) = (a, K) \restriction L^{\mathsf{a}}.$$

Proof. By 4.2.1.4, applied first with (L, L^{a}) instead of (L, L_0), and then with (K_{sep}, L) instead of (L, L_0). □

In the following Theorem 4.2.3 we highlight the simple but important fact that for unramified extensions L/K the norm residue symbol is essentially determined by the Frobenius automorphism $\varphi_{L/K}$.

Theorem 4.2.3. *Let L/K be a Galois extension, $G = \mathrm{Gal}(L/K)$ and $a \in K^\times$.*

1. Let $[L\colon K] = n$, $G = \langle\sigma\rangle$ and $m \in [0, n-1]$. Then

$$(a, L/K) = \sigma^m \quad \text{if and only if} \quad \mathrm{inv}_K(L, \sigma, a) = \frac{m}{n} + \mathbb{Z}.$$

2. If L/K is unramified, then $(a, L/K) = \varphi_{L/K}^{v_K(a)}$, and $(a, K_\infty/K) = \varphi_K^{v_K(a)}$.

3. Let $\chi \in \mathsf{X}(G)$, suppose that L_χ/K is unramified, and let $\varphi \in G$ be such that $\varphi \restriction L_\chi = \varphi_{L_\chi/K}$. Then $\langle a, \chi\rangle = v_K(a)\chi(\varphi)$.

Proof. 1. Suppose that $(a, L/K) = \sigma^m$, and let $\chi \in \mathsf{X}(G)$ be such that

$$\chi(\sigma) = \frac{1}{n} + \mathbb{Z} \in \mathbb{Q}/\mathbb{Z}.$$

Then $L_\chi = L$, $\sigma_\chi = \sigma$, and consequently

$$\mathrm{inv}_K(L, \sigma, a) = \mathrm{inv}_K(L_\chi, \sigma_\chi, a) = \langle a, \chi\rangle = \chi(a, L/K) = \frac{m}{n} + \mathbb{Z}.$$

The converse is now obvious.

2. Let L/K be unramified. If $[L\colon K] = n \in \mathbb{N}$, then $G = \langle\varphi_{L/K}\rangle$,

$$\mathrm{inv}_K(L, \varphi_{L/K}, a) = \frac{v_K(a)}{n} + \mathbb{Z}, \quad \text{and} \quad (a, L/K) = \varphi_{L/K}^{v_K(a)} \quad \text{by 1.}$$

Next, let $L = K_\infty$, and for $n \in \mathbb{N}$ let K_n be the unique unramified extension of K with $[K_n\colon K] = n$. Then

$$K_\infty = \bigcup_{n\in\mathbb{N}} K_n, \quad \mathrm{Gal}(K_\infty/K) = \overline{\langle\varphi_K\rangle},$$

and $(a, K_\infty/K) \restriction K_n = (a, K_n/K) = \varphi_{K_n/K}^{v_K(a)} = \varphi_K^{v_K(a)} \restriction K_n$ for all $n \in \mathbb{N}$. Hence $(a, K_\infty/K) = \varphi_K^{v_K(a)}$.

If finally L/K is any unramified extension, then $L \subset K_\infty$, and it follows that $(a, L/K) = (a, K)\restriction L = \varphi_K^{v_K(a)} \restriction L = \varphi_{L/K}^{v_K(a)}$.

3. Let $G_\chi = \mathrm{Gal}(L_\chi/K) = G/\mathrm{Ker}(\chi)$, and let $\chi^* \in \mathsf{X}(G_\chi)$ be the unique character satisfying $\chi^*(\sigma \restriction L_\chi) = \chi^*(\sigma\mathrm{Ker}(\chi)) = \chi(\sigma)$ for all $\sigma \in G$. By 2. and 4.2.1.4, it follows that

$$\begin{aligned}\langle a, \chi\rangle &= \chi(a, L/K) = \chi^*\big((a, L/K)\restriction L_\chi\big) = \chi^*(a, L_\chi/K) = \chi^*(\varphi_{L_\chi/K}^{v_K(a)}) \\ &= \chi^*(\varphi^{v_K(a)} \restriction L_\chi) = \chi(\varphi^{v_K(a)}) = v_K(a)\chi(\varphi).\end{aligned}$$

□

Theorem 4.2.4. *Let K'/K be a finite field extension, $b \in K'^{\times}$, and let L/K and L'/K' be Galois extensions such that $L \subset L'$. Let L^{a} be the largest over K abelian intermediate field of L/K and L'^{a} the largest over K' abelian intermediate field of L'/K'. Then $L^{\mathsf{a}} \subset L'^{\mathsf{a}}$, and*

$$(b, L'/K') \restriction L^{\mathsf{a}} = (\mathsf{N}_{K'/K}(b), L/K).$$

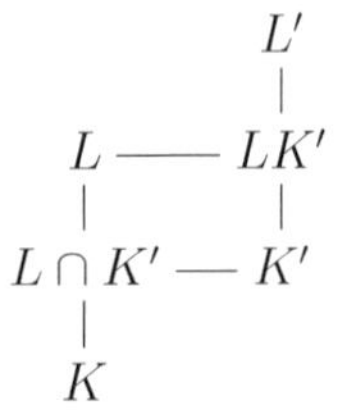

Proof. **Special Case**: $L = K_{\mathrm{ab}}$ (hence $L^{\mathsf{a}} = L$), and $L' = LK'$.
Let $G = \mathrm{Gal}(L/K)$ and $G' = \mathrm{Gal}(L'/K')$. By 1.8.4.1, the extension L'/K' is abelian, the restriction $\sigma \mapsto \sigma \restriction L$ yields $G' \xrightarrow{\sim} \mathrm{Gal}(L/L \cap K') \subset G'$, and if $\chi \in \mathsf{X}(G)$, then $\chi' = \chi \circ \rho \in \mathsf{X}(G')$. It suffices to prove:

A. If $\chi \in \mathsf{X}(G)$ and $\chi' = \chi \circ \rho$, then $\langle \mathsf{N}_{K'/K}(b), \chi \rangle_K = \langle b, \chi' \rangle_{K'}$.

Indeed, suppose that **A** holds. Then we obtain, for all $\chi \in \mathsf{X}(G)$ and $\chi' = \chi \circ \rho$,

$$\begin{aligned} \chi((b, L'/K') \restriction L) = \chi \circ \rho(b, L'/K') &= \chi'(b, L'/K') = \langle b, \chi' \rangle_{K'} \\ &= \langle \mathsf{N}_{K'/K}(b), \chi \rangle_K = \chi(\mathsf{N}_{K'/K}(b), L/K), \end{aligned}$$

and consequently $(b, L'/K') \restriction L) = (\mathsf{N}_{K'/K}(b), L/K)$ by 1.6.7.1(b).

Proof of **A.** Let $\chi \in \mathsf{X}(G)$, $\chi' = \chi \circ \rho$, and $|\mathsf{k}_{K'}| = q^f$, where $q = |\mathsf{k}_K|$ and $f = f(K'/K)$. Let $N \in \mathbb{N}$ and $k \in [0, N-1]$ be such that

$$\frac{1}{f v_{K'}(b)} \langle b, \chi' \rangle_{K'} = \frac{k}{N} + \mathbb{Z} \in \mathbb{Q}/\mathbb{Z}.$$

Since $K_\infty \subset L$, then exists a unique over K unramified intermediate field K_N of L/K such that $[K_N : K] = N$. Since $G_N = \mathrm{Gal}(K_N/K) = \langle \varphi_{K_N/K} \rangle$, there exists a character $\psi_0 \in \mathsf{X}(G_N)$ such that

$$\psi_0(\varphi_{K_N/K}) = \frac{k}{N} + \mathbb{Z} = \frac{1}{f v_{K'}(b)} \langle b, \chi' \rangle_{K'}.$$

Let $\psi \in \mathsf{X}(G)$ be defined by $\psi(\sigma) = \psi_0(\sigma \restriction K_N)$ for all $\sigma \in G$, and set $\psi' = \psi \circ \rho$. Then $\mathrm{Gal}(L/K_N) \subset \mathrm{Ker}(\psi)$, and consequently $L_\psi = L^{\mathrm{Ker}(\psi)} \subset K_N$. By definition, $\psi' \in \mathsf{X}(G')$, $\mathrm{Ker}(\psi') = \mathrm{Ker}(\psi) \cap G'$, and $L'_{\psi'} = L'^{\mathrm{Ker}(\psi')} = L_\psi K'$ by 1.8.4.1. As L_ψ/K is unramfied, $L'_{\psi'}/K'$ is unramified, too.

Let $\varphi \in G$ be such that $\varphi \restriction L_\psi = \varphi_{L_\psi/K}$, and let $\varphi' \in G'$ be such that $\varphi' \restriction L'_{\psi'} = \varphi_{L'_{\psi'}/K'}$. It follows that

$$\varphi' \restriction L_\psi = (\varphi' \restriction L'_{\psi'}) \restriction L_\psi = \varphi_{L'_{\psi'}/K'} \restriction L_\psi = \varphi^f_{L_\psi/K} = \varphi^f \restriction L_\psi,$$

and therefore $\psi'(\varphi') = \psi(\varphi' \restriction L) = \psi(\varphi^f) = f\psi(\varphi)$. Using 4.2.3.3 twice and observing that

$$\psi(\varphi) = \psi(\varphi \restriction L_\psi) = \psi(\varphi_{L_\psi/K}) = \psi_0(\varphi_{L_\psi/K} \restriction K_N) = \psi_0(\varphi_{K_N/K}),$$

we obtain

$$\begin{aligned}\langle b, \psi'\rangle_{K'} &= v_{K'}(b)\psi'(\varphi') = f v_{K'}(b)\psi(\varphi) = v_K(\mathsf{N}_{K'/K}(b))\psi(\varphi)\\ &= \langle \mathsf{N}_{K'/K}(b), \psi\rangle_K\\ &= f v_{K'}(b)\psi_0(\varphi_{K_N/K}) = \langle b, \chi'\rangle_{K'} \quad \text{by the very definition of } \psi_0.\end{aligned}$$

Let $\theta = \psi - \chi \in \mathsf{X}(G)$, $\theta' = \theta \circ \rho = \psi' - \chi' \in \mathsf{X}(G')$, $L_\theta = L^{\mathrm{Ker}(\theta)}$ and $L_{\theta'} = L^{\mathrm{Ker}(\theta')}$. Then $L_{\theta'} = L_\theta K'$, since $\mathrm{Gal}(L'/L_{\theta'}) = \mathrm{Ker}(\theta') = \mathrm{Ker}(\theta) \cap G'$.

Since $\langle b, \theta'\rangle_{K'} = \langle b, \psi'\rangle_{K'} - \langle b, \chi'\rangle_{K'} = 0$, it follows that $b = \mathsf{N}_{L_{\theta'}/K'}(c)$ for some $c \in L_{\theta'}^\times$ by 4.2.1.2, and

$$\mathsf{N}_{K'/K}(b) = \mathsf{N}_{K'/K} \circ \mathsf{N}_{L_{\theta'}/K'}(c) = \mathsf{N}_{L_\theta/K} \circ \mathsf{N}_{L_{\theta'}/L_\theta}(c) \in \mathsf{N}_{L_\theta/K}(L_\theta^\times).$$

Again by 4.2.1.2 we get

$$0 = \langle \mathsf{N}_{K'/K}(b), \theta\rangle_K = \langle \mathsf{N}_{K'/K}(b), \psi\rangle_K - \langle \mathsf{N}_{K'/K}(b), \chi\rangle_K$$

and therefore

$$\langle \mathsf{N}_{K'/K}(b), \chi\rangle_K = \langle \mathsf{N}_{K'/K}(b), \psi\rangle_K = \langle b, \chi'\rangle_{K'}. \qquad \square[\mathbf{A}.]$$

General Case: As $L^{\mathsf{a}}K'/K'$ is abelian, we obtain $L^{\mathsf{a}} \subset L'^{\mathsf{a}}$ and, using the special case and again 4.2.1.4, we obtain

$$\begin{aligned}(\mathsf{N}_{K'/K}(b), L/K) &= (\mathsf{N}_{K'/K}(b), L^{\mathsf{a}}/K) = (\mathsf{N}_{K'/K}(b), K_{\mathrm{ab}}/K) \restriction L^{\mathsf{a}}\\ &= (b, K_{\mathrm{ab}}K'/K') \restriction L^{\mathsf{a}}\\ &= (b, K'_{\mathrm{ab}}/K') \restriction L^{\mathsf{a}} = (b, L'^{\mathsf{a}}/K') \restriction L^{\mathsf{a}} = (b, L'/K') \restriction L^{\mathsf{a}}. \qquad \square\end{aligned}$$

Corollary 4.2.5. *Let L/K be a Galois extension, K' an over K finite intermediate field of L/K and $b \in K'^\times$. Let L^{a} be the largest over K abelian intermediate field of L/K and L'^{a} the largest over K' abelian intermediate field of L/K'. Then $L^{\mathsf{a}} \subset L'^{\mathsf{a}}$, and $(b, L/K') \restriction L^{\mathsf{a}} = (\mathsf{N}_{K'/K}(b), L/K)$.*

Proof. By 4.2.4, applied with $L = L'$. $\square$

Theorem 4.2.6. *Let L/K be a Galois extension, L^{a} the largest over K abelian intermediate field of L/K, $G = \mathrm{Gal}(L/K)$ and $G^{\mathsf{a}} = \mathrm{Gal}(L^{\mathsf{a}}/K)$.*

Let $\phi\colon L \stackrel{\sim}{\to} L'$ be a field isomorphism and $K' = \phi(K)$. Then K' is a local field, L'/K' is Galois, and $L'^{\mathsf{a}} = \phi(L^{\mathsf{a}})$ is the largest over K' abelian intermediate field of L'/K'. If $G' = \mathrm{Gal}(L'/K')$, then $G'^{\mathsf{a}} = \mathrm{Gal}(L'^{\mathsf{a}}/K')$, *and* ϕ *gives rise to a topological isomorphism* $\phi^*\colon G \to G'$ *such that* $\phi^*(\sigma) = \phi\circ\sigma\circ\phi^{-1}$ *for all* $\sigma \in G$. *If* $a \in K^\times$, *then*

$$(\phi(a), L'/K') = \phi\circ(a, L/K)\circ\phi^{-1}\restriction L'^{\mathsf{a}}. \qquad (*)$$

Proof. It suffices to prove $(*)$ (the other assertions are obvious), and we may assume that L/K is abelian. For $c \in Z^2(G, L^\times)$ let $c'\colon G'^2 \to L'^\times$ be defined by

$$c'(\phi\circ\sigma\circ\phi^{-1}, \phi\circ\tau\circ\phi^{-1}) = \phi(c(\sigma,\tau)) \quad \text{for all } (\sigma,\tau) \in G^2.$$

Then it follows that $c' \in Z^2(G', L'^\times)$, and there is an isomorphism $\phi^2\colon H^2(G, L^\times) \overset{\sim}{\to} H^2(G', L'^\times)$ such that $\phi^2([c]) = [c']$ for all $c \in Z^2(G, L^\times)$.

For $\chi \in \mathsf{X}(G)$ we set $\chi' = \chi\circ\phi^{*-1} \in \mathsf{X}(G')$. Then it apparently follows that $\mathrm{Ker}(\chi') = \phi^*(\mathrm{Ker}(\chi))$, $L'_{\chi'} = \phi(L_\chi)$, $n_\chi = [L_\chi : K] = [L'_{\chi'} : K']$ and $\sigma_{\chi'} = \phi^*(\sigma_\chi)$. If $a \in K^\times$, then

$$\begin{aligned}\langle a, \chi\rangle_K = \mathrm{inv}_K(L_\chi, \sigma_\chi, a) &= \frac{v_K(a)}{n_\chi} + \mathbb{Z} = \mathrm{inv}_K(L'_{\chi'}, \sigma_{\chi'}, \phi(a)) \\ &= \frac{v_{K'}(\phi(a))}{n_\chi} + \mathbb{Z} = \langle \phi(a), \chi'\rangle_{K'}.\end{aligned}$$

Hence $\chi(a, L/K) = \chi'(\phi(a), L'/K') = \chi\circ\phi^{*-1}(\phi(a), L'/K')$ for all $\chi \in \mathsf{X}(G)$, and consequently $(a, L/K) = \phi^{*-1}(\phi(a), L'/K')$ by 1.6.7.1(b). Therefore we finally obtain $(\phi(a), L'/K') = \phi^*(a, L/K) = \phi\circ(a, L/K)\circ\phi^{-1}$. $\square$

Theorem 4.2.7 (Local reciprocity law)**.** *Let* L/K *be a Galois extension, and let* $G = \mathrm{Gal}(L/K)$.

1. $\mathrm{Im}(\,\cdot\,, L/K)$ *is dense in* G^{a}.

2. *Assume that* $[L : K] < \infty$. *Then the norm residue symbol* $(\,\cdot\,, L/K)\colon K^\times \to G^{\mathsf{a}}$ *is surjective,* $\mathrm{Ker}(\,\cdot\,, L/K) = \mathsf{N}_{L/K}L^\times$, *and* $(\,\cdot\,, L/K)$ *induces an (equally denoted) isomorphism*

$$(\,\cdot\,, L/K)\colon K^\times/\mathsf{N}_{L/K}L^\times \overset{\sim}{\to} G^{\mathsf{a}}.$$

If L/K *is a finite abelian extension, then* $(K^\times : \mathsf{N}_{L/K}L^\times) = [L : K]$.

Proof. 1. Assume to the contrary that $U = \overline{\mathrm{Im}(\,\cdot\,, L/K)} \subsetneq G^{\mathsf{a}}$. By 1.6.7.1 there exists a character $\chi \in \mathsf{X}(G) = \mathsf{X}(G^{\mathsf{a}})$ such that $\chi \ne 0$ and $\chi \restriction U = 0$. If $L_\chi = L^{\mathrm{Ker}(\chi)}$, then L_χ/K is cyclic, and

$$(K^\times : \mathsf{N}_{L_\chi/K}L_\chi^\times) = [L_\chi : K] = \mathrm{ord}(\chi) > 1$$

by 4.1.10.1. If $a \in K^\times \setminus \mathsf{N}_{L_\chi/K}L_\chi^\times$, then $(a, L/K) \in U$ implies that $\langle a, \chi\rangle = \chi(a, L/K) = 0$, and therefore $a \in \mathsf{N}_{L_\chi/K}L_\chi^\times$ by 4.2.1.2, a contradiction.

2. By 1., the norm residue symbol $(\cdot, L/K)\colon K^\times \to G^{\mathsf{a}}$ is surjective. Since $\mathsf{N}_{L/K}L^\times \subset \operatorname{Ker}(\cdot, L/K)$ by 4.2.1.5, we must prove:

$$\text{If } a \in K^\times \text{ and } (a, L/K) = 1 \in G^{\mathsf{a}}, \text{ then } a \in \mathsf{N}_{L/K}L^\times. \qquad (*)$$

If L/K is cyclic, then there is an injective character $\chi \in \mathsf{X}(G)$. It satisfies $L = L_\chi$ and $0 = \chi(a, L/K) = \langle a, \chi\rangle$, hence $a \in \mathsf{N}_{L/K}L^\times$ by 4.2.1.2.

Thus we assume that L/K is not cyclic, and we proceed by induction on $[L:K]$. Let $a \in K^\times$ be such that $(a, L/K) = 1$. As L/K is solvable, there exists an over K cyclic intermediate field K' of L/K such that $K \subsetneq K' \subsetneq L$. Since $(a, K'/K) = (a, L/K)\restriction K' = 1$ by 4.2.1.4, it follows that $a \in \mathsf{N}_{K'/K}K'^\times$, say $a = \mathsf{N}_{K'/K}(b)$ for some $b \in K'^\times$.

If L/K is abelian, then $(b, L/K') = (\mathsf{N}_{K'/K}(b), L/K) = (a, L/K) = 1$ by 4.2.4, hence $b \in \mathsf{N}_{L/K'}L^\times$ by the induction hypothesis, and $a \in \mathsf{N}_{L/K}L^\times$. This completes the proof for abelian extensions L/K. As to the general case, it suffices to prove:

> **I.** There exists some $x \in K'^\times$ such that $(bx, L/K') = 1$ and $\mathsf{N}_{K'/K}(x) = 1$.

Indeed, if **I** is true, then $bx = \mathsf{N}_{L/K'}(c)$ for some $c \in L^\times$ by the induction hypothesis, and $\mathsf{N}_{L/K}(c) = \mathsf{N}_{K'/K}(bx) = a \in \mathsf{N}_{L/K}L^\times$.

Proof of **I.** Let L_0 be the largest over K' abelian intermediate field of L/K'. Then $L^{\mathsf{a}} \subset L_0$, and if $V = \operatorname{Gal}(L/K')$, then V is a normal subgroup of G, and $[V, V] = \operatorname{Gal}(L/L_0)$ is also a normal subgroup of G. Consequently L_0/K is Galois,

$$G_0 = G/[V, V] = \operatorname{Gal}(L_0/K) \text{ and } (a, L_0/K) = (a, L/K)\restriction L_0^{\mathsf{a}} = 1.$$

Let $U = \operatorname{Gal}(L_0/K')$. Since K'/K is cyclic, U is a normal subgroup of G_0, G_0/U is cyclic, and $G_0/[G_0, U]$ is abelian by the subsequent Lemma 4.2.8. Let $M = L_0^{[G_0, U]}$ be the fixed field of $[G_0, U]$. Then M/K is abelian, $\operatorname{Gal}(M/K) = G_0/[G_0, U]$, and

$$1 = (a, L/K)\restriction M = (a, M/K) = (\mathsf{N}_{K'/K}(b), M/K) = (b, L_0/K')\restriction M$$

by 4.2.4. Hence $(b, L_0/K') \in [G_0, U]$, and therefore

$$(b, L_0/K') = \prod_{i=1}^{k} \rho_i\sigma_i\rho_i^{-1}\sigma_i^{-1}$$

for some $k \in \mathbb{N}$, $\rho_1, \ldots, \rho_k \in G_0$ and $\sigma_1, \ldots, \sigma_k \in U$.
As U is abelian, the norm residue symbol $(\cdot, L_0/K')\colon K'^\times \to U$ is surjective, and therefore $\sigma_i = (x_i, L_0/K')$ for some $x_i \in K'^\times$. As $\rho_i(K') = K'$ for all

$i \in [1, k]$, it follows by 4.2.6 that

$$\begin{aligned}(b, L_0/K') &= \prod_{i=1}^{k} \rho_i(x_i, L_0/K')\rho_i^{-1}(x_i, L_0/K')^{-1} \\ &= \prod_{i=1}^{k} (x_i^{\rho_i}, L_0/K')(x_i^{-1}, L_0/K') \\ &= \prod_{i=1}^{k} (x_i^{\rho_i - 1}, L_0/K') = (x^{-1}, L_0/K'), \\ \text{where} \quad x &= \prod_{i=1}^{k} x_i^{1-\rho_i} \in K'^{\times}.\end{aligned}$$

Hence we obtain $(bx, L/K') = (bx, L_0/K') = 1$, and $\mathsf{N}_{K'/K}(x) = 1$, since $\mathsf{N}_{K'/K}(x_i^{\rho_i - 1}) = 1$ for all $i \in [1, k]$. □

Lemma 4.2.8. *Let G be a finite group. For subsets X, Y of G we set*

$$[X, Y] = \langle \{xyx^{-1}y^{-1} \mid x \in X,\ y \in Y\} \rangle.$$

Let U and V be normal subgroups of G. Then $[U, V]$ is a normal subgroup of G, and if G/U is cyclic, then $G/[G, U]$ is abelian.

Proof. If $\sigma \in G$, $u \in U$ and $v \in V$, then $\sigma[u, v]\sigma^{-1} = [\sigma u \sigma^{-1}, \sigma v \sigma^{-1}]$. Hence $[U, V]$ is a normal subgroup of G.

Let $\tau \in G$ be such that $G/U = \langle \tau U \rangle$. Then $G = \{\tau^k u \mid k \in \mathbb{Z},\ u \in U\}$. If $k \in \mathbb{Z}$ and $u \in U$, then $\tau^k u (u\tau^k)^{-1} = \tau^k u \tau^{-k} u^{-1} \in [G, U]$, and thus $\tau^k u \equiv u \tau^k \mod [G, U]$. Hence it follows that $[\tau^k u, \tau^l v] \in [G, U]$ for all k, $l \in \mathbb{Z}$ and u, $v \in U$. Consequently, $G/[G, U]$ is abelian. □

For unramified finite extensions we rediscover the following well-known result (see [18, Theorem 5.6.9]).

Corollary 4.2.9. *Let L/K be a finite unramified extension, $[L{:}K] = n$, and let π be a prime element of $\mathcal{O}_K$. Then $\mathsf{N}_{L/K} L^{\times} = \langle \pi^n \rangle {\cdot} U_K$.*

Proof. Let $a = \pi^k u \in K^{\times}$, where $k \in \mathbb{Z}$ and $u \in U_K$. Using 4.2.3.2, we obtain:

$$a \in \mathsf{N}_{L/K} L^{\times} \iff (a, L/K) = \varphi_{L/K}^{v(a)} = 1 \iff n \,|\, v(a) \iff a \in \langle \pi^n \rangle {\cdot} U_K.$$

□

Definition and Theorem 4.2.10 (Local norm theorem)**.** A subgroup B of $K^\times$ is called a **norm group** if $B = \mathsf{N}_{L/K}L^\times$ for some finite abelian extension L/K. In this case, L is called a **class field** to B. By 4.1.10.3, every norm group is an open subgroup of $K^\times$ of finite index. In 4.8.2 we shall prove that conversely every open subgroup of finite index in $K^\times$ is a norm group.

1. Let L/K be a finite field extension, and let L_0 be the largest over K abelian intermediate field of L/K. Then

$$\mathsf{N}_{L/K}L^\times = \mathsf{N}_{L/K}L_0^\times \quad \textit{and} \quad (K^\times : \mathsf{N}_{L/K}L^\times) = [L_0 : K].$$

In particular, $\mathsf{N}_{L/K}L^\times$ is a norm group.

2. Let L, L_1, L_2 be over K finite abelian extension fields of K.

(a) Let B be a subgroup of $K^\times$ such that $\mathsf{N}_{L/K}L^\times \subset B$. Then $B = \mathsf{N}_{K'/K}K'^\times$ for some intermediate field K' of L/K.

(b) $\mathsf{N}_{L_1L_2/K}(L_1L_2)^\times = \mathsf{N}_{L_1/K}L_1^\times \cap \mathsf{N}_{L_2/K}L_2^\times$.

(c) $L_1 \subset L_2$ if and only if $\mathsf{N}_{L_1/K}L_1^\times \supset \mathsf{N}_{L_2/K}L_2^\times$. In particular, $L_1 = L_2$ if and only if $\mathsf{N}_{L_1/K}L_1^\times = \mathsf{N}_{L_2/K}L_2^\times$, and every norm group has a unique class field.

(d) $\mathsf{N}_{L_1\cap L_2/K}(L_1 \cap L_2)^\times = (\mathsf{N}_{L_1/K}L_1^\times)(\mathsf{N}_{L_2/K}L_2^\times)$.

The assignment $L \mapsto \mathsf{N}_{L/K}L^\times$ defines an inclusion reversing lattice isomorphism between the set of all finite abelian extension fields of K (inside $K_{\rm ab}$) and the set of all norm groups in $K^\times$.

Proof. 1. It suffices to prove that $\mathsf{N}_{L_0/L}L_0^\times \subset \mathsf{N}_{L/K}L^\times$. Then equality holds, and $(K^\times : \mathsf{N}_{L_0/K}L_0^\times) = [L_0 : K]$ by the local reciprocity law 4.2.7.2.

CASE 1: L/K is separable. Let N be a Galois closure of L/K and N_0 the largest over K abelian intermediate field of N/K. Then $L_0 = L \cap N_0$, and there is an isomorphism $\mathrm{Gal}(N_0L/L) \to \mathrm{Gal}(N_0/L_0)$, given by $\sigma \mapsto \sigma \restriction N_0$.

Let $a \in \mathsf{N}_{L_0/K}L_0^\times$. Then $1 = (a, L_0/K) = (a, N/K) \restriction L_0$, and therefore it follows that $(a, N/K) \in \mathrm{Gal}(N_0/L_0)$. Let $\tau \in \mathrm{Gal}(N_0L/L)$ be such that $\tau \restriction N_0 = (a, N/K)$. Since $(\,\cdot\,, N_0L/L)\colon L^\times \to \mathrm{Gal}(N_0L/L)$ is surjective, there exists some $b \in L^\times$ such that $(b, N_0L/L) \restriction N_0 = (a, N/K)$, and we obtain

$$(a, N/K) = (b, N_0L/L) \restriction N_0 = (\mathsf{N}_{L/K}(b), N_0/K) = (\mathsf{N}_{L/K}(b), N/K).$$

Hence $(a\mathsf{N}_{L/K}(b)^{-1}, N/K) = 1$, $a\mathsf{N}_{L/K}(b)^{-1} \in \mathsf{N}_{N/K}N^\times \subset \mathsf{N}_{L/K}L^\times$ by the local reciprocity law 4.2.7.2, and $a \in \mathsf{N}_{L/K}L^\times$.

CASE 2: L/K is inseparable and $\mathrm{char}(K) = p > 0$. Let L' be the separable closure of K in L. Then $L_0 \subset L'$, $\mathsf{N}_{L'/K}L'^\times = \mathsf{N}_{L_0/K}L_0^\times$ by CASE 1, and L/L' is purely inseparable. If $[L : L'] = p^m$ for some $m \in \mathbb{N}_0$ and $L = \mathbb{F}_q((X))$ for some p-power q, then $L' = L^{p^m} = \mathbb{F}_q((X^{p^m}))$ (see [18, Theorem 1.4.7]). Since $\mathbb{F}_q^{p^m} = \mathbb{F}_q$ and $\mathsf{N}_{L/L'}(X) = X^{p^m}$, it follows that $L'^\times = \mathsf{N}_{L/L'}L^\times$, and consequently $\mathsf{N}_{L/K}L^\times = \mathsf{N}_{L'/K}(\mathsf{N}_{L/L'}L^\times) = \mathsf{N}_{L_0/K}L_0^\times$.

2. If $L_1 \subset L_2$, then $\mathsf{N}_{L_2/K}L_2^\times = \mathsf{N}_{L_1/K} \circ \mathsf{N}_{L_2/L_1}L_2^\times \subset \mathsf{N}_{L_1/L}L_1^\times$.

(a) Let $G' = (B, L/K) \subset G$ and $K' = L^{G'}$.

We assert that $B = \mathsf{N}_{K'/K}K'^\times$. If $a \in B$, then $(a, L/K) \in G'$, hence $(a, K'/K) = (a, L/K) \restriction K' = 1$ and $a \in \mathsf{N}_{K'/K}K'^\times$. Assume conversely that $a \in \mathsf{N}_{K'/K}K'^\times$. Then $1 = (a, K'/K) = (a, L/K) \restriction K'$, hence $(a, L/K) \in G'$, and there exists some $b \in B$ such that $(a, L/K) = (b, L/K)$. Then $(ab^{-1}, L/K) = 1$, hence $ab^{-1} \in \mathsf{N}_{L/K}L^\times$ and $a \in b\mathsf{N}_{L/K}L^\times \subset B$.

(b) For $i \in \{1, 2\}$, $L_i \subset L_1L_2$ implies $\mathsf{N}_{L_1L_2/K}(L_1L_2)^\times \subset \mathsf{N}_{L_i/K}L_i^\times$, and therefore $\mathsf{N}_{L_1L_2/K}(L_1L_2)^\times \subset \mathsf{N}_{L_1/K}L_1^\times \cap \mathsf{N}_{L_2/K}L_2^\times$. As to the converse, if $a \in \mathsf{N}_{L_1/K}L_1^\times \cap \mathsf{N}_{L_2/K}L_2^\times$, then $1 = (a, L_i/K) = (a, L_1L_1/K) \restriction L_i$ for $i \in \{1, 2\}$, hence $(a, L_1L_2/K) = 1$, and therefore $a \in \mathsf{N}_{L_1L_2/K}(L_1L_2)^\times$.

(c) Clearly, $L_1 \subset L_2$ implies $\mathsf{N}_{L_2/K}L_2^\times \subset \mathsf{N}_{L_1/K}L_1^\times$. Thus assume that $\mathsf{N}_{L_2/K}L_2^\times \subset \mathsf{N}_{L_1/K}L_1^\times$. Then (b) implies

$$\mathsf{N}_{L_1L_2/K}(L_1L_2)^\times = \mathsf{N}_{L_1/K}L_1^\times \cap \mathsf{N}_{L_2/K}L_2^\times = \mathsf{N}_{L_2/K}L_2^\times,$$

and since

$$[L_1L_2 : K] = (K^\times : \mathsf{N}_{L_1L_2/K}(L_1L_2)^\times) = (K^\times : \mathsf{N}_{L_2/K}L_2^\times) = [L_2 : K],$$

we obtain $L_1 \subset L_1L_2 = L_2$.

(d) $L_1 \cap L_2 \subset L_i$ implies $\mathsf{N}_{L_i/K}L_i^\times \subset \mathsf{N}_{L_1 \cap L_2/K}(L_1 \cap L_2)^\times$ for $i \in \{1, 2\}$ and therefore $(\mathsf{N}_{L_1/K}L_1^\times)(\mathsf{N}_{L_2/K}L_2^\times) \subset \mathsf{N}_{L_1 \cap N_2/K}(L_1 \cap L_2)^\times$. By (b), we have $\mathsf{N}_{L'/K}L'^\times = (\mathsf{N}_{L_1/K}L_1^\times)(\mathsf{N}_{L_2/K}L_2^\times)$ for some intermediate field L' of L_1/K, and therefore $\mathsf{N}_{L'/K}L'^\times \subset \mathsf{N}_{L_1 \cap L_2/K}(L_1 \cap L_2)^\times$. Hence $L_1 \cap L_2 \subset L'$ by (c). Since $\mathsf{N}_{L_i/K}L_i^\times \subset \mathsf{N}_{L'/K}L'^\times$ for $i \in \{1, 2\}$, we get $L' \subset L_1 \cap L_2$, and thus equality holds.

The final statement is now obvious. □

Theorem 4.2.11 (Characterization of the universal norm residue symbol). *Let L/K be a Galois extension, $K_\infty \subset L$ and $G = \mathrm{Gal}(L/K)$. Let $\psi \in G$ be such that $\psi \restriction K_\infty = \varphi_K$, and let L^ψ be the fixed field of ψ.*

1. $L = L^\psi K_\infty$, $L^\psi \cap K_\infty = K$, and the maps

$$G \to \mathrm{Gal}(K_\infty/K) \times \mathrm{Gal}(L^\psi/K), \quad \textit{defined by } \sigma \mapsto (\sigma \restriction K_\infty, \sigma \restriction L^\psi),$$

and

$$\mathrm{Gal}(L/L^\psi) \to \mathrm{Gal}(K_\infty/K), \quad \textit{defined by } \sigma \mapsto \sigma \restriction K_\infty,$$

are topological isomorphisms.

2. Let $f \colon K^\times \to G^{\mathsf{a}}$ a group homomorphism. Then $f = (\,\cdot\,, L/K)$ if and only if the following two conditions are fulfilled:

(a) *If* $a \in K^\times$, *then* $f(a)\restriction K_\infty = \varphi_K^{v_K(a)}$.

(b) *If* $\pi \in \mathcal{O}_K$ *is a prime element and* $L_\pi = L^{(\pi,L/K)}$ *is the fixed field of* $(\pi, L/K)$, *then* $f(\pi)\restriction L_\pi = \mathrm{id}_{L_\pi}$.

Proof. 1. Let K' be an over K finite intermediate field of $L^\psi \cap K_\infty/K$. Then $\psi\restriction K' = \varphi_K\restriction K' = \varphi_{K'/K}$ and $\psi\restriction K' = (\psi\restriction L^\psi)\restriction K' = 1$. Hence $K' = K$, and therefore $L^\psi \cap K_\infty = K$. By 1.8.3.1, $\mathrm{Gal}(L/L^\psi) = \overline{\langle\psi\rangle}$, and $\sigma \mapsto \sigma\restriction K_\infty L^\psi$ induces a continuous epimorphism $\theta\colon \mathrm{Gal}(L/L^\psi) \to \mathrm{Gal}(K_\infty L^\psi/L^\psi)$. By 1.8.4.1, it follows that the map $\rho\colon \mathrm{Gal}(K_\infty L^\psi/L^\psi) \overset{\sim}{\to} \mathrm{Gal}(K_\infty/K)$, defined by $\rho(\sigma) = \sigma\restriction K_\infty$, is a topological isomorphism, and by 1.8.10 there is a topological isomorphism $\Theta\colon \mathrm{Gal}(K_\infty/K) \overset{\sim}{\to} \widehat{\mathbb{Z}}$. In all, $\Theta\circ\rho\circ\theta\colon \overline{\langle\psi\rangle} \to \widehat{\mathbb{Z}}$ is a continuous epimorphism and thus an isomorphism by 1.5.6.2. Hence θ is also an isomorphism, and thus $L = K_\infty L^\psi$. By 1.8.4.2, the maps $G \to \mathrm{Gal}(K_\infty/K)\times\mathrm{Gal}(L^\psi/K)$ and $\mathrm{Gal}(L/L^\psi) \to \mathrm{Gal}(K_\infty/K)$ are topological isomorphisms.

2. Observe that $(a, L/K)\restriction K_\infty = (a, K_\infty/K) = \varphi_K^{v_K(a)}$ for all $a \in K^\times$ by 4.2.3.2 and $(\pi, L_\pi/K) = (\pi, L/K)\restriction L_\pi = \mathrm{id}_{L_\pi}$ for every prime element $\pi \in \mathcal{O}_K$. Hence $(\,\cdot\,, L/K)$ has the properties (a) and (b).

Let conversely $f\colon K^\times \to G$ be any homomorphism satisfying (a) and (b), and let $\pi \in \mathcal{O}_K$ be a prime element. Then

$$f(\pi)\restriction K_\infty = \varphi_K = (\pi, K_\infty/K) = (\pi, L/K)\restriction K_\infty$$

and

$$f(\pi)\restriction L_\pi = (\pi, L/K)\restriction L_\pi = \mathrm{id}_{L_\pi}.$$

By 1., applied with $\psi = (\pi, L/K)$, it follows that $L = L_\pi L_\infty$ and therefore $f(\pi) = (\pi, L/K)$. Since the prime elements of $\mathcal{O}_K$ generate $K^\times$, we obtain $f = (\,\cdot\,, L/K)$. □

4.3 Auxiliary results on complete discrete valued fields

In this section, we present additional results on the arithmetic of complete discrete valued fields, in particular with algebraically closed residue class fields. Such fields appear naturally in the theory of local fields as follows: If K is a local field, then its maximal unramified extension K_∞ is a discrete valued field satisfying $v_{K_\infty}\restriction K = v_K$ with an algebraically closed residue class field, and so is its completion $\widehat{K_\infty}$ which plays a prominent role in various aspects of local class field theory as we shall see in the course of this chapter.

We start with the investigation of fully ramified cyclic extensions of prime degree of complete discrete valued fields.

Let K be a complete discrete valued field, L/K a fully ramified cyclic extension of prime degree $e(L/K) = [L\colon K] = l$, and $G = \mathrm{Gal}(L/K) = \langle\sigma\rangle$. Let $\mathsf{k} = \mathsf{k}_L = \mathsf{k}_K$, $\mathrm{char}(\mathsf{k}) = p$, and let π be a prime element of $\mathcal{O}_L$. Then the number

$$t = t_{L/K} = v_L(\sigma\pi - \pi) - 1 \in \mathbb{N}_0$$

does not depend on π and is called the **ramification number** of L/K. If $(G_i)_{i\ge -1}$ is the Hilbert subgroup series of L/K, then $G_i = G$ if $i \le t$, and $G_i = \mathbf{1}$ if $i > t$. Moreover, there is a monomorphism

$$\Phi_t\colon G \to U_L^{(t)}/U_L^{(t+1)} \quad \text{such that} \quad \Phi_t(\sigma) = \frac{\sigma\pi}{\pi}\,U_L^{(t+1)}.$$

If $l \ne p$, then $t = 0$, and $l = p$, then $t \ge 1$ (see [18, Section 5.10]).

Lemma 4.3.1. *Let K be a complete discrete valued field and L/K a fully ramified cyclic extension of prime degee $l = [L:K]$. Let $t = t_{L/K}$ be the ramification number and $\mathfrak{D}_{L/K}$ the different of L/K.*

1. $\mathfrak{D}_{L/K} = \mathfrak{p}_L^{(t+1)(l-1)}$.
2. *If $n \in \mathbb{N}$, then*

$$\mathsf{Tr}_{L/K}(\mathfrak{p}_L^n) = \mathfrak{p}_L^r, \quad \textit{where} \;\; r = \Big\lfloor \frac{(t+1)(l-1)+n}{l} \Big\rfloor.$$

3. *If $n \in \mathbb{N}$ and $x \in \mathfrak{p}_L^n$, then*

$$\mathsf{N}_{L/K}(1+x) = 1 + \mathsf{Tr}_{L/K}(x) + \mathsf{N}_{L/K}(x) + \mathsf{Tr}_{L/K}(y) \quad \textit{for some } y \in \mathfrak{p}_L^{2n}.$$

Proof. 1. See [18, Theorem 5.10.1].

2. Clearly $\mathsf{Tr}_{L/K}(\mathfrak{p}^n)$ is an ideal of $\mathcal{O}_K$. If $k \in \mathbb{N}_0$, then

$$\begin{aligned}
\mathsf{Tr}_{L/K}(\mathfrak{p}_L^n) \subset \mathfrak{p}_K^k &\iff \mathfrak{p}_K^{-k}\mathsf{Tr}_{L/K}(\mathfrak{p}_L^n) = \mathsf{Tr}_{L/K}(\mathfrak{p}_L^{n-kl}) \subset \mathcal{O}_K \\
&\iff \mathfrak{p}_L^{n-kl} \subset \mathfrak{D}_{L/K}^{-1} = \mathfrak{p}_L^{-(t+1)(l-1)} \\
&\iff n - kl \ge -(t+1)(l-1) \\
&\iff k \le \frac{(t+1)(l-1)+n}{l}.
\end{aligned}$$

3. Let $G = \mathrm{Gal}(L(K) = \langle\sigma\rangle$. If $\alpha = a_0 + a_1\sigma + \ldots + a_{l-1}\sigma^{l-1} \in \mathbb{Z}[G]$ is such that $a_i \in \mathbb{N}_0$ for all $i \in [0, l-1]$ and $x \in L$, we set

$$x^\alpha = \prod_{i=0}^{l-1} \sigma^i(x)^{a_i}.$$

For $k \in [0, l]$ we denote by A_k the set of all sums $\sigma^{i_1} + \ldots + \sigma^{i_k} \in \mathbb{Z}[G]$, where $0 \le i_1 < \ldots < i_k < l$. In particular, $A_0 = \{0\}$, $A_1 = G$ and $A_n = \{N_G\}$. For $x \in K$, we obtain

$$\mathsf{N}_{L/K}(1 + x) = \prod_{\sigma \in G} (1 + x^\sigma) = \sum_{k=0}^{n} \sum_{\alpha \in A_k} x^\alpha$$

$$= 1 + \mathsf{Tr}_{L/K}(x) + \mathsf{N}_{L/K}(x) + \sum_{k=2}^{n-1} \sum_{\alpha \in A_k} x^\alpha.$$

If $k \in [2, n-1]$ and $\alpha \in A_k$, then $x^\alpha \in \mathfrak{p}_L^{2n}$, and $\alpha\sigma \ne \alpha$. Indeed, if $\alpha\sigma = \alpha$, then $\alpha(\sigma - 1) = 0$, and as $I_G = (\sigma - 1)\mathbb{Z}[G]$, it follows that $\alpha \in \operatorname{Ann}(I_G) = N_G\mathbb{Z}$ by 2.1.1, a contradiction. Hence there exist distinct $\alpha_1, \ldots, \alpha_s \in A_k$ such that

$$A_k = \biguplus_{i=1}^{s} \{\alpha_i, \alpha_i\sigma, \ldots, \alpha_i\sigma^{l-1}\},$$

and consequently

$$\sum_{\alpha \in A_k} x^\alpha = \sum_{i=1}^{s} \sum_{\nu=0}^{l-1} x^{\alpha_i \sigma^\nu} = \mathsf{Tr}_{L/K}\Big(\sum_{i=1}^{s} x^{\alpha_i}\Big) \in \mathsf{Tr}_{L/K}(\mathfrak{p}_L^{kn}) \subset \mathsf{Tr}_{L/K}(\mathfrak{p}_L^{2n}). \quad \square$$

For the next theorem, recall that for every discrete valued field K and a prime element $\pi \in \mathcal{O}_K$ there are isomorphisms

$$v_0 \colon U_K/U_K^{(1)} \xrightarrow{\sim} \mathsf{k}_K^\times \quad \text{and} \quad v_n^{(\pi)} \colon U_K^{(n)}/U_K^{(n+1)} \xrightarrow{\sim} \mathsf{k}_K \ \text{ for all } \ n \ge 1,$$

given by $v_0(uU_K^{(1)}) = u + \mathfrak{p}_K$ for all $u \in U_K$, and $v_n^{(\pi)}(1 + a\pi^n)U_K^{(n+1)} = a + \mathfrak{p}_K$ for all $a \in \mathcal{O}_K$ and $n \ge 1$ (see [18, Theorem 2.10.5], and note that $v_n^{(\pi)}$ depends on π).

Theorem 4.3.2. *Let K be a complete discrete valued field, and let L/K be a fully ramified cyclic extension of prime degee $l = [L\colon K]$ and ramification number $t = t_{L(K}$. Let $\psi = \psi_{L/K} \colon [-1, \infty) \to \mathbb{R}$ be the Herbrand function, given by*

$$\psi(x) = \begin{cases} x & \text{if } x \le t, \\ x + l(t - x) & \text{if } x \ge t \end{cases}$$

(see [18, Definition 5.10.5]). *Let $\mathsf{k} = \mathsf{k}_K = \mathsf{k}_L$, and $p = \operatorname{char}(\mathsf{k})$. Let π be a prime element of $\mathcal{O}_L$ and $\pi' = \mathsf{N}_{L/K}(\pi)$ (then π' is a prime element of $\mathcal{O}_K$).*

A. *If $n \in \mathbb{N}_0$, then $\mathsf{N}_{L/K}(U_L^{(\psi(n))}) \subset U_K^{(n)}$, $\mathsf{N}_{L/K}(U_L^{(\psi(n)+1)}) \subset U_K^{(n+1)}$, and there exist commutative diagrams*

$$\begin{array}{ccc} U_L/U_L^{(1)} & \xrightarrow{\overline{\mathsf{N}}_{L/K}^{(0)}} & U_K/U_K^{(1)} \\ {\scriptstyle v_0}\big\downarrow & & \big\downarrow{\scriptstyle v_0} \\ \mathsf{k}^\times & \xrightarrow{N_0} & \mathsf{k}^\times \end{array}, \quad \text{and} \quad \begin{array}{ccc} U_L^{(\psi(n))}/U_L^{(\psi(n)+1)} & \xrightarrow{\overline{\mathsf{N}}_{L/K}^{(n)}} & U_K^{(n)}/U_K^{(n+1)} \\ {\scriptstyle v_{\psi(n)}^{(\pi)}}\big\downarrow & & \big\downarrow{\scriptstyle v_n^{(\pi')}} \\ \mathsf{k} & \xrightarrow{N_n} & \mathsf{k} \end{array} \quad \text{for } n \ge 1,$$

where the vertical arrows are the canonical isomorphism given above, $\overline{\mathsf{N}}_{L/K}^{(n)}$ *is induced by* $\mathsf{N}_{L/K}$*, and the homomorphisms* N_n *which make the diagram commutative, are specified below.*

1. If $n=0$*, then* $N_0(\xi)=\xi^l$ *for all* $\xi\in\mathsf{k}^\times$*, and*

$$|\mathrm{Ker}(N_0)| = \begin{cases} l & \text{if } t=0, \\ 1 & \text{if } t\neq 0. \end{cases}$$

2. If $n\in[1,t-1]$*, then* N_n *is injective, and* $N_n(\xi)=\xi^p$ *for all* $\xi\in\mathsf{k}$*.*

3. If $n=t\geq 1$*, then* $|\mathrm{Ker}(N_n)|=l=p$*, and there exist some* $\beta\in\mathsf{k}^\times$ *such that* $N_n(\xi)=\xi^p+\beta\xi$ *for all* $\xi\in\mathsf{k}$*.*

4. If $n>t$*, then* N_n *is bijective, and there is some* $\beta_n\in\mathsf{k}^\times$ *such that* $N_n(\xi)=\beta_n\xi$ *for all* $\xi\in\mathsf{k}$*.*

B. $\mathsf{N}_{L/K}(U_L^{(\psi(m))})=U_K^{(m)}$ *for all* $m>t$*, and even for all* $m\geq 0$ *if* k *is algebraically closed.*

Proof. **A.** For $x\in\mathcal{O}_K$ let $\overline{x}=x+\mathfrak{p}_L\in\mathsf{k}$ be its residue class. Then $\overline{\sigma(x)}=\overline{x}$, and therefore $\overline{\mathsf{N}_{L/K}(x)}=\overline{x}^{\,l}$ for all $x\in\mathcal{O}_L$.

1. Let $n=0$. Clearly, $\psi(0)=0$, $\mathsf{N}_{L/K}(U_L)\subset U_K$ and $\mathsf{N}_{L/K}(U_L^{(1)})\subset U_K^{(1)}$. If $u\in U_L$, then $\overline{\mathsf{N}_{L/K}(u)}=\overline{u}^{\,l}$, and therefore $N_0(\xi)=\xi^l$ for all $\xi\in\mathsf{k}^\times$. If $t\neq 0$, then $l=p$, and N_0 is injective. If $t=0$, consider $\Phi_0\colon G\to U_L/U_L^{(1)}$. Since $\mathsf{N}_{L/K}(\sigma\pi/\pi)=1$, we obtain $\mathrm{Im}(\Phi_0)\subset\mathrm{Ker}(\overline{\mathsf{N}}_{L/K}^{(0)})$, and therefore $l\leq|\mathrm{Ker}(\overline{\mathsf{N}}_{L/K}^{(0)})|=|\mathrm{Ker}(N_0)|\leq l$. Hence $|\mathrm{Ker}(N_0)|=l$.

2. Let $n\in[1,t-1]$. Then $l=p$, $\psi(n)=n$, and $v_K\circ\mathsf{N}_{L/K}=v_L$ implies $\mathsf{N}_{L/K}(\mathfrak{p}_L^n)\subset\mathfrak{p}_K^n$. By 4.3.1.2, we obtain $\mathsf{Tr}_{L/K}(\mathfrak{p}_L^n)=\mathfrak{p}_K^r$, where

$$r=\Big\lfloor\frac{(t+1)(l-1)+n}{l}\Big\rfloor\geq\Big\lfloor t+1+\frac{n-t-1}{l}\Big\rfloor\geq\Big\lfloor n+2-\frac{2}{l}\Big\rfloor\geq n+1.$$

Hence $\mathsf{Tr}_{L/K}(\mathfrak{p}_L^n)\subset\mathfrak{p}_K^{n+1}$, and if $x\in\mathfrak{p}_L^n$, then $\mathsf{N}_{L/K}(1+x)\equiv 1+\mathsf{N}_{L/K}(x)$ mod $\mathfrak{p}_K^{n+1}$. Therefore we get $\mathsf{N}_{L/K}(U_L^{(n+\nu)})\subset U_K^{(n+\nu)}$ for $\nu\in\{0,1\}$. If now $u=1+\pi^n a\in U_L^{(n)}$, where $a\in\mathcal{O}_K$, then

$$\mathsf{N}_{L/K}(u)\equiv 1+\mathsf{N}_{L/K}(\pi^n a)\mod\mathfrak{p}_K^{n+1},\mathsf{N}_{L/K}(\pi^n a)=\pi'^n\mathsf{N}_{L(K}(a),$$
$$v_n^{(\pi)}(uU_L^{(n+1)})=\overline{a}\text{ and }v_n^{(\pi')}(\mathsf{N}_{L/K}(u)U_L^{(n+1)})=\overline{\mathsf{N}_{L/K}(a)}=\overline{a}^{\,l}.$$

Hence $N_n(\xi)=\xi^l=\xi^p$ for all $\xi\in\mathsf{k}$, and N_n is injective.

3. and 4. Let $n \ge t$, and $\psi(n) = t + l(n-t)$. If $\nu \in \{0,1\}$, then by 4.3.1.2 we get $\mathsf{Tr}_{L/K}(\mathfrak{p}^{\psi(n)+\nu}) = \mathfrak{p}_K^r$, where

$$r = \Big\lfloor \frac{(t+1)(l-1) + t + l(n-t) + \nu}{l} \Big\rfloor = \Big\lfloor n + 1 - \frac{\nu - 1}{l} \Big\rfloor = n + \nu,$$

and thus $\mathsf{Tr}_{L/K}(\mathfrak{p}_L^{\psi(n)+\nu}) = \mathfrak{p}_K^{n+\nu}$. If $x \in \mathfrak{p}_L^{\psi(n)+\nu}$, then $\mathsf{N}_{L/K}(x) \in \mathfrak{p}_K^{\psi(n)+\nu}$, hence $\mathsf{N}_{L/K}(1+x) \equiv 1 + \mathsf{Tr}_{L/K}(x) + \mathsf{N}_{L/K}(x) \mod \mathfrak{p}_K^{n+1}$ by 4.3.1.3, and consequently $\mathsf{N}_{L/K}(U_L^{(\psi(n)+\nu)}) \subset U_K^{(n+\nu)}$.

Let $u = 1 + a'\pi^{\psi(n)} \in U_L^{(\psi(n))}$, where $a' \in \mathcal{O}_L$, and let $a \in \mathcal{O}_K$ be such that $\overline{a} = \overline{a'}$. Then $uU_L^{(\psi(n)+1)} = (1 + a\pi^{\psi(n)})U_L^{(\psi(n)+1)}$, and $v_{\psi(n)}^{(\pi)}(uU_L^{(\psi(n)+1)}) = \overline{a}$. Since $\mathsf{Tr}_{L/K}(\pi^{\psi(n)}) = b\pi'^n$ for some $b \in \mathcal{O}_K$ and $\mathsf{N}_{L/K}(a\pi^{\psi(n)}) = \mathsf{N}_{L/K}(a)\pi'^{\psi(n)}$, we get

$$\begin{aligned}\mathsf{N}_{L/K}(1 + a\pi^{\psi(n)}) &\equiv 1 + \mathsf{Tr}_{L/K}(a\pi^{\psi(n)}) + \mathsf{N}_{L/K}(a\pi^{\psi(n)}) \\ &\equiv 1 + (ab + \mathsf{N}_{L/K}(a)\pi'^{(\psi(n)-n)})\pi'^n \mod \mathfrak{p}_K^{n+1},\end{aligned}$$

and therefore

$$v_n^{(\pi')}(\mathsf{N}_{L/K}(u)U_L^{(\psi(n)+1)}) = \overline{ab + \mathsf{N}_{L/K}(a)\pi'^{(\psi(n)-n)}} = \overline{a}\overline{b} + \overline{a^l\pi'^{(\psi(n)-n)}}.$$

CASE **a.** $t = n \ge 1$. Then $\psi(n) = n$, $l = p$, and

$$v_n^{(\pi')}(\mathsf{N}_{L/K}(u)U_L^{(n+1)})) = \overline{a}\overline{b} + \overline{a}^l, \quad \text{hence} \quad N_n(\xi) = \overline{b}\xi + \xi^p \ \text{ for all } \ \xi \in \mathsf{k}.$$

We consider the homomorphism $\Phi_t \colon G \to U_L^{(t)}/U_L^{(t+1)} = U_L^{(n)}/U_L^{(n+1)}$. Since apparently $\mathrm{Im}(\Phi_t) \subset \mathrm{Ker}(\overline{\mathsf{N}}_{L/K}^{(n)})$, we get $p \le |\mathrm{Ker}(\overline{\mathsf{N}}_{L/K}^{(n)}| = |\mathrm{Ker}(N_n)| \le p$, hence $|\mathrm{Ker}(N_n)| = p$, and thus $\overline{b} \ne 0$.

CASE **b.** $n > t$. Then $\psi(n) > n$ and

$$v_n^{(\pi')}(\mathsf{N}_{L/K}(u)U_L^{(\psi(n)+1)}) = \overline{a}\overline{b}, \quad \text{hence} \quad N_n(\xi) = \overline{b}\xi \ \text{ for all } \ \xi \in \mathsf{k}.$$

We must prove that $\overline{b} \ne 0$. Assume that $\overline{b} = 0$. Then $\mathsf{Tr}_{L/K}(\pi^{\psi(n)}) \in \mathfrak{p}_K^{n+1}$. If $x \in \mathfrak{p}_L^{\psi(n)}$, then $x = a_0\pi^{\psi(n)} + a_1$ for some $a_0 \in \mathcal{O}_K$ and $a_1 \in \mathfrak{p}_L^{\psi(n)+1}$, and we obtain $\mathsf{Tr}_{L/K}(x) = a_0\mathsf{Tr}_{L/K}(\pi^{\psi(n)}) + \mathsf{Tr}_{L/K}(a_1) \in \mathfrak{p}_K^{n+1}$. Hence $\mathsf{Tr}_{L/K}(\mathfrak{p}_L^{\psi(n)}) \subset \mathfrak{p}_K^{n+1}$, a contradiction.

B. By **A** the homomorphism $N_n \colon \mathsf{k} \to \mathsf{k}$ (and thus also the homomorphism $\overline{\mathsf{N}}_{L/K}^{(n)} \colon U_L^{(\psi(n))}/U_L^{(\psi(n)+1)} \to U_K^{(n)}/U_K^{(n+1)}$) is bijective for $n > t$, and if k is algebraically closed, then it is surjective for all $n \ge 0$. Since $\psi(n+1) \ge \psi(n) + 1$ for all $n \ge 0$, there exists a natural epimorphism $\rho_n \colon U_L^{(\psi(n))}/U_L^{(\psi(n+1))} \to U_L^{(\psi(n))}/U_L^{(\psi(n)+1)}$, and we consider the homomorphism

$$\begin{aligned}\widetilde{N}_n = \overline{\mathsf{N}}_{L/K}^{(n)} \circ \rho_n \colon U_L^{(\psi(n))}/U_L^{(\psi(n+1))} &\to U_K^{(n)}/U_K^{(n+1)}, \\ v_nU_L^{(\psi(n+1))} &\mapsto \mathsf{N}_{L/K}(v_n)U_K^{(n)}.\end{aligned}$$

$\widetilde{N}_n$ is surjective for all $n > t$, and even for all $n \geq 0$ if k is algebraically closed.

Now let $m \in \mathbb{N}$ such that $m > t$, and actually $m \geq 0$ if k is algebraically closed, and let $v \in U_K^{(m)}$.

We construct recursively sequences

$$(v_n \in U_L^{(\psi(n))})_{n \geq m} \text{ and } (u_n \in U_K^{(n)})_{n \geq m}$$

as follows. Set $v_m = v$; if $n \geq m$ and $v_n \in U_K^{(n)}$ is given, then by the surjectivity of $\widetilde{N}_n$ there exist elements $u_n \in U_L^{(\psi(n))}$ and $v_{n+1} \in U_K^{(n+1)}$ such that $\mathsf{N}_{L/K}(u_n) = v_n v_{n+1}^{-1}$. If $k \geq m$, then

$$u_m u_{m+1} \cdot \ldots \cdot u_{k+1} - u_m u_{m+1} \cdot \ldots \cdot u_k = u_m u_{m+1} \cdot \ldots \cdot u_k (u_{k+1} - 1) \in \mathfrak{p}_L^{\psi(k+1)},$$

and since $U_L^{(\psi(m))}$ is complete, there exists

$$u = \lim_{k \to \infty} u_m u_{m+1} \cdot \ldots \cdot u_k \in U_L^{(\psi(m))}.$$

As $N_{L/K}$ is continuous, it follows that

$$\mathsf{N}_{L/K}(u) = \lim_{k \to \infty} \prod_{n=m}^{k} \mathsf{N}_{L/K}(u_n) = \lim_{k \to \infty} \prod_{n=m}^{k} v_n v_{n+1}^{-1} = \lim_{k \to \infty} v_m v_{k+1}^{-1} = v_m = v.$$

□

Theorem 4.3.3. *Let L/K be a finite Galois extension of complete discrete valued fields such that k_K is algebraically closed. Then L/K is solvable,*

$$\mathsf{N}_{L/K} L^\times = K^\times \qquad \textit{and} \qquad \mathsf{N}_{L/K}(U_L) = U_K.$$

Proof. As k_K is algebraically closed, the extension L/K is fully ramified. If $G = \mathrm{Gal}(L/K)$ and $(G_i)_{i \geq -1}$ is the Hilbert subgroup series for L/K, then $G = G_0$, G_0/G_1 is cylcic, and G_i/G_{i+1} is an elementary abelian p-group (see [18, Section 5.10]). Hence G and thus L/K is solvable.

It suffices to prove that $\mathsf{N}_{L/K}(U_L) = U_K$. Indeed, if $\pi \in \mathcal{O}_L$ is a prime element, then $\mathsf{N}_{L/K}(\pi)$ is a prime element of $\mathcal{O}_K$, and $K^\times = \langle \mathsf{N}_{L/K}(\pi) \rangle \cdot U_K$. If $\mathsf{N}_{L/K}(U_L) = U_K$, then

$$\mathsf{N}_{L/K} L^\times = \mathsf{N}_{L/K}(\langle \pi \rangle \cdot U_L) = \langle \mathsf{N}_{L/K}(\pi) \rangle \cdot \mathsf{N}_{L/K}(U_L) = K^\times.$$

For the proof of $\mathsf{N}_{L/K}(U_L) = U_K$ we use induction on $[L:K]$. Since L/K is solvable, there exists an intermediate field M of L/K such that M/K is cyclic of prime degree. As $\mathsf{N}_{L/M}(U_L) = U_M$ by the induction hypothesis and $\mathsf{N}_{M/K}(U_M) = U_K$ by 4.3.2.**B**, it follows that $\mathsf{N}_{L/K}(U_L) = U_K$. □

Theorem 4.3.4. *Let L/K be a finite Galois extension of complete discrete valued fields such that k_K is algebraically closed, $G = \mathrm{Gal}(L/K)$, and consider the subgroup*

$$U_L^{I_G} = \langle \{u^{\sigma-1} \mid u \in U_L,\ \sigma \in G\} \rangle \subset U_L.$$

Let $\pi \in \mathcal{O}_L$ be a prime element, and let

$$\ell \colon G \to U_L/U_L^{I_G} \quad \text{be defined by} \quad \ell(\sigma) = \pi^{\sigma-1}U_L^{I_G}.$$

ℓ is a group homomorphism which does not depend on π and induces an (equally denoted) monomorphism $\ell\colon G^{\mathsf{a}} \to U_L/U_L^{I_G}$ which fits into the exact sequence

$$\mathbf{1} \longrightarrow G^{\mathsf{a}} \xrightarrow{\ \ell\ } U_L/U_L^{I_G} \xrightarrow{\overline{\mathsf{N}}_{L/K}} U_K \longrightarrow \mathbf{1}.$$

where $\overline{\mathsf{N}}_{L/K}(uU_L^{I_G}) = \mathsf{N}_{L/K}(u)$ for all $u \in U_L$.

Proof. Note that L/K is fully ramified, since k_K is algebraically closed.

If π' is another prime element of $\mathcal{O}_L$, then $\pi' = \pi\varepsilon$ for some $\varepsilon \in U_L$, and therefore $\pi'^{\sigma-1}U_L^{I_G} = \pi^{\sigma-1}\varepsilon^{\sigma-1}U_L^{I_G} = \pi^{\sigma-1}U_L^{I_G}$. Hence ℓ does not depend on π.

If $\sigma,\ \tau \in G$, then $\sigma\tau - 1 = (\sigma-1)(\tau-1) + (\sigma-1) + (\tau-1)$, and consequently

$$\begin{aligned}\ell(\sigma\tau) &= \pi^{\sigma\tau-1}U_L^{I_G} = \pi^{\sigma-1}\pi^{\tau-1}(\pi^{\tau-1})^{\sigma-1}U_L^{I_G} \\ &= (\pi^{\sigma-1}U_L^{I_G})(\pi^{\tau-1}U_L^{I_G}) = \ell(\sigma)\ell(\tau).\end{aligned}$$

Hence ℓ is a group homomorphism. Obviously, $\ell \restriction G^{(1)} = 1$, and thus ℓ induces an (equally denoted) homomorphism $\ell\colon G^{\mathsf{a}} \to U_L/U_L^{I_G}$. If L^{a} is the largest over K abelian intermediate field of L/K, then $G^{\mathsf{a}} = \mathrm{Gal}(L^{\mathsf{a}}/K$, and we obtain $\ell(\sigma) = \ell(\sigma \restriction L^{\mathsf{a}}$ for all $\sigma \in G$. Apparently, $\mathrm{Im}(\ell) \subset \mathrm{Ker}(\overline{\mathsf{N}}_{L/K})$, and $\overline{\mathsf{N}}_{L/K}$ is surjective. It remains to prove that $\mathrm{Ker}(\overline{\mathsf{N}}_{L/K}) \subset \mathrm{Im}(\ell)$ and that $\ell\colon G^{\mathsf{a}} \to U_L/U_L^{I_G}$ is injective.

SPECIAL CASE: L/K is cyclic, $[L\!:\!K] = n$, and $G = \langle\sigma\rangle$. Then we have $I_G = (\sigma-1)\mathbb{Z}[G]$ and $U_L^{I_G} = U_L^{\sigma-1}$.

$\mathrm{Ker}(\overline{\mathsf{N}}_{L/K}) \subset \mathrm{Im}(\ell)$: Let $\varepsilon \in U_L$ such that $\varepsilon U_L^{I_G} \in \mathrm{Ker}(\overline{\mathsf{N}}_{L/K})$. As $\mathsf{N}_{L/K}(\varepsilon) = 1$, we obtain $\varepsilon = a^{\sigma-1}$ for some $a \in L^\times$ by 2.6.4 (Hilbert's Theorem 90), and we may assume that $a \in \mathcal{O}_L$, say $a = \pi^m u$, where $m \in \mathbb{N}_0$ and $u \in U_L$. Then it follows that

$$\varepsilon U_L^{I_G} = (\pi^{\sigma-1})^m u^{\sigma-1}U_L^{I_G} = \ell(\sigma)^m = \ell(\sigma^m) \in \mathrm{Im}(l).$$

ℓ is injective. Let $m \in \mathbb{N}$ such that $\sigma^m \in \mathrm{Ker}(\ell)$. Then $\ell(\sigma^m) = \ell(\sigma)^m = 1$, and therefore $(\pi^{\sigma-1})^m \in U_L^{I_G} = U_L^{\sigma-1}$, say $(\pi^m)^{\sigma-1} = (\pi^{\sigma-1})^m = u^{\sigma-1}$ for some $u \in U_L$. Then $(\pi^m u^{-1})^{\sigma-1} = 1$, hence $\pi^m u^{-1} \in K$, and therefore $n \mid m$ and $\sigma^m = 1$.

GENERAL CASE: $\mathrm{Ker}(\overline{\mathsf{N}}_{L/K}) \subset \mathrm{Im}(\ell)$: We use induction on $[L:K]$. Since L/K is solvable (by 4.3.3), there exists an intermediate field M of L/K

such that M/K is cyclic, and $[L:M] < [L:K]$. Let $H = \mathrm{Gal}(L/M)$ and identify $G/H = \mathrm{Gal}(M/K)$ by means of $\sigma H = \sigma \upharpoonright M$ for all $\sigma \in G$. Then $\pi_M = \mathsf{N}_{L/M}(\pi)$ is a prime element of M, and let $\ell_M \colon G/H \to U_M/U_M^{I_{G/H}}$ be defined by $\ell_M(\tau H) = \pi_M^{\tau-1} U_M^{I_{G/H}}$ for all $\tau \in G$. By the SPECIAL CASE it follow that $\mathrm{Ker}(\overline{\mathsf{N}}_{M/K}) \subset \mathrm{Im}(\ell_M)$.

We apply 2.1.1.3(b) to the multiplicative G-module U_L. Since $\mathsf{N}_{L/M} = U_M$ (by 4.3.3), we obtain

$$\mathsf{N}_{L/M}(U_L^{I_G}) = (U_L^{I_G})^{N_H} = (U_L^{N_H})^{I_G} = \mathsf{N}_{L/M}(U_L)^{I_G} = U_M^{I_{G/H}}.$$

Now suppose that $\varepsilon \in U_L$ is such that $\varepsilon U_L^{I_G} \in \mathrm{Ker}(\overline{\mathsf{N}}_{L/K})$. Then we obtain $1 = \mathsf{N}_{L/K}(\varepsilon) = \mathsf{N}_{M/K}(\mathsf{N}_{L/M}(\varepsilon))$, and therefore $\mathsf{N}_{L/M}(\varepsilon) = \pi_M^{\tau-1}\eta$ for some $\tau \in G$ and $\eta \in U_M^{I_{G/H}} = \mathsf{N}_{L/M}(U_L^{I_G})$, say $\eta = \mathsf{N}_{L/K}(\xi)$ for some $\xi \in U_L^{I_G}$. Again by 2.1.1.3(b) we obtain

$$\mathsf{N}_{L/M}(\pi^{\tau-1}) = \pi^{N_H(\tau-1)} = \pi^{(\tau-1)N_H} = \mathsf{N}_{L/M}(\pi)^{\tau-1} = \pi_M^{\tau-1},$$

and therefore $\mathsf{N}_{L/M}(\varepsilon(\pi^{\tau-1}\xi)^{-1}) = \mathsf{N}_{L/M}(\varepsilon)(\pi_M^{\tau-1}\eta)^{-1}) = 1$. By the induction hypothesis it follows that $\varepsilon(\pi^{\tau-1}\xi)^{-1} = \pi^{\sigma-1}\rho$ for some $\sigma \in H$ and $\rho \in U_L^{I_H} \subset U_L^{I_G}$, and consequently

$$\varepsilon = \pi^{\sigma\tau-1}(\pi^{(\sigma-1)(\tau-1)})^{-1}\xi\rho \in \pi^{\sigma\tau-1}U_{I_G} = \ell(\sigma\tau) \in \mathrm{Im}(\ell).$$

$\ell \colon G^{\mathsf{a}} \to U_L/U_L^{I_G}$ is injective: Let $\sigma \in G$ such that $\sigma \upharpoonright L^{\mathsf{a}} \in \mathrm{Ker}(\ell)$, that is, $\ell(\sigma \upharpoonright L^{\mathsf{a}}) = \ell(\sigma) = \pi^{\sigma-1} \in U_L^{I_G}$. We must prove that $\sigma \upharpoonright L^{\mathsf{a}} = \mathrm{id}_{L^{\mathsf{a}}}$, and therefore it suffices to prove that $\sigma \upharpoonright M = \mathrm{id}_M$ for every over K cyclic intermediate field M of L/K. Thus let $\sigma \in G$, $\pi^{\sigma-1} \in U_L^{I_G}$ and M an over K cyclic intermediate field of L/K. Then $\mathsf{N}_{L/M}(\pi^{\sigma-1}) = \pi_M^{\sigma-1} \in \mathsf{N}_{L/M}(U_L^{I_G}) = U_M^{I_{G/H}}$. By the special case, the map $\ell_M \colon \mathrm{Gal}(M/K) \to U_M/U_M^{I_{G/H}}$ is injective, and since $\ell_M(\sigma \upharpoonright M) = \pi_M^{\sigma-1}U_M^{I_{G/H}} = U_M^{I_{G/H}}$, we obtain $\sigma \upharpoonright M = \mathrm{id}_M$. □

Corollary 4.3.5. *Let L/K be a finite Galois extension of complete discrete valued fields such that k_K is algebraically closed, $G = \mathrm{Gal}(L/K)$, $r \in \mathbb{N}$, $x_1, \ldots, x_r \in L^\times$ and $\sigma_1, \ldots, \sigma_r \in G$. Then*

$$\prod_{i=1}^r x_i^{\sigma_i - 1} = 1 \qquad \textit{implies} \qquad \sigma = \prod_{i=1}^r \sigma_i^{v_L(x_i)} \in G^{(1)}.$$

Proof. Let $\pi \in \mathcal{O}_L$ be a prime element, and for $i \in [1, r]$ let $x_i = \pi^{n_i}u_i$, where $n_i = v_L(x_i) \in \mathbb{Z}$ and $u_i \in U_L$. Then

$$\prod_{i=1}^r x_i^{\sigma_i-1} = \prod_{i=1}^r \pi^{\sigma_i-1} \prod_{i=1}^r u_i^{\sigma_i-1} = 1, \quad \text{and therefore} \quad \prod_{i=1}^r \pi^{\sigma_i-1} \in U_L^{I_G}.$$

By 4.3.4, the map $\ell\colon G \to U_L/U_L^{I_G}$, defined by $\ell(\tau) = \pi^{\tau-1}U_L^{I_G}$ for all $\tau \in G$, is a homomorphism with kernel $G^{(1)}$, and as

$$\ell(\sigma) = \prod_{i=1}^{r} \ell(\sigma_i^{n_i}) = \prod_{i=1}^{r} \pi^{\sigma_i - 1}U_L^{I_G} = U_L^{I_G},$$

it follows that $\sigma \in G^{(1)}$. □

Definitions and Remarks 4.3.6. Let K be a (non-Archimedian) local field, $\mathsf{k}_K = \mathbb{F}_q$, K_∞ its maximal unramified extension and φ_K the Frobenius automorphism over K (then follows that $\varphi_K(x) \equiv x^q \bmod \mathfrak{p}_{K_\infty}$, and $\mathrm{Gal}(K_\infty/K) = \overline{\langle\varphi_K\rangle} \cong \widehat{\mathbb{Z}}$).

1. The discrete valuation v_K has a unique extension to an (equally denoted) discrete valuation $v_K\colon K_\infty \to \mathbb{Z} \cup \{\infty\}$, K_∞ is a discrete valued field, $\mathsf{k}_{K_\infty} = \overline{\mathbb{F}}_q$ is the algebraic closure of $\mathbb{F}_q$, $K_\infty = K(\mu')$, where μ' is the group of roots of unity of order coprime to p in $\overline{K}$ (see [18, Theorem 5.6.7]), and K_∞ is not complete (see Theorem 4.3.7 below).

2. Let L/K be a finite Galois extension. Then $L_0 = L \cap K_\infty$ is the inertia field of L/K, $K_\infty = (L_0)_\infty$ and $L_\infty = L(\mu') = LK(\mu') = LK_\infty$. By 1.8.4, the restriction $\sigma \mapsto \sigma \restriction L$ defines an isomorphism $\mathrm{Gal}(L_\infty/K_\infty) \xrightarrow{\sim} \mathrm{Gal}(L/L_0)$, and as k_{K_∞} is algebraically closed, the extension L_∞/K_∞ is a fully ramified extension of degree $[L{:}L_0] = e(L/K)$.

3. Let $\widehat{K}_\infty$ be a completion of K_∞. Then $\widehat{K}_\infty$ is a complete discrete valued field, the discrete valuation v_K has a unique (equally denoted) extension to the discrete valuation of $\widehat{K}_\infty$, and $\mathsf{k}_{\widehat{K}_\infty} = \mathsf{k}_{K_\infty} = \overline{\mathbb{F}}_q$. In particular, if μ' denotes the group of all roots of unity of an order not divisible by p, then $R = \mu' \cup \{0\}$ is a multiplicative residue system of K_∞ and thus also of $\widehat{K}_\infty$.

The Frobenius automorphism $\varphi_K\colon K_\infty \to K_\infty$ is continuous (it even satisfies $v \circ \varphi_K = v$), and thus it has a unique (equally denoted) extension to an automorphism $\varphi_K \in \mathrm{Gal}(\widehat{K}_\infty/K)$ which we call the **Frobenius automorphism of** $\widehat{K}_\infty$. Its residue class homomorphism $\overline{\varphi}_K\colon \overline{\mathbb{F}}_q \to \overline{\mathbb{F}}_q$, defined by $\overline{\varphi}_K(x + \mathfrak{p}_{\widehat{K}_\infty}) = \varphi_K(x) + \mathfrak{p}_{\widehat{K}_\infty}$ satisfied $\overline{\varphi}_K(\xi) = \xi^q$ for all $\xi \in \overline{\mathbb{F}}_q$, and therefore $\overline{\varphi}_K = \varphi_{\mathbb{F}_q}$ is the Frobenius automorphism over $\mathbb{F}_q$.

4. Let L/K be a fully ramified Galois extension. Then $L_0 = K$, and the restriction $\sigma \mapsto \sigma \restriction L$ defines an isomorphism $\mathrm{Gal}(L_\infty/K_\infty) \xrightarrow{\sim} \mathrm{Gal}(L/K)$. Moreover, since L_∞/K_∞ is fully ramified, the restriction $\tau \mapsto \tau \restriction L_\infty$ defines an isomorphism $\mathrm{Gal}(\widehat{L}_\infty/\widehat{K}_\infty) \xrightarrow{\sim} \mathrm{Gal}(L_\infty/K_\infty)$, and the extension $\widehat{L}_\infty/\widehat{K}_\infty$ is fully ramified, too. The restriction $\mathrm{Gal}(\widehat{L}_\infty/\widehat{K}_\infty) \to \mathrm{Gal}(L/K)$ is an isomorphism; we identify.

Let $V = \mathrm{Gal}(L_\infty/L) = \overline{\langle\varphi\rangle} \cong \widehat{\mathbb{Z}}$. As $\varphi \restriction K_\infty$ is the Frobenius automorphism over K, it follows that $\mathrm{Gal}(K_\infty/K) = \overline{\langle\varphi \restriction K\rangle}$, and the restriction

$\varphi \mapsto \varphi \restriction K$ defines an isomorphism $V \stackrel{\sim}{\to} \mathrm{Gal}(K_\infty/K) \cong \widehat{\mathbb{Z}}$. By 1.8.4.2 we get $\mathrm{Gal}(L_\infty/K) = G{\cdot}V$.

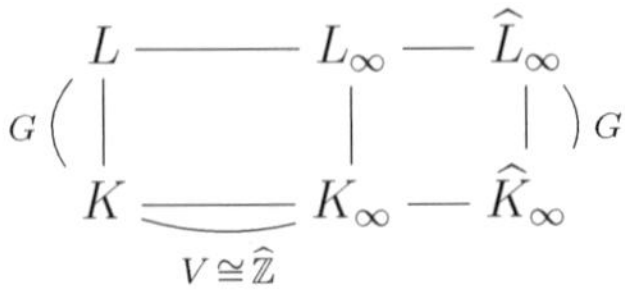

Theorem 4.3.7. *Let K be a local field. Then K_∞ is not complete.*

Proof. Assume that K_∞ is complete, let $\pi \in \mathcal{O}_K$ be a prime element, and let μ' be the group of all roots of unity of an order not divisible by p in $\overline{K}$. Since $K_\infty = K(\mu')$, there exists a sequence $(b_n)_{n\ge 0}$ in μ' such that $b_{n-1} \in K(b_n)$ and $[K(b_n){:}K(b_{n-1})] > n$ for all $n \in \mathbb{N}$. Then

$$c = \sum_{n=0}^{\infty} b_n\pi^n \in K_\infty, \quad \text{we set} \quad t = [K(c){:}K] \ \text{ and } \ c_t = \sum_{n=0}^{t} b_n\pi^n.$$

By definition, $v_K(c_t - c) \ge t+1$ and thus $v_K(\sigma c_t - \sigma c) \ge t+1$ for all $\sigma \in \mathrm{Gal}(K_\infty/K)$. As $[K(b_t) : K(b_{t-1})] > t$, there exits automorphisms $\sigma_1, \ldots, \sigma_{t+1} \in \mathrm{Gal}(K_\infty/K(b_{t-1}))$ such that the roots of unity $\sigma_1(b_t), \ldots, \sigma_{t+1}(b_t)$ are distinct, and since $K(b_t)/K$ is unramified, the residue classes $\overline{\sigma_1(b_t), \ldots, \sigma_{t+1}(b_t)} \in \mathsf{k}_{K(b_t)}$ are distinct, too. Hence, if $i,\ j \in [1, t+1]$ and $i \ne j$, then

$$v_K\big(\sigma_i(c_t) - \sigma_j(c_t)\big) = v_K\Big(\sum_{n=1}^{t} (\sigma_i(b_n) - \sigma_j(b_n))\pi^n\Big) = v_K\big(\sigma_i(b_t) - \sigma_j(b_t))\pi^t\big) = t,$$

and since $\sigma_i(c) - \sigma_j(c) = (\sigma_i(c) - \sigma_i(c_t)) + (\sigma_i(c_t) - \sigma_j(c_t)) + (\sigma_j(c_t) - \sigma_j(c))$, we obtain $v_K(\sigma_i(c) - \sigma_j(c)) = t$. In particular, $\sigma_1(t), \ldots, \sigma_{t+1}(t)$ are distinct, a contradiction, since $[K(c){:}K] = t$. □

Theorem 4.3.8. *Let K be a local field, $\widehat{K}_\infty$ a completion of K_∞, $\mathsf{k}_K = \mathbb{F}_q$, $\mathsf{k}_{K_\infty} = \mathsf{k}_{\widehat{K}_\infty} = \overline{\mathbb{F}}_q$ and φ the Frobenius automorphism of $\widehat{K}_\infty$. Let $\widehat{U} = U_{\widehat{K}_\infty}$ and $\widehat{U}^{(n)} = U^{(n)}_{\widehat{K}_\infty}$ for all $n \in \mathbb{N}$.*

1. *$K = \{x \in \widehat{K}_\infty \mid \varphi(x) = x\}$.*
2. *$(\widehat{U}^{(m)})^{\varphi-1} = \widehat{U}^{(m)}$ for all $m \in \mathbb{N}$.*
3. *$(\widehat{K}_\infty^\times)^{\varphi-1} = \widehat{U}^{\varphi-1} = \widehat{U}$*

Proof. Let $\pi \in \mathcal{O}_K$ be a prime element. For an element $a \in \mathcal{O}_{\widehat{K}}$ we denote by $\overline{a} = a + \mathfrak{p}_{\widehat{K}_\infty} \in \overline{\mathbb{F}}_q$ its residue class. If $\overline{\varphi} \in \mathrm{Gal}(\overline{\mathbb{F}}_q/\mathbb{F}_q)$ is the residue class automorphism of φ, then $\overline{\varphi}(\overline{a}) = \overline{a}^{\,q}$ for all $a \in \mathcal{O}_{\widehat{K}_\infty}$.

1. Let $x \in \widehat{K}_\infty^\times$. Then $x = \pi^n u$ for some $n \in \mathbb{Z}$ and $u \in \widehat{U}$, and therefore $\varphi(x) = \pi^n \varphi(u)$. If $R \subset K_\infty$ is a multiplicative residue system of K_∞ (and thus of $\widehat{K}_\infty$), then there exists a unique sequence $(c_n)_{n \ge 0}$ in R such that

$$u = \sum_{n=0}^{\infty} c_n \pi^n, \quad \text{and then} \quad \varphi(u) = \sum_{n=0}^{\infty} \varphi(c_n) \pi^n$$

(see [18, Theorem 5.3.5]). Since $\mathrm{Gal}(K_\infty/K) = \overline{\langle \varphi \rangle}$, we obtain:

$$\begin{aligned} \varphi(x) = x &\iff \varphi(u) = u \iff \varphi(c_n) = c_n \ \text{ for all } \ n \ge 0 \\ &\iff c_n \in K \ \text{ for all } \ n \ge 0 \iff u \in K \iff x \in K. \end{aligned}$$

2. For $n \in \mathbb{N}$ let $v_n = v_n^{(\pi)} \colon \widehat{U}^{(n)}/\widehat{U}^{(n+1)} \to \overline{\mathbb{F}}_q$ be the isomorphism given by $v_n((1 + a\pi^n)\widehat{U}^{(n+1)}) = \overline{a}$ for all $a \in \mathcal{O}_{\widehat{K}}$. If $a \in \mathcal{O}_{\widehat{K}_\infty}$, then

$$\begin{aligned} (1 + \pi^n a)^{\varphi - 1} &= \frac{1 + \varphi(a)\pi^n}{1 + a\pi^n} \\ &= \frac{(1 + \varphi(a)\pi^n)(1 - a\pi^n)}{1 - a^2\pi^{2n}} \in (1 + (\varphi(a) - a)\pi^n))\widehat{U}^{(n+1}, \end{aligned}$$

and therefore $v_n((1 + \pi^n a)^{\varphi - 1}) = \overline{\varphi(a) - a} = \overline{a}^{\,q} - \overline{a}$. Since obviously $(\widehat{U}^{(n)})^{\varphi - 1} \subset \widehat{U}^{(n)}$, we obtain a commutative diagram

$$\begin{array}{ccc} \widehat{U}^{(n)}/\widehat{U}^{(n+1)} & \xrightarrow{\ \widetilde{\varphi}\ } & \widehat{U}^{(n)}/\widehat{U}^{(n+1)} \\ {\scriptstyle v_n}\big\downarrow & & \big\downarrow{\scriptstyle v_n} \\ \overline{\mathbb{F}}_q & \xrightarrow{\ \varphi^*\ } & \overline{\mathbb{F}}_q\,, \end{array}$$

where $\widetilde{\varphi}(u\widehat{U}^{(n+1)}) = u^{\varphi - 1}\widehat{U}^{(n+1)}$ for all $u \in \widehat{U}^{(n)}$, and $\varphi^*(\xi) = \xi^q - \xi$. As φ^* is surjective, $\widetilde{\varphi}$ is surjective, too.

Now we assume that $m \in \mathbb{N}$ and prove that $(\widehat{U}^{(m)})^{\varphi - 1} = \widehat{U}^{(m)}$. Thus let $v \in \widehat{U}^{(m)}$. We construct recursively sequences

$$(u_n \in \widehat{U}^{(n)})_{n \ge m} \text{ and } (v_n \in \widehat{U}^{(n)})_{n \ge m}$$

as follows. Let $v_m = v$; if $n \ge m$ and v_n is given, then by the surjectivity of $\widehat{\varphi}$ there exist $u_n \in \widehat{U}^{(n)}$ and $v_{n+1} \in \widehat{U}^{(n+1)}$ such that $u_n^{\varphi - 1} = v_n v_{n+1}^{-1}$. If $k \ge m$, then

$$u_m u_{m+1} \cdot \ldots \cdot u_{k+1} - u_m u_{m+1} \cdot \ldots \cdot u_k = u_m u_{m+1} \cdot \ldots \cdot u_k(u_{k+1} - 1) \in \mathfrak{p}_{\widehat{K}_\infty}^{k+1},$$

and since $\widehat{U}^{(m)}$ is complete, there exists

$$u = \lim_{k\to\infty} u_m u_{m+1} \cdot \ldots \cdot u_k \in \widehat{U}^{(m)}.$$

As φ is a continuous homomorphism, we obtain

$$u^{\varphi-1} = \lim_{k\to\infty} \prod_{n=m}^{k} u_n^{\varphi-1} = \lim_{k\to\infty} \prod_{n=m}^{k} v_m v_{k+1}^{-1} = \lim_{k\to\infty} v v_{k+1}^{-1} = v.$$

3. If $x \in \widehat{K}_\infty^\times$, then $x = \pi^n u$ for some $n \in \mathbb{Z}$ and $u \in \widehat{U}$, and then $x^{\varphi-1} = u^{\varphi-1}$. Therefore it suffices to prove that $\widehat{U}^{\varphi-1} = \widehat{U}$. Let $u \in \widehat{U}$. Then $\overline{u} \in \overline{\mathbb{F}_q} = \overline{\mathbb{F}}_q^{q-1}$, and there exists some $u_1 \in \widehat{U}$ such that $\overline{u} = \overline{u}_1^{q-1} = \overline{u_1^{\varphi-1}}$. In particular, $u = u_1^{\varphi-1} u_1'$ for some $u_1' \in \widehat{U}^{(1)}$. By 2., there exists some $u_2 \in \widehat{U}^{(1)}$ such that $u_1' = u_2^{\varphi-1}$, and consequently $u = (u_1 u_2)^{\varphi-1} \in \widehat{U}^{\varphi-1}$. □

4.4 The reciprocity laws of Dwork and Neukirch

Let K be a local field, $\operatorname{char}(\mathsf{k}_K) = p$ and $\mathsf{k}_K = \mathbb{F}_q$

We present two alternative representations of the local norm residue symbol. The first one is true to B. Dwork [11] and was historically the first purely local proof of an explicit formula for the norm residue symbol of cyclotomic fields (here we shall do this by means of the Lubin-Tat theory in 4.9.2). The second one is due to J. Neukirch [47], where he used the underlying ideas for a new foundation of both local and global class field theory (see [48]). Here we discribe his approach within the frame of our algebra-theoretic foundation, which allows us to restrict ouselves to abelian extensions. Our presentation is based on [12] and [57].

In what follows we use the terminology of 4.3.6.

Theorem 4.4.1 (Theorem of Dwork). *Let L/K be a finite abelian fully ramified extension, $G = \operatorname{Gal}(L/K) = \operatorname{Gal}(\widehat{L}_\infty/\widehat{K}_\infty)$, and let φ be the Frobenius automorphism of $\widehat{L}_\infty$. Let $x \in K^\times$ and $y \in \widehat{L}_\infty$ such that $\mathsf{N}_{\widehat{L}_\infty/\widehat{K}_\infty}(y) = x$.*

1. *Let $r \in \mathbb{N}$, $\sigma_1, \ldots, \sigma_r \in G$ and $z_1, \ldots, z_r \in \widehat{L}_\infty^\times$.*

If $y^{\varphi-1} = \prod_{i=1}^{r} z_i^{\sigma_i - 1}$ *and* $\sigma = \prod_{i=1}^{r} \sigma_i^{v_L(z_i)}$, *then* $(x, L/K) = \sigma^{-1}$

(note that v_L has a unique extension to a discrete valuation of $\widehat{L}_\infty$).

2. If $\sigma \in G$ *and* $y^{\varphi-1} \equiv \pi^{\sigma-1} \bmod U_{\widehat{L}_\infty/\widehat{K}_\infty}$ (see 4.3.4), *then* $(x, L/K) = \sigma^{-1}$.

Proof. 1. Let $V = \mathrm{Gal}(L_\infty/L) = \overline{\langle\varphi\rangle} \cong \widehat{\mathbb{Z}}$. Then $\mathrm{Gal}(K_\infty/K) = \overline{\langle\varphi \restriction K_\infty\rangle} \cong V$ and $\mathrm{Gal}(L_\infty/K) = G{\cdot}V$.

Let $\pi \in \mathcal{O}_L$ be a prime element. Then $\pi_K = \mathsf{N}_{L/K}(\pi) = \mathsf{N}_{\widehat{L}_\infty/\widehat{K}_\infty}(\pi)$ is a prime element of $\mathcal{O}_K$, and we set $y = \pi^k y'$, where $k \in \mathbb{Z}$ and $y' \in U_{\widehat{L}_\infty}$. Then it follows that $x = \pi_K^k \mathsf{N}_{\widehat{L}_\infty/\widehat{K}_\infty}(y')$, $x' = \mathsf{N}_{\widehat{L}_\infty/\widehat{K}_\infty}(y') \in U_K$, and as $x = \mathsf{N}_{L/K}(\pi^k)x'$, we obtain $(x, L/K) = (x', L/K)$. Since also $y^{\varphi-1} = y'^{\varphi-1}$, we may replace (y, x) with (y', x'), and therefore in the sequel we assume that $y \in U_{\widehat{L}_\infty}$ and $x \in U_K$. Let

$$y^{\varphi-1} = \prod_{i=1}^{r} z_i^{\sigma_i-1}, \quad \sigma = \prod_{i=1}^{r} \sigma_i^{v_L(z_i)}, \quad \tau = (x, L/K) \in G \ \text{ and } \ L' = L_\infty^{\langle\tau\varphi\rangle}.$$

Then $L' \cap K_\infty = L_\infty^{\langle G, \tau\varphi\rangle} = L_\infty^{G\cdot V} = K$, $L'K_\infty = L_\infty^{G \cap \overline{\langle\tau\varphi\rangle}} = L_\infty$ (since $\overline{\langle\tau\varphi\rangle}$ contains no element of finite order except 1), and $\mathsf{N}_{\widehat{L}_\infty/\widehat{K}_\infty} \restriction L' = \mathsf{N}_{L'/K}$.

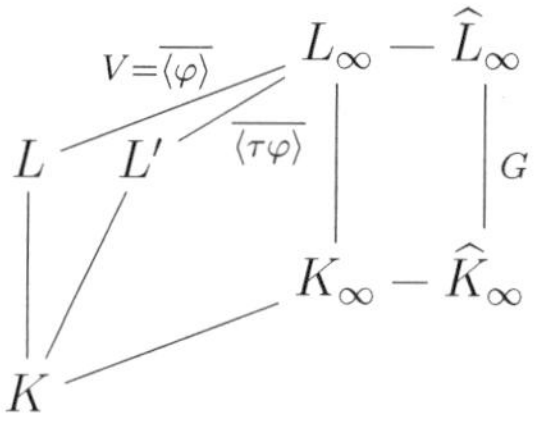

We assert that $\tau\varphi = (x\pi_K, L_\infty/K)$. Indeed, on the one hand,

$$(x\pi_K, L_\infty/K) \restriction L = (x, L/K) \circ (\pi_K, L/K) = \tau = \tau\varphi \restriction L,$$

since $\pi_K = \mathsf{N}_{L/K}(\pi)$ and $\varphi \restriction L = \mathrm{id}_L$. On the other hand,

$$(x\pi_K, L_\infty/K) \restriction K_\infty = (x, K_\infty/K) \circ (\pi_K, K_\infty/K) = \varphi \restriction K_\infty = \tau\varphi \restriction K_\infty,$$

since $(U_K, K_\infty/K) = 1$, $\varphi \restriction K_\infty = \varphi_K$ and $\tau \restriction K_\infty = \mathrm{id}_{K_\infty}$.

It follows that $(x\pi_K, L'/K) = \tau\varphi \restriction L' = \mathrm{id}_{L'}$. Therefore there exists some $z \in L'$ such that $\mathsf{N}_{\widehat{L}_\infty/\widehat{K}_\infty}(z) = \mathsf{N}_{L'/K}(z) = x\pi_K = \mathsf{N}_{\widehat{L}_\infty/\widehat{K}_\infty}(y\pi)$, and consequently we obtain $\mathsf{N}_{\widehat{L}_\infty/\widehat{K}_\infty}(z^{-1}y\pi) = 1$. In particular, $z^{-1}y\pi \in U_{\widehat{L}_\infty}$, and by 4.3.4 there is an exact sequence

$$G \xrightarrow{\ \ell\ } U_{\widehat{L}_\infty}/U_{\widehat{L}_\infty}^{I_G} \xrightarrow{\ \overline{N}\ } U_{\widehat{K}_\infty},$$

where $\ell(\theta) = \pi^{\theta-1}$ for all $\theta \in G$ and $\overline{N}(uU_{\widehat{L}_\infty^{I_G}}) = \mathsf{N}_{\widehat{L}_\infty/\widehat{K}_\infty}(u)$ for all $u \in U_{\widehat{L}_\infty}$.

Hence there exists some $\theta \in G$ such that $z^{-1}y\pi \in \pi^{\theta-1}U_{\widehat{L}_\infty}^{I_G}$. Consequently there exist $s \in \mathbb{N}$, $b_1, \ldots, b_s \in \widehat{L}_\infty^\times$ and $\theta_1, \ldots, \theta_s \in G$ such that

$$z^{-1}y\pi = \prod_{j=1}^{s} b_j^{\theta_j - 1}.$$

Since $z^{\tau\varphi-1} = 1$, $\pi^{\tau\varphi-1} = \pi^{\tau-1}$ and

$$y^{\tau\varphi-1} = y^{\tau-1}(y^{\varphi-1})^\tau = y^{\tau-1}\Big(\prod_{i=1}^{r} z_i^{\sigma_i-1}\Big)^\tau = y^{\tau-1}\prod_{i=1}^{r} z_i^{\tau(\sigma_i-1)},$$

we obtain

$$(z^{-1}y\pi)^{\tau\varphi-1} = \prod_{j=1}^{s} b_j^{(\tau\varphi-1)(\theta_j-1)} = y^{\tau-1}\pi^{\tau-1}\prod_{i=1}^{r} z_i^{\tau(\sigma_i-1)},$$

and consequently

$$y^{\tau-1}\pi^{\tau-1}\prod_{i=1}^{r} z_i^{\tau(\sigma_i-1)}\prod_{j=1}^{s} b_j^{(1-\tau\varphi)(\theta_j-1)} = 1.$$

By 4.3.5, it follows that

$$\tau^{v_L(y)}\tau^{v_L(\pi)}\prod_{i=1}^{r}\sigma_i^{v_L(z_i^\tau)}\prod_{j=1}^{s}\theta_j^{v_L(b_j^{1-\tau\varphi})} = 1.$$

As $v_L(y) = 0$, $v_L(\pi) = 1$, $v_L(z_i^\tau) = v_L(z_i)$ for all $i \in [1, r]$ and $v_L(b_j^{1-\tau\varphi}) = 0$ for all $j \in [1, s]$, it follows that $\tau\sigma = 1$, and therefore $\tau = (x, L/K) = \sigma^{-1}$.

2. Let $\sigma \in G$ and suppose that $y^{\varphi-1} \equiv \pi^{\sigma-1} \bmod U_{\widehat{L}_\infty/\widehat{K}_\infty}$. Then there exist $b_1, \ldots, b_r \in U_{\widehat{L}_\infty}$ and $\sigma_1, \ldots, \sigma_r \in G$ such that

$$y^{\varphi-1} = \pi^{\sigma-1}\prod_{i=1}^{r} b_i^{\sigma_i-1}. \quad \text{Then} \quad \sigma = \sigma^{v_L(\pi)}\prod_{i=1}^{r}\sigma_i^{v_L(b_i)},$$

and thus $(x, L/K) = \sigma^{-1}$ by 1. $\square$

Corollary 4.4.2. *Let L/K be a finite abelian fully ramified extension, and let $G = \mathrm{Gal}(L/K) = \mathrm{Gal}(\widehat{L}_\infty/\widehat{K}_\infty)$. Let φ be the Frobenius automorphism of $\widehat{L}_\infty$, $\pi \in \mathcal{O}_L$ a prime element and $\sigma \in G$.*

Then there exists some $y \in \widehat{L}_\infty^\times$ such that $y^{\varphi-1} = \pi^{\sigma-1}$, and for every such y the following holds: If $x = \mathsf{N}_{\widehat{L}_\infty/\widehat{K}_\infty}(y)$, then $x \in K^\times$, and $(x, L/K) = \sigma^{-1}$.

Proof. By 4.3.8.3, there exists some $y \in \widehat{L}_\infty^\times$ such that $y^{\varphi-1} = \pi^{\sigma-1}$. Thus let $y \in \widehat{L}_\infty^\times$ be such that $y^{\varphi-1} = \pi^{\sigma-1}$, and set $x = \mathsf{N}_{\widehat{L}_\infty/\widehat{K}_\infty}(y)$. Then

$$x^{\varphi-1} = \mathsf{N}_{\widehat{L}_\infty/\widehat{K}_\infty}(y^{\varphi-1}) = \mathsf{N}_{\widehat{L}_\infty/\widehat{K}_\infty}(\pi^{\sigma-1}) = \mathsf{N}_{L/K}(\pi^{\sigma-1}) = 1,$$

hence $x \in K^\times$ by 4.3.8.3, and the remaining assertion follow by 4.4.1.2. □

Definition 4.4.3. Let L/K be a finite abelian extension, $G = \mathrm{Gal}(L/K$, and let $\varphi = \varphi_K$ be the Frobenius automorphism over K.

1. We define

$$\mathrm{Frob}(L/K) = \{\theta \in \mathrm{Gal}(L_\infty/K) \mid \theta \restriction K_\infty = \varphi^n \text{ for some } n \in \mathbb{N}\}.$$

Obviously, $1 \notin \mathrm{Frob}(L/K)$, and if $\theta,\, \theta_1 \in \mathrm{Frob}(L/K)$, then $\theta\theta_1 \in \mathrm{Frob}(L/K)$, and $\theta^{-1} \notin \mathrm{Frob}(L/K)$.

2. For $\theta \in \mathrm{Frob}(L/K)$ we denote by $\Sigma_\theta = L_\infty^{\langle\theta\rangle}$ the fixed field of θ, and we set $\Sigma_\theta' = \Sigma_\theta \cap K_\infty$. Then $[\Sigma_\theta : K] < \infty$ (see 4.4.4.1 below). Let $\pi_\theta \in \mathcal{O}_{\Sigma_\theta}$ be a prime element, and define the **Neukirch map**

$$\Upsilon_{L/K}\colon \mathrm{Frob}(L/K) \to K^\times/\mathsf{N}_{L/K}(L^\times) \quad \text{by} \quad \Upsilon_{L/K}(\theta) = \mathsf{N}_{\Sigma_\theta/K}(\pi_\theta)\mathsf{N}_{L/K}L^\times.$$

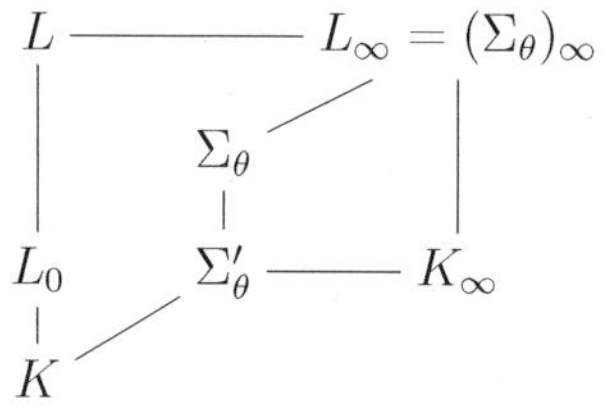

Below in 4.4.4.3 we shall see that the definition of $\Upsilon_{L/K}$ does not depend on the choice of π_θ.

Theorem 4.4.4. *Let L/K be a finite abelian extension, $G = \mathrm{Gal}(L/K)$, and let $\varphi = \varphi_K$ be the Frobenius automorphism over K.*

1. *Let $\theta \in \mathrm{Frob}(L/K)$, $n \in \mathbb{N}$ such that $\theta \restriction K_\infty = \varphi^n$, and set $\Sigma_\theta' = \Sigma_\theta \cap K_\infty$. Then $[\Sigma_\theta' : K] = n$, $[\Sigma_\theta : K] < \infty$, $(\Sigma_\theta)_\infty = L_\infty$, the restriction $\tau \mapsto \tau \restriction \Sigma_\theta$ defines an isomorphism*

$$\mathrm{Gal}(L_\infty/K_\infty) \to \mathrm{Gal}(\Sigma_\theta/\Sigma_\theta'),$$

and $\theta = \varphi_{\Sigma_\theta}$ is the Frobenius automorphism over Σ_θ.

2. $G = \{\theta \restriction L \mid \theta \in \mathrm{Frob}(L/K)\}$.

3. $\Upsilon_{L/K}$ *does not depend on choice of the prime element* π_θ.

4. Let L/K *be either fully ramified or unramified and suppose that* $\theta \in \mathrm{Frob}(L/K)$ *and* $\sigma = \theta \restriction L$.
If $x \in K^\times$ *and* $\Upsilon_{L/K}(\theta) = x\,\mathsf{N}_{L/K}L^\times$*, then* $(x, L/K) = \sigma$.
In particular, $\Upsilon_{L/K}$ *induces an isomorphism*

$$\overline{\Upsilon}_{L/K}\colon G \to K^\times/\mathsf{N}_{L/K}L^\times \ \textit{such that}\ \overline{\Upsilon}_{L/K}^{-1} = (\,\cdot\,, L/K).$$

Proof. Let $L_0 = L \cap K_\infty$ be the inertia field of L/K.

1. By 4.3.6.**2** it follows that $L_\infty = LK_\infty$, and since $K \subset \Sigma_\theta \subset L_\infty$, we obtain $K_\infty \subset (\Sigma_\theta)_\infty \subset L_\infty$. The restriction $\theta \mapsto \theta \restriction K_\infty = \varphi^n$ defines a continuous epimorphism $\overline{\langle\theta\rangle} \to \overline{\langle\varphi^n\rangle} \cong \widehat{\mathbb{Z}}$, and therefore we obtain $\mathrm{Gal}(L_\infty/\Sigma_\theta) = \overline{\langle\theta\rangle} \cong \widehat{\mathbb{Z}}$ by 1.5.6.2.

As $[L_\infty : K_\infty] < \infty$, it follows that $\mathrm{Gal}(L_\infty/(\Sigma_\theta)_\infty)$ is a finite closed subgroup of $\mathrm{Gal}(L_\infty/\Sigma_\theta)$. Hence $\mathrm{Gal}(L_\infty/(\Sigma_\theta)_\infty) = \mathbf{1}$ and $(\Sigma_\theta)_\infty = L_\infty$. The field $\Sigma'_\theta = \Sigma_\theta \cap K_\infty$ is the fixed field of $\theta \restriction K_\infty = \varphi^n$, and thus $[\Sigma'_\theta : K] = n$. Since $K_\infty = (L_0)_\infty = (\Sigma'_\theta)_\infty$, 4.3.6 shows that the restriction $\tau \mapsto \tau \restriction \Sigma_\theta$ defines an isomorphism $\mathrm{Gal}(L_\infty/K_\infty) \to \mathrm{Gal}(\Sigma_\theta/\Sigma'_\theta)$. In particular, we get $[\Sigma_\theta : \Sigma'_\theta] = [L_\infty : K_\infty] = [L : L_0]$, and therefore $[\Sigma_\theta : K] < \infty$. As Σ'_θ is the inertia field of Σ_θ/K, we obtain

$$\varphi_{\Sigma_\theta} = \varphi^{f(\Sigma_\theta/K)} = \varphi^{[\Sigma'_\theta:K]} = \varphi^n = \theta.$$

2. Let $\sigma \in G$. Then $\sigma \restriction L_0 = \varphi^m \restriction L_0$ for some $m \in \mathbb{N}$. Let $\overline{\varphi} \in \mathrm{Gal}(L_\infty/K)$ be such that $\overline{\varphi} \restriction K_\infty = \varphi$. Then $\sigma\overline{\varphi}^{-m} \restriction L_0 = \mathrm{id}_{L_0}$, and therefore we have $\sigma\overline{\varphi}^{-m} \restriction L \in \mathrm{Gal}(L/L_0)$. By 4.3.6.**4**, there exists some $\tau \in \mathrm{Gal}(L_\infty/K_\infty)$ such that $\tau \restriction L = \sigma\overline{\varphi}^{-m} \restriction L$. Then it follows that $\theta = \tau\overline{\varphi}^m \in \mathrm{Gal}(L_\infty/K)$, $\theta \restriction K_\infty = \varphi^m$, hence $\theta \in \mathrm{Frob}(L/K)$ and $\theta \restriction L = \sigma$.

3. Let π' be another prime element of $\mathcal{O}_{\Sigma_\theta}$, say $\pi' = \pi_\theta\varepsilon$ for some $\varepsilon \in U_{\Sigma_\theta}$. Since $L\Sigma_\theta \subset (\Sigma_\theta)_\infty$, the extension $L\Sigma_\theta/\Sigma_\theta$ is unramified, and therefore get $\varepsilon = \mathsf{N}_{L\Sigma_\theta/\Sigma_\theta}(\eta)$ for some $\eta \in L\Sigma_\theta$. Consequently we obtain

$$\begin{aligned}\mathsf{N}_{\Sigma_\theta/K}(\pi') &= \mathsf{N}_{\Sigma_\theta/K}(\pi_\theta)\mathsf{N}_{\Sigma_\theta/K}(\mathsf{N}_{L\Sigma_\theta/\Sigma_\theta}(\eta)) = \mathsf{N}_{\Sigma_\theta/K}(\pi_\theta)\mathsf{N}_{L/K}(\mathsf{N}_{L\Sigma_\theta/L}(\eta))\\ &\in \mathsf{N}_{\Sigma_\theta/K}(\pi_\theta)\mathsf{N}_{L/K}L^\times.\end{aligned}$$

4. CASE 1: L/K is fully ramified. Let $\theta \in \mathrm{Frob}(L/K)$, $n \in \mathbb{N}$ such that $\theta \restriction L_\infty = \varphi^n$, and $\theta \restriction L = \sigma$. According to 4.3.6.**4** we identify

$$G = \mathrm{Gal}(\widehat{L}_\infty/\widehat{K}_\infty) = \mathrm{Gal}(L_\infty/K_\infty),$$

and $\mathrm{Gal}(L_\infty/K) = G{\cdot}V$, where $V = \overline{\langle\varphi\rangle} = \mathrm{Gal}(L_\infty/L)$. Since $\theta \restriction L = \sigma$ we obtain (according to our identifications) $\theta = \sigma\varphi^n$.

Let $\pi \in \mathcal{O}_L$ and $\pi_\theta \in \mathcal{O}_{\Sigma_\theta}$ be a prime elements. Then both π and π_θ are prime elements of $\mathcal{O}_{L_\infty}$, hence $\pi = \pi_\theta\varepsilon$ for some $\varepsilon \in U_{L_\infty}$, and

$$\pi^{\sigma-1} = \pi^{\theta-1} = \varepsilon^{\theta-1} = \varepsilon^{\sigma\varphi^n-1}.$$

By definition, $\Upsilon_{L/K}(\theta) = \mathsf{N}_{\Sigma_\theta/K}(\pi_\theta)\,\mathsf{N}_{L/K}L^\times$, and thus we are going to calculate the norm. Recall that $\mathrm{Gal}(\Sigma'_\theta/K) = \langle \varphi \restriction \Sigma'_\theta\rangle$, and as $\Sigma_\theta/\Sigma'_\theta$ is fully ramified, 4.3.6.**4** implies

$$\mathrm{Gal}(\Sigma_\theta/\Sigma'_\theta) \cong \mathrm{Gal}((\Sigma_\theta)_\infty/(\Sigma'_\theta)_\infty) = \mathrm{Gal}(L_\infty/K_\infty) \cong \mathrm{Gal}(L/K),$$

and we identify. Now we obtain

$$\begin{aligned}\mathsf{N}_{\Sigma_\theta/K}(\pi_\theta) &= [\mathsf{N}_{\Sigma'_\theta/K}\circ\mathsf{N}_{\Sigma_\theta/\Sigma'_\theta}(\pi)]\,[\mathsf{N}_{\Sigma'_\theta/K}\circ\mathsf{N}_{\Sigma_\theta/\Sigma'_\theta}(\varepsilon]^{-1}\\ &= [\mathsf{N}_{\Sigma'_\theta/K}\circ\mathsf{N}_{L/K}(\pi)]\,[\mathsf{N}_{L_\infty/K_\infty}(\varepsilon)^{1+\varphi+\ldots+\varphi^{n-1}}]^{-1}\\ &= \mathsf{N}_{L/K}(\pi)^n\mathsf{N}_{L_\infty/K_\infty}(\varepsilon_1)^{-1} \in \mathsf{N}_{L_\infty/K_\infty}(\varepsilon_1^{-1})\mathsf{N}_{L/K}L^\times,\end{aligned}$$

where $\varepsilon_1 = \varepsilon^{1+\varphi+\ldots+\varphi^{n-1}}$. Hence we get $x = \mathsf{N}_{L_\infty/K_\infty}(\varepsilon_1)^{-1} \in K^\times$, and $\Upsilon_{L/K}(\theta) = x\,\mathsf{N}_{L/K}L^\times$.

Since $\sigma\varphi^n - 1 = \varphi^n - 1 + (\sigma-1)\varphi^n \equiv \varphi^n - 1 \mod I_G$, we get

$$\varepsilon_1^{\varphi-1} = \varepsilon^{\varphi^n-1} \equiv \varepsilon^{\sigma\varphi^n-1} = \pi^{\sigma-1} \mod U_{\widehat{L}_\infty}^{I_G}, \quad \text{and} \quad \mathsf{N}_{L/K}(\varepsilon_1) = x^{-1}.$$

Now 4.4.1.2 implies $(x^{-1}, L/K) = \sigma^{-1}$, and consequently $(x, L/K) = \sigma$.

CASE 2: L/K is unramified. Then $L \subset K_\infty = L_\infty$, and we set $m = [L\!:\!K]$. Let $\theta = \varphi^d$. Then $\sigma = \theta \restriction L = \varphi^d \restriction L$, and $\Sigma_\theta = K_\infty^{\langle\varphi^d\rangle} = K_d$ (the unramified extension of K of degree d). Let $\pi \in \mathcal{O}_K$ be a prime element. Then π is also a prime element of Σ_θ, and $\mathsf{N}_{\Sigma_\theta/K}(\pi) = \pi^d$. Hence $\Upsilon_{L/K}(\theta) = \pi^d\,\mathsf{N}_{L/K}L^\times$, and $(\pi^d, L/K) = \varphi^d \restriction L = \sigma$ by 4.2.3.2. □

Below in 4.4.7 we shall extend the assertion of 4.4.4.4 to arbitrary abelian extensions. As a preparation for this we investigate the functorial properties of $\Upsilon_{L/K}$.

Theorem 4.4.5. *Let K'/K be a finite field extension, and let L/K and L'/K' be finite abelian extensions such that $L \subset L'$. Then there is a commmutative diagram*

$$\begin{array}{ccc}\mathrm{Frob}(L'/K') & \xrightarrow{\Upsilon_{L'/K'}} & K'^\times/\mathsf{N}_{L'/K'}L'^\times\\ {\scriptstyle\iota}\big\downarrow & & \big\downarrow{\scriptstyle\nu}\\ \mathrm{Frob}(L/K) & \xrightarrow{\Upsilon_{L/K}} & K^\times/\mathsf{N}_{L/K}L^\times,\end{array}$$

where

$$\iota(\theta') = \theta' \restriction L_\infty \textit{ for all } \theta' \in \mathrm{Frob}(L'/K')$$

and

$$\nu(x'\mathsf{N}_{L'/K'}L'^\times) = \mathsf{N}_{K'/K}(x')\mathsf{N}_{L/K}L^\times \textit{ for all } x' \in K'^\times.$$

Proof. Let $\theta' \in \operatorname{Frob}(L'/K') \subset \operatorname{Gal}(L'_\infty/K')$, and set $\theta = \theta' \restriction L_\infty$. If then $\theta' \restriction K'_\infty = \varphi_{K'}^n$, where $n \in \mathbb{N}$, then $\theta \restriction K_\infty = \theta' \restriction K_\infty = \varphi_K^{f(K'/K)n}$, and therefore $\theta \in \operatorname{Frob}(L/K)$.

Let $\Sigma_{\theta'} = L'^{\langle\theta'\rangle}_\infty$ and $\Sigma_\theta = L^{\langle\theta\rangle}_\infty$. Since $(\Sigma_{\theta'})_\infty = L'_\infty$ by 4.4.4.1, it follows that

$$\Sigma_\theta = L_\infty^{\langle\theta'\restriction L_\infty\rangle} = L'^{\langle\theta'\rangle}_\infty \cap L_\infty = \Sigma_{\theta'} \cap L_\infty \subset \Sigma_{\theta'},$$

and therefore $\Sigma_{\theta'}/\Sigma_\theta$ is fully ramified. Let $\pi_{\theta'}$ be a pime element of $\mathcal{O}_{\Sigma_{\theta'}}$. Then $\pi_\theta = \mathsf{N}_{\Sigma_{\theta'}/\Sigma_\theta}(\pi_{\theta'})$ is a prime element of $\mathcal{O}_{\Sigma_\theta}$. If $x' = \mathsf{N}_{\Sigma_{\theta'}/K'}(\pi_{\theta'})$ and $x = \mathsf{N}_{\Sigma_\theta/K}(\pi_\theta)$, then $\Upsilon_{L'/K'}(\theta') = x'\mathsf{N}_{L'/K'}L'^\times$ and $\Upsilon_{L/K}(\theta) = x\mathsf{N}_{L/K}L'^\times$ (observe that by 4.4.4.3 it does not matter which prime element we choose). Since

$$x = \mathsf{N}_{\Sigma_\theta/K}\circ\mathsf{N}_{\Sigma_{\theta'}/\Sigma_\theta}(\pi_{\theta'}) = \mathsf{N}_{K'/K}\circ\mathsf{N}_{\Sigma_{\theta'}/K'}(\pi_{\theta'}) = \mathsf{N}_{K'/K}(x'),$$

we obtain

$$\Upsilon_{L/K}\circ\iota(\theta') = \Upsilon_{L/K}(\theta) = \mathsf{N}_{K'/K}(x')\mathsf{N}_{L/K}L^\times = \nu(x'\mathsf{N}_{L'/K'}L'^\times) = \nu\circ\Upsilon_{L'/K'}(\theta').$$

□

Corollary 4.4.6. *Let L/K be a finite abelian extension, and let M be an intermediate field of L/K. Then $\operatorname{Frob}(L/M) \subset \operatorname{Frob}(L/K)$, and there is the following commutative diagram in which ι is surjective and the bottom row is exact.*

$$\begin{array}{ccccccc}
\operatorname{Frob}(L/M) & \xrightarrow{i} & \operatorname{Frob}(L/K) & \xrightarrow{\iota} & \operatorname{Frob}(M/K) & \longrightarrow & \mathbf{1} \\
\Upsilon_{L/M}\downarrow & & \downarrow\Upsilon_{L/K} & & \Upsilon_{M/K}\downarrow & & \\
M^\times/\mathsf{N}_{L/M}L^\times & \xrightarrow{\nu} & K^\times/\mathsf{N}_{L/K}L^\times & \xrightarrow{j} & K^\times/\mathsf{N}_{M/K}M^\times & \longrightarrow & \mathbf{1}.
\end{array}$$

Here $i = (\operatorname{Frob}(L/M) \hookrightarrow \operatorname{Frob}(L/K))$, $\iota(\theta) = \theta\restriction M_\infty$ *for all* $\theta \in \operatorname{Frob}(L/K)$,

$$\nu(z\mathsf{N}_{L/M}L^\times) = \mathsf{N}_{M/K}(z)\mathsf{N}_{L/K}L^\times \text{ for all } z \in M^\times,$$

and

$$j(x\mathsf{N}_{L/K}L^\times) = x\mathsf{N}_{M/K}M^\times \text{ for all } x \in K^\times.$$

Proof. The left-hand square is 4.4.5, applied with $M = K' \subset L = L'$, and the right-hand square is 4.4.5, applied with $K = K'$ and (M, L) instead of (L, L'). By 4.2.10.2(c), j is surjective. If $\theta \in \operatorname{Frob}(M/K) \subset \operatorname{Gal}(M_\infty/K)$, then there exists some $\theta_1 \in \operatorname{Gal}(L_\infty/K)$ such that $\theta_1 \restriction M_\infty = \theta$, and since $\theta_1 \restriction K_\infty = \theta \restriction K_\infty$, it follows that $\theta_1 \in \operatorname{Frob}(L/K)$ and $\theta = \iota(\theta_1)$. Hence ι is surjective. □

Theorem 4.4.7 (Theorem of Neukirch). *Let L/K be a finite abelian extension and $G = \mathrm{Gal}(L/K)$. Let $\sigma \in G$, $\theta \in \mathrm{Frob}(L/K)$, $\theta \restriction L = \sigma$ and $x \in K^\times$ such that $\Upsilon_{L/K}(\theta) = x\,\mathsf{N}_{L/K}L^\times$. Then $(x, L/K) = \sigma$.*

In particular, $\Upsilon_{L/K}$ induces an isomorphism $\overline{\Upsilon}_{L/K} \colon G \to K^\times/\mathsf{N}_{L/K}L^\times$ which satisfies

$$\overline{\Upsilon}_{L/K}(\theta \restriction G) = \Upsilon_{L/K}(\theta) \text{ for all } \theta \in \mathrm{Frob}(L/K), \text{ and } \overline{\Upsilon}_{L/K}^{-1} = (\,\cdot\,,\, L/K).$$

Proof. We prove first:

> **A.** There exist a finite field extension $\overline{L}/L$ and intermediate fields L_1, L_2 of $\overline{L}/K$ such that $\overline{L} = L_1L_2$, $\overline{L}/K$ is abelian, L_1/K is fully ramified, and L_2/K is unramified.

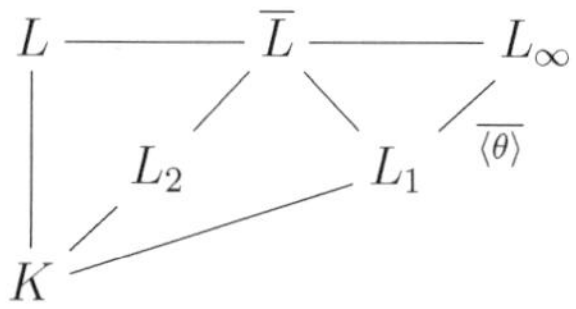

Proof of **A.** Recall that $L_\infty = LK_\infty$ by 4.3.6.**2**, and thus L_∞/K is abelian. Let $\theta \in \mathrm{Gal}(L_\infty/K)$ be such that $\theta \restriction K_\infty = \varphi_K$, set $L_1 = L_\infty^{\langle\theta\rangle}$. By 4.4.4.1 (note that $L_1 = \Sigma_\theta$), we obtain $[L_1 \colon K] < \infty$, $L_1 \cap K_\infty = K$, and $(L_1)_\infty = L_\infty$, L_∞/L_1 is unramified and L_1/K is fully ramified. We set $\overline{L} = LL_1 \subset L_\infty$ and denote by L_2 the inertia field of $\overline{L}/K$. Then $\overline{L}/K$ is abelian, $\overline{L}/L$ is unramified, $\overline{L}/L_1$ is unramified and $\overline{L}/L_2$ is fully ramified. As $\overline{L}/L_1L_2$ is both unramified and fully ramified, we get $\overline{L} = L_1L_2$. $\square$**A.**

Now we can do the very proof of Theorem 4.4.7. We repeatedly apply 4.4.6. Let $\theta \in \mathrm{Frob}(L/K)$, $\theta \restriction L = \sigma$, $x \in K^\times$ and $\Upsilon_{L/K}(\theta) = x\,\mathsf{N}_{L/K}L^\times$. Let $\overline{\theta} \in \mathrm{Frob}(\overline{L}/K)$ be such that $\overline{\theta} \restriction L_\infty = \theta$, and set $\overline{\sigma} = \overline{\theta} \restriction \overline{L}$ (then $\overline{\sigma} \restriction L = \sigma$). If $\overline{x} \in K^\times$ is such that $\Upsilon_{\overline{L}/K}(\overline{\theta}) = \overline{x}\,\mathsf{N}_{\overline{L}/K}\overline{L}^\times$, then $\overline{x}\,\mathsf{N}_{L/K}L^\times = x\,\mathsf{N}_{L/K}L^\times$.

For $i \in \{1, 2\}$ let

$$\theta_i = \overline{\theta} \restriction (L_i)_\infty, \sigma_i = \theta_i \restriction L_i, x_i \in K^\times, \text{ and } \Upsilon_{L_i/K} = x_i\,\mathsf{N}_{L_i/K}L_i^\times.$$

Then $\sigma_i = \overline{\sigma} \restriction L_i$ and $\overline{x}\,\mathsf{N}_{L_i/K}L_i^\times = x_i\,\mathsf{N}_{L_i/K}L_i^\times$. By 4.4.4.4, we obtain

$$\overline{\sigma} \restriction L_i = \sigma_i = (x_i, L_i/K) = (x_i, \overline{L}/K) \restriction L_i = (\overline{x}, \overline{L}/K) \restriction L_i,$$

and therefore $\overline{\sigma} = (\overline{x}, \overline{L}/K)$, since $\overline{L} = L_1L_2$. Now it eventually follows that

$$\sigma = \overline{\sigma} \restriction L = (\overline{x}, L/K) = (x, L/K). \qquad \square$$

4.5 Basics of formal groups

The remainder of the section is devoted to the theory of Lubin-Tate. This theory provides an explicit construction of the maximal abelian extension of a local field of characteristic 0. The main idea is to use division points of formal groups instead of roots of unity and to construct a sort of generalized cyclotomic fields with them. This construction has the essential advantage to deliver an explicit description of the norm residue symbol for these generalized cyclotomic fields.

As in the previous section, we describe this approach inside our algebra-theoretic foundation of local class field theory. A development of local class field theory by formal groups alone was presented by K. Iwasawa [26].

We start with preliminaries of the theory of formal power series.

Let R be a commutative ring. We consider power series rings $R[\![X]\!]$, $R[\![X,Y]\!]$, $R[\![X,Y,Z]\!]$, $R[\![T]\!]$ etc., and, more generally, $R[\![\boldsymbol{X}]\!]$, where $w\boldsymbol{X} = (X_1,\ldots,X_n)$ for some $n \in \mathbb{N}$. For a multiindex $\boldsymbol{i} = (i_1,\ldots,i_n) \in \mathbb{N}_0^n$ we set $|\boldsymbol{i}| = i_1 + \ldots + i_n$ and $\boldsymbol{X}^{\boldsymbol{i}} = X_1^{i_1} \cdot \ldots \cdot X_n^{i_n}$. We write power series $f \in R[\![X_1,\ldots,X_n]\!] = R[\![\boldsymbol{X}]\!]$ in the form

$$f = \sum_{i_1,\ldots,i_n=0}^{\infty} a_{i_1,\ldots,i_n} X_1^{i_1} \cdot \ldots \cdot X_n^{i_n} = \sum_{i_1,\ldots,i_n \geq 0} a_{i_1,\ldots,i_n} X_1^{i_1} \cdot \ldots \cdot X_n^{i_n} = \sum_{\boldsymbol{i} \in \mathbb{N}_0^n} a_{\boldsymbol{i}} \boldsymbol{X}^{\boldsymbol{i}},$$

where $a_{i_1,\ldots,i_n} = a_{\boldsymbol{i}} \in R$.

The assignment $f \mapsto f_{(0,\ldots,0)} = f_{\boldsymbol{0}} = f(\boldsymbol{0})$ defines a ring epimorphism $R[\![\boldsymbol{X}]\!] \to R$ with kernel

$$\mathfrak{M}_R(\boldsymbol{X}) = {}_{R[\![\boldsymbol{X}]\!]}\langle X_1,\ldots,X_n\rangle = \{f \in R[\![\boldsymbol{X}]\!] \mid f(\boldsymbol{0}) = 0\}.$$

For $d \in \mathbb{N}$, the power series $f \in \mathfrak{M}_R(\boldsymbol{X})^d$ are of the form

$$f = \sum_{\substack{\boldsymbol{i} \in \mathbb{N}_0^n \\ |\boldsymbol{i}| \geq d}} a_{\boldsymbol{i}} \boldsymbol{X}^{\boldsymbol{i}}, \quad \text{where } a_{\boldsymbol{i}} \in R.$$

In particular, $\mathfrak{M}_R(X)^d = X^d R[\![X]\!]$ for all $d \in \mathbb{N}$.

If $m, n \in \mathbb{N}$, $f \in R[\![X_1,\ldots,X_n]\!]$, $i \in [1,n]$, and $g \in \mathfrak{M}_R(Y_1,\ldots,Y_m)$, then we can replace X_i in f with g and obtain a power series

$$f(X_1,\ldots,X_{i-1},g(Y_1,\ldots,Y_m),X_{i+1},\ldots,X_n) \in R[\![X_1,\ldots,X_n,Y_1,\ldots,Y_m]\!],$$

In particular, every power series $f \in R[\![X]\!]$ defines an (equally denoted) map

$$f\colon \mathfrak{M}_R(\boldsymbol{X}) \to R[\![\boldsymbol{X}]\!], \quad \text{given by} \quad g \mapsto f(g) \ \text{ for all } \ g \in R[\![\boldsymbol{X}]\!].$$

For power series $f \in R[\![X]\!]$ and $g \in XR[\![X]\!]$ we shall often write $f \circ g$ instead of $f(g)$.

For a power series $f \in R[\![X]\!]$ we define its derivative in the usual way:

$$\text{If} \quad f = \sum_{i \ge 0} a_i X^i, \quad \text{then} \quad f' = \frac{df}{dX} = \sum_{i \ge 1} i a_i X^{i-1}.$$

The usual calculation rules hold: If $f,\ g \in R[\![X]\!]$ and $c \in R$, then $(cf)' = cf'$, $(f+g)' = f' + g'$, $(fg)' = f'g + fg'$, and $(f \circ g)' = (f' \circ g)g'$ if $g \in XR[\![X]\!]$. Moreover, $f \in R[\![X]\!]^\times$ if and only if $f(0) \in R^\times$, and then

$$\Big(\frac{g}{f}\Big)' = \frac{g'f - gf'}{f^2}.$$

If $\operatorname{char}(R) = 0$, then $f' = 0$ implies $f \in R$ (see [18, Theorem 5.3.8]). More general, for a power series $f \in R[\![X_1, \ldots, X_n]\!]$ we define its partial derivatives

$$\frac{\partial f}{\partial X_i} \in R[\![X_1, \ldots, X_n]\!] \quad \text{for} \quad i \in [1, n]$$

in the obvious way.

Definition and Theorem 4.5.1. *Let R be a commutative ring. For a power series $f \in XR[\![X]\!]$ the following assertions are equivalent:*

(a) There exists a (unique) power series $g \in XR[\![X]\!]$ such that

$$f \circ g = g \circ f = X.$$

(b) $f'(0) \in R^\times$ [i.e., $f \equiv aX \mod X^2 R[\![X]\!]$ for some $a \in R^\times$].

If these conditions hold, then f is called **composition-invertible**, and he power series $g = f^{-1}$ is called its **composition-inverse**.

Proof. Let $f \in XR[\![X]\!]$.

(a) $\Rightarrow$ (b) If $g \in XR[\![X]\!]$ and $g \circ f = X$, then $(g \circ f)' = (g' \circ f)f' = 1$, hence $(g' \circ f)(0)f'(0) = 1$, and thus $f'(0) \in R^\times$.

(b) $\Rightarrow$ (a) We show first:

A. For every $m \ge 0$ there exists a unique polynomial $g_m \in XR[X]$ such that $\partial(g_m) \le m$ and $f(g_m) \equiv X \mod X^{m+1} R[\![X]\!]$. In addition, it follows that $g_{m+1} \equiv g_m \mod X^m R[\![X]\!]$ for all $m \ge 0$.

Proof of **A.** Induction on m.

$m = 0$: $g_0 = 0$ is the only constant polynomial satisfying the requirements.

$m \ge 0$, $m \to m+1$: Let $g_m \in XR[X]$ be the unique polynomial which satisfies $\partial(g_m) \le m$ and $f(g_m) \equiv X \mod X^{m+1} R[\![X]\!]$. Then there exists a unique $b \in R$ such that $f(g_m) \equiv X + bX^{m+1} \bmod X^{m+2}R[X]$, and we set $g_{m+1} = h_m + cX^{m+1}$, where $h_m \in R[X]$, $\partial(h_m) \le m$ and $c \in R$. Hence

$$f(g_{m+1}) \equiv f(h_m) + cX^{m+1}f'(h_m) \equiv f(h_m) + cf'(0)X^{m+1} \mod X^{m+2}R[X],$$

and if $f(g_{m+1}) \equiv X \bmod X^{m+2}R[X]$, then $f(h_m) \equiv X \mod X^{m+1}R[X]$ and therefore $h_m = g_m$ by the induction hypothesis. If $h_m = g_m$, then

$$f(g_{m+1}) \equiv f(g_m) + cf'(0)X^{m+1} \equiv X + (b + cf'(0))X^{m+1} \mod X^{m+2}R[X].$$

Altogether we obtain $f(g_{m+1}) \equiv X \bmod X^{m+2}R[X]$ if and only if $h_m = g_m$ and $c = -f'(0)^{-1}b$. $\square$[**A.**]

Let $(g_m)_{m\ge 0}$ be as in **A**. Then there exists a power series $g \in R[\![X]\!]$ such that $g \equiv g_m \bmod X^{m+1}R[\![X]\!]$, hence $f(g) \equiv f(g_m) \equiv X \bmod X^{m+1}R[\![X]\!]$ for all $m \ge 0$, and consequently $f(g) = X$.

As to uniqueness, let $\widetilde{g} \in XR[\![X]\!]$ such that $f(\widetilde{g}) = X$. For $m \ge 0$ let $\widetilde{g}_m \in R[X]$ be the unique polynomial satisfying

$$\partial(\widetilde{g}) \le m \text{ and } \widetilde{g} \equiv \widetilde{g}_m \mod X^{m+1}R[\![X]\!].$$

Then $f(\widetilde{g}_m) \equiv f(\widetilde{g}) \equiv X \mod X^{m+1}R[\![X]\!]$, hence $\widetilde{g}_m \equiv g_m \mod X^{m+1}R[\![X]\!]$ by the uniqueness in **A**, and consequently $\widetilde{g} = g$.

If $g \in XR[\![X]\!]$ is such that $f(g) = X$, then $f'(g)g' = 1$, which implies that $f'(0)g'(0) = 1$, and $g'(0) \in R^\times$. Hence there exists a power series $h \in XR[X]$ such that $g(h) = X$. Then $h = f(g(h)) = f(X) = f$, $g(f) = X$, and f is composition-invertible. $\square$

Theorem 4.5.2. *Let K be a complete discrete valued field, $n \in \mathbb{N}$ and*

$$g = \sum_{\boldsymbol{i} \in \mathbb{N}_0^n} a_{\boldsymbol{i}} \boldsymbol{X}^{\boldsymbol{i}} \in \mathcal{O}_K[\![\boldsymbol{X}]\!], \quad \textit{where } \boldsymbol{X} = (X_1, \ldots, X_n).$$

For $\boldsymbol{x} = (x_1, \ldots, x_n)^{\mathrm{t}} \in \mathfrak{p}_K^n$ and $\boldsymbol{i} = (i_1, \ldots, i_n) \in \mathbb{N}_0^n$ we set $\boldsymbol{x}^{\boldsymbol{i}} = x_1^{i_1} \cdot \ldots \cdot x_n^{i_n}$. Then the series

$$g(\boldsymbol{x}) = \sum_{\boldsymbol{i} \in \mathbb{N}_0^n} a_{\boldsymbol{i}} \boldsymbol{x}^{\boldsymbol{i}} = \sum_{k=0}^{\infty} \sum_{\substack{\boldsymbol{i} \in \mathbb{N}_0^n \\ |\boldsymbol{i}| = k}} a_{\boldsymbol{i}} \boldsymbol{x}^{\boldsymbol{i}}$$

converges in $\mathcal{O}_K$, and the (equally denoted) function $g\colon \mathfrak{p}_K^n \to \mathcal{O}_K$, given by $\boldsymbol{x} \mapsto g(\boldsymbol{x})$, is conntinuous. Moreover, if $g \in \mathfrak{M}_K(\boldsymbol{X})$, then $g(\mathfrak{p}_K^n) \subset \mathfrak{p}_K$.

Proof. If $\boldsymbol{x} = (x_1, \ldots, x_n)^{\mathrm{t}} \in \mathfrak{p}_K^n$ then $|\boldsymbol{x}| = \max\{|x_1|, \ldots, |x_n|\} < 1$, and

$$\Bigl| \sum_{\substack{\boldsymbol{i} \in \mathbb{N}_0 \\ |\boldsymbol{i}| = k}} a_{\boldsymbol{i}} \boldsymbol{x}^{\boldsymbol{i}} \Bigr| \le |\boldsymbol{x}|^k \ \to 0 \ \text{ as } \ k \to \infty.$$

Hence the series converges in $\mathcal{O}_K$. If $(\boldsymbol{x}_\nu = (x_{1,\nu}, \ldots, x_{n,\nu})^{\mathrm{t}})_{\nu \ge 0}$ is a sequence in $\mathfrak{p}_K^n$ and $(\boldsymbol{x}_\nu)_{\nu \ge 0} \to \boldsymbol{x} = (x_1, \ldots, x_n)^{\mathrm{t}} \in \mathfrak{p}_K^n$, then $(\boldsymbol{x}_\nu^{\boldsymbol{i}})_{\nu \ge 0} \to \boldsymbol{x}$ for all $\boldsymbol{i} \in \mathbb{N}_0^n$, and

$$g(\boldsymbol{x}_\nu) - g(\boldsymbol{x}) = \sum_{k=0}^{\infty} \sum_{\substack{\boldsymbol{i} \in \mathbb{N}_0^n \\ |\boldsymbol{i}|=k}} a_{\boldsymbol{i}}(\boldsymbol{x}_\nu^{\boldsymbol{i}} - \boldsymbol{x}^{\boldsymbol{i}}) \quad \text{for all} \quad \nu \ge 0.$$

Let $\varepsilon \in \mathbb{R}_{>0}$. Suppose that $\mathfrak{p}_K = \pi\mathcal{O}_K$, let $k_0 \in \mathbb{N}$ be such that $|\pi|^{nk_0} < \varepsilon/2$, and let $\nu_0 \ge 0$ be such that $|a_{\boldsymbol{i}}(\boldsymbol{x}_\nu^{\boldsymbol{i}} - \boldsymbol{x}^{\boldsymbol{i}})| < \varepsilon/2$ for all $\boldsymbol{i} \in \mathbb{N}_0^n$ such that $|\boldsymbol{i}| \le k_0$. Then it follows that $|g(\boldsymbol{x}_\nu) - g(\boldsymbol{x})| < \varepsilon$, and thus $(g(\boldsymbol{x}_\nu))_{\nu \ge 0} \to g(\boldsymbol{x})$. If $g \in \mathfrak{M}_K(\boldsymbol{X})$, then $a_{\boldsymbol{0}} = 0$, hence $a_{\boldsymbol{i}}\boldsymbol{x}^{\boldsymbol{i}} \in \mathfrak{p}_K$ for all $\boldsymbol{i} \in \mathbb{N}_0^n$, and therefore $g(\boldsymbol{x}) \in \mathfrak{p}_K$. □

Definition and Remarks 4.5.3. Let R be a commutative ring.

1. A **formal group** (over R) is a power series $F = F(X,Y) \in R[\![X,Y]\!]$ with the following properties:

- $F(X,Y) \equiv X + Y \mod \mathfrak{M}_R(X,Y)^2$.
- $F(X,Y) = F(Y,X)$ (commutative law).
- $F(X, F(Y,Z)) = F(F(X,Y), Z)$ (associative law).

2. Let F, $G \in R[\![X,Y]\!]$ be formal groups. A power series $f \in XR[\![X]\!]$ is called a **morphism** from F to G (over R), and we write $f\colon F \to G$ if

$$f(F(X,Y)) = G(f(X), f(Y)).$$

If there is no doubt regarding the ring R, we shall omit the suffix "over R".

Suppose that $f\colon F \to G$ and $g\colon G \to H$ are morphisms of formal groups. Then $g \circ f\colon F \to H$ is a morphism.

For every formal group F, $1_F = X\colon F \to F$ is a morphism, called the **identity** of F. If $f\colon F \to G$ is a morphism of formal groups, then we have $f \circ 1_F = 1_G \circ f$, and f is called an **isomorphism** if it is composition-invertible (then $f^{-1}\colon G \to F$ is a morphism, $f^{-1} \circ f = 1_F$ and $f \circ f^{-1} = 1_G$).

3. A morphism $f\colon F \to F$ of a formal group F (over R) into itself is called an **endomorphism** of F (over R). We denote by $\mathrm{End}_R(F)$ the set of all endomorphisms of F over R.

4. We call $G_{\mathsf{a}} = X+Y$ the **additive formal group** and $G_{\mathsf{m}} = X+Y+XY$ the **multiplicative formal group**.

5. We consider the power series

$$E(X) = \sum_{n=0}^{\infty} \frac{1}{n!} X^n \in \mathbb{Q}[\![X]\!] \quad \text{and}$$

$$L(X) = \log(1+X) = \sum_{n=1}^{\infty} \frac{(-1)^n}{n} X^n \in \mathbb{Q}[\![X]\!].$$

The power series $\mathsf{e}(X) = E(X) - 1$ and $L(X)$ are mutually composition-inverse, i. e., $\mathsf{e} \circ L = L \circ \mathsf{e} = X$, and they satisfy the identities

$$E'(X) = E(X), \quad (1+X)L'(X) = 1, \quad E(X+Y) = E(X)E(Y),$$

$$\mathsf{e}(X+Y) = \mathsf{e}(X) + \mathsf{e}(Y) + \mathsf{e}(X)\mathsf{e}(Y) \quad \text{and} \quad L(X+Y+XY) = L(X) + L(Y)$$

(see [18, Theorem 5.3.8]). In particular, $\mathsf{e}\colon G_{\mathsf{a}} \to G_{\mathsf{m}}$ and $L\colon G_{\mathsf{m}} \to G_{\mathsf{a}}$ are mutually inverse isomorphisms of formal groups.

Theorem 4.5.4. *Let R be a commutative ring and $F \in R[\![X,Y]\!]$ a formal group.*

1. *$F(X,Y) = X+Y+XYF_1$ for some power series $F_1 \in R[\![X,Y]\!]$.*

2. *There exists a unique power series $j_F \in XR[\![X]\!]$ such that $F(X, j_F(X)) = 0$.*

Proof. 1. We set $F = XA + YB + XYF_1$, where $F_1 \in R[\![X,Y]\!]$, $A \in R[\![X]\!]$ and $B \in R[\![Y]\!]$. Now $F(X,Y) \equiv X + Y \bmod \mathfrak{M}_R(X,Y)^2$, and consequently we get $A(0) = B(0) = 1$. Therefore $F(X,0) = XA$ is not a zero divisor, and by the associative law we obtain

$$F(F(X,0),0) = F(X,0)A(F(X,0)) = F(X, F(0,0)) = F(X,0).$$

Hence it follows that $A(F(X,0)) = 1$.

If $A \neq 1$, then $A = 1 + X^n A_1$, where $n \in \mathbb{N}$ and $A_1 \in R[\![X]\!]$ is such that $A_1(0) \neq 0$. As $F(X,0) \equiv X \bmod X^2R[\![X]\!]$, we get

$$1 = A(F(X,0)) \equiv 1 + X^n A_1(0) \mod X^{n+1}R[\![X]\!],$$

and consequently $A_1(0) \equiv 0 \bmod XR[\![X]\!]$, a contradiction. Hence $A = 1$, and by the commutative law it follows that also $B = 1$.

2. We show first:

A. For every $m \geq 0$ there exists a unique polynomial $j_m \in XR[X]$ such that $\partial(j_m) \leq m$ and $F(X, j_m) \in X^{m+1}R[\![X]\!]$. Over and above, it follows that $j_{m+1} \equiv j_m \bmod X^{m+1}R[X]$ for all $m \in \mathbb{N}$.

Proof of **A.** By induction on m.

$m=0$: $j_0=0$ is the only constant polynomial satisfying the requirements.

$m\ge 0$, $m\to m+1$: Let $j_m\in XR[X]$ be the unique polynomial such that $\partial(j_m)\le m$ and $F(X,j_m)\in X^{m+1}R[\![X]\!]$, say $F(X,j_m)=X^{m+1}h$ for some $h\in R[\![X]\!]$. We set $j_{m+1}=k+cX^{m+1}$, where $k\in XR[X]$, $\partial(k)\le m$ and $c\in R$. If then $F(X,j_{m+1})\in X^{m+2}R[\![X]\!]$, then $F(X,k)\in X^{m+1}R[\![X]\!]$ and therefore $k=j_m$ by the induction hypothesis. If $j_{m+1}=j_m+cX^{m+1}$ for some $c\in R$, then

$$\begin{aligned}F(X,j_{m+1})&\equiv F(X,j_m)+cX^{m+1}\frac{\partial F}{\partial Y}(X,j_m)\\&\equiv X^{m+1}(h(0)+c)\bmod X^{m+2}R[\![X]\!],\end{aligned}$$

since

$$\frac{\partial F}{\partial Y}(X,j_m)=1+XF_1(X,j_m)+Xj_m\frac{\partial F_1}{\partial Y}(X,j_m)\equiv 1\bmod XR[\![X]\!].$$

Altogether we obtain $F(X,j_{m+1})\in X^{m+2}R[\![X]\!]$ if and only if $k=j_m$ and $c=-h(0)$. $\square$[**A.**]

Let $(j_m)_{m\ge 0}$ be as in **A**. Then there exists a power series $j_F\in XR[\![X]\!]$ such that $j_F\equiv j_m \bmod X^{m+1}R[\![X]\!]$, hence

$$F(X,j_F)\equiv F(X,j_m)\equiv 0\mod X^{m+1}R[\![X]\!]\text{ for all }m\in\mathbb{N},$$

and consequently $F(X,j_F)=0$.

As to uniqueness, let $\widetilde{j}\in R[\![X]\!]$ be any power series satisfying $F(X,\widetilde{j})=0$. For $m\ge 0$ let $\widetilde{j}_m\in R[X]$ be the unique polynomial such that $\partial(\widetilde{j}_m)\le m$ and $\widetilde{j}\equiv\widetilde{j}_m \bmod X^{m+1}R[\![X]\!]$. Then $F(X,\widetilde{j}_m)\equiv F(X,\widetilde{j})\equiv 0\bmod X^{m+1}R[\![X]\!]$, hence $\widetilde{j}_m\equiv j_m\bmod X^{m+1}R[\![X]\!]$ for all $m\ge 0$ by the uniqueness in **A**, and consequently $\widetilde{j}=j$. $\square$

Definition and Remarks 4.5.5. Let R be a commutative ring. Let $F\in R[\![X,Y]\!]$ be a formal group, $n\in\mathbb{N}$ and $\boldsymbol{X}=(X_1,\ldots,X_n)$. For power series $f,\,g\in\mathfrak{M}_R(\boldsymbol{X})$ we define their F-**sum** by

$$f+_F g=F(f,g)\in\mathfrak{M}_R(\boldsymbol{X}).$$

For $f,\,g,\,h\in\mathfrak{M}_R(\boldsymbol{X})$ it follows that

$$f+_F(g+_F h)=F(f,F(g,h))=F(F(f,g),h)=f+_F(g+_F h),$$

$$f+_F 0=F(f,0)=f\quad\text{and}\quad f+_F j_F(f)=F(f,j_F(f))=F(X,j_F)(f)=0$$

with $j_F\in XR[\![X]\!]$ as in 4.5.4.2. Hence

$$\mathfrak{M}_R(\boldsymbol{X})_F=(\mathfrak{M}_R(\boldsymbol{X}),+_F)$$

is an abelian group with zero $0 \in \mathfrak{M}_R(\boldsymbol{X})$, and for $f \in \mathfrak{M}_R(\boldsymbol{X})$ its negative is given by $-_F f = j_F(f)$.

Let F, $G \in R[\![X,Y]\!]$ be formal groups and $f \in XR[\![X]\!]$. By definition, $f(F(X,Y)) = G(f(X), f(Y))$ if and only if $f(x +_F y) = f(x) +_G f(y)$ for all $x, y \in \mathfrak{M}_R[\![\boldsymbol{X}]\!]$. Hence $f\colon F \to G$ is a morphism of formal groups if and only if $f\colon \mathfrak{M}_R(\boldsymbol{X})_F \to \mathfrak{M}_R(\boldsymbol{X})_G$ is a group homomorphism.

If f, $g\colon F \to G$ are morphisms of formal groups, then it is easily checked that

$$(f +_G g)(X +_F Y) = (f +_G g)(X) +_G (f +_G g)(Y).$$

Hence $f +_G g\colon F \to G$ is also a morphism of formal groups. If $e\colon E \to F$ and $h\colon G \to H$ are further morphisms of formal groups, then

$$(f +_F g)\circ e = f\circ e +_F g\circ e \text{ and } h\circ(f +_G g) = h\circ f +_H g\circ f.$$

In particular, $\mathrm{End}_R(F) = (\mathrm{End}_R(F), +_F, \circ)$ is a ring, called the **endomorphism ring** of F.

Definition and Remarks 4.5.6. Let R be a commutative ring.

1. A **formal R-module** $F = (F, [\,\cdot\,]_F)$ is comprised of a formal group $F \in R[\![X,Y]\!]$ together with a ring homomorphism

$$[\,\cdot\,]_F\colon R \to \mathrm{End}_R(F), \ a \mapsto [a]_F \in XR[\![X]\!],$$

such that $[a]_F \equiv aX \mod X^2R[\![X]\!]$ for all $a \in R$.

Let $(F, [\,\cdot\,]_F)$ be a formal R-module and a, $b \in R$. Then the following assertions follow immediately from the definition:

- If $[a]_F = 0$, then $a = 0$. Hence $[\,\cdot\,]_F\colon R \to \mathrm{End}_R(F)$ is a monomorphism.
- $[a+b]_F = [a]_F +_F [b]_F = F([a]_F, [b]_F) \equiv [a]_F + [b]_F \equiv (a+b)X \mod X^2R[\![X]\!]$
- $[ab]_F = [a]_F \circ [b]_F = [a]_F([b]_F) \equiv [a]_F(bX) \equiv abX \mod X^2R[\![X]\!]$.
- If $a \in R^\times$, then $[a]_F \in \mathrm{Aut}_R(F) = \mathrm{End}_R(F)^\times$, and $[a]_F^{-1} = [a^{-1}]_F$.

The additive formal group G_{a} is a formal R-module by means of $[a]_{G_{\mathsf{a}}} = aX$ for all $a \in R$.

2. Let F and G be formal R-modules. A **morphism** of formal R-modules is a morphism $f\colon F \to G$ of formal groups over R satisfying $f\circ[a]_F = [a]_G\circ f$ for all $a \in R$. If a morphism $f\colon F \to G$ of formal R-modules is an isomorphism of formal groups, then its composition-inverse $f^{-1}\colon G \to F$ is also a morphism of formal R-modules, and in this case F and G are called **isomorphic** formal R-modules.

4.6 Lubin-Tate formal groups

Let K be a local field, $\mathrm{char}(\mathsf{k}_K) = p$ *and* $\mathsf{k}_K = \mathbb{F}_q$

Definition 4.6.1. Let $\pi \in \mathcal{O}_K$ be a prime element.

1. A power series $e = e(X) \in X\mathcal{O}_K[\![X]\!]$ is called a **π-series** if

$$e \equiv \pi X \mod X^2\mathcal{O}_K[\![X]\!] \quad \text{and} \quad e \equiv X^q \mod \pi\mathcal{O}_K[\![X]\!]$$

[equivalently, $e = X^q + \pi X u$ for some power series $u \in 1 + X\mathcal{O}[\![X]\!]$]. A π-series lying in $X\mathcal{O}_K[X]$ is called a **π-polynomial**.

Let $e \in X\mathcal{O}_K[\![X]\!]$ be a π-series. For $n \in \mathbb{N}_0$ we define the **n-fold iterated π-series** $e_n \in X\mathcal{O}_K[\![X]\!]$ of e by $e_0 = X$ and $e_n = e_{n-1}\circ e = e\circ\ldots\circ e$ for all $n \geq 1$.

2. A **π-module** over $\mathcal{O}_K$ is a formal $\mathcal{O}_K$-module $F = (F, [\,\cdot\,]_F)$ such that the power series $[\pi]_F \in X\mathcal{O}_K[\![X]\!]$ is a π-series.

Lemma 4.6.2. *Let $\pi \in \mathcal{O}_K$ be a prime element.*

1. *If $g \in \mathcal{O}_K[\![X_1, \ldots, X_n]\!]$, then $g^q \equiv g(X_1^q, \ldots, X_n^q) \mod \pi\mathcal{O}_K[\![X_1, \ldots, X_n]\!]$.*

2. *Let $e \in X\mathcal{O}_K[\![X]\!]$ be a π-series, and let $g \in X\mathcal{O}_K[\![X]\!]$ be composition-invertible. Then $g\circ e\circ g^{-1}$ is a π-series.*

Proof. 1. Suppose that $\boldsymbol{X} = (X_1, \ldots, X_n)$ and

$$g = \sum_{\boldsymbol{i}\in\mathbb{N}_0^n} a_{\boldsymbol{i}}\boldsymbol{X}^{\boldsymbol{i}}, \quad \text{where } \boldsymbol{X}^{\boldsymbol{i}} = X_1^{i_1}\cdot\ldots\cdot X_n^{i_n} \text{ for all } \boldsymbol{i} = (i_1, \ldots, i_n) \in \mathbb{N}_0^n.$$

The power series converges in the $\mathfrak{M}_{\mathcal{O}_X}$-adic topology, the map $(x \mapsto x^q)$ is continuous, and $a^q \equiv a \mod \pi\mathcal{O}_K$ for all $a \in \mathcal{O}_K$. Hence it follows that

$$g^q \equiv \sum_{\boldsymbol{i}\in\mathbb{N}_0^n} a_{\boldsymbol{i}}^q\boldsymbol{X}^{\boldsymbol{i}q} \equiv \sum_{\boldsymbol{i}\in\mathbb{N}_0^n} a_{\boldsymbol{i}}\boldsymbol{X}^{\boldsymbol{i}q} \equiv g(\boldsymbol{X}^q) \mod \pi\mathcal{O}_K[\![\boldsymbol{X}]\!],$$

where $\boldsymbol{X}^q = (X_1^q, \ldots, X_n^q)$.

2. If $a \in U_K$ and $g \equiv aX \mod X^2\mathcal{O}_K[\![X]\!]$, then it follows that $g^{-1} \equiv a^{-1}X \mod X^2\mathcal{O}_K[\![X]\!]$,

$$g\circ e\circ g^{-1} \equiv g\circ e(a^{-1}X)) \equiv g(\pi a^{-1}X) \equiv a\pi a^{-1}X \equiv \pi X \mod X^2\mathcal{O}_K[\![X]\!],$$

and, using 1.,

$$g\circ e\circ g^{-1} = g\circ e(g^{-1}) \equiv g((g^{-1})^q) \equiv [g(g^{-1})]^q \equiv X^q \mod \pi\mathcal{O}_K[\![X]\!]. \qquad \square$$

Definition and Theorem 4.6.3. *Let p be a prime.*
For $a \in \mathbb{Z}_p$ and $n \in \mathbb{N}_0$ we define

$$\binom{a}{n} = \frac{a(a-1)\cdot\ldots\cdot(a-n+1)}{n!} \quad \text{and} \quad (1+X)^a = \sum_{n=0}^{\infty} \binom{a}{n} X^n \in \mathbb{Z}_p [\![X]\!].$$

(if $a \in \mathbb{Z}$, then $(1+X)^a \in \mathbb{Z}[\![X]\!]$ has the usual meaning). For $a \in \mathbb{Z}_p$, we define

$$[a]_{G_{\mathsf{m}}} = (1+X)^a - 1] \in \mathbb{Z}_p [\![X]\!].$$

Let $E(X)$ and $L(X) = \log(1+X)$ be as in 4.5.3.**5**.

1. *If $a \in \mathbb{Z}_p$, then $(1+X)^a = E(aL(X)) \in \mathbb{Z}_p [\![X]\!]$.*
2. *If $a,\, b \in \mathbb{Z}_p$ and $f,\, g \in X\mathbb{Z}_p [\![X]\!]$, then*
$$(1+f)^a(1+f)^b = (1+f)^{a+b},$$
$$(1+f)^a(1+g)^a = (1+f+g+fg)^a \quad \textit{and} \quad [(1+f)^a]^b = (1+f)^{ab}.$$
3. *The formal $\mathbb{Z}_p$-module $G_{\mathsf{m}} = (G_{\mathsf{m}}, [\,\cdot\,]_{G_{\mathsf{m}}})$ is a p-module.*

Proof. 1. If $a \in \mathbb{Z}_p$ and $(a_i)_{i\ge 0}$ is a sequence in $\mathbb{Z}$ such that $(a_i)_{i\ge 0} \to a$, then

$$\binom{a}{n} = \lim_{i\to\infty} \binom{a_i}{n} \in \mathbb{Z}_p \ \text{ for all } \ n \in \mathbb{N}_0, \quad \text{and therefore} \quad (1+X)^a \in \mathbb{Z}_p [\![X]\!].$$

If $f(X) = (1+X)^a$ and $g(X) = E(aL(X)) \in \mathbb{Q}_p [\![X]\!]$, then $f(0) = g(0) = 1$,

$$(1+X)f'(X) = (1+X)\sum_{n=1}^{\infty} n\binom{a}{n} X^{n-1} = \sum_{n=0}^{\infty} \Big[(n+1)\binom{k}{n+1} + n\binom{a}{n}\Big] X^n$$
$$= \sum_{n=0}^{\infty} a\binom{a}{n} X^n = af(X),$$

and $(1+X)g'(X) = E(aL(X))aL'(X)(1+X) = ag(X)$. Furthermore,

$$\frac{f}{g} \in \mathbb{Q}_p [\![X]\!], \quad \Big(\frac{f}{g}\Big)' = \frac{fg' - f'g}{g^2} = 0, \quad \text{hence } \ \frac{f}{g} = \frac{f}{g}(0) = 1,$$

and consequently $f = g$.

2. Let $a,\, b \in \mathbb{Z}_p$ and $f,\, g \in X\mathbb{Z}_p [\![X]\!]$. Then

$$(1+f)^a(1+f)^b = E(aL(f))E(bL(f)) = E((a+b)L(f)) = (1+f)^{a+b},$$

$$\begin{aligned}(1+f)^a(1+g)^a &= E(aL(f))E(aL(g)) = E(a(L(f)+L(g))) \\ &= E(aL(f+g+fg)) = (1+f+g+fg)^a,\end{aligned}$$

and since $L((1+f)^a-1) = L(E(aL(f))-1) = aL(f)$, it follows that

$$[(1+f)^a]^b = (1+[(1+f)^a-1])^b = E(bL((1+f)^a-1)) = E(abL(f)) = (1+f)^{ab}.$$

3. If $a \in \mathbb{Z}_p$, then $[a]_{G_{\mathsf{m}}} \in \mathrm{End}_{\mathbb{Z}_p}(G_{\mathsf{m}})$, since

$$\begin{aligned}[a]_{G_{\mathsf{m}}}(X +_{G_{\mathsf{m}}} Y) &= [a]_{G_{\mathsf{m}}}(X+Y+XY) = (1+X+Y+XY)^a - 1\\ &= (1+X)^a(1+Y)^a - 1\\ &= [(1+X)^a-1] + [(1+Y)^a-1] + [(1+X)^a-1]\,[(1+Y)^a-1]\\ &= [a]_{G_{\mathsf{m}}}(X) + [a]_{G_{\mathsf{m}}}(Y) + [a]_{G_{\mathsf{m}}}(X)[a]_{G_{\mathsf{m}}}(Y) = [a]_{G_{\mathsf{m}}}(X) +_{G_{\mathsf{m}}} [a]_{G_{\mathsf{m}}}(Y).\end{aligned}$$

If $a,\, b \in \mathbb{Z}_p$, then

$$\begin{aligned}[a+b]_{G_{\mathsf{m}}}(X) &= (1+X)^{a+b}-1] = (1+X)^a(1+X)^b - 1\\ &= [(1+X)^a-1] + [(1+X)^b-1] + [(1+X)^a-1][(1+X)^b-1]\\ &= [a]_{G_{\mathsf{m}}}(X) + [b]_{G_{\mathsf{m}}}(X) + [a]_{G_{\mathsf{m}}}(X)[b]_{G_{\mathsf{m}}}(X)\\ &= [a]_{G_{\mathsf{m}}}(X) +_{G_{\mathsf{m}}} [b]_{G_{\mathsf{m}}}(X)\end{aligned}$$

and

$$\begin{aligned}[ab]_{G_{\mathsf{m}}}(X) &= [(1+X)^{ab} - 1 = ((1+X)^b)^a - 1\\ &= \big\{\big[1 + \big((1+X)^b - 1\big)\big]^a - 1\big\} = [a]_{G_{\mathsf{m}}} \circ [b]_{G_{\mathsf{m}}}(X).\end{aligned}$$

Accordingly the map $[\,\cdot\,]_{G_{\mathsf{m}}} \colon \mathbb{Z}_p \to \mathrm{End}_{\mathbb{Z}_p}(G_{\mathsf{m}})$ is a ring homomorphism, and $G_{\mathsf{m}} = (G_{\mathsf{m}}, [\,\cdot\,]_{G_{\mathsf{m}}})$ is a formal $\mathbb{Z}_p$-module. Since

$$[p]_{G_{\mathsf{m}}}(X) = (1+X)^p - 1] \equiv \begin{cases} pX \mod X^2\mathcal{O}_K\,[\![X]\!],\\ X^p \mod p\mathcal{O}_K\,[\![X]\!],\end{cases}$$

the power series $[p]_{G_{\mathsf{m}}}(X) \in \mathbb{Z}_p\,[\![X]\!]$ is a p-series and therefor $G_{\mathsf{m}} = (G_{\mathsf{m}}, [\,\cdot\,]_{G_{\mathsf{m}}})$ is a p-module. □

In the sequel we consider the maximal unramified extension K_∞ of K and its completion $\widehat{K}_\infty$, and we use the notatios fixed in 4.3.6. In particular, we denote by $v_K \colon \widehat{K}_\infty \to \mathbb{Z} \cup \{\infty\}$ the discrete valuation and by $\varphi \in \mathrm{Gal}(\widehat{K}_\infty/K)$ the contiuous extension of the Frobenius automorphism $\varphi_K \in \mathrm{Gal}(K_\infty/K)$. We use exponential notation for the action of φ. Explicitly, for $x \in \widehat{K}_\infty$ we set $x^\varphi = \varphi(x)$, and consequenly $x^\varphi \equiv x^q \mod \mathfrak{p}_{\widehat{K}_\infty}$ for all $x \in \mathcal{O}_{\widehat{K}_\infty}$.

Let $n \in \mathbb{N}$, and $\boldsymbol{X} = (X_1, \ldots, X_n)$ an n-tuple of indeterminates. For a power series

$$f = \sum_{\boldsymbol{i} \in \mathbb{N}_0^n} a_{\boldsymbol{i}} \boldsymbol{X}^{\boldsymbol{i}} \in \mathcal{O}_{\widehat{K}_\infty}\,[\![\boldsymbol{X}]\!] \quad \text{we define} \quad f^\varphi = \sum_{\boldsymbol{i} \in \mathbb{N}_0^n} a_{\boldsymbol{i}}^\varphi \boldsymbol{X}^{\boldsymbol{i}},$$

and for a power series $e \in \mathcal{O}_{\widehat{K}_\infty}\,[\![X]\!]$ we define $(e, \ldots, e) \in \mathcal{O}_{\widehat{K}_\infty}\,[\![\boldsymbol{X}]\!]$ by

$$(e, \ldots, e)(\boldsymbol{X}) = (e(X_1), \ldots, e(X_n)).$$

The following Theorem 4.6.4 is basic for the theory of Lubin-Tate. It equally holds and will be applied for the local field K itself and for the completion $\widehat{K}_\infty$ of its maximal unramified extension. This is the reason for its seemingly strange formulation.

Theorem 4.6.4 (Existence theorem of Lubin-Tate). *Let either $\mathcal{O} = \mathcal{O}_{\widehat{K}_\infty}$, or $\mathcal{O} = \mathcal{O}_{K'}$ for an over K finite intermediate field K' of K_∞/K (in any case $\mathcal{O}$ is a complete subring of $\mathcal{O}_{\widehat{K}_\infty}$, and $\mathcal{O}^\varphi = \mathcal{O}$). Let π and $\overline{\pi}$ be prime elements of $\mathcal{O}$, $e \in \mathcal{O}[\![X]\!]$ a π-series and $\overline{e} \in \mathcal{O}[\![X]\!]$ a $\overline{\pi}$-series. Let $n \in \mathbb{N}$, $\boldsymbol{X} = (X_1, \ldots, X_n)$ and $L = a_1X_1 + \ldots + a_nX_n \in \mathcal{O}[\boldsymbol{X}]$ a linear form satisfying $\overline{\pi}L = \pi L^\varphi$. Then there exists a unique power series $F \in \mathcal{O}[\![\boldsymbol{X}]\!]$ such that*

$$F \equiv L \mod \mathfrak{M}_{\mathcal{O}}(\boldsymbol{X})^2 \quad \textit{and} \quad \overline{e} \circ F = F^\varphi \circ (e, \ldots, e) \in \mathcal{O}[\![\boldsymbol{X}]\!].$$

Proof. We set $\mathfrak{M} = \mathfrak{M}_{\mathcal{O}}(\boldsymbol{X})$. It suffices to prove the following assertion:

> **A.** For every $r \in \mathbb{N}$ there exists a unique polynomial $F_r \in \mathcal{O}[\boldsymbol{X}]$ such that

$$\partial(F_r) \le r, \quad F_r \equiv L \mod \mathfrak{M}^2. \quad \text{and} \quad \overline{e} \circ F_r \equiv F_r^\varphi \circ (e, \ldots, e) \mod \mathfrak{M}^{r+1}.$$

Indeed, assume that **A** holds, and let $(F_r)_{r \ge 1}$ be a sequence of polnyomials as in **A**. The uniqueness in **A** implies $F_{r+1} \equiv F_r \mod \mathfrak{M}^{r+1}$ for all $r \in \mathbb{N}$, and therefore there exists a power series $F \in \mathcal{O}[\![\boldsymbol{X}]\!]$ such that $F \equiv F_r \mod \mathfrak{M}^{r+1}$ for all $r \in \mathbb{N}$. It follows that

$$F \equiv L \mod \mathfrak{M}^2 \text{ and } \overline{e} \circ F \equiv \overline{e} \circ F_r \equiv F_r^\varphi \circ (e, \ldots, e) \equiv F^\varphi \circ (e, \ldots, e) \mod \mathfrak{M}^{r+1}$$

for all $r \in \mathbb{N}$, hence $\overline{e} \circ F = F^\varphi \circ (e, \ldots, e)$.

To prove uniqueness, let $\widetilde{F} \in \mathcal{O}[\![\boldsymbol{X}]\!]$ be such that $\widetilde{F} \equiv L \mod \mathfrak{M}^2$ and $\overline{e} \circ \widetilde{F} = \widetilde{F}^\varphi \circ (e, \ldots, e)$. For $r \in \mathbb{N}$ let $\widetilde{F}_r \in \mathcal{O}[\boldsymbol{X}]$ be the unique polynomial such that $\partial(\widetilde{F}_r) \le r$ and $\widetilde{F} \equiv \widetilde{F}_r \mod \mathfrak{M}^{r+1}$. Then it follows that $\widetilde{F}_r \equiv L \mod \mathfrak{M}^2$ and $\overline{e} \circ \widetilde{F}_r \equiv \widetilde{F}_r^\varphi \circ (e, \ldots, e) \mod \mathfrak{M}^{r+1}$, the uniqueness in **A** implies $\widetilde{F}_r = F_r$ for all $r \in \mathbb{N}$, and consequently $\widetilde{F} = F$.

Proof of **A.** We use induction on r.

$r = 1$: Obviously $F_1 = L$ is the only polynomial satisfying $\partial(F_1) \le 1$ and $F_1 \equiv L \mod \mathfrak{M}^2$, and by the very definition it follows that

$$\overline{e} \circ L \equiv \overline{\pi} L \equiv \pi L^\varphi \equiv \sum_{i=1}^n \pi a_i^\varphi X_i \equiv \sum_{i=1}^n a_i^\varphi e(X_i) \equiv L^\varphi \circ (e, \ldots, e) \mod \mathfrak{M}^2.$$

$r \ge 1$, $r \to r+1$: Let $F_r \in \mathcal{O}[\boldsymbol{X}]$ be the unique polynomial satisfying $\partial(F_r) \le r$, $F_r \equiv L \mod \mathfrak{M}^2$ and $\overline{e} \circ F_r \equiv F_r^\varphi \circ (e, \ldots, e) \mod \mathfrak{M}^{r+1}$. Let $\mathcal{O}(r+1)$ be the $(r+1)$-st homogenous component of $\mathcal{O}[\boldsymbol{X}]$ (consisting of 0 and all homogenous polynomials of degree $r+1$). If $F_{r+1} \in \mathcal{O}[\boldsymbol{X}]$ is such that $\partial(F_{r+1}) \le r+1$,

$$F_{r+1} \equiv L \mod \mathfrak{M}^2 \text{ and } \overline{e} \circ F_{r+1} \equiv F_{r+1}^\varphi \circ (e, \ldots, e) \mod \mathfrak{M}^{r+2},$$

then we set $F_{r+1} = H + E$, where $H \in \mathcal{O}[\boldsymbol{X}]$ is a polynomial such that $\partial(H) \le r$ and $E \in \mathcal{O}(r+1)$. Then we get $H \equiv L \bmod \mathfrak{M}^2$ and $\overline{e} \circ H \equiv H^{\varphi} \circ (e, \ldots, e) \bmod \mathfrak{M}^{r+1}$, and by the uniqueness of F_r it follows that $H = F_r$.

Thus we formulate $F_{r+1} = F_r + E$ with a suitable polynomial

$$E = \sum_{\substack{\boldsymbol{i} \in \mathbb{N}_0^n \\ |\boldsymbol{i}| = r+1}} a_{\boldsymbol{i}} \boldsymbol{X}^{\boldsymbol{i}} \in \mathcal{O}(r+1). \tag{†}$$

Then

$$F_{r+1} \in \mathcal{O}[\boldsymbol{X}], \quad \partial(F_{r+1}) \le r+1, \quad F_{r+1} \equiv L \mod \mathfrak{M}^2,$$

and as $\overline{e}' \equiv \overline{\pi} \bmod X\mathcal{O}[\![X]\!]$, we get

$$\overline{e} \circ F_{r+1} = \overline{e} \circ (F_r + E) \equiv \overline{e}(F_r) + \overline{e}'(F_r)E \equiv \overline{e} \circ F_r + \overline{\pi} E \mod \mathfrak{M}^{r+2}.$$

On the other hand, we obtain

$$\begin{aligned} E^{\varphi} \circ (e, \ldots, e) &= \sum_{\substack{\boldsymbol{i} = (i_1, \ldots, i_n) \in \mathbb{N}_0^n \\ |\boldsymbol{i}| = r+1}} a_{\boldsymbol{i}}^{\varphi} e(X_1)^{i_1} \cdot \ldots \cdot e(X_n)^{i_n} \\ &\equiv \sum_{\substack{\boldsymbol{i} = (i_1, \ldots, i_n) \in \mathbb{N}_0^n \\ |\boldsymbol{i}| = r+1}} a_{\boldsymbol{i}}^{\varphi} (\pi X_1)^{i_1} \cdot \ldots \cdot (\pi X_n)^{i_n} \\ &\equiv \pi^{r+1} E^{\varphi} \mod \mathfrak{M}^{r+2} \end{aligned}$$

and

$$\begin{aligned} F_{r+1}^{\varphi} \circ (e, \ldots, e) &= F_r^{\varphi} \circ (e, \ldots, e) + E^{\varphi} \circ (e, \ldots, e) \\ &\equiv F_r^{\varphi} \circ (e, \ldots, e) + \pi^{r+1} E^{\varphi} \mod \mathfrak{M}^{r+2}. \end{aligned}$$

Consequently, the congruence $\overline{e} \circ F_{r+1} \equiv F_{r+1}^{\varphi} \circ (e, \ldots, e) \bmod \mathfrak{M}^{r+2}$ holds if and only if

$$\overline{e} \circ F_r + \overline{\pi} E \equiv F_r^{\varphi} \circ (e, \ldots, e) + \pi^{r+1} E^{\varphi} \mod \mathfrak{M}^{r+2}.$$

We consider the polynomial

$$G = F_r^{\varphi} \circ (e, \ldots, e) - \overline{e} \circ F_r \in \mathcal{O}[\![\boldsymbol{X}]\!].$$

The induction hypothesis implies $G \in \mathfrak{M}^{r+1}$ and $\partial(G) \le r+1$, hence we obtain $G \in \mathcal{O}(r+1)$, and we must prove:

(∗) There exists a unique polynomial $E \in \mathcal{O}(r+1)$ satisfying

$$\overline{\pi} E - \pi^{r+1} E^{\varphi} \equiv G \mod \mathfrak{M}^{r+2}.$$

Proof of (∗) We prove first that $G \in \pi\mathcal{O}[\boldsymbol{X}]$. To do so, we set

$$F_r = \sum_{\boldsymbol{i}\in\mathbb{N}_0^n} c_{\boldsymbol{i}}\boldsymbol{X}^{\boldsymbol{i}} \in \mathcal{O}[\boldsymbol{X}].$$

Since $e \equiv \overline{e} \equiv X^q \mod \pi\mathcal{O}\,[\![X]\!]$, we obtain

$$\begin{aligned} F_r^{\varphi}\circ(e,\ldots,e) &= \sum_{\boldsymbol{i}=(i_1,\ldots,i_n)\in\mathbb{N}_0^n} c_{\boldsymbol{i}}^{\varphi}\, e(X_1)^{i_1}\cdot\ldots\cdot e(X_n)^{i_n} \equiv \sum_{\boldsymbol{i}\in\mathbb{N}_0^n} c_{\boldsymbol{i}}^{q}\boldsymbol{X}^{q\boldsymbol{i}} \equiv F_r^q \\ &\equiv \overline{e}\circ F_r \mod \pi\mathcal{O}\,[\![\boldsymbol{X}]\!], \end{aligned}$$

and therefore $G = F_r^{\varphi}\circ(e,\ldots,e) - \overline{e}\circ F_r \in \pi\mathcal{O}[\boldsymbol{X}]$.

We prove the existence and uniqueness of the polynomial $E \in \mathcal{O}(r+1)$ coefficient-wise. If E is as in (†), $\boldsymbol{i} \in \mathbb{N}_0^n$ with $|\boldsymbol{i}| = r+1$, then $\overline{\pi}a_{\boldsymbol{i}} - \pi^{r+1}a_{\boldsymbol{i}}^{\varphi}$ is the coefficient of $\boldsymbol{X}^{\boldsymbol{i}}$ in $\overline{\pi}E - \pi^{r+1}E^{\varphi}$. The coefficient of $\boldsymbol{X}^{\boldsymbol{i}}$ in G is of the form $\pi\beta$ for some $\beta \in \mathcal{O}$. Hence is suffices to prove:

For every $\beta \in \mathcal{O}$ there exists a unique $\alpha \in \mathcal{O}$ such that

$$\overline{\pi}\alpha - \pi^{r+1}\alpha^{\varphi} = \overline{\pi}\beta.$$

Suppose that $\beta \in \mathcal{O}$, set $\rho = \overline{\pi}^{-1}\pi^{r+1} \in \mathcal{O}$ and

$$\alpha = \beta + \sum_{i=0}^{\infty} \beta^{\varphi^{i+1}} \rho^{1+\varphi+\ldots+\varphi^i}.$$

Since $v_K(\beta^{\varphi^{i+1}} \rho^{1+\varphi+\ldots+\varphi^i}) \ge (i+1)v_K(\rho) = (i+1)r$, the series converges, we obtain $\alpha \in \mathcal{O}$,

$$\begin{aligned} \alpha - \rho\alpha^{\varphi} &= \beta + \sum_{i=0}^{\infty} \beta^{\varphi^{i+1}} \rho^{1+\varphi+\ldots+\varphi^i} - \rho\beta^{\varphi} - \sum_{i=0}^{\infty} \beta^{\varphi^{i+2}} \rho^{1+\varphi+\ldots+\varphi^{i+1}} \\ &= \beta + \sum_{i=0}^{\infty} \beta^{\varphi^{i+1}} \rho^{1+\varphi+\ldots+\varphi^i} - \rho\beta^{\varphi} - \sum_{i=1}^{\infty} \beta^{\varphi^{i+1}} \rho^{1+\varphi+\ldots+\varphi^i} \\ &= \beta + \beta^{\varphi}\rho - \rho\beta^{\varphi} = \beta \end{aligned}$$

and $\overline{\pi}\alpha - \pi^{r+1}\alpha^{\varphi} = \overline{\pi}(\alpha - \rho\alpha^{\varphi}) = \overline{\pi}\beta$. To prove uniqueness, let $\alpha,\ \alpha' \in \mathcal{O}$ be such that $\alpha - \rho\alpha^{\varphi} = \alpha' - \rho\alpha'^{\varphi}$. Then we obtain $\alpha - \alpha' = \rho(\alpha - \alpha')^{\varphi}$, and as $v_K(\alpha - \alpha') = v_K((\alpha - \alpha')^{\varphi})$ and $v_K(\rho) = r \ge 1$, it follows that $\alpha = \alpha'$. □

Theorem 4.6.5. *Let $\pi \in \mathcal{O}_K$ be a prime element, and let $e, \overline{e} \in X\mathcal{O}_K\,[\![X]\!]$ be π-series. Then there exist*

- *a unique power series $F_e \in \mathcal{O}_K\,[\![X,Y]\!]$ such that*

$$F_e(X,Y) \equiv X + Y \mod \mathfrak{M}_{\mathcal{O}_K}(X,Y)^2 \quad \textit{and} \quad e(F_e(X,Y)) = F_e(e(X),e(Y)),$$

- *for every* $a \in \mathcal{O}_K$ *a unique power series* $[a]_{e,\overline{e}} \in \mathcal{O}_K[\![X]\!]$ *such that*

$$[a]_{e,\overline{e}}(X) \equiv aX \mod X^2\mathcal{O}_K[\![X]\!] \quad \text{and} \quad e\circ[a]_{e,\overline{e}} = [a]_{e,\overline{e}}\circ\overline{e}.$$

For $a \in \mathcal{O}_K$ *we set* $[a]_e = [a]_{e,e}$. *Then the following assertions hold:*

1. $[\pi^n]_e = e_n = e\circ\ldots\circ e$ *for all* $n \geq 0$, *and* $F_e = (F_e, [\,\cdot\,]_e)$ *is a* π*-module over* $\mathcal{O}_K$.

2. *If* $a \in \mathcal{O}_K$, *then* $[a]_{e,\overline{e}}\colon F_{\overline{e}} \to F_e$ *is a morphism of* π*-modules, and if* $a \in U_K$, *then* $[a]_{e,\overline{e}}$ *is an isomorphism.*

Proof. From 4.6.4 (applied with $\mathcal{O} = \mathcal{O}_K$) we obtain the existence and uniqueness of the series F_e and $[a]_{e,\overline{e}}$. As to 1. and 2., it suffices to prove the following assertions for all a, $b \in \mathcal{O}_K$:

a. $F_e(X,Y) = F_e(Y,X)$.

b. $F_e(X,F_e(Y,Z)) = F_e(F_e(X,Y),Z)$.

c. $[a]_{e,\overline{e}}(F_{\overline{e}}(X,Y)) = F_e([a]_{e,\overline{e}}(X),[a]_{e,\overline{e}}(Y))$.

d. $[a+b]_{e,\overline{e}}(X) = F_e([a]_{e,\overline{e}}(X),[b]_{e,\overline{e}}(X))$.

e. $[ab]_e = [a]_{e,\overline{e}}\circ[b]_{\overline{e},e}$.

f. $[a]_{e,\overline{e}}\circ[b]_{\overline{e}} = [b]_e\circ[a]_{e,\overline{e}}$.

g. $[\pi^n]_e = e_n$ for all $n \geq 0$.

Indeed, if **a** to **g** hold, then:
a and **b** imply that F_e is a formal group. **c** implies that $[a]_{e,\overline{e}}\colon F_{\overline{e}} \to F_e$ is a morphism of formal groups, and that $[a]_e \in \operatorname{End}_{\mathcal{O}_K}(F_e)$. By **d** and **e** (applied with $e = \overline{e}$) it follows that $(F_e, [\,\cdot\,]_e)$ is a formal $\mathcal{O}_K$-module, and **f** implies that $[a]_{e,\overline{e}}\colon F_{\overline{e}} \to F_e$ is a morphism of formal $\mathcal{O}_K$-modules. If moreover $a \in U_K$, then $[a]_{e,\overline{e}}\circ[a^{-1}]_{\overline{e},e} = [1]_e = 1_{F_e}$ and $[a^{-1}]_{\overline{e},e}\circ[a]_{e,\overline{e}} = [1]_{\overline{e}} = 1_{F_{\overline{e}}}$ by **e**, and therefore $[a]_{e,\overline{e}}$ is an isomorphism. By **g** we obtain $[\pi^n]_e = e_n$ for all $n \geq 0$. In particular, $[\pi]_e = e \in \mathcal{O}_K[\![X]\!]$ is a π-series, and F_e is a π-module over $\mathcal{O}_K$.

Proof of the assertions **a** *to* **g**. We set $\mathfrak{M} = \mathfrak{M}_{\mathcal{O}_K}$.

a. We set $F_1(X,Y) = F_e(Y,X)$. Then we get $F_1 \equiv X+Y \mod \mathfrak{M}(X,Y)^2$ and $e(F_1(X,Y)) = e(F_e(Y,X)) = F_e(e(Y),e(X)) = F_1(e(X),e(Y))$. From the uniqueness of F_e it follows that $F_1 = F_e$.

b. We set $F_1(X,Y,Z) = F_e(X,F_e(Y,Z))$ and $F_2(X,Y,Z) = F_e(F_e(X,Y),Z)$. Then $F_1 \equiv F_2 \equiv X+Y+Z \mod \mathfrak{M}(X,Y,Z)^2$,

$$\begin{aligned} e(F_1(X,Y,Z)) &= e(F_e(X,F_e(Y,Z))) = F_e(e(X),e(F_e(Y,Z)) \\ &= F_e(e(X),F_e(e(Y),e(Z))) = F_1(e(X),e(Y),e(Z)) \end{aligned}$$

and

$$\begin{aligned} e(F_2(X,Y,Z)) &= e(F_e(F_e(X,Y),Z)) = F_e(e(F_e(X,Y)),e(Z)) \\ &= F_e(F_e(e(X),e(Y)),e(Z)) = F_2(e(X),e(Y),e(Z)). \end{aligned}$$

Hence the uniqueness in 4.6.4 implies $F_1 = F_2$.

c. We set $F_1(X,Y) = [a]_{e,\overline{e}}(F_{\overline{e}})$ and $F_2(X,Y) = F_e([a]_{e,\overline{e}}(X),[a]_{e,\overline{e}}(Y))$. Then

$$F_1(X,Y) \equiv aF_{\overline{e}}(X,Y) \equiv aX + aY \mod \mathfrak{M}(X,Y)^2,$$

$$F_2(X,Y) \equiv [a]_{e,\overline{e}}(X) + [a]_{e,\overline{e}}(Y) \equiv aX + aY \mod \mathfrak{M}(X,Y)^2,$$

$$\begin{aligned} e(F_1(X,Y)) &= e([a]_{e,\overline{e}}(F_{\overline{e}}(X,Y))) = [a]_{e,\overline{e}}(\overline{e}(F_{\overline{e}}(X,Y))) \\ &= [a]_{e,\overline{e}}(F_{\overline{e}}(\overline{e}(X),\overline{e}(Y))) = F_1(\overline{e}(X),\overline{e}(Y)) \end{aligned}$$

and

$$\begin{aligned} e(F_2(X,Y)) &= e(F_e([a]_{e,\overline{e}}(X)),[a]_{e,\overline{e}}(Y)) = F_e(e([a]_{e,\overline{e}}(X)),e([a]_{e,\overline{e}}(Y))) \\ &= F_e([a]_{e,\overline{e}}(\overline{e}(X)),[a]_{e,\overline{e}}(\overline{e}(Y))) \\ &= F_2(\overline{e}(X),\overline{e}(Y)). \end{aligned}$$

Hence the uniqueness in 4.6.4 implies $F_1 = F_2$.

d. We set $F_1(X) = [a+b]_{e,\overline{e}}(X)$ and $F_2(X) = F_e([a]_{e,\overline{e}}(X),[b]_{e,\overline{e}}(X))$. Then

$$F_1(X) \equiv (a+b)X \equiv aX + bX \mod X^2\mathcal{O}_K[\![X]\!],$$

$$F_2(X) \equiv [a]_{e,\overline{e}}(X) + [b]_{e,\overline{e}}(X) \equiv aX + bX \mod X^2\mathcal{O}_K[\![X]\!],$$

$$e(F_1(X)) = e([a+b]_{e,\overline{e}}(X)) = [a+b]_{e,\overline{e}}(\overline{e}(X)) = F_1(\overline{e}(X))$$

and

$$\begin{aligned} e(F_2(X)) = e(F_e([a]_{e,\overline{e}}(X),[b]_{e,\overline{e}}(X))) &= F_e(e([a]_{e,\overline{e}}(X)),e([b]_{e,\overline{e}}(X))) \\ &= F_e([a]_{e,\overline{e}}(\overline{e}(X)),[b]_{e,\overline{e}}(\overline{e}(X))) \\ &= F_2(\overline{e}(X)). \end{aligned}$$

Hence the uniqueness in 4.6.4 implies $F_1 = F_2$.

e. We set $F_1(X) = [ab]_e(X)$ and $F_2(X) = [a]_{e,\overline{e}} \circ [b]_{\overline{e},e}(X)$. Then

$$F_1(X) \equiv abX \mod X^2\mathcal{O}_K[\![X]\!],\ \ F_2(X) \equiv a[b]_{\overline{e},e}(X) \equiv abX \mod X^2\mathcal{O}_K[\![X]\!],$$

$$e(F_1(X) = e([ab]_e(X)) = [ab]_e(e(X)) = F_1(e(X))$$

and

$$\begin{aligned} e(F_2(X)) &= e([a]_{e,\overline{e}}([b]_{\overline{e},e}(X))) = [a]_{e,\overline{e}}(\overline{e}([b]_{\overline{e},e}(X))) \\ &= [a]_{e,\overline{e}}([b]_{\overline{e},e}(e(X))) = F_2(e(X)). \end{aligned}$$

Hence the uniqueness in 4.6.4 implies $F_1 = F_2$.

f. We set $F_1(X) = [a]_{e,\overline{e}} \circ [b]_{\overline{e}}(X)$ and $F_2(X) = [b]_e \circ [a]_{e,\overline{e}}(X)$. Then

$$F_1(X) \equiv F_2(X) \equiv abX \mod X^2 \mathcal{O}_K [\![X]\!],$$

$$e(F_1(X)) = e([a]_{e,\overline{e}} \circ [b]_{\overline{e}}(X)) = [a]_{e,\overline{e}}(\overline{e}([b]_{\overline{e}}(X))) = [a]_{e,\overline{e}} \circ [b]_{\overline{e}}(\overline{e}(X)) = F_1(\overline{e}(X))$$

and

$$e(F_2(X)) = e([b]_e \circ [a]_{e,\overline{e}}(X)) = [b]_e(e([a]_{e,\overline{e}}(X))) = [b]_e \circ [a]_{e,\overline{e}}(\overline{e}(X)) = F_2(\overline{e}(X)).$$

Hence the uniqueness in 4.6.4 implies $F_1 = F_2$.

g. By induction on n.

$n = 0$: Since $X \equiv [1]_e(X) \mod \mathcal{O}_K [\![X]\!]$ and $X \circ e = e \circ X$, we obtain $X = [1]_e$ by the uniqueness of $[1]_e$.

$n = 1$: Since $e \equiv [\pi]_e \equiv \pi X \mod X^2 \mathcal{O}_K [\![X]\!]$ and $e \circ e = e \circ e$, we obtain $[\pi]_e = e$ by the uniqueness of $[\pi]_e$.

$n \geq 1$, $n \to n+1$: If $[\pi^n]_e = e_n$, then

$$[\pi^{n+1}]_e = [\pi^n]_e \circ [\pi]_e = e_n \circ e = e_{n+1}. \qquad \square$$

Corollary 4.6.6. *Let $\pi \in \mathcal{O}_K$ be a prime element, and let $e, \overline{e} \in \mathcal{O}_K [\![X]\!]$ be π-series.*

1. *Let $(F, [\,\cdot\,]_F)$ be a π-module over $\mathcal{O}_K$ such that $[\pi]_F = e$. Then $F = F_e$.*
   We call F_e the π-module associated with e.

2. *Let $h \in X\mathcal{O}_K [\![X]\!]$ such that $\overline{e} \circ h = h \circ e$. Then $h \colon F_e \to F_{\overline{e}}$ is a morphism of π-modules. If $a \in \mathcal{O}_K$ and $h \equiv aX \mod X^2 \mathcal{O}_K [\![X]\!]$, then $h = [a]_{\overline{e},e}$.*

3. *For every $a \in \mathcal{O}_K$ there exists a unique morphism $h \colon F_e \to F_{\overline{e}}$ which satisfies $h \equiv aX \mod X^2 \mathcal{O}_K [\![X]\!]$. Further, h is an isomorphism if and only if $a \in U_K$, and then $\overline{e}_n = h \circ e_n \circ h^{-1}$ for all $n \in \mathbb{N}$.*

Proof. 1. Bytion,

$$F(X,Y) \equiv X + Y \mod \mathfrak{M}_{\mathcal{O}_K}(X,Y)^2 \quad \text{and} \quad e(F(X,Y)) = F(e(X), e(Y)).$$

Hence $F = F_e$ by 4.6.5.

2. If $a \in \mathcal{O}_K$ and $h \equiv aX \mod X^2 \mathcal{O}_K [\![X]\!]$, then $h = [a]_{\overline{e},e}$ and thus it is a morphism by 4.6.5.

3. Obvious by 2. and a simple induction on n. $\square$

Theorem 4.6.7. *Let $\pi \in \mathcal{O}_K$ be a prime element, and let $e = X^q + \pi X u$, where $u \in 1 + X\mathcal{O}_K[\![X]\!]$ (then e is a π-series). For $n \in \mathbb{N}_0$, suppose that $g_n = e_n^{q-1} + \pi u(e_n)$ (recall that $e_0 = X$ and $e_{n+1} = e(e_n)$ for all $n \ge 0$).*

1. *If $n \in \mathbb{N}$, then $e_n \equiv \pi^n X \bmod X^2\mathcal{O}_K[\![X]\!]$, $g_{n-1} \equiv X^{q^{n-1}(q-1)} \bmod \pi\mathcal{O}_K[\![X]\!]$, $g_{n-1}(0) = \pi$, and*

$$e_n = e_{n-1}g_{n-1} = X\prod_{i=0}^{n-1} g_i \equiv X^{q^n} \mod \pi\mathcal{O}_K[\![X]\!].$$

2. *Let $e = X^q + \pi X$ and $n \in \mathbb{N}$. Then g_n is an Eisenstein polynomial of degree $\partial(g_n) = q^n(q-1)$, e_n is separable of degree $\partial(e_n) = q^n$, and e_n has all its zeros in $\overline{\mathfrak{p}} = \{x \in K_{\text{sep}} \mid |x| < 1\}$.*

Proof. 1. By induction on n. For $n = 1$, there is nothing to do. Thus let $n \ge 1$, $e_n \equiv \pi^n X \bmod X^2\mathcal{O}_K[\![X]\!]$, $g_{n-1} \equiv X^{q^{n-1}(q-1)} \bmod \pi\mathcal{O}_K[\![X]\!]$, $g_{n-1}(0) = \pi$, and

$$e_n = e_{n-1}g_{n-1} = X\prod_{i=0}^{n-1} g_i \equiv X^{q^n} \mod \pi\mathcal{O}_K[\![X]\!].$$

Then $g_n(0) = \pi u(0) = \pi$,

$$\begin{aligned} e_{n+1} = e(e_n) = e_n^q + \pi e_n u(e_n) = e_n g_n &= X\prod_{i=0}^{n} g_i \\ &\equiv \pi^n X g_n(0) \equiv \pi^{n+1}X \mod X^2\mathcal{O}_K[\![X]\!], \end{aligned}$$

$$g_n \equiv e_n^{q-1} \equiv X^{q^n(q-1)} \mod \pi\mathcal{O}_K[\![X]\!],$$

and

$$e_{n+1} = e_n g_n \equiv X^{q^n + q^n(q-1)} \equiv X^{q^{n+1}} \mod \pi\mathcal{O}_K[\![X]\!].$$

2. As $\partial(e) = q$, we get $\partial(e_n) = q^n$ and $\partial(g_n) = \partial(e_{n+1}) - \partial(e_n) = q^n(q-1)$ for all $n \ge 1$. As $g_n \equiv X^{q^n(q-1)} \bmod \pi\mathcal{O}_K[X]$ and $g_n(0) = \pi$, it follows that g_n is an Eisenstein polynomial. Hence g_n is irreducible and has all its zeros in $\overline{\mathfrak{p}}$ (see [18, Theorem 2.12.8]). Since $e_n = X^{q^n} + \ldots + \pi^n X$, we obtain

$$\begin{aligned} g_n &= \frac{e_{n+1}}{e_n} = \frac{e(e_n)}{e_n} = \frac{e_n^q + \pi e_n}{e_n} = e_n^{q-1} + \pi = (X^{q^n} + \ldots + \pi^n X)^{q-1} + \pi \\ &= X^{q^n(q-1)} + \ldots + \pi^{n(q-1)}X^{q-1} + \pi, \end{aligned}$$

$g_n' = q^n(q-1)X^{q^n(q-1)-1} + \ldots + (q-1)\pi^{n(q-1)}X^{q-2} \neq 0$, and therefore g_n is separable. The polynomials $X, g_0, g_1, \ldots, g_n$ are irreducible, separable, distinct and have all its zeros in $\overline{\mathfrak{p}}$. Hence $e_n = Xg_0 \cdot \ldots \cdot g_{n-1}$ is separable and has all its zeros in $\overline{\mathfrak{p}}$, too. □

4.7 Lubin-Tate extensions

Let K be a local field, $\mathrm{char}(\mathsf{k}_K) = p$ *and* $\mathsf{k}_K = \mathbb{F}_q$.

Let $|\cdot|\colon K \to \mathbb{R}_{\ge 0}$ be an absolute of K belonging to v_K, tacitly extended to an (equally denoted) absolute value of $K_{\rm sep}$. Let $\overline{\mathcal{O}} = \{x \in K_{\rm sep} \mid |x| \le 1\}$ be the valuation domain of $K_{\rm sep}$ and by $\overline{\mathfrak{p}} = \{x \in K_{\rm sep} \mid |x| < 1\}$ its maximal ideal.

We start with some useful remarks and a lemma concerning continuous maps defined by power series and their zeros.

Remark 4.7.1.

1. Let $g \in \mathcal{O}_K[X]$ and $x \in \overline{\mathfrak{p}}$. Then $[K(x):K] < \infty$, hence $K(x)$ is a local field and $\mathfrak{p}_{K(x)} = \overline{\mathfrak{p}} \cap K(x)$. By 4.5.2, we obtain $g(x) \in \mathcal{O}_{K(x)} \subset \overline{\mathcal{O}}$, and therefore the power series g defines an (equally denoted) continuous map $g\colon \overline{\mathfrak{p}} \to \overline{\mathcal{O}}$. Moreover, if $g \in X\mathcal{O}_K[\![X]\!]$, then $g(\overline{\mathfrak{p}}) \subset \overline{\mathfrak{p}}$.

2. Let $h \in X\mathcal{O}_K[\![X]\!]$ be composition-invertible and $g = h^{-1} \in X\mathcal{O}_K[\![X]\!]$ its composition-inverse. Then $g \circ h(\alpha) = \alpha$ for all and $\alpha \in \overline{\mathfrak{p}}$. In particular, if $\alpha \in \overline{\mathfrak{p}}$ and $h(\alpha) = 0$, then $\alpha = 0$.

Lemma 4.7.2. *Let $f \in \mathcal{O}_K[\![X]\!]$ be a power series.*

1. *If $\alpha \in \mathfrak{p}_K$ and $f(\alpha) = 0$, then $f \in (X-\alpha)\mathcal{O}_K[\![X]\!]$.*

2. *Let $h \in \mathcal{O}_K[X]$, $n \in \mathbb{N}$ and $h = (X-\alpha_1)\cdot\ldots\cdot(X-\alpha_n)$ with distinct $\alpha_1, \ldots, \alpha_n \in \overline{\mathfrak{p}}$. If $f(\alpha_i) = 0$ for all $i \in [1,n]$, then $f \in h\mathcal{O}_K[\![X]\!]$.*

Proof. 1. Suppose that

$$f = \sum_{i=0}^{\infty} a_i X^i, \quad \alpha \in \mathfrak{p}_K, \quad f(\alpha) = 0, \quad \text{and for } j \ge 0 \text{ set } \beta_j = \sum_{\nu=0}^{\infty} a_{\nu+j+1}\alpha^{\nu} \in \mathcal{O}_K.$$

Then

$$\begin{aligned} f &= \sum_{i=0}^{\infty} a_i(X^i - \alpha^i) = (X-\alpha)\sum_{i=0}^{\infty} a_i \sum_{j=0}^{i-1} \alpha^{i-1-j}X^j \\ &= (X-\alpha)\sum_{j=0}^{\infty}\Big(\sum_{i=j+1}^{\infty} a_i\alpha^{i-1-j}\Big)X^j = (X-\alpha)\sum_{j=0}^{\infty}\Big(\sum_{\nu=0}^{\infty} a_{\nu+j+1}\alpha^{\nu}\Big)X^j \\ &= (X-\alpha)\sum_{j=0}^{\infty}\beta_j X^j. \end{aligned}$$

2. If $K' = K(\alpha_1,\ldots,\alpha_n)$, then K'/K is Galois. By a simple induction, using 1. for K' instead of K, we obtain a power series $g \in \mathcal{O}_{K'}[\![X]\!]$ such that $f = (X-\alpha_1)\cdot\ldots\cdot(X-\alpha_n)g = hg$. If $G = \mathrm{Gal}(K'/K)$ acts coefficient-wise on power series in $K'[\![X]\!]$, then $K'[\![X]\!]^G = K[\![X]\!]$, $f = hg = f^\sigma = h^\sigma g^\sigma = hg^\sigma$, hence $g^\sigma = g$ for all $\sigma \in G$ and therefore $g \in K[\![X]\!] \cap \mathcal{O}_{K'}[\![X]\!] = \mathcal{O}_K[\![X]\!]$. □

Definition 4.7.3. Let $\pi \in \mathcal{O}_K$ be a prime element, $e \in X\mathcal{O}_K[\![X]\!]$ a π-series and $F_e = (F_e, [\,\cdot\,]_e)$ the π-module associated with e (see 4.6.6.1). We define an addition $+_e \colon \overline{\mathfrak{p}}\times\overline{\mathfrak{p}} \to \overline{\mathfrak{p}}$ and a scalar multiplication $\cdot_e \colon \mathcal{O}_K\times\overline{\mathfrak{p}} \to \overline{\mathfrak{p}}$ by

$$x +_e y = F_e(x,y) \quad\text{and}\quad \alpha \cdot_e x = [\alpha]_e(x) \quad \text{for all } x,\, y \in \overline{\mathfrak{p}} \text{ and } \alpha \in \mathcal{O}_K.$$

It is easily checked that with these operations $\overline{\mathfrak{p}}$ is an $\mathcal{O}_K$-module (in the ordinary sense), denoted by $\overline{\mathfrak{p}}_e$.

For $n \in \mathbb{N}_0$, we consider the n-fold iterated series $e_n = [\pi^n]_e$, and we define

$$\mathsf{W}_e^n = \mathrm{Ker}([\pi^n]_e \colon \overline{\mathfrak{p}}_e \to \overline{\mathfrak{p}}_e) = \{\alpha \in \overline{\mathfrak{p}} \mid e_n(\alpha) = 0\} \subset \overline{\mathfrak{p}}_e \quad\text{and}\quad K_\pi^n = K(\mathsf{W}_e^n).$$

The elements $\alpha \in \mathsf{W}_e^n$ are called **π^n-torsion points** and K_π^n is called the **π^n-division field** defined by the π-series e over K. Moreover, we set

$$\mathsf{W}_e = (\overline{\mathfrak{p}}_e)_{\mathrm{tor}} = \{\alpha \in \overline{\mathfrak{p}} \mid [a]_e(\alpha) = 0 \text{ for some } a \in \mathcal{O}_K^\bullet\} \quad\text{and}$$
$$K_\pi = K(\mathsf{W}_e) \subset K_{\mathrm{sep}}.$$

By definition, W_e^n and W_e are $\mathcal{O}_K$-submodules of $\overline{\mathfrak{p}}_e$, and thus the valuation ideals $\mathfrak{p}_{K_\pi^n} = \overline{\mathfrak{p}} \cap K_\pi^n$ are also $\mathcal{O}_K$-submodules of $\overline{\mathfrak{p}}_e$. We shall see in 4.7.5.2 that the fields K_π^n and K_π actually only depend on π and not on e.

The fields K_π^n are called **generalized cyclotomic fields**. This terminology is justified by the following important Remark 4.7.4.

Remark 4.7.4. Let p be a prime and $K = \mathbb{Q}_p$. Then $\mathcal{O}_K = \mathbb{Z}_p$, $K_{\mathrm{sep}} = \overline{\mathbb{Q}}_p$, and if $|\cdot|_p \colon \overline{\mathbb{Q}}_p \to \mathbb{R}_{\ge 0}$ is the extension of the p-adic absolute value, then

$$\overline{\mathcal{O}} = \{x \in \overline{\mathbb{Q}}_p \mid |x|_p \le 1\}$$

is the valuation domain of $\overline{\mathbb{Q}}_p$ and $\overline{\mathfrak{p}} = \{x \in \overline{\mathbb{Q}}_p \mid |x|_p < 1\}$ is its maximal ideal. The polynomial $e = (1+X)^p - 1 = X^p + \ldots + p \in X\mathbb{Z}_p[X]$ is a p-polynomial and G_{m} is a p-module over $\mathbb{Z}_p$ (see 4.6.3). Since $[p]_e = e$, we get $[p^n]_e = e_n = (1+X)^{p^n} - 1$ for all $n \in \mathbb{N}$, and as $e(X +_{G_{\mathsf{m}}} Y) = e(X) +_{G_{\mathsf{m}}} e(Y)$, 4.6.6.1 implies that $F_e = G_{\mathsf{m}}$.

For $n \in \mathbb{N}$ let $\zeta_{p^n} \in \overline{\mathbb{Q}}_p$ be a primitive p^n-th root of unity, and let $\mu_{p^n} = \langle \zeta_{p^n} \rangle \subset \overline{\mathbb{Q}}_p^\times$. Then $\mathsf{W}_e^n = \{\alpha \in \overline{\mathfrak{p}} \mid (1+\alpha)^{p^n} - 1 = 0\} = \{\zeta - 1 \mid \zeta \in \mu_{p^n}\}$, and

$$(\mathbb{Q}_p)_p^n = \mathbb{Q}_p(\mathsf{W}_e^n) = \mathbb{Q}_p(\zeta_{p^n}) = \mathbb{Q}_p^{(p^n)} \quad \text{is the } p^n\text{-th cyclotomic field over } \mathbb{Q}_p.$$

Theorem 4.7.5. *Let $\pi \in \mathcal{O}_K$ be a prime element, and let $e, \overline{e} \in X\mathcal{O}_K[\![X]\!]$ be π-series. Let $h\colon F_e \to F_{\overline{e}}$ be a morphism of π-modules, and let $a \in \mathcal{O}_K$ be such that $h \equiv aX \mod X^2\mathcal{O}_K[\![X]\!]$* (see 4.6.6.3).

1. *The assignment $\alpha \mapsto h(\alpha)$ induces an (equally denoted) ordinary $\mathcal{O}_K$-module homomorphism $h\colon \overline{\mathfrak{p}}_e \to \overline{\mathfrak{p}}_{\overline{e}}$. If $a \in U_K$, then $h\colon \overline{\mathfrak{p}}_e \to \overline{\mathfrak{p}}_{\overline{e}}$ is an isomorphism, and $\overline{e} = h \circ e \circ h^{-1}$. In particular, $e = [\pi]_e \in \mathrm{End}_{\mathcal{O}_K}(\overline{\mathfrak{p}}_e)$.*

2. *Let $a \in U_K$. Then $h(\mathsf{W}_e^n) = \mathsf{W}_{\overline{e}}^n$ and $K(\mathsf{W}_e^n) = K(\mathsf{W}_{\overline{e}}^n)$ for all $n \ge 0$. Moreover, $|h(\alpha)| = |\alpha|$ for all $\alpha \in \overline{\mathfrak{p}}$.*
In particular, the fields K_π^n and K_π depend only on π, and the $\mathcal{O}_K$-modules W_e^n and W_e are up to isomorphisms uniquely determined by π.

Proof. 1. If $x, y \in \overline{\mathfrak{p}}$ and $\alpha \in \mathcal{O}_K$, then

$$h(x +_e y) = h \circ F_e(x, y) = F_{\overline{e}}(h(x), h(y)) = h(x) +_{\overline{e}} h(y)$$

and

$$h(\alpha \cdot_e x) = h \circ [\alpha]_e(x) = [\alpha]_{\overline{e}} \circ h(x) = \alpha \cdot_{\overline{e}} x.$$

Hence $h\colon \overline{\mathfrak{p}}_e \to \overline{\mathfrak{p}}_{\overline{e}}$ is an (ordinary) $\mathcal{O}_K$-module homomorphism. If $a \in U_K$, then h is composition-invertible, hence $h\colon \overline{\mathfrak{p}}_e \to \overline{\mathfrak{p}}_{\overline{e}}$ is an isomorphism, and $\overline{e} = h \circ e \circ h^{-1}$ by 4.6.6.3.

2. Let $a \in U_K$ and $x \in \overline{\mathfrak{p}}$. Then $h(x) \equiv ax \mod \overline{\mathfrak{p}}^2$, hence $h = aX(1 + Xg)$ for some $g \in \mathcal{O}_K[\![X]\!]$, and since $1 + xg(x) \in 1 + \overline{\mathfrak{p}}$, we obtain $|h(x)| = |ax| = |x|$. If $n \in \mathbb{N}_0$, then $h \circ [\pi^n]_e = [\pi^n]_{\overline{e}}$, and therefore

$$x \in \mathsf{W}_e^n \iff [\pi^n]_e(x) = 0 \iff h \circ [\pi^n]_e(x) = 0 \iff [\pi^n]_{\overline{e}} \circ h(x) = 0$$
$$\iff h(x) \in \mathsf{W}_{\overline{e}}^n.$$

It follows that $\mathsf{W}_{\overline{e}}^n = h(\mathsf{W}_e^n) \subset K(W_e^n)$, hence $K(\mathsf{W}_{\overline{e}}^n) \subset K(\mathsf{W}_e^n)$, and thus equality holds by symmetry. □

Theorem 4.7.6. *Let $\pi \in \mathcal{O}_K$ be a prime element and $e \in X\mathcal{O}_K[\![X]\!]$ a π-series.*

1. *If $a \in \mathcal{O}_K$ and $v_K(a) = n$, then $\mathrm{Ker}([a]_e) = \mathrm{Ker}([\pi^n]_e) = \mathsf{W}_e^n$. Moreover,*

$$\mathbf{0} = \mathsf{W}_e^0 \subset \mathsf{W}_e^n \subset \mathsf{W}_e^{n+1} \quad \text{and} \quad K = K_\pi^0 \subset K_\pi^n \subset K_\pi^{n+1} \quad \text{for all } n \in \mathbb{N}_0,$$

$$\mathsf{W}_e = \bigcup_{n \in \mathbb{N}} \mathsf{W}_e^n \quad \text{and} \quad K_\pi = \bigcup_{n \in \mathbb{N}} K_\pi^n.$$

2. If $n \in \mathbb{N}_0$ and $m \in [0, n]$ then $|\mathsf{W}_e^n| = q^n$,

$$e_m(\mathsf{W}_e^n) = \pi^m \cdot_e \mathsf{W}_e^n = \mathsf{W}_e^{n-m},$$

and if $m < n$, then

$$\pi^m \cdot_e (\mathsf{W}_e^n \setminus \mathsf{W}_e^{n-1}) = \mathsf{W}_e^{n-m} \setminus \mathsf{W}_e^{n-m-1}.$$

Proof. 1. Obviously, $\mathsf{W}_e^0 = \mathbf{0}$ and $K_\pi^0 = K$. If $a = \pi^n u$, where $n \in \mathbb{N}_0$ and $u \in U_K$, then $[a]_e = [u]_e \circ [\pi^n]_e$. Hence $\mathrm{Ker}([a]_e) = \mathrm{Ker}([\pi^n]_e) = \mathsf{W}_e^n$, since $[u]_e$ is an automorphism. As $e_{n+1} = e \circ e_n$, we get $\mathsf{W}_e^n \subset \mathsf{W}_e^{n+1}$, hence $K_\pi^n \subset K_\pi^{n+1}$, and obviously

$$\mathsf{W}_e = \bigcup_{a \in \mathcal{O}_K^\bullet} \mathrm{Ker}([a]_e) = \bigcup_{n \in \mathbb{N}} \mathsf{W}_e^n \qquad \text{and} \quad K_\pi = \bigcup_{n \in \mathbb{N}} K_\pi^n.$$

2. Let $\overline{e} = X^q + \pi X$. Then $|\mathsf{W}_{\overline{e}}^n| = |\{\alpha \in \overline{\mathfrak{p}} \mid \overline{e}_n(\alpha) = 0\}| = \partial(\overline{e}_n) = q^n$ by 4.6.7.2, and thus $|\mathsf{W}_e^n| = |\mathsf{W}_{\overline{e}}^n| = q^n$.

If $\alpha \in \mathsf{W}_e^n$ and $m \in [0, n]$, then $[\pi^{n-m}]_e \circ [\pi^m]_e(\alpha) = [\pi^n]_e(\alpha) = 0$, which implies $[\pi^m]_e(\alpha) \in \mathsf{W}_e^{n-m}$. In particular, $[\pi^m]_e \colon \mathsf{W}_e^n \to \mathsf{W}_e^{n-m}$ is an $\mathcal{O}_K$-homomorphism, and

$$|\mathrm{Im}([\pi^m]_e)| = \frac{|\mathsf{W}_e^n|}{|\mathrm{Ker}([\pi^m]_e)|} = \frac{|\mathsf{W}_e^n|}{|\mathsf{W}_e^m|} = q^{n-m} = |\mathsf{W}_e^{n-m}|.$$

Hence $[\pi^m]_e$ is surjective, and $\pi^m \cdot_e \mathsf{W}_e^n = [\pi^m]_e(\mathsf{W}_e^n) = \mathsf{W}_e^{n-m}$.

Suppose that $m < n$. Then we get $\pi^m \cdot_e \mathsf{W}_e^{n-1} = \mathsf{W}_e^{n-m-1}$, and if $\alpha \in \mathsf{W}_e^n$ and $\pi^m \cdot_e \alpha \in \mathsf{W}_e^{n-m-1}$, then $\pi^{n-1} \cdot_e \alpha = \pi^{n-m-1} \cdot_e (\pi^m \cdot_e \alpha) = 0$, hence $\alpha \in \mathsf{W}_e^{n-1}$. Therefore it follows that $\pi^m \cdot_e (\mathsf{W}_e^n \setminus \mathsf{W}_e^{n-1}) = \mathsf{W}_e^{n-m} \setminus \mathsf{W}_e^{n-m-1}$. $\square$

Theorem 4.7.7. *Let $\pi \in \mathcal{O}_K$ be a prime element, $e \in X\mathcal{O}_K[\![X]\!]$ a π-series, $n \in \mathbb{N}$ and $\alpha_n \in \mathsf{W}_e^n \setminus \mathsf{W}_e^{n-1}$. Then*

- *$\phi_n \colon \mathcal{O}_K/\pi^n\mathcal{O}_K \to \mathsf{W}_e^n$, defined by $\phi_n(a + \pi^n\mathcal{O}_K) = a \cdot_e \alpha_n$, is an $\mathcal{O}_K$-module isomorphism. In particular, $\mathsf{W}_e^n = \mathcal{O}_K \cdot_e \alpha_n$, $\mathsf{W}_e^n \setminus \mathsf{W}_e^{n-1} = U_K \cdot_e \alpha_n$, and $\mathrm{Ann}_{\mathcal{O}_K}(\mathsf{W}_n^e) = \mathrm{Ann}_{\mathcal{O}_K}(\alpha_n) = \pi^n\mathcal{O}_K$.*

- *$\psi_n \colon \mathcal{O}_K/\pi^n\mathcal{O}_K \to \mathrm{End}_{\mathcal{O}_K}(\mathsf{W}_e^n)$, defined by $\psi_n(a + \pi^n\mathcal{O}_K) = [a]_e$, is a ring isomorphism.*

- *$\psi_n^* \colon U_K/U_K^{(n)} \to \mathrm{Aut}_{\mathcal{O}_K}(\mathsf{W}_e^n)$, defined by $\psi_n^*(uU_K^{(n)}) = [u^{-1}]_e$, is a group isomorphism, and $|\mathrm{Aut}_{\mathcal{O}_K}(\mathsf{W}_e^n)| = q^{n-1}(q-1)$.*
  The occurrence of u^{-1} will become lucid in 4.8.1.

Proof. The map $\phi'_n\colon \mathcal{O}_K \to \mathsf{W}_e^n$, defined by $\phi'_n(a) = a \cdot_e \alpha_n$, is an $\mathcal{O}_K$-module homomorphism. Since $\pi^n \cdot_e \alpha = 0$ and $\pi^{n-1} \cdot_e \alpha \neq 0$, it follows that $\mathrm{Ker}(\phi'_n) = \pi^n\mathcal{O}_K$, ϕ'_n induces a monomorphism $\phi_n\colon \mathcal{O}_K/\pi^n\mathcal{O}_K \to \mathsf{W}_e^n$, and as $|\mathcal{O}_K/\pi^n\mathcal{O}_K| = q^n = |\mathsf{W}_e^n|$, ϕ_n is an isomorphism. In particular, $\mathsf{W}_e^n = \mathcal{O}_K \cdot_e \alpha_n$, and since $e \cdot_e \mathsf{W}_e^n = \mathsf{W}_e^{n-1}$ by 4.7.6.2, it follows that $\mathsf{W}_e^n \setminus \mathsf{W}_e^{n-1} = U_K \cdot_e \alpha_n$.

ϕ_n induces a ring isomorphism

$$\begin{aligned}\psi_n\colon \mathcal{O}_K/\pi^n\mathcal{O}_K &= \mathrm{End}_{\mathcal{O}_K/\pi^n\mathcal{O}_K}(\mathcal{O}_K/\pi^n\mathcal{O}_K)\\ &= \mathrm{End}_{\mathcal{O}_K}(\mathcal{O}_K/\pi^n\mathcal{O}_K) \overset{\sim}{\to} \mathrm{End}_{\mathcal{O}_K}(\mathsf{W}_e^n),\end{aligned}$$

given by $\psi_n(a+\pi^n\mathcal{O}_K) = [a]_e\colon \mathsf{W}_e^n \to \mathsf{W}_e^n$. As the map

$$U_K/U_K^{(n)} \to (\mathcal{O}_K/\pi^n\mathcal{O}_K)^\times, \ uU_K^{(n)} \mapsto (u^{-1}+\pi^n\mathcal{O}_K)$$

is an isomorphism (see [18, Theorem 2.10.5]), the ring isomorphism ψ_n induces a group isomorphism

$$\psi_n^*\colon U_K/U_K^{(n)}) \overset{\sim}{\to} (\mathcal{O}_K/\pi^n\mathcal{O}_K)^\times \overset{\sim}{\to} \mathrm{End}_{\mathcal{O}_K}(\mathsf{W}_e^n)^\times = \mathrm{Aut}_{\mathcal{O}_K}(\mathsf{W}_e^n),$$

given by $\psi_n^*(uU_K^{(n)}) = \psi_n(u^{-1}+\pi^n\mathcal{O}_K) = [u^{-1}]_e\colon \mathsf{W}_e^n \to \mathsf{W}_e^n$, and it follows that $|\mathrm{Aut}_{\mathcal{O}_K}(\mathsf{W}_e)| = |(\mathcal{O}_K/\pi^n\mathcal{O}_K)^\times| = q^{n-1}(q-1)$ (see [18, Theorem 2.14.1]). □

Theorem 4.7.8. *Let $\pi \in \mathcal{O}_K$ be a prime element, $e \in X\mathcal{O}_K[\![X]\!]$ a π-series and $n \in \mathbb{N}$.*

1. *The field extension K_π^n/K is abelian, fully ramified and of degree $[K_\pi^n : K] = q^{n-1}(q-1)$, and there is an isomorphism*

$$\Phi_n\colon U_K/U_K^{(n)} \overset{\sim}{\to} \mathrm{Gal}(K_\pi^n/K)$$

such that $\Phi_n(uU_K^{(n)})(\alpha) = [u^{-1}]_e(\alpha)$ for all $u \in U_K$ and $\alpha \in \mathsf{W}_e^n$. If $\alpha \in \mathsf{W}_e^n \setminus \mathsf{W}_e^{n-1}$, then $\mathsf{W}_e^n \setminus \mathsf{W}_e^{n-1} = U_K{\cdot}\alpha = \{[u^{-1}]_e(\alpha) \mid u \in U_K\}$ is the set of all conjugates of α over K. Moreover, $K_\pi^n = K(\alpha)$, $\mathcal{O}_{K_\pi^n} = \mathcal{O}_K[\alpha]$, α is a prime element of $\mathcal{O}_{K_\pi^n}$, and $(\pi, K_\pi^n/K) = 1$. In particular, if $\alpha \in (\mathsf{W}_e^1)^\bullet$, then $(\mathsf{W}_e^1)^\bullet$ is the set of all conjugates of α over K, and

$$\mathsf{N}_{K_\pi^1/K}(\alpha) = \prod_{\gamma \in (\mathsf{W}_e^1)^\bullet} \gamma.$$

2. *Let $m \in [1,n]$. Then $\Phi_n(U_K^{(m)}/U_K^{(n)}) = \mathrm{Gal}(K_\pi^n/K_\pi^m)$, and if $\alpha \in \mathsf{W}_e^n \setminus \mathsf{W}_e^{n-1}$, then $\{[u^{-1}]_e(\alpha) \mid u \in U_K^{(m)}\} = \alpha +_e \mathsf{W}_e^{n-m}$ is the set of all conjugates of α over K_π^m. In particular,*

$$\mathsf{N}_{K_\pi^n/K_\pi^m}(\alpha) = \prod_{\gamma \in \mathsf{W}_e^{n-m}} (\alpha +_e \gamma).$$

3. If $g \in \mathcal{O}_K[\![X]\!]$ *is such that* $g(\alpha) = 0$ *for all* $\alpha \in \mathsf{W}_e^n$*, then* $g \in e_n\mathcal{O}_K[\![X]\!]$.

The occurrence of u^{-1} in 2. and 3. will become lucid in 4.8.1.

Proof. Let $\overline{e} = X^q + \pi X \in \mathcal{O}_K[X]$ and $h = [1]_{\overline{e},e} \colon F_e \overset{\sim}{\to} F_{\overline{e}}$. By 4.7.5, the map $h \colon \overline{\mathfrak{p}}_e \overset{\sim}{\to} \overline{\mathfrak{p}}_{\overline{e}}$ is an $\mathcal{O}_K$-module isomorphism such that $|h(\alpha)| = |\alpha|$ for all $\alpha \in \overline{\mathfrak{p}}$, $h(\mathsf{W}_e^n) = \mathsf{W}_{\overline{e}}^n$ and $K_\pi^n = K(\mathsf{W}_e^n)) = K(\mathsf{W}_{\overline{e}}^n)$ for all $n \ge 0$.

1. By defintion, $\mathsf{W}_{\overline{e}}^n$ is the set of all zeros of $\overline{e}_n$ in $\overline{\mathfrak{p}}$, and we apply 4.6.7.2. First of all, $\overline{e}_n$ is separable, and thus K_π^n/K is Galois. Also, $\overline{e}_n = \overline{e}_{n-1}\overline{g}_{n-1}$, $\overline{g}_{n-1} \in \mathcal{O}_K[X]$ is an Eisenstein polynomial of degree $q^{n-1}(q-1)$, $\overline{g}_{n-1}(0) = \pi$, and $\mathsf{W}_{\overline{e}}^n \setminus \mathsf{W}_{\overline{e}}^{n-1}$ is the set of all zeros of $\overline{g}_{n-1}$. Thus, if $\overline{\alpha} \in \mathsf{W}_{\overline{e}}^n \setminus \mathsf{W}_{\overline{e}}^{n-1}$, then $K(\overline{\alpha})/K$ is fully ramified, $\overline{\alpha}$ is a prime element of $\mathcal{O}_{K(\overline{\alpha})}$, and $\mathcal{O}_{K(\overline{\alpha})} = \mathcal{O}_K[\overline{\alpha}]$ (see [18, Theorem 5.7.1]). Moreover, we get $\mathsf{N}_{K(\overline{\alpha})/K}(-\overline{\alpha}) = \overline{g}_{n-1}(0) = \pi$ and $[K(\overline{\alpha}) \colon K] = \partial(\overline{g}_{n-1}) = q^{n-1}(q-1)$.

If $\sigma \in \mathrm{Gal}(K_\pi^n/K)$, then $e_n(\sigma\alpha) = \sigma e_n(\alpha) = 0$ for all $\alpha \in \mathsf{W}_e^n$. Hence it follows that $\sigma \restriction \mathsf{W}_e^n \in \mathrm{Aut}_{\mathcal{O}_K}(\mathsf{W}_e^n)$, and the map

$$\Psi_n \colon \mathrm{Gal}(K_\pi^n/K) \to \mathrm{Aut}_{\mathcal{O}_K}(\mathsf{W}_e^n), \quad \text{defined by} \quad \Psi_n(\sigma) = \sigma \restriction \mathsf{W}_e^n,$$

is a group monomorphism. If $\alpha \in \mathsf{W}_e^n \setminus \mathsf{W}_e^{n-1}$, then $\overline{\alpha} = h(\alpha) \in \mathsf{W}_{\overline{e}}^n \setminus \mathsf{W}_{\overline{e}}^{n-1}$, $|\overline{\alpha}| = |\alpha|$ and $K(\alpha) = K(\overline{\alpha})$. Now we apply 4.7.7. Since

$$\begin{aligned} q^{n-1}(q-1) = |\mathrm{Aut}_{\mathcal{O}_K}(\mathsf{W}_e^n)| &\ge |\mathrm{Gal}(K_\pi^n/K)| = [K_\pi^n \colon K] \ge [K(\lambda) \colon K] \\ &= q^{n-1}(q-1), \end{aligned}$$

it follows that $K_\pi^n = K(\alpha)$, $|\mathrm{Gal}(K_\pi^n/K)| = |\mathrm{Aut}_{\mathcal{O}_K}(\mathsf{W}_e^n)|$, and Ψ_n is an isomorphism. As the map

$$\psi_n^* \colon U_K/U_K^{(n)} \to \mathrm{Aut}_{\mathcal{O}_K}(\mathsf{W}_e^n), \text{ defined by } \psi_n^*(uU_K^{(n)}) = [u^{-1}]_e,$$

is an isomorphism, it follows that the map

$$\Phi_n = \Psi_n^{-1} \circ \psi_n^* \colon U_K/U_K^{(n)} \to \mathrm{Gal}(K_\pi^n/K),$$

defined by $\Phi_n(uU_K^{(n)})(\alpha) = \Psi_n^{-1}([u^{-1}]_e)(\alpha)$ for all $\alpha \in \mathsf{W}_e^n$, is an isomorphism, too. If $\alpha \in \mathsf{W}_e^n \setminus \mathsf{W}_e^{n-1}$, then

$$\{\sigma\alpha \mid \sigma \in \mathrm{Gal}(K_\pi^n/K\} = \{[u^{-1}]_e(\alpha) \mid u \in U_K\}$$

is the set of all conjugates of α over K, and

$$\{[u^{-1}]_e(\alpha) \mid u \in U_K\} = U_K{\cdot}\alpha = W_e^n \setminus W_e^{n-1}$$

by 4.7.7.1.

$$\begin{array}{ccc} & & U_K/U_K^{(n)} \\ & \swarrow^{\Phi_n} & \downarrow{\scriptstyle \psi_n^*} \\ \mathrm{Gal}(K_\pi^n/K) & \xrightarrow{\Psi_n} & \mathrm{Aut}_{\mathcal{O}_K}(\mathsf{W}_e^n) \end{array}$$

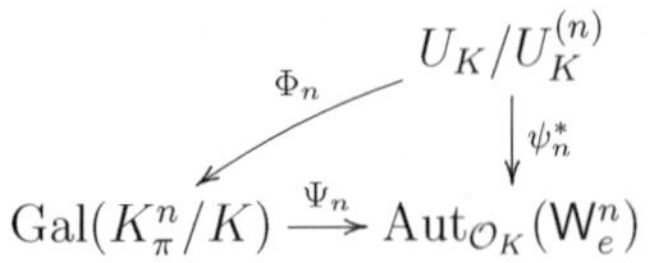

If $\alpha \in \mathsf{W}_e^n \setminus \mathsf{W}_e^{n-1}$ and $\overline{\alpha} = h(\alpha)$, then

$$\mathcal{O}_{K_\pi^n} = h^{-1}(\mathcal{O}_{K_{\overline{\pi}}^n}) = h^{-1}(\mathcal{O}_K[\overline{\alpha}]) = \mathcal{O}_K[\alpha],$$

and as $|\alpha| = |\overline{\alpha}|$, it follows that α is a prime element of $\mathcal{O}_{K_\pi^n}$.

If $\alpha \in (\mathsf{W}_e^1)^\bullet = \{\gamma \in \overline{\mathfrak{p}}^\bullet \mid e(\gamma) = 0\}$ and $\sigma \in \mathrm{Gal}(K_\pi^1/K)$, then we get $\sigma\alpha \in (\mathsf{W}_e^1)^\bullet$, and since $|(\mathsf{W}_e^1)^\bullet| = q-1 = [K_\pi^1 : K]$, $(\mathsf{W}_e^1)^\bullet$ is the set of all conjugates of α over K.

Finally, $\pi = \mathsf{N}_{K_\pi^n/K}(-\overline{\alpha}) \in \mathsf{N}_{K_\pi^n/K}(K_\pi^n)^\times$ implies $(\pi, K_\pi^n/K) = 1$ by 4.2.1.5.

2. Let $m \in [1,n]$. If $u \in U_K$, then $\Phi_n(uU_K^{(n)}) \in \mathrm{Gal}(K_\pi^n/K_\pi^m)$ if and only if $[u^{-1}]_e(\alpha) = \alpha$ for all $\alpha \in \mathsf{W}_e^m$, and this holds if and only if $\Phi_m(uU_K^{(m)}) = 1$, that is, if and only if $u \in U_K^{(m)}$. Hence $\Phi_n(U_K^{(m)}/U_K^{(n)}) = \mathrm{Gal}(K_\pi^n/K_\pi^m)$. If $\alpha \in \mathsf{W}_e^n \setminus \mathsf{W}_e^{n-1}$, then $K_\pi^n = K_\pi^m(\alpha)$, and therefore

$$\{\Phi_n(uU_K^{(n)})(\alpha) \mid u \in U_K^{(m)}\} = \{[u^{-1}]_e(\alpha) \mid u \in U_K^{(m)}\}$$

is the set of all conjugates of α over K_π^m. If $u \in U_K^{(m)}$, then $u^{-1} = 1 + \pi^m y$ for some $y \in \mathcal{O}_K$, and $[u^{-1}]_e(\alpha) = (1 +_e \pi^m y) \cdot_e \alpha = \alpha + (\pi^m y) \cdot_e \alpha$. Hence 4.7.7 implies

$$\{[u^{-1}]_e(\alpha) \mid u \in U_K^{(m)]}\} = \alpha +_e \pi^m \cdot_e (\mathcal{O}_K \cdot_e \alpha) = \alpha +_e \pi^m \cdot_e \mathsf{W}_e^n = \alpha +_e \mathsf{W}_e^{n-m}.$$

3. We continue to use the notions introduced in the proof of 1. above. Let $g \in \mathcal{O}_K [\![X]\!]$ be such that $g(\alpha) = 0$ for all $\alpha \in \mathsf{W}_e^n$. Then $g \circ h^{-1}(\overline{\alpha}) = 0$ for all $\overline{\alpha} \in \mathsf{W}_{\overline{e}}^n$, that is, for all zeros of $\overline{e}_n$. Hence 4.7.2.2 implies $g \circ h^{-1} \in \overline{e}_n \mathcal{O}_K [\![X]\!]$, say $g \circ h^{-1} = \overline{e}_n q$ for some $q \in \mathcal{O}_K [\![X]\!]$. Since $h = Xh_1$ for some $h_1 \in \mathcal{O}_K [\![X]\!]$, it follows that

$$g = g \circ h^{-1} \circ h = \overline{e}_n(h) q(h) = h(e_n) q(h) = e_n h_1(e_n) q(h) \in e_n \mathcal{O}_K [\![X]\!]. \quad \square$$

The following Theorem 4.7.9 relies anew on the Theorems 4.6.4 and 4.6.5. This time, we apply 4.6.4 for the completion $\widehat{K}_\infty$ of the maximal unramified extension of K, and as before we denote by $\varphi \in \mathrm{Gal}(\widehat{K}_\infty/K)$ the continuous extension of the Frobenius automorphism over K.

Theorem 4.7.9. *Let π and $\overline{\pi}$ be prime elements of $\mathcal{O}_K$, $u \in U_K$ such that $\overline{\pi} = \pi u$ and $\varepsilon \in U_{\widehat{K}_\infty}$ such that $\varepsilon^{\varphi - 1} = u$* (see 4.3.8,1). *Let $e \in X\mathcal{O}_K [\![X]\!]$ be a π-series and $\overline{e} \in X\mathcal{O}_K [\![X]\!]$ a $\overline{\pi}$-series. Then there exists a unique power series $T \in \mathcal{O}_{\widehat{K}_\infty} [\![X]\!]$ such that*

$$T \equiv \varepsilon X \bmod X^2 \mathcal{O}_{\widehat{K}_\infty} [\![X]\!] \quad \textit{and} \quad \overline{e} \circ T = T^\varphi \circ e.$$

This power series has the following properties:

(a) $T(F_e(X,Y)) = F_{\overline{e}}(T(X),T(Y))$.

(b) $[a]_{\overline{e}}\circ T = T\circ[a]_e$ *for all* $a \in \mathcal{O}_K$.

(c) $T^\varphi = T\circ[u]_e$.

(d) $T(\mathsf{W}_e^n) = \mathsf{W}_{\overline{e}}^n$ *for all* $n \in \mathbb{N}$, *and* $T(\mathsf{W}_e) = \mathsf{W}_{\overline{e}}$.

Proof. If $L = \varepsilon X \in \mathcal{O}_{\widehat{K}}[X]$, then $\overline{\pi}L = \pi u\varepsilon X = \pi\varepsilon^\varphi X = \pi L^\varphi$, and by 4.6.4 there exists a unique power series $T \in \mathcal{O}_{\widehat{K}_\infty}[\![X]\!]$ such that $T \equiv \varepsilon X \mod X^2\mathcal{O}_K[\![X]\!]$ and $\overline{e}\circ T = T^\varphi\circ e$.

(a) If $F_1 = T\circ F_e \in \mathcal{O}_{\widehat{K}_\infty}[\![X,Y]\!]$ and $F_2 = F_{\overline{e}}(T(X),T(Y)) \in \mathcal{O}_{\widehat{K}_\infty}[\![X,Y]\!]$, then

$$\overline{e}\circ F_1 = \overline{e}\circ T\circ F_e = T^\varphi\circ e\circ F_e = T^\varphi(F_e(e(X),e(Y))) = F_1^\varphi(e(X),e(Y)$$

and

$$\begin{aligned}\overline{e}\circ F_2 &= \overline{e}(F_{\overline{e}}(T(X),T(Y))) = F_{\overline{e}}(\overline{e}\circ T(X),\overline{e}\circ T(Y)) = F_{\overline{e}}(T^\varphi\circ e(X),T^\varphi\circ e(Y)) \\ &= F_2^\varphi(e(X),e(Y)).\end{aligned}$$

Hence the uniqueness in 4.6.4 implies $F_1 = F_2$.

(b) If $a \in \mathcal{O}_K$, $F_1 = [a]_{\overline{e}}\circ T \in \mathcal{O}_{\widehat{K}_\infty}[\![X]\!]$ and $F_2 = T\circ[a]_e \in \mathcal{O}_{\widehat{K}_\infty}[\![X]\!]$, then

$$\overline{e}\circ F_1 = \overline{e}\circ[a]_{\overline{e}}\circ T = [a]_{\overline{e}}\circ\overline{e}\circ T = [a]_{\overline{e}}\circ T^\varphi\circ e = F_1^\varphi\circ e.$$

and

$$\overline{e}\circ F_2 = \overline{e}\circ T\circ[a]_e = T^\varphi\circ e\circ[a]_e = T^\varphi\circ[a]_e\circ e = F_2^\varphi\circ e.$$

Again the uniqueness in 4.6.4 implies $F_1 = F_2$.

(c) If $T_1 = T^{\varphi^{-1}}\circ[u]_e \in \mathcal{O}_{\widehat{K}_\infty}[\![X]\!]$, then

$$T_1 \equiv \varepsilon^{\varphi^{-1}}uX \equiv \varepsilon X \mod X^2\mathcal{O}_{\widehat{K}_\infty}[\![X]\!]$$

and

$$\begin{aligned}\overline{e}\circ T_1 &= \overline{e}\circ T^{\varphi^{-1}}\circ[u]_e = (\overline{e}\circ T)^{\varphi^{-1}}\circ[u]_e = (T^\varphi\circ e)^{\varphi^{-1}}\circ[u]_e \\ &= T\circ e\circ[u]_e = T\circ[u]_e\circ e = T_1^\varphi\circ e.\end{aligned}$$

By the uniqueness of T we get $T = T_1 = T^{\varphi^{-1}}\circ[u]_e$ and thus $T^\varphi = T\circ[u]_e$.

(d) For $n \in \mathbb{N}$ and $\lambda \in \mathsf{W}_e^n$ we obtain

$$[\overline{\pi}^n]_{\overline{e}}(T(\lambda)) = [\overline{\pi}^n]_{\overline{e}}\circ T(\lambda) = T\circ[\overline{\pi}^n]_e(\lambda) = T\circ[u^{-n}]_e\circ[\pi^n]_e(\lambda) = 0,$$

and consequently $T(\lambda) \in \mathsf{W}_{\overline{e}}^n$. Hence $T(\mathsf{W}_e^n) \subset \mathsf{W}_{\overline{e}}^n$, but as $|\mathsf{W}_e^n)| = |\mathsf{W}_{\overline{e}}^n|$ and T is composition-invertible, equality holds. Now $T(\mathsf{W}_e) = \mathsf{W}_{\overline{e}}$ is obvious. □

Theorem 4.7.10. *Let $e \in X\mathcal{O}_K[\![X]\!]$ be a π-series.*

1. *K_π/K is abelian, and there exists a topological isomorphism*

$$\Phi\colon U_K \to \mathrm{Gal}(K_\pi/K)$$

such that $\Phi(u)(\alpha) = [u^{-1}]_e(\alpha)$ for all $u \in U_K$ and $\alpha \in \mathsf{W}_e$.

2. *The field $L = K_\infty K_\pi$ does not depend on π, the map*

$$\mathrm{Gal}(L/K) \to \mathrm{Gal}(K_\infty/K)\times\mathrm{Gal}(K_\pi/K), \quad \textit{defined by } \sigma \mapsto (\sigma\restriction K_\infty, \sigma\restriction K_\pi),$$

is a topological isomorphism, and $K_\pi = L^{(\pi,L/K)}$ is the fixed field of $(\pi, L/K)$.

The occurrence of u^{-1} will become lucid in 4.8.1, and there we shall also prove that $L = K_{\mathrm{ab}}$.

Proof. 1. We apply 4.7.8.1. If $n \in \mathbb{N}$, then K_π^n/K is abelian, and there is an isomorphism $\Phi_n\colon U_K/U_K^{(n)} \overset{\sim}{\to} \mathrm{Gal}(K_\pi^n/K)$ such that $\Phi_n(uU_K^{(n)})(\alpha) = [u^{-1}]_e(\alpha)$ for all $u \in U_K$ and $\alpha \in \mathsf{W}_e^n$. If $m, n \in \mathbb{N}$ and $m \ge n$, then there is a commutative diagram

$$\begin{array}{ccc} U_K/U_K^{(m)} & \xrightarrow{\Phi_n} & \mathrm{Gal}(K_\pi^m/K) \\ {\scriptstyle j}\big\downarrow & & \big\downarrow{\scriptstyle \rho} \\ U_K/U_K^{(n)} & \xrightarrow{\Phi_m} & \mathrm{Gal}(K_\pi^n/K) \end{array},$$

where we have $j(uU_K^{(m)}) = uU_K^{(n)}$ for all $u \in U_K$ and $\rho(\sigma) = \sigma\restriction K_\pi^n$ for all $\sigma \in \mathrm{Gal}(K_\pi^m/K)$.

As K_π is the union of the chain $(K_\pi^n)_{n\ge1}$, we obtain (using 1.5.4 and 1.8.2) a topological isomorphism

$$\Phi = \varprojlim \Phi_n\colon \varprojlim U_K/U_K^{(n)} = U_K \overset{\sim}{\to} \varprojlim \mathrm{Gal}(K_\pi^n/K) = \mathrm{Gal}(K_\pi/K)$$

such that $\Phi(u)\restriction K_\pi^n = \Phi_n(uU_K^{(n)})$ for all $u \in U_K$ and $n \in \mathbb{N}$. In particular, if $\alpha \in \mathsf{W}_e$ and $n \in \mathbb{N}$ is such that $\alpha \in \mathsf{W}_e^n$, then

$$\Phi(u)(\alpha) = \Phi_n(uU_K^{(n)})(\alpha) = [u^{-1}]_e(\alpha).$$

2. Let $\overline{\pi}$ be another prime element of $\mathcal{O}_K$ and $\overline{e} \in X\mathcal{O}_K[\![X]\!]$ a $\overline{\pi}$-series. By 4.7.9(d), there exists a composition-invertible power series $T \in \mathcal{O}_{\widehat{K}_\infty}[\![X]\!]$ such that $\mathsf{W}_{\overline{e}} = T(\mathsf{W}_e) \subset \widehat{K}_\infty(\mathsf{W}_e)$, and then

$$K_{\overline{\pi}} = K(\mathsf{W}_{\overline{e}}) \subset \widehat{K}_\infty(\mathsf{W}_e) = \widehat{K}_\infty K_\pi \subset \widehat{K_\infty K_\pi},$$

where $\widehat{K_\infty K_\pi}$ is a completion of $K_\infty K_\pi$. By definition, $K_{\overline{\pi}}/K$ is separable, hence $K_\infty K_\pi K_{\overline{\pi}}/K_\infty K_\pi$ is separable, too, but $K_\infty K_\pi K_{\overline{\pi}} \subset \widehat{K_\infty K_\pi}$. Hence the extension $K_\infty K_\pi K_{\overline{\pi}}/K_\infty K_\pi$ is purely inseparable (see [18, Corollary 5.5.5]). Therefore we obtain $K_\infty K_\pi K_{\overline{\pi}} = K_\infty K_\pi \supset K_\infty K_{\overline{\pi}}$, and consequently $K_\infty K_\pi = K_\infty K_{\overline{\pi}}$ by symmetry.

As $(\pi, K_\pi/K)\restriction K_\pi^n = (\pi, K_\pi^n/K) = 1$ for all $n \in \mathbb{N}$, we get $(\pi, K_\pi/K) = 1$, hence $(\pi, L/K)\restriction K_\pi = (\pi, K_\pi/K) = 1$ and thus $K_\pi \subset L^{(\pi,L/K)}$. By 4.2.3.2, we obtain $(\pi, L/K) \restriction K_\infty = (\pi, K_\infty/K) = \varphi_K$, hence 4.2.11.1 implies $L^{(\pi,L/K)} \cap K_\infty = K$ and therefore $K_\pi \cap K_\infty = K$.

Since $L = K_\infty K_\pi \subset K_\infty L^{(\pi,L/K)} \subset L$, equality holds, and by 1.8.4 the assignment $\sigma \mapsto \sigma\restriction K_\infty$ defines topological isomorphisms

$$\rho\colon \mathrm{Gal}(L/L^{(\pi,L/K)}) \xrightarrow{\sim} \mathrm{Gal}(K_\infty/K) \quad \text{and} \quad \overline{\rho}\colon \mathrm{Gal}(L/K_\pi) \xrightarrow{\sim} \mathrm{Gal}(K_\infty/K).$$

Since $\mathrm{Gal}(L/L^{(\pi,L/K)}) \subset \mathrm{Gal}(L/K_\pi)$ is closed and $\overline{\rho}\restriction \mathrm{Gal}(L/L^{(\pi,L/K)}) = \rho$, it follows that $\mathrm{Gal}(L/L^{(\pi,L/K)}) = \mathrm{Gal}(L/K_\pi)$, and thus $L^{(\pi,L/K)} = K_\pi$. □

4.8 The reciprocity law of Lubin-Tate

Let K be a local field, $\mathrm{char}(\mathsf{k}_K) = p$ *and* $\mathsf{k}_K = \mathbb{F}_q$.

We continue to use the terminology introduced in Section 4.7.

In 4.8.1, we present an explicit description of the local reciprocity law for Lubin-Tate extensions and, based on this, we derive the main results of local class field theory. In particular, we prove the equality $K_{\mathrm{ab}} = K_\infty K_\pi$ (announced in 4.7.10) and the characterization of norm groups (already announced after 4.2.10).

Theorem 4.8.1. *Let $\pi \in \mathcal{O}_K$ be a prime element, $e \in X\mathcal{O}_K[\![X]\!]$ a π-series, $L = K_\infty K_\pi$ and $G = \mathrm{Gal}(L/K)$.*

1. *Let $r_\pi\colon K^\times \to G$ be a homomorphism. Then $r_\pi = (\,\cdot\,, L/K)$ if and only if the following two conditions are fulfilled:*

   (a) $r_\pi(\pi) \restriction K_\pi = 1$ and $r_\pi(\pi) \restriction K_\infty = \varphi_K$ (the Frobenius automorphism).

   (b) $r_\pi(u)\restriction K_\infty = 1$ and $r_\pi(u)(\alpha) = [u^{-1}]_e(\alpha)$ for all $u \in U_K$ and $\alpha \in \mathsf{W}_e$.

2. The map $(\,\cdot\,, K_\pi/K)\restriction U_K\colon U_K \to \mathrm{Gal}(K_\pi/K)$ *is a topological isomorphism. If* $a = \pi^r u \in K^\times$, *where* $r \in \mathbb{Z}$ *and* $u \in U_K$, *then*

$$(a, K_\pi/K)(\alpha) = (u, K_\pi/K) = [u^{-1}]_e(\alpha) = u^{-1}\cdot_e\alpha \quad \textit{for all } \alpha \in \mathsf{W}_e.$$

3. Let $n \in \mathbb{N}$. *Then* $(\,\cdot\,, K_\pi^n/K)\restriction U_K\colon U_K \to \mathrm{Gal}(K_\pi^n/K)$ *is an epimorphism with kernel* $U_K^{(n)}$, *and* $(U_K^{(n)}, K_\pi/K) = \mathrm{Gal}(K_\pi/K_\pi^n)$. *If* $a = \pi^r u \in K^\times$, *where* $r = v_K(a)$ *and* $u \in U_K$, *then*

$$(a, K_\pi^n/K)(\alpha) = (u, K_\pi^n/K) = [u^{-1}]_e(\alpha) = u^{-1}\cdot_e \alpha \textit{ for all } \alpha \in \mathsf{W}_e^n.$$

If $m \in [0, n]$, *then* $(U_K^{(m)}, K_\pi^n/K) = \mathrm{Gal}(K_\pi^n/K_\pi^m)$, *and the norm residue symbol induces an isomorphism*

$$\Phi_{n,m}\colon U_K^{(m)}/U_K^{(n)} \stackrel{\sim}{\to} \mathrm{Gal}(K_\pi^n/K_\pi^m),$$

given by $\Phi_{n,m}(uU_K^{(n)}) = (u, K_\pi^n/K)$ *for all* $u \in U_K^{(m)}$.

Proof. 1. We prove first:

A. There exists a unique homomorphism $r_\pi\colon K^\times \to G$ such that (a) and (b) hold, and that is continuous.

Proof of **A.** By 4.7.10, there exist topological isomorphisms

$$\Phi\colon U_K \to \mathrm{Gal}(K_\pi/K) \quad\text{and}\quad \rho\colon \mathrm{Gal}(L/K) \to \mathrm{Gal}(K_\infty/K)\times\mathrm{Gal}(K_\pi/K)$$

satisfying $\Phi(u)(\alpha) = [u^{-1}]_e(\alpha)$ for all $u \in U_K$ and $\alpha \in \mathsf{W}_e$, and moreover $\rho(\sigma) = (\sigma\restriction K_\infty, \sigma\restriction K_\pi)$ for all $\sigma \in \mathrm{Gal}(L/K)$. Since $K^\times = \langle\pi\rangle \times U_K$ and $\mathrm{Gal}(K_\infty/K) = \overline{\langle\varphi_K\rangle}$, there exists a unique homomorphism $r_\pi\colon K^\times \to G$ such that (a) and (b) hold, and that is continuous. It is given by

$$r_\pi(\pi^n u)\restriction K_\pi = \Phi(u) \quad\text{and}\quad r_\pi(\pi^n u)\restriction K_\infty = \varphi_K^n$$

for all $n \in \mathbb{Z}$ and $u \in U_K$. $\square$[**A.**]

Now let $r_\pi\colon K^\times \to G$ be the unique (continuous) homomorphism satisfying (a) and (b). We use 4.2.11.2 to show that $r_\pi = (\,\cdot\,, L/K)$, and therefore we must prove:

a. $r_\pi(a)\restriction K_\infty = \varphi_K^{v_K(a)}$ for all $a \in K^\times$.

b. $r_\pi(\overline{\pi})\restriction L^{(\overline{\pi}, L/K)} = 1$ for every prime element $\overline{\pi}$ of $\mathcal{O}_K$.

Proof of **a.** If $a \in K^\times$, then $a = \pi^{v_K(a)}u$, where $u \in U_K$, and

$$r_\pi(a)\restriction K_\infty = (r_\pi(u)\restriction K_\infty)\circ(r_\pi(\pi)\restriction K_\infty)^{v_K(a)} = \varphi_K^{v_K(a)}.$$

Proof of **b.** Let $\Phi\colon U_K \to \mathrm{Gal}(K_\pi/K)$ be the topological isomorphism satisfying $\Phi(u)(\alpha) = [u^{-1}]_e(\alpha)$ for all $u \in U_K$ and $\alpha \in \mathsf{W}_e$. Let $\overline{\pi}$ be a prime element of $\mathcal{O}_K$, $u \in U_K$ such that $\overline{\pi} = \pi u$, and let $\overline{e} \in X\mathcal{O}_K[X]$ be a $\overline{\pi}$-series. By 4.7.10.2, we obtain $L^{(\overline{\pi},L/K)} = K_{\overline{\pi}} = K(\mathsf{W}_{\overline{e}})$, and therefore it suffices to prove that $r_\pi(\overline{\pi})(\alpha) = \alpha$ for all $\alpha \in \mathsf{W}_{\overline{e}}$. As before let $\varphi \in \mathrm{Gal}(\widehat{K}_\infty/K)$ be the continuous extension of the Frobenius automorphism over K, and in common with φ let also $r_\pi(\overline{\pi})\restriction K_\infty$ tacitly be continuously extended to $\widehat{K}_\infty$.

By 4.7.9, there exists a composition-invertible power series $T \in \mathcal{O}_{\widehat{K}_\infty}[\![X]\!]$ satisfying $T^\varphi = T\circ[u]_e$ and $T(\mathsf{W}_e) = \mathsf{W}_{\overline{e}}$. Let $\alpha \in \mathsf{W}_{\overline{e}}$ and $\beta \in \mathsf{W}_e$ such that $\alpha = T(\beta)$. By definition, $r_\pi(\overline{\pi})\restriction K_\pi = r_\pi(u)\circ r_\pi(\pi)\restriction K_\pi = r_\pi(u)\restriction K_\pi = \Phi(u)$, and since $r_\pi(\overline{\pi})\restriction K_\infty = r_\pi(\pi)\circ r_\pi(u)\restriction K_\infty = r_\pi(\pi)\restriction K_\infty = \varphi$, it follows that $r_\pi(\overline{\pi})(a) = a^\varphi$ for all $a \in \widehat{K}_\infty$. We set

$$T = \sum_{n=1}^\infty a_n T^n \in \mathcal{O}_{\widehat{K}_\infty}[\![X]\!]$$

and obtain

$$\begin{aligned} r_\pi(\overline{\pi})(\alpha) &= r_\pi(\overline{\pi})\Big(\sum_{n=1}^\infty a_n\beta^n\Big) = \sum_{n=1}^\infty a_n^\varphi r_\pi(\overline{\pi})(\beta)^n = T^\varphi(r_\pi(\overline{\pi})(\beta)) \\ &= T^\varphi(\Phi(u)(\beta)) = T^\varphi\circ[u^{-1}]_e(\beta) = T\circ[u]_e\circ[u^{-1}]_e(\beta) = T(\beta) = \alpha. \end{aligned}$$

2. If $u \in U_K$, then $(u, K_\pi/K) = (u, L/K)\restriction K_\pi = r_\pi(u)\restriction K_\pi = \Phi(u)$. Hence $(\,\cdot\,, K_\pi/K)\restriction U_K = \Phi\colon U_K \to \mathrm{Gal}(K_\pi/K)$ is a topological isomorphism by 4.7.10.1. If $a = \pi^r u \in K^\times$, where $r \in \mathbb{Z}$ and $u \in U$, and $\alpha \in \mathsf{W}_e$, then

$$\begin{aligned} (a, K_\pi/K)(\alpha) &= (u, K_\pi/K)\circ(\pi, K_\pi/K)^r(\alpha) = (u, K_\pi/K)(\alpha) \\ &= \Phi(u)(\alpha) = [u^{-1}]_e(\alpha). \end{aligned}$$

3. If $n \in \mathbb{N}$, then by 4.7.8.1 it follows that there is an isomorphism $\Phi_n\colon U_K/U_K^{(n)} \overset{\sim}{\to} \mathrm{Gal}(K_\pi^n/K)$ satisfying

$$\Phi_n(uU_K^{(n)})(\alpha) = [u^{-1}]_e(\alpha) = (u, K_\pi/K)(\alpha) \text{ for all } \alpha \in \mathsf{W}_e^n$$

by 2., and therefore

$$\Phi_n(uU_K^{(n)}) = (u, K_\pi/K)\restriction K_\pi^n = (u, K_\pi^n/K) \quad \text{for all} \quad n \in \mathbb{N}.$$

It follows that $(u, K_\pi/K)\restriction K_\pi^n = 1$ if and only if $u \in U_K^{(n)}$, and therefore we obtain $(U_K^{(n)}, K_\pi/K) = \mathrm{Gal}(K_\pi/K_\pi^n)$.

If $a = \pi^r u \in K^\times$, where $r = v_K(a)$ and $u \in U_K$, then it follows by 2. that $(a, K_\pi^n/K) = (a, K_\pi/K)\restriction K_\pi^n = (u, K_\pi/K)\restriction K_\pi^n = (u, K_\pi^n/K)$

If $m \in [0, n]$, then $(U_K^{(m)}, K_\pi^n/K) = (U_K^{(m)}, K_\pi/K)\restriction K_\pi^n = \mathrm{Gal}(K_\pi^n/K_\pi^m)$. If $u \in U_K^{(m)}$, then $(u, K_\pi^n/K) = 1$ if and only if $u \in U_K^{(n)}$, and thus then norm residue symbol induces an isomorphism $\Phi_{n,m}\colon U_K^{(m)}/U_K^{(n)} \overset{\sim}{\to} \mathrm{Gal}(K_\pi^n/K_\pi^m)$ as asserted. □

Theorem 4.8.2. *Let $\pi \in \mathcal{O}_K$ be a prime element, and for $m \in \mathbb{N}$ we denote by $K_m \subset K_\infty$ the (unique) unramified extension of degree m over K.*

1. If $m \in \mathbb{N}$ and $n \in \mathbb{N}_0$, then $\mathsf{N}_{K_m K_\pi^n/K}(K_m K_\pi^n)^\times = \langle \pi^m \rangle \cdot U_K^{(n)}$.

2. A subgroup of $K^\times$ is a norm group if and only if it is open and of finite index. Moreover,

$$K_{\mathrm{ab}} = K_\infty K_\pi = \bigcup_{m,n \in \mathbb{N}} K_m K_\pi^n, \qquad K_\pi = K_{\mathrm{ab}}^{(\pi,K)},$$

and the restriction $\rho\colon \mathrm{Gal}(K_{\mathrm{ab}}/K_\infty) \to \mathrm{Gal}(K_\pi/K)$, defined by $\rho(\sigma) = \sigma \restriction K_\pi$ is a topological isomorphism.

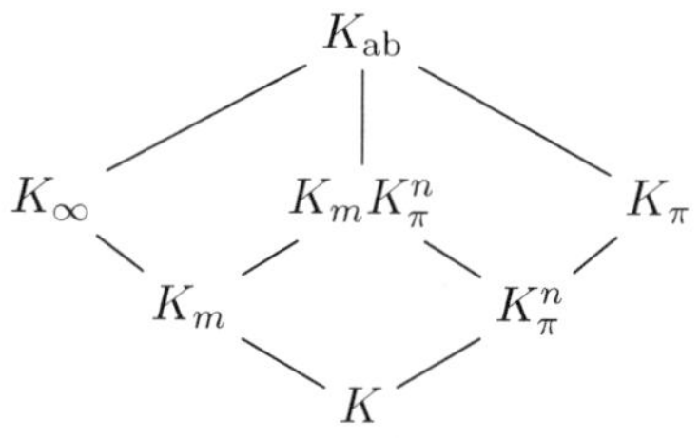

3. Let $B \subset K^\times$ be a norm group and $L \subset K_{\mathrm{ab}}$ the class field to B. Then L/K is unramified if and only if $U_K \subset B$.

4. Let L be an intermediate field of K_{sep}/K_π which is fully ramified over K, and let $\mathcal{E}(L/K)$ be the set of all over K finite intermediate fields of L/K. Then

$$\mathcal{N}(L/K) = \bigcap_{E \in \mathcal{E}(L/K)} \mathsf{N}_{E/K} E^\times = \langle \pi \rangle.$$

$\mathcal{N}(L/K)$ is called the group of **universal norms** of L/K.

Proof. Let $e \in \mathcal{O}_K [\![X]\!]$ be a π-series.

1. Let $m,\, n \in \mathbb{N}$ and $a = \pi^r u \in K^\times$, where $r = v_K(a)$ and $u \in U_K$. By 4.2.3.2 and the local reciprocity law 4.2.7.2, we get $(a, K_m/K) = \varphi_{K_m/K}^r = 1$ (and consequently $a \in \mathsf{N}_{K_m/K} K_m^\times$) if and only if $m \,|\, r$. Hence it follows that $\mathsf{N}_{K_m/K} K_m^\times = \langle \pi^m \rangle \cdot U_K$.

By 4.8.1.3, $(a, K_\pi^n/K) = (u, K_\pi^n/K) = 1$ (and thus $a \in \mathsf{N}_{K_\pi^n/K}(K_\pi^n)^\times)$ holds if and only if $u \in U_K^{(n)}$. Hence $\mathsf{N}_{K_\pi^n/K}(K_\pi^n)^\times = \langle \pi \rangle \cdot U_K^{(n)}$.

Finally, using 4.2.10.2(b), it follows that

$$\begin{aligned}\mathsf{N}_{K_m K_\pi^n/K}(K_m K_\pi^n)^\times &= \mathsf{N}_{K_m/K} K_m^\times \cap \mathsf{N}_{K_\pi^n/K}(K_\pi^n)^\times \\ &= (\langle \pi^m \rangle \cdot U_K) \cap (\langle \pi \rangle \cdot U_K^{(n)}) = \langle \pi^m \rangle \cdot U_K^{(n)}.\end{aligned}$$

2. By 4.1.10.3, every norm group in $K^\times$ is an open subgroup of finite index. Thus let B be an open subgroup of $K^\times$ of finite index. Then there exists some $m \in \mathbb{N}$ such that $\pi^m \in B$, and as $\{U^{(m)} \mid m \in \mathbb{N}\}$ is a fundamental system of neighbourhoods of 1 in $K^\times$, there exists some $n \in \mathbb{N}$ such that $U_K^{(n)} \subset B$. Hence using 1. it follows that $\langle \pi^m \rangle \cdot U_K^{(n)} = \mathsf{N}_{K_m K_\pi^n/K}(K_m K_\pi^n)^\times \subset B$, and by 4.2.10.2(a) there exists an intermediate field M of $K_m K_\pi^n/K$ such that $B = \mathsf{N}_{M/K} M^\times$.

By definition,

$$K_\infty K_\pi = \bigcup_{m,n \in \mathbb{N}} K_m K_\pi^n \subset K_{\mathrm{ab}},$$

and for the reverse inclusion it suffices to prove that every over K finite intermediate field of K_{ab}/K lies in $K_\infty K_\pi$. Thus let E be an over K finite intermediate field of K_{ab}/K. Then $\mathsf{N}_{E/K} E^\times \subset K^\times$ is an open subgroup of finite index by 4.1.10.3. As we have just seen, there exists an over K finite intermediate field M of $K_\infty K_\pi/K$ such that $\mathsf{N}_{E/K} E^\times = \mathsf{N}_{M/K} M^\times$, and 4.2.10.2(c) implies that $E = M \subset K_\infty K_\pi$. Since $K_\infty \cap K_\pi = K$, it follows from 1.8.4 that ρ is a topological isomorphism. Since $K_{\mathrm{ab}} = K_\infty K_\pi$, 4.7.10.2 (applied with $L = K_{\mathrm{ab}}$) implies $K_\pi = K_{\mathrm{ab}}^{(\pi, K_{\mathrm{ab}}/K)} = K_{\mathrm{ab}}^{(\pi,K)}$.

3. $B = \mathsf{N}_{L/K} L^\times \subset K^\times$ is a closed subgroup of finite index. If L/K is unramified and $m = [L\!:\!K]$, then $L = K_m$ and $\mathsf{N}_{L/K} L^\times = \langle \pi^m \rangle \times U_K \supset U_K$ by 1.

As to the converse, assume that $U_K \subset B$, and let $m \in \mathbb{N}$ be such that $\pi^m \in B$. Then $\mathsf{N}_{K_m/K} K_m^\times = \langle \pi^m \rangle \cdot U_K \subset B = \mathsf{N}_{L/K} L^\times$, hence $L \subset K_m$ and L/K is unramified.

4. By 1., we obtain

$$\mathcal{N}(L/K) \subset \mathcal{N}(K_\pi/K) = \bigcap_{n \in \mathbb{N}} \mathsf{N}_{K_\pi^n/K}(K_\pi^n)^\times = \bigcap_{n \in \mathbb{N}} \langle \pi \rangle \times U_K^n = \langle \pi \rangle.$$

Being fully ramified, $\mathcal{N}(L/K)$ contains a prime element (see [18, Theorem 5.7.8]), and therefore $\mathcal{N}(L/K) = \langle \pi \rangle$. □

Theorem 4.8.3.

1. The map $(\,\cdot\,, K) \restriction U_K \colon U_K \to \mathrm{Gal}(K_{\mathrm{ab}}/K_\infty)$ is a topological isomorphism.

2. The map $(\,\cdot\,, K) \colon K^\times \to G_K^{\mathsf{a}}$ is a continuous monomorphism, and $\mathrm{Im}(\,\cdot\,, K) = \{\sigma \in G_K^{\mathsf{a}} \mid \sigma \restriction K_\infty \in \langle \varphi_K \rangle\}$ is dense in G_K^{a}. If $m \in \mathbb{Z}$, $\sigma \in G_K^{\mathsf{a}}$ and $\sigma \restriction K_\infty = \varphi_K^m$, then there exists a unique $x \in K^\times$ such that $\sigma = (x, K)$, and then $m = v_K(x)$.

Proof. 1. The map $\rho\colon \mathrm{Gal}(K_{\mathrm{ab}}/K_\infty) \to \mathrm{Gal}(K_\pi/K)$, $\sigma \mapsto \sigma \restriction K_\pi$, is a topological isomorphism by 4.8.2.2, the map $(\,\cdot\,, K_\pi/K) \restriction U_K \colon U_K \to \mathrm{Gal}(K_\pi/K)$ is a topological isomorphism by 4.8.1.2, and if $u \in U_K$, then

$$(u, K_\pi/K) = (u, K) \restriction K_\pi = \rho \circ (u, K).$$

Hence

$$(\,\cdot\,, K) \restriction U_K = \rho^{-1} \circ (\,\cdot\,, K_\pi/K) \restriction U_K \colon U_K \to \mathrm{Gal}(K_{\mathrm{ab}}/K_\infty)$$

is a topological isomorphism.

2. Since U_K is an open subgroup of $K^\times$, the homomorphism $(\,\cdot\,, K)$ is continuous by 1., and its image is dense in G_K^{a} by 4.2.7.1.

If $x = \pi^m u \in K^\times$, where $\pi \in \mathcal{O}_K$ is a prime element, $m \in \mathbb{Z}$ and $u \in U_K$, then $(x, K) \restriction K_\infty = (x, K_\infty/K) = \varphi_K^m \in \langle \varphi_K \rangle$ by 4.8.1.1. It remains to prove that for every $\sigma \in G_K^{\mathsf{a}}$ with $\sigma \restriction K_\infty \in \langle \varphi_K \rangle$ there exists a unique $x \in K^\times$ such that $\sigma = (x, K)$.

Thus let $\sigma \in G_K^{\mathsf{a}}$ and $\sigma \restriction K_\infty = \varphi_K^m$ for some $m \in \mathbb{Z}$. By 4.8.1.2, there is a unique $u \in U_K$ such that $(u, K_\pi/K) = \sigma \restriction K_\pi$, and we set $x = \pi^m u \in K^\times$. Then

$$(x, K) \restriction K_\pi = (x, K_\pi/K) = (u, K_\pi/K) = \sigma \restriction K_\pi,$$

and

$$(x, K) \restriction K_\infty = (x, K_\infty/K) = \varphi_K^m = \sigma \restriction K_\infty.$$

Hence $(x, K) = \sigma$. □

Theorem 4.8.4. *Let K'/K be a finite field extension.*

1. *The transfer $\mathsf{V}_{K'/K}\colon G_K^{\mathsf{a}} \to G_{K'}^{\mathsf{a}}$ is a monomorphism, and*

$$(x, K') = \mathsf{V}_{K'/K}(x, K) \quad \textit{for all} \quad x \in K^\times.$$

2. *Let L/K be a Galois extension such that $K' \subset L$. Then*

$$(x, L/K') = \mathsf{V}^L_{K'/K}(x, L/K) \quad \textit{for all} \quad x \in K^\times.$$

Proof. The following proof, due to K. Iwasawa [25], is by far not the simplest one, but it provides a deep insight in what is going on.

1. Let $[K' \colon K] = n$. We proceed in 7 steps.

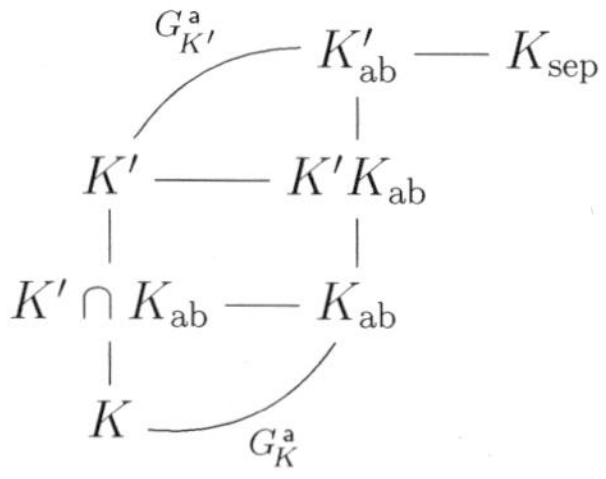

I. Assume that $(x,K') = \mathsf{V}_{K'/K}(x,K)$ for all $x \in K^\times$. Then $\mathsf{V}_{K'/K}$ is a monomorphism.

Proof of **I.** Let $\sigma \in \mathrm{Ker}(\mathsf{V}_{K'/K}) \subset G^{\mathsf{a}} = \mathrm{Gal}(K_{\mathrm{ab}}/K)$. Then 2.6.9.3 implies $1 = \mathsf{V}_{K'/K}(\sigma) \upharpoonright K_{\mathrm{ab}} = \sigma^n \upharpoonright K_{\mathrm{ab}}$, hence $\sigma^n \upharpoonright K_\infty = 1$ and thus $\sigma \upharpoonright K_\infty = 1$, since $\mathrm{Gal}(K_\infty/K) \cong \widehat{\mathbb{Z}}$. It follows that $\sigma \in \mathrm{Gal}(K_{\mathrm{ab}}/K_\infty)$, and by 4.8.3.1 there exists a unique $u \in U_K$ such that $\sigma = (u,K)$. Then

$$1 = \mathsf{V}_{K'/K}(u,K) = (u,K') \in G^{\mathsf{a}}_{K'}$$

and since $(\,\cdot\,,K') \upharpoonright U_{K'} \colon U_{K'} \xrightarrow{\sim} \mathrm{Gal}(K'_{\mathrm{ab}}/K'_\infty)$ is an isomorphism, we get $u = 1$, hence $\sigma = 1$. $\square$[**I.**]

II. Let K'' be an intermediate field of K'/K, and suppose that $(x',K'') = \mathsf{V}_{K''/K'}(x',K')$ for all $x' \in K'^\times$. Then we obtain, for all $x \in K^\times$,

$$(x,K'') = \mathsf{V}_{K''/K}(x,K) \quad \text{if and only if} \quad (x,K') = \mathsf{V}_{K'/K}(x,K).$$

Proof of **II.** Observe that $\mathsf{V}_{K''/K} = \mathsf{V}_{K''/K'} \circ \mathsf{V}_{K'/K}$ by 2.6.9.1 and that $\mathsf{V}_{K''/K'}$ is injective by **I**. Let $x \in K^\times$. If $(x,K'') = \mathsf{V}_{K''/K}(x,K)$, then

$$(x,K'') = \mathsf{V}_{K''/K'}(\mathsf{V}_{K'/K}(x,K)) = \mathsf{V}_{K''/K'}(x,K')$$

and therefore $(x,K') = \mathsf{V}_{K'/K}(x,K)$. Conversely, if $(x,K') = \mathsf{V}_{K'/K}(x,K)$, then

$$\mathsf{V}_{K''/K}(x,K) = \mathsf{V}_{K''/K'} \circ \mathsf{V}_{K'/K}(x,K) = \mathsf{V}_{K''/K'}(x,K') = (x,K''). \qquad \square[\textbf{II.}]$$

III. If K'/K is unramified, then $(x,K') = \mathsf{V}_{K'/K}(x,K)$ holds for all $x \in K^\times$.

Proof of **III.** Let K'/K be unramified. As $K^\times$ is generated by the prime elements of $\mathcal{O}_K$, it suffices to prove that $(\pi,K') = \mathsf{V}_{K'/K}(\pi,K)$ for all prime elements $\pi \in \mathcal{O}_K$. Note that K'/K is cyclic and

$$K \subset K' \subset K_\infty = K'_\infty \subset K_{\mathrm{ab}} \subset K'_{\mathrm{ab}} \subset K_{\mathrm{sep}}.$$

If $\theta \in G_K$, then $\theta(K') = K'$, hence $\theta(K'_{\mathrm{ab}}) = K'_{\mathrm{ab}}$, and therefore K'_{ab}/K is Galois.

Let $\pi \in \mathcal{O}_K$ be a prime element, and let $\psi \in \mathrm{Gal}(K'_{\mathrm{ab}}/K)$ such that $\psi \upharpoonright K_{\mathrm{ab}} = (\pi,K)$, and let $F = (K'_{\mathrm{ab}})^\psi$ be the fixed field of ψ. Then $\psi \upharpoonright K_\infty = (\pi, K_\infty/K) = \varphi_K$, and by 4.8.2.2 we obtain $K_\pi = K_{\mathrm{ab}}^{(\pi,K)} \subset F$. By 4.2.11.1 (applied with $L = K'_{\mathrm{ab}}$), it follows that $FK_\infty = K'_{\mathrm{ab}}$, $F \cap K_\infty = K$, $\mathrm{Gal}(K'_{\mathrm{ab}}/F) = \overline{\langle\psi\rangle} \cong \mathrm{Gal}(K_\infty/K)$, and as F/K is fully ramified, 4.8.2.4 implies $\mathcal{N}(F/K) = \langle\pi\rangle$.

We consider the field $F' = FK' \subset K'_{\rm ab}$. As $F \cap K' = K$, 1.8.4 implies $[F':F] = n$, and therefore $F' = (K'_{\rm ab})^{\psi^n}$. Moreover, $\psi^n \restriction K_\infty = \varphi_K^n = \varphi_{K'}$, and 4.8.3.2 implies $\psi^n = (\pi', K')$ for some prime element $\pi' \in \mathcal{O}_{K'}$. Hence $F' = (K'_{\rm ab})^{(\pi',K')} = K'_{\pi'}$, and since $F' \cap K_\infty = K'$, again 4.8.2.4 implies $\mathcal{N}(F'/K') = \langle \pi' \rangle$.

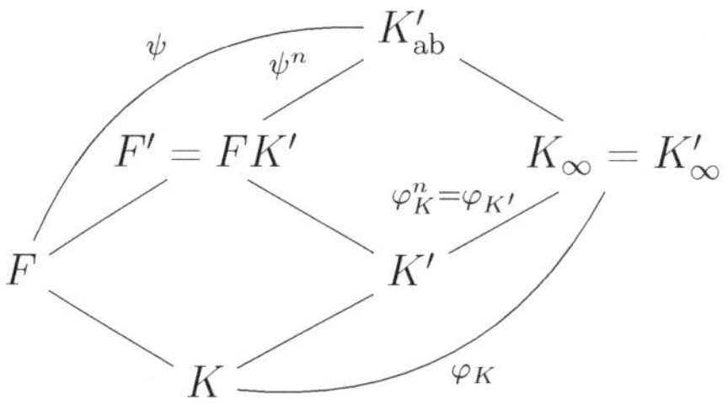

Next we prove that $\mathcal{N}(F/K) \subset \mathcal{N}(F'/K')$. Let $x \in \mathcal{N}(F/K)$. We must prove that $x \in \mathsf{N}_{E'/K'}E'^\times$ for every over K' finite intermediate field E' of F'/K'. Thus let E' be an over K' finite intermediate field of F'/K', and let E be an over K finite intermediate field of F/K such that $E' \subset EK'$. Then it suffices to prove that $x \in \mathsf{N}_{EK'/K'}(EK')^\times$. As K'/K is Galois, $E \cap K' = K$ and all extensions are separable, we get $[EK' : K'] = [E : K]$, and the assignment $\tau \mapsto \tau \restriction K$ defines a bijective map $\theta\colon \mathrm{Hom}^{K'}(EK', K'_{\rm sep}) \to \mathrm{Hom}^K(E, K_{\rm sep})$. If $y \in E$ is such that $x = \mathsf{N}_{E/K}(y)$, then

$$x = \prod_{\tau \in \mathrm{Hom}^K(E, K_{\rm sep})} \tau(y) = \prod_{\tau \in \mathrm{Hom}^{K'}(EK', K'_{\rm sep})} \tau(y)$$
$$= \mathsf{N}_{EK'/K'}(y) \in \mathsf{N}_{EK'/K'}(EK')^\times.$$

From $\langle \pi \rangle = \mathcal{N}(F/K) \subset \mathcal{N}(F'/K') = \langle \pi' \rangle$ we obtain $\pi = \pi'$. Let $\sigma \in G_K$ be such that $\sigma \restriction K'_{\rm ab} = \psi$. Then

$$\mathrm{Gal}(K'/K) = \langle \psi \restriction K' \rangle = \langle \sigma \restriction K' \rangle, \quad \sigma \restriction K_{\rm ab} = \psi \restriction K_{\rm ab} = (\pi, K),$$

and by 2.6.9.4 we obtain

$$\mathsf{V}_{K'/K}(\pi, K) = \mathsf{V}_{K'/K}(\sigma \restriction K_{\rm ab}) = \sigma^n \restriction K'_{\rm ab} = \psi^n = (\pi', K') = (\pi, K').$$

$\square$[**III.**]

IV. If K'/K is Galois and $x \in \mathsf{N}_{K'/K}K'^\times$, then it follows that $(x, K') = \mathsf{V}_{K'/K}(x, K)$.

Proof of **IV.** Let K'/K be Galois, and let R be a set if representatives for $G_{K'} \backslash G_K$ in G_K. Then $\mathrm{Gal}(K'/K) = \{\rho \restriction K' \mid \rho \in R\}$, and we let $x' \in K'^\times$ be such that

$$x = \mathsf{N}_{K'/K}(x') = \prod_{\rho \in R} \rho(x').$$

Let $\sigma \in G_{K'}$ be such that $\sigma \restriction K'_{\rm ab} = (x', K') \in G_{K'}^{\mathsf{a}}$. Then 4.2.4 implies

$$\sigma \restriction K_{\rm ab} = (x', K') \restriction K_{\rm ab} = (\mathsf{N}_{K'/K}(x'), K) = (x, K),$$

and by 2.6.9.2 we obtain

$$\mathsf{V}_{K'/K}(x,K) = \mathsf{V}_{K'/K}(\sigma \restriction K_{\rm ab}) = \prod_{\rho \in R} \rho\sigma\rho^{-1} \restriction K'_{\rm ab}.$$

If $\rho \in R$ and $\rho' = \rho \restriction K'_{\rm ab}$, then

$$\rho\sigma\rho^{-1} \restriction K'_{\rm ab} = \rho'(x',K')\rho'^{-1} = (\rho'(x'),K') = (\rho(x'),K')$$

by 4.2.6, and

$$\mathsf{V}_{K'/K}(x,K) = \prod_{\rho \in R} (\rho(x'),K') = \Big(\prod_{\rho \in R} \rho(x'),K'\Big) = (x,K'). \qquad \square[\textbf{IV.}]$$

V. If K'/K is abelian, then $(x,K') = \mathsf{V}_{K'/K}(x,K)$ for all $x \in K^\times$.

Proof of **V.** Let K'/K be abelian. It is sufficieent to prove that $(\pi,K') = \mathsf{V}_{K'/K}(\pi,K)$ for all prime elements π of $\mathcal{O}_K$. Let $\pi \in \mathcal{O}_K$ be a prime element, E the unique intermediate field of K_∞/K such that $[E:K] = n$, and set $E' = EK' \subset K_{\rm ab}$. If $\sigma \in G_K$ is such that $\sigma \restriction K_{\rm ab} = (\pi,K)$, then $\sigma \restriction E = (\pi,E/K) = \varphi_{E/K}$ and $\mathrm{Gal}(E/K) = \langle \sigma \restriction E\rangle$. Since π is a prime element of $\mathcal{O}_E$ and $E_\infty = K_\infty$, **III** implies $\mathsf{V}_{E/K}(\pi,K) = (\pi,E) = \varphi_E = \varphi_K^n$. Now $[K':K] = n$ implies $(\pi,E) \restriction K' = 1$. As obviously $(\pi,E) \restriction E = 1$, we obtain $(\pi,E) \restriction E' = (\pi,E'/E) = 1$ and consequently $\pi \in \mathsf{N}_{E'/E}E'^\times$ by the local reciprocity law 4.2.7.2. We apply **IV** for the extension E'/E and obtain

$$(\pi,E') = \mathsf{V}_{E'/E}(\pi,E) = \mathsf{V}_{E'/E} \circ \mathsf{V}_{E/K}(\pi,K) = \mathsf{V}_{E'/K}(\pi,K).$$

As E'/K' is unramified, **III** implies $(x',E') = \mathsf{V}_{E'/K'}(x',K')$ for all $x' \in E'^\times$, and **II** (applied with $K'' = E'$) shows that $(\pi,K') = \mathsf{V}_{K'/K}(\pi,K)$. $\square$[**V.**]

VI. If K'/K is Galois, then $(x,K') = \mathsf{V}_{K'/K}(x,K)$ for all $x \in K^\times$.

Proof of **VI.** Let $n > 1$ and proceed by induction on n. As K'/K is solvable, there exists an over K abelian intermediate field K'_1 of K'/K such that $[K':K'_1] < n$. Then $(x',K') = \mathsf{V}_{K'/K'_1}(x,K'_1)$ for all $x' \in K_1'^\times$ by the induction hypothesis, and $(x,K'_1) = \mathsf{V}_{K'_1/K}(x,K)$ for all $x \in K^\times$ by **V**. Hence the assertion follows using **II**. $\square$[**VI.**

VII. *End of proof.* Let $K'' \subset K_{\rm sep}$ be a Galois closure of K'/K. Then **VI** implies $(x,K'') = \mathsf{V}_{K''/K}(x,K)$ for all $x \in K$ and $(x',K'') = \mathsf{V}_{K''/K'}(x'.K')$ for all $x' \in K'$. Using **II**, we obtain $(x,K') = \mathsf{V}_{K'/K}(x,K)$ for all $x \in K$.

2. By 2.6.9.5 (applied with $L' = L \cap K'^{\mathsf{a}}$) we obtain

$$\begin{aligned}(x,L/K') &= (x,K') \restriction L'^{\mathsf{a}} = \mathsf{V}_{K'/K}(x,K) \restriction L'^{\mathsf{a}} = \mathsf{V}^L_{K'/K}((x,K) \restriction L^{\mathsf{a}}) \\ &= \mathsf{V}^L_{K'/K}(x,L/K).\end{aligned}$$

$\square$

In the following Theorems 4.8.5 and 4.8.6 we determine the higher ramification groups $G_i = \mathsf{G}_i(K_\pi^n/K)$ and calculate the different $\mathfrak{D}_{K_\pi^n/K}$ of the fully ramified extensions K_π^n/K. We tacitly use the results from local ramification theory (see [18, Section 5.11] or [57, Ch. IV]).

Theorem 4.8.5. *Let $\pi \in \mathcal{O}_K$ be a prime element, $n \in \mathbb{N}$, $G = \mathrm{Gal}(K_\pi^n/K)$, and for $i \ge -1$ let $G_i = \mathsf{G}_i(K_\pi^n/K)$ be the i-th ramification group of K_π^n/K. Then*

$$G_i = \begin{cases} G & \text{if } i \le 0, \\ \mathrm{Gal}(K_\pi^n/K_\pi^k) & \text{if } i \in [q^{k-1}, q^k - 1] \text{ for some } k \in [1, n-1], \\ \mathbf{1} & \text{if } i \ge q^{n-1}, \end{cases}$$

$$\mathfrak{D}_{K_\pi^n/K} = \mathfrak{p}_{K_\pi^n}^{n(q-1)q^{n-1}-q^{n-1}} = \pi^n \mathfrak{p}_{K_\pi^1}^{-1}\mathcal{O}_{K_\pi^n}, \quad \mathsf{Tr}_{K_\pi^n/K}(\mathcal{O}_{K_\pi^n}) \subset \pi^{n-1}\mathcal{O}_K,$$

and if $m \in [1, n]$, then

$$\mathfrak{D}_{K_\pi^n/K_\pi^m} = \pi^{n-m}\mathcal{O}_{K_\pi^n} \quad \text{and} \quad \mathsf{Tr}_{K_\pi^n/K_\pi^m}(\mathcal{O}_{K_\pi^n}) \subset \pi^{n-m}\mathcal{O}_{K_\pi^m}.$$

Proof. Let $e \in \mathcal{O}_K[\![X]\!]$ be a π-series and $\alpha \in \mathsf{W}_e^n \setminus \mathsf{W}_e^{n-1}$. By 4.7.8.1, K_π^n/K is fully ramified (hence $G_0 = G$), $[K_\pi^n : K] = q^{n-1}(q-1)$, and α is a prime element in $\mathcal{O}_{K_\pi^n}$. For $\sigma \in G$ we set $i_{K_\pi^n/K}(\sigma) = v_{K_\pi^n}(\alpha^\sigma - \alpha)$, and then

$$G_i = \{\sigma \in G \mid i_{K_\pi^n/K}(\sigma) \ge i+1\} \quad \text{for all } i \ge 0.$$

By 4.8.1.3, the norm residue symbol gives occasion to an isomorphism $(\,\cdot\,, K_\pi^n/K)\colon U_K/U_K^{(n)} \xrightarrow{\sim} G$, and $(\,\cdot\,, K_\pi^n/K)$ maps the p-Sylow subgroup $U_K^{(1)}/U_K^{(n)}$ of $U_K/U_K^{(n)}$ onto the p-Sylow subgroup G_1 of G.

Let $\sigma \in G_1$, say $\sigma = (u^{-1}, K_\pi^n/K) \in G_1$, where $u = 1 + \varepsilon\pi^m \in U_K^{(1)}$ for some $\varepsilon \in U_K$ and $m \in \mathbb{N}$. Then $\alpha^\sigma = [u]_e(\alpha)$ by 4.8.1.3, if $m \ge n$, then $u \in U_K^{(n)}$ and therefore $\sigma = 1$. Thus suppose that $m < n$. Then, using 4.7.6.2, it follows that $[\varepsilon\pi^m]_e(\alpha) = [\varepsilon]_e\circ[\pi^m]_e(\alpha) \in \pi^m\cdot_e(\mathsf{W}_e^n \setminus \mathsf{W}_e^{n-1}) = \mathsf{W}_e^{n-m}\setminus\mathsf{W}_e^{n-m-1}$, and $[\varepsilon\pi^m]_e(\alpha)$ is a prime element of $\mathcal{O}_{K_\pi^{n-m}}$. Since $e(K_\pi^n/K_\pi^{n-m}) = q^m$, we get $[\varepsilon\pi^m]_e(\alpha) = \alpha^{q^m}\eta$ for some $\eta \in U_{K_\pi^n}$, and consequently

$$\begin{aligned} \alpha^\sigma &= [1 + \varepsilon\pi^m]_e(\alpha) = [1]_e(\alpha) +_e [\varepsilon\pi^m]_e(\alpha) = \alpha +_e \alpha^{q^m}\eta = F_e(\alpha, \alpha^{q^m}\eta) \\ &\equiv \alpha + \alpha^{q^m}\eta \mod \alpha^{q^m+1}\mathcal{O}_{K_\pi^n}, \end{aligned}$$

which implies $i_{K_\pi^n/K}(\sigma) = v_{K_\pi^n}(\alpha^\sigma - \alpha) = q^m \le q^{n-1}$.

If $i \ge q^{n-1}$ and $\sigma \in G_i$, then $i_{K_\pi^n/K}(\sigma) \ge q^{n-1} + 1$, hence $i_{K_\pi^n/K}(\sigma) = \infty$, and therefore $\sigma = 1$, since $K_\pi^n = K(\alpha)$. Consequently, $G_i = \mathbf{1}$ for all $i \ge q^{n-1}$.

Let $k \in [1, n-1]$ and $i \in [q^{k-1}, q^k - 1]$. If $\sigma \in G_i$, then it follows that $i_{L/K}(\sigma) \ge i + 1 > q^{k-1}$, hence $m \ge k$, $u \in U_K^{(k)}$ and

$$\sigma \in (U_K^{(k)}, K_\pi^n/K) = \mathrm{Gal}(K_\pi^n/K_\pi^k) \text{ by } 4.8.1.3.$$

Conversely, if $\sigma \in \mathrm{Gal}(K_\pi^n/K_\pi^k)$, then $u \in U_K^{(k)}$, hence $m \geq k$, $q^m \geq q^k \geq i+1$ and therefore $\sigma \in G_i$. Altogether, $G_i = \mathrm{Gal}(K_\pi^n/K)$ follows. As

$$v_{K_\pi^n}(\mathfrak{D}_{K_\pi^n/K}) = \sum_{i \geq 0}(|G_i| - 1)$$

(see [18, Theorem 5.11.2]), we obtain

$$\begin{aligned} v_{K_\pi^n}(\mathfrak{D}_{K_\pi^n/K}) &= q^{n-1}(q-1) - 1 + \sum_{k=1}^{n-1}(q^k - q^{k-1})(q^{n-k} - 1) \\ &= q^{n-1}(q-1) - 1 + \sum_{k=1}^{n-1}[q^{n-1}(q-1) - q^{k-1}(q-1)] \\ &= q^{n-1}(q-1) - 1 + (n-1)(q-1)q^{n-1} - (q^{n-1} - 1) \\ &= n(q-1)q^{n-1} - q^{n-1}. \end{aligned}$$

Since $\pi\mathcal{O}_{K_\pi^n} = \mathfrak{p}_{K_\pi^n}^{q^{n-1}(q-1)}$ and $\mathfrak{p}_{K_\pi^1}\mathcal{O}_{K_\pi^n} = \mathfrak{p}_{K_\pi^n}^{q^{n-1}}$, it follows that

$$\begin{aligned} \mathfrak{D}_{K_\pi^n/K} = \mathfrak{p}_{K_\pi^n}^{n(q-1)q^{n-1} - q^{n-1}} &= \pi^n \mathfrak{p}_{K_\pi^1}^{-1}\mathcal{O}_{K_\pi^n}, \quad \mathcal{O}_{K_\pi^n} \\ &= \pi^n \mathfrak{p}_{K_\pi^1}^{-1}\mathfrak{D}_{K_\pi^n/K}^{-1} \subset \pi^{n-1}\mathfrak{D}_{K_\pi^n/K}^{-1}, \end{aligned}$$

and therefore $\mathsf{Tr}_{K_\pi^n/K}(\mathcal{O}_{K_\pi^n}) \subset \pi^{n-1}\mathcal{O}_K$.

If $m \in [1, n]$, then $\mathfrak{D}_{K_\pi^n/K_\pi^m} = \mathfrak{D}_{K_\pi^n/K}(\mathfrak{D}_{K_\pi^m/K}\mathcal{O}_{K_\pi^n})^{-1} = \pi^{n-m}\mathcal{O}_{K_\pi^n}$ by the different tower theorem (see [18, Theorem 5.10.3]), and therefore

$$\mathsf{Tr}_{K_\pi^n/K_\pi^m}(\mathcal{O}_{K_\pi^n}) = \pi^{n-m}\mathsf{Tr}_{K_\pi^n/K_\pi^m}(\mathfrak{D}_{K_\pi^n/K_\pi^m}^{-1}) \subset \pi^{n-m}\mathcal{O}_{K_\pi^m}. \qquad \square$$

In the following Theorem 4.8.6 we use the upper numbering of the ramification groups and the theory of Herbrand (see [18, Theorems 5.11.4, 5.11.5 and 5.11.6]). For a finite Galois extension L/K with $G = \mathrm{Gal}(L/K)$ and ramification groups G_s for $s \geq -1$, we consider the Herbrand function $\eta_{L/K}\colon [-1, \infty) \to \mathbb{R}$, defined by

$$\eta_{L/K}(s) = \int_0^\infty \frac{dx}{(G_0 : G_x)},$$

and its inverse $\psi = \psi_{K_\pi^n/K}\colon [-1, \infty) \to \mathbb{R}$. Then $G^s = G_{\psi(s)}$ for all $s \geq -1$.

Theorem 4.8.6. *Let L/K be a finite abelian extension. Then*

$$(U_K^{(m)}, L/K) = G^m \quad \textit{for all} \ \ m \geq 0,$$

where $G^m = \mathsf{G}^m(L/K)$ denotes the m-th ramification group of L/K in the upper numbering. In particular, $(U_K, L/K) = G_0$, and L/K is unramified if and only if $(U_K, L/K) = \mathbf{1}$.

Proof. SPECIAL CASE 1: $L = K_\pi^n$ for some prime element $\pi \in \mathcal{O}_K$ and $n \in \mathbb{N}$. For $i \geq 0$ we set $G_i = \mathsf{G}_i(K_\pi^n/K)$ and $G^i = \mathsf{G}^i(K_\pi^n/K)$.

If $m \in [1, n-1]$, then

$$\eta_{K_\pi^n/K}(q^m - 1) = \frac{1}{|G_0|}\sum_{j=0}^{q^m-1}|G_j| - 1 = \sum_{j=1}^{q^m-1}\frac{|G_j|}{|G_0|} = \sum_{k=1}^{m}\sum_{j=q^{k-1}}^{q^k-1}\frac{1}{[K_\pi^k\!:\!K]}$$
$$= \sum_{k=1}^{m}\sum_{j=q^{k-1}}^{q^k-1}\frac{1}{q^k(q-1)} = m.$$

Hence $\psi_{K_\pi^n/K}(m) = q^m - 1$, and

$$G^m = G_{q^m-1} = \mathrm{Gal}(K_\pi^n/K_\pi^m) = (U_K^{(m)}, K_\pi^n/K)$$

for all $m \in [1, n-1]$ by 4.8.5 and 4.8.1.3.

If $m \geq n$, then $\psi_{K_\pi^n/K}(m) > \psi_{K_\pi^n/K}(n-1) = q^{n-1} - 1$, and therefore we obtain $G^m = G_{\psi_{K_\pi^n/K}(m)} = \mathbf{1} = (U_K^{(m)}, K_\pi^n/K)$.

SPECIAL CASE 2: L/K is fully ramified. Let π_L be a prime element of $\mathcal{O}_L$. Then $\pi = \mathsf{N}_{L/K}(\pi_L)$ is a prime element of $\mathcal{O}_K$, and there exists some $n \in \mathbb{N}$ such that $\mathsf{N}_{L/K}L^\times \supset \langle\pi\rangle \times U_K^n = \mathsf{N}_{K_\pi^n/K}(K_\pi^n)^\times$. Hence it follows that $L \subset K_\pi^n$, and let $\rho\colon \mathrm{Gal}(K_\pi^n/K) \to \mathrm{Gal}(L/K)$ be defined by $\rho(\sigma) = \sigma \restriction L$. Then we obtain

$$G^m = \mathsf{G}^m(L/K) = \rho(\mathsf{G}^m(K_\pi^n/K)) = \rho(U_K^{(m)}, K_\pi^n/K) = (U_K^{(m)}, L/K).$$

GENERAL CASE: Let T be the inertia field of L/K. Then L/T if fully ramified, T/K is unramified, and $\mathsf{G}_i(L/T) = \mathsf{G}_i(L/K)$ for all $i \geq 0$. Since $\eta_{T/K}(s) = s$ for all $s \geq -1$, it follows that

$$\psi_{T/K}(s) = s \text{ and } \psi_{L/K} = \psi_{T/K}\circ\psi_{L/T}(s) = \psi_{L/T}(s) \text{ for all } s \geq -1.$$

Consequently, if $m \geq 0$, then

$$\mathsf{G}^m(L/K) = \mathsf{G}_{\psi_{L/K}(m)}(L/K) = \mathsf{G}_{\psi_{L/T}(m)}(L/K) = \mathsf{G}_{\psi_{L/T}(m)}(L/T)$$
$$= \mathsf{G}^m(L/T).$$

Since $U_K^{(m)} = \mathsf{N}_{T/K}(U_T^{(m)})$ (see [18, Theorem 5.6.9]), we obtain, using 4.2.4,

$$(U_K^{(m)}, L/K) = (\mathsf{N}_{T/K}U_T^{(m)}, L/K) = (U_T^{(m)}, L/T) = \mathsf{G}^m(L/T) = \mathsf{G}^m(L/K).$$

In particular, recall that $U_K^{(0)} = U_K$ and $G^0 = G_0$ is the ramification group of L/K. □

Corollary and Definition 4.8.7. *Let L/K be a finite abelian extension. Then for $n \in \mathbb{N}_0$ the following assertions are equivalent and hold for all $n \gg 1$:*

(a) $U_K^{(n)} \subset \mathsf{N}_{L/K}L^\times$; (b) $(U_K^{(n)}, L/K) = \mathbf{1}$; (c) $\mathsf{G}^n(L/K) = \mathbf{1}$.

If $n \in \mathbb{N}_0$ is the smallest integer for which these conditions are fulfilled, then we call $\mathfrak{f}_{L/K} = \mathfrak{p}_K^n$ the **conductor** of L/K.

1. *L/K is unramified if and only if $\mathfrak{f}_{L/K} = \mathbf{1}$.*

2. *Let $L_1, \ldots, L_m$ be intermediate fields of L/K, and suppose that $L = L_1 \cdot \ldots \cdot L_m$. Then $\mathfrak{f}_{L/K} = \mathrm{lcm}\,\{\mathfrak{f}_{L_j/K} \mid j \in [1,m]\}$.*

Proof. The equivalence of (a) and (b) holds by the local reciprocity law 4.2.7.2 and that of (b) and (c) by 4.8.6. By 4.1.10.3, (a) holds for $n \gg 1$.

1. By 4.8.2.3.

2. Let $\mathfrak{f}_{L/K} = \mathfrak{p}_K^n$, $\mathfrak{f}_{L_j/K} = \mathfrak{p}_K^{n_j}$ for all $j \in [1,m]$, and set $n^* = \max\{n_1, \ldots, n_m\}$. Then $\mathrm{lcm}\,\{\mathfrak{f}_{L_j/K} \mid j \in [1,m]\} = \mathfrak{p}_K^{n^*}$, and thus we must prove that $n = n^*$.

For $j \in [1,m]$, the inclusion $U_K^{(n)} \subset \mathsf{N}_{L/K}L^\times \subset \mathsf{N}_{L_j/K}L_j^\times$ implies $n_j \le n$, and thus $n^* \le n$. On the other hand, using 4.2.10.2(b), we obtain

$$\mathsf{N}_{L/K}L^\times = \bigcap_{j=1}^m \mathsf{N}_{L_j/K}L_j^\times \supset \bigcap_{j=1}^m U_K^{(n_j)} = U^{(n^*)},$$

and therefore $n \le n^*$. $\square$

The notion of a conductor is ubiquitous in algebraic number theory. Besides with field extensions it is also associated with (multiplicative) characters. Below in 4.8.8 we give a first instance of this concept.

Definition and Theorem 4.8.8. Recall from 1.7.10.1 that the Pontrjagin duality theorem holds for the locally compact group $K^\times$ and for the profinite group U_K.

If $\chi \in \mathsf{X}(K^\times)$, then $\chi(U_K)$ is a finite subgroup of $\mathbb{T}$ by 1.7.9.3, hence $\mathrm{Ker}(\chi) \cap U_K$ is an open subgroup of U_K, and thus there exists some $n \in \mathbb{N}_0$ such that $\chi \restriction U_K^{(n)} = 1$ (recall that $U_K^{(0)} = U_K$).

We call $f_\chi = \min\{n \in \mathbb{N}_0 \colon \chi \restriction U_K^{(n)} = 1\}$ the **conductor exponent** and $\mathfrak{f}_\chi = \mathfrak{p}_K^{f_\chi}$ the **conductor** of χ. We call χ **unramified** if $\mathfrak{f}_\chi = \mathbf{1}$ [equivalently, if $\chi \restriction U_K = 1$].

Let L/K be a finite abelian extension, $G = \mathrm{Gal}(L/K)$ and $\chi \in \mathsf{X}(G)$. Let $K_\chi = L^{\mathrm{Ker}(\chi)}$ be the fixed field of $\mathrm{Ker}(\chi)$, and

$$\widetilde{\chi} = \chi \circ (\,\cdot\,, L/K) \colon K^\times \to \mathbb{T}.$$

Then $\widetilde{\chi} \in \mathsf{X}(K^\times)$, K_χ is the class field to $\mathrm{Ker}(\chi)$, and $\mathfrak{f}_{\widetilde{\chi}} = \mathfrak{f}_{L_\chi/K}$.

Proof. If $a \in K^\times$, then

$$(a, K_\chi/K) = (a, L/K)\restriction K_\chi = 1 \iff (a, L/K) \in \mathrm{Ker}(\chi) \iff a \in \mathrm{Ker}(\widetilde{\chi}).$$

Hence K_χ is the class field to $\mathrm{Ker}(\chi)$. If $\chi \in \mathsf{X}(G)$, then clearly $\widetilde{\chi} \in \mathsf{X}(K^\times)$, and if $n \in \mathbb{N}_0$, then

$$\begin{aligned} \mathfrak{f}_{\widetilde{\chi}} \mid \mathfrak{p}_K^n &\iff \widetilde{\chi}\restriction U_K^{(n)} = 1 \iff (U_K^{(n)}, L/K) \subset \mathrm{Ker}(\chi) \\ &\iff (U_K^{(n)}, L_\chi/K) = 1 \iff \mathfrak{f}_{L_\chi/K} \mid \mathfrak{p}_K^n. \end{aligned}$$

Consequently, $\mathfrak{f}_{\widetilde{\chi}} = \mathfrak{f}_{L_\chi/K}$. □

4.9 Abelian extensions of $\mathbb{Q}_p$.

Let p be a prime and $\overline{\mathbb{Q}}_p$ an algebraic closure of $\mathbb{Q}_p$.

Theorem 4.9.1. *Let $W_p = \mu_{p^\infty}(\overline{\mathbb{Q}}_p)$ be the group of all roots of unity of p-power order and W_p' the group of all roots of unity of an order not divisible by p in $\overline{\mathbb{Q}}_p$.*

1. $(\mathbb{Q}_p)_\infty = \mathbb{Q}_p(W_p')$ is the maximal unramified extension, and $(\mathbb{Q}_p)_{\mathrm{ab}} = \mathbb{Q}_p(W_p \cup W_p')$ is the maximal abelian extension of $\mathbb{Q}_p$. In particular, every finite abelian extension of $\mathbb{Q}_p$ is contained in a cyclotomic extension of $\mathbb{Q}_p$.

2. $(p, \mathbb{Q}_p(W_p)/\mathbb{Q}_p) = 1$, and $(p, \mathbb{Q}_p(W_p')/\mathbb{Q}_p) = \varphi_p$ (the Frobenius automorphism over $\mathbb{Q}_p$).

3. $(\,\cdot\,, \mathbb{Q}_p(W_p)/\mathbb{Q}_p)\restriction \mathbb{Z}_p^\times : \mathbb{Z}_p^\times \xrightarrow{\sim} \mathrm{Gal}(\mathbb{Q}_p(W_p)/\mathbb{Q}_p)$ is a topological isomorphism, and

$$(u, \mathbb{Q}_p(W_p)/\mathbb{Q}_p)(\zeta) = \zeta^{u^{-1}} \quad \textit{for all } u \in \mathbb{Z}_p^\times \textit{ and } \zeta \in W_p.$$

Proof. 1. We recall 4.7.4 and consider the p-series $e = (1+X)^p - 1 \in \mathbb{Z}_p[X]$. If $n \in \mathbb{N}$, then $\mathsf{W}_e^n = \{\alpha \in \overline{\mathfrak{p}} \mid e_n(\alpha) = 0\} = \{\zeta - 1 \mid \zeta \in \overline{\mathbb{Q}}_p,\ \zeta^{p^n} = 1\}$, and therefore $(\mathbb{Q}_p)_e^n = \mathbb{Q}_p(\mathsf{W}_e^n) = \mathbb{Q}_p^{(p^n)}$ is the p^n-th cyclotomic field over $\mathbb{Q}_p$. Furthermore, $W_e = \{\zeta - 1 \mid \zeta \in W_p\}$, and $(\mathbb{Q}_p)_p = \mathbb{Q}_p(W_e) = \mathbb{Q}_p(W_p)$. It is well known that $(\mathbb{Q}_p)_\infty = \mathbb{Q}_p(W_p')$ (see [18, Theorem 5.6.7]), and therefore $(\mathbb{Q}_p)_{\mathrm{ab}} = \mathbb{Q}_p(W_p \cup W_p')$ by 4.8.2.2.

2. By 4.8.1.1.

3. By 4.8.1.2, the map $(\,\cdot\,, \mathbb{Q}_p(W_p)/\mathbb{Q}_p)\restriction\mathbb{Z}_p^\times : \mathbb{Z}_p^\times \to \mathrm{Gal}(\mathbb{Q}_p(W_p)/\mathbb{Q}_p)$ is a topological isomorphism. If $u \in \mathbb{Z}_p^\times$ and $\zeta \in W_p$, then

$$(u, \mathbb{Q}_p(W_p)/\mathbb{Q}_p)(\zeta - 1) = [u^{-1}]_e(\zeta - 1) = [1 + (\zeta - 1)]^{u^{-1}} - 1 = \zeta^{u^{-1}} - 1,$$

and therefore $(u, \mathbb{Q}_p(W_p)/\mathbb{Q}_p)(\zeta) = \zeta^{u^{-1}}$. □

Theorem 4.9.2. *Let $n \in \mathbb{N}$, $\mathbb{Q}_p^{(p^n)} = \mathbb{Q}_p(\zeta)$, where ζ is a primitive p^n-th root of unity, $\mathcal{O}_n = \mathcal{O}_{\mathbb{Q}_p^{(p^n)}}$ and $\mathfrak{p}_n = \mathfrak{p}_{\mathbb{Q}_p^{(p^n)}}$.*

1. $[\mathbb{Q}_p^{(p^n)} : \mathbb{Q}_p] = p^{n-1}(p-1)$, the extension $\mathbb{Q}_p^{(p^n)}/\mathbb{Q}_p$ is fully ramified, $\mathrm{Gal}(\mathbb{Q}_p^{(p^n)}/\mathbb{Q}_p) \cong (\mathbb{Z}/p^n\mathbb{Z})^\times$,

$$\mathcal{O}_n = \mathbb{Z}_p[\zeta] \text{ and } \mathfrak{p}_n = (\zeta - 1)\mathbb{Z}_p[\zeta].$$

2. $\mathfrak{D}_{\mathbb{Q}_p^{(p^n)}/\mathbb{Q}_p} = \mathfrak{p}_n^{p^{n-1}(pn-n-1)} = p^n\mathfrak{p}_1^{-1}\mathcal{O}_n$, and

$$\mathfrak{d}_{\mathbb{Q}_p^{(p^n)}/\mathbb{Q}_p} = p^{p^{n-1}(pn-n-1)}\mathbb{Z}_p.$$

In addition, if $m \in [1, n]$, then $\mathfrak{D}_{\mathbb{Q}_p^{(p^n)}/\mathbb{Q}_p^{(p^m)}} = p^{n-m}\mathcal{O}_n$.

3. $\mathsf{Tr}_{\mathbb{Q}_p^{(p^n)}/\mathbb{Q}_p}(\mathcal{O}_n) \subset p^{n-1}\mathbb{Z}_p$, and

$$\mathsf{Tr}_{\mathbb{Q}_p^{(p^n)}/\mathbb{Q}_p^{(p^m)}}(\mathcal{O}_n) \subset p^{n-m}\mathcal{O}_m \quad \text{for all } m \in [1, n].$$

4. $\mathsf{N}_{\mathbb{Q}_p^{(p^n)}/\mathbb{Q}_p}(\mathbb{Q}_p^{(p^n)})^\times = \langle p\rangle \cdot U_p^{(n)}$.

5. If $u \in \mathbb{Z}_p^\times$ and $k \in \mathbb{Z}$, then $(up^k, \mathbb{Q}_p^{(p^n)}/\mathbb{Q}_p)(\zeta) = \zeta^{u^{-1}}$.

Proof. By 4.7.4, $\mathbb{Q}_p^{(p^n)} = (\mathbb{Q}_p)_p^n$, and $\zeta - 1 \in \mathsf{W}_e^n \setminus \mathsf{W}_e^{n-1}$ (built with the p-polynomial $e = (1+X)^p - 1$).

1. By 4.7.8.1, $\mathbb{Q}_p^{(p^n)}/\mathbb{Q}_p$ is fully ramified,

$$\mathrm{Gal}(\mathbb{Q}_p^{(p^n)}/\mathbb{Q}_p) \cong U_p/U_p^{(n)} \cong (\mathbb{Z}/p^n\mathbb{Z})^\times,$$

$\mathbb{Q}_p^{(p^n)} : \mathbb{Q}_p] = p^{n-1})p - 1)$, $\mathcal{O}_n = \mathbb{Z}_p[\zeta - 1]$, $\zeta - 1$ is a prime element of $\mathcal{O}_n$, and thus $\mathfrak{p}_n = (\zeta - 1)\mathcal{O}_n = (\zeta - 1)\mathbb{Z}_p[\zeta]$.

2. and 3. By 4.8.5.

4. By 4.8.2.1.

5. By 4.9.1. (2 and 3), since $(\,\cdot\,, \mathbb{Q}_p^{(p^n)}/\mathbb{Q}_p) = (\,\cdot\,, \mathbb{Q}_p(W_p)/\mathbb{Q}_p)\restriction\mathbb{Q}_p^{(p^n)}$, and

$$(up^k, \mathbb{Q}_p^{(p^n)}/\mathbb{Q}_p) = (u, \mathbb{Q}_p^{(p^n)}/\mathbb{Q}_p)\circ(p, \mathbb{Q}_p^{(p^n)}/\mathbb{Q}_p)^k.$$ □

Corollary 4.9.3. *Let $p \neq 2$ be a prime and $L/\mathbb{Q}_p$ be a finite fully ramified p-extension. Then there exists some $n \in \mathbb{N}$ such that $L \subset \mathbb{Q}_p^{(p^n)}$. In particular, $L/\mathbb{Q}_p$ is cyclic.*

Proof. Assume first that $L/\mathbb{Q}_p$ is abelian. Since $p \in \mathsf{N}_{L/\mathbb{Q}_p} L^\times$, there exists some $n \in \mathbb{N}$ such that $\mathsf{N}_{L/\mathbb{Q}_p} L^\times \subset \langle p \rangle \cdot U_p^{(n)}$. Hence $L \subset \mathbb{Q}_p^{(p^n)}$ by 4.2.10.2(c), and as $\mathbb{Q}_p^{(p^n)}/\mathbb{Q}_p$ is cyclic, $L/\mathbb{Q}_p$ is cyclic, too.

Now let $L/\mathbb{Q}_p$ be a fully ramified p-extension, $G = \mathrm{Gal}(L/\mathbb{Q}_p)$, and let L_0 be the largest over $\mathbb{Q}_p$ abelian intermediate field of $L/\mathbb{Q}_p$. Then its Galois group $\mathrm{Gal}(L_0/\mathbb{Q}_p) = G^{\mathsf{a}}$ is cyclic, and therefore G is abelian and thus also cyclic. □

Corollary 4.9.4. *Let p be a prime, $n \in \mathbb{N}$ and $n = p^e m$, where $e = v_p(n)$, $m \in \mathbb{N}$, $p \nmid m$, and let $f \in \mathbb{N}$ be minimal such that $p^f \equiv 1 \bmod m$. Then*

$$\mathsf{N}_{\mathbb{Q}_p^{(n)}/\mathbb{Q}_p}(\mathbb{Q}_p^{(n)})^\times = \langle p^f \rangle \cdot U_p^{(e)} \quad \textit{and} \quad \mathfrak{f}_{\mathbb{Q}_p^{(n)}/\mathbb{Q}_p} = \begin{cases} \mathbb{Z}_2 & \textit{if } p = 2 \textit{ and } e = 1, \\ p^e \mathbb{Z}_p & \textit{otherwise} \end{cases}$$

Proof. As $\mathbb{Q}_p^{(n)} = \mathbb{Q}_p^{(p^e)} \mathbb{Q}_p^{(m)}$, we obtain

$$\mathsf{N}_{\mathbb{Q}_p^{(n)}/\mathbb{Q}_p}(\mathbb{Q}_p^{(n)})^\times = \mathsf{N}_{\mathbb{Q}_p^{(p^e)}/\mathbb{Q}_p}(\mathbb{Q}_p^{(p^e)})^\times \cap \mathsf{N}_{\mathbb{Q}_p^{(m)}/\mathbb{Q}_p}(\mathbb{Q}_p^{(m)})^\times = \langle p \rangle \cdot U_p^{(e)} \cap \langle p^f \rangle \cdot U_p$$

by 4.9.2.4 and 4.2.9, since $\mathbb{Q}_p^{(m)}/\mathbb{Q}_p$ is an unramified extension of degree f (see [18, Corollary 5.8.5]). By definition, $\mathfrak{f}_{\mathbb{Q}_p^{(p^n)}/\mathbb{Q}_p} = p^f \mathbb{Z}_p$, where

$$f = \min\{k \in \mathbb{N}_0 \mid U_p^{(k)} \subset \mathsf{N}_{\mathbb{Q}_p^{(m)}/\mathbb{Q}_p}(\mathbb{Q}_p^{(m)})^\times\} = \begin{cases} 0 & \text{if } p = 2 \text{ and } e = 1, \\ e & \text{otherwise} \end{cases}$$

(observe that $U_p/U_p^{(1)} \cong \mathbb{F}_p^\times$, and consequently $U_2 = U_2^{(1)}$). □

4.10 Hilbert symbols

Let K be a local field, $\mathrm{char}(\mathsf{k}_K) = p$ *and* $\mathsf{k}_K = \mathbb{F}_q$.

In this section, we describe the (local) norm residue symbol for Kummer extensions and Artin-Schreier extensions L/K using the defining data from the base field. For the basic algebraic properties of these extensions we refer to [18, Theorems 1.7.8 and 1.7.10], and for a presentation in a more general style to 2.6.6 and 2.6.7.

We start with Artin-Schreier extensions. Apart from the interest for themselves, the corresponding results will prove useful in the development of global class field theory.

Definition and Remarks 4.10.1. Let $K = \mathbb{F}_q((X))$ and $v_K = \mathsf{o}_X$. We set $\wp(X) = X^p - X \in \mathbb{F}_p[X]$, and for $a \in K$ let $\wp^{-1}a = \{x \in \overline{K} \mid \wp(x) = a\}$. If $x \in \wp^{-1}a$, then $\wp^{-1}a = x + \mathbb{F}_p \subset \overline{K}$. We consider the Artin-Schreier extension $K(x) = K(\wp^{-1}a)$, set $G_a = \mathrm{Gal}(K(x)/K)$ and define $\chi_a\colon G_a \to \mathbb{F}_p$ by $\chi_a(\sigma) = \sigma x - x$. Then χ_a is a group monomorphism which only depends on a (and not on x). If $a \notin \wp K$, then χ_a is an isomorphism and $K(\wp^{-1}a)/K$ is a cyclic extension of degree p. Moreover, every cyclic extension L/K of degree p is an Artin-Schreier extension as above.

We define the (local) p-**Hilbert symbol**

$$(\,\cdot\,,\,\cdot\,] = (\,\cdot\,,\,\cdot\,]_K\colon K^\times \times K \to \mathbb{F}_p$$

by $(b,a] = \chi_a(b, K(\wp^{-1}a)/K) = (b, K(\wp^{-1}a)/K)(x) - x$ for $a \in K$ and $b \in K^\times$, where $(\,\cdot\,, K(\wp^{-1}a)/K)\colon K^\times \to G_a$ denotes the norm residue symbol for the Artin-Schreier extensions $K(\wp^{-1}a)/K$.

Theorem 4.10.2. *Assume that* $\mathrm{char}(K) = p > 0$, $a,\ a' \in K$ *and* $b,\ b' \in K^\times$.

1. $(b, a+a'] = (b,a] + (b,a']$ *and* $(bb',a] = (b,a] + (b',a]$.
2. $(b,a] = 0$ *if and only if* $b \in \mathsf{N}_{K(\wp^{-1}a)/K}K(\wp^{-1}a)^\times$.
3. *If* $a \in K^\times$, *then* $(a,a] = 0$.
4. (a) $v_K(\wp(z)) \ne -1$ *for all* $z \in K$, *and consequently* $K \ne \wp K$.
   (b) $(K^\times, a] = \mathbf{0}$ *if and only if* $a \in \wp(K)$.
5. $(b, K] = \mathbf{0}$ *if and only if* $b \in K^{\times p}$.

Proof. Throughout, let $x,\ x' \in \overline{K}$ be such that $\wp(x) = a$ and $\wp(x') = a'$.

1. Observe that $\wp(x+x') = a + a'$, the extension $K(x,x')/K$ is abelian and $x + x' \in K(x,x')$. By 4.2.1.4 we obtain

$$\begin{aligned}(b, a+a'] &= (b, K(x+x')/K)(x+x') - (x+x')\\ &= (b, K(x,x')/K)(x+x') - (x+x')\\ &= [(b, K(x,x')/K)(x) - x] + [(b, K(x,x')/K)(x') - x']\\ &= [(b, K(x)/K)(x) - x] + [(b, K(x')/K)(x') - x'] = (b,a] + (b,a']\end{aligned}$$

and

$$\begin{aligned}(bb',a] &= \chi_a(bb', K(x)/K) = \chi_a((b, K(x)/K)\circ(b', K(x)/K))\\ &= \chi_a(b, K(x)/K) + \chi_a(b', K(x)/K) = (b,a] + (b',a].\end{aligned}$$

2. By definition and the local reciprocity law 4.2.7.2 we obtain:

$$(b,a]=0 \iff (b,K(\wp^{-1}a)/K)(x)=x \iff (b,K(\wp^{-1}a)/K)=1$$
$$\iff b\in \operatorname{Ker}(\,\cdot\,,K(x)/K)=\mathsf{N}_{K(x)/K}K(x)^\times .$$

3. By 2., it suffices to prove that $a\in \mathsf{N}_{K(x)/K}K(x)^\times$. If $x\in K$, there is nothing to do. If $x\notin K$, then $\wp(X)-a$ is the minimal polynomial of x over K and therefore $\mathsf{N}_{K(x)/K}(x)=a$.

4. (a) Let $z\in K$. Apparently, $v_K(z)\ge 0$ implies $v_K(\wp z)\ge 0$. If $v_K(z)<0$, then $v_K(z^p)=pv_K(z)<v_K(z)$, and therefore $v_K(\wp z)=v_K(z^p)\le -2$.

(b) If $a\in \wp K$, then $x\in K$ and therefore $(b,a]=0$ for all $b\in K^\times$. If $a\notin\wp K$, then 4.1.10.1 implies $(K^\times:\mathsf{N}_{K(x)/K}K(x)^\times)=[K(x):K]=p$, and thus $(b,a]\ne 0$ for all $b\in K^\times\setminus \mathsf{N}_{K(x)/K}K(x)^\times$ by 2.

5. If $b=b_0^p$ for some $b_0\in K^\times$, then $(b,z]=p(b_0,z]=0$ for all $z\in K$, and thus $(b,K]=\mathbf{0}$. To prove the converse, we set $M=\{\beta\in K^\times \mid (\beta,K]=0\}$. Then $K^{\times p}\subset M$, $M\subsetneq K^\times$ by 4., and M is a subgroup of $K^\times$ by 1. Since

$$0=(\alpha\beta,\alpha\beta]=(\alpha,\alpha\beta]+(\beta,\alpha\beta] \quad \text{for all}\ \ \alpha,\ \beta\in K^\times,$$

it follows that

$$\begin{aligned} M&=\{\beta\in K^\times \mid (\beta,\alpha\beta]=0 \ \text{ for all }\ \alpha\in K^\times\}\\ &=\{\beta\in K^\times \mid (\alpha,\alpha\beta]=0 \ \text{ for all }\ \alpha\in K^\times\}.\end{aligned}$$

If $\beta,\ \beta'\in M$ and $\beta+\beta'\ne 0$, then $(\alpha,\alpha(\beta+\beta')]=(\alpha,\alpha\beta]+(\alpha,\alpha\beta']=0$ for all $\alpha\in K$, and consequently $\beta+\beta'\in M$. Hence $M\cup\{0\}$ is a subfield of K such that $K^p\subset M$. Since $K=\mathbb{F}_q((X))$, it follows that $K^p=\mathbb{F}_q((X^p))$ and $[K\!:\!K^p]=p$. Hence $M=K^{\times p}$, and we are done. □

To obtain an explicit description of the p-Hilbert symbol, we use the calculus of residues as derived in [18, Section 6.7], and we use the terminology introduced there. For $K=\mathbb{F}_q((X))$ we consider the space Ω_K of local differentials, the universal local derivation $d=d_K\colon K\to\Omega_K$ and the residue map $\operatorname{Res}=\operatorname{Res}_K\colon \Omega_K\to\mathbb{F}_q$.

Theorem 4.10.3. *Let $K=\mathbb{F}_q((X))$ and $b\in K^\times$.*

1. *If $c\in\mathbb{F}_q$, then $(b,c]=v_K(b)\mathsf{Tr}_{\mathbb{F}_q/\mathbb{F}_p}(c)$.*

2. *If $a\in K$ and $t\in\mathcal{O}_K$ is a prime element, then*

$$(b,a]=\big(t,\operatorname{Res}(ab^{-1}db)\big]=\mathsf{Tr}_{\mathbb{F}_q/\mathbb{F}_p}\big(\operatorname{Res}(ab^{-1}db)\big).$$

Proof. Let $q=p^f$, where $f=[\mathbb{F}_q\!:\!\mathbb{F}_p]$. Recall that $v_K=\mathsf{o}_X$, $\mathcal{O}_K=\mathbb{F}_q[\![X]\!]$ and $\mathfrak{p}_K=X\mathbb{F}_q[\![X]\!]$.

1. The assignments $b \mapsto (b, c]$ and $b \mapsto v_K(b)\,\mathsf{Tr}_{\mathbb{F}_q/\mathbb{F}_p}(c)$ define homomorphisms $K^\times \to \mathbb{F}_p$, and $K^\times$ is generated by the prime elements of $\mathcal{O}_K$. Hence it suffices to prove that $(b, c] = \mathsf{Tr}_{\mathbb{F}_q/\mathbb{F}_p}(c)$ for all prime elements $b \in \mathcal{O}_K$ and all $c \in \mathbb{F}_q$.

Thus let $b \in \mathcal{O}_K$ be a prime element, hence $K = \mathbb{F}_q((b))$ (see [18, Theorem 5.3.7]), let $c \in \mathbb{F}_q$, $x \in \wp^{-1}c$ and $K_1 = K(\wp^{-1}c) = K(x) = K\mathbb{F}_q(x)$. Hence $K_1 = \mathbb{F}_q(x)((b))$, $\mathsf{k}_{K_1} = \mathbb{F}_q(x)$, and the extension K_1/K is unramified (see [18, Theorems 6.1.9 and 6.6.8]). By 4.2.3.2 it follows that $(b, K_1/K)$ is the Frobenius automorphism of K_1/K, hence $(b, K_1/K)(x) = x^q = x^{p^f}$. As $\mathrm{Gal}(\mathbb{F}_q/\mathbb{F}_p) = \langle x \mapsto x^p \rangle$, we obtain

$$\begin{aligned}\mathsf{Tr}_{\mathbb{F}/\mathbb{F}_p}(c) &= \sum_{i=0}^{f-1} c^{p^i} = \sum_{i=0}^{f-1} (x^p - x)^{p^i} = \sum_{i=0}^{f-1} (x^{p^{i+1}} - x^{p^i}) = x^{p^f} - x \\ &= (b, K_1/K)(x) - x = (b, c].\end{aligned}$$

2. It suffices to prove the first equality. Then the second one follows by 1., applied with $c = \mathrm{Res}(ab^{-1}db)$.

If $b_1, b_2 \in K^\times$,then $(b_1b_2)^{-1}d(b_1b_2) = b_1^{-1}db_1 + b_2^{-1}db_2$. Hence the assignment $b \mapsto (t, \mathrm{Res}(ab^{-1}db)]$ defines a homomorphism $K^\times \to \mathbb{F}_p$, and the same is true for the assignment $b \mapsto (b, a]$. As $K^\times$ is generated by the prime elements of $\mathcal{O}_K$, it suffices to prove that $(x, a] = (t, \mathrm{Res}(ax^{-1}dx)]$ for all prime elements t, $x \in \mathcal{O}_K$ and all $a \in K$.

Thus let t, $x \in \mathcal{O}_K$ be prime elements and $a \in K$. By 1. we obtain

$$(t, \mathrm{Res}(ax^{-1}dx)] = \mathsf{Tr}_{\mathbb{F}_q/\mathbb{F}_p}(\mathrm{Res}(ax^{-1}dx)) = (x, \mathrm{Res}(ax^{-1}dx)],$$

and therefore it suffices to prove that

$$(x, a] = (x, \mathrm{Res}(ax^{-1}dx)]. \tag{$*$}$$

As the assignments $a \mapsto (x, a]$ and $a \mapsto (x, \mathrm{Res}(ax^{-1}dx)]$ define homomorphisms $K \to \mathbb{F}_p$, it suffices to consider the following three cases: **1)** $a \in \mathbb{F}_q$; **2)** $v_K(a) \geq 1$; **3)** $a = ux^{-n}$ for some $u \in \mathbb{F}_q^\times$ and $n \in \mathbb{N}$.

1) If $a \in \mathbb{F}_q$, then $\mathrm{Res}(ax^{-1}dx) = a\mathrm{Res}_x(x^{-1}) = a$, and there is nothing to do.

2) If $v_K(a) \geq 1$, then

$$z = -\sum_{i=0}^{\infty} a^{p^i} \in K, \quad \text{and} \quad a = \wp z \in \wp K.$$

Hence $(x, a] = 0$, and $\mathrm{Res}(ax^{-1}dx) = \mathrm{Res}_x(ax^{-1}) = 0$ since $v_K(ax^{-1}) \geq 0$.

3) If $a = ux^{-n}$, where $u \in \mathbb{F}_q^\times$ and $n \in \mathbb{N}$, then we obtain $\mathrm{Res}(ax^{-1}dx) = \mathrm{Res}_x(ux^{-n-1}) = 0$, and we set $L = K(\wp^{-1}(ux^{-n}))$. We shall prove that $x \in \mathsf{N}_{L/K}L^\times$. Then 4.10.2.2 implies $(x, ux^{-n}] = 0$, and we are done.

If $L = K$, there is nothing to do. Thus let $[L:K] = p$ and proceed by induction on $v_p(n)$. Let $z \in \wp^{-1}(ux^{-n})$, hence $ux^{-n} = \mathsf{N}_{L/K}(z)$, and let $v \in \mathbb{F}_q$ such that $v^p = u$.

If $p \nmid n$, then there exist ν, $\mu \in \mathbb{Z}$ such that $1 = -n\nu + p\mu$, and we obtain $x = v^{-p\nu}(ux^{-n})^\nu (x^\mu)^p = \mathsf{N}_{L/K}(v^{-\nu} z^\nu x^\mu)$.

If $p \mid n$, then $n = pm$, where $m \in \mathbb{N}$. By the induction hypothesis there exists some $y \in L$ such that $\mathsf{N}_{L/K}(y) = vx^{-m}$, whence $ux^{-n} = \mathsf{N}_{L/K}(y^p)$. $\square$

Now we turn to the case of Kummer extensions.

Theorem and Definition 4.10.4. *Let* $|\mu_n(K)| = n \geq 2$, $L = K(\sqrt[n]{K^\times})$ *and* $G = \mathrm{Gal}(L/K)$.

1. *L/K is a finite abelian extension of exponent n, and* $\mathsf{N}_{L/K} L^\times = K^{\times n}$.

2. For a, $b \in K^\times$ we define the **Hilbert symbol** of exponent n for K by

$$(a,b) = (a,b)_{K,n} = \frac{(a, L/K)(\sqrt[n]{b})}{\sqrt[n]{b}} = \frac{(a, K(\sqrt[n]{b})/K)(\sqrt[n]{b})}{\sqrt[n]{b}} \in \mu_n(K).$$

Note that this definition does not depend on the choice of $\sqrt[n]{b}$.

The Hilbert symbol has the following properties:

Let a, b, a', $b' \in K^\times$.

(a) $(aa', b) = (a,b)(a',b)$ *and* $(a, bb') = (a,b)(a,b')$.

(b) $(a,b) = 1$ *if and only if* $a \in \mathsf{N}_{K(\sqrt[n]{b})/K} K(\sqrt[n]{b})^\times$.

(c) If $x \in K$ *and* $x^n \neq b$, *then* $(x^n - b, b) = 1$. *In particular,* $(a, -a) = 1$. *If* $a \neq 1$, *then* $(a, 1-a) = 1$, *and if n is odd, then* $(a,a) = 1$.

(d) $(a,b)(b,a) = 1$.

3. *The Hilbert symbol defines a non-degenerate pairing*

$$K^\times/K^{\times n} \times K^\times/K^{\times n} \to \mu_n(K) \quad by \quad (aK^{\times n}, bK^{\times n}) \mapsto (a,b).$$

Proof. 1. Recall that $|\mu_n(K)| = n$ implies $\mathrm{char}(K) \nmid n$ and $(K^\times : K^{\times n}) < \infty$ (see [18, Theorem 5.8.10]). By 2.6.6, L/K is a finite abelian extension of exponent n, and the map

$$\delta\colon K^\times/K^{\times n} \to \mathsf{X}(G) = \mathrm{Hom}(G, \mu_n(K)), \quad \text{defined by } \delta(aK^{\times n})(\sigma) = \sqrt[n]{a}^{\,\sigma-1},$$

is an isomorphism. According to the local reciprocity law 4.2.7.2, the norm residue symbol yields an isomorphism

$$(\,\cdot\,, L/K)\colon K^\times/\mathsf{N}_{L/K}L^\times \to G.$$

Hence $K^{\times n} \subset \mathsf{N}_{L/K}L^\times$, and equality holds, since

$$|G| = |\mathsf{X}(G)| = (K^\times : K^{\times n}) \le (K^\times : \mathsf{N}_{L/K}L^\times) = |G|.$$

2. (a) By the very definition,

$$\begin{aligned}(aa', b)\sqrt[n]{b} &= (aa', L/K)(\sqrt[n]{b}\,) = (a, L/K)\circ(a', L/K)(\sqrt[n]{b}\,) \\ &= (a, L/K)[(a', b)\sqrt[n]{b}\,] = (a', b)(a, L/K)(\sqrt[n]{b}\,) = (a', b)(a, b)\sqrt[n]{b}\end{aligned}$$

and

$$\begin{aligned}(a, bb')\sqrt[n]{bb'} &= (a, L/K)(\sqrt[n]{b}\,\sqrt[n]{b'}\,) = [(a, L/K))\sqrt[n]{b}\,)][(a, L/K)(\sqrt[n]{b'}\,)] \\ &= [(a, b)\sqrt[n]{b}\,][(a, b')\sqrt[n]{b'}\,] = (a, b)(a, b')\sqrt[n]{bb'}.\end{aligned}$$

Hence $(aa', b) = (a, b)(a', b)$ and $(a, bb') = (a, b)(a, b')$.

(b) By the local reciprocity law 4.2.7.2 we obtain:

$$\begin{aligned}(a, b) = 1 \iff (a, K(\sqrt[n]{b}\,)/K)(\sqrt[n]{b}\,) = \sqrt[n]{b} \iff (a, K(\sqrt[n]{b}\,)/K) = 1 \\ \iff a \in \mathsf{N}_{K(\sqrt[n]{b}\,)/K}K(\sqrt[n]{b}\,)^\times.\end{aligned}$$

(c) Let $\zeta \in \mu_n(K)$ be a primitive n-th root of unity, $\beta \in L$ such that $\beta^n = b$ and $[K(\beta) : K] = d$. Then $d\,|\,n$, say $n = dk$, where $k \in \mathbb{N}$, and $\mathrm{Gal}(K(\beta)/K) = \langle\sigma\rangle$, where $\sigma(\beta) = \zeta^k\beta$ (indeed, ζ^k is a primitive d-th root of unity). Let $x \in K$ be such that $x^n \ne b$. Then

$$\begin{aligned}x^n - b &= \prod_{i=0}^{n-1}(x - \zeta^i\beta) = \prod_{\nu=0}^{k-1}\prod_{j=0}^{d-1}(x - \zeta^{kj}(\zeta^\nu\beta)) = \prod_{\nu=0}^{k-1}\prod_{j=0}^{d-1}\sigma^j(x - \zeta^\nu\beta) \\ &= \prod_{\nu=0}^{k-1}\mathsf{N}_{K(\beta)/K}(x - \zeta^\nu\beta) = \mathsf{N}_{K(\beta)/K}\Big(\prod_{\nu=0}^{k-1}(x - \zeta^\nu\beta)\Big) \\ &\in \mathsf{N}_{K(\sqrt[n]{b}\,)/K}K(\sqrt[n]{b}\,)^\times,\end{aligned}$$

and therefore $(x^n - b, b) = 1$ by (b). If $x = 0$ and $b = -a$, then $(a, -a) = 1$, and if $x = 1$ and $b = 1 - a$ (for $a \ne 1$), then $(a, 1 - a) = 1$. If n is odd, then $1 = (a, -1)^n = (a, -1)$, and $(a, a) = (a, -a)(a, -1) = 1$.

(d) We have

$$\begin{aligned}(a, b)(b, a) &= (a, -a)(a, b)(b, a)(b, -b) \\ &= (a, -ab)(b, -ab) = (ab, -ab) = 1.\end{aligned}$$

3. If $a \in K^\times$, then $(a, y) = 1$ holds for all $y \in K^\times$ if and only if $a \in \mathsf{N}_{K(\sqrt[n]{y}\,)/K}K(\sqrt[n]{y}\,)^\times$ for all $y \in K^\times$, and by 4.2.10.2(b) this holds if and only if $a \in \mathsf{N}_{L/K}L^\times = K^{\times n}$.

If $b \in K^\times$, then $(x, b) = 1$ holds for all $x \in K^\times$ if and only if $K^\times = \mathsf{N}_{K(\sqrt[n]{b}\,)/K}K(\sqrt[n]{b}\,)^\times$, and this holds if and only if $K(\sqrt[n]{b}\,) = K$, that is, $b \in K^{\times n}$.

Therefore, and by 2.(a), the Hilbert symbol induces a non-degenerate pairing as asserted. □

If $(q,n)=1$, then $|\mu_n(K)|=n$ if and only if $q\equiv 1 \bmod n$. In this case, every $u\in U_K$ has a unique factorization $u=\omega(u)u_1$, where $\omega(u)\in\mu_{q-1}(K)$ and $u_1\in U_K^{(1)}$ (see [18, Theorem 5.8.10]). In particular, if $\zeta\in\mu_{q-1}(K)$, then $\omega(u)=\zeta$ if and only if $u\equiv\zeta \bmod \mathfrak{p}_K$.

If $(q,n)=1$, then the Hilbert symbol $(\,\cdot\,,\,\cdot\,)_{K,n}$ is called a **tame symbol**, otherwise a **wild symbol**.

In the following Theorem 4.10.5 we give explicit formulas for the tame symbol. Corresponding explicit formulas for the wild symbol were independently given by H. Brückner [3] and S. V. Vostokov [62]. A thorough presentation of these explicit formulas is beyond the scope of this volume, and we refer to [12].

Theorem 4.10.5. *Let* $|\mu_n(K)|=n\ge 2,\ \ q\equiv 1 \bmod n,\ \ a,\,b\in K^\times$, *and set* $\alpha=v_K(a)$ *and* $\beta=v_K(b)$. *Then*

$$(a,b)=(a,b)_{K,n}=\omega\big((-1)^{\alpha\beta}b^\alpha a^{-\beta}\big)^{(q-1)/n}\equiv\big((-1)^{\alpha\beta}b^\alpha a^{-\beta}\big)^{(q-1)/n} \mod \mathfrak{p}_K.$$

In particular, if $a,\,b\in U_K$, *then* $(a,b)=1$.

Proof. It suffices to prove that $(a,b)\equiv((-1)^{\alpha\beta}b^\alpha a^{-\beta})^{(q-1)/n} \bmod \mathfrak{p}_K$.

We assume first that $b\in U_K$. Then the extension $K(\sqrt[n]{b}\,)/K$ is unramified (see [18, Theorem 5.9.9]), and if $\varphi=\varphi_{K(\sqrt[n]{b})/K}$ is its Frobenius automorphism, then $\varphi(\sqrt[n]{b})=\zeta\sqrt[n]{b}$ for some $\zeta\in\mu_n(K)$, and $\varphi^\alpha(\sqrt[n]{b}\,)=\zeta^\alpha\sqrt[n]{b}$. Since $\varphi(\sqrt[n]{b})\equiv(\sqrt[n]{b})^q \bmod \mathfrak{p}_{K(\sqrt[n]{b})}$, we get $\zeta\equiv(\sqrt[n]{b})^{q-1} \bmod \mathfrak{p}_{K(\sqrt[n]{b})(K}$ and therefore $\zeta\equiv b^{(q-1)/n} \bmod \mathfrak{p}_K$ (since $\mathfrak{p}_K=\mathfrak{p}_{K(\sqrt[n]{b})}\cap K$ and $n\,|\,q-1$). By 4.2.3.2, we obtain

$$(a,b)\sqrt[n]{b}=(a,K(\sqrt[n]{b}\,)/K)(\sqrt[n]{b}\,)=\varphi^\alpha(\sqrt[n]{b}\,)=\zeta^\alpha\sqrt[n]{b},$$

and therefore

$$(a,b)=\zeta^\alpha\equiv b^{\alpha(q-1)/n} \mod \mathfrak{p}_K.$$

In particular, $(u,b)=1$ for all $u\in U_K$ and $(\pi,b)\equiv b^{(q-1)/n} \bmod \mathfrak{p}_K$ for all prime elements $\pi\in\mathcal{O}_K$.

Now we can do the general case. Suppose that $a=\pi^\alpha u$ and $b=\pi^\beta v$ for some prime element $\pi\in\mathcal{O}_K$ and $u,\,v\in U_K$. Observing $(\pi,-\pi)=1$, we obtain

$$\begin{aligned}(a,b)&=(\pi,b)^\alpha(u,b)=(\pi,\pi)^{\alpha\beta}(\pi,v)^\alpha(u,\pi)^\beta(u,v)\\&=(\pi,-\pi)^{\alpha\beta}(\pi,-1)^{\alpha\beta}(\pi,v)^\alpha(\pi,u)^{-\beta}(u,v)=(\pi,(-1)^{\alpha\beta}v^\alpha u^{-\beta})\\&\equiv\big((-1)^{\alpha\beta}v^\alpha u^{-\beta}\big)^{(q-1)/n} \mod \mathfrak{p}_K.\end{aligned}$$

□

Remark and Definition 4.10.6. Let $|\mu_n(K)| = n \geq 2$ and $q \equiv 1 \bmod n$. If $u \in U_K$ and $\pi \in \mathcal{O}_K$ is a prime element, then

$$\left(\frac{u}{\mathfrak{p}_K}\right)_n = (\pi, u)_{K,n} = \omega(u^{(q-1)/n}) \in \mu_n(K)$$

does not depend on π and is called the **n-th power residue symbol** for K. By definition,

$$\left(\frac{u}{\mathfrak{p}_K}\right)_n \equiv u^{(q-1)/n} \mod \mathfrak{p}_K$$

and since $\mathsf{k}_K^\times$ is cyclic of order $q-1$, we obtain the following equivalence which expounds the name power residue symbol:

$$\left(\frac{u}{\mathfrak{p}_K}\right)_n = 1 \iff u + \mathfrak{p}_K \in \mathsf{k}_K^{\times n}$$
$$\iff \text{there exists some } x \in U_K \text{ satisfying } x^n \equiv u \mod \mathfrak{p}_K.$$

In this case, u is called an **n-th power residue** modulo $\mathfrak{p}_K$.

We specialize our considerations to the simplest case $K = \mathbb{Q}_p$ for some prime p and $n = 2$ (then $U_K = U_p = \mathbb{Z}_p^\times$, and $\mathfrak{p}_K = p\mathbb{Z}_p$). We obtain the classical Hilbert symbol

$$\left(\frac{a, b}{p}\right) = (a, b)_{\mathbb{Q}_p,2} \in \{\pm 1\} \quad \text{for all} \quad a,\, b \in \mathbb{Q}_p^\times,$$

and the classical Legendre symbol:

$$\left(\frac{a}{p}\right) = \left(\frac{a}{p\mathbb{Z}_p}\right)_2 \in \{\pm 1\} \quad \text{for all} \quad a \in U_p.$$

Theorem 4.10.7 (Classical Hilbert and Legendre symbols).

1. *Let p be a prime and a, a', $b \in \mathbb{Q}_p^\times$. Then*

$$\left(\frac{a, b}{p}\right) = \left(\frac{b, a}{p}\right), \quad \left(\frac{aa', b}{p}\right) = \left(\frac{a, b}{p}\right)\left(\frac{a', b}{p}\right), \quad \left(\frac{a, a}{p}\right) = \left(\frac{a, -1}{p}\right),$$

and if $a \neq 1$, then

$$\left(\frac{a, 1-a}{p}\right) = 1.$$

2. *Let $p \neq 2$ be a prime.*

(a) If $u \in U_p$, then

$$\left(\frac{u}{p}\right) \equiv u^{(p-1)/2} \mod p.$$

(b) *Let* $a,\, b \in \mathbb{Q}_p^\times$, $a = p^\alpha a'$ *and* $b = p^\beta b'$, *where* $\alpha,\, \beta \in \mathbb{Z}$ *and* $a',\, b' \in U_p$. *Then*

$$\Big(\frac{a,b}{p}\Big) = \Big(\frac{(-1)^{\alpha\beta} a^\beta b^\alpha}{p}\Big) = \Big(\frac{-1}{p}\Big)^{\alpha\beta}\Big(\frac{a'}{p}\Big)^\beta\Big(\frac{b'}{p}\Big)^\alpha.$$

3. If $a,\, b \in U_2 = 1 + 2\mathbb{Z}_2$, *then*

$$\Big(\frac{2,a}{2}\Big) = (-1)^{(a^2-1)/8} \quad \textit{and} \quad \Big(\frac{a,b}{2}\Big) = (-1)^{(a-1)(b-1)/4}.$$

Proof. 1. By 4.10.4.2.

2. (a) Obvious by the definition.

(b) By 4.10.5 we obtain

$$\Big(\frac{a,b}{p}\Big) = (a,b)_{\mathbb{Q}_p,2} \equiv \Big((-1)^{\alpha\beta} b^\alpha a^{-\beta}\Big)^{(p-1)/2} \equiv \Big(\frac{(-1)^{\alpha\beta} b^\alpha a^{-\beta}}{p}\Big) \bmod p,$$

and consequently

$$\Big(\frac{(-1)^{\alpha\beta} b^\alpha a^{-\beta}}{p}\Big) = \Big(\frac{(-1)^{\alpha\beta} b'^\alpha a^{-\prime\beta}}{p}\Big) = \Big(\frac{-1}{p}\Big)^{\alpha\beta}\Big(\frac{a'}{p}\Big)^\beta\Big(\frac{b'}{p}\Big)^\alpha.$$

3. We define $\varepsilon,\, \eta \colon U_2 \to \mathbb{Z}/2\mathbb{Z}$ by

$$\varepsilon(a) = (-1)^{(a-1)/2} \text{ and } \eta(a) = (-1)^{(a^2-1)/8},$$

and we prove first that $\varepsilon(a_1a_2) = \varepsilon(a_1)\varepsilon(a_2)$ and $\eta(a_1a_2) = \eta(a_1)\eta(a_2)$ for all $a_1,\, a_2 \in U_2$.

For $i \in \{1,2\}$, we set $a_i = 1 + 2u_i$ for some $u_i \in \mathbb{Z}_2$. Then we obtain

$$a_1a_2 - 1 \equiv 2u_1 + 2u_2 \bmod 4 \quad \text{and} \quad a_1^2a_2^2 - 1 \equiv 4u_1^2 + 4u_1 + 4u_2^2 + 4u_2 \bmod 16,$$

$$\frac{a_1a_2-1}{2} \equiv u_1 + u_2 \equiv \frac{a_1-1}{2} + \frac{a_2-1}{2} \bmod 2,$$

and

$$\frac{a_1^2a_2^2-1}{8} \equiv \frac{u_1^2+u_1}{2} + \frac{u_2^2+u_2}{2} \equiv \frac{a_1^2-1}{8} + \frac{a_2^2-1}{8} \bmod 2.$$

Hence $\varepsilon(a_1a_2) = \varepsilon(a_1)\varepsilon(a_2)$ and $\eta(a_1a_2) = \eta(a_1)\eta(a_2)$.

Now we can do the proof itself. For $a,\, b \in U_2$ we set

$$(2,a) = \Big(\frac{2,a}{2}\Big) \quad \text{and} \quad (a,b) = \Big(\frac{a,b}{2}\Big).$$

Recall that $\{-1, 5\}$ is a set of representatives for U_2/U_2^2 (see [18, Theorem 5.8.11]). Therefore $\{2, -1, 5\}$ is a set of representatives for $\mathbb{Q}_2^\times/\mathbb{Q}_2^{\times 2}$, and due to 1. and the above congruences it suffices to prove that

$$(2, -1) = (5, -1) = 1 \quad \text{and} \quad (-1, -1) = (2, 5) = -1.$$

Since $2 = \mathsf{N}_{\mathbb{Q}_2(\sqrt{-1})/\mathbb{Q}_2}(1 + \sqrt{-1}\,)$ and $5 = \mathsf{N}_{\mathbb{Q}_2(\sqrt{-1})/\mathbb{Q}_2}(2 + \sqrt{-1}\,)$, it follows that $(2, -1) = (5, -1) = 1$.

If $(-1, -1) = 1$, then $(-1, a) = 1$ for all $a \in \mathbb{Q}_2^\times$ and therefore $-1 \in \mathbb{Q}_2^{\times 2}$, a contradiction. Similarly, if $(5, 2) = 1$, then $(5, a) = 1$ for all $a \in \mathbb{Q}_2^\times$ and $5 \in \mathbb{Q}_2^{\times 2}$, again a contradiction. $\square$

5

Global fields: Adeles, ideles and holomorphy domains

In this section, we provide the essential tools for global class field theory which we shall develop simultaneously for number fields and function fields. With this in mind, we introduce the concept of global fields, and we discuss the corresponding unified notation (valid both for number fields and function fields) in Section 5.1. Then we present the classical theory of adeles and ideles, essenially following [4], [40] and [48]. We investigate in detail ray class groups and ideal class groups of holomorphy domains, both in the idele-theoretic and divisor-theoretic setting. For these special topics we refer to [17], [13, Sec. 3.3], [15, Ch. 8.9], and [16]. Several of the technical results given there will be justified by their applications in Section 6. We close this section with a thorough discussion ray class charcters and their idelic interpretation.

5.1 Global fields

In this section, we sum up the unified terminology for global fields. For undefined items we refer to the chapter "Notation" at the beginning of this volume.

Definitions, Conventions and Remarks 5.1.1 (Global fields and places).

1. A **global field** is either an (algebraic) number field K (referred to as the "number field case") or an (algebraic) function field $K/\mathbb{F}_q$ such that $\mathrm{char}(K) = p > 0$, q is a p-power and $\mathbb{F}_q$ is the field of constants (referred to as the "function field case"). Our main reference for both cases is [18]. We start by recalling, amending and unifying our notation and conventions.

2. Let K be a global field. A **place** v of K is an equivalence class of absolute values of K (note that in the function field case we have $|\cdot|\restriction\mathbb{F}_q^\times = 1$ for all absolute values $|\cdot|$ of K). If the absolute values belonging to v are

DOI: 10.1201/9780429506574-5

Archimedian, we call v an **Archimedian** or **infinite place**, otherwise a **finite place**. We denote by M_K^0 the set of all finite places, by M_K^∞ the set of all Archimedian places and by $\mathsf{M}_K = \mathsf{M}_K^0 \cup \mathsf{M}_K^\infty$ the set of all places of K. In the function field case we have $\mathsf{M}_K^\infty = \emptyset$ and $\mathsf{M}_K^0 = \mathsf{M}_K$.

Let $v \in \mathsf{M}_K$, $|\cdot| \in v$ an absolute value and $\widehat{K} = (\widehat{K}, |\cdot|)$ a completion of the valued field $K = (K, |\cdot|)$. The locally compact field $\widehat{K}$ is (up to topological isomorphisms) and its absolute value is (up to equivalence) uniquely determined by v. We set $K_v = \widehat{K}$, and we call K_v a **completion of** K at v.

3. Let K be a number field, let $\sigma_1, \ldots, \sigma_{r_1} \colon K \to \mathbb{R}$ be the real embeddings and $\sigma_{r_1+1}, \overline{\sigma_{r_1+1}}, \ldots, \sigma_{r_1+r_2}, \overline{\sigma_{r_1+r_2}} \colon K \to \mathbb{C}$ the pairs of conjugate complex embeddings of K. Then $n = [K : \mathbb{Q}] = r_1 + 2r_2$, and we call (r_1, r_2) the **signature** of K. Throughout, we denote by $|\cdot|_\infty$ the ordinary absolute value of $\mathbb{C}$.

For $j \in [1, r_1]$ we set $l_j = 1$, for $j \in [r_1 + 1, r_2]$ we set $l_j = 2$, and for all $j \in [1, r_1 + r_2]$ we denote by $|\cdot|_{\infty,j} = |\cdot|_\infty^{l_j} \circ \sigma_j \colon K \to \mathbb{R}_{\ge 0}$ the normalized Archimedian absolute value of K.

If $v \in \mathsf{M}_K^\infty$, then there is a unique $j \in [1, r_1 + r_2]$ such that $|\cdot|_{\infty,j} \in v$; we set $|\cdot|_v = |\cdot|_{\infty,j}$ and $\sigma_v = \sigma_j \colon K \to \mathbb{C}$, we call $|\cdot|_v$ the **normalized v-adic absolute value** and σ_v the **embedding associated with** v. We call v **real** if $j \in [1, r_1]$ (then σ_v extends to a unique isomorphism $K_v \overset{\sim}{\to} \mathbb{R}$, and $|a|_v = |\sigma_j(a)|_\infty$ for all $a \in K$), and **complex** if $j \in [r_1 + 1, r_1 + r_2]$ (then σ_v extends to a unique isomorphism $K_v \overset{\sim}{\to} \mathbb{C}$, and $|a|_v = |\sigma_j(a)|_\infty^2$ for all $a \in K$).

Careful! While we deal with a singe $v \in \mathsf{M}_K^\infty$, we may identify K_v with $\mathbb{R}$ or $\mathbb{C}$ whenever this is convenient, but as soon as several Archimedian places are involved, we must consider the various distinct embeddings.

If $v \in \mathsf{M}_K^\infty$ is real and $x \in K$, then $|x - 1|_v < 1$ implies $\sigma_v(x) > 0$ and $|x + 1|_v < 1$ implies $\sigma_v(x) < 0$.

If $K = \mathbb{Q}$, then $|\cdot|_\infty$ is the unique normalized Archimedian absolute value, and we denote its place by ∞ (then $\mathsf{M}_\mathbb{Q}^\infty = \{\infty\}$ and $\mathbb{Q}_\infty = \mathbb{R}$).

4. Let K be a global field.

a. Let $K/\mathbb{F}_q$ is a function field. A **prime divisor** P of K is the maximal ideal of a discrete valuation domain $\mathcal{O}_P$ of K satisfying $\mathbb{F}_q \subset \mathcal{O}_P$. Let $\mathbb{P}_K$ be the set of all prime divisors of K. The divisor group $\mathbb{D}_K$ is the free (additive) abelian group and the monoid $\mathbb{D}'_K$ of effective divisors is the free (additive) abelian monoid with basis $\mathbb{P}_K$.

b. Let K be a number field. Then $\mathfrak{o}_K = \mathrm{cl}_K(\mathbb{Z})$ is a Dedekind domain. We set $\mathcal{P}_K = \mathcal{P}(\mathfrak{o}_K)$, $\mathcal{I}_K = \mathcal{I}(\mathfrak{o}_K)$, $\mathcal{I}'_K = \mathcal{I}'(\mathfrak{o}_K)$, $\mathcal{H}_K = \mathcal{H}(\mathfrak{o}_K)$, $\mathcal{H}'_K = \mathcal{H}_K \cap \mathcal{I}'_K$, and $\mathcal{C}_K = \mathcal{C}(\mathfrak{o}_K) = \mathcal{I}_K/\mathcal{H}_K$. For $\mathfrak{m} \in \mathcal{I}'_K$ we set $\mathcal{I}_K^\mathfrak{m} = \mathcal{I}^\mathfrak{m}(\mathfrak{o}_K)$, $\mathcal{I}_K'^\mathfrak{m} = \mathcal{I}'_K \cap \mathcal{I}_K^\mathfrak{m}$, $\mathcal{H}_K^\mathfrak{m} = \mathcal{H}_K \cap \mathcal{I}_K^\mathfrak{m}$ and $\mathcal{P}_K^\mathfrak{m} = \mathcal{P}_K \cap \mathcal{I}_K^\mathfrak{m}$. Recall that $\mathcal{I}_K'^\mathfrak{m}$ is the free (multiplicative) abelian monoid and $\mathcal{I}_K^\mathfrak{m}$ is the free (multiplicative) abelian group with basis $\mathcal{P}_K^\mathfrak{m}$.

c. Let anew K be an arbitrary global field. We introduce the following unifying notation for both cases. We set

$$\begin{cases} \mathcal{D}_K = \mathcal{J}_K, \ \mathcal{D}'_K = \mathcal{J}'_K, \ \text{and} \ \mathcal{P}_K = \mathcal{P}_K & \text{in the number field case,} \\ \mathcal{D}_K = \mathbb{D}_K, \ \mathcal{D}'_K = \mathbb{D}'_K \ \text{and} \ \mathcal{P}_K = \mathbb{P}_K & \text{in the function field case.} \end{cases}$$

We call $\mathcal{D}_K$ the (**abstract**) **divisor group** and its elements (**abstract**) **divisors** of K. The elements of $\mathcal{D}'_K$ are called (**abstract**) **integral divisors**. $\mathcal{D}'_K$ is the free monoid and $\mathcal{D}_K = \mathsf{q}(\mathcal{D}'_K)$ is the free abelian group with basis $\mathcal{P}_K$. The unit element of $\mathcal{D}_K$ is denoted by 1 and is called the (**abstract**) **unit divisor**.

We use the multiplicative notion for $\mathcal{D}_K$ also in the function field case, and as in every free monoid, we use in $\mathcal{D}_K$ the notions of divisibility, gcd, lcm, etc. However, if in the sequel we occasionally treat the number field case and the function field case separately for greater clarity, we shall return to the original notation as above in **a** and **b**, and which is consistent with the notation used in [18].

5. Let K be a global field and $v \in \mathsf{M}_K^0$. If $|\cdot| \in v$, then the absolute value $|\cdot|$ is discrete, and as the discrete valuation belonging to $|\cdot|$ is uniquely determined by v, we also denote it by v. In this way, we identify M_K^0 with the set of all discrete valuations of K. For $v \in \mathsf{M}_K^0$, we denote by $\mathcal{O}_v = \{x \in K \mid v(x) \ge 0\}$ the valuation domain, by $P_v = \{x \in K \mid v(x) > 0\}$ the valuation ideal, by $\mathsf{k}_v = \mathcal{O}_v/P_v$ the residue class field and by $\mathfrak{N}(v) = |\mathsf{k}_v|$ the **absolute norm** of v.

If $K = \mathbb{Q}$, then $\mathsf{M}_K^0 = \{v_p \mid p \in \mathbb{P}\}$, the set of all p-adic valuations, and if $p \in \mathbb{P}$, then $\mathcal{O}_{v_p} = \mathbb{Z}_{(p)}$, the domain of p-integral rational numbers, $\mathcal{P}_{v_p} = p\mathbb{Z}_{(p)}$ and $\mathsf{k}_{v_p} = \mathbb{F}_p$.

In the function field case, $\mathcal{P}_K = \{P_v \mid v \in \mathsf{M}_K\}$, and (according to the terminology of [18, Ch. 6]) we obtain $\mathcal{O}_v = \mathcal{O}_{P_v}$, $v = v_{P_v}$ and $\mathsf{k}_v = \mathsf{k}_{P_v}$. For $v \in \mathsf{M}_K$ we denote by $\deg(v) = \deg(P_v) = \dim_{\mathbb{F}_q}(\mathsf{k}_v)$ the **degree** of v, and we extend deg to an (equally denoted) group homomorphism $\deg\colon \mathcal{D}_K \to \mathbb{Z}$, and if $v \in \mathsf{M}_K$, then $|\mathsf{k}_v| = q^{\deg(v)}$. For every subset S of M_K we set $\partial_S = \gcd\{\deg(v) \mid v \in S\}$ (in particular, $\partial_\emptyset = 0$). Then $\partial_K = \partial_{\mathsf{M}_K} = 1$ (see [18, Theorem 6.9.3]) (a purely algebraic proof of the fact will be given in 6.1.7), and thus $\partial_S = 1$ for every sufficiently large finite subset S of M_K. If $v \in \mathsf{M}_K^0$, then $\mathfrak{N}(v) = \mathfrak{N}(P_v) = q^{\deg(v)}$, and for every $c \in \mathbb{R}_{\ge 1}$ the set $\{v \in \mathsf{M}_K \mid \deg(v) \le c\}$ is finite (see [18, Theorem 6.6.8]).

In the number field case we set $\mathfrak{p}_v = \mathcal{P}_v \cap \mathfrak{o}_K \in \mathcal{P}_K$. Then $\mathcal{P}_K = \{\mathcal{P}_v \cap \mathfrak{o}_K \mid v \in \mathsf{M}_K^0\}$, and if $v \in \mathsf{M}_K^0$, then according to the terminology of [18, Ch. 3] we obtain $\mathcal{O}_v = (\mathfrak{o}_K)_{\mathfrak{p}_v}$, $P_v = \mathfrak{p}_v(\mathfrak{o}_K)_{\mathfrak{p}_v}$, $v = v_{\mathfrak{p}_v}$ and $\mathsf{k}_v = \mathfrak{o}_K/\mathfrak{p}_v$. If p is a prime such that $\mathfrak{p}_v \cap \mathbb{Z} = p\mathbb{Z}$, then we write $v \downarrow p$, we say that v lies above p, we call p the underlying prime of v, we set $e_v = e(\mathfrak{p}_v/p\mathbb{Z})$, and $f_v = f(\mathfrak{p}_v/p\mathbb{Z}) = \dim_{\mathbb{F}_p} \mathsf{k}_v$; then $\mathbb{Q}_p \subset K_v$, and $\mathfrak{N}(v) = \mathfrak{N}(\mathfrak{p}_v) = p^{f_v}$. For every prime p there are at most $[K:\mathbb{Q}]$ prime ideals $\mathfrak{p} \in \mathcal{P}_K$ lying above p. Consequently, for every $c \in \mathbb{R}_{\ge 1}$ the set $\{v \in \mathsf{M}_K^0 \mid \mathrm{char}(\mathsf{k}_v) \le c\}$ is finite.

Let afresh K be an arbitrary global field. For $v \in \mathsf{M}_K^0$ we define

$$\mathsf{p}_v = \begin{cases} \mathfrak{p}_v & \text{in the number field case,} \\ P_v & \text{in the function field case.} \end{cases}$$

Consequently, $\mathcal{P}_K = \{\mathsf{p}_v \mid v \in \mathsf{M}_K^0\}$, $\mathcal{D}_K = \mathcal{P}_K^{(\mathbb{Z})}$ and $\mathcal{D}'_K = \mathcal{P}_K^{(\mathbb{N}_0)}$. The p_v-adic valuation of $\mathcal{D}_K$ is also denoted by v. Then $v\colon \mathcal{D}_K \to \mathbb{Z}$ is an epimorphism, and every $\mathsf{a} \in \mathcal{D}_K$ has the unique representation

$$\mathsf{a} = \prod_{v \in \mathsf{M}_K^0} \mathsf{p}_v^{v(\mathsf{a})}, \quad \text{and we set} \quad \operatorname{supp}(\mathsf{a}) = \{v \in \mathsf{M}_K^0 \mid v(\mathsf{a}) \neq 0\}.$$

Obviously, $\operatorname{supp}(\mathsf{a}) = \emptyset$ if and only if $\mathsf{a} = 1$, and in the number field case we obtain $v(\mathsf{a}) = \inf\{v(a) \mid a \in \mathsf{a}\}$.

Be aware that there is some ambiguity in this concept in the event that we compare it with [18]. If $\mathfrak{o}$ is a Dedekind domain and $\mathfrak{a} \in \mathcal{J}(\mathfrak{o})$, we there defined $\operatorname{supp}(\mathfrak{a}) = \{\mathfrak{p} \in \mathcal{P}(\mathfrak{o}) \mid v_\mathfrak{p}(\mathfrak{a}) \neq 0\}$, and if $K/\mathbb{F}_q$ is a function field and $D \in \mathbb{D}_K$, we defined there $\operatorname{supp}(D) = \{P \in \mathbb{P}_K \mid v_P(D) \neq 0\}$. We will not remedy this ambiguity and hope that from the context it will be lucid whether we mean places, ideals or divisors.

6. Let K be a global field. For $v \in \mathsf{M}_K^0$ we define the **normalized v-adic absolute value** $|\cdot|_v \colon K \to \mathbb{R}_{\geq 0}$ by $|a|_v = \mathfrak{N}(v)^{-v(a)}$ for all $a \in K$ and obtain the product formula

$$\prod_{v \in \mathsf{M}_K} |a|_v = 1 \quad \text{for all} \quad a \in K^\times,$$

which in the function field case is equivalent to

$$\sum_{v \in \mathsf{M}_K} v(a) \deg v = 0 \quad \text{for all} \quad a \in K^\times,$$

since $|a|_v = q^{-v(a)\deg(v)}$ for all $v \in \mathsf{M}_K$ and $a \in K^\times$.

7. Let K be a global field, $v \in \mathsf{M}_K^0$ and K_v a completion of K at v. Then every absolute value $|\cdot| \in v$ (and in particular $|\cdot|_v$ itself) has a unique extension to an (equally denoted) absolute value of K_v, $(K_v, |\cdot|_v)$ is complete, K is dense in K_v, and the discrete valuation $v\colon K \to \mathbb{Z} \cup \{\infty\}$ has a unique extension to an (equally denoted) discrete valuation of K_v. In particular, K_v is a non-Archimedean local field.

If $K = \mathbb{Q}$ and $v = v_p$, then $\mathbb{Q}_{v_p} = \mathbb{Q}_p$, the p-adic number field. If K is a number field and $v \downarrow p$, then K_v is a finite extension of $\mathbb{Q}_p$ (called a $\mathfrak{p}$-adic number field). If $K/\mathbb{F}_q$ is a function field and $t \in K$ is such that $v(t) = 1$, then $K_v = \mathsf{k}_v((t))$ is the field of formal Laurent series in t, and its discrete valation $v = \mathsf{o}_t \colon K_v \to \mathbb{Z} \cup \{\infty\}$ is the order valuation with respect to t.

Let anew K be an arbitrary glocal field and $v \in \mathsf{M}_K^0$. We denote by $\widehat{\mathcal{O}}_v$ the valuation domain, by $\widehat{\mathfrak{p}}_v$ the valuation ideal, by $U_v = U_v^{(0)} = \widehat{\mathcal{O}}_v^\times$ the unit group, and for $n \ge 1$ by $U_v^{(n)} = 1 + \widehat{\mathfrak{p}}_v^n$ the group of principal units of level n. Then $\mathcal{O}_v = \widehat{\mathcal{O}}_v \cap K$, $P_v = \widehat{\mathfrak{p}}_v \cap K$, $\mathsf{k}_v = \mathcal{O}_v/P_v = \widehat{\mathcal{O}}_v/\widehat{\mathfrak{p}}_v)$, $|x|_v = \mathfrak{N}(v)^{-v(x)}$ for all $x \in K_v^\times$, and in the number field case $\mathfrak{p}_v = \widehat{\mathfrak{p}}_v \cap \mathfrak{o}_K$. An element $\pi \in \widehat{\mathcal{O}}_v$ is a prime element if and only if $v(\pi) = 1$.

Recall that $\widehat{\mathcal{O}}_v = \{x \in K_v \mid |x|_v \le 1\} = \{x \in K_v \mid v(x) \ge 0\}$, and if $\pi \in \mathcal{O}_v$ is a prime element, then $\widehat{\mathfrak{p}}_v^n = \pi^n \widehat{\mathcal{O}}_v = \{x \in K_v \mid v(x) \ge n\}$ for all $n \in \mathbb{Z}$. By 1.5.4, $\widehat{\mathcal{O}}_v$ is a compact topological ring, and the groups U_v, $U_v^{(n)}$ for all $n \in \mathbb{N}$, and $\widehat{\mathfrak{p}}_v^n$ for all $n \in \mathbb{Z}$ are compact topological groups. $\{\widehat{\mathfrak{p}}_v^n \mid n \in \mathbb{N}\}$ is a fundamental system of neighbourhoods of 0 and $\{U_v^{(n)} \mid n \in \mathbb{N}\}$ is a fundamental system of neighbourhoods of 1 consisting of clopen subsets of K_v. In particular, the groups $\widehat{\mathcal{O}}_v$, U_v, $\widehat{\mathfrak{p}}_v^n$ and $U_v^{(n)}$ for all $n \in \mathbb{N}$ are profinite groups.

For $v \in \mathsf{M}_K^\infty$ we set $\widehat{\mathcal{O}}_v = K_v$, and

$$U_v = \begin{cases} K_v^\times = \mathbb{C}^\times & \text{if } v \text{ is complex,} \\ \mathbb{R}_{>0} & \text{if } v \text{ is real.} \end{cases}$$

If $K = \mathbb{Q}$, and $v = v_p$ and $\mathbb{Q}_{v_p} = \mathbb{Q}_p$, then $\widehat{\mathcal{O}}_{v_p} = \mathbb{Z}_p$ (the domaim of p-adic integers), $\widehat{\mathfrak{p}}_{v_p} = p\mathbb{Z}_p$, $U_{v_p} = U_p = \mathbb{Z}_p \setminus p\mathbb{Z}_p$ (the group of p-adic units), and $U_{v_p}^{(n)} = U_p^{(n)} = 1 + p^n\mathbb{Z}_p$ (the group of p-adic principal units of level n) for all $n \ge 1$.

8. Let $\phi\colon K \to K'$ be an isomorphism of global fields. If $|\cdot|\colon K \to \mathbb{R}_{\ge 0}$ is an absolute value of K, then ${}^\phi|\cdot| = |\cdot| \circ \phi^{-1}\colon K' \to \mathbb{R}_{\ge 0}$ is an absolute value of K', and ${}^\phi|\cdot|$ is normalized if and only if $|\cdot|$is normalized. Two absolute values $|\cdot|$, $|\cdot|'$ of K are equivalent if and only if ${}^\phi|\cdot|$, ${}^\phi|\cdot|'$ are equivalent absolute values of K'. For a place $v \in \mathsf{M}_K$ we denote by ϕv the set of all absolute values ${}^\phi|\cdot|$ with $|\cdot| \in v$. Then $\phi v \in \mathsf{M}_{K'}$, ${}^\phi|\cdot|_v = |\cdot|_{\phi v}$, and the assignment $v \mapsto \phi v$ induces a bijective map $\phi\colon \mathsf{M}_K \to \mathsf{M}_{K'}$. If $v \in \mathsf{M}_K$, then $\phi\colon (K, |\cdot|_v) \to (K', |\cdot|_{v'})$ is an isomorphism of valued fields and has a unique extension to an (equally denoted) isomorphism of valued fields $\phi\colon K_v \to K'_{v'}$.

Recall that we identify $v \in \mathsf{M}_K^0$ with the associated discrete valuation $v\colon K \to \mathbb{Z} \cup \{\infty\}$; thus we obtain $\phi v = v \circ \phi^{-1}\colon K' \to \mathbb{Z} \cup \{\infty\}$.

9. Let K be a global field. As element $a \in K^\times$ is called **totally positive** and we write $a \gg 0$ if $\sigma(a) > 0$ for real embeddings $\sigma\colon K \to \mathbb{R}$. We denote by K^+ the group of all totally positive elements of K. Note that $K^+ = K^\times$ if K has no real embeddings (in particular in the function field case), and $\mathbb{Q}^+ = \mathbb{Q}_{>0}$. In the number field case we obtain $(K^\times : K^+) = 2^{r_1}$.

For $\mathsf{a} \in \mathcal{D}_K$ we call

$$\mathfrak{N}(\mathsf{a}) = \prod_{v \in \mathrm{supp}(\mathsf{a})} \mathfrak{N}(v)^{v(\mathsf{a})}$$

the **absolute norm** of a. In particular, $\mathfrak{N}(\mathsf{p}_v) = \mathfrak{N}(v)$ for all $v \in \mathsf{M}_K^0$. We define

$$\varphi(\mathsf{a}) = \prod_{v \in \mathrm{supp}(\mathsf{a})} \mathfrak{N}(v)^{v(\mathsf{a})-1}(\mathfrak{N}(v)-1) \;=\; \mathfrak{N}(\mathsf{a}) \prod_{v \in \mathrm{supp}(\mathsf{a})} \Big(1 - \frac{1}{\mathfrak{N}(v)}\Big).$$

In the number field case, when $\mathsf{a} \in \mathcal{D}_K' = \mathcal{I}_K'$, it follows that $\mathfrak{N}(\mathsf{a}) = (\mathfrak{o}_K{:}\,\mathsf{a})$ and $\varphi(\mathsf{a}) = |(\mathfrak{o}_K/\mathsf{a})^\times|$. In the function field case, $\mathfrak{N}(\mathsf{a}) = q^{\deg(\mathsf{a})}$.

10. For $a \in K^\times$, we denote by

$$(a) = (a)^K = \prod_{v \in \mathsf{M}_K^0} \mathsf{p}_v^{v(a)} \in \mathcal{D}_K$$

the **principal divisor** of a; then $v(a) = v((a)^K)$ for all $a \in K^\times$ and $v \in \mathsf{M}_K^0$. We denote by $(K^\times) = \{(a) \mid a \in K^\times\} \subset \mathcal{D}_K$ the group of **principal divisors** and by $(K^+) = \{(a) \mid a \in K^+\}$ the group of **totally positive principal divisors**. The assignment $a \mapsto (a)^K$ defines a group homomorphism $(\,\cdot\,)^K \colon K^\times \to (K^\times)$ with kernel

$$U_K = \begin{cases} \mathfrak{o}_K^\times & \text{in the number field case,} \\ \mathbb{F}_q^\times & \text{in the function field case} \end{cases}$$

and also a group homomorphism $(\,\cdot\,) \restriction K^+ \colon K^+ \to (K^+)$ with kernel $U_K^+ = U_K \cap K^+$. We call $\mathcal{C}_K = \mathcal{D}_K/(K^\times)$ the **(divisor) class group** and $\mathcal{C}_K^+ = \mathcal{D}_K/(K^+)$ the **narrow (divisor) class group** of K. For $\mathsf{a} \in \mathcal{D}_K$ we denote by $[\mathsf{a}] = \mathsf{a}(K^\times) \in \mathcal{C}_K$ the divisor class and by $[\mathsf{a}]^+ = \mathsf{a}(K^+)$ the narrow divisor class of a. We summarize these facts in the exact sequences

$$\mathbf{1} \to U_K \to K^\times \overset{(\cdot)}{\to} \mathcal{D}_K \overset{[\cdot]}{\to} \mathcal{C}_K \to \mathbf{1} \quad \text{and} \quad \mathbf{1} \to U_K^+ \to K^+ \overset{(\cdot)}{\to} \mathcal{D}_K \overset{[\cdot]^+}{\to} \mathcal{C}_K^+ \to \mathbf{1}$$

In the number field case we recover well-known concepts from [18, Ch. 3]: If $a \in K^\times$, then $(a) = a\mathfrak{o}_K \in \mathcal{H}_K$ is the fractional principal ideal of $\mathfrak{o}_K$ generated by a, $(K^\times) = \mathcal{H}_K$ is the group of fractional principal ideals of $\mathfrak{o}_K$, and $(K^+) = \mathcal{H}_K^+$ is the group of totally positive principal ideal of $\mathfrak{o}_K$. The (ideal) class groups $\mathcal{C}_K = \mathcal{I}_K/\mathcal{H}_K$ and the narrow ideal class group $\mathcal{C}_K^+ = \mathcal{I}_K/\mathcal{H}_K^+$ are finite, $h_K = |\mathcal{C}_K|$ is the **class number** and $h_K^+ = |\mathcal{C}_K^+|$ is the **narrow class number** of K.

If $K/\mathbb{F}_q$ is a function field and $\mathsf{a} \in \mathcal{D}_K$, then its degree is given by

$$\deg(\mathsf{a}) = \sum_{v \in \mathsf{M}_K} v(\mathsf{a}) \deg(v). \quad \text{In particular,} \quad \deg(\mathsf{p}_v) = \deg(v) \quad \text{for all} \quad v \in \mathsf{M}_K.$$

$\deg \colon \mathcal{D}_K \to \mathbb{Z}$ is a group epimorphism, $\mathcal{D}_K^0 = \mathrm{Ker}(\deg)$, and since $(K^\times) \subset \mathcal{D}_K^0$ (see [18, Theorem 6.3.4]), deg induces an (equally denoted) homomorphism $\deg \colon \mathcal{C}_K \to \mathbb{Z}$. We call $\mathcal{C}_K^0 = \mathcal{D}_K^0/(K^\times) = \mathrm{Ker}(\deg \colon \mathcal{C}_K \to \mathbb{Z})$ the **0-divisor**

class group of K. It is finite, $h_K = |\mathcal{C}_K^0|$, $\mathcal{C}_K \cong \mathcal{C}_K^0 \times \mathbb{Z}$, and we obtain the exact sequence

$$1 \to \mathbb{F}_q \to K^\times \overset{(\cdot)}{\to} \mathcal{D}_K^0 \overset{[\cdot]}{\to} \mathcal{C}_K^0 \to 1.$$

5.2 Local direct products

Definition and Theorem 5.2.1. *Let $(X_i)_{i\in I}$ be a family of topological spaces, and for every $i \in I$ let O_i be an open subset of X_i.*
We define the **local direct product** X of the family $(X_i)_{i\in I}$ regarding the family $(O_i)_{i\in I}$ by

$$\begin{aligned} X &= \prod_{i\in I}(X_i \,|\, O_i) \\ &= \Big\{(x_i)_{i\in I} \in \prod_{i\in I} X_i \;\Big|\; x_i \in O_i \ \text{ for almost all } \ i \in I\Big\} \subset P = \prod_{i\in I} X_i. \end{aligned}$$

We endow P with the product topology, and we let

$$\mathfrak{B} = \Big\{\prod_{i\in S} U_i \times \prod_{i\in I\setminus S} O_i \;\Big|\; S \in \mathbb{P}_{\rm fin}(I), \ \ U_i \subset X_i \ \text{ open for all } \ i \in S\Big\}$$

be the basis of a (new) topology on X, called the **local topology**. For a finite subset S of I we set

$$X_S = \prod_{i\in S} X_i \times \prod_{i\in I\setminus S} O_i \subset X, \quad \text{and then}$$

$$X = \bigcup_{S\in\mathbb{P}_{\rm fin}(I)} X_S \ \text{ by the very definitio of } X.$$

Clearly, on X_S the subspace topology induced by P is the product topology of X_S.

1. The local topology on X is finer than the subspace topology induced by P. In particular, the local topology has the following properties:

- *For every $j \in I$ the projection $X \to X_j$, $(x_i)_{i\in I} \mapsto x_j$ is continuous.*
- *If all X_i are Hausdorff, then X is Hausdorff.*

2. If $S \in \mathbb{P}_{\text{fin}}(I)$, then the product topology on X_S coincides with the subspace topology induced by the local topology of X.

3. $(X_S)_{S\in\mathbb{P}_{\text{fin}}(I)}$ is a directed family of open subsets of X, and the local topology on X is the final topology induced by the family $(X_S)_{S\in\mathbb{P}_{\text{fin}}(I)}$.

4. For every $i \in I$ let O_i' be another open subset of X_i and let X' be the local direct product of the family $(X_i)_{i\in I}$ regarding the family $(O_i')_{i\in I}$. If $O_i = O_i'$ for almost all $i \in I$, then $X = X'$.

Proof. 1. Let $a = (a_i)_{i\in I} \in X$, and for $i \in I$ let $\mathfrak{U}_i$ be a fundamental system of neighbourhoods of a_i in X_i. Then the set $S(a) = \{i \in I \mid a_i \notin O_i\}$ is finite, and

$$\mathfrak{B}(a) = \Big\{\prod_{i\in S} U_i \times \prod_{i\in I\setminus S} O_i \;\Big|\; S \in \mathbb{P}_{\text{fin}}(I),\;\; S \supset S(a),\;\; U_i \in \mathfrak{U}_i \text{ for all } i \in S\Big\}$$

is a fundamental system of neighbourhoods of a in the local topology. However,

$$\mathfrak{U}(a) = \Big\{\prod_{i\in S} U_i \times \prod_{i\in I\setminus S} X_i \;\Big|\; S \in \mathbb{P}_{\text{fin}}(I),\;\; S \supset S(a),\;\; U_i \in \mathfrak{U}_i \text{ for all } i \in S\Big\}$$

is a fundamental system of neighbourhoods of a in P. For every $S \in \mathbb{P}_{\text{fin}}(I)$ we have

$$\Big(\prod_{i\in S} U_i \times \prod_{i\in S\setminus I} X_i\Big) \cap X \supset \prod_{i\in S} U_i \times \prod_{i\in S\setminus I} O_i,$$

and therefore every neighbourhood of a in the subspace topology induced by P is also a neighbourhood of a in the local topology. Hence the local topology is finer.

2. Let $S \in \mathbb{P}_{\text{fin}}(I)$. By 1., on X_S the subspace topology induced X is finer than the subspace topology induced by P. The subsets of X_S of the form

$$\prod_{i\in I} U_i,$$

where U_i is open in X_i for all $i \in I$, $U_i \subset O_i$ for all $i \in I \setminus S$, and $U_i = O_i$ for almost all $i \in I$, are a basis of the subspace topology induced by the local topology. Since all these sets are also open in the product topology of X_S, the two topologies coincide.

3. By definition, a subet U of X is open in the local topology if and only if $U \cap X_S$ is open for all finite subsets S of I.

4. For every finite subset S of I the definition of the local direct product implies

$$\prod_{i\in I}(X_i \mid O_i) = \prod_{i\in S} X_i \times \prod_{i\in I\setminus S} (X_i \mid O_i).$$

Hence the local direct product does not change if we modify finitely many O_i's. □

Theorem 5.2.2. *Let $(X_i)_{i\in I}$ be a family of locally compact topological spaces. For every $i \in I$ let O_i be an open subset of X_i such that O_i is compact for almost all $i \in I$. Then the local direct product*

$$X = \prod_{i\in I}(X_i \mid O_i)$$

is locally compact.

Proof. By 5.2.1.4, we may assume that all O_i are compact. Then X_S is locally compact for every finite subset S of I, and therefore X is locally compact by 5.2.1.3. □

We proceed with some elementary facts concerning local direct products of topological groups and rings.

Theorem 5.2.3. *Let $(X_i)_{i\in I}$ be a family of topological groups [rings]. For every $i \in I$ let O_i be an open subgroup [an open subring] of X_i. Then the local direct product*

$$X = \prod_{i\in I}(X_i \mid O_i),$$

endowed with component-wise composition and the local topology, is a topological group [a topological ring].

Proof. Obviously, X_S is a topological group [a topological ring] for every finite subset S of I. Therefore it suffices to prove: If $\mu\colon X\times X \to X$ is a map, such that $\mu(X_S\times X_S) \subset X_S$ and $\mu\restriction X_S\times X_S$ is continuous for all finite subsets S of I, then μ is continuous. But this is obvious since $X\times X$ carries the final topology induced by the ascending family $X_S\times X_S)_{S\in\mathbb{P}_{\text{fin}}(I)}$. □

Theorem 5.2.4. *Let $(A_i)_{i\in I}$ be a family of topological rings. For every $i \in I$ let O_i be an open subring of A_i, and*

$$A = \prod_{i\in I}(A_i \mid O_i).$$

For every $i \in I$ let $A_i^\times$ and $O_i^\times$ be equipped with the unit topology (so that $O_i^\times$ is an open subgroup of $A_i^\times$, see 1.2.4.2*). Then*

$$A^\times = \prod_{i\in I}(A_i^\times \mid O_i^\times),$$

and the local topology on $A^\times$ is the unit topology.

Proof. We prove first that algebraically

$$A^\times = \prod_{i\in I}(A_i^\times \mid O_i^\times) \subset A.$$

Let $a=(a_i)_{i\in I}\in A$. Then we have $a\in A^\times$ if and only if there exists a family $b=(b_i)_{i\in I}\in A$ such that $a_ib_i=1\in A_i$ for all $i\in I$. However, this holds if and only if $a_i\in A_i^\times$ for all $i\in I$ and $a_i\in O_i^\times$ for almost all $i\in I$.

As to the topology: Let $\varepsilon\colon A^\times\to A\times A$ be defined by $\varepsilon(a)=(a,a^{-1})$, and for $i\in I$ let $\varepsilon_i\colon A_i^\times\to A_i\times A_i$ be defined by $\varepsilon_i(a_i)=(a_i,a_i^{-1})$. From the associativity of direct products (both algebraically and topologically) it follows that

$$A\times A = \prod_{i\in I}(A_i \mid O_i)\times\prod_{i\in I}(A_i \mid O_i) = \prod_{i\in I}(A_i\times A_i \mid O_i\times O_i).$$

Hence $\varepsilon(a)=(\varepsilon(a_i))_{i\in I}$ for all $a=(a_i)_{i\in I}\in A$, and the sets

$$Y=\prod_{i\in S}(U_i\times V_i)\times\prod_{i\in I\setminus S}(O_i\times O_i)$$

(built with $S\in\mathbb{P}_{\rm fin}(I)$ and open subsets $U_i,\,V_i\subset O_i$ for all $i\in I$), are a basis of the topology of $A\times A$. For each such set Y we obtain

$$\varepsilon^{-1}(Y)=\prod_{i\in S}\varepsilon_i^{-1}(U_i\times V_i)\times\prod_{i\in I\setminus S}O_i^\times,$$

and these sets are a basis both of the unit topology and of the topology of the local direct product. □

Examples 5.2.5. We consider the local direct products

$$A=\prod_{p\in\mathbb{P}}(\mathbb{Q}_p \mid \mathbb{Z}_p) \quad\text{and}\quad A^\times=\prod_{p\in\mathbb{P}}(\mathbb{Q}_p^\times \mid \mathbb{Z}_p^\times).$$

1. The local topology on A is strictly finer than the subspace topology induced by the product topology. To see this, we construct a sequence in A which converges to 0 in the product topology but not in the local topology.

Für a prime q we define

$$x^{(q)}=(x_p^{(q)})_{p\in\mathbb{P}}\in A \quad\text{by}\quad x_p^{(q)}=\begin{cases}0 & \text{if } p\ne q,\\ q^{-1} & \text{if } p=q.\end{cases}$$

For every finite subset S of $\mathbb{P}$ and $n\in\mathbb{N}$ we set

$$W(S,n)=\prod_{p\in S}p^n\mathbb{Z}_p\times\prod_{p\notin S}\mathbb{Q}_p.$$

These sets $W(S,n)$ are a fundamental system of neighbourhoods of 0 in the product topology, and for a prime q we obtain $x^{(q)} \in W(S,n)$ if and only if $q \notin S$. Consequently, for every $S \in \mathbb{P}_{\text{fin}}(\mathbb{P})$ and $n \in \mathbb{N}$ almost all $x^{(q)}$ lie in $W(S,n)$. Hence $(x^{(q)})_{q\in\mathbb{P}} \to 0$ in the product topology.

The sets

$$V(S,n) = \prod_{p\in S} p^n\mathbb{Z}_p \times \prod_{p\notin S} \mathbb{Z}_p \quad \text{with} \quad S \in \mathbb{P}_{\text{fin}}(\mathbb{P}) \quad \text{and} \quad n \in \mathbb{N}$$

are a fundamental system of neighbourhoods of 0 in the local topology. If $q \in \mathbb{P}$, then $x^{(q)}$ lies in none of these sets. Hence $(x^{(q)})_{q\in\mathbb{P}} \not\to 0$ in the local topology.

2. The unit topology on $A^\times$ is strictly finer than the subspace topology induced by the local topology of A. We construct a sequence in $A^\times$ which converges to 1 in the local topology but not in the unit topology.

For a prime $q \in \mathbb{P}$ we define

$$x^{(q)} = (x_p^{(q)})_{p\in\mathbb{P}} \in A^\times \quad \text{by} \quad x_p^{(q)} = \begin{cases} 1 & \text{if } p \ne q, \\ q & \text{if } p = q. \end{cases}$$

For every finite subset S of $\mathbb{P}$ and $n \in \mathbb{N}$ we set

$$W_1(S,n) = \prod_{p\in S}(1+p^n\mathbb{Z}_p) \times \prod_{p\notin S} \mathbb{Z}_p.$$

These sets $W_1(S,n)$ are a fundamental system of neighbourhoods of 1 in the local topology of A, and for a prime q we obtain $x^{(q)} \in W_1(S,n)$ if and only if $q \notin S$. Consequently, for every $S \in \mathbb{P}_{\text{fin}}(\mathbb{P})$ and $n \in \mathbb{N}$ almost all $x^{(q)}$ lie in $W_1(S,n)$. Hence $(x^{(q)})_{q\in\mathbb{P}} \to 1$ in the local topology.

The sets

$$V_1(S,n) = \prod_{p\in S}(1+p^n\mathbb{Z}_p) \times \prod_{p\notin S} \mathbb{Z}_p^\times \quad \text{for} \quad S \in \mathbb{P}_{\text{fin}}(\mathbb{P}) \quad \text{and} \quad n \in \mathbb{N}$$

are a fundamental system of neighbourhoods of 1 in the unit topology. If $q \in \mathbb{P}$, then $x^{(q)}$ lies in none of these sets. Hence $(x^{(q)})_{q\in\mathbb{P}} \not\to 1$ in the unit topology.

5.3 Adeles and ideles of global fields

Let K be a global field.

Definition and Remarks 5.3.1.

1. The local direct product

$$\mathsf{A}_K = \prod_{v\in\mathsf{M}_K} (K_v \,|\, \widehat{\mathcal{O}}_v) \subset \prod_{v\in\mathsf{M}_K} K_v$$

is a locally compact topologica ring by 5.2.2 and 5.2.3; it is called the **adele ring** of K. For a family

$$\alpha = (\alpha_v)_{v\in\mathsf{M}_K} \in \prod_{v\in\mathsf{M}_K} K_v \quad \text{we set} \quad \|\alpha\| = \prod_{v\in\mathsf{M}_K} |\alpha_v|_v \in \mathbb{R}_{\ge 0}.$$

The local direct product

$$\mathsf{J}_K = \prod_{v\in\mathsf{M}_K} (K_v^\times \,|\, U_v) \subset \prod_{v\in\mathsf{M}_K} K_v^\times$$

is a locally compact topological group (again by 5.2.2 and 5.2.3); it is called the **idele group** of K. By 5.2.4, $\mathsf{J}_K = \mathsf{A}_K^\times$, and the local topology is the unit topology of $\mathsf{A}_K^\times$.

By definition, $\|\alpha\| \in \mathbb{R}_{>0}$ for all $\alpha \in \mathsf{J}_K$, and the map $\|\cdot\|\colon \mathsf{J}_K \to \mathbb{R}_{>0}$ is a group homomorphism with kernel

$$\mathsf{J}_K^0 = \{\alpha \in \mathsf{J}_K \mid \|\alpha\| = 1\}, \quad \text{called the } \textbf{idele kernel group}.$$

2. For a finite subset S of M_K we define the ring of **S-adeles** $\mathsf{A}_{K,S}$ and the group of **S-ideles** $\mathsf{J}_{K,S}$ by

$$\mathsf{A}_{K,S} = \prod_{v\in S} K_v \times \prod_{v\in\mathsf{M}_K\setminus S} \widehat{\mathcal{O}}_v \subset \mathsf{A}_K \quad \text{and}$$

$$\mathsf{J}_{K,S} = \prod_{v\in S} K_v^\times \times \prod_{v\in\mathsf{M}_K\setminus S} U_v = \mathsf{A}_{K,S}^\times \subset \mathsf{J}_K.$$

On $\mathsf{A}_{K,S}$ and $\mathsf{J}_{K,S}$ the product topology coincides with the subspace topology induced by the local topology (see 5.2.1.2). If S and S' are finite subsets of M_K such that $S \subset S'$, then $\mathsf{A}_{K,S} \subset \mathsf{A}_{K,S'}$ and $\mathsf{J}_{K,S} \subset \mathsf{J}_{K,S'}$. Hence the families $(\mathsf{A}_{K,S})_{S\in\mathbb{P}_{\rm fin}(\mathsf{M}_K)}$ and $(J_{K,S})_{S\in\mathbb{P}_{\rm fin}(\mathsf{M}_K)}$ are directed,

$$\mathsf{A}_K = \bigcup_{S\in\mathbb{P}_{\rm fin}(\mathsf{M}_K)} \mathsf{A}_{K,S} \quad \text{and} \quad \mathsf{J}_K = \bigcup_{S\in\mathbb{P}_{\rm fin}(\mathsf{M}_K)} \mathsf{J}_{K,S}.$$

The sets $\mathsf{A}_{K,S}$ (resp. $\mathsf{J}_{K,S}$) are open in A_K (resp. J_K), and the topology of A_K (resp. J_K) is the final topology induced by the directed family $(\mathsf{A}_{K,S})_{S\in\mathbb{P}_{\rm fin}(\mathsf{M}_K)}$ (resp. $(\mathsf{J}_{K,S})_{S\in\mathbb{P}_{\rm fin}(\mathsf{M}_K)}$.

3. We define the diagonal embeddings $K \to \mathsf{A}_K$ and $K^\times \to \mathsf{J}_K$ by the assignment $a \mapsto (a)_{v\in\mathsf{M}_K}$, and we identify: $K \subset \mathsf{A}_K$ and $K^\times \subset \mathsf{J}_K$. In this way, A_K is a K-algebra containing K, and $K^\times$ is a subgroup of J_K. If $a \in K^\times$, then the product formula implies $\|a\| = 1$, and consequently $K^\times \subset \mathsf{J}_K^0$.

We call $\mathsf{C}_K = \mathsf{J}_K/K^\times$ the **idele class group** and $\mathsf{C}_K^0 = \mathsf{J}_K^0/K^\times \subset \mathsf{C}_K$ the **idele class kernel group** of K. We endow C_K with the quotient topology and C_K^0 with the subspace topology induced by C_K (which coincides with the quotient topology induced by J_K^0, see 1.1.5.2(a)). Note that the residue class epimorphism $\mathsf{J}_K \to \mathsf{C}_K$ is continuous and open.

For a finite subset S of M_K we call $\mathsf{C}_{K,S} = \mathsf{J}_{K,S}K^\times/K^\times$ the ***S*-idele class group**. $(\mathsf{C}_{K,S})_{S\in\mathbb{P}_{\text{fin}}(\mathsf{M}_K)}$ is a directed family of open subgroups of C_K with union C_K, and the topology on C_K is the final topology induced by the family $(\mathsf{C}_{K,S})_{S\in\mathbb{P}_{\text{fin}}(\mathsf{M}_K)}$.

If S is finite subset of M_K, then

$$\mathsf{C}_K/\mathsf{C}_{K,S} = (\mathsf{J}_K/K^\times)(\mathsf{J}_{K,S}K^\times/K^\times) \cong \mathsf{J}_K/\mathsf{J}_{K,S}K^\times,$$

and we identify: $\mathsf{C}_K/\mathsf{C}_{K,S} = \mathsf{J}_K/\mathsf{J}_{K,S}K^\times$.

Since $K^\times \subset \mathsf{J}_K^0$, the homomorphism $\|\cdot\|\colon \mathsf{J}_K \to \mathbb{R}_{>0}$ induces an (equally denoted) homomorphism $\|\cdot\|\colon \mathsf{C}_K \to \mathbb{R}_{>0}$ with kernel $\mathsf{C}_K^0 = \mathsf{J}_K^0/K^\times$.

If $\alpha = (\alpha_v)_{v\in\mathsf{M}_K} \in \mathsf{J}_{K,S}$, then

$$\|\alpha\| = \prod_{v\in S\cup\mathsf{M}_K^\infty} |\alpha_v|_v.$$

Hence $\|\cdot\| \restriction \mathsf{J}_{K,S}\colon \mathsf{J}_{K,S} \to \mathbb{R}_{>0}$ is continuous, and since J_K carries the final topology induced by the family $(\mathsf{J}_{K,S})_{S\in\mathbb{P}_{\text{fin}}(\mathsf{M}_K)}$, the homomorphism $\|\cdot\|\colon \mathsf{J}_K \to \mathbb{R}_{>0}$ is continuous. Consequently the induced homomorphism $\|\cdot\|\colon \mathsf{C}_K \to \mathbb{R}_{>0}$ is continuous as well, $\mathsf{J}_K^0 \subset \mathsf{J}_K$ and $\mathsf{C}_K^0 \subset \mathsf{C}_K$ are closed subgroups. Since closed subgroups and factor groups of locally compact topological groups are locally compact, the groups J_K^0, C_K and C_K^0 are locally compact.

The group

$$\mathsf{E}_K = \mathsf{J}_{K,\mathsf{M}_K^\infty} = \prod_{v\in\mathsf{M}_K^\infty} K_v^\times \times \prod_{v\in\mathsf{M}_K^0} U_v \subset \mathsf{J}_K$$

is called the **idele unit group**, and $\mathcal{E}_K = \mathsf{E}_K K^\times/K^\times \subset \mathsf{C}_K$ is called the **idele class unit group** (the terminology will become lucid in 5.3.2 below). By definition, E_K is an open subgroup of J_K, and consequently $\mathcal{E}_K$ is an open subgoup of C_K.

In the function field case even more is true. As $\mathsf{M}_K^\infty = \emptyset$, it follows that

$$\mathsf{E}_K = \prod_{v\in\mathsf{M}_K} U_v \subset \mathsf{J}_K^0, \quad \mathcal{E}_K \subset \mathsf{C}_K^0, \quad \mathsf{C}_K^0/\mathcal{E}_K = \mathsf{J}_K^0/K^\times\mathsf{E}_K,$$

and therefore $\mathsf{J}_K^0 \subset \mathsf{J}_K$ and $\mathsf{C}_K^0 \subset \mathsf{C}_K$ are also open subgroups.

4. For a finite subset S of M_K we define

$$K_S = \mathsf{J}_{K,S} \cap K^\times = \{a \in K^\times \mid |a|_v = 1 \text{ for all } v \in \mathsf{M}_K \setminus S\}.$$

In particular,

$$K_{\mathsf{M}_K^\infty} = K^\times \cap \mathsf{E}_K = \{a \in K^\times \mid |a|_v = 1 \text{ for all } v \in \mathsf{M}_K^0\}$$
$$= U_K = \begin{cases} \mathfrak{o}_K^\times & \text{in the number field case,} \\ \mathbb{F}_q^\times & \text{in the function field case,} \end{cases}$$

and

$$K_\emptyset = \begin{cases} \mu(K) & \text{(the group of roots of unity in } K\text{) in the number field case,} \\ \mathbb{F}_q^\times & \text{in the function field case} \end{cases}$$

(see [18, Theorems 3.5.8 and 6.2.7]).

5. For $v \in \mathsf{M}_K$ we define the **local embedding** $j_v \colon K_v^\times \to \mathsf{J}_K$ by

$$j_v(a) = (j_v(a)_w)_{w \in \mathsf{M}_K} \in \mathsf{J}_K, \quad \text{where } j_v(a)_w = \begin{cases} a & \text{if } w = v, \\ 1 & \text{if } w \ne v, \end{cases}$$

and the **local embedding** $j_v^* \colon K_v^\times \to \mathsf{C}_K$ by $j_v^*(a) = j_v(a)K^\times$.

The local embeddings induce topological isomorphisms $j_v \colon K_v^\times \overset{\sim}{\to} j_v(K_v^\times)$ and $j_v^* \colon K_v^\times \overset{\sim}{\to} j_v^*(K_v^\times)$ for all $v \in \mathsf{M}_K$. Indeed, regarding the subspace topology of $j_v(K_v^\times)$ in J_K it is evidemt that the map $j_v \colon K_v^\times \to j_v(K_v^\times)$ is a topological isomorphism. Since $j_v(K_v^\times) \cap K^\times = \mathbf{1}$, the map $j_v^* \colon K_v^\times \to \mathsf{C}_K$ is a monomorphism, and since the residue class epimorphism $\mathsf{J}_K \to \mathsf{C}_K$ is continuous and open, it follows that $j_v^* \colon K_v^\times \to j_v^*(K_v^\times)$ is even a topological isomorphism.

Next we highlight the interrelationship between ideles and (abstract) divisors.

Definition and Theorem 5.3.2 (Ideles and divisors). We endow the groups $\mathcal{D}_K$ and $\mathcal{C}_K$ with the discrete topology and define the **divisor map**

$$\iota_K \colon \mathsf{J}_K \to \mathcal{D}_K \ \text{ by } \ \iota_K(\alpha) = \prod_{v \in \mathsf{M}_K^0} \mathfrak{p}_v^{v(\alpha_v)} \in \mathcal{D}_K \ \text{ for all } \ \alpha = (\alpha_v)_{v \in \mathsf{M}_K} \in \mathsf{J}_K.$$

1. ι_K is a continuous epimorphism with kernel E_K, and

$$\prod_{v \in \mathsf{M}_K^0} |\alpha|_v = \mathfrak{N}(\iota_K(\alpha))^{-1} \quad \textit{for all } \ \alpha \in \mathsf{J}_K.$$

In particular, ι_K induces isomorphisms

$$\mathsf{J}_K/\mathsf{E}_K \overset{\sim}{\to} \mathcal{D}_K \quad \textit{and} \quad \mathsf{J}_K/K^\times\mathsf{E}_K = \mathsf{C}_K/\mathcal{E}_K \overset{\sim}{\to} \mathcal{C}_K.$$

2. If $a \in K^\times$, then $\iota_K(a) = (a)^K$.

3. Let K be a number field.
If S is a subset of M_K such that $S \cap \mathsf{M}_K^\infty \neq \emptyset$, then $\mathsf{J}_K = \mathsf{J}_K^0 \mathsf{J}_{K,S}$. In particular, $\mathsf{J}_K = \mathsf{J}_K^0 \mathsf{E}_K$, and $\iota_K(\mathsf{J}_K^0) = \mathcal{D}_K$.

4. Let $K/\mathbb{F}_q$ be a function field.

(a) For an idel $\alpha = (\alpha_v)_{v \in \mathsf{M}_K} \in \mathsf{J}_K$ we define its **degree** by

$$\deg(\alpha) = \deg(\iota_K(\alpha)) = \sum_{v \in \mathsf{M}_K} v(\alpha_v) \deg(v) \in \mathbb{Z}.$$

Then $\|\alpha\| = q^{-\deg(\alpha)}$ for all $\alpha \in \mathsf{J}_K$, then $\deg\colon \mathsf{J}_K \to \mathbb{Z}$ is a continuous epimorphism with kernel J_K^0 and induces an (equally denoted) continuous epimorphism $\deg\colon \mathsf{C}_K \to \mathbb{Z}$ with kernel C_K^0.

(b) $\mathsf{E}_K \subset \mathsf{J}_K^0 = \iota_K^{-1}(\mathcal{D}_K^0)$, and ι_K induces isomorphisms

$$\mathsf{J}_K^0/\mathsf{E}_K \overset{\sim}{\to} \mathcal{D}_K^0, \quad \text{and} \quad \mathsf{C}_K^0/\mathcal{E}_K = \mathsf{J}_K^0/K^\times\mathsf{E}_K \overset{\sim}{\to} \mathcal{C}_K^0.$$

(c) If $S \subset \mathsf{M}_K$, then

$$\deg(\mathsf{J}_K^0\mathsf{J}_{K,S}) = \partial_S\mathbb{Z}, \text{ and } \mathsf{J}_K/\mathsf{J}_K^0\mathsf{J}_{K,S} \cong \mathbb{Z}/\partial_S\mathbb{Z}.$$

In particular, $\mathsf{J}_K = \mathsf{J}_K^0\mathsf{J}_{K,S}$ for every finite subset S of M_K satisfying $\partial_S = 1$.

The interrelationship between field elements, ideles and divisors is represented by the following (natural) commutative diagram with exact rows and columns.

$$\begin{array}{ccccccccc}
 & & 1 & & 1 & & 1 & & \\
 & & \downarrow & & \downarrow & & \downarrow & & \\
1 & \longrightarrow & U_K & \longrightarrow & \mathsf{E}_K & \longrightarrow & \mathcal{E}_K & \longrightarrow & 1 \\
 & & \downarrow & & \downarrow & & \downarrow & & \\
1 & \longrightarrow & K^\times & \longrightarrow & \mathsf{J}_K & \longrightarrow & \mathsf{C}_K & \longrightarrow & 1 \\
 & & \downarrow & & \downarrow & & \downarrow & & \\
1 & \longrightarrow & (K^\times) & \longrightarrow & \mathsf{C}_K & \xrightarrow{c} & \mathcal{C}_K & \longrightarrow & 1 \\
 & & \downarrow & & \downarrow & & \downarrow & & \\
 & & 1 & & 1 & & 1 & &
\end{array}$$

In the function field case, there is in addition the following modified commutative diagram with exact rows and columns.

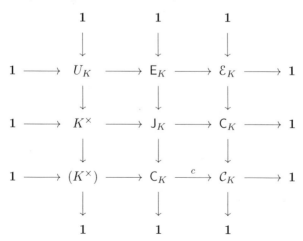

$$
\begin{array}{ccccccccc}
 & & \mathbf{1} & & \mathbf{1} & & \mathbf{1} & & \\
 & & \downarrow & & \downarrow & & \downarrow & & \\
\mathbf{1} & \longrightarrow & U_K & \longrightarrow & \mathsf{E}_K & \longrightarrow & \mathcal{E}_K & \longrightarrow & \mathbf{1} \\
 & & \downarrow & & \downarrow & & \downarrow & & \\
\mathbf{1} & \longrightarrow & K^\times & \longrightarrow & \mathsf{J}_K^0 & \longrightarrow & \mathsf{C}_K^0 & \longrightarrow & \mathbf{1} \\
 & & \downarrow & & \downarrow & & \downarrow & & \\
\mathbf{1} & \longrightarrow & (K^\times) & \longrightarrow & \mathsf{C}_K^0 & \xrightarrow{\ c\ } & \mathcal{C}_K^0 & \longrightarrow & \mathbf{1} \\
 & & \downarrow & & \downarrow & & \downarrow & & \\
 & & \mathbf{1} & & \mathbf{1} & & \mathbf{1} & &
\end{array}
$$

Proof. 1. By definition, ι_K is an epimorphism with kernel

$$\operatorname{Ker}(\iota_K) = \{\alpha \in \mathsf{J}_K \mid v(\alpha_v) = 0 \text{ for all } v \in \mathsf{M}_K^0\} = \prod_{v \in \mathsf{M}_K^\infty} K_v^\times \times \prod_{v \in \mathsf{M}_K^0} U_v = \mathsf{E}_K,$$

and therefore it induces the asserted isomorphisms. Since E_K is open in J_K, ι_K is continuous by 1.1.6.1.

If $\alpha = (\alpha_v)_{v \in \mathsf{M}_K} \in \mathsf{J}_K$, then

$$\prod_{v \in \mathsf{M}_K^0} |\alpha_v|_v = \prod_{v \in \mathsf{M}_K^0} \mathfrak{N}(v)^{-v(\alpha_v)} = \prod_{v \in \mathsf{M}_K^0} \mathfrak{N}(v)^{-v(\iota_K(\alpha))} = \mathfrak{N}(\iota_K(\alpha))^{-1}.$$

2. If $a \in K^\times$, then

$$\iota_K(a) = \prod_{v \in \mathsf{M}_K^0} \mathsf{p}_v^{v(a)} = (a)^K.$$

3. Let $\alpha \in \mathsf{J}_K$, fix some $w \in S \cap \mathsf{M}_K^\infty$, let $\beta_w \in K_w$ be such that $|\beta_w|_w = \|\alpha\|$, and for all $v \in \mathsf{M}_K \setminus \{w\}$ set $\beta_v = 1$. Then $\beta = (\beta_v)_{v \in \mathsf{M}_K} \in \mathsf{J}_{K,S}$ and $\|\beta\| = \|\alpha\|$, hence $\alpha\beta^{-1} \in \mathsf{J}_K^0$ and $\alpha = (\alpha\beta^{-1})\beta \in \mathsf{J}_K^0 \mathsf{J}_{K,S}$. In particular, $\mathsf{J}_K = \mathsf{J}_K^0 \mathsf{J}_{K,\mathsf{M}_K^\infty} = \mathsf{J}_K^0 \mathsf{E}_K$, and $\iota_K(\mathsf{J}_K^0) = \iota_K(\mathsf{J}_K^0 \mathsf{E}_K) = \iota_K(\mathsf{J}_K) = \mathcal{D}_K$.

4.(a) Clearly $\deg\colon \mathsf{J}_K \to \mathbb{Z}$ is an epimorphism, and if $\alpha = (\alpha_v)_{v \in \mathsf{M}_K} \in \mathsf{J}_K$, then

$$\|\alpha\| = \prod_{v \in \mathsf{M}_K} |\alpha_v|_v = \prod_{v \in \mathsf{M}_K} q^{-v(\alpha_v)\deg(v)} = q^{-\deg(\alpha)}.$$

Hence $\operatorname{Ker}(\deg) = \mathsf{J}_K^0$ is open in J_K, and thus deg is continuous. The remaining assertions follow, since $K^\times \subset \mathsf{J}_K^0$.

(b) Obvious by (a).

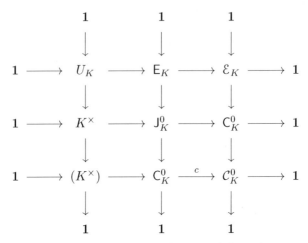

(c) By (a), deg induces an isomorphism $\mathsf{J}_K/\mathsf{J}_K^0\mathsf{J}_{K,S} \overset{\sim}{\to} \mathbb{Z}/\deg(\mathsf{J}_K^0\mathsf{J}_{K,S})$, and thus it suffices to prove that $\deg(\mathsf{J}_K^0\mathsf{J}_{K,S}) = \partial_S\mathbb{Z}$. This is obvious if $S = \emptyset$. Thus let $S \neq \emptyset$ and $(s_v)_{v\in S}$ a sequence in $\mathbb{Z}$ such that $s_v = 0$ for almost all $v \in S$ and

$$\partial_S = \sum_{v\in S} s_v \deg(v).$$

Let $\beta = (\beta_v)_{v\in\mathsf{M}_K} \in \mathsf{J}_{K,S}$ be an idele satisfying $v(\beta_v) = s_v$ if $v \in S$, and $\beta_v = 1$ if $v \notin S$. Then $\deg(\beta) = \partial_S \in \deg(\mathsf{J}_{K,S}) = \deg(\mathsf{J}_K^0\mathsf{J}_{K,S})$ and thus $\partial_S\mathbb{Z} \subset \deg(\mathsf{J}_K^0\mathsf{J}_{K,S})$. Conversely, if $\beta = \beta_0\gamma \in \mathsf{J}_K^0\mathsf{J}_{K,S}$, where $\beta_0 \in \mathsf{J}_K^0$ and $\gamma = (\gamma_v)_{v\in\mathsf{M}_K} \in \mathsf{J}_{K,S}$, then

$$\deg(\beta) = \deg(\beta_0) + \deg(\gamma) = \sum_{v\in S} v(\gamma_v)\deg(v) \in \partial_S\mathbb{Z}. \qquad \square$$

Theorem 5.3.3.

1. Let K be a number field, and $n = [K:\mathbb{Q}]$. We define the map $j\colon \mathbb{R}_{>0} \to \mathsf{J}_K$ by

$$j(\lambda)_v = \begin{cases} \lambda^{1/n} & \text{if } v \in \mathsf{M}_K^\infty, \\ 1 & \text{if } v \in \mathsf{M}_K^0 \end{cases} \quad \text{and set}$$

$$\Gamma = j(\mathbb{R}_{>0}) \subset \prod_{v\in\mathsf{M}_K^\infty} K_v^\times \times \prod_{v\in\mathsf{M}_K^0} \mathbf{1} \subset \mathsf{J}_K.$$

(a) $\|j(\lambda)\| = \lambda$ for all $\lambda \in \mathbb{R}_{>0}$, $j\colon \mathbb{R}_{>0} \to \Gamma$ is a topological isomorphism, and $\|\mathsf{J}_K\| = \|\mathsf{C}_K\| = \|\Gamma\| = \mathbb{R}_{>0}$.

(b) J_K and C_K decompose as topological inner direct products as $\mathsf{J}_K = \mathsf{J}_K^0 \cdot \Gamma$ and $\mathsf{C}_K = \mathsf{C}_K^0 \cdot \Gamma'$, where $\Gamma' = \Gamma K^\times/K^\times \subset \mathsf{C}_K$ and $\Gamma' \cong \Gamma \cong \mathbb{R}_{>0}$.

2. Let $K/\mathbb{F}_q$ be a function field, $\gamma \in \mathsf{J}_K$ such that $\deg(\gamma) = 1$, $\Gamma = \langle\gamma\rangle \subset \mathsf{J}_K$, $c = \gamma K^\times \in \mathsf{C}_K$, and $\Gamma' = \langle c\rangle = \Gamma K^\times/K^\times$.

(a) $\|\mathsf{J}_K\| = \|\mathsf{C}_K\| = q^{\mathbb{Z}}$.

(b) J_K and C_K decompose as topological inner direct products as

$$\mathsf{J}_K = \mathsf{J}_K^0{\cdot}\Gamma \quad \text{and} \quad \mathsf{C}_K = \mathsf{C}_K^0{\cdot}\Gamma', \quad \text{where} \quad \Gamma \cong \Gamma' \cong \mathbb{Z}.$$

Proof. 1. (a) If $\lambda \in \mathbb{R}_{>0}$, then

$$\|j(\lambda)\| = \prod_{v\in\mathsf{M}_K^\infty} |\lambda|_v^{1/n} = \prod_{v\in\mathsf{M}_K^\infty} \lambda^{l_v/n} = \lambda.$$

Apparently j is a continuous monomorphism. Assume now that $\varepsilon \in (0,1)$, $V = (1-\varepsilon, 1+\varepsilon) \subset \mathbb{R}_{>0}$ and $B_{v,\varepsilon} = \{x \in K_v \mid |x-1|_v^{1/l_v} < \varepsilon\}$ for all $v \in \mathsf{M}_K^\infty$. Then

$$j(V) = j(\mathbb{R}_{>0}) \cap \Big[\prod_{v \in \mathsf{M}_K^\infty} B_{v,\varepsilon} \times \prod_{v \in \mathsf{M}_K^0} (K_v^\times \mid U_v)\Big]$$

is relatively open in $j(\mathbb{R}_{>0})$. Hence $j\colon \mathbb{R}_{>0} \to j(\mathbb{R}_{>0})$ is a homeomorphism, and since $\mathbb{R}_{>0} = \|\Gamma\| \subset \|\mathsf{J}_K\| = \|\mathsf{C}_K\| \subset \mathbb{R}_{>0}$, we get $\|\mathsf{J}_K\| = \|\mathsf{C}_K\| = \|\Gamma\| = \mathbb{R}_{>0}$.

(b) Since $\operatorname{Ker}(\|\cdot\|) = \mathsf{J}_K^0$ and $\|\cdot\| \circ j = \mathrm{id}_{\mathbb{R}_{>0}}$, 1.1.8.3 implies $\mathsf{J}_K = \mathsf{J}_K^0 \cdot \Gamma$, and since $K^\times \subset \mathsf{J}_K^0$, it follows that $\mathsf{C}_K = (\mathsf{J}_K^0 \cdot \Gamma)/K^\times = \mathsf{C}_K^0 \cdot \Gamma'$, where $\Gamma' = \Gamma K^\times/K^\times \cong \Gamma \cong \mathbb{R}_{>0}$.

2. (a) Since $\deg(\Gamma) = \mathbb{Z} = \deg(\mathsf{J}_K) = \deg(\mathsf{C}_K)$, we get

$$q^{\mathbb{Z}} = \|\Gamma\| = \|\mathsf{J}_K\| = \|\mathsf{C}_K\|.$$

(b) Let $j\colon \mathbb{Z} \to \mathsf{J}_K$ be defined by $j(n) = \gamma^n$. Then we have $j(\mathbb{Z}) = \Gamma$, $\operatorname{Ker}(\deg) = \mathsf{J}_K^0$ and $\deg \circ j = \mathrm{id}_{\mathbb{Z}}$. Hence 1.1.8.3 implies $\mathsf{J}_K = \mathsf{J}_K^0 \cdot \Gamma$, and as $K^\times \subset \mathsf{J}_K^0$, it follows that $\mathsf{C}_K = (\mathsf{J}_K^0 \cdot \Gamma)/K^\times = \mathsf{C}_K^0 \cdot \Gamma'$, where

$$\Gamma' = \Gamma K^\times/K^\times = \langle c\rangle \cong \Gamma \cong \mathbb{Z}. \qquad \square$$

The following "Fundamental lemma" is the fundamentum for all quantitative assertions concerning global fields.

Theorem 5.3.4 (Fundamental lemma for global fields). *If K is a number field with signature (r_1, r_2), we set*

$$C = \Big(\frac{2}{\pi}\Big)^{r_2} \sqrt{|\mathsf{d}_K|} \quad \text{(the **Blichfeld constant**)},$$

and if $K/\mathbb{F}_q$ is a function field with genus g, we set $C = q^g$.

1. *Let $\alpha \in \mathsf{J}_K$ such that $\|\alpha\| > C$. Then there exists some $a \in K^\times$ satisfying $|a|_v \le |\alpha_v|_v$ for all $v \in \mathsf{M}_K$.*

2. *For all $v \in \mathsf{M}_K$ let $c_v', c_v'' \in \mathbb{R}$ such that $0 \le c_v' \le c_v''$, and suppose that $c_v'' \le 1$ for almost all $v \in \mathsf{M}_K$. Then the set*

$$C = \{\alpha \in \mathsf{A}_K \mid c_v' \le |\alpha_v|_v \le c_v''\}$$

is compact.

Proof. 1. CASE 1: K is a number field. As in 5.1.1.**3** let $\sigma_1, \ldots, \sigma_{r_1}\colon K \to \mathbb{R}$ be the real and $\sigma_{r_1+1}, \overline{\sigma_{r_1+1}}, \ldots, \sigma_{r_1+r_2}, \overline{\sigma_{r_1+r_2}}\colon K \to \mathbb{C}$ the pairs of conjugate complex embeddings. For $i \in [1, r_1]$ set $l_i = 1$, and for $i \in [r_1+1, r_1+r_2]$ set $l_i = 2$. For $i \in [1, r_1+r_2]$ let $v_i \in \mathsf{M}_K^\infty$ be the place satisfying $|x|_{v_i} = |\sigma_i(x)|_\infty^{l_i}$ for all $x \in K$, and set $c_i = |\alpha_{v_i}|_\infty$. Then 5.3.2.1 implies

$$\|\alpha\| = \prod_{v \in \mathsf{M}_K^\infty} |\alpha_v|_v \prod_{v \in \mathsf{M}_K^0} |\alpha_v|_v = \prod_{i=1}^{r_1} c_i \prod_{i=r_1+1}^{r_1+r_2} c_i^2\, \mathfrak{N}(\iota_K(\alpha))^{-1} > \Big(\frac{2}{\pi}\Big)^{r_2} \sqrt{|\mathsf{d}_K|},$$

and consequently

$$\prod_{i=1}^{r_1} c_i \prod_{i=r_1+1}^{r_1+r_2} c_i^2 > \Big(\frac{2}{\pi}\Big)^{r_2} \sqrt{|\mathsf{d}_K|\mathfrak{N}(\iota_K(\alpha))^2} = \Big(\frac{2}{\pi}\Big)^{r_2} \sqrt{\Delta(\iota_K(\alpha))},$$

where $\Delta(\iota_K(\alpha))$ denotes the discriminant of the lattice $\iota_K(\alpha)$. An application of Minkowski's convex body theorem yields the existence some $a \in \iota_K(\alpha)^\bullet$ such that $|\sigma_i(\alpha)| \le c_i$ hence $|a|_{v_i} \le c_i^{l_i} = |\alpha_{v_i}|_{v_i}$ for all $i \in [1, r_1+r_2]$ (see [18, Theorem 3.5.5]). If $v \in \mathsf{M}_K^0$, then $a \in \iota_K(\alpha)$ implies $v(a) \ge v(\iota_K(\alpha)) = v(\alpha_v)$ and therefore $|a|_v \le |\alpha_v|_v$.

CASE 2: $K/\mathbb{F}_q$ is a function field with genus g. Then

$$\|\alpha\| = q^{-\deg(\alpha)} > C = q^g \quad \text{and therefore} \quad -\deg(\alpha) = \deg(-\iota_K(\alpha)) \ge g.$$

By Riemann's inequality (see [18, Corollary 6.3.8]) it follows that

$$\dim(-\iota_K(\alpha)) \ge \deg(-\iota_K(\alpha)) - g + 1 \ge 1,$$

and thus there exists some $a \in K^\times$ such that $(a)^K \ge \iota_K(\alpha)$. If $v \in \mathsf{M}_K$, then $v(a) = v((a)^K) \ge v(\iota_K(\alpha)) = v(\alpha_v)$, and

$$|a|_v = q^{-\deg(v)v(a)} \le q^{-\deg(v)v(\alpha_v)} = |\alpha_v|_v.$$

2. For $v \in \mathsf{M}_K$ we set $C_v = \{\alpha_v \in K_v \mid c_v' \le |\alpha_v|_v \le c_v''\}$. By assumption, the set $S = \{v \in \mathsf{M}_K \mid c_v'' > 1\}$ is finite, and therefore

$$C = \prod_{v \in \mathsf{M}_K} C_v \subset \mathsf{A}_{K,S} = \prod_{v \in S} K_v \times \prod_{v \in \mathsf{M}_K \setminus S} \widehat{\mathcal{O}}_v.$$

As $\mathsf{A}_{K,S}$ carries the product topology, it suffices by Tychonov's theorem to prove that C_v is compact for all $v \in \mathsf{M}_K$. If $v \in \mathsf{M}_K^\infty$, then $C_v \subset \mathbb{C}$ is closed and bounded and thus compact.

Thus let $v \in \mathsf{M}_K^0$. Then $C_v \subset K_v$ is closed, since $|\cdot|_v \colon K_v \to \mathbb{R}_{\ge 0}$ is continuous. If $a \in C_v$, then $|a|_v = \mathfrak{N}(v)^{-v(a)} \le c_v''$, and therefore

$$v(a) \ge N = \Big\lceil -\frac{\log c_v''}{\log \mathfrak{N}(v)} \Big\rceil, \quad \text{which implies } a \in \mathfrak{p}_v^N \quad [= \mathbf{0} \text{ if } N = \infty].$$

It follows that $C_v \subset \widehat{\mathfrak{p}}_v^N$, and thus C_v is compact. □

Theorem 5.3.5 (Strong approximation theorem). *Let S be a finite subset of M_K, $v_0 \in \mathsf{M}_K \setminus S$, $\varepsilon \in \mathbb{R}_{>0}$ and $\alpha = (\alpha_v)_{v \in \mathsf{M}_K} \in \mathsf{A}_K$.*

1. There exists some $x \in K$ satisfying

$$|x - \alpha_v|_v < \varepsilon \ \text{for all } v \in S \quad \text{and} \quad |x|_v \le 1 \ \text{for all } v \in \mathsf{M}_K \setminus (S \cup \{v_0\}).$$

2. Assume that $\mathsf{M}_K^\infty \subset S \cup \{v_0\}$, *and for* $v \in S \cap \mathsf{M}_K^0$ *let* $m_v \in \mathbb{Z}$. *Then there exists some* $x \in K$ *satisfying*

$$v(x) = m_v \quad \text{for all} \quad v \in S \cap \mathsf{M}_K^0, \qquad |x - \alpha_v|_v < \varepsilon \quad \text{for all} \quad v \in S \cap \mathsf{M}_K^\infty,$$

and $|x|_v \le 1$ *for all* $v \in \mathsf{M}_K \setminus (S \cup \{v_0\})$.

3. A_K/K *is compact.*

Proof. We proceed in five steps.

I. Let K be a number field. There exists some $m \in \mathbb{R}_{>0}$ such that for every family $(\xi_v \in K_v)_{v \in \mathsf{M}_K^\infty}$ there exists some $d \in \mathfrak{o}_K$ such that $|d - \xi_v|_v \le m$ for all $v \in \mathsf{M}_K^\infty$.

Proof of **I.** Let (r_1, r_2) be the signature of K and $[K:\mathbb{Q}] = n = r_1 + 2r_2$. Let $\sigma_1, \ldots, \sigma_{r_1} : K \to \mathbb{R}$ be the real and $\sigma_{r_1+1}, \overline{\sigma_{r_1+1}}, \ldots, \sigma_{r_1+r_2}, \overline{\sigma_{r_1+r_2}} : K \to \mathbb{C}$ the pairs of conjugate complex embeddings. Let $\varphi \colon K \to \mathbb{R}^n$ be the field embedding, defined by

$$\varphi(x) = (\sigma_1(x), \ldots, \sigma_{r_1}(x), \Im\sigma_{r_1+1}(x), \ldots, \Im\sigma_{r_1+r_2}(x), \Re\sigma_{r_1+1}(x), \ldots, \Re\sigma_{r_1+r_2}(x))^{\mathsf{t}},$$

and let

$$j \colon \mathbb{R}^n \to \mathbb{R}^{r_1} \times \mathbb{C}^{r_2} = \prod_{v \in \mathsf{M}_K^\infty} K_v = K_\infty$$

be defined by

$$j((a_1, \ldots, a_n)^{\mathsf{t}}) = (a_1, \ldots, a_{r_1}, a_{r_1+r_2+1} + \mathrm{i}a_{r_2+1}, \ldots, a_{r_1+2r_2} + \mathrm{i}a_{r_1+r_2}).$$

Then j is an isomorphism of real vector spaces, and $j \circ \varphi \colon K \to K_\infty$ is the diagonal embedding of K. As $\varphi(\mathfrak{o}_K)$ is a complete lattice in $\mathbb{R}^n$ (see [18, Theorem 3.5.4]), it follows that $j \circ \varphi(\mathfrak{o}_K)$ is a complete lattice in K_∞. Let $\|\cdot\| \colon K_\infty \to \mathbb{R}_{\ge 0}$ be defined by $\|(\xi_v)_{v \in \mathsf{M}_K^\infty}\| = \max\{|\xi_v|_v \mid v \in \mathsf{M}_K^\infty\}$. Then $\|\cdot\|$ is a norm on K_∞, and there exists some $m \in \mathbb{N}$ such that for every $(\xi_v)_{v \in \mathsf{M}_K^\infty} \in K_\infty$ there exists some $d \in \mathfrak{o}_K$ satisfying

$$\|j \circ \varphi(d) - \xi\| = \max\{|d - \xi_v|_v \mid v \in \mathsf{M}_K^\infty\} \le m. \qquad \square[\mathbf{I.}]$$

II. For every $\xi = (\xi_v)_{v \in \mathsf{M}_K} \in \mathsf{A}_K$ there exists some $b \in K$ such that

$$|b - \xi_v|_v \le 1 \quad \text{for all} \quad v \in \mathsf{M}_K^0 \quad \text{and} \quad |b - \xi_v|_v \le m \quad \text{for all} \quad v \in \mathsf{M}_K^\infty,$$

where in the number field case m is as in **I**.

Proof of **II.** Let $\xi = (\xi_v)_{v \in \mathsf{M}_K} \in \mathsf{A}_K$, and $T = \{v \in \mathsf{M}_K^0 \mid |\xi_v|_v > 1\}$.
Let first K be a number field and m as in **I**. For $v \in T$ let $b_v \in \mathfrak{o}_K$ be such that $\mathfrak{p}_v^{h_K} = (b_v)$. As $|b_v|_v < 1$, there exists some $n \in \mathbb{N}$ such that $|b_v^n \xi_v|_v \le 1$

for all $v \in T$. For $v \in T$ let $a_v \in \mathfrak{o}_K$ be such that $a_v \equiv b_v^n \xi_v \mod \mathfrak{p}_v^{nh_K}$. Then we obtain

$$\left| \frac{a_v}{b_v^n} - \xi_v \right|_v \le 1, \quad |a_v|_v \le 1 \quad \text{and} \quad |b_v|_{v'} = 1 \quad \text{for all} \quad v,\, v' \in T, \quad v' \ne v. \qquad (\dagger)$$

Now let K be a function field and $v_0 \in \mathsf{M}_K \setminus T$. For $v \in T$, we set $\mathsf{c}_v = \mathsf{p}_v^{\deg(v_0)} \mathsf{p}_{v_0}^{-\deg(v)}$. Then $\deg(\mathsf{c}_v) = 0$, and therefore $\mathsf{c}_v^{h_K} = (b_v)$ for some $b_v \in K^\times$. As $|b_v|_v < 1$, there exists some $n \in \mathbb{N}$ be such that $|b_v^n \xi_v|_v \le 1$ for all $v \in T$. For $v \in T$, let $a_v \in K^\times$ such that $|a_v - b_v^n \xi_v|_v \le |b_v^n|_v$. Then, as in the number field case,

$$\left| \frac{a_v}{b_v^n} - \xi_v \right|_v \le 1, \quad |a_v|_v \le 1 \quad \text{and} \quad |b_v|_{v'} = 1 \quad \text{for all} \quad v,\, v' \in T, \quad v' \ne v. \qquad (\dagger)$$

Starting from $(\dagger)$, we proceed unified and set

$$a = \sum_{v \in T} \frac{a_v}{b_v^n}.$$

If $v \in T$, then

$$|a - \xi_v|_v = \left| \frac{a_v}{b_v^n} - \xi_v + \sum_{v' \in T \setminus \{v\}} \frac{a_{v'}}{b_{v'}^n} \right|_v$$
$$\le \max\left\{ \left| \frac{a_v}{b_v^n} - \xi_v \right|_v, \left| \frac{a_{v'}}{b_{v'}^n} \right|_v \;\middle|\; v' \in T \setminus \{v\} \right\} \le 1,$$

and if $v \in \mathsf{M}_K^0 \setminus T$, then

$$|a - \xi_v|_v = \left| \sum_{v' \in T} \frac{a_{v'}}{b_{v'}^n} - \xi_v \right|_v \le \max\left\{ \left| \frac{a_{v'}}{b_{v'}^n} \right|_v, |\alpha_v|_v \;\middle|\; v' \in T \right\} \le 1.$$

In the function field case we set $b = a$ and are done.

In the number field case we apply **I**. Let $d \in \mathfrak{o}_K$ such that $|a - \xi_v - d|_v \le m$ for all $v \in \mathsf{M}_K^\infty$, and set $b = a - d$. Then $|b - \xi_v|_v \le \max\{|a - \xi_v|_v, |d|_v\} \le 1$ for all $v \in \mathsf{M}_K^0$ and $|b - \xi_v|_v \le m$ for all $v \in \mathsf{M}_K^\infty$. $\square$[**II.**]

III. Proof of 3. Let m be as in **I**. The set

$$B = \{\beta = (\beta_v)_{v \in \mathsf{M}_K} \in \mathsf{A}_K \mid |\beta_v|_v \le 1 \text{ for } v \in \mathsf{M}_K^0 \text{ and } |\beta_v|_v \le m \text{ for } v \in \mathsf{M}_K^\infty\}$$

is compact by 5.3.4.2, and $B + K = \mathsf{A}_K$ by **II**. If $\pi \colon \mathsf{A}_K \to \mathsf{A}_K/K$ denotes the residue class homomorphism, then $\mathsf{A}_K/K = \pi(C)$ is compact. $\square$[**III.**

IV. Proof of 1. Let C be the Blichfeld constant given in 5.3.4. Let m be as in **I**, and let $(m_v)_{v \in \mathsf{M}_K} \in \mathsf{A}_K$ be such that $m_v = 1$ if $v \in \mathsf{M}_K^0$, and $m_v = m$ if $v \in \mathsf{M}_K^\infty$. Let $\gamma = (\gamma_v)_{v \in \mathsf{M}_K} \in \mathsf{J}_K$ be such that $|\gamma_v|_v < \varepsilon m_v^{-1}$ if $v \in S$, $|\gamma_v|_v \le m_v^{-1}$ if $v \in \mathsf{M}_K \setminus (S \cup \{v_0\})$, and let $\gamma_{v_0} \in K_{v_0}^\times$ be such that $\|\gamma\| > C$. By 5.3.4.1, there exists some $a \in K^\times$ satisfying $|a|_v \le |\gamma_v|_v$ for all $v \in \mathsf{M}_K$.

Let $\xi = (\xi_v)_{v \in \mathsf{M}_K} \in \mathsf{A}_K$ be such that $\xi_v = \alpha_v$ if $v \in S$ and $\xi_v = 0$ otherwise. Then $a^{-1}\xi \in \mathsf{A}_K$, and by **II** there exists some $b \in K$ such that $|b - a^{-1}\xi_v|_v \le m_v$ for all $v \in \mathsf{M}_K$. Set finally $x = ab \in K$. If $v \in S$, then

$$|x - \alpha_v|_v = |ab - \xi_v|_v = |a|_v |b - a^{-1}\xi_v|_v \le |\gamma_v|_v m_v < \varepsilon,$$

and if $v \in \mathsf{M}_K \setminus (S \cup \{v_0\})$, then

$$|x|_v = |ab - \xi_v|_v = |a|_v |b - a^{-1}\xi_v|_v \le |\gamma_v|_v m_v \le 1.$$

V. Proof of 2. For $v \in S \cap \mathsf{M}_K^0$ let $z_v \in K$ such that $v(z_v) = m_v$. By 1, there exist $x', x'' \in K$ such that

$$|x' - \alpha_v|_v < \mathfrak{N}(v)^{-m_v} \text{ for all } v \in S \cap \mathsf{M}_K^0, \quad |x' - \alpha_v|_v < \frac{\varepsilon}{2} \text{ for all } v \in S \cap \mathsf{M}_K^\infty,$$

and $|x'|_v \le 1$ for all $v \in \mathsf{M}_K \setminus (S \cup \{v_0\})$;

$$|x'' - z_v|_v < \mathfrak{N}(v)^{-m_v} \text{ for all } v \in S \cap \mathsf{M}_K^0, \quad |x''|_v < \frac{\varepsilon}{2} \text{ for all } v \in S \cap \mathsf{M}_K^\infty,$$

and $|x''|_v \le 1$ for all $v \in \mathsf{M}_K \setminus (S \cup \{v_0\})$.

We set $x = x' + x''$. If $v \in S \cap \mathsf{M}_K^0$, then $x - \alpha_v = (x' - \alpha_v) + (x'' - z_v) + z_v$, and since $|z_v|_v = \mathfrak{N}(v)^{-m_v}$, we get $|x - \alpha_v|_v = \mathfrak{N}(v)^{-\alpha_v}$ and thus $v(x - \alpha_v) = m_v$. If $v \in S \cap \mathsf{M}_K^\infty$, then $|x - \alpha_v|_v \le |x' - \alpha_v|_v + |x''|_v < \varepsilon$. If $v \in \mathsf{M}_K \setminus (S \cup \{v_0\})$, then $|x|_v \le \max\{|x'|_v, |x''|_v\} \le 1$, since $\mathsf{M}_K \setminus (S \cup \{v_0\}) \subset \mathsf{M}_K^0$.

The strong approximation theorem for function fields holds even for function fields with an arbitrary field of constants (see [18, Theorem 6.4.8]).

We proceed with a more precise investigation of the topology of J_K. As a consequence we shall see in 5.3.7 and 5.3.8 that the classical finiteness results of algebraic number theory concerning the ideal class group and the unit group have its origin in topological properties of the idele group.

Theorem 5.3.6 (Topology of J_K).

1. *The subsets K of A_K and $K^\times$ of J_K are discrete and closed.*

2. *J_K^0 is closed in A_K.*

3. *On J_K^0 the subspace topologies induced by A_K and by J_K coincide.*

Proof. 1. By 1.1.3.4 it suffices to prove discreteness. Let S be a finite non-empty subset of M_K, $\mathsf{M}_K^\infty \subset S$, and for $v \in S$ let $B_v = \{x \in K_v \mid |x|_v < 1\}$. Then

$$U = \prod_{v \in S} B_v \times \prod_{v \in \mathsf{M}_K \setminus S} \widehat{\mathcal{O}}_v \subset \mathsf{A}_K$$

is a neighbourhood of 0 in A_K such that $\|\alpha\| < 1$ for all $\alpha \in U$. Hence $U \cap K = \{0\}$, and K is discrete in A_K.

Now let V be a neighbourhood of 1 in A_K such that $V \cap K = \{1\}$. Then $V \cap \mathsf{J}_K$ is a neighbourhood of 1 in J_K (since the topology of J_K is finer than the subspace topology induced by A_K). Hence $K^\times$ is discrete in J_K.

2. Let $\alpha = (\alpha_v)_{v \in \mathsf{M}_K} \in \mathsf{A}_K \setminus \mathsf{J}_K^0$. For $\varepsilon \in \mathbb{R}_{>0}$ and a finite subset S of M_K satisfying $\mathsf{M}_K^\infty \cup \{v \in \mathsf{M}_K \mid |\alpha_v|_v > 1\} \subset S$ we set

$$W(S,\varepsilon) = \{\xi = (\xi_v)_{v \in \mathsf{M}_K} \in \mathsf{A}_K \mid \\ |\xi_v - \alpha_v|_v < \varepsilon \text{ for all } v \in S, \text{ and } |\xi_v|_v \le 1 \text{ for all } v \in \mathsf{M}_K \setminus S\}.$$

Then $W(S,\varepsilon)$ is a neighbourhood of α in A_K, and we show that a suitable choice of S and ε entails $W(S,\varepsilon) \cap \mathsf{J}_K^0 = \emptyset$.

CASE 1: $\|\alpha\| < 1$. Let S be a finite subset of M_K and $\varepsilon \in \mathbb{R}_{>0}$ such that

$$\mathsf{M}_K^\infty \cup \{v \in \mathsf{M}_K \mid |\alpha_v|_v > 1\} \subset S \quad \text{and} \quad \prod_{v \in S}(|\alpha_v|_v + \varepsilon) < 1.$$

If $\xi \in W(S,\varepsilon)$, then

$$\|\xi\| = \prod_{v \in \mathsf{M}_K} |\xi_v|_v \le \prod_{v \in S}(|\alpha_v|_v + \varepsilon) \prod_{v \notin S} |\xi_v|_v < 1, \quad \text{hence } \xi \notin \mathsf{J}_K^0,$$

and therefore $W(S,\varepsilon) \cap J_K^0 = \emptyset$.

CASE 2: $\|\alpha\| > 1$. Let S be a finite subset of M_K and $\varepsilon \in \mathbb{R}_{>0}$ such that

$$\mathsf{M}_K^\infty \cup \{v \in \mathsf{M}_K \mid |\alpha_v|_v > 1\} \cup \{v \in \mathsf{M}_K^0 \mid \mathfrak{N}(v) \le 2\|\alpha\|\} \subset S,$$

$$1 < \prod_{v \in S}(|\alpha_v|_v - \varepsilon) \quad \text{and} \quad \prod_{v \in S}(|\alpha_v|_v + \varepsilon) < 2\|\alpha\|,$$

and let $\xi \in W(S,\varepsilon)$.

CASE 2a: $|\xi_v|_v = 1$ for all $v \in \mathsf{M}_K \setminus S$. In this case we obtain

$$\|\xi\| = \prod_{v \in \mathsf{M}_K} |\xi_v|_v > \prod_{v \in S}(|\alpha_v|_v - \varepsilon) \prod_{v \in \mathsf{M}_K \setminus S} |\xi_v|_v > 1, \quad \text{hence } \xi \notin \mathsf{J}_K^0,$$

and consequently $W(S,\varepsilon) \cap J_K^0 = \emptyset$.

CASE 2b: There exists some $w \in \mathsf{M}_K \setminus S$ such that $|\xi_w|_w < 1$. Then $w \in \mathsf{M}_K^0$, $\mathfrak{N}(w) > 2\|\alpha\|$, and as $|\xi_w|_w = |\mathfrak{N}(w)|^{-w(\xi_w)} < 1$, it follows that $|\xi_w|_w \le \mathfrak{N}(w)^{-1} < (2\|\alpha\|)^{-1}$. Hence

$$\|\xi\| = |\xi_w|_w \prod_{v \in \mathsf{M}_K \setminus (S \cup \{w\})} |\xi_v|_v \prod_{v \in S} |\xi_v|_v < (2\|\alpha\|)^{-1} \prod_{v \in S}(|\alpha_v|_v + \varepsilon) < 1,$$

$\xi \notin \mathsf{J}_K^0$, and consequently $W(S,\varepsilon) \cap J_K^0 = \emptyset$.

3. Call the subspace topology on J_K^0 induced by A_K resp. J_K the A_K-topology resp. the J_K-topology. By 1.2.4.1, the topology of J_K (which is the unit topology induced by A_K) is finer than the subspace topology induced by A_K. Hence on J_K^0 the J_K-topology is finer than the A_K-topology, and thus it suffices to prove: If $\alpha \in \mathsf{J}_K^0$, then every J_K-neighbourhood of α in J_K^0 is an A_K-neighbourhood.

Let $\alpha \in \mathsf{J}_K^0$. For a finite subset S of M_K satisfying

$$\mathsf{M}_K^\infty \cup \{v \in \mathsf{M}_K \mid |\alpha_v|_v \neq 1\} \subset S$$

and $\varepsilon \in \mathbb{R}_{>0}$ we set

$$W(S,\varepsilon) = \{\xi = (\xi_v)_{v \in \mathsf{M}_K} \in \mathsf{A}_K \mid |\xi_v - \alpha_v|_v < \varepsilon \text{ for all } v \in S, \text{ and } |\xi_v|_v \le 1 \text{ for all } v \in \mathsf{M}_K \setminus S\}$$

and

$$W_1(S,\varepsilon) = \{\xi = (\xi_v)_{v \in \mathsf{M}_K} \in \mathsf{J}_K \mid |\xi_v - \alpha_v|_v < \varepsilon \text{ for all } v \in S, \text{ and } |\xi_v|_v = 1 \text{ for all } v \in \mathsf{M}_K \setminus S\}.$$

The sets $W(S,\varepsilon) \cap \mathsf{J}_K^0$ constitute a fundamental system of A_K-neighbourhoods, and the sets $W_1(S,\varepsilon) \cap \mathsf{J}_K^0$ constitute a fundamental system of J_K-neighbourhoods of α in J_K^0.

Let U be a J_K-neighbourhood of α in J_K^0. Then there exists a finite subset S of M_K such that $\mathsf{M}_K^\infty \cup \{v \in \mathsf{M}_K \mid |\alpha_v|_v \neq 1\} \subset S$ and some $\varepsilon_0 \in \mathbb{R}_{>0}$ with $W_1(S,\varepsilon_0) \cap \mathsf{J}_K^0 \subset U$. Since $\|\alpha\| = 1$, there exists some $\varepsilon \in (0, \varepsilon_0]$ such that

$$\prod_{v \in S} (|\alpha_v|_v + \varepsilon) < 2,$$

and it suffices to prove that

$$W(S,\varepsilon) \subset W_1(S,\varepsilon). \tag{$*$}$$

Indeed, $(*)$ implies $W(S,\varepsilon) \cap \mathsf{J}_K^0 \subset W_1(S,\varepsilon) \cap \mathsf{J}_K^0 \subset W_1(S,\varepsilon_0) \cap \mathsf{J}_K^0 \subset U$, and therefore U is an A_K-neighbourhood of α in J_K^0.

Proof of $(*)$. Assume the contrary and let, $\xi \in W(S,\varepsilon) \setminus W_1(S,\varepsilon)$. Then there exists some $w \in \mathsf{M}_K \setminus S$ such that $|\xi_w|_w < 1$, and as $w \in \mathsf{M}_K^0$, we obtain

$$|\xi_w|_w = \mathfrak{N}(w)^{-w(\xi_w)} \le \mathfrak{N}(w)^{-1} \le \frac{1}{2}, \text{ and therefore } \prod_{v \in \mathsf{M}_K \setminus S} |\xi_v|_v \le |\xi_w|_w \le \frac{1}{2}.$$

Hence it follows that

$$1 = \|\xi\| < \prod_{v \in \mathsf{M}_K \setminus S} |\xi_v|_v \prod_{v \in S} |\xi_v|_v < \frac{1}{2} \prod_{v \in S} (|\alpha_v|_v + \varepsilon) < 1,$$

a contradiction. □

We proceed with the announced finiteness results for class groups. Recall from 5.3.2 that

$$\mathsf{J}_K^0\mathsf{E}_K = \begin{cases} \mathsf{J}_K & \text{in the number field case,} \\ \mathsf{J}_K^0 & \text{in the function field case.} \end{cases}$$

Theorem 5.3.7 (Topology of J_K, continued).

1. C_K^0 is compact, $\mathsf{J}_K^0\mathsf{E}_K/K^\times\mathsf{E}_K$ is finite, and

$$\mathsf{J}_K^0\mathsf{E}_K/K^\times\mathsf{E}_K = \begin{cases} \mathsf{J}_K/K^\times\mathsf{E}_K = \mathsf{C}_K/\mathcal{E}_K \cong \mathcal{C}_K & \textit{in the number field case,} \\ \mathsf{J}_K^0/K^\times\mathsf{E}_K = \mathsf{C}_K^0/\mathcal{E}_K \cong \mathcal{C}_K^0 & \textit{in the function field case.} \end{cases}$$

In any case, $|\mathsf{J}_K^0\mathsf{E}_K/K^\times\mathsf{E}_K| = h_K$.

2. There exists a finite subset S of M_K such that $\mathsf{J}_K = \mathsf{J}_{K,S}K^\times$.

3. For every finite subset S of M_K the set $\mathsf{J}_{K,S} \cap \mathsf{J}_K^0/K_S$ is compact.

Proof. 1. Let C be the constant given in 5.3.4, let $\alpha \in \mathsf{J}_K$ be such that $\|\alpha\| > C$, and set

$$W = \{\xi = (\xi_v)_{v\in\mathsf{M}_K} \in \mathsf{A}_K \mid |\xi_v|_v \le |\alpha_v|_v \text{ for all } v \in \mathsf{M}_K\}.$$

Then W is compact by 5.3.4.2, and since J_K^0 is closed in A_K by 5.3.6.2, it follows that $W \cap \mathsf{J}_K^0$ is relatively closed in W and thus compact. We shall prove:

A. $(W \cap \mathsf{J}_K^0)K^\times = \mathsf{J}_K^0$.

Proof of **A.** Clearly, $(W \cap \mathsf{J}_K^0)K^\times \subset \mathsf{J}_K^0$. Thus let $\beta = (\beta_v)_{v\in\mathsf{M}_K} \in \mathsf{J}_K^0$. Then $\|\beta^{-1}\alpha\| = \|\alpha\| > C$, and by 5.3.4 there exists some $y \in K^\times$ satisfying $|y|_v \le |\beta_v^{-1}\alpha_v|_v$ for all $v \in \mathsf{M}_K$. It follows that $|y\beta_v|_v \le |\alpha_v|_v$ for all $v \in \mathsf{M}_K$, hence $y\beta \in W \cap \mathsf{J}_K^0$ and $\beta \in (W \cap \mathsf{J}_K^0)K^\times$. □[**A.**]

If $\pi\colon \mathsf{J}_K^0 \to \mathsf{J}_K^0/K^\times = \mathsf{C}_K^0$ denotes the residue class epimorphism, then **A** implies that $\mathsf{C}_K^0 = \pi(W \cap \mathsf{J}_K^0)$ is compact.

Together with E_K the group $K^\times\mathsf{E}_K$ is an open subgroup of J_K and thus of $\mathsf{J}_K^0\mathsf{E}_K$. Hence the factor group $\mathsf{J}_K^0\mathsf{E}_K/K^\times\mathsf{E}_K$ is discrete by 1.1.5.5. Since the residue class epimorphism $\mathsf{C}_K^0 = \mathsf{J}_K^0/K^\times \to \mathsf{J}_K^0\mathsf{E}_K/K^\times\mathsf{E}_K$ is continuous, it follows that $\mathsf{J}_K^0\mathsf{E}_K/K^\times\mathsf{E}_K$ is compact and thus finite.

In the number field case $\mathsf{J}_K^0\mathsf{E}_K/K^\times\mathsf{E}_K = \mathsf{J}_K/K^\times\mathsf{E}_K \cong \mathcal{C}_K$ by 5.3.2.1, in the function field case $\mathsf{J}_K^0\mathsf{E}_K/K^\times\mathsf{E}_K = \mathsf{J}_K^0/K^\times\mathsf{E}_K \cong \mathcal{C}_K^0$ by 5.3.2.4(b), and in any case $|\mathsf{J}_K^0\mathsf{E}_K/K^\times\mathsf{E}_K| = h_K$.

2. Let $\pi\colon \mathsf{J}_K^0 \to \mathsf{J}_K^0\mathsf{E}_K/K^\times\mathsf{E}_K$ be the residue class epimorphism. It follows that $\operatorname{Ker}(\pi) = \mathsf{J}_K^0 \cap K^\times\mathsf{E}_K$, and $h = (\mathsf{J}_K^0 : \operatorname{Ker}(\pi)) = |\mathsf{J}_K^0\mathsf{E}_K/K^\times\mathsf{E}_K| = h_K$.

Let $\{\alpha_1,\ldots,\alpha_h\}$ be a set of representatives for $\mathsf{J}_K^0/\mathrm{Ker}(\pi)$ in J_K^0, and let S be a finite subset of M_K such that $\mathsf{M}_K^\infty \subset S$, $\{\alpha_1,\ldots,\alpha_h\} \subset \mathsf{J}_{K,S}$, and $\partial_S = 1$ in the function field case. Then it follows by 5.3.2.3 and 5.3.2.4(c) that $\mathsf{J}_K = \mathsf{J}_K^0\mathsf{J}_{K,S}$, and since

$$\mathsf{J}_K^0 \subset \{\alpha_1,\ldots,\alpha_s\}\mathrm{Ker}(\pi) \subset \mathsf{J}_{K,S}\,\mathrm{Ker}(\pi) \subset \mathsf{J}_{K,S}K^\times\mathsf{E}_K = \mathsf{J}_{K,S}K^\times,$$

we obtain $\mathsf{J}_K = \mathsf{J}_K^0\mathsf{J}_{K,S} = \mathsf{J}_{K,S}K^\times$.

3. Let S be a finite subset of M_K. Let $\pi\colon \mathsf{J}_K^0 \to \mathsf{C}_K^0$ be the residue class epimorphism and $\phi = \pi \restriction \mathsf{J}_{K,S} \cap \mathsf{J}_K^0 \colon \mathsf{J}_{K,S} \cap \mathsf{J}_K^0 \to \mathsf{C}_K^0$. Since $\mathsf{J}_{K,S} \cap \mathsf{J}_K^0$ is relatively open in J_K^0, it follows that ϕ is continuous and open, and apparently $\mathrm{Ker}(\phi) = K^\times \cap \mathsf{J}_{K,S} \cap \mathsf{J}_K^0 = K_S$. Therefore $\Lambda = \phi(\mathsf{J}_{K,S} \cap \mathsf{J}_K^0)$ is an open subgroup of C_K^0, hence closed and thus compact by 1., and ϕ induces a topological isomorphism $\mathsf{J}_{K,S} \cap \mathsf{J}_K^0/K_S \overset{\sim}{\to} \Lambda$. Hence $\mathsf{J}_{K,S} \cap \mathsf{J}_K^0/K_S$ is compact. □

Theorem 5.3.8 (Generalized Dirichlet's unit theorem). *Let S be a finite subset of M_K, $s = |S| \ge 1$, and $S \cap \mathsf{M}_K^\infty \ne \emptyset$ in the number field case. Let*

$$\lambda\colon \mathsf{J}_{K,S} \to \mathbb{R}^S \ \ \textit{be defined by} \ \ \lambda(\alpha) = (\log|\alpha_v|_v)_{v\in S} \ \ \textit{for all}$$
$$\alpha = (\alpha_v)_{v\in\mathsf{M}_K} \in \mathsf{J}_{K,S},$$

and set

$$H = \Big\{(x_v)_{v\in S} \in \mathbb{R}^S \ \Big|\ \sum_{v\in S} x_v = 0\Big\}.$$

Then λ is a group homomorphism, $\mathrm{Ker}(\lambda \restriction K_S) = \mu(K)$, $\lambda(K_S)$ is a lattice in H and $\mathbb{R}\lambda(K_S) = H$. In particular,

$$K_S \cong \mu(K)\times\mathbb{Z}^{s-1}.$$

Proof. Obviously, λ is a group homomorphism, and we start with the following three assertions:

I. For any $c_1,\, c_2 \in \mathbb{R}$ with $0 \le c_1 \le c_2$ the set

$$C = \{a \in K_S \mid c_1 \le |a|_v \le c_2 \text{ for all } v \in S\}$$

is finite.

II. $\lambda(K_S)$ is a lattice in $\mathbb{R}^s$, and $\mathrm{Ker}(\lambda \restriction K_S) = \mu(K)$.

III. $\mathbb{R}\lambda(\mathsf{J}_{K,S} \cap \mathsf{J}_K^0) = H$.

Proof of **I.** Let $c_1,\, c_2 \in \mathbb{R}$ with $0 \le c_1 < c_2$. Then the set

$$\overline{C} = \{\alpha \in \mathsf{A}_K \mid c_1 \le |\alpha_v|_v \le c_2 \text{ for all } v \in S \text{ and } |\alpha_v|_v = 1 \text{ for all } v \in \mathsf{M}_K \setminus S\}.$$

is compact by 5.3.4.2, and by 5.3.6.1 the set

$$C = \{a \in K_S \mid c_1 \le |a|_v \le c_2 \text{ for all } v \in S\} = \overline{C} \cap K$$

is discrete and relatively closed in $\overline{C}$, hence compact and thus finite. $\square$[**I.**]

Proof of **II.** By **I**, $\mathrm{Ker}(\lambda \restriction K_S) = \{a \in K_S \mid |a|_v = 1 \text{ for all } v \in S\}$ is a finite subgroup of $K^\times$. Hence $\mathrm{Ker}(\lambda \restriction K_S) \subset \mu(K)$, and in fact equality holds. Indeed, if $\zeta \in \mu(K)$, then $|\zeta|_v = 1$ for all $v \in \mathsf{M}_K$, hence $\zeta \in K_S$ and $\lambda(\zeta) = 0$.

Let B be a bounded subset of $\lambda(K_S)$, say $B = \lambda(B')$ for some subset B' of K_S. Then there exists some $b \in \mathbb{R}_{>0}$ such that $|\log |a|_v| \le b$ and thus $\mathsf{e}^{-b} \le |a|_v \le \mathsf{e}^b$ for all $a \in B'$. Hence B' and thus B is finite by **I**. Since every bounded subset of $\lambda(K_S)$ is finite, $\lambda(K_S)$ is a discrete subgroup of $\mathbb{R}^S$, and therefore it is a lattice (see [18, Theorem 3.5.1]). $\square$[**II.**]

Proof of **III.** If $\alpha \in \mathsf{J}_{K,S} \cap \mathsf{J}_K^0$, then

$$\|\alpha\| = \prod_{v \in S} |\alpha_v|_v = 1, \quad \text{hence} \quad \sum_{v \in S} \log |\alpha_v|_v = 0, \quad \text{and} \quad \lambda(\alpha) \in H.$$

H is a hyperplane of $\mathbb{R}^S$, and $\mathbb{R}\lambda(\mathsf{J}_{K,S} \cap \mathsf{J}_K^0)$ is a subspace of H. Hence it suffices to prove that $\lambda(\mathsf{J}_{K,S} \cap \mathsf{J}_K^0)$ contains $s-1$ linearly independent vectors. Let $(\boldsymbol{e}_v)_{v \in S}$ be the canonical basis of $\mathbb{R}^S$ (defined by $\boldsymbol{e}_v = (\delta_{v,v'})_{v' \in S}$ for all $v \in S$). We fix some $v^* \in S$ such that $v^* \in \mathsf{M}_K^\infty$ in the number field case, we set $S^* = S \setminus \{v^*\}$, and for every $v \in S^*$ we construct an idele $\beta^{(v)} \in \mathsf{J}_{K,S} \cap \mathsf{J}_K^0$ such that $\lambda(\beta^{(v)}) = \log |\beta_v^{(v)}|_v (\boldsymbol{e}_v - \boldsymbol{e}_{v^*})$ (then $(\lambda(\beta^{(v)}))_{v \in S^*}$ is linearly independent and we are done).

Let $v \in S^*$, and define $\beta^{(v)} = (\beta_{v'}^{(v)})_{v' \in \mathsf{M}_K} \in \mathsf{J}_{K,S} \cap \mathsf{J}_K^0$ as follows.

- In the number field case
  let $\beta_v^{(v)} \in K^\times$ be arbitrary, $\beta_{v^*}^{(v)} \in K_{v^*}^\times$ such that $|\beta_{v^*}^{(v)}|_{v^*} = |\beta_v^{(v)}|_v^{-1}$, and $\beta_{v'}^{(v)} = 1$ for all $v' \in \mathsf{M}_K \setminus \{v, v^*\}$. Then $\lambda(\beta^{(v)}) = \log |\beta_v^{(v)}|_v (\boldsymbol{e}_v - \boldsymbol{e}_{v^*})$.
- In the function field case
  let $\beta_v^{(v)} \in K^\times$ be such that $\deg(v) v(\beta_v^{(v)}) \equiv 0 \mod \deg(v^*)$, let $\beta_{v^*}^{(v)} \in K^\times$ be such that $\deg(v^*) v^*(\beta_{v^*}^{(v)}) = -\deg(v) v(\beta_v^{(v)})$, and $\beta_{v'}^{(v)} = 1$ for all $v' \in \mathsf{M}_K \setminus \{v, v^*\}$. Then

  $$|\beta_{v^*}^{v)}|_{v^*} = q^{-\deg(v^*) v^*(\beta_{v^*}^{(v)})} = q^{\deg(v) v(\beta_v^{(v)})} = |\beta_v^{(v)}|_v^{-1},$$

  and consequently $\lambda(\beta^{(v)}) = \log |\beta_v^{(v)}|_v (\boldsymbol{e}_v - \boldsymbol{e}_{v^*})$. $\square$[**III.**]

Now we can finish the proof. Let $\lambda_1 \colon \mathsf{J}_{K,S} \cap \mathsf{J}_K^0 / K_S \to \lambda(\mathsf{J}_{K,S} \cap \mathsf{J}_K^0)/\lambda(K_S)$ be defined by $\lambda_1(\alpha K_S) = \lambda(\alpha)\lambda(K_S)$ for all $\alpha \in \mathsf{J}_{K,S} \cap \mathsf{J}_K^0$. As $\mathsf{J}_{K,S}$ carries the product topology, λ is continuous, and therefore λ_1 is a continuous epimorphism. By 5.3.7.3 the set $\mathsf{J}_{K,S} \cap \mathsf{J}_K^0 / K_S$ is compact, and therefore $\lambda(\mathsf{J}_{K,S} \cap \mathsf{J}_K^0)/\lambda(K_S)$ is compact, too. We obtain $\lambda(K_S) \subset \lambda(\mathsf{J}_{K,S} \cap \mathsf{J}_K^0) \subset H$, and **III** implies $\mathbb{R}\lambda(K_S) = \mathbb{R}\lambda(\mathsf{J}_{K,S} \cap \mathsf{J}_K^0) = H$.

Hence $\mathrm{rk}(\lambda(K_S)) = \dim_{\mathbb{R}}(H) = s-1$ and thus $\lambda(K_S) \cong \mathbb{Z}^{s-1}$. The exact sequence

$$\mathbf{1} \to \mu(K) \to K_S \to \mathbb{Z}^{s-1} \to \mathbf{0}$$

induced by $\lambda \restriction K_S$ splits, and therefore it follows that $K_S \cong \mu(K) \times \mathbb{Z}^{s-1}$. □

In the number field case with $S = \mathsf{M}_K^\infty$, Theorem 5.3.8 is the classical Dirichlet unit theorem (see [18, Theorem 3.5.8]).

5.4 Ideles in field extensions

Let L/K be a finite extension of global fields.

Definition and Remarks 5.4.1.

1. If $v \in \mathsf{M}_K$ and $\overline{v} \in \mathsf{M}_L$, then we say that $\overline{v}$ **lies above** v and we write $\overline{v} \,|\, v$ if $|\cdot| \in \overline{v}$ implies $|\cdot| \restriction K \in v$ for some (and then for every) absolute value $|\cdot|$ of L. If $\overline{v} \,|\, v$, then we call $\overline{v}$ an **extension** of v to L, v a **restriction** of $\overline{v}$ to K, and we assume that $K_v \subset L_{\overline{v}}$.

If $|\cdot|$ is an absolute value of L, then $|\cdot| \restriction K$ is an absolute value of K. Hence for every place $\overline{v} \in \mathsf{M}_L$ there exists a unique place $v \in \mathsf{M}_K$ such that $\overline{v} \,|\, v$, and we write $v = \overline{v} \downarrow K$. By [18, Theorem 5.5.7] the map $\rho \colon \mathsf{M}_L \to \mathsf{M}_K$, defined by $\rho(\overline{v}) = \overline{v} \downarrow K$, is surjective, $\rho^{-1}(v)$ is finite for all $v \in \mathsf{M}_K$, and apparently $\rho^{-1}(\mathsf{M}_K^0) = \mathsf{M}_L^0$ and $\rho^{-1}(\mathsf{M}_K^\infty) = \mathsf{M}_L^\infty$.

Let $v \in \mathsf{M}_K^\infty$. We call v **fully decomposed** in L if $|\rho^{-1}(v)| = [L\!:\!K]$, and then $L_{\overline{v}} = K_v$ for all $\overline{v} \in \rho^{-1}(v)$. If v is complex, then v is fully decomposed; if v is real, then v is fully decomposed if and only if all places of L lying above v are real.

If $\overline{v} \in \mathsf{M}_L^\infty$ is complex and $v = \overline{v} \downarrow K$ is real, then $\mathsf{N}_{L_{\overline{v}}/K_v} L_{\overline{v}}^\times = \mathsf{N}_{\mathbb{C}/\mathbb{R}}(\mathbb{C}^\times) = \mathbb{R}_{>0}$.

2. Let $v \in \mathsf{M}_K^0$ and $\overline{v} \in \mathsf{M}_L^0$. We apply [18, Theorem 5.5.9] to the extension $\mathfrak{o}_K \subset \mathfrak{o}_L$ in the number field case and to the extension $\mathcal{O}_v \subset \mathcal{O}'_v = \mathrm{cl}_L(\mathcal{O}_v)$ in the function field case (the latter together with [18, Theorem 6.2.5]). In either case it follows that $\overline{v} \,|\, v$ if and only if $\mathsf{p}_{\overline{v}} \cap K = \mathsf{p}_v$, and then we again say that $\mathsf{p}_{\overline{v}}$ **lies above** p_v and write $\mathsf{p}_{\overline{v}} \,|\, \mathsf{p}_v$.

If $\overline{v} \,|\, v$, we denote as usual by $e(\overline{v}/v)$ the **ramification index** and by $f(\overline{v}/v)$ the **inertia degree** of $\overline{v}/v$. Recall that

- $e(\overline{v}/v) = e(L_{\overline{v}}/K_v) = e(\mathsf{p}_{\overline{v}}/\mathsf{p}_v) = e(\widehat{\mathsf{p}}_{\overline{v}}/\widehat{\mathsf{p}}_v)$,
  and $\overline{v} \restriction K = e(\overline{v}/v)\, v \colon K \to \mathbb{Z} \cup \{\infty\}$;

- $f(\overline{v}/v) = f(L_{\overline{v}}/K_v) = f(\mathsf{p}_{\overline{v}}/\mathsf{p}_v) = f(\widehat{\mathsf{p}}_{\overline{v}}/\widehat{\mathsf{p}}_v) = [\mathsf{k}_{\overline{v}}\!:\!\mathsf{k}_v]$.
  and $v \circ \mathsf{N}_{L_{\overline{v}}(K_v} = f(\overline{v}/v)\, v \colon K_v \to \mathbb{Z} \cup \{\infty\}$.

(compare with [18, Theorems 2.12.2, 5.5.6, 5.5.9 and 6.2.5]).

We call $\overline{v}/v$ **ramified** [**unramified, tamely ramified** etc.] if the field extension $L_{\overline{v}}/K_v$ and thus $\mathsf{p}_{\overline{v}}/\mathsf{p}_v$ has this property. Corresponding, we call v in L **ramified** [**unramified, inert, undecomposed, fully decomposed** etc.] if p_v has this property. For precise definitions of these almost self-explaining notions we refer to [18, Definitions and Theorems 2.12.1, 2.12.5, 5.6.1, 5.7.4 and 6.2.6].

We denote by $\mathrm{Ram}(L/K)$ the set of all in L ramified places $v \in \mathsf{M}_K^0$ and by $\mathrm{Split}(L/K)$ the set of all in L fully decomposed places $v \in \mathsf{M}_K$. The extension L/K is called **unramified** if $\mathrm{Ram}(L/K) = \emptyset$.

Altogether we use the notions just introduced for places mutatis mutandis also for ideals, divisors and absolute values.

3. We define the **divisor embedding** $j_{L/K} \colon \mathcal{D}_K \to \mathcal{D}_L$ and the **divisor norm** $\mathcal{N}_{L/K} \colon \mathcal{D}_L \to \mathcal{D}_K$ by

$$j_{L/K}(\mathsf{a}) = \prod_{v \in \mathsf{M}_K^0} \prod_{\substack{\overline{v} \in \mathsf{M}_L \\ \overline{v} \mid v}} \mathsf{p}_{\overline{v}}^{e(\overline{v}/v)v(\mathsf{a})} \quad \text{for all } \mathsf{a} \in \mathcal{D}_K$$

and

$$\mathcal{N}_{L/K}(\mathsf{b}) = \prod_{v \in \mathsf{M}_K^0} \prod_{\substack{\overline{v} \in \mathsf{M}_L \\ \overline{v} \mid v}} \mathsf{p}_v^{f(\overline{v}/v)\overline{v}(\mathsf{b})} \quad \text{for all } \mathsf{b} \in \mathcal{D}_L.$$

For the corresponding notions in the classical case of number fields and funtion fields we refer to [18, Theorems 2.12.4, 6.5.2, 2.14.2 and 6.5.6].

4. Let L/K be separable. Recall the definition of the **different** $\mathsf{D}_{L/K} \in \mathcal{D}_L$ and of the **discriminant** $\mathsf{d}_{L/K} = \mathcal{N}_{L/K}(\mathsf{D}_{L/K}) \in \mathcal{D}_K$ (see [18, Definition 5.9.8] for the number field case, and [18, Definition 5.6.7] for the function field case). If $\mathfrak{D}_{L_{\overline{v}}/K_v}$ resp. $\mathfrak{d}_{L_{\overline{v}}/K_v}$ denotes the different resp. discim-inant of the local extension $L_{\overline{v}}/K_v$, then

$$\overline{v}(\mathsf{D}_{L/K}) = \overline{v}(\mathfrak{D}_{L_{\overline{v}}/K_v}) \quad \text{for all } \overline{v} \in \mathsf{M}_L^0 \text{ and } v = \overline{v} \downarrow K$$

and

$$v(\mathsf{d}_{L/K}) = \sum_{\substack{\overline{v} \in \mathsf{M}_L \\ \overline{v} \mid v}} v(\mathfrak{d}_{L_{\overline{v}}/K_v}) \quad \text{for all } v \in \mathsf{M}_K^0$$

(see [18, Theorem 5.9.8] for the number field case, and [18, Theorem 6.5.10] for the function field case).

In the following Theorem 5.4.2 we gather some properties of the divisor embedding, the divisor norm, the different and the discriminant. Instead of giving complete proofs, we carefully outline how the assertions follow from the corresponding results in [18].

Theorem 5.4.2.

1. *If* $v \in \mathsf{M}_K$*, then*

$$[L\colon K] = \sum_{\substack{\overline{v}\in\mathsf{M}_L\\ \overline{v}\,|\,v}} [L_{\overline{v}}\colon K_v] \quad \text{and} \quad \mathsf{N}_{L/K}(x) = \prod_{\substack{\overline{v}\in\mathsf{M}_L\\ \overline{v}\,|\,v}} \mathsf{N}_{L_{\overline{v}}/K_v}(x) \ \text{ for all } \ x \in L.$$

2. *If* $v \in \mathsf{M}_K^0$*, then*

$$v\circ\mathsf{N}_{L/K} = \sum_{\substack{\overline{v}\in\mathsf{M}_L\\ \overline{v}\,|\,v}} f(\overline{v}/v)\overline{v}\colon L \to \mathbb{Z}\cup\{\infty\} \quad \text{and}$$

$$\sum_{\substack{\overline{v}\in\mathsf{M}_L\\ \overline{v}\,|\,v}} e(\overline{v}/v)f(\overline{v}/v) = [L\colon K].$$

3. *Let* $\overline{v} \in \mathsf{M}_L$ *and* $v = \overline{v}\downarrow K$*. Then*

$$|\mathsf{N}_{L_{\overline{v}}/K_v}(x)|_v = |x|_{\overline{v}} \ \text{ for all}$$
$$x \in L_{\overline{v}}, \quad \text{and} \quad |x|_{\overline{v}} = |x|_v^{[L_{\overline{v}}:K_v]} \ \text{ for all } \ x \in K_v.$$

4. $j_{L/K}\colon \mathcal{D}_K \to \mathcal{D}_L$ *and* $\mathcal{N}_{L/K}\colon \mathcal{D}_L \to \mathcal{D}_K$ *are group homomorphisms,*

$$j_{L/K}((a)^K) = (a)^L \ \text{ for all } \ a \in K^\times$$

and

$$\mathcal{N}_{L/K}((b)^L) = (\mathsf{N}_{L/K}(b))^K \ \text{ for all } \ b \in L^\times.$$

5. *Let* $v \in \mathsf{M}_K$*. Then the diagonal embedding*

$$\delta\colon L \to A = \prod_{\substack{\overline{v}\in\mathsf{M}_L\\ \overline{v}\,|\,v}} L_{\overline{v}}, \quad x \mapsto (x)_{\overline{v}\,|\,v}$$

is a K*-linear monomorphism,* $\delta(L)$ *is dense in* A*, and* δ *induces a* K_v*-isomorphism* $\delta^*\colon K_v \otimes_K L \overset{\sim}{\to} A$ *such that* $\delta^*(x\otimes a) = \delta(x)a$ *for all* $x \in K_v$ *and* $a \in L$*.*

6. *Let* L/K *be separable.*

(a) *If* K' *is an intermediate field of* L/K*, then*

$$\mathsf{D}_{L/K} = \mathsf{j}_{L/K}(\mathsf{D}_{K'/K})\mathsf{D}_{L/K'} \quad \text{and} \quad \mathsf{d}_{L/K} = \mathsf{d}_{K'/K}^{[K':K]}\mathcal{N}_{L/K}(\mathsf{d}_{L/K'}).$$

(b) *If* $\overline{v} \in \mathsf{M}_L^0$ *and* $v = \overline{v} \downarrow K$, *then* $\overline{v}(\mathsf{D}_{L/K}) \ge e(\overline{v}/v) - 1$, *and equality holds if and only if* $\mathrm{char}(\mathsf{k}_v) \nmid e(\overline{v}/v)$. *In particular,* $\overline{v}/v$ *is unramified if and only if* $\overline{v}(\mathsf{D}_{L/K}) = 0$.

(c) $\mathrm{Ram}(L/K) = \mathrm{supp}(\mathsf{d}_{L/K})$. *In particular,* $\mathrm{Ram}(L/K)$ *is finite.*

Proof. 1. and 2. For $v \in \mathsf{M}_K^0$ the relation

$$\sum_{\substack{\overline{v} \in \mathsf{M}_L \\ \overline{v} \,|\, v}} e(\overline{v}/v) f(\overline{v}/v) = [L\!:\!K]$$

holds by [18, Theorem 5.5.9] in the number field case, and by [18, Theorem 6.5.3] in the function field case. By [18, Theorem 5.5.7] we obtain, for all $v \in \mathsf{M}_K$,

$$[L\!:\!K] = \sum_{\substack{\overline{v} \in \mathsf{M}_L \\ \overline{v} \,|\, v}} l(\overline{v}/v)[L_{\overline{v}}\!:\!K_v], \quad \mathsf{N}_{L/K}(x) = \prod_{\substack{\overline{v} \in \mathsf{M}_L \\ \overline{v} \,|\, v}} \mathsf{N}_{L_{\overline{v}}/K_v}(x)^{l(\overline{v}/v)} \text{ for all } x \in L,$$

and, if $v \in \mathsf{M}_K^0$, then (using [18, Theorem 5.5.6])

$$v \circ \mathsf{N}_{L/K}(x) = \sum_{\substack{\overline{v} \in \mathsf{M}_L \\ \overline{v} \,|\, v}} l(\overline{v}/v) v(\mathsf{N}_{L_{\overline{v}}/K_v}(x)) = \sum_{\substack{\overline{v} \in \mathsf{M}_L \\ \overline{v} \,|\, v}} l(\overline{v}/v) f(\overline{v}/v)\overline{v}(x) \text{ for all } x \in L.$$

where

$$l(\overline{v}/v) = \frac{[L\!:\!K]_{\mathsf{i}}}{[L_{\overline{v}}\!:\!K_v]_{\mathsf{i}}} \quad \text{for all } \overline{v} \in \mathsf{M}_L \text{ and } v = \overline{v} \downarrow K.$$

Hence it suffices to prove that $l(\overline{v}/v) = 1$ for all $\overline{v} \in \mathsf{M}_L$ and $v = \overline{v} \downarrow K$. In the number field case there is nothing to do. In the function field case, we use the relation

$$v \circ \mathsf{N}_{L/K}(x) = \sum_{\substack{\overline{v} \in \mathsf{M}_L \\ \overline{v} \,|\, v}} f(\overline{v}/v)\overline{v}(x) \quad \text{for all } x \in L \qquad (*)$$

(see [18, Theorem 6.5.6]). Let $\overline{v} \in \mathsf{M}_L$ and $v = \overline{v} \restriction K$. By the weak approximation theorem there exists some $x \in L$ such that $\overline{v}(x) \ne 0$ and $w(x) = 0$ for all $w \in \mathsf{M}_L \setminus \{\overline{v}\}$ satisfying $w \,|\, v$. Now $(*)$ implies

$$\mathsf{N}_{L/K}(x) = l(\overline{v}/v) f(\overline{v}/v)\overline{v}(x) = f(\overline{v}/v)\overline{v}(x),$$

and consequently $l(\overline{v}/v) = 1$.

3. Let $x \in L_{\overline{v}}$. If $\overline{v} \in \mathsf{M}_L^0$, then

$$|\mathsf{N}_{L_{\overline{v}}/K_v}(x)|_v = \mathfrak{N}(v)^{-v(\mathsf{N}_{L_{\overline{v}}/K_v}(x))} = \mathfrak{N}(v)^{-f(\overline{v}/v)\overline{v}(x)} = \mathfrak{N}(\overline{v})^{-\overline{v}(x)} = |x|_{\overline{v}}.$$

If $v \in \mathsf{M}_K^\infty$ and $L_{\overline{v}} = K_v$, then $|\cdot|_v = |\cdot|_{\overline{v}}$ and there is nothing to do. If $L_{\overline{v}} = \mathbb{C}$ and $K_v = \mathbb{R}$, then $|\mathsf{N}_{\mathbb{C}/\mathbb{R}}(x)|_v = |x\overline{x}|_\infty = |x|_\infty^2 = |x|_w$.

If $x \in K_v$, then it follows in any case that $|x|_v^{[L_{\overline{v}}:K_v]} = |\mathsf{N}_{L_{\overline{v}}/K_v}(x)|_v = |x|_{\overline{v}}$.

4. Apparently, $j_{L/K}$ and $\mathcal{N}_{L/K}$ are group homomorphisms. If $a \in K^\times$, then $v(a) = v((a)^K)$ for all $v \in \mathsf{M}_K^0$, and

$$j_{L/K}((a)^K) = \prod_{v \in \mathsf{M}_K^0} \prod_{\substack{\overline{v} \in \mathsf{M}_L \\ \overline{v} \,|\, v}} \mathsf{p}_{\overline{v}}^{e(\overline{v}/v)v(a)} = \prod_{\overline{v} \in \mathsf{M}_L^0} \mathsf{p}_{\overline{v}}^{\overline{v}(a)} = (a)^L.$$

If $b \in L^\times$, then $\overline{v}((b)^L) = \overline{v}(b)$, and

$$\mathcal{N}_{L/K}((b)^L) = \prod_{v \in \mathsf{M}_K^0} \prod_{\substack{\overline{v} \in \mathsf{M}_L \\ \overline{v} \,|\, v}} \mathsf{p}_v^{f(\overline{v}/v)\overline{v}(b)} = \prod_{v \in \mathsf{M}_K^0} \mathsf{p}_v^{v(\mathsf{N}_{L/K}(b))} = (\mathsf{N}_{L/K}(b))^K.$$

5. Obviously, δ is K-linear, and $\delta(L)$ is dense in A (see [18, Theorem 5.5.7]). As A is a vector space over K_v and the assignment $(x, a) \mapsto x\delta(a)$ defines a K-bilinear map $K_v \times L \to A$, there exist a unique K-linear map $\delta^*\colon K_v \otimes_K L \to A$ such that $\delta^*(x \otimes a) = x\delta(a)$ for all $x \in K_v$ and $a \in L$. Now δ^* is even K_v-linear, $\delta^*(K_v \otimes_K L)$ is a K_v-subspace of A, hence closed in A, and as $\delta(L) \subset \delta^*(K_v \otimes_K L) \subset A$, it follows that $\delta^*(K_v \otimes_K L)$ is dense in A. Hence δ^* is surjective, and since $\dim_{K_v}(A) = [L:K]$ (by 1.), δ^* is a K_v-isomorphism.

6. (a) This holds by [18, Theorem 5.9.8] in the number field case, and by [18, Theorem 6.5.9] in the function field case.

(b) Observe that $\overline{v}(\mathsf{D}_{L/K}) = \overline{v}(\mathfrak{D}_{L_{\overline{v}/v}/K_v})$, and apply [18, Theorem 5.9.6] for the extension $L_{\overline{v}}/K_v$.

(c) Again by [18, Theorem 5.9.6], we obtain:

$$\begin{aligned}\operatorname{Ram}(L/K) &= \{v \in \mathsf{M}_K^0 \mid L_{\overline{v}/K_v} \text{ is ramified for some } \overline{v} \in \mathsf{M}_L \text{ with } \overline{v} \,|\, v\} \\ &= \{v \in \mathsf{M}_K^0 \mid v(\mathfrak{d}_{L_{\overline{v}}/K_v}) \neq 0 \text{ for some } \overline{v} \in \mathsf{M}_L \text{ with } \overline{v} \,|\, v\} \\ &= \{v \in \mathsf{M}_K^0 \mid v(\mathsf{d}_{L/K}) \neq 0\} = \operatorname{supp}(\mathsf{d}_{L/K}).\end{aligned}$$ □

Definition and Remarks 5.4.3.

1. We define the **idele embedding** $i_{L/K}\colon \mathsf{J}_K \to \mathsf{J}_L$ by

$$i_{L/K}(\alpha)_{\overline{v}} = \alpha_v \quad \text{for all } \alpha \in \mathsf{J}_K,\ \overline{v} \in \mathsf{M}_L \text{ and } v = \overline{v} \downarrow K$$

(if $\overline{v} \in \mathsf{M}_L^0$ and $v = \overline{v} \downarrow K$, then $U_v \subset U_{\overline{v}}$; hence $i(\alpha)_{\overline{v}} \in U_{\overline{v}}$ for almost all $\overline{v} \in \mathsf{M}_L^0$ and thus indeed $i(\alpha) \in \mathsf{J}_L$).

Clearly $i_{L/K}\colon \mathsf{J}_K \to \mathsf{J}_L$ is a group monomorphism. If $\beta = (\beta_{\overline{v}})_{\overline{v} \in \mathsf{M}_L} \in \mathsf{J}_L$, then $\beta \in i_{L/K}(\mathsf{J}_K)$ if and only if $\beta_{\overline{v}} = \beta_{\overline{v}'} \in K_v$ for all $v \in \mathsf{M}_K$ and all $\overline{v}$, $\overline{v}' \in \mathsf{M}_L$ satisfying $\overline{v} \downarrow K = \overline{v}' \downarrow K = v$.

For $\alpha = (\alpha_v)_{v\in\mathsf{M}_K} \in \mathsf{J}_K$ we obtain, using 5.4.2.1,

$$\|i_{L/K}(\alpha)\| = \prod_{v\in\mathsf{M}_K} \prod_{\substack{\overline{v}\in\mathsf{M}_L \\ \overline{v}\,|\,v}} |\alpha_v|_{\overline{v}} = \prod_{v\in\mathsf{M}_K} \prod_{\substack{\overline{v}\in\mathsf{M}_L \\ \overline{v}\,|\,v}} |\alpha_v|_v^{[L_{\overline{v}}:K_v]}$$
$$= \prod_{v\in\mathsf{M}_K} |\alpha_v|_v^{[L:K]} = \|\alpha\|^{[L:K]},$$

and therefore $i_{L/K}(\mathsf{J}_K^0) \subset \mathsf{J}_L^0$.

Since $i_{L/K} \restriction K^\times = (K^\times \hookrightarrow L^\times)$, it follows that $i_{L/K}$ induces a group homomorphism $i_{L/K}^*\colon \mathsf{C}_K \to \mathsf{C}_L$ satisfying

$$i_{L/K}^*(\alpha K^\times) = i_{L/K}(\alpha)L^\times \in \mathsf{J}_L/L^\times = \mathsf{C}_L$$

for all $\alpha \in \mathsf{J}_K$, and clearly $i_{L/K}^*(\mathsf{C}_K^0) \subset \mathsf{C}_L^0$. If M is an intermediate field of L/K, then $i_{L/K} = i_{L/M} \circ i_{M/K}$ and $i_{L/K}^* = i_{L/M}^* \circ i_{M/K}^*$.

2. We define the **idele norm** $\mathsf{N}_{L/K}\colon \mathsf{J}_L \to \mathsf{J}_K$ by

$$\mathsf{N}_{L/K}(\beta)_v = \prod_{\substack{\overline{v}\in\mathsf{M}_L \\ \overline{v}\,|\,v}} \mathsf{N}_{L_{\overline{v}}/K_v}(\beta_{\overline{v}}) \quad \text{for all} \quad \beta = (\beta_{\overline{v}})_{\overline{v}\in\mathsf{M}_L} \in \mathsf{J}_L \quad \text{and} \quad v \in \mathsf{M}_K$$

(if $\overline{v} \in \mathsf{M}_L^0$ and $v = \overline{v} \downarrow K$, then $\mathsf{N}_{L_{\overline{v}}/K_v}(U_{\overline{v}}) \subset U_v$; hence $\mathsf{N}_{L/K}(\beta)_v \in U_v$ for almost all $v \in \mathsf{M}_K$ and thus indeed $\mathsf{N}_{L/K}(\beta) \in \mathsf{J}_K$).

Obviously, $\mathsf{N}_{L/K}\colon \mathsf{J}_L \to \mathsf{J}_K$ is a group homomorphism, and if M is an intermediate field of L/K, then $\mathsf{N}_{L/K} = \mathsf{N}_{M/K} \circ \mathsf{M}_{L/M}\colon \mathsf{J}_L \to \mathsf{J}_K$.

For $\beta = (\beta_{\overline{v}})_{\overline{v}\in\mathsf{M}_L} \in \mathsf{J}_L$ we obtain, using 5.4.2.3,

$$\|\mathsf{N}_{L/K}(\beta)\| = \prod_{v\in\mathsf{M}_K} |\mathsf{N}_{L/K}(\beta)_v|_v = \prod_{v\in\mathsf{M}_K} \Bigg| \prod_{\substack{\overline{v}\in\mathsf{M}_L \\ \overline{v}\,|\,v}} \mathsf{N}_{L_{\overline{v}}/K_v}(\beta_{\overline{v}}) \Bigg|_v$$
$$= \prod_{v\in\mathsf{M}_K} \prod_{\substack{\overline{v}\in\mathsf{M}_L \\ \overline{v}\,|\,v}} |\beta_{\overline{v}}|_{\overline{v}} = \|\beta\|$$

and therefore $\mathsf{N}_{L/K}(\mathsf{J}_L^0) \subset \mathsf{J}_K^0$. If $b \in L^\times$, then

$$\mathsf{N}_{L(K}(b) = \prod_{\substack{\overline{v}\in\mathsf{M}_L \\ \overline{v}\,|\,v}} \mathsf{N}_{L_{\overline{v}}/K_v}(b) \quad \text{for all} \quad v \in \mathsf{M}_K.$$

Hence $\mathsf{N}_{L/K} \restriction L^\times$ is the usual field-theoretic norm, and $\mathsf{N}_{L/K}$ induces an (equally denoted) homomorphism

$$\mathsf{N}_{L/K}\colon \mathsf{C}_L \to \mathsf{C}_K \quad \text{such that} \quad \mathsf{N}_{L/K}(\beta L^\times) = \mathsf{N}_{L/K}(\beta)K^\times \in \mathsf{C}_K = \mathsf{J}_K/K^\times$$

for all $\beta \in \mathsf{J}_L$. It follows that $\mathsf{N}_{L/K}(\mathsf{C}_L) = \mathsf{N}_{L/K}(\mathsf{J}_L)K^\times/K^\times$, and $\|\mathsf{N}_{L/K}(c)\| = \|c\|$ for all $c \in \mathsf{C}_L$. The factor group

$$\mathsf{C}_K/\mathsf{N}_{L/K}\mathsf{C}_L = (\mathsf{J}_K/K^\times)/(\mathsf{N}_{L/K}(\mathsf{J}_L)K^\times/K^\times) \cong \mathsf{J}_K/\mathsf{N}_{L/K}(\mathsf{J}_L)K^\times$$

is called the **norm residue group** of L/K.
We identify: $\mathsf{C}_K/\mathsf{N}_{L/K}\mathsf{C}_L = \mathsf{J}_K/\mathsf{N}_{L/K}(\mathsf{J}_L)K^\times$.

3. For a finite subset S of M_K we set $S_L = \{\overline{v} \in \mathsf{M}_L \mid \overline{v} \downarrow K \in S\}$. Then

$$\mathsf{J}_L = \bigcup_{S \in \mathbb{P}_{\rm fin}(\mathsf{M}_K)} \mathsf{J}_{L,S_L}, \quad i_{L/K}(\mathsf{J}_{K,S}) \subset \mathsf{J}_{L,S_L} \text{ and } \mathsf{N}_{L/K}(\mathsf{J}_{L,S_L}) \subset \mathsf{J}_{K,S}.$$

The maps $i_{L/K} \restriction \mathsf{J}_{K,S} \colon \mathsf{J}_{K,S} \to \mathsf{J}_{L,S_L}$ and $\mathsf{N}_{L/K} \restriction \mathsf{J}_{L,S_L} \colon \mathsf{J}_{L,S_L} \to \mathsf{J}_{K,S}$ are component-wise continuous and thus they are continuous (in the product topology). Since the topology of J_L is the final topology induced by the directed family $(\mathsf{J}_{L,S_L})_{S\in\mathbb{P}_{\rm fin}(\mathsf{M}_K)}$, the maps $i_{L/K} \colon \mathsf{J}_K \to \mathsf{J}_L$ and $\mathsf{N}_{L/K} \colon \mathsf{J}_L \to \mathsf{J}_K$ are continuous. Consequently, the maps $i^*_{L/K} \colon \mathsf{C}_K \to \mathsf{C}_L$ and $\mathsf{N}_{L/K} \colon \mathsf{C}_L \to \mathsf{C}_K$ are continuous, too.

Theorem 5.4.4.

1. The diagrams

$$\begin{array}{ccc} \mathsf{J}_L & \xrightarrow{\iota_L} & \mathcal{D}_L \\ {\scriptstyle i_{L/K}}\uparrow & & \uparrow{\scriptstyle j_{L/K}} \\ \mathsf{J}_K & \xrightarrow{\iota_K} & \mathcal{D}_K \end{array} \qquad \text{and} \qquad \begin{array}{ccc} \mathsf{J}_L & \xrightarrow{\iota_L} & \mathcal{D}_L \\ {\scriptstyle \mathsf{N}_{L/K}}\downarrow & & \downarrow{\scriptstyle \mathcal{N}_{L/K}} \\ \mathsf{J}_K & \xrightarrow{\iota_K} & \mathcal{D}_K \end{array}$$

are commutative.

2. $\mathsf{N}_{L/K}(\mathsf{J}_L)$ *and* $\mathsf{N}_{L/K}(\mathsf{C}_L)$ *are open subgroups of* J_K *resp.* C_K.

Proof. 1. If $\alpha = (\alpha_v)_{v\in\mathsf{M}_K} \in \mathsf{J}_K$, then $e(\overline{v}/v)v(\alpha_v) = \overline{v}(\alpha_v) = \overline{v}(i_{L/K}(\alpha)_{\overline{v}})$ for all $v \in \mathsf{M}_K^0$ and $\overline{v} \in \mathsf{M}_L^0$ such that $\overline{v}\,|\,v$, and therefore

$$j_{L/K}\circ\iota_K(\alpha)) = j_{L/K}\Big(\prod_{v\in\mathsf{M}_K^0} \mathsf{p}_v^{v(\alpha_v)}\Big) = \prod_{v\in\mathsf{M}_K^0}\prod_{\substack{\overline{v}\in\mathsf{M}_L\\ \overline{v}\,|\,v}} \mathsf{p}_{\overline{v}}^{e(\overline{v}/v)v(\alpha_v)}$$
$$= \prod_{\overline{v}\in\mathsf{M}_L^0} \mathsf{p}_{\overline{v}}^{\overline{v}(i_{L/K}(\alpha)_{\overline{v}})} = \iota_L\circ i_{L/K}(\alpha)).$$

If $\beta = (\beta_{\overline{v}})_{\overline{v}\in\mathsf{M}_L} \in \mathsf{J}_L$ and $v \in \mathsf{M}_K^0$, then

$$v(\mathsf{N}_{L/K}(\beta)_v) = v\Big(\prod_{\substack{\overline{v}\in\mathsf{M}_L\\ \overline{v}\,|\,v}} \mathsf{N}_{L_{\overline{v}}/K_v}(\beta_{\overline{v}})\Big) = \sum_{\substack{\overline{v}\in\mathsf{M}_L\\ \overline{v}\,|\,v}} f(\overline{v}/v)\overline{v}(\beta_{\overline{v}}),$$

and therefore

$$\iota_K(\mathsf{N}_{L/K}(\beta)) = \prod_{v\in\mathsf{M}_K^0} \mathsf{p}_v^{v(\mathsf{N}_{L/K}(\beta)_v)} = \prod_{v\in\mathsf{M}_K^0}\prod_{\substack{\overline{v}\in\mathsf{M}_L\\ \overline{v}\,|\,v}} \mathsf{p}_v^{f(\overline{v}/v)\overline{v}(\beta_{\overline{v}})}$$

$$= \prod_{v\in\mathsf{M}_K^0}\prod_{\substack{\overline{v}\in\mathsf{M}_L\\ \overline{v}\,|\,v}} \mathcal{N}_{L/K}(\mathsf{p}_{\overline{v}}^{\overline{v}(\beta_{\overline{v}})})$$

$$= \mathcal{N}_{L/K}\Big(\prod_{\overline{v}\in\mathsf{M}_L^0} \mathsf{p}_{\overline{v}}^{\overline{v}(\beta_{\overline{v}})}\Big) = \mathcal{N}_{L/K}(\iota_L(\beta)).$$

2. Let $S = \mathsf{M}_K^\infty \cup \mathrm{Ram}(L/K)$, and for $v \in \mathsf{M}_K$ we fix some $\overline{v} \in \mathsf{M}_L$ above v. Then $\mathsf{N}_{L_{\overline{v}}/K_v} L_{\overline{v}}^\times$ is open in K_v by 4.1.10.3 (apparently this also holds for $v \in \mathsf{M}_K^\infty$), and if $v \in \mathsf{M}_K \setminus S$, then $U_v = \mathsf{N}_{L_{\overline{v}}/K_v} U_{\overline{v}}$ (see [18, Theorem 5.6.9]). If

$$W = \prod_{w\in S_L} L_w^\times \times \prod_{w\in\mathsf{M}_L\setminus S_L} U_w,$$

then W is a subgroup of J_L, and

$$\mathsf{N}_{L/K}(W) = \prod_{v\in S} \mathsf{N}_{L_{\overline{v}}/K_v} L_{\overline{v}}^\times \times \prod_{v\in\mathsf{M}_K\setminus S} U_v$$

is an open subgroup of J_K contained in $\mathsf{N}_{L/K}\mathsf{J}_L$. Hence $\mathsf{N}_{L/K}(\mathsf{J}_L)$ is an open subgroup of J_K, and thus $\mathsf{N}_{L/K}(\mathsf{C}_L)$ is an open subgroup of C_K. □

Definition and Remarks 5.4.5. Let L/K be a finite Galois extension. Then $G = \mathrm{Gal}(L/K)$ operates on M_L by means of $|\cdot|_{\sigma\overline{v}} = |\cdot|_{\overline{v}}\circ\sigma^{-1}\colon L \to \mathbb{R}_{\ge 0}$ for all $\overline{v} \in \mathsf{M}_L$ and $\sigma \in G$ (see 5.1.1.**8**), and G acts on the discrete valuations $v \in \mathsf{M}_L^0$ by means of $\sigma\overline{v} = \overline{v}\circ\sigma^{-1}\colon L \to \mathbb{Z}\cup\{\infty\}$ for all $\overline{v} \in \mathsf{M}_L^0$ and $\sigma \in G$.

We recall and unify the results from [18, Section 2.13, Theorem 5.5.8 and Definition 6.5.1] adjusted to our present terminology.

1. Let $\overline{v} \in \mathsf{M}_L$, $v = \overline{v}\downarrow K$ and $\sigma \in G$. Then $\sigma\overline{v}\downarrow K = v$, and σ has a unique extension to an (equally denoted) K_v-isomorphism of valued fields $\sigma\colon L_{\overline{v}} \to L_{\sigma\overline{v}}$. We call the isotropy group $G_{\overline{v}} = \{\sigma \in G \mid \sigma\overline{v} = \overline{v}\}$ the **decomposition group** and its fixed field $L^{G_{\overline{v}}}$ the **decomposition field** of $\overline{v}$ over K. If $(G\colon G_{\overline{v}}) = r$ and $G/G_{\overline{v}} = \{\sigma_1 G_{\overline{v}}, \ldots, \sigma_r G_{\overline{v}}\}$, then $\sigma_1\overline{v}, \ldots, \sigma_r\overline{v}$ are the distinct places of M_L lying above v, and G operates transitively on them. The extension $L_{\overline{v}}/K_v$ is Galois, and the restriction $\tau \mapsto \tau\restriction L$ defines an isomorphism $\mathrm{Gal}(L_{\overline{v}}/K_v) \overset{\sim}{\to} G_{\overline{v}}$. We identify: $G_{\overline{v}} = \mathrm{Gal}(L_{\overline{v}}/K_v)$. Then $L^{G_{\overline{v}}} \subset K_v$, and in fact K_v is the completion of L^{G_v} with respect to $\overline{v}\downarrow L^{G_{\overline{v}}}$. If $\tau \in G$, then $G_{\tau\overline{v}} = \tau G_{\overline{v}}\tau^{-1}$. If $v \in \mathsf{M}_K^\infty$, then either $v \in \mathrm{Split}(L/K)$, or v is real and all places of L lying above v are complex. If $\overline{v} \in \mathsf{M}_L$ lies above v, then $G_{\overline{v}} = \mathbf{1}$ in the first case, and $G_{\overline{v}} = \langle\iota\rangle$ (the complex conjugation in $L_{\overline{v}} = \mathbb{C}$) in the second case.

2. Let $\overline{v} \in \mathsf{M}_L^0$, $v = \overline{v} \downarrow K$ and $\sigma \in G$. Then the K_v-isomorphism $\sigma\colon L_{\overline{v}} \overset{\sim}{\to} L_{\sigma\overline{v}}$ satisfies $\sigma\widehat{\mathcal{O}}_{\overline{v}} = \widehat{\mathcal{O}}_{\sigma\overline{v}}$, $\sigma\widehat{\mathfrak{p}}_{\overline{v}} = \widehat{\mathfrak{p}}_{\sigma\overline{v}}$, $\sigma U_{\overline{v}}^{(i)} = U_{\sigma\overline{v}}^{(i)}$ for all $i \ge 0$, and $\overline{v}\colon L^\times \to \mathbb{Z}$ is a G-homomorphism (with $\mathbb{Z}$ as trivial G-module). Apparently $\mathcal{O}_{\sigma\overline{v}} = \sigma\mathcal{O}_{\overline{v}}$, $P_{\sigma\overline{v}} = \sigma P_{\overline{v}}$, $\mathsf{p}_{\sigma\overline{v}} = \sigma\mathsf{p}_{\overline{v}}$, $e(\overline{v}/v) = e(\sigma\overline{v}/v)$, $f(\overline{v}/v) = f(\sigma\overline{v}/v)$, and σ induces a k_v-isomorphism $\sigma_{\overline{v}}\colon \mathsf{k}_{\overline{v}} \to \mathsf{k}_{\sigma\overline{v}}$ such that $\overline{\sigma}(a+P_{\overline{v}}) = \sigma a + P_{\sigma\overline{v}}$ for all $a \in \mathcal{O}_{\overline{v}}$. The extension $\mathsf{k}_{\overline{v}}/\mathsf{k}_v$ is cyclic (as a finite extension of finite fields), and the assignment $\sigma \mapsto \sigma_{\overline{v}}$ defines an epimorphism $G_{\overline{v}} \to \mathrm{Gal}(\mathsf{k}_{\overline{v}}/\mathsf{k}_v)$. Its kernel $I_{\overline{v}} = I_{\widehat{\mathfrak{p}}_{\overline{v}}} = \{\sigma \in G_{\overline{v}} \mid \sigma_{\overline{v}} = 1\}$ is called the **inertia group** and its fixed field $L^{I_{\overline{v}}}$ the **inertia field** of $\overline{v}$ over K. Recall that $G_{\overline{v}}$ is the decomposition group, $L^{G_{\overline{v}}}$ is the decomposition field, $I_{\overline{v}}$ is the inertia group and $L^{I_{\overline{v}}}$ is the inertia field of $\mathsf{p}_{\overline{v}}$ over K (as defined in [18, Section 2.13 and Definition 6.5.4]). Moreover, $T_{\overline{v}} = L_{\overline{v}}^{I_{\overline{v}}}$ is the inertia field of the extension $L_{\overline{v}}/K_v$ (see [18, Definition 5.6.5]). By definition, $K \subset L^{G_{\overline{v}}} \subset L^{I_{\overline{v}}} \subset L$, and we refer to [18, Theorem 2.13.4 and Corollary 2.13.6] for the relevance of these intermediate fields for the decomposition properties of v in L. In particular, v is unramified in L if and only if $I_{\overline{v}} = \mathbf{1}$; in this case, $G_{\overline{v}} \cong \mathrm{Gal}(\mathsf{k}_{\overline{v}}/\mathsf{k}_v)$ is cyclic, and v is fully decomposed in $L^{G_{\overline{v}}}$.

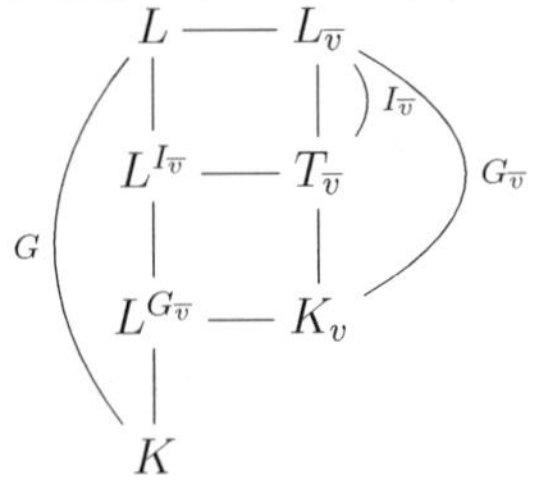

Assume now that $\overline{v}, \overline{v}' \in \mathsf{M}_L$ such that $\overline{v} \downarrow K = \overline{v}' \downarrow K = v \in \mathsf{M}_K$. Then $\overline{v}' = \tau\overline{v}$ for some $\tau \in G$, and τ has a unique extension to an (equally denoted) K_v-isomorphism $\tau\colon L_{\overline{v}} \to L_{\overline{v}'}$. Moreover, $G_{\overline{v}'} = \tau G_{\overline{v}}\tau^{-1}$, and if $v \in \mathsf{M}_K^0$, then $I_{\overline{v}'} = \tau I_{\overline{v}}\tau^{-1}$ (see [18, Theorems 2.13.2 and 5.5.8]).

Altogether we stress once more that we use mutatis mutandis for places all ways of speaking introduced in [18] for ideals, absolute values and divisors.

3. We recall the concept of Frobenius automorphisms. Here we do not restrict to unramified extensions (as it was done in [18, Definition 5.6.7 and Corollary 5.6.8]), but we rather go back to the more general concept presented in [18, Definition 2.14.5 and Theorem 2.14.6] for extensions of Dedekind domains, and reformulate the corresponding notions in the present terminology for global fields.

Let $\overline{v} \in \mathsf{M}_L^0$ and $v = \overline{v} \downarrow K$. An automorphism $\varphi \in G$ is called a **Frobenius automorphism** for $\overline{v}$ (or for $\mathsf{p}_{\overline{v}}$) over K if

$$\overline{v}(\varphi(x) - x^{\mathfrak{N}(v)}) \ge 1 \quad \text{for all } x \in \widehat{\mathcal{O}}_{\overline{v}}.$$

Then φ is a Frobenius automorphism for $\widehat{\mathfrak{p}}_{\overline{v}}$ over K_v and of $\mathsf{p}_{\overline{v}}$ over K in the number field case. Recall that there exists a Frobenius automorphisms φ for $\overline{v}$ over K and that it has the following properties (see [18, Theorem 2.14.6]):

- $\varphi \in G_{\overline{v}}$, and $\varphi I_{\overline{v}}$ is the set of all Frobenius automorphisms for $\overline{v}$ over K.
- If $\sigma \in G$, then $\sigma\varphi\sigma^{-1}$ is a Frobenius automorphism for $\sigma\overline{v}$ over K.
- Let K' is an intermediate field of L/K and $v' = \overline{v} \downarrow K'$. Then:
  - $\varphi^{f(v'/v)}$ is a Frobenius automorphism for $\overline{v}$ over K'.
  - If K'/K is Galois, then $\varphi \restriction K'$ is a Frobenius automorphism for v' over K.
- If $\overline{v}/v$ is unramified, then φ is the Frobenius automorphism for $L_{\overline{v}}/K_v$ (according to [18, Definition 5.6.7]), and it is denoted by

$$\begin{aligned}\varphi_{L_{\overline{v}}/K_v} &= (\overline{v}, L/K) = (\widehat{\mathfrak{p}}_{\overline{v}}, L_{\overline{v}}/K_v) \\ &= (\mathsf{p}_v, L/K) = (P_v, L/K) = \Big(\frac{L/K}{\overline{v}}\Big) \in G_{\overline{v}}\end{aligned}$$

in the various cases.

If L/K is abelian, then $(\overline{v}, L/K)$ only depends on v, we set

$$\varphi_{L_{\overline{v}}/K_v} = (v, L/K) = (\mathsf{p}_v, L/K) = (P_v, L/K) = \Big(\frac{L/K}{v}\Big).$$

If L/K is unramified, then 4.2.3.2. implies

$$(\alpha_v, L_{\overline{v}}/K_v) = (v, L/K)^{v(\alpha_v)} \quad \text{for all} \quad \alpha_v \in K_v.$$

4. G operates on $\mathcal{D}_K$ by means of

$$\sigma\mathsf{b} = \prod_{\overline{v} \in \mathsf{M}_L^0} \mathsf{p}_{\sigma\overline{v}}^{\overline{v}(\mathsf{b})} = \prod_{\overline{v} \in \mathsf{M}_L^0} \mathsf{p}_{\overline{v}}^{(\sigma^{-1}\overline{v})(\mathsf{b})} \quad \text{for all} \quad \mathsf{b} \in \mathcal{D}_L;$$

hence $\overline{v}(\sigma\mathsf{b}) = (\sigma^{-1}\overline{v})(\mathsf{b})$. If $b \in L^\times$, then $(\sigma b)^L = \sigma(b)^L$, and thus G operates on $\mathcal{C}_L$ by means of $\sigma[\mathsf{b}] = [\sigma\mathsf{b}]$ for all $\mathsf{b} \in \mathcal{D}_L$. In particular,

$$\mathbf{1} \to U_L \to L^\times \overset{(\cdot)_L}{\to} \mathcal{D}_L \to \mathcal{C}_L \to \mathbf{1}$$

is an exact sequence of G-modules.

In the function field case it follows that $\deg(\sigma\mathsf{b}) = \deg(\mathsf{b})$ for all $\mathsf{b} \in \mathcal{D}_L$. Hence G operates on $\mathcal{D}_L^0$ and $\mathcal{C}_L^0$, and

$$\mathbf{1} \to U_L \to L^\times \overset{(\cdot)_L}{\to} \mathcal{D}_L^0 \to \mathcal{C}_L^0 \to \mathbf{1}$$

is an exact sequence of G-modules, too.

5. For $\sigma \in G$ and an idele $\beta = (\beta_{\overline{v}})_{\overline{v} \in \mathsf{M}_L} \in \mathsf{J}_L$ we define $\sigma\beta \in \mathsf{J}_L$ by $(\sigma\beta)_{\overline{v}} = \sigma\beta_{\sigma^{-1}\overline{v}} \in L_{\overline{v}}$ for all $\overline{v} \in \mathsf{M}_L$ (as $\sigma U_{\sigma^{-1}\overline{v}} = U_{\overline{v}}$, we have $(\sigma\beta)_{\overline{v}} \in U_{\overline{v}}$ for almost all $\overline{v} \in \mathsf{M}_L^0$, and thus in fact $\sigma\beta \in \mathsf{J}_L$).

If S is a finite subset of M_L, then we obviously have $\sigma(\mathsf{J}_{L,S}) = \mathsf{J}_{L,\sigma S}$, and $\sigma \restriction \mathsf{J}_{L,S} \colon \mathsf{J}_{L,S} \to \mathsf{J}_{L,\sigma S}$ is component-wise (and thus also in the product topology) a topological isomorphism. If $\sigma \in G$, then $\sigma(\mathsf{J}_{L,S_L}) = \mathsf{J}_{L,S_L}$ for all finite subsets S of M_K, and as J_L carries the final topology induced by the directed family $(\mathsf{J}_{L,S_L})_{S \in \mathbb{P}_{\text{fin}}(M_K)}$ it follows that $\sigma \colon \mathsf{J}_L \to \mathsf{J}_L$ is a topological isomorphism. By definition, $\sigma \restriction L^\times$ has the usual meaning, and thus σ induces an (equally denoted) topological isomorphism $\sigma \colon \mathsf{C}_L \to \mathsf{C}_L$ satisfying $\sigma(\beta\mathsf{J}_L) = (\sigma\beta)\mathsf{J}_L \in \mathsf{C}_L$ for all $\beta \in \mathsf{J}_L$.

$\sigma,\, \tau \in G$ and $\beta,\, \beta' \in \mathsf{J}_L$, then $(\sigma\tau)\beta = \sigma(\tau\beta)$ and $\sigma(\beta\beta') = (\sigma\beta)(\sigma\beta')$. Therefore J_L is a (multiplicative) G-module, $L^\times$ is a G-submodule of J_L, and consequently $\mathsf{C}_L = \mathsf{J}_L/L^\times$ is a G-module, too. In particular,

$$\mathbf{1} \to L^\times \to \mathsf{J}_L \to \mathsf{C}_L \to \mathbf{1}$$

is an exact sequence of G-modules.

Theorem 5.4.6 (Galois operation on ideles). *Let L/K be a finite Galois extension and $G = \mathrm{Gal}(L/K)$.*

1. *If $\sigma \in G$, then $\sigma \circ \iota_L = \iota_L \circ \sigma \colon \mathsf{J}_L \to \mathcal{D}_L$.*
2. *If $\beta \in \mathsf{J}_L$, then*
$$i_{L/K} \circ \mathsf{N}_{L/K}(\beta) = \prod_{\sigma \in G} \sigma\beta.$$
3. *$\mathsf{J}_L^G = i_{L/K}(\mathsf{J}_K)$.*
4. *For every $v \in \mathsf{M}_K^0$ let $\overline{v} \in \mathsf{M}_L$ be an arbitrary place lying above v. Then*
$$\mathcal{D}_L^G / j_{L/K}(\mathcal{D}_K) \;\cong\; \prod_{v \in \mathsf{M}_K^0} \mathbb{Z}/e(\overline{v}/v)\mathbb{Z}.$$

Proof. 1. If $\beta = (\beta_{\overline{v}})_{\overline{v} \in \mathsf{M}_L} \in \mathsf{J}_L$, then

$$\begin{aligned}
\sigma \circ \iota_L(\beta) &= \sigma\Big(\prod_{\overline{v} \in \mathsf{M}_L^0} \mathsf{p}_{\overline{v}}^{\overline{v}(\beta_{\overline{v}})} \Big) = \prod_{\overline{v} \in \mathsf{M}_L^0} \mathsf{p}_{\sigma\overline{v}}^{\overline{v}(\beta_{\overline{v}})} \\
&= \prod_{\overline{v} \in \mathsf{M}_L^0} \mathsf{p}_{\overline{v}}^{(\sigma^{-1}\overline{v})(\beta_{\sigma^{-1}\overline{v}})} = \prod_{\overline{v} \in \mathsf{M}_L^0} \mathsf{p}_{\overline{v}}^{\overline{v} \circ \sigma(\beta_{\sigma^{-1}\overline{v}})} \\
&= \prod_{\overline{v} \in \mathsf{M}_L^0} \mathsf{p}_{\overline{v}}^{\overline{v}((\sigma\beta)_{\overline{v}})} = \iota_L \circ \sigma(\beta).
\end{aligned}$$

2. Let $\overline{v} \in \mathsf{M}_L$, $v = \overline{v} \downarrow K$, $(G\!:\!G_{\overline{v}}) = r$ and $G/G_{\overline{v}} = \{\sigma_1 G_{\overline{v}}, \ldots, \sigma_r G_{\overline{v}}\}$. Then $G = \{\tau\sigma_i^{-1} \mid \tau \in G_{\overline{v}},\ i \in [1,r]\}$, $\{\sigma_1\overline{v}, \ldots, \sigma_r\overline{v}\} = \{w \in \mathsf{M}_L \mid w \,|\, v\}$, $\mathrm{Gal}(L_{\sigma_i\overline{v}}/K_v) = \sigma_i G_{\overline{v}} \sigma_i^{-1}$ for all $i \in [1,r]$, and $\tau\overline{v} = \overline{v}$ for all $\tau \in G_{\overline{v}}$. Hence we obtain

$$\begin{aligned}\Big(\prod_{\sigma\in G}\sigma\beta\Big)_{\overline{v}} &= \prod_{\sigma\in G}(\sigma\beta)_{\overline{v}} = \prod_{\sigma\in G}\sigma\beta_{\sigma^{-1}\overline{v}} = \prod_{i=1}^{r}\prod_{\tau\in G_{\overline{v}}}\tau\sigma_i^{-1}\beta_{\sigma_i\overline{v}} \\ &= \prod_{i=1}^{r}\sigma_i^{-1}\prod_{\tau\in G_{\overline{v}}}\sigma_i\tau\sigma_i^{-1}\beta_{\sigma_i\overline{v}} \\ &= \prod_{i=1}^{r}\sigma_i^{-1}\mathsf{N}_{L_{\sigma_i\overline{v}}/K_v}(\beta_{\sigma_i\overline{v}}) = \prod_{i=1}^{r}\mathsf{N}_{L_{\sigma_i\overline{v}}/K_v}(\beta_{\sigma_i\overline{v}}) \\ &= \mathsf{N}_{L/K}(\beta)_v = i_{L/K}\circ\mathsf{N}_{L/K}(\beta)_{\overline{v}},\end{aligned}$$

and consequently

$$i_{L/K}\circ\mathsf{N}_{L/K}(\beta) = \prod_{\sigma\in G}\sigma\beta.$$

3. Let first $\beta \in i_{L/K}(\mathsf{J}_K)$, $\sigma \in G$, $\overline{v} \in \mathsf{M}_L$ and $v = \overline{v} \downarrow K$. Then $\beta_{\overline{v}} = \beta_{\sigma\overline{v}} \in K_v$, hence $(\sigma\beta)_{\overline{v}} = \sigma(\beta_{\sigma^{-1}\overline{v}}) = \beta_{\sigma^{-1}\overline{v}} = \beta_{\overline{v}}$ for all $\sigma \in G$ and $\overline{v} \in \mathsf{M}_L$, and consequently $\beta \in \mathsf{J}_L^G$.

Now let $\beta \in \mathsf{J}_L^G$. If $\overline{v} \in \mathsf{M}_L$, $v = \overline{v} \downarrow K$ and $\tau \in G_{\overline{v}} = \mathrm{Gal}(L_{\overline{v}}/K_v)$, then $\tau^{-1}\overline{v} = \overline{v}$, hence $\beta_{\overline{v}} = (\tau\beta)_{\overline{v}} = \tau(\beta_{\tau^{-1}\overline{v}}) = \tau(\beta_{\overline{v}})$, and consequently $\beta_{\overline{v}} \in K_v$. If $\overline{v}, \overline{v}' \in \mathsf{M}_L$ are such that $\overline{v} \downarrow K = \overline{v}' \downarrow K = v \in \mathsf{M}_K$, then $\overline{v}' = \sigma\overline{v}$ for some $\sigma \in G$. As $\beta \in K_v$, this implies

$$\beta_{\overline{v}} = \sigma(\beta_{\overline{v}}) = \sigma(\beta_{\sigma^{-1}\overline{v}'}) = (\sigma\beta)_{\overline{v}'} = \beta_{\overline{v}'}$$

and consequently $\beta \in i_{L/K}(\mathsf{J}_K)$.

4. Let $\mathsf{b} \in \mathcal{D}_L$, say

$$\mathsf{b} = \prod_{v\in\mathsf{M}_K^0}\prod_{\substack{\overline{v}\in\mathsf{M}_L\\ \overline{v}\,|\,v}}\mathsf{p}_{\overline{v}}^{a_{\overline{v}}} \quad\text{with}\quad a_{\overline{v}} \in \mathbb{Z}.$$

For all $\sigma \in G$ we obtain

$$\sigma\mathsf{b} = \prod_{v\in\mathsf{M}_K^0}\prod_{\substack{\overline{v}\in\mathsf{M}_L\\ \overline{v}\,|\,v}}\mathsf{p}_{\sigma\overline{v}}^{a_{\overline{v}}} = \prod_{v\in\mathsf{M}_K^0}\prod_{\substack{\overline{v}\in\mathsf{M}_L\\ \overline{v}\,|\,v}}\mathsf{p}_{\overline{v}}^{a_{\sigma^{-1}\overline{v}}}.$$

Hence $\mathsf{b} \in \mathcal{D}_L^G$ if and only if $a_{\sigma\overline{v}} = a_{\overline{v}}$ for all $\overline{v} \in \mathsf{M}_L^0$ and all $\sigma \in G$, and this holds if and only if

$$\mathsf{b} = \prod_{v\in\mathsf{M}_K^0}\Big(\prod_{\substack{\overline{v}\in\mathsf{M}_L\\ \overline{v}\,|\,v}}\mathsf{p}_{\overline{v}}\Big)^{a_v} \quad\text{with}\quad a_v \in \mathbb{Z}.$$

On the other hand, $\mathsf{b} \in j_{L/K}(\mathcal{D}_K)$ holds if and only if

$$\mathsf{b} = j_{L/K}\Big(\prod_{v \in \mathsf{M}_K^0} \mathsf{p}_v^{a_v}\Big) = \prod_{v \in \mathsf{M}_K^0} \Big(\prod_{\substack{\overline{v} \in \mathsf{M}_L \\ \overline{v} \,|\, v}} \mathsf{p}_{\overline{v}}\Big)^{e(\overline{v}/v)a_v} \quad \text{with} \quad a_v \in \mathbb{Z}.$$

Hence it follows that

$$\mathcal{D}_L^G / j_{L/K}(\mathcal{D}_K) \cong \prod_{v \in \mathsf{M}_K^0} \mathbb{Z}/e(\overline{v}/v)\mathbb{Z}. \qquad \square$$

Theorem 5.4.7.

1. *The homomorphism* $i^*_{L/K} \colon \mathsf{C}_K \to \mathsf{C}_L$ *is injective.*

2. $i_{L/K}(\mathsf{J}_K)$ *and* $i^*_{L/K}(\mathsf{C}_K)$ *are closed subsets of* J_L *resp.* C_L*, and the induced maps*

$$i_{L/K} \colon \mathsf{J}_K \overset{\sim}{\to} i_{L/K}(\mathsf{J}_K) \;\; \textit{and} \;\; i^*_{L/K} \colon \mathsf{C}_K \overset{\sim}{\to} i^*_{L/K}(\mathsf{C}_K)$$

are topological isomorphisms.

We identify: $\mathsf{J}_K = i_{L/K}(\mathsf{J}_K)$ and $\mathsf{C}_K = i^*_{L/K}(\mathsf{C}_K)$. Then $\mathsf{J}_K \subset \mathsf{J}_L$ and $\mathsf{C}_K \subset \mathsf{C}_L$ are closed subgroups, $\mathsf{J}_K^0 = \mathsf{J}_K \cap \mathsf{J}_L^0$, $\mathsf{C}_K^0 = \mathsf{C}_K \cap \mathsf{C}_L^0$ and $K^\times = \mathsf{J}_K \cap L^\times$.

Proof. 1. Since

$$\operatorname{Ker}(i^*_{L/K} \colon \mathsf{C}_K \to \mathsf{C}_L) = \{\alpha K^\times \in \mathsf{C}_K \mid \alpha \in \mathsf{J}_K,\ i_{L/K}(\alpha) \in L^\times\},$$

it suffices to prove: If $\alpha \in \mathsf{J}_K$ is such that $i_{L/K}(\alpha) \in L^\times$, then $\alpha \in K^\times$. Thus let $\alpha = (\alpha_v)_{v \in \mathsf{M}_K} \in \mathsf{J}_K$ be such that $i_{L/K}(\alpha) = b \in L^\times$.

CASE 1: L/K is separable. Let N/K be a Galois closure of L/K and $G = \operatorname{Gal}(N/K)$. Then $i_{L/K}(\alpha) = b \in N^G = K$, and thus $\alpha \in K^\times$.

CASE 2: L/K is purely inseparable. Assume that $b \notin K^\times$. As $b = i_{L/K}(\alpha)$, it follows that $b = \alpha_v \in K_v$ for all $v \in \mathsf{M}_K$. However, being purely inseparable, $K(b)/K$ is fully ramified (see [18, Theorem 6.5.5]), and therefore $K_v(b) \supsetneq K_v$ for all $v \in \mathsf{M}_K$, a contradiction.

GENERAL CASE: Let M be the separable closure of K in L. Then M/L is purely inseparable. Since $b = i_{L/K}(\alpha) = i_{M/L} \circ i_{M/K}(\alpha) \in L^\times$, we get $i_{M/K}(\alpha) \in M^\times$ by CASE 2, and then $\alpha \in K^\times$ by CASE 1.

2. For $v \in \mathsf{M}_K$ we consider the diagonal embeddings

$$\delta \colon K_v^\times \to \prod_{\substack{\overline{v} \in \mathsf{M}_L \\ \overline{v} \,|\, v}} L_{\overline{v}}^\times \quad \text{and} \quad \delta' \colon U_v \to \prod_{\substack{\overline{v} \in \mathsf{M}_L \\ \overline{v} \,|\, v}} U_{\overline{v}}.$$

Their images are sequentially complete and thus closed by 1.4.4.1(b). For every finite subset S of M_K containing M_K^∞ it follows that

$$\begin{aligned} i_{L/K}(\mathsf{J}_K) \cap \mathsf{J}_{L,S_L} &= \prod_{v\in S} \delta(K_v^\times) \times \prod_{v\in \mathsf{M}_K\setminus S} \delta(U_v) \ \text{ is closed in } \ \mathsf{J}_{L,S_L} \\ &= \prod_{\overline{v}\in S_L} L_{\overline{v}}^\times \times \prod_{\overline{v}\in \mathsf{M}_L\setminus S_L} U_{\overline{v}}. \end{aligned}$$

Now we can prove that $\mathsf{J}_L \setminus i_{L/K}(\mathsf{J}_K)$ is open. Let $\alpha \in \mathsf{J}_L \setminus i_{L/K}(\mathsf{J}_K)$, and let S be a finite subset of M_K containing M_K^∞ such that $\alpha \in \mathsf{J}_{L,S_L}$. Then $U = \mathsf{J}_{L,S_L} \setminus i_{L/K}(\mathsf{J}_K)$ is relatively open in J_{L,S_L} hence open in J_L, and since $U \cap i_{L/K}(\mathsf{J}_K) = \emptyset$ and $\alpha \in U$, the assertion follows.

For a finite subset S of M_K containing M_K^∞ and $\varepsilon \in \mathbb{R}_{>0}$ we set

$$B(S,\varepsilon) = \prod_{v\in S} \{x \in K_v \mid |x-1|_v < \varepsilon\} \times \prod_{v\in \mathsf{M}_K\setminus S} U_v.$$

The sets $B(S,\varepsilon)$ constitute a fundamental system of neighbourhoods of 1 in J_K, and

$$i_{L/K}(B(S,\varepsilon)) = \Big(\prod_{\overline{v}\in S_L} \{x \in L_{\overline{v}} \mid |x-1|_{\overline{v}}^{[L_{\overline{v}}:K_v]} < \varepsilon\} \times \prod_{\overline{v}\in \mathsf{M}_L\setminus S_L} U_{\overline{v}} \Big) \cap i_{L/K}(\mathsf{J}_K)$$

is an open neighbourhood of 1 in $i_{L/K}(\mathsf{J}_L)$. Therefore $i_{L/K} \colon \mathsf{J}_K \to i_{L/K}(\mathsf{J}_K)$ is an open isomorphism by 1.1.6.2, and as it is continuous, it is a topological isomorphism.

The epimorphism

$$i \colon \mathsf{J}_K^0 \to i_{L/K}(\mathsf{J}_K^0)L^\times/L^\times = i_{L/K}^*(\mathsf{C}_K^0) \subset \mathsf{C}_L, \ \text{ defined by } i(\alpha) = i_{L/K}(\alpha)L^\times,$$

is continuous, and since $\mathrm{Ker}(i) = K^\times$, it follows that i induces a continuous isomorphism $i^* = i_{L/K}^* \restriction \mathsf{C}_K^0 \colon \mathsf{C}_K^0 \to i_{L/K}^*(\mathsf{C}_K^0)$ which is even a topological isomorphism since C_K^0 is compact (see 5.3.7.1).

By 5.3.3.2(b), C_K is a topological direct product $\mathsf{C}_K = \mathsf{C}_K^0 \cdot \Gamma'$, where $\Gamma' \cong \mathbb{R}_{>0}$ in the number field case and $\Gamma' \cong \mathbb{Z}$ in the function field case. If $\alpha \in \mathsf{J}_K$, then $\|i_{L/K}(\alpha)\| = \|\alpha\|^n$, where $n = [L\colon K]$. Hence $i_{L/K}^*(\alpha b) = i^*(\alpha)b^n$ for all $\alpha \in \mathsf{J}_K^0$ and $b \in \Gamma'$, and consequently $i_{L/K}^*(\mathsf{C}_K) = i^*(\mathsf{C}_K^0)\cdot\Gamma'^n \subset \mathsf{C}_L^0\cdot\Gamma'$. The map $\Gamma' \to \Gamma'^n$, $x \to x^n$ is a topological isomorphism, and since we have already seen that $i^* \colon \mathsf{C}_K^0 \stackrel{\sim}{\to} i^*(\mathsf{C}_K^0)$ is a topological isomorphism, it follows that $i_{L/K}^* \colon \mathsf{C}_K \to i_{L/K}^*(\mathsf{C}_K)$ is a topological isomorphism, too.

Being compact, $i^*(C_K^0)$ is closed in C_L^0, and thus $i_{L/K}^*(\mathsf{C}_K) = i^*(\mathsf{C}_K^0)\cdot\Gamma'^n$ is closed in $\mathsf{C}_L = \mathsf{C}_L^0 \cdot \Gamma'$, too. □

Theorem 5.4.8. *Let L/K be a finite Galois extension of global fields with Galois group G. Then $\mathsf{J}_L^G = \mathsf{J}_K$ and $\mathsf{C}_L^G = \mathsf{C}_K$.*

Proof. The equality $\mathsf{J}_L^G = \mathsf{J}_K$ holds by 5.4.6.3. Since $L^{\times G} = K^\times$ and $\mathsf{J}_L^G = \mathsf{J}_K$, the exact sequence of G-modules

$$\mathbf{1} \to L^\times \to \mathsf{J}_L \to \mathsf{C}_L \to \mathbf{1}$$

yields (using 2.4.1 and 2.6.3) the exact sequence

$$\mathbf{1} \;\to\; K^\times \;\to\; \mathsf{J}_K \;\to\; \mathsf{C}_L^G \;\to\; H^1(G, L^\times) = \mathbf{1},$$

which shows that $\mathsf{C}_L^G = \mathsf{J}_K/K^\times = \mathsf{C}_K$. □

5.5 S-class groups and holomorphy domains

Let K be a global field.

Definition and Remarks 5.5.1. We generalize the concept of class groups. Let S be a finite subset of M_K^0. We call

$$\mathcal{D}_{K,S} = \{\mathsf{a} \in \mathcal{D}_K \mid \operatorname{supp}(\mathsf{a}) \cap S = \emptyset\} = \{\mathsf{a} \in \mathcal{D}_K \mid v(\mathsf{a}) = 0 \text{ for all } v \in S\}$$

the **S-divisor group** of K; its elements are called **S-divisors**, and we denote by $\mathcal{D}'_{K,S} = \mathcal{D}_{K,S} \cap \mathcal{D}'_K$ the monoid of all integral S-divisors.

For $a \in K^\times$ we define the **S-principal divisor** of a by

$$(a)_S = (a)_S^K = \prod_{v \in \mathsf{M}_K^0 \setminus S} \mathsf{p}_v^{v(a)} \in \mathcal{D}_{K,S}.$$

We denote by $(K^\times)_S$ the group of S-principal divisors of K. The assignment $a \mapsto (a)_S$ defines an epimorphism $(\,\cdot\,)_S \colon K^\times \to (K^\times)_S$ with kernel

$$U_{K,S} = K_{S \cup \mathsf{M}_K^\infty} = \mathsf{J}_{K, S \cup \mathsf{M}_K^\infty} \cap K^\times = \{a \in K^\times \mid v(a) = 0 \text{ for all } v \in \mathsf{M}_K^0 \setminus S\}$$

and cokernel

$$\mathcal{C}_{K,S} = \mathcal{D}_{K,S}/(K^\times)_S.$$

Hence there are exact sequences

$$\begin{array}{ccccccccccc}
\mathbf{1} & \longrightarrow & U_{K,S} & \longrightarrow & K^\times & & \longrightarrow & & \mathcal{D}_{K,S} & \longrightarrow & \mathcal{C}_{K,S} \longrightarrow \mathbf{1} \\
 & & & & & \searrow & & \nearrow & & & \\
 & & & & & & (K^\times)_S & & & & \\
 & & & & & \nearrow & & \searrow & & & \\
 & & & & \mathbf{1} & & & & \mathbf{1} & &
\end{array}$$

Obviously, $(K^\times) \subset \mathcal{D}_{K,\mathsf{M}_K^0\setminus S}(K^\times)_S$, and

$$\mathcal{K}_S = \mathcal{D}_{K,\mathsf{M}_K^0\setminus S}(K^\times)_S/(K^\times) \quad \text{is a subgroup of } \mathcal{C}_K.$$

We call $U_{K,S}$ the **S-unit group** and $\mathcal{C}_{K,S}$ the **S-class group** of K. For $\mathsf{a} \in \mathcal{D}_{K,S}$ we denote by $[\mathsf{a}]_S \in \mathcal{C}_{K,S}$ the S-ideal class containing a. For $S = \emptyset$ we retrieve the original concepts presented in 5.1.1.**10** above.

We set $(K^+)_S = \{(a)_S \mid a \in K^+\}$, $U_{K,S}^+ = K^+ \cap U_{K,S}$, and we call $\mathcal{C}_{K,S}^+ = \mathcal{D}_{K,S}/(K^+)_S$ the **narrow S-class group**. For $\mathsf{a} \in \mathcal{D}_{K,S}$ we denote by $[\mathsf{a}]_S^+ \in \mathcal{C}_{K,S}^+$ the narrow S-class containing a. The ordinary and the narrow S-class groups are connected by the exact sequence

$$\mathbf{1} \to U_{K,S}/U_{K,S}^+ \to K^\times/K^+ \overset{\tau}{\to} \mathcal{C}_{K,S}^+ \to \mathcal{C}_{K,S} \to \mathbf{1},$$

where $\tau(aK^+) = [(a)]_S^+$
which shows that

$$h_{K,S}^+ = \frac{2^{r_1}}{(U_{K,S} : U_{K,S}^+)}\, h_{K,S},$$

where r_1 is the number of real embeddings of K. If K has no real embeddings (and in particular in the function field case) we obviously have $\mathcal{C}_{K,S} = \mathcal{C}_{K,S}^+$.

An ideal-theoretic description of S-class groups will be given below in Theorem 5.5.6. In the following Theorem 5.5.2 we connect the S-class groups with the ordinary class groups and show their finiteness.

Theorem 5.5.2. *Let S be a finite non-empty subset of M_K^0. Then*

$$\mathcal{C}_K = \mathcal{D}_{K,S}(K^\times)/(K^\times),$$

and there is an exact sequence

$$\mathbf{1} \to \mathcal{K}_S \to \mathcal{C}_K \overset{\tau}{\to} \mathcal{C}_{K,S} \to \mathbf{1}, \quad \textit{where}$$

$$\tau([\mathsf{a}]) = \Big[\prod_{v \in \mathsf{M}_K^0\setminus S} \mathsf{p}_v^{v(\mathsf{a})}\Big]_S \quad \textit{for all } \mathsf{a} \in \mathcal{D}_K,$$

and $\mathcal{K}_S = \mathcal{D}_{K,\mathsf{M}_K^0\setminus S}(K^\times)/(K^\times) \subset \mathcal{C}_K$. In the function field case, there is the additional exact sequence

$$\mathbf{1} \to \mathcal{K}_S \cap \mathcal{C}_K^0 \to \mathcal{C}_K^0 \overset{\tau^0}{\to} \mathcal{C}_{K,S} \to \mathbb{Z}/\partial_S\mathbb{Z} \to \mathbf{0}, \quad \textit{where } \tau^0 = \tau \restriction \mathcal{C}_K^0$$

In any case, $\mathcal{C}_{K,S}$ is a finite group, and

$$|\mathcal{C}_{K,S}| \ \textit{divides} \ \begin{cases} h_K & \textit{in the number field case,} \\ h_K\partial_S & \textit{in the function field case.} \end{cases}$$

Proof. Let $\mathfrak{a} = [\mathsf{a}] \in \mathcal{C}_K$ for some $\mathsf{a} \in \mathcal{D}_K$. By the weak approximation theorem there exists some $a \in K^\times$ such that $v(a) = v(\mathsf{a})$ for all $v \in S$, hence $a^{-1}\mathsf{a} \in \mathcal{D}_{K,S}$ and $\mathfrak{a} = [a^{-1}\mathsf{a}] \in \mathcal{D}_{K,S}(K^\times)/(K^\times)$.

As $\mathcal{D}_K = \mathcal{D}_{K,S} \cdot \mathcal{D}_{K,\mathsf{M}_K^0 \setminus S}$, every $\mathsf{a} \in \mathcal{D}_K$ has a unique factorization $\mathsf{a} = \mathsf{a}_S \mathsf{a}_{S'}$, where $\mathsf{a}_S \in \mathcal{D}_{K,S}$ and $\mathsf{a}_{S'} \in \mathcal{D}_{K,\mathsf{M}_K^0 \setminus S}$. We define $\tau_0 \colon \mathcal{D}_K \to \mathcal{C}_{K,S}$ by $\tau_0(\mathsf{a}) = [\mathsf{a}_S]_S$. Then $\mathrm{Ker}(\tau_0) = \mathcal{D}_{K,\mathsf{M}_K^0 \setminus S}(K^\times)$, and since

$$\mathcal{D}_K/\mathcal{D}_{K,\mathsf{M}_K^0 \setminus S}(K^\times) \cong \mathcal{D}_K/(K^\times)/\mathcal{D}_{K,\mathsf{M}_K^0 \setminus S}(K^\times)/(K^\times) = \mathcal{C}_K/\mathcal{K}_S,$$

we arrive at the asserted exact sequence, where τ is induced by τ_0.

Now let $K/\mathbb{F}_q$ be a function field. Since $\deg(\mathcal{K}) = \deg(\mathcal{D}_{K,\mathsf{M}_K^0 \setminus S}) = \partial_S\mathbb{Z}$, the homomorphism $\tau^0 = \tau \upharpoonright \mathcal{C}_K^0$ fits into the following commutative diagram with exact rows and columns.

$$\begin{array}{ccccccccc}
 & & \mathbf{1} & & \mathbf{1} & & \mathbf{1} & & \\
 & & \downarrow & & \downarrow & & \downarrow & & \\
\mathbf{1} & \longrightarrow & \mathcal{K}_S \cap \mathcal{C}_K^0 & \longrightarrow & \mathcal{C}_K^0 & \longrightarrow & \tau(\mathcal{C}_K^0) & \longrightarrow & \mathbf{0} \\
 & & \downarrow & & \downarrow & & \downarrow & & \\
\mathbf{1} & \longrightarrow & \mathcal{K}_S & \longrightarrow & \mathcal{C}_K & \longrightarrow & \mathcal{C}_{K,S} & \longrightarrow & \mathbf{0} \\
 & & {\scriptstyle\deg}\downarrow & & {\scriptstyle\deg}\downarrow & & & & \\
\mathbf{0} & \longrightarrow & \partial_S\mathbb{Z} & \longrightarrow & \mathbb{Z} & \longrightarrow & \mathbb{Z}/\partial_S\mathbb{Z} & \longrightarrow & \mathbf{0} \\
 & & \downarrow & & \downarrow & & \downarrow & & \\
 & & \mathbf{0} & & \mathbf{0} & & \mathbf{0} & &
\end{array}$$

A simple diagram chasing shows that this diagram can be completed by an epimorphism $\mathcal{C}_{K,S} \to \mathbb{Z}/\partial_S\mathbb{Z}$, which implies the asserted exact sequence.

As we already know that $\mathcal{C}_K$ in the number field case and $\mathcal{C}_K^0$ in the function field case are finite groups, the exact sequences show that $\mathcal{C}_{K,S}$ is finite and has the asserted divisibility property. $\square$

Definition 5.5.3. For a subset S of M_K^0 we define

$$\mathfrak{o}_{K,S} = \bigcap_{v \in \mathsf{M}_K^0 \setminus S} \mathcal{O}_v = \{z \in K \mid v(z) \ge 0 \ \text{ for all } \ v \in \mathsf{M}_K^0 \setminus S\}.$$

$\mathfrak{o}_{K,S}$ is a subring of K, and if $v \in \mathsf{M}_K^0 \setminus S$, then $\mathfrak{p}_{S,v} = P_v \cap \mathfrak{o}_{K,S}$ is a prime ideal of $\mathfrak{o}_{K,S}$. In particular, $\mathfrak{o}_{K,\mathsf{M}_K^0} = K$, and $\mathfrak{o}_{K,\mathsf{M}_K^0 \setminus \{v\}} = \mathcal{O}_v$ for all $v \in \mathsf{M}_K^0$.

A subring $\mathfrak{o}$ of K is called a **holomorphy domain** in K if $\mathfrak{o} = \mathfrak{o}_{K,S}$ for some subset S of M_K^0.

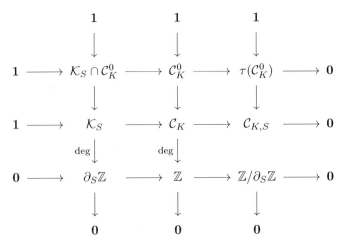

In the number field case

$$\mathfrak{o}_{K,S} = \bigcap_{\substack{\mathfrak{p}\in\mathcal{P}_K \\ v_\mathfrak{p}\in\mathsf{M}_K^0\setminus S}} (\mathfrak{o}_K)_\mathfrak{p}, \quad \text{and} \quad \mathfrak{o}_{K,\emptyset} = \mathfrak{o}_K \text{ (see [18, Theorem 2.7.5]).}$$

In the function field case, $\mathfrak{o}_{K,\emptyset} = \mathbb{F}_q$ (see [18, Theorem 6.2.7]).

Theorem 5.5.4. *Let S and T be subsets of M_K^0 and $v \in \mathsf{M}_K^0$. Then we have $\mathfrak{o}_{K,S} \subset \mathfrak{o}_{K,T}$ if and only if $S \subset T$, and $\mathfrak{o}_{K,S} \subset \mathcal{O}_v$ if and only if $v \in \mathsf{M}_K^0 \setminus S$.*

Proof. Obviously, $S \subset T$ implies $\mathsf{M}_K^0 \setminus T \subset \mathsf{M}_K^0 \setminus S$ and therefore $\mathfrak{o}_{K,S} \subset \mathfrak{o}_{K,T}$. Thus assume that $S \not\subset T$ and let $v \in S \setminus T$.

In the number field case we consider the principal ideal $\mathfrak{p}_v^{h_K} = b\mathfrak{o}_K$, where $b \in \mathfrak{o}_K^\bullet$. Since $v(b) > 0$ and $v'(b) = 0$ for all $v' \in \mathsf{M}_K^0 \setminus \{v\}$, it follows that $b^{-1} \in \mathfrak{o}_{K,S} \setminus \mathfrak{o}_{K,T}$. In the function field case there exists some $z \in K$ such that v is the only place of K satisfying $v(z) < 0$ (see [18, Corollary 6.4.6]), and then $z \in \mathfrak{o}_{K,S} \setminus \mathfrak{o}_{K,T}$.

The second assertion follows with $T = \mathsf{M}_K^0 \setminus \{v\}$ since $\mathfrak{o}_{K,\mathsf{M}_K^0\setminus\{v\}} = \mathcal{O}_v$. $\square$

Theorem 5.5.5. *Let $S \subsetneq \mathsf{M}_K^0$ and $S \neq \emptyset$ in the function field case.*

1. *If $v \in \mathsf{M}_K^0 \setminus S$, then $\mathfrak{p}_{S,v}$ is a maximal ideal of $\mathfrak{o}_{K,S}$, $(\mathfrak{o}_{K,S})_{\mathfrak{p}_{S,v}} = \mathcal{O}_v$, and the embedding $\mathfrak{o}_{K,S} \hookrightarrow \mathcal{O}_v$ induces an isomorphism*

$$\phi\colon \mathfrak{o}_{K,S}/\mathfrak{p}_{S,v} \overset{\sim}{\to} \mathsf{k}_v \quad \textit{satisfying} \quad \phi(x+\mathfrak{p}_{S,v}) = x+P_v \quad \textit{for all} \quad x \in \mathfrak{o}_{K,S}.$$

We identify: $\mathfrak{o}_{K,S}/\mathfrak{p}_{S,v} = \mathsf{k}_v$.

2. *$\mathfrak{o}_{K,S}$ is a Dedekind domain with finite residue class rings and quotient field K. The map*

$$\Phi\colon \mathsf{M}_K^0\setminus S \to \mathcal{P}(\mathfrak{o}_{K,S}), \quad \textit{defined by} \quad \Phi(v) = \mathfrak{p}_{S,v} \quad \textit{for all} \quad v \in \mathsf{M}_K^0\setminus S,$$

is bijective, and if $v \in \mathsf{M}_K^0\setminus S$, then $v\colon K \to \mathbb{Z}\cup\{\infty\}$ is the $\mathfrak{p}_{S,v}$-adic valuation. In particular, if $\mathfrak{a} \in \mathcal{J}(\mathfrak{o}_{K,S})$, then

$$\mathfrak{a} = \prod_{v\in\mathsf{M}_K^0\setminus S} \mathfrak{p}_{S,v}^{v(\mathfrak{a})},$$

where $v(\mathfrak{a}) = v_{\mathfrak{p}_{S,v}}(\mathfrak{a}) = \min\{v(x) \mid x \in \mathfrak{a}\}$ for all $v \in \mathsf{M}_K^0 \setminus S$.
In an obvious way, we define $\operatorname{supp}(\mathfrak{a}) = \{v \in \mathsf{M}_K^0 \setminus S \mid v(\mathfrak{a}) \neq 0\}$.

3. Suppose that $S \subset S' \subsetneq \mathsf{M}_K^0$. Then

$$\mathfrak{a}\mathfrak{o}_{K,S'} = \prod_{v \in \mathsf{M}_K^0 \setminus S'} \mathfrak{p}_{S',v}^{v(\mathfrak{a})} \quad \textit{for all} \quad \mathfrak{a} \in \mathcal{I}(\mathfrak{o}_{K,S}).$$

In particular, the ideal extension $\mathcal{I}(\mathfrak{o}_{K,S}) \to \mathcal{I}(\mathfrak{o}_{K,S'})$, $\mathfrak{a} \mapsto \mathfrak{a}\mathfrak{o}_{K,S'}$, is an epimorphism, and if $v \in \mathsf{M}_K^0 \setminus S'$, then $\mathfrak{p}_{S',v} = \mathfrak{p}_{S,v}\mathfrak{o}_{K,S'}$ and $v(\mathfrak{a}\mathfrak{o}_{K,S'}) = v(\mathfrak{a})$ for all $\mathfrak{a} \in \mathcal{I}(\mathfrak{o}_{K,S})$.

Proof. 1. Let $v \in \mathsf{M}_K^0 \setminus S$, and let $\varphi \colon \mathfrak{o}_{K,S} \to \mathsf{k}_v = \mathcal{O}_v/P_v$ be defined by $\varphi(x) = x + P_v$. Then φ is a ring homomorphism with kernel $P_v \cap \mathfrak{o}_{K,S} = \mathfrak{p}_{S,v}$, and we prove that φ is surjective. For that let $\xi = x + P_v \in \mathsf{k}_v$ with $x \in \mathcal{O}_v$. By 5.3.5, there exists some $z \in K$ satisfying $v(z - x) > 0$, and $v'(z) \ge 0$ for all $v' \in \mathsf{M}_K^0 \setminus (S \cup \{v\})$. Since $v(z) = v(x + (z-x)) \ge \min\{v(x), v(z-x)\} \ge 0$, it follows that $z \in \mathfrak{o}_{K,S}$, and $\varphi(z) = \xi$. Therefore φ induces an isomorphism $\phi \colon \mathfrak{o}_{K,S}/\mathfrak{p}_{S,v} \overset{\sim}{\to} \mathsf{k}_v$ satisfying $\phi(x + \mathfrak{p}_{S,v}) = x + P_v$ for all $x \in \mathfrak{o}_{K,S}$, and thus $\mathfrak{p}_{S,v}$ is a maximal ideal of $\mathfrak{o}_{K,S}$.

Since $\mathfrak{o}_{K,S} \subset \mathcal{O}_v$ and $\mathfrak{p}_{S,v} = \mathfrak{o}_{K,S} \cap P_v$, we obtain $(\mathfrak{o}_{K,S})_{\mathfrak{p}_{S,v}} \subset \mathcal{O}_v$. To prove the reverse inclusion, let $x \in \mathcal{O}_v^\bullet$. By 5.3.5, there exists some $a \in K$ such that $v(a) = v(x)$ and $v'(a) \ge \max\{v'(x), 0\}$ for all $v' \in \mathsf{M}_K^0 \setminus (S \cup \{v\})$. It follows that $v'(a) \ge 0$ and $v'(x^{-1}a) \ge 0$ for all $v' \in \mathsf{M}_K^0 \setminus S$, and $v(x^{-1}a) = 0$. Hence $a \in \mathfrak{o}_{K,S}$, $x^{-1}a \in \mathfrak{o}_{K,S} \setminus \mathfrak{p}_{S,v}$, and thus $x = (x^{-1}a)^{-1}a \in (\mathfrak{o}_{K,S})_{\mathfrak{p}_{S,v}}$.

2. Since $\mathcal{O}_v = (\mathfrak{o}_{K,S})_{\mathfrak{p}_{S,v}}$ for all $v \in \mathsf{M}_K^0 \setminus S$, we get $\mathsf{q}(\mathfrak{o}_{K,S}) = \mathsf{q}(\mathcal{O}_v) = K$. If K is a number field, we set $\mathfrak{o} = \mathfrak{o}_K$. If $K/\mathbb{F}_q$ is a function field, let $x \in \mathfrak{o}_{K,S} \setminus \mathbb{F}_q$, and set $\mathfrak{o} = \mathrm{cl}_K(\mathbb{F}_q[x])$. In any case $\mathfrak{o}$ is a Dedekind domain and $K = \mathsf{q}(\mathfrak{o})$. If $v \in \mathsf{M}_K^0 \setminus S$, then $x \in \mathcal{O}_v$ implies $\mathfrak{o} \subset \mathcal{O}_v$, since $\mathcal{O}_v$ is integrally closed, and therefore $\mathfrak{o} \subset \mathfrak{o}_{K,S}$. Being an overring of a Dedekind domain, $\mathfrak{o}_{K,S}$ is itself a Dedekind domain (see [18, Theorem 2.10.3]).

If $v, v' \in \mathsf{M}_K^0 \setminus S$ are such that $\mathfrak{p}_{S,v} = \mathfrak{p}_{S,v'}$, then it follows that $\mathcal{O}_v = (\mathfrak{o}_{K,S})_{\mathfrak{p}_{S,v}} = (\mathfrak{o}_{K,S})_{\mathfrak{p}_{S,v'}} = \mathcal{O}_{v'}$ and therefore $v = v'$ (see [18, Theorem 2.10.4]). Hence Φ is injective. To prove that Φ is surjective, let $\mathfrak{p} \in \mathcal{P}(\mathfrak{o}_{K,S})$ and $v = v_\mathfrak{p} \in \mathsf{M}_K^0$. Then $(\mathfrak{o}_{K,S})_\mathfrak{p} = \mathcal{O}_v$, $\mathfrak{p}(\mathfrak{o}_{K,S})_\mathfrak{p} = P_v$, and $\Phi(v) = \mathfrak{p}(\mathfrak{o}_{K,S})_\mathfrak{p} \cap \mathfrak{o}_{K,S} = \mathfrak{p}$.

If $v \in \mathsf{M}_K^0 \setminus S$, then $|\mathfrak{o}_{K,S}/\mathfrak{p}_{S,v}| = \mathfrak{N}(v) < \infty$, and thus $\mathfrak{o}_{K,S}$ has finite residue class rings. Since $(\mathfrak{o}_{K,S})_{\mathfrak{p}_{S,v}} = \mathcal{O}_v$ for all $v \in \mathsf{M}_K^0 \setminus S$, we obtain $v_{\mathfrak{p}_{S,v}} = v$, and the implication of the remaining assertions is well known (see [18, Theorem 2.8.3]).

3. Let $\mathfrak{a} \in \mathcal{I}(\mathfrak{o}_{K,S})$. Then $\mathfrak{a} = a\mathfrak{o}_{K,S} + b\mathfrak{o}_{K,S}$ for some $a, b \in \mathfrak{o}_{K,S}$ (see [18, Corollary 2.9.3]), and consequently $\mathfrak{a}\mathfrak{o}_{K,S'} = a\mathfrak{o}_{K,S'} + b\mathfrak{o}_{K,S'}$. For all $v \in \mathsf{M}_K^0 \setminus S'$ we obtain

$$\begin{aligned} v(\mathfrak{a}\mathfrak{o}_{K,S'}) &= \min\{v(a\mathfrak{o}_{K,S'}), v(b\mathfrak{o}_{K,S'})\} = \min\{v(a), v(b)\} \\ &= \min\{v(a\mathfrak{o}_{K,S}), v(b\mathfrak{o}_{K,S}\} = v(\mathfrak{a}), \end{aligned}$$

and therefore

$$\mathfrak{a}\mathfrak{o}_{K,S'} = \prod_{v \in \mathsf{M}_K^0 \setminus S'} \mathfrak{p}_{S',v}^{v(\mathfrak{a}\mathfrak{o}_{K,S'})} = \prod_{v \in \mathsf{M}_K^0 \setminus S'} \mathfrak{p}_{S',v}^{v(\mathfrak{a})}. \qquad \square$$

Next we investigate the unit group $\mathfrak{o}_{K,S}^\times$ of a holomorphy domain and the interrelationship between its ideal class group $\mathcal{C}(\mathfrak{o}_{K,S}) = \mathcal{J}(\mathfrak{o}_{K,S})/\mathcal{H}(\mathfrak{o}_{K,S})$ and the S-divisor class group $\mathcal{C}_{K,S}$ of K. For a fractional ideal $\mathfrak{a} \in \mathcal{J}(\mathfrak{o}_{K,S})$ we denote by $[\mathfrak{a}] \in \mathcal{C}(\mathfrak{o}_{K,S})$ its ideal class.

Theorem 5.5.6. *Let S be a finite subset of M_K^0 and $S \neq \emptyset$ in the function field case.*

1. $\mathfrak{o}_{K,S}^\times = U_{K,S} \cong \mu(K) \times \mathbb{Z}^{s-1}$, *where* $s = |S \cup \mathsf{M}_K^\infty|$.
2. *The map*

$$\theta_S \colon \mathcal{D}_{K,S} \to \mathcal{J}(\mathfrak{o}_{K,S}), \quad \textit{defined by} \quad \theta_S(\mathsf{a}) = \prod_{v \in \mathsf{M}_K^0 \setminus S} \mathfrak{p}_{S,v}^{v(\mathsf{a})},$$

is an isomorphism, $\theta_S((a)_S) = a\mathfrak{o}_{K,S}$ *for all* $a \in K^\times$, θ_S *induces an (equally denoted) isomorphism*

$$\theta_S \colon \mathcal{C}_{K,S} \overset{\sim}{\to} \mathcal{C}(\mathfrak{o}_{K,S}) \quad \textit{satisfying} \quad \theta_S([\mathsf{a}]_S) = [\theta_S(\mathsf{a})] \quad \textit{for all} \quad \mathsf{a} \in \mathcal{D}_{K,S},$$

and there is the commutative diagram with exact rows

$$\begin{array}{ccccccccc}
\mathbf{1} & \longrightarrow & \mathfrak{o}_{K,S}^\times & \longrightarrow & K^\times & \xrightarrow{(\cdot)_S} & \mathcal{D}_{K,S} & \xrightarrow{[\cdot]_S} & \mathcal{C}_{K,S} & \longrightarrow & \mathbf{1} \\
 & & \| & & \| & & \downarrow{\scriptstyle \theta_S} & & \downarrow{\scriptstyle \theta_S} & & \\
\mathbf{1} & \longrightarrow & U_{K,S} & \longrightarrow & K^\times & \xrightarrow{\delta} & \mathcal{J}(\mathfrak{o}_{K,S}) & \xrightarrow{[\cdot]} & \mathcal{C}(\mathfrak{o}_{K,S}) & \longrightarrow & \mathbf{1}
\end{array},$$

where $\delta(a) = a\mathfrak{o}_{K,S}$ *for all* $a \in K^\times$.

3. *If* $\mathsf{a} \in \mathcal{D}_{K,S}$, *then*

$$\mathfrak{o}_{K,S}/\theta_S(\mathsf{a}) \cong \prod_{v \in \mathrm{supp}(\mathsf{a})} \widehat{\mathcal{O}}_v/\widehat{\mathfrak{p}}_v^{v(\mathfrak{a})} \quad \textit{and} \quad (\mathfrak{o}_{K,S}/\theta_S(\mathsf{a}))^\times \cong \prod_{v \in \mathrm{supp}(\mathsf{a})} U_v/U_v^{(v(\mathsf{a}))}.$$

In particular, $(\mathfrak{o}_{K,S} : \theta_S(\mathsf{a})) = \mathfrak{N}(\mathsf{a})$, *and*

$$\varphi(\mathsf{a}) = |(\mathfrak{o}_{K,S}/\theta_S(\mathsf{a}))^\times| = \prod_{v \in \mathrm{supp}(\mathsf{a})} \mathfrak{N}(v)^{v(\mathsf{a})-1}(\mathfrak{N}(v) - 1).$$

Proof. 1. By definition,

$$\mathfrak{o}_{K,S}^\times = \{x \in K^\times \mid v(x) = 0 \text{ for all } v \in \mathsf{M}_K^0 \setminus S\} = U_{K,S} = K_{S \cup \mathsf{M}_K^\infty},$$

and the assertion follows by 5.3.8.

2. By definition, $\mathcal{D}_{K,S}$ is the free abelian group with basis $\{\mathsf{p}_v \mid v \in \mathsf{M}_K^0 \backslash S\}$, and as $\mathfrak{o}_{K,S}$ is a Dedekind domain by 5.5.5.2 it follows that $\mathcal{I}(\mathfrak{o}_{K,S})$ is the free abelian group with bases $\{\mathfrak{p}_{S,v} \mid v \in \mathsf{M}_K^0\}$. Moreover, $\theta_S \colon \mathcal{D}_{K,S} \to \mathcal{I}(\mathfrak{o}_{K,S})$ is an isomorphism, since it is the unique group homomorphism satisfying $\theta_S(\mathsf{p}_v) = \mathfrak{p}_{S,v}$ for all $v \in \mathsf{M}_K^0 \setminus S$.

If $a \in K^\times$, then

$$\theta_S((a)_S) = \theta_S\Big(\prod_{v \in \mathsf{M}_K^0 \backslash S} \mathsf{p}_v^{v(a)}\Big) = \prod_{v \in \mathsf{M}_K^0 \backslash S} \mathfrak{p}_{S,v}^{v(a)} = a\mathfrak{o}_{K,S} = \delta(a),$$

and θ_S induces the equally denoted isomorphism $\theta_S \colon \mathcal{D}_{K,S} \to \mathcal{I}(\mathfrak{o}_{K,S})$ making the diagram with obviously exact rows commutative.

3. Let $\mathsf{a} \in \mathcal{D}_{K,S}$. Then

$$\mathsf{a} = \prod_{v \in \mathrm{supp}(\mathsf{a})} \mathsf{p}_v^{v(\mathsf{a})}$$

implies

$$\theta_S(\mathsf{a}) = \prod_{v \in \mathrm{supp}(\mathsf{a})} \mathfrak{p}_{S,v}^{v(\mathsf{a})} \quad \text{and} \quad \mathfrak{o}_{K,S}/\theta_S(\mathsf{a}) \cong \prod_{v \in \mathrm{supp}(\mathsf{a})} \mathfrak{o}_{K,S}/\mathfrak{p}_{S,v}^{v(\mathsf{a})}.$$

Since $(\mathfrak{o}_{K,S})_{\mathfrak{p}_{S,v}} = \mathcal{O}_v$, we get

$$\mathfrak{o}_{K,S}/\theta_S(\mathsf{a}) \cong \prod_{v \in \mathrm{supp}(\mathsf{a})} \widehat{\mathcal{O}}_v/\widehat{\mathfrak{p}}_v^{v(\mathsf{a})}$$

and

$$(\mathfrak{o}_{K,S}/\theta_S(\mathsf{a}))^\times \cong \prod_{v \in \mathrm{supp}(\mathsf{a})} (\widehat{\mathcal{O}}_v/\widehat{\mathfrak{p}}_v^{v(\mathsf{a})})^\times \cong \prod_{v \in \mathrm{supp}(\mathsf{a})} U_v/U_v^{(v(\mathsf{a}))}$$

(see [18, Theorems 5.3.4 and 2.10.5]). Moreover,

$$(\mathfrak{o}_{K,S} \colon \theta_S(\mathsf{a})) = \prod_{v \in \mathrm{supp}(\mathsf{a})} \mathfrak{N}(v)^{v(\mathsf{a})} = \mathfrak{N}(\mathsf{a})$$

and

$$|(\mathfrak{o}_{K,S}/\theta_S(\mathfrak{a}))^\times| = \prod_{v \in \mathrm{supp}(\mathfrak{a})} |\widehat{\mathcal{O}}_v/\widehat{\mathfrak{p}}^{v(\mathsf{a})})^\times| = \prod_{v \in \mathrm{supp}(\mathsf{a})} |U_v/U_v^{(v(\mathsf{a}))}|$$

$$= \prod_{v \in \mathrm{supp}(\mathsf{a})} \mathfrak{N}(v)^{v(\mathsf{a})-1}(\mathfrak{N}(v) - 1) = \varphi(\mathsf{a}). \qquad \square$$

Theorem 5.5.7 (Characterization of holomorphy domains).

1. Let $S' \subset S \subsetneq \mathsf{M}_K^0$ and $S' \ne \emptyset$ in the function field case. Then $\mathfrak{o}_{K,S} = T^{-1}\mathfrak{o}_{K,S'}$, where $T = \mathfrak{o}_{K,S}^\times \cap \mathfrak{o}_{K,S'}$.
In the number fied case we even obtain $\mathfrak{o}_{K,S} = T^{-1}\mathfrak{o}_K$, where $T = \mathfrak{o}_{K,S}^\times \cap \mathfrak{o}_K$.

2. Let $\mathfrak{o}$ be a subring of K such that $K = \mathsf{q}(\mathfrak{o})$, and $\mathbb{F}_q \subset \mathfrak{o}$ in the function field case. Then its integral closure $\mathfrak{o}' = \mathrm{cl}_K(\mathfrak{o})$ is a holomorphy domain in K. Explicitly, if $S(\mathfrak{o}) = \{v \in \mathsf{M}_K^0 \mid \mathfrak{o} \subset \mathcal{O}_v\}$, then $\mathfrak{o}' = \mathfrak{o}_{K,\mathsf{M}_K^0\setminus S(\mathfrak{o})}$.

Proof. 1. This follows from [18, Theorem 2.7.2], since $\mathcal{C}(\mathfrak{o}_{K,S'}) \cong \mathcal{C}_{K,S'}$ is finite (see 5.5.6.2 and 5.5.2). The special assertion in the number field case follows with $S' = \emptyset$.

2. In the number field case we set $\mathfrak{O} = \mathfrak{o}_K$; then $\mathfrak{O} = \mathrm{cl}_K(\mathbb{Z}) \subset \mathrm{cl}_K(\mathfrak{o}) = \mathfrak{o}'$. In the function field case let $x \in \mathfrak{o} \setminus \mathbb{F}_q$, and set $\mathfrak{O} = \mathrm{cl}_K(\mathbb{F}_q[x])$; then again $\mathfrak{O} \subset \mathrm{cl}_K(\mathfrak{o}) = \mathfrak{o}'$. In either case $\mathfrak{O}$ is a Dedekind domain and $K = \mathsf{q}(\mathfrak{O})$, and thus $\mathfrak{o}'$ is a Dedekind domain (see [18, Theorem 2.10.3]). If $v \in \mathsf{M}_K^0$, then $\mathfrak{o} \subset \mathcal{O}_v$ if and only if $\mathfrak{o}' \subset \mathcal{O}_v$, and this holds if and only if $v = v_\mathfrak{p}$ for some $\mathfrak{p} \in \mathcal{P}(\mathfrak{o}')$, and then $\mathcal{O}_v = \mathfrak{o}'_\mathfrak{p}$ (see [18, Theorem 2.10.7]). Hence we obtain

$$\mathfrak{o}' = \bigcap_{\mathfrak{p}\in\mathcal{P}(\mathfrak{o}')} \mathfrak{o}'_\mathfrak{p} = \bigcap_{v\in S(\mathfrak{o})} \mathcal{O}_v = \mathcal{O}_{\mathsf{M}_K^0\setminus S(\mathfrak{o})}. \qquad \square$$

Theorem 5.5.8 (Extension of holomorphy domains). *Let L/K be a finite field extension. Let S be a subset of M_K such that $S \ne \emptyset$ in the function field case, and $S_L = \{\overline{v} \in \mathsf{M}_L \mid \overline{v}\downarrow K \in S\}$.*

1. $\mathfrak{o}_{L,S_L} = \mathrm{cl}_L(\mathfrak{o}_{K,S})$.
If $v \in \mathsf{M}_K^0$ and $\overline{v} \in \mathsf{M}_L$, then $\overline{v}\,|\,v$ if and only if $\mathfrak{p}_{S_L,\overline{v}} \cap K = \mathfrak{p}_{S,v}$; if this holds, then $f(\mathfrak{p}_{S_L,\overline{v}}/\mathfrak{p}_{S,v}) = f(\overline{v}/v)$ and $e(\mathfrak{p}_{S_L,\overline{v}}/\mathfrak{p}_{S,v}) = e(\overline{v}/v)$.

2. There exist commutative diagrams

$$\begin{array}{ccc} \mathcal{D}_{L,S_L} & \xrightarrow{\theta_{S_L}} & \mathcal{I}(\mathfrak{o}_{L,S_L}) \\ {\scriptstyle i}\uparrow & & \uparrow{\scriptstyle \nu} \\ \mathcal{D}_{K,S} & \xrightarrow{\theta_S} & \mathcal{I}(\mathfrak{o}_{K,S}) \end{array} \quad \text{and} \quad \begin{array}{ccc} \mathcal{D}_{L,S_L} & \xrightarrow{\theta_{S_L}} & \mathcal{I}(\mathfrak{o}_{L,S_L}) \\ {\scriptstyle \mathcal{N}_{L/K}}\downarrow & & \downarrow{\scriptstyle \mathcal{N}_{L/K,S}} \\ \mathcal{D}_{K,S} & \xrightarrow{\theta_S} & \mathcal{I}(\mathfrak{o}_{K,S}) \end{array},$$

where $i = i_{L/K} \restriction \mathcal{D}_{K,S} \colon \mathcal{D}_{K,S} \to \mathcal{D}_{L,S_L}$, $j(\mathfrak{a}) = \mathfrak{a}\mathfrak{o}_{L,S_L}$ for all $\mathfrak{a} \in \mathcal{I}(\mathfrak{o}_{K,S}$, $\mathcal{N}_{L/K,S} = \mathcal{N}_{\mathfrak{o}_{L,S_L}/\mathfrak{o}_{K,S}}$ and the horizontal arrows given by 5.5.6.2.

Proof. 1. For $v \in \mathsf{M}_K^0$ we set $\mathcal{O}'_v = \mathrm{cl}_L(\mathcal{O}_v)$. Then $\mathcal{O}'_v$ is a semilocal Dedekind domain satisfying $\mathsf{q}(\mathcal{O}'_v) = K$ (see [18, Theorem 2.12.4]). We assert:

A. If $v \in \mathsf{M}_K^0$, then $\mathcal{P}(\mathcal{O}'_v) = \{P_{\overline{v}} \cap \mathcal{O}'_v \mid \overline{v} \in \mathsf{M}_L,\ \overline{v}\,|\,v\}$, $(\mathcal{O}'_v)_{P_{\overline{v}} \cap \mathcal{O}'_v} = \mathcal{O}_{\overline{v}}$ for all $\overline{v} \in \mathsf{M}_L$ with $\overline{v}\,|\,v$, and

$$\mathcal{O}'_v = \bigcap_{\substack{\overline{v} \in \mathsf{M}_L \\ \overline{v}\,|\,v}} \mathcal{O}_{\overline{v}}.$$

Proof of **A.** Let $v \in \mathsf{M}_K^0$, and let $\overline{v} \in \mathsf{M}_L$ such that $\overline{v}\,|\,v$. Then $\mathcal{O}_v \subset \mathcal{O}_{\overline{v}}$ implies $\mathcal{O}'_v \subset \mathcal{O}_{\overline{v}}$, and there exists some $\mathfrak{P} \in \mathcal{P}(\mathcal{O}'_v)$ such that $\mathcal{O}_{\overline{v}} = (\mathcal{O}'_v)_{\mathfrak{P}}$ (see [18, Theorem 2.10.7]). Hence $P_{\overline{v}} = \mathfrak{P}(\mathcal{O}'_v)_{\mathfrak{P}}$ and $\mathfrak{P} = P_{\overline{v}} \cap \mathcal{O}'_v$.

Conversely, if $\mathfrak{P} \in \mathcal{P}(\mathcal{O}'_v)$, then $(\mathcal{O}'_v)_{\mathfrak{P}}$ is a discrete valuation domain (containing $\mathbb{F}_q$ in the function field case). Hence $\mathfrak{P}(\mathcal{O}'_v)_{\mathfrak{P}} = P_{\overline{v}}$ for some $\overline{v} \in \mathsf{M}_L^0$, $\mathcal{O}_{\overline{v}} = (\mathcal{O}'_v)_{\mathfrak{P}}$, $\mathfrak{P} = P_{\overline{v}} \cap \mathcal{O}'_v$, and $\mathcal{O}_{\overline{v}} \supset \mathcal{O}_v$ implies $\overline{v}\,|\,v$.

Since $\mathcal{P}(\mathcal{O}'_v) = \{P_{\overline{v}} \cap \mathcal{O}'_v \mid \overline{v} \in \mathsf{M}_L,\ \overline{v}\,|\,v\}$, it follows that

$$\mathcal{O}'_v = \bigcap_{\mathfrak{P} \in \mathcal{P}(\mathcal{O}'_v)} (\mathcal{O}'_v)_{\mathfrak{P}} = \bigcap_{\substack{\overline{v} \in \mathsf{M}_L \\ \overline{v}\,|\,v}} (\mathcal{O}'_v)_{P_{\overline{v}} \cap \mathcal{O}'_v} = \bigcap_{\substack{\overline{v} \in \mathsf{M}_L \\ \overline{v}\,|\,v}} \mathcal{O}_{\overline{v}}. \qquad \square[\mathbf{A.}]$$

Apparently $\mathfrak{o}_{K,S} \subset \mathfrak{o}_{L,S_L}$ implies $\mathrm{cl}_L(\mathfrak{o}_{K,S}) \subset \mathfrak{o}_{L,S_L}$. To prove the reverse inclusion, let $x \in \mathfrak{o}_{L,S_L}$ and $f \in K[X]$ the minimal polynomial of x over K. Since

$$\mathfrak{o}_{L,S_L} = \bigcap_{\overline{v} \in \mathsf{M}_L^0 \setminus S_L} \mathcal{O}_{\overline{v}} = \bigcap_{v \in \mathsf{M}_K^0 \setminus S} \ \bigcap_{\substack{\overline{v} \in \mathsf{M}_L \\ \overline{v}\,|\,v}} \mathcal{O}_{\overline{v}} = \bigcap_{v \in \mathsf{M}_K^0 \setminus S} \mathcal{O}'_v,$$

it follows that $f \in \mathcal{O}'_v[X]$ for all $v \in \mathsf{M}_K^0 \setminus S$, hence $f \in \mathfrak{o}_{K,S}[X]$ and thus $x \in \mathrm{cl}_L(\mathfrak{o}_{K,S})$.

If $v \in \mathsf{M}_K$ and $\overline{v} \in \mathsf{M}_L$, then

$$\overline{v}\,|\,v \iff P_{\overline{v}} \subset P_v \iff \mathfrak{p}_{S,v} \subset \mathfrak{p}_{S_L,\overline{v}} \iff \mathfrak{p}_{S,v} = \mathfrak{p}_{S_L,\overline{v}} \cap K.$$

If $\overline{v}\,|\,v$, then $\overline{v}$ is the $\mathfrak{p}_{S_L,\overline{v}}$-adic valuation of L and v is the $\mathfrak{p}_{S,v}$-adic valuation of K. Hence it follows that $f(\mathfrak{p}_{S_L,\overline{v}}/\mathfrak{p}_{S,v}) = f(L_{\overline{v}}/K_v) = f(\overline{v}/v)$ and $e(\mathfrak{p}_{S_L,\overline{v}}/\mathfrak{p}_{S,v}) = e(L_{\overline{v}}/K_v) = e(\overline{v}/v)$.

2. If $v \in \mathsf{M}_K^0 \setminus S$, then

$$\theta_{S_L} \circ i(\mathsf{p}_v) = \theta_{S_L}\Big(\prod_{\substack{\overline{v} \in \mathsf{M}_L \\ \overline{v}\,|\,v}} \mathsf{p}_{\overline{v}}^{e(\overline{v}/v)}\Big) = \prod_{\substack{\overline{v} \in \mathsf{M}_L \\ \overline{v}\,|\,v}} \mathfrak{p}_{S_L,\overline{v}}^{e(\mathfrak{p}_{S_L,\overline{v}}/\mathfrak{p}_{S,v})} = \mathfrak{p}_{S,v}\mathfrak{o}_{L,S_L} = \nu \circ \theta_S(\mathsf{p}_v),$$

and if $\overline{v} \in \mathsf{M}_L^0 \setminus S_L$, then

$$\theta_S \circ \mathcal{N}_{L/K}(\mathsf{p}_{\overline{v}}) = \theta_{K,S}(\mathsf{p}_v^{f(\overline{v}/v)}) = \mathfrak{p}_{S,v}^{f(\mathfrak{p}_{S_L,v}/\mathfrak{p}_{S,v})} = \mathcal{N}_{L/K,S} \circ \theta_{S_L}(\mathsf{p}_{\overline{v}}). \qquad \square$$

5.6 Ray class groups 1: Ideal- and divisor-theoretic approach

Let K be a global field.

We continue to use the notions and definitions introduced hitherto in this chapter. We introduce ray class groups in a thorough way. We aim at a complete characterization from the ideal-theoretic, the divisor-theoretic and the idel-theoretic point of view.

Definition 5.6.1. Let S be a finite subset of M_K^0 and $\mathsf{m} \in \mathcal{D}'_{K,S}$ an integral S-divisor. In the sequel, whenever we omit the suffix S, we mean $S = \emptyset$.

A. We set

- $\mathcal{D}_{K,S}^{\mathsf{m}} = \mathcal{D}_{K,S\cup\mathrm{supp}(\mathsf{m})}$, $\mathcal{D}_K^{\mathsf{m}} = \mathcal{D}_{K,\emptyset}^{\mathsf{m}}$ and $\mathcal{D}_K'^{\mathsf{m}} = \mathcal{D}_K^{\mathsf{m}} \cap \mathcal{D}'_K$; hence
  $\mathcal{D}_K^{\mathsf{m}} = \{\mathsf{a} \in \mathcal{D}_K \mid \mathrm{supp}(\mathsf{a}) \cap \sup(\mathsf{m}) = \emptyset\}$, $\mathcal{D}_{K,S} = \mathcal{D}_{K,S}^1$,
  $\mathcal{D}_{K,S}^{\mathsf{m}} = \{\mathsf{a} \in \mathcal{D}_K^{\mathsf{m}} \mid \mathrm{supp}(\mathsf{a}) \cap S = \emptyset\}$;
  $\mathcal{D}_{K,S}^{\mathsf{m},0} = \mathcal{D}_{K,S}^{\mathsf{m}} \cap \mathcal{D}_K^0$ and $\mathcal{D}_K^{\mathsf{m},0} = \mathcal{D}_K^{\mathsf{m}} \cap \mathcal{D}_K^0$ in the function field case.
- $K(\mathsf{m}) = \{a \in K^\times \mid (a) \in \mathcal{D}_K^{\mathsf{m}}\} = \{a \in K^\times \mid v(a) = 0 \text{ for all } v \in \mathrm{supp}(\mathsf{m})\}$;
- $K^+(\mathsf{m}) = K(\mathsf{m}) \cap K^+$;
  hence $\mathcal{D}_K^{\mathsf{m}} \cap (K^\times) = (K(\mathsf{m}))$ and $\mathcal{D}_K^{\mathsf{m}} \cap (K^+) = (K^+(\mathsf{m}))$.
- $K^{\circ\mathsf{m}} = \{a \in K^\times \mid v(a-1) \geq v(\mathsf{m}) \text{ for all } v \in \mathrm{supp}(\mathsf{m})\}$ and
  $K^{\mathsf{m}} = K^{\circ\mathsf{m}} \cap K^+$;
  in particular, $K(1) = K^{\circ 1} = K^\times$ and $K^+(1) = K^1 = K^+$.

For a subset T of $K^\times$, we set $(T)_S = \{(a)_S \mid a \in T\} \subset \mathcal{D}_{K,S}$ (see 5.5.1). We call

- $\mathcal{C}_{K,S}^{\mathsf{m}} = \mathcal{D}_{K,S}^{\mathsf{m}}/(K^{\mathsf{m}})_S$ the **S-ray class group** modulo m;
- $\mathcal{C}_{K,S}^{\circ\mathsf{m}} = \mathcal{D}_{K,S}^{\mathsf{m}}/(K^{\circ\mathsf{m}})_S$ the **small S-ray class group** modulo m;
- $\mathcal{C}_K^{\mathsf{m}} = \mathcal{C}_{K,\emptyset}^{\mathsf{m}} = \mathcal{D}_K^{\mathsf{m}}/(K^{\mathsf{m}})$ the **ray class group** modulo m;
- $\mathcal{C}_K^{\circ\mathsf{m}} = \mathcal{D}_K^{\mathsf{m}}/(K^{\circ\mathsf{m}})$ the **small ray class group** modulo m.

In particular:

- $\mathcal{C}_{K,S}^{\circ 1} = \mathcal{C}_{K,S} = \mathcal{D}_{K,S}/(K^\times)_S$ (the S-class group);
- $\mathcal{C}_{K,S}^1 = \mathcal{C}_{K,S}^+ = \mathcal{D}_{K,S}/(K^+)_S$ (the narrow S-class group);

- $\mathcal{C}_K^{\circ 1} = \mathcal{C}_K = \mathcal{D}_K/(K^\times)$ (the class group);
- $\mathcal{C}_K^{1} = \mathcal{C}_K^{+} = \mathcal{D}_K/(K^+)$ (the narrow class group).

Recall that in the number field case with $\mathsf{m} = \mathfrak{m} \in \mathcal{I}_K = \mathcal{D}_K$ we have $\mathcal{H}_K^{\mathfrak{m}} = (K^{\mathfrak{m}})$, $\mathcal{H}_K^{\circ\mathfrak{m}} = (K^{\circ\mathfrak{m}})$, $\mathcal{H}_K^{+} = \mathcal{H}_K^{1} = (K^+)$, $\mathcal{H}_K = \mathcal{H}_K^{\circ 1} = (K^\times)$, and correspondingly $\mathcal{C}_K^{\mathfrak{m}} = \mathcal{I}_K^{\mathfrak{m}}/\mathcal{H}_K^{\mathfrak{m}}$, $\mathcal{C}_K^{\circ\mathfrak{m}} = \mathcal{I}_K^{\mathfrak{m}}/\mathcal{H}_K^{\circ\mathfrak{m}}$, $\mathcal{C}_K = \mathcal{I}_K/\mathcal{H}_K$, $\mathcal{C}_K^{+} = \mathcal{I}_K/\mathcal{H}_K^{+}$ etc.

For $\mathsf{a} \in \mathcal{D}_{K,S}^{\mathsf{m}}$ we set $[\mathsf{a}]_S^{\mathsf{m}} = \mathsf{a}(K^{\mathsf{m}})_S \in \mathcal{C}_{K,S}^{\mathsf{m}}$ and $[\mathsf{a}]_S^{\circ\mathsf{m}} = \mathsf{a}(K^{\circ\mathsf{m}})_S \in \mathcal{C}_{K,S}^{\circ\mathsf{m}}$;
for $\mathsf{a} \in \mathcal{D}_K^{\mathsf{m}}$ we set $[\mathsf{a}]^{\mathsf{m}} = \mathsf{a}(K^{\mathsf{m}}) \in \mathcal{C}_K^{\mathsf{m}}$ and $[\mathsf{a}]^{\circ\mathsf{m}} = \mathsf{a}(K^{\circ\mathsf{m}}) \in \mathcal{C}_K^{\circ\mathsf{m}}$;
for $\mathsf{a} \in \mathcal{D}_{K,S}$ we set $[\mathsf{a}]_S = [\mathsf{a}]_S^{\circ 1} = \mathsf{a}(K^\times)_S$ and $[\mathsf{a}]_S^{+} = [\mathsf{a}]_S^{1} = \mathsf{a}(K^+)_S$;
for $\mathsf{a} \in \mathcal{D}_K$ let $[\mathsf{a}] = [\mathsf{a}]^{\circ 1} = \mathsf{a}(K^\times) \in \mathcal{C}_K$ and $[\mathsf{a}]^{+} = [\mathsf{a}]^{1} = \mathsf{a}(K^+) \in \mathcal{C}_K^{+}$.

B. Suppose that $S \ne \emptyset$ in the function field case, let $\theta_S \colon \mathcal{D}_{K,S} \to \mathcal{I}(\mathfrak{o}_{K,S})$ be the isomorphism defined in 5.5.6 and $\mathfrak{m} = \theta_S(\mathsf{m}) \in \mathcal{I}'(\mathfrak{o}_{K,S})$. We denote by

- $\mathcal{I}^{\mathfrak{m}}(\mathfrak{o}_{K,S})$ the group of fractional ideals of $\mathfrak{o}_{K,S}$ coprime to $\mathfrak{m}$;
- $\mathcal{H}^{\mathfrak{m}}(\mathfrak{o}_{K,S}) = \{a\mathfrak{o}_{K,S} \mid a \in K^{\mathsf{m}}\}$ the **principal ray** modulo $\mathfrak{m}$ in $\mathfrak{o}_{K,S}$;
- $\mathcal{H}^{\circ\mathfrak{m}}(\mathfrak{o}_{K,S}) = \{a\mathfrak{o}_{K,S} \mid a \in K^{\circ\mathsf{m}}\}$ the **large principal ray** modulo $\mathfrak{m}$ in $\mathfrak{o}_{K,S}$;
- $\mathcal{C}^{\mathfrak{m}}(\mathfrak{o}_{K,S}) = \mathcal{I}^{\mathfrak{m}}(\mathfrak{o}_{K,S})/\mathcal{H}^{\mathfrak{m}}(\mathfrak{o}_{K,S})$ the **narrow ray class group** modulo $\mathfrak{m}$ of $\mathfrak{o}_{K,S}$;
- $\mathcal{C}^{\circ\mathfrak{m}}(\mathfrak{o}_{K,S}) = \mathcal{I}^{\mathfrak{m}}(\mathfrak{o}_{K,S})/\mathcal{H}^{\circ\mathfrak{m}}(\mathfrak{o}_{K,S})$ the **(ordinary) ray class group** modulo $\mathfrak{m}$ of $\mathfrak{o}_{K,S}$;

In particular:

- $\mathcal{H}^{\circ 1}(\mathfrak{o}_{K,S}) = \mathcal{H}(\mathfrak{o}_{K,S})$ is the group of fractional principal ideals of $\mathfrak{o}_{K,S}$;
- $\mathcal{H}^{1}(\mathfrak{o}_{K,S}) = \mathcal{H}^{+}(\mathfrak{o}_{K,S})$ is the group of totally positive fractional principal ideals of $\mathfrak{o}_{K,S}$;
- $\mathcal{C}^{\circ 1}(\mathfrak{o}_{K,S}) = \mathcal{C}(\mathfrak{o}_{K,S}) = \mathcal{I}(\mathfrak{o}_{K,S})/\mathcal{H}(\mathfrak{o}_{K,S})$ is the (ideal) class group of $\mathfrak{o}_{K,S}$;
- $\mathcal{C}^{1}(\mathfrak{o}_{K,S}) = \mathcal{C}^{+}(\mathfrak{o}_{K,S}) = \mathcal{I}(\mathfrak{o}_{K,S})/\mathcal{H}^{+}(\mathfrak{o}_{K,S})$ is the narrow (ideal) class group of $\mathfrak{o}_{K,S}$.

C. Let $K/\mathbb{F}_q$ be a function field. Then $\deg(\mathcal{D}_K^{\mathsf{m}}) = \mathbb{Z}$. Indeed, let $\mathsf{c} \in \mathcal{D}_K$ be such that $\deg(\mathsf{c}) = 1$ and $a \in K^\times$ such that $v(a) = v(\mathsf{c})$ for all $v \in \operatorname{supp}(\mathsf{m})$ (by the weak approximation theorem); then $a^{-1}\mathsf{c} \in \mathcal{D}_K^{\mathsf{m}}$ and $\deg(a^{-1}\mathsf{c}) = 1$.

Hence deg induces an (equally denoted) epimorphism $\deg\colon \mathcal{C}_K^{\mathsf{m}} \to \mathbb{Z}$ satisfying $\deg([\mathsf{a}]^{\mathsf{m}}) = \deg(\mathsf{a})$ for all $\mathsf{a} \in \mathcal{D}_K^{\mathsf{m}}$, and we denote by

$$\mathcal{C}_K^{\mathsf{m},0} = \{\mathfrak{c} \in \mathcal{C}_K^{\mathsf{m}} \mid \deg(\mathfrak{c}) = 0\} = \mathcal{D}_K^{\mathsf{m},0}/(K^{\mathsf{m}})$$

the 0-**ray class group modulo** m of K. If $\mathfrak{c} \in \mathcal{C}_K^{\mathsf{m}}$ is any class satisfying $\deg(\mathfrak{c}) = 1$, then

$$\mathcal{C}_K^{\mathsf{m}} = \mathcal{C}_K^{\mathsf{m},0} \cdot \langle \mathfrak{c} \rangle \cong \mathcal{C}_K^{\mathsf{m},0} \times \mathbb{Z}.$$

Theorem 5.6.2. *Let S be a finite subset of M_K^0 such that $S \ne \emptyset$ in the function field case. Let $\theta_S\colon \mathcal{D}_{K,S} \to \mathcal{I}(\mathfrak{o}_{K,S})$ be the isomorphism defined in* 5.5.6 *and $\mathsf{m} \in \mathcal{D}'_{K,S}$. Then $\theta_S(\mathsf{m}) \in \mathcal{I}'(\mathfrak{o}_{K,S})$,*

$$\theta_S(\mathcal{D}_{K,S}^{\mathsf{m}}) = \mathcal{I}^{\mathsf{m}}(\mathfrak{o}_{K,S}), \quad \theta_S((K^{\mathsf{m}})_S) = \mathcal{H}^{\mathsf{m}}(\mathfrak{o}_{K,S}), \quad \theta_S((K^{\circ\mathsf{m}})_S) = \mathcal{H}^{\circ\mathsf{m}}(\mathfrak{o}_{K,S}),$$

and θ_S induces (equally denoted) isomorphisms

$$\theta_S\colon \mathcal{C}_{K,S}^{\mathsf{m}} \to \mathcal{C}^{\mathsf{m}}(\mathfrak{o}_{K,S}) \quad \text{and} \quad \theta_S\colon \mathcal{C}_{K,S}^{\circ\mathsf{m}} \to \mathcal{C}^{\circ\mathsf{m}}(\mathfrak{o}_{K,S}).$$

Proof. Obvious by the definitions. □

We proceed with a generalization of the simple observation that every ray class contains divisors coprime to a given one. This is the abstract counterpart to the corresponding well-known result for Dedekind domains (see [18, Theorems 2.9.2 and 2.9.9]).

Theorem 5.6.3. *Let S be a finite subset of M_K^0 and $\mathsf{m} \in \mathcal{D}'_{K,S}$.*

1. *Let S' be a finite subset of M_K^0 such that $S \subset S'$, and suppose that $S' \cap \operatorname{supp}(\mathsf{m}) = \emptyset$. Then*

$$\mathcal{C}_{K,S}^{\mathsf{m}} = \{[\mathsf{a}]_S^{\mathsf{m}} \mid \mathsf{a} \in \mathcal{D}_{K,S'}^{\mathsf{m}}\} = \mathcal{D}_{K,S'}/(K^{\mathsf{m}})_S \cap \mathcal{D}_{K,S'}^{\mathsf{m}}, \quad \textit{and}$$
$$\mathcal{C}_{K,S}^{\circ\mathsf{m}} = \mathcal{D}_{K,S'}/(K^{\circ\mathsf{m}})_S \cap \mathcal{D}_{K,S'}^{\mathsf{m}}.$$

In particular:
$\mathcal{C}_K^{\mathsf{m}} = \{[\mathsf{a}]^{\mathsf{m}} \mid \mathsf{a} \in \mathcal{D}_{K,S}^{\mathsf{m}}\} = \mathcal{D}_{K,S}^{\mathsf{m}}/\mathcal{D}_{K,S}^{\mathsf{m}} \cap (K^{\mathsf{m}})$ *and* $\mathcal{C}_K^{\circ\mathsf{m}} = \mathcal{D}_{K,S}^{\mathsf{m}}/\mathcal{D}_{K,S}^{\mathsf{m}} \cap (K^{\circ\mathsf{m}})$.

2. *Let $\mathsf{m}' \in \mathcal{D}'_{K,S}$ be such that $\mathsf{m}' \mid \mathsf{m}$. Then*

$$\mathcal{C}_{K,S}^{\mathsf{m}} = \{[\mathsf{a}]_S^{\mathsf{m}} \mid \mathsf{a} \in \mathcal{D}_{K,S}^{\mathsf{m}'}\} = \mathcal{D}_{K,S}^{\mathsf{m}'}/(K^{\mathsf{m}})_S \cap \mathcal{D}_{K,S}^{\mathsf{m}'}, \quad \textit{and}$$
$$\mathcal{C}_{K,S}^{\circ\mathsf{m}} = \mathcal{D}_{K,S}^{\mathsf{m}'}/(K^{\circ\mathsf{m}})_S \cap \mathcal{D}_{K,S}^{\mathsf{m}'}.$$

$$\mathcal{C}_{K,S}^{\mathsf{m}'} = \{[\mathsf{a}]_S^{\mathsf{m}'} \mid \mathsf{a} \in \mathcal{D}_{K,S}^{\mathsf{m}}\} = \mathcal{D}_{K,S}^{\mathsf{m}}/(K^{\mathsf{m}'})_S \cap \mathcal{D}_{K,S}^{\mathsf{m}}, \quad \textit{and}$$
$$\mathcal{C}_{K,S}^{\circ\mathsf{m}'} = \mathcal{D}_{K,S}^{\mathsf{m}}/(K^{\circ\mathsf{m}'})_S \cap \mathcal{D}_{K,S}^{\mathsf{m}}.$$

In particular:
$\mathcal{C}_{K,S}^{+} = \{[\mathsf{a}]_S^{+} \mid \mathsf{a} \in \mathcal{D}_{K,S}^{\mathsf{m}}\} = \mathcal{D}_{K,S}^{\mathsf{m}}/(K^{+})_S \cap \mathcal{D}_{K,S}^{\mathsf{m}}$, *and*
$\mathcal{C}_{K,S} = \mathcal{D}_{K,S}^{\mathsf{m}}/(K^{\times})_S \cap \mathcal{D}_{K,S}^{\mathsf{m}}$.

3. $\mathcal{C}_K^+ = \{[\mathsf{a}]^+ \mid \mathsf{a} \in \mathcal{D}_K^{\mathsf{m}}\} = \mathcal{D}_K^{\mathsf{m}}/(K^+(\mathsf{m}))$, *and* $\mathcal{C}_K = \mathcal{D}_K^{\mathsf{m}}/(K(\mathsf{m}))$.

4. Assume that $S \ne \emptyset$ *in the function field case, and let* $\theta_S \colon \mathcal{D}_{K,S} \to \mathcal{I}(\mathfrak{o}_{K,S})$ *be the isomorphism defined in* 5.5.6.
If $\mathfrak{m} = \theta_S(\mathsf{m})$, *then*

$$\mathcal{C}^+(\mathfrak{o}_{K,S}) = \mathcal{I}^{\mathfrak{m}}(\mathfrak{o}_{K,S})/\mathcal{H}^{\mathfrak{m}}(\mathfrak{o}_{K,S}), \quad \textit{and} \quad \mathcal{C}(\mathfrak{o}_{K,S}) = \mathcal{I}^{\mathfrak{m}}(\mathfrak{o}_{K,S})/\mathcal{H}^{\circ\mathfrak{m}}(\mathfrak{o}_{K,S}).$$

Proof. 1. Let $\mathfrak{c} = [\mathsf{c}]_S^{\mathsf{m}} \in \mathcal{C}_{K,S}^{\mathsf{m}} = \mathcal{D}_{K,S}^{\mathsf{m}}/(K^{\mathsf{m}})_S$, where $\mathsf{c} \in \mathcal{D}_{K,S}^{\mathsf{m}}$. By the weak approximation theorem there exists some $a \in K^\times$ such that $v(a-1) \ge v(\mathsf{m})$ for all $v \in \operatorname{supp}(\mathsf{m})$, $v(a) = v(\mathsf{c})$ for all $v \in S'$ and $|a-1|_v < 1$ for all real $v \in \mathsf{M}_K^\infty$. Then $a \in K^{\mathsf{m}}$, and $v(a^{-1}\mathsf{c}) = 0$ for all $v \in S' \cup \operatorname{supp}(\mathsf{m})$, hence $a^{-1}\mathsf{c} \in \mathcal{D}_{K,S'}^{\mathsf{m}}$, and $\mathfrak{c} = [a^{-1}\mathsf{c}]^{\mathsf{m}}$.

Consequenly $\mathcal{C}_{K,S}^{\mathsf{m}} = \{[\mathsf{a}]^{\mathsf{m}} \mid \mathsf{a} \in \mathcal{D}_{K,S'}^{\mathsf{m}}\} = \mathcal{D}_{K,S'}^{\mathsf{m}}/\mathcal{D}_{K,S'}^{\mathsf{m}} \cap (K^{\mathsf{m}})_S$, and the assertion concerning $\mathcal{C}_K^{\circ\mathsf{m}}$ follows if we disregard the real places.

The particular assertion follows with $(\emptyset, S)$ instead of (S, S').

2. Let $\mathfrak{c} = [\mathsf{c}]_S^{\mathsf{m}} \in \mathcal{C}_{K,S}^{\mathsf{m}} = \mathcal{D}_{K,S}^{\mathsf{m}}/(K^{\mathsf{m}})_S$, where $\mathsf{c} \in \mathcal{D}_{K,S}^{\mathsf{m}}$. By the weak approximation theorem there exists some $a \in K^\times$ such that $|a-1|_v < 1$ for all real $v \in \mathsf{M}_K^\infty$, $v(a-1) \ge v(\mathsf{m})$ for all $v \in \operatorname{supp}(\mathsf{m})$ and $v(a) = v(\mathsf{c})$ for all $v \in S \cup \operatorname{supp}(\mathsf{m}') \setminus \operatorname{supp}(\mathsf{m})$. Then $a \in K^{\mathsf{m}}$, $a^{-1}\mathsf{c} \in \mathcal{D}_{K,S}^{\mathsf{m}'}$, and $\mathfrak{c} = [a^{-1}\mathsf{c}]_S^{\mathsf{m}}$.

Consequently, $\mathcal{C}_{K,S}^{\mathsf{m}} = \{[\mathsf{a}]_S^{\mathsf{m}} \mid \mathsf{a} \in \mathcal{D}_{K,S}^{\mathsf{m}'}\} = \mathcal{D}_{K,S}^{\mathsf{m}'}/(K^{\mathsf{m}})_S \cap \mathcal{D}_{K,S}^{\mathsf{m}'}$, and the assertion concerning $\mathcal{C}_{K,S}^{\circ\mathsf{m}}$ follows if we disregard the real places.

The particular assertion follows with $(1, \mathsf{m})$ instead of $(\mathsf{m}, \mathsf{m}')$.

3. This follows by the particular case in 2., applied with $S = \emptyset$, since $\mathcal{D}_K^{\mathsf{m}} \cap (K^\times) = (K(\mathsf{m}))$ and $\mathcal{D}_K^{\mathsf{m}} \cap (K^+) = (K^+(\mathsf{m}))$.

4. By 5.6.2 and the particular case in 2. we obtain

$$\begin{aligned}\mathcal{C}^+(\mathfrak{o}_{K,S}) &= \theta_S(\mathcal{C}_{K,S}^+) = \theta_S(\mathcal{D}_{K,S}^{\mathsf{m}})/\theta_S(\mathcal{D}_{K,S}^{\mathsf{m}}) \cap \theta_S((K^+)_S) \\ &= \mathcal{I}^{\mathfrak{m}}(\mathfrak{o}_{K,S})/\mathcal{I}^{\mathfrak{m}}(\mathfrak{o}_{K,S}) \cap \mathcal{H}^+(\mathfrak{o}_{K,S}) = \mathcal{I}^{\mathfrak{m}}(\mathfrak{o}_{K,S})/\mathcal{H}^{\mathfrak{m}}(\mathfrak{o}_{K,S})\end{aligned}$$

and

$$\begin{aligned}\mathcal{C}(\mathfrak{o}_{K,S}) &= \theta_S(\mathcal{C}_{K,S}) = \theta_S(\mathcal{D}_{K,S}^{\mathsf{m}})/\theta_S(\mathcal{D}_{K,S}^{\mathsf{m}}) \cap \theta_S((K^\times)_S) \\ &= \mathcal{I}^{\mathfrak{m}}(\mathfrak{o}_{K,S})/\mathcal{I}^{\mathfrak{m}}(\mathfrak{o}_{K,S}) \cap \mathcal{H}(\mathfrak{o}_{K,S}) = \mathcal{I}^{\mathfrak{m}}(\mathfrak{o}_{K,S})/\mathcal{H}^{\circ\mathfrak{m}}(\mathfrak{o}_{K,S}).\end{aligned}$$ □

Remark 5.6.4 (Ray class groups in $\mathbb{Q}$). Recall that $\mathsf{M}_{\mathbb{Q}}^\infty = \{\infty\}$, $\mathbb{Q}_\infty = \mathbb{R}$, $\mathbb{Q}^+ = \mathbb{Q}_{>0} = \mathsf{q}(\mathbb{N})$ and $\mathfrak{o}_{\mathbb{Q}} = \mathbb{Z}$. The map $\mathbb{Q}^+ \to \mathcal{D}_{\mathbb{Q}} = \mathcal{I}_{\mathbb{Q}}$, $a \mapsto \mathbb{Z}a$, is a group isomorphism, and we identify: $\mathcal{D}_{\mathbb{Q}} = \mathbb{Q}^+$. Then $\mathcal{D}'_{\mathbb{Q}} = \mathbb{N}$, $(\mathbb{Q}^\times) = \mathbb{Q}^+$ is the set of principal divisors and $\mathcal{C}_{\mathbb{Q}} = \mathbf{1}$.

1. For $m \in \mathbb{N}$ we denote by $\mathbb{Z}(m) = \{a \in \mathbb{Z} \mid (a, m) = 1\}$ the monoid of all integers coprime to m, and by

$$\mathbb{Q}(m) = \mathsf{q}(\mathbb{Z}(m)) = \{a^{-1}b \mid a \in \mathbb{N},\ b \in \mathbb{Z},\ a \equiv 1 \mod m,\ (b, m) = 1\}$$

its quotient group. By virtue of quotients we extended the residue class epimorphism $\pi_m\colon \mathbb{Z}(m) \to (\mathbb{Z}/m\mathbb{Z})^\times$ to a (uniquely determined and equally denoted) group epimorphism $\pi_m\colon \mathbb{Q}(m) \to (\mathbb{Z}/m\mathbb{Z})^\times$ with kernel

$$\mathbb{Q}^{\circ m} = \mathsf{q}(1+m\mathbb{Z}) = \{a^{-1}b \mid a \in \mathbb{N},\ b \in \mathbb{Z},\ a \equiv b \equiv 1 \bmod m\}.$$

Indeed, if $x \in \mathbb{Q}(m)$ and $\pi_m(x) = 1 + m\mathbb{Z}$, then $x = a^{-1}b$, where $a,\, b \in \mathbb{Z}(m)$ and $a \equiv b \bmod m$, hence $x \in \mathsf{q}(1+m\mathbb{Z})$. The reverse inclusion is obvious.

In particular, π_m induces an isomorphism $\mathbb{Q}(m)/\mathbb{Q}^{\circ m} \overset{\sim}{\to} (\mathbb{Z}/m\mathbb{Z})^\times$

Now let $m \in \mathbb{N} = \mathcal{D}'_{\mathbb{Q}}$. By virtue of the identification $\mathcal{D}_{\mathbb{Q}} = \mathbb{Q}^+$ we obtain

$$\mathcal{D}_{\mathbb{Q}}^m = \mathbb{Q}^+(m) = \mathbb{Q}(m) \cap \mathbb{Q}^+, \quad (\mathbb{Q}^m) = \mathbb{Q}^m = \mathbb{Q}^{\circ m} \cap \mathbb{Q}^+,$$

and

$$(\mathbb{Q}^{\circ m}) = \mathbb{Q}_{\pm}^m = \{a \in \mathbb{Q}(m) \cap \mathbb{Q}^+ \mid a \equiv \pm 1 \bmod m\}.$$

The residue class epimorphism $\pi_m\colon \mathbb{Q}(m) \to (\mathbb{Z}/m\mathbb{Z})^\times$ satisfies

$$\pi_m(\mathbb{Q}^+(m)) = (\mathbb{Z}/m\mathbb{Z})^\times, \quad \pi_m(\mathbb{Q}^m) = 1 + m\mathbb{Z} \quad \text{and} \quad \pi_m(\mathbb{Q}_{\pm}^m) = \pm 1 + m\mathbb{Z}.$$

Hence π_m induces isomorphisms

$$\mathcal{C}_{\mathbb{Q}}^m = \mathbb{Q}^+(m)/\mathbb{Q}^m \overset{\sim}{\to} (\mathbb{Z}/m\mathbb{Z})^\times \quad \text{and}$$
$$\mathcal{C}_{\mathbb{Q}}^{\circ m} = \mathbb{Q}^+(m)/\mathbb{Q}_{\pm}^m \overset{\sim}{\to} (\mathbb{Z}/m\mathbb{Z})^\times/\{\pm 1 + m\mathbb{Z}\}.$$

We identify: $\mathcal{C}_{\mathbb{Q}}^m = (\mathbb{Z}/m\mathbb{Z})^\times$ and $\mathcal{C}_{\mathbb{Q}}^{\circ m} = (\mathbb{Z}/m\mathbb{Z})^\times/\{\pm1+m\mathbb{Z}\}$. In particular, $|\mathcal{C}_{\mathbb{Q}}^m| = \varphi(m)$ and $|\mathcal{C}_{\mathbb{Q}}^{\circ m}| = \varphi(m)/2$

2. Let $S = \{p_1, \ldots, p_k\}$ be a finite set of primes not dividing m, and suppose that $s = p_1 \cdot \ldots \cdot p_k$. Then $\mathcal{D}_{\mathbb{Q},S}^m = \mathbb{Q}^+(sm)$,

$$\pi_m(\mathbb{Q}^+(sm)) = \mathbb{Q}^+(sm)/\mathbb{Q}^m \cap \mathbb{Q}^+(sm) = \{a + m\mathbb{Z} \mid (a, sm) = 1\} = (\mathbb{Z}/m\mathbb{Z})^\times,$$

hence $\mathcal{C}_{\mathbb{Q}}^m = \mathbb{Q}^+(sm)/\mathbb{Q}^m \cap \mathbb{Q}^+(sm)$, and $\mathcal{C}_{\mathbb{Q},S}^m = \mathbb{Q}^+(sm)/(\mathbb{Q}^m)_S$. We assert that

$$(\mathbb{Q}^m)_S = \langle p_1, \ldots, p_k\rangle \mathbb{Q}^m \cap \mathbb{Q}^+(sm). \tag{†}$$

Proof of (†). If $c = (a)_S$, where $a \in \mathbb{Q}^m$, then

$$c = \prod_{p \in \mathbb{P} \setminus S} p^{v_p(a)} = \prod_{i=1}^k p_i^{-v_{p_i}(a)} a \in \langle p_1, \ldots, p_k\rangle \mathbb{Q}^m \cap \mathbb{Q}^+(sm).$$

Conversely, suppose that $c = p_1^{\nu_1} \cdot \ldots \cdot p_k^{\nu_k} a \in \langle p_1, \ldots, p_k\rangle \mathbb{Q}^m \cap \mathbb{Q}^+(sm)$, where $\nu_1, \ldots, \nu_k \in \mathbb{Z}$ and $a \in \mathbb{Q}^m$. Then $a = p_1^{-\nu_1} \cdot \ldots \cdot p_k^{-\nu_k} u$, where $(u, s) = 1$, and $c = (a)_S$. $\square$[(†)].

Since $\mathbb{Q}^m \cap \mathbb{Q}^+(sm) \subset \langle p_1, \ldots, p_k\rangle \mathbb{Q}^m \cap \mathbb{Q}^+(sm) \subset \mathbb{Q}^+(sm)$, we obtain a (natural) epimorphism

$$\mathcal{C}_{\mathbb{Q}}^m = \mathbb{Q}^+(sm)/\mathbb{Q}^m \cap \mathbb{Q}^+(sm) \to \mathbb{Q}^+(sm)/\langle p_1, \ldots, p_k\rangle \mathbb{Q}^m \cap \mathbb{Q}^+(sm) = \mathcal{C}_{\mathbb{Q},S}^m$$

with kernel

$$\langle p_1,\ldots,p_k\rangle\mathbb{Q}^m\cap\mathbb{Q}^+(sm)/\mathbb{Q}^m\cap\mathbb{Q}^+(sm)\ \cong\ \pi_m(\langle p_1,\ldots,p_k\rangle)\subset(\mathbb{Z}/m\mathbb{Z})^\times.$$

It follows that

$$|\mathcal{C}^m_{\mathbb{Q},S}|\ =\ \frac{\varphi(m)}{|\langle\pi_m(p_1),\ldots,\pi_m(p_k)\rangle|},$$

and similarly

$$|\mathcal{C}^{\circ m}_{\mathbb{Q},S}|\ =\ \frac{\varphi(m)}{|\langle\pi_m(-1),\pi_m(p_1),\ldots,\pi_m(p_k)\rangle|}\ =\ \frac{|\mathcal{C}^m_{\mathbb{Q},S}|}{|\langle\pi_m(-1)\rangle|}.$$

Let us finally look at the corresponding holomorphy domain

$$\mathfrak{o}_{\mathbb{Q},S}\ =\ \mathbb{Z}_S\ =\ \langle S\rangle^{-1}\mathbb{Z}\ =\ \mathbb{Z}\Big[\frac{1}{s}\Big].$$

The isomorphism $\theta_S\colon\mathbb{Q}^+(ms)\ \tilde{\to}\ \mathcal{J}^m(\mathbb{Z}_S)$ is given by $\theta_S(a)=a\mathbb{Z}_S$ for all $a\in\mathbb{Q}^+(ms)$.

3. Let K be an algebraic number field. If $\mathfrak{a}\in\mathcal{D}_K=\mathcal{J}_K$, then it follows that $\mathcal{N}_{K/\mathbb{Q}}\in\mathcal{J}_\mathbb{Q}=\mathbb{Q}_{>0}$, and consequently $\mathcal{N}_{K/\mathbb{Q}}(\mathfrak{a})=\mathfrak{N}(\mathfrak{a})$.

We continue with an investigation of the (small) ray class group modulo m (Theorem 5.6.6), and afterwards and we connect the (small) S-ray class group module m with the (small) ray class group modulo m (Theorem 5.6.7).

Definition and Remarks 5.6.5. Let $\sigma_1,\ldots,\sigma_{r_1}:K\to\mathbb{R}$ be the real embeddings of K and $s\colon K^\times\to\{\pm1\}^{r_1}$ the signature map, defined by

$$s(a)=(\operatorname{sgn}\sigma_1(a),\ldots,\operatorname{sgn}\sigma_{r_1}(a)).$$

Then $s\colon K^\times\to\{\pm1\}$ is an epimorphism with kernel K^+ (see [18, Theorem 2.9.7]). For a non-empty subgroup U of $K^\times$ we obtain

$$|U/K^+\cap U|=|K^+U/K^+|=|s(U)|=2^{\sigma(U)},\quad\text{where }\ \sigma(U)\in[0,r_1],$$

and $U/K^+\cap U$ is an elementary-abelian 2-group of rank $\sigma(U)$. We call $\sigma(U)$ the **signature rank** of U.

Theorem 5.6.6. *Let* $\mathsf{m}\in\mathcal{D}'_K$. *We set* $U_K^{\circ\mathsf{m}}=U_K\cap K^{\circ\mathsf{m}}$, $U_K^{\mathsf{m}}=U_K\cap K^{\mathsf{m}}$ *and* $U_K^+=U_K^1=U_K\cap K^+$.

1. *There exists an exact sequence*

$$\mathbf{1}\ \to\ K^{\circ\mathsf{m}}/U_K^{\circ\mathsf{m}}K^{\mathsf{m}}\ \overset{\delta}{\to}\ \mathcal{C}_K^{\mathsf{m}}\ \overset{\rho}{\to}\ \mathcal{C}_K^{\circ\mathsf{m}}\ \to\ \mathbf{1},$$

where $\delta(xU_K^{\circ\mathsf{m}}K^{\mathsf{m}})=[(x)]^{\mathsf{m}}$ *for all* $x\in K^{\circ\mathsf{m}}$ *and* $\rho([\mathsf{a}]^{\mathsf{m}})=[\mathsf{a}]^{\circ\mathsf{m}}$. *The group* $K^{\circ\mathsf{m}}/U_K^{\circ\mathsf{m}}K^{\mathsf{m}}$ *is an elementary-abelian 2-group of rank* $\sigma(K^{\circ\mathsf{m}})-\sigma(U^{\circ\mathsf{m}})$.

2. There exist (natural) exact sequences

$$\mathbf{1} \to U_K^+/U_K^{\mathsf{m}} \to K^+(\mathsf{m})/K^{\mathsf{m}} \to \mathcal{C}_K^{\mathsf{m}} \to \mathcal{C}_K^+ \to \mathbf{1},$$

$$\mathbf{1} \to U_K/U_K^{\circ\mathsf{m}} \to K(\mathsf{m})/K^{\circ\mathsf{m}} \to \mathcal{C}_K^{\circ\mathsf{m}} \to \mathcal{C}_K \to \mathbf{1},$$

$$(K^+(\mathsf{m})\colon K^{\mathsf{m}}) = (K(\mathsf{m})\colon K^{\circ\mathsf{m}}) = \prod_{v\in\mathrm{supp}(\mathsf{m})} \mathfrak{N}(v)^{v(\mathsf{m})-1}(\mathfrak{N}(v)-1),$$

and in the function field case there exists further the exact sequence

$$\mathbf{1} \to U_K/U_K^{\mathsf{m}} \to K(\mathsf{m})/K^{\mathsf{m}} \to \mathcal{C}_K^{\mathsf{m},0} \to \mathcal{C}_K^0 \to \mathbf{1}.$$

In particular, the groups $\mathcal{C}_K^{\mathsf{m}}$ and $\mathcal{C}_K^{\circ\mathsf{m}}$ in the number field case and $\mathcal{C}_K^{\mathsf{m},0}$ in the function field case are finite.

Proof. 1. Since $\mathcal{C}_K^{\mathsf{m}} = \mathcal{D}_K^{\mathsf{m}}/(K^{\mathsf{m}})$, $\mathcal{C}_K^{\circ\mathsf{m}} = \mathcal{D}_K^{\mathsf{m}}/(K^{\circ\mathsf{m}})$ and $K^{\mathsf{m}} \subset K^{\circ\mathsf{m}}$, there exists an epimorphism $\rho\colon \mathcal{C}_K^{\mathsf{m}} \to \mathcal{C}_K^{\circ\mathsf{m}}$ such that $\rho([\mathsf{a}]^{\mathsf{m}}) = [\mathsf{a}]^{\circ\mathsf{m}}$ for all $\mathsf{a} \in \mathcal{D}_K^{\mathsf{m}}$, and $\mathrm{Ker}(\rho) = (K^{\circ\mathsf{m}})/(K^{\mathsf{m}})$. The natural epimorphism $K^{\circ\mathsf{m}} \to (K^{\circ\mathsf{m}})/(K^{\mathsf{m}})$ has the kernel $U_K^{\circ\mathsf{m}}K^{\mathsf{m}}$, and consequently we obtain an exact sequence as announced.

Since

$$K^{\circ\mathsf{m}}/U_K^{\circ\mathsf{m}}K^{\mathsf{m}} \cong (K^{\circ\mathsf{m}}/K^{\mathsf{m}})/(U_K^{\circ\mathsf{m}}K^{\mathsf{m}}/K^{\mathsf{m}}), \;\; K^{\circ\mathsf{m}}/K^{\mathsf{m}} = K^{\circ\mathsf{m}}/K^+ \cap K^{\circ\mathsf{m}}$$

and $U_K^{\circ\mathsf{m}}K^{\mathsf{m}}/K^{\mathsf{m}} \cong U_K^{\circ\mathsf{m}}/U_K^{\circ\mathsf{m}} \cap K^{\mathsf{m}} = U_K^{\circ\mathsf{m}}/K^+ \cap U_K^{\circ\mathsf{m}}$, it follows that $K^{\circ\mathsf{m}}/U_K^{\circ\mathsf{m}}K^{\mathsf{m}}$ is an elementary-abelian 2-group of rank $\sigma(K^{\circ\mathsf{m}}) - \sigma(U^{\circ\mathsf{m}})$.

2. As $\mathcal{C}_K^+ = \mathcal{D}_K^{\mathsf{m}}/(K^+(\mathsf{m}))$ by 5.6.3.2 and $\mathcal{C}_K^{\mathsf{m}} = \mathcal{D}_K^{\mathsf{m}}/(K^{\mathsf{m}})$, there exists an epimorphism $\alpha\colon \mathcal{C}_K^{\mathsf{m}} \to \mathcal{C}_K^+$ such that $\alpha([\mathsf{a}]^{\mathsf{m}}) = [\mathsf{a}]^+$ for all $\mathsf{a} \in \mathcal{D}_K^{\mathsf{m}}$, and appartently $\mathrm{Ker}(\alpha) = (K^+(\mathsf{m}))/(K^{\mathsf{m}})$. Now we define

$$\beta\colon K^+(\mathsf{m})/K^{\mathsf{m}} \to (K^+(\mathsf{m}))/(K^{\mathsf{m}}) \text{ by } \beta(aK^{\mathsf{m}}) = (a)(K^{\mathsf{m}}).$$

Then β is an epimorphism, and

$$\mathrm{Ker}(\beta) = K^{\mathsf{m}}U_K^+/K^{\mathsf{m}} = U_K^+/U_K^+ \cap K^{\mathsf{m}} = U_K^+/U_K^{\mathsf{m}}.$$

This gives the first exact sequence.

As $\mathcal{C}_K = \{[\mathsf{a}] \mid \mathsf{a} \in \mathcal{D}_K^{\mathsf{m}}\} = \mathcal{D}_K^{\mathsf{m}}/(K^\times) \cap \mathcal{D}_K^{\mathsf{m}} = \mathcal{D}_K^{\mathsf{m}}/(K(\mathsf{m}))$ (by 1., applied with $\mathsf{m} = 1$), and $\mathcal{C}_K^{\circ\mathsf{m}} = \mathcal{D}_K^{\mathsf{m}}/K^{\circ\mathsf{m}}$, there exists an epimorphism $\alpha^\circ\colon \mathcal{C}_K^{\circ\mathsf{m}} \to \mathcal{C}_K$ such that $\alpha^\circ([\mathsf{a}]^{\circ\mathsf{m}}) = [\mathsf{a}]$ for all $\mathsf{a} \in \mathcal{D}_K^{\mathsf{m}}$, and apparently $\mathrm{Ker}(\alpha^\circ) = (K(\mathsf{m}))/(K^{\circ\mathsf{m}})$. We define $\beta^\circ\colon K(\mathsf{m})/K^{\circ\mathsf{m}} \to (K(\mathsf{m}))/(K^{\circ\mathsf{m}})$ by $\beta^\circ(aK^{\circ\mathsf{m}}) = (a)(K^{\circ\mathsf{m}})$ for all $a \in K(\mathsf{m})$. Then β° is an epimorphism, and $\mathrm{Ker}(\beta^\circ) = K^{\circ\mathsf{m}}U_K/K^{\circ\mathsf{m}} = U_K/U_K \cap K^{\circ\mathsf{m}} = U_K/U_K^{\circ\mathsf{m}}$. This gives the second exact sequence.

To calculate $(K(\mathsf{m})\colon K^{\circ\mathsf{m}})$, we consider the homomorphism

$$\theta_0\colon K(\mathsf{m}) \to \mathsf{U}(\mathsf{m}) = \prod_{v\in\mathrm{supp}(\mathsf{m})} U_v/U_v^{v(\mathsf{m})}, \quad \text{defined by}$$

$$\theta_0(a) = (aU_v^{(v(\mathsf{m})})_{v\in\mathrm{supp}(\mathsf{m})}.$$

θ_0 is surjective. Indeed if $(uU_v^{v(\mathsf{m})})_{v\in\mathrm{supp}(\mathsf{m})} \in \mathsf{U}(\mathsf{m})$, then by the weak approximation theorem there exists some $a \in K^\times$ such that $v(a - u_v) \geq v(\mathsf{m})$ for all $v \in \mathrm{supp}(\mathsf{m})$. Then $a \in K(\mathsf{m})$, and since $v(au_v^{-1} - 1) \geq v(\mathsf{m})$, it follows that $aU_v^{(v(\mathsf{m})} = u_vU_v^{(v(\mathsf{m})}$ for all $v \in \mathrm{supp}(\mathsf{m})$. Obviously, $\mathrm{Ker}(\theta_0) = K^{\circ\mathsf{m}}$, θ_0 induces an isomorphism $\theta\colon K(\mathsf{m})/K^{\circ\mathsf{m}} \to \mathsf{U}(\mathsf{m})$, and

$$(K(\mathsf{m})\colon K^{\circ\mathsf{m}}) = |\mathsf{U}(\mathsf{m})| = \prod_{v\in\mathrm{supp}(\mathsf{m})} |U_v/U_v^{(v(\mathsf{m})}| = \prod_{v\in\mathrm{supp}(\mathsf{m})} \mathfrak{N}(v)^{v(\mathsf{m})-1}(\mathfrak{N}(v)-1)$$

by 5.5.6.3.

Next we prove that $K^+(\mathsf{m})K^{\circ\mathsf{m}} = K(\mathsf{m})$. Clearly, $K^+(\mathsf{m})K^{\circ\mathsf{m}} \subset K(\mathsf{m})$. Thus let $a \in K(\mathsf{m})$. By the weak approximation theorem there exists some $x \in K^\times$ such that $v(x-1) \geq v(\mathsf{m})$ for all $v \in \mathrm{supp}(\mathsf{m})$, and $|x-a|_v < |a|_v$ for all real $v \in \mathsf{M}_K^\infty$. Then $|xa^{-1} - 1|_v < 1$ for all real $v \in \mathsf{M}_K^\infty$, hence $xa^{-1} \in K^+(\mathsf{m})$, and therefore we obtain $ax^{-1} \in K^+(\mathsf{m})$, $x \in K^{\circ m}$ and $a = (ax^{-1})x \in K^+(\mathsf{m})K^{\circ\mathsf{m}}$.

Now it follows that

$$K^+(\mathsf{m})/K^{\mathsf{m}} = K^+(\mathsf{m})/K^+(\mathsf{m}) \cap K^{\circ\mathsf{m}} \cong K^+(\mathsf{m})K^{\circ\mathsf{m}}/K^{\circ\mathsf{m}} = K(\mathsf{m})/K^{\circ\mathsf{m}},$$

and consequently $(K^+(\mathsf{m})\colon K^{\mathsf{m}}) = (K(\mathsf{m})\colon K^{\circ\mathsf{m}})$.

In the function field case, the restriction to divisor classes of degree 0 yields the asserted exact sequence. Since $K^+(\mathsf{m})/K^{\mathsf{m}}$ is finite, the finiteness of $\mathcal{C}_K^{\mathsf{m}}$ in the number field case follows from the finiteness of $\mathcal{C}_K^+$, and that of $\mathcal{C}_K^{\mathsf{m},0}$ in the function field case follows from the finiteness of $\mathcal{C}_K^0$. □

Theorem 5.6.7. *Let S be a finite subset of M_K^0 and $\mathsf{m} \in \mathcal{D}'_{K,S}$.*

1. $(K^{\mathsf{m}})_S = \mathcal{D}_{K,S}^{\mathsf{m}} \cap (K^{\mathsf{m}})\mathcal{D}_{K,\mathsf{M}_K^0\setminus S}$, *there exists an epimorphism*

$$\Theta\colon \mathcal{C}_K^{\mathsf{m}} \to \mathcal{C}_{K,S}^{\mathsf{m}} \text{ such that } \Theta([\mathsf{a}]^{\mathsf{m}}) = [\mathsf{a}]_S^{\mathsf{m}} \text{ for all } \mathsf{a} \in \mathcal{D}_{K,S}^{\mathsf{m}},$$

and $\mathrm{Ker}(\Theta) = (K^{\mathsf{m}})_S/\mathcal{D}_{K,S}^{\mathsf{m}} \cap (K^{\mathsf{m}})$.

2. *Suppose that $S \neq \emptyset$.*

(a) *There exists the (natural) exact sequence*

$$\mathbf{1} \to K^\times/K^+\mathfrak{o}_{K,S}^\times \to \mathcal{C}_{K,S}^+ \to \mathcal{C}_{K,S} \to \mathbf{1},$$

and $K^\times/K^+\mathfrak{o}_{K,S}^\times \cong (\mathbb{Z}/2\mathbb{Z})^t$, where $t = r_1 - \sigma(\mathfrak{o}_{K,S}^\times)$. In particular,

$$|\mathcal{C}_{K,S}^+| = |\mathcal{C}_{K,S}|2^t < \infty.$$

If $r_1 = 0$, then $\mathcal{C}_{K,S} = \mathcal{C}_{K,S}^+$. If $r_1 \geq 1$, then $t \in [0, r_1 - 1]$.

(b) Let $\theta_S\colon \mathcal{D}_{K,S} \to \mathcal{J}(\mathfrak{o}_{K,S})$ be the isomorphism defined in 5.5.6 and $\mathfrak{m} = \theta_S(\mathsf{m})$. There exists the (natural) exact sequence

$$\mathbf{1} \to \mathfrak{o}_{K,S}^\times \cap K^+/\mathfrak{o}_{K,S}^\times \cap K^{\mathsf{m}} \to (\mathfrak{o}_{K,S}/\mathfrak{m})^\times \to \mathcal{C}_{K,S}^{\mathsf{m}} \to \mathcal{C}_{K,S}^+ \to \mathbf{1},$$

and the group $\mathcal{C}_{K,S}^{\mathsf{m}}$ is finite.

Proof. 1. By definition, $(K^{\mathsf{m}})_S \subset \mathcal{D}_{K,S}^{\mathsf{m}}$. Thus let $\mathsf{a} \in \mathcal{D}_{K,S}^{\mathsf{m}}$. If $\mathsf{a} = (a)_S \in (K^{\mathsf{m}})_S$, where $a \in K^{\mathsf{m}}$, then

$$\mathsf{a} = \prod_{v \in \mathsf{M}_K^0 \setminus S} \mathsf{p}_v^{v(a)} = (a) \prod_{v \in S} \mathsf{p}_v^{-v(a)} \in (K^{\mathsf{m}})\mathcal{D}_{K,\mathsf{M}_K^0 \setminus S}.$$

Conversely, let $\mathsf{a} = (a)\mathsf{c} \in (K^{\mathsf{m}})\mathcal{D}_{K,\mathsf{M}_K^0 \setminus S}$, where $a \in K^{\mathsf{m}}$ and $\mathsf{c} \in \mathcal{D}_{K,\mathsf{M}_K^0 \setminus S}$. By definition, $(a) = (a)_S\mathsf{b}$, where $(a)_S \in \mathcal{D}_{K,S}^{\mathsf{m}}$ and $\mathsf{b} \in \mathcal{D}_{K,\mathsf{M}_K^0 \setminus S}^{\mathsf{m}}$. Hence we obtain

$$(a) = \mathsf{a}\mathsf{c}^{-1} = (a)_S\mathsf{b}, \quad \text{where} \quad \mathsf{a},\, (a)_S \in \mathcal{D}_{K,S}^{\mathsf{m}}, \quad \text{and} \quad \mathsf{c}^{-1},\, \mathsf{b} \in \mathcal{D}_{K,\mathsf{M}_K^0 \setminus S}^{\mathsf{m}}.$$

Since $\mathcal{D}_K^{\mathsf{m}} = \mathcal{D}_{K,S}^{\mathsf{m}} \cdot \mathcal{D}_{K,\mathsf{M}_K^0 \setminus S}^{\mathsf{m}}$, it follows that $\mathsf{a} = (a)_S \in (K^{\mathsf{m}})_S$.

Recall that $\mathcal{C}_K^{\mathsf{m}} = \mathcal{D}_{K,S}^{\mathsf{m}}/\mathcal{D}_{K,S}^{\mathsf{m}} \cap (K^{\mathsf{m}})$ (by 5.6.3.1) and that $\mathcal{C}_{K,S}^{\mathsf{m}} = \mathcal{D}_{K,S}^{\mathsf{m}}/(K^{\mathsf{m}})_S$. Since $\mathcal{D}_{K,S}^{\mathsf{m}} \cap (K^{\mathsf{m}}) \subset (K^{\mathsf{m}})_S$, there exists a (natural) epimorphism $\Theta\colon \mathcal{C}_K^{\mathsf{m}} \to \mathcal{C}_{K,S}^{\mathsf{m}}$ such that $\Theta([\mathsf{a}]^{\mathsf{m}}) = [\mathsf{a}]_S^{\mathsf{m}}$ for all $\mathsf{a} \in \mathcal{D}_{K,S}^{\mathsf{m}}$, and $\operatorname{Ker}(\Theta) = (K^{\mathsf{m}})_S/\mathcal{D}_{K,S}^{\mathsf{m}} \cap (K^{\mathsf{m}})$.

2. (a) The epimorphism

$$\mathcal{C}^+(\mathfrak{o}_{K,S}) \to \mathcal{C}(\mathfrak{o}_{K,S}), \quad \mathfrak{a}\mathcal{H}^+(\mathfrak{o}_{K,S}) \mapsto \mathfrak{a}\mathcal{H}^+(\mathfrak{o}_{K,S})$$

has the kernel $\mathcal{H}(\mathfrak{o}_{K,S})/\mathcal{H}^+(\mathfrak{o}_{K,S})$, and the epimorphism $K^\times \to \mathcal{H}(\mathfrak{o}_{K,S})/\mathcal{H}^+(\mathfrak{o}_{K,S})$, given by $a \mapsto (a\mathfrak{o}_{K,S})\mathcal{H}^+(\mathfrak{o}_{K,S}$, has the kernel $K^+\mathfrak{o}_{K,S}^\times$. This yields the exact sequence

$$\mathbf{1} \to K^\times/K^+\mathfrak{o}_{K,S}^\times \to \mathcal{C}^+(\mathfrak{o}_{K,S}) \to \mathcal{C}(\mathfrak{o}_{K,S} \to \mathbf{1}.$$

Since $\mathcal{C}^+(\mathfrak{o}_{K,S}) \cong \mathcal{C}_{K,S}^+$ and $\mathcal{C}(\mathfrak{o}_{K,S}) \cong \mathcal{C}_{K,S}$ by 5.6.2, the asserted exact sequence follows.

By means of the isomorphisms $K^\times/K^+\mathfrak{o}_{K,S}^\times \cong (K^\times/K^+)/(K^+\mathfrak{o}_{K,S}^\times/K^+)$ and $K^+\mathfrak{o}_{K,S}^\times/K^+ = \mathfrak{o}_{K,S}^\times/K^+ \cap \mathfrak{o}_{K,S}^\times$, it follows that $K^\times/K^+\mathfrak{o}_{K,S}^\times$ is an elementary abelian 2-group of rank $t = r_1 - \sigma(\mathfrak{o}_{K,S}^\times)$. If $r_1 = 0$, the clearly $t = 0$ and $\mathcal{C}_{K,S}^+ = \mathcal{C}_{K,S}$. If $r_1 \geq 1$, then $\sigma(\mathfrak{o}_{K,S}^\times) \geq 1$, since $\{1, -1\} \subset \mathfrak{o}_{K,S}^\times$, and consequently $t \in [0, r_1 - 1]$.

By 5.5.2, the group $\mathcal{C}_{K,S}$ is finite, and therefore $|\mathcal{C}_{K,S}^+| = |\mathcal{C}_{K,S}|2^t < \infty$.

(b) Since $\mathcal{C}^{\mathsf{m}}(\mathfrak{o}_{K,S}) \cong \mathcal{C}_{K,S}^{\mathsf{m}}$ and $\mathcal{C}^+(\mathfrak{o}_{K,S}) \cong \mathcal{C}_{K,S}^+$ by 5.6.2, it suffices to prove the exactness of the (natural) exact sequence

$$\mathbf{1} \to \mathfrak{o}_{K,S}^\times \cap K^+/\mathfrak{o}_{K,S}^\times \cap K^{\mathsf{m}} \to (\mathfrak{o}_{K,S}/\mathfrak{m})^\times \to \mathcal{C}^{\mathsf{m}}(\mathfrak{o}_{K,S}) \to \mathcal{C}^+(\mathfrak{o}_{K,S}) \to \mathbf{1}.$$

We only outline the main ideas of a proof and refer to [18, Theorem 2.9.9] for more details. Since $\mathcal{C}^+(\mathfrak{o}_{K,S}) = \mathcal{J}^{\mathfrak{m}}(\mathfrak{o}_{K,S})/\mathcal{H}^{\mathfrak{m}}(\mathfrak{o}_{K,S})$ by 5.6.3.4, we get an epimorphism $\mathcal{C}^{\mathfrak{m}}(\mathfrak{o}_{K,S}) \to \mathcal{C}^+(\mathfrak{o}_{K,S})$ with kernel $\mathcal{H}^+(\mathfrak{o}_{K,S})/\mathcal{H}^{\mathfrak{m}}(\mathfrak{o}_{K,S})$. Let $\pi_{\mathfrak{m}}\colon K^+(\mathfrak{m}) \to (\mathfrak{o}_{K,S}/\mathfrak{m})^\times$ be the residue class epimorphism, and define $\psi\colon (\mathfrak{o}_{K,S}/\mathfrak{m})^\times \to \mathcal{H}^+(\mathfrak{o}_{K,S})/\mathcal{H}^{\mathfrak{m}}(\mathfrak{o}_{K,S})$ by $\psi(\pi_{\mathfrak{m}}(a)) = (a\mathfrak{o}_{K,S})\mathcal{H}^{\mathfrak{m}}(\mathfrak{o}_{K,S})$. Then ψ is an epimorphism, it does no depend on a, $\pi_{\mathfrak{m}}^{-1}(\mathrm{Ker}(\psi)) = \mathfrak{o}_{K,S}^\times \cap K^+$, and $\mathrm{Ker}(\pi_{\mathfrak{m}} \restriction \mathfrak{o}_{K,S}^\times \cap K^+) = \mathfrak{o}_{K,S}^\times \cap K^{\mathfrak{m}}$. □

5.7 Ray class groups 2: Idelic approach

Let K be a global field.

We continue to use the notations of the preceding sections.

Definition 5.7.1. Let S be a finite subset of M_K^0 and $\mathsf{m} \in \mathcal{D}'_{K,S}$.

1. We define

$$\mathsf{J}_{K,S}^{\mathsf{m}} = \prod_{v \in \mathsf{M}_K^\infty} U_v \times \prod_{v \in S} K_v^\times \times \prod_{v \in \mathsf{M}_K^0 \setminus S} U_v^{(v(\mathsf{m}))}, \quad \mathsf{J}_{K,S}^{\circ\mathsf{m}} = \prod_{v \in \mathsf{M}_K^\infty \cup S} K_v^\times \times \prod_{v \in \mathsf{M}_K^0 \setminus S} U_v^{(v(\mathsf{m}))}$$

and

$$\mathsf{J}_{K,S}^{(\mathsf{m})} = \prod_{v \in \mathsf{M}_K^\infty} U_v \times \prod_{v \in S} K_v^\times \times \prod_{v \in \mathrm{supp}(\mathsf{m})} U_v^{(v(\mathsf{m}))} \times \prod_{v \in \mathsf{M}_K^0 \setminus (S \cup \mathrm{supp}(\mathsf{m}))} (K_v^\times \mid U_v).$$

$\mathsf{J}_{K,S}^{\mathsf{m}}$, $\mathsf{J}_{K,S}^{\circ\mathsf{m}}$ and $\mathsf{J}_{K,S}^{(\mathsf{m})}$ are open subgroups of J_K, $\mathsf{J}_{K,S}^{\mathsf{m}} = \mathsf{J}_{K,S}^{(\mathsf{m})} \cap \mathsf{J}_{K,S \cup \mathsf{M}_K^\infty}$, $\mathsf{J}_{K,S}^{(\mathsf{m})} \cap K^\times = K^{\mathsf{m}}$, $\mathsf{J}_{K,S}^{1} = \mathsf{J}_{K,S}$, and $\mathsf{J}_{K,S}^{\circ 1} = \mathsf{J}_{K,S \cup \mathsf{M}_K^\infty}$. Note that $\mathsf{J}_{K,S}^{\circ\mathsf{m}} = \mathsf{J}_{K,S}^{\mathsf{m}}$ if K admits no real embeddings (and in particular in the function field case).

The groups $\mathsf{C}_{K,S}^{\mathsf{m}} = \mathsf{J}_{K,S}^{\mathsf{m}} K^\times / K^\times$ and $\mathsf{C}_{K,S}^{\circ\mathsf{m}} = \mathsf{J}_{K,S}^{\circ\mathsf{m}} K^\times / K^\times$ are open subgroups of C_K. We call $\mathsf{C}_K/\mathsf{C}_{K,S}^{\mathsf{m}}$ the **(idelic) S-ray class group** and $\mathsf{C}_K/\mathsf{C}_{K,S}^{\circ\mathsf{m}}$ the **small (idelic) S-ray class group** modulo m. As before in all cases, if $S = \emptyset$ we omit the suffix $\emptyset$. Then $\mathsf{J}_K^{\mathsf{m}} \subset \mathsf{J}_K^{\circ\mathsf{m}} \subset \mathsf{J}_K^{\circ 1} = \mathsf{E}_K$ and $\mathsf{J}_K^{\mathsf{m}} \subset \mathsf{J}_K^1$, hence $\mathsf{C}_K^{\mathsf{m}} \subset \mathsf{C}_K^{\circ\mathsf{m}} \subset \mathsf{C}_K^{\circ 1} = \mathcal{E}_K$ and $\mathsf{C}_K^{\mathsf{m}} \subset \mathsf{C}_K^1$.

In addition, in the function field case we get $\mathsf{J}_K^{\mathsf{m}} \subset \mathsf{J}_K^1 = \mathsf{E}_K \subset \mathsf{J}_K^0$ and thus also $\mathsf{C}_K^{\mathsf{m}} \subset \mathsf{C}_K^1 = \mathcal{E}_K \subset \mathsf{C}_K^0$.

2. We define

$$\iota_S\colon \mathsf{J}_K \to \mathcal{D}_{K,S} \quad \text{by} \quad \iota_S(\alpha) = \prod_{v \in \mathsf{M}_K^0 \setminus S} \mathsf{p}_v^{v(\alpha_v)} \quad \text{if } \alpha = (\alpha_v)_{v \in \mathsf{M}_K} \in \mathsf{J}_K,$$

and we set

$$\iota_S^{(\mathsf{m})} = \iota_S \restriction \mathsf{J}_{K,S}^{(\mathsf{m})}\colon \mathsf{J}_{K,S}(\mathsf{m}) \to \mathcal{D}_{K,S}^{\mathsf{m}}.$$

Then ι_S and $\iota_S^{(\mathsf{m})}$ are epimorphisms, $\operatorname{Ker}(\iota_S) = \mathsf{J}_{K,S\cup\mathsf{M}_K^\infty}$, $\operatorname{Ker}(\iota_S^{(\mathsf{m})}) = \mathsf{J}_{K,S}^{\mathsf{m}}$, and $\iota_S^{(\mathsf{m})}(a\alpha) = (a)_S\iota_S^{\mathsf{m}}(\alpha)$ for all $a \in K(\mathsf{m})$ and $\alpha \in \mathsf{J}_{K,S}^{(\mathsf{m})}$. If we endow $\mathcal{D}_{K,S}^{\mathsf{m}}$ with the discrete topology, then $\iota_S^{(\mathsf{m})}$ is continuous.
If $S = \emptyset$, then $\iota^{(\mathsf{m})} = \iota_\emptyset^{(\mathsf{m})} = \iota_K \restriction \mathsf{J}_K^{(\mathsf{m})} \colon \mathsf{J}_K^{(\mathsf{m})} \to \mathcal{D}_K^{\mathsf{m}}$.

Theorem 5.7.2. *Let S be a finite subset of M_K^0 and $\mathsf{m} \in \mathcal{D}'_{K,S}$.*

1. $\mathsf{J}_{K,S}^{(\mathsf{m})} \cap \mathsf{J}_{K,S}^{\mathsf{m}}K^\times = \mathsf{J}_{K,S}^{\mathsf{m}}K^{\mathsf{m}}$, $\mathsf{J}_K = \mathsf{J}_{K,S}^{(\mathsf{m})}K^\times$, *and there exist* (*natural*) *isomorphisms* $\mathsf{C}_K \cong \mathsf{J}_{K,S}^{(\mathsf{m})}/K^{\mathsf{m}}$, $\mathsf{C}_{K,S}^{\mathsf{m}} \cong \mathsf{J}_{K,S}^{\mathsf{m}}K^{\mathsf{m}}/K^{\mathsf{m}}$,

$$\mathsf{C}_K/\mathsf{C}_{K,S}^{\mathsf{m}} \cong \mathsf{J}_K/\mathsf{J}_{K,S}^{\mathsf{m}}K^\times \cong \mathsf{J}_{K,S}^{(\mathsf{m})}/\mathsf{J}_{K,S}^{\mathsf{m}}K^{\mathsf{m}},$$

$\mathsf{C}_K/\mathsf{C}_{K,S}^{\circ\mathsf{m}} \cong \mathsf{J}_K/\mathsf{J}_{K,S}^{\circ\mathsf{m}}K^\times$ *and* $\mathsf{C}_{K,S}^{\circ\mathsf{m}}/\mathsf{C}_{K,S}^{\mathsf{m}} \cong \mathsf{J}_{K,S}^{\circ\mathsf{m}}K^\times/\mathsf{J}_{K,S}^{\mathsf{m}}K^\times$. We identify.

2. *There exists a unique isomorphism*

$$\iota_S^{\mathsf{m}} \colon \mathsf{C}_K/\mathsf{C}_{K,S}^{\mathsf{m}} = \mathsf{J}_{K,S}^{(\mathsf{m})}/\mathsf{J}_{K,S}^{\mathsf{m}}K^{\mathsf{m}} \overset{\sim}{\to} \mathcal{C}_{K,S}^{\mathsf{m}}$$

satisfying $\iota_S^{\mathsf{m}}(\alpha\mathsf{J}_{K,S}^{\mathsf{m}}K^{\mathsf{m}}) = [\iota_S^{(\mathsf{m})}(\alpha)]_S^{\mathsf{m}} \in \mathcal{C}_{K,S}^{\mathsf{m}}$ *for all* $\alpha \in \mathsf{J}_{K,S}^{(\mathsf{m})}$ (*where* $\iota_S^{(\mathsf{m})}$ *is given in* 5.7.1).

(a) *If $S \ne \emptyset$ in the function field case, then $\mathsf{C}_{K,S}^{\mathsf{m}}$ is an open subgroup of C_K of finite index* $(\mathsf{C}_K : \mathsf{C}_{K,S}^{\mathsf{m}}) = |\mathcal{C}_{K,S}^{\mathsf{m}}| < \infty$.

(b) *Let $K/\mathbb{F}_q$ be a function field. The epimorphism* $\deg\colon \mathsf{C}_K \to \mathbb{Z}$ (see 5.3.2.4) *induces an* (*equally denoted*) *epimorphism*

$$\deg\colon \mathsf{C}_K/\mathsf{C}_K^{\mathsf{m}} \to \mathbb{Z} \quad \textit{with kernel} \quad \mathsf{C}_K^0/\mathsf{C}_K^{\mathsf{m}}.$$

If $c \in \mathsf{C}_K$, *then* $\deg(c\mathsf{C}_K^{\mathsf{m}}) = \deg(c) = \deg(\iota^{\mathsf{m}}(c\mathsf{C}_K^{\mathsf{m}}))$. *In particular,* $\mathsf{C}_K/\mathsf{C}_K^{\mathsf{m}} \cong \mathsf{C}_K^0/\mathsf{C}_K^{\mathsf{m}} \times \mathbb{Z}$, $\iota^{\mathsf{m}}(\mathsf{C}_K^0/\mathsf{C}_K^{\mathsf{m}}) = \mathcal{C}_K^{\mathsf{m},0}$, *and* $|\mathsf{C}_K^0/\mathsf{C}_K^{\mathsf{m}}| = |\mathcal{C}_K^{\mathsf{m},0}| < \infty$.

3. *Let S' be a finite subset of M_K^0 and $\mathsf{m}' \in \mathcal{D}'_{K,S'}$ such that $S' \supset S$, $\mathsf{m}' \mid \mathsf{m}$ and $\mathsf{m} \in \mathcal{D}'_{K,S'}$. Then* $\mathsf{J}_{K,S}^{\mathsf{m}} \subset \mathsf{J}_{K,S'}^{\mathsf{m}'}$, $\mathsf{C}_{K,S}^{\mathsf{m}} \subset \mathsf{C}_{K,S'}^{\mathsf{m}'}$, *and there exist* (*natural*) *epimorphisms* $\varpi\colon \mathsf{C}_K/\mathsf{C}_{K,S}^{\mathsf{m}} \to \mathsf{C}_K/\mathsf{C}_{K,S'}^{\mathsf{m}'}$ *and* $\pi\colon \mathcal{C}_{K,S}^{\mathsf{m}} \to \mathcal{C}_{K,S'}^{\mathsf{m}'}$ *such that the following diagram commutes.*

$$\begin{array}{ccc}
\mathsf{C}_K/\mathsf{C}_{K,S}^{\mathsf{m}} & \xrightarrow{\ \varpi\ } & \mathsf{C}_K/\mathsf{C}_{K,S'}^{\mathsf{m}'} \\
\Big\downarrow{\scriptstyle \iota_S^{\mathsf{m}}} & & \Big\downarrow{\scriptstyle \iota_{S'}^{\mathsf{m}'}} \\
\mathcal{C}_{K,S}^{\mathsf{m}} & \xrightarrow{\ \pi\ } & \mathcal{C}_{K,S'}^{\mathsf{m}'} \\
\Big\| & & \Big\| \\
\mathcal{D}_{K,S'}^{\mathsf{m}}/(K^{\mathsf{m}})_{S'} \cap \mathcal{D}_{K,S'}^{\mathsf{m}} & & \mathcal{D}_{K,S'}^{\mathsf{m}}/(K^{\mathsf{m}'})_{S'} \cap \mathcal{D}_{K,S'}
\end{array}$$

Explicitly, $\varpi(\mathsf{c}\mathsf{C}^{\mathsf{m}}_{K,S}) = \mathsf{c}\mathsf{C}^{\mathsf{m}'}_{K,S'}$ *for all* $\mathsf{c} \in \mathsf{C}_K$, $\pi([\mathsf{a}]^{\mathsf{m}}_S) = [\mathsf{a}]^{\mathsf{m}'}_{S'}$ *for all* $\mathsf{a} \in \mathcal{D}^{\mathsf{m}}_{K,S'}$, $\operatorname{Ker}(\varpi) = \mathsf{C}^{\mathsf{m}'}_{K',S'}/\mathsf{C}^{\mathsf{m}}_{K,S}$,

$$Ker(\pi) = (K^{\mathsf{m}'})_{S'} \cap \mathcal{D}^{\mathsf{m}}_{K,S'}/(K^{\mathsf{m}})_{S'} \cap \mathcal{D}^{\mathsf{m}}_{K,S'},$$

and ι^{m}_S *induces an isomorphism*

$$\iota^{\mathsf{m}}_S \restriction \cdots : \mathsf{C}^{\mathsf{m}'}_{K',S'}/\mathsf{C}^{\mathsf{m}}_{K,S} \overset{\sim}{\to} (K^{\mathsf{m}'})_{S'} \cap \mathcal{D}^{\mathsf{m}}_{K,S'}/(K^{\mathsf{m}})_{S'} \cap \mathcal{D}^{\mathsf{m}}_{K,S'}.$$

In particular, if $S = S' = \emptyset$, *then we obtain an isomorphism*

$$\iota^{\mathsf{m}} \restriction \cdots : \mathsf{C}^{\mathsf{m}'}_K/\mathsf{C}^{\mathsf{m}}_K \overset{\sim}{\to} (K^{\mathsf{m}'}) \cap \mathcal{D}^{\mathsf{m}}_K/(K^{\mathsf{m}})$$

4. Let K *be a number field with at least one real embedding. Then*

$$\iota^{\mathsf{m}}_S(\mathsf{C}^{\circ\mathsf{m}}_{K,S}/\mathsf{C}^{\mathsf{m}}_{K,S}) = (K^{\circ\mathsf{m}})_S/(K^{\mathsf{m}})_S,$$

and ι^{m}_S *induces an isomorpism* $\iota^{\circ\mathsf{m}}_S \colon \mathsf{C}_K/\mathsf{C}^{\circ\mathsf{m}}_{K,S} \overset{\sim}{\to} \mathcal{C}^{\circ\mathsf{m}}_{K,S}$.

Proof. 1. $\mathsf{J}^{\mathsf{m}}_{K,S} \subset \mathsf{J}^{(\mathsf{m})}_{K,S}$ implies $\mathsf{J}^{(\mathsf{m})}_{K,S} \cap \mathsf{J}^{\mathsf{m}}_{K,S}K^\times = \mathsf{J}^{\mathsf{m}}_{K,S}(\mathsf{J}^{(\mathsf{m})}_{K,S} \cap K^\times) = \mathsf{J}^{\mathsf{m}}_{K,S}K^{\mathsf{m}}$.

For the proof of $\mathsf{J}_K = \mathsf{J}^{(\mathsf{m})}_{K,S}K^\times$ let $\alpha = (\alpha_v)_{v\in\mathsf{M}_K} \in \mathsf{J}_K$. By the weak approximation theorem let $a \in K^\times$ such that $v(a - \alpha_v) \ge v(\alpha_v) + v(\mathsf{m})$ for all $v \in \operatorname{supp}(\mathsf{m})$, and $|a - \alpha_v|_v < |\alpha_v|_v$ for all real $v \in \mathsf{M}^\infty_K$. If $v \in \operatorname{supp}(\mathsf{m})$, then $v(\alpha_v^{-1}a - 1) \ge v(\mathsf{m})$, and consequently $\alpha_v^{-1}a \in U_v^{(v(\mathsf{m}))}$. If $v \in \mathsf{M}^\infty_K$ is real, then $|\alpha_v^{-1}a - 1|_v < 1$, hence $\alpha_v^{-1}a > 0$, and consequently $\alpha_v^{-1}a \in U_v$. It follows that $\alpha^{-1}a \in \mathsf{J}^{(\mathsf{m})}_{K,S}$, hence $\alpha a^{-1} \in \mathsf{J}^{(\mathsf{m})}_{K,S}$ and $\alpha \in \mathsf{J}^{(\mathsf{m})}_{K,S}K^\times$.

Now we obtain

$$\mathsf{C}_K = \mathsf{J}_K/K^\times = \mathsf{J}^{(\mathsf{m})}_{K,S}K^\times/K^\times \cong \mathsf{J}^{(\mathsf{m})}_{K,S}/K^\times \cap \mathsf{J}^{(\mathsf{m})}_{K,S} = \mathsf{J}^{(\mathsf{m})}_{K,S}/K^{\mathsf{m}}$$

and

$$\mathsf{C}^{\mathsf{m}}_{K,S} = \mathsf{J}^{\mathsf{m}}_{K,S}K^\times/K^\times \cong \mathsf{J}^{\mathsf{m}}_{K,S}/\mathsf{J}^{\mathsf{m}}_{K,S} \cap K^\times = \mathsf{J}^{\mathsf{m}}_{K,S}/\mathsf{J}^{\mathsf{m}}_{K,S} \cap K^{\mathsf{m}} \cong \mathsf{J}^{\mathsf{m}}_{K,S}K^{\mathsf{m}}/K^{\mathsf{m}}.$$

The remaining isomorphisms follow by an application of the second isomorphism theorem.

2. Uniqueness is clear. We define $j \colon \mathsf{J}^{(\mathsf{m})}_{K,S} \to \mathcal{C}^{\mathsf{m}}_{K,S}$ by $j(\alpha) = [\iota^{(\mathsf{m})}_S(\alpha)]^{\mathsf{m}}_S$. Then j is an epimorphism, and it suffices to prove that $\operatorname{Ker}(j) = K^{\mathsf{m}}\mathsf{J}^{\mathsf{m}}_{K,S}$.

Let $\alpha \in \mathsf{J}^{(\mathsf{m})}_{K,S}$. Since $(K^{\mathsf{m}})_S = \mathcal{D}^{\mathsf{m}}_{K,S} \cap (K^{\mathsf{m}})\mathcal{D}_{K,\mathsf{M}^0_K\setminus S}$ (by 5.6.7.1), $\operatorname{Im}(\iota^{(\mathsf{m})}_S) \subset \mathcal{D}^{\mathsf{m}}_{K,S}$ and $\mathcal{D}^{\mathsf{m}}_{K,S} \cap \mathcal{D}_{K,\mathsf{M}^0_K\setminus S} = \mathbf{0}$, we obtain:

$$\begin{aligned} \alpha \in \operatorname{Ker}(j) &\iff \iota^{(\mathsf{m})}_S(\alpha) \in (K^{\mathsf{m}})_S \iff \iota^{(\mathsf{m})}_S(\alpha) \in (K^{\mathsf{m}})\mathcal{D}_{K,\mathsf{M}^0_K\setminus S} \\ &\iff (a)^{-1}\iota^{(\mathsf{m})}_S(\alpha) = \iota^{(\mathsf{m})}_S(a^{-1}\alpha) \in \mathcal{D}_{K,\mathsf{M}^0_K\setminus S} \text{ for some } a \in K^{\mathsf{m}} \\ &\iff a^{-1}\alpha \in \operatorname{Ker}(\iota^{(\mathsf{m})}_S) = \mathsf{J}^{\mathsf{m}}_{K,S} \text{ for some } a \in K^{\mathsf{m}} \\ &\iff \alpha \in K^{\mathsf{m}}\mathsf{J}^{\mathsf{m}}_{K,S}. \end{aligned}$$

(a) $\mathsf{C}^{\mathsf{m}}_{K,S}$ is an open subgroup of C_K, since $\mathsf{J}^{\mathsf{m}}_{K,S}$ is an open subgroup of J_K. Apparently, $(\mathsf{C}_K\!:\!\mathsf{C}^{\mathsf{m}}_{K,S}) = |\mathcal{C}^{\mathsf{m}}_{K,S}|$, and $\mathcal{C}^{\mathsf{m}}_{K,S}$ is finite by 5.6.7.2(b).

(b) As C^0_K is the kernel of $\deg\colon \mathsf{C}_K \to \mathbb{Z}$ and $\mathsf{C}^{\mathsf{m}}_K \subset \mathsf{C}^0_K$, deg induces an epimorphism $\deg\colon \mathsf{C}_K/\mathsf{C}^{\mathsf{m}}_K \to \mathbb{Z}$ with kernel $\mathsf{C}^0_K/\mathsf{C}^{\mathsf{m}}_K$. It splits, and consequently $\mathsf{C}_K/\mathsf{C}^{\mathsf{m}}_K \cong \mathsf{C}^0_K/\mathsf{C}^{\mathsf{m}}_K\times\mathbb{Z}$. If $c = \alpha K^{\mathsf{m}} \in \mathsf{C}_K = \mathsf{J}^{(\mathsf{m})}_K/K^{\mathsf{m}}$, where $\alpha \in \mathsf{J}^{(\mathsf{m})}_K$, then, by definition and 5.3.2.4(b), we obtain

$$\deg(\mathsf{c}\mathsf{C}^{\mathsf{m}}_K) = \deg(\mathsf{c}) = \deg(\alpha) = \deg(\iota_K(\alpha)) = \deg(\iota^{(\mathsf{m})}(\alpha)) = \deg(\iota^{\mathsf{m}}(\mathsf{c}\mathsf{C}^{\mathsf{m}}_K)).$$

In particular, $\iota^{\mathsf{m}}(\mathsf{C}^0_K/\mathsf{C}^{\mathsf{m}}_K) = \mathcal{C}^{\mathsf{m},0}_K$, and the latter group is finite by 5.6.6.2.

3. Let $\alpha = (\alpha_v)_{v\in\mathsf{M}_K} \in \mathsf{J}^{\mathsf{m}}_{K,S}$. Then $\alpha_v \in U_v$ for all $v \in \mathsf{M}^\infty_K$, $\alpha_v \in U_v^{(v(\mathsf{m}))}$ for all $v \in \mathsf{M}^0_K \setminus S$ and thus $\alpha_v \in U_v^{(v(\mathsf{m}'))}$ for all $v \in \mathsf{M}^0_K \setminus S'$, which holds since $\mathsf{M}^0_K \setminus S \supset \mathsf{M}^0_K \setminus S'$ and $v(\mathsf{m}') \le v(\mathsf{m})$ for all $v \in \mathsf{M}^0_K$. It follows that $\alpha \in \mathsf{J}^{\mathsf{m}'}_{K,S'}$, and therefore $\mathsf{J}^{\mathsf{m}}_{K,S} \subset \mathsf{J}^{\mathsf{m}'}_{K,S'}$. Hence $\mathsf{C}^{\mathsf{m}}_{K,S} \subset \mathsf{C}^{\mathsf{m}'}_{K,S'}$, and we obtain the residue class epimorphism

$$\varpi\colon \mathsf{C}_K/\mathsf{C}^{\mathsf{m}}_{K,S} = \mathsf{J}_K/\mathsf{J}^{\mathsf{m}}_{K,S}K^\times \to \mathsf{J}_K/\mathsf{J}^{\mathsf{m}'}_{K,S'} = \mathsf{C}_K/\mathsf{C}^{\mathsf{m}'}_{K,S'}$$

with kernel $\operatorname{Ker}(\varpi) = \mathsf{C}^{\mathsf{m}'}_{K,S'}/\mathsf{C}^{\mathsf{m}}_{K,S}$.

Since $K^{\mathsf{m}} \subset K^{\mathsf{m}'}$, and $(a)_S \cap \mathcal{D}^{\mathsf{m}}_{K,S'} = (a)_{S'} \in (K^{\mathsf{m}'})_{S'}$ for all $a \in K^{\mathsf{m}}$, it follows that $(K^{\mathsf{m}})_S \cap \mathcal{D}^{\mathsf{m}}_{K,S'} \subset (K^{\mathsf{m}'})_{S'} \cap \mathcal{D}^{\mathsf{m}}_{K,S'}$, and using 5.6.3 we obtain the residue class epimorphism

$$\pi\colon \mathcal{C}^{\mathsf{m}}_{K,S} = \mathcal{D}^{\mathsf{m}}_{K,S'}/(K^{\mathsf{m}})_S \cap \mathcal{D}^{\mathsf{m}}_{K,S'} \to \mathcal{D}^{\mathsf{m}}_{K,S'}/(K^{\mathsf{m}'})_{S'} \cap \mathcal{D}^{\mathsf{m}}_{K,S'} = \mathcal{C}^{\mathsf{m}'}_{K,S'}$$

with kernel $\operatorname{Ker}(\pi) = (K^{\mathsf{m}'}) \cap \mathcal{D}^{\mathsf{a}}_K/(K^{\mathsf{m}})$.

By the very definiton of ι^{m}_S and $\iota^{\mathsf{m}'}_{S'}$, the diagram is commutative, and therefore ι^{m}_S induces an isomorhism $\operatorname{Ker}(\varpi) \overset{\sim}{\to} \operatorname{Ker}(\pi)$.

4. Let $c \in \mathsf{C}^{\circ\mathsf{m}}_{K,S}/\mathsf{C}^{\mathsf{m}}_{K,S} = \mathsf{J}^{\circ\mathsf{m}}_{K,S}K^\times/\mathsf{J}^{\mathsf{m}}_{K,S}K^\times$, say $c = \alpha\mathsf{J}^{\mathsf{m}}_{K,S}K^\times$ for some $\alpha \in \mathsf{J}^{\circ\mathsf{m}}_{K,S}$. By the weak approximation theorem there exists some $x \in K$ such that $|x - \alpha_v^{-1}|_v < |\alpha_v^{-1}|_v$ for all real $v \in \mathsf{M}^\infty_K$ and $v(x-1) \ge v(\mathsf{m})$ for all $v \in \operatorname{supp}(\mathsf{m})$. Then $x \in K^{\circ\mathsf{m}}$, and if $v \in \operatorname{supp}(\mathsf{m})$, then $\alpha_v \in U_v^{(v(\mathsf{m}))}$, hence $x\alpha_v \in U_v^{(v(\mathsf{m}))}$. If $v \in \mathsf{M}^\infty_K$ is real, then $|x\alpha_v - 1|_v < 1$ implies $x\alpha_v \in U_v$, and altogether we obtain $x\alpha \in \mathsf{J}^{(\mathsf{m})}_{K,S}$. Hence $c = x\alpha\mathsf{J}^{\mathsf{m}}_{K,S}K^\times$ and $\iota^{\mathsf{m}}_S(c) = [\iota^{(\mathsf{m})}_S(x\alpha)]^{\mathsf{m}}_S$. Since

$$\iota^{(\mathsf{m})}_S(x\alpha) = \prod_{v\in\mathsf{M}^0_K\setminus S} \mathsf{p}_v^{v(x\alpha_v)} = \prod_{v\in\mathsf{M}^0_K\setminus S} \mathsf{p}_v^{v(x)} = (x)_S \in (K^{\circ\mathsf{m}})_S,$$

it follows that $\iota^{\mathsf{m}}_S(c) = (x)_S(K^{\mathsf{m}})_S \in (K^{\circ\mathsf{m}})_S/(K^{\mathsf{m}})_S$.

As to the reverse inclusion, let $\xi = (x)_S(K^{\mathsf{m}})_S \in (K^{\circ\mathsf{m}})_S(K^{\mathsf{m}})_S$, where $x \in K^{\circ\mathsf{m}}$, and define $\alpha \in \mathsf{J}^{(m)}_{K,S}$ by

$$\alpha_v = \begin{cases} x^{-1} & \text{if} \quad v \in \mathsf{M}^\infty_K \text{ is real}\,, \\ 1 & \text{otherwise.} \end{cases}$$

Then $x\alpha \in \mathsf{J}^{\circ\mathsf{m}}_{K,S}$, hence $c = x\alpha\mathsf{J}^{\mathsf{m}}_{K,S}K^\times \in \mathsf{C}^{\circ\mathsf{m}}_{K,S}/\mathsf{C}^{\mathsf{m}}_{K,S}$ and $\iota^{\mathsf{m}}_S(c) = \xi$.

Now we consider the following commutative diagram with exact rows:

$$\begin{array}{ccccccccc} \mathbf{1} & \longrightarrow & \mathsf{C}^{\circ\mathfrak{m}}_{K,S}/\mathsf{C}^{\mathfrak{m}}_{K,S} & \longrightarrow & \mathsf{C}_K/\mathsf{C}^{\mathfrak{m}}_{K,S} & \longrightarrow & \mathsf{C}_K/\mathsf{C}^{\circ\mathfrak{m}}_{K,S} & \longrightarrow & \mathbf{1} \\ & & {\scriptstyle \iota^{\mathfrak{m}}_S\restriction\ldots}\big\downarrow & & \big\downarrow{\scriptstyle \iota^{\mathfrak{m}}_S} & & & & . \\ \mathbf{1} & \longrightarrow & (K^{\circ\mathfrak{m}})_S/(K^{\mathfrak{m}})_S & \longrightarrow & \mathcal{D}^{\mathfrak{m}}_{K,S}/(K^{\mathfrak{m}})_S & \longrightarrow & \mathcal{C}^{\circ\mathfrak{m}}_{K,S} & \longrightarrow & \mathbf{1} \end{array}$$

There exists a unique isomorphism $\iota^{\circ\mathfrak{m}}\colon \mathsf{C}_K/\mathsf{C}^{\circ\mathfrak{m}}_{K,S} \overset{\sim}{\to} \mathcal{C}^{\circ\mathfrak{m}}_{K,S}$ which completes the commutative diagram. □

Next we investigate finite extensions L/K of global fields and recall the corresponding notations from 5.4.3 and 5.5.8. In particular,

$$\mathsf{N}_{L/K}(\mathsf{C}_L) = \mathsf{N}_{L/K}(\mathsf{J}_L)K^\times/K^\times \quad \text{and} \quad \mathsf{C}_K/\mathsf{N}_{L/K}\mathsf{C}_L = \mathsf{J}_K/\mathsf{N}_{L/K}(\mathsf{J}_L)K^\times.$$

Theorem 5.7.3. *Let S be a finite subset of M^0_K and $\mathfrak{m} \in \mathcal{D}'_{K,S}$. Let L/K a finite field extension and $\overline{\mathfrak{m}} = j_{L/K}(\mathfrak{m}) \in \mathcal{D}'_{L,S_L}$.*

1. $\mathsf{N}_{L/K}(\mathsf{J}^{(\overline{\mathfrak{m}})}_{L,S_L}) \subset \mathsf{J}^{(\mathfrak{m})}_{K,S}$, *and there exist the following commutative diagrams (where for the right-hand squares we assume that $S \ne \emptyset$ in the function field case, $\mathfrak{m} = \theta_S(\mathfrak{m})$ and $\overline{\mathfrak{m}} = \theta_{S_L}(\overline{\mathfrak{m}})$) :*

$$\begin{array}{ccccc} \mathsf{J}^{(\overline{\mathfrak{m}})}_{L,S_L} & \xrightarrow{\iota^{(\overline{\mathfrak{m}})}_{S_L}} & \mathcal{D}^{(\overline{\mathfrak{m}})}_{L,S_L} & \xrightarrow{\theta_{S_L}} & \mathcal{J}^{\overline{\mathfrak{m}}}(\mathfrak{o}_{L,S_L}) \\ \big\downarrow{\scriptstyle \mathsf{N}_{L/K}\restriction\ldots} & & \big\downarrow{\scriptstyle \mathcal{N}_{L/K}\restriction\ldots} & & \big\downarrow{\scriptstyle \mathcal{N}_{L/K,S}\restriction\ldots} \\ \mathsf{J}^{(\mathfrak{m})}_{K,S} & \xrightarrow{\iota^{(\mathfrak{m})}_{S}} & \mathcal{D}^{\mathfrak{m}}_{K,S} & \xrightarrow{\theta_S} & \mathcal{J}^{\mathfrak{m}}(\mathfrak{o}_{K,S}) \end{array}$$

and

$$\begin{array}{ccccc} \mathsf{J}^{(\overline{\mathfrak{m}})}_{L,S_L} & \xrightarrow{\iota^{(\overline{\mathfrak{m}})}_{S_L}} & \mathcal{D}^{(\overline{\mathfrak{m}})}_{L,S_L} & \xrightarrow{\theta_{S_L}} & \mathcal{J}^{\overline{\mathfrak{m}}}(\mathfrak{o}_{L,S_L}) \\ \big\uparrow{\scriptstyle i_{L/K}\restriction} & & \big\uparrow{\scriptstyle j_{L/K}\restriction\ldots} & & \big\uparrow{\scriptstyle \nu} \\ \mathsf{J}^{(\mathfrak{m})}_{K,S} & \xrightarrow{\iota^{(\mathfrak{m})}_{S}} & \mathcal{D}^{\mathfrak{m}}_{K,S} & \xrightarrow{\theta_S} & \mathcal{J}^{\mathfrak{m}}(\mathfrak{o}_{K,S}) \end{array} ,$$

where $\nu(\mathfrak{a}) = \mathfrak{a}\mathfrak{o}_{L,S_L}$ for all $\mathfrak{a} \in \mathcal{J}^{\mathfrak{m}}(\mathfrak{o}_{K,S})$.
Moreover, there exist (natural) isomorphisms

$$\mathsf{N}_{L/K}\mathsf{C}_L \cong [\mathsf{N}_{L/K}(\mathsf{J}_L)K^\times \cap \mathsf{J}^{(\mathfrak{m})}_{K,S}]/K^{\mathfrak{m}},$$

$$\mathsf{C}_K/\mathsf{N}_{L/K}\mathsf{C}_L \cong \mathsf{J}^{(\mathfrak{m})}_{K,S}/[\mathsf{N}_{L/K}(\mathsf{J}_L)K^\times \cap \mathsf{J}^{(\mathfrak{m})}_{K,S}].$$

and, if $\mathsf{C}^{\mathfrak{m}}_{K,S} \subset \mathsf{N}_{L/K}\mathsf{C}_L$ (or, equivalently, $\mathsf{J}^{\mathfrak{m}}_{K,S} \subset \mathsf{N}_{L/K}(\mathsf{J}_L)K^\times$), then

$$\mathsf{N}_{L/K}\mathsf{C}_L/\mathsf{C}^{\mathfrak{m}}_{K,S} \cong \mathsf{N}_{L/K}(\mathsf{J}_L)K^\times \cap \mathsf{J}^{(\mathfrak{m})}_{K,S}/\mathsf{J}^{\mathfrak{m}}_{K,S}K^{\mathfrak{m}}.$$

We identify.

2. We set $\mathcal{N}^{\mathsf{m}}_{L/K,S} = \mathcal{N}_{L/K}(\mathcal{D}^{\overline{\mathsf{m}}}_{L,S_L})(K^{\mathsf{m}})_S$. *Then*

$$\iota_S^{(\mathsf{m})}\big(\mathsf{N}_{L/K}(\mathsf{J}_L)K^\times \cap \mathsf{J}^{(\mathsf{m})}_{K,S}\big) = \mathcal{N}^{\mathsf{m}}_{L/K,S},$$

and if $\mathsf{C}^{\mathsf{m}}_{K,S} \subset \mathsf{N}_{L/K}\mathsf{C}_L$, *then* $\iota_S^{(\mathsf{m})}$ *induces the isomorphisms*

$$\iota^{\mathsf{m}}_{S,L/K}\colon \mathsf{C}_K/\mathsf{N}_{L/K}\mathsf{C}_L \stackrel{\sim}{\to} \mathcal{D}^{\mathsf{m}}_{K,S}/\mathcal{N}^{\mathsf{m}}_{L/K,S}$$

and

$$\iota^{\mathsf{m}}_S\restriction_{\ldots}\colon \mathsf{N}_{L/K}\mathsf{C}_L/\mathsf{C}^{\mathsf{m}}_{K,S} \stackrel{\sim}{\to} \mathcal{N}^{\mathsf{m}}_{L/K,S}/(K^{\mathsf{m}})_S.$$

Proof. 1. Let $\beta = (\beta_{\overline{v}})_{\overline{v}\in\mathsf{M}_L} \in \mathsf{J}^{(\overline{\mathsf{m}})}_{L,S_L}$ and $v \in \mathsf{M}_K$. Then

$$\mathsf{N}_{L/K}(\beta)_v = \prod_{\substack{\overline{v}\in\mathsf{M}_L\\ \overline{v}\,|\,v}} \mathsf{N}_{L_{\overline{v}}/K_v}(\beta_{\overline{v}}),$$

and we fix some $\overline{v} \in \mathsf{M}_L$ such that $\overline{v}\,|\,v$. Let first $v \in \mathsf{M}_K^\infty$ be real. Then either $\overline{v}$ is real and $\mathsf{N}_{L_{\overline{v}}/K_v}(\beta_{\overline{v}}) = \beta_{\overline{v}} > 0$, or $\overline{v}$ is complex and $\mathsf{N}_{L_{\overline{v}}/K_v}(\beta_{\overline{v}}) = \overline{\beta}_{\overline{v}}\beta_{\overline{v}} > 0$. In any case $\mathsf{N}_{L/K}(\beta)_v > 0$. Now let $v \in \mathrm{supp}(\mathsf{m})$ and $e = e(\overline{v}/v)$. Then $\overline{v}(\overline{\mathsf{m}}) = ev(\mathsf{m})$, hence $\beta_{\overline{v}} \in U_{\overline{v}}^{(ev(\mathsf{m}))}$ and $\mathsf{N}_{L_{\overline{v}}/K_v}(\beta_{\overline{v}}) \in U_v^{(v(\mathsf{m}))}$ (see [18, Theorem 5.5.6]). Consequently $\mathsf{N}_{L/K}(\beta)_v \in U_v^{(v(\mathsf{m}))}$, and altogether it follows that $\mathsf{N}_{L/K}(\beta) \in \mathsf{J}^{(\mathsf{m})}_{K,S}$.

The diagrams are commutative by 5.4.4.1 and 5.5.8.2.

As to the isomorphisms, we obtain, using 5.7.2.1,

$$\begin{aligned}
\mathsf{N}_{L/K}\mathsf{C}_L &= \mathsf{N}_{L/K}(\mathsf{J}_L)K^\times/K^\times = \mathsf{N}_{L/K}(\mathsf{J}_L)K^\times \cap \mathsf{J}^{(\mathsf{m})}_{K,S}K^\times/K^\times\\
&= [\mathsf{N}_{L/K}(\mathsf{J}_L)K^\times \cap \mathsf{J}^{(\mathsf{m})}_{K,S}]K^\times/K^\times\\
&\cong [\mathsf{N}_{L/K}(\mathsf{J}_L)K^\times \cap \mathsf{J}^{(\mathsf{m})}_{K,S}]/[K^\times \cap \mathsf{N}_{L/K}(\mathsf{J}_L)K^\times \cap \mathsf{J}^{(\mathsf{m})}_{K,S}]\\
&= \mathsf{N}_{L/K}(\mathsf{J}_L)K^\times \cap \mathsf{J}^{(\mathsf{m})}_{K,S}/K^{\mathsf{m}}
\end{aligned}$$

and

$$\begin{aligned}
\mathsf{C}_K/\mathsf{N}_{L/K}\mathsf{C}_L &\cong (\mathsf{J}^{(\mathsf{m})}_{K,S}/K^{\mathsf{m}})/(\mathsf{N}_{L/K}(\mathsf{J}_L)K^\times \cap \mathsf{J}^{(\mathsf{m})}_{K,S}/K^{\mathsf{m}})\\
&\cong \mathsf{J}^{(\mathsf{m})}_{K,S}/[\mathsf{N}_{L/K}(\mathsf{J}_L)K^\times \cap \mathsf{J}^{(\mathsf{m})}_{K,S}].
\end{aligned}$$

By definition we have $\mathsf{C}^{\mathsf{m}}_{K,S} \subset \mathsf{N}_{L/K}\mathsf{C}_L$ if and only if $\mathsf{J}^{\mathsf{m}}_{K,S} \subset \mathsf{N}_{L/K}(\mathsf{J}_L)K^\times$, and then

$$\begin{aligned}
\mathsf{N}_{L/K}\mathsf{C}_L/\mathsf{C}^{\mathsf{m}}_{K,S} &= [\mathsf{N}_{L/K}(\mathsf{J}_L)K^\times \cap \mathsf{J}^{(\mathsf{m})}_{K,S}/K^{\mathsf{m}}]/[\mathsf{J}^{\mathsf{m}}_{K,S}K^{\mathsf{m}}/K^{\mathsf{m}}]\\
&\cong \mathsf{N}_{L/K}(\mathsf{J}_L)K^\times \cap \mathsf{J}^{(\mathsf{m})}_{K,S}/\mathsf{J}^{\mathsf{m}}_{K,S}K^{\mathsf{m}}.
\end{aligned}$$

2. $\subset$: Let $\beta \in \mathsf{J}_L$ and $a \in K^\times$ such that $\mathsf{N}_{L/K}(\beta)a \in \mathsf{J}_{K,S}^{(\mathsf{m})}$. As $\mathsf{J}_L = \mathsf{J}_{L,S_L}^{(\overline{\mathsf{m}})} L^\times$, we obtain $\beta = \gamma b$ for some $\gamma \in \mathsf{J}_{L,S_L}^{(\overline{\mathsf{m}})}$ and $b \in L^\times$. Hence $\mathsf{N}_{L/K}(\beta)a = \mathsf{N}_{L/K}(\gamma)\mathsf{N}_{L/K}(b)a$, and as $\mathsf{N}_{L/K}(\gamma) \in \mathsf{J}_{K,S}^{(\mathsf{m})}$, we get $\mathsf{N}_{L/K}(b)a = \mathsf{N}_{L/K}(\gamma)^{-1}\mathsf{N}_{L/K}(\beta)a \in \mathsf{J}_{K,S}^{(\mathsf{m})} \cap K^\times = K^{\mathsf{m}}$. Using 1., we eventually obtain

$$
\begin{aligned}
\iota_S^{(\mathsf{m})}(\mathsf{N}_{L/K}(\beta)a) &= \iota_S^{(\mathsf{m})}(\mathsf{N}_{L/K}(\gamma))\iota_S^{(\mathsf{m})}(\mathsf{N}_{L/K}(b)a) \\
&= \mathcal{N}_{L/K}(\iota_{S_L}^{(\overline{\mathsf{m}})}(\gamma))(\mathsf{N}_{L/K}(b)a)_S \in \mathcal{N}_{L/K,S}^{\mathsf{m}}.
\end{aligned}
$$

$\supset$: Let $\mathsf{a} \in \mathcal{N}_{L/K,S}^{\mathsf{m}}$, say $\mathsf{a} = \mathcal{N}_{L/K}(\mathsf{b})(a)_S$, where $\mathsf{b} \in \mathcal{D}_{L,S_L}^{\overline{\mathsf{m}}}$ and $a \in K^{\mathsf{m}}$. Then $\mathsf{b} = \iota_{S_L}^{(\overline{\mathsf{m}})}(\beta)$ for some $\beta \in \mathsf{J}_{L,S_L}^{(\overline{\mathsf{m}})}$, $\mathsf{N}_{L/K}(\beta)a \in \mathsf{J}_{K,S}^{(\mathsf{m})}K^{\mathsf{m}} = \mathsf{J}_{K,S}^{(\mathsf{m})}$ by 1., and $\mathsf{a} = \mathcal{N}_{L/K}(\iota_{S_L}^{(\overline{\mathsf{m}})}(\beta))(a)_S = \iota_S^{(\mathsf{m})}(\mathsf{N}_{L/K}(\beta)a) \in \iota_S^{(\mathsf{m})}(\mathsf{N}_{L/K}(\mathsf{J}_L)K^\times \cap \mathsf{J}_{K,S}^{(\mathsf{m})})$.

Let now $\mathsf{C}_{K,S}^{\mathsf{m}} \subset \mathsf{N}_{L/K}\mathsf{C}_L$ (i. e., $\mathrm{Ker}(\iota_S^{(\mathsf{m})}) = \mathsf{J}_{K,S}^{\mathsf{m}} \subset \mathsf{N}_{L/K}(\mathsf{J}_L)K^\times$). Then $\iota_S^{(\mathsf{m})}$ induces the following commutative diagram (in which the vertical arrows are inclusions).

$$
\begin{array}{ccc}
\mathsf{J}_{K,S}^{(\mathsf{m})} & \xrightarrow{\iota_S^{(\mathsf{m})}} & \mathcal{D}_{K,S}^{\mathsf{m}} \\
\uparrow & & \uparrow \\
\mathsf{N}_{L/K}(\mathsf{J}_L)K^\times \cap \mathsf{J}_{K,S}^{(\mathsf{m})} & \xrightarrow{\iota_S^{(\mathsf{m})}} & \mathcal{N}_{L/K,S}^{\mathsf{m}} \\
\uparrow & & \uparrow \\
\mathsf{J}_{K,S}^{\mathsf{m}}K^{\mathsf{m}} & \xrightarrow{\iota_S^{(\mathsf{m})}} & (K^{\mathsf{m}})_S
\end{array}
.
$$

Since $(\iota_S^{(\mathsf{m})})^{-1}((K^{\mathsf{m}})_S) = \mathsf{J}_{K,S}^{\mathsf{m}}K^{\mathsf{m}}$, this diagram induces the isomorphisms

$$
\iota_{S,L/K}^{\mathsf{m}} \colon \mathsf{C}_K/\mathsf{N}_{L/K}\mathsf{C}_L = \mathsf{J}_{K,S}^{(\mathsf{m})}/[\mathsf{N}_{L/K}(\mathsf{J}_L)K^\times \cap \mathsf{J}_{K,S}^{(\mathsf{m})}] \overset{\sim}{\to} \mathcal{D}_{K,S}^{\mathsf{m}}/\mathcal{N}_{L/K,S}^{\mathsf{m}}
$$

and

$$
\iota_S^{\mathsf{m}} \restriction \ldots \colon \mathsf{N}_{L/K}\mathsf{C}_L/\mathsf{C}_{K,S}^{\mathsf{m}} = [\mathsf{N}_{L/K}(\mathsf{J}_L)K^\times \cap \mathsf{J}_{K,S}^{(\mathsf{m})}]/\mathsf{J}_{K,S}^{\mathsf{m}})K^{\mathsf{m}} \overset{\sim}{\to} \mathcal{N}_{L/K,S}^{\mathsf{m}}/(K^{\mathsf{m}})_S.
$$

□

Corollary 5.7.4. *Let S be a finite subset of M_K^0 and L/K a finite field extension. Then the following diagrams are commutative.*

$$
\begin{array}{ccccc}
\mathsf{J}_L & \xrightarrow{\iota_{S_L}} & \mathcal{D}_{L,S_L} & \xrightarrow{\theta_{S_L}} & \mathfrak{J}(\mathfrak{o}_{L,S_L}) \\
{\scriptstyle i_{L/K}}\uparrow\ \downarrow{\scriptstyle \mathsf{N}_{L/K}} & & {\scriptstyle j_{L/K}}\uparrow\ \downarrow{\scriptstyle \mathcal{N}_{L/K}} & & {\scriptstyle \nu}\uparrow\ \downarrow{\scriptstyle \mathcal{N}_{L/K,S}} \\
\mathsf{J}_K & \xrightarrow{\iota_S} & \mathcal{D}_{K,S} & \xrightarrow{\theta_S} & \mathfrak{J}(\mathfrak{o}_{K,S})
\end{array}
,
$$

where $\nu(\mathfrak{a}) = \mathfrak{a}\mathfrak{o}_{L,S_L}$ for all $\mathfrak{a} \in \mathfrak{o}_{K,S}$.

Proof. By 5.7.3.1 with $\mathsf{m} = 1$, since $\iota_{S_L}^{(\overline{1})} = \iota_{S_L} \restriction \mathsf{J}_{L,S_L}^{(\overline{1})}$ and $\iota_S^{(1)} = \iota_S \restriction \mathsf{J}_{K,S}^{(1)}$. □

Definition and Theorem 5.7.5. *Let L/K be a finite field extension.*

A positive (abstract) divisor $\mathsf{m} \in \mathcal{D}'_K$ is called a **modulus of definition** for L/K if $\mathsf{J}_K^{\mathsf{m}} \subset \mathsf{N}_{L/K}(\mathsf{J}_L)K^\times$ [equivalently, $\mathsf{C}_K^{\mathsf{m}} \subset \mathsf{N}_{L/K}(\mathsf{C}_L)$].

1. Let $\mathsf{m} \in \mathcal{D}'_K$ be such that $\operatorname{supp}(\mathsf{m}) \supset \operatorname{Ram}(L/K)$*, and let S be a finite subset of* $\operatorname{Split}(L/K)$*. Then there exists some $m \in \mathbb{N}$ such that $\mathsf{J}_{K,S}^{\mathsf{m}^m} \subset \mathsf{N}_{L/K}\mathsf{J}_L$, and therefore m^m is a modulus of definition for L/K.*

2. Let $\mathsf{m},\, \mathsf{m}' \in \mathcal{D}'_K$.

(a) If m is a modulus of definition for L/K and $\mathsf{m} \,|\, \mathsf{m}'$, then m' is a modulus of definition for L/K, too.

(b) If m and m' are moduli of definition for L/K, then $\gcd(\mathsf{m},\mathsf{m}')$ *is a modulus of definition for L/K, too.*

The smallest modulus of definition (in the sense of divisibility) for L/K is called the **conductor** of L/K and is denoted by $\mathsf{f}_{L/K}$. *In particular, L/K has a conductor, and if m is any modulus of definition of L/K, then $\mathsf{f}_{L/K} \,|\, \mathsf{m}$.*

3. Let m be a modulus of definition for L/K, $\overline{\mathsf{m}} = j_{L/K}(\mathsf{m})$, let $\iota^{(\mathsf{m})} \colon \mathsf{J}_K^{(\mathsf{m})} \to \mathcal{D}_K^{\mathsf{m}}$ be the epimorphism defined in 5.7.1.**2** *and $\mathcal{N}_{L/K}^{\mathsf{m}} = \mathcal{N}_{L/K,\emptyset}^{\mathsf{m}} = \mathcal{N}_{L/K}(\mathcal{D}_L^{\overline{\mathsf{m}}})(K^{\mathsf{m}}) \subset \mathcal{D}_K^{\mathsf{m}}$. Then ι^{m} induces an isomorphism $\iota_{L/K}^{\mathsf{m}} \colon \mathsf{C}_K/\mathsf{N}_{L/K}\mathsf{C}_L \overset{\sim}{\to} \mathcal{D}_K^{\mathsf{m}}/\mathcal{N}_{L/K}^{\mathsf{m}}$.*

Proof. 1. By 4.1.10.3, there exists some $m \in \mathbb{N}$ satisfying $U_v^{(m)} \subset \mathsf{N}_{L_{\overline{v}}/K_v}U_{\overline{v}}$ for all $v \in \operatorname{supp}(\mathsf{m})$ and $\overline{v} \in \mathsf{M}_L$ such that $\overline{v} \,|\, v$. We show $\mathsf{J}_{K,S}^{\mathsf{m}^m} \subset \mathsf{N}_{L/K}(\mathsf{J}_L)$ component-wise. Let $v \in \mathsf{M}_K$ and $\overline{v} \in \mathsf{M}_L$ such that $\overline{v} \,|\, v$. If $v \in \mathsf{M}_K^\infty$, then clearly $U_v \subset \mathsf{N}_{L_{\overline{v}}/K_v}L_{\overline{v}}^\times$. If $v \in S$, then $L_{\overline{v}} = K_v$ and thus $\mathsf{N}_{L_{\overline{v}}/K_v}L_{\overline{v}}^\times = K_v^\times$. If $v \in \mathsf{M}_K^0 \setminus \operatorname{supp}(\mathsf{m})$, then $L_{\overline{v}}/K_v$ is unramified and $U_v = \mathsf{N}_{L_{\overline{v}}/K_v}U_{\overline{v}}$ (see [18, Theorem 5.6.9]). Thus assume after all that $v \in \operatorname{supp}(\mathsf{m})$. Then $U_v^{(v(\mathsf{m}^m))} = U_v^{(mv(\mathsf{m}))} \subset U_v^{(m)} \subset \mathsf{N}_{L_{\overline{v}}/K_v}U_{\overline{v}}$. Overall it follows that

$$\mathsf{J}_{K,S}^{\mathsf{m}^m} = \prod_{v \in \mathsf{M}_K^\infty} U_v \times \prod_{v \in S} K_v^\times \times \prod_{v \in \mathsf{M}_K^0 \setminus S} U_v^{(v(\mathsf{m}^m))}$$
$$\subset \mathsf{N}_{L/K}\Bigl(\prod_{\overline{v} \in \mathsf{M}_L^\infty \cup S_L} L_{\overline{v}}^\times \times \prod_{\overline{v} \in \mathsf{M}_L^0 \setminus S_L} U_{\overline{v}} \Bigr) \subset \mathsf{N}_{L/K}(\mathsf{J}_L)K^\times.$$

2. (a) If $\mathsf{m} \,|\, \mathsf{m}'$, then $v(\mathsf{m}) \le v(\mathsf{m}'$ for all $v \in \mathsf{M}_K^0$ and therefore $\mathsf{J}_K^{\mathsf{m}'} \subset \mathsf{J}_K^{\mathsf{m}}$. Thus if m is a modulus of definition for L/K, then so is m'.

(b) Since $U_v^{(v(\mathsf{m}))}U_v^{(v(\mathsf{m}'))} = U_v^{(\min\{v(\mathsf{m}),v(\mathsf{m}')\})} = U_v^{(v(\gcd(\mathsf{m},\mathsf{m}'))}$ holds for all $v \in \mathsf{M}_K^0$, it follows that $\mathsf{J}_K^{\mathsf{m}}\mathsf{J}_M^{\mathsf{m}'} = \mathsf{J}_K^{\gcd(\mathsf{m},\mathsf{m}')}$. Consequently, if m and m' are moduli of definition for L/K, then so is $\gcd(\mathsf{m},\mathsf{m}')$.

If m is any modulus of definition for L/K, then $\gcd(\mathsf{m},\mathsf{f}_{L/K})$ is also a modulus of definition for L/K, and therefore $\gcd(\mathsf{m},\mathsf{f}_{L/K}) = \mathsf{f}_{L/K} \,|\, \mathsf{m}$.

3. By 5.7.3.2. □

Remark 5.7.6. Let L/K be a finite abelian extension and m a modulus of definition for L/K. In 6.5.6.2 we shall prove that $\mathrm{Ram}(L/K) = \mathrm{supp}(\mathsf{f}_{L/K})$ and consequently $\mathrm{Ram}(L/K) \subset \mathrm{supp}(\mathsf{m})$. Note that in 5.7.5.1 above we have proved a partial converse: If $\mathsf{m} \in \mathcal{D}'_K$ and $\mathrm{Ram}(L/K) \subset \mathrm{supp}(\mathsf{m})$, then some power of m is a modulus of definition for L/K.

Now let m be a modulus of definition for L/K and $\mathrm{supp}(\mathsf{m}) \supset \mathrm{Ram}(L/K)$. Then 5.7.5.3, together with the analytically proved normindex inequality (see [18, Theorem 4.4.1]), implies the so-called **second inequality of class field theory** in the number field case:

$$(\mathsf{C}_K : \mathsf{N}_{L/K}\mathsf{C}_L) = (\mathcal{I}_K^{\mathsf{m}} : \mathcal{N}_{L/K}^{\mathsf{m}}) \le [L:K].$$

That here in fact equality holds, is a main statement of global class field theory (see 6.4.7).

One of the main results in local class field theory is the characterization of norm groups as open subgroups of finite index in $K^\times$ (see 4.2.10 and 4.8.2). In global class field we shall prove the analogous result for C_K instead of $K^\times$ (see 6.5.3). We proceed with a preliminary result on open subgroups of finite index of the idele class group.

Theorem 5.7.7 (Subgroups of finite index of the idele class group)**.**

1. If H is an open subgroup of J_K, then there exists an integral divisor $\mathsf{m} \in \mathcal{D}'_K$ such that $\mathsf{J}_K^{\mathsf{m}} \subset H$.

2. For a subgroup N of C_K the following assertions are equivalent:

(a) N is an open subgroup of finite index.

(b) There exists an open subgroup H of J_K such that $(\mathsf{J}_K : H) < \infty$, $K^\times \subset H$ and $N = H/K^\times$.

(c) There exists an integral divisor $\mathsf{m} \in \mathcal{D}'_K$ such that

- *$\mathsf{C}_K^{\mathsf{m}} \subset N$ in the number field case,*
- *$\mathsf{C}_K^{\mathsf{m}} \cdot \langle c_1^d \rangle \subset N$ in the function field case for some $d \in \mathbb{N}$, where $c_1 \in \mathsf{C}_K$ is any idele class of degree 1.*

Proof. 1. Let H be an open subgroup J_K. Then H is a neighbourhood of 1, and thus there exist a finite subset S of M_K such that $\mathsf{M}_K^\infty \subset S$, a family $(n_v)_{v \in S \cap \mathsf{M}_K^0}$ of exponents in $\mathbb{N}$ and some $\varepsilon \in \mathbb{R}_{>0}$ such that

$$H \supset W = \prod_{v \in \mathsf{M}_K^\infty} \{\alpha_v \in K_v \mid |\alpha_v - 1| < \varepsilon\} \times \prod_{v \in S \cap \mathsf{M}_K^0} U_v^{(n_v)} \times \prod_{v \in \mathsf{M}_K \setminus S} U_v,$$

and then

$$H \supset \langle W \rangle = \prod_{v \in \mathsf{M}_K^\infty} U_v \times \prod_{v \in \mathsf{M}_K^0} U_v^{(v(\mathsf{m}))} = \mathsf{J}_K^{\mathsf{m}}, \quad \text{where} \quad \mathsf{m} = \prod_{v \in S \cap \mathsf{M}_K^0} \mathsf{p}_v^{n_v}.$$

2. Let $\pi \colon \mathsf{J}_K \to \mathsf{C}_K$ be the residue class epimorphism.

(a) $\Rightarrow$ (b) $H = \pi^{-1}(N) \subset \mathsf{J}_K$ is an open subgroup of J_K satisfying $(\mathsf{J}_K \colon H) < \infty$, $K^\times \subset H$ and $N = H/K^\times$.

(b) $\Rightarrow$ (a) Obvious.

(a) and (b) $\Rightarrow$ (c) By 2., there exists an integral divisor $\mathsf{m} \in \mathcal{D}'_K$ such that $\mathsf{J}_K^{\mathsf{m}} K^\times \subset H$, and then $\mathsf{C}_K^{\mathsf{m}} = \mathsf{J}_K^{\mathsf{m}} K^\times / K^\times \subset H/K^\times = N$. In the function field case, if $c_1 \in C$ is any idele class of degree 1, then $c_1^d \in N$ for some $d \in \mathbb{N}$, and consequently $\mathsf{C}_K^{\mathsf{m}} \cdot \langle c_1^d \rangle \subset N$.

(c) $\Rightarrow$ (a) Let $\mathsf{m} \in \mathcal{D}'_K$. Then $\mathsf{C}_K^{\mathsf{m}}$ is an open subgroup of C_K, and in the number fied case it is of finite index by 5.7.2.2(a). In the function field case let $c_1 \in \mathsf{C}_K$ be any idele class of degree 1 and $d \in \mathbb{N}$ such that $\mathsf{C}_K^{\mathsf{m}} \cdot \langle c_1^d \rangle \subset N$. Then N is open in C_K, and since $\mathsf{C}_K = \mathsf{C}_K^0 \cdot \langle c_1 \rangle$ and $\mathsf{C}_K^{\mathsf{m}} \subset \mathsf{C}_K^0$, it follows that

$$(\mathsf{C}_K \colon \mathsf{C}_K^{\mathsf{m}} \cdot \langle c_1^d \rangle) = (\mathsf{C}_K^0 \colon \mathsf{C}_K^{\mathsf{m}})(\langle c_1 \rangle \colon \langle c_1^d \rangle) < \infty$$

by 5.7.2.2(b). Hence in either case N is an open subgroup of finite index in C_K. □

In the function field case, the assertion of 5.7.7.2(c) is not as precise as it could be, since $\mathsf{C}_K^{\mathsf{m}} \cdot \langle c^d \rangle \subset N$ merely implies $\deg(N) \supset d\mathbb{Z}$. More precisely, the following holds:

Corollary 5.7.8. *Let $K/\mathbb{F}_q$ be a function field, N an open subgroup of C_K of finite index and $N^0 = N \cap \mathsf{C}_K^0$. Then $\deg(N) = d\mathbb{Z}$ for some $d \in \mathbb{N}$. If $c \in N$ is such that $\deg(c) = d$, then $N = N^0 \cdot \langle c \rangle$, and there exists some $\mathsf{m} \in \mathcal{D}'_K$ such that $\mathsf{C}_K^{\mathsf{m}} \subset N^0$.*

Proof. As $N \not\subset \mathsf{C}_K^0$, there exists some $d \in \mathbb{N}$ such that $\deg(N) = d\mathbb{Z}$. If $c \in N$ is such that $\deg(c) = d$, then $N^0 = N \cap \mathsf{C}_K^0$ is an open subgroup of finite index of C_K^0. Hence there exists some $\mathsf{m} \in \mathcal{D}'_K$ such that $\mathsf{C}_K^{\mathsf{m}} \subset N^0$, and then $\mathsf{C}_K^{\mathsf{m}} \cdot \langle c \rangle \subset N^0 \cdot \langle c \rangle \subset N$. Conversely, if $a \in N$, then $\deg(a) = dk$ for some $k \in \mathbb{Z}$, hence $ac^{-k} \in N^0$ and thus $a \in N_0 \cdot \langle c \rangle$. □

We close this section with a short preview of class field theory. We recall the main notions and results of class field theory in the ideal-theoretic version (as given in [18, Section 3.7]) and write them up in the idelic language.

Remarks 5.7.9 (Reformulation of the classical class field theory). Let K be an algebraic number field (and $S = \emptyset$). We tacitly assume that all algebraic extensions of K are inside a fixed algebraic closure $\overline{K}$.

1. We recall the notations introduced in 5.1.1.**4**, 5.4.1.**3** and 5.6.1.**B**:

- $\mathfrak{o}_K$, $\mathcal{P}_K = \mathcal{P}(\mathfrak{o}_K)$, $\mathcal{I}_K = \mathcal{I}(\mathfrak{o}_K)$, $\mathcal{I}'_K = \mathcal{I}'(\mathfrak{o}_K)$;
- if $\mathfrak{m} \in \mathcal{I}'_K$, then $\mathcal{I}_K^{\mathfrak{m}} = \mathcal{I}^{\mathfrak{m}}(\mathfrak{o}_K)$, $\mathcal{P}_K^{\mathfrak{m}} = \mathcal{P}^{\mathfrak{m}}(\mathfrak{o}_K)$, $\mathcal{H}_K^{\mathfrak{m}} = \mathcal{H}^{\mathfrak{m}}(\mathfrak{o}_K)$, and $\mathcal{C}_K^{\mathfrak{m}} = \mathcal{I}_K^{\mathfrak{m}}/\mathcal{H}_K^{\mathfrak{m}}$;
- if L/K is a finite field extension, then $\mathcal{N}_{L/K}\colon \mathcal{I}_L \to \mathcal{I}_K$ is the ideal norm; it satisfies $\mathcal{N}_{L/K}(\mathcal{I}_L^{\mathfrak{m}\mathfrak{o}_L}) \subset \mathcal{I}_K^{\mathfrak{m}}$ (see 5.7.3.1).
  Recall that we denote by $\mathrm{Split}(L/K)$ the set of all in L fully decomposed and $\mathrm{Ram}(L/K)$ the set of all in L ramified prime ideals $\mathfrak{p} \in \mathcal{P}_K$.

2. (Class field theory, ideal-theoretic version) Let L/K be a finite abelian extension, $G = \mathrm{Gal}(L/K)$, and $\mathfrak{m} \in \mathcal{I}'_K$ such that $\mathrm{Ram}(L/K) \subset \mathrm{supp}(\mathfrak{m})$.

For $\mathfrak{p} \in \mathcal{P}_K^{\mathfrak{m}}$ we denote by $(\mathfrak{p}, L/K) \in G$ the Frobenius automorphis of $\mathfrak{p}$ (see 5.4.5.**3**), and for $\mathfrak{a} \in \mathcal{I}_K^{\mathfrak{m}}$ we define the **Artin symbol** by

$$(\mathfrak{a}, L/K) = \prod_{\mathfrak{p} \in \mathcal{P}_K^{\mathfrak{m}}} (\mathfrak{p}, L/K)^{v_{\mathfrak{p}}(\mathfrak{a})} \in G.$$

We call $(\,\cdot\,, L/K)\colon \mathcal{I}_K^{\mathfrak{m}} \to G$ the **Artin map**.

Then $\mathcal{N}_{L/K}(\mathcal{I}_L^{\mathfrak{m}\mathfrak{o}_L}) \subset \mathrm{Ker}(\,\cdot\,, L/K)$, and a prime ideal $\mathfrak{p} \in \mathcal{P}_K^{\mathfrak{m}}$ is fully decomposed in L if and only if $(\mathfrak{p}, L/K) = 1$ (see [18, Theorems 2.14.7 and 2.14.8]). Now the main theorems of classical class field theory read as follows.

Theorem and Definition 5.7.10 (Classical class field theory).
Let K be an algebraic number field.

> *1.* (Artin reciprocity law) *Let L/K be a finite abelian extension. Then there exists an ideal $\mathfrak{m} \in \mathcal{I}'_K$ such that* $\mathrm{Ram}(L/K) \subset \mathrm{supp}(\mathfrak{m})$ *and the Artin map $(\,\cdot\,, L/K)\colon \mathcal{I}_K^{\mathfrak{m}} \to G$ is an epimorphism with kernel*
>
> $$\mathcal{N}_{L/K}^{\mathfrak{m}} = \mathcal{N}_{L/K}(\mathcal{I}_L^{\mathfrak{m}\mathfrak{o}_L})\mathcal{H}_K^{\mathfrak{m}}.$$
>
> *In particular,* $\mathcal{P}_K^{\mathfrak{m}} \cap \mathrm{Split}(L/K) = \mathcal{P}_K^{\mathfrak{m}} \cap \mathcal{N}_{L/K}^{\mathfrak{m}}$.
>
> *2.* (Existence theorem) *Let $\mathfrak{m} \in \mathcal{I}'_K$ be an ideal and $\mathcal{G}$ a subgroup of $\mathcal{I}_K^{\mathfrak{m}}$ such that $\mathcal{H}_K^{\mathfrak{m}} \subset \mathcal{G}$. Then there exists a unique over K finite abelian extension field $K[\mathcal{G}]$ such that $\mathcal{G} = \mathcal{N}_{K[\mathcal{G}]/K}^{\mathfrak{m}}$. Furthermore,* $\mathcal{P}_K \cap \mathcal{G} = \mathcal{P}_K^{\mathfrak{m}} \cap \mathrm{Split}(K[\mathcal{G}]/K)$ *and* $\mathrm{Ram}(K[\mathcal{G}]/K) \subset \mathrm{supp}(\mathfrak{m})$.
>
> We call $K[\mathcal{G}]$ the **class field** to $\mathcal{G}$.
> The Artin map $(\,\cdot\,, K[\mathcal{G}]/K)\colon \mathcal{I}_K^{\mathfrak{m}} \to \mathrm{Gal}(K[\mathcal{G}]/K)$ induces an (equally denoted) isomorphism
>
> $$(\,\cdot\,, K[\mathcal{G}]/K)\colon \mathcal{I}_K^{\mathfrak{m}}/\mathcal{G} \xrightarrow{\sim} \mathrm{Gal}(K[\mathcal{G}]/K),$$
>
> called the **Artin isomorphism**.

The class field $K[\mathfrak{m}] = K[\mathcal{H}_K^{\mathfrak{m}}]$ to $\mathcal{H}_K^{\mathfrak{m}}$ is called the **ray class field** modulo $\mathfrak{m}$.

The assignment $\mathcal{G} \mapsto K[\mathcal{G}]$ defines an inclusion-reversing bijection from the set of all subgroups $\mathcal{G}$ of $\mathcal{J}_K^{\mathfrak{m}}$ containing $\mathcal{H}_K^{\mathfrak{m}}$ onto the set of all intermediate fields of $K[\mathfrak{m}]/K$.

3. Let L/K be a finite abelian field extension and $\mathfrak{m} \in \mathcal{J}_K'$ the product of all in L ramified prime ideals. Then there exists some $m \in \mathbb{N}$ such that $L \subset K[\mathfrak{m}^m]$.

In the classical literature the field $K[\mathcal{H}_K^{\circ\mathfrak{m}}]$ is often called ray class field modulo $\mathfrak{m}$, while the field $K[\mathcal{H}_K^{\mathfrak{m}}]$ ic called ray class field modulo $\mathfrak{m}\infty$. Our terminology coincides with that of [48].

3. (Idele-theoretic reformulation) Let Y be an open subgroup of C_K and $\mathfrak{m} \in \mathcal{J}_K'$ such that $(\mathsf{C}_K : Y) < \infty$ and $\mathsf{C}_K^{\mathfrak{m}} \subset Y$ (see 5.7.7.2). Let

$$\iota^{\mathfrak{m}} \colon \mathsf{C}_K/\mathsf{C}_K^{\mathfrak{m}} \overset{\sim}{\to} \mathcal{C}_K^{\mathfrak{m}} = \mathcal{J}_K^{\mathfrak{m}}/\mathcal{H}_K^{\mathfrak{m}}$$

be the isomorphism given in 5.7.2.2. Then $\iota^{\mathfrak{m}}(Y/\mathsf{C}_K^{\mathfrak{m}}) = \mathcal{G}/\mathcal{H}_K^{\mathfrak{m}}$ for some subgroup $\mathcal{G}$ of $\mathcal{J}_K^{\mathfrak{m}}$ such that $\mathcal{H}_K^{\mathfrak{m}} \subset \mathcal{G}$. Let $K[\mathfrak{m}]$ be the ray class field modulo $\mathfrak{m}$ of K and $L = K[\mathcal{G}]$ the class field to $\mathcal{G}$ as defined in 5.7.10 above. It follows that $L \subset K[\mathfrak{m}]$, $\mathcal{G} = \mathcal{N}_{L/K}^{\mathfrak{m}} = \mathcal{N}_{L/K}(\mathcal{J}_L^{\mathfrak{m}\mathfrak{o}_L})\mathcal{H}_K^{\mathfrak{m}}$, there exists the Artin isomorphism

$$\theta_{\mathcal{G}} \colon \mathcal{J}_K^{\mathfrak{m}}/\mathcal{G} \overset{\sim}{\to} \mathrm{Gal}(L/K), \quad \text{defined by} \quad \theta_{\mathcal{G}}([\mathfrak{p}]) = (\mathfrak{p}, L/K) \quad \text{for all} \quad \mathfrak{p} \in \mathcal{P}_K \cap \mathcal{J}_K^{\mathfrak{m}},$$

and we call L the class field to Y.

Let $\mathfrak{m}$ be a modulus of definition for L/K. Then $\mathsf{C}_K^{\mathfrak{m}} \subset \mathsf{N}_{L/K}\mathsf{C}_L$, and using 5.7.3.2 we obtain the following commutative diagram in which the vertical arrows are inclusions and the horizontal arrows are isomorphisms.

$$\begin{array}{ccc} \mathsf{C}_K/\mathsf{C}_K^{\mathfrak{m}} & \xrightarrow{\iota^{\mathfrak{m}}} & \mathcal{C}_K^{\mathfrak{m}} \\ \uparrow & & \uparrow \\ \mathsf{N}_{L/K}\mathsf{C}_L/\mathsf{C}_K^{\mathfrak{m}} & \longrightarrow & \mathcal{N}_{L/K}^{\mathfrak{m}}/\mathcal{H}_K^{\mathfrak{m}}. \end{array}$$

We obtain $\iota^{\mathfrak{m}}(Y/\mathsf{C}_K^{\mathfrak{m}}) = \mathcal{G}/\mathcal{H}_K^{\mathfrak{m}} = \mathcal{N}_{L/K}^{\mathfrak{m}}/\mathcal{H}_K^{\mathfrak{m}} = \iota^{\mathfrak{m}}(\mathsf{N}_{L/K}\mathsf{C}_L/\mathsf{C}_K^{\mathfrak{m}})$, it follows that $Y/\mathsf{C}_K^{\mathfrak{m}} = \mathsf{N}_{L/K}\mathsf{C}_L/\mathsf{C}_K^{\mathfrak{m}}$, hence $Y = \mathsf{N}_{L/K}\mathsf{C}_L$, and we get the idele-theoretic Artin isomorphism

$$(\,\cdot\,, L/K) \colon \mathsf{C}_K/\mathsf{N}_{L/K}\mathsf{C}_L \overset{\sim}{\to} \mathcal{J}_K^{\mathfrak{m}}/\mathcal{G} \overset{\theta_{\mathcal{G}}}{\to} \mathrm{Gal}(L/K).$$

The assignment $L \mapsto \mathsf{N}_{L/K}\mathsf{C}_L$ defines a bijective map from the set of all finite abelian extensions of K onto the set of all open subgroups of finite index in C_K.

5.8 Ray class characters

We recall and unify the various concepts of generalized Dirichlet characters used in algebraic number theory and investigated in [18]. To do so, we introduce the more general concept of Hecke characters. Although we will not use this concept in its full generality, we present it here in order to highlight the connection between the ideal-theoretic and the idelic point of view.

Definition and Remarks 5.8.1. Let K be a global field and $\mathsf{m} \in \mathcal{D}'_K$.

A **ray class character** modulo m is a character $\chi \in \mathsf{X}(\mathcal{C}_K^{\mathsf{m}})$. Every ray class character $\chi \in \mathsf{X}(\mathcal{C}_K^{\mathsf{m}})$ induces an (equally denoted) homomorphism $\chi\colon \mathcal{D}_K^{\mathsf{m}} \to \mathbb{T}$ satisfying $\chi \restriction (K^{\mathsf{m}}) = 1$ by means of $\chi(\mathsf{a}) = \chi([\mathsf{a}]^{\mathsf{m}})$ for all $\mathsf{a} \in \mathcal{D}_K^{\mathsf{m}}$, and we set $\chi(\mathsf{a}) = 0$ for all $\mathsf{a} \in \mathcal{D}_K \setminus \mathcal{D}_K^{\mathsf{m}}$. Conversely, every homomorphism $\chi\colon \mathcal{D}_K^{\mathsf{m}} \to \mathbb{T}$ satisfying $\chi \restriction (K^{\mathsf{m}}) = 1$ is induced by a ray class character modulo m and is itself called a **ray class character** modulo m.

In the number field case the ray class groups $\mathcal{C}_K^{\mathsf{m}}$ are finite, and thus every ray class character is of finite order.

If $\mathsf{m}_1 \in \mathcal{D}'_K$ and $\mathsf{m}_1 \,|\, \mathsf{m}$, then $\mathcal{D}_K^{\mathsf{m}} \subset \mathcal{D}_K^{\mathsf{m}_1}$ and $(K^{\mathsf{m}}) \subset (K^{\mathsf{m}_1}) \cap \mathcal{D}_K^{\mathsf{m}}$. Consequently, if $\chi_1\colon \mathcal{D}_K^{\mathsf{m}_1} \to \mathbb{T}$ is a ray class character modulo m_1, then $\chi = \chi_1 \restriction \mathcal{D}_K^{\mathsf{m}}$ is a ray class character modulo m. In this case, we say that χ **is induced by** χ_1, and we call m_1 a **modulus of definition** for χ. If furthermore $\mathsf{m}_2 \in \mathcal{D}'_K$, $\mathsf{m}_2 \,|\, \mathsf{m}_1$ and χ_1 is induced by a ray class character χ_2 modulo m_2, the χ is also induced by χ_2. In particular, every modulus of definition for χ_1 is also a modulus of definition for χ.

Let $\chi\colon \mathcal{D}_K^{\mathsf{m}} \to \mathbb{T}$ be a ray class character modulo m. If m_1 and m_2 are moduli of definition for χ, then $\mathsf{m}^* = \gcd(\mathsf{m}_1, \mathsf{m}_2)$ is also a modulus of definition for χ, since $\mathcal{D}_K^{\mathsf{m}^*} = \mathcal{D}_K^{\mathsf{m}_1} \cap \mathcal{D}_K^{\mathsf{m}_2}$ and $K^{\mathsf{m}^*} = K^{\mathsf{m}_1} \cap K^{\mathsf{m}_2}$. Consequently, there exists a smallest (in the sense of divisibility) modulus of definition for χ, which is called the **conductor** of χ and is denoted by $\mathsf{f}(\chi)$. The ray class character χ is called **primitive** if $\mathsf{m} = \mathsf{f}(\chi)$ [equivalently, m is the smallest modulus of definition for χ].

If K is a number field, then the definition of a ray class character coincides with that given in [18, Definition 4.2.2]. If $K = \mathbb{Q}$ and $\mathsf{m} = m\mathbb{Z}$ for some $m \in \mathbb{N}$, then $\mathcal{C}_{\mathbb{Q}}^{\mathsf{m}} = (\mathbb{Z}/m\mathbb{Z})^\times$ by 5.6.4, and thus ray class characters modulo $m\mathbb{Z}$ are just Dirichlet characters modulo m.

However, we must point out already now that there is another way to look at ray class characters of $\mathbb{Q}$. Below in 5.8.7 we shall assoiate with a Dirichlet character a special Hecke character and associate it with an idele class character.

Theorem and Definition 5.8.2. *Let K be a global field, and let $\mathsf{m}, \mathsf{m}_1 \in \mathcal{D}'_K$ such that $\mathsf{m}_1 \,|\, \mathsf{m}$. Let φ be a ray class character modulo m and m_1 a modulus* *of definition for φ. Then there exists a unique ray class character φ_1 modulo m_1 such that φ is induced by φ_1.*

In particular, if $\mathsf{f} = \mathsf{f}(\varphi)$, then there exists a unique ray class character φ_0 modulo f such that φ is induced by φ_0, and φ_0 is primitive.

φ_0 is called the **associated primitive ray class character** of φ.

Proof. By definition, it suffices to prove uniqueness. Note that φ is induced by a ray class character φ_1 modulo m_1 if and only if there is a commutative diagram

$$\begin{array}{ccc} \mathcal{C}_K^{\mathsf{m}} = \mathcal{D}_K^{\mathsf{m}}/(K^{\mathsf{m}}) & \xrightarrow{\ \varphi\ } & \mathbb{T} \\ {\scriptstyle \pi}\big\downarrow & \nearrow{\scriptstyle \varphi_1} & \\ \mathcal{C}_K^{\mathsf{m}_1} = \mathcal{D}_K^{\mathsf{m}_1}/(K^{\mathsf{m}_1}) & & \end{array},$$

where π is the epimorphisms defined in 5.7.2.3. Therefore φ_1 is uniquely determined by φ.

If $\mathsf{f} = \mathsf{f}(\varphi)$ and φ is induced by the ray class character φ_0 modulo f, then φ_0 is primitive. Indeed, if $\mathsf{f}_0 = \mathsf{f}(\varphi_0)$, then $\mathsf{f}_0 \mid \mathsf{f}$, and f_0 is a modulus of definition for φ. Hence $\mathsf{f}_0 = \mathsf{f}$. □

Definition and Remarks 5.8.3. Let K be a global field.

1. We set

$$K_\infty^\times = \prod_{v \in \mathsf{M}_K^\infty} K_v^\times; \quad \text{in particular, } K_\infty^\times = \mathbf{1} \text{ in the function field case.}$$

In the number field case we endow $K_\infty^\times$ with the product topology, and we embed $K^\times$ into $K_\infty^\times$ by identifying $a \in K^\times$ with $(\sigma_v(a))_{v \in \mathsf{M}_K^\infty}$.

2. A **Hecke character modulo** m is a homomorphism $\chi\colon \mathcal{D}_K^{\mathsf{m}} \to \mathbb{T}$ such that there exists a (continuous) character $\chi_\infty \in \mathsf{X}(K_\infty^\times)$ satisfying $\chi((a)^K) = \chi_\infty(a)$ for all $a \in K^{\circ\mathsf{m}}$. Consequently, in the function field case a Hecke character is nothing but a ray class character modulo m.

3. Let K be a number field. By 1.7.8 we obtain

$$\mathsf{X}(K_\infty^\times) = \prod_{v \in \mathsf{M}_K^\infty} \mathsf{X}(K_v^\times),$$

and if $\psi = (\psi_v)_{v \in \mathsf{M}_K^\infty} \in \mathsf{X}(K_\infty^\times)$ and $\alpha = (\alpha_v)_{v \in \mathsf{M}_K^\infty}$, then

$$\psi(\alpha) = \prod_{v \in \mathsf{M}_K^\infty} \psi_v(\alpha_v).$$

We will mainly be concerned with characters of finite order.

Theorem 5.8.4. *Let K be a number field and (r_1, r_2) its signature.*

1. $(K_\infty^\times : K_\infty^{\times 2}) = 2^{r_1}$, and for $\psi \in \mathsf{X}(K_\infty^\times)$ the following assertions are equivalent:

(a) ψ is of finite order.

(b) $\psi = (\psi_v)_{v \in \mathsf{M}_K^\infty}$ such that, for all $v \in \mathsf{M}_K^\infty$, either $\psi_v = 1$, or v is real and $\psi_v = \mathrm{sgn}\colon K_v^\times \to \{\pm 1\} \subset \mathbb{T}$.

(c) $\psi^2 = 1$.

2. Let $\mathfrak{m} \in \mathcal{J}'_K$.

(a) $K^{\circ\mathfrak{m}}$ is dense in $K_\infty^\times$.

(b) The embedding $K^\times \hookrightarrow K_\infty^\times$ induces an isomorphism

$$j_\mathfrak{m}\colon K^{\circ\mathfrak{m}}/K^\mathfrak{m} \overset{\sim}{\to} K_\infty^\times/K_\infty^{\times 2}.$$

(c) A homomorphism $\chi\colon \mathcal{J}_K^\mathfrak{m} \to \mathbb{T}$ is a ray class character modulo $\mathfrak{m}$ if and only if χ is a Hecke character modulo $\mathfrak{m}$ such that χ_∞ is of finite order.

Proof. 1. If $v \in \mathsf{M}_K^\infty$, then $K_v^{\times 2} = K_v^+ = \mathbb{R}_{>0}$ if v is real, and $K_v^{\times 2} = K_v^\times = \mathbb{C}^\times$ if v is complex. Hence

$$K_\infty^\times/K_\infty^{\times 2} = \prod_{v \in \mathsf{M}_K^\infty} K_v^\times/K_v^{\times 2} \cong (\mathbb{R}^\times/\mathbb{R}_{>0})^{r_1}, \quad \text{and} \quad (K_\infty^\times : K_\infty^{\times 2}) = 2^{r_1}.$$

(a) $\Rightarrow$ (b) $\psi = (\psi_v)_{v \in \mathsf{M}_K^\infty}$ is of finite order if and only if all ψ_v are of finite order. Hence the assertion follows by 1.7.11.

(b) $\Rightarrow$ (c) $\Rightarrow$ (a) Obvious.

2. (a) Let $(x_v)_{v \in \mathsf{M}_K^\infty} \in K_\mathbb{R}^\times$ and $\varepsilon \in \mathbb{R}_{>0}$. Then there exists some $z \in K$ such that $|z - x_v|_v < \varepsilon/2$ for all $v \in \mathsf{M}_K^\infty$ (see 5.4.2.5), and by the weak approximation theorem there exists some $x \in K$ such that $v(x-1) \ge v(\mathfrak{m})$ for all $v \in \mathrm{supp}(\mathfrak{m})$, and $|x - z|_v < \varepsilon/2$ for all $v \in \mathsf{M}_K^\infty$. Then $x \in K^{\circ\mathfrak{m}}$ and $|x - x_v|_v < \varepsilon$ for all $v \in \mathsf{M}_K^\infty$.

(b) Let $j'_\mathfrak{m}\colon K^{\circ\mathfrak{m}} \to K_\infty^\times/K_\infty^{\times 2}$ be defined by $j'_\mathfrak{m}(a) = aK_\infty^{\times 2}$. Then

$$\begin{aligned}\mathrm{Ker}(j'_\mathfrak{m}) &= \{a \in K^{\circ\mathfrak{m}} \mid \sigma_v(a) \in K_v^{\times 2} \ \text{ for all real } v \in \mathsf{M}_K^\infty\} \\ &= \{a \in K^{\circ\mathfrak{m}} \mid \sigma_v(a) > 0 \ \text{ for all real } v \in \mathsf{M}_K^\infty\} = K^\mathfrak{m}.\end{aligned}$$

By (a), $\mathrm{Im}(j'_\mathfrak{m})$ is dense in $K_\infty^\times/K_\infty^{\times 2}$. Since $K_\infty^\times/K_\infty^{\times 2}$ is finite, $j'_\mathfrak{m}$ is surjective and induces an isomorphism as asserted.

(c) Let first $\chi\colon \mathcal{J}_K^\mathfrak{m} \to \mathbb{T}$ be a ray class character modulo $\mathfrak{m}$. For $\alpha \in K_\infty^\times$ let $a \in K^{\circ\mathfrak{m}}$ be such that $j_\mathfrak{m}(aK^\mathfrak{m}) = \alpha K_\infty^{\times 2}$, and set $\chi_\infty(\alpha) = \chi(a\mathfrak{o}_K)$. Then it follows that $\chi_\infty \in \mathsf{X}(K_\infty^\times)$, $\chi_\infty^2 = 1$, and $\chi(a\mathfrak{o}_K) = \chi_\infty(a)$ for all $a \in K^{\circ\mathfrak{m}}$. Hence χ is a Hecke character modulo $\mathfrak{m}$.

Let now $\chi\colon \mathfrak{J}_K^{\mathfrak{m}} \to \mathbb{T}$ be a Hecke character modulo $\mathfrak{m}$ such that χ_∞ is of finite order (then $\chi_\infty^2 = 1$ by 1.). If $a \in K^{\mathfrak{m}}$, then $a \in K_\infty^{\times 2}$, and therefore $\chi(a\mathfrak{o}_K) = \chi_\infty(a) = 1$. Hence χ is a ray class character modulo $\mathfrak{m}$. □

We shall now interpret Hecke characters (and in particular ray class characters) in terms of ideles. To that end we return to a common treatment of number fields and function fields.

Definition and Remarks 5.8.5 (Hecke characters vs. idele class characters).

Let K be a global field.

1. An **idele class character** is a continuous homomorphism $\chi\colon \mathsf{J}_K \to \mathbb{T}$ such that $\chi \restriction K^\times = 1$ [then χ induces an (equally denoted) continuous homomorphism $\chi\colon \mathsf{C}_K \to \mathbb{T}$ which we also call an idele class character].

2. For $\mathsf{m} \in \mathcal{D}'_K$ we set

$$\mathsf{U}_K^{\mathsf{m}} = \prod_{v \in \mathsf{M}_K^\infty} 1 \times \prod_{v \in \mathsf{M}_K^0} U_v^{(v(\mathsf{m}))} \subset \mathsf{U}_K = \mathsf{U}_K^1 = \prod_{v \in \mathsf{M}_K^\infty} 1 \times \prod_{v \in \mathsf{M}_K^0} U_v \subset \mathsf{J}_K^0 \cap \mathsf{E}_K.$$

Then U_K carries the product topology (see 5.3.6.3). Hence U_K is a compact topological group, and $\{\mathsf{U}_K^{\mathsf{m}} \mid \mathsf{m} \in \mathcal{D}'_K\}$ is a fundamental system of neighbourhoods of 1 in U_K.

Let χ be an idele class character. Then $\chi(\mathsf{U}_K)$ is a compact subgroup of $\mathbb{T}$ with subgroup topology. By 1.6.1.2, it is finite, $\mathrm{Ker}(\chi) \cap \mathsf{U}_K$ is an open subgroup of U_K, and thus there exists some $\mathsf{m} \in \mathcal{D}'_K$ such that $\mathsf{U}_K^{\mathsf{m}} \subset \mathrm{Ker}(\chi)$.

An integral divisor $\mathsf{m} \in \mathcal{D}'_K$ satisfying $\mathsf{U}_K^{\mathsf{m}} \subset \mathrm{Ker}(\chi)$ is called a **modulus of definition** for χ. If m' and m'' are moduli of definition for χ, then $\mathsf{U}_K^{\gcd(\mathsf{m}',\mathsf{m}'')} = \mathsf{U}_K^{\mathsf{m}'}\mathsf{U}_K^{\mathsf{m}''} \subset \mathrm{Ker}(\chi)$, and therefore $\gcd(\mathsf{m}',\mathsf{m}'')$ is a modulus of definition for χ, too. Consequently, there exists a unique smallest modulus of definition for χ (in the sense of divisibility). It is called the **conductor** of χ and denoted by $\mathsf{f}(\chi)$. An integral divisor $\mathsf{m} \in \mathcal{D}'_K$ is a modulus of definition for χ if and only if $\mathsf{f}(\chi) \mid \mathsf{m}$.

3. Let χ be an idele class character, $v \in \mathsf{M}_K$ and $j_v\colon K_v^\times \to \mathsf{J}_K$ the local embedding (see 5.3.1.5). Then $\chi_v = \chi \circ j_v \in \mathsf{X}(K_v^\times)$ is called the (**local**) v-**component** of χ (see 4.8.8). Its conductor exponent $f_v(\chi) = f(\chi_v)$ is called the (**local**) **conductor exponent** of χ at v.

If $\alpha = (\alpha_v)_{v \in \mathsf{M}_K} \in \mathsf{J}_K$, let S be a finite subset of M_K such that

$$\alpha \in \mathsf{J}_{K,S} = \prod_{v \in S} K_v^\times \times \prod_{v \in \mathsf{M}_K \setminus S} U_v.$$

Then 1.7.8 implies $\chi_v(\alpha_v) = 1$ for almost all $v \in \mathsf{M}_K$,

$$\chi(\alpha) = \prod_{v \in \mathsf{M}_K} \chi_v(\alpha_v), \quad \text{and we assert that} \quad \mathsf{f}(\chi) = \prod_{v \in \mathsf{M}_K^0} \mathsf{p}_v^{f_v(\chi)}.$$

Indeed, it $v \in \mathsf{M}_K^0$ and $n \in \mathbb{N}_0$, then

$$\mathsf{p}_v^n \,|\, \mathsf{f}(\chi) \iff j_v(U_v^{(n)}) \subset \mathrm{Ker}(\chi) \iff U_v^{(n)} \subset \mathrm{Ker}(\chi_v) \iff n \ge f_v(\chi).$$

4. Let $\chi\colon \mathsf{J}_K \to \mathbb{T}$ be an idele class character and $\mathsf{m} \in \mathcal{D}_K'$ a modulus of definition for χ. We shall construct a Hecke character χ' modulo m.

For $\mathsf{a} \in \mathcal{D}_K^{\mathsf{m}}$ we choose an idele $\alpha = (\alpha_v)_{v \in \mathsf{M}_K} \in \mathsf{J}_K$ such that $\mathsf{a} = \iota_K(\alpha)$, and we define

$$\chi'(\mathsf{a}) = \overline{\chi}(\alpha) \prod_{v \in \mathrm{supp}(\mathsf{m}) \cup \mathsf{M}_K^\infty} \chi_v(\alpha_v) = \prod_{v \in \mathsf{M}_K^0 \setminus \mathrm{supp}(\mathsf{m})} \overline{\chi}_v(\alpha_v).$$

This definition is independent of the choice of α. Indeed, if $\alpha' = (\alpha_v')_{v \in \mathsf{M}_K} \in \mathsf{J}_K$ is another idele such that $\mathsf{a} = \iota_K(\alpha')$, then $\alpha_v' = \alpha_v u_v$ for some $u_v \in U_v$ for all $v \in \mathsf{M}_K^0$, but if $v \in \mathsf{M}_K^0 \setminus \mathrm{supp}(\mathsf{m})$, then $\chi_v \restriction U_v = 1$, and consequently $\chi_v(\alpha_v) = \chi_v(\alpha_v')$.

Obviously, $\chi'\colon \mathcal{D}_K^{\mathsf{m}} \to \mathbb{T}$ is a homomorphism. If $a \in \mathcal{D}_K^{\mathsf{m}}$, then $(a)^K = \iota_K(a)$ and $a \in U_v$ for all $v \in \mathrm{supp}(\mathfrak{m})$. Hence it follows that

$$\chi'((a)^K) = \overline{\chi}(a) \prod_{v \in \mathrm{supp}(\mathsf{m}) \cup \mathsf{M}_K^\infty} \chi_v(a) = \prod_{v \in \mathrm{supp}(\mathsf{m}) \cup \mathsf{M}_K^\infty} \chi_v(a).$$

We define $\chi_\infty'\colon K_\infty^\times \to \mathbb{T}$ by

$$\chi_\infty'((\alpha_v)_{v \in \mathsf{M}_K^\infty}) = \prod_{v \in \mathsf{M}_K^\infty} \chi_v(\alpha_v).$$

Then $\chi_\infty' \in \mathsf{X}(K_\infty^\times)$, and $\chi_{\infty,v}' = \chi_v$ for all $v \in \mathsf{M}_K^\infty$. If $a \in K^{\circ\mathsf{m}}$, then $a \in U_v^{(v(\mathsf{m}))}$, hence $\chi_v(a) = 1$ for all $v \in \mathrm{supp}(\mathsf{m})$, and consequently

$$\chi'((a)^K) = \prod_{v \in \mathsf{M}_K^\infty} \chi_v(a) = \chi_\infty'(a).$$

Therefore χ' is a Hecke character, and it is a ray class character if and only if χ_v is of finite order for all $v \in \mathsf{M}_K^\infty$.

5. Now we proceed in the opposite direction. Let $\mathsf{m} \in \mathcal{D}_K'$, $\chi\colon \mathcal{D}_K^{\mathsf{m}} \to \mathbb{T}$ a Hecke character modulo m, and $\chi_\infty = (\chi_{\infty,v})_{v \in \mathsf{M}_K^\infty} \in \mathsf{X}(K_\infty^\times)$. We consider the group

$$\overline{\mathsf{J}}_K^{(\mathsf{m})} = \prod_{v \in \mathsf{M}_K^\infty} K_v^\times \times \prod_{v \in \mathrm{supp}(\mathsf{m})} U_v^{(v(\mathsf{m}))} \times \prod_{v \in \mathsf{M}_K^0 \setminus \mathrm{supp}(\mathsf{m})} (K_v^\times \mid U_v) \subset \mathsf{J}_K.$$

$\overline{\mathsf{J}}_K^{(\mathsf{m})}$ is an open subgroup of J_K satisfying $\mathsf{J}_K^{(\mathsf{m})} \subset \overline{\mathsf{J}}_K^{(\mathsf{m})}$, $\overline{\mathsf{J}}_K^{(\mathsf{m})} K^\times = \mathsf{J}_K$, and $\iota_K(\overline{\mathsf{J}}_K^{(\mathsf{m})}) = \mathcal{D}_K^{\mathsf{m}}$. For $\alpha = (\alpha_v)_{v \in \mathsf{M}_K} \in \overline{\mathsf{J}}_K^{(\mathsf{m})}$ we set

$$\chi^*(\alpha) = \overline{\chi}(\iota_K(\alpha))\chi_\infty((\alpha_v)_{v \in \mathsf{M}_K^\infty}) = \overline{\chi}(\iota_K(\alpha)) \prod_{v \in \mathsf{M}_K^\infty} \chi_{\infty,v}(\alpha_v).$$

Then $\chi^* \colon \overline{\mathsf{J}}_K^{(\mathsf{m})} \to \mathbb{T}$ is a continuous homomorphism. If $a \in K^{\circ\mathsf{m}}$, then $a \in \overline{\mathsf{J}}_K^{(\mathsf{m})}$, $\iota_K(a) = (a)^K$, and therefore thus $\chi^*(a) = \chi((a)^K)^{-1}\chi_\infty(a) = 1$. Since $\mathsf{J}_K = \overline{\mathsf{J}}_K^{(\mathsf{m})} K^\times$, χ^* has a unique extension to an (equally denoted) homomorphism $\chi^* \colon \mathsf{J}_K \to \mathbb{T}$ such that $\chi^* \restriction K^\times = 1$, and as $\overline{\mathsf{J}}_K^{(\mathsf{m})}$ is open in J_K, it follows that χ is continuous. If $\alpha \in \mathsf{U}_K^{\mathsf{m}}$, then $\iota_K(\alpha) = 1$ and thus $\chi^*(\alpha) = 1$. Consequently, χ^* is an idele class character and m is a modulus of definition for χ^*. If $v \in \mathsf{M}_K^\infty$ and $\alpha_v \in K_v^\times$, then $\iota_K(j_v(\alpha_v)) = 1$, and consequently

$$\chi_v^*(\alpha_v) = \chi^*(j_v(\alpha_v)) = \chi_{\infty,v}(\alpha_v).$$

Theorem and Conventions 5.8.6. *Let K be a global field and $\mathsf{m} \in \mathcal{D}_K'$. The assignments $\chi \mapsto \chi'$ and $\chi \mapsto \chi^*$ constructed in* 5.8.5.(**4** and **5**) *define mutually inverse bijective maps between the set of all idele class characters admitting m as a modulus of definition and the set of all Hecke characters modulo m.*

For a Hecke character χ modulo $\mathfrak{m}$ we denote the associated idele class character likewise by χ. Then we have

$$\chi((a)^K) = \prod_{v \in \mathrm{supp}(\mathfrak{m}) \cup \mathsf{M}_K^\infty} \chi_v(a) \quad \text{for all } a \in K(\mathfrak{m}).$$

Proof. Let $\mathsf{m} \in \mathcal{D}_K'$, $\alpha = (\alpha_v)_{v \in \mathsf{M}_K} \in \overline{\mathsf{J}}_K^{(\mathsf{m})}$ and $\mathsf{a} = \iota_K(\alpha) \in \mathcal{D}_K^{\mathsf{m}}$.

Let first $\chi \colon \mathsf{J}_K \to \mathbb{T}$ be an idele class character and m a modulus of definition for χ. If $v \in \mathrm{supp}(\mathsf{m})$, then $\alpha_v \in U_v^{(v(\mathsf{m}))}$ and therefore $\chi_v(\alpha_v) = 1$. Hence we obtain

$$\chi'(\mathsf{a}) = \overline{\chi}(\alpha) \prod_{v \in \mathsf{M}_K^\infty} \chi_v(\alpha_v) \quad \text{and} \quad \chi'_{\infty,v} = \chi_v \quad \text{for all } v \in \mathsf{M}_K^\infty.$$

It follows that

$$\chi'^*(\alpha) = \overline{\chi}'(\mathsf{a}) \prod_{v \in \mathsf{M}_K^\infty} \chi'_{\infty,v}(\alpha_v) = \chi(\alpha) \prod_{v \in \mathsf{M}_K^\infty} \overline{\chi}_v(\alpha_v) \prod_{v \in \mathsf{M}_K^\infty} \chi_v(\alpha_v) = \chi(\alpha).$$

Hence $\chi \restriction \overline{\mathsf{J}}_K^{\mathsf{m}} = \chi'^* \restriction \overline{\mathsf{J}}_K^{\mathsf{m}}$, and since $\chi \restriction K^\times = \chi'^* \restriction K^\times = 1$, it follows that $\chi = \chi'^*$.

Now let $\chi \colon \mathcal{D}_K^{\mathsf{m}} \to \mathbb{T}$ be a Hecke character modulo m. Then

$$\chi^*(\alpha) = \overline{\chi}(\mathsf{a}) \prod_{v \in \mathsf{M}_K^\infty} \chi_{\infty,v}(\alpha_v) \quad \text{and} \quad \chi_v^* = \chi_{\infty,v} \quad \text{for all } v \in \mathsf{M}_K^\infty.$$

It follows that

$$\chi^{*\prime}(\mathsf{a}) = \overline{\chi}^*(\alpha) \prod_{v\in \mathsf{M}_K^\infty} \chi_v^*(\alpha_v) = \chi(\mathsf{a}) \prod_{v\in \mathsf{M}_K^\infty} \overline{\chi}_{\infty,v}(\alpha_v) \prod_{v\in \mathsf{M}_K^\infty} \chi_{\infty,v}(\alpha_v) = \chi(\mathsf{a}),$$

hence $\chi^{*\prime} = \chi$. □

Finally we compare (as already announced) the Hecke characters of finite order with the Dirichlet characters of elementary number theory.

Remark 5.8.7. Let $m \in \mathbb{N}$, $m \geq 3$, and let $\varphi \in \mathsf{X}(m) = \mathsf{X}((\mathbb{Z}/m\mathbb{Z})^\times)$ be a Dirichlet character modulo m. We view φ as a homorphism $\varphi\colon \mathbb{Q}^{\circ m} \to \mathbb{T}$ satisfying $\varphi \upharpoonright \mathbb{Q}^{\circ m} = 1$. Let $\chi\colon \mathcal{J}_{\mathbb{Q}}^m \to \mathbb{T}$ be defined by $\chi(a\mathbb{Z}) = \varphi(a)\mathrm{sgn}(a)$ for all $a \in \mathbb{Q}(m)$. Then it follows that χ is a Hecke character modulo $m\mathbb{Z}$ and $\chi_\infty = \mathrm{sgn}\colon \mathbb{R}^\times \to \{\pm 1\} \subset \mathbb{T}$. We identify it with the associated idele class character. We obtain

$$\chi(a\mathbb{Z}) = \prod_{p\in\mathbb{P}} \chi_p(a)\,\mathrm{sgn}(a) \quad \text{for all} \quad a \in \mathbb{Q}(m),$$

and it is easily checked that χ_p is just the p-component of the Dirichlet character φ as defined in [18, Theorem 3.3.6]: $\chi_p = \varphi_p$ for all primes p.

6

Global class field theory

Our development of global class field theory and its connection with central simple algebras is heavily influenced by [40]. I put emphasis on the theory of ray class fields and their connection with the ideal theory of holomorphy domains. Thereby results an ideal-theoretic formulation of class field theory valid both for number fields and function fields. General references for this chapter (besides [40]) are [6], [46], [48], [60], [1], [33], [38], [64].

Throughout this chapter, let K be a global field.

Let $\overline{K}$ be an algebraic closure of K, $K_{\rm sep}$ the separable closure of K in $\overline{K}$, and let $G_K = \text{Gal}(K_{\rm sep}/K) = \text{Gal}(\overline{K}/K)$ be the absolute Galois group of K. Let $K_{\rm ab}$ be the maximal abelian extension of K inside $K_{\rm sep}$. Then $\text{Gal}(K_{\rm sep}/K_{\rm ab}) = G_K^{(1)}$ and $\text{Gal}(K_{\rm ab}/K) = G_K/G_K^{(1)} = G_K^{\mathsf{a}}$ (see 1.8.5.2). We assume that all algebraic extensions of K are contained in $\overline{K}$. We tacitly use notation and results of Chapter 5.

6.1 Cohomology of the idele groups

As a first step towards global class field theory, we study the Galois cohomology of the idele group and the idele class group.

Theorem 6.1.1. *Let L/K be a finite Galois extension, $G = \text{Gal}(L/K)$, and let S be a finite subset of M_K such that $\mathsf{M}_K^\infty \cup \text{Ram}(L/K) \subset S$. For each $v \in \mathsf{M}_K$ we fix some $\overline{v} \in \mathsf{M}_L$ lying above v, we set $n_v = [L_{\overline{v}} : K_v]$, $G_{\overline{v}} = \text{Gal}(L_{\overline{v}}/K_v) \subset G$, and we denote by R_v a set of representatives for $G/G_{\overline{v}}$ in G such that $1 \in R_v$ (then $n_v = [L_w\!:\!K_v]$ and $G_{\overline{v}} \cong \text{Gal}(L_w/K_v)$ for all $w \in \mathsf{M}_L$ above v, and $\{w \in \mathsf{M}_L \mid w\,|\,v\} = \{\sigma\overline{v} \mid \sigma \in R_v\}$).*

DOI: 10.1201/9780429506574-6

1. For all $q \geq -1$ there exists a (natural) commutative diagram

$$\begin{array}{ccc} \theta_S^q\colon H^q(G, \mathsf{J}_{L,S_L}) & \xrightarrow{(\theta_v^q)_{v\in S}} & \bigoplus_{v\in S} H^q(G_{\overline{v}}, L_{\overline{v}}^\times) \\ {\scriptstyle H^q(j)}\big\downarrow & & \big\downarrow{\scriptstyle i} \\ \theta^q\colon\ H^q(G, \mathsf{J}_L) & \xrightarrow{(\theta_v^q)_{v\in \mathsf{M}_K}} & \bigoplus_{v\in \mathsf{M}_K} H^q(G_{\overline{v}}, L_{\overline{v}}^\times), \end{array}$$

where $H^q(j)$ is induced by the inclusion $j = (\mathsf{J}_{L,S_L} \hookrightarrow \mathsf{J}_L)$ and i is the embedding of a subsum. The horizontal arrows are isomorphisms whose local components $\theta_v^q\colon H^q(G, \mathsf{J}_L) \to H^q(G_{\overline{v}}, L_{\overline{v}}^\times)$ are given as follows:

(a) Let $q \geq 1$ and $x\colon G^q \to \mathsf{J}_L$ a defining q-cocycle of $\xi \in H^q(G, \mathsf{J}_L)$. If $x = (x_v\colon G^q \to L_v^\times)_{v\in\mathsf{M}_K}$, then

$$\theta_v^q x = x_v \restriction G_{\overline{v}}^q\colon G_{\overline{v}}^q \to L_{\overline{v}}^\times$$

is a defining q-cocycle of $\theta_v^q(\xi)$ for all $v \in \mathsf{M}_K$.

(b) Let $q = 0$, and let $\xi = \alpha\mathsf{N}_{L/K}\mathsf{J}_L \in \mathsf{J}_K/\mathsf{N}_{L/K}\mathsf{J}_L = H^0(G, \mathsf{J}_L)$ for some $\alpha = (\alpha_v)_{v\in\mathsf{M}_K} \in \mathsf{J}_K$. Then

$$\theta_v^0\xi = \alpha_v\mathsf{N}_{L_{\overline{v}}/K_v}L_{\overline{v}}^\times \in K_v^\times/\mathsf{N}_{L_{\overline{v}}/K_v}L_{\overline{v}}^\times = H^0(G_{\overline{v}}, L_{\overline{v}}^\times).$$

In particular, the isomorphism

$$\theta^0\colon \mathsf{J}_K/\mathsf{N}_{L/K}\mathsf{J}_L \to \bigoplus_{v\in\mathsf{M}_K} K_v^\times/\mathsf{N}_{L_{\overline{v}}/K_v}L_{\overline{v}}^\times$$

is given by

$$\theta^0(\alpha\mathsf{N}_{L/K}\mathsf{J}_L) = (\alpha_v\mathsf{N}_{L_{\overline{v}}/K_v}L_{\overline{v}}^\times)_{v\in\mathsf{M}_K}$$

for all $\alpha = (\alpha_v)_{v\in\mathsf{M}_K}$, and the following local-global-principle holds:

$$\alpha \in \mathsf{N}_{L/K}(\mathsf{J}_L) \iff \alpha_v \in \mathsf{N}_{L_{\overline{v}}/K_v}L_{\overline{v}}^\times \ \text{ for all } \ v \in \mathsf{M}_K.$$

(c) Let $q = -1$ and $\xi = \beta I_G\mathsf{J}_L \in {}_{N_G}\mathsf{J}_L/I_G\mathsf{J}_L = H^{-1}(G, \mathsf{J}_L)$ for some $\beta = (\beta_{\overline{v}})_{\overline{v}\in\mathsf{M}_L} \in {}_{N_G}\mathsf{J}_L$. Then

$$\theta_v^{-1}\xi = \eta I_{G_{\overline{v}}}L_{\overline{v}}^\times, \quad \text{where} \quad \eta = \sum_{\rho\in R_v} \rho^{-1}\beta_{\rho\overline{v}} \in {}_{N_{G_{\overline{v}}}}L_{\overline{v}}^\times.$$

2. $H^1(G, \mathsf{J}_{L,S_L}) = H^1(G, \mathsf{J}_L) = \mathbf{1}$.

3. Let L/K be cyclic. Then $H^{-1}(G, \mathsf{J}_{L,S_L}) = H^{-1}(G, \mathsf{J}_L) = \mathbf{1}$,

$$H^0(G, \mathsf{J}_L) \cong H^0(G, \mathsf{J}_{L,S_L}) \cong \bigoplus_{v \in S} \mathbb{Z}/n_v\mathbb{Z} = \bigoplus_{v \in \mathsf{M}_K} \mathbb{Z}/n_v\mathbb{Z},$$

and the Herbrand quotients are given by

$$\mathsf{h}(G, J_L) = \mathsf{h}(G, J_{L,S_L}) = \prod_{v \in S} n_v.$$

Proof. 1. If $v \in \mathsf{M}_K$, then G operates transitively on the places of L lying above v (see 5.4.5.1, and G operates on

$$\prod_{\substack{w \in \mathsf{M}_L \\ w \mid v}} L_w^\times \quad \text{as follows: If } \beta = (\beta_w)_{w \mid v}, \text{ then } (\sigma\beta)_w = \sigma\beta_{\sigma^{-1}w} \text{ for all } \sigma \in G.$$

In this way,

$$\prod_{\substack{w \in \mathsf{M}_L \\ w \mid v}} L_w^\times = \prod_{\sigma \in R_v} \sigma L_{\overline{v}}^\times \text{ if } v \in S, \quad \text{and} \quad \prod_{\substack{w \in \mathsf{M}_L \\ w \mid v}} U_w = \prod_{\sigma \in R_v} \sigma U_{\overline{v}} \text{ if } v \in \mathsf{M}_K \setminus S.$$

are G-modules,

$$\mathsf{J}_{L,S_L} = \prod_{v \in S} \prod_{\substack{w \in \mathsf{M}_L \\ w \mid v}} L_w^\times \times \prod_{v \in \mathsf{M}_K \setminus S} \prod_{\substack{w \in \mathsf{M}_L \\ w \mid v}} U_w,$$

is their direct product, and 2.3.1.2 yields for every $q \ge -1$ an isomorphism

$$(\overline{\theta}_v^q)_{v \in S} \colon H^q(G, \mathsf{J}_{L,S_L}) \overset{\sim}{\to} \bigoplus_{v \in S} H^q\Big(G, \prod_{\substack{w \in \mathsf{M}_L \\ w \mid v}} L_w^\times\Big) \oplus \prod_{v \in \mathsf{M}_K \setminus S} H^q\Big(G, \prod_{\substack{w \in \mathsf{M}_L \\ w \mid v}} U_w\Big),$$

where for each $v \in S$ the homomorphism $\overline{\theta}_v^q$ is induced by the projection of J_{L,S_L} onto the v-component.

By means of 2.4.4 and 2.2.7.1 we obtain the isomorphisms

$$\theta_v'^q \colon H^q\Big(G, \prod_{\substack{w \in \mathsf{M}_L \\ w \mid v}} L_w^\times\Big) \overset{\sim}{\to} H^q(G_{\overline{v}}, L_{\overline{v}}^\times), \quad \text{for } v \in S,$$

and

$$H^q\Big(G, \prod_{\substack{w \in \mathsf{M}_L \\ w \mid v}} U_w\Big) \overset{\sim}{\to} H^q(G_{\overline{v}}, U_{\overline{v}}) \quad \text{for } v \in \mathsf{M}_K \setminus S.$$

If $v \in \mathsf{M}_K \setminus S$, then $L_{\overline{v}}/K_v$ is unramified, $H^0(G_{\overline{v}}, U_{\overline{v}}) = H^1(G_{\overline{v}}, U_{\overline{v}}) = \mathbf{0}$ by 4.1.10 and consequently $H^q(G_{\overline{v}}, U_{\overline{v}}) = \mathbf{0}$ for all $q \geq -1$ by 2.4.5.1, since $L_{\overline{v}}/K_v$ is unramified and thus cyclic. For $v \in S$, we set

$$\theta_v^q = \theta_v'^q \circ \overline{\theta}_v^q \colon H^q(G, \mathsf{J}_{L,S_L}) \to H^q(G_{\overline{v}}, L_{\overline{v}}^\times),$$

and then we obtain the isomorphism

$$\theta_S^q = (\theta_v^q)_{v \in S} \colon H^q(G, \mathsf{J}_{L,S_L}) \overset{\sim}{\to} \bigoplus_{v \in S} H^q(G_{\overline{v}}, L_{\overline{v}}^\times).$$

Regarding the explicit description of $\theta_v'^q$ in 2.4.4 and 2.2.7, the assertions (a), (b) and (c) follow.

Let $(\mathfrak{S}, \subset)$ be the directed set of all finite subsets S of M_K satifying $\mathsf{M}_K^\infty \cup \mathrm{Ram}(L/K) \subset S$, ordered by inclusion. If S, $S' \in \mathfrak{S}$ and $S \subset S'$ then the natural inclusions induce the commutative diagram

$$\begin{array}{ccc} \theta_S^q \colon H^q(G, \mathsf{J}_{L,S_L}) & \xrightarrow{(\theta_v^q)_{v \in S}} & \bigoplus_{v \in S} H^q(G_{\overline{v}}, L_{\overline{v}}^\times) \\ {\scriptstyle H^q(j_{S,S'})} \big\downarrow & & \big\downarrow {\scriptstyle i_{S,S'}} \\ \theta_{S'}^q \colon H^q(G, \mathsf{J}_{L,S'}) & \xrightarrow{(\theta_v^q)_{v \in S'}} & \bigoplus_{v \in S'} H^q(G_{\overline{v}}, L_{\overline{v}}^\times), \end{array}$$

in which the vertical arrows constitute inductive systems. Passing to the inductive limits, we obtain

$$\varinjlim_{\mathfrak{S}} H^q(G, \mathsf{J}_{L,S_L}) = H^q(G, \mathsf{J}_L) \quad \text{by 2.3.3},$$

$$\varinjlim_{\mathfrak{S}} \bigoplus_{v \in S} H^q(G_{\overline{v}}, L_{\overline{v}}^\times) = \bigoplus_{v \in \mathsf{M}_K} H^q(G_{\overline{v}}, L_{\overline{v}}^\times) \quad \text{by 1.3.2.}\mathbf{2}.,$$

and the commutative diagram

$$\begin{array}{ccc} H^q(G, \mathsf{J}_{L,S_L}) & \xrightarrow{\theta_S^q} & \bigoplus_{v \in S} H^q(G_{\overline{v}}, L_{\overline{v}}^\times) \\ {\scriptstyle H^q(j)} \big\downarrow & & \big\downarrow {\scriptstyle i} \\ H^q(G, \mathsf{J}_L) & \xrightarrow{\theta^q} & \bigoplus_{v \in \mathsf{M}_K} H^q(G_{\overline{v}}, L_{\overline{v}}^\times), \end{array}$$

where $\theta^q = \varinjlim_{\mathfrak{S}} \theta_S^q$ is an isomorphism by 1.3.3.2.

2. If $v \in \mathsf{M}_K$, then $H^1(G_{\overline{v}}, L_{\overline{v}}^\times) = \mathbf{1}$ by the 2.6.3, and thus it follows that $H^1(G, \mathsf{J}_{L,S_L}) = H^1(G, \mathsf{J}_L) = \mathbf{1}$ by 1.

3. By 2. and 2.4.5.1, it follows that $H^{-1}(G, \mathsf{J}_{L,S_L}) = H^{-1}(G, \mathsf{J}_L) = \mathbf{1}$.

If $v \in \mathsf{M}_K^0$, then $G_{\overline{v}}$ is cyclic (it is a subgroup of G), and therefore we obtain $H^0(G_{\overline{v}}, L_{\overline{v}}^\times) \cong \mathbb{Z}/n_v\mathbb{Z}$ by 4.1.10.1.

If $v \in \mathsf{M}_K^\infty$, then either $K_v = L_{\overline{v}}$ and $H^0(G_{\overline{v}}, L_{\overline{v}}^\times) = \mathbf{1}$, or v is real, $\overline{v}$ is complex, and $H^0(G_{\overline{v}}, L_{\overline{v}}^\times) = \mathbb{R}^\times/\mathsf{N}_{\mathbb{C}/\mathbb{R}}\mathbb{C}^\times \cong \mathbb{Z}/2\mathbb{Z}$. Therefore and by the very definition of the Herbrand quotient (see 2.4.5.2) the assertions follow. □

Theorem 6.1.2 (Main lemma of Artin and Tate). *Let G be a finite cyclic group operating on a non-empty finite set I by means of $(\sigma, i) \mapsto \sigma(i)$ for all $(\sigma, i) \in G \times I$. Let $V_1, \ldots, V_k$ be the distinct orbits in I of this operation. For $j \in [1, k]$, let $\iota_j \in V_j$, and denote by G_j the isotropy group of ι_j. Let U be a vector space over $\mathbb{R}$, $(u_i)_{i \in I}$ an $\mathbb{R}$-basis of U, and define a G-module structure on U by*

$$\sigma \sum_{i \in I} c_i u_i = \sum_{i \in I} c_i u_{\sigma(i)} \quad \text{for all} \quad \sigma \in G \quad \text{and} \quad (c_i)_{i \in I} \in \mathbb{R}^I.$$

Let Λ be a G-submodule of U which is a lattice in U satisfying $\mathbb{R}\Lambda = U$. Then

$$\mathsf{h}(G, \Lambda) = \prod_{j=1}^{k} |G_j|.$$

Proof. By definition, U is a G-module. We prove first:

> **A.** There exist a sublattice Λ' of Λ and a basis $(w_i)_{i \in I}$ of Λ' such that $\sigma(w_i) = w_{\sigma(i)}$ for all $\sigma \in G$ and $i \in I$.

Proof of **A.** Let $\|\cdot\| \colon U \to \mathbb{R}_{\ge 0}$ be the sum norm regarding the basis $(u_i)_{i \in I}$, i. e.,

$$\Big\| \sum_{i \in I} c_i u_i \Big\| = \sum_{i \in I} |c_i| \quad \text{for all} \quad (c_i)_{i \in I} \in \mathbb{R}^I.$$

Then $\|\sigma u\| = \|u\|$ for all $\sigma \in G$ and $u \in U$, and there exists some $b \in \mathbb{R}_{>0}$ such that for every $x \in U$ there exists some $\gamma \in \Lambda$ with $\|x - \gamma\| < b$.

Let $t \in \mathbb{R}$ such that $t > |G|\, b$, and for $j \in [1, k]$ let $\gamma_j \in \Lambda$ such that $\|t u_{\iota_j} - \gamma_j\| < b$. For $j \in [1, k]$ and $i \in V_j$ we set

$$G(i) = \{\sigma \in G \mid \sigma(\iota_j) = i\} \quad \text{and} \quad w_i = \sum_{\tau \in G(i)} \tau \gamma_j \in \Lambda.$$

If $\sigma \in G$ and $\tau \in G(i)$, then $\sigma\tau(\iota_j) = \sigma(i)$, hence $\sigma\tau \in G(\sigma(i))$. If $j \in [1, k]$, $i \in V_j$, and $\sigma \in G$, then

$$\sigma w_i = \sum_{\tau \in G(i)} \sigma\tau\gamma_j = \sum_{\rho \in G(\sigma(i))} \rho\gamma_j = w_{\sigma(i)},$$

and therefore $\sigma w_i = w_{\sigma(i)}$ for all $\sigma \in G$ and $i \in I$. Now it suffices to prove that $(w_i)_{i \in I}$ is linearly independent, for then

$$\Lambda' = \bigoplus_{i \in I} \mathbb{Z} w_i$$

fulfills our requirements.

We suppose to the contrary that

$$\sum_{i\in I} c_i w_i = 0 \quad \text{for some sequence} \quad (c_i)_{i\in I} \in \mathbb{R}^I \setminus \{\mathbf{0}\}.$$

We may assume that $|c_i| \le 1$ for all $i \in I$, and $c_\kappa = 1$ for some $\kappa \in I$. Then

$$\begin{aligned} 0 &= \sum_{j=1}^k \sum_{i\in V_j} c_i \sum_{\tau\in G(i)} \tau[tu_{\iota_j} - (tu_{\iota_j} - \gamma_j)] \\ &= t\sum_{j=1}^k \sum_{i\in V_j} c_i \sum_{\tau\in G(i)} \tau u_{\iota_j} - \sum_{j=1}^k \sum_{i\in I} c_i \sum_{\tau\in G(i)} (tu_{\iota_j} - \gamma_j), \end{aligned}$$

and consequently

$$t \le t|G(\kappa)| \le t\sum_{i\in I} |G(i)||c_i| = \Big\| t\sum_{i\in I} |G(i)|c_i u_i \Big\| = \Big\| t\sum_{j=1}^k \sum_{i\in V_j} c_i \sum_{\tau\in G(i)} \tau u_{\iota_j} \Big\|$$

$$= \Big\| \sum_{j=1}^k \sum_{i\in V_j} c_i \sum_{\tau\in G(i)} \tau(tu_{\iota_j} - \gamma_j) \Big\| \le \sum_{i\in I} |c_i|\,|G(i)|\,b \le |G|\,b,$$

a contradiction. $\square$[**A.**]

From **A** we obtain

$$\Lambda' = \bigoplus_{j=1}^k \bigoplus_{i\in V_j} \mathbb{Z}w_i.$$

For $j \in [1,k]$ let R_j be a set of representatives for G/G_j containing 1 in G. By means of 2.1.9 it follows that

$$\bigoplus_{i\in V_j} \mathbb{Z}w_i = \bigoplus_{\sigma\in R_j} \mathbb{Z}w_{\sigma(\iota_j)} = \bigoplus_{\sigma\in R_j} \sigma(\mathbb{Z}w_{\iota_j}) \cong \mathcal{M}_G^{G_j}(\mathbb{Z}w_{\iota_j}),$$

observing that $\mathbb{Z}w_{\iota_j}$ is a trivial G_j-module isomorphis to $\mathbb{Z}$. By 2.2.7.1 and 2.4.5.3 we obtain, for $\nu \in \{-1,0\}$,

$$H^\nu\Big(G, \bigoplus_{i\in V_j} \mathbb{Z}w_i\Big) \cong H^\nu(G_j, \mathbb{Z}w_{\iota_j}) \cong H^\nu(G_j, \mathbb{Z}) = \begin{cases} \mathbb{Z}/|G_j|\mathbb{Z} & \text{if } \nu = 0, \\ 0 & \text{if } \nu = -1. \end{cases}$$

By 2.3.1 it follows that

$$H^{-1}(G,\Lambda') = \mathbf{0} \quad \text{and} \quad |H^0(G,\Lambda')| = \prod_{j=1}^k |H^0(G_j,\mathbb{Z})| = \prod_{j=1}^k |G_j| = h(G,\Lambda').$$

By 2.4.5, the exact sequence $\mathbf{0} \to \Lambda' \hookrightarrow \Lambda \to \Lambda/\Lambda' \to \mathbf{0}$ together with $|\Lambda/\Lambda'| < \infty$ implies

$$\mathsf{h}(G,\Lambda) = \mathsf{h}(G,\Lambda')\mathsf{h}(G,\Lambda/\Lambda') = \mathsf{h}(G,\Lambda') = \prod_{j=1}^k |G_j|. \qquad \square$$

Theorem 6.1.3 (First inequality).
Let L/K be a cyclic extension, $G = \mathrm{Gal}(L/K)$ and $n = [L{:}K]$.

1. *Let S be a finite subset of M_K such that $\mathsf{M}_K^\infty \subset S$. For $v \in S$ and $\overline{v} \in \mathsf{M}_L$ such that $\overline{v}\,|\,v$ we set $n_v = [L_{\overline{v}}{:}K_v]$. Then the S_L-unit group*

$$L_{S_L} = \mathsf{J}_{L,S_L} \cap L^\times = \{x \in L^\times \mid |x|_w = 1 \text{ for all } w \in \mathsf{M}_L \setminus S_L\}$$

is a G-submodule of $L^\times$, and

$$\mathsf{h}(G, L_{S_L}) = \frac{1}{n} \prod_{v \in S} n_v.$$

2. *$H^0(G, \mathsf{C}_L)$ and $H^{-1}(G, \mathsf{C}_L)$ are finite,*

$$\mathsf{h}(G, \mathsf{C}_L) = n, \quad \text{and} \quad (\mathsf{C}_K : \mathsf{N}_{L/K}\mathsf{C}_L) \geq n.$$

Proof. 1. If $x \in L_{S_L}$, $\sigma \in G$ and $v \in \mathsf{M}_K \setminus S_L$, then $\sigma^{-1}v \in \mathsf{M}_K \setminus S_L$ and thus $|\sigma a|_v = |x|_{\sigma^{-1}v}| = 1$. Hence L_{S_L} is a G-submodule of $L^\times$.

For $v \in \mathsf{M}_K$ we denote by V_v the set of all places of L lying above v. By 5.4.5.1, G operates on the finite set S_L, and $\{V_v \mid v \in S\}$ is the set of orbits of this operation. If $w \in V_v$, then $G_w = \mathrm{Gal}(L_w/K_v)$ is the isotropy group of w and $|G_w| = n_v$. Let $\{\boldsymbol{e}_w \mid w \in S_L\}$ be the canonical basis of $\mathbb{R}^{S_L}$, that is, $\boldsymbol{e}_w = (\delta_{w,v})_{v \in S_L}$ for all $w \in S_L$. Then $\mathbb{R}^{S_L}$ is a G-module by means of the action

$$\sigma \sum_{w \in S_L} c_w \boldsymbol{e}_w = \sum_{w \in S_L} c_w \boldsymbol{e}_{\sigma w} \quad \text{for all } (c_w)_{w \in S_L} \in \mathbb{R}^{S_L},$$

$$H = \Big\{ (x_w)_{w \in S_L} \Bigm| \sum_{w \in S_L} x_w = 0 \Big\}$$

is a G-invariant hyperplane in $\mathbb{R}^{S_L}$, and $\mathbb{R}^{S_L} = H \oplus \mathbb{R}\boldsymbol{e}^*$, where

$$\boldsymbol{e}^* = \sum_{w \in S_L} \boldsymbol{e}_w \quad \text{and} \quad \sigma \boldsymbol{e}^* = \boldsymbol{e}^* \quad \text{for all } \sigma \in G.$$

Let $\lambda \colon L^{S_L} \to \mathbb{R}^{S_L}$ be defined by

$$\lambda(a) = (\log |a|_w)_{w \in S_L} = \sum_{w \in S_L} \log |a|_w \boldsymbol{e}_w \quad \text{for all } a \in L^{S_L}.$$

By 5.3.8, λ is a group homomorphism, $\mathrm{Ker}(\lambda) = \mu_L$, $\lambda(L_{S_L})$ is a lattice in H and $\mathbb{R}\lambda(L_{S_L}) = H$. If $a \in L_{S_L}$ and $\sigma \in G$, then

$$\lambda(\sigma a) = \sum_{w \in S_L} \log |\sigma a|_w \boldsymbol{e}_w = \sum_{w \in S_L} \log |a|_{\sigma^{-1}w} \boldsymbol{e}_w = \sum_{w \in S_L} \log |a|_w \boldsymbol{e}_{\sigma w} = \sigma\lambda(a).$$

Hence λ is a G-homomorphism, and $\Lambda = \lambda(L_{S_L}) \oplus \mathbb{Z}e^*$ is a G-submodule of $\mathbb{R}^{S_L}$ which is a lattice in $\mathbb{R}^{S_L}$ satisfying $\mathbb{R}\Lambda = \mathbb{R}^{S_L}$. Now we apply 6.1.2 and obtain

$$\mathsf{h}(G, \lambda(L_{S_L}) \oplus \mathbb{Z}e^*) = \prod_{v \in S} n_v.$$

We calculate this Herbrand quotient by means of 2.4.5. From the exact sequences

$$\mathbf{0} \to \mathbb{Z} \to \lambda(L_{S_L}) \oplus \mathbb{Z}e^* \to \lambda(L_{S_L}) \to \mathbf{0} \quad \text{and} \quad \mathbf{0} \to \mu_L \to L_{S_L} \to \lambda(L_{S_L}) \to \mathbf{0}$$

together with $\mathsf{h}(G, \mu_L) = 1$ it follows that

$$\mathsf{h}(G, L_{S_L}) = (G, \lambda(L_{S_L}) = \frac{\mathsf{h}(G, \lambda(L_{S_L}) \oplus \mathbb{Z}e^*)}{\mathsf{h}(G, \mathbb{Z})} = \frac{1}{n} \prod_{v \in S} n_v.$$

2. By 5.3.7.2, there exists a finite subset S of M_K such that $\mathsf{J}_L = \mathsf{J}_{L,S_L} L^\times$, and we may assume that $\mathsf{M}_K^\infty \cup \mathrm{Ram}(L/K) \subset S$. Then it follows that $\mathsf{C}_L = \mathsf{J}_{L,S_L} L^\times / L^\times \cong \mathsf{J}_{L,S_L} / L_{S_L}$. By 1., 6.1.1.3, 2.4.5.2(b) and the exact sequence

$$\mathbf{1} \to L_{S_L} \to \mathsf{J}_{L,S_L} \to \mathsf{C}_L \to \mathbf{1}$$

we see that the Herbrand quotient $\mathsf{h}(G, \mathsf{C}_L)$ is defined (i. e., the groups $H^0(G, \mathsf{C}_L)$ and $H^{-1}(G, \mathsf{C}_L)$ are finite) and

$$\mathsf{h}(G, \mathsf{C}_L) = \frac{\mathsf{h}(G, \mathsf{J}_{L,S_L})}{\mathsf{h}(G, L_{S_L})} = n.$$

In particular, it follows that

$$(\mathsf{C}_K : \mathsf{N}_{L/K} \mathsf{C}_L) = |H^0(G, \mathsf{C}_L)| \geq \frac{|H^0(G, \mathsf{C}_L)|}{|H^{-1}(G, \mathsf{C}_L)|} = h(G, \mathsf{C}_L) = n. \qquad \square$$

Remark 6.1.4. Traditionally, the inequality $(C_K : \mathsf{N}_{L/K} C_L) \geq [L:K]$ which we just proved for cyclic extensions L/K is called the **first inequality of class field theory**. In 5.7.6 we already met the second inequality $(C_K : \mathsf{N}_{L/K} C_L) \leq [L:K]$ for every Galois extension of number fields (a proof using analytical tools is in [18, Theorem 4.4.1]). For cyclic extensions of prime degree $n = [L:K]$ we shall prove it in 6.1.10, provided that K contains the n-th roots of unity. Later, in 6.4.4 we shall prove that $(\mathsf{C}_K : \mathsf{N}_{L/K} \mathsf{C}_L) = [L:K]$ and $H^1(G, \mathsf{C}_L) = \mathbf{1}$ for all Galois extensions L/K of global fields (Theorem of Artin and Tate) which will turn out to be a cornerstone of the theory.

We proceed with several consequences of 6.1.3 which may be viewed as preparations for the proof of the theorem of Artin and Tate.

Theorem 6.1.5. *Let L/K be a finite separable extension and $L \neq K$.*

1. *There exist infinitely many places of K which are not fully decomposed in L.*

2. *Let L/K be cyclic of prime power degree. Then there exist infinitely many places of K which are inert in L.*

Proof. Let $S_{L/K}$ be the set of all places of K which are not fully decomposed in L and $S^*_{L/K}$ the set of all places of K which are inert in L. We prove both assertions at a time and proceed in three steps.

a. Let L/K be cyclic, and assume to the contrary that $S_{L/K}$ is finite. For every $v \in \mathsf{M}_K$ we fix som $\overline{v} \in \mathsf{M}_L$ lying above f. Then $L_{\overline{v}} \neq K_v$ for all $v \in S_{L/K}$, and we shall prove that $\mathsf{N}_{L/K}\mathsf{C}_L = \mathsf{C}_K$, which implies $L = K$ by the first inequality 6.1.3.2, a contradiction.

Let $x = \alpha K^\times \in \mathsf{C}_K$, where $\alpha = (\alpha_v)_{v\in\mathsf{M}_K} \in \mathsf{J}_K$. If $v \in S_{L/K}$, then $\mathsf{N}_{L_{\overline{v}}/K_v}L_{\overline{v}}^\times$ is open in $K_v^\times$ (note that for $v \in \mathsf{M}_K^\infty$ this is obvious, since $\mathsf{N}_{L_{\overline{v}}/K_v}L_{\overline{v}}^\times = \mathsf{N}_{\mathbb{C}/\mathbb{R}}\mathbb{C}^\times = \mathbb{R}_{>0}$, and for $v \in \mathsf{M}_K^0$ this follows from 4.1.10.3). Consequently there exists some $\varepsilon \in \mathbb{R}_{>0}$ such that

$$\{x \in K_v \mid |x-1|_v < \varepsilon\} \subset \mathsf{N}_{L_{\overline{v}}/K_v}L_{\overline{v}}^\times \text{ for all } v \in S_{L/K}.$$

By the weak approximation theorem there exists some $a \in K^\times$ such that $|a-\alpha_v|_v < \varepsilon|\alpha_v|_v$ for all $v \in S_{L/K}$, hence $|\alpha_v^{-1}a-1|_v < \varepsilon$, $\alpha_v^{-1}a \in \mathsf{N}_{L_{\overline{v}}/K_v}L_{\overline{v}}^\times$ and thus $\alpha_v a^{-1} \in \mathsf{N}_{L_{\overline{v}}/K_v}L_{\overline{v}}^\times$. If $v \in \mathsf{M}_K \setminus S_{L/K}$ and $\overline{v} \in \mathsf{M}_L$ lies above v, then $L_{\overline{v}} = K_v$ and therefore $\alpha_v a^{-1} \in \mathsf{N}_{L_{\overline{v}}/K_v}L_{\overline{v}}^\times$. Now 6.1.1.1(b) implies $\alpha a^{-1} \in \mathsf{N}_{L/K}(\mathsf{J}_L)$ and therefore

$$x = \alpha K^\times \in \mathsf{N}_{L/K}(\mathsf{J}_L)K^\times/K^\times = \mathsf{N}_{L/K}\mathsf{C}_L.$$

b. Let L/K be cyclic of prime power degree p^n, and let L_0 be the (unique) intermediate field of L/K such that $[L_0\!:\!K] = p$. Then $S_{L_0/K}$ is infinite by **a**, and thus it suffices to prove that $S_{L_0/K} \cap \mathsf{M}_K^0 \subset S^*_{L/K}$. Let $v \in \mathsf{S}_{L_0/K} \cap \mathsf{M}_K^0$, $\overline{v} \in \mathsf{M}_L$ such that $\overline{v} \,|\, v$ and $G_{\overline{v}}$ be the decomposition group of $\overline{v}$ over K. Then $L_0 \not\subset L^{G_{\overline{v}}}$ (see [18, Corollary 2.13.6]), hence $G_{\overline{v}} = G$, and therefore $v \in S^*_{L/K}$.

c. Let L/K be separable, N a Galois closure of L/K, $G = \mathrm{Gal}(N/K)$, $\tau \in G \setminus \{1\}$ and $L_1 = N^{\langle\tau\rangle}$. Then a place $v \in \mathsf{M}_K^0$ is fully decomposed in L if and only if it is fully decomposed in N (see [18, Theorem 2.13.5]), and if this is the case, then every place $v_1 \in \mathsf{M}_{L_1}$ above v is fully decomposed in N. Consequently, since S_{N/L_1} is infinite by **a** it follows that $S_{L/K}$ is infinite, too. □

Corollary 6.1.6. *Let L/K and L'/K be distinct cyclic extensions of the same prime degree (inside a fixed algebraic closure). Then there exist infinitely many places $v \in \mathsf{M}_K^0$ which are (fully) decomposed in L and undecomposed in L'.*

Proof. Let $[L\!:\!K] = [L'\!:\!K] = p$ and $\overline{L} = LL'$. Then there is an isomorphism

$$G = \mathrm{Gal}(\overline{L}/K) \xrightarrow{\sim} \mathrm{Gal}(L/K) \times \mathrm{Gal}(L'/K) \cong \mathbb{Z}/p\mathbb{Z} \times \mathbb{Z}/p\mathbb{Z},$$

given by $\sigma \mapsto (\sigma\restriction L, \sigma\restriction L')$. For a place $v \in \mathsf{M}_K^0$ we fix a place $\overline{v} \in \mathsf{M}_{\overline{L}}$ above v and denote by $G_{\overline{v}}$ the decomposition group of $\overline{v}$ over K. Let W be the set of all places $v \in \mathsf{M}_K^0$ which are unramified in $\overline{L}$ such that $\overline{v}{\downarrow}L$ is inert in $\overline{L}$. Then W is infinite by 6.1.5. If $v \in W$, then $\mathrm{Gal}(\overline{L}/L) \cap G_{\overline{v}}$ is the decomposition group of $\overline{v}$ over L, hence $\mathrm{Gal}(\overline{L}/L) \cap G_{\overline{v}} = \mathrm{Gal}(\overline{L}/L) \subset G_{\overline{v}} \subsetneq G$ (as $G_{\overline{v}}$ is cyclic), and therefore $G_{\overline{v}} = \mathrm{Gal}(\overline{L}/L)$. Consequently, $\mathbf{1} = G_{\overline{v}} \cap \mathrm{Gal}(\overline{L}/L')$ is the decomposition group of $\overline{v}$ over L'. Hence $\overline{v}{\downarrow}L'$ is decomposed in $\overline{L}$, and as v cannot be fully decomposed in $\overline{L}$, it follows that v is inert in L'. $\square$

Before we proceed, we use 6.1.5 to give an algebraic proof of F. K. Schmidt's Theorem $\partial_K = \gcd\{\deg(v) \mid v \in \mathsf{M}_K\} = 1$ in the function field case (the tricky original analytic proof is in [18, Theorem 6.9.3]).

Corollary 6.1.7. *Let $K/\mathbb{F}_q$ be a function field. Then $\partial_K = 1$.*

Proof. Let $\partial = \partial_K$, and let $L = K\mathbb{F}_{q^\partial}$ be the constant field extension of degree ∂. Then $[L:K] = \partial$, and as $\deg(v) \equiv 0 \bmod \partial$ for all $v \in \mathsf{M}_K$, all $v \in \mathsf{M}_K$ are fully decomposed in L (see [18, Theorem 6.6.8]). Hence $L = K$ and thus $\partial = 1$ by 6.1.5. $\square$

The aim of the following two theorems is to prove that $(\mathsf{C}_K : \mathsf{N}_{L/K}\mathsf{C}_L) = n$ and $H^1(G, \mathsf{C}_L) = \mathbf{1}$ for certain cyclic Kummer extensions L/K of degree n. For this we must assume that $|\mu_n(K)| = n$ and thus $\mathrm{char}(K) \nmid n$.

If $\mathrm{char}(K) = 0$, then the set $\{v \in \mathsf{M}_K^0 \mid \mathrm{char}(\mathsf{k}_v) \,|\, n\}$ is finite by 5.1.1.**5**, and if $\mathrm{char}(K) = p > 0$, then $\mathrm{char}(\mathsf{k}_v) = p \nmid n$ for all $v \in \mathsf{M}_K^0$.

In any case, if $v \in \mathsf{M}_K^0$, $\mathrm{char}(\mathsf{k}_v) \nmid n$ and $|\mu_n(K_v)| = |\mu_n(K)| = n$, then it follows that $\mathfrak{N}(v) \equiv 1 \bmod n$, $U_v/U_v^n \cong \mathbb{Z}/n\mathbb{Z}$ and $K_v^\times/K_v^{\times n} \cong \mathbb{Z}/n\mathbb{Z} \times \mathbb{Z}/n\mathbb{Z}$ (see [18, Theorem 5.8.10]).

The case of p-extensions of global fields of characteristic p requires different methods and will be postponed until Section 6.3.

In the following two theorems we observe that empty direct products equal $\mathbf{1}$, and 2.6.6 is applied again and again.

Theorem 6.1.8. *Let $n \in \mathbb{N}$, $|\mu_n(K)| = n$, and let S be a finite subset of M_K such that $\mathsf{M}_K^\infty \cup \{v \in \mathsf{M}_K^0 \mid \mathrm{char}(\mathsf{k}_v) \,|\, n\} \subset S$ and $\mathsf{J}_K = \mathsf{J}_{K,S}K^\times$ (see 5.3.6.2). Let T be a finite (maybe empty) subset of M_K such that $T \cap S = \emptyset$, and assume that the diagonal homomorphism*

$$\theta \colon K_S \to \prod_{v \in T} U_v/U_v^n, \ \textit{defined by} \ \ \theta(a) = (aU_v^n)_{v \in T} \ \ \textit{for all} \ \ a \in K_S,$$

is surjective. Let $\Delta = \mathrm{Ker}(\theta)$ (hence $\Delta = K_S$ if $T = \emptyset$), $s = |S|$,

$$\mathsf{J}_K^{(n)}(S,T) = \prod_{v \in S} K_v^{\times n} \times \prod_{v \in T} K_v^\times \times \prod_{v \in \mathsf{M}_K \setminus (S \cup T)} U_v, \quad \textit{and}$$

$$\mathsf{C}_K^{(n)}(S,T) = \mathsf{J}_K^{(n)}(S,T)K^\times/K^\times.$$

1. $K_S/K_S^n \cong (\mathbb{Z}/n\mathbb{Z})^s$, $K_S^n \subset \Delta$, $(\Delta : K_S^n) = n^r$ *and* $|T| = s - r$ *for some* $r \in [0, s]$.

2. $\mathsf{J}_K^{(n)}(S,T) \cap K^\times = (K_{S\cup T})^n$.

3. Let $L = K(\sqrt[n]{\Delta})$. *Then* $(\mathsf{C}_K : \mathsf{C}_K^{(n)}(S,T)) = [L:K] = n^r$,

$$\mathsf{C}_K^{(n)}(S,T) \subset \mathsf{N}_{L/K}\mathsf{C}_L \quad \textit{and} \quad (\mathsf{C}_K : \mathsf{N}_{L/K}\mathsf{C}_L) \le [L:K].$$

Proof. 1. As $K_S \cong \mu(K) \times \mathbb{Z}^{s-1}$ (see 5.3.8) and $|\mu(K)| \equiv 0 \bmod n$, we obtain $K_S/K_S^n \cong (\mathbb{Z}/n\mathbb{Z})^s$. Obviously, $K_S^n \subset \Delta$, and θ induces an isomorphism

$$K_S/\Delta \overset{\sim}{\to} \prod_{v \in T} U_v/U_v^n.$$

It follows that $n^s = (K_S : K_S^n) = (K_S : \Delta)(\Delta : K_S^n) = n^{|T|}(\Delta : K_S^n)$ and therefore $(\Delta : K_S^n) = n^{s-|T|} = n^r$, where $r = s - |T| \in [0, s]$.

2. Clearly, $(K_{S\cup T})^n \subset \mathsf{J}_K^{(n)}(S,T) \cap K^\times$. Thus let $a \in \mathsf{J}_K^{(n)}(S,T) \cap K^\times$ and set $K' = K(\sqrt[n]{a})$. We shall prove that $\mathsf{C}_K = \mathsf{N}_{K'/K}\mathsf{C}_{K'}$. Then 6.1.3.2 implies $K' = K$ and thus $a \in \mathsf{J}_K^{(n)}(S,T) \cap K^{\times n} = (K_{S\cup T})^n$. As $\mathsf{C}_K/\mathsf{N}_{K'/K}\mathsf{C}_{K'} = \mathsf{J}_K/\mathsf{N}_{K'/K}(\mathsf{J}_{K'})K^\times$ and $\mathsf{J}_K = \mathsf{J}_{K,S}K^\times$, it suffices to prove that $\mathsf{J}_{K,S} \subset \mathsf{N}_{K'/K}(\mathsf{J}_{K'})K^\times$.

Let $\alpha = (\alpha_v)_{v \in \mathsf{M}_K} \in \mathsf{J}_{K,S}$. Since $S \cap T = \emptyset$, we obtain $\alpha_v \in U_v$ for all $v \in T$, and as θ is surjective, there exists some $x \in K_S$ such that $\alpha_v = xu_v^n$ with $u_v \in U_v$ for all $v \in T$. It follows that $\beta = \alpha x^{-1} \in \mathsf{J}_{K,S}$, and we shall prove that $\beta \in \mathsf{N}_{K'/K}(\mathsf{J}_{K'})$ (then $\alpha = \beta x \in \mathsf{N}_{K'/K}(\mathsf{J}_{K'})K^\times$). By 6.1.1.1(b), it suffices to prove that $\beta_v \in \mathsf{N}_{K'_{v'}/K_v}K'^\times_{v'}$ for all $v \in \mathsf{M}_K$ (where $v' \in \mathsf{M}_{K'}$ is any place above v). So let $v \in \mathsf{M}_K$ and $v' \in \mathsf{M}_{K'}$ such that $v' \downarrow K = v$. Then $K'_{v'} = K'K_v = K_v(\sqrt[n]{a})$ and thus $[K'_{v'} : K_v] \mid n$. Now we have to distinguish several cases.

- If $v \in T$, then $\beta_v = \alpha_v x^{-1} = u_v^n \in K_v^{\times n} \subset \mathsf{N}_{K'_{v'}/K_v}K'^\times_{v'}$.
- If $v \in S$, then $a \in K_v^{\times n}$, hence $K'_{v'} = K_v$, and there is nothing to do.
- If $v \in \mathsf{M}_K \setminus (S \cup T)$, then $\beta_v = \alpha_v x^{-1} \in U_v$ and $a \in U_v$. Hence $K'_{v'}/K_v$ is unramified and $\beta_v \in \mathsf{N}_{K'_{v'}/K_v}K'^\times_{v'}$ (see [18, Theorems 5.9.9 and 5.6.9]).

3. Since $\mathsf{J}_K^{(n)}(S,T)K^\times \cap \mathsf{J}_{K,S\cup T} = \mathsf{J}_K^{(n)}(S,T)K_{S\cup T}$, we obtain

$$\begin{aligned}
\mathsf{C}_K/\mathsf{C}_K^{(n)}(S,T) &= (\mathsf{J}_{K,S\cup T}K^\times/K^\times)/(\mathsf{J}_K^{(n)}(S,T)K^\times/K^\times) \\
&\cong \mathsf{J}_{K,S\cup T}K^\times/\mathsf{J}_K^{(n)}(S,T)K^\times \\
&\cong \mathsf{J}_{K,S\cup T}/\mathsf{J}_K^{(n)}(S,T)K^\times \cap \mathsf{J}_{K,S\cup T} = \mathsf{J}_{K,S\cup T}/\mathsf{J}_K^{(n)}(S,T)K_{S\cup T}.
\end{aligned}$$

Let $\eta\colon \mathsf{J}_{K,S\cup T}/\mathsf{J}_K^{(n)}(S,T) \to \mathsf{J}_{K,S\cup T}/\mathsf{J}_K^{(n)}(S,T)K_{S\cup T} \xrightarrow{\sim} \mathsf{C}_K/C_K^{(n)}(S,T)$ be the natural epimorphism. Then

$$\begin{aligned}\mathrm{Ker}(\eta) &= \mathsf{J}_K^{(n)}(S,T)K_{S\cup T}/\mathsf{J}_K^{(n)}(S,T) \cong K_{S\cup T}/\mathsf{J}_K^{(n)}(S,T)\cap K_{S\cup T} \\ &= K_{S\cup T}/\mathsf{J}_K^{(n)}(S,T)\cap K^\times = K_{S\cup T}/(K_{S\cup T})^n \quad \text{by 2.,}\end{aligned}$$

and by 1. we obtain

$$\begin{aligned}(\mathsf{C}_K : \mathsf{C}_K^{(n)}(S,T)) &= \frac{(\mathsf{J}_{K,S\cup T} : \mathsf{J}_K^{(n)}(S,T))}{(K_{S\cup T} : (K_{S\cup T})^n)} = \frac{1}{n^{|S\cup T|}} \prod_{v\in S} (K_v^\times : K_v^{\times n}) \\ &= \frac{n^{2s}}{n^{2s-r}} = n^r.\end{aligned}$$

Since $\Delta \subset K_S$, it follows that $\Delta \cap K^{\times n} = K_S^n$, and using 1. we obtain

$$n^r = (\Delta : K_S^n) = (\Delta : \Delta \cap K^{\times n}) = (\Delta K^{\times n} : K^{\times n}) = [L:K].$$

It is now sufficient to prove that

$$\mathsf{J}_K^{(n)}(S,T) \subset \mathsf{N}_{L/K}\mathsf{J}_L \tag{$*$}$$

then $(\mathsf{C}_K^{(n)}(S,T) = \mathsf{J}_K^{(n)}(S,T)K^\times/K^\times \subset \mathsf{N}_{L/K}(\mathsf{J}_L)K^\times/K^\times = \mathsf{N}_{L/K}\mathsf{C}_L)$, and consequently $(\mathsf{C}_K : \mathsf{N}_{L/K}\mathsf{C}_L) \le (\mathsf{C}_K : \mathsf{C}_K^{(n)}(S,T)) = n^r = [L:K]$.

For the proof of $(*)$ let $\alpha = (\alpha_v)_{v\in \mathsf{M}_K} \in \mathsf{J}_K^{(n)}(S,T)$. For each $v \in \mathsf{M}_K$ we fix some $\overline{v} \in \mathsf{M}_L$ such that $\overline{v}\,|\,v$. By 6.1.1.1(b) it suffices to prove that $\alpha_v \in \mathsf{N}_{L_{\overline{v}}/K_v} L_{\overline{v}}^\times$ for all $v \in \mathsf{M}_K$.

Thus let $v \in \mathsf{M}_K$. We distinguish several cases:

- Let $v \in \mathsf{M}_K^\infty$. If $L_{\overline{v}} = K_v$, there is nothing to do. If $(K_v, L_{\overline{v}}) = (\mathbb{R}, \mathbb{C})$, then $2\,|\,n$, hence $K_v^{\times n} = \mathbb{R}_{>0} \in \mathsf{N}_{\mathbb{C}/\mathbb{R}}\mathbb{C}^\times$.

- If $v \in S \cap \mathsf{M}_K^0$, then $\alpha_v \in K_v^{\times n}$. By the local reciprocity law 4.2.7.2 the norm residue symbol $(\,\cdot\,, L_{\overline{v}}/K_v)\colon K_v^\times \to \mathrm{Gal}(L_{\overline{v}}/K_v)$ is an epimorphism with kernel $\mathsf{N}_{L_{\overline{v}}/K_v} L_{\overline{v}}^\times$. Since $L_{\overline{v}} = LK_v = K_v(\sqrt[n]{\Delta}\,)$, it follows that $\sigma^n = 1$ for all $\sigma \in \mathrm{Gal}(L/K)$. Hence $(\alpha_v, L_{\overline{v}}/K_v) = 1$ and thus $\alpha_v \in \mathsf{N}_{L_{\overline{v}}/K_v} L_{\overline{v}}^\times$.

- If $v \in T$, then $\Delta \subset U_v \cap K_v^{\times n}$. Hence $L_{\overline{v}} = K_v$ and thus $\mathsf{N}_{L_{\overline{v}}/K_v} L_{\overline{v}}^\times = K_v^\times$.

- If $v \in \mathsf{M}_K \setminus (S \cup T)$, then $\alpha_v \in U_v$ and $\Delta \subset U_v$. The field $K_v(\sqrt[n]{\Delta})$ is the compositum of the fields $K_v(\sqrt[n]{\delta})$ for $\delta \in \Delta$. Hence $K_v(\sqrt[n]{\Delta})/K_v$ is unramified (see [18, Theorems 5.9.9 and 5.6.4]), and $\alpha_v \in \mathsf{N}_{L_{\overline{v}}/K_v} L_{\overline{v}}^\times$ (see [18, Theorem 5.6.9]). □

Corollary 6.1.9 (Local-global principle for n-th powers). *Let $a \in K^\times$, $n \in \mathbb{N}$ and $|\mu_n(K)| = n$. If $a \in K_v^{\times n}$ for all $v \in \mathsf{M}_K$, then $a \in K^{\times n}$.*

Proof. Let $a \in K_v^{\times n}$ for all $v \in \mathsf{M}_K$ and S be a finite subset of M_K such that $\mathsf{M}_K^\infty \cup \{v \in \mathsf{M}_K^0 \mid \mathrm{char}(\mathsf{k}_v) \mid n\} \cup \{v \in \mathsf{M}_K^0 \mid v(a) \ne 0\} \subset S$ and $\mathsf{J}_K = \mathsf{J}_{K,S}K^\times$. Using the terminology of 6.1.8, we obtain $a \in \mathsf{J}_K^{(n)}(S,\emptyset) \cap K^\times = K_S^n \subset K^{\times n}$ by 6.1.8.2. □

Theorem 6.1.10. *Let $n \in \mathbb{N}$ be a prime power, $|\mu_n(K)| = n$, $r \in \mathbb{N}$, and let L/K be a Galois extension such that $\mathrm{Gal}(L/K) \cong (\mathbb{Z}/n\mathbb{Z})^r$. Let S be a finite subset of M_K such that $\mathsf{M}_K^\infty \cup \{v \in \mathsf{M}_K^0 \mid \mathrm{char}(\mathsf{k}_v) \mid n\} \cup \mathrm{Ram}(L/K) \subset S$ and $\mathsf{J}_K = \mathsf{J}_{K,S}K^\times$. Let $|S| = s$ and $\overline{L} = K(\sqrt[n]{K_S}\,)$.*

1. *$L = K(\sqrt[n]{L^{\times n} \cap K_S}\,) \subset \overline{L}$, $[L:K] = (L^{\times n} \cap K_S : K_S^n)$, $\mathrm{Gal}(\overline{L}/K) \cong (\mathbb{Z}/n\mathbb{Z})^s$, $\mathrm{Gal}(\overline{L}/L) \cong (\mathbb{Z}/n\mathbb{Z})^t$ for some $t \in [0,s]$, and $\mathrm{Ram}(\overline{L}/K) \subset S$.*

2. *There exists a finite subset T of $\mathsf{M}_K^0 \setminus S$ such that the diagonal homomorphism*

$$\eta\colon K_S \to \prod_{v \in T} U_v/U_v^n, \quad \textit{defined by} \quad \eta(a) = (aU_v^n)_{v\in T},$$

is surjective and $\mathrm{Ker}(\eta) = L^{\times n} \cap K_S$.

3. *If L/K is cyclic, then $H^1(G, \mathsf{C}_L) = \mathbf{1}$ and $(\mathsf{C}_K : \mathsf{N}_{L/K}\mathsf{C}_L) = n$.*

Proof. 1. By 2.6.6, $L = K(\sqrt[n]{L^{\times n} \cap K^\times}\,)$. Clearly $L \supset K(\sqrt[n]{L^{\times n} \cap K_S}\,)$, and to prove of equality, we must show that $L^{\times n} \cap K^\times \subset (L^{\times n} \cap K_S)K^{\times n}$. Thus let $a \in L^{\times n} \cap K^\times$. If $v \in \mathsf{M}_K \setminus S$ and $\overline{v} \in \mathsf{M}_L$ lies above v, then $K_v(\sqrt[n]{a}) \subset L_{\overline{v}}$, hence $K_v(\sqrt[n]{a})/K_v$ is unramified, and $a = u_v\alpha_v^n$ for some $u_v \in U_v$ and $\alpha_v \in K_v^\times$ (see [18, Theorem 5.9.9]). For $v \in S$ we set $\alpha_v = 1$. Then $\alpha = (\alpha_v)_{v\in\mathsf{M}_K} \in \mathsf{J}_K = \mathsf{J}_{K,S}K^\times$, and therefore we get $\alpha = \beta b$, where $\beta = (\beta_v)_{v\in\mathsf{M}_K} \in \mathsf{J}_{K,S}$ and $b \in K^\times$. If $v \in \mathsf{M}_K \setminus S$, then $a = u_v\alpha_v^n = u_v\beta_v^n b^n$, hence $ab^{-n} = u_v\beta_v^n \in U_v$ and $ab^{-n} \in \mathsf{J}_{K,S} \cap K^\times = K_S$. Since $ab^{-n} \in L^{\times n} \cap K_S$, it follows that $a \in (L^{\times n} \cap K_S)K^{\times n}$.
As $L = K(\sqrt[n]{L^{\times n} \cap K_S}\,) = K(\sqrt[n]{(L^{\times n} \cap K_S)K^{\times n}}\,)$, we obtain

$$[L:K] = ((L^{\times n} \cap K_S)K^{\times n} : K^{\times n}) = (L^{\times n} \cap K_S : K^{\times n} \cap K_S) = (L^{\times n} \cap K_S : K_S^n).$$

Since $L = K(\sqrt[n]{L^{\times n} \cap K_S}\,) \subset K(\sqrt[n]{K_S}\,) = \overline{L}$ and $K_S/(K_S)^n \cong (\mathbb{Z}/n\mathbb{Z})^s$ by 6.1.8.1, it follows that $G = \mathrm{Gal}(\overline{L}/K) \cong \mathsf{X}((\mathbb{Z}/n\mathbb{Z})^s) \cong (\mathbb{Z}/n\mathbb{Z})^s$. In particular, $s \ge r$, and $\overline{L} = L(\sqrt[n]{K_S}\,)$ implies $\mathrm{Gal}(\overline{L}/L) \cong (\mathbb{Z}/n\mathbb{Z})^t$, where $t = s - r \in [0,s]$.

If $v \in \mathsf{M}_K^0 \setminus S$, then $K_v(\sqrt[n]{K_S})$ is the compositum of the fields $K_v(\sqrt[n]{\delta})$ for $\delta \in K_S$. Hence $\delta \in U_v$, and $K_v(\sqrt[n]{K_S})/K_v$ is unramified (see [18, Theorems 5.9.9 and 5.6.4]). Since $K_v(\sqrt[n]{K_S}\,) = \overline{L}_{\overline{v}}$ for some $\overline{v} \in \mathsf{M}_{\overline{L}}$ above v, this implies $v \notin \mathrm{Ram}(\overline{L}/K)$, and consequently $\mathrm{Ram}(\overline{L}/K) \subset S$.

2. Let $\mathrm{Gal}(\overline{L}/L) = \langle \sigma_1, \ldots, \sigma_t \rangle$. For $i \in [1,t]$ we set $G_i = \langle \sigma_i \rangle \cong \mathbb{Z}/n\mathbb{Z}$ and $L_i = \overline{L}^{G_i}$. Then $\overline{L}/L_i$ is cyclic of prime power degree n, and $L = L_1 \cap \ldots \cap L_t$.

By 6.1.5.2, there exist infinitely many places of L_i which are inert in $\overline{L}$. Therefore for every $i \in [1,t]$ there exists some place $w_i \in \mathsf{M}^0_{L_i}$ which is inert in $\overline{L}$ such that $v_i = w_i \downarrow K \notin S$, and $v_1, \ldots, v_t$ are distinct. Now we set $T = \{v_1, \ldots, v_t\}$. Then $T \cap S = \emptyset$ and $|T| = t$. For $i \in [1,t]$ let $\overline{v}_i \in \mathsf{M}^0_{\overline{L}}$ be the (unique) place of $\overline{L}$ above w_i. Let

$$G_{\overline{v}_i} = \{\sigma \in G \mid \sigma\overline{v}_i = \overline{v}_i\} = \mathrm{Gal}(\overline{L}_{\overline{v}_i}/K_{v_i})$$

be the decomposition group of $\overline{v}_i$ over K (see 5.4.5.1). Then $G_i \subset G_{\overline{v}_i}$, and $G_{\overline{v}_i}$ is cyclic, since v_i is unramified in $\overline{L}$. Hence $G_i = G_{\overline{v}_i}$, L_i is the decomposition field of v_i in $\overline{L}$, v_i is fully decomposed in L_i, hence in L, and thus $K_{v_i} = (L_i)_{w_i} = L_{w_i \downarrow L}$.

Now we investigate the homomorphism

$$\eta\colon K_S \to \prod_{i=1}^t U_{v_i}/U_{v_i}^n$$

and show first that $\mathrm{Ker}(\eta) = L^{\times n} \cap K_S$. If $a \in L^{\times n} \cap K_S$ and $i \in [1,t]$, then $a \in U_{v_i}$ and $K_{v_i}(\sqrt[n]{a}) \subset L_{w_i \downarrow L} = K_{v_i}$, hence $a \in K_{v_i}^{\times n} \cap U_{v_i} = U_{v_i}^n$ and thus $a \in \mathrm{Ker}(\eta)$. As to the converse, let $a \in \mathrm{Ker}(\eta) \subset K_S$. If $i \in [1,t]$, then $a \in U_{v_i}^n$, hence $K_{v_i} = K_{v_i}(\sqrt[n]{a})$, v_i is fully decomposed in $K(\sqrt[n]{a})$ and thus $K(\sqrt[n]{a}) \subset L_i$. Consequenetly, it follows that $\sqrt[n]{a} \in L_1 \cap \ldots \cap L_t = L$ and $a \in L^{\times n} \cap K_S$.

As $\mathrm{Ker}(\eta) = L^{\times n} \cap K_S$, we obtain (using 1.),

$$|\mathrm{Im}(\eta)| = (K_S \colon L^{\times n} \cap K_S) = \frac{(K_S \colon K_S^n)}{(L^{\times n} \cap K_S \colon K_S^n)} = \frac{n^s}{[L \colon K]} = n^t = \Big|\prod_{v \in T} U_v/U_v^n\Big|.$$

Hence η is surjective.

3. Let $r = 1$. By 2., the assumptions of 6.1.8 are fulfilled, and by 6.1.8.3 we obtain $(\mathsf{C}_K \colon \mathsf{N}_{L/K}\mathsf{C}_L) \le [L \colon K]$. On the other hand, $(\mathsf{C}_K \colon \mathsf{N}_{L/K}\mathsf{C}_L) \ge [L \colon K]$ and $h(G, \mathsf{C}_L) = [L \colon K]$ by 6.1.3.2, hence $(\mathsf{C}_K \colon \mathsf{N}_{L/K}\mathsf{C}_L) = [L \colon K]$ and (using 2.4.5.1) $H^1(G, \mathsf{C}_L) = H^{-1}(G, \mathsf{C}_L) = \mathbf{1}$. □

6.2 The global norm residue symbol

Definitions and Remarks 6.2.1. Let L/K be a finite Galois extension, L^{a} be the largest over K abelian intermediate field of L/K, $G = \mathrm{Gal}(L/K)$ and $G^{\mathsf{a}} = \mathrm{Gal}(L^{\mathsf{a}}/K)$. For every place $v \in \mathsf{M}_K$ we fix some $\overline{v} \in \mathsf{M}_L$ above v, we assume that $K_v \subset L_{\overline{v}}$ and denote by $G_{\overline{v}} = \mathrm{Gal}(L_{\overline{v}}/K_v) \subset G$ the decomposition group of $\overline{v}$ over K. Let $L^{\mathsf{a}}_{\overline{v}}$ be the largest over K_v abelian intermediate field of $L_{\overline{v}}/K_v$ and $G^{\mathsf{a}}_{\overline{v}} = \mathrm{Gal}(L^{\mathsf{a}}_{\overline{v}}/K_v)$. By 1.8.4, the extension

$L^{\mathsf a}K_v/K_v$ is abelian and thus $L^{\mathsf a}K_v \subset L_{\overline v}^{\mathsf a}$. The embedding $G_{\overline v} \hookrightarrow G$ induces a homomorphism $\iota_{\overline v}\colon G_{\overline v}^{\mathsf a} \to G^{\mathsf a}$, and if $\sigma \in G_{\overline v}^{\mathsf a}$, then $\iota_{\overline v}(\sigma) = \sigma \restriction L^{\mathsf a}$. For $v \in \mathsf M_K^0$ let $(\,\cdot\,, L_{\overline v}/K_v)\colon L_v^\times \to G_{\overline v}^{\mathsf a}$ be the local norm residue symbol as introduced in 4.2.1.

For $v \in \mathsf M_K^\infty$ we define the (local) **Archimedean norm rest symbol** by

$$(\,\cdot\,, \mathbb C/\mathbb C) = (\,\cdot\,, \mathbb R/\mathbb R) = 1, \quad \text{and} \quad (a, \mathbb C/R) = \begin{cases} \iota & \text{if } a \in \mathbb R_{<0}, \\ 1 & \text{if } a \in \mathbb R_{>0}, \end{cases}$$

where $\iota\colon \mathbb C \to \mathbb C$ is the complex conjugation. Based on this definition, for every $v \in \mathsf M_K$ the local norm residue symbol induces an isomorphism

$$(\,\cdot\,, L_{\overline v}/K_v)\colon K_v^\times/\mathsf N_{L_{\overline v}/K_v}L_{\overline v}^\times \overset{\sim}{\to} G_{\overline v}^{\mathsf a} = \mathrm{Gal}(L_{\overline v}^{\mathsf a}/K_v)$$

regardless whether v is finite or Archimedean (see 4.2.7.2). Moreover we prove:

> **A.** Let $v \in \mathsf M_K$ and $a \in K_v^\times$. Then $(a, L_{\overline v}/K_v) \restriction L^{\mathsf a} \in \mathrm{Gal}(L^{\mathsf a}/K)$ only depends on a and v (and not on the choice of $\overline v$).

Proof of **A.** If $v \in \mathsf M_K^\infty$, then either $K_v = L_{\overline v}$ and $(a, L_{\overline v}/K_v) = 1$ for all $a \in K_v$, or $(K_v, L_{\overline v}) = (\mathbb R, \mathbb C)$, and then $(a, L_{\overline v}/K_v) = (a, \mathbb C/\mathbb R)$ does by the very definition only depend on $a \in K_v$.

Let $v \in \mathsf M_K^0$, and let $\overline v$, $\overline v' \in \mathsf M_L$ be places above v such that $\overline v \ne \overline v'$. Then there exists some $\sigma \in G$ such that $\overline v' = \sigma\overline v$, and σ induces an (equally denoted) K_v-isomorphism $\sigma\colon L_{\overline v}^{\mathsf a} \to L_{\overline v'}^{\mathsf a}$. By 4.2.6, $(a, L_{\overline v'}/K_v) = \sigma\circ(a, L_{\overline v}/K_v)\circ\sigma^{-1}$ for all $a \in K_v^\times$ and consequently

$$(a, L_{\overline v'}/K_v) \restriction L^{\mathsf a} = \overline\sigma\circ(a, L_{\overline v}/K_v)\circ\overline\sigma^{-1} \restriction L^{\mathsf a} = (a, L_{\overline v}/K_v) \restriction L^{\mathsf a}. \qquad \square[\mathbf{A.}]$$

We define the (global) **norm residue symbol** $(\,\cdot\,, L/K)\colon \mathsf J_K \to G^{\mathsf a}$ by

$$(\alpha, L/K) = \prod_{v \in \mathsf M_K} (\alpha_v, L_{\overline v}/K_v) \restriction L^{\mathsf a} \in G^{\mathsf a} \quad \text{for all } \alpha = (\alpha_v)_{v\in\mathsf M_K} \in \mathsf J_K. \qquad (*)$$

If $\alpha = (\alpha_v)_{v\in\mathsf M_K} \in \mathsf J_K$, then $\alpha_v \in U_v$ and $L_{\overline v}/K_v$ is unramified for almost all $v \in \mathsf M_K^0$. If $v \in \mathsf M_K^0$, $L_{\overline v}/K_v$ is unramified and $\alpha_v \in U_v$, then $\alpha_v \in \mathsf N_{L_{\overline v}/K_v}L_{\overline v}^\times$ (see [18, Theorem 5.6.9]), and therefore $(\alpha_v, L_{\overline v}/K_v) = 1$. Consequently, the product in $(*)$ is finite, and as $G^{\mathsf a}$ is abelian, it does not depend on the order of the factors. In all, $(a, L/K)$ is well defined.

For $v \in \mathsf M_K$ let $j_v\colon K_v^\times \to \mathsf J_K$ be the local embedding (see 5.3.1.**5**). Then the definition of the norm residue symbol yields the commutative diagram

$$\begin{array}{ccccc} \mathsf K_v^\times & \xrightarrow{(\,\cdot\,, L_{\overline v}/K_v)} & G_{\overline v}^{\mathsf a} & = & \mathrm{Gal}(L_{\overline v}^{\mathsf a}/K_v) \\ {\scriptstyle j_v}\downarrow & & \downarrow{\scriptstyle \iota_{\overline v}} & & \\ \mathsf J_K & \xrightarrow{(\,\cdot\,, L/K)} & G^{\mathsf a} & = & \mathrm{Gal}(L^{\mathsf a}/K) \end{array} \qquad \text{where } \iota_{\overline v}(\sigma) = \sigma \restriction L^{\mathsf a},$$

i. e., $(j_v(x), L/K) = (x, L_{\overline v}/K_v) \restriction L^{\mathsf a}$ for all $x \in K_v^\times$.

Theorem 6.2.2. *Let L/K be a finite Galois extension and $G = \mathrm{Gal}(L/K)$.*

1. $(\,\cdot\,, L/K)\colon \mathsf{J}_L \to G^{\mathsf{a}}$ *is a group homomorphism, and*

$$(\mathsf{N}_{L/K}(\beta),\, L/K) = \mathrm{id}_{L^{\mathsf{a}}} \quad \textit{for all} \quad \beta \in \mathsf{J}_L.$$

2. *Let K'/K be a finite and L'/K' a finite Galois extension such that $L \subset L'$. Let L^{a} be the largest over K abelian intermediate field of L/K and L'^{a} the largest over K' abelian intermediate field of L'/K'. Then $L^{\mathsf{a}} \subset L'^{\mathsf{a}}$, and*

$$(\alpha', L'/K') \restriction L^{\mathsf{a}} = (\mathsf{N}_{K'/K}(\alpha'), L/K) \quad \textit{for all} \;\; \alpha' \in \mathsf{J}_{K'}.$$

$$\begin{array}{ccccc}
\mathsf{J}_{K'} & \xrightarrow{(\,\cdot\,, L'/K')} & G^{\mathsf{a}} & = & \mathrm{Gal}(L^{\mathsf{a}}/K) \\
{\scriptstyle \mathsf{N}_{K'/K}}\big\downarrow & & \big\downarrow{\scriptstyle \rho} & & \\
\mathsf{J}_K & \xrightarrow{(\,\cdot\,, L/K)} & G^{\mathsf{a}} & = & \mathrm{Gal}(L^{\mathsf{a}}/K)
\end{array} \qquad \textit{where} \;\; \rho(\sigma) = \sigma \restriction L^{\mathsf{a}}$$

3. *Let K' be an over K Galois intermediate field of L/K, L^{a} the largest over K abelian intermediate field of L/K and K'^{a} the largest over K abelian intermediate field of K'/K. Then $K'^{\mathsf{a}} \subset L^{\mathsf{a}}$,*

$$(\alpha, K'/K) = (\alpha, L/K) \restriction K'^{\mathsf{a}} \quad \textit{and} \quad (\alpha, L^{\mathsf{a}}/K) = (\alpha, L/K) \;\; \textit{for all} \;\; \alpha \in \mathsf{J}_K.$$

$$\begin{array}{ccccc}
\mathsf{J}_K & \xrightarrow{(\,\cdot\,, L/K)} & G^{\mathsf{a}} & = & \mathrm{Gal}(L^{\mathsf{a}}/K) \\
\big\| & & \big\downarrow{\scriptstyle \rho} & & \\
\mathsf{J}_K & \xrightarrow{(\,\cdot\,, K'/K)} & G'^{\mathsf{a}} & = & \mathrm{Gal}(K'^{\mathsf{a}}/K)
\end{array} \qquad \textit{where} \;\; \rho(\sigma) = \sigma \restriction K'^{\mathsf{a}}$$

4. *Let $\phi\colon L \to L'$ be an isomorphism of global fields, $K' = \phi(K)$ and L^{a} the largest over K abelian intermediate field of L/K. Then $L'^{\mathsf{a}} = \phi(L^{\mathsf{a}})$ is the largest over K' abelian intermediate field of L'/K', and*

$$(\phi(\alpha), L'/K') = \phi \circ (\alpha, L/K) \circ \phi^{-1} \restriction L'^{\mathsf{a}} \quad \textit{for all} \;\; \alpha \in \mathsf{J}_K.$$

$$\begin{array}{ccccc}
\mathsf{J}_K & \xrightarrow{(\,\cdot\,, L/K)} & G^{\mathsf{a}} & = & \mathrm{Gal}(L^{\mathsf{a}}/K) \\
{\scriptstyle \phi}\big\downarrow & & \big\downarrow{\scriptstyle \phi^*} & & \\
\mathsf{J}_{K'} & \xrightarrow{(\,\cdot\,, L'/K')} & G'^{\mathsf{a}} & = & \mathrm{Gal}(L'^{\mathsf{a}}/K')
\end{array}$$

where $\phi^*(\sigma) = \phi \circ \sigma \circ \phi^{-1}$.

5. *Let K' be an intermediate field of L/K and $G' = \mathrm{Gal}(L/K')$. Let L^{a} the largest over K abelian intermediate field of L/K and*

L'^{a} the largest over K' abelian intermediate field of L/K'. Then $L^{\mathsf{a}} \subset L'^{\mathsf{a}}$, and

$$(\alpha, L/K') = \mathsf{V}^L_{K'/K}(\alpha, L/K) \quad \textit{for all} \quad \alpha \in \mathsf{J}_K.$$

$$\begin{array}{ccccc} \mathsf{J}_K & \xrightarrow{(\,\cdot\,, L/K)} & G^{\mathsf{a}} & = & \mathrm{Gal}(L^{\mathsf{a}}/K) \\ {\scriptstyle i}\big\downarrow & & \big\downarrow {\scriptstyle \mathsf{V}^L_{K'/K} = \mathsf{V}_{G \to G'}} & & \\ \mathsf{J}_{K'} & \xrightarrow{(\,\cdot\,, L/K')} & G'^{\mathsf{a}} & = & \mathrm{Gal}(L'^{\mathsf{a}}/K') \end{array}$$

where $i = (\mathsf{J}_K \hookrightarrow \mathsf{J}_{K'})$

Proof. 1. Let $\alpha = (\alpha_v)_{v \in \mathsf{M}_K}$, $\beta = (\beta_v)_{v \in \mathsf{M}_K} \in \mathsf{J}_K$, and for each $v \in \mathsf{M}_K$ we fix some $\overline{v} \in \mathsf{M}_L$ above v. For all $v \in \mathsf{M}_K$, $(\,\cdot\,, L_{\overline{v}}/K_v)\colon K_v^\times \to G^{\mathsf{a}}_{\overline{v}}$ is a homomorphism, and therefore

$$\begin{aligned} (\alpha\beta, L/K) &= \prod_{v \in \mathsf{M}_K} (\alpha_v\beta_v, L_{\overline{v}}/K_v) \restriction L^{\mathsf{a}} = \prod_{v \in \mathsf{M}_K} (\alpha_v, L_{\overline{v}}/K_v) \circ (\beta_v, L_{\overline{v}}/K_v) \restriction L^{\mathsf{a}} \\ &= \prod_{v \in \mathsf{M}_K} (\alpha_v, L_{\overline{v}}/K_v) \restriction L^{\mathsf{a}} \circ \prod_{v \in \mathsf{M}_K} (\beta_v, L_{\overline{v}}/K_v) \restriction L^{\mathsf{a}} \\ &= (\alpha, L/K) \circ (\beta, L/K). \end{aligned}$$

If $\beta = (\beta_{\overline{v}})_{\overline{v} \in \mathsf{M}_L} \in \mathsf{J}_L$, then 4.2.1.5 implies

$$\begin{aligned} (\mathsf{N}_{L/K}(\beta), L/K) &= \prod_{v \in \mathsf{M}_K} (\mathsf{N}_{L/K}(\beta)_v, L_{\overline{v}}/K_v) \restriction L^{\mathsf{a}} \\ &= \prod_{v \in \mathsf{M}_K} \prod_{\substack{\overline{v} \in \mathsf{M}_L \\ \overline{v} \mid v}} (\mathsf{N}_{L_{\overline{v}}/K_v}(\beta_{\overline{v}}), L_{\overline{v}}/K_v) \restriction L^{\mathsf{a}} = \mathrm{id}_{L^{\mathsf{a}}}. \end{aligned}$$

2. If $v \in \mathsf{M}_K$, $v' \in \mathsf{M}_{K'}$ and $\overline{v}' \in \mathsf{M}_{L'}$ are such that $\overline{v}' \mid v' \mid v$ and $\overline{v} = \overline{v}' \downarrow L$, then $\overline{v} \mid v$, and we obtain the following field diagrams.

$$\begin{array}{ccc} L & - & L' \\ | & & | \\ K & - & K' \end{array} \qquad\qquad \begin{array}{ccc} L_{\overline{v}} & - & L'_{\overline{v}'} \\ | & & | \\ K_v & - & K'_{v'} \end{array}$$

Let $L^{\mathsf{a}}_{\overline{v}}$ be the largest over K_v abelian intermediate field of $L_{\overline{v}}/K_v$ and $L'^{\mathsf{a}}_{\overline{v}'}$ the largest over $K'_{v'}$ abelian intermediate field of $L'_{\overline{v}'}/K'_{v'}$. Since $K'L^{\mathsf{a}} \subset L'$,

$K_vL^{\mathsf a} \subset L_{\overline{v}}$ and $K'_{v'}L^{\mathsf a}_{\overline{v}} \subset L'_{\overline{v}'}$, and since the extensions $K'L^{\mathsf a}/K'$, $K_vL^{\mathsf a}/K_v$ and $K'_{v'}L^{\mathsf a}_{\overline{v}}/K'_{v'}$ are abelian by 1.8.4.1, it follows that $L^{\mathsf a} \subset L'^{\mathsf a}$, $L^{\mathsf a} \subset L^{\mathsf a}_{\overline{v}}$ and $L^{\mathsf a}_{\overline{v}} \subset L'^{\mathsf a}_{\overline{v}'}$.

Let $\alpha = (\alpha_{v'})_{v' \in \mathsf{M}_{K'}} \in \mathsf{J}_{K'}$. If $v \in \mathsf{M}^0_K$, then 4.2.4 implies

$$(\alpha_{v'}, L'_{\overline{v}'}/K'_{v'}) \restriction L^{\mathsf a}_{\overline{v}} = (\mathsf{N}_{K'_{v'}/K_v}(\alpha_{v'}), L_{\overline{v}}/K_v) \in \mathrm{Gal}(L^{\mathsf a}_{\overline{v}}/K_v),$$

and we asseret that this also holds for $v \in \mathsf{M}^\infty_K$. Indeed, there is nothing to do if $L_{\overline{v}} = \mathbb{R}$ or $K_v = \mathbb{C}$. Thus let $L_{\overline{v}} = \mathbb{C}$ (then $L'_{\overline{v}'} = \mathbb{C}$, too) and $K_v = \mathbb{R}$. In this case, the assertion reduces to the obviously valid relations $(\alpha_{v'}, \mathbb{C}/\mathbb{R}) = (\alpha_{v'}, \mathbb{C}/\mathbb{R})$ if $K'_{v'} = \mathbb{R}$, and $1 = (\mathsf{N}_{\mathbb{C}/\mathbb{R}}(\alpha_{v'}), \mathbb{C}/\mathbb{R})$ if $K'_{v'} = \mathbb{C}$.

Now we obtain

$$\begin{aligned}
(\mathsf{N}_{K'/K}(\alpha), L/K) &= \prod_{v \in \mathsf{M}_K} (\mathsf{N}_{K'/K}(\alpha)_v, L_{\overline{v}}/K_v) \restriction L^{\mathsf a} \\
&= \prod_{v \in \mathsf{M}_K} \prod_{\substack{v' \in \mathsf{M}_{K'} \\ v' \mid v}} (\mathsf{N}_{K'_{v'}/K_v}(\alpha_{v'}), L_{\overline{v}}/K_v) \restriction L^{\mathsf a} \\
&= \Big[\prod_{v \in \mathsf{M}_K} \prod_{\substack{v' \in \mathsf{M}_{K'} \\ v' \mid v}} (\alpha_{v'}, L'_{\overline{v}'}/K'_{v'}) \restriction L^{\mathsf a}_{\overline{v}} \Big] \restriction L^{\mathsf a} \\
&= \Big[\prod_{v' \in \mathsf{M}_{K'}} (\alpha_{v'}, L'_{\overline{v}'}/K'_{v'}) \restriction L'^{\mathsf a} \Big] \restriction L^{\mathsf a} = (\alpha, L'/K') \restriction L^{\mathsf a}.
\end{aligned}$$

3. We apply 2. with (K, K, K', L) instead of (K, K', L, L'). Then it follows that $(\alpha, K'/K) = (\alpha, L/K) \restriction K'^{\mathsf a}$. If $K' = L^{\mathsf a}$, then $(\alpha, L^{\mathsf a}/K) = (\alpha, L/K)$.

4. Apparently, $L'^{\mathsf a} = \phi(L^{\mathsf a})$ is the largest over K' abelian intermediate field of L'/K'. ϕ induces (equally denoted) bijective maps $\phi\colon \mathsf{M}_L \to \mathsf{M}_{L'}$ and $\phi\colon \mathsf{M}_K \to \mathsf{M}_{K'}$. If $v \in \mathsf{M}_K$, $\overline{v} \in \mathsf{M}_L$ and $\overline{v} \mid v$, then $\phi\overline{v} \mid \phi v$, and ϕ has a unique extension to a topological isomorphism $\phi_v\colon K_v \to K'_{\phi v}$. Moreover, ϕ induces an (equally denoted) isomorphism $\phi\colon \mathsf{J}_K \to \mathsf{J}_{K'}$ satisfying $\phi(\alpha)_{\phi v} = \phi_v(\alpha_v)$ for all $\alpha \in \mathsf{J}_K$ and $v \in \mathsf{M}_K$. An application of these isomorphisms together with 4.2.6 yields, for all $\alpha \in \mathsf{J}_K$,

$$\begin{aligned}
(\phi(\alpha), L'/K') &= \prod_{v \in \mathsf{M}_{K'}} ((\phi(\alpha)_{\phi v}, (\phi L)_{\phi\overline{v}}/(\phi K)_{\phi v}) \restriction \phi(L^{\mathsf a}) \\
&= \prod_{v \in \mathsf{M}_K} \phi \circ (\alpha_v, L_{\overline{v}}/K_v) \circ \phi^{-1} \restriction \phi(L^{\mathsf a} \\
&= \phi \circ \Big(\prod_{v \in \mathsf{M}_K} (\alpha_v, L_{\overline{v}}/K_v) \restriction L^{\mathsf a} \Big) \circ \phi^{-1} = \phi \circ (\alpha, L/K) \circ \phi^{-1}.
\end{aligned}$$

5. Since $K'L^{\mathsf a}/K'$ is abelian, it follows that $L^{\mathsf a} \subset L'^{\mathsf a}$.

For $v \in \mathsf{M}_K$ we fix some $\overline{v} \in \mathsf{M}_L$ above v, and we let $G_{\overline{v}} = \mathrm{Gal}(L_{\overline{v}}/K_v)$

be the decomposition group of $\overline{v}$ over K. Let $\mathsf{T}_v \subset G$ be a set of representatives for the double coset space $G'\backslash G/G_{\overline{v}}$ and consider the set $W_v = \{v' \in \mathsf{M}_{K'} \mid v'\!\downarrow\! K = v\}$.

A. The map $\theta\colon \mathsf{T}_v \to W_v$, defined by $\theta(\tau) = \tau\overline{v}\!\downarrow\! K'$, is bijective.

Proof of **A.** Since $\{\tau\overline{v} \mid \tau \in G\}$ is the set of all places of L above v, it follows that $W_v = \{\tau\overline{v}\!\downarrow\! K' \mid \tau \in G\}$. Hence it suffices to prove: If $\tau,\ \tau' \in G$, then $\tau\overline{v}\!\downarrow\! K' = \tau'\!\downarrow\! K'$ if and only if $G'\tau G_{\overline{v}} = G'\tau' G_{\overline{v}}$.

Thus let $\tau,\ \tau' \in G$. If $G'\tau G_{\overline{v}} = G'\tau' G_{\overline{v}}$, then $\tau' = \sigma'\tau\sigma$, where $\sigma' \in G'$ and $\sigma \in G_{\overline{v}}$. Hence it follows that $\tau'\overline{v}\!\downarrow\! K' = \sigma'\tau\sigma\overline{v}\!\downarrow\! K' = \tau\!\downarrow\! K'$. Conversely, if $\tau\overline{v}\!\downarrow\! K' = \tau\!\downarrow\! K'$, then $\tau'\overline{v} = \rho\tau\overline{v}$ for some $\rho \in G'$. Thus we get $(\rho\tau)^{-1}\tau' \in G_{\overline{v}}$, hence $\tau' \in \rho\tau G_{\overline{v}} \subset G'\tau G_{\overline{v}}$, and consequently $G'\tau' G_{\overline{v}} = G'\tau G_{\overline{v}}$. $\square$[**A**]

If $\tau \in \mathsf{T}_v$ and $v \in \mathsf{M}_K$, then

$$\tau G_{\overline{v}}\tau^{-1} = \mathrm{Gal}(L_{\tau\overline{v}}/K_v), \quad \tau G_{\overline{v}}\tau^{-1} \cap G' = \mathrm{Gal}(L_{\tau\overline{v}}/K'_{\tau\overline{v}\downarrow K'}),$$

and $(\alpha_v, L_{\tau\overline{v}}/K_v) = \tau(\alpha_v, L_{\overline{v}}/K_v)\tau^{-1}$ for all $\alpha_v \in K_v$ by 4.2.6. Hence, using **A** and 4.8.4.2 we obtain, for all $\alpha = (\alpha_v)_{v\in\mathsf{M}_K} \in \mathsf{J}_K$,

$$\begin{aligned}(\alpha, L/K') &= \prod_{v\in\mathsf{M}_K}\prod_{\tau\in\mathsf{T}_v} (\alpha_v, L_{\tau\overline{v}}/K'_{\tau\overline{v}\downarrow K'})\!\restriction\! L'^{\mathsf{a}} \\ &= \prod_{v\in\mathsf{M}_K}\prod_{\tau\in\mathsf{T}_v} \mathsf{V}^{L_{\tau\overline{v}}}_{K'_{\tau\overline{v}\downarrow K'}/K_v}(\alpha_v, L_{\tau\overline{v}}/K_v)\!\restriction\! L'^{\mathsf{a}} \\ &= \prod_{v\in\mathsf{M}_K}\prod_{\tau\in\mathsf{T}_v} \mathsf{V}_{\tau G_{\overline{v}}\tau^{-1}\to\tau G_{\overline{v}}\tau^{-1}\cap G'}(\tau(\alpha_v, L_{\overline{v}}/K_v)\tau^{-1})\!\restriction\! L'^{\mathsf{a}}.\end{aligned}$$

On the other hand,

$$\begin{aligned}\mathsf{V}^L_{K'/K}(\alpha, L/K) &= \mathsf{V}_{G\to G'}\Big(\prod_{v\in\mathsf{M}_K}(\alpha_v, L_{\overline{v}}/K_v)\!\restriction\! L^{\mathsf{a}}\Big) \\ &= \prod_{v\in\mathsf{M}_K}\mathsf{V}_{G\to G'}\big((\alpha_v, L_{\overline{v}}/K_v)\!\restriction\! L^{\mathsf{a}}\big),\end{aligned}$$

since $\mathsf{V}_{G\to G'}$ is a homomorphism. Hence is sufficient to prove that, for all $v \in \mathsf{M}_K$,

$$\mathsf{V}_{G\to G'}\big((\alpha_v, L_{\overline{v}}/K_v)\!\restriction\! L^{\mathsf{a}}\big) = \prod_{\tau\in\mathsf{T}_v}\mathsf{V}_{\tau G_{\overline{v}}\tau^{-1}\to\tau G_{\overline{v}}\tau^{-1}\cap G'}(\tau(\alpha_v, L_{\overline{v}}/K_v)\tau^{-1})\!\restriction\! L'^{\mathsf{a}}.$$

For $v \in \mathsf{M}_K$ and $\tau \in \mathsf{T}_v$ let $\Lambda_\tau \subset G_{\overline{v}}$ be a set of representatives for the cosets $(G_{\overline{v}} \cap \tau^{-1}G'\tau)\backslash G_{\overline{v}}$. Then we obtain

$$\begin{aligned}G = \biguplus_{\tau\in\mathsf{T}_v} G'\tau G_{\overline{v}} &= \biguplus_{\tau\in\mathsf{T}_v}\biguplus_{\lambda\in\Lambda_\tau} G'\tau(G_{\overline{v}} \cap \tau^{-1}G'\tau)\lambda \\ &= \biguplus_{\tau\in\mathsf{T}_v}\biguplus_{\lambda\in\Lambda_\tau} G'(\tau G_{\overline{v}}\tau^{-1} \cap G')\tau\lambda \\ &= \biguplus_{\tau\in\mathsf{T}_v}\biguplus_{\lambda\in\Lambda_\tau} G'\tau\lambda.\end{aligned}$$

Let $\omega \in G_{\overline{v}}$ be such that $\omega \restriction L_{\overline{v}}^{\mathsf{a}} = (\alpha_v, L_{\overline{v}}/K_v)$, and suppose that $\lambda\omega = \omega_\lambda\lambda_\omega$, where $\omega_\lambda \in G_{\overline{v}} \cap \tau^{-1}G'\tau$ and $\lambda_\omega \in \Lambda_\tau$. Then $\tau\lambda\omega = (\tau\omega_\lambda\tau^{-1})(\tau\lambda_\omega)$, and since $\tau\omega_\lambda\tau^{-1} \in \tau G_{\overline{v}}\tau^{-1} \cap G' \subset G'$, it follows that

$$\mathsf{V}_{G\to G'}(\omega \restriction L^{\mathsf{a}}) = \prod_{\tau\in\mathsf{T}_v}\prod_{\lambda\in\Lambda_\tau} \tau\omega_\lambda\tau^{-1} \restriction L'^{\mathsf{a}},$$

observing that $\sigma G^{(1)} = \sigma \restriction L^{\mathsf{a}}$ for all $\sigma \in G$, and $\sigma' G'^{(1)} = \sigma' \restriction L'^{\mathsf{a}}$ for all $\sigma' \in G'$. Hence we get

$$\mathsf{V}_{G\to G'}\big((\alpha_v, L_{\overline{v}}/K_v) \restriction L^{\mathsf{a}}\big) = \prod_{\tau\in\mathsf{T}_v}\prod_{\lambda\in\Lambda_\tau} \tau\omega_\lambda\tau^{-1} \restriction L'^{\mathsf{a}}.$$

On the other hand, if $\tau \in \mathsf{T}_v$, then

$$G_{\overline{v}} = \biguplus_{\lambda\in\Lambda_\tau} (G_{\overline{v}} \cap \tau^{-1}G'\tau)\lambda \quad \text{implies} \quad \tau G_{\overline{v}}\tau^{-1} = \biguplus_{\lambda\in\Lambda_\tau} (\tau G_{\overline{v}}\tau^{-1} \cap G')(\tau\lambda\tau^{-1}),$$

and since $(\tau\lambda\tau^{-1})(\tau\omega\tau^{-1}) = \tau\lambda\omega\tau^{-1} = (\tau\omega_\lambda\tau^{-1})(\tau\lambda_\omega\tau^{-1})$, we obtain

$$\mathsf{V}_{\tau G_{\overline{v}}\tau^{-1}\to\tau G_{\overline{v}}\tau^{-1}\cap G'}(\tau\omega\tau^{-1}) \restriction L'^{\mathsf{a}} = \prod_{\lambda\in\Lambda_\tau} \tau\omega_\lambda\tau^{-1} \restriction L'^{\mathsf{a}}. \qquad \square$$

Definition 6.2.3. Let L/K be a (not necessarily finite) Galois extension, L^{a} the largest over K abelian intermediate field of L/K and $\mathcal{E}(L/K)$ the set of all over K finite Galois intermediate fields of L/K. Let $G = \mathrm{Gal}(L/K)$, hence $G^{\mathsf{a}} = \mathrm{Gal}(L^{\mathsf{a}}/K)$, and for $E \in \mathcal{E}(L/K)$ let $E^{\mathsf{a}} = E \cap L^{\mathsf{a}}$ be the largest over K abelian intermediate field of E/K. Then

$$L = \bigcup_{E\in\mathcal{E}(L/K)} E, \qquad L^{\mathsf{a}} = \bigcup_{E\in\mathcal{E}(L/K)} E^{\mathsf{a}},$$

$$G = \varprojlim_{E\in\mathcal{E}(L/K} \mathrm{Gal}(E/K) \quad \text{and} \quad G^{\mathsf{a}} = \varprojlim_{E\in\mathcal{E}(L/K} \mathrm{Gal}(E/K)^{\mathsf{a}}.$$

If $E,\, E' \in \mathcal{E}(L/K)$ and $E \subset E'$, then $E^{\mathsf{a}} \subset E'^{\mathsf{a}}$ and $(\,\cdot\,, E/K) = (\,\cdot\,, E'/K) \restriction E^{\mathsf{a}}$ by 6.2.2.3. Hence the map

$$(\,\cdot\,, L/K) = \varprojlim_{E\in\mathcal{E}(L/K)} (\,\cdot\,, E/K)\colon \mathsf{J}_K \to G^{\mathsf{a}}$$

is a homomorphism satisfying $(\alpha, L/K) \restriction E^{\mathsf{a}} = (\alpha, E/K)$ for all $E \in \mathcal{E}(L/K)$ and $\alpha \in \mathsf{J}_K$. By definition, $(\,\cdot\,, L/K) = (\,\cdot\,, L^{\mathsf{a}}/K)\colon \mathsf{J}_K \to G^{\mathsf{a}}$. We call $(\,\cdot\,, L/K)$ the **norm residue symbol** for L/K. In particular,

$$(\,\cdot\,, K) = (\,\cdot\,, K_{\mathrm{sep}}/K) = (\,\cdot\,, K_{\mathrm{ab}}/K)\colon \mathsf{J}_K \to G_K^{\mathsf{a}}$$

is called the **universal norm residue symbol** for K.

Theorem 6.2.4. *Let L/K be a (not necessarily finite) Galois extension, and let $G = \mathrm{Gal}(L/K)$. Then the norm residue symbol $(\,\cdot\,, L/K)\colon \mathsf{J}_K \to G^{\mathsf{a}}$ is a continuous homomorphism and its image is dense in G^{a}.*

In particular, if $[L\!:\!K] < \infty$, then $(\,\cdot\,, L/K)\colon \mathsf{J}_K \to G^{\mathsf{a}}$ is surjective.

Proof. Since $(\,\cdot\,, L/K) = (\,\cdot\,, L^{\mathsf{a}}/K)$, we may assume that L/K is abelian.

Assume first that $[L:K] < \infty$. Then $\mathsf{N}_{L/K}(\mathsf{J}_L)$ is an open subgroup of J_K by 5.4.4, and $\mathsf{N}_{L/K}(\mathsf{J}_L) \subset \mathrm{Ker}(\,\cdot\,, L/K)$ by 6.2.2.1. Hence $\mathrm{Ker}(\,\cdot\,, L/K)$ is open, too, and thus $(\,\cdot\,, L/K)$ is continuous. Let $G' = \mathrm{Im}(\,\cdot\,, L/K) \subset G$, and let $K' = L^{G'}$ be the fixed field of G'. Then $(\alpha, K'/K) = (\alpha, L/K) \restriction K' = 1$ for all $\alpha \in \mathsf{J}_K$. If $v \in \mathsf{M}_K$ and $v' \in \mathsf{M}_{K'}$ lies above v, then the local extension $K'_{v'}/K_v$ is also abelian. If $x \in K_v^\times$ and $j_v\colon K_v^\times \to \mathsf{J}_K$ is the local embedding, then $(j_v(x), K'/K) = (x, K'_{v'}/K_v) \restriction K' = 1$, and therefore $(x, K'_{v'}/K_v) = 1$ (observe that $\sigma \mapsto \sigma \restriction K'$ defines the embedding $\mathrm{Gal}(K'_{v'}/K_v) \hookrightarrow \mathrm{Gal}(K'/K)$). As this holds for all $x \in K_v^\times$, the local reciprocity law 4.2.7.2 impies $K_v = K'_{v'}$, and therefore v is fully decomposed in K'. Since this holds for all $v \in \mathsf{M}_K$, it follows by 6.1.5.1 that $K' = K$, and consequently $G' = G$.

Now let $[L\!:\!K] = \infty$ and $\mathcal{E}(L/K)$ the set of all over K finite intermediate fields of L/K. As $(\,\cdot\,, L/K) = \varprojlim_{E \in \mathcal{E}(K)} (\,\cdot\,, E/K)$ and all homomorphisms $(\,\cdot\,, E/K)$ are continuous, it follows that $(\,\cdot\,, L/K)$ is continuous.

Let finally $\tau \in G$ and U a neighbourhood of τ in G. By definition of the Krull topology, there exists a field $E \in \mathcal{E}(L/K)$ such that $\tau\mathrm{Gal}(L/E) \subset U$, and, as we have just proved, there exists some $\alpha \in \mathsf{J}_K$ such that

$$\tau \restriction E = (\alpha, E/K) = (\alpha, L/K) \restriction E,$$

and consequently $(\alpha, L/K) \in \tau\mathrm{Gal}(L/E) \subset U$. □

We close this section with an algebra-theoretic interpretation of the global norm residue symbol which supplies the connection between the local Hasse invariants and the global norm residue symbol. Later in 6.4.5 this connection will come out essentially in the proof of the main results of class field theory.

Definition and Theorem 6.2.5. *Let A be a central simple K-algebra.* For $v \in \mathsf{M}_K$ the central simple K_v-algebra $A_v = A_{K_v} = A \otimes_K K_v$ is called the **completion** of A, and $\mathrm{inv}_v(A) = \mathrm{inv}_{K_v}(A_v) \in \mathbb{Q}/\mathbb{Z}$ is called the **local Hasse invariant** of A at v (see 4.1.5 and 4.1.7). Note that $\mathrm{inv}_v(A) = 0$ if and only if $A_v \sim K_v$ (see 4.1.6.2).

Let L/K be a finite Galois extension, $G = \mathrm{Gal}(L/K)$, and suppose that $\chi \in \mathsf{X}(G) = \mathrm{Hom}^{\mathsf{c}}(G, \mathbb{Q}/\mathbb{Z})$. Let $L_\chi = L^{\mathrm{Ker}(\chi)}$ be the fixed field of χ, and

$$\mathrm{Gal}(L_\chi/K) = \langle \sigma_\chi \rangle, \quad \textit{where} \quad \chi(\sigma_\chi) = \frac{1}{[L_\chi\!:\!K]} + \mathbb{Z} \in \mathbb{Q}/\mathbb{Z}.$$

If $a \in K^\times$, then

$$\chi(a, L/K) = \sum_{v \in \mathsf{M}_K} \mathrm{inv}_v(L_\chi, \sigma_\chi, a),$$

Proof. For $v \in \mathsf{M}_K$ let $\overline{v} \in \mathsf{M}_L$ be a place above v, $G_{\overline{v}} = \mathrm{Gal}(L_{\overline{v}}/K_v) \subset G$, $\chi_{\overline{v}} = \chi \restriction G_{\overline{v}}$, $(L_{\overline{v}})_{\chi_{\overline{v}}} = L_{\overline{v}}^{\mathrm{Ker}(\chi_{\overline{v}})}$, and $v_\chi = \overline{v} \downarrow L_\chi$. Then it follow that $\mathrm{Ker}(\chi_{\overline{v}}) = \mathrm{Ker}(\chi) \cap G_{\overline{v}}$, hence $(L_{\overline{v}})_{\chi_{\overline{v}}} = L_\chi K_v$, and therefore $(L_{\overline{v}})_{\chi_{\overline{v}}} = (L_\chi)_{v_\chi}$. The extension L_χ/K is cyclic of degree $[L_\chi : K] = n(\chi) = \mathrm{ord}(\chi)$, and

$$\mathrm{Gal}(L_\chi/K) = \langle \sigma_\chi \rangle \ \text{ for some } \ \sigma_\chi \in G \ \text{ such that } \ \chi(\sigma_\chi) = \frac{1}{n(\chi)} + \mathbb{Z}.$$

The extension $(L_{\overline{v}})_{\chi_{\overline{v}}}/K_v$ is cyclic of degree $[(L_{\overline{v}})_{\chi_{\overline{v}}} : K_v] = n_v(\chi) = \mathrm{ord}(\chi_{\overline{v}})$, and

$$\mathrm{Gal}((L_{\overline{v}})_{\chi_{\overline{v}}}/K_v) = \langle \sigma_{\chi_{\overline{v}}} \rangle \ \text{ for some } \ \sigma_{\chi_{\overline{v}}} \in G, \ \text{ such that } \ \chi_{\overline{v}}(\sigma_{\chi_{\overline{v}}}) = \frac{1}{n_v(\chi)} + \mathbb{Z}.$$

As $(L_{\overline{v}})_{\chi_{\overline{v}}} = L_\chi K_v$, 1.8.4.1 implies

$$\mathrm{Gal}((L_{\overline{v}})_{\chi_{\overline{v}}}/K_v) \cong \mathrm{Gal}(L_\chi/L_\chi \cap K_v) \subset \mathrm{Gal}(L_\chi/K),$$

hence $n_v(\chi) \,|\, n(\chi)$ and $\mathrm{Gal}(L_\chi K_v/K_v) = \mathrm{Gal}((L_\chi)_{v_\chi}/K_v) = \langle \sigma_\chi^{n(\chi)/n_v(\chi)} \rangle$. Since

$$\chi\big(\sigma_\chi^{n(\chi)/n_v(\chi)}\big) = \frac{n(\chi)}{n_v(\chi)} \Big(\frac{1}{n(\chi} + \mathbb{Z} \Big) = \frac{1}{n_v(\chi)} + \mathbb{Z} = \chi_{\overline{v}}(\sigma_{\chi_{\overline{v}}}),$$

it follows that $\sigma_{\chi_{\overline{v}}} = \sigma_\chi^{n(\chi)/n_v(\chi)}$.

For an idele $\alpha = (\alpha_v)_{v \in \mathsf{M}_K} \in \mathsf{J}_K$ we obtain, using 4.2.1.3,

$$\begin{aligned}
\chi(\alpha, L/K) &= \chi\Big(\prod_{v \in \mathsf{M}_K} (\alpha_v, L_{\overline{v}}/K_v) \restriction L^{\mathsf{a}} \Big) = \sum_{v \in \mathsf{M}_K} \chi\big((\alpha_v, L_{\overline{v}}/K_v) \restriction L^{\mathsf{a}}\big) \\
&= \sum_{v \in \mathsf{M}_K} \chi(\alpha_v, L_{\overline{v}}/K_v) = \sum_{v \in \mathsf{M}_K} \chi_{\overline{v}}(\alpha_v, L_{\overline{v}}/K_v) \\
&= \sum_{v \in \mathsf{M}_K} \mathrm{inv}_{K_v}((L_{\overline{v}})_{\chi_{\overline{v}}}, \sigma_{\chi_{\overline{v}}}, \alpha_v).
\end{aligned}$$

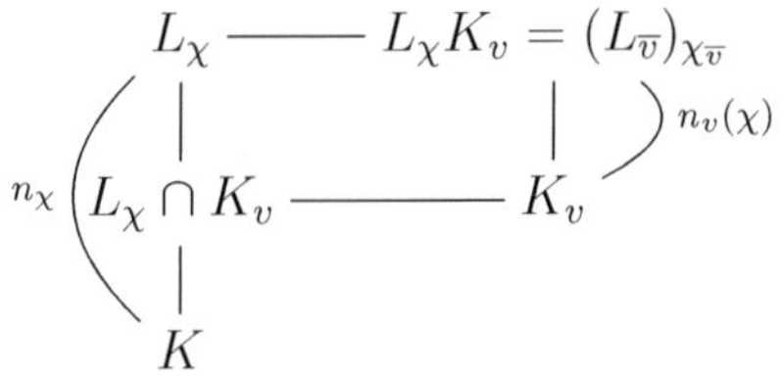

If $a \in K^\times$, then 3.6.4 implies

$$(L_\chi, \sigma_\chi, a) \otimes_K K_v \sim ((L_{\overline{v}})_{\chi_{\overline{v}}}, \sigma_\chi^{n(\chi)/n_v(\chi)}, a) = ((L_{\overline{v}})_{\chi_{\overline{v}}}, \sigma_{\chi_{\overline{v}}}, a),$$

and consequently

$$\begin{aligned}
\chi(a, L_{\overline{v}}/K_v) &= \mathrm{inv}_{K_v}((L_{\overline{v}})_{\chi_{\overline{v}}}, \sigma_{\chi_{\overline{v}}}, a) = \mathrm{inv}_{K_v}\big((L_\chi, \sigma_\chi, a) \otimes_K K_v\big) \\
&= \mathrm{inv}_v(L_\chi, \sigma_\chi, a).
\end{aligned}$$

Hence we obtain

$$\chi(a, L/K) = \sum_{v \in \mathsf{M}_K} \mathrm{inv}_v(L_\chi, \sigma_\chi, a). \qquad \square$$

6.3 p-extensions in characteristic p

Let $K/\mathbb{F}_q$ be a function field and $\mathrm{char}(K) = p > 0$.

This section is attended to preparations for a proof of the theorem of Artin-Tate in the function field case. It fills the gap left in Section 6.1. In the following we use notation and results from [18, Ch. 6] and complement them for our purposes.

We consider the adele ring

$$\begin{aligned}\mathsf{A}_K &= \prod_{v \in \mathsf{M}_K} (K_v \mid \widehat{\mathcal{O}}_v) \\ &= \Big\{ (\alpha_v)_{v \in \mathsf{M}_K} \in \prod_{v \in \mathsf{M}_K} K_v \;\Big|\; v(\alpha_v) \ge 0 \text{ for almost all } v \in \mathsf{M}_K \Big\}\end{aligned}$$

and its subring $\mathbb{A}_K = \mathsf{A}_K \cap K^{\mathsf{M}_K}$, consisting of all families $(\alpha_v)_{v \in \mathsf{M}_K}$ in K such that $v(\alpha_v) \ge 0$ for almost all $v \in \mathsf{M}_K$. Note that $\mathbb{A}_K$ is just the ring of repartitions of K (as defined in [18, Definition 6.3.1]). As in [18], we now use additive notation for the group $\mathbb{D}_K$ of divisors of K and note that $\mathbb{D}_K$ is the free abelian group with basis $\mathbb{P}_K = \{P_v \mid v \in \mathsf{M}_K\}$. In particular, every divisor $D \in \mathbb{D}_K$ has a unique representation

$$D = \sum_{v \in \mathsf{M}_K} n_v P_v, \quad \text{where } (n_v)_{v \in \mathsf{M}_K} \in \mathbb{Z}^{(\mathsf{M}_K)},$$

and we set $v(D) = n_v$ for all $v \in \mathsf{M}_K$.

We investigate the embedding $\mathbb{A}_K \hookrightarrow \mathsf{A}_K$, and we aim to extend Weil differentials (as defined in [18, Section 6.4]) to A_K.

For a divisor $D \in \mathbb{D}_K$ we set

$$\begin{aligned}\mathsf{A}_K(D) &= \{\alpha = (\alpha_v)_{v \in \mathsf{M}_K} \in \mathsf{A}_K \mid v(\alpha_v) \ge -v(D) \text{ for all } v \in \mathsf{M}_K\} \\ &= \prod_{v \in \mathsf{M}_K} \widehat{\mathfrak{p}}_v^{-v(D)} = \prod_{\substack{v \in \mathsf{M}_K \\ v(D) \ne 0}} \widehat{\mathfrak{p}}_v^{-v(D)} \times \prod_{\substack{v \in \mathsf{M}_K \\ v(D) = 0}} \widehat{\mathcal{O}}_v.\end{aligned}$$

Then $\mathsf{A}_K(D) \cap \mathbb{A}_K = \mathbb{A}_K(D)$ for all $D \in \mathbb{D}_K$, and $\{\mathsf{A}_K(D) \mid D \in \mathbb{D}_K\}$ is a fundamental system of open neighbourhoods of 0 in A_K.

We endow $\mathbb{F}_q$ with the discrete topology. A homomorphism $\varphi\colon \mathsf{A}_K \to \mathbb{F}_q$ is continuous if and only if $\varphi \restriction \mathsf{A}_K(D) = 0$ for some $D \in \mathbb{D}_K$. Consequently, a homomorphism $\sigma\colon \mathbb{A}_K \to \mathbb{F}_q$ is continuous (in the relative topology) if and only if $\sigma \restriction \mathbb{A}_K(D) = 0$ for some divisor $D \in \mathbb{D}_K$. Hence a Weil differential is just a continuous homomorphism $\sigma\colon \mathbb{A}_K \to \mathbb{F}_q$ satisfying $\sigma \restriction K = 0$.

Lemma 6.3.1. *$\mathbb{A}_K$ is dense in A_K, and if $C \in \mathbb{D}_K$, then $\mathbb{A}_K(C)$ is dense in $\mathsf{A}_K(C)$.*

Proof. Let $\alpha = (\alpha_v)_{v\in\mathsf{M}_K} \in \mathsf{A}_K$. For every $D \in \mathbb{D}_K$ there exists some repartition $\xi = (\xi_v)_{v\in\mathsf{M}_K} \in \mathbb{A}_K$ such that $v(\xi_v) \ge v(\alpha_v) - v(D)$ for all $v \in \mathsf{M}_K$, and consequently $\xi \in (\alpha + \mathsf{A}_K(D)) \cap \mathbb{A}_K$. Hence $\mathbb{A}_K$ is dense in A_K. If $C \in \mathbb{D}_K$, then $\mathsf{A}_K(C)$ is open in A_K, and therefore $\mathbb{A}_K(C) = \mathsf{A}_K(C) \cap \mathbb{A}_K$ is dense in $\mathsf{A}_K(C)$. □

Definition and Remarks 6.3.2.

1. We review the relevant concepts from [18, Sections 6.7 and 6.8].

Let Σ_K be the space of Weil differentials of K and $\mathrm{Der}_{\mathbb{F}_q}(K)$ the space of $\mathbb{F}_q$-derivations. Its dual space $\Omega_K = \mathrm{Hom}_K(\mathrm{Der}_{\mathbb{F}_q}(K), K)$ is the (global) differential module, and the map $d = d_K\colon K \to \Omega_K$, defined by $dz(D) = D(z)$ for all $z \in K$ and $D \in \mathrm{Der}_{\mathbb{F}_q}(K)$, is the universal derivation of K. Recall that Σ_K, $\mathrm{Der}_{\mathbb{F}_q}(K)$ and Ω_K are one-dimensional vector spaces over K, and if z is any separating element of $K/\mathbb{F}_q$, then $\Omega_K = K dz$.

Let $v \in \mathsf{M}_K$ and $t \in K$ such that $v(t) = 1$. Then $\Omega_K = K dt$, $K_v = \mathsf{k}_v((t))$, we denote by $\delta_t\colon \mathsf{k}_v((t)) \to \mathsf{k}_v((t))$ the differentiation with respect to t and by $\mathrm{Res}_t\colon \mathsf{k}_v((t)) \to \mathsf{k}_v$ the residue map (for a Laurent series $f \in \mathsf{k}_v((t))$ we let $\mathrm{Res}_t(f)$ be the coefficient of t^{-1}). A local differential at v is an equivalence class $f d_v u$ of a pair $(f, u) \in K_v^2$ with respect to the equivalence relation $(f, u) \sim (g, v)$ if and only if $f\delta_t(u) = g\delta_t(v)$. The set $\Omega_v = K_v d_v t$ of all local differentials at t is a one-dimensional vector space over K_v, and for a local differential $\omega = f d_v t \in \Omega_v$ (where $f \in K_v$) let $\mathrm{Res}_v(\omega) = \mathrm{Res}_t(f) \in \mathsf{k}_v$ be its residuum; it does not depend on t. The map $d_v\colon K_v \to \Omega_v$, $u \mapsto d_v u$, is the local universal derivation, and $\iota_v\colon \Omega_K \to \Omega_v$, defined by $\iota_v(f dt) = f d_v t$, is the local embedding of differentials.

Now let $t, f \in K$, $v \in \mathsf{M}_K$, $v(t) = 1$ and $\omega = f dt \in \Omega_K$. Then we call $\mathrm{Res}_v(\omega) = \mathrm{Res}_v(\iota_v(\omega)) = \mathrm{Res}_t(f) \in \mathsf{k}_v$ the residue of ω at v, we denote by, $v(\omega) = v(f) = \mathsf{o}_t(f) \in \mathbb{Z} \cup \{\infty\}$ the value of ω at v, and

$$(\omega) = \sum_{v\in\mathsf{M}_K} v(\omega) P_v \in \mathbb{D}_K$$

the divisor of ω.

2. By a **complete Weil differential** we mean a continuous $\mathbb{F}_q$-linear map $\theta\colon \mathsf{A}_K \to \mathbb{F}_q$ satisfying $\theta \restriction K = 0$. We denote by $\overline{\Sigma}_K$ the space of complete Weil differentials of K. For $c \in K$ and $\theta \in \overline{\Sigma}_K$ we define $c\theta\colon \mathsf{A}_K \to \mathbb{F}_q$ by

$c\theta(\alpha) = \theta(c\alpha)$; then $c\theta \in \overline{\Sigma}_K$. With this scalar multiplication and pointwise addition, $\overline{\Sigma}_K$ is a vector space over K.

A continuous map $\theta\colon \mathsf{A}_K \to \mathbb{F}_q$ belongs to $\overline{\Sigma}_K$ if and only if $\theta \restriction \mathbb{A}_K \in \Sigma_K$. Moreover, if $\theta \in \overline{\Sigma}_K$ and $\theta \neq 0$, then $\theta \restriction \mathsf{A}_K \neq 0$ (since θ is continuous and $\mathbb{A}_K$ is dense in A_K), and then we call $(\theta) = (\theta \restriction \mathbb{A}_K) \in \mathbb{D}_K$ the **divisor** of θ. If $D \in \mathbb{D}_K$ and $\theta \in \overline{\Sigma}_K$, then $\theta \restriction \mathsf{A}_K(D) = 0$ if and only if $\theta \restriction \mathbb{A}_K(D) = 0$, and this holds if and only if $D \le (\theta)$ (see [18, Theorem 6.4.3]).

3. For a differential $\omega \in \Omega_K$ we define $\omega^{\#}\colon \mathsf{A}_K \to \mathbb{F}_q$ by

$$\omega^{\#}(\alpha) = \sum_{v \in \mathsf{M}_K} \mathsf{Tr}_{\mathsf{k}_v/\mathbb{F}_q}(\mathrm{Res}_v(\alpha_v \iota_v(\omega)) \quad \text{for all} \quad \alpha = (\alpha_v)_{v \in \mathsf{M}_K} \in \mathsf{A}_K.$$

Then $\omega^{\#} \restriction \mathbb{A}_K = \sigma_\omega \in \Sigma_K$, $(\omega) = (\sigma_\omega) \in \mathbb{D}_K$, and the assignment $\omega \mapsto \sigma_\omega$ defines a K-isomorphism $\Omega_K \overset{\sim}{\to} \Sigma_K$ (see [18, Theorem 6.8.5]). As the map $\mathrm{Res}_t\colon \mathsf{k}_v\,((t)) \to \mathbb{F}_q$ is continuous (see [18, Theorem 6.7.8]), it follows that $\omega^{\#}$ is continuous. Consequently, $\omega^{\#} \in \overline{\Sigma}_K$, and $(\omega^{\#}) = (\omega) \in \mathbb{D}_K$.

Theorem 6.3.3.

1. The maps

$$\Theta\colon \begin{cases} \Omega_K & \to \ \overline{\Sigma}_K \\ \omega & \mapsto \ \omega^{\#} \end{cases} \qquad \text{and} \qquad \rho\colon \begin{cases} \overline{\Sigma}_K & \to \ \Sigma_K \\ \theta & \mapsto \ \theta \restriction \mathbb{A}_K \end{cases}$$

are K-isomorphisms. In particular, $\dim_K \overline{\Sigma}_K = 1$, and if x is a separating element for $K/\mathbb{F}_q$, then $\overline{\Sigma}_K = K(dx)^{\#}$.

2. If $\theta \in \overline{\Sigma}{}_K^{\bullet}$ and $\alpha \in \mathsf{A}_K$ are such that $\theta(\xi\alpha) = 0$ for all $\xi \in \mathsf{A}_K$, then $\alpha = 0$.

Proof. 1. Apparently, Θ and ρ are $\mathbb{F}_q$-linear, and by definition ρ is even K-linear. We have already seen that $\mathrm{Ker}(\rho) = \mathbf{0}$ and that $\rho \circ \Theta$ is a K-isomorphism. Hence Θ and ρ are both K-isomorphisms.

2. Let $\theta = \omega^{\#} \in \overline{\Sigma}{}_K^{\bullet}$, where $\omega \in \Omega_K^{\bullet}$, and suppose at the contrary that there is some $\alpha = (\alpha_v)_{v \in \mathsf{M}_K} \in \mathsf{A}_K^{\bullet}$ such that $\theta(\xi\alpha) = 0$ for all $\xi \in \mathsf{A}_K$. Let $v \in \mathsf{M}_K$ be such that $\alpha_v \neq 0$. For $u \in K_v$ let $\overline{u} \in \mathsf{A}_K$ be defined by $\overline{u}_v = u_v$ and $\overline{u}_{v'} = 0$ for all $v' \in \mathsf{M}_K \setminus \{v\}$. Then it follows that

$$0 = \omega^{\#}(\overline{u}\alpha) = \mathsf{Tr}_{\mathsf{k}_v/\mathbb{F}_q}(\mathrm{Res}_v(\alpha_v u\, \iota_v(\omega)) \quad \text{for all} \quad u \in K_v.$$

If $u \in K_v$, then $0 = \mathsf{Tr}_{\mathsf{k}_v/\mathbb{F}_q}(\mathrm{Res}_v(cu\iota_v(\omega)) = \mathsf{Tr}_{\mathsf{k}_v/\mathbb{F}_q}(c\mathrm{Res}_v(u\iota_v(\omega))$ for all $c \in \mathsf{k}_v$, and as $\mathsf{Tr}_{\mathsf{k}_v/\mathbb{F}_q} \neq 0$, we finally obtain $\mathrm{Res}_v(u\iota_v(\omega)) = 0$ for all $u \in K_v$.

Let $t \in K$ such that $v(t) = 1$ and $\omega = g dt$, where $g \in K^{\times}$. If $u \in K_v$ is such that $v(u) = -1 - v(g)$, then $v(u\iota_v(\omega) = v(u) + v(g) = -1$, and consequently $\mathrm{Res}_v(u\iota_v(\omega)) \neq 0$, a contradiction. □

Now we investigate abelian extensions of exponent p of K and define (in the style of 4.10.1 and 6.2.1) a global p-Hilbert symbol. We fix an algebraic closure $\overline{K}$ of K and consider algebraic extensions of K inside $\overline{K}$. We refer to 2.6.7 for the general theory of Artin-Schreier extensions and to Section 4.10 for the local theory.

Definition 6.3.4. For $v \in \mathsf{M}_K$ let $(\,\cdot\,,\,\cdot\,]_v = (\,\cdot\,,\,\cdot\,]_{K_v} \colon K_v^\times \times K_v \to \mathbb{F}_p$ be the local p-Hilbert symbol, defined in 4.10.1 by

$$(\beta_v, a]_v = (\beta_v, K_v(\wp^{-1}a)/K_v)(x) - x \quad \text{for } x \in \wp^{-1}a.$$

Now we define the **(global) p-Hilbert symbol** $(\,\cdot\,,\,\cdot\,] \colon \mathsf{J}_K \times K \to \mathbb{F}_p$ by

$$(\beta, a] = \sum_{v \in \mathsf{M}_K} (\beta_v, a]_v \quad \text{for all } \beta = (\beta_v)_{v \in \mathsf{M}_K} \in \mathsf{J}_K \text{ and } a \in K.$$

Let $a \in K$ and $x \in \wp^{-1}a$. If $v \in \mathsf{M}_K$ is unramified in $K(\wp^{-1}a)$, then the extension $K_v(\wp^{-1}a)/K_v$ is unramified, hence $U_v \subset \mathsf{N}_{K_v(\wp^{-1}a)/K_v} K_v(\wp^{-1}a)^\times$, and if $\beta_v \in U_v$, then $(\beta_v, a]_v = (\beta_v, K_v(\wp^{-1}a)/K_v)(x) - x = 0$. Consequently, if $\beta = (\beta_v)_{v \in \mathsf{M}_K} \in \mathsf{J}_K$, then $\beta_v \in U_v$ for almost all $v \in \mathsf{M}_K$, and as almost all $v \in \mathsf{M}_K$ are unramified in K, in the above definition of $(\beta, a]$ only finitely many summands are different from 0. Hence the p-Hilber symbol $(\beta, a]$ is well-defined.

Theorem 6.3.5 (Properties of the global p-Hilbert symbol)**.**

1. *If β, $\beta' \in \mathsf{J}_K$ and a, $a' \in K$, then*

$$(\beta\beta', a] = (\beta, a] + (\beta', a], \quad (\beta, a + a'] = (\beta, a] + (\beta, a'],$$

and

$$(\beta^p, a] = (\beta, \wp a] = 0.$$

2. *If $\beta = (\beta_v)_{v \in \mathsf{M}_K} \in \mathsf{J}_K$ and $a \in K$, then*

$$(\beta, a] = \sum_{v \in \mathsf{M}_K} \mathsf{Tr}_{\mathsf{k}_v/\mathbb{F}_p}(\mathrm{Res}_v(a\beta_v^{-1} d_v \beta_v))\Big).$$

3. *If $b \in K^\times$ and $a \in K$, then $(b, a] = 0$.*

Proof. According to the definition, 1. follows by 4.10.2.1 and 2. by 4.10.3.

3. If $b \in K^\times$ and $a \in K$, then

$$\begin{aligned}(b, a] &= \sum_{v \in \mathsf{M}_K} (b, a]_v = \sum_{v \in \mathsf{M}_K} \mathsf{Tr}_{\mathsf{k}_v/\mathbb{F}_p}(\mathrm{Res}_{\mathsf{k}_v/\mathbb{F}_p}(ab^{-1} d_v b)) \\ &= \mathsf{Tr}_{\mathbb{F}_q/\mathbb{F}_p}\Big(\sum_{v \in \mathsf{M}_K} \mathsf{Tr}_{\mathsf{k}_v/\mathbb{F}_q}(\mathrm{Res}_{\mathsf{k}_v/\mathbb{F}_p}(ab^{-1} d_v b))\Big) = 0\end{aligned}$$

by 2. and the residue theorem (see [18, Theorem 6.8.4]). □

In the following definition we introduce the concept of an adelic derivation. We shall essentially use this concept for the proof that the global p-Hilbert symbol is non-degerate, which is the essential step in the proof of the global reciprocity law in characteristic p.

Definition 6.3.6. Let $t \in K$ be a separating element for $K/\mathbb{F}_q$ and for $v \in \mathsf{M}_K$ let $\delta_{t,v} \colon K_v \to K_v$ be the unique derivation satisfying

$$\delta_{t,v} \restriction K = D_t \in \mathrm{Der}_{\mathbb{F}_q}(K), \text{ the unique } \mathbb{F}_q\text{-derivation satisfying } D_t(t) = 1.$$

If $t_v \in K_v$ is any element satisfying $v(t_v) = 1$, then $\delta_{t,v}(x) = \delta_{t_v}(x)\delta_{t_v}(t)^{-1}$ for all $x \in K_v$ (see [18, Theorem 6.7.9]). We define

$$\vartheta_t \colon \mathsf{A}_K \to \mathsf{A}_K \quad \text{by} \quad \vartheta_t(\alpha) = (\delta_{t,v}(\alpha_v))_{v \in \mathsf{M}_K} \quad \text{for all} \quad \alpha = (\alpha_v)_{v \in \mathsf{M}_K} \in \mathsf{A}_K.$$

It is easily checked (component-wise) that ϑ_t is a derivation, called the **adelic derivation** induced by t.

Theorem 6.3.7. *The global p-Hilbert symbol* $(\,\cdot\,,\cdot\,] \colon \mathsf{J}_K \times K \to \mathbb{F}_p$ *defines a non-degenerate pairing*

$$\mathsf{J}_K/K^\times \mathsf{J}_K^p \times K/\wp K \;\to\; \mathbb{F}_p, \quad [\beta K^\times \mathsf{J}_K^p, a + \wp K) \;\mapsto\; (\beta, a].$$

in particular:

(a) If $a \in K$*, then* $[\beta, a) = 0$ *for all* $\beta \in \mathsf{J}_K$ *if and only if* $a \in \wp K$*.*

(b) If $\beta \in \mathsf{J}_K$*, then* $[\beta, a) = 0$ *for all* $a \in K$ *if and only if* $\beta \in K^\times \mathsf{J}_K^p$*.*

Proof. By 6.3.5.1, the p-Hilbert symbol induces a pairing $\mathsf{J}_K \times K \to \mathbb{F}_p$. Therefore it suffices to prove the assertions (a) and (b).

(a) If $a \in \wp K$, then $(\beta, a] = 0$ for all $\beta \in \mathsf{J}_K$ by 6.3.5.1. Thus let $a \in K$ be such that $(\beta, a] = 0$ for all $\beta \in \mathsf{J}_K$. If $v \in \mathsf{M}_K$, then $0 = (j_v(b), a] = (b, a]_v$ for all $b \in K_v^\times$, hence $a \in \wp K_v$ by 4.10.2.4 and $K_v(\wp^{-1}a) = K_v$ für all $v \in \mathsf{M}_K$. Therefore all $v \in \mathsf{M}_K$ are fully decomposed in $K(\wp^{-1}a)$, and by 6.1.5.1 it follows that $K(\wp^{-1}a) = K$, that is, $a \in \wp K$.

(b) If $\beta = b\alpha^p \in K^\times \mathsf{J}_K^p$, where $b \in K^\times$ and $\alpha \in \mathsf{J}_K$, then it follows that $(\beta, a] = (b, a] + p(\beta, a] = 0$ by 6.3.5.3. Thus let $\beta \in \mathsf{J}_K$ be such that $(\beta, a] = 0$ for all $a \in K$. In particular, by 6.3.4.2 we obtain, for all $a \in K$ and $c \in \mathbb{F}_q$,

$$\begin{aligned} 0 = (\beta, ac] &= \sum_{v \in \mathsf{M}_K} \mathsf{Tr}_{\mathsf{k}_v/\mathbb{F}_p}(\mathrm{Res}_v(ca\beta_v^{-1} d_v \beta_v)) \\ &= \mathsf{Tr}_{\mathbb{F}_q/\mathbb{F}_p}\Big(c \sum_{v \in \mathsf{M}_K} \mathsf{Tr}_{\mathsf{k}_v/\mathbb{F}_q}(\mathrm{Res}_v(a\beta_v^{-1} d_v \beta_v))\Big). \end{aligned}$$

As $\mathsf{Tr}_{\mathbb{F}_q/\mathbb{F}_p} \neq 0$, it follows that

$$\sum_{v \in \mathsf{M}_K} \mathsf{Tr}_{\mathsf{k}_v/\mathbb{F}_q}(\mathrm{Res}_v(a\beta_v^{-1}d_v\beta_v)) = 0 \quad \text{for all} \quad a \in K. \tag{$\dagger$}$$

We investigate this sum by means of an adelic derivation. For every $v \in \mathsf{M}_K$ let $t_v \in K$ be such that $v(t_v) = 1$. We fix some $w \in \mathsf{M}_K$ and set $t = t_w$. Then t is a separating element for $K/\mathbb{F}_q$, and we consider the adelic derivation $\vartheta_t \colon \mathsf{A}_K \to \mathsf{A}_K$. Let

$$\beta^* = \beta^{-1}\vartheta_t(\beta) \in \mathsf{A}_K, \quad \text{and define} \quad \theta \colon \mathsf{A}_K \to \mathbb{F}_q \quad \text{by} \quad \theta(\xi) = (dt)^{\#}(\xi\beta^*).$$

We shall prove:

A. θ is a complete Weil differential.

Proof of **A.** As $(dt)^{\#} \in \overline{\Sigma}_K$, θ is continuous, and thus it suffices to prove that $\theta(a) = 0$ for al $a \in K$. First we calculate

$$\begin{aligned} \beta_v^*\iota_v(dt) &= \beta_v^{-1}\delta_{t,v}(\beta_v)\iota_v(dt) = [\beta_v^{-1}\delta_{t_v}(\beta_v)\delta_{t_v}(t)^{-1}]\,[\delta_{t_v}(t)d_vt_v] \\ &= \beta_v^{-1}\delta_{t_v}(\beta_v)d_vt_v = \beta_v^{-1}d_v\beta_v \quad \text{for all} \quad v \in \mathsf{M}_K, \end{aligned}$$

and then we obtain, for all $a \in K$,

$$\begin{aligned} \theta(a) = (dt)^{\#}(a\beta^*) &= \sum_{v \in \mathsf{M}_K} \mathsf{Tr}_{\mathsf{k}_v/\mathbb{F}_q}(\mathrm{Res}_v(a\beta_v^*\iota_v(dt)) \\ &= \sum_{v \in \mathsf{M}_K} \mathsf{Tr}_{\mathsf{k}_v/F}(a\beta_v^{-1}d_v\beta_v) = 0 \end{aligned}$$

by ($\dagger$). $\square$[**A.**]

As $\theta \in \overline{\Sigma}_K = K(dt)^{\#}$, there exists some $y \in K$ such that $\theta = y(dt)^{\#}$. Then

$$0 = (\theta - y(dt)^{\#})(\xi) = (dt)^{\#}(\xi(\beta^* - y)) \quad \text{for all} \quad \xi \in \mathsf{A}_K,$$

and therefore $y = \beta^* = \beta^{-1}\vartheta_t(\beta)$ by 6.3.3.3. In particular, since $t_w = t$, it follows that

$$y = \beta_w^* = \beta_w^{-1}\delta_{t_w,w}(\beta_w) = \beta_w^{-1}\delta_{t_w}(\beta_w)\delta_{t_w}(t_w)^{-1} = \beta_w^{-1}\delta_{t_w}(\beta_w).$$

The derivation $\delta_t \colon K_w \to K_w$ satisfies $\delta_t \restriction K = D_t \colon K \to K$, $\delta_t^p = 0$ (see [18, Theorem 6.7.8]), and $\beta_w^{-1}\delta_t(\beta_w) = y \in K$. But then there exists even some $b \in K^\times$ such that $y = b^{-1}\delta_t(b) = b^{-1}D_t(b)$ (see [18, Theorem 6.7.3]).

Now we consider the idele $\gamma = b^{-1}\beta$. Since

$$b^{-1}\vartheta_t(b) = (b^{-1}\delta_{t,v}(b))_{v \in \mathsf{M}_K} = b^{-1}D_t(b) = y = \beta^{-1}\delta_t(\beta),$$

it follows that

$$\vartheta_t(\gamma) = -b^{-2}\vartheta_t(b)\beta + b^{-1}\vartheta_t(\beta) = -b^{-1}\delta_t(\beta) + b^{-1}\delta_t(\beta) = 0.$$

For every $v \in \mathsf{M}_K$ this implies $0 = \vartheta_t(\gamma)_v = \delta_{t,v}(\gamma_v) = \delta_{t_v}(\gamma_v)\delta_{t_v}(t)^{-1}$, and therefore $\delta_{t_v}(\gamma_v) = 0$, which entails $\gamma_v \in K_v^p$ (see [18, Theorem 6.7.4]). Eventually this implies $\gamma \in \mathsf{J}_K^p$ and $\beta = b\gamma \in K^\times \mathsf{J}_K^p$. □

In the following Theorem 6.3.8 we prove the global reciprocity law for the maximal abelian extension of exponent p (see 2.6.7.3). We tacitly use the results of Section 6.2. We consider the groups

$$\mathsf{C}_K^p = \mathsf{J}_K^p K^\times / K^\times \quad \text{and} \quad \mathsf{C}_K/\mathsf{C}_K^p \cong \mathsf{J}_K/\mathsf{J}_K^p K^\times.$$
$$\text{We identify: } \mathsf{C}_K/\mathsf{C}_K^p = \mathsf{J}_K/\mathsf{J}_K^p K^\times.$$

By 6.3.7, the global p-Hilbert symbol $(\,\cdot\,,\cdot\,]$ induces a non-degenerate pairing

$$\mathsf{C}_K/C_K^p \times K/\wp K \;\to\; \mathbb{F}_p.$$

Theorem 6.3.8. *Let $N = K(\wp^{-1}K)$ be the largest over K abelian extension of exponent p and $G = \mathrm{Gal}(N/K)$.*

1. The norm residue symbol $(\,\cdot\,, N/K) \colon \mathsf{J}_K \to G$ is surjective with kernel $J_K^p K^\times$, and it induces an (equally denoted) topological isomorphism

$$(\,\cdot\,, N/K) \colon \mathsf{C}_K/\mathsf{C}_K^p = \mathsf{J}_K/\mathsf{J}_K^p K^\times \;\to\; G.$$

In particular, the group $\mathsf{C}_K/\mathsf{C}_K^p$ is compact.

2. Let L be an intermediate field of N/K such that $[L:K] = p$. Then

$$(\mathsf{C}_K \!:\! \mathsf{N}_{L/K}\mathsf{C}_L) = p \quad \textit{and} \quad H^1(G, \mathsf{C}_L) = H^{-1}(G, \mathsf{C}_L) = \mathbf{1}.$$

Proof. 1. If $a \in K$, $x \in \wp^{-1}a \varepsilon N$ and $\beta \in \mathsf{J}_K$, then 6.2.3 and 6.2.1 yield

$$(\beta, N/K)(x) = (\beta, K(x)/K)(x) = \Big[\prod_{v \in \mathsf{M}_K} (\beta_v, K_v(x)/K) \restriction K(x)\Big](x).$$

If $v \in \mathsf{M}_K$, then $(\beta_v, K_v(x)/K_v)(x) = x + (\beta_v, a]_v$ by the very definition of the local p-Hilbert symbol. Hence

$$(\beta, N/K)(x) = x + \sum_{v \in \mathsf{M}_K} (\beta_v, a]_v = x + (\beta, a],$$

and consequently

$$\begin{aligned} \mathrm{Ker}(\,\cdot\,, N/K) &= \{\beta \in \mathsf{J}_K \mid (\beta, N/K)(x) = x \ \text{ for all } \ x \in N\} \\ &= \{\beta \in \mathsf{J}_K \mid (\beta, a] = 0 \ \text{ for all } \ a \in K\} = \mathsf{J}_K^p K^\times \quad \text{by 6.3.7.(b)}. \end{aligned}$$

By 6.2.4, the homomorphism $(\,\cdot\,, N/K)\colon \mathsf{J}_K \to G$ is continuous and its image is dense in G. Hence it induces an (equally denoted) continuous monomorphism

$$(\,\cdot\,, N/K)\colon \mathsf{J}_K/\mathsf{J}_K^p K^\times = \mathsf{C}_K/\mathsf{C}_K^p \;\to\; G$$

having a dense image. The topological isomorphism $\mathsf{C}_K \cong \mathsf{C}_K^0 \times \mathbb{Z}$ (see 5.3.3.2(b)) induces a topological isomorphism $\mathsf{C}_K/\mathsf{C}_K^p \cong \mathsf{C}_K^0/(\mathsf{C}_K^0)^p \times \mathbb{Z}/p\mathbb{Z}$, and C_K^0 is compact by 5.3.7.1. Hence $\mathsf{C}_K/\mathsf{C}_K^p$ is compact, too, and therefore $(\,\cdot\,, N/K)\colon \mathsf{C}_K/\mathsf{C}_K^p \to G$ is a topological isomorphism.

2. It suffices to prove that $|H^0(G, \mathsf{C}_L)| = (\mathsf{C}_K : \mathsf{N}_{L/K}\mathsf{C}_L) = p$. Indeed, since $h(G, \mathsf{C}_L) = p$ by 6.1.3.2, we then obtain $H^{-1}(G, \mathsf{C}_L) = 1$ and thus also $H^1(G, \mathsf{C}_L) = 1$ by 2.4.5.1.

By 1., $\eta = (\,\cdot\,, N/K)\colon \mathsf{C}_K/\mathsf{C}_K^p \overset{\sim}{\to} G$ is a topological isomorphism. Apparently $\mathsf{C}_K^p \subset \mathsf{N}_{L/K}\mathsf{C}_L$, and as $\mathsf{N}_{L/K}\mathsf{C}_L$ is open in C_K by 5.4.4.2, it follows that $\mathsf{N}_{L/K}\mathsf{C}_L/\mathsf{C}_K^p$ is open in $\mathsf{C}_K/\mathsf{C}_K^p$ and thus closed. Therefore $H = \eta(\mathsf{N}_{L/K}\mathsf{C}_L/\mathsf{C}_K^p)$ is closed in G, and if $E = N^H$, then $H = \mathrm{Gal}(N/E)$. Hence η induces an isomorphism

$$\overline{\eta}\colon \mathsf{J}_K/K^\times\mathsf{N}_{L/K}\mathsf{J}_L = \mathsf{C}_K/\mathsf{N}_{L/K}\mathsf{C}_L \;\to\; G/H = \mathrm{Gal}(E/K)$$

satisfying $\overline{\eta}(\beta K^\times \mathsf{N}_{L/K}\mathsf{J}_L) = (\beta, E/K)$ for all $\beta \in \mathsf{J}_K$. In particular, $(\beta, E/K) = 1$ for all $\beta \in \mathsf{N}_{L/K}\mathsf{J}_L$. Since $(\mathsf{C}_K : \mathsf{N}_{L/K}\mathsf{C}_L) = [E:K]$, it suffices to prove that $E = L$.

We assume to the contrary that $E \ne L$. Since E/K is abelian of exponent p, there exists an intermediate field L' of E/K such that $L' \ne L$ and $[L' : K] = p$. By 6.1.6, there exists some place $w \in \mathsf{M}_K$ which is decomposed in L and inert in L'. Now let $\beta = (\beta_v)_{v \in \mathsf{M}_K} \in \mathsf{J}_K$ be such that $w(\beta_w) = 1$ and $\beta_v = 1$ for all $v \in \mathsf{M}_K \setminus \{w\}$. Let $\overline{w} \in \mathsf{M}_L$ be a place above w and $\overline{\beta} \in \mathsf{M}_L$ defined by $\overline{\beta}_{\overline{w}} = \beta_v$ and $\overline{\beta}_{w'} = 1$ for all $w' \in \mathsf{M}_L \setminus \{\overline{w}\}$. Since $L_{\overline{w}} = K_w$, we get $\beta = \mathsf{N}_{L/K}(\overline{\beta})$, hence $(\beta, E/K) = 1$ and therefore $(\beta, L'/K) = (\beta, E/K) \restriction L' = 1$. For every $v \in \mathsf{M}_K$ we fix a place $v' \in \mathsf{M}_{L'}$ above v. Then $L'_{w'} \ne K_w$, $L'_{w'}/K_w$ is inert and thus $(\beta_w, L'_{w'}/K_w) = \varphi_{L_{w'}/K_w} \ne 1$ by 4.2.3.2. Therefore we obtain

$$(\beta, L'/K) = \prod_{v \in \mathsf{M}_K} (\beta_v, L'_{v'}/K_v) = (\beta_w, L'_{w'}/K_w) = \varphi_{L'_{w'}/K_w} \ne 1,$$

a contradiction. □

6.4 The global reciprocity law

Theorem 6.4.1. *Let L/K be a cyclic extension and $G = \mathrm{Gal}(L/K)$.*

1. $H^1(G, \mathsf{C}_L) = H^{-1}(G, \mathsf{C}_L) = 1$, *and* $(\mathsf{C}_K : \mathsf{N}_{L/K}\mathsf{C}_L) = [L:K]$.

2. (Hasse norm theorem) *The map*

$$\theta\colon K^\times/\mathsf{N}_{L/K}L^\times \to \bigoplus_{v\in\mathsf{M}_K} K_v^\times/\mathsf{N}_{L_{\overline{v}}/K_v}L_{\overline{v}}^\times,$$
$$a\mathsf{N}_{L/K}L^\times \mapsto (a\mathsf{N}_{L_{\overline{v}}/K_v}K_{\overline{v}}^\times)_{v\in\mathsf{M}_K}$$

is a group monomorphism. In particular the following local-global principle for norms holds for all $a \in K^\times$:

$$a \in \mathsf{N}_{L/K}L^\times \quad\Longleftrightarrow\quad a \in \mathsf{N}_{L_{\overline{v}}/K_v}L_{\overline{v}}^\times \ \text{ for all } \ v\in\mathsf{M}_K.$$

Proof. 1. Recall that $h(G,\mathsf{C}_L) = 1$ by 6.1.3.2 and $H^1(G,\mathsf{C}_L) \cong H^{-1}(G,\mathsf{C}_L)$ by 2.4.5.1. Hence it suffices to prove that $H^1(G,\mathsf{C}_L) = \mathbf{1}$, and for this we use induction on $[L\!:\!K]$. Thus assume that $[L\!:\!K] > 1$ and the assertion holds for all extensions of smaller degree.

CASE 1: $[L:K] = p$ is a prime. If $\operatorname{char}(K) = p$, the assertion follows by 6.3.8.2. Thus let $\operatorname{char}(K) \ne p$, $K' = K(\zeta)$ and $L' = L(\zeta) = K'L$, where $\zeta \in \overline{K}$ is a primitive p-th root of unity. Then L'/K is Galois, K'/K and L'/K' are cyclic, $[L':K'] = p$ and $[K':K] < p$. By 5.4.8 it follows that $\mathsf{C}_{K'} = \mathsf{C}_{L'}^{\mathrm{Gal}(L'/K')}$, and $\mathsf{C}_L = \mathsf{C}_{L'}^{\mathrm{Gal}(L'/L)}$. Hence 2.5.4.3 yields the exact sequences

$$\mathbf{1} \to H^1(\mathrm{Gal}(K'/K),\mathsf{C}_{K'}) \overset{\mathrm{Inf}^1}{\to} H^1(\mathrm{Gal}(L'/K),\mathsf{C}_{L'}) \overset{\mathrm{Res}^1}{\to} H^1(\mathrm{Gal}(L'/K'),\mathsf{C}_{L'})$$

and

$$\mathbf{1} \to H^1(G,\mathsf{C}_L) \overset{\mathrm{Inf}^1}{\to} H^1(\mathrm{Gal}(L'/K),\mathsf{C}_{L'}).$$

Since as $H^1(\mathrm{Gal}(L'/K'),\mathsf{C}_{L'}) = \mathbf{1}$ by 6.1.10.3, and $H^1(\mathrm{Gal}(K'/K),\mathsf{C}_{K'}) = \mathbf{1}$ by the induction hypothesis, the first exact sequence shows that $H^1(\mathrm{Gal}(L'/K),\mathsf{C}_{L'}) = \mathbf{1}$, and the second one implies $H^1(G,\mathsf{C}_L) = \mathbf{1}$.

CASE 2: $[L\!:\!K] = pm$ for some prime p and $m \in \mathbb{N}$. Then G has a subgroup G' of order p, and we set $L' = L^{G'}$. Then $G/G' = \mathrm{Gal}(L'/K)$, $\mathsf{C}_{L'} = \mathsf{C}_L^{G'}$ by 5.4.8 and 2.5.4.3 yields the exact sequence

$$\mathbf{1} \to H^1(G/G',\mathsf{C}_{L'}) \overset{\mathrm{Inf}^1}{\to} H^1(G,\mathsf{C}_L) \overset{\mathrm{Res}^1}{\to} H^1(G',\mathsf{C}_L).$$

Since $H^1(G',\mathsf{C}_L) = \mathbf{1}$ by CASE 1 and $H^1(G/G',\mathsf{C}_{L'}) = \mathbf{1}$ by the induction hypothesis, it follows that $H^1(G,\mathsf{C}_L) = \mathbf{1}$.

2. By 2.4.2 and 1. the exact sequence $\mathbf{1} \to L^\times \hookrightarrow \mathsf{J}_L \to \mathsf{C}_L \to \mathbf{1}$ yields the exact sequence $\mathbf{1} = H^{-1}(G,\mathsf{C}_L) \to H^0(G,L^\times) \to H^0(G,\mathsf{J}_L)$ and thus a monomorphism

$$i\colon K^\times/\mathsf{N}_{L/K}L^\times \to \mathsf{J}_K/\mathsf{N}_{L/K}(\mathsf{J}_L) \quad\text{such that}\quad i(a\mathsf{N}_{L/K}L^\times) = a\,\mathsf{N}_{L/K}\mathsf{J}_L.$$

By 6.1.1.1(b) there exists an isomorphism

$$\theta^0 \colon \mathsf{J}_K/\mathsf{N}_{L/K}\mathsf{J}_L \to \bigoplus_{v\in\mathsf{M}_K} K_v^\times/\mathsf{N}_{L_{\overline{v}}/K_v}L_{\overline{v}}^\times,$$

given by $\theta^0(\alpha\mathsf{N}_{L/K}\mathsf{J}_L) = (\alpha_v\mathsf{N}_{L_{\overline{v}}/K_v}L_{\overline{v}}^\times)_{v\in\mathsf{M}_K}$. Hence the monomorphism

$$\theta = \theta^0\circ i \colon K^\times/\mathsf{N}_{L(K}L^\times \to \bigoplus_{v\in\mathsf{M}_K} K_v^\times/\mathsf{N}_{L_{\overline{v}}/K_v}L_{\overline{v}}^\times$$

satisfies $\theta(a\mathsf{N}_{L/K}L^\times) = (a\mathsf{N}_{L_{\overline{v}}/K_v}L_{\overline{v}}^\times$ for all $a \in K^\times$, which implies the assertion. □

Next we investigate the influence of the Hasse norm theorem on the splitting properties of central simple K-algebras. The main result is the splitting theorem of Hasse - Brauer - Noether 6.4.3, which is a local-global principle for central simple algebras. Based on it, we shall prove the theorem of Artin - Tate 6.4.4, and from all this we shall derive the global reciprocity law in 6.4.7.

Remarks 6.4.2. Let L/K be a finite Galois extension, $G = \mathrm{Gal}(L/K)$ and $v \in \mathsf{M}_K$. Then 3.5.6.1, applied with $K' = K_v$, yields the following commutative diagram in which i denotes the inclusion:

$$\begin{array}{ccccc} \mathrm{Br}(K) & \xleftarrow{\;i\;} & \mathrm{Br}(L/K) & \xrightarrow{\Theta_{L/K}} & H^2(G, L^\times) \\ \downarrow{\scriptstyle \mathrm{res}_{K_v/K}} & & \downarrow{\scriptstyle \mathrm{res}_{L/K}^{L_{\overline{v}}/K_v}} & & \downarrow{\scriptstyle \mathrm{Res}^2_{L/K\to L_{\overline{v}}/K_v}} \\ \mathrm{Br}(K_v) & \xleftarrow{\;i\;} & \mathrm{Br}(L_{\overline{v}}/K_v) & \xrightarrow{\Theta_{L_{\overline{v}}/K_v}} & H^2(G_{\overline{v}}, L_{\overline{v}}^\times) \end{array}$$

Let A be a central simple K-algebra which splits over L. Let $v \in \mathsf{M}_K$, and denote by $A_v = A\otimes_K K_v$ the completion of A at v (see 6.2.5). Then $[A] \in \mathrm{Br}(L/K)$ and $[A_v] = \mathrm{Res}_{L/K}^{L_{\overline{v}}/K_v}([A]) \in \mathrm{Br}(L_{\overline{v}}/K_v)$. Recall from 3.5.4 and 3.5.5 that $A = (L, G, \gamma)$, where $\gamma = \gamma_L(A) = \Theta_{L/K}([A]) \in H^2(G, L^\times)$. Then

$$A_v = (L_{\overline{v}}, G_{\overline{v}}, \gamma_v), \quad \text{where} \quad \gamma_v = \mathrm{Res}_{L/K\to L_{\overline{v}}/K_v}(\gamma) = \Theta_{L_{\overline{v}}/K_v}([A_v]).$$

If $\gamma = [c]$ for some $c \in C^2(G, L)$ and

$$c_v = (G_{\overline{v}} \hookrightarrow G \overset{c}{\to} L^\times \hookrightarrow L_{\overline{v}}^\times) \in C^2(G_{\overline{v}}, L_{\overline{v}}^\times),$$

then $\gamma_v = [c_v]$.

Let eventually $G = \langle\tau\rangle$ be cyclic, $a \in K^\times$ and $A = (L, \tau, a)$. Then 3.6.4 (applied with $K' = K_v$) implies $A_v = (L_{\overline{v}}, \tau_v, a)$, where $\tau_v = \tau^{(G:G_{\overline{v}})}$, since $G = \langle\tau\rangle$ implies $G_{\overline{v}} = \langle\tau^{(G:G_{\overline{v}})}\rangle$. By 3.6.2.2(b) we obtain $A \sim K$ if and only if $a \in \mathsf{N}_{L/K}L^\times$, and $A_v \sim K_v$ if and only if $a \in \mathsf{N}_{L_{\overline{v}}/K_v}L_{\overline{v}}^\times$. Hence the Hasse norm theorem 6.4.1.2 implies:

$$A \sim K \iff A_v \sim K_v \quad \text{for all} \quad v \in \mathsf{M}_K.$$

Theorem 6.4.3 (Splitting theorem of Hasse - Brauer - Noether).
Let A be a central simple K-algebra.

1. *For almost all $v \in \mathsf{M}_K$ the algebra A splits over K_v.*

2. (Local-global principle for central simple algebras) *The map*

$$(\mathrm{res}_{K_v/K})_{v\in\mathsf{M}_K} : \mathrm{Br}(K) \to \bigoplus_{v\in\mathsf{M}_K} \mathrm{Br}(K_v), \qquad [A] \mapsto ([A_v])_{v\in\mathsf{M}_K},$$

is a monomorphism.
In particular, $A \sim K$ if and only if $A_v \sim K_v$ for all $v \in \mathsf{M}_K$.

Proof. 1. Let L/K be a finite Galois extension such that A splits over L, and let $G = \mathrm{Gal}(L/K)$. Then 3.5.5.1 implies that $A \sim (L, G, \gamma)$ for some $\gamma \in H^2(G, L^\times)$, and we fix a γ defining normalized factor system $c\colon G^2 \to L^\times$. For $v \in \mathsf{M}_K$ we obtain $A_v \sim (L_{\overline{v}}, G_{\overline{v}}, \gamma_v)$, where $\gamma_v \in H^2(G_{\overline{v}}, L_{\overline{v}}^\times)$, and $c_v = (G_{\overline{v}} \hookrightarrow G \overset{c}{\to} L^\times \hookrightarrow L_{\overline{v}}^\times)$ is a defining normalized factor system of γ_v.

Almost all $v \in \mathsf{M}_K$ are unramified in L, and for almost all $v \in \mathsf{M}_K$ we have $c(\sigma, \rho) \in U_{\overline{v}}$ for all σ, $\rho \in G$. Hence it suffices to prove:

> **A.** If $v \in \mathsf{M}_K$ is unramified in L and $c(\sigma, \rho) \in U_{\overline{v}}$ for all σ, $\rho \in G$, then $A_v \sim K_v$.

Proof of **A.** Let $v \in \mathsf{M}_K$ be unramified in L and $c(\sigma, \rho) \in U_{\overline{v}}$ for all σ, $\rho \in G$. Then the extension $L_{\overline{v}}/K_v$ is unramified, hence cyclic, and 3.6.1.3 implies $A_v \sim (L_{\overline{v}}, \tau_v, a_v)$, where

$$\tau_v = \tau^{(G:G_{\overline{v}})} \quad \text{and} \quad a_v = \prod_{\nu=1}^{|G_{\overline{v}}|-1} c_v(\tau_v^\nu, \tau_v) = \prod_{\nu=1}^{|G_{\overline{v}}|-1} c(\tau_v^\nu, \tau_v) \in U_{\overline{v}}.$$

Hence $a_v \in \mathsf{N}_{L_{\overline{v}}/K_v} L_{\overline{v}}^\times$ (see [18, Theorem 5.6.9]), and thus $A_v \sim K_v$ by 3.6.2.2(b).

2. Apparently $A \sim K$ implies $A_v \sim K_v$ for all $v \in \mathsf{M}_K$. For the converse we assume that L/K (with variable K) is a finite Galois extension of smallest degree such that there exists a central simple K-algebra A which splits over L and over all completions K_v with $v \in \mathsf{M}_K$, but not over K. Then L/K is not cyclic by 6.4.2. Suppose that $A \cong \mathsf{M}_n(D)$ for some central K-division algebra D and $n \in \mathbb{N}$. Let $\dim_K(D) = d^2 > 1$, let p be a prime dividing d, G' a p-Sylow subgroup of G and $K' = L^{G'}$ its fixed field. Then $d \nmid [K' : K]$, and therefore D (and thus also A) does not split over K' by 3.4.8.2. If $v' \in \mathsf{M}_{K'}$ and $v = v' \restriction K$, then it follows that $(A_{K'})_{v'} = (A_v)_{K'_{v'}} \sim (K_v)_{K'_{v'}} = K'_{v'}$. Since $A_{K'} \not\sim K'$, the minimal choice of L/K shows that $K' = K$, and therefore G is a p-group.

Being a p-group, G has a normal subgroup G_1 of index p. If $K_1 = L^{G_1}$, then L/K_1 is cyclic, hence $K_1 \neq K$, and as above $(A_{K_1})_{v_1} \sim (K_1)_{v_1}$ for all $v_1 \in \mathsf{M}_{K'}$. By the minimal choice of L/K it follows that $A_{K_1} \not\sim K_1$, but this contradicts 6.4.2, since L/K' is cyclic and $A_L \sim L$. □

Theorem 6.4.4 (Theorem of Artin - Tate)**.** *If L/K is a finite Galois extension, then*

$$H^1(G, \mathsf{C}_L) = \mathbf{1}.$$

Proof. We consider the homomorphism

$$\theta\colon H^2(G, L^\times) \stackrel{i}{\to} H^2(G, \mathsf{J}_L) \stackrel{\theta^2}{\to} \bigoplus_{v \in \mathsf{M}_K} H^2(G_{\overline{v}}, L_{\overline{v}}^\times),$$

where i is induced by the inclusion $K^\times \hookrightarrow \mathsf{J}_K$, and θ^2 is the isomorphism given in 6.1.1.1(a). If $\xi \in H^2(G, L^\times)$ and $x\colon G^2 \to L^\times$ is a ξ defining factor system, then $\theta x = (G_{\overline{v}}^2 \hookrightarrow G^2 \stackrel{x}{\to} L^\times \hookrightarrow L_{\overline{v}}^\times)_{v \in \mathsf{M}_K}$ is a $\theta(\xi)$ defining factor system, and consequently $\theta = (\mathrm{Res}^2_{L/K \to L_{\overline{v}}/K_v})_{v \in \mathsf{M}_K}$ by 2.6.8.

From the exact sequence $\mathbf{1} \to L^\times \to \mathsf{J}_L \to \mathsf{C}_L \to \mathbf{1}$ together with 6.1.1.2, the commutative diagram in 6.4.2 and the homomorphism θ defined above we receive the following commutative diagram with an exact row (note that $H^1(G, \mathsf{J}_L) = \mathbf{1}$):

$$\begin{array}{ccccccc}
\mathbf{1} & \longrightarrow & H^1(G, \mathsf{C}_L) & \longrightarrow & H^2(G, L^\times) & \stackrel{\theta}{\longrightarrow} & \bigoplus_{v \in \mathsf{M}_K} H^2(G_{\overline{v}}, L_{\overline{v}}^\times) \\
 & & & & {\scriptstyle \mathsf{f}_{L/K}}\uparrow & & \uparrow{\scriptstyle (\mathsf{f}_{L_{\overline{v}}/K_v})_{v \in \mathsf{M}_K}} \\
 & & & & \mathrm{Br}(L/K) & \xrightarrow{(\mathrm{res}_{L/K}^{L_{\overline{v}}/K_v})_{v \in \mathsf{M}_K}} & \bigoplus_{v \in \mathsf{M}_K} \mathrm{Br}(L_{\overline{v}}/K_v) \\
 & & & & {\scriptstyle j}\downarrow & & \downarrow{\scriptstyle j} \\
 & & & & \mathrm{Br}(K) & \xrightarrow{(\mathrm{res}_{K_v/K})_{v \in \mathsf{M}_K}} & \bigoplus_{v \in \mathsf{M}_K} \mathrm{Br}(K_v)
\end{array}$$

In this diagram, j denotes the inclusions, the maps $\mathsf{f}_{L/K}$ and $(\mathsf{f}_{L_{\overline{v}}/K_v})_{v \in \mathsf{M}_K}$ are the isomorphisms given in 3.5.4, and $(\mathrm{res}_{K_v/K})_{v \in \mathsf{M}_K}$ is injective by 6.4.3. Hence θ is injective and thus $H^1(G, \mathsf{C}_L) = \mathbf{1}$. □

Theorem and Definition 6.4.5 (Sum and product formulas)**.**

1. *Let A be a central simple K-algebra. Then* $\mathrm{inv}_v(A) = 0$ *for almost all* $v \in \mathsf{M}_K$*, and*

$$\sum_{v \in \mathsf{M}_K} \mathrm{inv}_v(A) = 0 \in \mathbb{Q}/\mathbb{Z}.$$

2. *Let L/K be a (not necessarily finite) Galois extension, and let* $G = \mathrm{Gal}(L/K)$*. If* $a \in K^\times$*, then*

$$(a, L/K) = 1 \in G^{\mathsf{a}}.$$

In particular, the global norm residue symbol $(\,\cdot\,, L/K)\colon \mathsf{J}_K \to G^{\mathsf{a}}$ *induces an (equally denoted and equally named) continuous homomorphism*

$$(\,\cdot\,, L/K)\colon \mathsf{C}_K \to G^{\mathsf{a}}$$

satisfying $(\alpha K^\times, L/K) = (\alpha, L/K)$ *for all* $\alpha \in \mathsf{J}_K$. *Its image*

$$(\mathsf{C}_K, L/K) = \{(a, L/K) \mid a \in \mathsf{C}_K\} = \{(\alpha, L/K) \mid \alpha \in \mathsf{J}_K\}$$

is dense in G^{a}, *and if* $[L:K] < \infty$, *then* $(\,\cdot\,, L/K)\colon \mathsf{C}_K \to G^{\mathsf{a}}$ *is surjective.*

Proof. We start our investigations with a special case of 2.

I. Let $m \in \mathbb{N}$ and $\mathbb{Q}^{(m)} = \mathbb{Q}(\zeta)$ for some primitive m-th root of unity ζ. Then $(a, \mathbb{Q}^{(m)}/\mathbb{Q}) = 1$ for all $a \in \mathbb{Q}^\times$.

Proof of **I.** We may assume that $m > 2$ and set $m = q_1^{e_1} \cdot \ldots \cdot q_s^{e_s}$, where $q_1, \ldots, q_s$ are distinct primes and $e_1, \ldots, e_s \in \mathbb{N}$. Then the map

$$\mathrm{Gal}(\mathbb{Q}^{(m)}/\mathbb{Q}) \to \prod_{i=1}^{s} \mathrm{Gal}(\mathbb{Q}^{(q_i^{e_i})}/\mathbb{Q}), \quad \text{defined by}$$

$$\tau \mapsto (\tau \restriction \mathbb{Q}^{(q_1^{e_1})}, \ldots, \tau \restriction \mathbb{Q}^{(q_s^{e_s})}),$$

is an isomorphism, and if $a \in \mathbb{Q}^\times$, then $(a, \mathbb{Q}^{(m)}/\mathbb{Q}) \restriction \mathbb{Q}^{(q_i^{e_i})} = (a, \mathbb{Q}^{(q_i^{e_i})}/\mathbb{Q})$ for all $i \in [1, s]$ by 6.2.2.3. Hence it suffices to prove that $(a, \mathbb{Q}^{(q^e)}/\mathbb{Q}) = 1$ for all $a \in \mathbb{Q}^\times$ and all prime powers $q^e > 2$.

Thus let q be a prime, $e \in \mathbb{N}$ and $q^e > 2$. Since $(\,\cdot\,, \mathbb{Q}^{(q^e)}/\mathbb{Q})$ is a homomorphism, it suffices to prove that $(-1, \mathbb{Q}^{(q^e)}/\mathbb{Q}) = (t, \mathbb{Q}^{(q^e)}/\mathbb{Q}) = 1$ for all primes t. If $a \in \mathbb{Q}^\times$, then

$$\begin{aligned}(a, \mathbb{Q}^{(q^e)}/\mathbb{Q}) &= \prod_{v \in \mathsf{M}_{\mathbb{Q}}} (a_v, \mathbb{Q}_v^{(q^e)}/\mathbb{Q}_v) \restriction \mathbb{Q}^{(q^e)} \\ &= (a, \mathbb{C}/\mathbb{R}) \restriction \mathbb{Q}^{(q^e)} \circ \prod_{p \in \mathbb{P}} (a, \mathbb{Q}_p^{(q^e)}/\mathbb{Q}_p) \restriction \mathbb{Q}^{(q^e)}.\end{aligned}$$

Let $\zeta \in \mathbb{Q}^{(q^e)}$ be a primitive p^e-th root of unity. Then it suffices to prove that $(-1, \mathbb{Q}^{(p^e)}/\mathbb{Q})(\zeta) = (t, \mathbb{Q}^{p^e}/\mathbb{Q})(\zeta) = \zeta$ for all primes t.

Apparently, $(-1, \mathbb{C}/\mathbb{R})(\zeta) = \overline{\zeta} = \zeta^{-1}$, and $(t, \mathbb{C}/\mathbb{R}) = 1$ for all primes t.

Let p be a prime and $p \neq q$. Then $\mathbb{Q}_p^{(q^e)}/\mathbb{Q}_p$ is unramified (see [18, Corollary 5.8.5]), and consequently $(b, \mathbb{Q}_p^{(q^e)}/\mathbb{Q}_p) = \varphi_{\mathbb{Q}_p^{(q^e)}/\mathbb{Q}_p}^{v_p(b)}$ for all $b \in \mathbb{Q}_p^\times$ by 4.2.3.2. Hence it follows that $(-1, \mathbb{Q}_p^{(q^e)}/\mathbb{Q}_p) = (t, \mathbb{Q}_p^{(q^e)}/\mathbb{Q}_p) = 1$ for all primes $t \neq p$, and $(p, \mathbb{Q}_p^{(q^e)}/\mathbb{Q}_p)(\zeta) = \varphi_{\mathbb{Q}^{(p^e)}/\mathbb{Q}}(\zeta) = \zeta^p$ (see [18, Theorem 5.6.7]).

Since $q = \mathsf{N}_{\mathbb{Q}_q^{(q^e)}/\mathbb{Q}_q}(1 - \zeta)$ (see [18, Theorem 3.1.11]), we get $(q, \mathbb{Q}_q^{(q^e)}/\mathbb{Q}_q) = 1$ by 6.2.2.1, and by 4.9.1.3 we obtain $(b, \mathbb{Q}_q^{(q^e)}/\mathbb{Q}_q)(\zeta) = \zeta^{b^{-1}}$ for all $b \in U_p = \mathbb{Z}_p^\times$.

Summarizing, we obtain

$$\begin{aligned}(-1, \mathbb{Q}^{(q^e)}/\mathbb{Q})(\zeta) &= (-1, \mathbb{Q}_q^{(q^e)}/\mathbb{Q}_q)\circ(-1, \mathbb{C}/\mathbb{R})(\zeta) \\ &= (-1, \mathbb{Q}_q^{(q^e)}/\mathbb{Q}_q)(\zeta^{-1}) = (\zeta^{-1})^{-1} = \zeta,\end{aligned}$$

$(q, \mathbb{Q}^{(q^e)}/\mathbb{Q})(\zeta) = \zeta$, and

$$\begin{aligned}(p, \mathbb{Q}^{(q^e)}/\mathbb{Q})(\zeta) &= (p, \mathbb{Q}_q^{(q^e)}/\mathbb{Q}_q)\circ(p, \mathbb{Q}_p^{(q^e)}/\mathbb{Q}_p)(\zeta) = (p, \mathbb{Q}_q^{(q^e)}/\mathbb{Q}_q)(\zeta^p) \\ &= (p, \mathbb{Q}_q^{(q^e)}/\mathbb{Q}_q)(\zeta)^p = (\zeta^{p^{-1}})^p = \zeta\end{aligned}$$

for all primes $p \neq q$. □[**I.**]

Next we prove:

II. If $m \in \mathbb{N}$ and S is a finite subset of M_K^0, then there exists a cyclic extension L/K with the following properties:

- $m \,|\, [L_{\overline{v}} \colon K_v]$ for all $v \in S$.
- $L_{\overline{v}} = \mathbb{C}$ for all $\overline{v} \in \mathsf{M}_L^\infty$.
- $(a, L/K) = 1$ for all $a \in K^\times$.

Proof of **II.** Let $m \in \mathbb{N}$ and S be a finite subset of M_K^0.

CASE 1: K is a number field.

For $n \in \mathbb{N}$, $n \geq 2$, and a prime $p \neq 2$, let $\mathbb{Q}(p^n)$ be the over $\mathbb{Q}$ cyclic subfield of $\mathbb{Q}^{(p^n)}$ satisfying $[\mathbb{Q}(p^n)\colon\mathbb{Q}] = p^{n-1}$. For $n \geq 2$ let $\mathbb{Q}(2^n) = \mathbb{Q}(\zeta - \zeta^{-1})$, where ζ is a primitive 2^n-th root of unity ζ; then $\mathbb{Q}(2^n)/\mathbb{Q}$ is cyclic of degree 2^{n-2} (see [18, Corollary 1.7.6]).

For every prime p, $n \in \mathbb{N}$ and $v \in \mathsf{M}_K^0$, the extension $K_v\mathbb{Q}(p^n)/K_v$ is cyclic of degree

$$[K_v\mathbb{Q}(p^n)\colon K_v] = [\mathbb{Q}(p^n)\colon\mathbb{Q}(p^n) \cap K_v] = \frac{[\mathbb{Q}(p^n)\colon\mathbb{Q}]}{[\mathbb{Q}(p^n) \cap K_v \colon \mathbb{Q}]}.$$

The field $\mathbb{Q}(p^n) \cap K_v$ is a maximal over $\mathbb{Q}$ cyclic subfield of $\mathbb{Q}^{(p^n)} \cap K_v$ of p-power degree. Since $\mathbb{Q}^{(p^n)} \cap K_v = \mathbb{Q}^{(p^{n+1})} \cap K_v$ for $n \gg 1$, it follows that $[\mathbb{Q}(p^n) \cap K_v : \mathbb{Q}] = p^{\lambda(p,v)}$ does for $n \gg 1$ not depend on n, and therefore $[K_v\mathbb{Q}(p^n)\colon K_v] = p^{n-\lambda(p,v)} \to \infty$ for $n \to \infty$.

Let $p_1, \ldots, p_s$ be the primes dividing $2m$, $n \gg 1$,

$$L = K\mathbb{Q}(p_1^n)\mathbb{Q}(p_2^n) \cdot \ldots \cdot \mathbb{Q}(p_s^n)$$

and $\lambda \in \mathbb{N}$ such that $\lambda \geq \lambda(p_j, v)$ for all $j \in [1, s]$ and $v \in S$. For $n \gg 1$, all $j \in [1, s]$, and all $v \in S$ it follows that $p_j^{n-\lambda} \mid [K_v\mathbb{Q}(p_j^n)\colon K_v] \mid [L_{\overline{v}}\colon K_v]$, and consequently $m \,|\, [L_{\overline{v}}\colon K_v]$ for all $v \in S$.

Moreover, $\mathbb{Q}(2^n) \subset L$ implies $L_w = \mathbb{C}$ for all $w \in \mathsf{M}_L^\infty$, and, by construction, $L \subset K(\zeta)$ for some root of unity ζ. By 1.8.4, the map

$$\mathrm{Gal}(L/K) \to \prod_{i=1}^{s} \mathrm{Gal}(\mathbb{Q}(p^n)/\mathbb{Q}), \quad \text{defined by } \tau \mapsto (\tau \restriction \mathbb{Q}(p_1^n), \ldots, \tau \restriction \mathbb{Q}(p_s^n))$$

is a monomorphism. Hence L/K is cyclic.

Now let $a \in K^\times$. By means of 6.2.2 and **I** we gain

$$(a, L/K) = (a, K(\zeta)/K) \restriction L \quad \text{and}$$
$$(a, K(\zeta)/K) \restriction \mathbb{Q}(\zeta) = (\mathsf{N}_{K/\mathbb{Q}}(a), \mathbb{Q}(\zeta)/\mathbb{Q}) = 1.$$

Since the map $\mathrm{Gal}(K(\zeta)/K) \to \mathrm{Gal}(\mathbb{Q}(\zeta)/\mathbb{Q})$, $\tau \mapsto \tau \restriction \mathbb{Q}(\zeta)$ is a monomorphism, it follows that $(a, K(\zeta)/K) = 1$ and therefore $(a, L/K) = 1$.

CASE 2: $K/\mathbb{F}_q$ is a function field. For $n \ge 2$ let $L = K\mathbb{F}_{q^n}$ be the constant field extension of degree n. Then L/K is cyclic, and if $v \in \mathsf{M}_K$ and $\overline{v} \in \mathsf{M}_L$ lies above v, then

$$[L_{\overline{v}} : K_v] = [K_v\mathbb{F}_{q^n} : K_v] = [\mathbb{F}_{q^n} : \mathbb{F}_{q^n} \cap K_v] = \frac{[\mathbb{F}_{q^n} : \mathbb{F}_q]}{[\mathbb{F}_{q^n} \cap K_v : \mathbb{F}_q]}.$$

Since $\mathbb{F}_{q^n} \cap K_v = \mathbb{F}_{q^{n+1}} \cap K_v$ for $n \gg 1$, it follows that $\lambda(v) = [\mathbb{F}_{q^n} \cap K_v : \mathbb{F}_q]$ does for $n \gg 1$ not depend on n, and therefore $[L_{\overline{v}} : K_v] = n/\lambda(v) \to \infty$ for $n \to \infty$. In particular, there exists some $n \in \mathbb{N}$ such that $\lambda(v)m \mid n$ and thus $m \mid [L_{\overline{v}} : K_v]$ for all $v \in S$. By construction, $L = K(\zeta)$ for some root of unity ζ, and all $v \in \mathsf{M}_K$ are unramified in L (see [18, Theorem 6.6.7]). If $v \in \mathsf{M}_K$, then $L_{\overline{v}}/K_v$ is unramified, and if $a \in K_v$, then by 4.2.3.2 we achieve

$$(a, L_{\overline{v}}/K_v)(\zeta) = \varphi_{L_{\overline{v}}/K_v}^{v(a)}(\zeta) = \zeta^{\mathfrak{N}(v)^{v(a)}} = \zeta^{q^{v(a)\deg(v)}}.$$

Consequently, if $a \in K$, then

$$(a, L/K)(\zeta) = \prod_{v \in \mathsf{M}_K} (a, L_{\overline{v}}/K_v) = \prod_{v \in \mathsf{M}_K} \varphi_{L_{\overline{v}}/K_v}^{v(a)}(\zeta) = \zeta^{q^s},$$

where

$$s = \sum_{v \in \mathsf{M}_K} v(a)\deg(v) = \deg((a)^K) = 0 \quad \text{(see [18, Theorem 6.3.4])},$$

and therefore $(a, L/K) = 1$. $\square$[**II.**]

After these preparations we can do the real proof of the theorem.

1. Let A be a central simple K-algebra. Then there exists a finite subset S of M_K such that $A_v \sim K_v$ and thus $\mathrm{inv}_v(A) = 0$ for all $v \in \mathsf{M}_K \setminus S$ (see 6.4.3.1 and 4.1.6.2). Hence there exists some $m \in \mathbb{N}$ such that $m\,\mathrm{inv}_v(A) = 0$ for all $v \in \mathsf{M}_K$. Let L/K be a cyclic extension such that the properties stated

in **II** above are fulfilled with the exceptional set $S \cap \mathsf{M}_K^0$. Let $[L\!:\!K] = n$ and $\mathrm{Gal}(L/K) = \langle\sigma\rangle$. If $v \in \mathsf{M}_K^0$, then 4.1.8.1 implies

$$\mathrm{inv}_{\overline{v}}((A_L)_{\overline{v}}) = [L_{\overline{v}}\!:\!K_v]\,\mathrm{inv}_v(A) = 0$$

and thus $(A_L)_{\overline{v}} \sim L_{\overline{v}}$. If $\overline{v} \in \mathsf{M}_L^\infty$, then $L_{\overline{v}} = \mathbb{C}$ and thus likewise $(A_L)_{\overline{v}} \sim L_{\overline{v}}$. Consequently, $(A_L)_{\overline{v}} \sim L_{\overline{v}}$ for all $\overline{v} \in \mathsf{M}_L$, hence $A_L \sim L$ by 6.4.3.2, and $A \sim (L, \sigma, a)$ for some $a \in K^\times$ by 3.6.1.3.

Let $\chi \in \mathsf{X}(G)$ be the character satisfying

$$\chi(\sigma) = \frac{1}{n} + \mathbb{Z} \in \mathbb{Q}/\mathbb{Z}.$$

Then $\mathsf{X}(G) = \langle\chi\rangle$, $L = L_\chi$, $\sigma = \sigma_\chi$, and following 6.2.5 and **II**, we achieve

$$\sum_{v \in \mathsf{M}_K} \mathrm{inv}_v(A) = \sum_{v \in \mathsf{M}_K} \mathrm{inv}_v(L, \sigma, a) = \chi(a, L/K) = 0 \in \mathbb{Q}/\mathbb{Z}.$$

2. By 6.2.3 it suffices to prove that $(a, L/K) = 1$ if L/K is finite. Thus let L/K be finite. For all $\chi \in \mathsf{X}(G)$ we obtain, using 1. and 6.2.5,

$$\chi(a, L/K) = \sum_{v \in \mathsf{M}_K} \mathrm{inv}_v(L_\chi, \sigma_\chi, a) = 0 \in \mathbb{Q}/\mathbb{Z},$$

and therefore $(a, L/K) = 1 \in G^{\mathsf{a}}$ by 1.6.7.1.

By 6.2.4 the homomorphism $(\,\cdot\,, L/K)\colon \mathsf{C}_K \to G^{\mathsf{a}}$ is continuous, its image is dense in G^{a}, and if $[L\!:\!K] < \infty$, then $(\,\cdot\,, L/K)\colon \mathsf{C}_K \to G^{\mathsf{a}}$ is surjective. □

Definition and Remarks 6.4.6.

1. Let L/K be a finite Galois extension, and let $G = \mathrm{Gal}(L/K)$. The calculation rules stated in 6.2.2 for the global norm residue symbol remain valid word for word if we replace J_K with C_K.

As $(\mathsf{N}_{L/K}(\beta), L/K) = 1$ for all $\beta \in \mathsf{J}_L$, we get $(\mathsf{N}_{L/K}(c), L/K) = 1$ for all $c \in \mathsf{C}_L$. Therefore $\mathsf{N}_{L/K}\mathsf{C}_L$ is contained in the kernel of the epimorphism $(\,\cdot\,, L/K)\colon \mathsf{C}_K \to G^{\mathsf{a}}$ (see 6.4.5.2), and thus the norm residue symbol for L/K induces an equally denoted and equally named continuous epimorphism

$$(\,\cdot\,, L/K)\colon \mathsf{C}_K/\mathsf{N}_{L/K}\mathsf{C}_L \to G^{\mathsf{a}}.$$

In the subsequent Theorem 6.4.7 we shall prove the global reciprocity law which states that $(\,\cdot\,, L/K)\colon \mathsf{C}_K/\mathsf{N}_{L/K}\mathsf{C}_L \to G^{\mathsf{a}}$ is even an isomorphism. This is a perfect counterpart of the local reciprocity law stated in 4.2.7 if we replace the multiplicative group $K^\times$ with the idele class group C_K. As a consequence of the reciprocity law we shall prove the global norm theorem 6.4.8 which is also a perfect counterpart of the corresponding local theorem 4.2.10. Surprisingly, the proofs for the global and the local theorems are very similar. This is one of the reasons to look for an abstract class field theory valid in both cases. Such abstract class field theories are presented in [48] and [1, Ch. XIV].

2. By 6.4.5.2, the universal norm residue symbol $(\,\cdot\,,K)\colon \mathsf{J}_K \to G_K^{\mathsf{a}}$ introduced in 6.2.3 induces an (equally denoted and equally named) continuous homomorphism

$$(\,\cdot\,,K) = (\,\cdot\,,K_{\rm sep}/K)\colon \mathsf{C}_K \to G_K^{\mathsf{a}},$$

whose image $(\mathsf{C}_K, K) = \{(a,K) \mid a \in \mathsf{C}_K\} = \{(\alpha,K) \mid \alpha \in \mathsf{J}_K\}$ is dense in G_K^{a}. Its kernel

$$\mathsf{D}_K = \operatorname{Ker}(\,\cdot\,,\,K) = \{c \in \mathsf{C}_K \mid (c,K) = 1 \in G_K^{\mathsf{a}}\}$$

is called the **group of universal norms**. Obviouly, the universal norm residue symbol induces an (equally denoted and equally named) continuous momorphism

$$(\,\cdot\,,K)\colon \mathsf{C}_K/\mathsf{D}_K \to G_K^{\mathsf{a}} \quad \text{satisfying} \quad (a\mathsf{D}_K,K) = (a,K) \quad \text{for all} \quad a \in \mathsf{C}_K,$$

whose image is dense in G_K^{a}.
We shall prove:

- In the number field case:
  $\mathsf{D}_K \ne \mathbf{1}$, the universal norm residue symbol $(\,\cdot\,,K)\colon \mathsf{C}_K \to G_K^{\mathsf{a}}$ is surjective, and it induces a topological isomorphism $\mathsf{C}_K/\mathsf{D}_K \overset{\sim}{\to} G_K^{\mathsf{a}}$ (see 6.4.10.3 and 6.5.1).

- In the function field case:
  $\mathsf{D}_K = \mathbf{1}$, and the universal norm residue symbol $(\,\cdot\,,K)\colon \mathsf{C}_K \to G_K^{\mathsf{a}}$ is not surjective (see 6.5.2.1).

3. Let $\mathcal{E}(K)$ be the set of all finite Galois extensions of K. If $L \in \mathcal{E}(K)$, then by the global reciprocity law $\mathsf{N}_{L/K}\mathsf{C}_L$ is the kernel of the norm residue symbol $(\,\cdot\,,L/K)\colon \mathsf{C}_K \to \operatorname{Gal}(L^{\mathsf{a}}/K)$. Hence it follows that

$$\mathsf{D}_K = \bigcap_{L \in \mathcal{E}(K)} \mathsf{N}_{L/K}\mathsf{C}_L.$$

A subgroup B of C_K is called a **norm group** if there exists a finite abelian extension L/K such that $B = \mathsf{N}_{L/K}\mathsf{C}_L$. In this case, we call L a **class field** to B (in 6.4.8 we shall prove that L is uniquely determined by B).

For every finite field extension L/K the group $\mathsf{N}_{L/K}\mathsf{C}_L$ is an open subgroup of C_K by 5.4.4.2), and in 6.4.8, we shall prove that it is actually a norm group of finite index. In 6.5.3, we shall prove that conversely every open subgroup of C_K of finite index is a norm group. This is again a perfect counterpart of the corresponding local Theorem 4.8.2.2.

4. Sometimes it is more convenient to work with ideles instead of idele classes, and for this reason we adopt our definitions for ideles. Let $\pi\colon \mathsf{J}_K \to \mathsf{C}_K$

be the residue class epimorphism. A subgroup B_0 of J_K is called a **norm group** if $B_0 = \pi^{-1}(B)$ for some norm group $B \subset \mathsf{C}_K$ [or, equivalently, $B_0 = \mathsf{N}_{L/K}(\mathsf{J}_L)K^\times$ for some finite abelian extension L/K].

Theorem 6.4.7 (Global reciprocity law)**.** *Let L/K be a finite Galois extension and $G = \mathrm{Gal}(L/K)$. Then the global norm residue symbol*

$$(\,\cdot\,, L/K)\colon \mathsf{C}_K/\mathsf{N}_{L/K}\mathsf{C}_L \stackrel{\sim}{\to} G^{\mathsf{a}}$$

is an isomorphism. In particular:

(a) If L/K is abelian, then $[L\!:\!K] = (\mathsf{C}_K\!:\!\mathsf{N}_{L/K}\mathsf{C}_L)$.

(b) If B is a subgroup of C_K, then it follows that L is a class field to B if and only if B is the kernel of the global norm residue symbol $(\,\cdot\,, L/K)\colon \mathsf{C}_K \to G$.

(c) If B_0 is a subgroup of J_K, then it follows that L is a class field to B_0 if and only if B_0 is the kernel of the norm residue symbol $(\,\cdot\,, L/K)\colon \mathsf{J}_K \to G$.

(d) If $c \in \mathsf{C}_K$, then $c \in \mathsf{N}_{L/K}\mathsf{C}_L$ if and only if $(c, L/K) = 1$.

Proof. It suffices to prove that $(\,\cdot\,, L/K)\colon \mathsf{C}_K/\mathsf{N}_{L/K}\mathsf{C}_L \to G^{\mathsf{a}}$ is an isomorphism, for then (a), (b), (c) and (d) are evident. By 6.2.4 and 6.4.6, the global norm residue symbol $(\,\cdot\,, L/K)\colon \mathsf{C}_K/\mathsf{N}_{L/K}\mathsf{C}_L \to G^{\mathsf{a}}$ is an epimorphism. If L/K is even cyclic, then $(\mathsf{C}_K\!:\!\mathsf{N}_{L/K}\mathsf{C}_L) = [L\!:\!K]$ by 6.4.1.1, and thus there is nothing to do.

As to the general case, it suffices to prove:

$$\text{If } a \in \mathsf{C}_K \text{ and } (a, L/K) = 1 \in G^{\mathsf{a}}, \text{ then } a \in \mathsf{N}_{L/K}\mathsf{C}_L. \qquad (*)$$

We prove $(*)$ by induction on $[L\!:\!K]$ and argue essentially as in the proof of 4.2.7. Assume that $[L\!:\!K] > 1$ and the assertion holds for all extensions of global fields of smaller degree. Let L^{a} be the largest over K abelian intermediate field of L/K and $G^{\mathsf{a}} = \mathrm{Gal}(L^{\mathsf{a}}/K)$. We freely use the calculation rules given in 6.2.2, and (as stated in 6.4.6) in doing so we may replace J_K with C_K.

CASE 1: $L^{\mathsf{a}} \ne K$. Then there exists an over K cyclic intermediate field K' of L^{a}/K such that $[K'\!:\!K] > 1$. Since $(a, K'/K) = (a, L/K) \restriction K' = 1$, there exists some $b \in K'^\times$ such that $a = \mathsf{N}_{K'/K}(b)$, and it suffices to prove:

I. There exists some $x \in \mathsf{C}_{K'}$ such that $(bx, L/K') = 1$ and $\mathsf{N}_{K'/K}(x) = 1$.

Indeed, **I** implies $bx = \mathsf{N}_{L/K'}(c)$ for some $c \in L^\times$ by the induction hypothesis, and then we obtain $\mathsf{N}_{L/K}(c) = \mathsf{N}_{K'/K}(bx) = a \in \mathsf{N}_{L/K}L^\times$.

Proof of **I.** Let L_0 be the largest over K' abelian intermediate field of L/K'. Then $L^{\mathsf{a}} \subset L_0$, and if $V = \mathrm{Gal}(L/K')$, then V is a normal subgroup of G, and $[V,V] = \mathrm{Gal}(L/L_0)$ is also a normal subgroup of G. Consequently L_0/K is Galois, $G_0 = G/[V,V] = \mathrm{Gal}(L_0/K)$, and

$$(a, L_0/K) = (a, L/K) \restriction L_0^{\mathsf{a}} = 1.$$

Let $U = \mathrm{Gal}(L_0/K')$. Since K'/K is cyclic, U is a normal subgroup of G_0, G_0/U is cyclic, and $G_0/[G_0,U]$ is abelian by 4.2.8. Let $M = L_0^{[G_0,U]}$ be the fixed field of $[G_0,U]$. Then M/K is abelian, $\mathrm{Gal}(M/K) = G_0/[G_0,U]$, and

$$1 = (a, L/K) \restriction M = (a, M/K) = (\mathsf{N}_{K'/K}(b), M/K) = (b, L_0/K') \restriction M$$

by 6.2.2.2. Hence $(b, L_0/K') \in [G_0,U]$, and therefore

$$(b, L_0/K') = \prod_{i=1}^{k} \rho_i\sigma_i\rho_i^{-1}\sigma_i^{-1}$$

for some $k \in \mathbb{N}$, $\rho_1,\ldots,\rho_k \in G_0$ and $\sigma_1,\ldots,\sigma_k \in U$.

As U is abelian, the norm residue symbol $(\,\cdot\,, L_0/K') \colon \mathsf{C}_{K'} \to U$ is surjective, and therefore $\sigma_i = (x_i, L_0/K')$ for some $x_i \in C_{K'}$. As $\rho_i(K') = K'$ for all $i \in [1,k]$, it follows that

$$\begin{aligned}\rho_i\sigma_i\rho_i^{-1}\sigma_i^{-1} &= \rho_i(x_i, L_0/K')\rho_i^{-1}(x_i, L_0/K')^{-1} = (x_i^{\rho_i}, L_0/K')(x_i^{-1}, L_0/K') \\ &= (x_i^{\rho_i-1}, L_0/K')\end{aligned}$$

by 6.2.2.4. If

$$x = \Big(\prod_{i=1}^{k} x_i^{\rho_i-1}\Big)^{-1} \in \mathsf{C}_{K'}, \quad \text{then} \quad (x, L_0/K')$$
$$= \prod_{i=1}^{k} (x_i^{\rho_i-1}, L_0/K')^{-1} = (b, L_0/K)^{-1},$$

and therefdore $(bx, L/K') = (bx, L_0/K') = 1$, and $\mathsf{N}_{K'/K}(x) = 1$, since $\mathsf{N}_{K'/K}(x_i^{\rho_i-1}) = 1$ for all $i \in [1,k]$. □[**I.**]

CASE 2: $L^{\mathsf{a}} = K$. Then $G^{\mathsf{a}} = \mathbf{1}$, and we must prove that $\mathsf{C}_K = \mathsf{N}_{L/K}\mathsf{C}_L$. Let $p_1,\ldots,p_k$ be the distinct prime divisors of $|G|$. For $i \in [1,k]$ let G_i be a p_i-Sylow group of G, $K_i = L^{G_i}$ and $n_i = (G:G_i) = [K_i:K]$. Since $G^{\mathsf{a}} = \mathbf{1}$, it follows that G is not a p-group, hence $k \ge 2$, $K_i \ne K$ for all $i \in [1,k]$, and $\gcd(n_1,\ldots,n_k) = 1$. Let $a \in \mathsf{C}_K$. If $i \in [1,k]$, then $(a, L/K_i) = \mathsf{V}^L_{K_i/K}(a, L/K) = 1$, hence $a \in \mathsf{N}_{L/K_i}\mathsf{C}_L$ by the induction hypothesis, and $a^{n_i} = \mathsf{N}_{K_i/K}(a) \in \mathsf{N}_{L/K}\mathsf{C}_L$. Since $\gcd(n_1,\ldots,n_k) = 1$, it follows that $a \in \langle a^{n_1},\ldots a^{n_k}\rangle \subset \mathsf{N}_{L/K}\mathsf{C}_L$. □

Theorem 6.4.8 (Global norm theorem).

1. Let L/K be a finite extension and L_0 the largest over K abelian intermediate field of L/K. Then

$$\mathsf{N}_{L/K}\mathsf{C}_L = \mathsf{N}_{L_0/K}\mathsf{C}_{L_0} \quad \text{and} \quad (\mathsf{C}_K : \mathsf{N}_{L/K}\mathsf{C}_L) = [L_0 : K].$$

In particular, $\mathsf{N}_{L/K}\mathsf{C}_L$ is a norm group, L_0 is a class field to $\mathsf{N}_{L/K}\mathsf{C}_L$, and

$$\mathsf{D}_K = \bigcap_{E/K \text{ finite abelian}} \mathsf{N}_{E/K}\mathsf{C}_E = \bigcap_{E/K \text{ finite}} \mathsf{N}_{E/K}\mathsf{C}_E$$

is the intersection of all norm groups in C_K.

2. Let L, L_1, L_2 be finite extension fields of K.

(a) Let B be a subgroup of C_K such that $\mathsf{N}_{L/K}\mathsf{C}_L \subset B$. Then $B = \mathsf{N}_{K'/K}\mathsf{C}_{K'}$ for some intermediate field K' of L/K.

(b) $\mathsf{N}_{L_1L_2/K}\mathsf{C}_{L_1L_2} = \mathsf{N}_{L_1/K}\mathsf{C}_{L_1} \cap \mathsf{N}_{L_2/K}\mathsf{C}_{L_2}$.

(c) $L_1 \subset L_2$ if and only if $\mathsf{N}_{L_1/K}\mathsf{C}_{L_1} \supset \mathsf{N}_{L_2/K}\mathsf{C}_{L_2}$. In particular, $L_1 = L_2$ if and only if $\mathsf{N}_{L_1/K}\mathsf{C}_{L_1} = \mathsf{N}_{L_2/K}\mathsf{C}_{L_2}$, and every norm group has a unique class field.

(d) $\mathsf{N}_{L_1\cap L_2/K}\mathsf{C}_{L_1\cap L_2} = (\mathsf{N}_{L_1/K}\mathsf{C}_{L_1})(\mathsf{N}_{L_2/K}C_{L_2})$.

The assignment $L \mapsto \mathsf{N}_{L/K}\mathsf{C}_L$ defines an inclusion reverting lattice isomorphism between the set of all finite abelian extension fields of K and the set of all norm groups in C_K.

3. Let B be a norm group in C_K and L the class field to B. Let K_0 be a subfield of K such that K/K_0 is finite Galois, $G_0 = \mathrm{Gal}(K/K_0)$ and $\sigma B = B$ for all $\sigma \in G_0$. Then L/K_0 is Galois.

Proof. 1. It suffices to prove that $\mathsf{N}_{L_0/K}\mathsf{C}_{L_0} \subset \mathsf{N}_{L/K}\mathsf{C}_L$. Indeed, then equality holds, and by 6.4.7 we obtain $(\mathsf{C}_K : \mathsf{N}_{L_0/K}\mathsf{C}_{L_0}) = [L_0 : K]$.

CASE 1: L/K is separable. Let N be a Galois closure of L/K and N_0 the largest over K abelian intermediate field of N/K. Then $L_0 = L \cap N_0$, and the map $\mathrm{Gal}(N_0L/L) \to \mathrm{Gal}(N_0/L_0)$, $\sigma \mapsto \sigma \restriction L_0$, is an isomorphism.

Now let $a \in \mathsf{N}_{L_0/K}\mathsf{C}_{L_0}$. Then we obtain $1 = (a, L_0/K) = (a, N/K) \restriction L_0$, hence $(a, N/K) \in \mathrm{Gal}(N_0/L_0)$ and therefore $(a, N/K) = \tau \restriction N_0$ for some $\tau \in \mathrm{Gal}(N_0L/L)$. Since $(\,\cdot\,, N_0L/L) \colon \mathsf{C}_L \to \mathrm{Gal}(N_0L/L)$ is surjective, there exists some $b \in \mathsf{C}_L$ such that

$$(a, N/K) = \tau \restriction N_0 = (b, N_0L/L) \restriction N_0 = (\mathsf{N}_{L/K}(b), N_0/K) = (\mathsf{N}_{L/K}(b), N/K).$$

Hence $(a\mathsf{N}_{L/K}(b)^{-1}, N/K) = 1$, $a\mathsf{N}_{L/K}(b)^{-1} \in \mathsf{N}_{N/K}\mathsf{C}_N \subset \mathsf{N}_{L/K}\mathsf{C}_L$ and $a \in \mathsf{N}_{L/K}\mathsf{C}_L$.

CASE 2: L/K is inseparable and $\operatorname{char}(K) = p > 0$. We use induction on $[L:K]$. Let L' be the separable closure of K in L and $L_1 = L^p$. Then $L' \subset L_1 \subset L$ and $[L:L_1] = p$ (see [18, Theorem 6.1.6]). Clearly, $L_0 \subset L'$, and $\mathsf{N}_{L_1/K}\mathsf{C}_{L_1} = \mathsf{N}_{L_0/K}\mathsf{C}_{L_0}$ either by the induction hypothesis (if L_1/K is inseparable) or by CASE 1 (if L_1/K is separable).

Hence it suffices to prove that $\mathsf{N}_{L/L_1}\mathsf{C}_L = \mathsf{C}_{L_1}$. If $v_1 \in \mathsf{M}_{L_1}$ and $v \in \mathsf{M}_L$ lies above v, then L_v/L_{1v_1} is purely inseparable and $\mathsf{N}_{L_v/L_{1v_1}} L_v^\times = L_{1v_1}^\times$ by 4.2.10.1. It follows that $\mathsf{N}_{L_v/L_{1v_1}}(\mathsf{J}_L) = \mathsf{J}_{L_1}$ and thus $\mathsf{N}_{L_v/L_{1v_1}}(\mathsf{C}_L) = \mathsf{C}_{L_1}$.

2. Clearly, $L_1 \subset L_2$ implies $\mathsf{N}_{L_2/K}\mathsf{C}_{L_2} = \mathsf{N}_{L_1/K} \circ \mathsf{N}_{L_2/L_1}\mathsf{C}_{L_2} \subset \mathsf{N}_{L_1/L}\mathsf{C}_{L_1}$.

(a) Let $G' = (B, L/K) \subset G$ and $K' = L^{G'}$. We assert that $B = \mathsf{N}_{K'/K}\mathsf{C}_{K'}$. If $a \in B$, then $(a, L/K) \in G'$, hence $(a, K'/K) = (a, L/K) \restriction K' = 1$ and $a \in \mathsf{N}_{K'/K}\mathsf{C}_{K'}$. As to the reverse inclusion, let $a \in \mathsf{N}_{K'/K}\mathsf{C}_{K'}$. Then $1 = (a, K'/K) = (a, L/K) \restriction K'$, hence $(a, L/K) \in G'$, and there exists some $b \in B$ such that $(a, L/K) = (b, L/K)$. It follows that $(ab^{-1}, L/K) = 1$, hence $ab^{-1} \in \mathsf{N}_{L/K}\mathsf{C}_L$ and $a \in b\mathsf{N}_{L/K}\mathsf{C}_L \subset B$.

(b) Clearly, $L_i \subset L_1L_2$ implies $\mathsf{N}_{L_1L_2/K}\mathsf{C}_{L_1L_2} \subset \mathsf{N}_{L_i/K}\mathsf{C}_{L_i}$ for $i \in \{1,2\}$, and therefore $\mathsf{N}_{L_1L_2/K}\mathsf{C}_{L_1L_2} \subset \mathsf{N}_{L_1/K}\mathsf{C}_{L_1} \cap \mathsf{N}_{L_2/K}\mathsf{C}_{L_2}$. To prove the reverse inclusion, let $a \in \mathsf{N}_{L_1/K}\mathsf{C}_{L_1} \cap \mathsf{N}_{L_2/K}\mathsf{C}_{L_2}$. Then it follows that $1 = (a, L_i/K) = (a, L_1L_1/K) \restriction L_i$ for $i \in \{1,2\}$, hence $(a, L_1L_2/K) = 1$ and $a \in \mathsf{N}_{L_1L_2/K}\mathsf{C}_{L_1L_2}$.

(c) Clearly $L_1 \subset L_2$ implies $\mathsf{N}_{L_2/K}\mathsf{C}_{L_2} \subset \mathsf{N}_{L_1/K}\mathsf{C}_{L_1}$. Thus assume conversely that $\mathsf{N}_{L_2/K}\mathsf{C}_{L_2} \subset \mathsf{N}_{L_1/K}\mathsf{C}_{L_1}$. Then

$$\mathsf{N}_{L_1L_2/K}\mathsf{C}_{L_1L_2} = \mathsf{N}_{L_1/K}\mathsf{C}_{L_1} \cap \mathsf{N}_{L_2/K}\mathsf{C}_{L_2} = \mathsf{N}_{L_2/K}\mathsf{C}_{L_2}$$

by (b), and since

$$[L_1L_2:K] = (\mathsf{C}_K : \mathsf{N}_{L_1L_2/K}\mathsf{C}_{L_1L_2}) = (\mathsf{C}_K : \mathsf{N}_{L_2/K}\mathsf{C}_{L_2}) = [L_2:K],$$

we obtain $L_1 \subset L_1L_2 = L_2$.

(d) For $i \in \{1,2\}$ we have $\mathsf{N}_{L_i/K}\mathsf{C}_{L_i} \subset \mathsf{N}_{L_1\cap L_2/K}\mathsf{C}_{L_1\cap L_2}$ and therefore $(\mathsf{N}_{L_1/K}\mathsf{C}_{L_1})(\mathsf{N}_{L_2/K}\mathsf{C}_{L_2}) \subset \mathsf{N}_{L_1\cap L_2/K}\mathsf{C}_{L_1\cap L_2}$. By (a) there exists an intermediate field L' of L_1/K such that

$$\mathsf{N}_{L'/K}\mathsf{C}_{L'} = (\mathsf{N}_{L_1/K}\mathsf{C}_{L_1})(\mathsf{N}_{L_2/K}\mathsf{C}_{L_2}) \subset \mathsf{N}_{L_1\cap L_2/K}\mathsf{C}_{L_1\cap L_2},$$

and (b) implies $L_1 \cap L_2 \subset L'$. For $i \in \{1,2\}$, $\mathsf{N}_{L_i/K}\mathsf{C}_{L_i} \subset \mathsf{N}_{L'/K}\mathsf{C}_{L'}$ implies $L' \subset L_i$, hence $L' \subset L_1 \cap L_2$ and therefore $L' = L_1 \cap L_2$.

3. Since K/K_0 and L/K are separable, the extension L/K_0 is separable, too. If $\sigma \in \operatorname{Gal}(\overline{K}/K_0)$ and $b \in \mathsf{C}_K$, then $(\sigma b, \sigma L/K) = \sigma(b, L/K)\sigma^{-1}$ by 6.2.2.4. It follows from the global reciprocity law 6.4.7 that $\sigma b \in \mathsf{N}_{\sigma L/K}\mathsf{C}_{\sigma L}$ if and only if $b \in \mathsf{N}_{L/K}\mathsf{C}_L$. Hence σL is the class field to $\sigma B = B$. Consequently $\sigma L = L$, and L/K_0 is Galois. □

As an application of the global norm theorem we investigate the conductor of a compositum of finite abelian extensions of global fields and obtain a counterpart to the corresponding local behavior investigated in 4.8.7.2. Recall from 5.7.5.2 that the conductor $\mathsf{f}_{L/K}$ of a finite extension L/K is the smallest divisor (in the sense of divisibility) satisfying $\mathsf{C}_K^{\mathsf{f}_{L/K}} \subset \mathsf{N}_{L/K}\mathsf{C}_L$.

Corollary 6.4.9. *Let L/K be a finite abelian extension, and let $L_1, \ldots, L_m$ be intermediate fields of L/K such that $L = L_1 \cdot \ldots \cdot L_m$. Then*

$$\mathsf{f}_{L/K} = \mathrm{lcm}\{\mathsf{f}_{L_j/K} \mid j \in [1,m]\,\}.$$

Proof. Let $\mathsf{f}^* = \mathrm{lcm}\{\mathsf{f}_{L_j/K} \mid j \in [1,m]\,\}$. Then it follows that

$$\mathsf{C}_K^{\mathsf{f}^*} = \bigcap_{j=1}^{m} \mathsf{C}_K^{\mathsf{f}_{L_j/K}} \subset \bigcap_{j=1}^{m} \mathsf{N}_{L_j/K}\mathsf{C}_{L_j} = \mathsf{N}_{L/K}\mathsf{C}_L$$

by 6.4.8.2(b) and thus $\mathsf{f}_{L/K} \,|\, \mathsf{f}^*$. Otherwise $\mathsf{C}_K^{\mathsf{f}_{L/K}} \subset \mathsf{N}_{L/K}\mathsf{C}_L \subset \mathsf{N}_{L_j/K}\mathsf{C}_{L_j}$ implies $\mathsf{f}_{L_j/K} \,|\, \mathsf{f}_{L/K}$ for all $j \in [1,m]$, hence $\mathsf{f}^* \,|\, \mathsf{f}_{L/K}$, and eventually $\mathsf{f}_{L/K} = \mathsf{f}^*$. □

Theorem 6.4.10 (Structure of the group of universal norms).

1. *For all $n \in \mathbb{N}$ the group C_K^n is a closed subgroup of C_K, and*

$$\mathsf{D}_K = \bigcap_{n \in \mathbb{N}} \mathsf{C}_K^n.$$

2. *If $K/\mathbb{F}_q$ is a function field, then $\mathsf{D}_K = 1$, and C_K is totaly disconnected.*

3. *Let K be a number field. Let $\mathsf{C}_K = \mathsf{C}_K^0 \cdot \Gamma'$, where $\Gamma' \cong \mathbb{R}_{>0}$ as in* 5.3.3.1(b), *and set*

$$\mathsf{J}_K^+ = \prod_{\substack{v \in \mathsf{M}_K^\infty \\ v \text{ real}}} \mathbb{R}_{>0} \times \prod_{\substack{v \in \mathsf{M}_K^\infty \\ v \text{ complex}}} \mathbb{C}^\times \times \prod_{v \in \mathsf{M}_K^0} 1.$$

Then $\Gamma' \subset \mathsf{D}_K$, and $\mathsf{D}_K = \overline{K^\times \mathsf{J}_K^+}/K^\times$ is the component of 1 in C_K. Moreover, if $\mathsf{D}_K^0 = \mathsf{D}_K \cap \mathsf{C}_K^0$, then $\mathsf{D}_K = \mathsf{D}_K^0 {\cdot} \Gamma'$, and $\mathsf{C}_K/\mathsf{D}_K$ is compact.

Proof. By 5.3.3 we have $\mathsf{J}_K = \mathsf{J}_K^0 \cdot \Gamma$ and $\mathsf{C}_K = \mathsf{C}_K^0 \cdot \Gamma'$, where $\Gamma \cong \Gamma' \cong \mathbb{R}_{>0}$ in the number field case, and $\Gamma \cong \Gamma' \cong \mathbb{Z}$ in the function field case.

1. By 5.3.7.1, C_K^0 is compact. Hence $(\mathsf{C}_K^0)^n$ is also compact and thus closed in C_K^0. Since Γ'^n is closed in Γ', it follows that $\mathsf{C}_K^n = (\mathsf{C}_K^0)^n \cdot \Gamma'^n$ is closed in $\mathsf{C}_K = \mathsf{C}_K^0 \cdot \Gamma'$.

If B is a norm group in C_K, then $(\mathsf{C}_K : B) < \infty$, and thus there exists some $n \in \mathbb{N}$ such that $\mathsf{C}_K^n \subset B$. Consequently, we obtain

$$\bigcap_{n\in\mathbb{N}} \mathsf{C}_K^n \subset \mathsf{D}_K,$$

and it remains to prove the reverse inclusion.

We prove first:

I. If E/K is a finite field extensions, then $\mathsf{D}_K \subset \mathsf{N}_{E/K}\mathsf{D}_E$.

Proof of **I.** Let E/K be a finite extension, $x \in \mathsf{D}_K$, and let $\mathcal{E}$ be the set of all over E finite abelian extension fields of E. For $L \in \mathcal{E}$, we set

$$\mathsf{X}_L = \mathsf{N}_{E/K}^{-1}(x) \cap \mathsf{N}_{L/E}\mathsf{C}_L \subset \mathsf{C}_E, \quad \text{and} \quad \mathsf{X} = \bigcap_{L\in\mathcal{E}} \mathsf{X}_L.$$

If $y \in \mathsf{X}$, then $y \in \mathsf{D}_E$ and $x = \mathsf{N}_{E/K}(y) \in \mathsf{N}_{E/K}\mathsf{D}_E$. Hence it suffices to prove that $\mathsf{X} \neq \emptyset$, and considering that we show:

If $L \in \mathcal{E}$, then X_L is compact and not empty, and the family $(\mathsf{X}_L)_{L\in\mathcal{E}}$ has the finite intersection property.

Let $L \in \mathcal{E}$. As $x \in \mathsf{D}_K$, we obtain $x = \mathsf{N}_{L/K}(y) = \mathsf{N}_{E/K}\circ\mathsf{N}_{L/E}(y)$ for some $y \in \mathsf{C}_L$ and therefore $\mathsf{N}_{L/E}(y) \in \mathsf{X}_L$. By 5.4.3, the map $\mathsf{N}_{E/K} \colon \mathsf{C}_E \to \mathsf{C}_K$ is continuous, and $\|\mathsf{N}_{E/K}(z)\| = \|z\|$ for all $z \in \mathsf{C}_E$. If $z_0 \in \mathsf{N}_{E/K}^{-1}(x)$, then $\|z_0\| = \|x\| = \|z\|$ for all $z \in \mathsf{N}_{E/K}^{-1}(x)$, and therefore $\mathsf{N}_{E/K}^{-1}(x) \subset z_0\mathsf{C}_E^0$. Being a closed subset of the compact set $z_0\mathsf{C}_E^0$, the set $\mathsf{N}_{E/K}^{-1}(x)$ is compact. By 5.4.4.2, the set $\mathsf{N}_{L/E}\mathsf{C}_L$ is open and thus closed in C_E, and therefore X_L is compact.

To prove the finite intersection property, let $n \in \mathbb{N}$ and $L_1, \ldots, L_n \in \mathcal{E}$. Then $L = L_1 \cdot \ldots \cdot L_n \in \mathcal{E}$, and if $y \in \mathsf{X}_L$, then $\mathsf{N}_{E/K}(y) = x$, and 6.4.8.2(b) implies

$$y \in \mathsf{N}_{L/K}\mathsf{C}_L = \bigcap_{i=1}^n \mathsf{N}_{L_i/E}\mathsf{C}_{L_i} \quad \text{hence} \quad y \in \bigcap_{i=1}^n \mathsf{X}_{L_i}, \quad \text{and}$$

$$\emptyset \neq \mathsf{X}_L \subset \bigcap_{i=1}^n \mathsf{X}_{L_i}. \qquad \square[\mathbf{I.}]$$

Our next assertion comprises the core of the proof.

A. Let $n \in \mathbb{N}$ and $\mathrm{char}(K) \nmid n$. For a finite subset S of M_K containing M_K^∞ we set

$$\mathsf{J}_{K,S}^{(n)} = \prod_{v\in S} K_v^{\times n} \times \prod_{v\in \mathsf{M}_K\setminus S} U_v \quad \text{and} \quad \mathsf{C}_{K,S)}^{(n)} = \mathsf{J}_{K,S}^{(n)}K^\times/K^\times.$$

a. For sufficiently large S, the group $\mathsf{C}_{K,S}^{(n)}$ is a norm group.

b. $\mathsf{D}_K \subset \mathsf{C}_K^n$.

Proof of **A. a.** Assume first that $\mu_n \subset K$, and set $L = K(\sqrt[n]{K_S})$. If S is sufficiently large, then 6.1.8.3 (applied with $T = \emptyset$ and $\mathsf{C}_K^{(n)}(S,\emptyset) = \mathsf{C}_{K,S}^{(n)}$) shows that $\mathsf{C}_{K,S}^{(n)} \subset \mathsf{N}_{L/K}\mathsf{C}_L$ and $(\mathsf{C}_K : \mathsf{C}_{K,S}^{(n)}) = [L:K]$.

Since $(\mathsf{C}_K : \mathsf{N}_{L/K}\mathsf{C}_L) = [L:K]$ by the global reciprocity law 6.4.7, it follows that $\mathsf{C}_{K,S}^{(n)} = \mathsf{N}_{L/K}\mathsf{C}_L$ is a norm group.

Now we do the general case. We set $K' = K(\mu_n)$ and $S' = S_{K'}$. Together with S also S' is sufficiently large such that $\mathsf{C}_{K',S'}^{(n)}$ is a norm group. If L' is the class field to $C_{K',S'}^{(n)}$, then

$$\mathsf{N}_{L'/K}\mathsf{C}_{L'} = \mathsf{N}_{K'/K} \circ \mathsf{N}_{L'/K'}(\mathsf{C}_{L'}) = \mathsf{N}_{K'/K}\mathsf{C}_{K',S'}^{(n)} \subset \mathsf{C}_{K,S}^{(n)}.$$

Hence $\mathsf{C}_{K,S}^{(n)}$ is a norm group by 6.4.8.2(a). □[**a.**]

b. Assume to the contrary that there is some $a \in \mathsf{D}_K \setminus \mathsf{C}_K^n$. As C_K^n is closed in C_K by 1., there exists a neighbourhood V of 1 in J_K such that $aVK^\times \cap \mathsf{C}_K^n = \emptyset$. We may assume that

$$V = \prod_{v \in S} V_v \times \prod_{v \in \mathsf{M}_K \setminus S} U_v$$

for a sufficiently large finite subset S of M_K containing M_K^∞ and sufficiently small neighbourhoods V_v of 1 in $K_v^\times$ for all $v \in S$. Explicitly, we choose S so large that $\mathsf{C}_{K,S}^{(n)}$ is a norm group, and for $v \in S$ w choose V_v so small that $V_v \subset K_v^{\times n}$ (note that $K_v^{\times n}$ is open in $K_v^\times$, see [18, Theorem 5.8.10]). Then it follows that $\mathsf{D}_K \subset \mathsf{C}_{K,S}^{(n)}$, $V \subset \mathsf{J}_{K,S}^{(n)}$ and $VK^\times \subset \mathsf{C}_{K,S}^{(n)}$. Let $a = \alpha K^\times$ for some $\alpha \in \mathsf{J}_{K,S}^{(n)}$, and define $\beta = (\beta_v)_{v \in \mathsf{M}_K} \in \mathsf{J}_K$ by $\beta_v = 1$ if $v \in S$, and $\beta_v = \alpha_v^{-1}$ if $v \notin S$. It follows that $\beta \in V$, $(\alpha\beta)_v = \alpha_v \in K_v^{\times n}$ if $v \in S$, and $(\alpha\beta)_v = 1$ if $v \notin S$. Hence $\alpha\beta \in \mathsf{J}_K^n \cap \alpha V$ and $\alpha\beta K^\times \in \mathsf{C}_K^n \cap aVK^\times$, a contradiction. □[**b.**]

By **A** it follows that

$$\mathsf{D}_K \subset \bigcap_{\substack{n \in \mathbb{N} \\ \mathrm{char}(K) \nmid n}} \mathsf{C}_K^n,$$

and therefore it remains to prove:

B. If $\mathrm{char}(K) = p > 0$, then $\mathsf{D}_K \subset \mathsf{C}_K^{p^j}$ for all $j \in \mathbb{N}$.

Proof of **B**. Let $\mathrm{char}(K) = p > 0$. It suffices to prove that $\mathsf{D}_K \subset \mathsf{D}_K^p$. Indeed, then a simple induction shows that $\mathsf{D}_K \subset \mathsf{D}_K^{p^j} \subset \mathsf{C}_K^{p^j}$ for all $j \geq 0$.

Let $x \in \mathsf{D}_K$. Then $(x,K)\restriction K(\wp^{-1}z) = (x, K(\wp^{-1}z)/K) = 1$ for all $z \in K$, since $x \in \mathsf{N}_{K(\wp^{-1}z)/K}K(\wp^{-1}z)^\times$. Hence

$$(x, K(\wp^{-1}K)/K) = (x,K)\restriction K(\wp^{-1}K) = 1,$$

and by 6.3.8.1 there exists some $y \in \mathsf{C}_K$ such that $x = y^p$. We assert that $y \in \mathsf{N}_{L/K}\mathsf{C}_L$ for every finite extension L/K (then it follows that $y \in \mathsf{D}_K$ and $x \in \mathsf{D}_K^p$). If L/K is a finite field extension, then $x = \mathsf{N}_{L/K}(z)$ for some $z \in \mathsf{D}_L$ by **I**, and as above it follows that $z = u^p$ for some $u \in \mathsf{C}_L$. But then $x = y^p = \mathsf{N}_{L/K}(u^p) = \mathsf{N}_{L/K}(u)^p$ and therefore $y = \mathsf{N}_{L/K}(u) \in \mathsf{N}_{L/K}\mathsf{C}_L$.

2. Let $K/\mathbb{F}_q$ be a function field. Then $\mathsf{M}_K = \mathsf{M}_K^0$, and thus J_K has subgroup topology by 5.2.3. Hence the groups C_K, J_K^0 and $\mathsf{C}_K^0 = \mathsf{J}_K^0/K^\times$ have subgroup topology, too. In particular, C_K is totally disconnected by 1.1.10.3.

If U is an open subgroup of C_K^0, then $(\mathsf{C}_K^0 : U) < \infty$ (by 1.1.3.3, since C_K^0 is compact), hence $(\mathsf{C}_K^0)^n \subset U$ for some $n \in \mathbb{N}$. Consequently

$$\bigcap_{n\in\mathbb{N}} (\mathsf{C}_K^0)^n = \mathbf{1}, \quad \text{and therefore} \quad \bigcap_{n\in\mathbb{N}} \mathsf{C}_K^n = \bigcap_{n\in\mathbb{N}} [(\mathsf{C}_K^0)^n \cdot \Gamma'^n] = \mathbf{1},$$

since $\Gamma' \cong \mathbb{Z}$.

3. Apparently, J_K^+ is a connected subgroup of J_K, and

$$\mathsf{J}_K/\mathsf{J}_K^+ = \prod_{\substack{v\in\mathsf{M}_K^\infty \\ v\,\text{real}}} \{\pm 1\} \times \prod_{\substack{v\in\mathsf{M}_K^\infty \\ v\,\text{complex}}} \mathbf{1} \times \prod_{v\in\mathsf{M}_K^0} (K_v^\times \mid U_v)$$

is a locally compact topological group with subgroup topology by 5.2.2 and 5.2.3. Being a continuous image of J_K^+, the group $\mathsf{J}_K^+K^\times/K^\times$ is a connected subgroup of C_K. Using 1.1.5, it follows that $\mathsf{C}_K^+ = \overline{\mathsf{J}_K^+K^\times/K^\times} = \overline{\mathsf{J}_K^+K^\times}/K^\times$ is a connected closed subgroup of C_K, $\mathsf{C}_K/\mathsf{C}_K^+$ is Hausdorff, and being a factor group of $\mathsf{J}_K/\mathsf{J}_K^+$, it has subgroup topology. Hence $\mathsf{C}_K/\mathsf{C}_K^+$ is totally disconnected, and therefore C_K^+ is the component of 1 in C_K by 1.1.10.2.

Since $\Gamma \subset \mathsf{J}_K^+K^\times$, it follows that $\Gamma' = \Gamma K^\times/K^\times \subset \mathsf{C}_K^+$, and we infer $\mathsf{C}_K^+ = W \cdot \Gamma'$, where $W = \mathsf{C}_K^+ \cap \mathsf{C}_K^0$. Indeed, if $c \in \mathsf{C}_K^+$, then c has a unique factorization $c = c_0 z$, where $c_0 \in \mathsf{C}_K^0$ and $z \in \Gamma'$, but $\Gamma' \subset \mathsf{C}_K^+$ implies $c_0 = cz^{-1} \in \mathsf{C}_K^0 \cap \mathsf{C}_K^+ = W$.

As W is a closed subgroup of C_K^0, it follows that $\mathsf{C}_K/\mathsf{C}_K^+ = \mathsf{C}_K^0 \cdot \Gamma'/W \cdot \Gamma' = \mathsf{C}_K^0/W$ is compact. Being compact and totally disconnected, $\mathsf{C}_K/\mathsf{C}_K^+$ if profinite, and 1.5.1.2 implies

$$\mathbf{1} = \bigcap_{n\in\mathbb{N}} (\mathsf{C}_K/\mathsf{C}_K^+)^n = \bigcap_{n\in\mathbb{N}} \mathsf{C}_K^n\mathsf{C}_K^+/\mathsf{C}_K^+ = \Big[\bigcap_{n\in\mathbb{N}} \mathsf{C}_K^n\mathsf{C}_K^+\Big]\Big/\mathsf{C}_K^+,$$

and consequently

$$\mathsf{D}_K = \bigcap_{n\in\mathbb{N}} \mathsf{C}_K^n \subset \bigcap_{n\in\mathbb{N}} \mathsf{C}_K^n K^+ = \mathsf{C}_K^+.$$

If $n \in \mathbb{N}$, then $\overline{\mathsf{J}_K^+ K^\times/K^\times} = \mathsf{J}_K^{+n} K^\times/K^\times \subset \mathsf{C}_K^n$, therefore it follows that $\mathsf{C}_K^+ = \overline{\mathsf{J}_K^+ K^\times/K^\times} \subset \mathsf{C}_K^n$, and consequently

$$\mathsf{C}_K^+ \subset \bigcap_{n\in\mathbb{N}} \mathsf{C}_K^n = \mathsf{D}_K.$$

Hence $\mathsf{D}_K = \mathsf{C}_K^+$, $W = \mathsf{D}_K \cap \mathsf{C}_K^0 = \mathsf{D}_K^0$, and $\mathsf{C}_K/\mathsf{D}_K = \mathsf{C}_K^0 \cdot \Gamma'/\mathsf{D}_K^0 \cdot \Gamma' \cong \mathsf{C}_K^0/\mathsf{D}_K^0$ (both algebraically and topologically). Consequently $\mathsf{C}_K/\mathsf{D}_K$ is compact. □

6.5 Global class fields

In the sequel we aim again at an utmost unified treatment of number fields and function fields. For that purpose we use once more the notion of an abstract divisor group $\mathcal{D}_K$ as introduced in 5.1.1, and we freely use all further concepts introduced as a result of this in Chapter 5, and all concepts and results provided hitherto in the current chapter.

We start with an investigation of the universal norm residue symbol in the number field case.

Theorem 6.5.1 (Universal norm residue symbol for number fields). *Let K be a number field. Then $(\mathsf{C}_K, K) = (\mathsf{C}_K^0, K) = G_K^{\mathsf{a}}$, and the universal norm residue symbol $(\,\cdot\,, K)\colon \mathsf{C}_K/\mathsf{D}_K \xrightarrow{\sim} G^{\mathsf{a}}$ is a topological isomorphism.*

Proof. By 6.4.6.**2** the universal norm residue symbol $(\,\cdot\,, K)\colon \mathsf{C}_K/\mathsf{D}_K \to G_K^{\mathsf{a}}$ is a continuous monomorphism with dense image $U = (\mathsf{C}_K/\mathsf{D}_K, K) \subset G_K^{\mathsf{a}}$. By 6.4.10.3 the group $\mathsf{C}_K/\mathsf{D}_K$ is compact. Hence U is compact and thus closed in G_K^{a}, which implies that $U = G_K^{\mathsf{a}}$, $(\,\cdot\,, K)\colon \mathsf{C}_K/\mathsf{D}_K \to G_K^{\mathsf{a}}$ is a continuous bijection, and as $\mathsf{C}_K/\mathsf{D}_K$ is compact, it is even a topological isomorphism. Since $\mathsf{C}_K = \mathsf{C}_K^0 \cdot \Gamma'$ and $\Gamma' \subset \mathsf{D}_K$, it follows that

$$G_K^{\mathsf{a}} = (\mathsf{C}_K/\mathsf{D}_K, K) = (\mathsf{C}_K, K) = (\mathsf{C}_K^0, K). \qquad \square$$

We proceed with the class field theory of constant field extensions in the function field case. We recall some terminology. Let $K/\mathbb{F}_q$ be a function field, L/K a finite field extension and $\mathbb{F}_{q^r}$ the field of constants of L. We call L/K a

- constant field extension if $L = K\mathbb{F}_{q^r}$,
- geometric extension if $r = 1$.

Theorem 6.5.2 (Universal global norm residue symbol). *Let $K/\mathbb{F}_q$ be a function field, $\overline{\mathbb{F}}_q$ an algebraic closure of $\mathbb{F}_q$ and $K_\infty = K\overline{\mathbb{F}}_q$ the largest constant field extension of K. Then the restriction map*

$$\mathrm{Gal}(K_\infty/K) \to \mathrm{Gal}(\overline{\mathbb{F}}_q/\mathbb{F}_q), \quad \sigma \mapsto \sigma \restriction \overline{\mathbb{F}}_q,$$

is a topological isomorphism, and

$$\mathrm{Gal}(K_\infty/K) = \overline{\langle\phi\rangle} \cong \widehat{\mathbb{Z}},$$

where $\phi \in \mathrm{Gal}(K_\infty/K)$ is the unique automorphism such that $\phi \restriction \overline{\mathbb{F}}_q = \phi_q$ is the Frobenius automorphism over $\mathbb{F}_q$ (see 1.8.7).
Let $c \in \mathsf{C}_K$ be such that $\deg(c) = 1$ (then $\mathsf{C}_K = \mathsf{C}_K^0{\cdot}\langle c\rangle$, see 5.3.3). For $r \in \mathbb{N}$ let $K_r = K\mathbb{F}_{q^r}$ be the constant field extension of degree r (recall that then $\mathrm{Gal}(K_r/K) = \langle\phi \restriction K_r\rangle$, and the restriction map

$$\mathrm{Gal}(K_r/K) \to \mathrm{Gal}(\mathbb{F}_{q^r}/\mathbb{F}_q), \ \sigma \mapsto \sigma \restriction \mathbb{F}_{q^r}$$

is an isomorphism satisfying $(\phi \restriction K_r) \restriction \mathbb{F}_{q^r} = \phi_q \restriction \mathbb{F}_{q^r}$).

1. *Let $a \in \mathsf{C}_K$ and $r \in \mathbb{N}$. Then*

$$(a, K_r/K) = \phi^{\deg(a)} \restriction K_r \quad \text{and} \quad \mathsf{N}_{K_r/K}\mathsf{C}_{K_r} = \mathsf{C}_K^0{\cdot}\langle c^r\rangle.$$

In particular, $(a, K_\infty/K) = \phi^{\deg(a)}$, $(\mathsf{C}_K, K_\infty/K) = \langle\phi\rangle$ is a dense proper subgroup of $\mathrm{Gal}(K_\infty/K)$, and $(\,\cdot\,, K)\colon \mathsf{C}_K \to G_K^{\mathsf{a}}$ is not surjective.

2. *$(\,\cdot\,, K) \restriction \mathsf{C}_K^0 \colon \mathsf{C}_K^0 \to \mathrm{Gal}(K_{\mathrm{ab}}/K_\infty)$ is a topological isomorphism.*

3. *Let L/K be a finite abelian extension, $B = \mathsf{N}_{L/K}\mathsf{C}_L$ and let $\mathbb{F}_{q^r}$ be the field of constants of L. Then $\deg(B) = r\mathbb{Z}$, and $(\mathsf{C}_K^0, L/K) = \mathrm{Gal}(L/K_r)$. In particular:*

- *L/K is a constant field extension if and only if $\mathsf{C}_K^0 \subset B$.*
- *L/K is a geometric extension if and only if $(\mathsf{C}_K^0, L/K) = G$.*

Proof. By 1.8.4, the map $\mathrm{Gal}(K_\infty/K) \to \mathrm{Gal}(\overline{\mathbb{F}}_q/\mathbb{F}_q)$, $\tau \mapsto \tau \restriction \overline{\mathbb{F}}_q$ is a topological isomorphism, and by 1.8.7, $\mathrm{Gal}(\overline{\mathbb{F}}_q/\mathbb{F}_q) = \overline{\langle\phi_q\rangle}$, where $\phi_q = (x \mapsto x^q)$ is the Frobenius automorphism over $\mathbb{F}_q$. Hence there exists a unique automorphism $\phi \in \mathrm{Gal}(K_\infty/K)$ such that $\phi \restriction \overline{\mathbb{F}}_q = \phi_q$, and then $\mathrm{Gal}(K_\infty/K) = \overline{\langle\phi\rangle}$.

If $r \in \mathbb{N}$, then $\mathrm{Gal}(K_r/K) = \langle\phi \restriction K_r\rangle$, and there is an isomorphism $\mathrm{Gal}(K_r/K) \overset{\sim}{\to} \mathrm{Gal}(\mathbb{F}_{q^r}/\mathbb{F}_q)$ such that $\phi \restriction K_r \mapsto \phi \restriction \mathbb{F}_{q^r}$.

1. Let $r \in \mathbb{N}$ and $a = \alpha K^\times$, where $\alpha = (\alpha_v)_{v\in\mathsf{M}_K} \in \mathsf{J}_K$. If $v \in \mathsf{M}_K$, $v_r \in \mathsf{M}_{K_r}$ and $v_r \,|\, v$, then v_r/v and thus $(K_r)_{v_r}/K_v$ is unramified (see [18,

Theorem 6.6.7]). Let φ_v be the Frobenius automorphism of $(K_r)_{v_r}/K_v$. If $x \in K_r$, then $\varphi_v(x) = x^{q^{\deg(v)}}$, and

$$\begin{aligned}(a, K_r/K)(x) = (\alpha, K_r/K)(x) &= \prod_{v \in \mathsf{M}_K} (\alpha_v, (K_r)_{v_r}/K_v)(x) = \prod_{v \in \mathsf{M}_K} \varphi_v^{v(\alpha_v)}(x) \\ &= x^{q^s} = \phi^s(x), \quad \text{where} \quad s = \sum_{v \in \mathsf{M}_K} \deg(v)v(\alpha_v) \\ &= \deg(\alpha) = \deg(c).\end{aligned}$$

Hence $(a, K_r/K) = \phi^{\deg(a)} \restriction K_r$, and as

$$\begin{aligned}K_\infty = \bigcup_{r \in \mathbb{N}} K_r \quad \text{and} \quad (a, K_\infty/K) \restriction K_r &= (a, K_r/K) \\ &= \phi^{\deg(a)} \restriction K_r \text{ for all } r \in \mathbb{N},\end{aligned}$$

it follows that $(a, K_\infty/K) = \phi^{\deg(a)}$. In particular, $(\mathsf{C}_K, K_\infty/K) = \langle \phi \rangle$ is a dense proper subgroup of $\mathrm{Gal}(K_\infty/K)$.

For $a \in \mathsf{C}_K$ we obtain by the global reciprocity law 6.4.7:

$$\begin{aligned}a \in \mathsf{N}_{K_r/K}\mathsf{C}_{K_r} &\iff (a, K_r/K) = \phi^{\deg(a)} \restriction K_r = 1 \iff r \mid \deg(a) \\ &\iff a \in \mathsf{C}_K^0 \cdot \langle c^r \rangle.\end{aligned}$$

The restriction $\rho\colon G_K \to \mathrm{Gal}(K_\infty/K)$, defined by $\rho(\sigma) = \sigma \restriction K_\infty$, is an epimorphism, and $(\,\cdot\,, K_\infty/K) = \rho \circ (\,\cdot\,, K)\colon \mathsf{C}_K \to \mathrm{Gal}(K_\infty/L)$ is not surjective. Hence the universal norm residue symbol $(\,\cdot\,, K)\colon \mathsf{C}_K \to G_K^{\mathsf{a}}$ is not surjective.

2. Since $\mathsf{D}_K = \mathrm{Ker}(\,\cdot\,, K) = \mathbf{1}$ by 6.4.10.2, the universal norm residue symbol $(\,\cdot\,, K)\colon \mathsf{C}_K \to G_K^{\mathsf{a}}$ is a continuous monomorphism by 6.4.5.2. If now $a \in \mathsf{C}_K^0$, then 1. implies $(a, K) \restriction K_\infty = (a, K_\infty/K) = 1$, and therefore $U = (\mathsf{C}_K^0, K) \subset \mathrm{Gal}(K_{\mathrm{ab}}/K_\infty)$. As U is compact and thus closed in $\mathrm{Gal}(K_{\mathrm{ab}}/K_\infty)$, we get $U = \mathrm{Gal}(K_{\mathrm{ab}}/K_{\mathrm{ab}}^U)$ by 1.8.3.1. Therefore it suffices to prove that $K_{\mathrm{ab}}^U \subset K_\infty$ (for then equality holds, and $U = \mathrm{Gal}(K_{\mathrm{ab}}/K_\infty)$.

Let L be an over K finite intermediate field of K_{ab}^U/K and $[L{:}K] = r \in \mathbb{N}$. If $a \in \mathsf{C}_K^0$, then $(a, K) \in U$, hence $(a, L/K) = (a, K) \restriction L = 1$, and consequently $\mathsf{C}_K^0 \subset \mathsf{N}_{L/K}\mathsf{C}_L \subset \mathsf{C}_K = \mathsf{C}_K^0 \cdot \langle c \rangle$. Since $(\mathsf{C}_K : \mathsf{N}_{L/K}\mathsf{C}_L) = r$ by the global reciprocity law 6.4.7, it follows that $\mathsf{N}_{L/K}\mathsf{C}_L = \mathsf{C}_K^0 \cdot \langle c^r \rangle = \mathsf{N}_{K_r/K}\mathsf{C}_{K_r}$ by 1. By 6.4.8.2, we eventually obatin $L = K_r \subset K_\infty$, and consequently, $K_{\mathrm{ab}}^U \subset K_\infty$.

3. For any $n \in \mathbb{N}$ we obtain, using 1. and 6.4.8.2(c):

$$\begin{aligned}\deg(B) \subset n\mathbb{Z} &\iff \mathsf{N}_{L/K}\mathsf{C}_L = B \subset \mathsf{C}_K^0 \cdot \langle a^n \rangle = \mathsf{N}_{K_n/K}\mathsf{C}_{K_n} \\ &\iff K_n \subset L \iff \mathbb{F}_{q^n} \subset L \iff n \,|\, r \iff r\mathbb{Z} \subset n\mathbb{Z}.\end{aligned}$$

Hence it follows that $\deg(B) = r\mathbb{Z}$.

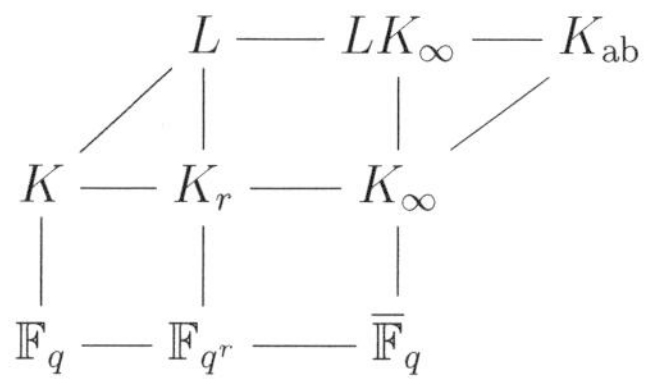

Let $G = \mathrm{Gal}(L/K)$, and define $\rho\colon \mathrm{Gal}(K_{\mathrm{ab}}/K) \to G$ by $\rho(\sigma) = \sigma \restriction L$. Then $(\mathsf{C}_K^0, L/K) = \rho(\mathsf{C}_K^0, K) = \rho(\mathrm{Gal}(K_{\mathrm{ab}}/K_\infty) = \mathrm{Gal}(L/L \cap K_\infty) = \mathrm{Gal}(L/K_r)$.

□

Theorem 6.5.3. *A subgroup of C_K is a norm group if and only if it is an open subgroup of finite index.*

Proof. By 5.4.4.2, norm groups are open, and by 6.4.8.1 they are of finite index. Thus let H be an open subgroup of finite index of C_K.

CASE 1: K is a number field. If $(\mathsf{C}_K : H) = n$, then $\mathsf{C}_K^n \subset H$, and consequently $\mathsf{D}_K \subset H$ by 6.4.10.1. Since $(\,\cdot\,, K)\colon \mathsf{C}_K/\mathsf{D}_K \overset{\sim}{\to} G_K^{\mathsf{a}}$ is a topological isomorphism by 6.4.10.3(b) it follows that $U = (H, K) = (H/\mathsf{D}_K, K)$ is an open subgroup of G_K^{a}. Hence $U = \mathrm{Gal}(K_{\mathrm{ab}}/L)$ for some over K finite intermediate field L of K_{ab}/K, and we obtain:

$$c \in H \iff (c\mathsf{D}_K, K) = (c, K) \in U \iff (c, K) \restriction L = (c, L/K) = 1$$
$$\iff c \in \mathsf{N}_{L/K}\mathsf{C}_L.$$

Hence $H = \mathsf{N}_{L/K}\mathsf{C}_L$ is a norm group.

CASE 2: $K/\mathbb{F}_q$ is a function field. We set $H^0 = H \cap \mathsf{C}_K^0$, and (as in 5.3.3.2) $\mathsf{C}_K = \mathsf{C}_K^0 \cdot \Gamma'$, where $\Gamma' = \langle c \rangle \cong \mathbb{Z}$ for some $c \in \mathsf{C}_K$ satisfying $\deg(c) = 1$. Then H^0 is an open subgroup of finite index in C_K^0, and the universal global norm residue symbol $(\,\cdot\,, K)\colon \mathsf{C}_K^0 \overset{\sim}{\to} \mathrm{Gal}(K_{\mathrm{ab}}/K_\infty)$ is a topological isomorphism by 6.5.2.2. Hence (H^0, K) is an open subgroup of $\mathrm{Gal}(K_{\mathrm{ab}}/K_\infty)$, and therefore $(H^0, K) = \mathrm{Gal}(K_{\mathrm{ab}}/M)$ for some over K_∞ finite intermediate field M of K_{ab}/K_∞. Let L be an over K finite intermediate field of M/K such that $M = LK_\infty$. If $c \in \mathsf{C}_K^0 \cap \mathsf{N}_{L/K}\mathsf{C}_L$, then $(c, K) \restriction K_\infty = 1$ and $(c, K) \restriction L = 1$, hence $(c, K) \restriction M = 1$ and therefore $c \in H^0$. It follows that $\mathsf{C}_K^0 \cap \mathsf{N}_{L/K}\mathsf{C}_L \subset H^0 = H \cap \mathsf{C}_K^0$, and we assert that

$$[(\mathsf{N}_{L/K}\mathsf{C}_L \cap H)\mathsf{C}_K^0] \cap \mathsf{N}_{L/K}\mathsf{C}_L \subset H. \qquad (*)$$

Proof of $(*)$. Suppose that $c = c_1c_0 \in [(\mathsf{N}_{L/K}\mathsf{C}_L \cap H)\mathsf{C}_K^0] \cap \mathsf{N}_{L/K}\mathsf{C}_L$, where $c_1 \in \mathsf{N}_{L/K}\mathsf{C}_L \cap H$ and $c_0 \in \mathsf{C}_K^0$. Then it follows that

$$c_0 = cc_1^{-1} \in \mathsf{C}_K^0 \cap \mathsf{N}_{L/K}\mathsf{C}_L \subset H^0,$$

and consequently $c = c_0c_1 \in H^0(\mathsf{N}_{L/K}\mathsf{C}_L \cap H) \subset H$. □[$(*)$]

By 6.4.8.2(a), it suffices to prove that $\mathsf{N}_{L/K}\mathsf{C}_L \cap [(\mathsf{N}_{L/K}\mathsf{C}_L \cap H)\mathsf{C}_K^0]$ is a norm group, and as $\mathsf{N}_{L/K}\mathsf{C}_L$ is a norm group, it suffices to prove that $(\mathsf{N}_{L/K}\mathsf{C}_L \cap H)\mathsf{C}_K^0$ is a norm group (see 6.4.8.2(b)). Since $\mathsf{N}_{L/K}\mathsf{C}_L$ and H are subgroups of finite index of C_K, it follows that $\mathsf{C}_K^0(\mathsf{N}_{L/K} \cap H)$ is a subgroup of C_K of finite index. Hence it follows that $c^r \in \mathsf{C}_K^0(\mathsf{N}_{L/K} \cap H)$ for some $r \in \mathbb{N}$, and then $\mathsf{C}_K^0 \cdot \langle c^r \rangle \subset \mathsf{C}_K^0(\mathsf{N}_{L/K} \cap H)$. However, $\mathsf{C}_K^0 \cdot \langle a^r \rangle$ is a norm group by 6.5.2.1, and therefore also $(\mathsf{N}_{L/K}\mathsf{C}_L \cap H)\mathsf{C}_K^0$ is a norm group (again by 6.4.8.2(a)). □

For $v \in \mathsf{M}_K$ we denote by $j_v \colon K_v^\times \to \mathsf{J}_K$ and $j_v^* \colon K_v^\times \to \mathsf{C}_K$ the local embeddings. Recall that $j_v \colon K_v^\times \overset{\sim}{\to} j_v(K_v^\times)$ and $j_v^* \colon K_v^\times \overset{\sim}{\to} j_v^*(K_v^\times)$ are topological isomorphisms (see 5.3.1.**5**).

Theorem 6.5.4. *Let L/K be a finite abelian extension, $G = \mathrm{Gal}(L/K)$, $v \in \mathsf{M}_K$, $\overline{v} \in \mathsf{M}_L$, $\overline{v}\,|\,v$ and $G_{\overline{v}} = \mathrm{Gal}(L_{\overline{v}}/K_v) \subset G$.*

1. There is a commutative diagram

$$\begin{array}{ccccc} (\,\cdot\,, L_{\overline{v}}/K_v) : K_v^\times & \xrightarrow{\ =\ } & K_v^\times & \xrightarrow{(\,\cdot\,,L_{\overline{v}}/K_v)} & G_{\overline{v}} \\ {\scriptstyle j_v}\big\downarrow & & {\scriptstyle j_v^*}\big\downarrow & & \big\downarrow{\scriptstyle \rho}, \\ (\,\cdot\,, L/K) : \ \mathsf{J}_K & \longrightarrow & \mathsf{C}_K & \xrightarrow{(\,\cdot\,,L/K)} & G \end{array} \qquad \textit{where } \rho = (G_{\overline{v}} \hookrightarrow G).$$

Moreover,

$$\mathsf{N}_{L/K}\mathsf{J}_L \cap j_v(K_v^\times) = j_v(\mathsf{N}_{L_{\overline{v}}/K_v}L_{\overline{v}}^\times) \ \textit{ and } \ \mathsf{N}_{L/K}\mathsf{C}_L \cap j_v^*(K_v^\times) = j_v^*(\mathsf{N}_{L_{\overline{v}}/K_v}L_{\overline{v}}^\times).$$

In particular, if $a \in K_v^\times$, then

$$a \in \mathsf{N}_{L_{\overline{v}}/K_v}L_{\overline{v}}^\times \iff j_v(a) \in \mathsf{N}_{L/K}\mathsf{J}_L \iff j_v^*(a) \in \mathsf{N}_{L/K}\mathsf{C}_L.$$

2. $$v \in \mathrm{Split}(L/K) \iff j_v(K_v^\times) \subset \mathsf{N}_{L/K}\mathsf{J}_L \iff j_v^*(K_v^\times) \subset \mathsf{N}_{L/K}\mathsf{C}_L.$$

3. Let $v \in \mathsf{M}_K^0$. Then

$$v \textit{ is unramified in } L \iff j_v(U_v) \subset \mathsf{N}_{L/K}\mathsf{J}_L \iff j_v^*(U_v) \subset \mathsf{N}_{L/K}\mathsf{C}_L.$$

4. Let $v \in \mathsf{M}_K^0$, $\overline{v} \in \mathsf{M}_L$ and $\overline{v}\,|\,v$. Let L_0 be the decomposition field and L_1 the inertia field of $\overline{v}$ over K (see 5.4.5). *Then*

$$\mathsf{N}_{L_0/K}\mathsf{C}_{L_0} = (\mathsf{N}_{L/K}\mathsf{C}_L)j_v^*(K_v^\times) \quad \textit{and} \quad \mathsf{N}_{L_1/K}\mathsf{C}_{L_1} = (\mathsf{N}_{L/K}\mathsf{C}_L)j_v^*(U_v).$$

If $\overline{v}/v$ is unramified, $\pi_v \in K$ and $v(\pi_v) = 1$, then

$$f(\overline{v}/v) = \min\{f \in \mathbb{N} \mid j_v^*(\pi_v^f) \in \mathsf{N}_{L/K}\mathsf{C}_L\}.$$

Proof. 1. If $a \in K_v^\times$, then

$$(j_v^*(a), L/K) = (j_v(a), L/K) = \prod_{w \in \mathsf{M}_K} (j_v(a)_w, L_{\overline{w}}/K_w)$$
$$= (a, L_{\overline{v}}/K_v) \in G_{\overline{v}} \subset G$$

by the very definition of the global norm residue symbol.

Now let $a \in K_v^\times$. By 6.1.1.1(b), it follows that $j_v(a) \in \mathsf{N}_{L/K}\mathsf{J}_L$ if and only if $a \in \mathsf{N}_{L_{\overline{v}}/K_v} L_{\overline{v}}^\times$. Using the global reciprocitiy law 6.4.7 and the local reciprocity law 4.2.7.2, we obtain

$$j_v^*(a) \in \mathsf{N}_{L/K}\mathsf{C}_L \iff (j_v^*(a), L/K) = (a, L_{\overline{v}}/K_v) = 1 \iff a \in \mathsf{N}_{L_{\overline{v}}/K_v} L_{\overline{v}}^\times.$$

2. We use the local reciprocity law 4.2.7.2, and for $v \in \mathsf{M}_K^\infty$ we recall the Archimedian norm residue symbol defined in 6.2.1. Then, using 1.,

$$v \in \mathrm{Split}(L/K) \iff L_{\overline{v}} = K_v \iff K_v^\times = \mathsf{N}_{L_{\overline{v}}/K_v} L_{\overline{v}}^\times$$
$$\iff j_v(K_v^\times) \subset \mathsf{N}_{L/K}\mathsf{J}_L \iff j_v^*(K_v^\times) \subset \mathsf{N}_{L/K}\mathsf{C}_L.$$

3. Using 4.8.2.3 we obtain, again using 1.,

$$v \text{ is unramified in } L \iff L_{\overline{v}}/K_v \text{ is unramified} \iff U_v \subset \mathsf{N}_{L_{\overline{v}}/K_v} L_{\overline{v}}^\times$$
$$\iff j_v(U_v) \subset \mathsf{N}_{L/K}\mathsf{J}_L \iff j_v^*(U_v) \subset \mathsf{N}_{L/K}\mathsf{C}_L.$$

4. As v is fully decomposed in L_0, we obtain $j_v^*(K_v^\times) \subset \mathsf{N}_{L_0/K}\mathsf{C}_{L_0}$ by 2., and thus $(\mathsf{N}_{L/K}\mathsf{C}_L)j_v^*(K_v^\times) \subset \mathsf{N}_{L_0/K}\mathsf{C}_{L_0}$. By 6.4.8.2(a) there exists an intermediate field L_0' of L/K such that

$$\mathsf{N}_{L_0'/K}\mathsf{C}_{L_0'} = (\mathsf{N}_{L/K}\mathsf{C}_L)j_v^*(K_v^\times) \subset \mathsf{N}_{L_0/K}\mathsf{C}_{L_0},$$

and 6.4.8.2(c) implies $L_0 \subset L_0'$. By 2. it follows that v is fully decomposed in L_0', and the maximality of L_0 implies $L_0 = L_0'$ (see [18, Corollary 2.13.6]).

As v is unramified in L_1, we obtain $j_v^*(U_v) \subset \mathsf{N}_{L_1/K}C_{L_1}$ by (a), and thus also $(\mathsf{N}_{L/K}\mathsf{C}_L)j_v^*(U_v) \subset \mathsf{N}_{L_1/K}\mathsf{C}_{L_1}$. By 6.4.8.2(a), there exists an intermediate field L_1' of L/K such that $\mathsf{N}_{L_1'/K}\mathsf{C}_{L_1'} = (\mathsf{N}_{L/K}\mathsf{C}_L)j_v^*(U_v) \subset \mathsf{N}_{L_1/K}\mathsf{C}_{L_1}$, and 6.4.8.2(c) implies $L_1 \subset L_1'$. By (a), it follows that v is unramified in L_1', and the maximality of L_1 implies $L_1 = L_1'$ (see again [18, Corollary 2.13.6]).

Let after all $\overline{v}/v$ be unramified, $\pi_v \in K$ such that $v(\pi_v) = 1$ and φ_v the Frobenius automorphism of $L_{\overline{v}}/K_v$. Then $\mathrm{ord}(\varphi_v) = [L_{\overline{v}}{:}K_v] = f(\overline{v}/v)$, and for $f \in \mathbb{N}$ we obtain:

$$j_v^*(\pi^f) \in \mathsf{N}_{L/K}\mathsf{C}_L \iff (j_v^*(\pi_v^f), L/K) = (\pi_v^f, L_{\overline{v}}/K_v) = \varphi_v^f = 1$$
$$\iff f(\overline{v}/v) \mid f.$$

Hence $f(\overline{v}/v) = \min\{f \in \mathbb{N} \mid j_v^*(\pi^f) \in \mathsf{N}_{L/K}\mathsf{C}_L$. □

Corollary 6.5.5. *Let B be an open subgroup of C_K of finite index, S a subset of M_K and*

$$B_S = B \prod_{v \in S} j_v^*(K_v^\times).$$

Let L be the class field to B and L_S the class field to B_S. Then L_S is the largest intermediate field of L/K in which all places of S are fully decomposed.

Proof. By 6.5.4.2 and 6.4.8.2(b), a finite abelian extension L_0 of K is contained in L and all places form S are fully decomposed in L_0 if and only if $\mathsf{N}_{L_0/K}\mathsf{C}_{L_0} \subset \mathsf{N}_{L/K}\mathsf{C}_L$ and $j_v^*(K_v^\times) \subset \mathsf{N}_{L_0/K}\mathsf{C}_{L_0}$ for all $v \in S$. Hence L_S is the largest field with these properties. □

Let S be a finite subset of M_K^0 such that $S \neq \emptyset$ in the function field case, and let $\mathfrak{m} \in \mathcal{D}'_{K,S}$ be an integral S-divisor. Recall from 5.7.1 that

$$\mathsf{J}_{K,S}^{\mathfrak{m}} = \prod_{v \in \mathsf{M}_K^\infty} U_v \times \prod_{v \in S} K_v^\times \times \prod_{v \in \mathsf{M}_K^0 \setminus S} U_v^{(v(\mathfrak{m}))} \quad \text{and} \quad \mathsf{C}_{K,S}^{\mathfrak{m}} = \mathsf{J}_{K,S}^{\mathfrak{m}} K^\times / K^\times,$$

By 5.7.2.2(a), the group $\mathsf{C}_{K,S}^{\mathfrak{m}}$ is an open subgroup of finite index in C_K. In the subsequent Theorem 6.5.6 0, we investigate its role as a norm group, and as a consequence we obtain a local-global principle for conductors of abelian extensions. We briefly recall their definitions. The (global) conductor $\mathfrak{f}_{L/K}$ is the smallest (abstract) divisor $\mathfrak{m} \in \mathcal{D}'_K$ satisfying $\mathsf{C}_K^{\mathfrak{m}} \subset \mathsf{N}_{L/K}\mathsf{C}_L$ (see 5.7.5.2). If $v \in \mathsf{M}_K^0$ and $\overline{v} \in \mathsf{M}_L$ lies above v, then the (local) conductor is defined by $\mathfrak{f}_{L_{\overline{v}}/K_v} = \widehat{\mathfrak{p}}_v^n$, where $n \in \mathbb{N}_0$ is the smallest integer satisfying $U_v^{(n)} \subset \mathsf{N}_{L_{\overline{v}}/K_v} L_{\overline{v}}^\times$ (see 4.8.7).

Theorem 6.5.6. *Let L/K be a finite abelian extension. For each $v \in \mathsf{M}_K$ we fix some $\overline{v} \in \mathsf{M}_L$ lying above v.*

1. *Let S be a finite subset of M_K^0 such that $S \neq \emptyset$ in the function field case, and let $\mathfrak{m} \in \mathcal{D}'_{K,S}$. Then $\mathsf{C}_{K,S}^{\mathfrak{m}} \subset \mathsf{N}_{L/K}\mathsf{C}_L$ if and only if $S \subset \mathrm{Split}(L/K)$, and $U_v^{(v(\mathfrak{m}))} \subset \mathsf{N}_{L_{\overline{v}}/K_v} L_{\overline{v}}^\times$ for all $v \in \mathsf{M}_K^0$.*
2. *If $v \in \mathsf{M}_K^0$, then $v(\mathfrak{f}_{L/K}) = v(\mathfrak{f}_{L_{\overline{v}}/K_v})$.*
3. $\mathrm{Ram}(L/K) = \mathrm{supp}(\mathfrak{f}_{L/K})$.

Proof. 1. By definition, $\mathsf{C}_{K,S}^{\mathfrak{m}} \subset \mathsf{N}_{L/K}\mathsf{C}_L$ if and only if $j_v^*(K_v^\times) \subset \mathsf{N}_{L/K}\mathsf{C}_L$ for all $v \in S$, and $j_v^*(U_v^{(v(\mathfrak{m}))}) \subset \mathsf{N}_{L/K}\mathsf{C}_L$ for all $v \in \mathsf{M}_K^0$. If $v \in S$, then 6.5.4.2 implies that $j_v^*(K_v^\times) \subset \mathsf{N}_{L/K}\mathsf{C}_L$ if and only if $v \in \mathrm{Split}(L/K)$. If $v \in \mathsf{M}_K^0$, then 6.5.4.1 implies that $j_v^*(U_v^{(v(\mathfrak{m}))}) \subset \mathsf{N}_{L/K}\mathsf{C}_L$ if and only if $U_v^{(v(\mathfrak{m}))} \subset \mathsf{N}_{L_{\overline{v}}/K_v} L_{\overline{v}}^\times$.

2. For $v \in \mathsf{M}_K^0$ we set $\mathfrak{f}_{L_{\overline{v}}/K_v} = \widehat{\mathfrak{p}}_v^{n_v}$, where $n_v \in \mathbb{N}_0$. For $\mathfrak{m} \in \mathcal{D}'_K$, we shall prove that $\mathfrak{f}_{L/K} \,|\, \mathfrak{m}$ if and only if $v(\mathfrak{m}) \geq n_v$ for all $v \in \mathsf{M}_K^0$ (then $v(\mathfrak{f}_{L/K}) = v(\mathfrak{f}_{L_{\overline{v}}/K_v})$ follows).

Thus let $\mathsf{m} \in \mathcal{D}'_K$. By 5.7.5.2, we get $\mathsf{f}_{L/K} \mid \mathsf{m}$ if and only if m is a modulus of definition for L/K, that is, $\mathsf{C}_K^{\mathsf{m}} \subset \mathsf{N}_{L/K}\mathsf{C}_L$, and by 1. this holds if and only if $U_v^{(v(\mathsf{m}))} \subset \mathsf{N}_{L_{\overline{v}}/K_v} L_{\overline{v}}^\times$ and thus $v(\mathsf{m}) \geq n_v$ for all $v \in \mathsf{M}_K^0$ (note that $S = \emptyset$).

3. If $v \in \mathsf{M}_K^0$, then

$$\begin{aligned} v \in \mathrm{Ram}(L/K) &\iff L_{\overline{v}}/K_v \text{ is ramified } \iff \mathfrak{f}_{L_{\overline{v}}/K_v} \neq \mathbf{1} \\ &\iff v(\mathsf{f}_{L/K}) = v(\mathfrak{f}_{L_{\overline{v}}/K_v}) > 0 \\ &\iff v \in \mathrm{supp}(\mathsf{f}_{L/K}). \end{aligned}$$ □

Our next result, the Artin reciprocity law, is the main result of class field theory. The most important special case occurs for $S = \emptyset$ (both in the number field and in the function field case). The classical formulation with ray class fields will be dealt with in the next section (see 6.6.2). There we shall present it in the generalized form with holomorphy domains (which requires the assumption $S \neq \emptyset$ in the function field case).

Theorem and Definition 6.5.7 (Artin reciprocity law). *Let S be a finite subset of M_K^0 and $\mathsf{m} \in \mathcal{D}'_{K,S}$.*

1. *Let L/K be a finite abelian extension, $G = \mathrm{Gal}(L/K)$, let $\mathsf{C}_{K,S}^{\mathsf{m}} \subset \mathsf{N}_{L/K}\mathsf{C}_L$ and $\overline{\mathsf{m}} = j_{L/K}(\mathsf{m})$. Then $\mathrm{supp}(\mathsf{a}) \cap \mathrm{Ram}(L/K) = \emptyset$ for all $\mathsf{a} \in \mathcal{D}_K^{\mathsf{m}}$.*

   For a divisor $\mathsf{a} \in \mathcal{D}_K^{\mathsf{m}}$ we define the **Artin symbol** $(\mathsf{a}, L/K)$ by

   $$(\mathsf{a}, L/K) = \prod_{v \in \mathrm{supp}(\mathsf{a})} (v, L/K)^{v(\mathsf{a})} \in G,$$

   where $(v, L/K)$ denotes the Frobenius automorphism of $L_{\overline{v}}/K_v$ for some place $\overline{v} \in \mathsf{M}_L$ above v (see 5.4.5.**3**).

   The Artin symbol defines an (equally denoted) epimorphism

   $$(\,\cdot\,, L/K) \colon \mathcal{D}_{K,S}^{\mathsf{m}} \to G, \quad \text{called the } \textbf{Artin map},$$

   with kernel $\mathcal{N}_{L/K,S}^{\mathsf{m}} = \mathcal{N}_{L/K}(\mathcal{D}_{L,S_L}^{\overline{\mathsf{m}}})(K^{\mathsf{m}})_S$ (see 5.7.3.2), and a (likewise equally denoted) isomorphism

   $$(\,\cdot\,, L/K) \colon \mathcal{D}_{K,S}^{\mathsf{m}}/\mathcal{N}_{L/K,S}^{\mathsf{m}} \xrightarrow{\sim} G, \quad \text{called the } \textbf{Artin isomorphism}.$$

   *If $\alpha \in \mathsf{J}_{K,S}^{(\mathsf{m})}$, then $(\alpha, L/K) = (\iota_S^{(\mathsf{m})}(\alpha), L/K)$ (see 5.7.1.**2**), and there is a commutative diagram*

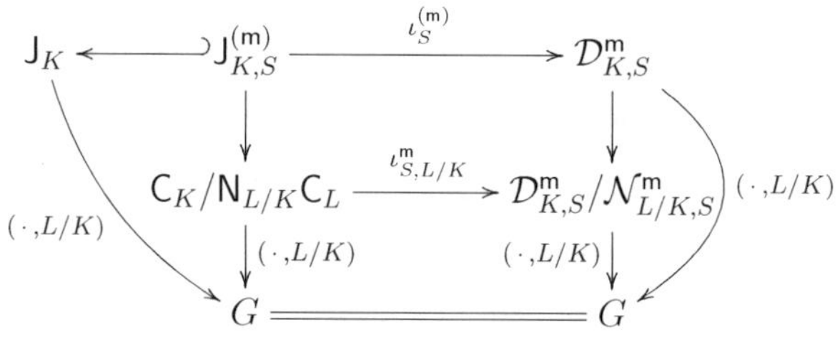

in which all arrows of the bottom square are isomorphisms. On the left-hand side, $(\,\cdot\,, L/K)$ *denotes the norm residue symbol, and on the right-hand side,* $(\,\cdot\,, L/K)$ *denotes the Artin map resp. the Artin isomorphism. The isomorphism* $\iota_{S,L/K}^{\mathsf{m}}$ *is given by* 5.7.3.2 *and fits into the following commutative diagram with (natural) exact rows in which the vertical arrows are the isomorphisms given in* 5.7.3.2.

$$\begin{array}{ccccccccc}
1 & \longrightarrow & \mathsf{N}_{L/K}\mathsf{C}_L/\mathsf{C}_{K,S}^{\mathsf{m}} & \longrightarrow & \mathsf{C}_K/\mathsf{C}_{K,S}^{\mathsf{m}} & \longrightarrow & \mathsf{C}_K/\mathsf{N}_{L/K}\mathsf{C}_L & \longrightarrow & 1 \\
 & & \downarrow{\scriptstyle \iota_S^{\mathsf{m}}\restriction\ldots} & & \downarrow{\scriptstyle \iota_S^{\mathsf{m}}} & & \downarrow{\scriptstyle \iota_{S,L/K}^{\mathsf{m}}} & & \\
1 & \longrightarrow & \mathcal{N}_{L/K,S}^{\mathsf{m}}/(K^{\mathsf{m}})_S & \longrightarrow & \mathcal{D}_{K,S}^{\mathsf{m}}/(K^{\mathsf{m}})_S & \longrightarrow & \mathcal{D}_{K,S}^{\mathsf{m}}/\mathcal{N}_{L/K,S}^{\mathsf{m}} & \longrightarrow & 1. \\
 & & & & \| & & & & \\
 & & & & \mathcal{C}_{K,S}^{\mathsf{m}} & & & &
\end{array}$$

2. Let $\mathcal{G}$ *be a subgroup of* $\mathcal{D}_{K,S}^{\mathsf{m}}$ *of finite index and* $(K^{\mathsf{m}})_S \subset \mathcal{G}$*. Then there exists a unique finite abelian extension* L/K *such that* $\mathcal{G} = \mathcal{N}_{L/K,S}^{\mathsf{m}}$*. If* $v \in \mathsf{M}_K^0 \setminus \operatorname{supp}(\mathsf{m})$*, then* $v \in \operatorname{Split}(L/K)$ *if and only if either* $v \in S$ *or* $\mathsf{p}_v \in \mathcal{G}$*.*

We call L the **class field** to $\mathcal{G}$.

Proof. For each $v \in \mathsf{M}_K$ we fix some $\overline{v} \in \mathsf{M}_L$ above v.

1. Let $\alpha = (\alpha_v)_{v \in \mathsf{M}_K} \in \mathsf{J}_{K,S}^{(\mathsf{m})}$ and $\mathsf{a} = \iota_S^{(\mathsf{m})}(\alpha) \in \mathcal{D}_{K,S}^{\mathsf{m}}$. Then 6.5.6.2 implies $\operatorname{Ram}(L/K) \subset \operatorname{supp}(\mathsf{m})$ and thus $\operatorname{supp}(\mathsf{a}) \cap \operatorname{Ram}(L/K) = \emptyset$. If $v \in \operatorname{supp}(\mathsf{a})$, then v is unramified in L, and therefore

$$(\alpha_v, L_{\overline{v}}/K_v) = (v, L/K)^{v(\alpha_v)} = (v, L/K)^{v(\mathsf{a})}$$

by 4.2.3.2. We assert that $(\alpha_v, L_{\overline{v}}/K_v) = 1$ for all $v \in \mathsf{M}_K \setminus \operatorname{supp}(\mathsf{a})$.

- If $v \in \mathsf{M}_K^\infty$, then $\alpha_v \in U_v \subset \mathsf{N}_{L_{\overline{v}}/K_v} L_{\overline{v}}^\times$ and therefore $(\alpha_v, L_{\overline{v}}/K_v) = 1$.
- If $v \in \mathsf{M}_K^0 \setminus (\operatorname{supp}(\mathsf{a}) \cup \operatorname{supp}(\mathsf{m}))$, then v is unramified in L and $v(\alpha_v) = 0$, hence $(\alpha_v, L_{\overline{v}}/K_v) = (v, L/K)^{v(\alpha_v)} = 1$.
- If $v \in \operatorname{supp}(\mathsf{m})$, then it follows that $\alpha_v \in U_v^{(v(\mathsf{m}))}$, and $\mathsf{f}_{L/K} \,|\, \mathsf{m}$ implies $v(\mathsf{m}) \ge v(\mathsf{f}_{L/K}) = v(\mathfrak{f}_{L_{\overline{v}}/K_v})$ by 6.5.6.2, hence $U_v^{(v(\mathsf{m}))} \subset \mathsf{N}_{L_{\overline{v}}/K_v} L_{\overline{v}}^\times$, and therefore $(\alpha_v, L_{\overline{v}}/K_v) = 1$.

All in all, it follows that

$$(\alpha, L/K) = \prod_{v \in \mathsf{M}_K} (\alpha_v, L_{\overline{v}}/K_v) = \prod_{v \in \operatorname{supp}(\mathsf{a})} (v, L/K)^{v(\mathsf{a})} = (\mathsf{a}, L/K).$$

By 5.7.3.2, the isomorphism $\iota_{S,L/K}^{\mathsf{m}} \colon \mathsf{C}_K/\mathsf{N}_{L/K}\mathsf{C}_L \overset{\sim}{\to} \mathcal{D}_{K,S}^{\mathsf{m}}/\mathcal{N}_{L/K,S}^{\mathsf{m}}$ is induced by the epimorphism $\iota_S^{(\mathsf{m})} \colon \mathsf{J}_{K,S}^{(\mathsf{m})} \to \mathcal{D}_{K,S}^{\mathsf{m}}$ considering the equality

$$\mathsf{C}_K/\mathsf{N}_{L/K}\mathsf{C}_L = \mathsf{J}_{K,S}^{(\mathsf{m})}/\mathsf{N}_{L/K}(\mathsf{J}_L)K^\times \cap \mathsf{J}_{K,S}^{(\mathsf{m})}.$$

Hence the upper square of the first diagram commutes by the very definition. As to the bottom square, we consider the isomorphism

$$\phi = (\,\cdot\,, L/K)\circ(\iota^{\mathsf{m}}_{S,L/K})^{-1} \colon \mathcal{D}^{\mathsf{m}}_{K,S}/\mathcal{N}^{\mathsf{m}}_{L/K,S} \to G$$

and prove that $\phi(\mathsf{a}\mathcal{N}^{\mathsf{m}}_{L/K,S}) = (\mathsf{a}, L/K)$ for all $\mathsf{a} \in \mathcal{D}^{\mathsf{m}}_{K,S}$. If $\mathsf{a} \in \mathcal{D}^{\mathsf{m}}_{K,S}$ and $\mathsf{a} = \iota^{(\mathsf{m})}_S(\alpha)$ for some $\alpha \in \mathsf{J}^{(\mathsf{m})}_{K,S}$, then $\mathsf{a}\mathcal{N}^{\mathsf{m}}_{L/K,S} = \iota^{\mathsf{m}}_{S,L/K}((\alpha K^{\mathsf{m}})\mathsf{N}_{L/K}\mathsf{C}_L)$ by 5.7.3.2, and

$$\phi\circ\iota^{\mathsf{m}}_{S,L/K}((\alpha K^{\mathsf{m}})\mathsf{N}_{L/K}\mathsf{C}_L) = ((\alpha K^{\mathsf{m}})\mathsf{N}_{L/K}\mathsf{C}_L), L/K) = (\alpha, L/K) = (\mathsf{a}, L/K).$$

The second commutative diagram follows immediately from 5.7.2.2 and 5.7.3.2.

2. Let $\iota^{\mathsf{m}}_S \colon \mathsf{C}_K/\mathsf{C}^{\mathsf{m}}_{K,S} \overset{\sim}{\to} \mathcal{C}^{\mathsf{m}}_{K,S}$ be the isomorphism given in 5.7.2.2, and observe that $\mathcal{G}/(K^{\mathsf{m}})_S \subset \mathcal{D}^{\mathsf{m}}_{K,S}/(K^{\mathsf{m}})_S = \mathcal{C}^{\mathsf{m}}_{K,S}$. Hence there exists a subgroup N of C_K such that $\mathsf{C}^{\mathsf{m}}_{K,S} \subset N$ and $\iota^{\mathsf{m}}_S(N/\mathsf{C}^{\mathsf{m}}_{K,S}) = \mathcal{G}/(K^{\mathsf{m}})_S$, N is a norm group by 6.5.3, and thus there exists a finite abelian extension L/K such that $N = \mathsf{N}_{L/K}(\mathsf{C}_L)$. By 1., it follows that

$$\mathcal{G}/(K^{\mathsf{m}})_S = \iota^{\mathsf{m}}_S(\mathsf{N}_{L/K}\mathsf{C}_L/\mathsf{C}^{\mathsf{m}}_{K,S}) = \mathcal{N}^{\mathsf{m}}_{L/K,S}/(K^{\mathsf{m}})_S,$$

and consequently $\mathcal{G} = \mathcal{N}^{\mathsf{m}}_{L/K,S}$.

Let $v \in \mathsf{M}^0_K \setminus \operatorname{supp}(\mathsf{m})$. Then v is unramified in L. If $v \in S$, then v is fully decomposed in L by 6.5.4.1. If $v \notin S$, then $\mathsf{p}_v \in \mathcal{D}^{\mathsf{m}}_{K,S}$, and $v \in \operatorname{Split}(L/K)$ if and only if $(v, L/K) = 1$. By 1., it follows that $(v, L/K) = (\mathsf{p}_v, L/K) = 1$ if and only if $\mathsf{p}_v \in \mathcal{N}^{\mathsf{m}}_{L/K,S} = \mathcal{G}$. □

In the following theorem we consider a finite abelian extension and associate the characters of its Galois group with the ray class characters of its norm residue group as investigated in Section 5.8.

Theorem 6.5.8. *Let L/K be a finite abelian extension, $G = \operatorname{Gal}(L/K)$, and let $\mathsf{m} \in \mathcal{D}'_K$ be such that $\mathsf{f}_{L/K} \,|\, \mathsf{m}$. Let $\chi \in \mathsf{X}(G)$ be a character of G and $K_\chi = L^{\operatorname{Ker}(\chi)}$ the fixed field of $\operatorname{Ker}(\chi)$.*

*1. $\widetilde{\chi} = \chi\circ(\,\cdot\,, L/K) \colon \mathcal{D}^{\mathsf{m}}_K \to \mathbb{T}$ is a ray class character modulo m of finite order, $\mathsf{f}(\widetilde{\chi}) = \mathsf{f}_{K_\chi/K}$, and K_χ is the class field to $\operatorname{Ker}(\widetilde{\chi})$. If $\widehat{\chi} = \widetilde{\chi}^* \colon \mathsf{J}_K \to \mathbb{T}$ denotes the idele class character associated with $\widetilde{\chi}$ (see 5.8.5.**5**), then there is a commutative diagram*

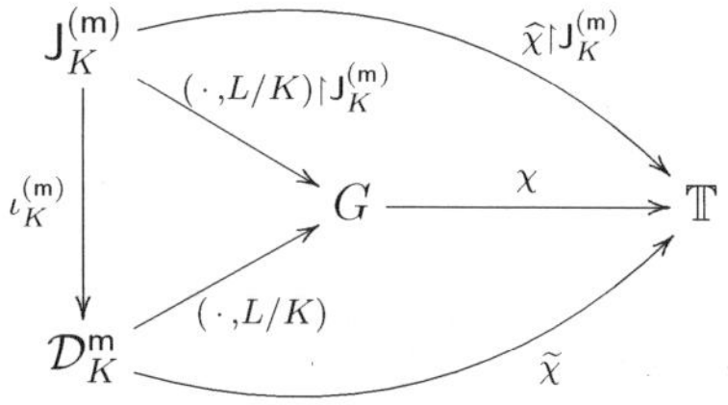

where the upper $(\,\cdot\,, L/K)$ *denotes the global norm residue symbol and the lower* $(\,\cdot\,, L/K)$ *denotes the Artin symbol.*

2. Let $K/\mathbb{F}_q$ *be a function field. Then* K_χ/K *is a constant field extension if and only if* $\widetilde{\chi}\restriction\mathcal{D}_K^0 = 1$.

Proof. 1. By definition, $\widetilde{\chi}$ is a homomorphism of finite order. As $\mathfrak{f}_{L/K} \mid \mathsf{m}$, it follows that $((K^{\mathsf{m}}), L/K) = \mathbf{1}$, and therefore $\widetilde{\chi}\restriction(K^{\mathsf{m}}) = 1$. Hence $\widetilde{\chi}$ is a ray class character modulo m of finite order, and $\mathsf{f}(\widetilde{\chi}) \mid \mathsf{m}$. For $\alpha \in \mathsf{J}_K$ we obtain

$$\alpha \in \mathrm{Ker}(\widetilde{\chi}) \iff (\alpha, L/K) \in \mathrm{Ker}(\chi) \iff (\alpha, K_\chi/K) = (\alpha, L/K)\restriction K_\chi = 1.$$

Hence K_χ is a class field to $\mathrm{Ker}(\widetilde{\chi})$.

Next we consider the diagram. Let $\widehat{\chi} = \widetilde{\chi}^*$ be the idele class character associated with $\widetilde{\chi}$. If $\alpha \in \mathsf{J}_K^{(\mathsf{m})}$, then $\widetilde{\chi}(\iota_K^{(\mathsf{m})}(\alpha)) = \widehat{\chi}(\alpha)$ by 5.8.5.**5** (observe that $\widetilde{\chi}_{\infty,v}(\alpha_v) = 1$ for all $v \in \mathsf{M}_K^\infty$)), and 6.5.7.1 implies

$$(\iota_K^{(\mathsf{m})}(\alpha), L/K) = (\alpha K^\times, L/K) = (\alpha, L/K).$$

Hence the diagram commutes.

It remains to prove that $\mathfrak{f}_{K_\chi/K} = \mathsf{f}(\widetilde{\chi})$. We first assume that χ is injective (hence $K_\chi = L$), and we consider $\widehat{\chi}\colon \mathsf{J}_K \to \mathbb{T}$. As $\mathsf{f}(\widetilde{\chi}) = \mathsf{f}(\widehat{\chi})$ by 5.8.6.2, it suffices to prove that $\mathsf{f}(\widehat{\chi}) = \mathfrak{f}_{L/K}$, and we show that $\mathsf{f}(\widehat{\chi}) \mid \mathsf{n}$ if and only if $\mathfrak{f}_{L/K} \mid \mathsf{n}$ for all $\mathsf{n} \in \mathcal{D}'_K$. Thus let $\mathsf{n} \in \mathcal{D}'_K$, apply 5.8.5.2, 4.8.7 and 6.5.6, and recall that

$$\mathsf{U}_K^{\mathsf{n}} = \prod_{v \in \mathsf{M}_K^\infty} 1 \times \prod_{v \in \mathsf{M}_K^0} U_v^{(v(\mathsf{n}))}.$$

Then we obtain

$$\begin{aligned}
\mathsf{f}(\widehat{\chi}) \mid \mathsf{n} &\iff \mathsf{U}_K^{\mathsf{n}} \subset \mathrm{Ker}(\widehat{\chi}) = \mathrm{Ker}(\,\cdot\,, L/K) \\
&\iff U_v^{(v(\mathsf{n}))} \subset \mathrm{Ker}(\,\cdot\,, L_{\overline{v}}/K_v) \ \text{ for all } v \in \mathsf{M}_K^0 \text{ and } \overline{v} \in \mathsf{M}_L \text{ above } v \\
&\iff v(\mathsf{n}) \ge v(\mathfrak{f}_{L_{\overline{v}}/K_v}) = v(\mathfrak{f}_{L/K}) \ \text{ for all } v \in \mathsf{M}_K^0 \text{ and } \overline{v} \in \mathsf{M}_L \\
&\qquad \text{above } v \iff \mathfrak{f}_{L/K} \mid \mathsf{n}\,.
\end{aligned}$$

Now let χ be arbitrary, $G' = G/\mathrm{Ker}(\chi) = \mathrm{Gal}(K_\chi/K)$, and let $\chi'\colon G' \to \mathbb{T}$ be defined by $\chi'(\sigma\mathrm{Ker}(\chi)) = \chi(\sigma \restriction K_\chi) = \chi(\sigma)$. Then χ' is injective, hence $\mathsf{f}(\widetilde{\chi}') = \mathfrak{f}_{K_\chi/K}$, and since

$$\widetilde{\chi}' = \chi' \circ (\,\cdot\,, K_\chi/K) = \chi' \circ [(\,\cdot\,, L/K)\restriction K_\chi] = \chi \circ (\,\cdot\,, L/K) = \widetilde{\chi},$$

it follows that $\mathsf{f}(\widetilde{\chi}) = \mathsf{f}(\widetilde{\chi}') = \mathfrak{f}_{K_\chi/K}$.

2. Let $G_\chi = G/\mathrm{Ker}(\chi) = \mathrm{Gal}(K_\chi/K)$. By 6.5.2.3, the extension K_χ/K is a constant field extension if and only if $(\mathsf{C}_K^0, K_\chi/K) = \mathbf{1}$. If this is the case, then K_χ/K is unramified and $\mathfrak{f}_{K_\chi/K} = \mathsf{f}(\widetilde{\chi}) = 1$. On the other hand, if $\widetilde{\chi}\restriction\mathcal{D}_K^0 = 1$, then $\widetilde{\chi}\restriction(K^\times) = 1$, and therefore again $\mathfrak{f}_{K_\chi/K} = \mathsf{f}(\widetilde{\chi}) = 1$.

Thus suppose that $\mathsf{f}_{K_\chi/K} = 1$. A slight modification of 6.5.7.1 (for $S = \emptyset$ and $\mathsf{m} = 1$) yields the commutative diagram

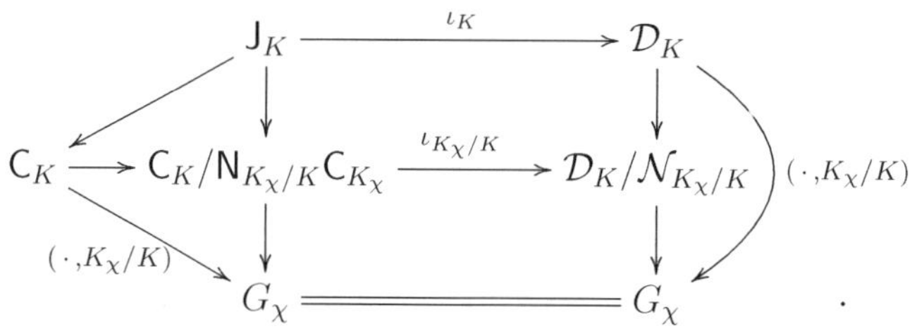

Let $c = \gamma K^\times \in \mathsf{C}_K$, where $\gamma \in \mathsf{J}_K$, and $\mathsf{c} = \iota_K(\gamma) \in \mathcal{D}_K$. Then we have $\deg(c) = \deg(\mathsf{c})$, and the above commutative diagram shows that $(c, K_\chi/K) = (\mathsf{c}, K_\chi/K)$. Since ι_K is an epimorphism, it follows that $(\mathsf{C}_K^0, K_\chi/K) = (\mathcal{D}_K^0, K_\chi/K)$, and consequently $(\mathsf{C}_K^0, K_\chi/K) = \mathbf{1}$ if and only if $\widetilde{\chi} \restriction \mathcal{D}_K^0 = 1$. □

6.6 Special class fields and decomposition laws

Definition and Remarks 6.6.1. Let S be a finite subset of M_K^0 such that $S \neq \emptyset$ in the function field case, and $\mathsf{m} \in \mathcal{D}'_{K,S}$. Then $\mathsf{C}_{K,S}^{\mathsf{m}}$ is an open subgroup of finite index of C_K, hence a norm group, and there exists a unique class field $K[\mathsf{m}, S]$ to $\mathsf{C}_{K,S}^{\mathsf{m}}$, which is called the S**-ray class field modulo** m.

Let S and S' be finite subsets of M_K^0 such that $S \subset S'$ and $S \neq \emptyset$ in the function field case, and let $\mathsf{m}, \mathsf{m}' \in \mathcal{D}'_K(S')$ be such that $\mathsf{m}' \mid \mathsf{m}$. Then $\mathsf{C}_{K,S}^{\mathsf{m}} \subset \mathsf{C}_{K,S'}^{\mathsf{m}'}$ by 5.7.2.3, hence $K[\mathsf{m}', S'] \subset K[\mathsf{m}, S]$ by 6.4.8.2(b), and clearly $\mathsf{f}_{K[\mathsf{m}',S']/K} \mid \mathsf{f}_{K[\mathsf{m},S]/K} \mid \mathsf{m}$. In particular, $K[1, S] \subset K[\mathsf{m}, S]$ for all $\mathsf{m} \in \mathcal{D}'_K$. The field $K[1, S] = \mathsf{H}_{K,S}$ is called the **large Hilbert** S**-class field**.

Now let K be a number field and $\mathsf{m} = \mathfrak{m} \in \mathcal{J}'_K$ an integral ideal. For $v \in \mathsf{M}_K$ let $j_v \colon K_v^\times \to \mathsf{J}_K$ and $j_v^* \colon K_v^\times \to \mathsf{C}_K$ be the local embeddings as defined if 6.5.4. By 5.7.1 we obtain

$$\mathsf{J}_{K,S}^{\circ\mathsf{m}} = \mathsf{J}_{K,S}^{\mathsf{m}} \Big(\prod_{v \in \mathsf{M}_K^\infty} K_v^\times \times \prod_{v \in \mathsf{M}_K^0} 1 \Big) = \mathsf{J}_{K,S}^{\mathsf{m}} \prod_{\substack{v \in \mathsf{M}_K^\infty \\ v \text{ real}}} j_v(K_v^\times) \supset \mathsf{J}_{K,S}^{\mathsf{m}},$$

and consequently

$$\mathsf{C}_{K,S}^{\circ\mathsf{m}} = \mathsf{J}_{K,S}^{\circ\mathsf{m}} K^\times / K^\times = \mathsf{C}_{K,S}^{\mathsf{m}} \prod_{\substack{v \in \mathsf{M}_K^\infty \\ v \text{ real}}} j_v^*(K_v^\times) \supset \mathsf{C}_{K,S}^{\mathsf{m}}.$$

Hence $\mathsf{C}_{K,S}^{\circ\mathfrak{m}}$ is a norm group in C_K. Its class field $K[\mathfrak{m},S]^\circ$ is called the **small S-ray class field modulo $\mathfrak{m}$**. The field $K[1,S]^\circ = \mathsf{H}_{K,S}^\circ$ is called the **small Hilbert S-class field** of K. We call

- $K[\mathfrak{m}] = K[\mathfrak{m},\emptyset]$ the **large ray class field modulo** $\mathfrak{m}$,
- $K[\mathfrak{m}]^\circ = K[\mathfrak{m},\emptyset]^\circ$ the **small ray class field modulo** $\mathfrak{m}$,
- $\mathsf{H}_K = \mathsf{H}_{K,\emptyset} = K[1]$ the **large Hilbert class field**, and
- $\mathsf{H}_K^\circ = \mathsf{H}_{K,\emptyset}^\circ = K[1]^\circ$ the **small Hilbert class field** of K.

By definition $K[\mathfrak{m},S] \subset K[\mathfrak{m}]$ and $\mathfrak{f}_{K[\mathfrak{m},S]/K} \,|\, \mathfrak{f}_{K[\mathfrak{m}]/K} \,|\, \mathfrak{m}$ for all $\mathfrak{m} \in \mathcal{I}_K'$ and all finite subsets S of M_K^0. Equality need not hold here (e. g., $\mathbb{Q}[2] = \mathbb{Q}$, see 6.6.6).

Of course, if K admits no real embeddings (and in particular in the function field case) we get $K[\mathsf{m},S]^\circ = K[\mathsf{m},S]$ and then we omit the prefixes "large" and "small".

For algebraic number fields, our terminology differs from that used in most books.

What we call	is mostly called
large ray class field modulo $\mathfrak{m}$	ray class field modulo $\mathfrak{m}\infty$
small ray class field modulo $\mathfrak{m}$	ray class field modulo $\mathfrak{m}$
large Hilbert class field	ray class field modulo ∞
small Hilbert class field	Hilbert class field

In particular, the small Hilbert class field is just the classical Hilbert class field occuring in most books. Our terminology is influenced by [48]. It is motivated by our opinion that an infinite place is always unramified and that the cyclotomic field $\mathbb{Q}^{(m)}$ is the ray class field modulo m, whereas in the traditional opinion an undecomposed infinite place is viewed as being ramified and the greatest real subfield $(\mathbb{Q}^{(m)})^+$ of the cyclotomic field is the ray class field modulo m. These different points of view are made explicit in Corollary 6.6.5 and Theorem 6.6.6 below.

In the following two tables we collocate the various types of class fields together with their norm groups.

Table 1: Let S be a finite non-empty subset of M_K^0 such that $S \neq \emptyset$ in the function field case, and $\mathsf{m} \in \mathcal{D}_{K,S}'$. Let L/K be a finite abelian extension, $\overline{\mathsf{m}} = j_{L/K}(\mathsf{m})$, $G = \mathrm{Gal}(L/K)$,

$$\mathcal{N}_{L/K,S}^{\mathsf{m}} = \mathcal{N}_{L/K}(\mathcal{D}_{L,S_L}^{\overline{\mathsf{m}}})(K^{\mathsf{m}})_S \subset \mathcal{D}_{K,S}^{\mathsf{m}}$$

and

$$\mathcal{C}_{L/K,S}^{\mathsf{m}} = \mathcal{D}_{K,S}^{\mathsf{m}}/\mathcal{N}_{L/K,S}^{\mathsf{m}} = \iota_{S,L/K}^{\mathsf{m}}(\mathsf{C}_K/\mathsf{N}_{L/K}\mathsf{C}_L)$$

(see 5.7.3.2). Having this in mind and considering 5.7.2, we arrive at the tables below.

Class field	L	$\mathsf{N}_{L/K}\mathsf{C}_L$	$\mathcal{N}^{\mathsf{m}}_{L/K,S}$	$\mathcal{C}^{\mathsf{m}}_{L/K,S}$
S-ray class field modulo m	$K[\mathsf{m},S]$	$\mathsf{C}^{\mathsf{m}}_{K,S}$	$(K^{\mathsf{m}})_S$	$\mathcal{C}^{\mathsf{m}}_{K,S}$
Small S-ray class field modulo m	$K[\mathsf{m},S]^\circ$	$\mathsf{C}^{\circ\mathsf{m}}_{K,S}$	$(K^{\circ\mathsf{m}})_S$	$\mathcal{C}^{\circ\mathsf{m}}_{K,S}$
Large Hilbert S-class field	$K[1,S]=\mathsf{H}_{K,S}$	$\mathsf{C}^{1}_{K,S}$	$(K^{+})_S$	$\mathcal{C}^{+}_{K,S}$
Small Hilbert S-class field	$K[1,S]^\circ=\mathsf{H}^\circ_{K,S}$	$\mathsf{C}^{\circ 1}_{K,S}$	$(K^{\times})_S$	$\mathcal{C}_{K,S}$

Table 2: We specialize the entries of Table 1 for a number field K with $\mathsf{m}=\mathfrak{m}\in\mathcal{J}'_K$ and $S=\emptyset$. In this case, $\mathcal{N}^{\mathfrak{m}}_{L/K}=\mathcal{N}_{L/K}(\mathcal{J}^{\mathfrak{m}\mathfrak{o}_L}_L)\mathcal{H}^{\mathfrak{m}}_K\subset\mathcal{J}^{\mathfrak{m}}_K$ and $\mathcal{C}^{\mathfrak{m}}_{L/K}=\mathcal{J}^{\mathfrak{m}}_K/\mathcal{N}^{\mathfrak{m}}_{L/K}$.

Class field	L	$\mathsf{N}_{L/K}\mathsf{C}_L$	$\mathcal{N}^{\mathfrak{m}}_{L/K}$	$\mathcal{C}^{\mathfrak{m}}_{L/K}$
Ray class field modulo $\mathfrak{m}$	$K[\mathfrak{m}]$	$\mathsf{C}^{\mathfrak{m}}_K$	$\mathcal{H}^{\mathfrak{m}}_K$	$\mathcal{C}^{\mathfrak{m}}_K$
Small ray class field modulo $\mathfrak{m}$	$K[\mathfrak{m}]^\circ$	$\mathsf{C}^{\circ\mathfrak{m}}_K$	$\mathcal{H}^{\circ\mathfrak{m}}_K$	$\mathcal{C}^{\circ\mathfrak{m}}_K$
Large Hilbert class field	$K[1]=\mathsf{H}_K$	C^{1}_K	$\mathcal{H}^{+}_K$	$\mathcal{C}^{+}_K$
Small Hilbert class field	$K[1]^\circ=\mathsf{H}^\circ_K$	$\mathsf{C}^{\circ 1}_K=\mathcal{E}_K$	$\mathcal{H}_K$	$\mathcal{C}_K$

Recall that $[\mathsf{H}_K:K]=|\mathsf{C}^+_K|=h^+_K$ is the narrow class number and that $[\mathsf{H}^\circ_K:K]=|\mathsf{C}_K|=h_K$ is the class number of K.

We proceed with an ideal-theoretic reformulation of 6.5.7 using holomorphy domains.

Theorem and Definition 6.6.2 (Artin reciprocity law for ideals). *Let S be a finite subset of M^0_K such that $S\neq\emptyset$ in the function field case. Let $\mathsf{m}\in\mathcal{D}'_{K,S}$, $K[\mathsf{m},S]$ the S-ray class field modulo m, and $\mathfrak{m}=\theta_S(\mathsf{m})\in\mathcal{J}'(\mathfrak{o}_{K,S})$.*

1. *Let L be an intermediate field of $K[\mathsf{m},S]/K$ with Galois group $G=\mathrm{Gal}(L/K)$.*
If $\mathfrak{a}\in\mathcal{J}^{\mathfrak{m}}(\mathfrak{o}_{K,S})$, then $\mathrm{supp}(\mathfrak{a})\cap\mathrm{Ram}(L/K)=\emptyset$ and we definie the **Artin symbol** *$(\mathfrak{a},L/K)$ by*

$$(\mathfrak{a},L/K)=\prod_{v\in\mathrm{supp}(\mathfrak{a})}(v,L/K)^{v(\mathfrak{a})}\in G,$$

where $(v,L/K)$ denotes the Frobenius automorphism of $L_{\overline{v}}/K_v$ (see 5.4.5.**3**).

The Artin symbol defines an epimorphism $(\,\cdot\,,L/K)\colon\mathcal{J}^{\mathfrak{m}}(\mathfrak{o}_{K,S})\to G$ with kernel

$$\overline{\mathcal{N}}^{\mathfrak{m}}_{L/K,S}=\mathcal{N}_{L/K,S}(\mathcal{J}^{\overline{\mathfrak{m}}}(\mathfrak{o}_{L,S_L}))\mathcal{H}^{\mathfrak{m}}(\mathfrak{o}_{K,S}),$$

where $\overline{\mathfrak{m}}=\mathfrak{m}\mathfrak{o}_{L,S_L}\in\mathcal{J}'(\mathfrak{o}_{L,S_L})$ and $\mathcal{N}_{L/K,S}=\mathcal{N}_{\mathfrak{o}_{L,S_L}/\mathfrak{o}_{K,S}}$ (see 5.5.8 and 5.7.3).

If $\alpha \in \mathsf{J}_{K,S}^{(\mathfrak{m})}$*, then* $(\alpha, L/K) = (\iota_S^{(\mathfrak{m})}(\alpha), L/K)$ *(see* 5.7.1.**2**)*, and there is a commutative diagram*

$$\begin{array}{ccc} \mathsf{C}_K/\mathsf{N}_{L/K}\mathsf{C}_L & \xrightarrow{\ \iota^{\mathfrak{m}}_{S,L/K}\ } & \mathcal{J}^{\mathfrak{m}}(\mathfrak{o}_{K,S})/\overline{\mathcal{N}}^{\mathfrak{m}}_{L/K,S} \\ & {\scriptstyle (\cdot,L/K)}\searrow \quad \swarrow{\scriptstyle (\cdot,L/K)} & \\ & G & \end{array}$$

in which all arrows are isomorphisms. The left down arrow is the global norm residue symbol, the right down arrow is the **Artin isomorphism** *induced by the Artin symbol, and* $\iota^{\mathfrak{m}}_{S,L/K}$ *is the isomorphism given in* 5.7.3.2. *Explicitly, if* $\alpha \in \mathsf{J}_{K,S}^{(\mathfrak{m})}$, $a = \alpha K^\times \in \mathsf{C}_K$ *and* $\mathsf{a} = \iota_S^{(\mathfrak{m})}(\alpha) \in \mathcal{D}_K$*, then*

$$\iota^{\mathfrak{m}}_{S,L/K}(a\mathsf{N}_{L/K}\mathsf{C}_L) = \mathsf{a}\overline{\mathcal{N}}^{\mathfrak{m}}_{L/K,S} \quad \text{and} \quad (a, L/K) = (\mathsf{a}, L/K).$$

The isomorphism $\iota^{\mathfrak{m}}_{S,L/K}$ *fits into the following commutative diagram with (natural) exact columns in which the horizontal arrows are the isomorphisms given in* 5.7.3.2.

$$\begin{array}{ccccc} \mathbf{1} & & \mathbf{1} & & \\ \downarrow & & \downarrow & & \\ \mathsf{N}_{L/K}\mathsf{C}_L/\mathsf{C}^{\mathfrak{m}}_{K,S} & \xrightarrow{\ \iota^{\mathfrak{m}}_S\restriction\ldots\ } & \overline{\mathcal{N}}^{\mathfrak{m}}_{L/K,S}/\mathcal{H}^{\mathfrak{m}}(\mathfrak{o}_{K,S}) & & \\ \downarrow & & \downarrow & & \\ \mathsf{C}_K/\mathsf{C}^{\mathfrak{m}}_{K,S} & \xrightarrow{\ \iota^{\mathfrak{m}}_S\ } & \mathcal{J}^{\mathfrak{m}}(\mathfrak{o}_{K,S})/\mathcal{H}^{\mathfrak{m}}(\mathfrak{o}_{K,S}) & = & \mathcal{C}^{\mathfrak{m}}(\mathfrak{o}_{K,S}) \\ \downarrow & & \downarrow & & \\ \mathsf{C}_K/\mathsf{N}_{L/K}\mathsf{C}_L & \xrightarrow{\ \iota^{\mathfrak{m}}_{S,L/K}\ } & \mathcal{J}^{\mathfrak{m}}(\mathfrak{o}_{K,S})/\overline{\mathcal{N}}^{\mathfrak{m}}_{L/K,S} & & \\ \downarrow & & \downarrow & & \\ \mathbf{1} & & \mathbf{1} & & \end{array}$$

2. Let $\mathcal{G}$ *be a subgroup of* $\mathcal{J}^{\mathfrak{m}}(\mathfrak{o}_{K,S})$ *satisfying* $\mathcal{H}^{\mathfrak{m}}(\mathfrak{o}_{K,S}) \subset \mathcal{G}$*. Then there exists a unique intermediate field* L *of* $K[\mathfrak{m}, S]/K$ *such that* $\mathcal{G} = \overline{\mathcal{N}}^{\mathfrak{m}}_{L/K,S}$.
If $v \in \mathsf{M}_K^0 \setminus \mathrm{supp}(\mathfrak{m})$*, then* $v \in \mathrm{Split}(L/K)$ *if and only if either* $v \in S$ *or* $\mathfrak{p}_{S,v} \in \mathcal{G}$.
We call L the **class field** to $\mathcal{G}$.

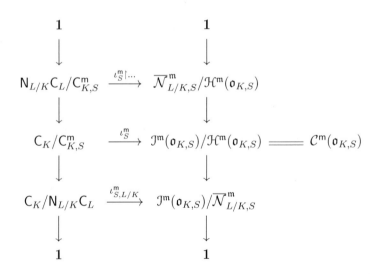

Proof. This is merely a reformulation of 6.5.7 by means of 5.6.2 conidering the terminology of 6.6.1 and the finiteness of $\mathcal{C}_{K,S}^{\mathsf{m}}$ (see 5.6.6.2). □

Theorem 6.6.3. *Let L/K be a finite abelian extension, S a finite subset of M_K^0 such that $S \neq \emptyset$ in the function field case, and $\mathsf{m} \in \mathcal{D}'_{K,S}$.*

1. *$L \subset K[\mathsf{m}, S]$ if and only if $S \subset \mathrm{Split}(L/K)$ and $\mathfrak{f}_{L/K} \mid \mathsf{m}$.*

2. *L is contained in $K[\mathsf{m}, S]^\circ$ if and only if $L \subset K[\mathsf{m}, S]$ and all real places of K are fully decomposed in L. In particular, $K[\mathsf{m}, S]^\circ$ is the largest subfield of $K[\mathsf{m}, S]$ in which all real places of K are fully decomposed.*

Proof. 1. By 6.4.8.2(c) we obtain $L \subset K[\mathsf{m}, S]$ if and only if $\mathsf{C}_{K,S}^{\mathsf{m}} \subset \mathsf{N}_{L/K}\mathsf{C}_L$, and by 6.5.6.1 this holds we have $S \subset \mathrm{Split}(L/K)$ and $j_v^*(U_v^{(v(\mathsf{m}))}) \subset \mathsf{N}_{L/K}\mathsf{C}_L$ for all $v \in \mathsf{M}_K^0$.

If $v \in \mathsf{M}_K^0$, then (by 6.5.4.1 and 6.5.6.2) we it follows that $j_v^*(U_v^{(v(\mathsf{m}))}) \subset \mathsf{N}_{L/K}\mathsf{C}_L$ $U_v^{(v(\mathsf{m}))} \subset \mathsf{N}_{L_{\overline{v}}/K_v}L_{\overline{v}}^\times$, that is, if and only if $v(\mathsf{m}) \geq v(\mathfrak{f}_{L_{\overline{v}}/K_v}) = v(\mathfrak{f}_{L/K})$. However, this holds for all $v \in \mathsf{M}_K^0$ if and only if $\mathfrak{f}_{L/K} \mid \mathsf{m}$.

2. By 6.5.5, it follows that $K[\mathsf{m}, S]^\circ$ is the largest subfield of $K[\mathsf{m}, S]$ in which all real places of K are fully decomposed. Consequently, if L is a subfield of $K[\mathsf{m}, S]$, then $L \subset K[\mathsf{m}, S]^\circ$ if and only if all real places of K are fully decomposed in L. □

Corollary 6.6.4. *Let S be a finite subset of M_K^0 such that $S \neq \emptyset$ in the function field case.*

1. *The large Hilbert S-class field $\mathsf{H}_{K,S}$ is the largest over K unramified finite abelian extension field L satisfying $S \subset \mathrm{Split}(L/K)$.*

2. *The Hilbert S-class field $\mathsf{H}_{K,S}^\circ$ is the largest over K unramified finite abelian extension field L satisfying $S \cup \mathsf{M}_K^\infty \subset \mathrm{Split}(L/K)$.*

Proof. Since $\mathsf{H}_{K,S} = K[1, S]$ and $\mathsf{H}_{K,S}^\circ = K[1, S]^\circ$, the assertions follow by 6.6.3 (observe that $\mathfrak{f}_{\mathsf{H}_{K,S}/K} = 1$ implies $\mathrm{Ram}(\mathsf{H}_{K,S}/K) = \emptyset$ by 6.5.6.2). □

In the following corollary we specialize the assertions of 6.6.3 and 6.5.7 to the number field case if $S = \emptyset$.

Corollary 6.6.5. *Let K be a number field and $\mathfrak{m} \in \mathcal{I}'_K$.*

1. *Let L/K be a finite abelian extension.*

(a) $L \subset K[\mathfrak{m}]$ if and only if $\mathfrak{f}_{L/K} \mid \mathfrak{m}$.

(b) For $v \in \mathsf{M}_K^0$ the field $K[\mathfrak{m}, \{v\}]$ is the decomposition field of v in $K[\mathfrak{m}]$, and if S is a finite non-empty subset of M_K, then

$$K[\mathfrak{m}, S] = \bigcap_{v \in S} K[\mathfrak{m}, \{v\}].$$

In particular, every finite abelian extension field L of K is contained in a ray class field.

2. The large Hilbert class field H_K of K is the largest abelian extension of K in which all finite places of K are unramified, and the Hilbert class field H_K° is the largest intermediate field of H_K/K in which all real places of K are fully decomposed.

3. Let L be an intermediate field of $K[\mathfrak{m}]/K$. Then the Artin map $(\,\cdot\,, L/K)\colon \mathcal{J}_K^{\mathfrak{m}} \to G$ is an epimorphism with kernel $\mathcal{N}_{L/K}^{\mathfrak{m}}$. If $\alpha \in \mathsf{J}_K^{(\mathfrak{m})}$ and $\mathfrak{a} = \iota^{(\mathfrak{m})}(\alpha) \in \mathcal{J}_K^{\mathfrak{m}}$, then the global norm residue symbol $(\alpha, L/K)$ coincides with the Artin symbol $(\mathfrak{a}, L/K)$.

4. Let $\mathcal{G}$ be a subgroup of $\mathcal{J}_K^{\mathfrak{m}}$ such that $\mathcal{H}_K^{\mathfrak{m}} \subset \mathcal{G}$. Then there exists a unique intermediate field L of $K[\mathfrak{m}]/K$ such that $\mathcal{G} = \mathcal{N}_{L/K}^{\mathfrak{m}}$.

Proof. 1. (a) holds by 6.6.3.1, and (b) follows by 6.5.5.1, since

$$\mathsf{C}_{K,\{v\}}^{\mathfrak{m}} = \mathsf{C}_K^{\mathfrak{m}}\, j_v^*(K_v^\times) \quad \text{and} \quad \mathsf{C}_{K,S}^{\mathfrak{m}} = \prod_{v \in S} \mathsf{C}_{K,\{v\}}^{\mathfrak{m}}.$$

2. Obvious by 6.6.4.
3. and 4. hold by 6.6.2. □

For $m \in \mathbb{N}$, let $\zeta_m \in \mathbb{C}$ be a primitive m-th root of unity, $\mathbb{Q}^{(m)} = \mathbb{Q}(\zeta_m)$ and $\mathbb{Q}[m]$ the ray class field modulo m of $\mathbb{Q}$. Already in [18, Theorem 3.7.9] it was proved that $\mathbb{Q}[m] = \mathbb{Q}^{(m)}$, at that time using the splitting properties of primes referring to analytic methods. Now we give an algebraic proof using class field theory in the idelic language.

Theorem 6.6.6.

1. If $m \in \mathbb{N}$, then $\mathbb{Q}^{(m)} = \mathbb{Q}[m]$, i. e., the m-th cyclotomic field is the ray class field modulo m. In particular, $\mathbb{Q}[m]^\circ$ is the greatest real subfield of $\mathbb{Q}^{(m)}$. If $m = 2m_0 \equiv 2 \bmod 4$, then $\mathbb{Q}[m] = \mathbb{Q}[m_0]$.

2. (Kronecker - Weber). *Every abelian number field is contained in some $\mathbb{Q}^{(m)}$.*

Proof. 1. Let $m \in \mathbb{N}$. If $m = 2m_0 \equiv 2 \bmod 4$, then apparently $\mathbb{Q}^{(m)} = \mathbb{Q}^{(m_0)}$ and $\mathsf{C}_{\mathbb{Q}}^m = \mathsf{C}_{\mathbb{Q}}^{m_0}$ (since $U_2^{(0)} = U_2 = U_2^{(1)}$). Therefore we may assume that $m \not\equiv 2 \bmod 4$.

By 6.5.6 and 4.9.4 it follows that $v_p(f_{\mathbb{Q}^{(m)}/\mathbb{Q}}) = v_p(\mathfrak{f}_{\mathbb{Q}_p^{(m)}/\mathbb{Q}_p}) = v_p(m)$ for all primes p, hence $f_{\mathbb{Q}^{(m)}/\mathbb{Q}} = m$, and therefore $\mathbb{Q}^{(m)} \subset \mathbb{Q}[m]$ by 6.6.5.1(a). Since $[\mathbb{Q}^{(m)}\!:\!\mathbb{Q}] = \varphi(m) = [\mathbb{Q}[m]\!:\!\mathbb{Q}]$, equality holds.

By 6.6.3.2 $\mathbb{Q}[m]^\circ$ is the largest subfield of $\mathbb{Q}[m] = \mathbb{Q}^{(m)}$ in which the unique Archimedean place of $\mathbb{Q}$ is fully decomposed, and thus it is its greatest real subfield of $\mathbb{Q}^{(m)}$.

2. Obvious by 1. and 6.6.5.1. $\square$

In the function field case, the restriction to $S \neq \emptyset$ in the definition of S-ray class fields was essential since the (ordinary) ray class groups are infinite. In 6.6.7 we will show how to modify the notion in the function field case to obtain a satisfactory substitute for the concept of ray class fields.

Definitions and Remarks 6.6.7. Let $K/\mathbb{F}_q$ be a function field, and let $\mathsf{m} \in \mathcal{D}'_K$. Then $\mathcal{C}_K^{\mathsf{m}} \cong \mathcal{C}_K^{\mathsf{m},0} \times \mathbb{Z}$ is infinite, but $h_{\mathsf{m}} = (\mathsf{C}_K^0 : \mathsf{C}_K^{\mathsf{m}}) = |\mathcal{C}_K^{\mathsf{m},0}| < \infty$ by 5.7.2.2(b). Observe that there exist arbitrarily large cyclic extensions L/K such that $\mathsf{f}_{L/K} \mid \mathsf{m}$. Indeed, if $d \in \mathbb{N}$, then $\mathrm{Gal}(K\mathbb{F}_{q^d}/K) \cong \mathrm{Gal}(\mathbb{F}_{q^d}/\mathbb{F}_q)$ is cyclic, $\mathrm{Ram}(K\mathbb{F}_{q^d}/K) = \emptyset$ (see [18, Theorem 6.6.7]), hence $\mathsf{f}_{K\mathbb{F}_{q^d}/K} = 1 \mid \mathsf{m}$ and $[K\mathbb{F}_{q^d} : K] = d$. We get over this difficulty if we restrict to geometric extensions.

Let $b \in \mathsf{C}_K$ be any idele class of degree 1. Then $B = \mathsf{C}_K^{\mathsf{m}} \cdot \langle b \rangle$ is an open subgroup of C_K, and $\mathsf{C}_K/B = \mathsf{C}_K^0 \cdot \langle b \rangle / \mathsf{C}_K^{\mathsf{m}} \cdot \langle b \rangle) \cong \mathsf{C}_K^0/\mathsf{C}_M^{\mathsf{m}} \cong \mathcal{C}_K^{\mathsf{m},0}$, and $(\mathsf{C}_K : B) = h_{\mathsf{m}}$. Let K_1 be the class field to B. Then K_1/K is a geometric extension, $\mathrm{Gal}(K_1/K) \cong \mathcal{C}_K^{\mathsf{m},0}$, $[K_1\!:\!K] = h_{\mathsf{m}}$, and $\mathsf{f}_{K_1/K} = \mathsf{m}$.

Let $b' \in \mathsf{C}_K$ be another idele class of degree 1, $b' = bc$ for some $c \in \mathsf{C}_K^0$, and $\mathsf{C}_K^{\mathsf{m}} \cdot \langle b \rangle = \mathsf{C}_K^{\mathsf{m}} \cdot \langle b' \rangle$ if and only if $c \in \mathsf{C}_K^{\mathsf{m}}$. Let $\{u_1 = 1, u_2, \ldots, u_{h_{\mathsf{m}}}\} \subset \mathsf{C}_K^0$ be a set of representatives for $\mathsf{C}_K^0/\mathsf{C}_K^{\mathsf{m}}$. For $i \in [1, h_{\mathsf{m}}]$ let $B_i = \mathsf{C}_K^{\mathsf{m}} \cdot \langle bu_i \rangle$ and K_i the class field to B_i. Then (as above) K_i/K is a geometric extension, $\mathrm{Gal}(K_i/K) \cong \mathcal{C}_K^{\mathsf{m},0}$, $[K_i : K] = h_{\mathsf{m}}$, $\mathsf{f}_{K_i/K} = \mathsf{m}$, and the fields $K_1, \ldots, K_{h_{\mathsf{m}}}$ are distinct. We assert that $K_1, \ldots, K_{h_{\mathsf{m}}}$ are all fields with this property. More precisely, we claim:

> **A.** Let K'/K be any geometric extension satisfying $\mathsf{f}_{K'/K} \mid \mathsf{m}$. Then there exists some $i \in [1, h_{\mathsf{m}}]$ such that $K' \subset K_i$.

Proof of **A.** Let $B' = \mathsf{N}_{K'/K} C_{K'}$. Then $\mathsf{C}_K^{\mathsf{m}} \subset B'$, and by 6.5.2.3 there exists some $b' \in B'$ such that $\deg(b') = 1$. Then $b'b^{-1} \in \mathsf{C}_K^0$, and there exists some $i \in [1, h]$ such that $b'b^{-1} \in u_i \mathsf{C}_K^{\mathsf{m}}$. It follows that

$$B' \supset \mathsf{C}_K^{\mathsf{m}} \cdot \langle b' \rangle = \mathsf{C}_K^{\mathsf{m}} \cdot \langle bu_i \rangle = B_i$$

and therefore $K' \subset K_i$. $\square$[**A.**]

Set eventually $h = h_{\mathsf{m}}$, and let $L = K\mathbb{F}_{q^h}$ be the constant field extension of K of degree h. Then L is the class field to $\mathsf{C}_K^0 \cdot \langle b^h \rangle$ by 6.5.2.1. For $i \in [1, h]$ we set $b_i = bu_i$, and we contend that

$$\mathsf{C}_K^0 \cdot \langle b^h \rangle \cap \mathsf{C}_K^{\mathsf{m}} \cdot \langle b_i \rangle = \mathsf{C}_K^{\mathsf{m}} \cdot \langle b_i^h \rangle. \qquad (\#)$$

Proof of (#). Since $b_i^h = b^h u_i^h \in \mathsf{C}_K^0\cdot\langle b^h\rangle$, we get $\mathsf{C}_K^{\mathfrak{m}}\cdot\langle b_i^h\rangle \subset \mathsf{C}_K^0\cdot\langle b^h\rangle \cap \mathsf{C}_K^{\mathfrak{m}}\cdot\langle b_i\rangle$. As to the reverse inclusion, let $a = c_0 b^{hk} = c_1 b_i^l \in \mathsf{C}_K^0\cdot\langle b^h\rangle \cap \mathsf{C}_K^{\mathfrak{m}}\cdot\langle b_i\rangle$, where $c_0 \in \mathsf{C}_K^0$, $c_1 \in \mathsf{C}_K^{\mathfrak{m}}$, and $k, l \in \mathbb{Z}$. Then $\deg(a) = hk = l$, and therefore $a \in \mathsf{C}_K^{\mathfrak{m}}\cdot\langle b_i^h\rangle$. □[(#)].

By (#) it follows that $\mathsf{C}_K^0\cdot\langle b^h\rangle \cap \mathsf{C}_K^{\mathfrak{m}}\cdot\langle b_i^h\rangle = \mathsf{C}_K^0\cdot\langle b^h\rangle \cap \mathsf{C}_K^{\mathfrak{m}}\cdot\langle b_j^h\rangle$ and therefore $L^* = LK_i = LK_j$ for all $i, j \in [1, h]$. Hence $L^* = LK_1 \cdot \ldots \cdot K_h$, $[L^* : K] = h^2$, and L^* is the compositum of all geometric abelian extensions of K whose conductor divides $\mathfrak{m}$. Therefore we may regard it as "the geometric ray class field modulo $\mathfrak{m}$".

We close this section with an up to date version of two of the oldest main theorems of class field theory, the decomposition law and the principal ideal theorem.

Theorem 6.6.8 (Decomposition law of class field theory). *Let L/K be a finite abelian extension of number fields, $G = \mathrm{Gal}(L/K)$ and $\mathfrak{m} \in \mathcal{I}'_K$ such that $L \subset K[\mathfrak{m}]$. Let $\mathcal{G} = \mathcal{N}_{L/K}^{\mathfrak{m}} = \mathcal{N}_{L/K}(\mathcal{I}_L^{\mathfrak{m}\mathfrak{o}_L})\mathcal{H}_K^{\mathfrak{m}}$, $\mathfrak{p} \in \mathcal{P}_K$ and $\mathfrak{P} \in \mathcal{P}_L$ such that $\mathfrak{P} \,|\, \mathfrak{p}$.*

1. *If $\mathfrak{p} \nmid \mathfrak{m}$, then $\mathfrak{p}$ is unramified in L, and $f(\mathfrak{P}/\mathfrak{p})$ is the smallest integer $f \in \mathbb{N}$ satisfying $\mathfrak{p}^f \in \mathcal{G}$. In particular, $\mathfrak{p} \in \mathrm{Split}(L/K)$ if and only if $\mathfrak{p} \in \mathcal{G}$.*

2. *Assume that $\mathfrak{m} = \mathfrak{p}^t\mathfrak{m}_0$, where $t \in \mathbb{N}$, $\mathfrak{m}_0 \in \mathcal{I}'_K$ and $\mathfrak{p} \nmid \mathfrak{m}_0$. Then the inertia field of $\mathfrak{p}$ in L is the class field to $\mathcal{G}_0 = \mathcal{G}(\mathcal{H}_K^{\mathfrak{m}_0} \cap \mathcal{I}_K^{\mathfrak{m}})$. If $e = (\mathcal{G}_0 : \mathcal{G})$ and $f \in \mathbb{N}$ is minimal such that $\mathfrak{p}^f \in \mathcal{G}_0$, then $e = e(\mathfrak{P}/\mathfrak{p})$ and $f = f(\mathfrak{P}/\mathfrak{p})$.*

Proof. 1. Let $\varphi_{\mathfrak{p}} = (\mathfrak{p}, L/K) \in G$ be the Frobenius automorphism of $\mathfrak{p}$. Then $f(\mathfrak{P}/\mathfrak{p})$ is the order of $\varphi_{\mathfrak{p}}$ in G, and as $(\,\cdot\,, L/K) \colon \mathcal{I}_K^{\mathfrak{m}}/\mathcal{G} \to G$ is an isomorphism, it follows that $f = f(\mathfrak{P}/\mathfrak{p})$ is the smallest $f \in \mathbb{N}$ with $\mathfrak{p}^f \in \mathcal{G}$.

2. Let $w \in \mathsf{M}_K^0$ be such that $\mathfrak{p}_w = \mathfrak{p}$. By 6.5.4.3 the inertia field of w (and thus of $\mathfrak{p}$) in L_1 is the class field to $(\mathsf{N}_{L/K}\mathsf{C}_L)j_w^*(U_w)$. Since $\mathsf{C}_K^{\mathfrak{m}} \subset \mathsf{N}_{L/K}\mathsf{C}_L$, it follows that

$$\mathsf{N}_{L_1/K}\mathsf{C}_{L_1} = (\mathsf{N}_{L/K}\mathsf{C}_L)j_w^*(U_w) = (\mathsf{N}_{L/K}\mathsf{C}_L)\mathsf{C}_K^{\mathfrak{m}_0}.$$

Using 5.7.2.3, we obtain

$$\begin{aligned}\mathcal{N}_{L_1/K}^{\mathfrak{m}}/\mathcal{H}_K^{\mathfrak{m}} &= \iota^{\mathfrak{m}}(\mathsf{N}_{L_1/K}\mathsf{C}_{L_1}/\mathsf{C}_K^{\mathfrak{m}}) = \iota^{\mathfrak{m}}(\mathsf{N}_{L/(K}(\mathsf{C}_L)\mathsf{C}_K^{\mathfrak{m}_0}/\mathsf{C}_K^{\mathfrak{m}}) \\ &= \iota^{\mathfrak{m}}(\mathsf{N}_{L/K}\mathsf{C}_L/\mathsf{C}_K^{\mathfrak{m}})\,\iota^{\mathfrak{m}}(\mathsf{C}_K^{\mathfrak{m}_0}/\mathsf{C}_K^{\mathfrak{m}}) = (\mathcal{G}/\mathcal{H}_K^{\mathfrak{m}})(\mathcal{H}_K^{\mathfrak{m}_0} \cap \mathcal{I}_K^{\mathfrak{m}}/\mathcal{H}_K^{\mathfrak{m}}) \\ &= \mathcal{G}(\mathcal{H}_K^{\mathfrak{m}_0} \cap \mathcal{I}_K^{\mathfrak{m}})/\mathcal{H}_K^{\mathfrak{m}},\end{aligned}$$

hence $\mathcal{N}_{L_1/K}^{\mathfrak{m}} = \mathcal{G}(\mathcal{H}_K^{\mathfrak{m}_0} \cap \mathcal{I}_K^{\mathfrak{m}}$, and L_1 is the class field to $\mathcal{G}_0 = \mathcal{G}(\mathcal{H}_K^{\mathfrak{m}_0} \cap \mathcal{I}_K^{\mathfrak{m}})$. By the very definition of the inertia field we obtain $e(\mathfrak{P}/\mathfrak{p}) = [L : L_1] = (\mathcal{G}_0 : \mathcal{G})$, and 1. implies $f = f(\mathfrak{P} \cap L_1/\mathfrak{p}) = f(\mathfrak{P}/\mathfrak{p})$. □

Theorem 6.6.9 (Principal ideal theorem). *Let K be a global field and S a finite subset of M_K^0 such that $S \neq \emptyset$ in the function field case. Let $L = \mathsf{H}_{K,S}^\circ$ be the small Hilbert S-class field of K, and define the embedding of S-ideal classes by*

$$\nu_{L/K}^* \colon \mathcal{C}_{K,S} \to \mathcal{C}_{L,S_L}, \quad \textit{where} \quad \nu_{L/K}^*([\mathfrak{a}]_S) = [\mathfrak{a}\mathfrak{o}_{L,S_L}]_{S_L}] \quad \textit{for all} \quad \mathfrak{a} \in \mathcal{J}(\mathfrak{o}_{K,S}).$$

Then ν^* is trivial [e. g., every ideal of $\mathfrak{o}_{K,S}$ becomes principal in $\mathfrak{o}_{L,S_L}$].

Proof. Let $L_1 = \mathsf{H}_{L,S_L}^\circ$ be the small Hilbert S_L-class field of L. Then L_1 is the class field to $\mathsf{C}_{L,S_L}^{\circ 1} = \mathsf{J}_{L,S_L}^{\circ 1}/L^\times = \mathsf{J}_{L,S_L \cup \mathsf{M}_L^\infty}/L^\times$, and since obviously $\sigma(S_L \cup \mathsf{M}_L^\infty) = S_L \cup \mathsf{M}_L^\infty$ for all $\sigma \in \mathrm{Gal}(L/K)$, we get $\sigma \mathsf{C}_{L,S_L}^{\circ 1} = \mathsf{C}_{L,S_L}^{\circ 1}$. Hence L_1/K is Galois by 6.4.8.3, and we set $G = \mathrm{Gal}(L_1/K)$. By 6.6.4.2, the extensions L/K and L_1/L are unramified, all real places of K are fully decomposed in L, and all places of L are fully decomposed in L_1. Hence L_1/K is unramified and all real places of K are fully decomposed in L_1. By the maximality property of the Hilbert class field it follows that L is the largest over K abelian intermediate field of L_1/K, and therefore $\mathrm{Gal}(L_1/L) = G^{(1)}$ is the commutator subgroup of G. By 6.2.2.5 and 2.5.8.3 it follows that $(\alpha, L_1/L) = \mathsf{V}_{L/K}^{L_1}(\alpha, L_1/K) = \mathsf{V}_{G \to G^{(1)}}(\alpha, L_1/K) = 1$ for all $\alpha \in \mathsf{J}_K$.

Let $\mathfrak{a} \in \mathcal{J}(\mathfrak{o}_{K,S})$ and $\alpha \in \mathsf{J}_K$ such that $\mathfrak{a} = \theta_S \circ \iota_S(\alpha)$; then we have $\mathfrak{a}\mathfrak{o}_{L,S_L} = \theta_{S_L} \circ \iota_{S_L}(\alpha)$ by 5.7.4. By 6.5.7.**A**.1, $(\mathfrak{a}\mathfrak{o}_{L,S_L}, L_1/L) = (\alpha, L_1/L) = 1$, and consequently $\mathfrak{a}\mathfrak{o}_{L,S_L} \in \mathcal{N}_{L_1/L,S_L}(\mathcal{J}(\mathfrak{o}_{L_1,S_{L_1}})) \subset \mathcal{H}(\mathfrak{o}_{L,S_L})$. □

Corollary 6.6.10 (Classical principal ideal theorem). *Let K be an algebraic number field and $K^1 = \mathsf{H}_K^\circ$ its small Hilbert class field. Then $\mathfrak{a}\mathfrak{o}_{K_1}$ is a principal ideal for all ideals $\mathfrak{a} \in \mathcal{J}_K'$.*

Proof. By 6.6.9. □

6.7 Power residues

Definition and Theorem 6.7.1. *Let $n \in \mathbb{N}_{\geq 2}$, $|\mu_n(K)| = n$, and let $a, b \in K^\times$.*
If $v \in \mathsf{M}_K^0$, then $(\mathfrak{N}(v), n) = 1$ implies $\mathfrak{N}(v) \equiv 1 \bmod n$ (see [18, Theorem 5.8.10]). For $v \in \mathsf{M}_K$ we recall from 4.10.4 the definition of the local Hilbert symbol of exponent n for K_v. We slightly change the notation in order to investigate the interplay of all local symbols in the light of a global point of view (for $K = \mathbb{Q}$ see 4.10.6).

For $v \in \mathsf{M}_K$ we define

$$\left(\frac{a,b}{v}\right)_n = (a,b)_{K_v,n} = \frac{(a, K_v(\sqrt[n]{b})/K_v)(\sqrt[n]{b})}{\sqrt[n]{b}}.$$

In particular, if $v \in \mathsf{M}_K^\infty$, then

$$\left(\frac{a,b}{v}\right)_n = \begin{cases} -1 & \text{if } v \text{ is real, } n \equiv 0 \bmod 2, \ \sigma_v(a) < 0 \text{ and } \sigma_v(b) < 0, \\ 1 & \text{otherwise.} \end{cases}$$

1. *If* $v \in \mathsf{M}_K$ *and* a', $b' \in K^\times$, *then*

$$\left(\frac{aa',b}{v}\right)_n = \left(\frac{a,b}{v}\right)_n\left(\frac{a',b}{v}\right)_n, \quad \left(\frac{a,bb'}{v}\right)_n = \left(\frac{a,b}{v}\right)_n\left(\frac{a,b'}{v}\right)_n,$$
$$\left(\frac{a,b}{v}\right) = \left(\frac{b,a}{v}\right)^{-1},$$

and

$$\left(\frac{a,b}{v}\right)_n = 1 \quad \textit{if and only if} \quad a \in \mathsf{N}_{K_v(\sqrt{b})/K_v}K_v(\sqrt{b})^\times.$$

2. (Product formula)

$$\left(\frac{a,b}{v}\right)_n = 1 \ \textit{ for almost all } \ v \in \mathsf{M}_K, \quad \textit{and} \quad \prod_{v\in\mathsf{M}_K}\left(\frac{a,b}{v}\right)_n = 1.$$

3. *If* $v \in \mathsf{M}_K^0$, $\mathfrak{N}(v) \equiv 1 \bmod n$ *and* $v(a) = v(b) = 0$, *then*

$$\left(\frac{a,b}{v}\right)_n = 1.$$

Proof. 1. For $v \in \mathsf{M}_K^0$ the assertions hold by 4.10.4, and for $v \in \mathsf{M}_K^\infty$ they are easily checked.

2. By 6.2.1 and the definition of the local Hilbert symbols it follows that

$$\left(\frac{a,b}{v}\right)_n = 1 \ \text{ for almost all } \ v \in \mathsf{M}_K.$$

Let $L = K(\sqrt[n]{b})$, and for $v \in \mathsf{M}_K$ we fix some $\overline{v} \in \mathsf{M}_L$ above v. Then we have $L_{\overline{v}} = K_v(\sqrt[n]{b})$ and we set $\tau_v = (a, L_{\overline{v}}/K) \in \mathrm{Gal}(L_{\overline{v}}/K_v) \subset \mathrm{Gal}(L/K)$. Then

$$\tau_v(\sqrt[n]{b}) = \left(\frac{a,b}{v}\right)_n\sqrt[n]{b}, \ \text{ and } 1 = (a, L/K) = \prod_{v\in\mathsf{M}_K}(a, K_v(\sqrt[n]{b})/K_v) = \prod_{v\in\mathsf{M}_K}\tau_v$$

by 6.4.5.2. Hence

$$\sqrt[n]{b} = \Big[\prod_{v\in\mathsf{M}_K}\tau_v\Big](\sqrt[n]{b}) = \Big[\prod_{v\in\mathsf{M}_K}\left(\frac{a,b}{v}\right)_n\Big]\sqrt[n]{b}, \ \text{ and therefore } \prod_{v\in\mathsf{M}_K}\left(\frac{a,b}{v}\right)_n = 1.$$

3. By 4.10.5. □

Definition 6.7.2. Let $n \in \mathbb{N}$, $n \geq 2$ and $|\mu_n(K)| = n$. For $a \in K^\times$ we set $\text{supp}(a) = \{v \in \mathsf{M}_K^0 \mid v(a) \neq 0\}$

1. We recall from 4.10.6 the definition of the local power residue symbol, and again we slightly change the notation in order to investigate the interplay of all local symbols.

Let $v \in \mathsf{M}_K^0$ be such that $\mathfrak{N}(v) \equiv 1 \bmod n$. For $a \in \mathcal{O}_v^\times$ (that is, $a \in K$ and $v(a) = 0$) we define the **n-th power residue symbol modulo** v by

$$\left(\frac{a}{v}\right)_n = \left(\frac{a}{\widehat{\mathfrak{p}}_v}\right)_n \in \mu_n(K).$$

From 4.10.6 we obtain:
If $\pi \in K$ is such that $v(\pi) = 1$, then

$$\left(\frac{a}{v}\right)_n = \left(\frac{\pi, a}{v}\right)_n \equiv a^{(\mathfrak{N}(v)-1)/n} \bmod P_v,$$

and

$$\begin{aligned}\left(\frac{a}{v}\right)_n = 1 &\iff a + P_v \in \mathsf{k}_v^{\times n} \\ &\iff \text{there exists some } x \in \mathcal{O}_v \text{ such that } x^n \equiv 1 \bmod P_v.\end{aligned}$$

In this case we call a an **n-th power residue modulo** P_v.

2. Let $a, b \in K^\times$ be such that $\text{supp}(a) \cap \text{supp}(b) = \emptyset$ and $\mathfrak{N}(v) \equiv 1 \bmod n$ for all $v \in \text{supp}(b)$. Then we define the **n-th generalized Jacobi symbol** by

$$\left(\frac{a}{b}\right)_n = \prod_{v \in \text{supp}(b)} \left(\frac{a}{v}\right)_n^{v(b)}.$$

Theorem 6.7.3 (Reciprocity law for power residues)**.** *Let $n \in \mathbb{N}_{\geq 2}$, and assume that $|\mu_n(K)| = n$. Let $a, b \in K^\times$, $\text{supp}(a) \cap \text{supp}(b) = \emptyset$, and suppose that $\mathfrak{N}(v) \equiv 1 \bmod n$ for all $v \in \text{supp}(a) \cup \text{supp}(b)$. Then*

$$\left(\frac{a}{b}\right)_n \left(\frac{b}{a}\right)_n^{-1} = \prod_{\substack{v \in \mathsf{M}_K^0 \\ (n, \mathfrak{N}(v)) \neq 1}} \left(\frac{a, b}{v}\right)_n \prod_{v \in \mathsf{M}_K^\infty} \left(\frac{a, b}{v}\right)_n.$$

Proof. We use 6.7.1 and suppress the index n in all power series and Hilbert symbols.

Assume first that $v \in \mathsf{M}_K^0 \setminus \text{supp}(a)$ and $(n, \mathfrak{N}(v)) = 1$. Let $\pi \in K$ be such that $v(\pi) = 1$ and $b = \pi^{v(b)} u$, where $u \in K^\times$ and $v(u) = 0$. Then

$$\left(\frac{a}{v}\right)^{v(b)} = \left(\frac{\pi, a}{v}\right)^{v(b)} = \left(\frac{\pi^{v(b)}, a}{v}\right)\left(\frac{u, a}{v}\right) = \left(\frac{b, a}{v}\right),$$

and alike, for all $v \in \mathsf{M}_K^0 \setminus \operatorname{supp}(b)$ satisfying $(n, \mathfrak{N}(v)) = 1$,

$$\Big(\frac{b}{v}\Big)^{v(a)} = \Big(\frac{a,b}{v}\Big).$$

Now it follows that

$$\begin{aligned}
\Big(\frac{a}{b}\Big)\Big(\frac{b}{a}\Big)^{-1} &= \prod_{v\in\operatorname{supp}(b)} \Big(\frac{a}{v}\Big)^{v(b)} \prod_{v\in\operatorname{supp}(a)} \Big(\frac{b}{v}\Big)^{-v(a)} \\
&= \prod_{v\in\operatorname{supp}(b)} \Big(\frac{b,a}{v}\Big) \prod_{v\in\operatorname{supp}(a)} \Big(\frac{a,b}{v}\Big)^{-1} \\
&= \prod_{v\in\operatorname{supp}(a)\cup\operatorname{supp}(b)} \Big(\frac{b,a}{v}\Big) = \prod_{\substack{v\in\mathsf{M}_K \\ v\notin\operatorname{supp}(a)\cup\operatorname{supp}(b)}} \Big(\frac{b,a}{v}\Big)^{-1} \\
&= \prod_{\substack{v\in\mathsf{M}_K \\ v\notin\operatorname{supp}(a)\cup\operatorname{supp}(b)}} \Big(\frac{a,b}{v}\Big).
\end{aligned}$$

If $v \in \mathsf{M}_K^0 \setminus [\operatorname{supp}(a) \cup \operatorname{supp}(b)]$ and $(n, \mathfrak{N}(v)) = 1$, then

$$\Big(\frac{a,b}{v}\Big) = 1,$$

and consequently

$$\Big(\frac{a}{b}\Big)\Big(\frac{b}{a}\Big)^{-1} = \prod_{\substack{v\in\mathsf{M}_K^0 \\ (n,\mathfrak{N}(v))\neq 1}} \Big(\frac{a,b}{v}\Big) \prod_{v\in\mathsf{M}_K^\infty} \Big(\frac{a,b}{v}\Big). \qquad \square$$

We finally return to the case $K = \mathbb{Q}$ and set

$$\Big(\frac{a,b}{p}\Big) = \Big(\frac{a,b}{v_p}\Big) \quad \text{for all primes } p \text{ and } \quad \Big(\frac{a,b}{\infty}\Big) = \Big(\frac{a,b}{v_\infty}\Big)$$

Corollary 6.7.4 (Gauss' reciprocity law). *Let $a,\, b \in \mathbb{Q}^\times$, $v_2(a) = v_2(b) = 0$ and $\operatorname{supp}(a) \cap \operatorname{supp}(b) = \emptyset$. Then the generalized quadratic Jacobi symbol satisfies*

- *the quadratic reciprocity law*

$$\Big(\frac{a}{b}\Big)\Big(\frac{b}{a}\Big) = (-1)^{\frac{a-1}{2}\frac{b-1}{2}}\,\varepsilon(a,b),$$

where

$$\varepsilon(a,b) = \Big(\frac{a,b}{\infty}\Big) = \begin{cases} -1 & \text{if } a<0 \text{ and } b<0, \\ 1 & \text{otherwise.} \end{cases}$$

• *and the two supplementary laws*

$$\Big(\frac{-1}{b}\Big) = (-1)^{(b-1)/2}\,\mathrm{sgn}(b) \quad \textit{and} \quad \Big(\frac{2}{b}\Big) = (-1)^{(b^2-1)/8}.$$

Proof. Using 6.7.3 and 4.10.7 we obtain

$$\Big(\frac{a}{b}\Big)\Big(\frac{b}{a}\Big) = \Big(\frac{a}{b}\Big)\Big(\frac{b}{a}\Big)^{-1} = \Big(\frac{a,b}{2}\Big)\Big(\frac{a,b}{\infty}\Big) = (-1)^{\frac{a-1}{2}\frac{b-1}{2}}\varepsilon(a,b),$$

and therefore

$$\Big(\frac{-1}{b}\Big) = \Big(\frac{-1}{b}\Big)\Big(\frac{b}{-1}\Big) = (-1)^{(b-1)/2}\mathrm{sgn}(b).$$

By definition,

$$\Big(\frac{2}{b}\Big) = \prod_{\substack{p\in\mathbb{P}\\ v_p(b)\neq 0}} \Big(\frac{2}{p}\Big)^{v_p(b)} = \prod_{p\in\mathbb{P}\setminus\{2\}} \Big(\frac{p,2}{p}\Big)^{v_p(b)}.$$

For a prime $p \neq 2$ we obtain

$$1 = \prod_{q\in\mathbb{P}\cup\{\infty\}} \Big(\frac{p,2}{q}\Big) = \Big(\frac{p,2}{2}\Big)\Big(\frac{p,2}{p}\Big),$$

hence

$$\Big(\frac{p,2}{p}\Big) = \Big(\frac{p,2}{2}\Big) = \Big(\frac{2,p}{2}\Big) = (-1)^{(p^2-1)/8}$$

by 4.10.7. Moreover, if $p = 2k_p + 1$, then $p^2 - 1 \equiv 4k_p^2 + 4k_p \mod 16$ and

$$b^2 - 1 = \prod_{p\in\mathbb{P}\setminus\{2\}} (2k_p+1)^{2v_p(b)} - 1 \equiv \sum_{p\in\mathbb{P}\setminus\{2\}} (4k_p^2 + 4k_p)v_p(b) \mod 16,$$

hence

$$\frac{b^2-1}{8} \equiv \sum_{p\in\mathbb{P}\setminus\{2\}} \frac{(k_p^2+k_p)v_p(b)}{2} \equiv \sum_{p\in\mathbb{P}\setminus\{2\}} \frac{(p^2-1)v_p(b)}{8} \mod 2$$

and

$$\Big(\frac{2}{b}\Big) = \prod_{p\in\mathbb{P}\setminus\{2\}} \Big(\frac{p,2}{p}\Big)^{v_p(b)} = \prod_{p\in\mathbb{P}\setminus\{2\}} (-1)^{(p^2-1)v_p(b)/8} = (-1)^{(b^2-1)/8}. \qquad \square$$

For $a,\, b \in \mathbb{Q}^\times$ we compare the generalized quadratic Jacobi symbol with the classical Kronecker symbol as investigated in [18, Theorem 3.4.3]. To do so, we denote the classical Kronecker symbol by

$$\Big(\frac{a}{b}\Big)_{\mathrm{Kr}}.$$

Both symbols are defined for all $a,\, b \in \mathbb{Z}$ satisfying $(a,b) = 1$. For a prime $p \neq 2$

$$\left(\frac{a}{p}\right)_{\mathrm{Kr}} = \left(\frac{a}{p}\right) \quad \text{is the Legendre symbol (see [18, Section 3.4]);}$$

by 6.7.4 and [18, Theorem 3.4.3] it follows that

$$\left(\frac{2}{b}\right)_{\mathrm{Kr}} = \left(\frac{2}{b}\right) = (-1)^{(b^2-1)/8}.$$

Eventually we obtain, for all $a,\, b \in \mathbb{Z}$ satisfying $(a,b) = 1$,

$$\left(\frac{a}{b}\right)_{\mathrm{Kr}} = \left(\frac{a}{\mathrm{sgn}(b)}\right)_{\mathrm{Kr}} \prod_{p \in \mathbb{P}} \left(\frac{a}{p}\right)_{\mathrm{Kr}}^{v_p(b)} = \varepsilon(a,b) \prod_{p \in \mathbb{P}} \left(\frac{a}{p}\right)^{v(b)} = \varepsilon(a,b)\left(\frac{a}{b}\right).$$

7

Functional equations and Artin L functions

In this final chapter we investigate the analytic continuation and the functional equation for Dirichlet and Artin L functions for global fields in the classical setting renouncing Hecke L functions and harmonic analysis. As to the farther-reaching theory of Hecke L functions we refer to [44] and [51] for a presentation using harmonic analysis, and to [48] for a presentation using the classical methods of Hecke.

First we investigate Dirichlet L functions, separately for number fields (in the first three sections) and function fields (in the fourth section). In doing so, we depart from the unified language for global fields and take recourse to the classical terminology for number fields and funcion fields as in [18]. Our main references for these sections are [20], [41], [19], [23], [21], [22], [35], [43] and [61].

The subsequent two sections are devoted to a thorough presentation of the tools from representation theory, and there we mainly refer to [37], [56] and [7].

Only in the final three sections we present the theory of Artin conductors, Artin L functions and arithmetical applications, now again in the unified language for all global fields. Here our main references are [41], [19], [48], [55] and [31].

7.1 Gauss sums and L functions of number fields

Overall assumptions for the sections 7.1, 7.2 *and* 7.3. Let K be an algebraic number field. Let $\mathfrak{o}_K$ be its ring of integers, (r_1, r_2) its signature, $r = r_1 + r_2$ and $[K:\mathbb{Q}] = n = r_1 + 2r_2$. Let $\sigma_1, \ldots, \sigma_n \colon K \to \mathbb{C}$ be the embeddings of K, such that $\sigma_i(K) \subset \mathbb{R}$, $l_i = 1$ for $i \in [1, r_1]$, and $\sigma_{r+i} = \overline{\sigma_{r_1+i}} \ne \sigma_{r_1+i}$, $l_i = 2$ for $i \in [1, r_2]$. For $i \in [1, r]$ and $a \in K$ let $a^{(i)} = \sigma_i(a) \in \mathbb{C}$, and let $v_i \in \mathsf{M}_K^\infty$ be the place such that σ_i is the embedding associated with v_i, i. e., $\sigma_{v_i} = \sigma_i$. Then the normalized absolute value associated with v_i is given by $|a|_{v_i} = |a^{(i)}|^{l_i}$ for all $a \in K$ (here $|\cdot| = |\cdot|_\infty$ is the ordinary absolute value of $\mathbb{C}$). A number $a \in K^\times$ is called totally positive, $a \gg 0$, if $a^{(i)} > 0$ for all $i \in [1, r_1]$; $K^+ = \{x \in K^\times \mid x \gg 0\}$.

DOI: 10.1201/9780429506574-7

Let d_K be the discriminant and $\mathfrak{D} = \mathfrak{D}_{K/\mathbb{Q}} \in \mathcal{J}'_K$ the different of K. Note that $|\mathsf{d}_K| = \mathfrak{N}(\mathfrak{D})$ (see [18, Theorem 5.9.8]). Let R_K be the regulator, h_K the class number and $w_K = |\mu(K)|$ the number of roots of unity in K.

In the sequel we rather use the language of ideals instead of those of divisors (we refer to the introductory chapter 'Notation', and in particular to 5.1.1.**4**). For $\mathfrak{m} \in \mathcal{J}'_K$ we denote by $\mathcal{J}_K^{\mathfrak{m}} = \{\mathfrak{a} \in \mathcal{J}_K \mid \operatorname{supp}(\mathfrak{a}) \cap \operatorname{supp}(\mathfrak{m}) = \emptyset\}$ the group of fractional ideals coprime to $\mathfrak{m}$, by $\mathcal{J}_K'^{\mathfrak{m}} = \mathcal{J}_K^{\mathfrak{m}} \cap \mathcal{J}'_K$ the set of all ideals coprime to $\mathfrak{m}$, by $K(\mathfrak{m}) = \{a \in K^\times \mid a\mathfrak{o}_K \in \mathcal{J}_K^{\mathfrak{m}}\}$ the group of all numbers coprime to $\mathfrak{m}$, and by $K^{\mathfrak{m}}$ the set of all totally positive $a \in K^\times$ satisfying $v(a-1) \ge v(\mathfrak{m})$ for all $v \in \operatorname{supp}(\mathfrak{m})$. For a non-Archimedian place $v \in \mathsf{M}_K^0$ we denote the associated discrete valuation again by v.

We briefly recall the concepts of ray class characters and associated idele class charactes from Section 5.8. Let $\mathfrak{f} \in \mathcal{J}'_K$ be an ideal, $\mathcal{C}_K^{\mathfrak{f}}$ the ray class group modulo $\mathfrak{f}$ and $\chi \in \mathsf{X}(\mathcal{C}_K^{\mathfrak{f}})$ a primitive ray class character modulo $\mathfrak{f}$. We denote the associated idele class character likewise by χ (see 5.8.6). For a place $v \in \mathsf{M}_K$, let $\chi_v \in \mathsf{X}(K_v^\times)$ be its local v-component, and for $v \in \mathsf{M}_K^0$ let $f_v = f(\chi_v) = \min\{n \in \mathbb{N}_0 \mid \chi_v \restriction U_v^{(n)} = 1\} = v(\mathfrak{f})$ be its conductor exponent.

If $\mathfrak{a} = \iota_K(\alpha) \in \mathcal{J}_K^{\mathfrak{f}}$ for some $\alpha = (\alpha_v)_{v \in \mathsf{M}_K} \in \mathsf{J}_K$, then

$$\chi(\mathfrak{a}) = \overline{\chi}(\alpha) \prod_{v \in \operatorname{supp}(\mathfrak{f}) \cup \mathsf{M}_K^\infty} \chi_v(\alpha_v) = \prod_{v \in \mathsf{M}_K^0 \setminus \operatorname{supp}(\mathfrak{f})} \overline{\chi}_v(\alpha_v),$$

and if $\mathfrak{a} \in \mathcal{J}_K \setminus \mathcal{J}_K^{\mathfrak{f}}$, then $\chi(\mathfrak{a}) = 0$.

If $v \in \mathsf{M}_K^\infty$, then either $\chi_v = 1$, or v is real and $\chi_v = \operatorname{sgn}\colon \mathbb{R} \to \{\pm 1\}$. We suppose that $\{i \in [1, r_1] \mid \chi_{v_i} \ne 1\} = [1, q]$ for some $q = q_\chi \in [0, r_1]$. Then we obtain, for all $a \in K^\times$,

$$\chi_{v_i}(a) = \begin{cases} \operatorname{sgn}(a^{(i)}) & \text{if } i \in [1, q], \\ 1 & \text{if } i \in [q+1, r], \end{cases} \qquad \chi_\infty(a) = \prod_{i=1}^{q_\chi} \operatorname{sgn}(a^{(i)}),$$

$$\chi(a\mathfrak{o}_K) = \prod_{v \in \operatorname{supp}(\mathfrak{f}) \cup \mathsf{M}_K^\infty} \chi_v(a) \text{ if } a \in K(\mathfrak{f}), \quad \text{and} \quad \chi(a\mathfrak{o}_K) = 0 \text{ if } a \notin K(\mathfrak{f}).$$

In particular, if $a \in K(\mathfrak{f})$ is totally positive, then

$$\chi_\infty(a) = 1 \quad \text{and} \quad \chi(a\mathfrak{o}_K) = \prod_{v \in \operatorname{supp}(\mathfrak{f})} \chi_v(a).$$

We start with local considerations.

Definition and Theorem 7.1.1 (Local Gauss sums). Let $\mathfrak{f} \in \mathcal{J}'_K$ be an ideal and $\chi \in \mathsf{X}(\mathcal{C}_K^{\mathfrak{f}})$ a primitive ray class character modulo $\mathfrak{f}$. Let $v \in \mathsf{M}_K^0$ and $v \downarrow p$ so that $p = \operatorname{char}(\mathsf{k}_v)$ and $\mathbb{Q}_p \subset K_v$ (see 5.1.1). Let $\chi_v \in \mathsf{X}(K_v^\times)$ be the v-component of χ and $f_v = v(\mathfrak{f})$ its character exponent.

Let $\lambda_v\colon K_v \to \mathbb{T}$ be the additive standard character of K_v (see 1.7.2.2(c)). If $x \in K_v$ and $\mathsf{Tr}_{K_v/\mathbb{Q}_p}(x) = p^{-m}a + b$ where $m \in \mathbb{N}$, $a \in [0, p^m - 1]$ and $b \in \mathbb{Z}_p$, then $\lambda_v(x) = \mathsf{e}^{2\pi \mathsf{i}\, a/p^m}$. Consequently,

$$\operatorname{Ker}(\lambda_v) = \{x \in K_v \mid \mathsf{Tr}_{K_v/\mathbb{Q}_p}(x) \in \mathbb{Z}_p\} = \mathfrak{D}_{K_v/\mathbb{Q}_p}.$$

Let $\{a_1, \ldots, a_m\} \subset U_v$ be a set of representatives for $U_v/U_v^{(f_v)}$, $d,\, c \in K_v$ such that $v(d) = v(\mathfrak{D}\mathfrak{f})$ and $v(c) \geq -v(d)$. We define

$$\tau_v(\chi) = \sum_{j=1}^{m} \chi_v(d^{-1}a_j)\lambda_v(d^{-1}a_j) \quad \text{and} \quad F_v(\chi, c) = \sum_{j=1}^{m} \chi_v(a_j)\lambda_v(a_jc).$$

We call $\tau_v(\chi)$ the (**local**) **Gauss sum** for χ at v.

1. The numbers $\tau_v(\chi)$ and $F_v(\chi, c)$ do not depend on the choice of $\{a_1, \ldots, a_m\}$ and d, and $F_v(\chi, c) = M_v(\chi, c)\tau_v(\chi)$, where

$$M_v(\chi, c) = \begin{cases} \overline{\chi}_v(c) & \text{if} \;\; v(c) = -v(d), \\ \chi_v(d) & \text{if} \;\; v(c) > -v(d) \;\; \text{and} \;\; f_v = 0, \\ 0 & \text{if} \;\; v(c) > -v(d) \;\; \text{and} \;\; f_v > 0. \end{cases}$$

2. If $f_v = 0$, then $\tau_v(\chi) = \overline{\chi}_v(d)$ and $F_v(\chi, c) = 1$. In particular, if $v(\mathfrak{D}\mathfrak{f}) = 0$ or $\chi_v = 1$, then $\tau_v(\chi) = 1$.

3. $\overline{\tau_v(\chi)} = \chi_v(-1)\tau_v(\overline{\chi})$, and $|\tau_v(\chi)|^2 = \mathfrak{N}(v)^{f_v}$.

Proof. Since $v(\mathfrak{D}) = v(\mathfrak{D}_{K_v/\mathbb{Q}_p})$ (see [18, Theorem 5.9.8]), we get $\lambda_v(y) = 1$ for all $y \in K_v$ satisfying $v(y) \geq -v(\mathfrak{D})$.

1. We prove first that the definition of $F_v(\chi, c)$ does not depend on the choice of $\{a_1, \ldots, a_m\}$. Let $\{a_1', \ldots, a_m'\} \subset U_v$ be another set of representatives for $U_v/U_v^{(f_v)}$. As the definition of $F_v(\chi, c)$ does not depend on the order of $a_1, \ldots, a_m$, we may assume that $a_j' = a_ju_j$ for some $u_j \in U_v^{(f_v)}$. Then we get $\chi_v(a_j') = \chi_v(a_j)\chi_v(u_j) = \chi_v(a_j)$, and since

$$v(a_j'c - a_jc) = v(u_j - 1) + v(a_j) + v(c) \geq f_v - v(d) = -v(\mathfrak{D}),$$

it follows that $\lambda_v(a_j'c - a_jc) = \lambda_v(a_j'c)\lambda_v(a_jc)^{-1} = 1$, and eventually $\chi_v(a_j')\lambda_v(a_j'c) = \chi_v(a_j)\lambda_v(a_jc)$ for all $j \in [1, m]$. Consequently, $F_v(\chi, c)$ does not depend on the choice of $\{a_1, \ldots, a_m\}$.

Now let $\{b_1, \ldots, b_m\} \subset U_v$ be any set of representatives for $U_v/U_v^{(f_v)}$, and let $c,\, d \in K_v$ be arbitrary such that $-v(c) = v(d) = v(\mathfrak{D}\mathfrak{f})$. Then it follows that $cd \in U_v$, hence $\{(cd)^{-1}a_1, \ldots, (cd)^{-1}a_m\} \subset U_v$ is again a set of representatives for $U_v/U_v^{(f_v)}$, and consequently

$$F_v(\chi, c) = \sum_{j=1}^{m} \chi_v((cd)^{-1}a_j)\lambda_v((cd)^{-1}a_jc) = \overline{\chi}_v(c)\sum_{j=1}^{m} \chi_v(d^{-1}a_j)\lambda_v(d^{-1}a_j)$$
$$= \overline{\chi}_v(c)\tau_v(\chi).$$

Thus the definition of $\tau_v(\chi)$ likewise does not depend on the choice of $\{a_1, \ldots, a_m\}$, and it also does not depend on d.

It remains to prove: If $v(c) > -v(d)$, then

$$F_v(\chi, c) = \begin{cases} \chi_v(d)\tau_v(\chi) & \text{if } f_v = 0, \\ 0 & \text{if } f_v > 0. \end{cases}$$

CASE 1: $v(c) \geq -v(\mathfrak{D})$. If $j \in [1, m]$, then $v(a_j c) = v(c) \geq -v(\mathfrak{D})$, hence $\lambda_v(a_j c) = 1$, and therefore

$$F_v(\chi, c) = \sum_{j=1}^{m} \chi_v(a_j) = \sum_{\omega \in U_v/U_v^{(f_v)}} \chi'_v(\omega),$$

where $\chi'_v \in \mathsf{X}(U_v/U_v^{(f_v)})$ is defined by $\chi'_v(uU_v^{(f_v)}) = \chi_v(u)$ for all $u \in U_v$. If $f_v > 0$, then $\chi'_v \neq 1$ and $F_v(\chi, c) = 0$. If $f_v = 0$, then $U_v^{(f_v)} = U_v$, $m = 1$, and with $a_1 = 1$ we obtain get $F_v(\chi, c) = 1$ and $\tau_v(\chi) = \chi_v(d^{-1}) = \chi_v(d)^{-1}$.

CASE 2: $v(c) < -v(\mathfrak{D})$. Then $f_v = v(\mathfrak{f}) = v(\mathfrak{D}\mathfrak{f}) - v(\mathfrak{D}) > d + v(c)$, and therefore $1 \leq f' = -v(\mathfrak{D}) - v(c) < f_v$. Let $\{a'_1, \ldots, a'_{m'}\} \subset U_v$ be a set of representatives for $U_v/U_v^{(f')}$ and $\{a''_1, \ldots, a''_{m''}\} \subset U_v^{(f')}$ a set of representatives for $U_v^{(f')}/U_v^{(f_v)}$. Then $\{a'_j a''_k \mid j \in [1, m'],\ k \in [1, m'']\}$ is a set of representatives for $U_v/U_v^{(f_v)}$, and

$$F_v(\chi, c) = \sum_{j=1}^{m'} \sum_{k=1}^{m''} \chi_v(a'_j)\chi_v(a''_k)\lambda_v(a'_j a''_k c).$$

Since

$$\begin{aligned} v(a'_j a''_k c - a'_j c) &= v(a'_j c) + v(a''_k - 1) \\ &= v(c) + v(a''_k - 1) \geq -f' - v(\mathfrak{D}) + f' = -v(\mathfrak{D}), \end{aligned}$$

we obtain $1 = \lambda_v(a'_j a''_k c - a'_j c) = \lambda_v(a'_j a''_k c)\lambda_v(a'_j c)^{-1}$, and

$$F_v(\chi, c) = \sum_{j=1}^{m'} \sum_{k=1}^{m''} \chi_v(a'_j)\chi_v(a''_k)\lambda_v(a'_j c) = \sum_{j=1}^{m'} \chi_v(a'_j)\lambda_v(a'_j c) \sum_{k=1}^{m''} \chi_v(a''_k).$$

Let $\chi'_v \in \mathsf{X}(U_v^{(f')}/U_v^{(f_v)})$ be defined by $\chi'_v(uU_v^{(f_v)}) = \chi_v(u)$ for all $u \in U_v^{(f')}$, then $\chi_v \restriction U_v^{(f')} \neq 1$ (since χ is primitive) implies $\chi'_v \neq 1$, hence

$$\sum_{k=1}^{m''} \chi_v(a''_k) = \sum_{\omega \in U_v^{(f')}/U_v^{(f_v)}} \chi'_v(\omega) = 0,$$

and consequently $F_v(\chi, c) = 0$.

2. If $f_v = 0$, then $\chi_v \restriction U_v = 1$ and $v(c) \geq -v(\mathfrak{D})$. By the very definitions of $\tau_v(\chi)$ and $F_v(\chi, c)$ (with $m = 1$ and $a_1 = 1$) we get $\tau_v(\chi) = \chi_v(d^{-1})$ and $F_v(\chi, c) = 1$. If additionally $v(\mathfrak{D}) = v(d) = 0$, then $\chi_v(d^{-1}) = 1$ and thus $\tau_v(\chi) = 1$.

3. As $\chi_v(-1) \in \{\pm 1\}$, $\overline{\lambda}_v(y) = \lambda_v(-y)$ for all $y \in K$, and $\{-a_1, \ldots, -a_m\} \subset U_v$ is also a set of representatives for $U_v/U_v^{(f_v)}$, we obtain

$$\overline{\tau_v(\chi)} = \sum_{j=1}^{m} \overline{\chi}_v(d^{-1}a_j)\overline{\lambda}_v(d^{-1}a_j) = \sum_{j=1}^{m} \overline{\chi}_v(-1)\overline{\chi}_v(-d^{-1}a_j)\lambda_v(-d^{-1}a_j)$$
$$= \chi_v(-1)\tau_v(\overline{\chi}).$$

If $f_v = 0$, then $|\tau_v(\chi)| = |\chi_v(d^{-1})| = 1$ by 2. Thus let $f_v > 0$, let $\{b_1, \ldots, b_\rho\} \subset \mathfrak{o}_K$ be a set of representative for $\mathfrak{o}_K/\mathfrak{p}_v^{f_v}$ and $c \in K$ such that $v(c) = -v(d) = -v(\mathfrak{D}\mathfrak{f})$. Then 1. implies

$$\sum_{j=1}^{\rho} |F_v(\chi, b_j c)|^2 = |\tau_v(\chi)|^2 \sum_{j=1}^{\rho} |M_v(\chi, b_j c)|^2 = |\tau_v(\chi)|^2 s, \tag{1}$$

where

$$s = |\{j \in [1, \rho] \mid v(b_j c) = -v(d)\}| = |\{j \in [1, \rho] \mid v(b_j) = 0\}| = |(\mathfrak{o}_K/\mathfrak{p}_v^{f_v})^\times|.$$

On the other hand,

$$\sum_{j=1}^{\rho} |F_v(\chi, b_j c)|^2 = \sum_{j=1}^{\rho} F_v(\chi, b_j c)\overline{F_v(\chi, b_j c)}$$
$$= \sum_{k,l=1}^{m} \chi_v(a_k)\overline{\chi}_v(a_l) \sum_{j=1}^{\rho} \lambda_v(a_k b_j c)\overline{\lambda}_v(a_l b_j c).$$

Let $k,\, l \in [1, m]$. If $j \in [1, \rho]$, then $\lambda_v(a_k b_j c)\overline{\lambda}_v(a_l b_j c) = \lambda_v((a_k - a_l)b_j c)$, and we consider the homomorphism $\theta\colon \mathfrak{o}_K \to \mathbb{T}$, defined by $\theta(b) = \lambda_v((a_k - a_l)bc)$. If $b \in \mathfrak{p}_v^{f_v}$, then $v((a_k - a_l)bc) \geq v(b) + v(c) \geq f_v - v(d) = -v(\mathfrak{D})$, then $\theta(b) = \lambda_v((a_k - a_l)bc) = 1$. Hence θ induces a character $\theta' \in \mathsf{X}(\mathfrak{o}_K/\mathfrak{p}_v^{f_v})$ satisfying $\theta'(b + \mathfrak{p}_v^{f_v}) = \theta(b)$ for all $b \in \mathfrak{o}_K$, and sinc $(\mathfrak{o}_K : \mathfrak{p}_v^{f_v}) = \mathfrak{N}(v)^{f_v}$, we obtain

$$\sum_{j=1}^{\rho} \lambda_v(a_k b_j c)\overline{\lambda}_v(a_l b_j c) = \sum_{\beta \in \mathfrak{o}_K/\mathfrak{p}_v^{f_v}} \theta(\beta) = \begin{cases} \mathfrak{N}(v)^{f_v} & \text{if } \theta = 1, \\ 0 & \text{if } \theta \neq 1. \end{cases}$$

By definition,

$$\theta = 1 \iff \lambda_v((a_k - a_l)bc) = 1 \ \text{ for all } \ b \in \widehat{\mathcal{O}}_v$$
$$\iff v(a_k - a_l) + v(c) \geq -v(\mathfrak{D}) \iff v(a_k - a_l) \geq f_v \iff k = l.$$

Hence it follows that

$$\sum_{j=1}^{\rho}|F_v(\chi,b_jc)|^2=\sum_{k=1}^{m}|\chi_v(a_k)|^2\,\mathfrak{N}(v)^{f_v}=m\,\mathfrak{N}(v)^{f_v}=(U_v\!:\!U_v^{(f_v)})\,\mathfrak{N}(v)^{f_v}. \tag{2}$$

Since $(U_v:U_v^{(f_v)})=|(\widehat{\mathcal{O}_v}/\widehat{\mathfrak{p}_v}^{f_v})^\times|=|(\mathfrak{o}_K/\mathfrak{p}_v^{f_v})^\times|$ (see [18, Theorems 2.10.5 and 5.3.4]), the comparison of (1) and (2) reveals $|\tau_v(\chi)|^2=\mathfrak{N}(v)^{f_v}$. □

The following lemma is an essential tool for the globalization of the notion of Gauss sums.

Lemma 7.1.2. *If $a\in K$, then $\lambda_v(a)=1$ for almost all $v\in\mathsf{M}_K^0$, and*

$$\mathsf{e}^{2\pi\mathsf{i}\,\mathsf{Tr}_{K/\mathbb{Q}}(a)}=\prod_{v\in\mathsf{M}_K^0}\lambda_v(a).$$

Proof. Let $a\in K$. Then we use the uniquely detemined expansion into partial fractions (see [18, Theorem 2.3.1])

$$\mathsf{Tr}_{K/\mathbb{Q}}(a)=\sum_{p\in\mathbb{P}}\frac{b_p}{p^{n_p}}+b,$$

where $b\in\mathbb{Z}$, $(b_p,n_p)=(0,0)$ for almost all $p\in\mathbb{P}$, and for all $p\in\mathbb{P}$ we have either $(b_p,n_p)=(0,0)$ or $n_p\in\mathbb{N}$, $b_p\in[1,p^n-1]$ and $p\nmid b_p$.

For $p\in\mathbb{P}$ and all $v\in\mathsf{M}_K^0$ lying above p we set

$$\mathsf{Tr}_{K_v/\mathbb{Q}_p}(a)=\frac{a_v}{p^{m_v}}+c_v,\quad\text{and therefore}\quad\lambda_v(a)=\mathsf{e}^{2\pi\mathsf{i}\,a_v/p^{m_v}},$$

where $m_v\in\mathbb{N}$, $a_v\in[0,p^{m_v}-1]$ and $c_v\in\mathbb{Z}_p$. Using [18, Theorem 5.5.9], it follows that

$$\mathsf{Tr}_{K/\mathbb{Q}}(a)=\sum_{\substack{v\in\mathsf{M}_K^0\\ v\downarrow p}}\mathsf{Tr}_{K_v/\mathbb{Q}_p}(a)=\sum_{\substack{v\in\mathsf{M}_K^0\\ v\downarrow p}}\frac{a_v}{p^{m_v}}+\sum_{\substack{v\in\mathsf{M}_K^0\\ v\downarrow p}}c_v=\frac{b_p}{p^{n_p}}+\sum_{q\in\mathbb{P}\setminus\{p\}}\frac{b_q}{q^{n_q}}+b,$$

and therefore

$$\frac{b_p}{p^{n_p}}=\sum_{\substack{v\in\mathsf{M}_K^0\\ v\downarrow p}}\frac{a_v}{p^{m_v}}+a,\quad\text{where (for some }N\in\mathbb{N})\ a\in\mathbb{Z}_p\cap p^{-N}\mathbb{Z}=\mathbb{Z}.$$

Hence we eventually get

$$\mathsf{e}^{2\pi\mathsf{i}\,\mathsf{Tr}_{K/\mathbb{Q}}(a)}=\prod_{p\in\mathbb{P}}\mathsf{e}^{2\pi\mathsf{i}b_p/p^{n_p}}=\prod_{p\in\mathbb{P}}\prod_{\substack{v\in\mathsf{M}_K^0\\ v\downarrow p}}\mathsf{e}^{2\pi\mathsf{i}a_v/p^{m_v}}=\prod_{v\in\mathsf{M}_K^0}\lambda_v(a).$$ □

Definition 7.1.3. Let $\mathfrak{f} \in \mathcal{J}'_K$ and $\chi \in \mathsf{X}(\mathcal{C}_K^{\mathfrak{f}})$ be a primitive ray class character. We shall repeatedly use the following argument for the existence of totally positive numbers. Let $\mathfrak{a}$ be a complete module in K (i. e., an additive submodule of K which contains a basis of K). Then for every $a \in K$ the coset $a + \mathfrak{a}$ contains a totally positive number (see [18, Theorem 3.1.3]).

1. Let $\{a_1, \ldots, a_\rho\} \subset (\mathfrak{D}\mathfrak{f})^{-1}$ be a set of totally positive representatives for $(\mathfrak{D}\mathfrak{f})^{-1}/\mathfrak{D}^{-1}$. We define the **Gauss sum** attached to χ by

$$\tau(\chi) = \sum_{j=1}^{\rho} \chi(a_j\mathfrak{D}\mathfrak{f})\mathsf{e}^{2\pi \mathsf{i}\,\mathsf{Tr}_{K/\mathbb{Q}}(a_j)}.$$

2. Let $\{b_1, \ldots, b_\rho\} \subset \mathfrak{o}$ be a set of totally positive representatives for $\mathfrak{o}/\mathfrak{f}$. For $c \in (\mathfrak{D}\mathfrak{f})^{-1}$ we define

$$F(\chi, c) = \sum_{j=1}^{\rho} \chi(b_j\mathfrak{o}_K)\mathsf{e}^{2\pi \mathsf{i}\,\mathsf{Tr}_{K/\mathbb{Q}}(b_j c)}.$$

It is easily checked that the definitions of $\tau(\chi)$ and $F(\chi, c)$ do not depend on the respective representatives. Yet we will not do this, because it follows immediately from the subsequent Theorem 7.1.4, and there we will also present a more general formula for $F(\chi, c)$.

3. The number

$$\mathsf{W}(\chi) = \frac{\mathsf{i}^{-q_\chi}\tau(\chi)}{\sqrt{\mathfrak{N}(\mathfrak{f})}}$$

is called the **Artin root number** of χ.

By the very definitions, we get $\tau(1) = F(1, c) = \mathsf{W}(1) = 1$.

Theorem 7.1.4 (Local-global principle for Gauss sums). *Let $\mathfrak{f} \in \mathcal{J}'_K$ be an ideal, $c \in (\mathfrak{D}\mathfrak{f})^{-1}$ and $\chi \in \mathsf{X}(\mathcal{C}_K^{\mathfrak{f}})$ a primitive ray class character.*

1. $\tau_v(\chi) = F_v(\chi, c) = 1$ *for all* $v \in \mathsf{M}_K^0 \setminus \operatorname{supp}(\mathfrak{D}\mathfrak{f})$,

$$\tau(\chi) = \prod_{v \in \mathsf{M}_K^0} \tau_v(\chi) \quad \text{and} \quad F(\chi, c) = \prod_{v \in \mathsf{M}_K^0} F_v(\chi, c)$$

2. $F(\chi, c) = \tau(\chi)M(\chi, c)$, *where*

$$M(\chi, c) = \begin{cases} \overline{\chi}(c\mathfrak{D}\mathfrak{f})\chi_\infty(c) & \text{if } c \neq 0, \\ \overline{\chi}(\mathfrak{D}\mathfrak{f}) & \text{if } c = 0. \end{cases}$$

3. $\chi_\infty(-1) = (-1)^{q_\chi}$, $\tau(\overline{\chi}) = (-1)^{q_\chi}\,\overline{\tau(\chi)}$ *and* $|\tau(\chi)|^2 = \mathfrak{N}(\mathfrak{f})$.

4. $\mathsf{W}(\overline{\chi}) = \overline{\mathsf{W}(\chi)}$ *and* $|\mathsf{W}(\chi)| = 1$.

5. $\tau(1) = F(1, c) = \mathsf{W}(1) = 1$.

Proof. 1. If $v \in \mathsf{M}_K^0 \setminus \mathrm{supp}(\mathfrak{D}\mathfrak{f})$, then $\tau_v(\chi) = F_v(\chi, c) = 1$ by 7.1.1.2. Now we proceed in 3 steps.

I. Let $\{b_1, \ldots, b_\rho\} \subset \mathfrak{o}$ be a set of totally positive representatives for $\mathfrak{o}/\mathfrak{f}$, and let $m \in [1, \rho]$ be such that $\{b_1, \ldots, b_m\} = \{b_j \mid j \in [1, \rho],\ (b_j, \mathfrak{f}) = 1\}$. Then $\{b_1, \ldots, b_m\}$ is a set of representatives for $(\mathfrak{o}/\mathfrak{f})^\times$.

For $v \in \mathrm{supp}(\mathfrak{f})$, there is an isomorphism $(\mathfrak{o}_K/\mathfrak{p}_v^{f_v})^\times \overset{\sim}{\to} U_v/U_v^{(f_v)}$, given by the assignment $a + \mathfrak{p}_v^{f_v} \mapsto aU_v^{(f_v)}$ for all $a \in \mathfrak{o}_K(\mathfrak{f})$. We identify. Let $\{b_1^{(v)}, \ldots, b_{m_v}^{(v)}\} \subset \mathfrak{o}_K$ be a set of totally positive representatives for $(\mathfrak{o}_K/\mathfrak{p}_v^{f_v})^\times$ satisfying $b_j^{(v)} \equiv 1 \mod \mathfrak{p}_{v'}^{f_{v'}}$ for all $v' \in \mathrm{supp}(\mathfrak{f}) \setminus \{v\}$ and all $j \in [1, m_v]$. By our identification, $\{b_1^{(v)}, \ldots, b_{m_v}^{(v)}\}$ is also a set of representaves for $U_v/U_v^{(f)}$.

With this choice of $b_j^{(v)}$ we obtain a bijective map

$$[1, m] \;\to \prod_{v \in \mathrm{supp}(\mathfrak{f})} [1, m_v], \quad j \mapsto (j_v)_{v \in \mathrm{supp}(\mathfrak{f})}$$

such that

$$b_j \;=\; \prod_{v \in \mathrm{supp}(\mathfrak{f})} b_{j_v}^{(v)} b \ \text{ with } \ b \in K^{\mathfrak{f}} \ \text{ for all } \ j \in [1, m],$$

and $b_{j_v}^{(v)} \equiv b_j \mod \mathfrak{p}_v^{f_v}$ for all $j \in [1, m]$ and $v \in \mathrm{supp}(\mathfrak{f})$.

II. Proof of the product formula for $F(\chi, c)$.
Let again $\{b_1, \ldots, b_\rho\} \subset \mathfrak{o}_K$ be a set of totally positive representatives for $\mathfrak{o}_K/\mathfrak{f}$, and suppose that $\{j \in [1, \rho] \mid (b_j, \mathfrak{f}) = 1\} = [1, m]$ for some $m \in [1, \rho]$. Then

$$F(\chi, c) = \sum_{j=1}^{\rho} \chi(b_j \mathfrak{o}_K) \mathsf{e}^{2\pi \mathsf{i}\, \mathsf{Tr}_{K/\mathbb{Q}}(b_j c)} = \sum_{j=1}^{m} \chi(b_j \mathfrak{o}_K) \mathsf{e}^{2\pi \mathsf{i}\, \mathsf{Tr}_{K/\mathbb{Q}}(b_j c)}.$$

For $j \in [1, m]$ and $v \in \mathrm{supp}(\mathfrak{f})$ let $m_v \in \mathbb{N}$ and $b_j^{(v)} \in \mathfrak{o}_K$ for $j \in [1, m_v]$ be as in **I**. Then $b_j \equiv b_{j_v}^{(v)} \mod \mathfrak{p}_v^{f_v}$, hence $\chi_v(b_j) = \chi_v(b_{j_v}^{(v)})$, and

$$v(b_j c - b_{j_v}^{(v)} c) \ge f_v - v(\mathfrak{D}\mathfrak{f}) = -v(\mathfrak{D}) \quad \text{implies} \quad \lambda_v(b_j c) = \lambda_v(b_{j_v}^{(v)} c).$$

Using 7.1.2, we finally obtain

$$\begin{aligned} F(\chi, c) &= \sum_{j=1}^{m} \chi(b_j \mathfrak{o}_K) \mathsf{e}^{2\pi \mathsf{i}\, \mathsf{Tr}_{K/\mathbb{Q}}(b_j c)} = \sum_{j=1}^{m} \prod_{v \in \mathrm{supp}(\mathfrak{f})} \chi_v(b_j) \lambda_v(b_j c) \\ &= \sum_{j=1}^{m} \prod_{v \in \mathrm{supp}(\mathfrak{f})} [\chi_v(b_{j_v}^{(v)}) \lambda_v(b_{j_v}^{(v)} c)] = \prod_{v \in \mathrm{supp}(\mathfrak{f})} \sum_{j_v=1}^{m_v} \chi_v(b_{j_v}^{(v)}) \lambda_v(b_{j_v}^{(v)} c) \\ &= \prod_{v \in \mathrm{supp}(\mathfrak{f})} F_v(\chi, c) = \prod_{v \in \mathsf{M}_K^0} F_v(\chi, c). \end{aligned}$$

III. Proof of the product formula for the Gauss sums.
Let $\{a_1,\ldots,a_\rho\}\subset(\mathfrak{D}\mathfrak{f})^{-1}$ be a set of totally positive representatives for $(\mathfrak{D}\mathfrak{f})^{-1}/\mathfrak{D}^{-1}$, and suppose that $\{j\in[1,\rho]\mid(a_j\mathfrak{D}\mathfrak{f},\mathfrak{f})=1\}=[1,m]$ for some $m\in[1,\rho]$. Then

$$\tau(\chi)=\sum_{j=1}^{\rho}\chi(a_j\mathfrak{D}\mathfrak{f})\mathrm{e}^{2\pi\mathrm{i}\,\mathrm{Tr}_{K/\mathbb{Q}}(a_j)}=\sum_{j=1}^{m}\chi(a_j\mathfrak{D}\mathfrak{f})\mathrm{e}^{2\pi\mathrm{i}\,\mathrm{Tr}_{K/\mathbb{Q}}(a_j)}.$$

Let $\gamma\in\mathsf{J}_K$ be such that $\gamma_v=1$ for all $v\in\mathsf{M}_K^\infty$ and $\iota_K(\gamma)=\mathfrak{D}\mathfrak{f}$ (see 5.3.2). If $j\in[1,\rho]$, then $v(a_j\gamma_v)=v(a_j\mathfrak{D}\mathfrak{f})=v(a_j)+v(\mathfrak{D}\mathfrak{f})\geq 0$ for all $v\in\mathsf{M}_K^0$, hence there exists some totally positive $b_j\in\mathfrak{o}_K$ such that $b_j\equiv a_j\gamma_v \bmod \mathfrak{p}_v^{f_v}$ for all $v\in\mathrm{supp}(\mathfrak{D}\mathfrak{f})$, and $(b_j,\mathfrak{f})=1$ if and only if $(a_j\mathfrak{D}\mathfrak{f},\mathfrak{f})=1$, that is, if and only if $j\in[1,m]$.

We assert that $\{b_1,\ldots,b_\rho\}$ is a set of representatives for $\mathfrak{o}_K/\mathfrak{f}$ (then it follows that $\{b_1,\ldots,b_m\}$ is a set of representatives for $(\mathfrak{o}_K/\mathfrak{f})^\times$). Since $(\mathfrak{D}\mathfrak{f})^{-1}/\mathfrak{D}^{-1}\cong\mathfrak{o}_K/\mathfrak{f}$ (see [18, Corollary 2.9.3]), it is sufficient to prove that $b_j\not\equiv b_k \bmod \mathfrak{f}$ for all $j,\,k\in[1,\rho]$ such that $j\neq k$. Thus let $j,\,k\in[1,\rho]$ and $b_j\equiv b_k \bmod \mathfrak{f}$. Then

$$0=v(b_j-b_k)=v(a_j-a_k)+v(\gamma_v)\geq v(\mathfrak{f})$$

and thus $v(a_j-a_k)\geq v(\mathfrak{f})-v(\gamma_v)=v(\mathfrak{D}^{-1})$ for all $v\in\mathrm{supp}(\mathfrak{D}\mathfrak{f})$, hence $a_j-a_k\in\mathfrak{D}^{-1}$ and therefore $j=k$.

For $j\in[1,m]$ and $v\in\mathrm{supp}(\mathfrak{f})$ let $m_v\in\mathbb{N}$ and $b_j^{(v)}\in\mathfrak{o}_K$ for $j\in[1,m_v]$ be as in **I**. Then $a_j\gamma_v\equiv b_j\equiv b_{j_v}^{(v)} \bmod \mathfrak{p}_v^{f_v}$, hence $\chi_v(a_j\gamma_v)=\chi_v(b_j)=\chi_v(b_{j_v}^{(v)})$, and as $\gamma_v=1$ for all $v\in\mathsf{M}_K^\infty$, we obtain

$$\chi(a_j\mathfrak{D}\mathfrak{f})=\chi(\iota_K(a_j\gamma))=\overline{\chi}(a_j\gamma)\prod_{v\in\mathrm{supp}(\mathfrak{f})}\chi_v(a_j\gamma_v)=\overline{\chi}(\gamma)\prod_{v\in\mathrm{supp}(\mathfrak{f})}\chi_v(b_{j_v}^{(v)})$$

$$=\overline{\chi}(\gamma)\prod_{v\in\mathrm{supp}(\mathfrak{f})}\chi_v\Big(\frac{b_{j_v}^{(v)}}{\gamma_v}\Big)\chi_v(\gamma_v)=\prod_{v\in\mathsf{M}_K^0\setminus\mathrm{supp}(\mathfrak{f})}\overline{\chi}_v(\gamma_v)\prod_{v\in\mathrm{supp}(\mathfrak{f})}\chi_v\Big(\frac{b_{j_v}^{(v)}}{\gamma_v}\Big).$$

If $v\in\mathrm{supp}(\mathfrak{f})$, then

$$v\Big(\frac{b_j^{(v)}}{\gamma_v}-a_j\Big)=v(b_{j_v}^{(v)}-a_j\gamma_v)-v(\gamma_v)\geq f_v-v(\mathfrak{D}\mathfrak{f})=-v(\mathfrak{D})=v(\mathfrak{D}_{K_v/\mathbb{Q}_p}^{-1}),$$

and therefore

$$\lambda_v(a_j)=\lambda_v\Big(\frac{b_j^{(v)}}{\gamma_v}\Big).$$

If $v \in \mathsf{M}_K^0 \setminus \operatorname{supp}(\mathfrak{f})$, then $v(a_j) \geq -v(\mathfrak{D}\mathfrak{f}) = -v(\mathfrak{D}_{K_v/\mathbb{Q}_p})$, hence $\lambda_v(a_j) = 0$, and $\overline{\chi}_v(\gamma_v) = \tau_v(\chi)$ by 7.1.1.2. Allowing for all this and using 7.1.2 we obtain

$$\begin{aligned}
\tau(\chi) &= \sum_{j=1}^m \chi(a_j\mathfrak{D}\mathfrak{f})\mathsf{e}^{2\pi\mathsf{i}\,\mathsf{Tr}_{K/\mathbb{Q}}(a_j)} \\
&= \sum_{j=1}^m \prod_{v\in\mathsf{M}_K^0\setminus\operatorname{supp}(\mathfrak{f})} \overline{\chi}_v(\gamma_v) \prod_{v\in\operatorname{supp}(\mathfrak{f})} \chi_v\Big(\frac{b_{j_v}^{(v)}}{\gamma_v}\Big) \prod_{v\in\mathsf{M}_K^0} \lambda_v(a_j) \\
&= \prod_{v\in\mathsf{M}_K^0\setminus\operatorname{supp}(\mathfrak{f})} \tau_v(\chi) \sum_{j=1}^m \prod_{v\in\operatorname{supp}(\mathfrak{f})} \chi_v\Big(\frac{b_{j_v}^{(v)}}{\gamma_v}\Big)\lambda_v\Big(\frac{b_{j_v}^{(v)}}{\gamma_v}\Big) \\
&= \prod_{v\in\mathsf{M}_K^0\setminus\operatorname{supp}(\mathfrak{f})} \tau_v(\chi) \prod_{v\in\operatorname{supp}(\mathfrak{f})} \sum_{j_v}^{m_v} \chi_v\Big(\frac{b_{j_v}^{(v)}}{\gamma_v}\Big)\lambda_v\Big(\frac{b_{j_v}^{(v)}}{\gamma_v}\Big) \\
&= \prod_{v\in\mathsf{M}_K^0\setminus\operatorname{supp}(\mathfrak{f})} \tau_v(\chi) \prod_{v\in\operatorname{supp}(\mathfrak{f})} \tau_v(\chi) = \prod_{v\in\mathsf{M}_K^0} \tau_v(\chi).
\end{aligned}$$

2. By 1. and 7.1.1.1 we obtain

$$F(\chi,c) = \prod_{v\in\mathsf{M}_K^0} F_v(\chi,c) = \prod_{v\in\mathsf{M}_K^0} [\tau_v(\chi)M_v(\chi,c)] = \tau(\chi) \prod_{v\in\mathsf{M}_K^0} M_v(\chi,c).$$

CASE 1: There exists some $v \in \operatorname{supp}(\mathfrak{f})$ such that $v(c) > -v(\mathfrak{D}\mathfrak{f})$. As $f_v > 0$, 7.1.1.1 implies $M_v(\chi,c) = 0$ and consequently $F(\chi,c) = 0$. Since $\overline{\chi}(\mathfrak{f}\mathfrak{D}) = \overline{\chi}(c\mathfrak{D}\mathfrak{f}) = 0$, we obtain $M(\chi,c) = 0$, too.

CASE 2: $v(c) = -v(\mathfrak{D}\mathfrak{f})$ for all $v \in \operatorname{supp}(\mathfrak{f})$. For each $v \in \mathsf{M}_K^0$ fix some $d_v \in K$ such that $v(d_v) = v(\mathfrak{D}\mathfrak{f})$.

CASE 2a: $c = 0$. By assumption, $\operatorname{supp}(\mathfrak{f}) = \emptyset$ and thus $\mathfrak{f} = \mathbf{1}$. Let $\gamma \in \mathsf{J}_K$ be defined by $\gamma_v = d_v$ for all $v \in \mathsf{M}_K^0$ and $\gamma_v = 1$ for all $v \in \mathsf{M}_K^\infty$. Then $\iota_K(\gamma) = \mathfrak{D}$, and 7.1.1.1 implies

$$\prod_{v\in\mathsf{M}_K^0} M_v(\chi,0) = \prod_{v\in\mathsf{M}_K^0} \chi_v(d_v) = \chi(\gamma) = \overline{\chi}(\mathfrak{D}) = M(\chi,0).$$

CASE 2b: $c \neq 0$. Then $v(c\mathfrak{D}\mathfrak{f}) = 0$ for all $v \in \operatorname{supp}(\mathfrak{f})$, and therefore $(c\mathfrak{D}\mathfrak{f},\mathfrak{f}) = 1$ (note that apparently this also holds if $\mathfrak{f} = 1$). Let $\gamma \in \mathsf{J}_K$ be defined by

$$\gamma_v = \begin{cases} c^{-1} & \text{if } v \in \operatorname{supp}(\mathfrak{f}), \\ d_v & \text{if } v \in \mathsf{M}_K^0 \setminus \operatorname{supp}(\mathfrak{f}), \\ 1 & \text{if } v \in \mathsf{M}_K^\infty. \end{cases}$$

Then $v(c\gamma_v) = v(c\mathfrak{D}\mathfrak{f})$ for all $v \in \mathsf{M}_K^0$, hence $\iota_K(c\gamma) = c\mathfrak{D}\mathfrak{f}$ and

$$\begin{aligned}
\chi(c\mathfrak{D}\mathfrak{f}) = \overline{\chi}(c\gamma) \prod_{v\in\operatorname{supp}(\mathfrak{f})\cup\mathsf{M}_K^\infty} \chi_v(c\gamma_v) = \overline{\chi}(\gamma)\Big[\prod_{v\in\operatorname{supp}(\mathfrak{f})} \chi_v(c\gamma_v)\Big]\chi_\infty(c) \\
= \overline{\chi}(\gamma)\chi_\infty(c).
\end{aligned}$$

By 7.1.1.1, we eventually obtain

$$\prod_{v\in \mathsf{M}_K^0} M_v(\chi,c) = \prod_{v\in \mathsf{M}_K^0} \chi_v(\gamma_v) = \chi(\gamma) = \overline{\chi}(c\mathfrak{D}\mathfrak{f})\chi_\infty(c) = M(\chi,c).$$

3. By definition $\chi_\infty(-1) = (-1)^{q_\chi}$. By 1. and 7.1.1.3 it follows that

$$\tau(\overline{\chi}) = \prod_{v\in \mathsf{M}_K^0} \tau_v(\overline{\chi}) = \prod_{v\in \mathsf{M}_K^0} \Big[\chi_v(-1)\overline{\tau_v(\chi)}\Big] = \prod_{v\in \mathsf{M}_K^\infty} \chi_v(-1)\,\overline{\prod_{v\in \mathsf{M}_K^0} \tau_v(\chi)}$$
$$= (-1)^{q_\chi}\overline{\tau(\chi)}$$

and

$$|\tau(\chi)|^2 = \prod_{v\in \mathsf{M}_K^0} |\tau_v(\chi)|^2 = \prod_{v\in \mathsf{M}_K^0} \mathfrak{N}(\mathfrak{p}_v^{f_v}) = \mathfrak{N}(\mathfrak{f}).$$

4. By the very definition of $\mathsf{W}(\chi)$, 2. implies (since $q_\chi = q_{\overline{\chi}}$)

$$\overline{\mathsf{W}(\chi)} = \frac{\mathrm{i}^{q_\chi}\,\overline{\tau(\chi)}}{\sqrt{\mathfrak{N}(\mathfrak{f})}} = \frac{\mathrm{i}^{q_\chi}(-1)^{q_\chi}\tau(\overline{\chi})}{\sqrt{\mathfrak{N}(\mathfrak{f})}} = \mathsf{W}(\overline{\chi}) \quad\text{and}\quad |\mathsf{W}(\chi)| = \frac{|\tau(\chi)|}{\sqrt{\mathfrak{N}(\mathfrak{f})}} = 1.$$

5. Obvious by the definitions. □

In the following analytical considerations we denote for $\vartheta\in\mathbb{R}$ by

$$\mathcal{H}_\vartheta = \{s\in\mathbb{C} \mid \Re(s) > \vartheta\}$$

the open right half-plane.

Before we proceed with the functional equation of L-functions we recall the definition and the main properties of the Gamma function.

Definition and Theorem 7.1.5. For $s\in\mathcal{H}_0$ the Gamma function is defined by the absolutely convergent improper integral

$$\Gamma(s) = \int_0^\infty t^{s-1}\mathrm{e}^{-t}dt.$$

1. The Gamma function has an extension to a zero-free holomorphic function in $\mathbb{C}\setminus(-\mathbb{N}_0)$*. If* $n\in\mathbb{N}_0$*, then* Γ *has a simple pole at* $-n$*, and*

$$\mathrm{Res}\{\Gamma;-n\} = \frac{(-1)^n}{n!}.$$

2. If $s\in\mathbb{C}$*, then*

$$\Gamma(s+1) = s\Gamma(s), \quad \Gamma(s)\Gamma(1-s) = \frac{\pi}{\sin\pi s},$$

$$\Gamma(s)\Gamma\big(s+\frac{1}{2}\big) = 2^{1-2s}\sqrt{\pi}\,\Gamma(2s), \quad \textit{and} \quad \Gamma\big(\frac{1}{2}\big) = \sqrt{\pi}.$$

3. $\Gamma(k+1) = k!$ *for all* $k \in \mathbb{N}_0$.

Proof. [52, Ch. A.2] □

Definition and Remarks 7.1.6. Let $\mathfrak{f} \in \mathcal{J}'_K$ be an ideal and $\chi \colon \mathcal{J}_K^{\mathfrak{f}} \to \mathbb{T}$ a ray class character modulo $\mathfrak{f}$. Then the **(generalized) Dirichlet L function** built with χ is defined by the (for $s \in \mathcal{H}_1$ convergent) Dirichlet series

$$L(s,\chi) = \sum_{\mathfrak{a} \in \mathcal{J}'_K} \frac{\chi(\mathfrak{a})}{\mathfrak{N}(\mathfrak{a})^s}.$$

The following properties of $L(s,\chi)$ are well known (see [18, Section 4.2]):

- The function $L(\,\cdot\,,\chi)$ has an extension to a meromorphic function in the half-plane $\mathcal{H}_{1-1/n}$, and there

$$L(s,\chi) = \frac{\rho^*(\mathfrak{f})}{s-1}\varepsilon(\chi) + G(s,\chi),$$

  where $G(\,\cdot\,,\chi)$ is holomorphic in $\mathcal{H}_{1-1/n}$,

$$\rho^*(\mathfrak{f}) = \frac{2^{r_1}(2\pi)^{r_2} h_K \mathsf{R}_K}{w_K\sqrt{|\mathsf{d}_K|}} \prod_{\mathfrak{p} \in \mathrm{supp}(\mathfrak{f})} \Big(1 - \frac{1}{\mathfrak{N}(\mathfrak{p})^s}\Big) \quad \text{and} \quad \varepsilon(\chi) = \begin{cases} 1 & \text{if } \chi = 1, \\ 0 & \text{if } \chi \neq 1. \end{cases}$$

- If $\Re(s) > 1$, then

$$L(s,\chi) = \prod_{\mathfrak{p} \in \mathcal{P}_K} \frac{1}{1 - \chi(\mathfrak{p})\mathfrak{N}(\mathfrak{p})^{-s}} \neq 0.$$

- Let $\mathfrak{f}_0 = \mathfrak{f}(\chi)$ be the conductor of χ and $\chi_0 \colon \mathcal{J}_K^{\mathfrak{f}_0} \to \mathbb{T}$ the associated primitive ray class character (see 5.8.2). Then

$$L(s,\chi) = L(s,\chi_0) \prod_{\mathfrak{p} \in \mathrm{supp}(\mathfrak{f})} (1 - \chi_0(\mathfrak{p})\mathfrak{N}(\mathfrak{p})^{-s}).$$

In the sequel we shall limit our investigations to the analytical properties of $L(s,\chi)$ for primitive ray class characters.

Thus let $L(s,\chi)$ be the (generalized) Dirichlet L series built with the primitive ray class character χ with conductor $\mathfrak{f} = \mathfrak{f}(\chi)$ and $q = q_\chi$. For an ideal class $C \in \mathcal{C}_K$ and $s \in \mathcal{H}_1$ we set

$$L(s,\chi,C) = \sum_{\mathfrak{a} \in \mathcal{J}'_K \cap C} \frac{\chi(\mathfrak{a})}{\mathfrak{N}(\mathfrak{a})^s}$$

and

$$\Lambda(s,\chi,C) = 2^{r_2(1-s)}\pi^{-(q+ns)/2}\mathfrak{N}(\mathfrak{D}\mathfrak{f})^{s/2}\gamma(s,\chi)L(s,\chi,C),$$

where

$$\gamma(s,\chi) = \Gamma\Big(\frac{s+1}{2}\Big)^q\Gamma\Big(\frac{s}{2}\Big)^{r_1-q}\Gamma(s)^{r_2}.$$

It follows that

$$L(s,\chi) = \sum_{C\in\mathcal{C}_K} L(s,\chi,C), \quad \text{and we set} \quad \Lambda(s,\chi) = \sum_{C\in\mathcal{C}_K} \Lambda(s,\chi,C).$$

Then

$$\Lambda(s,\chi) = 2^{r_2(1-s)}\pi^{-(q+ns)/2}\mathfrak{N}(\mathfrak{D}\mathfrak{f})^{s/2}\Gamma\Big(\frac{s+1}{2}\Big)^q\Gamma\Big(\frac{s}{2}\Big)^{r_1-q}\Gamma(s)^{r_2}L(s,\chi),$$

and $L(s,\chi) = A(s,\chi)\Lambda(s,\chi)$, where $A(\,\cdot\,,\chi)$ is an entire function without zeros.
The function $\Lambda(\,\cdot\,,\chi)$ is called the **extended Dirichlet L function** built with the ray class character χ.

Theorem 7.1.7. *Let $\mathfrak{f}\in\mathcal{J}'_K$ be an ideal, $\chi\colon \mathcal{J}^{\mathfrak{f}}_K \to \mathbb{T}$ a primitive ray class character modulo $\mathfrak{f}$ and $q = q_\chi$. Let $C\in\mathcal{C}_K$ and $\overline{C} = C^{-1}[\mathfrak{D}\mathfrak{f}]\in\mathcal{C}_K$.*

1. *The function $\Lambda(\,\cdot\,,\chi,C)$ has an extension to a meromorphic function in $\mathbb{C}$ with (simple) poles at most at 0 and 1, and*

$$\Lambda(s,\chi,C) = \mathsf{W}(\chi)\Lambda(1-s,\overline{\chi},\overline{C}) \quad \textit{for all} \;\; s\in\mathbb{C}.$$

2. *If $\mathfrak{f}\neq\mathbf{1}$ or $q\neq 0$, then $\Lambda(\,\cdot\,,\chi,C)$ is holomorphic in $\mathbb{C}$.*

3. *If $\mathfrak{f}=\mathbf{1}$ and $q=0$, then $\Lambda(\,\cdot\,,\chi,C)$ has simple poles at 0 and 1 with residues*

$$\mathrm{Res}\{\Lambda(s,\chi,C);s=0\} = -\frac{2^r\mathsf{R}_K}{w_K}\chi(C)$$

and

$$\mathrm{Res}\{\Lambda(s,\chi,C);s=1\} = \frac{2^r\mathsf{R}_K}{w_K}\mathsf{W}(\chi)\overline{\chi}(\overline{C}).$$

The proof of 7.1.7 will occupy the whole next section. We close the present section with some important consequences and special cases.

Theorem 7.1.8 (Functional equation of the L function)**.** *Let $\mathfrak{f}\in\mathcal{J}'_K$ be an ideal, $\chi\colon\mathcal{J}^{\mathfrak{f}}_K\to\mathbb{T}$ a primitive ray class character modulo $\mathfrak{f}$ and $q=q_\chi$.*

1. The function $\Lambda(\,\cdot\,,\chi)$ *has an extension to a meromorphic function in* $\mathbb{C}$ *with poles at most at* $s=0$ *and* $s=1$*, and*

$$\Lambda(s,\chi) = \mathsf{W}(\chi)\Lambda(1-s,\overline{\chi}) \quad \textit{for all} \;\; s \in \mathbb{C}.$$

2. If $\chi \neq 1$*, then the function* $\Lambda(\,\cdot\,,\chi)$ *is holomorphic in* $\mathbb{C}$.

3. The function $\Lambda(\,\cdot\,,1)$ *has simple poles at* $s=0$ *and* $s=1$ *with residues*

$$\operatorname{Res}\{\Lambda(s,1); s=1\} \;=\; -\operatorname{Res}\{\Lambda(s,1); s=0\} \;=\; \frac{2^r \mathsf{R}_K h_K}{w_K}.$$

Proof using 7.1.7. By 7.1.7 the function

$$\Lambda(\,\cdot\,,\chi) = \sum_{C \in \mathcal{C}_K} \Lambda(\,\cdot\,,\chi,C)$$

has an extension to a meromorphic function in $\mathbb{C}$ with poles at most at $s=0$ and $s=1$.

If $\mathfrak{f} \neq \mathbf{1}$ or $q \neq 0$, then all $\Lambda(\,\cdot\,,\chi,C)$ and thus also $\Lambda(\,\cdot\,,\chi)$ are holomorphic in $\mathbb{C}$.

If $\mathfrak{f} = \mathbf{1}$ and $q=0$, then $\chi \in \mathsf{X}(\mathcal{C}_K)$, and we obtain

$$\operatorname{Res}\{\Lambda(s,\chi); s=0\} = \sum_{C \in \mathcal{C}_K} \operatorname{Res}\{\Lambda(s,\chi,C)\; s=0\} = -\frac{2^r \mathsf{R}_K}{w_K} \sum_{C \in \mathcal{C}_K} \chi(C)$$

and

$$\operatorname{Res}\{\Lambda(s,\chi); s=1\} = \sum_{C \in \mathcal{C}_K} \operatorname{Res}\{\Lambda(s,\chi,C)\; s=1\} = \frac{2^r \mathsf{R}_K}{w_K} \mathsf{W}(\chi) \sum_{C \in \mathcal{C}_K} \overline{\chi}(\overline{C})$$

The assertion follows since

$$\sum_{C \in \mathcal{C}_K} \chi(C) = \mathsf{W}(\chi) \sum_{C \in \mathcal{C}_K} \overline{\chi}(\overline{C}) = \begin{cases} h_K & \text{if } \chi = 1, \\ 0 & \text{if } \chi \neq 1. \end{cases}$$

In particular, if $\chi \neq 1$, then $\Lambda(s,\chi)$ is holomorphic in $\mathbb{C}$. □

Corollary 7.1.9 (Functional equation of the L function, explicit form).

1. Let $\mathfrak{f} \in \mathcal{I}_K'$ *be an ideal,* $\chi\colon \mathcal{I}_K^{\mathfrak{f}} \to \mathbb{T}$ *a primitive ray class character modulo* $\mathfrak{f}$ *and* $q = q_\chi$*. Then the Dirichlet L function* $L(\cdot,\chi)$ *satisfies for all* $s \in \mathbb{C}$ *the functional equation*

$$\frac{L(1-s,\overline{\chi})}{L(s,\chi)}$$
$$= \mathsf{i}^{-q}\tau(\overline{\chi})[2(2\pi)^{-s}\Gamma(s)]^n |\mathsf{d}_K|^{s-1/2}\mathfrak{N}(\mathfrak{f})^{s-1}\Big(\cos\frac{\pi s}{2}\Big)^{r-q}\Big(\sin\frac{\pi s}{2}\Big)^{r_2+q}.$$

2. Let $f \in \mathbb{N}$, $f \ge 3$ and $\varphi \in \mathsf{X}(n)$ a Dirichlet character modulo f and $q \in \{0,1\}$ such that $\varphi(-1) = (-1)^q$. Then the (classical) Dirichlet L function and the (classical) Gauss sum attached with φ are given by

$$L(s,\varphi) = \sum_{n=1}^{\infty} \frac{\varphi(n)}{n^s} \quad \text{for } s \in \mathcal{H}_1, \quad \text{and} \quad \tau(\varphi) = \sum_{\substack{a=1 \\ (a,f)=1}}^{f} \varphi(a)\mathrm{e}^{2\pi \mathrm{i} a/f}.$$

$L(\,\cdot\,,\varphi)$ has an extension to a meromorphis function on $\mathbb{C}$, and satisfies for all $s \in \mathbb{C}$ the functional equation

$$\frac{L(1-s,\overline{\varphi})}{L(s,\varphi)} = \frac{2f^{s-1}\mathrm{i}^q\tau(\overline{\chi})}{(2\pi)^s}\,\Gamma(s)\cos\frac{\pi(s+q)}{2}.$$

Proof. 1. We start with the extended Dirichlet L function

$$\Lambda(s,\chi) = 2^{r_2(1-s)}\pi^{-(q+ns)/2}\mathfrak{N}(\mathfrak{D}\mathfrak{f})^{s/2}\Gamma\Big(\frac{s+1}{2}\Big)^q\Gamma\Big(\frac{s}{2}\Big)^{r_1-q}\Gamma(s)^{r_2}L(s,\chi),$$

and its functional equation $\Lambda(s,\chi) = \mathsf{W}(\chi)\Lambda(1-s,\overline{\chi})$ (see 7.1.8.1). We rewrite it for $L(s,\chi)$ and obtain

$$L(s,\chi) = \mathsf{W}(\chi)2^{(2s-1)r_2}\pi^{n(2s-1)/2}\mathfrak{N}(\mathfrak{D}\mathfrak{f})^{(1-2s)/2}\Gamma^*(s)L(1-s,\overline{\chi}),$$

where

$$\Gamma^*(s) = \Big(\frac{\Gamma(1-\frac{s}{2})}{\Gamma(\frac{s+1}{2})}\Big)^q\Big(\frac{\Gamma(\frac{1-s}{2})}{\Gamma(\frac{s}{2})}\Big)^{r_1-q}\Big(\frac{\Gamma(1-s)}{\Gamma(s)}\Big)^{r_2}.$$

We calculate the Gamma factors by means of 7.1.5 as follows.

$$\frac{\Gamma(1-\frac{s}{2})}{\Gamma(\frac{s+1}{2})} = \frac{\pi}{\Gamma(\frac{s}{2})\Gamma(\frac{s+1}{2})\sin\frac{\pi s}{2}} = \frac{2^{s-1}\sqrt{\pi}}{\Gamma(s)\sin\frac{\pi s}{2}},$$

$$\frac{\Gamma(\frac{1-s}{2})}{\Gamma(\frac{s}{2})} = \frac{\Gamma(\frac{1-s}{2})\Gamma(\frac{1+s}{2})}{\Gamma(\frac{s}{2})\Gamma(\frac{1+s}{2})} = \frac{\Gamma(\frac{1-s}{2})\Gamma(1-\frac{1-s}{2})}{\Gamma(\frac{s}{2})\Gamma(\frac{1+s}{2})} = \frac{\pi}{\sin\frac{\pi(1-s)}{2}}\cdot\frac{1}{2^{1-s}\sqrt{\pi}\Gamma(s)}$$
$$= \frac{2^{s-1}\sqrt{\pi}}{\Gamma(s)\cos\frac{\pi s}{2}},$$

and

$$\frac{\Gamma(1-s)}{\Gamma(s)} = \frac{\pi}{\Gamma(s)^2\sin\pi s}.$$

This amounts to

$$\Gamma^*(s) = \frac{2^{(s-1)r_1-r_2}\pi^{n/2}}{\Gamma(s)^n\big(\cos\frac{\pi s}{2}\big)^{r_1+r_2-q}\big(\sin\frac{\pi s}{2}\big)^{q+r_2}}$$

and

$$L(s,\chi) = \frac{\mathsf{W}(\chi)2^{s-1)n}\pi^{ns}\mathfrak{N}(\mathfrak{D}\mathfrak{f})^{(1-2s)/2}}{\Gamma(s)^n\big(\cos\frac{\pi s}{2}\big)^{r_1+r_2-q}\big(\sin\frac{\pi s}{2}\big)^{q+r_2}}\, L(1-s,\overline{\chi})$$

Since

$$\mathsf{W}(\chi) = \frac{\sqrt{\mathfrak{N}(\mathfrak{f})}}{\mathsf{i}^{-q}\tau(\overline{\chi})} \quad \text{and} \quad \mathfrak{N}(\mathfrak{D}\mathfrak{f}) = |\mathsf{d}_K|\mathfrak{N}(\mathfrak{f}),$$

it eventually follows that

$$\frac{L(1-s,\overline{\chi})}{L(s,\chi)} = \mathsf{i}^{-q}\tau(\overline{\chi})[2(2\pi)^{-s}\Gamma(s)]^n|\mathsf{d}_K|^{s-1/2}\mathfrak{N}(\mathfrak{f})^{s-1}\Big(\cos\frac{\pi s}{2}\Big)^{r-q}\Big(\sin\frac{\pi s}{2}\Big)^{r_2+q}.$$

2. Let χ be the (primitive) ray class character modulo $f\mathbb{Z}$ with associated Dirichlet character φ. Then

$$\sum_{n=1}^{\infty}\frac{\varphi(n)}{n^s} = \sum_{n=1}^{\infty}\frac{\chi(n\mathbb{Z})}{\mathfrak{N}(n\mathbb{Z})^s} = L(s,\chi),$$

and we may apply 1. Since $f^{-1}[1,f]$ is a set of representatives for $f^{-1}\mathbb{Z}/\mathbb{Z}$, it follows that

$$\tau(\chi) = \sum_{a=1}^{f}\chi(a\mathbb{Z})\mathsf{e}^{2\pi\mathsf{i}a/f} = \sum_{\substack{a=1\\(a,f)=1}}^{f}\varphi(a)\mathsf{e}^{2\pi\mathsf{i}a/f} = \tau(\varphi).$$

Hence we obtain

$$\frac{L(1-s,\overline{\varphi})}{L(s,\varphi)} = \frac{2f^{s-1}\mathsf{i}^{-q}\tau(\overline{\varphi})}{(2\pi)^s}\Gamma(s)\big(\cos\frac{\pi s}{2}\big)^{1-q}\big(\sin\frac{\pi s}{2}\big)^{q}.$$

If $q=0$, this is the assertion. If $q=1$, the assertion follows, since

$$\sin\frac{\pi s}{2} = -\cos\frac{\pi(s+1)}{2}. \qquad \square$$

To conclude, we formulate our results for the Dedekind and the Riemann zeta function.

Corollary 7.1.10.

1. *The Dedekind zeta function*

$$\zeta_K(s) = \sum_{\mathfrak{a}\in\mathcal{J}_K'}\frac{1}{\mathfrak{N}(\mathfrak{a})^s} \quad \textit{for } s\in\mathcal{H}_1$$

has an extension to a holomorphic function in $\mathbb{C}\setminus\{1\}$ *with a simple pole at* 1. *It satisfies the functional equation*

$$\zeta_K(1-s) = [2(2\pi)^{-s}\Gamma(s)]^n|\mathsf{d}_K|^{s-1/2}\Big(\cos\frac{\pi s}{2}\Big)^{r}\Big(\sin\frac{\pi s}{2}\Big)^{r_2}\zeta_K(s),$$

$$\operatorname{Res}\{\zeta_K(s); s=1\} = \frac{2^{r_1}(2\pi)^{r_2}\mathsf{R}_K h_K}{w_K\sqrt{|\mathsf{d}_K|}}, \quad \textit{and} \quad \lim_{s\to 0}\frac{\zeta_K(s)}{s^{r-1}} = -\frac{\mathsf{R}_K h_K}{w_K}.$$

2. The Riemann zeta function

$$\zeta(s) = \sum_{n=1}^{\infty}\frac{1}{n^s} \quad \textit{for} \quad s \in \mathcal{H}_1$$

has an extension to a holomorphic function in $\mathbb{C}\setminus\{1\}$ *with a simple pole at* $s=1$*. It satisfies the functional equation*

$$\zeta(1-s) = 2(2\pi)^{-s}\Gamma(s)\cos\frac{\pi s}{2}\,\zeta(s), \quad \operatorname{Res}\{\zeta(s); s=1\} = 1 \ \textit{ and } \ \zeta(0) = -\frac{1}{2}.$$

Proof. 1. The functional equation follows by 7.1.9.1 with $\chi = 1$. As to the behavior at $s=0$ and $s=1$, we set

$$Z_K(s) = \Lambda(s,1) = 2^{r_2(1-s)}\pi^{-ns/2}\sqrt{|\mathsf{d}_K|}\,\Gamma\Big(\frac{s}{2}\Big)^{r_1}\Gamma(s)^{r_2}$$

and apply 7.1.8.3 to obtain

$$\begin{aligned}\frac{2^r\mathsf{R}_K h_K}{w_K} &= \operatorname{Res}\{Z_K(s); s=1\}\\ &= \lim_{s\to 1}(s-1)Z_K(s) = \pi^{-n/2}\sqrt{|\mathsf{d}_K|}\,\pi^{r_1/2}\lim_{s\to 1}(s-1)\zeta_K(s),\end{aligned}$$

and consequently

$$\operatorname{Res}\{\zeta_K(s); s=1\} = \lim_{s\to 1}(s-1)\zeta_K(s) = \frac{2^{r_1}(2\pi)^{r_2}\mathsf{R}_K h_K}{w_K\sqrt{|\mathsf{d}_K|}}.$$

On the other hand,

$$\begin{aligned}-\frac{2^r\mathsf{R}_K h_K}{w_K} &= \operatorname{Res}\{Z_K(s); s=0\} = \lim_{s\to 0} sZ_K(s)\\ &= 2^r\lim_{s\to 0}\Big[\Big(\frac{s}{2}\Gamma\big(\frac{s}{2}\big)\Big)^{r_1}\big(s\Gamma(s)\big)^{r_2}\frac{\zeta_K(s)}{s^{r-1}}\Big] = 2^r\lim_{s\to 0}\frac{\zeta_K(s)}{s^{r-1}},\end{aligned}$$

and therefore

$$\lim_{s\to 0}\frac{\zeta_K(s)}{s^{r-1}} = -\frac{\mathsf{R}_K h_K}{w_K}.$$

2. Obvious by 1. □

7.2 Further analytic tools

Before we proceed with the proof of the functional equation, we provide three essential analytic tools: Fourier transforms, the Poisson summation formula, and the elementary properties of theta series. For the frequent interchange of integration and differentiation throughout this section we use [10, Theorem 13.8].

Definitions and Remarks 7.2.1. Let $n \in \mathbb{N}$ and $\|\cdot\|$ be the Euclidean norm on $\mathbb{R}^n$. A function $f\colon \mathbb{R}^n \to \mathbb{C}$ is called

- **rapidly decreasing** if

$$\lim_{\|\boldsymbol{x}\|\to\infty} \|\boldsymbol{x}\|^m f(\boldsymbol{x}) = 0 \quad \text{for all} \quad m \in \mathbb{N}_0$$

  (equivalently, for each $m \in \mathbb{N}$ the function $\boldsymbol{x} \mapsto \|\boldsymbol{x}\|^m f(\boldsymbol{x})$ is bounded);

- a **Schwartz function** if it is a C^∞-function such that the function itself and all its (higher) partial derivatives are rapidly decreasing (and therefore they are themselves Schwartz functions).

The set $\boldsymbol{S}(n)$ of all Schwartz functions $f\colon \mathbb{R}^n \to \mathbb{C}$ is a vector space over $\mathbb{C}$. If $f \in \boldsymbol{S}(n)$, then there is some $C \in \mathbb{R}_{>0}$ such that

$$|f(\boldsymbol{x})| \le \frac{C}{(1+x_1^2)\cdot\ldots\cdot(1+x_n^2)} \quad \text{for all} \quad \boldsymbol{x} = (x_1,\ldots,x_n)^{\mathrm{t}} \in \mathbb{R}^n.$$

Hence f is integrable, and

$$\int_{\mathbb{R}^n} f(\boldsymbol{x})d\boldsymbol{x} = \int_{-\infty}^{\infty}\cdots\int_{-\infty}^{\infty} f(x_1,\ldots,x_n)dx_1\ldots dx_n.$$

For $j \in [1,n]$ we define the functions $D_j,\ M_j\colon \boldsymbol{S}(n) \to \boldsymbol{S}(n)$ by

$$D_j f(\boldsymbol{x}) = \frac{\partial f}{\partial x_j}(\boldsymbol{x}) \quad \text{and} \quad M_j f(\boldsymbol{x}) = x_j f(\boldsymbol{x}).$$

For $p = (p_1,\ldots,p_n) \in \mathbb{N}_0^n$ we set

$$|p| = p_1 + \ldots + p_n, \quad D^p = D_1^{p_1}\ldots D_n^{p_n} \quad \text{and} \quad M^p = M_1^{p_1}\ldots M_n^{p_n}.$$

For a function $f \in \boldsymbol{S}(n)$ we define its **Fourier transform** $\hat{f}\colon \mathbb{R}^n \to \mathbb{C}$ by

$$\hat{f}(\boldsymbol{y}) = \int_{\mathbb{R}^n} f(\boldsymbol{x})\mathrm{e}^{-2\pi\mathrm{i}\langle\boldsymbol{x},\boldsymbol{y}\rangle}d\boldsymbol{x}.$$

Theorem 7.2.2. *Let $n \in \mathbb{N}$. The assignment $f \mapsto \hat{f}$ defines a $\mathbb{C}$-linear map $\boldsymbol{S}(n) \to \boldsymbol{S}(n)$. If $f \in \boldsymbol{S}(n)$ and $p \in \mathbb{N}_0^n$, then*

$$D^p \hat{f} = (-2\pi\mathsf{i})^{|p|}(M^p f)\hat{}\quad \textit{and} \quad M^p \hat{f} = (2\pi\mathsf{i})^{-|p|}(D^p f)\hat{}. \qquad (\dagger)$$

Proof. Let $f \in \boldsymbol{S}(n)$. Then obviously $D^p f \in \boldsymbol{S}(n)$ and $M^p f \in \boldsymbol{S}(n)$ for all $p \in \mathbb{N}_0^n$. For every $\boldsymbol{y} \in \mathbb{R}^n$ the function $\boldsymbol{x} \mapsto f(\boldsymbol{x})\mathsf{e}^{-2\pi\mathsf{i}\langle \boldsymbol{x},\boldsymbol{y}\rangle}$ and its partial derivatives

$$\frac{\partial}{\partial y_j} f(\boldsymbol{x})\mathsf{e}^{-2\pi\mathsf{i}\langle \boldsymbol{x},\boldsymbol{y}\rangle} = -2\pi\mathsf{i} x_j f(\boldsymbol{x})\mathsf{e}^{-2\pi\mathsf{i}\langle \boldsymbol{x},\boldsymbol{y}\rangle} \quad \text{for all} \quad j \in [1,n]$$

are Schwartz functions. Consequently $\hat{f}$ is a C^∞-function.

Next we prove $(\dagger)$ by induction on $|p|$, and therefore it suffices to show that for $f \in \boldsymbol{S}(n)$ and all $j \in [1,n]$

$$D_j \hat{f} = -2\pi\mathsf{i}(M_j f)\hat{}\quad \text{and} \quad M_j \hat{f} = \frac{1}{2\pi\mathsf{i}}(D_j f)\hat{}.$$

Thus let $f \in \boldsymbol{S}(n)$ and $j \in [1,n]$. Then

$$\begin{aligned} D_j \hat{f}(\boldsymbol{y}) &= \frac{\partial}{\partial y_j}\int_{\mathbb{R}^n} f(\boldsymbol{x})\mathsf{e}^{-2\pi\mathsf{i}\langle \boldsymbol{x},\boldsymbol{y}\rangle}\boldsymbol{x} = -2\pi\mathsf{i}\int_{\mathbb{R}^n} x_j f(\boldsymbol{x})\mathsf{e}^{-2\pi\mathsf{i}\langle \boldsymbol{x},\boldsymbol{y}\rangle}d\boldsymbol{x} \\ &= -2\pi\mathsf{i}(M_j f)\hat{}(\boldsymbol{y}), \end{aligned}$$

and consequently $D_j \hat{f} = -2\pi\mathsf{i}(M_j f)\hat{}$. On the other hand,

$$M_j \hat{f}(\boldsymbol{y}) = y_j \int_{\mathbb{R}^n} f(\boldsymbol{x})\mathsf{e}^{-2\pi\mathsf{i}\langle \boldsymbol{x},\boldsymbol{y}\rangle}d\boldsymbol{x} = \int_{\mathbb{R}^n} f(\boldsymbol{x})\frac{\partial}{\partial x_j}\Big(\frac{-1}{2\pi\mathsf{i}}\mathsf{e}^{-2\pi\mathsf{i}\langle \boldsymbol{x},\boldsymbol{y}\rangle}\Big)d\boldsymbol{x},$$

and integration regarding x_j yields

$$\begin{aligned} &\int_{-\infty}^{\infty} f(\boldsymbol{x})\frac{\partial}{\partial x_j}\Big(\frac{-1}{2\pi\mathsf{i}}\mathsf{e}^{-2\pi\mathsf{i}\langle \boldsymbol{x},\boldsymbol{y}\rangle}\Big)dx_j \\ &\quad = \Big[\frac{-1}{2\pi\mathsf{i}}\mathsf{e}^{-2\pi\mathsf{i}\langle \boldsymbol{x},\boldsymbol{y}\rangle}f(\boldsymbol{x})\Big]_{-\infty}^{\infty} + \frac{1}{2\pi\mathsf{i}}\int_{-\infty}^{\infty}\frac{\partial}{\partial x_j}f(\boldsymbol{x})\mathsf{e}^{-2\pi\mathsf{i}\langle \boldsymbol{x},\boldsymbol{y}\rangle}dx_j \\ &\quad = \frac{1}{2\pi\mathsf{i}}\int_{-\infty}^{\infty} D_j f(\boldsymbol{x})\mathsf{e}^{-2\pi\mathsf{i}\langle \boldsymbol{x},\boldsymbol{y}\rangle}dx_j. \end{aligned}$$

We insert this into the n-fold integral and obtain

$$M_j \hat{f}(\boldsymbol{y}) = \frac{1}{2\pi\mathsf{i}}(D_j f)\hat{}. \qquad \Box[(\dagger)]$$

If $f \in \boldsymbol{S}(n)$ and $p \in \mathbb{N}_0^n$, then $M^p f \in \boldsymbol{S}(n)$, and therefore the function $D^p \hat{f} = (-2\pi\mathsf{i})^{|p|}(M^p f)$ is rapidly decreasing. As $\hat{f}$ is a C^∞-function, it follows that $\hat{f} \in \boldsymbol{S}(n)$. Apparently $(f+g)\hat{} = \hat{f} + \hat{g}$, and $(\lambda f)\hat{} = \lambda\hat{f}$ for all $f, g \in \boldsymbol{S}(n)$ and $\lambda \in \mathbb{C}$. Hence $(f \mapsto \hat{f})$ is a $\mathbb{C}$-linear map $\boldsymbol{S}(n) \to \boldsymbol{S}(n)$. $\Box$

For a matrix $A = (a_{i,j})_{i,j\in[1,n]} \in \mathsf{M}_n(\mathbb{C})$ we denote by $\overline{A} = (\overline{a}_{i,j})_{i,j\in[1,n]}$ the conjugate-complex matrix. The following special Fourier transform plays a prominent role in our theory.

Theorem 7.2.3. *Let $B \in \mathsf{GL}_n(\mathbb{R})$ be a symmetric positive definite matrix, and let $Q\colon \mathbb{R}^n \to \mathbb{R}$ be the positive definite quadratic form defined by $Q(\boldsymbol{z}) = \boldsymbol{z}^{\mathrm{t}}B\boldsymbol{z}$ for all $\boldsymbol{z} \in \mathbb{R}^n$. Let $\boldsymbol{w} \in \mathbb{R}^n$, and define $g\colon \mathbb{R}^n \to \mathbb{R}$ by $g(\boldsymbol{z}) = \mathsf{e}^{-\pi Q(\boldsymbol{z}+\boldsymbol{w})}$. Then*

$$\hat{g}(\boldsymbol{y}) = \frac{\mathsf{e}^{2\pi\mathsf{i}\langle\boldsymbol{w},\boldsymbol{y}\rangle}}{|\det(A)|}\,\mathsf{e}^{-\pi Q^*(\boldsymbol{y})}, \quad \textit{where}\ \ Q^*(\boldsymbol{y}) = \boldsymbol{y}^{\mathrm{t}}B^{-1}\boldsymbol{y}.$$

In particular, if $g_0\colon \mathbb{R} \to \mathbb{R}$ is defined by $g_0(z) = \mathsf{e}^{-\pi z^2}$, then $\hat{g}_0 = g_0$.

Proof. Let first $g_0\colon \mathbb{R} \to \mathbb{R}$ be defined by $g_0(z) = \mathsf{e}^{-\pi z^2}$. Then differentiation under the integral sign and integration by parts yields

$$\begin{aligned}
\frac{d\hat{g}(y)}{dy} &= \frac{d}{dy}\int_{-\infty}^{\infty} \mathsf{e}^{-\pi z^2 - 2\pi\mathsf{i}yz}\,dz = \int_{-\infty}^{\infty} -2\pi\mathsf{i}z\mathsf{e}^{-\pi z^2}\mathsf{e}^{-2\pi\mathsf{i}yz}\,dz \\
&= \mathsf{i}\int_{-\infty}^{\infty} \mathsf{e}^{-2\pi\mathsf{i}yz}\frac{d}{dz}\mathsf{e}^{-\pi z^2}\,dz \\
&= \mathsf{i}\Big[\mathsf{e}^{-\pi z^2 - 2\pi\mathsf{i}yz}\Big|_{-\infty}^{\infty} - \int_{-\infty}^{\infty} \mathsf{e}^{-\pi z^2}(-2\pi\mathsf{i}y)\mathsf{e}^{-2\pi\mathsf{i}yz}\,dz\Big] \\
&= -2\pi y\int_{-\infty}^{\infty} \mathsf{e}^{-\pi z^2 - 2\pi\mathsf{i}yz}\,dz \\
&= -2\pi y\hat{g}_0(y).
\end{aligned}$$

From this differential equation we obtain

$$\hat{g}(y) = C\mathsf{e}^{-\pi y^2} \quad\text{and}\quad C = \hat{g}(0) = \int_{-\infty}^{\infty} \mathsf{e}^{-\pi z^2}\,dz = 1.$$

Next we consider the function $g\colon \mathbb{R}^n \to \mathbb{R}$, defined by $g(\boldsymbol{z}) = \mathsf{e}^{-\pi\langle\boldsymbol{z},\boldsymbol{z}\rangle}$. Putting $\boldsymbol{z} = (z_1, \ldots, z_n)$ and $\boldsymbol{y} = (y_1, \ldots, y_n)$, we obtain

$$g(\boldsymbol{z}) = \prod_{j=1}^{n} g_0(z_j),$$

and

$$\begin{aligned}
\hat{g}(\boldsymbol{y}) &= \int_{\mathbb{R}^n} \mathsf{e}^{-\pi\langle\boldsymbol{z},\boldsymbol{z}\rangle - 2\pi\mathsf{i}\langle\boldsymbol{y},\boldsymbol{z}\rangle}\,d\boldsymbol{z} = \int_{-\infty}^{\infty} \cdots \\
&\int_{-\infty}^{\infty} \mathsf{e}^{-\pi(z_1^2+\ldots+z_n^2) - 2\pi\mathsf{i}(y_1z_1+\ldots+y_nz_n)}\,dz_1\ldots dz_n \\
&= \prod_{j=1}^{n}\int_{-\infty}^{\infty} \mathsf{e}^{-\pi z_j^2 - 2\pi\mathsf{i}y_jz_j}\,dz_j = \prod_{j=1}^{n}\hat{g}_0(y_j) = \prod_{j=1}^{n} g_0(y_j) = g(\boldsymbol{y}).
\end{aligned}$$

Now we can do the general case. As B is positive definite, there exists some matrix $D \in \mathsf{GL}_n(\mathbb{R})$ such that $B = D^{\mathrm{t}}D$. Then $Q(\boldsymbol{z}) = \langle D\boldsymbol{z}, D\boldsymbol{z}\rangle$ for all $\boldsymbol{z} \in \mathbb{R}^n$, and

$$\hat{g}(\boldsymbol{y}) = \int_{\mathbb{R}^n} \mathrm{e}^{-\pi Q(\boldsymbol{z}+\boldsymbol{w})-2\pi\mathrm{i}\langle \boldsymbol{z},\boldsymbol{y}\rangle} d\boldsymbol{z} = \int_{\mathbb{R}^n} \mathrm{e}^{-\pi\langle D(\boldsymbol{z}+\boldsymbol{w}),D(\boldsymbol{z}+\boldsymbol{w})\rangle-2\pi\mathrm{i}\langle \boldsymbol{z},\boldsymbol{y}\rangle} d\boldsymbol{z}.$$

We set $\boldsymbol{z}' = D(\boldsymbol{z}+\boldsymbol{w})$. Then $\boldsymbol{z} = D^{-1}\boldsymbol{z}' - \boldsymbol{w}$,

$$\langle \boldsymbol{z}, \boldsymbol{y}\rangle = \langle D^{-1}\boldsymbol{z}', \boldsymbol{y}\rangle - \langle \boldsymbol{w}, \boldsymbol{y}\rangle = \langle \boldsymbol{z}', (D^{-1})^{\mathrm{t}}\boldsymbol{y}\rangle - \langle \boldsymbol{w}, \boldsymbol{y}\rangle, \quad \frac{\partial \boldsymbol{z}}{\partial \boldsymbol{z}'} = \det(D^{-1}),$$

and consequently

$$\begin{aligned}\hat{g}(\boldsymbol{y}) &= \frac{\mathrm{e}^{2\pi\mathrm{i}\langle \boldsymbol{w},\boldsymbol{y}\rangle}}{|\det(D)|}\int_{\mathbb{R}^n} \mathrm{e}^{-\pi\langle \boldsymbol{z}',\boldsymbol{z}'\rangle-2\pi\mathrm{i}\langle \boldsymbol{z}',(D^{-1})^{\mathrm{t}}\boldsymbol{y}\rangle} d\boldsymbol{z}' \\ &= \frac{\mathrm{e}^{2\pi\mathrm{i}\langle \boldsymbol{w},\boldsymbol{y}\rangle}}{|\det(D)|}\mathrm{e}^{-\pi\langle (D^{-1})^{\mathrm{t}}\boldsymbol{y},(D^{-1})^{\mathrm{t}}\boldsymbol{y}\rangle} = \frac{\mathrm{e}^{2\pi\mathrm{i}\langle \boldsymbol{w},\boldsymbol{y}\rangle}}{|\det(B)|}\mathrm{e}^{-\pi Q^*(\boldsymbol{y})},\end{aligned}$$

since $\langle (D^{-1})^{\mathrm{t}}\boldsymbol{y}, (D^{-1})^{\mathrm{t}}\boldsymbol{y}\rangle = \boldsymbol{y}^{\mathrm{t}}D^{-1}(D^{-1})^{\mathrm{t}}\boldsymbol{y} = \boldsymbol{y}^{\mathrm{t}}B^{-1}\boldsymbol{y} = Q^*(\boldsymbol{y})$. □

Definition 7.2.4. A function $f\colon \mathbb{R}^n \to \mathbb{C}$ is called **periodic** if

$$f(\boldsymbol{x}+\boldsymbol{k}) = f(\boldsymbol{x}) \text{ for all } \boldsymbol{x} \in \mathbb{R}^n \text{ and } \boldsymbol{k} \in \mathbb{Z}^n.$$

For a periodic C^∞-function $f\colon \mathbb{R}^n \to \mathbb{C}$ and $\boldsymbol{k} \in \mathbb{Z}^n$ we define the $\boldsymbol{k}$-th **Fourier coefficient** by

$$c_{\boldsymbol{k}} = \int_{[0,1]^n} f(\boldsymbol{x})\mathrm{e}^{-2\pi\mathrm{i}\langle \boldsymbol{k},\boldsymbol{x}\rangle} d\boldsymbol{x}, \quad \text{and we call} \sum_{\boldsymbol{k}\in\mathbb{Z}^n} c_{\boldsymbol{k}}\mathrm{e}^{2\pi\mathrm{i}\langle \boldsymbol{k},\boldsymbol{x}\rangle}$$

the **Fourier series** of f.

Theorem 7.2.5. *Let $f\colon \mathbb{R}^n \to \mathbb{C}$ be a periodic C^∞-function. For every $d \in \mathbb{N}$ there exists some $C \in \mathbb{R}_{>0}$ such that $|c_{\boldsymbol{k}}| \le C\|\boldsymbol{k}\|^{-d}$ for all $\boldsymbol{k} \in \mathbb{Z}^n \setminus \{\boldsymbol{0}\}$. In particular, the Fourier series of f converges normally to f.*

Proof. We may assume that $\|\cdot\|$ is the maximum norm on $\mathbb{R}^n$. Suppose that $\boldsymbol{k} = (k_1, \dots, k_n) \in \mathbb{Z}^n \setminus \{\boldsymbol{0}\}$ and $j \in [1, n]$ is such that $\|\boldsymbol{k}\| = k_j$. Then partial integration re x_j yields, for all $\boldsymbol{x}' = (x_1, \dots, x_{j-1}, x_{j+1}, \dots x_n) \in [0,1]^{n-1}$,

$$\begin{aligned}\int_0^1 f(\boldsymbol{x})\mathrm{e}^{-2\pi\mathrm{i}\langle \boldsymbol{k},\boldsymbol{x}\rangle} dx_j &= \int_0^1 f(\boldsymbol{x})\frac{\partial}{\partial x_j}\Big(-\frac{1}{2\pi\mathrm{i}k_j}\mathrm{e}^{-2\pi\mathrm{i}\langle \boldsymbol{k},\boldsymbol{x}\rangle}\Big) dx_j \\ &= -\frac{1}{2\pi\mathrm{i}k_j}\Big[f(\boldsymbol{x})\mathrm{e}^{-2\pi\mathrm{i}\langle \boldsymbol{k},\boldsymbol{x}\rangle}\Big|_{x_j=0}^{x_j=1} \\ &\quad - \int_0^1 \frac{\partial f}{\partial x_j}(\boldsymbol{x})\mathrm{e}^{-2\pi\mathrm{i}\langle \boldsymbol{k},\boldsymbol{x}\rangle} dx_j\Big] \\ &= \frac{1}{2\pi\mathrm{i}k_j}\int_0^1 \frac{\partial f}{\partial x_j}(\boldsymbol{x})\mathrm{e}^{-2\pi\mathrm{i}\langle \boldsymbol{k},\boldsymbol{x}\rangle} dx_j,\end{aligned}$$

and by a simple induction on d we obtain

$$\int_0^1 f(\boldsymbol{x})\mathrm{e}^{-2\pi\mathrm{i}\langle\boldsymbol{k},\boldsymbol{x}\rangle}d\boldsymbol{x} = \frac{1}{(2\pi\mathrm{i}k_j)^d}\int_0^1 \frac{\partial^d f}{\partial x_j^d}(\boldsymbol{x})\mathrm{e}^{-2\pi\mathrm{i}\langle\boldsymbol{k},\boldsymbol{x}\rangle}d\boldsymbol{x} \quad \text{for all} \quad d \in \mathbb{N}.$$

Integration over ther remaining variables eventually yields the estimate

$$\begin{aligned}
|c_{\boldsymbol{k}}| &= \Big|\int_{[0,1]^n} f(\boldsymbol{x})\mathrm{e}^{-2\pi\mathrm{i}\langle\boldsymbol{k},\boldsymbol{x}\rangle}d\boldsymbol{x}\Big| \\
&= \Big|\int_{[0,1]^{n-1}} \frac{1}{(2\pi\mathrm{i}k_j)^d}\int_0^1 \frac{\partial^d f}{\partial x_j^d}(\boldsymbol{x})\mathrm{e}^{-2\pi\mathrm{i}\langle\boldsymbol{k},\boldsymbol{x}\rangle}dx_j d\boldsymbol{x}'\Big| \\
&\le \frac{1}{(2\pi k_j)^d}\int_{[0,1]^n}\Big|\frac{\partial^d f}{\partial x_j^d}(\boldsymbol{x})\Big|d\boldsymbol{x} \\
&\le \frac{1}{(2\pi)^d\|\boldsymbol{k}\|^d}\sup\Big\{\Big|\frac{\partial^d f}{\partial x_\nu^d}(\boldsymbol{x})\Big| \;\Big|\; \nu \in [1,n],\ \boldsymbol{x} \in \mathbb{R}^n\Big\} \\
&= C\|k\|^{-d} \quad \text{(say)}.
\end{aligned}$$

Hence the Fourier series converges normally, and we must prove that

$$f(\boldsymbol{x}) = \sum_{\boldsymbol{k}\in\mathbb{Z}^n}\int_0^1 f(\boldsymbol{y})\mathrm{e}^{2\pi\mathrm{i}\langle\boldsymbol{k},\boldsymbol{x}-\boldsymbol{y}\rangle}d\boldsymbol{y}.$$

The one-dimensional case is in textbooks on elementary analysis, e. g. [59, Ch. 16]. The general case follows easily by induction on n using Fubini's theorem. □

Theorem 7.2.6 (Poisson summation formula). *If $n \in \mathbb{N}$ and $f \in \boldsymbol{S}(n)$, then*

$$\sum_{\boldsymbol{m}\in\mathbb{Z}^n} f(\boldsymbol{m}) = \sum_{\boldsymbol{m}\in\mathbb{Z}^n}\hat{f}(\boldsymbol{m}).$$

Proof. We define $g\colon \mathbb{R}^n \to \mathbb{C}$ by

$$g(\boldsymbol{x}) = \sum_{\boldsymbol{k}\in\mathbb{Z}^n} f(\boldsymbol{x}+\boldsymbol{k}),$$

and we assert that this series converges locally normally. To show this, let B be a compact subset of $\mathbb{R}^n$, $M = \max\{\|\boldsymbol{x}\| \mid \boldsymbol{x} \in B\}$, $m \in \mathbb{N}$ and $C \in \mathbb{R}_{>0}$ such that $\|\boldsymbol{x}\|^m|f(\boldsymbol{x})| \le C$ for all $\boldsymbol{x} \in \mathbb{R}^n$. If $\boldsymbol{k} \in \mathbb{Z}$ is such that $\|\boldsymbol{k}\| \ge 2M$, then

$$C \ge |f(\boldsymbol{x}+\boldsymbol{k})|\,\|\boldsymbol{x}+\boldsymbol{k}\|^m \ge |f(\boldsymbol{x}+\boldsymbol{k})|(\|\boldsymbol{k}\|-\|\boldsymbol{x}\|)^m \ge |f(\boldsymbol{x}+\boldsymbol{k})|\Big(\frac{\|\boldsymbol{k}\|}{2}\Big)^m,$$

and therefore

$$|f(\boldsymbol{x}+\boldsymbol{k})| \le \frac{2^m C}{\|\boldsymbol{k}\|^m} \quad \text{for all} \quad \boldsymbol{x} \in B \quad \text{and} \quad \boldsymbol{k} \in \mathbb{Z}^n \quad \text{such that} \quad \|\boldsymbol{k}\| \ge 2M.$$

Hence the series converges normally on B, and therefore g is a C^∞-function. Apparently g is periodic, and for $\boldsymbol{m} \in \mathbb{Z}^n$ let $c_{\boldsymbol{m}}$ be its $\boldsymbol{m}$-th Fourier coefficient. Then 7.2.5 implies

$$\sum_{\boldsymbol{m}\in\mathbb{Z}^n} c_{\boldsymbol{m}} = g(0) = \sum_{\boldsymbol{k}\in\mathbb{Z}^n} f(\boldsymbol{k}).$$

By definition, and since all series are absolutely and uniformly convergent and all functions are integrable, it follows that

$$\begin{aligned} c_{\boldsymbol{m}} &= \int_{[0,1]^n} g(\boldsymbol{x})\mathrm{e}^{-2\pi\mathrm{i}\langle \boldsymbol{m},\boldsymbol{x}\rangle} d\boldsymbol{x} = \sum_{\boldsymbol{k}\in\mathbb{Z}^n} \int_{[0,1]^n} f(\boldsymbol{x}+\boldsymbol{k})\mathrm{e}^{-2\pi\mathrm{i}\langle \boldsymbol{m},\boldsymbol{x}\rangle} d\boldsymbol{x} \\ &= \sum_{\boldsymbol{k}\in\mathbb{Z}^n} \int_{[0,1]^n} f(\boldsymbol{x}+\boldsymbol{k})\mathrm{e}^{-2\pi\mathrm{i}\langle \boldsymbol{m},\boldsymbol{x}+\boldsymbol{k}\rangle} d\boldsymbol{x} = \int_{\mathbb{R}^n} f(\boldsymbol{x})\mathrm{e}^{-2\pi\mathrm{i}\langle \boldsymbol{m},\boldsymbol{x}\rangle} d\boldsymbol{x} = \hat{f}(\boldsymbol{m}), \end{aligned}$$

and consequently

$$\sum_{\boldsymbol{m}\in\mathbb{Z}^n} f(\boldsymbol{m}) = \sum_{\boldsymbol{m}\in\mathbb{Z}^n} c_{\boldsymbol{m}} = \sum_{\boldsymbol{m}\in\mathbb{Z}^n} \hat{f}(\boldsymbol{m}). \qquad \square$$

Definition and Theorem 7.2.7. A **theta series** is a series of the form

$$\sum_{\boldsymbol{a}\in\Gamma} f(\boldsymbol{x},\boldsymbol{a})\mathrm{e}^{-Q(\boldsymbol{a})+\ell(\boldsymbol{x},\boldsymbol{a})},$$

where Γ is a lattice in $\mathbb{R}^n$, $\boldsymbol{a} = (a_1,\ldots,a_n)$, $\boldsymbol{x} = (x_1,\ldots,x_n) \in \mathbb{C}^n$,

$$f(\boldsymbol{x},\boldsymbol{a}) = \sum_{\nu_1,\ldots,\nu_n=0}^{N} f_{\nu_1,\ldots,\nu_n}(\boldsymbol{x})a_1^{\nu_1}\cdot\ldots\cdot a_n^{\nu_n}$$

for some $N \in \mathbb{N}$ is a polynomial in $\boldsymbol{a}$ whose coefficients $f_{\nu_1,\ldots,\nu_n} : \mathbb{C}^n \to \mathbb{C}$ are C^∞-functions, $Q\colon \mathbb{R}^n \to \mathbb{R}$ is a positive definite quadratic form and

$$\ell(\boldsymbol{x},\boldsymbol{a}) = \sum_{i=1}^{n} g_i(\boldsymbol{x})a_i$$

is a linear form in $\boldsymbol{a}$ whose coefficients $g_i\colon \mathbb{C}^n \to \mathbb{C}$ are C^∞-functions.

A theta series as above as well as all its term-wise higher partial derivatives are locally normally convergent. The function

$$F\colon \mathbb{C}^n \to \mathbb{C}, \quad \textit{defined by} \quad F(\boldsymbol{x}) = \sum_{\boldsymbol{a}\in\Gamma} f(\boldsymbol{x},\boldsymbol{a})\mathrm{e}^{-Q(\boldsymbol{a})+\ell(\boldsymbol{x},\boldsymbol{a})},$$

is a C^∞-function and all its higher partial derivatives can be calculated term by term.

Proof. Let $\|\cdot\|$ be the maximum norm on $\mathbb{R}^n$. Let B be a compact subset of $\mathbb{C}^n$ and $C \in \mathbb{R}_{>0}$ such that $|f_{\nu_1,\ldots,\nu_n}(\boldsymbol{x})| \le C$ and $|g_i(\boldsymbol{x})| \le C$ for all $(\nu_1,\ldots,\nu_n) \in [0,N]^n$, $i \in [1,n]$, and all $\boldsymbol{x} \in B$. As Q is a positive definite quadratic form, there exists some $C_0 \in \mathbb{R}_{>0}$ such that $|Q(\boldsymbol{a})| \ge C_0\|\boldsymbol{a}\|^2$ for all $\boldsymbol{a} \in \Gamma$. Hence we obtain

$$|f(\boldsymbol{x},\boldsymbol{a})| \le (N+1)^n C\|\boldsymbol{a}\|^N \quad \text{and} \quad \mathsf{e}^{-Q(\boldsymbol{a})+\ell(\boldsymbol{x},\boldsymbol{a})} \le \mathsf{e}^{-C_0\|\boldsymbol{a}\|^2+Cn\|\boldsymbol{a}\|}$$

for all $\boldsymbol{a} \in \Gamma$ and $\boldsymbol{x} \in B$, and therefore (with new constants C_1, C_2, $C_3 \in \mathbb{R}_{>0}$)

$$|f(\boldsymbol{x},\boldsymbol{a})\mathsf{e}^{-Q(\boldsymbol{a})+\ell(\boldsymbol{x},\boldsymbol{a})}| \le M(\boldsymbol{a}) = C_1\|\boldsymbol{a}\|^N\mathsf{e}^{-C_2\|\boldsymbol{a}\|^2+C_3\|\boldsymbol{a}\|} \quad \text{for all} \quad \boldsymbol{x} \in B \ \text{and} \ \boldsymbol{a} \in \Gamma.$$

Assume that $\mathrm{rk}(\Gamma) = m \in [1,n]$, let $\boldsymbol{u} = (\boldsymbol{u}_1,\ldots,\boldsymbol{u}_m)$ be a basis of Γ, and define the $\boldsymbol{u}$-norm $\|\cdot\|_{\boldsymbol{u}} \colon \mathbb{R}\Gamma \to \mathbb{R}_{\ge 0}$ by

$$\|z_1\boldsymbol{u}_1+\ldots+z_m\boldsymbol{u}_m\|_{\boldsymbol{u}} = \max\{|z_1|,\ldots|z_m|\} \quad \text{for all} \quad (z_1,\ldots,z_m)^{\mathrm{t}} \in \mathbb{R}^m.$$

Since $\|\cdot\|\restriction\mathbb{R}\Gamma$ is also a norm on $\mathbb{R}\Gamma$, there exist constants γ_0, $\gamma_1 \in \mathbb{R}_{>0}$ such that $\gamma_0\|\boldsymbol{z}\|_{\boldsymbol{u}} \le \|\boldsymbol{z}\| \le \gamma_1\|\boldsymbol{z}\|_{\boldsymbol{u}}$ for all $\boldsymbol{z} \in \mathbb{R}\Gamma$, and consequently

$$M(\boldsymbol{a}) \le C_1\gamma_1^N\|\boldsymbol{a}\|_{\boldsymbol{u}}^N\,\mathsf{e}^{-C_2\gamma_0^2\|\boldsymbol{a}\|_{\boldsymbol{u}}^2-C_3\gamma_1\|\boldsymbol{a}\|_{\boldsymbol{u}}} = B_1\|\boldsymbol{a}\|_{\boldsymbol{u}}^N\,\mathsf{e}^{-B_2\|\boldsymbol{a}\|_{\boldsymbol{u}}^2+B_3\|\boldsymbol{a}\|d_{\boldsymbol{u}}}$$

for all $\boldsymbol{a} \in \Gamma$ and some (new) constants B_1, B_2, $B_3 \in \mathbb{R}_{>0}$.

If $k \in \mathbb{N}$, then

$$|\{\boldsymbol{a} \in \Gamma \mid \|\boldsymbol{a}\|_{\boldsymbol{u}} = k\}| \le |\{\boldsymbol{a} \in \Gamma \mid \|\boldsymbol{a}\|_{\boldsymbol{u}} \le k\}| = (k+1)^m \le (2k)^m,$$

and consequently

$$\sum_{\boldsymbol{a}\in\Gamma} M(\boldsymbol{a}) \le \sum_{k=1}^{\infty}(2k)^m B_1 k^N\mathsf{e}^{-B_2k^2+B_3k} = 2^m B_1\sum_{k=1}^{\infty}k^{N+m}\mathsf{e}^{-B_2k^2+B_3k} < \infty,$$

since $k^{N+m}\mathsf{e}^{-B_2k^2} \ll k^{-2}$.

Therefore the theta series is normally convergent in B and thus locally normally convergent. As all series of term-wise higher partial derivatives are of the same type, they are locally normally convergent, too. Hence $F\colon C^n \to \mathbb{C}$ is a C^∞-function, and all of its higher partial derivatives can be calculatet term by term. □

We close this section with the application of the Poisson summation formula to a theta series associated with an algebraic number field in a form which goes back to E. Hecke.

We use the notations introduced in the *Overall assumptions* at the beginning of Section 7.1.

Theorem 7.2.8 (Hecke's theta-transformation).
Let $u \in K^\times$, $x_1, \ldots, x_n \in \mathbb{R}_{>0}$ *and* $\mathfrak{a} \in \mathcal{J}_K$. *Then*

$$\sum_{a\in\mathfrak{a}} \prod_{j=1}^{q} [(a^{(j)} + u^{(j)})x_j] \exp\Big\{-\pi \sum_{j=1}^{n} |a^{(j)} + u^{(j)}|^2 x_j\Big\}$$
$$= \mathsf{i}^{-q}\Big(\mathfrak{N}(\mathfrak{a})^2 |\mathsf{d}_K| \prod_{j=1}^{n} x_j\Big)^{-1/2} \sum_{b\in(\mathfrak{a}\mathfrak{D})^{-1}} \prod_{j=1}^{q} b^{(j)}$$
$$\exp\Big\{-\pi \sum_{j=1}^{n} |b^{(j)}|^2 x_j^{-1} + 2\pi\mathsf{i}\,\mathsf{Tr}_{K/\mathbb{Q}}(bu)\Big\}.$$

Proof. It suffices to prove the formula

$$\sum_{a\in\mathfrak{a}} \exp\Big\{-\pi \sum_{j=1}^{n} |a^{(j)} + u^{(j)}|^2 x_j\Big\}$$
$$= \Big(\mathfrak{N}(\mathfrak{a})^2 |\mathsf{d}_K| \prod_{j=1}^{n} x_j\Big)^{-1/2} \sum_{b\in(\mathfrak{a}\mathfrak{D})^{-1}} \exp\Big\{-\pi \sum_{j=1}^{n} |b^{(j)}|^2 x_j^{-1} + 2\pi\mathsf{i}\,\mathsf{Tr}_{K/\mathbb{Q}}(bu)\Big\} \qquad (*)$$

Indeed, suppose that $(*)$ holds for all $u \in K$. Let $K_\mathbb{R}$ be the real vector space consisting of all $(\xi_1, \ldots, \xi_n) \in \mathbb{C}^n$ such that $\xi_j \in \mathbb{R}$ for $j \in [1, r_1]$ and $\xi_{r_1+r_2+j} = \overline{\xi}_{r_1+j}$ for all $j \in [1, r_2]$, and consider the map $\delta\colon K \to K_\mathbb{R}$, defined by $\delta(u) = (u^{(1)}, \ldots, u^{(n)})$. Then $\delta(K)$ is dense in $K_\mathbb{R}$.

Let $(\xi_1, \ldots, \xi_n) \in K_\mathbb{R}$, and let $(u_\lambda)_{\lambda\in\mathbb{N}}$ b e a sequence in K such that

$$(\xi_1, \ldots, \xi_n) = \lim_{\lambda\to\infty} (u_\lambda^{(1)}, \ldots, u_\lambda^{(n)}).$$

Since $(*)$ holds for u_λ for all $\lambda \in \mathbb{N}$, we obtain, as $\lambda \to \infty$,

$$\sum_{a\in\mathfrak{a}} \exp\Big\{-\pi \sum_{j=1}^{n} |a^{(j)} + \xi_j|^2 x_j\Big\}$$
$$= \Big(\mathfrak{N}(\mathfrak{a})^2 |\mathsf{d}_K| \prod_{j=1}^{n} x_j\Big)^{-1/2} \sum_{b\in(\mathfrak{a}\mathfrak{D})^{-1}} \exp\Big\{-\pi \sum_{j=1}^{n} |b^{(j)}|^2 x_j^{-1} + 2\pi\mathsf{i} \sum_{j=1}^{n} b^{(j)}\xi_j\Big\}. \qquad (**)$$

Both sides of $(**)$ are theta series in $(\xi_1, \ldots, \xi_q) \in \mathbb{R}^q$ in the sense of 7.2.7. Therefore we can apply the partial differential operator

$$\Big(\frac{-1}{2\pi}\Big)^q \frac{\partial^q}{\partial\xi_1 \ldots \partial\xi_q}$$

term by term to both sides of $(**)$ and obtain, for all $(\xi_1,\ldots,\xi_n)\in K_\mathbb{R}$,

$$\sum_{a\in\mathfrak{a}}\prod_{j=1}^{q}[(a^{(j)}+\xi_j)x_j]\exp\Big\{-\pi\sum_{j=1}^{n}|a^{(j)}+\xi_j|^2x_j\Big\}$$
$$=\mathsf{i}^{-q}\Big(\mathfrak{N}(\mathfrak{a})^2|\mathsf{d}_K|\prod_{j=1}^{n}x_j\Big)^{-1/2}\sum_{b\in(\mathfrak{a}\mathfrak{D})^{-1}}\prod_{j=1}^{q}b^{(j)}$$
$$\exp\Big\{-\pi\sum_{j=1}^{n}|b^{(j)}|^2x_j^{-1}+2\pi\mathsf{i}\sum_{j=1}^{n}b^{(j)}\xi_j\Big\}.$$

If $u\in K$, we may insert $(u^{(1)},\ldots,u^{(n)})$ instead of $(\xi_1,\ldots,\xi_n)$ to obtain our assertion.

Proof of $(*)$. Let $(a_1,\ldots,a_n)$ be a ($\mathbb{Z}$-)basis of $\mathfrak{a}$. Then $(a_1,\ldots,a_n)$ is a basis of K, and thus there exists a unique $\boldsymbol{w}=(w_1,\ldots,w_n)^{\mathsf{t}}\in\mathbb{R}^n$ such that $u=w_1a_1+\ldots+w_na_n$, and

$$\mathfrak{a}=\{m_1a_1+\ldots+m_na_n\mid \boldsymbol{m}=(m_1,\ldots,m_n)^{\mathsf{t}}\in\mathbb{Z}^n\}.$$

Hence we obtain

$$\sum_{a\in\mathfrak{a}}\exp\Big\{-\pi\sum_{j=1}^{n}|a^{(j)}+u^{(j)}|^2x_j\Big\}=\sum_{\boldsymbol{m}\in\mathbb{Z}^n}\exp\Big\{-\pi\sum_{j=1}^{n}\Big|\sum_{k=1}^{n}a_k^{(j)}(m_k+w_k)\Big|^2x_j\Big\}$$
$$=\sum_{\boldsymbol{m}\in\mathbb{Z}^n}\mathsf{e}^{-\pi Q(\boldsymbol{m}+\boldsymbol{w})},$$

where $Q\colon\mathbb{R}^n\to\mathbb{R}^n$ is defined by

$$Q(\boldsymbol{z})=\sum_{j=1}^{n}\Big|\sum_{k=1}^{n}a_k^{(j)}z_k\Big|^2x_j=\sum_{j=1}^{n}\sum_{k,k'=1}^{n}a_k^{(j)}\overline{a_{k'}^{(j)}}x_jz_kz_{k'}=\boldsymbol{z}^{\mathsf{t}}A^{\mathsf{t}}\overline{A}\boldsymbol{z}$$

for all $\boldsymbol{z}=(z_1,\ldots,z_n)^{\mathsf{t}}\in\mathbb{R}^n$, where $A=(x_j^{1/2}a_k^{(j)})_{j,k\in[1,n]}$.

We assert that $A^{\mathsf{t}}\overline{A}\in\mathsf{GL}_n(\mathbb{R})$ is a positive definite symmetric matrix. Indeed, by its definition, A is of the form

$$A=\begin{pmatrix}A_1\\A_2\\\overline{A}_2\end{pmatrix},\quad\text{where } A_1\in\mathsf{M}_{r_1,n}(\mathbb{R}) \text{ and } A_2\in\mathsf{M}_{r_2,n}(\mathbb{C}),$$

and therefore

$$A^{\mathsf{t}}\overline{A}=(A_1^{\mathsf{t}}\ A_2^{\mathsf{t}}\ \overline{A}_2^{\mathsf{t}})\begin{pmatrix}A_1\\\overline{A}_2\\A_2\end{pmatrix}=A_1^{\mathsf{t}}A_1+A_2^{\mathsf{t}}\overline{A}_2+\overline{A}_2^{\mathsf{t}}A_2$$
$$=A_1^{\mathsf{t}}A_1+2\,\Re(A_2^{\mathsf{t}}\overline{A}_2)\in\mathsf{GL}_n(\mathbb{R}).$$

Moreover, $(A^{\rm t}\overline{A})^{\rm t} = \overline{A}^{\rm t}A = \overline{A^{\rm t}\overline{A}} = A^{\rm t}\overline{A}$, hence $A^{\rm t}\overline{A}$ is symmetric, and as $Q(\boldsymbol{z}) \geq 0$ for all $\boldsymbol{z} \in \mathbb{R}^n$, the matrix $A^{\rm t}\overline{A}$ is positive definite.

We apply the Poisson summation formula 7.2.6 to $g\colon \mathbb{R}^n \to \mathbb{R}$, defined by $g(\boldsymbol{z}) = \mathsf{e}^{-\pi Q(\boldsymbol{z}+\boldsymbol{w})}$, whose Fourier transform $\hat{g}\colon \mathbb{R}^n \to \mathbb{R}$ is by 7.2.3 given by

$$\hat{g}(\boldsymbol{y}) = \frac{\mathsf{e}^{2\pi\mathsf{i}\langle \boldsymbol{w},\boldsymbol{y}\rangle}}{|\det(A)|}\,\mathsf{e}^{-\pi \boldsymbol{y}^{\rm t}A^{-1}(\overline{A}^{-1})^{\rm t}\boldsymbol{y}} \quad \text{for all } \ \boldsymbol{y} \in \mathbb{R}^n.$$

The dual basis $(b_1,\ldots,b_n)$ of $(a_1,\ldots,a_n)$ with respect to $\mathsf{Tr}_{K/\mathbb{Q}}$ is a $(\mathbb{Z})$-basis of the Dedekind complementary module $\mathfrak{C}_{\mathfrak{a}/\mathbb{Z}} = (\mathfrak{a}\mathfrak{D})^{-1}$ (see [18, Theorems 5.9.2 and 5.9.3]). Hence

$$\sum_{j=1}^n a_k^{(j)} b_{k'}^{(j)} = \delta_{k,k'} \quad \text{for all } \ k,\, k' \in [1,n],$$

which implies that $(A^{-1})^{\rm t} = (x_j^{-1/2} b_k^{(j)})_{j,k\in[1,n]}$. For all $\boldsymbol{y} = (y_1,\ldots,y_n) \in \mathbb{R}^n$ we get

$$\boldsymbol{y}^{\rm t}A^{-1}(\overline{A}^{-1})^{\rm t}\boldsymbol{y} = \sum_{j=1}^n \sum_{k,k'=1}^n b_k^{(j)}\overline{b_{k'}^{(j)}}\,x_j^{-1} y_k y_{k'} = \sum_{j=1}^n \Bigl|\sum_{k=1}^n b_k^{(j)} y_k\Bigr|^2 x_j^{-1}$$

and

$$\begin{aligned}\langle \boldsymbol{w},\boldsymbol{y}\rangle &= \sum_{j=1}^n y_k w_k = \sum_{k,\,k'=1}^n y_k w_{k'} \sum_{j=1}^n b_k^{(j)} a_{k'}^{(j)} = \sum_{j=1}^n \Bigl(\sum_{k=1}^n b_k^{(j)} y_k\Bigr)\Bigl(\sum_{k'=1}^n a_{k'}^{(j)} w_{k'}\Bigr) \\ &= \sum_{j=1}^n \Bigl(\sum_{k=1}^n b_k^{(j)} y_k\Bigr) u^{(j)}.\end{aligned}$$

Since $\det(a_k^{(j)})^2_{j,k\in[1,n]} = d_{K/\mathbb{Q}}(a_1,\ldots,a_n) = \Delta(\mathfrak{a}) = \mathfrak{N}(\mathfrak{a})^2\mathsf{d}_K$ (see [18, Theorem 1.6.9 and Corollary 3.1.4]), it follows that

$$\det(A)^2 = \det(a_k^{(j)})^2_{j,k\in[1,n]} \prod_{j=1}^n x_j = \mathfrak{N}(\mathfrak{a})^2\mathsf{d}_K \prod_{j=1}^n x_j.$$

The Poisson summation formula yields

$$\sum_{a\in\mathfrak{a}} \exp\Big\{-\pi\sum_{j=1}^{n}|a^{(j)}+u^{(j)}|^2x_j\Big\} = \sum_{\boldsymbol{m}\in\mathbb{Z}^n} \mathsf{e}^{-\pi Q(\boldsymbol{m}+\boldsymbol{w})}$$

$$= \sum_{\boldsymbol{y}\in\mathbb{Z}^n} \frac{\mathsf{e}^{2\pi\mathsf{i}\langle\boldsymbol{w},\boldsymbol{y}\rangle}}{|\det(A)|} \exp\Big\{-\pi\sum_{j=1}^{n}\Big|\sum_{k=1}^{n} b_k^{(j)}y_k\Big|^2 x_j^{-1}\Big\}$$

$$= \Big(\mathfrak{N}(\mathfrak{a})^2|\mathsf{d}_K|\prod_{j=1}^{n}x_j\Big)^{-1/2}\sum_{\boldsymbol{y}\in\mathbb{Z}^n}\exp\Big\{-\pi\sum_{j=1}^{n}\Big|\sum_{k=1}^{n} b_k^{(j)}y_k\Big|^2 x_j^{-1}$$

$$+ 2\pi\mathsf{i}\sum_{j=1}^{n}\Big(\sum_{k=1}^{n} b_k^{(j)}y_k\Big)u^{(j)}\Big\}$$

$$= \Big(\mathfrak{N}(\mathfrak{a})^2|\mathsf{d}_K|\prod_{j=1}^{n}x_j\Big)^{-1/2}\sum_{b\in(\mathfrak{a}\mathfrak{D})^{-1}}\exp\Big\{-\pi\sum_{j=1}^{n}|b^{(j)}|^2x_j^{-1}+2\pi\mathsf{i}\,\mathsf{Tr}_{K/\mathbb{Q}}(bu)\Big\}. \;\square$$

7.3 Proof of the functional equation for L functions of number fields

In this section, we perform the proof of 7.1.8. For this, we maintain the *Overall assumptions* made at the beginning of Section 7.1.

We fix some $s \in \mathcal{H}_1$ and an ideal $\mathfrak{a} \in C^{-1}\cap\mathcal{J}_K$. Then it follows that $C\cap\mathcal{J}_K' = \{a\mathfrak{a}^{-1} \mid a \in \mathfrak{a}^\bullet\}$, and if $a,\ a' \in \mathfrak{a}^\bullet$, then $a\mathfrak{a}^{-1} = a'\mathfrak{a}^{-1}$ if and only if $a' \in aU_K$ (that is, a and a' are associated). Let $\mathfrak{a}^\#$ be a set of representatives for the classes of associates in $\mathfrak{a}^\bullet$. Then

$$L(s,\chi,C) = \mathfrak{N}(\mathfrak{a})^s\sum_{a\in\mathfrak{a}^\#}\frac{\chi(a\mathfrak{a}^{-1})}{|\mathsf{N}_{K/\mathbb{Q}}(a)|^s}.$$

We aim at an integral representaion of $\Lambda(s,\chi)$ which shows its extension to a meromorphic function in the whole plane. We start with a suitable representation of the Γ-factors. If $z \in \mathcal{H}_0$ and $t \in \mathbb{R}_{>0}$, then

$$\Gamma\Big(\frac{z}{2}\Big) = \int_0^\infty \mathsf{e}^{-x}x^{z/2-1}dx = t^z\int_0^\infty \mathsf{e}^{-x}\Big(\frac{x}{t^2}\Big)^{z/2-1}\frac{dx}{t^2}$$

$$\underset{[x=t^2y]}{=} t^z\int_0^\infty \mathsf{e}^{-t^2y}y^{z/2-1}dy. \qquad (*)$$

Let $a \in \mathfrak{a}^\#$.

For $j \in [1, q]$ we apply $(*)$ with $z = s+1,\ \ t = |a^{(j)}|\sqrt{\pi}$ and obtain

$$\Gamma\Big(\frac{s+1}{2}\Big) = \big(|a^{(j)}|\sqrt{\pi}\big)^{s+1} \int_0^\infty \exp\{-\pi|a^{(j)}|^2 y_j\}\, y_j^{(s-1)/2}\, dy_j.$$

For $j \in [q+1, r_1]$ we we apply $(*)$ with $z = s,\ \ t = |a^{(j)}|\sqrt{\pi}$ and obtain

$$\Gamma\Big(\frac{s}{2}\Big) = \big(|a^{(j)}|\sqrt{\pi}\big)^{s} \int_0^\infty \exp\{-\pi|a^{(j)}|^2 y_j\}\, y_j^{s/2-1}\, dy_j.$$

For $j \in [r_1+1, r]$ we apply $(*)$ with $z = 2s,\ \ t = |a^{(j)}|^2\sqrt{2\pi}$ and obtain

$$\Gamma(s) = \big(2\pi|a^{(j)}|^2\big)^{s} \int_0^\infty \exp\{-2\pi|a^{(j)}|^2 y_j\}\, y_j^{s-1}\, dy_j.$$

We multiply the above expressions, set $\boldsymbol{y} = (y_1, \ldots, y_r)^{\rm t} \in \mathbb{R}^r_{>0}$ and observe that all integrands are L^1-functions. Then

$$\begin{aligned}\chi(a\mathfrak{a}^{-1})\gamma(s,\chi) &= \chi(a\mathfrak{a}^{-1})\Gamma\Big(\frac{s+1}{2}\Big)^q\Gamma\Big(\frac{s}{2}\Big)^{r_1-q}\Gamma(s)^{r_2} \\ &= 2^{r_2 s}\pi^{(q+ns)/2}|\mathsf{N}_{K/\mathbb{Q}}(a)|^s \int_{\mathbb{R}^r_{>0}} f_{\chi,a,s}(\boldsymbol{y})d\boldsymbol{y},\end{aligned}$$

where

$$f_{\chi,a,s}(\boldsymbol{y}) = \chi(a\mathfrak{a}^{-1}) \prod_{j=1}^{q} \big(|a^{(j)}|\, y_j^{1/2}\big) \exp\Big\{-\pi\sum_{j=1}^{r} l_j|a^{(j)}|^2 y_j\Big\} \prod_{j=1}^{r} y_j^{l_j s/2-1}.$$

Hence we obtain

$$\begin{aligned}\Lambda(s,\chi,C) &= 2^{r_2(1-s)}\pi^{-(q+ns)/2}\mathfrak{N}(\mathfrak{D}\mathfrak{f})^{s/2}\gamma(s,\chi)\mathfrak{N}(\mathfrak{a})^s \sum_{a\in\mathfrak{a}^\#}\frac{\chi(a\mathfrak{a}^{-1})}{|\mathsf{N}_{K/\mathbb{Q}}(a)|^s} \\ &= 2^{r_2}\mathfrak{N}(\mathfrak{a}^2\mathfrak{D}\mathfrak{f})^{s/2}\sum_{a\in\mathfrak{a}^\#}\frac{\chi(a\mathfrak{a}^{-1})\gamma(s,\chi)}{2^{r_2 s}\pi^{(q+ns)/2}|\mathsf{N}_{K/\mathbb{Q}}(a)|^s} \\ &= 2^{r_2}\mathfrak{N}(\mathfrak{a}^2\mathfrak{D}\mathfrak{f})^{s/2}\sum_{a\in\mathfrak{a}^\#}\int_{\mathbb{R}^r_{>0}} f_{\chi,a,s}(\boldsymbol{y})d\boldsymbol{y}\,.\end{aligned}$$

If $\sigma = \Re(s) > 1$, then

$$\begin{aligned}&\sum_{a\in\mathfrak{a}^\#}\int_{\mathbb{R}^r_{>0}} |f_{\chi,a,s}(\boldsymbol{y})|d\boldsymbol{y} \\ &= \sum_{a\in\mathfrak{a}^\#}\int_{\mathbb{R}^r_{>0}} f_{1,a,\sigma}(\boldsymbol{y})d\boldsymbol{y} = \Lambda(\sigma,1,C)2^{-r_2}\mathfrak{N}(\mathfrak{D}\mathfrak{f}\mathfrak{a}^2)^{-\sigma/2} < \infty.\end{aligned}$$

Therefore we may interchange sum and integral, and we obtain

$$\Lambda(s,\chi,C) = 2^{r_2}\mathfrak{N}(\mathfrak{a}^2\mathfrak{D}\mathfrak{f})^{s/2}\int_{\mathbb{R}^r_{>0}}\Big(\sum_{a\in\mathfrak{a}^\#} f_{\chi,a,s}(\boldsymbol{y})\Big)d\boldsymbol{y}.$$

To get rid of the set of representatives, we introduce new coordinates which take account of the units of K. Let $\{e_1,\ldots,e_{r-1}\}$ be a system of fundamental units of K (see [18, Definition 3.5.10]).

For $\boldsymbol{z}=(z_1,\ldots,z_{r-1})^{\mathsf t}\in\mathbb{R}^{r-1}$, $t\in\mathbb{R}_{>0}$ and $j\in[1,r]$ we set

$$y_j=y_j(t,\boldsymbol{z})=t^{1/n}\prod_{k=1}^{r-1}|e_k^{(j)}|^{2z_k},$$

we define

$$G\colon\mathbb{R}_{>0}\times\mathbb{R}^{r-1}\to\mathbb{R}^r_{>0}\quad\text{by}\quad G(t,\boldsymbol{z})=(y_1(t,\boldsymbol{z}),\ldots,y_r(t,\boldsymbol{z})),$$

and we assert that G is a diffeomorphism.

G is a surjective C^1-map, and as it is a group homomorphism it suffices to prove that its kernel is trivial. Let $t\in\mathbb{R}_{>0}$ and $\boldsymbol{z}=(z_1,\ldots,z_{r-1})^{\mathsf t}\in\mathbb{R}^{r-1}$ be such that $G(t,\boldsymbol{z})=\mathbf{1}$. For $j\in[1,r]$ this implies

$$0=\log y_j(t,\boldsymbol{z})=\frac{1}{n}\log t+2\sum_{k=1}^{r-1}z_k\log|e_k^{(j)}|,$$

and

$$-\log t=\sum_{j=1}^{r}2l_j\sum_{k=1}^{r-1}z_k\log|e_k^{(j)}|=2\sum_{k=1}^{r-1}z_k\sum_{j=1}^{r}l_j\log|e_k^{(j)}|$$

$$=2\sum_{k=1}^{r-1}z_k\log|\mathsf{N}_{K/\mathbb{Q}}(e_k)|=0.$$

Hence $t=1$, and

$$\sum_{k=1}^{r-1}z_k\log|e_k^{(j)}|=0\quad\text{for all}\quad j\in[1,r].$$

Since $\operatorname{rk}\big((\log|e_k^{(j)}|)_{j\in[1,r],\,k\in[1,r-1]}\big)=r-1$ (see [18, Theorem 3.5.10]), it follows that $\boldsymbol{z}=\mathbf{0}$.

Next we calculate the Jacobian $\mathsf{J}(G)(t,\boldsymbol{z})$. If $j\in[1,r]$ and $k\in[1,r-1]$, then

$$\frac{\partial y_j}{\partial t}=\frac{y_j}{nt}\qquad\text{and}\qquad\frac{\partial y_j}{\partial z_k}=2y_j\log|e_k^{(j)}|,$$

and therefore

$$\begin{aligned}\det\mathsf{J}(G)(t,\boldsymbol{z})&=\det\Big(\frac{y_j}{nt},\ 2y_j\log|e_1^{(j)}|,\ldots,2y_j\log|e_{r-1}^{(j)}|\Big)_{j\in[1,r]}\\&=\frac{2^{r-1}}{nt}\prod_{j=1}^{r}y_j\,\det\big(1,\ \log|e_1^{(j)}|,\ldots,\log|e_{r-1}^{(j)}|\big)_{j\in[1,r]}\\&=\frac{2^{r-1-r_2}}{nt}\prod_{j=1}^{r}y_j\,\det\big(l_j,\ l_j\log|e_1^{(j)}|,\ldots,l_j\log|e_{r-1}^{(j)}|\big)_{j\in[1,r]}.\end{aligned}$$

We replace the last line of this determinant with the sum of all lines and obtain

$$
\begin{aligned}
&|\det \mathsf{J}(G)(t,\boldsymbol{z})| \\
&= \frac{2^{r-1-r_2}}{nt}\prod_{j=1}^{r} y_j \left|\det\begin{pmatrix} l_1 & l_1\log|e_1^{(1)}| & \ldots & l_1\log|e_{r-1}^{(1)}| \\ \cdot & \cdot & \ldots & \cdot \\ \cdot & \cdot & \ldots & \cdot \\ l_{r-1} & l_{r-1}\log|e_1^{(r-1)}| & \ldots & l_{r-1}\log|e_{r-1}^{(r-1)}| \\ n & 0 & \ldots & 0 \end{pmatrix}\right| \\
&= \frac{2^{r_1-1}\mathsf{R}_K}{t}\prod_{j=1}^{r} y_j \qquad \text{(see [18, Theorem 3.5.10]).}
\end{aligned}
$$

The transformation formula yields (with $y_j = y_j(t,\boldsymbol{z})$ for all $j \in [1,r]$)

$$
\begin{aligned}
\Lambda(s,\chi,C) &= 2^{r_2}\mathfrak{N}(\mathfrak{a}^2\mathfrak{D}\mathfrak{f})^{s/2}\int_{\mathbb{R}_{>0}\times\mathbb{R}^{r-1}}\Big(\sum_{a\in\mathfrak{a}^\#} f_{\chi,a,s}(y_1,\ldots,y_r)\Big) \\
&\qquad |\det \mathsf{J}(G)(t,\boldsymbol{z})|\,d(t,\boldsymbol{z}) \\
&\qquad = 2^{r-1}\mathsf{R}_K\,\mathfrak{N}(\mathfrak{a}^2\mathfrak{D}\mathfrak{f})^{s/2}\int_{\mathbb{R}_{>0}\times\mathbb{R}^{r-1}}\Big(\sum_{a\in\mathfrak{a}^\#} F_{\chi,a,s}(t,\boldsymbol{z})\Big)d(t,\boldsymbol{z})
\end{aligned}
$$

and

$$
\sum_{a\in\mathfrak{a}^\#}\int_{\mathbb{R}_{>0}\times\mathbb{R}^{r-1}} |F_{\chi,a,s}(t,\boldsymbol{z})|\,d(t,\boldsymbol{z}) < \infty,
$$

where

$$
\begin{aligned}
F_{\chi,a,s}(t,\boldsymbol{z}) &= f_{\chi,a,s}(y_1,\ldots,y_r)t^{-1}\prod_{j=1}^{r} y_j \\
&= \chi(a\mathfrak{a}^{-1})\prod_{j=1}^{q}(|a^{(j)}|\,y_j^{1/2})\exp\Big\{-\pi\sum_{j=1}^{r} l_j|a^{(j)}|^2 y_j\Big\}\Big[t^{-1}\prod_{j=1}^{r} y_j^{l_j s/2}\Big].
\end{aligned}
$$

The last factor simplifies to

$$
\begin{aligned}
t^{-1}\prod_{j=1}^{r} y_j^{l_j s/2} &= t^{-1}\prod_{j=1}^{r}\Big(t^{1/n}\prod_{k=1}^{r-1}|e_k^{(j)}|^{2k}\Big)^{l_j s/2} \\
&= t^{s/2-1}\prod_{k=1}^{r-1}\Big(\prod_{j=1}^{r}|e_k^{(j)}|^{l_j}\Big)^{ks} = t^{s/2-1},
\end{aligned}
$$

since

$$
\prod_{j=1}^{r}|e_k^{(j)}|^{l_j} = \prod_{j=1}^{n}|e_k^{(j)}| = |\mathsf{N}_{K/\mathbb{Q}}(e_k)| = 1 \quad \text{for all} \quad k \in [1, r-1].
$$

For $j \in [1, r_2]$ and $(t, \boldsymbol{z}) \in \mathbb{R}_{>0} \times \mathbb{R}^{r-1}$ we set $y_{r+j}(t, \boldsymbol{z}) = y_{r_1+j}(t, \boldsymbol{z})$, and then

$$F_{\chi,a,s}(t, \boldsymbol{z}) = \chi(a\mathfrak{a}^{-1}) \prod_{j=1}^{q} \left(|a^{(j)}| y_j(t, \boldsymbol{z})^{1/2}\right) \exp\Big\{-\pi \sum_{j=1}^{n} |a^{(j)}|^2 y_j(t, \boldsymbol{z})\Big\} t^{s/2-1} \tag{1}$$

for all $(t, \boldsymbol{z}) \in \mathbb{R}_{>0} \times \mathbb{R}^{r-1}$.

If $\boldsymbol{m} = (m_1, \ldots, m_{r-1})^{\mathsf{t}} \in \mathbb{Z}^{r-1}$, $\boldsymbol{z} = (z_1, \ldots, z_{r-1})^{\mathsf{t}} \in \mathbb{R}^{r-1}$ and $j \in [1, n]$, then

$$y_j(t, \boldsymbol{z} + \boldsymbol{m}) = y_j(t, \boldsymbol{z}) \prod_{k=1}^{r-1} |e_k^{(j)}|^{2m_k},$$

and therefore

$$F_{\chi,a,s}(t, \boldsymbol{z} + \boldsymbol{m}) = F_{\chi, a e_1^{m_1} \cdot \ldots \cdot e_{r-1}^{m_{r-1}}, s}(t, \boldsymbol{z}).$$

If $c \in \mu(K)$ is a root of unity, then $F_{\chi,ac,s} = F_{\chi,a,s}$. A twofold application of Lebesgue's dominated convergence theorem yields

$$\begin{aligned}
&\int_{\mathbb{R}_{>0} \times \mathbb{R}^{r-1}} \Big(\sum_{a \in \mathfrak{a}^{\#}} F_{\chi,a,s}(t, \boldsymbol{z})\Big) d(t, \boldsymbol{z}) \\
&= \sum_{\boldsymbol{m} \in \mathbb{Z}^{r-1}} \int_{\mathbb{R}_{>0} \times [-\frac{1}{2}, \frac{1}{2}]^{r-1}} \Big(\sum_{a \in \mathfrak{a}^{\#}} F_{\chi,a,s}(t, \boldsymbol{z} + \boldsymbol{m})\Big) d(t, \boldsymbol{z}) \\
&= \int_{\mathbb{R}_{>0} \times [-\frac{1}{2}, \frac{1}{2}]^{r-1}} \Big(\sum_{\boldsymbol{m} \in \mathbb{Z}^{r-1}} \sum_{a \in \mathfrak{a}^{\#}} F_{\chi,a,s}(t, \boldsymbol{z} + \boldsymbol{m})\Big) d(t, \boldsymbol{z}) \\
&= \frac{1}{w_K} \int_{\mathbb{R}_{>0} \times [-\frac{1}{2}, \frac{1}{2}]^{r-1}} \Big(\sum_{\boldsymbol{m} \in \mathbb{Z}^{r-1}} \sum_{c \in \mu(K)} \sum_{a \in \mathfrak{a}^{\#}} F_{\chi, ace_1^{m_1} \cdot \ldots \cdot e_{r-1}^{m_{r-1}}, s}(t, \boldsymbol{z})\Big) d(t, \boldsymbol{z}) \\
&= \frac{1}{w_K} \int_{\mathbb{R}_{>0} \times [-\frac{1}{2}, \frac{1}{2}]^{r-1}} \Big(\sum_{a \in \mathfrak{a}^{\bullet}} F_{\chi,a,s}(t, \boldsymbol{z})\Big) d(t, \boldsymbol{z}).
\end{aligned}$$

Thus we got rid of the representatives for conjugate elements.

The main tool for the proof of the functional equation is a transformation formula for Hecke's theta series which we announce in the subsequent Theorem 7.3.3. To ensure a smooth formulation, we make the following definition.

Definition 7.3.1. Let $\mathfrak{a} \in \mathfrak{J}_K$, $a \in \mathfrak{a}$ and $\boldsymbol{x} = (x_1, \ldots, x_n)^{\mathsf{t}} \in \mathbb{R}^n_{>0}$. We set

$$\gamma(\mathfrak{a}, \chi) = \begin{cases} \overline{\chi}(\mathfrak{a}) & \text{if } \mathfrak{f} = 1 \text{ and } q = 0, \\ 0 & \text{otherwise,} \end{cases}$$

$$\psi(\mathfrak{a}, a, \chi) = \begin{cases} \chi(\mathfrak{a}^{-1}a)\chi_\infty(a) & \text{if } a \neq 0, \\ \gamma(\mathfrak{a}, \chi) & \text{if } a = 0, \end{cases}$$

and observe that for $a \in \mathfrak{a}^\bullet$ we have

$$\prod_{j=1}^{q} |a^{(j)}| = \chi_\infty(a) \prod_{j=1}^{q} a^{(j)}.$$

Then we define the Hecke theta series by

$$\begin{aligned}\theta(\boldsymbol{x}, \mathfrak{a}, \chi) &= \sum_{a\in\mathfrak{a}} \psi(\mathfrak{a}, a, \chi) \prod_{j=1}^{q} \big(a^{(j)} x_j^{1/2}\big) \exp\Big\{-\pi \sum_{j=1}^{n} |a^{(j)}|^2 x_j\Big\} \\ &= \sum_{a\in\mathfrak{a}^\bullet} \chi(\mathfrak{a}^{-1}a) \prod_{j=1}^{q} \big(|a^{(j)}| x_j^{1/2}\big) \exp\Big\{-\pi \sum_{j=1}^{n} |a^{(j)}|^2 x_j\Big\} + \gamma(\mathfrak{a}, \chi).\end{aligned}$$

With this terminology we summarize the results obtained hitherto as follows.

Theorem 7.3.2. *Let $C \in \mathcal{C}_K$, $\mathfrak{a} \in C^{-1}$ and $a \in \mathfrak{a}$. For $j \in [1, n]$, $t \in \mathbb{R}_{>0}$ and $\boldsymbol{z} = (z_1, \ldots, z_{r-1}) \in \mathbb{R}^{r-1}$ we set*

$$\begin{aligned}\boldsymbol{y}(t, \boldsymbol{z}) = (y_1(t, \boldsymbol{z}), \ldots, y_n(t, \boldsymbol{z})), \quad &\textit{where} \ \ y_j(t, \boldsymbol{z}) \\ &= t^{1/n} \prod_{k=1}^{r-1} |e_k^{(j)}|^{2z_k} \ \ \textit{for all} \ \ j \in [1, n].\end{aligned}$$

Then

$$\begin{aligned}\Lambda(s, \chi, C) = \frac{2^{r-1} \mathsf{R}_K \mathfrak{N}(\mathfrak{a}^2\mathfrak{D}\mathfrak{f})^{s/2}}{w_K} \\ \int_{\mathbb{R}_{>0}\times[-\frac{1}{2},\frac{1}{2}]^{r-1}} \big[\theta(\boldsymbol{y}(t, \boldsymbol{z}), \mathfrak{a}, \chi) - \gamma(\mathfrak{a}, \chi)\big]\, t^{s/2-1} d(t, \boldsymbol{z}).\end{aligned}$$

Theorem 7.3.3 (Transformation formula for theta series). *Let $\mathfrak{a} \in \mathfrak{J}_K$. Suppose that $\boldsymbol{x} = (x_1, \ldots, x_n)^{\mathsf{t}} \in \mathbb{R}^n_{>0}$ and $\boldsymbol{x}^{-1} = (x_1^{-1}, \ldots, x_n^{-1})^{\mathsf{t}}$. Then*

$$\theta(\boldsymbol{x}, \mathfrak{a}, \chi) = \mathsf{W}(\chi) \mathfrak{N}(\mathfrak{a}^2\mathfrak{D}\mathfrak{f})^{-1/2} \prod_{j=1}^{n} x_j^{-1/2}\, \theta(\boldsymbol{x}^{-1}, (\mathfrak{a}\mathfrak{D}\mathfrak{f})^{-1}, \overline{\chi}).$$

Proof of Theorem 7.3.3. We show first:

> **A.** If $\boldsymbol{x} = (x_1, \ldots, x_n)^{\mathsf{t}} \in \mathbb{R}^n_{>0}$, $\mathfrak{a} \in \mathfrak{J}_K$ and $c \in K^+$, then $\theta(\boldsymbol{x}, c\mathfrak{a}, \chi) = \theta(c^2\boldsymbol{x}, \mathfrak{a}, \chi)$, where $c^2\boldsymbol{x} = (c^{(1)2}x_1, \ldots, c^{(n)2}x_n)^{\mathsf{t}}$.

Proof of **A.** Let $\boldsymbol{x} = (x_1, \ldots, x_n)^{\mathsf{t}} \in \mathbb{R}^n_{>0}$, $\mathfrak{a} \in \mathfrak{J}_K$ and $c \in K^+$. If $a \in \mathfrak{a}$, then $\psi(c\mathfrak{a}, ca, \chi) = \psi(\mathfrak{a}, a, \chi)$ (this is obvious unless $a = 0$, $\mathfrak{f} = 1$ and $q = 0$, and in this exceptional case it follows since $\chi \in \mathsf{X}(\mathcal{C}_K)$ and thus $\overline{\chi}(c\mathfrak{a}) = \overline{\chi}(\mathfrak{a})$). Now we obtain

$$\begin{aligned}\theta(\boldsymbol{x}, c\mathfrak{a}, \chi) &= \sum_{a\in\mathfrak{a}} \psi(c\mathfrak{a}, ca, \chi) \prod_{j=1}^{q} (c^{(j)} a^{(j)} x_j^{1/2}) \exp\Big\{-\pi \sum_{j=1}^{n} |c^{(j)} a^{(j)}|^2 x_j\Big\} \\ &= \sum_{a\in\mathfrak{a}} \psi(\mathfrak{a}, a, \chi) \prod_{j=1}^{q} \big(a^{(j)} (c^{(j)2} x_j)^{1/2}\big) \exp\Big\{-\pi \sum_{j=1}^{n} |a^{(j)}|^2 (c^{(j)2} x_j)\Big\} \\ &= \theta(c^2\boldsymbol{x}, \mathfrak{a}, \chi).\end{aligned}$$

□[**A.**]

By **A** it suffices to prove the transformation formula for ideals $\mathfrak{a} \in \mathcal{J}_K'^{\mathfrak{f}}$. Indeed, suppose that this is done, and let $\mathfrak{a} \in \mathcal{J}_K$ and $\boldsymbol{x} = (x_1, \ldots, x_n) \in \mathbb{R}^n_{>0}$. Then the narrow ideal class $[\mathfrak{a}]^+$ of $\mathfrak{a}$ contains an ideal $\mathfrak{a}_0 \in \mathcal{J}_K'^{\mathfrak{f}}$ (see [18, Theorem 2.11.5]), that is, $\mathfrak{a} = c\mathfrak{a}_0$ for some $c \in K^+$, and **A** implies

$$\begin{aligned}\theta(\boldsymbol{x}, \mathfrak{a}, \chi) &= \theta(\boldsymbol{x}, c\mathfrak{a}_0, \chi) = \theta(c^2\boldsymbol{x}, \mathfrak{a}_0, \chi) \\ &= \mathsf{W}(\chi)\mathfrak{N}(\mathfrak{a}_0^2\mathfrak{D}\mathfrak{f})^{-1/2} \prod_{j=1}^{n} (c^{(j)2} x_j)^{-1/2} \theta(c^{-2}\boldsymbol{x}^{-1}, (\mathfrak{a}_0\mathfrak{D}\mathfrak{f})^{-1}, \overline{\chi}) \\ &= \mathsf{W}(\chi)\mathfrak{N}(\mathfrak{a}^2\mathfrak{D}\mathfrak{f})^{-1/2} \prod_{j=1}^{n} x_j^{-1/2}\, \theta(\boldsymbol{x}^{-1}(\mathfrak{a}\mathfrak{D}\mathfrak{f})^{-1}, \overline{\chi}).\end{aligned}$$

Now we start with the actual proof of Theorem 7.3.3. Let $\mathfrak{a} \in \mathcal{J}_K'^{\mathfrak{f}}$, and let $\{b_1, \ldots, b_\rho\}$ be a set of totally positive representatives for $\mathfrak{a}/\mathfrak{a}\mathfrak{f}$ in $\mathfrak{a}$.

I. Let $a_1 \in \mathfrak{a}$, $k \in [1, \rho]$ and $a_1 \equiv b_k \bmod \mathfrak{a}\mathfrak{f}$. Then it follows that $\psi(\mathfrak{a}, a_1, \chi) = \overline{\chi}(\mathfrak{a})\chi(b_k\mathfrak{o}_K)$ unless $a_1 = 0$, $\mathfrak{f} = \mathbf{1}$ and $q > 0$.

Proof of **I.** If $a_1 \neq 0$, then

$$\begin{aligned}\psi(\mathfrak{a}, a_1, \chi) &= \chi(\mathfrak{a}^{-1}a_1)\chi_\infty(a_1) = \overline{\chi}(\mathfrak{a})\chi(a_1\mathfrak{o}_K)\chi_\infty(a_1) = \overline{\chi}(\mathfrak{a})\chi_{\mathsf{fin}}(a_1) \\ &= \overline{\chi}(\mathfrak{a})\chi_{\mathsf{fin}}(b_k) = \overline{\chi}(\mathfrak{a})\chi(b_k\mathfrak{o}_K).\end{aligned}$$

If $a_1 = 0$ and $\mathfrak{f} \neq \mathbf{1}$, then $\psi(\mathfrak{a}, a_1, \chi) = 0$ and $b_k \in \mathfrak{f}$, hence $\chi(b_k\mathfrak{o}_K) = 0$. If $a_1 = 0$, $\mathfrak{f} = \mathbf{1}$ and $q = 0$, then $\psi(\mathfrak{a}, a_1, \chi) = \overline{\chi}(\mathfrak{a})$, and since $\chi \in \mathsf{X}(\mathcal{C}_K)$, it follows that $\chi(b_k\mathfrak{o}_K) = 1$. □[**I.**]

Using **I**, we obtain

$$\theta(\boldsymbol{x},\mathfrak{a},\chi) = \sum_{a_1\in\mathfrak{a}} \psi(\mathfrak{a},a_1,\chi)\prod_{j=1}^{q}\bigl(a_1^{(j)}x_j^{1/2}\bigr)\exp\Bigl\{-\pi\sum_{j=1}^{n}|a_1^{(j)}|^2x_j\Bigr\}$$

$$= \sum_{k=1}^{\rho}\sum_{a\in\mathfrak{af}} \overline{\chi}(\mathfrak{a})\chi(b_k\mathfrak{o}_K)\prod_{j=1}^{q}\bigl[(b_k^{(j)}+a^{(j)})x_j^{1/2}\bigr]\exp\Bigl\{-\pi\sum_{j=1}^{n}|b_k^{(j)}+a^{(j)}|^2x_j\Bigr\}$$

$$= \sum_{k=1}^{\rho}\overline{\chi}(\mathfrak{a})\chi(b_k\mathfrak{o}_K)\prod_{j=1}^{q}x_j^{-1/2}\Bigl[\sum_{a\in\mathfrak{af}}\prod_{j=1}^{q}\bigl[x_j(b_k^{(j)}+a^{(j)})\bigr]$$

$$\exp\Bigl\{-\pi\sum_{j=1}^{n}|b_k^{(j)}+a^{(j)}|^2x_j\Bigr\}\Bigr],$$

and we apply 7.2.8 to each summand with $\mathfrak{af}$ instead of $\mathfrak{a}$ and b_k instead of u. Then

$$\theta(\boldsymbol{x},\mathfrak{a},\chi) = \sum_{k=1}^{\rho}\overline{\chi}(\mathfrak{a})\chi(b_k\mathfrak{o}_K)\Bigl[\prod_{j=1}^{q}x_j^{-1/2}\Bigr]\,\mathsf{i}^{-q}\Bigl[\mathfrak{N}(\mathfrak{af})^2|\mathsf{d}_K|\prod_{j=1}^{n}x_j\Bigr]^{-1/2}\cdot$$

$$\cdot\sum_{b\in(\mathfrak{aDf})^{-1}}\prod_{j=1}^{q}b^{(j)}\exp\Bigl\{-\pi\sum_{j=1}^{n}|b^{(j)}|^2x_j^{-1}+2\pi\mathsf{i}\,\mathsf{Tr}_{K/\mathbb{Q}}(bb_k)\Bigr\}$$

$$= \overline{\chi}(\mathfrak{a})\mathsf{i}^{-q}\Bigl[\prod_{j=1}^{q}x_j\prod_{j=1}^{n}x_j\Bigr]^{-1/2}\mathfrak{N}(\mathfrak{a}^2\mathfrak{D}\mathfrak{f}^2)^{-1/2}.$$

$$\cdot\sum_{b\in(\mathfrak{aDf})^{-1}}\prod_{j=1}^{q}b^{(j)}\exp\Bigl\{-\pi\sum_{j=1}^{n}|b^{(j)}|^2x_j^{-1}\Bigr\}\sum_{k=1}^{\rho}\chi(b_k\mathfrak{o}_K)\mathsf{e}^{2\pi\mathsf{i}\mathsf{Tr}_{K/\mathbb{Q}}(bb_k)}.$$

Note that $\{b_1,\ldots,b_\rho\}$ is also a set of totally positive representatives for $\mathfrak{o}_K/\mathfrak{f}$. Indeed, $|\mathfrak{o}_K/\mathfrak{f}| = |\mathfrak{a}/\mathfrak{af}|$, and if $i,\,j\in[1,\rho]$ and $b_i\equiv b_j \bmod \mathfrak{f}$, then we obtain $b_i-b_j\in\mathfrak{a}\cap\mathfrak{f}=\mathfrak{af}$. Therefore we may apply 7.1.3 and 7.1.4 to obtain

$$\sum_{k=1}^{\rho}\chi(b_k\mathfrak{o}_K)\mathsf{e}^{2\pi\mathsf{i}\,\mathsf{Tr}_{K/\mathbb{Q}}(bb_k)} = F(\chi,b) = \tau(\chi)M(\chi,b) = \mathsf{i}^q\mathfrak{N}(\mathfrak{f})^{1/2}\mathsf{W}(\chi)M(\chi,b),$$

and consequently

$$\theta(\boldsymbol{x},\mathfrak{a},\chi) = \overline{\chi}(\mathfrak{a})\mathsf{W}(\chi)\Bigl[\mathfrak{N}(\mathfrak{a}^2\mathfrak{D}\mathfrak{f})\prod_{j=1}^{q}x_j\prod_{j=1}^{n}x_j\Bigr]^{-1/2}\cdot$$

$$\cdot\sum_{b\in(\mathfrak{afD})^{-1}}M(\chi,b)\prod_{j=1}^{q}b^{(j)}\exp\Bigl\{-\pi\sum_{j=1}^{n}|b^{(j)}|^2x_j^{-1}\Bigr\}.$$

On the other hand, by the very definition we have

$$\mathsf{W}(\chi)\mathfrak{N}(\mathfrak{a}^2\mathfrak{D}\mathfrak{f})^{-1/2}\Big[\prod_{j=1}^{n} x_j\Big]^{-1/2}\theta(\boldsymbol{x}^{-1},(\mathfrak{a}\mathfrak{D}\mathfrak{f})^{-1},\overline{\chi})$$
$$= \mathsf{W}(\chi)\Big[\mathfrak{N}(\mathfrak{a}^2\mathfrak{D}\mathfrak{f})\prod_{j=1}^{q} x_j \prod_{j=1}^{n} x_j\Big]^{-1/2}\cdot$$
$$\cdot \sum_{b\in(\mathfrak{a}\mathfrak{D}\mathfrak{f})^{-1}} \psi((\mathfrak{a}\mathfrak{D}\mathfrak{f})^{-1},b,\overline{\chi})\prod_{j=1}^{q} b^{(j)}\exp\Big\{-\pi\sum_{j=1}^{n}|b^{(j)}|^2x_j^{-1}\Big\},$$

and therefore it remains to prove that

$$\psi((\mathfrak{a}\mathfrak{D}\mathfrak{f})^{-1},b,\overline{\chi}) = \overline{\chi}(\mathfrak{a})M(\chi,b) \text{ for all } b\in(\mathfrak{a}\mathfrak{D}\mathfrak{f})^{-1} \text{ unless } b=0 \text{ and } q>0.$$

Let $b\in(\mathfrak{a}\mathfrak{D}\mathfrak{f})^{-1}$. If $b\neq 0$, then

$$\psi((\mathfrak{a}\mathfrak{D}\mathfrak{f})^{-1},b,\overline{\chi}) = \overline{\chi}(b\mathfrak{a}\mathfrak{D}\mathfrak{f})\chi_\infty(b) = \overline{\chi}(\mathfrak{a})\overline{\chi}(b\mathfrak{D}\mathfrak{f})\chi_\infty(b) = \overline{\chi}(\mathfrak{a})M(\chi,b).$$

If $b=0$, then $\overline{\chi}(\mathfrak{a})M(\chi,b) = \overline{\chi}(\mathfrak{a})\overline{\chi}(\mathfrak{D}\mathfrak{f}) = \overline{\chi}(\mathfrak{a}\mathfrak{D}\mathfrak{f})$, and

$$\psi((\mathfrak{a}\mathfrak{D}\mathfrak{f})^{-1},0,\overline{\chi}) = \gamma((\mathfrak{a}\mathfrak{D}\mathfrak{f})^{-1},\overline{\chi}) = \begin{cases}\overline{\chi}(\mathfrak{a}\mathfrak{D}\mathfrak{f}) & \text{if } \mathfrak{f}=1 \text{ and } q=0,\\ 0 & \text{otherwise.}\end{cases}$$

Hence $\psi((\mathfrak{a}\mathfrak{D}\mathfrak{f})^{-1},b,\overline{\chi}) = \overline{\chi}(\mathfrak{a})M(\chi,b)$ unless $q>0$. □

We proceed with the proof of the analytic continuation and the functional equation of $\Lambda(s,\chi)$ (see 7.1.8) and start with the formula given in 7.3.2. We evaluate the responsible integral using Theorem 7.3.3. We split

$$\int_{\mathbb{R}_{>0}\times[-\frac12,\frac12]^{r-1}} [\theta(\boldsymbol{y}(t,\boldsymbol{z}),\mathfrak{a},\chi)-\gamma(\mathfrak{a},\chi)]\,t^{s/2-1}dt$$
$$= \int_{[-\frac12,\frac12]^{r-1}}\Big\{\int_0^1+\int_1^\infty\Big\}[\theta(\boldsymbol{y}(t,\boldsymbol{z}),\mathfrak{a},\chi)-\gamma(\mathfrak{a},\chi)]t^{s/2-1}dtd\boldsymbol{z},$$

and we evaluate the first integral by means of the substitution $t\mapsto t^{-1}$, $\boldsymbol{z}\mapsto -\boldsymbol{z}$. Since $(t^{-1})^{s/2-1}dt^{-1} = -t^{-s/2-1}dt$ and

$$\boldsymbol{y}(t^{-1},-\boldsymbol{z}) = \boldsymbol{y}(t,\boldsymbol{z})^{-1} = (y_1(t,\boldsymbol{z})^{-1},\ldots,y_n(t,\boldsymbol{z})^{-1}),$$

we get

$$\int_0^1 [\theta(\boldsymbol{y}(t,\boldsymbol{z}),\mathfrak{a},\chi)-\gamma(\mathfrak{a},\chi)]\,t^{s/2-1}dt$$
$$= \int_1^\infty [\theta(\boldsymbol{y}(t,\boldsymbol{z})^{-1},\mathfrak{a},\chi)-\gamma(\mathfrak{a},\chi)]\,t^{-s/2-1}dt.$$

By 7.3.3 and the very definition of $\boldsymbol{y}(t,\boldsymbol{z})$, we obtain

$$\begin{aligned}\theta(\boldsymbol{y}(t,\boldsymbol{z})^{-1},\mathfrak{a},\chi) &= \mathsf{W}(\chi)\mathfrak{N}(\mathfrak{a}^2\mathfrak{D}\mathfrak{f})^{-1/2}\Big[\prod_{j=1}^{n} y_j(t,\boldsymbol{z})\Big]^{1/2}\theta(\boldsymbol{y}(t,\boldsymbol{z}),(\mathfrak{a}\mathfrak{D}\mathfrak{f})^{-1},\overline{\chi})\\ &= \mathsf{W}(\chi)\mathfrak{N}(\mathfrak{a}^2\mathfrak{D}\mathfrak{f})^{-1/2}\theta(\boldsymbol{y}(t,\boldsymbol{z}),(\mathfrak{a}\mathfrak{D}\mathfrak{f})^{-1},\overline{\chi})\,t^{1/2},\end{aligned}$$

and therefore

$$\begin{aligned}&\int_1^\infty [\theta(\boldsymbol{y}(t,\boldsymbol{z})^{-1},\mathfrak{a},\chi)-\gamma(\mathfrak{a},\chi]\,t^{-s/2-1}dt\\ &= \mathsf{W}(\chi)\mathfrak{N}(\mathfrak{a}^2\mathfrak{D}\mathfrak{f})^{-1/2}\int_1^\infty [\theta(\boldsymbol{y}(t,\boldsymbol{z}),(\mathfrak{a}\mathfrak{D}\mathfrak{f})^{-1},\overline{\chi})-\gamma((\mathfrak{a}\mathfrak{D}\mathfrak{f})^{-1},\overline{\chi})]\,t^{-(s+1)/2}dt\\ &+\mathsf{W}(\chi)\mathfrak{N}(\mathfrak{a}^2\mathfrak{D}\mathfrak{f})^{-1/2}\gamma((\mathfrak{a}\mathfrak{D}\mathfrak{f})^{-1},\overline{\chi})\int_1^\infty t^{-(1+s)/2}dt-\gamma(\mathfrak{a},\chi)\int_1^\infty t^{-s/2-1}dt.\end{aligned}$$

We insert this formula in our expression for $\Lambda(s,\chi,C)$ given in 7.3.2, observe that

$$\int_{[-\frac12,\frac12]^{r-1}\times(1,\infty)} t^{-(1+s)/2}d(\boldsymbol{z},t) = -\frac{2}{1-s},$$

$$\int_{[-\frac12,\frac12]^{r-1}\times(1,\infty)} t^{-s/2-1}d(\boldsymbol{z},t) = -\frac{2}{s},$$

we set $\Delta_{\mathfrak{a}}=\mathfrak{N}(\mathfrak{a}^2\mathfrak{D}\mathfrak{f})$ and $\overline{\mathfrak{a}}=(\mathfrak{a}\mathfrak{D}\mathfrak{f})^{-1}$. Then we arrive at the following final formula for $\Lambda(s,\chi,C)$.

$$\begin{aligned}\Lambda(s,\chi,C) &= -\frac{2^r\mathsf{R}_K}{w_K}\Big[\Delta_{\mathfrak{a}}^{s/2}\frac{\gamma(\mathfrak{a},\chi)}{s}+\Delta_{\mathfrak{a}}^{(s-1)/2}\frac{\mathsf{W}(\chi)\gamma(\overline{\mathfrak{a}},\overline{\chi})}{1-s}\Big]\\ &+\frac{2^{r-1}\mathsf{R}_K}{w_K}\Big[\Delta_{\mathfrak{a}}^{s/2}\int_{[-\frac12,\frac12]^{r-1}\times(1,\infty)}[\theta(\boldsymbol{y}(t,\boldsymbol{z}),\mathfrak{a},\chi)-\gamma(\mathfrak{a},\chi)]\,t^{s/2-1}d(\boldsymbol{z},t)\\ &\quad+\mathsf{W}(\chi)\Delta_{\mathfrak{a}}^{(s-1)/2}\int_{[-\frac12,\frac12]^{r-1}\times(1,\infty)}[\theta(\boldsymbol{y}(t,\boldsymbol{z}),\overline{\mathfrak{a}},\overline{\chi})\\ &\quad-\gamma(\overline{\mathfrak{a}},\overline{\chi})]\,t^{-(1+s)/2}d(\boldsymbol{z},t)\Big].\end{aligned} \qquad (\#)$$

We assert that both integrals in (#) represent integral functions. We postpone the proof of this purely analytic fact. For the time being, we accept it and finish the proof of 7.1.8.

The formula (#) represents an extension of $\Lambda(\,\cdot\,,\chi,C)$ to a meromorphic function in $\mathbb{C}$ with (simple) poles at most at $s=0$ and $s=1$, with residues

$$-\frac{2^r\mathsf{R}_K}{w_K}\gamma(\mathfrak{a},\chi)\quad\text{at}\quad s=0\quad\text{and}\quad\frac{2^r\mathsf{R}_K}{w_K}\mathsf{W}(\chi)\gamma(\overline{\mathfrak{a}},\overline{\chi})\quad\text{at}\quad s=1.$$

Since $[\mathfrak{a}]=C^{-1}$, it follows that $[\overline{\mathfrak{a}}]=[(\mathfrak{a}\mathfrak{D}\mathfrak{f})^{-1}]=(C^{-1}[\mathfrak{D}\mathfrak{f}])^{-1}=\overline{C}^{-1}$.

If $\mathfrak{f} \neq 1$ or $q \neq 0$, then $\gamma(\mathfrak{a}, \chi) = \gamma(\overline{\mathfrak{a}}, \overline{\chi}) = 0$, and $L(\,\cdot\,, \chi, C)$ is an entire function. Thus let $\mathfrak{f} = 1$, $q = 0$, and consequently $\chi \in \mathsf{X}(\mathcal{C}_K)$. Then it follows that $\gamma(\mathfrak{a}, \chi) = \overline{\chi}(\mathfrak{a}) = \chi(C)$ and $\gamma(\overline{\mathfrak{a}}, \overline{\chi}) = \chi(\overline{\mathfrak{a}}) = \overline{\chi}(\overline{C})$.

Since $\overline{\mathfrak{a}} \in \overline{C}^{-1}$, $\overline{\overline{\chi}} = \chi$, $\overline{\overline{\mathfrak{a}}} = \mathfrak{a}$, $\Delta_{\overline{\mathfrak{a}}} = \mathfrak{N}(\mathfrak{a}^2\mathfrak{D}\mathfrak{f})^{-1}$ and $\mathsf{W}(\chi)\mathsf{W}(\overline{\chi}) = 1$, (#) implies the functional equation $\Lambda(s, \chi, C) = \mathsf{W}(\chi)\Lambda(1-s, \overline{\chi}, \overline{C})$ for all $s \in \mathbb{C}$.

It remains to prove that the integrals in (#) are entire function. For this, we apply the following holomorphy criterion (see [10, Theorem 13.8.6]).

Holomorphy criterion. Let $D \neq \emptyset$ be an open subset of $\mathbb{C}$, $p \in \mathbb{N}$, and let $f\colon D \times \mathbb{R}^p \to \mathbb{C}$ be a function with the following properties:

(a) For all $s \in D$, the function $\boldsymbol{x} \mapsto f(s, \boldsymbol{x})$ is integrable;

(b) For all $\boldsymbol{x} \in \mathbb{R}^p$, the function $s \mapsto f(s, \boldsymbol{x})$ is holomorphic in D;

(c) There exists an integrable function $g\colon \mathbb{R}^p \to \mathbb{R}$ such that we have $|f(s, \boldsymbol{x})| \leq g(\boldsymbol{x})$ for all $(s, \boldsymbol{x}) \in D \times \mathbb{R}^p$.

Then the function
$$s \mapsto \int_{\mathbb{R}^p} f(s, \boldsymbol{x}) d\boldsymbol{x}$$
is holomorphic in D.

We apply the holomorphy criterion to the functions
$$(s, \boldsymbol{x}) \mapsto [\theta(\boldsymbol{y}(t, \boldsymbol{z}), \mathfrak{a}, \chi) - C(\mathfrak{a}, \chi)]\, t^{s/2-1}$$
and
$$(s, \boldsymbol{x}) \mapsto [\theta(\boldsymbol{y}(t, \boldsymbol{z}), \overline{\mathfrak{a}}, \overline{\chi}) - C(\overline{\mathfrak{a}}, \overline{\chi})]\, t^{-(1+s)/2},$$
defined for
$$\boldsymbol{x} = (\boldsymbol{z}, t) \in \Big[-\frac{1}{2}, \frac{1}{2}\Big]^{r-1} \times (1, \infty).$$
It suffices to prove that the integrals represent holomorphic functions in each strip $\{s \in \mathbb{C} \mid \Re(s) \in [u, u']\}$, where $u,\, u' \in \mathbb{R}$ and $u < u'$. Condition (b) of the holomorphy criterion is obviously fulfilled, and as the functions are continuous in $\boldsymbol{x}$, condition (a) is a consequence of (c). If $u,\, u' \in \mathbb{R}$, $u < \Re(s) < u'$ and $t \in \mathbb{R}_{>1}$, then $|t^{s/2-1}| \leq t^{u/2-1}$ and $|t^{-(1+s)/2}| \leq t^{-(u'+1)/2}$. Hence it suffices to prove:

Lemma 7.3.4. *Let $u \in \mathbb{R}$, and let*
$$g\colon \Big[-\frac{1}{2}, \frac{1}{2}\Big]^{r-1} \times (1, \infty) \to \mathbb{C} \quad \textit{be defined by}$$
$$g(\boldsymbol{z}, t) = |\theta(\boldsymbol{y}(t, \boldsymbol{z}), \mathfrak{a}, \chi) - C(\mathfrak{a}, \chi)|t^u.$$
Then
$$\int_{[-\frac{1}{2}, \frac{1}{2}]^{r-1} \times (1, \infty)} g(\boldsymbol{z}, t) d(\boldsymbol{z}, t) < \infty.$$

Proof. Let $(\boldsymbol{z},t)\in\left[-\frac{1}{2},\frac{1}{2}\right]^{r-1}\times(1,\infty)$. By definition,

$$g(\boldsymbol{z},t)\le\sum_{a\in\mathfrak{a}}\prod_{j=1}^{q}\left(|a^{(j)}|y_j^{1/2}\right)\exp\Big\{-\pi\sum_{j=1}^{n}|a^{(j)}|^2y_j\Big\}t^u.$$

For every $j\in[1,n]$ there exist (by continuity) bounds $m_j,\,M_j\in\mathbb{R}_{>0}$ such that

$$m_j\le\prod_{k=1}^{r-1}|e_k^{(j)}|^{2z_k}\le M_j,\ \text{ hence }\ m_jt^{1/n}\le y_j(t,\boldsymbol{z})\le M_jt^{1/n}$$
$$\text{for all }\ \boldsymbol{z}\in\Big[-\frac{1}{2},\frac{1}{2}\Big]^{r-1}.$$

Let $(a_1,\ldots,a_n)$ be a $\mathbb{Z}$-basis of $\mathfrak{a}$, and for $j\in[1,n]$ and $\boldsymbol{x}=(x_1,\ldots,x_n)^{\mathsf{t}}\in\mathbb{Z}^n$ we set

$$L(\boldsymbol{x})=\sum_{k=1}^{n}a_kx_k\quad\text{and}\quad L^{(j)}(\boldsymbol{x})=\sum_{k=1}^{n}a_k^{(j)}x_k.$$

Then $\mathfrak{a}^{\bullet}=\{L(\boldsymbol{x})\mid\boldsymbol{0}\ne\boldsymbol{x}\in\mathbb{Z}^n\}$, and thus we obtain

$$g(\boldsymbol{z},t)\le\sum_{\boldsymbol{0}\ne\boldsymbol{x}\in\mathbb{Z}^n}\prod_{j=1}^{q}\left(|L^{(j)}(\boldsymbol{x})|M_j^{1/2}t^{1/2n}\right)\exp\Big\{-\pi\sum_{j=1}^{n}|L^{(j)}(\boldsymbol{x})|^2m_jt^{1/n}\Big\}t^u.$$

We prove:

A. For every $N\in\mathbb{N}$ there exists some $B\in\mathbb{R}_{>0}$ such that

$$\prod_{j=1}^{q}\Big(\frac{M_j}{m_j}\Big)^{1/2}\prod_{j=1}^{q}\eta_j\exp\Big\{-\pi\sum_{j=1}^{n}\eta_j^2\Big\}\le B\Big(\sum_{j=1}^{n}\eta_j^2\Big)^{-N}$$

for all $(\eta_1,\ldots,\eta_n)\in\mathbb{R}^n$.

Proof of **A.** Let $(\eta_1,\ldots,\eta_n)^{\mathsf{t}}\in\mathbb{R}^n$ and $\eta=\eta_1^2+\ldots+\eta_n^2$. Since

$$\prod_{j=1}^{q}\eta_j=\Big(\prod_{j=1}^{q}\eta_j^2\Big)^{1/2}\le\Big(\frac{1}{q}\sum_{j=1}^{q}\eta_j^2\Big)^{q/2}\le q^{-q/2}\eta^{q/2}\quad\text{and}$$
$$\lim_{\eta\to\infty}\eta^{q/2+N}\mathsf{e}^{-\pi\eta}=0,$$

the assertion follows. $\square$[**A.**]

By Minkowski's inequality, the map

$$\boldsymbol{x}\ \mapsto\ \Big(\sum_{j=1}^{n}m_j|L^{(j)}(\boldsymbol{x})|^2\Big)^{1/2}$$

is a norm on $\mathbb{R}^n$, and thus there exists some $N \in \mathbb{N}$ such that

$$-\frac{N}{n} + u < -1 \quad \text{and} \quad M = \sum_{\mathbf{0} \neq \boldsymbol{x} \in \mathbb{Z}^n} \Big(\sum_{j=1}^{n} m_j |L^{(j)}(\boldsymbol{x})|^2\Big)^{-N} < \infty.$$

Let $B \in \mathbb{R}_{>0}$ be such that **A** holds. For $t \in (1, \infty)$, $\mathbf{0} \neq \boldsymbol{x} \in \mathbb{Z}^n$ and $j \in [1, n]$ we set $\eta_j = |L^{(j)}(\boldsymbol{x})| m_j^{1/2} t^{1/2n}$ and obtain

$$\begin{aligned} g(\boldsymbol{z}, t) &\leq \sum_{\mathbf{0} \neq \boldsymbol{x} \in \mathbb{Z}^n} \prod_{j=1}^{q} \Big(\frac{M_j}{m_j}\Big)^{1/2} \prod_{j=1}^{q} \eta_j \exp\Big\{-\pi \sum_{j=1}^{n} \eta_j^2\Big\} t^u \leq B \Big(\sum_{j=1}^{n} \eta_j^2\Big)^{-N} t^u \\ &= B t^{-N/n+u} \Big(\sum_{j=1}^{n} m_j |L^{(j)}(\boldsymbol{x})|^2\Big)^{-N} = M B t^{-N/n+u}. \end{aligned}$$

Now it follows that

$$\begin{aligned} \int_{[-\frac{1}{2}, \frac{1}{2}]^{r-1} \times (1,\infty)} g(\boldsymbol{z}, t) d(\boldsymbol{z}, t) &\leq M B \int_{[-\frac{1}{2}, \frac{1}{2}]^{r-1}} \int_{1}^{\infty} t^{-N/n+u} dt\, d\boldsymbol{z} \\ &= \frac{MB}{N/n - u + 1} < \infty. \qquad \square \end{aligned}$$

7.4 The functional equation for L functions of function fields

Let q be a prime power, $K/\mathbb{F}_q$ a function field, g its genus and $h = |\mathcal{C}_K^0|$ it class number.

The main goal of this section is to prove the counterpart of 7.1.8 for function fields. For this purpose we use the standard terminology for function fields as introduced in [18, Ch. 6]. We recall the basic concepts from [18, Definition 6.9.1] (see also 5.1.1).

Let $\mathbb{P}_K$ be the set of prime divisors, $\mathbb{D}_K = \mathbb{Z}^{(\mathbb{P}_K)}$ the (now additively written) group of divisors, $\mathbb{D}_K' = \mathbb{N}_0^{(\mathbb{P}_K)} = \{D \in \mathbb{D}_K \mid D \geq 0\}$ the monoid of effective divisors, $\mathsf{0}$ the zero divisor and $(K^\times) = \{(a) \mid a \in K^\times\}$ the group of principal divisors. For $P \in \mathbb{P}_K$ let v_P be the P-adic valuation, and for a divisor $D \in \mathbb{D}_K$ let $\operatorname{supp}(D) = \{P \in \mathbb{P}_K \mid v_P(D) \neq 0\}$ be its support. Let $\mathcal{C}_K = \mathbb{D}_K/(K^\times)$ be the divisor class group, and for a divisor D we denote by $[D] = D + (K^\times)$ its divisor class. For divisors D, D' we set $D \sim D'$ if $[D] = [D']$, and $D \leq D'$ if $v_P(D) \leq v_P(D')$ for all $P \in \mathbb{P}_K$. For $D \in \mathbb{D}_K$, we denote by $\deg(D) = \deg([D])$ its degree, we set

$\mathbb{D}_K^0 = \{D \in \mathbb{D}_K \mid \deg(D) = 0\}$, observe that $K^\times \subset \mathbb{D}_K^0$, that $\mathcal{C}_K^0 = \mathbb{D}_K^0/(K^\times)$ is finite, and we set $h_K = |\mathcal{C}_K^0|$.

For $M \in \mathbb{D}'_K$ we denote by $\mathbb{D}_K^M = \{D \in \mathbb{D}_K \mid \operatorname{supp}(D) \cap \operatorname{supp}(M) = \emptyset\}$ the group of all divisors coprime to M. We set $\mathbb{D}'^M_K = \mathbb{D}'_K \cap \mathbb{D}_K^M$,

$$K(M) = \{a \in K^\times \mid (a) \in \mathbb{D}_K^M\} = \{a \in K^\times \mid v_P(a) = 0 \text{ for all } P \in \operatorname{supp}(M)\},$$

$$K^M = \{a \in K^\times \mid v_P(a-1) \geq v_P(M) \text{ for all } P \in \operatorname{supp}(M)\},$$

and we call

$$\mathcal{C}_K^M = \mathbb{D}_K^M/(K^M) \quad \text{the } \textbf{ray class group modulo } M.$$

For $D \in \mathbb{D}_K^M$, we denote by $[D]^M = D + (K^M) \in \mathcal{C}_K^M$ the ray class of D modulo M. In particular, for $M = 0$ we obtain $\mathbb{D}_K^0 = \mathbb{D}_K$, $K^0 = K^\times$ and $\mathcal{C}_K^0 = \mathcal{C}_K = \mathbb{D}_K/(K^\times)$.

If $D, D' \in \mathbb{D}_K^M$, then $[D]^M = [D']^M$ implies $D \sim D'$, and we define the degree homomorphism $\deg\colon \mathcal{C}_K^M \to \mathbb{Z}$ by $\deg([D]^M) = \deg(D)$ for all $D \in \mathbb{D}_K^M$.

We proceed (as already announced) with the investigation of L functions associated with ray class characters and recall the relevant definitions (see [18, Definition 6.9.1]).

Remark 7.4.1. Let $M \in \mathbb{D}'_K$. A character $\chi \in \mathsf{X}(\mathcal{C}_K^M)$ is called a **ray class character modulo** M. Note that contrary to the number field case a ray class character need not be of finite order. A ray class character $\chi \in \mathsf{X}(\mathcal{C}_K^M)$ induces an (equally denoted) homomorphism $\chi\colon \mathbb{D}_K^M \to \mathbb{T}$ satisfying $\chi \restriction (K^M) = 1$ (where $\chi(D) = \chi([D]^M)$ for all $D \in \mathbb{D}_K^M$), and we set $\chi(D) = 0$ for all $D \in \mathbb{D}_K \setminus \mathbb{D}_K^M$. Conversely every homomorphism $\chi\colon \mathbb{D}_K^M \to \mathbb{T}$ with $\chi \restriction (K^M) = 1$ is induced by a ray class character modulo M and is itself called a ray class character modulo M.

If $\chi\colon \mathbb{D}_K^M \to \mathbb{T}$ is a ray class character modulo M, then the L **function** built with χ is given by the Dirichlet series

$$L(s,\chi) = \sum_{D \in \mathbb{D}'_K} \frac{\chi(D)}{\mathfrak{N}(D)^s} = \sum_{D \in \mathbb{D}'_K} \chi(D) q^{-s\deg(D} \quad \text{for } \ s \in \mathcal{H}_1,$$

and there it has an Euler product

$$L(s,\chi) = \prod_{P \in \mathbb{P}_K} \frac{1}{1 - \chi(P)\mathfrak{N}(P)^{-s}}.$$

In particular, $L(\,\cdot\,,\chi)\colon \mathcal{H}_1 \to \mathbb{C}$ is a holomorphic function without zeros.

We set $t = q^{-s}$ and represent the L function as a power series in t. Then

$$\mathcal{Z}(t,\chi) = L(s,\chi) = \sum_{D \in \mathbb{D}'_K} \chi(D) t^{\deg(D)} = \prod_{P \in \mathbb{P}_K} \frac{1}{1 - \chi(P) t^{\deg(P)}} \quad \text{for } \ |t| < q^{-1},$$

and $\mathcal{Z}(\,\cdot\,,\chi)$ is a holomorphic function without zeros in the disc $|t| < q^{-1}$ (see [18, Theorem 6.9.2]).

Let $F = F(\chi)$ be the conductor and $\chi_0 \colon \mathbb{D}_K^F \to \mathbb{T}$ is the associated primitive character of χ (see 5.8.2). Then it follows that

$$\mathcal{Z}(t,\chi) = \mathcal{Z}(t,\chi_0) \prod_{P \in \operatorname{supp}(M)} \bigl(1 - \chi_0(P)t^{\deg(P)}\bigr) \quad \text{for} \quad |t| < q^{-1}.$$

Hence it suffices to study $\mathcal{Z}(t,\chi)$ for primitive ray class characters χ. The main result is as follows.

Theorem 7.4.2. *Let $M \in \mathbb{D}'_K$, and let $\chi \in \mathsf{X}(\mathcal{C}_K^M)$ be a primitive ray class character modulo M.*

1. *If $\chi \restriction \mathcal{C}_K^0 = 1$, then $M = 0$, and there is some $\eta \in \mathbb{T}$ such that $\chi(C) = \eta^{\deg(C)}$ for all $C \in \mathcal{C}_K$. In this case, $\mathcal{Z}(t,\chi)$ is a rational function in t having only simple poles at $\overline{\eta}$ and $q^{-1}\overline{\eta}$ with residues*

$$\operatorname{Res}\{\mathcal{Z}(\,\cdot\,,\chi);\overline{\eta}\} = \frac{\overline{\eta}h}{q-1} \quad \textit{and} \quad \operatorname{Res}\{\mathcal{Z}(\,\cdot\,,\chi);q^{-1}\overline{\eta}\} = -\frac{q^{-g}\overline{\eta}h}{q-1}.$$

2. *If $\chi \restriction \mathcal{C}_K^0 \neq 1$, then $\mathcal{Z}(t,\chi)$ is a polynomial in t of degree*

$$2g - 2 + \deg(M).$$

3. *In any case, $\mathcal{Z}(t,\chi)$ satisfies a functional equation*

$$\mathcal{Z}(t,\chi) = \mathsf{W}(\chi)\bigl(t\sqrt{q}\bigr)^{2g-2+\deg(M)}\,\mathcal{Z}\Bigl(\frac{1}{qt},\overline{\chi}\Bigr) \quad \textit{for all} \;\; t \in \mathbb{C},$$

where $\mathsf{W}(\chi) \in \mathbb{C}$, $\mathsf{W}(\overline{\chi}) = \overline{\mathsf{W}(\chi)}$ and $|\mathsf{W}(\chi)| = 1$. More precisely, if $M = 0$, then $\mathsf{W}(\chi) = \chi(\boldsymbol{w})$, where $\boldsymbol{w}$ is the canonical class.

Although the result for $M = 0$ can be derived from [18, Theorem 6.9.3], we give an alternative fresh proof here. Note that $\chi \restriction \mathcal{C}_K^0 = 1$ implies $\chi \restriction (K^\times) = 1$, hence $M = 0$. If $\chi = 1$, then $\mathcal{Z}_K(t) = \mathcal{Z}(t,1)$ is the congruence zeta function which is investigated in detail in [18, Section 6.9].

Before we start with the proof, we reformulate Theorem 7.4.2 for $L(s,\chi)$ as a function in s.

Corollary 7.4.3. *Let $M \in \mathbb{D}'_K$, and let $\chi \in \mathsf{X}(\mathcal{C}_K^M)$ a primitive ray class character modulo M.*

1. *If $\chi \restriction \mathcal{C}_K^0 \neq 1$, then $L(s,\chi)$ is a polynomial in q^{-s} of degree $2g - 2 + \deg(M)$. In particular, $L(\,\cdot\,,\chi)$ is an entire function.*

2. If $\chi \restriction \mathcal{C}_K^0 = 1$ and $\chi(C) = \mathsf{e}^{\mathsf{i}\delta}$ for all $C \in \mathcal{C}_K^1$, then $L(s,\chi)$ has an extension to a meromorphic function in $\mathbb{C}$ with simple poles at

$$s = \mathsf{i}\Big(\delta + \frac{2k\pi}{\log q}\Big) \quad \text{and} \quad s = 1 + \mathsf{i}\Big(\delta + \frac{2k\pi}{\log q}\Big) \quad \text{for all } k \in \mathbb{Z}.$$

If χ is of finite order and $\chi \ne 1$, then $\mathcal{L}(s,\chi)$ is holomorphic in 0 and 1.

3. The extended L function

$$\Lambda(s,\chi) = q^{s[g-1+\deg(M)/2]} L(s,\chi)$$

satisfies the functional equation

$$\Lambda(s,\chi) = \mathsf{W}(\chi)\Lambda(1-s,\overline{\chi}),$$

where $\mathsf{W}(\chi)$ is as in 7.4.2.

Proof. By 7.4.2. If $s \in \mathbb{C}$ and $\mathcal{L}(\,\cdot\,,\chi)$ has a pole at q^{-s}, then $q^{-s} = q^{-e}\mathsf{e}^{-\mathsf{i}\delta}$, where $e \in \{0,1\}$, and then

$$s = e + \mathsf{i}\Big(\delta + \frac{2k\pi}{\log q}\Big).$$

In particular, if $\chi \ne 1$ and χ is of finite order, then $\delta = \pi r$ for some $r \in \mathbb{Q}^\times$, and consequently $s \notin \{0,1\}$. □

Proof of Theorem 7.4.2 if $M = 0$. Let $\chi \in \mathcal{C}_K$, fix a class $\boldsymbol{c}_1 \in \mathcal{C}_K$ such that $\deg(\boldsymbol{c}_1) = 1$ (see 6.1.7), and let $\chi(\boldsymbol{c}_1) = \eta \in \mathbb{T}$. Then

$$\mathcal{C}_K = \{\boldsymbol{c} + n\boldsymbol{c}_1 \mid \boldsymbol{c} \in \mathcal{C}_K^0,\ n \in \mathbb{Z}\},$$

and if $\boldsymbol{c} \in \mathcal{C}_K^0$ and $n \in \mathbb{Z}$, then $\deg(\boldsymbol{c} + n\boldsymbol{c}_1) = n$ and $\chi(\boldsymbol{c} + n\boldsymbol{c}_1) = \chi(\boldsymbol{c})\eta^n$. We set

$$\varepsilon(\chi) = \sum_{\boldsymbol{c} \in \mathcal{C}_K^0} \chi(\boldsymbol{c}) = \begin{cases} h & \text{if } \chi \restriction \mathcal{C}_K^0 = 1, \\ 0 & \text{if } \chi \restriction \mathcal{C}_K^0 \ne 1. \end{cases}$$

In order to avoid problems with convergence, we consider the power series ring $\mathbb{C}\,[\![T,T^{-1}]\!] = \mathbb{C}\,[\![T,X]\!]/(TX-1)$ and prove the results for the formal L function

$$\mathcal{Z}(T,\chi) = \sum_{D \in \mathbb{D}'_K} \chi(D) T^{\deg(D)} = \sum_{\boldsymbol{c} \in \mathcal{C}_K} |\boldsymbol{c} \cap \mathbb{D}'_K| \chi(\boldsymbol{c}) T^{\deg(\boldsymbol{c})} \in \mathbb{C}\,[\![T,T^{-1}]\!].$$

With the aid of [18, Theorem 6.6.8] it follows that

$$\mathcal{Z}(T,\chi) = \frac{1}{q-1} \sum_{\boldsymbol{c} \in \mathcal{C}_K} \chi(\boldsymbol{c}) T^{\deg(\boldsymbol{c})}(q^{\dim(\boldsymbol{c})} - 1) = \frac{1}{q-1} \sum_{\boldsymbol{c} \in \mathcal{C}_K} \chi(\boldsymbol{c}) T^{\deg(\boldsymbol{c})} q^{\dim(\boldsymbol{c})},$$

since

$$\sum_{\boldsymbol{c}\in\mathcal{C}_K}\chi(\boldsymbol{c})T^{\deg(\boldsymbol{c})}=\sum_{n\in\mathbb{Z}}\sum_{\boldsymbol{c}\in\mathcal{C}_K^0}\chi(\boldsymbol{c})\eta^nT^n=\varepsilon(\chi)\Big[\sum_{n=0}^{\infty}(\eta T)^n+\sum_{n=1}^{\infty}(\eta T)^{-n}\Big]$$
$$=\varepsilon(\chi)\Big[\frac{1}{1-\eta T}+\frac{1}{\eta T-1}\Big]=0.$$

Now we can prove the functional equation. Let $\boldsymbol{w}\in\mathcal{C}_K$ be the canonical class, and observe that $\mathcal{C}_K=\{\boldsymbol{c}-\boldsymbol{w}\mid\boldsymbol{c}\in\mathcal{C}_K\}$. Then

$$\mathcal{Z}\Big(\frac{1}{qT},\overline{\chi}\Big)=\frac{1}{q-1}\sum_{\boldsymbol{c}\in\mathcal{C}_K}\overline{\chi}(\boldsymbol{c}-\boldsymbol{w})\Big(\frac{1}{qT}\Big)^{\deg(\boldsymbol{c}-\boldsymbol{w})}q^{\dim(\boldsymbol{c}-\boldsymbol{w})},$$

and the Riemann-Roch theorem implies

$$\deg(\boldsymbol{w}-\boldsymbol{c})=2g-2-\deg(\boldsymbol{c})\ \text{ and }\ \dim(\boldsymbol{c}-\boldsymbol{w})-\deg(\boldsymbol{c}-\boldsymbol{w})=-g+1+\dim(\boldsymbol{c}).$$

Hence

$$\mathcal{Z}\Big(\frac{1}{qT},\overline{\chi}\Big)=\frac{\overline{\chi}(\boldsymbol{w})q^{-g+1}T^{2-2g}}{q-1}\sum_{\boldsymbol{c}\in\mathcal{C}_K}\chi(\boldsymbol{c})T^{\deg(C)}q^{\dim(C)}$$
$$=\overline{\chi}(\boldsymbol{w})q^{-g+1}T^{2-2g}\mathcal{Z}(T,\chi),$$

and consequently

$$\mathcal{Z}(T,\chi)=\chi(\boldsymbol{w})q^{g-1}T^{2g-2}\mathcal{Z}\Big(\frac{1}{qT},\overline{\chi}\Big).$$

As to the remaining assertions, observe that for $\boldsymbol{c}\in\mathcal{C}_K$ the following holds by [18, Theorems 6.2.10 and 6.4.5]: If $\deg(\boldsymbol{c})<0$, then $\dim(\boldsymbol{c})=0$, and if $\deg(\boldsymbol{c})\ge 2g-1$, then $\dim(\boldsymbol{c})=\deg(\boldsymbol{c})-g+1$. Therefore we obtain

$$\mathcal{Z}(T,\chi)=\frac{1}{q-1}\sum_{\boldsymbol{c}\in\mathcal{C}_K}\chi(\boldsymbol{c})T^{\deg(\boldsymbol{c})}q^{\dim(\boldsymbol{c})}$$
$$=\frac{1}{q-1}\sum_{n\in\mathbb{Z}}\sum_{\boldsymbol{c}\in\mathcal{C}_K^0}\chi(\boldsymbol{c})\eta^nT^nq^{\dim(\boldsymbol{c}+n\boldsymbol{c}_1)}=P(T)+Q(T),$$

where

$$P(T)=\frac{1}{q-1}\sum_{n=0}^{2g-2}\sum_{\boldsymbol{c}\in\mathcal{C}_K^0}\chi(\boldsymbol{c})\eta^n[q^{\dim(\boldsymbol{c}+n\boldsymbol{c}_1)}-q^{n-g+1}]T^n\quad[=0\ \text{ if }\ g=0],$$

and

$$Q(T)=\frac{\varepsilon(\chi)}{q-1}\Big[\sum_{n=1}^{\infty}\eta^{-n}T^{-n}+\sum_{n=0}^{\infty}\eta^nq^{n-g+1}T^n\Big]=\frac{\varepsilon(\chi)}{q-1}\Big[\frac{-1}{1-\eta T}+\frac{q^{1-g}}{1-q\eta T}\Big].$$

Hence $\mathcal{Z}(T,\chi)$ is a rational function. If $\chi\restriction\mathcal{C}_K^0=1$, then $\varepsilon(\chi)=h$, and $\mathcal{Z}(T,\chi)$ has only simple poles at $\overline{\eta}$ and $q^{-1}\overline{\eta}$ with residues as asserted. If $\chi\restriction\mathcal{C}_K^0\ne 1$, then $\mathcal{Z}(T,\chi) = P(T)$ is a polynomial of degree at most $2g-2$, and the functional equation shws that it is of precise degree $2g-2$. □

The remainder of this section is devoted to the proof of Theorem 7.4.2 for $M\ne 0$. After all we shall reformulate the theorem and give a precise value for $\mathsf{W}(\chi)$ in Theorem 7.4.9.

We start the proof of 7.4.2 with necessary algebraic arrangements. The concepts which we present in 7.4.4 below are fundamental for the investigation of the L functions in the case $M\ne 0$. The notation introduced there will be retained throughout this chapter.

Definition and Remarks 7.4.4. Let

$$M=e_1P_1+\ldots+e_rP_r\in\mathbb{D}_K',$$

where $r,e_1,\ldots,e_r\in\mathbb{N}$, $P_1,\ldots,P_r\in\mathbb{P}_K$ are distinct, and we set $S=\mathsf{M}_K\setminus\{P_1,\ldots,P_r\}$. Then $\mathbb{D}_K^M=\{D\in\mathbb{D}_K\mid \operatorname{supp}(D)\subset S\})$, the holomorphy domain

$$\mathfrak{o}_{K,S}=\bigcap_{i=1}^r\mathcal{O}_{P_i}$$

is a semilocal principal ideal domain, and if $\mathfrak{p}_i=P_i\cap\mathfrak{o}_{K,S}$ for all $i\in[1,r]$, then $\mathcal{P}(\mathfrak{o}_{K,S})=\{\mathfrak{p}_1,\ldots,\mathfrak{p}_r\}$ (see 5.5.3 and 5.5.5). For $i\in[1,r]$, we set $\mathfrak{p}_i=t_i\mathfrak{o}_{K,S}$; then $v_{P_i}(t_i)=1$, $K_{P_i}=\mathsf{k}_{P_i}(\!(t_i)\!)$ is a completion of (K,v_{P_i}), and $\widehat{\mathcal{O}}_{P_i}=\mathsf{k}_{P_i}[\![t_i]\!]$ is its valuation domain (see [18, Theorem 6.2.8]). We consider the $\mathbb{F}_q$-algebra

$$R_M=\bigoplus_{i=1}^r\mathsf{k}_{P_i}[\![t_i]\!]/t_i^{e_i}\mathsf{k}_{P_i}[\![t_i]\!]=\bigoplus_{i=1}^r\mathsf{k}_{P_i}[\mathsf{t}_i],$$

where $\mathsf{t}_i=t_i+t_i^{e_i}\mathsf{k}_{P_i}[\![t_i]\!]\in R_M$, $\mathsf{t}_i^{e_i}=0$ and $\mathsf{t}_i\mathsf{t}_j=0$ for all $i,\,j\in[1,r]$ such that $i\ne j$. We identify $x\in\mathbb{F}_q$ with the constant vector $(x,\ldots,x)\in R_M$. Then $\mathbb{F}_q$ is a one-dimensional subspace of R_M. Obviously,

$$\dim_{\mathbb{F}_q}(R_M)=\sum_{i=1}^r e_i\dim_{\mathbb{F}_q}(\mathsf{k}_{P_i})=\sum_{i=1}^r e_i\deg(P_i)=\deg(M).$$

With M we associate the ideal

$$\mathfrak{a}_M=\prod_{i=1}^r\mathfrak{p}_i^{e_i}=\{x\in K\mid v_{P_i}(x)\ge e_i\text{ for all }i\in[1,r]\}\ \text{ of }\ \mathfrak{o}_{K,S},$$

and we consider the (natural) ring isomorphisms

$$\mathfrak{o}_{K,S}/\mathfrak{a}_M\ \overset{\sim}{\to}\ \bigoplus_{i=1}^r\mathfrak{o}_{K,S}/\mathfrak{p}_i^{e_i}\ \overset{\sim}{\to}\ \bigoplus_{i=1}^r\mathcal{O}_{P_i}/P_i^{e_i}\ \overset{\sim}{\to}\ \bigoplus_{i=1}^r\mathsf{k}_{P_i}[\mathsf{t}_i]=R_M.$$

We call R_M the abstract residue class ring modulo M and in the following identify: $\mathfrak{o}_{K,S}/\mathfrak{a}_M = R_M$. Every $x \in R_M$ has a unique representation

$$x = (x_1, \ldots, x_r), \quad \text{where} \quad x_i = \sum_{j=0}^{e_i-1} x_{i,j}\mathsf{t}_i^j \quad \text{and} \quad x_{i,j} \in \mathsf{k}_{P_i} \quad \text{for all} \quad i \in [1, r]. \tag{\#}$$

If $\xi \in \mathfrak{o}_{K,S}$, then $\xi + \mathfrak{a}_M = x \in \mathfrak{o}_{K,S}/\mathfrak{a}_M = R_M$ if and only if

$$\xi \equiv \sum_{j=0}^{e_i-1} x_{i,j}t_i^j \bmod t_i^{e_i}\mathsf{k}_{P_i}[\![t_i]\!] \quad \text{for all} \quad i \in [1, r].$$

Let $x \in R_M$ be given by (#). Then $x \in R_M^\times$ if and only if $x_{i,0} \neq 0$ for all $i \in [1, r]$, x is a zero divisor of R_M if and only if $x \notin R_M^\times$, and we define the **trace** of x by

$$\mathsf{Tr}_M(x) = \sum_{i=1}^{r} \mathsf{Tr}_{\mathsf{k}_{P_i}/\mathbb{F}_q}(x_{i,e_i-1}).$$

Apparently, $\mathsf{Tr}_M \colon M \to \mathbb{F}_q$ is $\mathbb{F}_q$-linear and surjective. For an $\mathbb{F}_q$-subspace X of R_M we define its **dual** $X^\perp$ by

$$X^\perp = \{y \in R_M \mid \mathsf{Tr}_M(xy) = 0\}.$$

Let $e^\# \colon \mathbb{F}_q \to \mathbb{C}$ be any function satisfying

$$e^\#(0) = 1 \quad \text{and} \quad \sum_{a \in \mathbb{F}_q} e^\#(a) = 0.$$

Theorem 7.4.5. *Let $M = e_1P_1 + \ldots + e_rP_r \in \mathbb{D}_K'$ be as in 7.4.4.*

1. Let $D \in \mathbb{D}_K^M$. Then $\mathcal{L}(D) \subset \mathfrak{o}_{K,S}$, $\mathcal{L}(D - M) = \mathcal{L}(D) \cap \mathfrak{a}_M$, and the inclusion $\mathcal{L}(D) \hookrightarrow \mathfrak{o}_{K,S}$ induces an $\mathbb{F}_q$-monomorphism

$$\mathcal{L}(D)/\mathcal{L}(D - M) \to \mathfrak{o}_{K,S}/\mathfrak{a}_M = R_M.$$

We identify and set

$$X_D = \mathcal{L}(D)/\mathcal{L}(D - M) \subset R_M.$$

Then X_D is an $\mathbb{F}_q$-subspace of R_M, and

$$\dim_{\mathbb{F}_q}(X_D) = \dim(D) - \dim(D - M).$$

2. The map $\tau \colon R_M \times R_M \to \mathbb{F}_q$, defined by $\tau(x, y) = \mathsf{Tr}_M(xy)$, is a non-degenerate symmetric $\mathbb{F}_q$-bilinear form.

3. Let X be an $\mathbb{F}_q$-subspace of R_M, Then

$$\dim_{\mathbb{F}_q}(X) + \dim_{\mathbb{F}_q}(X^\perp) = \deg(M), \quad X^{\perp\perp} = X,$$

and if $y \in R_M$, then

$$\sum_{x \in X} e^{\#}(\mathsf{Tr}_M(xy)) = \begin{cases} |X| & \text{if } y \in X^\perp, \\ 0 & \text{if } y \notin X^\perp. \end{cases}$$

Proof. 1. Let $x \in \mathcal{L}(D)$. Then $v_{P_i}(x) \ge -v_{P_i}(D) = 0$ for all $i \in [1, r]$, and therefore $x \in \mathfrak{o}_{K,S}$. Moreover, $x \in \mathfrak{a}_M$ if and only if $v_{P_i}(x) \ge e_i = v_{P_i}(M)$ for all $i \in [1, r]$, and this holds if and only if $x \in \mathcal{L}(D-M)$. Hence the inclusion $\mathcal{L}(D) \hookrightarrow \mathfrak{o}_{K,S}$ yields an $\mathbb{F}_q$-monomorphism $\mathcal{L}(D)/\mathcal{L}(D-M) \to \mathfrak{o}_{K,S}/\mathfrak{a}_M = R_M$. We identify $X_D = \mathcal{L}(D)/\mathcal{L}(D-M) \subset \mathcal{O}_S/\mathfrak{a}_M = R_K$, and then the remaining assertions follow from the very definitions.

2. Apparently, τ is $\mathbb{F}_q$-bilinear and symmetric, and we must prove that for every $x \in R_M^\bullet$ there exists some $y \in R_M$ such that $\mathsf{Tr}_M(xy) \ne 0$. Thus let $x \in R_M^\bullet$ be given by (#), let $i \in [1, r]$ be such that $x_{i,j} \ne 0$ for some $j \in [0, e_i - 1]$, and let $j \in [0, e_i - 1]$ be minimal such that $x_{i,j} \ne 0$. Let $a_i \in \mathsf{k}_{P_i}$ be such that $\mathsf{Tr}_{\mathsf{k}_{P_i}/\mathbb{F}_q}(a_i) \ne 0$, and let $y = (y_1, \ldots, y_r) \in R_M$ be such that $y_i = a_i a_{i,j}^{-1} \mathsf{t}_i^{e_i - 1 - j}$ and $y_k = 0$ for all $k \in [1, r] \setminus \{i\}$. Then $(xy)_i = a_i \mathsf{t}_i^{e_i - 1}$, $(xy)_k = 0$ for all $k \in [1, r] \setminus \{i\}$, and $\mathsf{Tr}_M(xy) = \mathsf{Tr}_{\mathsf{k}_{P_i}/\mathbb{F}_q}(a_i) \ne 0$.

3. For $x \in R_M$, let $\varphi_x \colon R_M \to \mathbb{F}_q$ be defined by $\varphi_x(y) = \mathsf{Tr}_M(xy)$. Then φ_x is $\mathbb{F}_q$-linear, and the map $\varphi \colon R_M \to R_M^* = \mathrm{Hom}_{\mathbb{F}_q}(R_M, \mathbb{F}_q)$, defined by $\varphi(x) = \varphi_x$ for all $x \in R_M$, is $\mathbb{F}_q$-linear, too. Since τ is nondegenerate, it follows that

$$\varphi(X^\perp) = \{\lambda \in R_M^* \mid \lambda \restriction X = 0\} \cong (R_M/X)^*,$$

and consequently

$$\begin{aligned} \dim_{\mathbb{F}_q}(X^\perp) &= \dim_{\mathbb{F}_q}(R_M/X)^* = \dim_{\mathbb{F}_q}(R_M/X) = \dim_{\mathbb{F}_q}(R_M) - \dim_{\mathbb{F}_q}(X) \\ &= \deg(M) - \dim_{\mathbb{F}_q}(X). \end{aligned}$$

Since obviously $X \subset X^{\perp\perp}$ and

$$\begin{aligned} \dim_{\mathbb{F}_q}(X^{\perp\perp}) &= \deg(M) - \dim_{\mathbb{F}_q})(X^\perp) = \deg(M) - [\deg(M) - \dim_{\mathbb{F}_q}(X)] \\ &= \dim_{\mathbb{F}_q}(X), \end{aligned}$$

equality holds.

If $y \in X^\perp$, then $\mathsf{Tr}_M(xy) = 0$ for all $x \in X$, and consequently

$$\sum_{x \in X} e^{\#}(\mathsf{Tr}_M(xy)) = |X|.$$

Thus let $y \notin X^\perp$, and let $(x_1, \ldots, x_m)$ be an $\mathbb{F}_q$-basis of X. Then we have

$\mathsf{Tr}_M(x_jy) \neq 0$ for some $j \in [1, m]$, and we assume that $\mathsf{Tr}_M(x_1y) \neq 0$. Then

$$\sum_{x \in X} e(\mathsf{Tr}_M(xy)) = \sum_{(a_1,\ldots,a_m) \in \mathbb{F}_q^m} e^{\#}\Big(\mathsf{Tr}_M\Big(\sum_{j=1}^m a_j x_j y\Big)\Big)$$
$$= \sum_{(a_2,\ldots,a_m) \in \mathbb{F}_q^{m-1}} \sum_{a_1 \in \mathbb{F}} e^{\#}\Big(a_1 \mathsf{Tr}_M(x_1y) + \sum_{j=2}^m a_j \mathsf{Tr}_M(x_jy)\Big).$$

For fixed $(a_2, \ldots, a_m) \in \mathbb{F}_q^{m-1}$, the assignment

$$a_1 \mapsto a_1 \mathsf{Tr}_M(x_1y) + \sum_{j=2}^m a_j \mathsf{Tr}_M(x_jy)$$

defines a bijective map $\mathbb{F}_q \to \mathbb{F}_q$, which implies

$$\sum_{a_1 \in \mathbb{F}} e^{\#}\Big(a_1 \mathsf{Tr}_M(x_1y) + \sum_{j=2}^m a_j \mathsf{Tr}_M(a_jy)\Big) = 0 \quad \text{for all} \quad (a_2, \ldots, a_m) \in \mathbb{F}_q^{m-1}$$

by the very definition of the function $e^{\#}\colon \mathbb{F}_q \to \mathbb{C}$, and therefore

$$\sum_{x \in X} e^{\#}(\mathsf{Tr}_M(xy)) = 0. \qquad \square$$

Definition 7.4.6. Let $M = e_1P_1 + \ldots + e_rP_r \in \mathbb{D}_K'$ be as in 7.4.4.

1. A **primitive character** of R_M is a map $\varphi\colon R_M \to \mathbb{C}$ with the following properties:

- $\varphi \upharpoonright \mathbb{F}_q^\times = 1$;
- $\varphi(xy) = \varphi(x)\varphi(y)$ for all x, $y \in R_M$,
- If $x \in R_M$, then $\varphi(x) = 0$ if and only if $x \notin R_M^\times$.
- For every $i \in [1, r]$ there exists some $x \in R_M^\times$ such that $\varphi(x) \neq 1$, $x_k = 1$ for all $k \in [1, r] \setminus \{i\}$, and if $e_i > 1$, then $x_i = 1 + at_i^{e_i - 1}$ for some $a \in \mathsf{k}_{P_i}$.

If φ is a primitive character of R_M, then so is $\overline{\varphi}$.

2. Let φ be a primitive character of R_M. For $x \in R_M$ we define the **Gauss sum** $G_\varphi(x)$ attached to φ by

$$G_\varphi(x) = \sum_{y \in R_M} \varphi(y) e^{\#}(\mathsf{Tr}_M(xy)), \quad \text{and we set} \quad Y(\varphi) = G_\varphi(1)$$
$$= \sum_{z \in R_M} \varphi(z) e^{\#}(\mathsf{Tr}_M(z)).$$

For an $\mathbb{F}_q$-subspace X of R_M we set

$$\Lambda_\varphi(X) = \sum_{x\in X} \varphi(x).$$

Theorem 7.4.7. *Let* $M = e_1P_1 + \ldots + e_rP_r \in \mathbb{D}'_K$ *be as in* RefM, *and let* φ *be a primitive character of* R_M.

1. $G_\varphi(x) = \overline{\varphi}(x)Y(\varphi)$ *for all* $x \in R_M$.
2. *Let* X *be an* $\mathbb{F}_q$*-subspace of* R_M*. Then*

$$|X|\,\Lambda_\varphi(X^\perp) = Y(\varphi)\Lambda_{\overline{\varphi}}(X) \quad \textit{and} \quad |X^\perp|\,\Lambda_{\overline{\varphi}}(X) = Y(\overline{\varphi})\Lambda_\varphi(X^\perp).$$

3. $Y(\overline{\varphi}) = \overline{Y(\varphi)}$ *and* $|Y(\varphi)|^2 = q^{\deg(M)}$.

Proof. 1. If $x \in R_M^\times$, then

$$\begin{aligned} G_\varphi(x) &= \sum_{y\in R_M} \varphi(y)e^{\#}(\mathsf{Tr}_M(xy)) \underset{[z=xy]}{=} \sum_{z\in R_M} \varphi(x^{-1}z)e^{\#}(\mathsf{Tr}_M(z)) \\ &= \overline{\varphi}(x) \sum_{z\in R_M} \varphi(z)e^{\#}(\mathsf{Tr}_M(z)) = \overline{\varphi}(x)Y(\varphi). \end{aligned}$$

Thus let $x \in R_M \setminus R_M^\times$ be given by (#), and let $i \in [1,r]$ be such that $x_{i,0} = 0$. Let $z = (z_1,\ldots,z_r) \in R_M^\times$ be such that $z_i = 1 + a_i\mathsf{t}_i^{e_i-1}$ for some $a_i \in \mathsf{k}_{P_i}$, $z_k = 1$ for all $k \in [1,r] \setminus \{i\}$ and $\varphi(z) \ne 1$. Then $xz = x$ and

$$\begin{aligned} \overline{\varphi}(z)G_\varphi(x) &= \sum_{y\in R_M} \varphi(z^{-1}y)e^{\#}(\mathsf{Tr}_M(xy)) \underset{[y'=z^{-1}y]}{=} \sum_{y'\in R_M} \varphi(y')e^{\#}(\mathsf{Tr}_M(xzy')) \\ &= \sum_{y'\in R_M} \varphi(y')e^{\#}(\mathsf{Tr}_M(xy')) = G_\varphi(x). \end{aligned}$$

Hence $G_\varphi(x) = 0$, and $\varphi(x) = 0$, since $x \notin^\times_M$. Hence $\overline{\varphi}(x)Y(\varphi) = 0$, too.

2. By 7.4.5.3 we obtain

$$\begin{aligned} \sum_{y\in R_M} \varphi(y) \sum_{x\in X} e^{\#}(\mathsf{Tr}_M(xy)) &= \sum_{y\in X^\perp} \varphi(y)|X| = |X|\,\Lambda_\varphi(X^\perp) \\ &= \sum_{x\in X}\sum_{y\in R_M} \varphi(y)e^{\#}(\mathsf{Tr}_M(xy)) = \sum_{x\in X} G_\varphi(x) = \sum_{x\in X} \overline{\varphi}(x)Y(\varphi) \\ &= Y(\varphi)\Lambda_{\overline{\varphi}}(X), \end{aligned}$$

and therefore $|X|\Lambda_\varphi(X^\perp) = Y(\varphi)\Lambda_{\overline{\varphi}}(X)$. Applied with $(X^\perp, \overline{\varphi})$ instead of (X, φ), it follows that $|X^\perp|\Lambda_{\overline{\varphi}}(X) = Y(\overline{\varphi})\Lambda_\varphi(X^\perp)$.

3. We apply 2. with $X = \mathbb{F}_q \subset R_M$. Since $|\mathbb{F}_q| = q$, $|\mathbb{F}_q^\perp| = q^{\det(M)-1}$ and $\Lambda_\varphi(\mathbb{F}_q) = q-1$, we obtain

$$q\Lambda_\varphi(\mathbb{F}_q^\perp) = Y(\varphi)(q-1) \quad \text{and} \quad q^{\deg(M)-1}(q-1) = Y(\overline{\varphi})\Lambda_\varphi(\mathbb{F}_q^\perp).$$

From the first equation we infer

$$Y(\varphi) = \frac{q}{q-1}\Lambda_\varphi(\mathbb{F}_q^\perp) \quad \text{and therefore} \quad Y(\overline{\varphi}) = \overline{Y(\varphi)}.$$

The second equation shows that $\Lambda_\varphi(\mathbb{F}_q^\perp) \neq 0$. The product of the two equations yields

$$q^{\deg(M)}(q-1)\Lambda_\varphi(\mathbb{F}_q^\perp) = Y(\varphi)Y(\overline{\varphi})(q-1)\Lambda_\varphi(\mathbb{F}_q^\perp),$$

hence $Y(\varphi)Y(\overline{\varphi}) = |Y(\varphi)|^2 = q^{\deg(M)}$. □

In the following Theorem 7.4.8 we substantially use [18, Section 6.8].

Theorem 7.4.8. *Let $M = e_1P_1 + \ldots + e_rP_r \in \mathbb{D}'_K$ be as in 7.4.4. Then there exists a differential $\omega \in \Omega_K$ such that, for all $i \in [1, r]$,*

$$\iota_{P_i}(\omega) = (t_i^{-e_i} + \alpha_i)dt_i, \quad \textit{where} \quad \alpha_i \in \mathsf{k}_{P_i}[\![t_i]\!].$$

It has the following properties:

(a) If $\alpha \in \mathfrak{o}_{K,S}$ and $a = \alpha + \mathfrak{a}_M \in \mathcal{O}_S/\mathfrak{a}_M = R_M$, then

$$\mathsf{Tr}_M(a) = \sum_{i=1}^{r} \mathsf{Tr}_{\mathsf{k}_{P_i}/\mathbb{F}_q}(\mathrm{Res}_{P_i}(\alpha\omega)).$$

(b) If $D \in \mathbb{D}_K^M$ and $W = (\omega)$, then $D^* = W + M - D \in \mathbb{D}_K^M$ and $X_{D^*} = X_D^\perp$.
We call W a **special canonical divisor** for M.

(c) If φ is a primitive character of R_M and

$$\deg(D) \notin [0, 2g-2+\deg(M)],$$

then $\Lambda_\varphi(X_D) = 0$.

Proof. We show first:

A. Let $i \in [1, r]$, $a \in \mathsf{k}_{P_i}^\times$ and $d \in \mathbb{N}$. Then there exists some $\omega \in \Omega_K$ such that

$$\iota_{P_i}(\omega) = at_i^{-d} + t_i^{-d+1}\beta, \quad \text{where} \quad \beta \in \mathsf{k}_{P_i}[\![t_i]\!],$$

and $v_{P_k}(\omega) \geq 0$ for all $k \in [1, r] \setminus \{i\}$.

Proof of **A.** Let $\omega' \in \Omega_K$ be any differential such that $(\omega') \geq 0$, say

$$(\omega') = \sum_{k=1}^{r} s_k P_k + N,$$

where $s_k \in \mathbb{N}_0$ for all $k \in [1,r]$, $N \in \mathbb{D}_K'^M$, and $\iota_{P_i}(\omega') = (a' t_i^{s_i} + t_i^{s_i+1}\beta')dt_i$ for some $a' \in \mathsf{k}_{P_i}^\times$ and $\beta' \in \mathsf{k}_{P_i}[\![t_i]\!]$.

Let $D \in \mathbb{D}_K'^M$ be such that $\deg(D) \gg 1$. Then the Riemann-Roch theorem yields

$$\begin{aligned}\dim_{\mathbb{F}_q} \mathcal{L}(D+sP_i)/\mathcal{L}(D+(s-1)P_i) &= \dim(D+sP_i) - \dim(D+(s-1)P_i)\\ &= \deg(P_i)\end{aligned}$$

for all $s \in \mathbb{N}$. Let $L = \mathcal{L}(D+(d+s_i)P_i) \setminus \mathcal{L}(D+(d+s_i-1)P_i) \subset K$. If $f \in L$, then $v_{P_i}(f) = -d - s_i$, $v_{P_k}(f) \geq 0$ for all $k \in [1,r] \setminus \{i\}$, and $af \in L$ for all $a \in \mathsf{k}_{P_i}^\times$. Hence there exists some $f^* \in L$ such that $f^* = a'^{-1} a t_i^{-d-s_i} + t_i^{-d-s_i+1}\beta^*$, where $\beta^* \in \mathsf{k}_{P_i}[\![t_i]\!]$, and we set $\omega = f^*\omega'$. Then

$$\iota_{P_i}(\omega) = (a'^{-1} a t_i^{-d-s_i} + t_i^{-d-s_i+1}\beta^*)(a' t_i^{s_i} + t_i^{s_i+1}\beta')dt_i = (a t_i^{-d} + t_i^{-d+1}\beta)dt_i,$$

and $\beta \in \mathsf{k}_{P_i}[\![t_i]\!]$. If $k \in [1,r] \setminus \{i\}$, then $v_{P_k}(\omega) = v_{P_k}(f^*) + v_{P_k}(\omega') \geq s_k \geq 0$. $\square$[**A.**]

Next we tighten **A** as follows.

> **B.** Let $i \in [1,r]$. Then there exists some $\omega \in \Omega_K$ such that $\iota_{P_i}(\omega) = (t_i^{-e_i} + \beta)dt_i$ for some $\beta \in \mathsf{k}_{P_i}[\![t_i]\!]$ and $v_{P_k}(\omega) \geq 0$ for all $k \in [1,r] \setminus \{i\}$.

Proof of **B.** By **A** there exists some $\omega' \in \Omega_K$ with the property that $\iota_{P_i}(\omega') = (t_i^{-e_i} + t_i^{-d}\beta')dt_i$ for some $d \in [0, e_i - 1]$ and $\beta' \in \mathsf{k}_{P_i}[\![t_i]\!]$ such that $v_{P_k}(\omega') \geq 0$ for all $k \in [1,r] \setminus \{i\}$. We may assume that d is minimal with this property, and if $d = 0$, then we are done. Thus suppose that $d \geq 1$. Then it follows that $\iota_{P_i}(\omega') = t_i^{-e_i} + b t_i^{-d} + t_i^{-d+1}\beta_1'$, where $b \in \mathsf{k}_{P_i}^\times$ and $\beta_1' \in \mathsf{k}_{P_i}[\![t_i]\!]$. By **A**, there exists some $\omega'' \in \Omega_K$ such that $\iota_{P_i}(\omega'') = -b t_i^{-d} + t_i^{-d+1}\beta''$, where $\beta'' \in \mathsf{k}_{P_i}[\![t_i]\!]$ and $v_{P_k}(\omega'') \geq 0$ for all $k \in [1,r] \setminus \{i\}$. Then $\omega = \omega' + \omega''$ has the asserted properties. $\square$[**B.**]

Now we can do the real proof. For $i \in [1,r]$, **B** implies the existence of some $\omega_i \in \Omega_K$ such that $\iota_{P_i}(\omega_i) = (t_i^{-e_i} + \beta_i)dt_i$ with $\beta_i \in \mathsf{k}_{P_i}[\![t_i]\!]$ and $v_{P_k}(\omega_i) \geq 0$ for all $k \in [1,r] \setminus \{i\}$. Then the differential

$$\omega = \sum_{i=1}^{r} \omega_i \in \Omega_K$$

has the required properties.

(a) Let $\alpha \in \mathfrak{o}_{K,S}$ and $\alpha + \mathfrak{a}_M = a = (a_1, \ldots, a_r) \in R_M$. If $i \in [1, r]$ and

$$\alpha \equiv \sum_{j=0}^{e_i-1} a_{i,j} t_i^j \mod t_i^{e_i} \mathsf{k}_{P_i} [\![t_i]\!], \quad \text{where} \quad a_{i,1}, \ldots, a_{i,e_i-1} \in \mathsf{k}_{P_i}, \quad \text{then}$$

$$a_i = \sum_{j=0}^{e_i-1} a_{i,j} \mathsf{t}_i^j,$$

and consequently

$$\iota_{P_i}(\alpha\omega) = \Big(\sum_{j=0}^{e_i-1} a_{i,j} t_i^j + t_i^{e_i} \alpha_i' \Big) (t_i^{-e_i} + \alpha_i) dt_i = \Big(\sum_{j=0}^{e_i-1} a_{i,j} t_i^{j-e_i} + \beta_i \Big) dt_i,$$

where α_i', $\beta_i \in \mathsf{k}_{P_i} [\![t_i]\!]$. Hence it follows that

$$\sum_{i=1}^{r} \mathsf{Tr}_{\mathsf{k}_{P_i}/\mathbb{F}_q}(\mathrm{Res}_{P_i}(\alpha\omega)) = \sum_{i=1}^{r} \mathsf{Tr}_{\mathsf{k}_{P_i}/\mathbb{F}_q}(a_{i,e_i-1}) = \mathsf{Tr}_M(a).$$

(b) Let $D \in \mathbb{D}_K^M$, $W = (\omega)$ and $D^* = W + M - D$. If $i \in [1, r]$, then we obtain $v_{P_i}(D^*) = v_{P_i}(W) + v_{P_i}(M) - v_{P_i}(D) = -e_i + e_i = 0$ and thus $D^* \in \mathbb{D}_K^M$. It suffices to prove:

a. $\dim_{\mathbb{F}_q}(X_D) + \dim_{\mathbb{F}_q}(X_{D^*}) = \deg(M)$.

b. If $x \in X_D$ and $y \in X_{D^*}$, then $\mathsf{Tr}_M(xy) = 0$.

Indeed, $X_{D^*} \subset X_D^\perp$ by **b**, and

$$\dim_{\mathbb{F}_q}(X_{D^*}) = \deg(M) - \dim_{\mathbb{F}_q}(X_D = \dim_{\mathbb{F}_q}(X_D^\perp)$$

by **a** and 7.4.5.3. Hence equality holds.

Proof of **a.** By 7.4.5.1 and the Riemann-Roch theorem we obtain

$$\begin{aligned} &\dim_{\mathbb{F}_q}(X_D) + \dim_{\mathbb{F}_q}(X_{D^*}) \\ &\quad = \dim(D) - \dim(D - M) + \dim(W + M - D) - \dim(W - D) \\ &\quad = [\dim(D) - \dim(W - D)] - [\dim(D - M) - \dim(W + M - D)] \\ &\quad = [\deg(D) + 1 - g] - [\deg(D - M) + 1 - g] = \deg(M). \qquad \square[\mathbf{a}.] \end{aligned}$$

Proof of **b.** Let $x = \xi + \mathfrak{a}_M \in X_D$, where $\xi \in \mathcal{L}(D)$ and $y = \eta + \mathfrak{a}_M \in X_{D^*}$, where $\eta \in \mathcal{L}(D^*)$. Then ξ, $\eta \in \mathfrak{o}_{K,S}$ by 7.4.5.1, and

$$(\xi\eta\omega) = (\xi) + (\eta) + W \geq -D - D^* + W = -D - (W + M - D) + W = -M$$

and therefore $\mathrm{Res}_P(\xi\eta\omega) = 0$ for all $P \in \mathbb{P}_K \setminus \{P_1, \ldots, P_r\}$. By the residue theorem (see [18, Theorem 6.8.4]) and (a), we obtain

$$0 = \sum_{P \in \mathbb{P}_K} \mathsf{Tr}_{\mathsf{k}_P/\mathbb{F}_q}(\mathrm{Res}_P(\xi\eta\omega)) = \sum_{i=1}^{r} \mathsf{Tr}_{\mathsf{k}_{P_i}/\mathbb{F}_q}(\mathrm{Res}_{P_i}(\xi\eta\omega)) = \mathsf{Tr}_M(xy). \quad \square[\mathbf{b}.]$$

(c) If $\deg(D) < 0$, then $\mathcal{L}(D) = \mathbf{0}$, hence $X_D = \mathbf{0}$ and thus $\Lambda_\varphi(X_D) = 0$. If $\deg(D) > 2g-2+\deg(M)$, then $\deg(D^*) = 2g-2+\deg(M)-\deg(D) < 0$, hence $X_D^\perp = X_{D^*} = \mathbf{0}$ and thus

$$\Lambda_\varphi(X_D) = \frac{|X_D|}{Y(\overline{\varphi})}\Lambda_{\overline{\chi}}(X_{D^\perp}) = 0. \qquad \square$$

Now we are in a position to formulate and prove the main result. For this we maintain all notations and definitions introduced hitherto in the current section.

Theorem 7.4.9. *Let $M = e_1P_1 + \ldots + e_rP_r \in \mathbb{D}'_K$ be as in 7.4.4. Let W be a special canonical divisor for M, and let $\chi \in \mathsf{X}(\mathcal{C}_K^M)$ be a primitive ray class character modulo M. For $\xi \in K^\times$ we denote by $(\xi)^K$ the principal divisor of ξ.*

1. *There exists a unique primitive character φ of R_M such that*

$$\varphi(\xi + \mathfrak{a}_M) = \chi((\xi)^K) \quad \textit{for all} \quad \xi \in \mathfrak{o}^\bullet_{K,S}.$$

We call φ the primitive character of R_M associated with χ.

2. *Let φ be the primitive character of R_M associated with χ, and set*

$$\mathsf{W}(\chi) = \frac{\chi(W)Y(\varphi)}{q^{\deg(M)/2}} \in \mathbb{C}.$$

Then $\overline{\mathsf{W}(\chi)} = \mathsf{W}(\overline{\chi})$, and $|\mathsf{W}(\chi)| = 1$.
$\mathsf{W}(\chi)$ is called the **root number** attached with χ.

3. *$\mathcal{Z}(t,\chi)$ is a polynomial in t of degree $2g-2+\deg(M)$ and satisfies a functional equation*

$$\mathcal{Z}(t,\chi) = \mathsf{W}(\chi)\big(t\sqrt{q}\big)^{2g-2+\deg(M)}\mathcal{Z}\Big(\frac{1}{qt},\overline{\chi}\Big) \quad \textit{for all} \quad t \in \mathbb{C},$$

where $\mathsf{W}(\chi) \in \mathbb{C}$, $\mathsf{W}(\overline{\chi}) = \overline{\mathsf{W}(\chi)}$ and $|\mathsf{W}(\chi)| = 1$.

Proof. 1. Uniqueness is obvious. As to the existence, we prove first:

A. If $\xi,\ \xi' \in \mathfrak{o}^\bullet_{K,S}$ and $\xi \equiv \xi' \bmod \mathfrak{a}_M$, then $\chi((\xi)^K) = \chi((\xi')^K)$.

Proof of **A.** Let $\xi,\ \xi' \in \mathfrak{o}_{K,S}$ be such that $\xi \equiv \xi' \bmod \mathfrak{a}_M$. If $\xi \notin K(M)$, then $v_{P_i}(\xi) > 0$ for some $i \in [1,r]$, and since $v_{P_i}(\xi - \xi') \ge e_i > 0$ it follows that $v_{P_i}(\xi') > 0$ and therefore $\chi((\xi)^K) = \chi((\xi')^K) = 0$. Thus suppose that $\xi \in K(M)$. Then

$$v_{P_i}(\xi^{-1}\xi' - 1) = v_{P_i}(\xi' - \xi) - v_{P_i}(\xi) = v_{P_i}(\xi' - \xi) \geq e_i \quad \text{for all} \quad i \in [1, r],$$

hence $\xi^{-1}\xi' \in K^M$ and thus $\chi((\xi)^K) = \chi((\xi')^K)$. □[**A.**]

We define $\varphi\colon R_M = \mathfrak{o}_{K,S}/\mathfrak{a}_M \to \mathbb{C}$ by

$$\varphi(\xi + \mathfrak{a}_M) = \begin{cases} \chi((\xi)^K) & \text{if } \xi \neq 0, \\ 0 & \text{if } \xi = 0 \end{cases} \quad \text{for all} \quad \xi \in \mathfrak{o}_{K,S}.$$

Then $\varphi \restriction \mathbb{F}_q^\times = 1$, $\varphi(xy) = \varphi(x)\varphi(y)$ for all $x, y \in R_M$ and $\varphi(x) = 0$ if and only if $x \notin R_M^\times$. In order to prove that φ is a primitive character of R_M, let $i \in [1, r]$ and $M_i = M - P_i$. Then M_i is not a modulus of definition of χ, and thus there exists some $\xi \in K(M) \cap K^{M_i} \subset \mathfrak{o}_{K,S}$ such that $\chi((\xi)^K) \neq 1$. If $x = (x_1, \ldots, x_r) = \xi + \mathfrak{a}_M \in R_M$, then $x \in R_M^\times$, $\varphi(x) \neq 1$, $x_k = 1$ for all $k \in [1, r] \setminus \{i\}$, and if $e_i > 0$, then $x_i = 1 + at^{e_i - 1}$ for some $a \in \mathsf{k}_{P_i}$.

2. By 7.4.7.3, we obtain

$$|\mathsf{W}(\chi)| = \frac{|Y(\varphi)|}{q^{\deg(M)/2}} = 1 \quad \text{and} \quad \mathsf{W}(\overline{\chi}) = \frac{\overline{\chi}(W)Y(\overline{\varphi})}{q^{\deg(M)/2}} = \overline{\frac{\chi(W)Y(\varphi)}{q^{\deg(M)/2}}} = \overline{\mathsf{W}(\chi)}.$$

3. Let φ be the primitive character of R_M associated with χ. By 5.6.3.3, there exists a set of representatives $\mathfrak{D}$ for $\mathcal{C}_K$ in $\mathbb{D}_K^M$. If $t \in \mathbb{C}$ is such that $|t| < q^{-1}$, then

$$\mathcal{Z}(t, \chi) = \sum_{D \in \mathbb{D}'_K} \chi(D)t^{\deg(D)} = \sum_{D \in \mathfrak{D}} \Theta_\chi(D)t^{\deg(D)}, \quad \text{where}$$

$$\Theta_\chi(D) = \sum_{\substack{D' \in \mathbb{D}'^M_K \\ D' \sim D}} \chi(D').$$

If $D \in \mathfrak{D}$, then $\{D' \in \mathbb{D}'^M_K \mid D' \sim D\} = \{D + (\xi)^K \mid \xi \in \mathcal{L}(D) \cap K(M)\}$, and if $\xi \in \mathcal{L}(D) \cap K(M)$, then $\chi(D + (\xi)^K) = \chi(D)\chi((\xi)^K)$. If $\xi, \xi' \in \mathcal{L}(D) \cap K(M)$, then $(\xi)^K = (\xi')^K$ if and only if $\xi' \in \xi\mathbb{F}_q^\times$. Moreover, if $\xi - \xi' \in \mathcal{L}(D - M)$, then $x = \xi + \mathfrak{a}_M = \xi' + \mathfrak{a}_M \in X_D$, and $\varphi(x) = \chi((\xi)^K)$. Hence for $D \in \mathfrak{D}$ we obtain (using 7.4.5.1)

$$\begin{aligned} \Theta_\chi(D) &= \frac{\chi(D)}{q-1} \sum_{\xi \in \mathcal{L}(D) \cap K(M)} \chi((\xi)^K) = \frac{\chi(D)}{q-1} \sum_{\xi \in \mathcal{L}(D)} \varphi(\xi + \mathfrak{a}_M) \\ &= \frac{\chi(D)}{q-1} \sum_{x \in X_D} \sum_{\substack{\xi \in \mathcal{L}(D) \\ \xi + \mathcal{L}(D-M) = x}} \varphi(x) = \frac{\chi(D)}{q-1} q^{\dim(D-M)} \Lambda_\varphi(X_D), \end{aligned}$$

and 7.4.8(c) implies $\Theta_\chi(D) = 0$ if $\deg(D) \notin [0, 2g - 2 + \deg(M)]$. Consequently, $\mathcal{Z}(t, \chi)$ is a polynomial in t of degree at most $2g - 2 + \deg(M)$.

For $D \in \mathbb{D}_K$ we set $D^* = W + M - D$. Then $\{D^* \mid D \in \mathfrak{D}\}$ is also a set of representatives for $\mathcal{C}_K$ in $\mathbb{D}_K^M$, and as above it follows that

$$\mathcal{Z}\Big(\frac{1}{qt}, \overline{\chi}\Big) = \sum_{D \in \mathfrak{D}} \Theta_{\overline{\chi}}(D^*)\Big(\frac{1}{qt}\Big)^{\deg(D^*)},$$

where

$$\Theta_{\overline{\chi}}(D^*) = \frac{\overline{\chi}(D^*)}{q-1}\, q^{\dim(D^*-M)} \Lambda_{\overline{\varphi}}(X_{D^*}).$$

Observe that $X_{D^*} = X_D^{\perp}$ by 7.4.8(b), $\Lambda_{\overline{\varphi}}(X_D^{\perp}) = |X_D|^{-1} Y(\overline{\varphi}) \Lambda_{\varphi}(X_D)$ by 7.4.7.2, $|X_D| = q^{\dim_{\mathbb{F}_q}(X_D)} = q^{\dim(D)-\dim(D-M)}$ by 7.4.5.1, and that $\overline{\chi}(D^*) = \overline{\chi}(W)\chi(D)$. Then

$$\begin{aligned}
\Theta_{\overline{\chi}}(D^*) &= \frac{\overline{\chi}(W)\chi(D)}{q-1}\, q^{\dim(W-D)+\dim(D-M)-\dim(D)} Y(\chi)\Lambda_{\varphi}(X_D) \\
&= \overline{\chi}(W) Y(\chi) q^{\dim(W-D)-\dim(D)} \Theta_{\chi}(D) \\
&= \overline{\chi}(W) Y(\chi) q^{-\deg(D)+g-1} \Theta_{\chi}(D),
\end{aligned}$$

since $\dim(W-D) - \dim(D) = -\deg(D) + g - 1$ by the Riemann-Roch theorem. Now it follows that

$$\begin{aligned}
\mathcal{Z}\Big(\frac{1}{qt}, \overline{\chi}\Big) &= \frac{\overline{\chi}(W) Y(\chi)}{q^{\deg(M)/2}} \big(t\sqrt{q}\big)^{-2g+2-\deg(M)} \mathcal{Z}(t, \chi) \\
&= \mathsf{W}(\overline{\chi}) \big(t\sqrt{q}\big)^{-2g+2-\deg(M)} \mathcal{Z}(t, \chi)
\end{aligned}$$

for all $t \in \mathbb{C}$, and finally

$$\mathcal{Z}(t, \chi) = \mathsf{W}(\chi) \big(t\sqrt{q}\big)^{2g-2+\deg(M)} \mathcal{Z}\Big(\frac{1}{qt}, \overline{\chi}\Big).$$

In particular, the functional equation implies that $\mathcal{Z}(t, \chi)$ is a polynomial of precise degree $2g - 2 + \deg(M)$ in t. □

The functional equation for $\mathcal{Z}(t, \chi)$ expresses a duality in Grothendieck's cohomological interpretation, see [8].

7.5 Representation theory 1: Basic concepts

Let G be a finite group.

In the following two sections we provide the tools from representation theory needed for the subsequent theory of Artin L functions.

We consider the group algebra

$$\mathbb{C}[G] = \bigoplus_{\sigma\in G} \mathbb{C}\sigma.$$

By defnition, G is a $\mathbb{C}$-basis of $\mathbb{C}[G]$, and we identify $1 = 1_G = 1 \in \mathbb{C} \subset \mathbb{C}[G]$.

Let V be a $\mathbb{C}[G]$-module. Then V is a vector space over $\mathbb{C}$ whose $\mathbb{C}[G]$-module structure is induced by a $\mathbb{C}$-algebra homomorphism

$$\rho = \rho_V : \mathbb{C}[G] \to \mathrm{End}_{\mathbb{C}} V$$

by means of $\rho(\alpha)(v) = \alpha v$ for all $\alpha \in \mathbb{C}[G]$ and $v \in V$. Occasionally we shall write (V, ρ) instead of V to emphasize the $\mathbb{C}[G]$-module structure. Note that a $\mathbb{C}[G]$-module is a G-module in the sense of 2.1.3 such that the action of G is $\mathbb{C}$-linear. A submodule of a $\mathbb{C}[G]$-module is a subspace which is G-invariant, and an $\mathbb{C}[G]$-module homomorphism is a $\mathbb{C}$-linear map which is a G-homomorphism.

If V_1 and V_2 are $\mathbb{C}[G]$-modules, then we endow the vector space $V_1 \otimes_{\mathbb{C}} V_2$ with the unique $\mathbb{C}[G]$-module structure satisfying $\sigma(v_1 \otimes v_2) = \sigma v_1 \otimes \sigma v_2$ for all $\sigma \in G,\ v_1 \in V_1$ and $v_2 \in V_2$.

Recall from 3.1.1 that a $\mathbb{C}[G]$-module V is called **simple** if $V \neq \mathbf{0}$ and V has no non-zero proper $\mathbb{C}[G]$-submodules.

Definition 7.5.1. A (linear) **representation** $V = (V, \rho_V)$ of G is a non-zero finitely generated $\mathbb{C}[G]$-module. A representation V of G is called **irreducible** if V is a simple $\mathbb{C}[G]$-module.

For a representation V of G we define

$$\theta_V : G \to \mathbb{C} \quad \text{by} \quad \theta_V(\sigma) = \mathsf{Tr}(\rho_V(\sigma)) \quad \text{for all} \quad \sigma \in G.$$

(where for an endomorphism $\psi \in \mathrm{End}_{\mathbb{C}}(V)$ we denote by $\mathsf{Tr}(\psi) \in \mathbb{C}$ its trace). We call θ_V the **character** of V, and we tacitly extend it to a $\mathbb{C}$-linear map $\theta_V : \mathbb{C}[G] \to \mathbb{C}$; explicitly,

$$\text{if} \ \ \alpha = \sum_{\sigma\in G} c_\sigma \sigma \in \mathbb{C}[G], \ \ \text{then} \ \ \theta_V(\alpha) = \sum_{\sigma\in G} c_\sigma \theta_V(\sigma) = \mathsf{Tr}(\rho_V(\alpha)).$$

By definition, $\theta_V(1) = \dim_{\mathbb{C}} V$. Below in 7.5.10 we shall prove that a representation of G is (up to isomorphisms) uniquely determined by its character. With this in mind, we make already now the following definition.

A **character** of G is a map $\psi : G \to \mathbb{C}$ such that $\psi = \theta_V$ is the character of some representation V of G, and again it is tacitly extended to a function $\psi : \mathbb{C}[G] \to \mathbb{C}$. For a character ψ of G we call $\psi(1)$ its **degree**. We call ψ **linear** if $\psi(1) = 1$, and we call ψ **irreducible** if it is the character of an irreducible representation. We denote by $\mathrm{Irr}(G)$ the set of all irreducible characters of G.

A one-dimensional representation of G (i. e., a representation V with $\dim_{\mathbb{C}} V = 1$) is (up to isomorphisms) nothing but a $\mathbb{C}[G]$-module structure on

$V = \mathbb{C}$, given by a $\mathbb{C}$-algebra homomorphism $\rho\colon \mathbb{C}[G] \to \mathrm{End}_{\mathbb{C}}\mathbb{C}$. For $c \in \mathbb{C}$ we define $\mu_c \in \mathrm{End}_{\mathbb{C}}\mathbb{C}$ by $\mu_c(z) = cz$ for all $z \in \mathbb{C}$. Then $\mathsf{Tr}(\mu_c) = \det(\mu_c) = c$, and the assignment $c \mapsto \mu_c$ defines an isomorphism $\mathbb{C} \overset{\sim}{\to} \mathrm{End}_{\mathbb{C}}\mathbb{C}$. We identify. Then a one-dimensional representation of G is just a homomorphism $\rho\colon \mathbb{C}[G] \to \mathbb{C}$, $\chi = \rho \restriction G \in \mathsf{X}(G)$ is its (linear) character, and the action on $\mathbb{C}$ is given by $\alpha v = \chi(\alpha)v$ for all $\alpha \in \mathbb{C}[G]$ and $v \in \mathbb{C}$. Consequently, in this context the dual group $\mathsf{X}(G)$ is just the set of all linear characters of G.

A representation V of G is called **trivial** if $\sigma v = v$ for all $v \in V$ and $\sigma \in G$ [equivalently, $\rho_V(\sigma) = \mathrm{id}_V$ for all $\sigma \in G$]. If V is trivial and $d = \dim_{\mathbb{C}} V$, then $\theta_V(\sigma) = d$ for all $\sigma \in G$, hence V is irreducible if and only if $d = 1$, and then $\theta_V = 1_G$ is the **unit character** of G.

The $\mathbb{C}[G]$-module $\mathbb{C}[G]$ itself is called the **regular representation** of G. Its character $r_G\colon G \to \mathbb{C}$ is given by

$$r_G(\sigma) = \begin{cases} |G| & \text{if } \sigma = 1, \\ 0 & \text{otherwise} \end{cases}$$

Indeed, $\boldsymbol{u} = (\tau)_{\tau\in G}$ is a basis of $\mathbb{C}[G]$, and $\rho = \rho_{\mathbb{C}[G]}\colon \mathbb{C}[G] \to \mathrm{End}_{\mathbb{C}}\mathbb{C}[G]$ is given by $\rho(\alpha)(\beta) = \alpha\beta$ for all $\alpha,\ \beta \in \mathbb{C}[G]$. If $\sigma \in G$, then the diagonal of the $\rho(\sigma)$ representing matrix with respect to $\boldsymbol{u}$ is $(1,\ldots,1)$ if $\sigma = 1$, and $(0,\ldots,0)$ otherwise.

Let $V = (V,\rho)$ be a representation of G, $d = \dim_{\mathbb{C}} V$ and $\boldsymbol{v} = (v_1,\ldots,v_d)$ a basis of V. For $\sigma \in G$ let $R(\sigma) = (R_{i,j}(\sigma))_{i,j\in[1,d]}$ be the matrix of $\rho(\sigma)$ with respect to $\boldsymbol{v}$, explicitly

$$\sigma v_i = \sum_{j=1}^{d} R_{j,i}(\sigma)v_j \quad \text{for all } \ i \in [1,d].$$

Then $R\colon G \to \mathsf{GL}_d(\mathbb{C})$, $\sigma \mapsto R(\sigma)$, is a group homomorphism, called the **matrix representation** associated with (V,ρ) with respect to $\boldsymbol{v}$. Its character $\theta_R = \theta_V$ is given by

$$\theta_R(\sigma) = \mathsf{Tr}(\rho(\sigma)) = \mathrm{trace}(R(\sigma)) = \sum_{i=1}^{d} R_{i,i}(\sigma) \in \mathbb{C} \quad \text{for all } \ \sigma \in G,$$

and it is called the character of the matrix representation R.

More generally, any group homomorphism $R\colon G \to \mathsf{GL}_d(\mathbb{C})$ is called a (d-dimensional) matrix representation of G. In particular, the one-dimensional matrix representations are just the one-dimensional characters $R\colon G \to \mathbb{C}^\times$. A matrix representation $R\colon G \to \mathsf{GL}_d(\mathbb{C})$ has a unique extension to an (equally denoted) $\mathbb{C}$-algebra homomorphism $R\colon \mathbb{C}[G] \to \mathsf{M}_n(\mathbb{C})$, and then $\rho\colon \mathbb{C}[G] \to \mathrm{End}(\mathbb{C}^d)$, defined by $\rho(\sigma)(\boldsymbol{v}) = R(\sigma)\boldsymbol{v}$ for all $\sigma \in G$ and $\boldsymbol{v} \in \mathbb{C}^d$, is a representation of G, and R is the matrix representation of G associated with $(\mathbb{C}^d,\rho)$ with respect to the canonical basis.

Theorem 7.5.2. *Let $|G| = n$, ζ a primitive n-th root of unity, and $\psi\colon G \to \mathbb{C}$ a character of G. If $\sigma \in G$, then $\psi(\sigma) \in \mathbb{Z}[\zeta]$, $\psi(\sigma^{-1}) = \overline{\psi(\sigma)}$, and the map $\overline{\psi}\colon G \to \mathbb{C}$, defined by $\overline{\psi}(\sigma) = \overline{\psi(\sigma)}$, is also a character of G.*

Proof. Let ψ be the character of a matrix representation $R\colon G \to \mathsf{GL}_d(\mathbb{C})$, and for $\sigma \in G$ let $(\lambda_1(\sigma), \ldots, \lambda_d(\sigma))$ be the vector of eigenvalues of $R(\sigma)$ (listed with multiplicities). As $R(\sigma)^n = R(\sigma^n) = R(1) = I$ (the unit matrix), the eigenvalues $\lambda_i(\sigma)$ are n-th roots of unity, and $\psi(\sigma) = \lambda_1(\sigma)+\ldots+\lambda_n(\sigma) \in \mathbb{Z}[\zeta]$. Since $R(\sigma^{-1}) = R(\sigma)^{-1}$ and $\lambda_i(\sigma)^{-1} = \overline{\lambda_i(\sigma)}$ for $i \in [1, n]$ are the eigenvales of $R(\sigma)^{-1}$, we obtain

$$\chi(\sigma^{-1}) = \sum_{i=1}^{n} \overline{\lambda_i(\sigma)} = \overline{\chi(\sigma)}.$$

Let $\widetilde{R}\colon G \to \mathsf{GL}_d(\mathbb{C})$ be defined by $\widetilde{R}(\sigma) = (R(\sigma)^{-1})^{\mathrm{t}}$ for all $\sigma \in G$. Then $\widetilde{R}$ is a matrix representation of G, and the numbes $\lambda_i(\sigma)^{-1} = \overline{\lambda_i(\sigma)}$ for $i \in [1, d]$ are the eigenvalues of $\widetilde{R}(\sigma)$. If $\widetilde{\psi}$ is the character of $\widetilde{R}$, then

$$\widetilde{\psi}(\sigma) = \sum_{i=1}^{n} \overline{\lambda_i(\sigma)} = \overline{\psi(\sigma)} = \overline{\psi}(\sigma) \quad \text{for all} \quad \sigma \in G,$$

and therefore $\overline{\psi}$ is a character of G. □

Lemma 7.5.3. *Let V, V_1, V_2 be representations of G.*

1. *If $V = V_1 \oplus V_2$, then $\theta_V = \theta_{V_1} + \theta_{V_2}$.*
2. *If $V = V_1 \otimes_{\mathbb{C}} V_2$, then $\theta_V = \theta_{V_1}\theta_{V_2}$.*
3. *If $V_1 \cong V_2$, then $\theta_{V_1} = \theta_{V_2}$.*
4. *Let ψ be a character of G. Then we have $\psi(\sigma^{-1}x\sigma) = \psi(x)$ for all x, $\sigma \in G$.*

Proof. 1. If $\sigma \in G$, then $\rho_V(\sigma) = \rho_{V_1}(\sigma) \oplus \rho_{V_2}(\sigma)$, and consequently

$$\theta_V(\sigma) = \mathsf{Tr}(\rho_V(\sigma)) = \mathsf{Tr}(\rho_{V_1}(\sigma)) + \mathsf{Tr}(\rho_{V_2}(\sigma)) = \theta_{V_1}(\sigma) + \theta_{V_2}(\sigma) = (\theta_{V_1} + \theta_{V_2})(\sigma).$$

2. If $\sigma \in G$, then $\rho_V(\sigma) = \rho_{V_1}(\sigma) \otimes \rho_{V_2}(\sigma)$, and consequently

$$\theta_V(\sigma) = \mathsf{Tr}(\rho_V(\sigma)) = \mathsf{Tr}(\rho_{V_1}(\sigma))\mathsf{Tr}(\rho_{V_2}(\sigma)) = \theta_{V_1}(\sigma)\theta_{V_2}(\sigma) = (\theta_{V_1}\theta_{V_2})(\sigma).$$

3. If $f\colon V_1 \to V_2$ is a G-isomorphism and $\sigma \in G$, then $\theta_{V_2}(\sigma) \circ f = f \circ \theta_{V_1}(\sigma)$, hence $\rho_{V_2}(\sigma) = f \circ \rho_{V_1}(\sigma) \circ f^{-1}$, and $\theta_{V_2}(\sigma) = \mathsf{Tr}(\rho_{V_2}(\sigma)) = \mathsf{Tr}(\rho_{V_1}(\sigma)) = \theta_{V_1}(\sigma)$.

4. Let ψ be the character of the matrix representation $R\colon G \to \mathsf{GL}_d(\mathbb{C})$. If x, $\sigma \in G$, then $\psi(\sigma^{-1}x\sigma) = \mathrm{trac}(R(\sigma)^{-1}R(x)R(\sigma)) = \mathrm{trace}(R(x)) = \psi(x)$. □

Theorem 7.5.4 (Maschke). *Let V be an $\mathbb{C}[G]$-module.*

1. *Every $\mathbb{C}[G]$-submodule of V is a direct summand.*

2. *If $\dim_{\mathbb{C}} V < \infty$, then V is the direct sum of finitely many simple $\mathbb{C}[G]$-submodules.*

3. *Let $V = V_1 \oplus \ldots \oplus V_m$, where $V_1, \ldots, V_m$ are simple $\mathbb{C}[G]$-submodules of V. Then every simple $\mathbb{C}[G]$-submodule of V is isomorphic to some V_j.*

Proof. 1. Let W be a $\mathbb{C}[G]$-submodule of V and $p \in \mathrm{End}_{\mathbb{C}}(V)$ a projection onto W (i. e., $\mathrm{Im}(v) = W$ and $p \upharpoonright W = \mathrm{id}_W$). Then $V = W \oplus \mathrm{Ker}(p)$, but $\mathrm{Ker}(p)$ need not be G-invariant. We define

$$p^{\#} : V \to V \quad \text{by} \quad p^{\#}(v) = \frac{1}{|G|} \sum_{\sigma \in G} \sigma p(\sigma^{-1} v) \quad \text{for all} \quad v \in V.$$

Then $p^{\#}$ is a G-homomorphism. Indeed, if $v \in V$ and $\tau \in G$, then

$$p^{\#}(\tau v) = \frac{1}{|G|} \sum_{\sigma \in G} \sigma p(\sigma^{-1} \tau v) \underset{[\sigma' = \sigma^{-1}\tau]}{=} \frac{1}{|G|} \sum_{\sigma' \in G} \tau \sigma'^{-1} p(\sigma' v) = \tau p^{\#}(v).$$

Apparently, $\mathrm{Im}(p^{\#}) = W$ and $p^{\#} \upharpoonright W = \mathrm{id}_W$. Hence $W' = \mathrm{Ker}(p')$ is a $\mathbb{C}[G]$-submodule of V, and $V = W \oplus W'$.

2. By induction on $\dim(V)$. If V is simple, there is nothing to. Otherwise, there exists a non-zero proper $\mathbb{C}[G]$-submodule V' of V, and by 1. we obtain $V = V' \oplus V''$ for some $\mathbb{C}[G]$-submodule V'' of V. By the induction hypothesis both V' and V'' are direct summands of simple submodules, and thus also V has this property.

3. Let W be a simple $\mathbb{C}[G]$-submodule of V. Then

$$W = \sum_{j=1}^{m} W \cap V_j,$$

and thus $W \cap V_j \neq \mathbf{0}$ for some $j \in [1, m]$. Then $W \cap V_j = W \subset V_j$, and thus $W = V_j$. □

Theorem 7.5.5 (Schur's Lemma). *Let $f \colon V_1 \to V_2$ be G-homomorphism of simple $\mathbb{C}[G]$-modules.*

1. *If $f \neq 0$, then f is an isomorphism.*

2. *If $V_1 = V_2$ and $f \neq 0$, then there exists some $\lambda \in \mathbb{C}$ such that $f = \lambda\, \mathrm{id}_{V_1}$.*

Proof. 1. If $f \neq 0$, then $\mathrm{Ker}(f)$ is a proper $\mathbb{C}[G]$-submodule of V_1 and $\mathrm{Im}(f)$ is a non-zero $\mathbb{C}[G]$-submodule of V_2. Hence $\mathrm{Ker}(f) = \mathbf{0}$, $\mathrm{Im}(f) = V_2$, and f is an isomorphism.

2. By 1., f is an isomorphism. If $\lambda \in \mathbb{C}$ is an eigenvalue of f, then it follows that $\mathrm{Ker}(f - \lambda\,\mathrm{id}_{V_1}) \neq \mathbf{0}$, hence $f - \lambda\,\mathrm{id}_{V_1} = 0$ and $f = \lambda\,\mathrm{id}_{V_1}$. □

Theorem 7.5.6. *Let G be abelian. Then every irreduciblbe represenation of G is one-dimensional. In particular,* $\mathrm{Irr}(G) = \mathsf{X}(G)$ *(i. e., every irreducible character of G is linear).*

Proof. Let (V,ρ) be an irreducible representation of G. For all $\sigma, \tau \in G$ and $v \in V$ we have $\rho(\sigma)(\tau v) = \sigma\tau v = \tau\sigma v = \tau\rho(\sigma)(v)$. Hence $\rho(\sigma)\colon V \to V$ is a G-isomorphism and by 7.5.5.2 it follows that $\rho(\sigma) = \lambda_\sigma\,\mathrm{id}_V$ for some λ_σ in $\mathbb{C}^\times$. Hence every subspace of V is G-invariant, and as V is irreducible, it follows that $\dim_{\mathbb{C}} V = 1$. □

Theorem 7.5.7. *Let L be an irreducible left ideal of $\mathbb{C}[G]$ and E an irreducible $\mathbb{C}[G]$-module. If E is not G-isomorphic to L, then $LE = \mathbf{0}$.*

Proof. Since $\mathbb{C}[G]LE = LE$, it follows that LE is a $\mathbb{C}[G]$-submodule of E, hence $LE = \mathbf{0}$ or $LE = E$. If $LE = E$, let $y \in E$ be such that $Ly \neq \mathbf{0}$. Then $Ly = E$, and the map $\varphi\colon L \to E$, defined by $\varphi(\lambda) = \lambda y$, is a non-zero G-homomorphism and thus an isomorphism. □

Now we fix some notations which we will retain during this and the following sections.

Definitions and Remarks 7.5.8. By 7.5.4.3 it follows that

$$\mathbb{C}[G] = L_1 \oplus \ldots \oplus L_m, \quad \text{where } L_1, \ldots, L_m \text{ are simple left ideals of } \mathbb{C}[G],$$

and every simple left ideal of $\mathbb{C}[G]$ is isomorphic to some L_j. We assume that there is some $h \in [1,m]$ such that $\{L_1, \ldots, L_h\}$ is a maximal set of mutually non-isomorphic left ideals among $L_1, \ldots, L_m$. For $j \in [1,h]$ let $\chi_j = \theta_{L_j}$ be the character of L_j, $d_j = \chi_j(1) = \dim_{\mathbb{C}} L_j$, $m_j = |\{i \in [1,m] \mid L_i \cong L_j\}$, and

$$A_j = \bigoplus_{\substack{i=1 \\ L_i \cong L_j}}^{m_j} L_i \;\cong\; L_j^{m_j}.$$

Then every irreducible G-submodule of A_j is isomorphic to L_j by 7.5.4.3, and

$$\mathbb{C}[G] = \bigoplus_{j=1}^{h} A_j.$$

Below in 7.5.11 we shall see that $m_j = d_j$ and thus $\dim_{\mathbb{C}} A_j = d_j^2$.

If $j,\ k \in [1,h]$ and $j \neq k$, then $L_jL_k = \mathbf{0}$ by 7.5.7, hence $A_jA_k = \mathbf{0}$, and $A_j \subset A_j\mathbb{C}[G] = A_j(A_1 + \ldots + A_h) = A_jA_j \subset A_j$. Thus $A_j = A_jA_j = A_j\mathbb{C}[G]$, and A_j is a (two-sided) ideal of $\mathbb{C}[G]$.

Every $x \in \mathbb{C}[G]$ has a decomposition $x = x_1 + \ldots + x_h$ with uniquely determined $x_j \in A_j$ for all $j \in [1,h]$. In particular,

$$1 = \sum_{i=1}^{h} 1_i, \quad \text{where } 1_i \in A_i \text{ for all } i \in [1,h],$$

and if $x \in \mathbb{C}[G]$, then

$$x = \sum_{j=1}^{h} x_j = 1x = \sum_{j=1}^{h} 1_jx = x1_A = \sum_{j=1}^{h} x1_j, \quad \text{and therefore } x_j = 1_jx = x1_j.$$

Consequently, if $j \in [1,h]$ and $x_j \in A_j$, then $1_jx_j = x_j1_j$, and therefore A_j is a $\mathbb{C}$-algebra with unit element 1_j. Moreover, $A_j = \mathbb{C}[G]\,1_j = 1_j\mathbb{C}[G]$, every A_j-module is a $\mathbb{C}[G]$-module, and every simple left ideal of A_j is G-isomorphic to L_j.

Note that, contrary to what we did in Chapter 3, we do not identify $\mathbb{C}1_j$ with $\mathbb{C}$ since we consider the algebras $A_1, \ldots, A_h$ simultaneously.

We call $A_1, \ldots, A_h$ the **isotypic components** and $1_1, \ldots, 1_h$ the **orthogonal idempotents** of $\mathbb{C}[G]$.

Theorem and Definition 7.5.9. *Let $L_1, \ldots, L_h$ be the (non-isomorphic) simple left ideals and $A_1, \ldots, A_h$ the isotypic components of $\mathbb{C}[G]$. Let V be a representation of G and $\psi = \theta_V \colon G \to \mathbb{C}$ its character. Then*

$$V = \bigoplus_{j=1}^{h} V_j, \quad \textit{where } V_j = A_jV = 1_jV \textit{ for all } j \in [1,h].$$

For every simple $\mathbb{C}[G]$-submodule E of V there exists a unique $j \in [1,h]$ such that $E \cong L_j$, and then $E \subset V_j$. In particular:

$V_j \cong L_j^{n_j}$ for some $n_j \in \mathbb{N}_0$, $\theta_{V_j} = n_j\chi_j$ for all $j \in [1,h]$, and

$$\theta_V = \sum_{j=1}^{h} \theta_{V_j} = \sum_{j=1}^{h} n_j\chi_j \colon G \to \mathbb{C}.$$

For $j \in [1,h]$ we call V_j the **isotypic component** of V associated with χ_j.

Proof. For $j \in [1,h]$, let W_j be the sum of all simple $\mathbb{C}[G]$-submodules of V which are isomorphis to L_j. If E is any simple $\mathbb{C}[G]$-submodule of V, then

$$E = \mathbb{C}[G]E = \sum_{i=1}^{h} L_i E.$$

Hence there exists some $i \in [1,h]$ such that $E = L_iE$, and then $E \cong L_i$ by 7.5.7. In particular, there exists some $j \in [1,h]$ such that $E \cong L_j$, and thus $E \subset W_j$. Consequently $W_j \cong L_j^{n_j}$ for some $n_j \in \mathbb{N}_0$ by 7.5.4.2, $1_jw = w$ for all $w \in W_j$, $A_jW_j = W_j$, and if $k \in [1,h] \setminus \{j\}$, then $1_jW_k \subset A_jW_k = \mathbf{0}$, since $A_jL_k = \mathbf{0}$. Since every irreducible submodule of V is contained in some W_j, it follows that $V = W_1 + \ldots + W_h$. If $j \in [1,h]$ and

$$x \in W_j \cap \sum_{\substack{i=1\\ i\neq j}}^{h} W_i, \quad \text{then } x = 1_jx \in \sum_{\substack{i=1\\ i\neq j}}^{h} 1_jW_i = \mathbf{0}, \quad \text{and thus } V = \bigoplus_{j=1}^{h} W_j.$$

Moreover, if $j \in [1,h]$, then apparently $V_j = A_jV = 1_j\mathbb{C}[G]V = 1_jV$, and

$$V_j = 1_jV = \sum_{i=1}^{h} 1_jW_i = 1_jW_j = W_j.$$

The assertions concerning the characters are now obvious. □

Theorem 7.5.10. *Let $L_1, \ldots, L_h$ be the (non-isomorphic) simple left ideals and $1_1, \ldots, 1_h$ the orthogonal idempotens of $\mathbb{C}[G]$. For $j \in [1,h]$ let $\chi_j = \theta_{L_j}$ be the character of L_j, and let $d_j = \chi_j(1) = \dim_{\mathbb{C}} L_j$.*

1. *$\mathrm{Irr}(G) = \{\chi_1, \ldots, \chi_h\}$ is the set of all irreducible characters of G.*

2. *$\mathrm{Irr}(G)$ is linearly independent over $\mathbb{C}$, and $\chi_j(1_k) = \delta_{j,k}d_k$ for all $j,\ k \in [1,h]$.*

3. *Two representations V and V' of G are isomorphic if and only if $\theta_V = \theta_{V'}$.*

Proof. 1. By 7.5.9, every simple $\mathbb{C}[G]$-module is isomorphis to L_j for a unique $j \in [1,h]$, and therefore $\mathrm{Irr}(G) = \{\chi_1, \ldots, \chi_h\}$.

2. If $j,\ k \in [1,h]$, then $\rho_{L_j}(1_k) \in \mathrm{End}_{\mathbb{C}}(L_j)$. If $j \neq k$ and $x \in L_j$, then

$$\rho_{L_j}(1_k)(x) = 1_kx \in L_kL_j = \mathbf{0},$$

hence $\rho_{L_j}(1_k) = 0$ and thus $\chi_j(1_k) = \mathsf{Tr}(\rho_{L_j}(x_k)) = 0$. By definition,

$$d_k = \chi_k(1) = \sum_{i=1}^{h} \chi_k(1_i) = \chi_k(1_k).$$

If $c_1, \ldots, c_h \in \mathbb{C}$ and $c_1\chi_1 + \ldots + c_h\chi_h = 0$, then it follows that

$$(c_1\chi_1 + \ldots + c_h\chi_h)(1_i) = c_i d_i = 0,$$

and thus $c_i = 0$ for all $i \in [1, h]$. Hence $\mathrm{Irr}(G)$ is linearly independent over $\mathbb{C}$.

3. Let $\psi\colon G \to \mathbb{C}$ be any character of G, and let V be a representation such that $\psi = \theta_V$. If, as in 7.5.9,

$$V = \bigoplus_{j=1}^{h} V_j \quad \text{and} \quad V_j \cong L_j^{n_j} \text{ for all } j \in [1, h], \quad \text{then} \quad \psi = \sum_{j=1}^{h} n_j \chi_j,$$

and $n_i = \psi(1_i)$ for all $i \in [1, h]$. Hence $n_1, \ldots, n_h$ and thus also V (up to isomorphisms) are uniquely determined by ψ. □

Finally, we investigate the algebra $\mathbb{C}[G]$ itself from an algebra-theoretic point of view.

Theorem 7.5.11. *Let $A_1, \ldots, A_h$ be the isotypic components and $1_1, \ldots, 1_h$ the orthogonal idempotents of $\mathbb{C}[G]$. Let again $\mathrm{Irr}(G) = \{\chi_1, \ldots, \chi_h\}$, and for $j \in [1, h]$ let $d_j = \chi_j(1)$.*

If $j \in [1, h]$, then A_j is a central simple $\mathbb{C}$-algebra with center $\mathsf{c}(A_j) = \mathbb{C}1_j$,

$$A_j \cong L_j^{d_j}, \quad \mathsf{c}(\mathbb{C}[G]) = \bigoplus_{j=1}^{h} \mathbb{C}1_j, \quad r_G = \sum_{j=1}^{h} d_j\chi_j, \quad \text{and} \quad |G| = \sum_{j=1}^{h} d_j^2.$$

In particular, $\dim_{\mathbb{C}} \mathsf{c}(\mathbb{C}[G]) = h$.

Proof. Let $j \in [1, h]$. We prove first:

A. A_j is simple (that is, it has no non-zero proper ideals).

Proof of **A.** It suffices to prove that $LA_j = A_j$ for every simple left ideal L of A_j. Let L be a simple left ideal of A_j. As A_j is the sum of all simple left ideals, it suffices to prove that every simple left ideal of A_j is contained in LA_j. Thus let M be any simple left ideal of A_j. As all simple left ideals of A_j are isomorphic to L_j. Hence there is a G-isomorphism $\psi\colon L \to M$, and as L is a direct summand of A_j, there is a projection $p\colon A_j \to L$. As $\alpha\colon A_j \overset{p}{\to} L \overset{\psi}{\to} M \hookrightarrow A_j$ is a G-endomorphism, it is an A_j-endomorphism, and therefore $\alpha(x) = x\alpha(1_j)$ for all $x \in A_j$. Hence $L\alpha(1_j) = \mathrm{Im}(\alpha) = M \subset LA_j$. □[**A.**]

By 3.3.3.2 and 3.1.6, it follows that $A_j \cong \mathsf{M}_{n_j}(\mathbb{C})$ for some $n_j \in \mathbb{N}$, and A_j is the direct sum of n_j irreducible left ideals. Hence $A_j \cong L_j^{n_j}$, and $n_j^2 = \dim_{\mathbb{C}} A_j = n_j d_j$, which eventually implies $n_j = d_j = \chi_j(1)$. It follows that

$$r_G = \sum_{j=1}^{h} d_j\chi_j \quad \text{and} \quad |G| = \sum_{j=1}^{h} d_j^2.$$

Moreover, $\mathsf{c}(A_j) = \mathbb{C}1_j$, hence

$$\mathsf{c}(\mathbb{C}[G]) = \bigoplus_{j=1}^{h} \mathsf{c}(A_j) = \bigoplus_{j=1}^{h} \mathbb{C}1_j, \quad \text{and} \quad \dim_{\mathbb{C}} \mathsf{c}(\mathbb{C}[G]) = h. \qquad \square$$

7.6 Representation theory 2: Class functions and induced characters

We maintain all notations from the preceding section.

Two elements $\sigma,\ \sigma' \in G$ are called **conjugate**, $\sigma \sim \sigma'$ if $\sigma' = \tau^{-1}\sigma\tau$ for some $\tau \in G$. For $\sigma \in G$, we call $\Gamma_\sigma = \{\sigma' \in G \mid \sigma' \sim \sigma\}$ its **conjugacy class**, and

$$\gamma_\sigma = \sum_{\rho \in \Gamma_\sigma} \rho \quad \text{its } \textbf{conjugacy sum}.$$

Theorem 7.6.1. *The set* $\{\gamma_\sigma \mid \sigma \in G\}$ *of all conjugacy sums is a* $\mathbb{C}$*-basis of* $\mathsf{c}(\mathbb{C}[G])$.

Proof. Obviously, $\mathsf{c}(\mathbb{C}[G]) = \{\alpha \in \mathbb{C}[G] \mid \tau^{-1}\alpha\tau = \alpha \text{ for all } \tau \in G\}$. If $\sigma \in G$, then $\tau^{-1}\Gamma_\sigma\tau = \Gamma_\sigma$ and thus $\tau^{-1}\gamma_\sigma\tau = \gamma_\sigma$ for all $\tau \in G$. Hence $\gamma_\sigma \in \mathsf{c}(\mathbb{C}[G])$. Conversely, if $\alpha \in \mathbb{C}[G]$, say

$$\alpha = \sum_{\sigma \in G} a_\sigma \sigma, \quad \text{then} \quad \tau^{-1}\alpha\tau = \sum_{\sigma \in G} a_\sigma \tau^{-1}\sigma\tau = \sum_{\sigma \in G} a_\sigma \sigma \quad \text{for all} \quad \tau \in G.$$

Hence $a_\sigma = a_{\tau\sigma\tau^{-1}}$ for all $\tau \in G$, and therefore $\alpha \in {}_{\mathbb{C}}\langle\{\gamma_\sigma \mid \sigma \in G\}\rangle$. Since apparently $\{\gamma_\sigma \mid \sigma \in G\}$ is linearly independent over $\mathbb{C}$, the assertion follows. $\square$

Definition 7.6.2. A map $f\colon G \to \mathbb{C}$ is called a **class function** if

$$f(\tau^{-1}\sigma\tau) = f(\sigma) \quad \text{for all} \quad \sigma,\ \tau \in G.$$

Every class function $f\colon G \to \mathbb{C}$ has a unique (equally denoted) extension to a $\mathbb{C}$-linear map $f\colon \mathbb{C}[G] \to \mathbb{C}$ satisfying $f(\alpha\beta) = f(\beta\alpha)$ for all $\alpha,\ \beta \in \mathbb{C}[G]$. By 7.5.3.4, every character of G is a class function.

Let $\mathcal{X}(G)$ be the $\mathbb{C}$-algebra of all class functions (with pointwise addition and multiplication). For $f,\ g \in \mathcal{X}(G)$, we define

$$\langle f, g\rangle = \langle f, g\rangle_G = \frac{1}{|G|} \sum_{\sigma \in G} f(\sigma)\overline{g(\sigma)}.$$

Apparently, $\langle\cdot,\cdot\rangle$ is an Hermitian scalar product on $\mathcal{X}(G)$, and in 7.6.4 below we shall see that it is positive definite.

Theorem 7.6.3. *The map*

$$\beta\colon \mathcal{X}(G)\times \mathsf{c}(\mathbb{C}[G]) \to \mathbb{C}, \quad (f,\alpha)\mapsto f(\alpha)$$

is $\mathbb{C}$*-bilinear and non-degenerate,* $\mathcal{X}(G)\cong \operatorname{Hom}_{\mathbb{C}}(\mathsf{c}(\mathbb{C}[G]),\mathbb{C})$*, and* $\operatorname{Irr}(G)$ *is a basis of* $\mathcal{X}(G)$.

Proof. Clearly β is $\mathbb{C}$-bilinear. Let $f\in\mathcal{X}(G)$ and $f(\alpha)=0$ for all $\alpha\in\mathsf{c}(\mathbb{C}[G])$. If $\sigma\in G$ and $k=|\Gamma_\sigma|$, then $0=f(\gamma_\sigma)=kf(\sigma)$ and thus $f(\sigma)=0$. Hence $f=0$. Now let $\alpha\in\mathsf{c}(\mathbb{C}[G])$ and $f(\alpha)=0$ for all $f\in\mathcal{X}(G)$. By 7.5.11, $\alpha=c_1 1_1+\ldots+c_h 1_h$ for some $c_1,\ldots,c_h\in\mathbb{C}$. Therefore it follows that

$$\chi_j(\alpha)=\sum_{i=1}^h c_i\chi_j(1_i)=c_jd_j=0, \ \text{ hence } \ c_j=0$$
$$\text{for all } \ j\in[1,h], \ \text{ and thus } \ \alpha=0.$$

In particular, $\mathcal{X}(G)\cong\operatorname{Hom}_{\mathbb{C}}(\mathsf{c}(\mathbb{C}[G]),\mathbb{C})$, $\dim_{\mathbb{C}}\mathcal{X}(G)=h=|\operatorname{Irr}(G)|$, and as $\operatorname{Irr}(G)$ is linearly independent over $\mathbb{C}$, it is a basis of $\mathcal{X}(G)$. □

Theorem 7.6.4.

1. *If* $i\in[1,h]$*, then*

$$1_i=\sum_{\sigma\in G}a_\sigma\sigma \quad \textit{implies} \quad a_\tau=\frac{1}{|G|}r_G(1_i\tau^{-1})=\frac{d_i}{|G|}\chi_i(\tau^{-1}) \ \textit{ for all } \ \tau\in G.$$

2. $\langle\chi_j,\chi_i\rangle=\delta_{i,j}$ *for all* $i,j\in[1,h]$.

3. *If* $f\in\mathcal{X}(G)$*, then*

$$f=\sum_{j=1}^h\langle f,\chi_j\rangle\chi_j=\sum_{\chi\in\operatorname{Irr}(G)}\langle f,\chi\rangle\chi \quad \textit{and} \quad \langle f,f\rangle=\sum_{i=1}^h|c_i|^2.$$

Notably the Hermitean scalar product $\langle\cdot,\cdot\rangle\colon\mathcal{X}(G)\times\mathcal{X}(G)\to\mathbb{C}$ *is positive definite.*

4. *Let* C *be a conjugacy class of* G, $\sigma\in C$ *and* $\tau\in G$*. Then*

$$\sum_{\chi\in\operatorname{Irr}(G)}\overline{\chi}(\sigma)\chi(\tau)=\begin{cases}|G|/|C| & \textit{if } \tau\in C,\\ 0 & \textit{if } \tau\notin C.\end{cases}$$

Proof. 1. Let $\tau \in G$. On the one hand,

$$r_G(1_i\tau^{-1}) = r_G\Bigl(\sum_{\sigma\in G} a_\sigma \sigma\tau^{-1}\Bigr) = \sum_{\sigma\in G} a_\sigma r_G(\sigma\tau^{-1}) = |G|\, a_\tau,$$

and on the other hand (by 7.5.11)

$$r_G(1_i\tau^{-1}) = \sum_{j=1}^{h} d_j\chi_j(1_i\tau^{-1}) = d_i\chi_i(1_i\tau^{-1}) = d_i\chi_i(\tau^{-1}),$$

since

$$\chi_i(\tau^{-1}) = \chi_i\Bigl(\sum_{j=1}^{h} 1_j\tau^{-1}\Bigr) = \chi_i(1_i\tau^{-1}).$$

Hence it follows that

$$a_\tau = \frac{1}{|G|} r_G(1_i\tau^{-1}) = \frac{d_i}{|G|}\chi_i(\tau^{-1}).$$

2. If $i,\ j \in [1,h]$, then 7.5.10.2 and 1. yield

$$\begin{aligned}\delta_{i,j}d_i = \chi_j(1_i) = \chi_j\Bigl(\sum_{\sigma\in G}\frac{d_i}{|G|}\chi_i(\sigma^{-1})\sigma\Bigr) &= \sum_{\sigma\in G}\frac{d_i}{|G|}\chi_i(\sigma^{-1})\psi_j(\sigma)\\ &= \frac{d_i}{|G|}\sum_{\sigma\in G}\psi_j(\sigma)\overline{\psi_i(\sigma)} = d_i\langle\chi_j,\psi_i\rangle,\end{aligned}$$

and thus $\langle\chi_j,\chi_i\rangle = \delta_{i,j}$.

3. By 7.6.3, there exist $c_1,\ldots,c_h \in \mathbb{C}$ such that $f = c_1\chi_1+\ldots+c_h\chi_h$. An application of 2. yields

$$\langle f,\chi_i\rangle = \sum_{j=1}^{h} c_j\langle\chi_j,\psi_i\rangle = c_i \quad\text{for all}\quad i\in[1,h],$$

and therefore

$$\langle f,f\rangle = \sum_{i,j=1}^{h} = c_i\overline{c}_j\langle\chi_i,\chi_j\rangle = \sum_{i=1}^{h}|c_i|^2.$$

4. Let $1_C\colon G\to\mathbb{C}$ be the characteristic function of C, i. e.,

$$1_C(\rho) = \begin{cases}1 & \text{if}\ \ \rho\in C,\\ 0 & \text{if}\ \ \rho\notin C.\end{cases}$$

Then 1_C is a class function, and therefore $1_C = c_1\chi_1 + \ldots + c_h\chi_h$ for some $c_1, \ldots, c_h \in \mathbb{C}$ by 7.6.3. By 3. we obtain

$$c_j = \langle 1_C, \psi_j \rangle = \frac{1}{|G|} \sum_{\rho \in G} 1_C(\rho)\overline{\chi}_j(\rho) = \frac{|C|}{|G|}\overline{\chi}_j(\sigma), \quad \text{hence}$$

$$1_C(\tau) = \sum_{i=1}^{h} \frac{|C|}{|G|}\overline{\chi}_i(\sigma)\chi_i(\tau),$$

and consequently

$$\sum_{i=1}^{h} \overline{\chi}_i(\sigma)\chi_i(\tau) = \frac{|G|}{|C|} 1_C(\tau). \qquad \square$$

Definition and Remark 7.6.5. Let H be a subgroup of G.

Let $V = (V, \rho)$ be a $\mathbb{C}[G]$-module and $\psi = \theta_V : G \to \mathbb{C}$ its character. Then (by restriction of scalars) $V_H = (V, \rho \restriction \mathbb{C}[H])$ is a $\mathbb{C}[H]$-module, the restriction $\psi \restriction H = \theta_{V_H} : H \to \mathbb{C}$ is its character, and we define

$$\psi(H) = \frac{1}{|H|} \sum_{\tau \in H} \psi(\tau) = \langle \psi \restriction H, 1_H \rangle_H.$$

V^H is a $\mathbb{C}[H]$-submodule of V, and if $d = \dim_{\mathbb{C}}(V^H)$, then $\theta_{V^H} = d1_H$. Indeed, if $\tau \in H$, then $\rho(\tau) \restriction V^H = \mathrm{id}_{V^H}$ and $\theta_{V^H}(\tau) = \mathsf{Tr}(\mathrm{id}_{V^H}) = d$.

The $\mathbb{C}[H]$-homomorphism

$$p = \frac{1}{|H|} \sum_{\tau \in H} \tau : V \to V^H, \quad \text{defined by} \quad p(v) = \frac{1}{|H|} \sum_{\tau \in H} \tau v,$$

is the projection of V onto V^H, and $V' = \mathrm{Ker}(p)$ is a $\mathbb{C}[H]$-submodule of $V = V^H \oplus V'$. Since V' contains no subspace on which H acts trivially, we obtain $\langle \theta_{V'}, 1_H \rangle_H = 0$. Hence

$$\psi \restriction H = d1_H + \theta_{V'}, \psi(H) = \langle \psi \restriction H, 1_H \rangle_H = d\langle 1_H, 1_H \rangle_H + \langle \theta_{V'}, 1_H \rangle_H = d,$$

and therefore

$$\dim_{\mathbb{C}} V^H = \psi(H).$$

For later use we note that in terms of representations

$$\frac{1}{|H|} \sum_{\tau \in H} \tau \in \mathbb{C}[H], \quad \text{and} \quad p = \rho\Big(\frac{1}{|H|} \sum_{\tau \in H} \tau\Big) = \frac{1}{|H|} \sum_{\tau \in H} \rho(\tau) \in \mathrm{End}_{\mathbb{C}}(V).$$

Definition and Theorem 7.6.6. *Let H be a subgroup of G.*
For a class function $f \in \mathcal{X}(H)$ we define

$$f^* = \mathrm{Ind}_G^H(f) : G \to \mathbb{C} \quad \text{by}$$

$$f^*(\sigma) = \frac{1}{|H|} \sum_{\substack{\tau \in G \\ \tau^{-1}\sigma\tau \in H}} f(\tau^{-1}\sigma\tau) = \frac{1}{|H|} \sum_{\tau \in G} \dot{f}(\tau^{-1}\sigma\tau),$$

where $\dot{f}\colon G \to \mathbb{C}$ is given by $\dot{f}\restriction H = f$ and $\dot{f}\restriction G \setminus H = 0$. We say that f^* is **induced** by f from H.

1. If $f \in \mathcal{X}(H)$, then $f^* = \mathrm{Ind}_G^H f \in \mathcal{X}(G)$, and the map $\mathrm{Ind}_G^H\colon \mathcal{X}(H) \to \mathcal{X}(G)$ is $\mathbb{C}$-linear.

2. (Frobenius reciprocity law) *If $f \in \mathcal{X}(H)$ and $g \in \mathcal{X}(G)$, then*

$$\langle f^*, g\rangle_G = \langle f, g\restriction H\rangle_H \qquad \text{and} \qquad f^*g = (fg\restriction H)^*.$$

3. Let $f \in \mathcal{X}(H)$, $(G:H) = r$ and $G/H = \{\omega_1 H, \ldots, \omega_r H\}$, where $\omega_1 = 1$. Then

$$f^*(\sigma) = \sum_{i=1}^{r} \dot{f}(\omega_i^{-1}\sigma\omega_i) = \sum_{\substack{i=1 \\ \omega_i^{-1}\sigma\omega_i \in H}}^{r} f(\omega_i^{-1}\sigma\omega_i) \quad \text{for all } \sigma \in G.$$

4. If $H = \mathbf{1}$ and $f = 1_{\mathbf{1}}$ (the unit character of $H = \mathbf{1}$) , then $f^* = r_G$

Proof. 1. If $f \in \mathcal{X}(H)$ and σ, $\nu \in G$, then

$$f^*(\nu^{-1}\sigma\nu) = \frac{1}{|H|}\sum_{\tau\in G} \dot{f}(\tau^{-1}\nu^{-1}\sigma\nu\tau) \underset{\tau'=\nu\tau}{=} \frac{1}{|H|}\sum_{\tau'\in G} \dot{f}(\tau'^{-1}\sigma\tau') = f^*(\sigma),$$

and therefore $f^* \in \mathcal{X}(G)$. Obviously, the map $\mathrm{Ind}_G^H\colon \mathcal{X}(H) \to \mathcal{X}(G)$ is $\mathbb{C}$-linear.

2. By definition,

$$\begin{aligned}
\langle f^*, g\rangle_G &= \frac{1}{|G|\,|H|}\sum_{\sigma\in G}\sum_{\substack{\tau\in G \\ \tau^{-1}\sigma\tau\in H}} f(\tau^{-1}\sigma\tau)\overline{g(\sigma)} \\
&= \frac{1}{|G|\,|H|}\sum_{\sigma\in G}\sum_{\tau\in G} \dot{f}(\tau^{-1}\sigma\tau)\overline{g(\tau^{-1}\sigma\tau)} \\
&\underset{[\tau^{-1}\sigma\tau=\theta]}{=} \frac{1}{|G|\,|H|}\sum_{\theta\in G}\sum_{\tau\in G} \dot{f}(\theta)\overline{g(\theta)} \\
&= \frac{1}{|H|}\sum_{\theta\in H} f(\theta)\overline{g(\theta)} = \langle f, g\restriction H\rangle_H.
\end{aligned}$$

If $\sigma \in G$, then

$$\begin{aligned}
(f^*g)(\sigma) &= \frac{1}{|H|}\sum_{\substack{\tau\in G \\ \tau^{-1}\sigma\tau\in H}} f(\tau^{-1}\sigma\tau)g(\sigma) \\
&= \frac{1}{|H|}\sum_{\substack{\tau\in G \\ \tau^{-1}\sigma\tau\in H}} (fg)(\tau^{-1}\sigma\tau) = (fg\restriction H)^*(\sigma).
\end{aligned}$$

3. If $\sigma \in G$, then

$$f^*(\sigma) = \frac{1}{|H|} \sum_{\tau \in H} \sum_{i=1}^{r} \dot{f}(\tau^{-1}\omega_i^{-1}\sigma\omega_i\tau)$$
$$= \frac{1}{|H|} \sum_{\tau \in H} \sum_{i=1}^{r} \dot{f}(\omega_i^{-1}\sigma\omega_i) = \sum_{i=1}^{r} \dot{f}(\omega_i^{-1}\sigma\omega_i).$$

4. If $H = \mathbf{1}$ and $f = 1_{\mathbf{1}}$, then

$$\dot{f}(\sigma) = \begin{cases} 1 & \text{if } \sigma = 1, \\ 0 & \text{if } \sigma \neq 1, \end{cases} \quad \text{and} \quad f^*(\sigma) = \sum_{\tau \in G} \dot{f}(\sigma) = \begin{cases} |G| & \text{if } \sigma = 1, \\ 0 & \text{if } \sigma \neq 1. \end{cases}$$

Hence $f^* = r_G$. □

Definition and Theorem 7.6.7. *Let H be a subgroup of G, let $(G{:}H) = r$ and $G/H = \{\omega_1 H, \ldots \omega_r H\}$ with $\omega_1 = 1$. Then $\mathbb{C}[G]$ is a free right $\mathbb{C}[H]$-module with basis $\omega_1, \ldots, \omega_r$, and G acts transitively on G/H by means of $\sigma\omega_i = \omega_{\sigma(i)}\sigma_i$ for all $i \in [1, r]$, where $\sigma_i \in H$ and $i \mapsto \sigma(i)$ is a permutation of $[1, r]$.*

For a $\mathbb{C}[H]$-module V we define a $\mathbb{C}[G]$-module

$$V^* = \operatorname{Ind}_G^H V \quad \text{by} \quad V^* = \omega_1 V \oplus \ldots \oplus \omega_r V,$$

where $\omega_i V \cong V$ as an abelian group for all $i \in [1, r]$, and the $\mathbb{C}[G]$-action on V^* is given by

$$c\sigma\omega_i v = \omega_{\sigma(i)}\sigma_i c v \quad \text{for all} \quad c \in \mathbb{C},\ \sigma \in G,\ i \in [1, r], \text{ and linear extension.}$$

In other words if $V = (V, \rho)$ is a representation of H, then $V^* = (V^*, \rho^*)$ is a representation of G, given by $\rho^*(\sigma)(\omega_i v) = \omega_{\sigma(i)}\rho(\sigma_i)(v)$ for all $\sigma \in G$, $i \in [1, r]$ and $v \in V$. We call $V^* = (V^*, \rho^*)$ the representation of G induced from H by $V = (V, \rho)$.

From 2.1.9 it follows that $V^* \cong \mathcal{M}_G^H(V)$ as a G-module and thus (up to $\mathbb{C}[G]$-isomorphisms) V^* is the induced G-module by V from H. Hence it is (up to G-isomorphisms) uniquely determined by V.

1. *Let V be a representation of H and $V^* = \operatorname{Ind}_G^H V$ the induced representation of G. Let $\boldsymbol{v} = (v_1, \ldots, v_d)$ be a basis of V, and let $R\colon H \to \mathsf{GL}_d(\mathbb{C})$ be the associated matrix representation. Then $\boldsymbol{v}^* = (\omega_\rho v_i)_{\rho \in [1,r],\, i \in [1,d]}$ is a basis of V^*, and if $R^*\colon G \to \mathsf{GL}_{rd}(\mathbb{C})$ is the associated matrix representation of G, then*

$$R^*(\sigma) = \begin{pmatrix} \dot{R}(\omega_1^{-1}\sigma\omega_1) & \dot{R}(\omega_1^{-1}\sigma\omega_2) & \ldots & \dot{R}(\omega_1^{-1}\sigma\omega_r) \\ \dot{R}(\omega_2^{-1}\sigma\omega_1) & \dot{R}(\omega_2^{-1}\sigma\omega_2) & \ldots & \dot{R}(\omega_2^{-1}\sigma\omega_r) \\ \cdot & \cdot & \ldots & \cdot \\ \cdot & \cdot & \ldots & \cdot \\ \dot{R}(\omega_r^{-1}\sigma\omega_1) & \dot{R}(\omega_r^{-1}\sigma\omega_2) & \ldots & \dot{R}(\omega_r^{-1}\sigma\omega_r) \end{pmatrix} \quad \textit{for all } \sigma \in G.$$

where $\dot{R}(\tau) \in \mathsf{M}_d(\mathbb{C})$ *is given by*

$$\dot{R}(\tau) = \begin{cases} R(\tau) & \text{if } \tau \in H, \\ \mathbf{0} & \text{if } \tau \in G \setminus H, \end{cases}$$

and

$$\theta_{V^*} = \operatorname{Ind}_G^H(\theta_V) \qquad \text{(see 7.6.6).}$$

2. Let K *be a subgroup of* H *and* W *a* $\mathbb{C}[K]$*-module. Then*

$$\operatorname{Ind}_G^K W \cong \operatorname{Ind}_G^H \operatorname{Ind}_H^K W.$$

If ψ *is a character of* K*, then* $\operatorname{Ind}_G^K(\psi) = \operatorname{Ind}_G^H \operatorname{Ind}_H^K(\psi)$.

Proof. 1. Let V be a representation of H. Then obviously V^* is a representation of G, and if $\boldsymbol{v} = (v_1, \ldots, v_d)$ is a basis of V, then $\boldsymbol{v}^* = (\omega_\nu v_i)_{\nu \in [1,r],\, i \in [1,d]}$ is a basis of V^*. For all $\sigma \in G$, $\nu \in [1, r]$ and $i \in [1, d]$ we obtain

$$\sigma\omega_\nu v_i = \omega_{\sigma(\nu)}\sigma_\nu v_i = \omega_{\sigma(\nu)} \sum_{j=1}^{d} R_{j,i}(\sigma_\nu) v_j = \sum_{j=1}^{d} R_{j,i}(\sigma_\nu)\omega_{\sigma(\nu)} v_j$$
$$= \sum_{j=1}^{d}\sum_{k=1}^{r} \dot{R}_{j,i}(\omega_k^{-1}\sigma\omega_\nu)\omega_k v_j = \sum_{(k,j)\in[1,r]\times[1,d]} R^*_{(k,j),(\nu,i)}(\sigma)\omega_k v_j.$$

Indeed, if $\sigma \in G$, $\nu, k \in [1, r]$ and $i, j \in [1, d]$, then $\omega_k^{-1}\sigma\omega_\nu \in H$ if and only if $k = \sigma(\nu)$, and then $\omega_k^{-1}\sigma\omega_\nu = \sigma_\nu$. In particular,

$$\theta_{V^*}(\sigma) = \operatorname{trace} R^*(\sigma) = \sum_{i=1}^{r} \operatorname{trace} \dot{R}(\omega_i^{-1}\sigma\omega_i)$$
$$= \sum_{i=1}^{r} \dot{\theta}_V(\omega_i^{-1}\sigma\omega_i) = \operatorname{Ind}_G^H(\theta_V)(\sigma),$$

where $\dot{\theta}_V \restriction H = \theta_V$ and $\dot{\theta}_V \restriction G \setminus H = 0$.

2. By 2.1.6.2 we have $\operatorname{Ind}_G^K W \cong \operatorname{Ind}_G^H \operatorname{Ind}_H^K W$. Let $\psi = \theta_W$, $V = \operatorname{Ind}_H^K W$ and $V^* = \operatorname{Ind}_G^H V$. It follows from 1. that $\operatorname{Ind}_H^K(\psi) = \theta_V$, $\operatorname{Ind}_G^K(\psi) = \theta_{V^*}$, and alike $\operatorname{Ind}_G^H(\operatorname{Ind}_G^H(\psi)) = \operatorname{Ind}_G^H(\theta_V) = \theta_{V^*}$. Hence $\operatorname{Ind}_G^H(\operatorname{Ind}_G^H(\psi)) = \operatorname{Ind}_G^K(\psi)$. □

We close this section by quoting the theorems of Aramata and Brauer. Although these two theorems play an important role in the theory of Artin L functions, we renounce to prove them, and refer to the expositions in S. Lang's book.

Theorem 7.6.8 (Aramata). *There exists a (finite) family of linear characters $(\chi_j \in \mathsf{X}(A_j))_{j\in[1,m]}$ of cyclic subgroups A_j of G such that*

$$|G|(r_G - 1_G) = \sum_{j=1}^{m} c_j\chi_j^*,$$

where $c_j \in \mathbb{N}$ and $\chi_j^* = \mathrm{Ind}_G^{A_j}(\chi_j)$ for all $j \in [1, m]$.

Proof. [37, Chap. XVIII, Theorem 8.4]. □

Theorem 7.6.9 (Theorem of Brauer). *Every character ψ of G has a representation*

$$\psi = \sum_{j=1}^{m} c_j\chi_j^*,$$

where $c_j \in \mathbb{Z}$, $\chi_j \in \mathsf{X}(H_j)$ for some subgroup H_j of G, and $\chi_j^* = \mathrm{Ind}_G^{H_j}(\chi_j)$ for all $j \in [1, m]$.

Proof. [37, Chap. XVIII, Theorem 10.13]. □

7.7 Artin conductors

In this section, we substantially use the concepts and results from [18, Section 5.10]. We recall the basic defintions.

Let L/K be a finite Galois extension of local fields, $G = \mathrm{Gal}(L/K)$, and recall that $i_{L/K}\colon G \to \mathbb{N}_0 \cup \{\infty\}$ is defined by

$$i_{L/K}(\sigma) = \inf\{v_L(\sigma a - a) \mid a \in \mathcal{O}_L\} = v_L(\sigma\pi - \pi), \quad \text{where } \mathfrak{p}_L = \pi\mathcal{O}_L,$$

$i_{L/K}(\sigma) = \infty$ if and only if $\sigma = 1 (= \mathrm{id}_L \in G)$, and $i_{L/K}(\tau^{-1}\sigma\tau) = i_{L/K}(\sigma)$ for all σ, $\tau \in G$. For a real number $s \ge 1$ let

$$G_s = \mathsf{G}_s(L/K) = \{\sigma \in G \mid v_L(\sigma a - a) \ge s + 1 \text{ for all } a \in \mathcal{O}_L\} = G_{\lceil s\rceil},$$

and for an integer $j \ge -1$ let $g_j = |G_j|$. Note that $G_{-1} = G$, G_0 is the inertia group of L/K, $g_j = 0$ for all $j \gg 1$, and $G = G_{-1} \supset G_0 \supset G_1 \supset \ldots$ is the Hilbert subgroup series of L/K. In particular, $(G : G_0) = f(L/K)$, $g_0 = e(L/K)$, $G_{j-1} \setminus G_j = \{\sigma \in G \mid i_{L/K}(\sigma) = j\}$ for all $j \in \mathbb{N}_0$, and $G_s = \{\sigma \in G \mid i_{L/K}(\sigma) \ge s + 1\}$ for all $s \ge -1$.

Definition and Theorem 7.7.1. *Let L/K be a finite Galois extension of local fields and $G = \mathrm{Gal}(L/K)$.*
Let $\psi \in \mathcal{X}(G)$ be a class function of G. For a non-empty subset X of G we set

$$\psi(X) = \frac{1}{|X|}\sum_{\sigma\in X}\psi(\sigma) \quad\text{and}\quad m_{L/K}(\psi) = \sum_{j=0}^{\infty}\frac{g_j}{g_0}[\psi(1) - \psi(G_j)].$$

We call $m_{L/K}(\psi)$ the **Artin exponent** of ψ.
By definition, $m_{L/K}(\psi) \in \mathbb{C}$, $m_{L/K}(\psi) = 0$ if L/K is unramified, $m_{L/K}(1_G) = 0$, and the function $m_{L/K}\colon \mathcal{X}(G) \to \mathbb{C}$ is $\mathbb{C}$-linear. Below in 7.7.6 we shall see that $m_{L/K}(\psi) \in \mathbb{N}_0$ if ψ is a character of G.

1. *Let $\psi \in \mathcal{X}(G)$ be a class function. Then*

$$m_{L/K}(\psi) = \frac{1}{e(L/K)}\sum_{\sigma\in G\setminus\{1\}} i_{L/K}(\sigma)[\psi(1) - \psi(\sigma)].$$

2. *Let K' be an intermediate field of L/K and $H = \mathrm{Gal}(L/K')$.*

(a) *Let K'/K be Galois, $G' = G/H = \mathrm{Gal}(K'/K)$, $\psi' \in \mathcal{X}(G')$, and let $\psi \in \mathcal{X}(G)$ be defined by $\psi(\sigma) = \psi'(\sigma\restriction K')$ for all $\sigma \in G$. Then*

$$m_{L/K}(\psi) = m_{K'/K}(\psi').$$

(b) *Let $\psi \in \mathcal{X}(H)$ be a class function and $\psi^* = \mathrm{Ind}_G^H(\psi) \in \mathcal{X}(G)$. Then*

$$m_{L/K}(\psi^*) = \psi(1)v_K(\mathfrak{d}_{K'/K}) + f(K'/K)\,m_{L/K'}(\psi).$$

Proof. 1. By definition,

$$\begin{aligned} m_{L/K}(\psi) &= \sum_{j=0}^{\infty}\frac{g_j}{g_0}[\psi(1) - \psi(G_j)] = \frac{1}{g_0}\sum_{j=0}^{\infty}\Big[g_j\psi(1) - \sum_{\sigma\in G_j}\psi(\sigma)\Big] \\ &= \frac{1}{e(L/K)}\sum_{j=0}^{\infty}\sum_{\sigma\in G_j}[\psi(1) - \psi(\sigma)] \\ &= \frac{1}{e(L/K)}\sum_{\sigma\in G\setminus\{1\}} |\{j \ge 0 \mid \sigma \in G_j\}|\,[\psi(1) - \psi(\sigma)] \\ &= \frac{1}{e(L/K)}\sum_{\sigma\in G\setminus\{1\}} |\{j \ge 0 \mid i_{L/K}(\sigma) \ge j+1\}|\,[\psi(1) - \psi(\sigma)] \\ &= \frac{1}{e(L/K)}\sum_{\sigma\in G\setminus\{1\}} i_{L/K}(\sigma)[\psi(1) - \psi(\sigma)]. \end{aligned}$$

2. We apply 1. and [18, Theorem 5.10.2].
(a) We set $1' = \mathrm{id}_{K'} = 1 \restriction K'$. Then

$$\begin{aligned}
m_{K'/K}(\psi') &= \frac{1}{e(K'/K)} \sum_{\sigma' \in G' \setminus \{1'\}} i_{K'/K}(\sigma')[\psi'(1') - \psi'(\sigma')] \\
&= \frac{1}{e(K'/K)} \sum_{\sigma' \in G' \setminus \{1'\}} \Big(\frac{1}{e(L/K')} \sum_{\substack{\sigma \in G \\ \sigma \restriction K' = \sigma'}} i_{L/K}(\sigma) \Big) [\psi'(1') - \psi'(\sigma')] \\
&= \frac{1}{e(L/K)} \sum_{\sigma \in G \setminus \{1\}} i_{L/K}(\sigma)[\psi(1) - \psi(\sigma)] = m_{L/K}(\psi)
\end{aligned}$$

(if $\sigma \in H$, then $\psi(1) - \psi(\sigma) = 0$).
(b) We define $\dot\psi \colon G \to \mathbb{C}$ by $\dot\psi \restriction H = \psi$ and $\dot\psi \restriction G \setminus H = 0$. Then

$$\begin{aligned}
m_{L/K}(\psi^*) &= \frac{1}{e(L/K)} \sum_{\sigma \in G \setminus \{1\}} i_{L/K}(\sigma)[\psi^*(1) - \psi^*(\sigma)] \\
&= \frac{1}{e(L/K)} \Big[\sum_{\sigma \in G \setminus \{1\}} \psi^*(1) i_{L/K}(\sigma) - \sum_{\sigma \in G \setminus \{1\}} i_{L/K}(\sigma) \psi^*(\sigma) \Big] \\
&= \frac{1}{e(L/K)} \Big[\frac{|G|}{|H|} \psi(1) \sum_{\sigma \in G \setminus \{1\}} i_{L/K}(\sigma) \\
&\quad - \sum_{\sigma \in G \setminus \{1\}} i_{L/K}(\sigma) \frac{1}{|H|} \sum_{\tau \in G} \dot\psi(\tau^{-1}\sigma\tau) \Big],
\end{aligned}$$

and

$$\begin{aligned}
&\sum_{\sigma \in G \setminus \{1\}} i_{L/K}(\sigma) \frac{1}{|H|} \sum_{\tau \in G} \dot\psi(\tau^{-1}\sigma\tau) \\
&\quad = \frac{1}{|H|} \sum_{\tau \in G} \sum_{\sigma \in G \setminus \{1\}} i_{L/K}(\tau^{-1}\sigma\tau) \dot\psi(\tau^{-1}\sigma\tau) \\
&\quad = \frac{1}{|H|} \sum_{\tau \in G} \sum_{\sigma \in G \setminus \{1\}} i_{L/K}(\sigma) \dot\psi(\sigma) \\
&\quad = \frac{|G|}{|H|} \sum_{\sigma \in G \setminus \{1\}} i_{L/K}(\sigma) \dot\psi(\sigma) = [K' : K] \sum_{\sigma \in H \setminus \{1\}} i_{L/K'}(\sigma) \psi(\sigma),
\end{aligned}$$

since $i_{L/K}\upharpoonright H = i_{L/K'}$. Hence it follows that

$$\begin{aligned} m_{L/K}(\psi^*) &= \frac{[K':K]}{e(L/K)}\Big[\psi(1)\sum_{\sigma\in G\setminus 1} i_{L/K}(\sigma) - \sum_{\sigma\in H\setminus\{1\}} i_{L/K'}(\sigma)\psi(\sigma)\Big] \\ &= \frac{[K':K]}{e(L/K)}\Big[\psi(1)\sum_{\sigma\in G\setminus H} i_{L/K}(\sigma) + \sum_{\sigma\in H\setminus\{1\}} i_{L/K'}(\sigma)[\psi(1)-\psi(\sigma)]\Big] \\ &= \frac{[K':K]}{e(L/K)}\Big[\psi(1)\sum_{\sigma\in G\setminus H} i_{L/K}(\sigma) + e(L/K')m_{L/K'}(\psi)\Big]. \end{aligned}$$

By [18, Theorems 5.9.8 and 5.10.1] we obtain

$$\begin{aligned} \sum_{\sigma\in G\setminus H} i_{L/K}(\sigma) &= \sum_{\sigma\in G\setminus\{1\}} i_{L/K}(\sigma) - \sum_{\sigma\in H\setminus\{1\}} i_{L/K}(\sigma) \\ &= v_L(\mathfrak{D}_{L/K}) - v_L(\mathfrak{D}_{L/K'}) \\ &= e(L/K')v_{K'}(\mathfrak{D}_{K'/K}) \end{aligned}$$

and thus finally

$$\begin{aligned} m_{L/K}(\psi^*) &= \frac{[K':K]}{e(L/K)}\,[\psi(1)e(L/K')v_{K'}(\mathfrak{D}_{K'/K}) + e(L/K')m_{L/K'}(\psi)] \\ &= f(K'/K)[\psi(1)v_{K'}(\mathfrak{D}_{K'/K}) + m_{L/K'}(\psi)] \\ &= \psi(1)v_K(\mathfrak{d}_{K'/K}) + f(K'/K)m_{L/K'}(\psi), \end{aligned}$$

since $f(K'/K)v_{K'}(\mathfrak{D}_{K'/K}) = v_K\circ\mathcal{N}_{K'/K}(\mathfrak{D}_{K'/K}) = v_K(\mathfrak{d}_{K'/K})$. □

Theorem 7.7.2. *Let L/K be a finite Galois extension of local fields, and let $G=\mathrm{Gal}(L/K)$. Let $\chi\in\mathsf{X}(G)$ be a linear character, $K_\chi = L^{\mathrm{Ker}(\chi)}$ the fixed field of* $\mathrm{Ker}(\chi)$ *and $\mathfrak{f}_{K_\chi/K}$ its conductor* (see 4.8.7).

Let $\eta_{L/K}\colon[-1,\infty)\to\mathbb{R}$ be the Herbrand function (see [18, Theorem 5.10.3]), *and*

$$r = \begin{cases} \max\{j\ge -1 \mid \chi\upharpoonright G_j\ne 1\} & \text{if } \chi\ne 1_G, \\ -1 & \text{if } \chi = 1_G. \end{cases}$$

Then $m_{L/K}(\chi) = \eta_{L/K}(r)+1\in\mathbb{N}_0$, and $\mathfrak{f}_{K_\chi/K} = \mathfrak{p}_K^{m_{L/K}(\chi)}$.

Proof. If $\chi = 1_G$, there is nothing to do, since $m_{L/K}(1_G) = 0$, $K_\chi = K$, $r=-1$, and $\eta_{L/K}(-1) = -1$. Thus suppose that $\chi\ne 1_G$. Since

$$\chi(G_j) = \begin{cases} 0 & \text{if } j\le r, \\ 1 & \text{if } j> r, \end{cases}$$

it follows that

$$m_{L/K}(\chi) = \sum_{j=0}^{r} \frac{g_j}{g_0} = \eta_{L/K}(r) + 1.$$

Now let $H = \mathrm{Ker}(\chi) = \mathrm{Gal}(L/K_\chi)$ and identify as usual $G/H = \mathrm{Gal}(K_\chi/K)$. Then $G_rH/H \neq \mathbf{1}$ and $G_{r+\delta}H/H = \mathbf{1}$ for all $\delta \in \mathbb{R}_{>0}$. By [18, Theorems 5.10.4 and 5.10.5] we obtain

$$G_jH/H = \mathsf{G}_{j'}(K_\chi/K) = \mathsf{G}^{j''}(K_\chi/K) \quad \text{for all } j \geq -1,$$

where $j' = \eta_{L/K_\chi}(j)$ and $j'' = \eta_{K(\chi)/K}(j') = \eta_{K(\chi)/K} \circ \eta_{L/K(\chi)}(j) = \eta_{L/K}(j)$.

If $r' = \eta_{L/K_\chi}(r)$, then $\mathsf{G}_{r'}(K_\chi/K) \neq \mathbf{1}$ and $\mathsf{G}_{r'+\delta}(K_\chi/K) = \mathbf{1}$ for all $\delta \in \mathbb{R}_{>0}$. If $t = \eta_{L/K}(r) = \eta_{K_\chi/K}(r')$, then $\mathsf{G}^t(K_\chi/K) \neq \mathbf{1}$ and $\mathsf{G}^{t+\varepsilon}(K_\chi/K) = \mathbf{1}$ for all $\varepsilon \in \mathbb{R}_{>0}$, since η_{L/K_χ} is strictly monotonically increasing and continuous. Hence t is a jump in the upper numbering of the ramification groups of K_χ/K, and thus $t \in \mathbb{Z}$ by the Hasse-Arf theorem, since K_χ/K is abelian (see [18, Theorem 5.10.7]). Therefore we obtain $\eta_{L/K}(r) = \max\{t \in \mathbb{Z} \mid \mathsf{G}^t(K_\chi/K) \neq \mathbf{1}\} \in \mathbb{Z}_{\geq -1}$, and consequently

$$m_{L/K}(\chi) = \eta_{L/K}(r) + 1 = \min\{t \in \mathbb{N}_0 \mid \mathsf{G}^t(K_\chi/K) = \mathbf{1}\} = v(\mathfrak{f}_{K(\chi)/K}) \in \mathbb{N}_0$$

(by 1. and 4.8.7). □

Theorem and Definition 7.7.3. *Let L/K be a finite Galois extension of local fields and $G = \mathrm{Gal}(L/K)$.*

1. *If ψ is a character of G, then $m_{L/K}(\psi) \in \mathbb{N}_0$.*
   The ideal $\mathfrak{f}_{L(K}(\psi) = \mathfrak{p}_K^{m_{L/K}(\psi)}$ is called the **(local) Artin conductor** of ψ. By definition, $\mathfrak{f}_{L/K}(1_G) = \mathbf{1}$, and if L/K is unramified, then $\mathfrak{f}_{L/K}(\psi) = \mathbf{1}$.

2. *Let $\chi \in \mathsf{X}(G)$ be a linear character and $K_\chi = L^{\mathrm{Ker}(\chi)}$ the fixed field of* $\mathrm{Ker}(\chi)$. *Then $\mathfrak{f}_{L/K}(\chi) = \mathfrak{f}_{K_\chi/K}$, and if K_χ/K is unramified, then $\mathfrak{f}_{L/K}(\chi) = \mathbf{1}$.*

3. *If ψ_1 and ψ_2 are characters of G, then*

$$\mathfrak{f}_{L/K}(\psi_1 + \psi_2) = \mathfrak{f}_{L/K}(\psi_1)\mathfrak{f}_{L/K}(\psi_2).$$

4. *Let K' be an intermediate field of L/K and $H = \mathrm{Gal}(L/K')$.*

(a) *Let K'/K be Galois, $G' = G/H = \mathrm{Gal}(K'/K)$, ψ' a character of G', and let $\psi\colon G \to \mathbb{C}$ be defined by $\psi(\sigma) = \psi'(\sigma \restriction K')$ for all $\sigma \in G$. Then $\mathfrak{f}_{L/K}(\psi) = \mathfrak{f}_{K'/K}(\psi')$.*

(b) *Let ψ be a character of H and $\psi^* = \mathrm{Ind}_G^H(\psi)$. Then*

$$\mathfrak{f}_{L/K}(\psi^*) = \mathfrak{d}_{K'/K}^{\psi(1)} \mathcal{N}_{K'/K}(\mathfrak{f}_{L/K'}(\psi)).$$

In particular,

$$\mathfrak{f}_{L/K}(1_H^*) = \mathfrak{d}_{K'/K} \quad \textit{and} \quad \mathfrak{d}_{L/K} = \prod_{\chi \in \mathrm{Irr}(G)} \mathfrak{f}_{L/K}(\psi)^{\chi(1)}.$$

Proof. 1. Let ψ be a character of G. By Brauer's theorem 7.6.9,

$$\psi = \sum_{i=1}^{r} c_i \,\mathrm{Ind}_G^{H_i}(\chi_i)$$

where $r \in \mathbb{N}$, H_i is a subgroup of G and $\chi_i \in \mathsf{X}(H_i)$ for all $i \in [1,r]$. If $K_i = L^{H_i}$, then $m_{L/K_i}(\chi_i) \in \mathbb{N}_0$ by 7.7.2. As $m_{L/K}$ is $\mathbb{C}$-linear, 7.7.1.2(b) implies

$$m_{L/K}(\psi) = \sum_{i=1}^{r} c_i \big[\chi_i(1) v_K(\mathfrak{d}_{K_i/K}) + f(K_i/K) m_{L/K_i}(\chi_i)\big] \in \mathbb{Z}.$$

If V is a representation of G such that $\psi = \theta_V$, then

$$\psi(1) - \psi(G_j) = \dim_{\mathbb{C}} V - \dim_{\mathbb{C}} V^{G_j} \geq 0 \quad \text{for all} \quad j \geq 0 \quad \text{by } 7.6.5,$$

hence

$$m_{L/K}(\psi) = \frac{1}{e(L/K)} \sum_{j=0}^{\infty} \frac{g_j}{g_0} [\psi(1) - \psi(G_j)] \geq 0,$$

and therefore $m_{L/K}(\psi) \in \mathbb{N}_0$.

2. By 7.7.2, $\mathfrak{f}_{K_\chi/K} = \mathfrak{p}_K^{m_{L/K}(\chi)} = \mathfrak{f}_{L/K}(\chi)$, and if L/K is unramified, then we get $m_{L/K}(\chi) = 0$.

3. Obvious, since $m_{L/K}(\psi_1 + \psi_2) = m_{L/K}(\psi_1) + m_{L/K}(\psi_2)$.

4.(a) Obvious, since $m_{L/K}(\psi) = m_{K'/K}(\psi')$ by 7.7.1.2(a).

(b) By definition and 7.7.1.2(b) we obtain

$$\begin{aligned} v_K(\mathfrak{f}_{L/K}(\psi^*)) = m_{L/K}(\psi^*) &= \psi(1) v_K(\mathfrak{d}_{K'/K}) + f(K'/K) v_{K'}(\mathfrak{f}_{L/K'}(\psi)) \\ &= v_K(\mathfrak{d}_{K'/K}^{\psi(1)}) + v_K(\mathcal{N}_{K'/K}(\mathfrak{f}_{L/K'}(\psi)) \\ &= v_K\big(\mathfrak{d}_{K'/K}^{\psi(1)} \mathcal{N}_{K'/K}(\mathfrak{f}_{K'/K}(\psi))\big), \end{aligned}$$

and consequently $\mathfrak{f}_{L/K}(\psi^*) = \mathfrak{d}_{K'/K}^{\psi(1)} \mathcal{N}_{K'/K}(\mathfrak{f}_{K'/K}(\psi))$.

Since $\mathfrak{f}_{L/K'}(1_H) = \mathbf{1}$, we get $\mathfrak{f}_{L/K}(1_H^*) = \mathfrak{d}_{K'/K}$. In particular, if $H = \mathbf{1}$, then 7.6.6.4 and 7.5.11 imply

$$1_1^* = r_G = \sum_{\chi \in \mathrm{Irr}(G)} \chi(1)\chi,$$

and consequently

$$\mathfrak{d}_{L/K} = \mathfrak{f}_{L/K}(r_G) = \prod_{\chi \in \mathrm{Irr}(G)} \mathfrak{f}_{L/K}(\chi)^{\chi(1)}. \qquad \square$$

Corollary 7.7.4 (Hasse's conductor-discriminant formula, local case). *Let L/K be a finite abelian extension of local fields and $G = \mathrm{Gal}(L/K)$. Then*

$$\mathfrak{f}_{L/K} = \mathrm{lcm}\{\mathfrak{f}_{L/K}(\chi) \mid \chi \in \mathsf{X}(G)\} \quad \textit{and} \quad \mathfrak{d}_{L/K} = \prod_{\chi \in \mathsf{X}(G)} \mathfrak{f}_{L/K}(\chi).$$

If L/K is cyclic of prime degree l, then $\mathfrak{d}_{L/K} = \mathfrak{f}_{L/K}^{l-1}$.

Proof. Note that $\mathrm{Irr}(G) = \mathsf{X}(G)$ by 7.5.6. Hence the product formula for the discriminant follows by 7.7.3.4(b). For $\chi \in \mathsf{X}(G)$ let $K_\chi = L^{\mathrm{Ker}(\chi)}$ be the fixed field of $\mathrm{Ker}(\chi)$ and L' the compositum of $\{K_\chi \mid \chi \in \mathsf{X}(G)\}$. Then

$$\mathrm{Gal}(L/L') = \bigcap_{\chi \in \mathsf{X}(G)} \mathrm{Ker}(\chi) = \mathbf{1} \qquad \text{(see [18, Theorem 3.3.1])},$$

and thus $L' = L$. By 4.8.7.2 and 7.7.3.2 it follows that

$$\mathfrak{f}_{L/K} = \mathrm{lcm}\{\mathfrak{f}_{K_\chi/K} \mid \chi \in \mathsf{X}(G)\} = \mathrm{lcm}\{\mathfrak{f}_{L/K}(\chi) \mid \chi \in \mathsf{X}(G)\}.$$

Now let L/K be cyclic of prime degree l. Then $\mathsf{X}(G)$ is a cyclic group of order l, and we set $\mathsf{X}(G) = \langle \chi_0 \rangle = \{\chi_0^j \mid j \in [0, l-1]\}$. If $j \in [1, l-1]$, then $\mathrm{Ker}(\chi_0^j) = \mathrm{Ker}(\chi_0)$, hence $K_{\chi_0^j} = K_{\chi_0}$, and consequently

$$\mathfrak{f}_{L/K}(\chi_0^j) = \mathfrak{f}_{K_{\chi_0^j}/K} = \mathfrak{f}_{K_{\chi_0}/K} = \mathfrak{f}_{L/K}(\chi_0).$$

Since $\mathfrak{f}_{L(K}(1_G) = \mathbf{1}$, we obtain

$$\mathfrak{f}_{L/K} = \mathrm{lcm}\{\mathfrak{f}_{K_{\chi_0^j}/K} \mid j \in [0, l-1]\} = \mathfrak{f}_{L/K}(\chi_0)$$

and

$$\mathfrak{d}_{L/K} = \prod_{j=0}^{l-1} \mathfrak{f}_{K_{\chi_0^j}/K} = \mathfrak{f}_{L/K}(\chi_0)^{l-1} = \mathfrak{f}_{L/K}^{l-1}. \qquad \square$$

Now we turn to the theory of global fields and use the unified terminology for global fields introduced in 5.1.

Definition 7.7.5. Let L/K be a finite Galois extension of global fields, and let $G = \mathrm{Gal}(L/K)$. For every $v \in \mathsf{M}_K$ we fix a place $\overline{v} \in \mathsf{M}_L$ above v, and we denote by $G_{\overline{v}}$ the decomposition group of $\overline{v}$ over K. Then $G_{\overline{v}} = \mathrm{Gal}(L_{\overline{v}}/K_v)$ (see 5.4.5. **1**). For a character χ of G and $v \in \mathsf{M}_K^0$ we call

$$m_{L/K,v}(\chi) = m_{L_{\overline{v}}/K_v}(\chi \restriction G_{\overline{v}})$$

the **local Artin exponent** of χ at v. Then $m_{L/K,v}(\chi) = 0$ if v is unramified in L (hence for almost all $v \in \mathsf{M}_K^0$), and we call

$$\mathfrak{f}_{L/K}(\chi) = \prod_{v \in \mathsf{M}_K^0} \mathfrak{p}_v^{m_{L/K,v}(\chi)} \in \mathcal{D}'_K \qquad \text{the (**global**) **Artin conductor** of } \chi.$$

By definition, $m_{L/K,v}(\chi) = v(\mathfrak{f}_{L/K}(\chi)) = v(\mathfrak{f}_{L_{\overline{v}}/K_v}(\chi \restriction G_{\overline{v}}))$ for all $v \in \mathsf{M}_K^0$.

Theorem 7.7.6. *Let L/K be a finite Galois extension of global fields, and let $G = \mathrm{Gal}(L/K)$.*

1. *Let $\chi \in \mathsf{X}(G)$ be a linear character and $K_\chi = L^{\mathrm{Ker}(\chi)}$. Then $\mathfrak{f}_{L/K}(\chi) = \mathfrak{f}_{K_\chi/K}$. Notably if χ is injective, then $\mathfrak{f}_{L/K}(\chi) = \mathfrak{f}_{L/K}$.*
2. *If ψ_1 and ψ_2 are characters of G, then*

$$\mathfrak{f}_{L/K}(\psi_1 + \psi_2) = \mathfrak{f}_{L/K}(\psi_1)\mathfrak{f}_{L/K}(\psi_2).$$

3. *Let K' be an intermediate field of L/K and $H = \mathrm{Gal}(L/K')$.*

(a) Let K'/K be Galois and $G' = G/H = \mathrm{Gal}(K'/K)$. Let ψ' be a character of G', and define $\psi\colon G \to \mathbb{C}$ by $\psi(\sigma) = \psi'(\sigma \restriction K')$ for all $\sigma \in G$. Then $\mathfrak{f}_{L/K}(\psi) = \mathfrak{f}_{K'/K}(\psi')$.

(b) Let ψ be a character of H and $\psi^* = \mathrm{Ind}_G^H(\psi)$. Then

$$\mathfrak{f}_{L/K}(\psi^*) = \mathfrak{d}_{K'/K}^{\psi(1)} \mathcal{N}_{K'/K}(\mathfrak{f}_{L/K'}(\psi)) \quad \text{and} \quad \mathfrak{d}_{L/K} = \prod_{\chi \in \mathrm{Irr}(G)} \mathfrak{f}_{L/K}(\chi)^{\chi(1)}.$$

(c) If $1_G^H = \mathrm{Ind}_G^H(1_H)$, then $\mathfrak{f}_{L/K}(1_G^H) = \mathfrak{d}_{K'/K}$.

Proof. 1. It suffices to prove that $v(\mathfrak{f}_{L/K}(\chi)) = v(\mathfrak{f}_{K_\chi/K})$ for all $v \in \mathsf{M}_K^0$. Let $v \in \mathsf{M}_K^0$, $v' \in \mathsf{M}_{K_\chi}$ and $\overline{v} \in \mathsf{M}_L$ such that $\overline{v}\,|\,v'\,|\,v$. Let $G_{\overline{v}}$ be the decomposition group of $\overline{v}$ over K and $G'_{\overline{v}}$ the decomposition group of $\overline{v}$ over K_χ. Then $G_{\overline{v}} = \mathrm{Gal}(L_{\overline{v}}/K_v)$,

$$G'_{\overline{v}} = G_{\overline{v}} \cap \mathrm{Ker}(\chi) = \mathrm{Gal}(K_{\overline{v}}/(K_\chi)_{v'}) = \mathrm{Ker}(\chi \restriction G_{\overline{v}})$$

and $(K_\chi)_{v'} = (K_v)_{\chi \restriction G_{\overline{v}}}$. By 7.7.5, 7.7.3.2 and 6.5.6.2 we obtain

$$v(\mathfrak{f}_{L/K}(\chi)) = v(\mathfrak{f}_{L_{\overline{v}}/K_v}(\chi \restriction G_{\overline{v}})) = v(\mathfrak{f}_{(K_\chi)_{v'}/K_v}) = v(\mathfrak{f}_{K_\chi/K}).$$

If χ is injective, then $K_\chi = L$.

2. Obvious, since $m_{L/K,v}(\psi_1 + \psi_2) = m_{L/K,v}(\psi_1) + m_{L/K,v}(\psi_2)$ for all $v \in \mathsf{M}_K^0$.

3.(a) It suffices to prove that $m_{L/K,v}(\psi) = m_{K'/K,v}(\psi')$ for all $v \in \mathsf{M}_K^0$. Let $v \in \mathsf{M}_K^0$, $v' \in \mathsf{M}_{K'}$ and $\overline{v} \in \mathsf{M}_L$ such that $\overline{v}\,|\,v'\,|\,v$. Let $G_{\overline{v}}$ be the decomposition group of $\overline{v}$ over K and $H_{\overline{v}}$ the decomposition group of $\overline{v}$ over K'. Then $G'_{v'} = G_{\overline{v}}/H_{\overline{v}}$ is the decomposition group of v' over K (see [18, Theorem 2.13.3]), and if $\sigma \in G_{\overline{v}}$, then $(\psi \restriction G_{\overline{v}})(\sigma) = (\psi' \restriction G'_{v'})(\sigma \restriction K'_{v'})$. By 7.7.1.2(a) we obtain

$$m_{L/K,v}(\psi) = m_{L_{\overline{v}}/K_v}(\psi \restriction G_{\overline{v}}) = m_{K'_{v'}/K_v}(\psi' \restriction G'_{v'}) = m_{K'/K,v}(\psi').$$

3.(b) We first calculate $m_{L/K,v}(\psi^*)$. Let $v \in \mathsf{M}_K^0$, let $v'_1, \ldots, v'_t \in \mathsf{M}_{K'}$ be the places above v, and for each $j \in [1,t]$ let $\overline{v}_j \in \mathsf{M}_L$ be a place above v'_j. We set $\overline{v} = \overline{v}_1$, for $j \in [1,t]$ fix some $\sigma_j \in G$ be such that $\sigma_j\overline{v} = \overline{v}_j$, and we set $\sigma_1 = 1$. For $j \in [1,t]$ let $G_{\overline{v}_j}$ be the decomposition group of $\overline{v}_j$ over K. Then $H_{\overline{v}_j} = G_{\overline{v}_j} \cap H$ is the decomposition group of $\overline{v}_j$ over K', hence $G_{\overline{v}_j} = \mathrm{Gal}(L_{\overline{v}_j}/K_v)$ and $H_{\overline{v}_j} = \mathrm{Gal}(L_{\overline{v}_j}/K_{v'_j})$.

Let $n_j = [K'_{v'_j} : K_v] = (G_{\overline{v}_j} : H_{\overline{v}_j})$, and let $\{\gamma_{j,1}, \ldots, \gamma_{j,n_j}\} \subset G_{\overline{v}_j}$ be a set of representatives for $G_{\overline{v}_j}/H_{\overline{v}_j}$. We assert that

$$\{\sigma_j^{-1}\gamma_{j,\nu} \mid j \in [1,t],\ \nu \in [1,n_j]\} \quad \text{is a set of representatives for } G/H. \quad (*)$$

Proof of $(*)$. Since $(G:H) = [K':K] = n_1 + \ldots + n_t$, it suffices to prove that $\sigma_j^{-1}\gamma_{j,\nu}H = \sigma_k^{-1}\gamma_{k,\mu}H$ implies $(j,\nu) = (k,\mu)$ for all indices in question. Thus let $j,\ k \in [1,t]$, $\nu \in [1,n_j]$ and $\mu \in [1,n_k]$ be such that $\theta = \gamma_{k,\mu}^{-1}\sigma_k\sigma_j^{-1}\gamma_{j,\nu} \in H$. Then $\theta v'_j = v'_j$ and $\theta\overline{v}_j = \overline{v}_k$, hence $k = j$, and $\gamma_{j,\mu}^{-1}\gamma_{j,\nu} = \theta \in G_{\overline{v}_j} \cap H = H_{\overline{v}_j}$ implies $\nu = \mu$. $\square[(*)]$

If $\sigma \in G_{\overline{v}}$, then 7.6.6.3 implies

$$(\psi^* \restriction G_{\overline{v}})(\sigma) = \psi^*(\sigma) = \sum_{j=1}^{t}\sum_{\nu=1}^{n_j} \dot{\psi}(\gamma_{j,\nu}^{-1}\sigma_j\sigma\sigma_j^{-1}\gamma_{j,\nu}),$$

where $\dot{\psi} \restriction H = \psi$ and $\dot{\psi} \restriction G \setminus H = 0$.

For $j \in [1,t]$, we set $(\psi \restriction H_{\overline{v}_j})^* = \mathrm{Ind}_{G_{\overline{v}_j}}^{H_{\overline{v}_j}}(\psi \restriction H_{\overline{v}_j})$, and if $\sigma' \in G_{\overline{v}_j}$, then

$$(\psi \restriction H_{\overline{v}_j})^*(\sigma') = \sum_{\substack{\nu=1 \\ \gamma_{j,\nu}^{-1}\sigma'\gamma_{j,\nu} \in H_{\overline{v}_j}}}^{n_j} (\psi \restriction H_{\overline{v}_j})(\gamma_{j,\nu}^{-1}\sigma'\gamma_{j,\nu}) = \sum_{\nu=1}^{n_j} \dot{\psi}(\gamma_{j,\nu}^{-1}\sigma'\gamma_{j,\nu}),$$

since $H_{\overline{v}_j} = G_{\overline{v}_j} \cap H$.

If $\sigma \in G_{\overline{v}}$, then $\sigma_j\sigma\sigma_j^{-1} \in G_{\overline{v}_j}$, hence

$$(\psi^* \restriction G_{\overline{v}})(\sigma) = \sum_{j=1}^{t} (\psi \restriction H_{\overline{v}_j})^*(\sigma_j\sigma\sigma_j^{-1}),$$

and therefore

$$\begin{aligned}
m_{L/K,v}(\psi^*) &= m_{L_{\overline{v}}/K_v}(\psi^* \restriction G_{\overline{v}}) \\
&= \frac{1}{e(\overline{v}/v)} \sum_{\sigma \in G_{\overline{v}} \setminus \{1\}} i_{L_{\overline{v}}/K_v}(\sigma)\big[(\psi^* \restriction G_{\overline{v}})(1) - (\psi^* \restriction G_{\overline{v}})(\sigma)\big] \\
&= \frac{1}{e(\overline{v}/v)} \sum_{\sigma \in G_{\overline{v}} \setminus \{1\}} i_{L_{\overline{v}}/K_v}(\sigma) \sum_{j=1}^{t} \big[(\psi \restriction H_{\overline{v}_j})^*(1) - (\psi \restriction H_{\overline{v}_j})^*(\sigma_j \sigma \sigma_j^{-1})\big] \\
&= \frac{1}{e(\overline{v}/v)} \sum_{j=1}^{t} \sum_{\sigma \in G_{\overline{v}_j} \setminus \{1\}} i_{L_{\overline{v}}/K_v}(\sigma_j^{-1} \sigma \sigma_j)\big[(\psi \restriction H_{\overline{v}_j})^*(1) - (\psi \restriction H_{\overline{v}_j})^*(\sigma)\big]
\end{aligned}$$

We assert that $i_{L_{\overline{v}}/K_v}(\sigma_j^{-1}\sigma\sigma_j) = i_{L_{\overline{v}_j}/K_v}(\sigma)$ for all $j \in [1,t]$ and $\sigma \in G_{\overline{v}_j}$. Indeed, let $j \in [1,t]$ and $\sigma \in G_{\overline{v}_j}$. Then $\overline{v}_j = \overline{v} \circ \sigma_j^{-1} \colon L \to \mathbb{Z} \cup \{\infty\}$, and if $\pi \in L$ is such that $\overline{v}_j(\pi) = 1$, then $\overline{v}(\sigma_j^{-1}(\pi)) = 1$, and consequently

$$\begin{aligned}
i_{L_{\overline{v}_j}/K_v}(\sigma) = \overline{v}_j(\sigma\pi - \pi) = \overline{v} \circ \sigma_j^{-1}(\sigma\pi - \pi) &= \overline{v}(\sigma_j^{-1}\sigma\sigma_j(\sigma_j^{-1}\pi) - \sigma_j^{-1}\pi) \\
&= i_{L_{\overline{v}}/K_v}(\sigma_j^{-1}\sigma\sigma_j).
\end{aligned}$$

Since furthermore $e(\overline{v}_j/v) = e(\overline{v}/v)$, it follows by 7.7.1.3(b) that

$$\begin{aligned}
m_{L/K,v}(\psi^*) &= \sum_{j=1}^{t} \sum_{\sigma \in G_{\overline{v}_j} \setminus \{1\}} \frac{1}{e(\overline{v}_j/v)} i_{L_{\overline{v}_j}/K_v}(\sigma)\big[(\psi \restriction H_{\overline{\sigma}_j})^*(1) - (\psi \restriction H_{\overline{v}_j})^*(\sigma)\big] \\
&= \sum_{j=1}^{t} m_{L_{\overline{v}_j}/K_v}((\psi \restriction H_{\overline{v}_j})^*) \\
&= \sum_{j=1}^{t} \big[\psi(1) v(\mathfrak{d}_{K'_{v'_j}/K_v}) + f(v'_j/v) m_{L_{\overline{v}_j}/K'_{v'_j}}(\psi \restriction H_{\overline{v}_j})\big]
\end{aligned}$$

By 5.4.1.**4** we obtain

$$\sum_{j=1}^{t} \psi(1) v(\mathfrak{d}_{K'_{v'_j}/K_v}) = \psi(1) v\Big(\prod_{j=1}^{t} \mathfrak{d}_{K'_{v'_j}/K_v}\Big) = \psi(1) v(\mathsf{d}_{K'/K}) = v(\mathsf{d}_{K'/K}^{\psi(1)}),$$

and by the very definition it follows that

$$\begin{aligned}
\sum_{j=1}^{t} f(v'_j/v) m_{L_{\overline{v}_j}/K'_{v'_j}}(\psi \restriction H_{\overline{v}_j}) &= \sum_{j=1}^{t} f(v'_j/v) m_{L/K',v'_j}(\psi) \\
&= \sum_{j=1}^{t} f(v'_j/v) v'_j(\mathsf{f}_{L/K'}(\psi)) \\
&= v(\mathcal{N}_{K'/K}(\mathsf{f}_{L/K'}(\psi)).
\end{aligned}$$

Putting all together, we get

$$v(\mathsf{f}_{L/K}(\psi^*)) = m_{L/K,v}(\psi^*) = v(\mathsf{d}_{K'/K}^{\psi(1)}\mathcal{N}_{K'/K}(\mathsf{f}_{L/K'}(\psi))) \quad \text{for all} \quad v \in \mathsf{M}_K^0,$$

and consequently $\mathsf{f}_{L/K}(\psi^*) = \mathsf{d}_{K'/K}^{\psi(1)}\mathcal{N}_{K'/K}(\mathsf{f}_{L/K'}(\psi))$.

Since $\psi(1_H) = 1$ and $\mathsf{f}_{L/K'}(1_H) = \mathbf{1}$, we get $\mathsf{f}_{L/K}(1_H^*) = \mathsf{d}_{K'/K}$. If moreover $H = 1$, then 7.6.6.4 and 7.5.11 imply

$$\psi^* = r_G = \sum_{\chi \in \mathrm{Irr}(G)} \chi(1)\chi \quad \text{and thus} \quad \mathsf{d}_{L/K} = \mathsf{f}_{L/K}(r_G) = \prod_{\chi \in \mathrm{Irr}(G)} \mathsf{f}_{L/K}(\chi)^{\chi(1)}.$$

(c) Obvious since $\mathsf{f}_{L/K'}(1_H) = 1$. □

Corollary 7.7.7 (Hasse's conductor-discriminant formula). *Let L/K be a finite abelian extension of global fields and $G = \mathrm{Gal}(L/K)$. Then*

$$\mathsf{f}_{L/K} = \mathrm{lcm}\{\mathsf{f}_{L/K}(\chi) \mid \chi \in \mathsf{X}(G)\} \quad \text{and} \quad \mathsf{d}_{L/K} = \prod_{\chi \in \mathsf{X}(G)} \mathsf{f}_{L/K}(\chi).$$

If L/K is cyclic of prime degree l, then $\mathsf{d}_{L/K} = \mathsf{f}_{L/K}^{l-1}$.

Proof. The argument are word for word the same as those in the proof of 7.7.4 if we use 7.7.6 instead of 7.7.3 and 6.4.9 instead of 4.8.7. □

7.8 Artin L functions

Let K be a global field.

We recall the definitions of Dirichlet L and zeta functions from 7.1.6 in the number field case and from 7.4.1 in the function field case (for more detailed information we refer to [18, Sections 4.2 and 6.9]).

In the present and in the following section we use again the unified terminology for global fields introduced in 5.1, and we start with a replication of the basic concepts in this terminology.

For an integral divisor $\mathsf{m} \in \mathcal{D}'_K$ and a ray class character $\chi\colon \mathcal{D}_K^{\mathsf{m}} \to \mathbb{T}$ of finite order the L function $L(s,\chi)$ and the zeta function $\zeta_K(s)$ are defined by

$$L(s,\chi) = \sum_{\mathsf{a} \in \mathcal{D}'_K} \frac{\chi(\mathsf{a})}{\mathfrak{N}(\mathsf{a})^s} \quad \text{and} \quad \zeta_K(s) = L(s,1) \quad \text{for } s \in \mathcal{H}_1.$$

If $\sigma \in \mathbb{R}_{>1}$ and $s \in \mathbb{C}$ such that $\Re(s) \geq \sigma$, then

$$\sum_{\mathfrak{a}\in\mathcal{D}'_K} \left|\frac{\chi(\mathfrak{a})}{\mathfrak{N}(\mathfrak{a})^s}\right| \leq \sum_{\mathfrak{a}\in\mathcal{D}'_K} \frac{1}{\mathfrak{N}(\mathfrak{a})^\sigma} < \infty.$$

Hence the series defining $L(s,\chi)$ converge absolutely and locally uniformly in the half-plane $\mathcal{H}_1$, and there it has a (convergent) product representation

$$L(s,\chi) = \prod_{v\in\mathsf{M}_K^0} \frac{1}{1-\chi(\mathfrak{p}_v)\mathfrak{N}(v)^{-s}}.$$

Consequently $L(\,\cdot\,,\chi)$ is a holomorphic function without zeroes in $\mathcal{H}_1$. In 7.1.8 (for the number field case) and in 7.4.2 and 7.4.3 (for the function field case) we proved that $L(\,\cdot\,,\chi)$ has an extension to a meromorphic function in $\mathbb{C}$ and satisfies a functional equation.

Dirichlet L series (and more general Hecke L series) were built with ray class characters or idele class characters of a global field. Contrary to that, Artin L functions depend on characters of the Galois group of finite Galois extensions, and in the case of abelian extensions they are connected with the Dirichlet L functions by means of the formalism of global class field theory.

Definition and Remark 7.8.1. Let L/K be a finite Galois extension, and let $G = \mathrm{Gal}(L/K)$ be its Galois group. For a character ψ of G and $s \in \mathcal{H}_1$ we define the Artin L function $L(s,\psi)$ as follows.

Let $V = (V,\rho)$ be a representation of G such that $\psi = \theta_V$. Recall that V is a $\mathbb{C}[G]$-module, $\rho\colon \mathbb{C}[G] \to \mathrm{End}_{\mathbb{C}}(V)$ is a $\mathbb{C}$-algebra homomorphism, and $\alpha v = \rho(\alpha)(v)$ for all $\alpha \in \mathbb{C}[G]$ and $v \in V$.

For $v \in \mathsf{M}_K^0$ we fix a place $\overline{v} \in \mathsf{M}_L$ above v and a Frobenius automorphism $\varphi = \varphi_{\overline{v}} \in G_{\overline{v}}$ for $\overline{v}$ over K. We use the properties of Frobenius automorphisms as itemized in 5.4.5.**3**. Let $I_{\overline{v}}$ be the inertia group of $\overline{v}$ over K and consider the subspace

$$V^{I_{\overline{v}}} = \{x \in V \mid \sigma x = x \ \text{ for all } \ \sigma \in I_{\overline{v}}\}.$$

We assert that $\sigma V^{I_{\overline{v}}} = V^{I_{\overline{v}}}$ for all $\sigma \in G_{\overline{v}}$. Indeed, let $\sigma \in G_{\overline{v}}$, $v \in V^{I_{\overline{v}}}$ and $\tau \in I_{\overline{v}}$. Since $I_{\overline{v}}$ is a normal subgroup of $G_{\overline{v}}$, it follows that $\sigma^{-1}\tau\sigma \in I_{\overline{v}}$, hence $\sigma^{-1}\tau\sigma v = v$, and $\tau\sigma v = \sigma v$, which shows that $\sigma v \in V^{I_{\overline{v}}}$. In particular, $V^{I_{\overline{v}}}$ is a $\mathbb{C}[G_{\overline{v}}]$-module, and we consider the representation $V^{I_{\overline{v}}} = (V^{I_{\overline{v}}}, \rho\restriction G_{\overline{v}})$ of $G_{\overline{v}}$, given by $\alpha v = \rho(\alpha)(v)$ for all $\alpha \in \mathbb{C}[G_{\overline{v}}]$ and $v \in V^{I_{\overline{v}}}$.

If $\overline{v}/v$ is unramified, then $I_{\overline{v}} = \mathbf{1}$ and $V^{I_{\overline{v}}} = V$. Consequently $V^{I_{\overline{v}}} = V$ for almost all $v \in \mathsf{M}_K^0$.

A. If $s \in \mathcal{H}_1$, then $\phi = [\mathrm{id}_V - \mathfrak{N}(v)^{-s}\rho(\varphi)] \restriction V^{I_{\overline{v}}}$ is an automorphism of $V^{I_{\overline{v}}}$ which does not depend on the choice of $\overline{v}$, and whose determinant only depends on ψ, v and s (and not on $\overline{v}$).

Proof of **A.** Since $1-\mathfrak{N}(v)^{-s}\varphi \in \mathbb{C}[G_{\overline{v}}]$, it follows that

$$\phi = \rho(1-\mathfrak{N}(v)^{-s}\varphi)\restriction V^{I_{\overline{v}}} = [\mathrm{id}_V - \mathfrak{N}(v)^{-s}\rho(\varphi)]\restriction V^{I_{\overline{v}}} \in \mathrm{End}_{\mathbb{C}}(V^{I_{\overline{v}}}).$$

If $x \in \mathrm{Ker}(\phi)$, then $\rho(\varphi)(x) = \mathfrak{N}(v)^s x$, hence $\mathfrak{N}(v)^s$ is an eigenvalue of $\rho(\varphi)$, and thus it is an $f(\overline{v}/v)$-th roots of unity, which is impossible since $s \in \mathcal{H}_1$. Therefore ϕ is an automorphism of $V^{I_{\overline{v}}}$.

If φ' is another Frobenius automorphism for $\overline{v}$ over K, then $\varphi' \in \varphi I_{\overline{v}}$, and if $x \in V^{I_{\overline{v}}}$, then $\varphi x = \varphi' x$, hence

$$\begin{aligned}(x) &= [1-\mathfrak{N}(v)^{-s}\varphi']x = [1-\mathfrak{N}(v)^{-s}\varphi]x \\ &= [\mathrm{id}_V - \mathfrak{N}(v)^{-s}\rho(\varphi)](x),\end{aligned}$$

and therefore $[\mathrm{id}_V - \mathfrak{N}(v)^{-s}\varphi']\restriction V^{I_{\overline{v}}} = [\mathrm{id}_V - \mathfrak{N}(v)^{-s}\varphi]\restriction V^{I_{\overline{v}}}$.

Let now $\overline{v}' \in \mathsf{M}_L$ be another place above v and φ' a Frobenius automorphism for $\overline{v}'$ over K. Then $I_{\overline{v}'} = \tau I_{\overline{v}}\tau^{-1}$,

$$\varphi' = \tau\varphi\tau^{-1}, 1-\mathfrak{N}(v)^{-s}\varphi' = \tau(1-\mathfrak{N}(v)^{-s}\varphi)\tau^{-1},$$

and $V^{I_{\overline{v}'}} = \tau V^{I_{\overline{v}}}$. Hence $\rho(\tau)(V^{I_{\overline{v}}}) = V^{I_{\overline{v}'}}$, $\rho'(\tau) = \rho(\tau)\restriction V^{I_{\overline{v}}}\colon V^{I_{\overline{v}}} \to V^{I_{\overline{v}'}}$ is a $\mathbb{C}$-isomorphism,

$$[\mathrm{id}_V - \mathfrak{N}(v)^{-s}\rho(\varphi')]\restriction V^{I_{\overline{v}'}} = \rho'(\tau)\circ\Big([\mathrm{id}_V - \mathfrak{N}(v)^{-s}\rho(\varphi)]\restriction V^{I_{\overline{v}}}\Big)\circ\rho'(\tau)^{-1},$$

and consequently

$$\det\big([\mathrm{id}_V - \mathfrak{N}(v)^{-s}\rho(\varphi')]\restriction V^{I_{\overline{v}}}\big) = \det\big([\mathrm{id}_V - \mathfrak{N}(v)|^{-s}\rho(\varphi)]\restriction V^{I_{\overline{v}}}\big).$$

Assume finally that $\psi = \theta_{V'}$ for another representation $V' = (V', \rho')$ of G. Then there is a $\mathbb{C}[G]$-isomorphism $\Phi\colon V \to V'$. It apparently follows that $\Phi(V^{I_{\overline{v}}}) = V'^{I_{\overline{v}}}$, $\Phi' = \Phi\restriction V^{I_{\overline{v}}}\colon V^{I_{\overline{v}}} \to V'^{I_{\overline{v}}}$ is a $\mathbb{C}$-isomorphism,

$$[\mathrm{id}_{V'} - \mathfrak{N}(v)^{-s}\rho'(\varphi)]\restriction V'^{I_{\overline{v}}} = \Phi'\circ\Big([\mathrm{id}_V - \mathfrak{N}(v)^{-s}\rho(\varphi)]\restriction V^{I_{\overline{v}}}\Big)\circ\Phi'^{-1},$$

and consequently

$$\det\big([\mathrm{id}_V - \mathfrak{N}(v)^{-s}\varphi']\restriction V^{I_{\overline{v}}}\big) = \det\big([\mathrm{id}_{V'} - \mathfrak{N}(v)^{-s}\varphi]\restriction V'^{I_{\overline{v}}}\big). \qquad \square[\mathbf{A.}]$$

B. We denote by

$$\phi_{\overline{v}/v} = \frac{1}{e(\overline{v}/v)}\sum_{\tau\in I_{\overline{v}}}\varphi_{\overline{v}}\tau \in \mathbb{Q}[G_{\overline{v}}]$$

the arithmetic mean of all Frobeius automorphisms for $\overline{v}$ over K (so that $\phi_{\overline{v}/v} = \varphi_{\overline{v}}$ if $\overline{v}/v$ is unramified). Then

$$\det\big([\mathrm{id}_V - \mathfrak{N}(v)^{-s}\rho(\varphi_{\overline{v}})]\restriction V^{I_{\overline{v}}}\big) = \det\big(\mathrm{id}_V - \mathfrak{N}(v)^{-s}\rho(\phi_{\overline{v}/v})\big).$$

Proof of **B.** By 7.6.5, the endomorphism

$$p = \frac{1}{e(\overline{v}/v)}\rho\Big(\sum_{\tau\in I_{\overline{v}}}\tau\Big) = \frac{1}{e(\overline{v}/v)}\sum_{\tau\in I_{\overline{v}}}\rho(\tau) \in \mathrm{End}_{\mathbb{C}}V$$

is a projection of V onto $V^{I_{\overline{v}}}$. Hence $p \restriction V^{I_{\overline{v}}} = \mathrm{id}_{V^{I_{\overline{v}}}}$, and $V = V^{I_{\overline{v}}} \oplus V'$, where $V' = \mathrm{Ker}(p)$. Since

$$\rho(\Phi_{\overline{v}/v}) = \rho\Big(\frac{1}{e(\overline{v}/v)} \sum_{\tau \in I_{\overline{v}}} \varphi_{\overline{v}}\tau\Big) = \rho(\varphi_{\overline{v}})\rho\Big(\frac{1}{e(\overline{v}/v)} \sum_{\tau \in I_{\overline{v}}}\Big) = \rho(\varphi_{\overline{v}})\circ p,$$

we obtain $\rho(\Phi_{\overline{v}/v}) \restriction V' = 0$ and $\rho(\Phi_{\overline{v}/v}) \restriction V^{I_{\overline{v}}} = \rho(\varphi_{\overline{v}}) \restriction V^{I_{\overline{v}}}$. Now it follows that

$$\begin{aligned}\mathrm{id}_V - \mathfrak{N}(v)^{-s}\rho(\phi_{\overline{v}/v}) &= (\mathrm{id}_{V^{I_{\overline{v}}}} \oplus \mathrm{id}_{V'}) - (\mathfrak{N}(v)^{-s}\rho(\varphi_{\overline{v}}) \restriction V^{I_{\overline{v}}} \oplus 0) \\ &= (\mathrm{id}_V - \mathfrak{N}(v)^{-s}\rho(\varphi_{\overline{v}})) \restriction V^{I_{\overline{v}}} \oplus \mathrm{id}_{V'},\end{aligned}$$

and therefore $\det\big([\mathrm{id}_V - \mathfrak{N}(v)^{-s}\rho(\varphi_{\overline{v}})] \restriction V^{I_{\overline{v}}}\big) = \det\big(\mathrm{id}_V - \mathfrak{N}(v)^{-s}\rho(\phi_{\overline{v}/v})\big)$.
□[**B.**]

Now we define the **Artin L function** for L/K built with the character ψ of G for $s \in \mathcal{H}_1$ by means of the Euler product

$$\mathcal{L}_{L/K}(s,\psi) = \prod_{v \in \mathsf{M}_K^0} \mathcal{L}_{L/K,v}(s,\psi),$$

whose local factors for $v \in \mathsf{M}_K^0$ are given by

$$\mathcal{L}_{L/K,v}(s,\chi) = \det\big([\mathrm{id}_V - \mathfrak{N}(v)^{-s}\rho(\varphi_{\overline{v}})] \restriction V^{I_{\overline{v}}}\big)^{-1} = \det\big(\mathrm{id}_V - \mathfrak{N}(v)^{-s}\rho(\phi_{\overline{v}/v})\big)^{-1}.$$

C. The infinite product defining $\mathcal{L}_{L/K}(s,\psi)$ converges absolutely and locally uniformly in the half-plane $\mathcal{H}_1$, and consequently $\mathcal{L}_{L/K}(\,\cdot\,,\psi)$ is a holomorphic function without zeroes in $\mathcal{H}_1$.

Proof of **C.** Let $\dim_{\mathbb{C}} V = m$, for $v \in \mathsf{M}_K^0$ let $\dim_{\mathbb{C}} V^{I_{\overline{v}}} = m_v$, and denote by $\lambda_{v,1}, \ldots \lambda_{v,m_v}$ the eigenvalues of $\rho(\varphi_{\overline{v}}) \restriction V^{I_{\overline{v}}}$ (they are $f(\overline{v}/v)$-th roots of unity). Then

$$\mathcal{L}_{L/K}(s,\psi) = \prod_{v \in \mathsf{M}_K^0} \prod_{j=1}^{m_v} \frac{1}{1 - \mathfrak{N}(v)^{-s}\lambda_{v,j}},$$

and this product converges absolutely and locally uniformly if and only if the series

$$\sum_{v \in \mathsf{M}_K^0} \sum_{j=1}^{m_v} \mathfrak{N}(v)^{-s}\lambda_{v,j}$$

has this property. If $\sigma \in \mathbb{R}_{>1}$, $s \in \mathbb{C}$ and $\Re(s) \ge \sigma$, then

$$\sum_{v \in \mathsf{M}_K^0} \sum_{j=1}^{m_v} \big|\,|\mathfrak{N}(v)^{-s}\lambda_{v,j}\big| \le m \sum_{v \in \mathsf{M}_K^0} \mathfrak{N}(v)^{-\sigma} \le m \sum_{\mathsf{a} \in \mathcal{D}'_K} \mathfrak{N}(\mathsf{a})^{-\sigma} < \infty,$$

and the assertion follows by the Weierstrass criterion. □[**C.**]

Theorem 7.8.2. *Let L/K be a finite Galois extension, K' an intermediate field of L/K such that K'/K is Galois, $G = \mathrm{Gal}(L/K)$ and $H = \mathrm{Gal}(L/K')$. Let $V = (V, \rho)$ be a representation of H and $V^* = \mathrm{Ind}_G^H(V) = (V^*, \rho^*)$ the induced representation of G. Let $v \in \mathsf{M}_K^0$, and let $v_1', \ldots, v_t' \in \mathsf{M}_{K'}$ be the places above v. For $j \in [1, t]$ let $\overline{v}_j \in \mathsf{M}_L$ be a place above v_j', $f_j = f(v_j'/v)$, and set $\overline{v} = \overline{v}_1$. Let $\phi_{\overline{v}/v}$ be the arithmetic mean of the Frobenius automorphisms for $\overline{v}$ over K, and for $j \in [1, t]$ let $\phi_{\overline{v}_j/v_j'}$ be the arithmetic mean of the Frobenius automorphisms for $\overline{v}_j$ over K'. Then*

$$\det\bigl(\mathrm{id}_{V^*} - X\rho^*(\phi_{\overline{v}/v})\bigr) = \prod_{j=1}^{t} \det\bigl(\mathrm{id}_V - X^{f_j}\rho(\phi_{\overline{v}_j/v_j'})\bigr) \in \mathbb{C}[X].$$

Moreover, if $\psi = \theta_V$ is the character of (V, ρ) and $\psi^* = \theta_{V^*} = \mathrm{Ind}_G^H(\psi)$ is the character of (V^*, ρ^*), then

$$\psi^*(\phi_{\overline{v}/v}^m) = \sum_{\substack{j=1 \\ f_j \mid m}}^{t} f_j \psi(\phi_{\overline{v}_j/v_j'}^{m/f_j}) \quad \textit{for all} \ \ m \in \mathbb{N}.$$

Proof. Let $G' = \mathrm{Gal}(K'/K) = G/H$, for $j \in [1, t]$ let $\sigma_j \in G$ such that $\overline{v}_j = \sigma_j \overline{v}$, and set $\sigma_1 = 1$. Let $G_{\overline{v}_j}$ be the decomposition group and $I_{\overline{v}_j}$ the inertia group of $\overline{v}_j$ over K. Then $G_{\overline{v}_j} = \sigma_j G_{\overline{v}} \sigma_j^{-1}$ and $I_{\overline{v}_j} = \sigma_j I_{\overline{v}} \sigma_j^{-1}$, $H_{\overline{v}_j} = G_{\overline{v}_j} \cap H$ is the decomposition group of $\overline{v}_j$ and $I_{\overline{v}_j} \cap H$ is the inertia group of $\overline{v}_j$ over K'. Let $\varphi_{\overline{v}}$ be a Frobenius automorphism for $\overline{v}$ over K. Then $\varphi_{\overline{v}_j} = \sigma_j \varphi_{\overline{v}} \sigma_j^{-1}$ is a Frobenius automorphism for $\overline{v}_j$ over K and $\varphi_{\overline{v}_j}' = \varphi_{\overline{v}_j}^{f_j}$ is a Frobenius automorphism for $\overline{v}_j$ over K'. Let $e = e(\overline{v}/v)$, $e_j = e(v_j'/v)$, $n_j = [K_{v_j'}' : K_v] = (G_{\overline{v}_j} : H_{\overline{v}_j}) = e_j f_j$, and let $\{\gamma_{j,1}, \ldots, \gamma_{j,n_j}\} \subset G_{\overline{v}_j}$ be a set of representatives for $G_{\overline{v}_j}/H_{\overline{v}_j}$. Then

$$\{\sigma_j^{-1}\gamma_{j,\nu} \mid (j,\nu) \in J\}, \quad \text{where} \ \ J = \{(j,\nu) \mid j \in [1,t], \ \nu \in [1, n_j]\}$$

is a set of representatives for G/H as we have shown in course of the proof of 7.7.6.3(b).

Let $d = \dim_{\mathbb{C}} V$, $n = [K' : K] = (G : H) = n_1 + \ldots + n_t$ and $R \colon H \to \mathsf{GL}_d(\mathbb{C})$ a matrix representation associated with (V, ρ). Then the map

$$R^* \colon G \to \mathsf{GL}_{nd}(\mathbb{C}), \quad \text{defined by} \quad R^*(\sigma) = \bigl(\dot{R}(\gamma_{j,\nu}^{-1}\sigma_j \sigma \sigma_k^{-1}\gamma_{k,\mu})\bigr)_{(j,\nu),(k,\mu) \in J}$$

is a matrix representation associated with (V^*, ρ^*) by 7.6.7.2, with $\dot{R} \restriction H = R$ and $\dot{R} \restriction G \setminus H = 0$. More detailed, if $\sigma \in G$, then $R^*(\sigma) = (A_{j,k}(\sigma))_{j,k \in [1,t]}$, where $A_{j,k}(\sigma) = (\dot{R}(\gamma_{j,\nu}^{-1}\sigma_j \sigma \sigma_k^{-1}\gamma_{k,\mu}))_{\nu \in [1,n_j],\, \mu \in [1,n_k]}$ for all $(j,k) \in [1,t]$.

We must prove that

$$\det\bigl(I_{nd} - XR^*(\phi_{\overline{v}/v})\bigr) = \prod_{j=1}^{t} \det\bigl(I_d - X^{f_j} R(\phi_{\overline{v}_j/v'_j})\bigr)$$

(where $I_\bullet \in \mathsf{M}_\bullet(\mathbb{C})$ denotes the unit matrix).

Let $\tau \in I_{\overline{v}}$. If $(j,\nu),\ (k,\mu) \in J$, then $\gamma_{j,\nu}^{-1}\sigma_j\varphi_{\overline{v}}\tau\sigma_k^{-1}\gamma_{k,\mu}\overline{v}_k = \overline{v}_j$. Hence, if $k \ne j$, then $\gamma_{j,\nu}^{-1}\sigma_j\varphi_{\overline{v}}\tau\sigma_k^{-1}\gamma_{k,\mu} \notin H$ and therefore $A_{j,k}(\varphi_{\overline{v}}\tau) = 0$. It follows that

$$R^*(\varphi_{\overline{v}}\tau) = \operatorname{diag}(A_1(\tau), \ldots, A_t(\tau)), \quad \text{where} \quad A_j(\tau) = A_{j,j}(\varphi_{\overline{v}}\tau) \in \mathsf{GL}_{n_jd}(\mathbb{C}).$$

Our next task is to construct for each $j \in [1,t]$ a suitable set of representatives for $G_{\overline{v}_j}/H_{\overline{v}_j}$. Thus let $j \in [1,t]$, observe that

$$(I_{\overline{v}_j} : I_{\overline{v}_j} \cap H) = \frac{e(\overline{v}_j/v)}{e(\overline{v}_j/\overline{v}'_j)} = e(v'_j/v) = e_j,$$

and let $\{\tau_{j,a} \mid a \in [1,e_j]\}$ be a set of representatives for $I_{\overline{v}_j}/I_{\overline{v}_j} \cap H$. We assert that

$$\{\varphi_{\overline{v}_j}^l \tau_{j,a} \mid l \in [0, f_j-1],\ a \in [1,e_j]\} \quad \text{is a set of representatives for} \quad G_{\overline{v}_j}/H_{\overline{v}_j}. \tag{$*$}$$

Proof of $(*)$. Since $(G_{\overline{v}_j} : H_{\overline{v}_j}) = [K'_{v'_j} : K_v] = e_jf_j$ and $H_{\overline{v}_j} = G_{\overline{v}_j} \cap H$, it is sufficient to prove that $\varphi_{\overline{v}_j}^l\tau_{j,a}H = \varphi_{\overline{v}_j}^k\tau_{j,b}H$ implies $(a,l) = (b,k)$ for all indices in question. Therefore let $(j,a),\ (k,b) \in [1,e_j] \times [0,f_j-1]$ and $\theta = \tau_{j,b}^{-1}\varphi_{\overline{v}_j}^{-k}\varphi_{\overline{v}_j}^{l}\tau_{j,a} \in H$. The residue class automorphism $\overline{\theta} \in \operatorname{Gal}(\mathsf{k}_{\overline{v}_j}/\mathsf{k}_v)$ causes exponentiation with $\mathfrak{N}(v)^{l-k}$ and induces the identity on $\mathsf{k}_{v'_j}$. Since $\mathfrak{N}(v'_j) = \mathfrak{N}(v)^{f_j}$ it follows that $k - l \equiv 0 \bmod f_j$ and thus $k = l$. Then we obtain $\tau_{j,b}^{-1}\tau_{j,a} \in H \cap I_{\overline{v}_j} = I'_{\overline{v}_j}$ and thus $a = b$. $\square[(*)]$

Now we get, for all $j \in [1,t]$ and $\tau \in I_{\overline{v}}$,

$$A_j(\tau) = \bigl(\dot{R}(\tau_{j,a}^{-1}\varphi_{\overline{v}_j}^{-l}\sigma_j\varphi_{\overline{v}}\tau\sigma_j^{-1}\varphi_{\overline{v}_j}^{m}\tau_{j,b})\bigr)_{(a,l),(b,m)\in[1,e_j]\times[0,f_j-1]} \in \mathsf{GL}_{n_jd}(\mathbb{C}),$$

and therefore

$$R^*(\phi_{\overline{v}/v}) = \sum_{\tau\in I_{\overline{v}}} \frac{1}{e} R^*(\varphi_{\overline{v}}\tau) = \frac{1}{e}\operatorname{diag}(A_1, \ldots, A_t) \in \mathsf{GL}_{nd}(\mathbb{C}),$$

where, for all $j \in [1,t]$,

$$\begin{aligned} A_j &= \sum_{\tau\in I_{\overline{v}}} A_j(\tau) \\ &= \Bigl(\sum_{\tau\in I_{\overline{v}}} \dot{R}(\tau_{j,a}^{-1}\varphi_{\overline{v}_j}^{-l}\sigma_j\varphi_{\overline{v}}\tau\sigma_j^{-1}\varphi_{\overline{v}_j}^{m}\tau_{j,b})\Bigr)_{(a,l),(b,m)\in[1,e_j]\times[0,f_j-1]} \in \mathsf{M}_{n_jd}(\mathbb{C}). \end{aligned}$$

Let now $j \in [1, t]$. For $(a, l), (b, m) \in [1, e_j] \times [0, f_j - 1]$ we obtain (observing that $I_{\overline{v}_j} = \sigma_j I_{\overline{v}} \sigma_j^{-1}$)

$$\begin{aligned}
&\sum_{\tau \in I_{\overline{v}}} \dot{R}(\tau_{j,a}^{-1} \varphi_{\overline{v}_j}^{-l} \sigma_j \varphi_{\overline{v}} \tau \sigma_j^{-1} \varphi_{\overline{v}_j}^{m} \tau_{j,b}) \\
&\underset{[\tau' = \sigma_j \tau \sigma_j^{-1}]}{=} \sum_{\tau' \in I_{\overline{v}_j}} \dot{R}(\tau_{j,a}^{-1} \varphi_{\overline{v}_j}^{-l} \sigma_j \varphi_{\overline{v}} \sigma_j^{-1} \tau' \sigma_j \sigma_j^{-1} \varphi_{\overline{v}_j}^{m} \tau_{j,b}) \\
&= \sum_{\tau' \in I_{\overline{v}_j}} \dot{R}(\tau_{j,a}^{-1} \varphi_{\overline{v}_j}^{-l+1} \tau' \varphi_{\overline{v}_j}^{m} \tau_{j,b}) \underset{(*)}{=} \sum_{\tau \in I_{\overline{v}_j}} \dot{R}(\varphi_{\overline{v}_j}^{m+1-l} \tau),
\end{aligned}$$

where $(*)$ arises by the substitution

$$\tau' = \varphi_{\overline{v}_j}^{l-1} \tau_{j,a} \varphi_{\overline{v}_j}^{m+1-l} \tau \tau_{j,b}^{-1} \varphi_{\overline{v}_j}^{-m} = (\varphi_{\overline{v}_j}^{l-1} \tau_{j,a}) \varphi_{\overline{v}_j}^{m} (\tau \tau_{j,b}^{-1}) \varphi_{\overline{v}_j}^{-m}$$

(note that $\tau \mapsto \tau'$ defines a bijective map $I_{\overline{v}_j} \to I_{\overline{v}_j}$, since $I_{\overline{v}_j}$ is a normal subgroup of $G_{\overline{v}_j}$). Now it follows that

$$\begin{aligned}
A_j &= \sum_{\tau \in I_{\overline{v}_j}} \big(\dot{R}(\varphi_{\overline{v}_j}^{m+1-l} \tau)\big)_{(a,l),(b,m) \in [1,e_j] \times [0, f_j - 1]} \\
&= \begin{pmatrix} B_j & \dots & B_j \\ \cdot & \dots & \cdot \\ \cdot & \dots & \cdot \\ B_j & \dots & B_j \end{pmatrix} \in \mathsf{M}_{e_j}(\mathsf{M}_{f_j d}(\mathbb{C})),
\end{aligned}$$

where

$$B_j = \sum_{\tau \in I_{\overline{v}_j}} \big(\dot{R}(\varphi_{\overline{v}_j}^{m+1-l} \tau)\big)_{l,m \in [0, f_j - 1]} \in \mathsf{M}_{f_j d}(\mathbb{C}).$$

To calculate B_j, we assert:

> **A.** Let l, $m \in [0, f_j - 1]$ and $\tau \in I_{\overline{v}_j}$. Then $\varphi_{\overline{v}_j}^{m+1-l} \tau \in H$ if and only if one of the following two assertions hold:
>
> (a) $l = m + 1$ and $\varphi_{\overline{v}_j}^{m+1-l} \tau = \tau \in H \cap I_{\overline{v}_j}$;
>
> (b) $l = 0$, $m = f_j - 1$, $\tau \in H \cap I_{\overline{v}_j}$ and $\varphi_{\overline{v}_j}^{m+1-l} \tau = \varphi'_{\overline{v}_j} \tau$.

Proof of **A.** If (a) or (b) holds, then clearly $\varphi_{\overline{v}_j}^{m+1-l} \tau = \tau \in H \cap I_{\overline{v}_j}$ (observe that $\varphi_{\overline{v}_j}^{f_j} = \varphi'_{\overline{v}_j} \in H \cap I_{\overline{v}_j}$).

Thus suppose that $\theta = \varphi_{\overline{v}_j}^{m+1-l} \tau \in H$. Then the residue class automorphism $\overline{\theta} \in \mathrm{Gal}(\mathsf{k}_{\overline{v}_j}/\mathsf{k}_v)$ causes exponentiation with $\mathfrak{N}(v)^{m+1-l}$ and satisfies $\overline{\theta} \restriction \mathsf{k}_{v'_j} = \mathrm{id}_{\mathsf{k}_{v'_j}}$. Since $\mathfrak{N}(v'_j) = \mathfrak{N}(v)^{f_j}$, this implies $m + 1 - l \equiv 0 \bmod f_j$. Since l, $m \in [0, f_j - 1]$, it follows that either $l = m + 1$ (and we are done), or $l = 0$ and $m = f_j - 1$. In the latter case $\theta = \varphi_{\overline{v}_j}^{f_j} \tau = \varphi'_{\overline{v}_j} \tau \in H$ and therefore $\tau \in H$. □[**A.**]

By **A** we obtain

$$B_j = \begin{pmatrix} 0 & 0 & \dots & 0 & D_j \\ C_j & 0 & \dots & 0 & 0 \\ 0 & C_j & \dots & 0 & 0 \\ \cdot & \cdot & \cdot & \cdot & \cdot \\ 0 & 0 & \dots & C_j & 0 \end{pmatrix} \in \mathsf{M}_{f_j}(\mathsf{M}_d(\mathbb{C})),$$

where

$$C_j = \sum_{\tau \in H \cap I_{\overline{v}_j}} R(\tau) \in \mathsf{M}_d(\mathbb{C}) \quad \text{and}$$

$$D_j = \sum_{\tau \in H \cap I_{\overline{v}_j}} R(\varphi'_{\overline{v}_j}\tau) = R(\varphi'_{\overline{v}_j})C_j \in \mathsf{M}_d(\mathbb{C}).$$

Now we start with the final calculation. The following determinant equations marked by (1) and (2) will be justified by suitable matrix identities at the end of the proof.

$$\begin{aligned} \det(I_{nd} - XR^*(\phi_{\overline{v}/v})) &= \det\left[\,\mathrm{diag}(I_{n_1 d} - e^{-1}XA_1, \ldots, I_{n_t d} - e^{-1}XA_t\right] \\ &= \prod_{j=1}^{t} \det(I_{n_j d} - e^{-1}XA_j], \end{aligned}$$

and, for $j \in [1, t]$,

$$\begin{aligned} &\det(I_{n_j d} - e^{-1}XA_j) \\ &= \det \begin{pmatrix} I_{f_j d} - e^{-1}XB_j & -e^{-1}XB_j & \dots & -e^{-1}XB_j & -e^{-1}XB_j \\ -e^{-1}XB_j & I_{f_j d} - e^{-1}XB_j & \dots & -e^{-1}XB_j & -e^{-1}XB_j \\ \cdot & \cdot & \cdot & \cdot & \cdot \\ -e^{-1}XB_j & -e^{-1}XB_j & \dots & I_{f_j d} - e^{-1}XB_j & -e^{-1}XB_j \\ -e^{-1}XB_j & -e^{-1}XB_j & \dots & -e^{-1}XB_j & I_{f_j d} - e^{-1}XB_j \end{pmatrix} \\ &\underset{(1)}{=} \det(I_{f_j d} - e_j e^{-1}XB_j). \end{aligned}$$

If $f_j = 1$, then $B_j = D_j$ and $\det(I_d - e^{-1}XA_j) = \det(I_d - e_jXD_j)$. Thus suppose that $f_j \ge 2$. Then

$$\begin{aligned} \det(I_{n_j d} - e^{-1}XA_j) &= \det(I_{f_j d} - e_j e^{-1}XB_j) \\ &= \det \begin{pmatrix} I_d & 0 & 0 & \dots & 0 & -e_j e^{-1}XD_j \\ -e_j e^{-1}XC_j & I_d & 0 & \dots & 0 & 0 \\ 0 & -e_j e^{-1}XC_j & I_d & \dots & 0 & 0 \\ \cdot & \cdot & \cdot & \dots & \cdot & \cdot \\ 0 & 0 & 0 & \dots & I_d & 0 \\ 0 & 0 & 0 & \dots & -e_j e^{-1}XC_j & I_d \end{pmatrix} \\ &\underset{(2)}{=} \det\left[\,I - (e_j e^{-1}XD_j)(e_j e^{-1}XC_j)^{f_j - 1}\right]. \end{aligned}$$

In particular, for every $f_j \geq 1$ we obtain

$$\det(I_{n_jd} - e^{-1}XA_j) = \det\left[\, I_d - (e_je^{-1}XD_j)(e_jXC_j)^{f_j-1}\right].$$

Now observe that $|H \cap I_{\overline{v}_j}| = e(\overline{v}_j/v'_j) = e_j^{-1}e$, and that

$$\frac{e_j}{e} \sum_{\tau \in H \cap I_{\overline{v}_j}} R(\tau)$$

is idempotent, since it is the projection of V onto $V^{H\cap I_{\overline{v}_j}}$ by 7.6.5. Therefore

$$\begin{aligned}(e_je^{-1}XD_j)(e_je^{-1}XC_j)^{f_j-1} &= X^{f_j}R(\varphi'_{\overline{v}_j})\Big(\frac{e_j}{e}\sum_{\tau\in H\cap I_{\overline{v}_j}} R(\tau)\Big)^{f_j} \\ &= X^{f_j}R\Big(\frac{e_j}{e}\sum_{\tau\in H\cap I_{\overline{v}_j}} \varphi_{\overline{v}_j}\tau\Big) = X^{f_j}R(\phi_{\overline{v}_j/v'_j}).\end{aligned}$$

Altogether it follows that

$$\det(I - XR^*(\phi_{\overline{v}/v})) = \prod_{j=1}^{t}\det(I - e^{-1}XA_j) = \prod_{j=1}^{t}\det(I - X^{f_j}R(\phi_{\overline{v}_j/v'_j})).$$

It remains to prove the relation between the character values. Let $\lambda_1, \ldots, \lambda_{nd} \in \mathbb{C}$ be the eigenvalues of $R^*(\phi_{\overline{v}/v})$, and for $j \in [1, t]$ let $\lambda_{j,1}, \ldots, \lambda_{j,d} \in \mathbb{C}$ be the eigenvalues of $R(\phi_{\overline{v}_j/v'_j})$, both counted with their multiplicities. If $m \in \mathbb{N}$, then

$$\psi^*(\phi^m_{\overline{v}/v}) = \sum_{\nu=1}^{nd}\lambda_\nu^m, \qquad \psi(\phi^m_{\overline{v}_j/v'_j}) = \sum_{\nu=1}^{d}\lambda_{j,\nu}^m$$

and

$$\det(I - XR^*(\phi_{\overline{v}/v})) = \prod_{\nu=1}^{nd}(1-\lambda_\nu X) = \prod_{j=1}^{t}\prod_{\nu=1}^{d}(1-\lambda_{j,\nu}X^{f_j}) \in 1 + X\mathbb{C}[\![X]\!].$$

We apply the homomorphism $\mathrm{Log}\colon 1 + X\mathbb{C}[\![X]\!] \to \mathbb{C}[\![X]\!]$, defined by

$$\mathrm{Log}(1{+}x) = \sum_{k=1}^{\infty}\frac{(-1)^{k-1}}{k}x^k \quad \text{for all } x \in X\mathbb{C}[\![X]\!] \quad \text{(see [18, Theorem 5.3.8])}$$

and obtain

$$-\sum_{\nu=1}^{nd}\sum_{k=1}^{\infty}\frac{\lambda_\nu^kX^k}{k} = -\sum_{j=1}^{t}\sum_{\nu=1}^{d}\sum_{n=1}^{\infty}\frac{\lambda_{j,\nu}^kX^{f_jk}}{k}.$$

We compare the coefficients of X^m and obtain

$$\psi^*(\phi^m_{\overline{v}/v}) = \sum_{\nu=1}^{nd} \lambda^m_\nu = \sum_{\substack{j=1 \\ f_j \mid m}}^{t} \sum_{\nu=1}^{d} \lambda^{m/f_j}_{j,\nu} = \sum_{\substack{j=1 \\ f_j \mid m}}^{t} f_j \psi(\phi^{s/f_j}_{\overline{v}_j/v'_j}). \qquad \square$$

We add the promised matrix identities in $\mathsf{M}_{e_j}(\mathsf{M}_{f_j d}(\mathbb{C}))$ justifying the determinant calculations.

$$(1) \quad \begin{pmatrix} I & 0 & \dots & 0 & -I \\ 0 & I & \dots & 0 & -I \\ \cdot & \cdot & \dots & \cdot & \cdot \\ 0 & 0 & \dots & I & -I \\ -B & -B & \dots & -B & I+(e_j-1)B \end{pmatrix} \begin{pmatrix} I+B & B & \dots & B & B \\ B & I+B & \dots & B & B \\ \cdot & \cdot & \dots & \cdot & \cdot \\ B & B & \dots & I+B & B \\ B & B & \dots & B & I+B \end{pmatrix}$$

$$= \begin{pmatrix} I & 0 & \dots & 0 & -I \\ 0 & I & \dots & 0 & -I \\ \cdot & \cdot & \dots & \cdot & \cdot \\ 0 & 0 & \dots & I & -I \\ 0 & 0 & \dots & 0 & I+e_jB \end{pmatrix}$$

and

$$\begin{pmatrix} I & 0 & \dots & 0 & 0 \\ 0 & I & \dots & 0 & 0 \\ \cdot & \cdot & \dots & \cdot & \cdot \\ 0 & 0 & \dots & I & 0 \\ -B & -B & \dots & -B & I \end{pmatrix} \begin{pmatrix} I & 0 & \dots & 0 & -I \\ 0 & I & \dots & 0 & -I \\ \cdot & \cdot & \dots & \cdot & \cdot \\ 0 & 0 & \dots & I & -I \\ 0 & 0 & \dots & 0 & I \end{pmatrix}$$

$$= \begin{pmatrix} I & 0 & \dots & 0 & -I \\ 0 & I & \dots & 0 & -I \\ \cdot & \cdot & \dots & \cdot & \cdot \\ 0 & 0 & \dots & I & -I \\ -B & -B & \dots & -B & I+(e_j-1)B \end{pmatrix}.$$

$$
(2)\quad \begin{pmatrix} I & 0 & 0 & \ldots & 0 & -D \\ -C & I & 0 & \ldots & 0 & 0 \\ 0 & -C & I & \ldots & 0 & 0 \\ \cdot & \cdot & \cdot & \ldots & \cdot & \cdot \\ 0 & 0 & 0 & \ldots & I & 0 \\ 0 & 0 & 0 & \ldots & -C & I \end{pmatrix} \begin{pmatrix} I & 0 & 0 & \ldots & 0 & D \\ 0 & I & 0 & \ldots & 0 & CD \\ 0 & 0 & I & \ldots & 0 & C^2D \\ \cdot & \cdot & \cdot & \ldots & \cdot & \cdot \\ 0 & 0 & 0 & \ldots & I & C^{f-2}D \\ 0 & 0 & 0 & \ldots & 0 & I \end{pmatrix}
$$

$$
= \begin{pmatrix} I & 0 & 0 & \ldots & 0 & 0 \\ -C & I & 0 & \ldots & 0 & 0 \\ 0 & -C & I & \ldots & 0 & 0 \\ \cdot & \cdot & \cdot & \ldots & \cdot & \cdot \\ 0 & 0 & 0 & \ldots & I & 0 \\ 0 & 0 & 0 & \ldots & -C & I-DC^{f-1} \end{pmatrix}.
$$

Theorem 7.8.3. *Let L/K be a finite Galois extension, $G = \mathrm{Gal}(L/K)$, and let $v \in \mathsf{M}_K^0$ and $s \in \mathcal{H}_1$.*

1. Let ψ_1 and ψ_2 be characters of G. Then

$$\mathcal{L}_{L/K,v}(s, \psi_1 + \psi_2) = \mathcal{L}_{L/K,v}(s, \psi_1)\mathcal{L}_{L/K,v}(s, \psi_2).$$

2. Let K' be an intermediate field of L/K and $H = \mathrm{Gal}(L/K')$.

(a) Let K'/K be Galois, $G' = G/H = \mathrm{Gal}(K'/K)$, ψ' a character of G', and let $\psi\colon G \to \mathbb{C}$ be defined by $\psi(\sigma) = \psi'(\sigma \restriction K')$ for all $\sigma \in G$. Then

$$\mathcal{L}_{L/K,v}(s, \psi) = \mathcal{L}_{K'/K,v}(s, \psi').$$

(b) Let ψ be a character of H and $\psi^* = \mathrm{Ind}_G^H(\psi)$. Then

$$\mathcal{L}_{L/K,v}(s, \psi^*) = \prod_{\substack{v' \in \mathsf{M}_{K'} \\ v' \mid v}} \mathcal{L}_{L/K',v'}(s, \psi).$$

Proof. 1. Let $\overline{v} \in \mathsf{M}_L$ be a place above v and $\phi = \phi_{\overline{v}/v}$ the arithmetic mean of the Frobenius automorphisms for $\overline{v}$ over K. For $i \in \{1,2\}$ let $V_i = (V_i, \rho_i)$ be a representation of G with character ψ_i. Then $(V_1 \oplus V_2, \rho_1 \oplus \rho_2)$ is a representation of G with character $\psi_1 + \psi_2$ and we obtain

$$
\begin{aligned}
\mathcal{L}_{L/K,v}(s, \psi) &= \det\bigl(\mathrm{id}_{V_1 \oplus V_2} - \mathfrak{N}(v)^{-s}(\rho_1 \oplus \rho_2)(\phi)\bigr)^{-1} \\
&= \det\bigl(\bigl[\mathrm{id}_{V_1} - \mathfrak{N}(v)^{-s}\rho_1(\phi)\bigr] \oplus [\mathrm{id}_{V_2} - \mathfrak{N}(v)^{-s}\rho_2(\phi)]\bigr)^{-1} \\
&= \det(\mathrm{id}_{V_1} - \mathfrak{N}(v)^{-s}\rho_1(\phi))^{-1} \det(\mathrm{id}_{V_2} - \mathfrak{N}(v)^{-s}\rho_2(\phi))^{-1} \\
&= \mathcal{L}_{L/K,v}(s, \psi_1)\mathcal{L}_{L/K,v}(s, \psi_2).
\end{aligned}
$$

2. (a) Let (V, ρ') be a representation of G' with character ψ', and let $\pi\colon G \to G'$ be the residue class map, defined by $\pi(\sigma) = \sigma \restriction K'$. Then $\rho\colon G \to \mathrm{Aut}_{\mathbb{C}}(V)$, defined by $\rho = \rho' {\circ} \pi$, is a representation of G with character $\psi = \psi' {\circ} \pi$. Let $\overline{v} \in \mathsf{M}_L$ and $v' \in \mathsf{M}_{K'}$ be such that $\overline{v} \,|\, v' \,|\, v$. Then $V^{I_{\overline{v}}} = V^{I_{v'}}$ since $I_{v'} = \pi(I_{\overline{v}})$ (see [18, Theorem 2.13.3]). Let $\varphi_{\overline{v}}$ be a Frobenius automorphism for $\overline{v}$ over K. Then $\varphi_{\overline{v}} \restriction K'$ is a Frobenius automorphism for v' over K, and it follows that

$$[\mathrm{id}_V - \mathfrak{N}(v)^{-s}\rho(\varphi_{\overline{v}})] \restriction V^{I_{\overline{v}}} = [\mathrm{id}_V - \mathfrak{N}(v)^{-s}\rho'(\varphi_{\overline{v}} \restriction K')] \restriction V^{I_{v'}},$$

and consequently

$$\mathcal{L}_{L/K,v}(s, \psi) = \mathcal{L}_{K'/K,v}(s, \psi').$$

(b) Let (V, ρ) be a representation of H with character ψ and (V^*, ρ^*) the induced representation of G with character ψ^*. Let $v_1', \ldots, v_t'$ be the places of K' above v, for $j \in [1, t]$ let $\overline{v}_j \in \mathsf{M}_L$ such that $\overline{v}_j \,|\, v_j'$, and set $\overline{v} = \overline{v}_1$. Let $\phi_{\overline{v}/v}$ be the arithmetic mean of the Frobenious automorphisms of $\overline{v}$ over K, and for $j \in [1, t]$ let $\phi_{\overline{v}_j/v_j'}$ be the arithmetic mean of the Frobenius automorphisms of $\overline{v}_j$ over K'. Then 7.8.2 implies

$$\begin{aligned}
\mathcal{L}_{L/K,v}(s, \psi^*) &= \det(\mathrm{id}_V - \mathfrak{N}(v)^{-s}\rho^*(\phi_{\overline{v}/v})))^{-1} \\
&= \prod_{j=1}^{t} \det(\mathrm{id}_V - \mathfrak{N}(v)^{-sf_j}\rho(\phi_{\overline{v}_j/v_j'}))^{-1} \\
&= \prod_{j=1}^{t} \det(\mathrm{id}_V - \mathfrak{N}(v_j')^{-s}\rho(\phi_{\overline{v}_j/v_j'}))^{-1} = \prod_{j=1}^{t} \mathcal{L}_{L/K',v_j'}(s, \psi) \\
&= \prod_{\substack{v' \in \mathsf{M}_{K'} \\ v' | v}} \mathcal{L}_{L/K',v'}(s, \psi).
\end{aligned}$$

□

Corollary 7.8.4. *Let L/K be a finite Galois extension, $G = \mathrm{Gal}(L/K)$ and $s \in \mathcal{H}_1$.*

1. *Let ψ_1 and ψ_2 be characters of G. Then*

$$\mathcal{L}_{L/K}(s, \psi_1 + \psi_2) = \mathcal{L}_{L/K}(s, \psi_1)\mathcal{L}_{L/K}(s, \psi_2).$$

2. *Let K' be an intermediate field of L/K and $H = \mathrm{Gal}(L/K')$.*

(a) *Let K'/K be Galois, $G' = G/H = \mathrm{Gal}(K'/K)$, ψ' a character of G', and let $\psi\colon G \to \mathbb{C}$ be defined by $\psi(\sigma) = \psi'(\sigma \restriction K')$ for all $\sigma \in G$. Then*

$$\mathcal{L}_{L/K}(s, \psi) = \mathcal{L}_{K'/K}(s, \psi').$$

(b) Let ψ be a character of H and $\psi^* = \mathrm{Ind}_G^H(\psi)$. Then

$$\mathcal{L}_{L/K}(s,\psi^*) = \mathcal{L}_{L/K'}(s,\psi).$$

Proof. 1. and 2.(a) are obvious by 7.8.3. As to 2.(b) we obtain

$$\mathcal{L}_{L/K}(s,\psi^*) = \prod_{v\in \mathsf{M}_K^0} \mathcal{L}_{L/K,v}(s,\psi^*) = \prod_{v\in \mathsf{M}_K^0} \prod_{\substack{v'\in \mathsf{M}_{K'}\\ v'|v}} \mathcal{L}_{L/K',v'}(s,\psi)$$

$$= \prod_{v'\in \mathsf{M}_{K'}} \mathcal{L}_{L/K',v'}(s,\psi) = \mathcal{L}_{L/K'}(s,\psi). \qquad \square$$

We denote by $\mathcal{I}(G)$ the free abelian group with basis $\mathrm{Irr}(G)$; its elements are called **virtual characters**. By definition $\mathcal{I}(G)$ is a subgroup of $\mathsf{X}(G)$. We extend the definition of Artin L functions to virtual charactes.

Theorem and Definition 7.8.5. *Let L/K be a finite Galois extension. Suppose that $G = \mathrm{Gal}(L/K)$ and $s \in \mathcal{H}_1$.*
Let $\mathrm{Irr}(G) = \{\chi_1,\ldots,\chi_h\}$ and $\psi = c_1\chi_1 + \ldots + c_h\chi_h \in \mathcal{I}(G)$, where $c_1,\ldots,c_h \in \mathbb{Z}$. Then we define

$$\mathcal{L}_{L/K}(s,\psi) = \prod_{j=1}^h \mathcal{L}_{L/K}(s,\chi_j)^{c_j}.$$

If ψ is a proper character of G, this definition gives nothing new. Indeed, suppose that ψ is a character of G. Then

$$\psi + \sum_{\substack{j=1\\ c_j<0}}^h |c_j|\chi_j = \sum_{\substack{j=1\\ c_j>0}}^h c_j\chi_j,$$

and an easy induction using 7.8.4.1 yields

$$\mathcal{L}_{L/K}(s,\psi) \prod_{\substack{j=1\\ c_j<0}}^h \mathcal{L}_{L/K}(s,\chi_j)^{|c_j|} = \prod_{\substack{j=1\\ c_j>0}}^h \mathcal{L}_{L/K}(s,\chi_j)^{c_j},$$

and consequently

$$\mathcal{L}_{L/K}(s,\psi) = \prod_{j=1}^h \mathcal{L}_{L/K}(s,\chi_j)^{c_j}.$$

1. Let $r \in \mathbb{N}$, $\psi, \psi_1,\ldots,\psi_r \in \mathcal{I}(G)$ and $c_1,\ldots,c_r \in \mathbb{Z}$. If

$$\psi = \sum_{i=1}^r c_i\psi_i, \quad \text{then} \quad \mathcal{L}_{L/K}(s,\psi) = \prod_{i=1}^r \mathcal{L}_{L/K}(s,\psi_i)^{c_i}.$$

2. Let K' be an intermediate field of L/K and $H = \mathrm{Gal}(L/K')$.

(a) Let K'/K be Galois, $G' = G/H = \mathrm{Gal}(K'/K)$ and $\psi' \in \mathcal{I}(G')$. If $\psi \in \mathcal{I}(G)$ is defined by $\psi(\sigma) = \psi'(\sigma\restriction K')$ for all $\sigma \in G$, then

$$\mathcal{L}_{L/K}(s,\psi) = \mathcal{L}_{K'/K}(s,\psi').$$

(b) If $\psi \in \mathcal{I}(H)$ and $\psi^* = \mathrm{Ind}_G^H(\psi)$, then

$$\mathcal{L}_{L/K}(s,\psi^*) = \mathcal{L}_{L/K'}(s,\psi).$$

Proof. 1. For $i \in [1,r]$ let $\psi_i = c_{i,1}\chi_1 + \ldots + c_{i,h}\chi_h$, where $c_{i,1},\ldots,c_{i,h} \in \mathbb{Z}$. Then

$$\psi = \sum_{j=1}^{h}\Bigl(\sum_{i=1}^{r} c_ic_{i,j}\Bigr)\chi_j,$$

and consequently

$$\prod_{i=1}^{r}\mathcal{L}_{L/K}(\psi_i)^{c_i} = \prod_{i=1}^{r}\prod_{j=1}^{h}\mathcal{L}_{L/K}(s,\chi_i)^{c_ic_{i,j}}$$
$$= \prod_{j=1}^{h}\mathcal{L}_{L/K}(s,\chi_j)^{c_1c_{1,j}+\ldots+c_hc_{h,j}} = \mathcal{L}_{L/K}(s,\psi).$$

2.(a) Let $\mathrm{Irr}(G') = \{\chi_1',\ldots,\chi_{h'}'\}$ and $\psi = c_1\chi_1' + \ldots + c_{h'}\chi_{h'}'$, where $c_1,\ldots,c_{h'} \in \mathbb{Z}$. For $j \in [1,h']$ let $\psi_j\colon G \to \mathbb{C}$ be defined by $\psi_j(\sigma) = \chi_j'(\sigma\restriction K)$. Then $\psi_1,\ldots,\psi_{h'}$ are characters of G, and $\psi = c_1\psi_1 + \ldots + c_{h'}\psi_{h'}$. Hence $\psi \in \mathcal{I}(G)$, and by 1. and 7.8.4.2(a) we obtain

$$\mathcal{L}_{L/K}(s,\psi) = \prod_{j=1}^{h'}\mathcal{L}_{L/K}(s,\psi_j)^{c_j} = \prod_{j=1}^{h'}\mathcal{L}_{K'/K}(s,\chi_j')^{c_j} = \mathcal{L}_{K'/K}(s,\psi').$$

(b) Let $\mathrm{Irr}(H) = \{\chi_1',\ldots,\chi_{h'}'\}$ and $\psi = c_1\chi_1' + \ldots + c_{h'}\chi_{h'}'$, where $c_1,\ldots,c_{h'} \in \mathbb{Z}$. For $j \in [1,h']$ let $\chi_j'^* = \mathrm{Ind}_G^H(\chi_j')$. Then it follows that $\psi^* = c_1\chi_1'^* + \ldots + c_{h'}\chi_{h'}'^*$, and $\chi_1'^*,\ldots,\chi_{h'}'^*$ are characters of G. Hence $\psi^* \in \mathcal{I}(G)$, and by 1. and 7.8.4.2(b) we obtain

$$\mathcal{L}_{L/K}(s,\psi^*) = \prod_{j=1}^{m}\mathcal{L}_{L/K}(s,\chi_j^*)^{c_j} = \prod_{j=1}^{m}\mathcal{L}_{L/K'}(s,\chi_j)^{c_j} = \mathcal{L}_{L/K'}(s,\psi). \quad \square$$

Now we highlight the already announced connection between Artin L functions and Dirichlet L functions by means of class field theory.

Theorem 7.8.6. *Let L/K be a finite Galois extension and $G = \mathrm{Gal}(L/K)$. Let L^{a} be the largest over K abelian intermediate field of L/K let $\mathsf{f} = \mathsf{f}_{L^{\mathsf{a}}/K} \in \mathcal{D}'_K$ its conductor and $(\,\cdot\,, L^{\mathsf{a}}/K)\colon \mathcal{D}_K^{\mathsf{f}} \to G^{\mathsf{a}}$ the Artin map (as defined in* 6.5.7.1). *Let $\chi \in \mathsf{X}(G) = \mathsf{X}(G^{\mathsf{a}})$ be a linear character, and let $\widetilde{\chi} = \chi \circ (\,\cdot\,, L/K)\colon \mathcal{D}_K^{\mathsf{f}} \to \mathbb{T}$ be the associated ray class character modulo f of finite order* (see 6.5.8). *Then $\mathcal{L}_{L/K}(s,\chi)$ is a Dirichlet L function. Explicitly, if $s \in \mathcal{H}_1$, then*

$$\mathcal{L}_{L/K}(s,\chi) = \prod_{v \in \mathsf{M}_K^0} (1 - \widetilde{\chi}(\mathsf{p}_v)\mathfrak{N}(v)^{-s})^{-1} = L(s, \widetilde{\chi}).$$

In particular:

- *$\mathcal{L}_{L/K}(\,\cdot\,, \chi)$ has an extension to a meromorphic function on $\mathbb{C}$.*
- *If $\chi = 1_G$, then $\widetilde{\chi} = 1$, and $\mathcal{L}_{L/K}(s, 1_G) = \mathcal{L}_{K/K}(s, 1_1) = \zeta_K(s)$.*
- *If $\chi \ne 1_G$, then $\widetilde{\chi} \ne 1$, and $\mathcal{L}_{L/K}(\,\cdot\,, \chi)$ has an extension to a holomorphic function on $\mathbb{C}$, provided that in the function field case K_χ/K is not a constant field extension.*
- *Suppose that we are in the function field case and K_χ/K is a constant field extension. If $\chi \ne 1_G$, then $\mathcal{L}_{L/K}(\,\cdot\,, \chi)$ is holomorphic in 0 and 1.*

Proof. Apparently, $\widetilde{\chi}$ is a group homomorphism, and as $\widetilde{\chi} \restriction (K^{\mathsf{f}}) = 1$, it is a ray class character modulo f of finite order. As to the remaining assertions, we do a special case first.

SPECIAL CASE: χ is injective. Then we have $K = K_\chi = K^{\mathsf{a}}$ and $\mathsf{f}(\widetilde{\chi}) = \mathsf{f}_{L/K} = \mathsf{f}_{L/K}(\chi)$ by 7.7.6.1. By 7.5.1 we may assume that $\chi = \chi_V$, where $V = \mathbb{C}$, and the structural homomorphism ρ_V is given by $\rho_V(\alpha)(z) = \chi(\alpha)z$ for all $\alpha \in \mathbb{C}[G]$ and $z \in \mathbb{C}$. If $v \in \mathsf{M}_K^0$ and $\overline{v} \in \mathsf{M}_L$ lies above v, then

$$V^{I_{\overline{v}}} = \begin{cases} \mathbb{C} & \text{if } \chi \restriction I_{\overline{v}} = 1, \\ \mathbf{0} & \text{if } \chi \restriction I_{\overline{v}} \ne 1. \end{cases}$$

If $v \notin \mathrm{supp}(\mathsf{f})$, then v is unramified in L, $V^{I_{\overline{v}}} = \mathbb{C}$, and if $\varphi_{\overline{v}}$ is a Frobenius automorphism for $\overline{v}$ over K, then

$$\chi(\varphi_{\overline{v}}) = \chi(v, L^{\mathsf{a}}/K) = \chi(\mathsf{p}_v, L^{\mathsf{a}}/K) = \widetilde{\chi}(\mathsf{p}_v).$$

Hence the corresponding local factor is given by

$$\mathcal{L}_{L/K,v}(s,\chi) = \det\bigl([\mathrm{id}_V - \mathfrak{N}(v)^{-s}\rho(\varphi_{\overline{v}})] \restriction V^{I_{\overline{v}}}\bigr)^{-1} = \frac{1}{1 - \widetilde{\chi}(\mathsf{p}_v)\mathfrak{N}(v)^{-s}}.$$

If $v \in \text{supp}(\mathsf{f})$, then v is ramified in L, hence $I_{\overline{v}} \neq \mathbf{1}$ and therefore $\chi \restriction I_{\overline{v}} \neq 1$ (since χ is injective) which implies $V^{I_{\overline{v}}} = \mathbf{0}$ and $\mathcal{L}_{L/K,v}(s,\chi) = 1$. Moreover, as $\mathsf{f} = \mathsf{f}(\widetilde{\chi})$, it follows that $\widetilde{\chi}(\mathsf{p}_v) = 0$. Altogether we obtain

$$\mathcal{L}_{L/K}(s,\chi) = \prod_{v \in \mathsf{M}_K^0} \frac{1}{1 - \widetilde{\chi}(\mathsf{p}_v)\mathfrak{N}(v)^{-s}} = L(s,\widetilde{\chi}) \quad \text{for all} \quad s \in \mathcal{H}_1.$$

GENERAL CASE: Let $K_\chi = L^{\text{Ker}(\chi)} \subset L^{\mathsf{a}}$ be the fixed field of $\text{Ker}(\chi)$, $G_\chi = G/\text{Ker}(\chi) = \text{Gal}(K_\chi/K)$, and define $\chi' \in \mathsf{X}(G_\chi)$ by $\chi'(\sigma \restriction K_\chi) = \chi(\sigma)$ for all $\sigma \in G$. Then χ' is injective and

$$\widetilde{\chi}' = \chi' \circ (\,\cdot\,, K_\chi/K) = \chi' \circ (\,\cdot\,, L^{\mathsf{a}}/K) \restriction K_\chi = \chi \circ (\,\cdot\,, L^{\mathsf{a}}/K) = \widetilde{\chi}.$$

By the special case it follows that $\mathsf{f}(\widetilde{\chi}) = \mathsf{f}(\widetilde{\chi}') = \mathsf{f}_{K_\chi/K} = \mathsf{f}_{L/K}(\chi)$ by 7.7.6.1. Also by the special case and 7.8.5.2(a) we obtain

$$\mathcal{L}_{L/K}(s,\chi) = \mathcal{L}_{K_\chi/K}(s,\chi') = L(s,\widetilde{\chi}') = L(s,\widetilde{\chi}).$$

If $\chi = 1_G$, then clearly $\widetilde{\chi} = 1$ and $\mathcal{L}_{K/K}(s,1_1) = \mathcal{L}_{L/K}(s,1_G) = \zeta_K(s)$ by the very definition.

The assertions concerning the meromorphic continuation follow from 7.1.8.2 in the number field case, and from 7.4.3 in the function field case.

Thus suppose that we are in the function field case. By 7.4.3, the function $L(s,\widetilde{\chi}) = \mathcal{L}_{L/K}(s,\chi)$ is an entire function if and only if $\widetilde{\chi} \restriction \mathcal{D}_K^0 \neq 1$, and by 6.5.8.2 this holds if and only if K_χ/K is not a constant field extension. Moreover, if $\widetilde{\chi} \restriction \mathcal{D}_K^0 = 1$ but $\widetilde{\chi} \neq 1$, then $L(s,\widetilde{\chi}) = \mathcal{L}_{L/K}(s,\chi)$ has no poles at 0 and 1 by 7.4.3.2. □

Corollary 7.8.7. *Let L/K be a finite abelian extension, $G = \text{Gal}(L/K)$, $\chi \neq 1_G$ an irreducible character of G, and suppose that L/K is geometric in the function field case. Then $\mathcal{L}_{L/K}(s,\chi)$ is holomorphic in $\mathbb{C}$.*

Proof. If L/K is geometric, then K_χ/K is geometric, too, and the assertion follows by 7.8.6. □

The famous (and hitherto in its generality for number fields unsolved) **Artin conjecture** asserts that for a finite Galois extension L/K the Artin L function $\mathcal{L}_{L/K}(s,\chi)$ is holomorphic in $\mathbb{C}$ for every irreducible character $\chi \neq 1_G$ of $G = \text{Gal}(L/K)$. For function fields the Artin conjecture is true for geometric extensions due to A. Weil, see [63]. Here we can only prove the exceedingly modest result 7.8.10.

Theorem 7.8.6 also implies that every Dirichlet L function is in fact an Artin L function. In the following Remark 7.8.8 we give details.

Remark 7.8.8. Let $\mathfrak{m} \in \mathcal{D}'_K$ and $\psi\colon \mathcal{D}_K^{\mathfrak{m}} \to \mathbb{T}$ a ray class character. Then the Dirichlet L function $L(s,\psi)$ is an Artin L function in the following way.

The group $\mathcal{G} = \mathrm{Ker}(\psi)$ is a subgroup of finite index of $\mathcal{D}_K^{\mathfrak{m}}$, and we consider the class field L to $\mathcal{G}$ (see 6.5.7.2).

By the Artin reciprocity law, L/K is a finite abelian extension, $\mathcal{G} = \mathcal{N}_{L/K}^{\mathfrak{m}}$, and the Artin map $(\,\cdot\,, L/K)\colon \mathcal{D}_K^{\mathfrak{m}} \to G = \mathrm{Gal}(L/K)$ induces the Artin isomorphism $\mathcal{D}_K^{\mathfrak{m}}/\mathcal{G} \overset{\sim}{\to} G$. Hence there exists a unique character $\chi \in \mathsf{X}(G)$ such that $\widetilde{\chi} = \chi\circ(\,\cdot\,, L/K) = \psi$, and then 7.8.6 shows that

$$L(s,\psi) = \mathcal{L}_{L/K}(s,\chi).$$

Theorem 7.8.9. *Let L/K be a finite Galois extension and $G = \mathrm{Gal}(L/K)$. Then*

$$\zeta_L(s) = \zeta_K(s) \prod_{\chi\in\mathrm{Irr}(G)\setminus\{1\}} \mathcal{L}_{L/K}(s,\chi) \quad \textit{for all} \quad s\in\mathbb{C},$$

and if in the function field case L/K is a geometric extension, then

$$\frac{\zeta_L}{\zeta_K} \quad \textit{is an entire function.}$$

Proof. We apply 7.8.5.2(b) with $H = \mathbf{1}$, $\chi = 1_1$, $G' = G$ and $K' = L$. Then

$$1_1^* = r_G = \sum_{\chi\in\mathrm{Irr}(G)} \chi(1)\chi$$

by 7.6.6.4 and 7.5.11. Since $\mathcal{L}_{L/L}(s,1_1) = \zeta_L(s)$ and $\mathcal{L}_{L/K}(s,1_1) = \zeta_K(s)$ by 7.8.6, we obtain

$$\zeta_L(s) = \mathcal{L}_{L/L}(s,1_1) = \mathcal{L}_{L/K}(s,1_1^*) = \prod_{\chi\in\mathrm{Irr}(G)} \mathcal{L}_{L/K}(s,\chi)$$
$$= \zeta_K(s) \prod_{\chi\in\mathrm{Irr}(G)\setminus\{1\}} \mathcal{L}_{L/K}(s,\chi).$$

Let $[L\colon K] = n$. By Aramata's theorem 7.6.8 we obtain

$$n(r_G - 1_G)) = \sum_{j=1}^m c_j\chi_j^*,$$

where $m\in\mathbb{N}$, $\chi_j \in \mathsf{X}(A_j)$ for a cyclic subgroup A_j of G, $\chi_j^* = \mathrm{Ind}_G^{A_j}(\chi_j)$ and $c_j\in\mathbb{N}$ for all $j\in[1,m]$. By 7.8.5.2(b) it follows that

$$\Big(\frac{\zeta_L(s)}{\zeta_K(s)}\Big)^n = \prod_{\chi\in\mathsf{X}(G)\setminus\{1_G\}} \mathcal{L}_{L/K}(s,\chi)^n = \mathcal{L}_{L/K}(n(r_G - 1_G))$$
$$= \prod_{j=1}^m \mathcal{L}_{L/K}(s,\chi_j^*)^{c_j} = \prod_{j=1}^m \mathcal{L}_{L/K_j}(s,\chi_j)^{c_j} \quad \text{where } K_j = L^{A_j}.$$

Hence 7.8.7 implies that

$$\Big(\frac{\zeta_L}{\zeta_K}\Big)^n \quad \text{and thus also} \quad \frac{\zeta_L}{\zeta_K} \quad \text{is an entire function}$$

(observe that we already know that ζ_L/ζ_K is a meromorpic function). □

Theorem 7.8.10. *Let L/K be a finite Galois extension, $G = \mathrm{Gal}(L/K)$, and let $\chi \ne 1_G$ be an irreducible character of G. Then the Artin L function $\mathcal{L}_{L/K}(\,\cdot\,,\chi)$ has an extension to a meromorphic function on $\mathbb{C}$ which is holomorphic and non-zero in 1.*

Proof. SPECIAL CASE: $\chi \in \mathsf{X}(G)$ is a linear character. Then $\mathcal{L}_{L/K}(\,\cdot\,,\chi)$ has an extension to a meromorphic function on $\mathbb{C}$ by 7.8.6. Let $K_\chi = L^{\mathrm{Ker}(\chi)}$ be the fixed field of $\mathrm{Ker}(\chi)$, $G_\chi = G/\mathrm{Ker}(\chi) = \mathrm{Gal}(K_\chi/K)$, $h = \mathrm{ord}(\chi)$, $\zeta_h = \mathsf{e}^{2\pi\mathsf{i}/h}$ and $\sigma \in G$ such that $\chi(\sigma) = \zeta_h$. Let $\chi' \in \mathsf{X}(G_\chi)$ be defined by $\chi'(\tau \upharpoonright K_\chi) = \chi(\tau)$ for all $\tau \in G$. Then $\chi'(\sigma \upharpoonright K_\chi) = \chi(\sigma) = \zeta_h$, hence $\mathsf{X}(G_\chi) = \langle \chi' \rangle = \{\chi'^j \mid j \in]0, h-1]\}$, and $\chi'^j(\tau \upharpoonright K_\chi) = \chi^j(\tau)$ for all $\tau \in G$ and $j \in [0, h-1]$. Now 7.8.9 together with 7.8.5.2(a) implies

$$\zeta_{K_\chi}(s) = \zeta_K(s) \prod_{j=1}^{h-1} \mathcal{L}_{K_\chi/K}(s, \chi'^j) = \zeta_K(s) \prod_{j=1}^{h-1} \mathcal{L}_{L/K}(s, \chi^j) \quad \text{for all } \ s \in \mathcal{H}_1,$$

hence for all $s \in \mathbb{C}$, since all involved functions are meromorphic in $\mathbb{C}$. The functions $\mathcal{L}_{L/K}(\,\cdot\,,\chi^j)$ are holomorphic in 1, and the functions ζ_K and ζ_{K_χ} have a simple pole at 1. Comparing the orders yields $\mathcal{L}_{L/K}(1,\chi^j) \ne 0$ for all $j \in [1, h-1]$.

GENERAL CASE: By 7.6.9 we obtain

$$\chi = \sum_{j=1}^r c_j \chi_j^*,$$

where $r \in \mathbb{N}$, $c_j \in \mathbb{Z}$, H_j is a subgroup of G, $\chi_j^* = \mathrm{Ind}_G^{H_j}(\chi_j)$ for some $\chi_j \in \mathsf{X}(H_j)$, and we set $K_j = L^{H_j}$ for all $j \in [1, r]$. By 7.8.5 we get

$$\mathcal{L}_{L/K}(s,\chi) = \prod_{j=1}^r \mathcal{L}_{L/K}(s, \chi_j^*)^{c_j} = \prod_{j=1}^r \mathcal{L}_{L/K_j}(s, \chi_j)^{c_j},$$

and as the functions $\mathcal{L}_{L/K_j}(\,\cdot\,,\chi_j)$ have extensions to meromorphic functions on $\mathbb{C}$, the same is true for $\mathcal{L}_{L/K}(\,\cdot\,,\chi)$. By 7.8.6 and the SPECIAL CASE above we achieve

$$\mathrm{ord}\{\mathcal{L}_{L/K_j}(s,\chi_j); s = 1\} = \begin{cases} -1 & \text{if } \ \chi_j = 1_{H_j}, \\ 0 & \text{if } \ \chi_j \ne 1_{H_j}, \end{cases}$$

and consequently

$$\mathrm{ord}\{\mathcal{L}_{L/K}(s,\chi);\, s=1\} \;=\; -\sum_{\substack{j=1\\ \chi_j=1_{H_j}}}^{r} c_j.$$

On the other hand, using 7.6.6.2,

$$0=\langle\chi,1_G\rangle=\sum_{j=1}^{r}c_j\langle\chi_j^*,1_G\rangle=\sum_{j=1}^{r}c_j\langle\chi_j,1_{H_j}\rangle=\sum_{\substack{j=1\\ \chi_j=1_{H_j}}}^{r}c_j.$$

It follows that $\mathrm{ord}\{\mathcal{L}_{L/K}(s,\chi);\, s=1\}=0$, and therefore $\mathcal{L}_{L/K}(\,\cdot\,,\chi)$ is holomorphis and non-zero in 1. □

We close this section with a proof of the functional equation for Artin L functions in the number field case. To do so, we must add suitable Gamma factors for the Archimedian places.

Definition 7.8.11. Let L/K be a finite Galois extension of algebraic number fields, $G=\mathrm{Gal}(L/K)$ and $V=(V,\rho)$ a representation of G and $\psi=\theta_V$. For $s\in\mathcal{H}_1$ we define

$$\gamma(s)=\pi^{-s/2}\Gamma\Big(\frac{s}{2}\Big).$$

For a complex place $v\in\mathsf{M}_K^\infty$ we define

$$\mathcal{L}_{L/K,v}(s,\psi)=[\gamma(s)\gamma(s+1)]^{\psi(1)}.$$

Let $v\in\mathsf{M}_K^\infty$ be real, let $\overline{v}\in\mathsf{M}_L$ be a place above v and $\mathrm{Gal}(L_{\overline{v}}/K_v)=\langle\varphi_{\overline{v}}\rangle$, where $\varphi_{\overline{v}}\in G$, $\varphi_{\overline{v}}=1$ if $\overline{v}$ is real, and $\mathrm{ord}(\varphi_{\overline{v}})=2$ if $\overline{v}$ is complex (see 5.4.5.1). Then $\rho(\varphi_{\overline{v}})\in\mathrm{End}_{\mathbb{C}}V$ has at most the eigenvalues ±1, and we denote by $V_{\overline{v},\psi}^{\pm}$ the corresponding eigenspaces. Then $V=V_{\overline{v},\psi}^{+}\oplus V_{\overline{v},\psi}^{-}$, and if $x\in V$, then

$$x=x^++x^-,\quad \text{where}\quad x^{\pm}=\frac{1}{2}(x\pm\varphi_{\overline{v}}x)\in V_{\overline{v},\psi}^{\pm}.$$

Since

$$\psi(\varphi_{\overline{v}})=\mathsf{Tr}(\rho(\overline{\varphi}))=\mathsf{Tr}(\rho(\overline{\varphi})\restriction V_{\overline{\varphi},\psi}^{+})+\mathsf{Tr}(\rho(\overline{\varphi})\restriction V_{\overline{\varphi},\psi}^{-})=\dim_{\mathbb{C}}V_{\overline{\varphi},\psi}^{+}-\dim_{\mathbb{C}}V_{\overline{\varphi},\psi}^{-}$$

and $\psi(1)=\dim_{\mathbb{C}}V=\dim_{\mathbb{C}}V_{\overline{\varphi},\psi}^{+}+\dim_{\mathbb{C}}V_{\overline{\varphi},\psi}^{-}$, we obtain

$$n_{\overline{v},\psi}^{+}=\dim_{\mathbb{C}}V_{\overline{v},\psi}^{+}=\frac{1}{2}\big(\psi(1)+\psi(\varphi_{\overline{v}})\big)\quad\text{and}\quad n_{\overline{v},\psi}^{-}=\dim_{\mathbb{C}}V_{\overline{v},\psi}^{-}=\frac{1}{2}\big(\psi(1)-\psi(\varphi_{\overline{v}})\big).$$

We define

$$\mathcal{L}_{L/K,v}(s,\psi)=\gamma(s)^{n_{\overline{v},\psi}^{+}}\,\gamma(s+1)^{n_{\overline{v},\psi}^{-}}.$$

Now we define the **extended Artin L-function** by

$$\mathbf{\Lambda}_{L/K}(s,\psi) = \left[\mathfrak{N}(\mathfrak{f}_{L/K}(\psi))|\mathsf{d}_K|^{\psi(1)}\right]^{s/2}\Big[\prod_{v\in\mathsf{M}_K^\infty}\mathcal{L}_{L/K,v}(s,\psi)\Big]\mathcal{L}_{L/K}(s,\psi).$$

Theorem 7.8.12. *Let L/K be a finite Galois extension of algebraic number fields, $G=\mathrm{Gal}(L/K)$ and $s\in\mathcal{H}_1$.*

1. Let ψ_1 and ψ_2 be characters of G. Then

$$\mathbf{\Lambda}_{L/K}(s,\psi_1+\psi_2) = \mathbf{\Lambda}_{L/K}(s,\psi_1)\mathbf{\Lambda}_{L/K}(s,\psi_2).$$

2. Let K' be an intermediate field of L/K and $H=\mathrm{Gal}(L/K')$.

(a) Let K'/K be Galois, $G'=G/H=\mathrm{Gal}(K'/K)$, ψ' a character of G', and let $\psi\colon G\to\mathbb{C}$ be defined by $\psi(\sigma)=\psi'(\sigma\restriction K')$ for all $\sigma\in G$. Then

$$\mathbf{\Lambda}_{L/K}(s,\psi) = \mathbf{\Lambda}_{K'/K}(s,\psi').$$

(b) Let ψ be a character of H and $\psi^*=\mathrm{Ind}_G^H(\psi)$. Then

$$\mathbf{\Lambda}_{L/K}(s,\psi^*) = \mathbf{\Lambda}_{L/K'}(s,\psi).$$

3. Let $\psi\in\mathsf{X}(G)$ be a character of degree 1, and suppose that $\widetilde{\psi}=\psi\circ(\,\cdot\,,L/K)\colon\mathcal{D}_K^{\mathfrak{f}_{L/K}}\to\mathbb{C}$, where $(\,\cdot\,,L/K)\colon\mathcal{D}_K^{\mathfrak{f}_{L/K}}\to G$ denotes the Artin map. Then

$$\mathbf{\Lambda}_{L/K}(s,\psi) = \Lambda(s,\widetilde{\psi}),$$

where $\Lambda(\,\cdot\,,\widetilde{\psi})$ denotes the extended Dirichlet L function (see 7.1.6*).*

Proof. 1. Observe that $\mathcal{L}_{L/K}(s,\psi_1+\psi_2)=\mathcal{L}_{L/K}(s,\psi_1)\mathcal{L}_{L/K}(s,\psi_2)$ by 7.8.5.1, $\mathfrak{N}(\mathfrak{f}_{L/K}(\psi_1+\psi_2))|\mathsf{d}_K|^{(\psi_1+\psi_2)(1)}=\mathfrak{N}(\mathfrak{f}_{L/K}(\psi_1))|\mathsf{d}_K|^{\psi_1(1)}\mathfrak{N}(\mathfrak{f}_{L/K}(\psi_2))$ $\mathsf{d}_K|^{\psi_2(1)}$ by 7.7.6.2 and $\mathcal{L}_{L/K,v}(s,\psi_1+\psi_2)=\mathcal{L}_{L/K,v}(s,\psi_1)\mathcal{L}_{L/K,v}(s,\psi_2)$ for all $v\in\mathsf{M}_K^\infty$ by the very definition. Hence altogether we obtain $\mathbf{\Lambda}_{L/K}(s,\psi_1+\psi_2)=\mathbf{\Lambda}_{L/K}(s,\psi_1)\mathbf{\Lambda}_{L/K}(s,\psi_2)$.

2.(a) Since $\psi(1)=\psi'(1)$, 7.8.5.2(a) implies $\mathcal{L}_{L/K}(s,\psi)=\mathcal{L}_{K'/K}(s,\psi')$, and by 7.7.6.3(a) we get $\mathfrak{N}(\mathfrak{f}_{L/K}(\psi))|\mathsf{d}_K|^{\psi(1)}=\mathfrak{N}(\mathfrak{f}_{K'/K}(\psi'))|\,\mathsf{d}_K|^{\psi'(1)}$.

If $v\in\mathsf{M}_K^\infty$ is complex, then apparently $\mathcal{L}_{L/K,v}(s,\psi)=\mathcal{L}_{L/K,v}(s,\psi')$. Let finally $v\in\mathsf{M}_K^\infty$ be real, $v'\in\mathsf{M}_{K'}$ and $\overline{v}\in\mathsf{M}_L$ such that $\overline{v}\,|\,v'\,|\,v$. If $\mathrm{Gal}(L_{\overline{v}}/K_v)=\langle\varphi_{\overline{v}}\rangle$, then $\mathrm{Gal}(K'_{v'}/K_v)=\langle\varphi_{\overline{v}}\restriction K'\rangle$, hence $\varphi_{\overline{v}}\restriction K'=\varphi_{v'}$ and $\psi'(\varphi'_v)=\psi(\varphi_{\overline{v}})$. Hence

$$n^{\pm}_{v',\psi'}=\frac{1}{2}\big(\psi'(1)\pm\psi'(\varphi_{v'})\big)=\frac{1}{2}\big(\psi(1)\pm\psi(\varphi_{\overline{v}})\big)=n^{\pm}_{v,\psi},$$

therefore $\mathcal{L}_{L/K,v}(s,\psi) = \mathcal{L}_{K'/K,v}(s,\psi')$, and $\mathbf{\Lambda}_{L/K}(s,\psi) = \mathbf{\Lambda}_{K'/K}(s,\psi')$ follows.

2.(b) Let $n = [K' : K]$. By means of 7.7.6.3(b) we obtain (using [18, Theorem 5.9.8 and Corollary 2.14.3])

$$\begin{aligned}\mathfrak{N}(\mathfrak{f}_{L/K}(\psi^*))|\mathsf{d}_K|^{\psi^*(1)} &= \mathfrak{N}(\mathfrak{d}_{K'/K}^{\psi(1)}\mathcal{N}_{K'/K}(\mathfrak{f}_{L/K'}(\psi)))|\mathsf{d}_K|^{n\psi(1)} \\ &= \mathfrak{N}(\mathfrak{f}_{L/K'}(\psi))\big[|\mathsf{d}_K|^n\mathfrak{N}(\mathfrak{d}_{K'/K})\big]^{\psi(1)} \\ &= \mathfrak{N}(\mathfrak{f}_{L/K'}(\psi))|\mathsf{d}_{K'}|^{\psi(1)}.\end{aligned}$$

For the proof of $\mathbf{\Lambda}_{L/K}(s,\psi^*) = \mathbf{\Lambda}_{L/K'}(s,\psi)$ it is now sufficient to show that

$$\mathcal{L}_{L/K,v}(s,\psi^*) = \prod_{\substack{v'\in\mathsf{M}_{K'}\\ v'\,|\,v}} \mathcal{L}_{L/K',v'}(s,\psi) \quad \text{for all } \ v\in\mathsf{M}_K^\infty. \tag{$*$}$$

Proof of $(*)$. If $v\in\mathsf{M}_K^\infty$ is complex, then all $v'\in\mathsf{M}_{K'}$ above v are complex, and therefore

$$\prod_{\substack{v'\in\mathsf{M}_{K'}\\ v'\,|\,v}} \mathcal{L}_{L/K',v'}(s,\psi) = [\,\gamma(s)\gamma(s+1)\,]^{n\psi(1} = [\,\gamma(s)\gamma(s+1)\,]^{\psi^*(1} = \mathcal{L}_{L/K,v}(s,\psi^*).$$

Now let v be real, suppose that $H\backslash G = \{H\tau_1,\ldots,H\tau_n\}$, and for $i\in[1,n]$ let $v_i' = \tau_i\overline{v}\downarrow K'\in\mathsf{M}_{K'}$. Then $v_1',\ldots,v_n'$ are precisely the places of K' above v, where the complex ones are counted twice. We may assume that $n = r_1+2r_2$ with $r_1,\,r_2\in\mathbb{N}_0$ such that $v_1',\ldots,v_{r_1}'$ are real, $v_{r_1+1}',\ldots,v_{r_1+r_2}'$ are complex, and $v_{r_1+r_2+i}' = v_{r_1+i}'$ for all $i\in[1,r_2]$. Suppose that $\mathrm{Gal}(L_{\tau_i\overline{v}}/K_v) = \langle\varphi_{\tau_i\overline{v}}\rangle$ and $\mathrm{Gal}(L_{\tau_i\overline{v}}/K_{v_i'}') = \langle\varphi_{\tau_i\overline{v}}'\rangle$. Then we get $\varphi_{\tau_i\overline{v}} = \tau_i\varphi_{\overline{v}}\tau_i^{-1}$, $\varphi_{\tau_i\overline{v}}' = \varphi_{\tau_i\overline{v}}$ if $i\in[1,r_1]$ and $\varphi_{\tau_i\overline{v}}\notin H$ if $i\in[r_1+1,n]$.

As $\{\tau_1^{-1},\ldots,\tau_n^{-1}\}$ is a set of representatives for G/H, it follows that

$$\psi^*(\varphi_{\overline{v}}) = \sum_{i=1}^n \dot\psi(\tau_i\varphi_{\overline{v}}\tau_i^{-1}) = \sum_{i=1}^n \dot\psi(\varphi_{\tau_i\overline{v}}) = \sum_{i=1}^{r_1}\psi(\varphi_{\tau_i\overline{v}}'),$$

where $\dot\psi\restriction H = \psi$ and $\dot\psi\restriction G\setminus H = 0$. Hence we obtain

$$\mathcal{L}_{L/K,v}(s,\psi^*) = \gamma(s)^{A^+}\gamma(s+1)^{A^-},$$

where

$$A^\pm = \frac12\Big[\psi^*(1)\pm\psi^*(\varphi_{\overline{v}})\Big] = \frac12\Big[n\psi(1)\pm\sum_{i=1}^{r_1}\psi(\varphi_{\tau_i\overline{v}}')\Big].$$

On the other hand,

$$\prod_{\substack{v'\in \mathsf{M}_{K'}\\ v'\mid v}} \mathcal{L}_{L/K',v'}(s,\psi) = [\,\gamma(s)\gamma(s+1)\,]^{r_2\psi(1)}$$

$$\prod_{i=1}^{r_1} \gamma(s)^{[\psi(1)+\psi(\varphi'_{\tau_i\overline{v}})]/2}\gamma(s+1)^{[\psi(1)-\psi(\varphi'_{\tau_i\overline{v}})]/2}$$

$$= \gamma(s)^{A^+}\gamma(s+1)^{A^-} = \mathcal{L}_{L/K,v}(s,\psi^*).$$

3. It suffices to prove the assertion if χ is injective. Indeed, suppose that this is done. Then we set $K_\chi = L^{\mathrm{Ker}(\chi)}$, $G_\chi = \mathrm{Gal}(K_\chi/K) = G/\mathrm{Ker}(\chi)$ and define $\chi'\colon K_\chi \to G$ by $\chi'(\sigma\restriction K_\chi) = \chi(\sigma)$ for all $\sigma \in G$. Then χ' is injective,

$$\widetilde{\chi}' = \chi'\circ(\,\cdot\,,K_\chi/K) = \chi'\circ(\,\cdot\,,L/K)\restriction K_\chi = \chi\circ(\,\cdot\,,L/K) = \widetilde{\chi},$$

and by 2.(a) we obtain

$$\mathbf{\Lambda}_{L/K}(s,\chi) = \mathbf{\Lambda}_{K_\chi/K}(s,\chi') = \Lambda(s,\widetilde{\chi}') = \Lambda(s,\widetilde{\chi}).$$

Assume now that χ is injective. Let r_1 be the number of real and r_2 the number of complex places of K, and let $q \in [0,r_1]$ be the number of real places v such that $\widetilde{\chi}_v \ne 1$. Then

$$\Lambda(s,\widetilde{\chi}) = 2^{r_2(1-s)}\pi^{-(q+ns)/2}\mathfrak{N}(\mathfrak{D}\mathfrak{f}(\widetilde{\chi}))^{s/2}\Gamma\Big(\frac{s+1}{2}\Big)^q\Gamma\Big(\frac{s}{2}\Big)^{r_1-q}\Gamma(s)^{r_2}L(s,\widetilde{\chi})$$

by 7.1.6,

and

$$\mathbf{\Lambda}_{L/K}(s,\chi) = [\,\mathfrak{N}(\mathfrak{f}_{L/K}(\chi))|\mathsf{d}_K|\,]^{s/2}\Big[\prod_{v\in\mathsf{M}_K^\infty}\mathcal{L}_{L/K,v}(s,\chi)\Big]\mathcal{L}_{L/K}(s,\chi) \quad \text{by 7.8.11.}$$

Since $\mathfrak{f}_{L/K}(\chi) = \mathfrak{f}(\widetilde{\chi})$ by 7.7.6.1 and $\mathcal{L}_{L/K}(s,\chi) = L(s,\widetilde{\chi})$ by 7.8.6, it follows that $\mathfrak{N}(\mathfrak{D}\mathfrak{f}(\widetilde{\chi}))^{s/2} = [\,\mathfrak{N}(\mathfrak{f}_{L/K}(\chi))|\mathsf{d}_K|\,]^{s/2}$ (see [18, Theorem5.9.8]).

For any $v \in \mathsf{M}_K^\infty$ fix a place $\overline{v} \in \mathsf{M}_L$ above v and set $\mathrm{Gal}(L_{\overline{v}}/K_v) = \langle\varphi_{\overline{v}}\rangle$. Then $\varphi_{\overline{v}} = 1$ unless v is real and $\overline{v}$ is complex, and in the latter case $\mathrm{ord}(\varphi_{\overline{v}}) = 2$ and $\chi(\varphi_{\overline{v}}) = -1$. Hence

$$\mathcal{L}_{L/K,v}(s,\chi) = \begin{cases} \gamma(s)\gamma(s+1) & \text{if } v \text{ is complex,}\\ \gamma(s) & \text{if } \overline{v} \text{ is real,}\\ \gamma(s+1) & \text{if } v \text{ is real and } \overline{v} \text{ is complex.}\end{cases}$$

We view $\widetilde{\chi}$ as an idele class character, $\widetilde{\chi}\colon \mathsf{J}_K \to \mathbb{T}$, and for $v \in \mathsf{M}_K^\infty$ we consider its local component $\widetilde{\chi}_v = \widetilde{\chi}\circ j_v\colon K_v^\times \to \mathbb{T}$. We shall prove:

A. Let $v \in \mathsf{M}_K^\infty$ be real. Then $\overline{v}$ is complex if and only if $\widetilde{\chi}_v \ne 1$.

Proof of **A.** Recall that $\widetilde{\chi}_v = \widetilde{\chi}\circ j_v = \chi\circ(\,\cdot\,, L/K)\circ j_v = \chi\circ(\,\cdot\,, L_{\overline{v}}/K_v)$. Since χ is injective, we obtain, using 6.5.4.1,

$$\widetilde{\chi}_v = 1 \iff (\,\cdot\,, L_{\overline{v}}/K_v) = 1 \iff K_v^\times \subset \mathsf{N}_{L_{\overline{v}}/K_v} L_{\overline{v}}^\times$$
$$\iff \overline{v} \text{ is real.} \qquad \square[\mathbf{A}.]$$

By **A** q is the number of real places $v \in \mathsf{M}_K$ for which $\overline{v}$ is complex. Consequently

$$\begin{aligned}\prod_{v\in\mathsf{M}_K^\infty} \mathcal{L}_{L/K,v}(s,\chi) &= \gamma(s)^{r_1-q+r_2}\gamma(s+1)^{r_2+q}\\ &= \pi^{-(q+ns)/2}\Gamma\Big(\frac{s+1}{2}\Big)^q\Gamma\Big(\frac{s}{2}\Big)^{r_1-q}\Big[\pi^{-1/2}\Gamma\Big(\frac{s}{2}\Big)\Gamma\Big(\frac{s+1}{2}\Big)\Big]^{r_2}\\ &= \pi^{-(q+ns)/2}\Gamma\Big(\frac{s+1}{2}\Big)^q\Gamma\Big(\frac{s}{2}\Big)^{r_1-q}[2^{1-s}\Gamma(s)]^{r_2}\\ &= 2^{r_2(1-s}\pi^{-(q+ns)/2}\Gamma\Big(\frac{s+1}{2}\Big)^q\Gamma\Big(\frac{s}{2}\Big)^{r_1-q}\Gamma(s)^{r_2}.\end{aligned}$$

Altogether we achieve the desired equality $\mathbf{\Lambda}_{L/K}(s,\chi) = \Lambda(s,\widetilde{\chi})$.

Theorem 7.8.13. *Let L/K be a finite Galois extension of algebraic number fields, $G = \mathrm{Gal}(L/K)$ χ a character of G and $s \in \mathcal{H}_1$. Then the extended Artin L function $\mathbf{\Lambda}_{L/K}(s,\chi)$ has an extension to a meromorphic function on $\mathbb{C}$ and satisfies a functional equation of the form*

$$\mathbf{\Lambda}(s,\chi) = \mathbf{W}(\chi)\mathbf{\Lambda}(1-s,\overline{\chi}),$$

where $\mathbf{W}(\chi) \in \mathbb{C}$, $|\mathbf{W}(\chi)| = 1$ *and* $\mathbf{W}(\overline{\chi}) = \overline{\mathbf{W}(\chi)}$.

Proof. By 7.6.9

$$\chi = \sum_{j=1}^r c_j\chi_j^*,$$

where $r \in \mathbb{N}$, $c_1,\ldots,c_r \in \mathbb{Z}$, $H_1,\ldots,H_r$ are a subgroups of G and $\chi_j \in \mathsf{X}(H_j)$, $\chi_j^* = \mathrm{Ind}_G^{H_j}(\chi_j)$ for all $j \in [1,r]$. By 7.8.5 and 7.8.6 it follows that

$$\mathbf{\Lambda}_{L/K}(s,\chi) = \prod_{j=1}^r \mathbf{\Lambda}_{L/K}(s,\chi_j^*)^{c_j} = \prod_{j=1}^s \mathbf{\Lambda}_{L/K_j}(s,\chi_j)^{c_j} = \prod_{j=1}^r \Lambda(s,\widetilde{\chi}_j)^{c_j}$$

where $K_j = L^{H_j}$, and $\widetilde{\chi}_j = \chi_j\circ(\,\cdot\,, L/K_j)\colon \mathsf{J}_{K_j} \to \mathbb{T}$ is a ray class character. By 7.1.8 the functions $\Lambda(s,\widetilde{\chi}_j)^{c_j}$ can be extended to meromorphic functions in $\mathbb{C}$ and satisfy a functional equation

$$\mathbf{\Lambda}(s,\widetilde{\chi}_j) = \mathsf{W}_j(\widetilde{\chi}_j)\mathbf{\Lambda}(1-s,\overline{\widetilde{\chi}}_j) = \mathsf{W}_j(\widetilde{\chi}_j)\Lambda(1-s,\widetilde{\overline{\chi}}_j),$$

where, for all $j \in [1, r]$, $\mathsf{W}_j(\widetilde{\chi}_j) \in \mathbb{C}$, $|\mathsf{W}_j(\widetilde{\chi}_j)| = 1$ and $\overline{\mathsf{W}_j(\widetilde{\chi}_j)} = \mathsf{W}_j(\widetilde{\overline{\chi}}_j)$ by 7.1.4.4 and 7.1.8.1, since obviously $\overline{\widetilde{\chi}}_j = \widetilde{\overline{\chi}}_j$. In particular, $\mathbf{\Lambda}_{L/K}(s,\chi)$ has an extension to a meromorphic function on $\mathbb{C}$. Since

$$\overline{\chi} = \sum_{j=1}^{r} c_j \overline{\chi_j^*} = \sum_{j=1}^{r} c_j \overline{\chi}_j{}^*,$$

we obtain

$$\begin{aligned}\mathbf{\Lambda}_{L/K}(1-s,\overline{\chi}) &= \prod_{j=1}^{s} \mathbf{\Lambda}_{L/K}(1-s,\overline{\chi}_j{}^*)^{c_j} = \prod_{j=1}^{s} \mathbf{\Lambda}_{L/K_j}(1-s,\overline{\chi})^{c_j} \\ &= \prod_{j=1}^{r} \Lambda(1-s,\widetilde{\overline{\chi}}_j),\end{aligned}$$

and consequently

$$\mathbf{\Lambda}_{L/K}(s,\chi) = \prod_{j=1}^{r} \Lambda(s,\widetilde{\chi}_j)^{c_v} = \prod_{j=1}^{r} \mathsf{W}_j(\widetilde{\chi}_j)^{c_j} \prod_{j=1}^{r} \Lambda(1-s,\widetilde{\overline{\chi}}_j) = \mathbf{W}(\chi)\Lambda(1-s,\overline{\chi}),$$

where

$$\mathbf{W}(\chi) = \prod_{j=1}^{r} \mathsf{W}_j(\widetilde{\chi}_j)^{c_j}.$$

It is now clear, that $\mathbf{W}(\chi) \in \mathbb{C}$, $|\mathbf{W}(\chi)| = 1$ and $\overline{\mathbf{W}}(\chi) = \mathbf{W}(\overline{\chi})$. □

7.9 Prime decomposition and density results

Let K be a global field.

In [18, Section 4.3] we investigated densities of sets of prime ideals in algebraic number fields. Now we return to this subject in the general frame of global fields using the formalism of Artin L functions which opens the door for fresh proofs and further applications.

For a subset D of $\mathbb{C}$ we denote by $\mathcal{O}(D)$ the set of all holomorphic functions in some open neighbourhood of D.

Theorem and Definition 7.9.1. *If $n \in \mathbb{N}$, then*

$$A_n = |\{v \in \mathsf{M}_K^0 \mid \mathfrak{N}(v) \le n\}| \ll n \quad \textit{for all } n \ge 1, \tag{1}$$

and, for all $s \in \mathcal{H}_1$,

$$\sum_{v \in \mathsf{M}_K} \frac{1}{\mathfrak{N}(v)^s} = \log \frac{1}{s-1} + g(s) \quad \textit{where} \quad g \in \mathcal{O}(\mathcal{H}_1 \cup \{1\}). \tag{2}$$

In particular,

$$\sum_{v \in \mathsf{M}_K^0} \frac{1}{\mathfrak{N}(v)^\sigma} < \infty \ \textit{for all} \ \ \sigma \in \mathbb{R}_{>1}, \quad \textit{and} \quad \lim_{\sigma \to 1+} \sum_{v \in \mathsf{M}_K^0} \frac{1}{\mathfrak{N}(v)^\sigma} = \infty.$$

For functions $f,\, g \in \mathcal{O}(\mathcal{H}_1)$, we write $f \sim g$, if $f - g \in \mathcal{O}(\mathcal{H}_1 \cup \{1\})$. Then

$$\sum_{v \in \mathsf{M}_K^0} \frac{1}{\mathfrak{N}(v)^s} \sim \log \frac{1}{s-1} \quad \text{for} \quad s \in \mathcal{H}_1,$$

and for a subset T of M_K^0 we define its **Dirichlet density** by

$$d(T) = \lim_{\sigma \to 1+} \frac{\sum_{v \in \mathsf{M}_K^0} \mathfrak{N}(v)^{-\sigma}}{\log \frac{1}{\sigma - 1}},$$

provided this limit exists. A subset T of M_K^0 is called **regular** with density $c \in [0,1]$ if

$$\sum_{v \in T} \frac{1}{\mathfrak{N}(v)^s} \sim c \log \frac{1}{s-1} \quad \text{for} \quad s \in \mathcal{H}_1.$$

Obviously, M_K^0 is regular with density 1, and if a set T is regular with density $c \in [0,1]$, then $c = d(T)$ is its Dirichlet density. More about the Dirichlet density and its connection with the natural density is in [18, Sect. 4.3].

Proof of (1). In the number field case,

$$A_n = \{\mathfrak{p} \in \mathcal{P}_K \mid \mathfrak{N}(\mathfrak{p}) \le n\} \le [K \,{:}\, \mathbb{Q}]\, |\{p \in \mathbb{P} \mid p \le n\}| \ll n.$$

Thus let $K/\mathbb{F}_q$ be a function field. Then the prime divisor theorem (see [18, Theorem 6.9.9]) implies

$$|\{v \in \mathsf{M}_K \mid \deg(v) = d\}| = |\mathbb{P}_K^d| = \frac{q^d}{d} + \left(\frac{q^{d/2}}{d}\right) \ll q^d \quad \text{for all} \ \ d \in \mathbb{N}.$$

If $n \in \mathbb{N}$ and $d \in \mathbb{N}$ is such that $q^{d-1} \le n < q^d$, then

$$A_n = |\{v \in \mathsf{M}_K \mid \mathfrak{N}(v) \le n\}| \ll \sum_{k=1}^{d-1} q^k = \frac{q^d - 1}{q - 1} \ll n. \qquad \square$$

Proof of (2). We use the zeta function, given by a Dirichlet series and an Euler product for $s \in \mathcal{H}_1$ by

$$\zeta_K(s) = \sum_{\mathsf{a} \in \mathcal{D}_K'^{\bullet}} \mathfrak{N}(\mathsf{a})^{-s} = \prod_{v \in \mathsf{M}_K^0} \frac{1}{1 - \mathfrak{N}(P)^{-s}}.$$

It has an extension to a holomorphic function without zeros in $\mathcal{H}_1 \cup U \setminus \{1\}$ for some small disc $U = B_\varepsilon(1)$ with a simple pole at 1 (see [18, Sec. 4.2] for the number field case, and [18, Sec. 6.9] for the function field case). Hence

$$\zeta_K(s) = \frac{g(s)}{s-1} \quad \text{for some zero-free function } \ g \in \mathcal{O}(\mathcal{H}_1 \cup U),$$

which there has there a holomorphic logarithm $\log g(s)$, given by

$$\log \zeta_K(s) = \log \frac{1}{s-1} + \log g(s) \sim \log \frac{1}{s-1} \quad \text{in } \mathcal{H}_1.$$

On the other hand, for $s \in \mathcal{H}_1$ the Euler product gives

$$\log \zeta_K(s) = \sum_{v \in \mathsf{M}_K^0} -\log(1 - \mathfrak{N}(v)^{-s}) = \sum_{v \in \mathsf{M}_K^0} \sum_{k=1}^{\infty} \frac{1}{k\mathfrak{N}(v)^{ks}} \sim \sum_{v \in \mathsf{M}_K^0} \frac{1}{\mathfrak{N}(v)^s},$$

since, for $\sigma \in \mathbb{R}_{>1}$,

$$\sum_{v \in \mathsf{M}_K^0} \sum_{k=2}^{\infty} \frac{1}{k\mathfrak{N}(v)^{k\sigma}} < \sum_{v \in \mathsf{M}_K^0} \frac{1}{\mathfrak{N}(v)^{2\sigma}(1 - \mathfrak{N}(v)^{-1})} \ll \zeta_K(2\sigma) < \infty. \qquad \square$$

We start with a fresh proof of Čebotarev's density theorem (see [18, Theorem 4.4.6]). For more precise and effective versions we refer to [34], [27] and [13, Ch.6].

Theorem 7.9.2 (Čebotarev's density theorem)**.** *Let L/K be a finite Galois extension, $G = \mathrm{Gal}(L/K)$, $\sigma \in G$ and $\mathsf{M}_{L/K}(\sigma)$ the set of all $v \in \mathsf{M}_K^0$ which are unramified in L such that σ is the Frobenius automorphism over K of some place $\overline{v} \in \mathsf{M}_L$ lying above v. Then*

$$d(\mathsf{M}_{L/K}(\sigma)) = \frac{|\Gamma_\sigma|}{[L:K]}, \quad \text{where } \Gamma_\sigma \text{ denotes the conjugacy class of } \sigma \text{ in } G.$$

Note: If v is unramified in L, then σ is the Frobenius automorphism over K of some place $\overline{v} \in \mathsf{M}_L$ lying above v if and only if Γ_σ is the set of all Frobenius automorphisms over K of places of M_L lying above v (see 5.4.5.**3**).

Proof. We denote by $\mathsf{M}_{L/K}$ the set of all in L unramified places $v \in \mathsf{M}_K^0$. For each $v \in \mathsf{M}_{L/K}$ we fix a place $\overline{v} \in \mathsf{M}_L$ above v and denote by $\varphi_{\overline{v}} \in G$ its Frobenius automorphism over K. Let χ be an irreducible character of G, $\chi = \theta_V$ for some representation $V = (V, \rho)$, and $d = \dim_{\mathbb{C}} V = \chi(1)$.

We define the truncated Artin L function for $s \in \mathcal{H}_1$ by

$$\mathcal{L}_{L/K}^*(s,\chi) = \prod_{v \in \mathsf{M}_{L/K}} \mathcal{L}_{L/K,v}(s,\chi) = \prod_{v \in \mathsf{M}_{L/K}} \det\bigl[1_V - \mathfrak{N}(v)^{-s}\rho(\varphi_{\overline{v}})\bigr]^{-1}.$$

For $v \in \mathsf{M}_{L/K}$ we denote by $\lambda_{\overline{v},1}, \ldots \lambda_{\overline{v},d}$ the eigenvalues of $\rho(\varphi_{\overline{v}})$ (counted with multiplicity). They are roots of unity,

$$\det\big[1_V - \mathfrak{N}(v)^{-1}\rho(\varphi_{\overline{v}})\big] = \prod_{\nu=1}^{d}(1-\mathfrak{N}(v)^{-s}\lambda_{\overline{v},\nu}), \quad \text{and} \quad \chi(\varphi_{\overline{v}}) = \sum_{\nu=1}^{d}\lambda_{\overline{v},\nu}.$$

It follows from 7.8.1 that $\mathcal{L}^*_{L/K}(s,\chi)$ is holomorphic and non-zero in $\mathcal{H}_1$, and thus it possesses there a holomorphic logarithm, given by

$$\begin{aligned}\log \mathcal{L}^*_{L/K}(s,\chi) &= \sum_{v\in\mathsf{M}_{L/K}}\sum_{\nu=1}^{d}\sum_{n=1}^{\infty}\frac{\lambda^n_{\overline{v},\nu}}{n\mathfrak{N}(v)^{sn}}\\ &= \sum_{v\in\mathsf{M}_{L/K}}\frac{\chi(\varphi_{\overline{v}})}{\mathfrak{N}(v)^s} + \sum_{v\in\mathsf{M}_{L/K}}\sum_{n=2}^{\infty}\frac{1}{n\mathfrak{N}(v)^{sn}}\sum_{\nu=1}^{d}\lambda^n_{\overline{v},\nu}.\end{aligned}$$

If $\sigma = \Re(s) > 1/2$, then

$$\Big|\sum_{v\in\mathsf{M}_{L/K}}\sum_{n=2}^{\infty}\frac{1}{n\mathfrak{N}(v)^{sn}}\sum_{\nu=1}^{d}\lambda^n_{\overline{v},\nu}\Big| \le d$$

$$\sum_{v\in\mathsf{M}^0_K}\frac{1}{2\mathfrak{N}(v)^{2\sigma}(1-\mathfrak{N}(v)^{-\sigma})} \ll \zeta_K(2\sigma) < \infty,$$

and consequently

$$\log \mathcal{L}^*_{L/K}(s,\chi) \sim \sum_{v\in\mathsf{M}_{L/K}}\frac{\chi(\varphi_{\overline{v}})}{\mathfrak{N}(v)^s}.$$

By definition, $\mathcal{L}^*_{L/K}(s,1_G) = A(s)\zeta_K(s)$, and $\mathcal{L}_{L/K}(s,\chi) = A(s,\chi)\mathcal{L}_{L/K}(s,\chi)$ for all $\chi \in \mathrm{Irr}(G)$, where $A(s)$ and $A(s,\chi)$ are entire functions without zeros. Hence

$$\log \mathcal{L}^*_{L/K}(s,1) \sim \log\zeta_K(s) \sim \log\frac{1}{s-1},$$

and if $\chi \in \mathrm{Irr}(G)\setminus\{1_G\}$, then $\mathcal{L}^*_{L/K}(s,\chi) \sim \mathcal{L}_{L/K}(s,\chi) \sim 0$, since by 7.8.10 $\mathcal{L}_{L/K}(s,\chi)$ has an extension to a meromorphic function in $\mathbb{C}$ which is holomorphic and non-zero in 1. It follows that

$$\sum_{\chi\in\mathrm{Irr}(G)}\overline{\chi}(\sigma)\log\mathcal{L}^*_{L/K}(s,\chi) \sim \log\frac{1}{s-1}.$$

On the other hand, using 7.6.4.4, we obtain

$$\sum_{\chi\in\mathrm{Irr}(G)} \overline{\chi}(\sigma)\log\mathcal{L}^*_{L/K}(s,\chi) \sim \sum_{\chi\in\mathrm{Irr}(G)} \overline{\chi}(\sigma) \sum_{v\in\mathsf{M}_{L/K}} \frac{\chi(\varphi_{\overline{v}})}{\mathfrak{N}(v)^s}$$
$$= \sum_{v\in\mathsf{M}_{L/K}} \frac{1}{\mathfrak{N}(v)^s} \sum_{\chi\in\mathrm{Irr}(G)} \overline{\chi}(\sigma)\chi(\varphi_{\overline{v}})$$
$$= \frac{|G|}{|\Gamma_\sigma|} \sum_{v\in\mathsf{M}_{L/K}(\sigma)} \frac{1}{\mathfrak{N}(v)^s},$$

hence

$$\sum_{v\in\mathsf{M}_{L/K}(\sigma)} \frac{1}{\mathfrak{N}(v)^s} \sim \frac{|\Gamma_\sigma|}{[L:K]}\log\frac{1}{s-1}, \quad \text{and thus} \quad d(\mathsf{M}_{L/K}(\sigma)) = \frac{|\Gamma_\sigma|}{[L:K]}.$$

□

In the following corollary we apply Čebotarev's density theorem to obtain results about primes in S-ray classes and in ray classes of holomorphy domains.

Corollary 7.9.3. *Let S be a finite subset of M_K^0 such that $S \neq \emptyset$ in the function field case, and let $\theta_S\colon \mathcal{D}_{K,S} \overset{\sim}{\to} \mathcal{J}(\mathfrak{o}_{K,S})$ be the isomorphism defined in 5.5.6. Let $\mathsf{m} \in \mathcal{D}'_{K,S}$ be an integral divisor, $\mathfrak{m} = \theta_S(\mathsf{m}) \in \mathcal{J}'(\mathfrak{o}_{K,S})$, and $\theta_S\colon \mathcal{C}^{\mathsf{m}}_{K,S} \overset{\sim}{\to} \mathcal{C}^{\mathfrak{m}}(\mathfrak{o}_{K,S})$ the (equally denoted) induced isomorphism of the class groups (see 5.6.2).*

Let $C \in \mathcal{C}^{\mathsf{m}}_{K,S}$ and $\mathfrak{c} = \theta_S(C) \in \mathcal{C}^{\mathfrak{m}}(\mathfrak{o}_{K,S})$. Then

$$d(\mathcal{P}_K \cap C) = d(\mathcal{P}(\mathfrak{o}_{K,S}) \cap \mathfrak{c}) = \frac{1}{[K[\mathsf{m},S]:K]},$$

where $K[\mathsf{m},S]$ denotes the S-ray class field modulo m of K.

In particular, every S-ray class modulo m in K contains infinitely many prime divisors, and every (narrow) ray class modulo $\mathfrak{m}$ in $\mathfrak{o}_{K,S}$ contains infinitely many prime ideals.

Proof. By definition, $\theta_S(\mathcal{P}_K \cap \mathcal{D}^{\mathsf{m}}_{K,S}) = \mathcal{P}(\mathfrak{o}_{K,S}) \cap \mathcal{J}^{\mathfrak{m}}(\mathfrak{o}_{K,S})$, and the isomorphism $\theta_S\colon \mathcal{C}^{\mathsf{m}}_{K,S} \overset{\sim}{\to} \mathcal{C}^{\mathfrak{m}}(\mathfrak{o}_{K,S})$ is given by $\theta_S([\mathsf{a}]^{\mathsf{m}}_S) = [\theta_S(\mathsf{a})]^{\mathfrak{m}} \in \mathcal{C}^{\mathfrak{m}}(\mathfrak{o}_{K,S})$ for all $\mathsf{a} \in \mathcal{D}^{\mathsf{m}}_{K,S}$. Hence $\theta_S(\mathcal{P}_K \cap C) = \mathcal{P}(\mathfrak{o}_{K,S}) \cap \mathfrak{c}$, and since $\mathfrak{N}(\mathsf{p}) = \mathfrak{N}(\theta_S(\mathsf{p}))$ for all $\mathsf{p} \in \mathcal{P}_K$, it follows that $d(\mathcal{P}_K \cap C) = d(\mathcal{P}(\mathfrak{o}_{K,S}) \cap \mathfrak{c})$.

By 6.6.1 and 6.6.2 the Artin symbol induces an isomorphism

$$(\,\cdot\,, K[\mathsf{m},S]/K)\colon \mathcal{C}^{\mathsf{m}}_{K,S} \overset{\sim}{\to} \mathrm{Gal}(K[\mathsf{m},S]/K).$$

If $C \in \mathcal{C}^{\mathsf{m}}_{K,S}$ and $\sigma = (C, K[\mathsf{m},S]/K)$, then

$$\mathcal{P}_K \cap C = \{\mathsf{p} \in \mathcal{P}_K \cap \mathcal{D}^{\mathsf{m}}_{K,S} \mid (\mathsf{p}, K[\mathsf{m}]/K) = \sigma\},$$

and thus

$$d(\mathcal{P}_K \cap C) = \frac{1}{[K[\mathsf{m},S]:K]} \qquad \text{by 7.9.2.}$$

□

In the number field case, we could use the Tauberian theorem (see [18, Theorem 4.3.10]) to derive asymptotic results from 7.9.3. To do the same in the function field case one needs a special Tauber theorem for function fields. We will not do this here and refer to [15, Theorem 8.7.11].

The role of the Frobenius automorphism for the decomposition of unramified primes in finite extensions was already addressed in [18, Theorem 2.14.9]. Now we can do this more explicitly for global fields, including ramification. The following results in its sharp form (including ramified primes) are due to Klingen (see [31, Ch. II and III]). We start with some additional terminology.

Definition and Remark 7.9.4. Let G be a finite group and H a subgroup of G. We set

$$H^G = \bigcup_{\tau\in G} \tau H\tau^{-1}, \quad H_G = \bigcap_{\tau\in G} \tau H\tau^{-1} \quad \text{and} \quad 1_G^H = \mathrm{Ind}_G^H 1_H \in \mathsf{X}_G.$$

Apparently, $H_G \subset H \subset H^G$. If $(G\colon H) = n$, $G/H = \{\omega_1 H, \ldots, \omega_n H\}$ with $\omega_1 = 1$ and $\sigma \in G$, then

$$1_G^H(\sigma) = \frac{1}{|H|}\big|\{\tau\in G \mid \tau^{-1}\sigma\tau \in H\}\big| = \big|\{i\in[1,n] \mid \omega_i^{-1}\sigma\omega_i \in H\}\big| \in [0,n].$$

In particular, for all $\sigma \in G$ we obtain:

$$\sigma\in H_G \iff \sigma\in\tau H\tau^{-1} \text{ for all } \tau\in G \iff 1_G^H(\sigma) = n\,;$$
$$\sigma\in H^G \iff \sigma\in\tau H\tau^{-1} \text{ for some } \tau\in G \iff 1_G^H(\sigma) > 0.$$

Theorem 7.9.5. *Let L/K be a finite Galois extension, K' an intermediate field of L/K, $G = \mathrm{Gal}(L/K)$, $H = \mathrm{Gal}(L/K')$ and $1_G^H = \mathrm{Ind}_G^H(1_H)$. Let $v \in \mathsf{M}_K^0$, $\{v_1', \ldots, v_t'\} = \{v' \in \mathsf{M}_{K'} \mid v'\!\restriction\! K = v\}$, and for $d \in \mathbb{N}$ let*

$$\psi(d) = |\{j\in[1,t] \mid f(v_j'/v) = d\}|.$$

Let $\overline{v} \in \mathsf{M}_L$ be a place above v and $\phi_{\overline{v}/v}$ the arithmetic mean of all Frobenius automorphisms of $\overline{v}$ over K. If $s \in \mathbb{N}$, then

$$1_G^H(\phi_{\overline{v}/v}^s) = \sum_{1\le d\,|\,s} \psi(d) \quad \textit{and} \quad \psi(s) = \sum_{1\le d\,|\,s} \mu\Big(\frac{s}{d}\Big) 1_G^H(\phi_{\overline{v}/v}^d).$$

In particular,

$$1_G^H(\phi_{\overline{v}/v}) = |\{j\in[1,t] \mid f(v_j'/v) = 1\}|.$$

Proof. We apply the second formula in 7.8.2 with $\chi = 1_H$. As $1_H(\phi_{\overline{v}_j/v_j'}^{s/f_j}) = 1$ for all $j \in [1,t]$ we obtain, for any $s \in \mathbb{N}$,

$$1_G^H(\phi_{\overline{v}/v}^s) = \sum_{\substack{j=1\\ f(v_j'/v)\,|\,s}}^{t} f(v_j'/v) = \sum_{1\le d\,|\,s} \psi(d),$$

and the second equality follows by Möbius inversion. □

For a finite extension L/K of global fields we consider the **splitting set**

$$\mathrm{Split}(L/K) = \{v \in \mathsf{M}_K^0 \mid v \text{ is fully decomposed in } L\}$$

and the **Kronecker set**

$$\mathrm{Kron}(L/K) = \{v \in \mathsf{M}_K^0 \mid f(v'/v) = 1 \text{ for some } v' \in \mathsf{M}_L \text{ lying above } v\}.$$

Note that in the definition of the Kronecker set ramified primes are not excluded.

Theorem 7.9.6. *Let L/K be a finite Galois extension, K' an intermediate field of L/K, $[K' : K] = n$, $G = \mathrm{Gal}(L/K)$ and $H = \mathrm{Gal}(L/K')$. Let $v \in \mathsf{M}_K^0$, $\overline{v} \in \mathsf{M}_L$ a place lying above v and $\phi_{\overline{v}/v}$ the arithmetic mean of all Frobenius automorphisms of $\overline{v}$ over K.*

1. *For $v \in \mathsf{M}_K^0$ the following assertions are equivalent:*

   (a) $v \in \mathrm{Kron}(K'/K)$.

   (b) $1_G^H(\phi_{\overline{v}/v}) > 0$.

   (c) $1_G^H(\varphi) > 0$ *for some Frobenius automorphism* φ *of* $\overline{v}$ *over* K.

   (d) $\varphi \in H^G$ *for some Frobenius automorphism* φ *of* $\overline{v}$ *over* K.

2. *For $v \in \mathsf{M}_K^0$ the following assertions are equivalent:*

   (a) $v \in \mathrm{Split}(L/K)$.

   (b) $1_G^H(\phi_{\overline{v}/v}) = n$.

   (c) $1_G^H(\varphi) = n$ *for all Frobenius automorphisms* φ *of* $\overline{v}$ *over* K.

   (d) $\varphi \in H^G$ *for all Frobenius automorphisms* φ *of* $\overline{v}$ *over* K.

Note that in both assertions (a) does not depend on the choice of $\overline{v}$, and thus (b), (c) and (d) hold for all $\overline{v} \in \mathsf{M}_L$ lying above v.

Proof. Let $F(\overline{v}/v)$ be the set of all Frobenius automorphisms of $\overline{v}$ over K.

1. (a) $\Leftrightarrow$ (b) By definition, $v \in \mathrm{Kron}(K'/K)$ if and only if $f(v'/v) = 1$ for some $v' \in \mathsf{M}_{K'}$ lying above v, and by 7.9.5 this holds if and only if $1_G^H(\phi_{K'/K}) > 0$.

(b) $\Leftrightarrow$ (c) The assertion follows since

$$1_G^H(\phi_{\overline{v}/v}) = \frac{1}{e(\overline{v}/v)} \sum_{\varphi \in F(\overline{v}/v)} 1_G^H(\varphi).$$

(c) $\Leftrightarrow$ (d) By 7.9.4.

2. (a) $\Leftrightarrow$ (b) By 7.9.5, since

$$v \in \mathrm{Split}(L/K \iff 1_G^H(\phi_{\overline{v}/v}) = |\{v' \in \mathsf{M}_{K'} \mid v' \downarrow L = v,\ f(v'/v) = 1\}| = n.$$

(b) $\Leftrightarrow$ (c) Note that $|F(\overline{v}/v)| = e(\overline{v}/v)$ by definition, and $1_G^H(\varphi) \in [0, n]$ for all $\varphi \in F(\overline{v}/v)$ by 7.6.6.3. Hence

$$e(\overline{v}/v) 1_G^H(\phi_{\overline{v}/v}) = \sum_{\varphi \in F(\overline{v}/v)} 1_G^H(\varphi) \le e(\overline{v}/v)n.$$

Consequently, $1_G^H(\phi_{\overline{v}/v}) = n$ if and only if $1_G^H(\varphi) = n$ for all $\varphi \in F(\overline{v}/v)$.

(c) $\Leftrightarrow$ (d) By 7.9.4. □

The following result is a generalization of the theorem of Bauer (see [18, Theorem 4.4.9]).

Theorem 7.9.7. *Let K_1 and K_2 be intermediate fields of a finite Galois extension L/K. Let $G = \mathrm{Gal}(L/K)$, $H_1 = \mathrm{Gal}(L/K_1)$ and $H_2 = \mathrm{Gal}(L/K_2)$. Then the following assertions are equivalent:*

(a) $\mathrm{Kron}(K_1/K) \subset \mathrm{Kron}(K_2/K) \cup T$ *for some subset T of M_{K_2} of Dirichlet density* 0.

(b) $\mathrm{Kron}(K_1/K) \subset \mathrm{Kron}(K_2/K)$.

(c) $H_1^G \subset H_2^G$.

If K_2/K is Galois, then these conditions are also equivalent to

(d) $K_2 \subset K_1$ (Theorem of Bauer).

Proof. (b) $\Leftrightarrow$ (c) By the equivalence of (a) and (d) in 7.9.6.1.

(b) $\Rightarrow$ (a) Obvious.

(a) $\Rightarrow$ (c) Assume to the contrary that there exists some $\sigma \in H_1^G \setminus H_2^G$, and let Γ_σ be the conjugacy class of σ. As H_1^G and H_2^G consist of full conjugacy classes, it follows that $\Gamma_\sigma \subset H_1^G$ and $\Gamma_\sigma \cap H_2^G = \emptyset$. Let S be the set of all $v \in \mathsf{M}_K^0$ which are unramified in L such that Γ_σ is the set of all Frobenius automorphisms of places of L lying above v. Then 7.9.2 implies

$$d(S) = \frac{|\Gamma_\sigma|}{|G|} > 0,$$

and by 7.9.6.1 it follows that $S \subset \mathrm{Kron}(K_1/K) \setminus \mathrm{Kron}(K_2/K) \subset T$, a contradiction.

(d) $\Rightarrow$ (b) Obvious, even without assuming that K_2/K is Galois.

(c) $\Rightarrow$ (d) Let K_2/K be Galois. Then H_2 is a normal subgroup of G, and

$$H_1 \subset H_1^G \subset H_2^G = H_2 \quad \text{implies} \quad K_2 \subset K_1. \qquad \square$$

Corollary 7.9.8. *Let K_1/K and K_2/K be Galois extensions of global fields (inside a fixed algebraic closure). Then $K_1 = K_2$ if and only if $\mathrm{Kron}(K_1/K)$ and $\mathrm{Kron}(K_2/K)$ only differ by a set of Dirichlet density 0.*

Proof. Obvious by 7.9.7. □

Two finite extension fields K_1 and K_2 of a global field K are called **Kronecker equivalent** (over K) if $\mathrm{Kron}(K_1/K) = \mathrm{Kron}(K_2/K)$ (and by 7.9.7 this holds already if $\mathrm{Kron}(K_1/K)$ and $\mathrm{Kron}(K_2/K)$ only differ by a set of Dirichlet density 0). By Corollary 7.9.8, two finite Galois extensions inside a fixed algebraic closure are Kronecker equivalent if and only if they are equal. This is no longer true for non-Galois extensions. For a thorough study of Kronecker equivalence classes we refer to [31, Ch. II].

We close with the investigation of arithmetical equivalence which is stronger than Kronecker equivalence. We need some terminology.

Let L/K be a finite extension of global fields and $\boldsymbol{f} = (f_1, \ldots, f_r)$ a finite sequence of integers such that $1 \le f_1 \le \ldots \le f_r$. We say that a place $v \in \mathsf{M}_K^0$ has **decomposition type $\boldsymbol{f}$** in L if $\{\overline{v} \in \mathsf{M}_L \mid \overline{v} \,|\, v\} = \{\overline{v}_1, \ldots, \overline{v}_r\}$ such that $f(\overline{v}_i/v) = f_i$ for all $i \in [1, r]$, and we denote by $A_{L/K}(\boldsymbol{f})$ the set of all $v \in \mathsf{M}_K^0$ which have decomposition type $\boldsymbol{f}$ in L.

Two finite extension fields K_1 and K_2 of a global field K are called **arithmetically equivalent** (over K) if $A_{K_1/K}(\boldsymbol{f}) = A_{K_2/K}(\boldsymbol{f})$ for all possible decomposition types $\boldsymbol{f}$.

Theorem 7.9.9. *Let K_1, K_2 be intermediate fields of a finite Galois extension L/K of global fields. Let $G = \mathrm{Gal}(L/K)$, $H_1 = \mathrm{Gal}(L/K_1)$ and $H_2 = \mathrm{Gal}(L/K_2)$. Then the following assertions are equivalent:*

(a) K_1 and K_2 are arithmetically equivalent over K.

(b) For all decomposition types $\boldsymbol{f}$ the sets $A_{K_1/K}(\boldsymbol{f})$ and $A_{K_2/K}(\boldsymbol{f})$ only differ by a set of Dirichlet density 0.

(c) $1_G^{H_1} = 1_G^{H_2}$.

(d) $|\Gamma \cap H_1| = |\Gamma \cap H_2|$ for every conjugacy class Γ in G.

(e) All semilocal zeta functions of K_1 and K_2 coincide, i. e., for all $v \in \mathsf{M}_K^0$ and $s \in \mathcal{H}_1$ we have

$$\zeta_{K_1}^{(v)} = \prod_{\substack{v_1 \in \mathsf{M}_{K_1} \\ v_1 \,|\, v}} \frac{1}{1 - \mathfrak{N}(v_1)^{-s}} = \prod_{\substack{v_2 \in \mathsf{M}_{K_2} \\ v_2 \,|\, v}} \frac{1}{1 - \mathfrak{N}(v_2)^{-s}} = \zeta_{K_2}^{(v)}.$$

Proof. (a) ⇒ (b) Obvious.

(b) ⇒ (c) Let S be the set of all $v \in \mathsf{M}_K^0$ such that $A_{K_1/K}(\boldsymbol{f}) \ne A_{K_2/K}(\boldsymbol{f})$ for some decomposition type $\boldsymbol{f}$. Since $A_{K_1/K}(\boldsymbol{f}) = A_{K_2/K}(\boldsymbol{f}) = \emptyset$ for almost all decomposition types, it follows that S is the union of finitely many sets of Dirichlet density 0 and thus $d(S) = 0$. Let $\sigma \in G$, and let $\mathsf{M}_{L/K}(\sigma)$ be the set of all $v \in \mathsf{M}_K^0$ which are unramified in L such that σ is the Frobenius automorphism over K of some place $\overline{v} \in \mathsf{M}_L$ lying above v. Since $d(\mathsf{M}_{L/K}(\sigma)) > 0$ by 7.9.2, there exists some $v \in \mathsf{M}_{L/K}(\sigma) \setminus S$, and 7.9.5 implies

$$\begin{aligned} 1_G^{H_1}(\sigma) &= |\{v_1 \in \mathsf{M}_{K_1} \mid v_1 \restriction K = v \text{ and } f(v_1/v) = 1\}| \\ &= |\{v_2 \in \mathsf{M}_{K_2} \mid v_2 \restriction K = v \text{ and } f(v_2/v) = 1\}| = 1_G^{H_2}(\sigma). \end{aligned}$$

(c) $\Leftrightarrow$ (d) We prove first:

A. For any subgroup H of G and $\sigma \in G$ we have

$$1_G^H(\sigma) = \frac{|\Gamma_\sigma \cap H|\,|\mathsf{c}_G(\sigma)|}{|H|},$$

where $\mathsf{c}_G(\sigma)$ denotes the centralizer of σ in G. In particular,

$$1_G^H(1) = \frac{|G|}{|H|}.$$

Proof of **A.** Let H be a subgroup of G and $\sigma \in G$. By 7.9.4 we obtain

$$1_G^H(\sigma) = \frac{1}{|H|}\,|\{\tau \in G \mid \tau^{-1}\sigma\tau \in H\}|.$$

The map $\phi\colon \{\tau \in G \mid \tau^{-1}\sigma\tau \in H\} \to \Gamma_\sigma \cap H$, defined by $\phi(\tau) = \tau^{-1}\sigma\tau$, is an epimorphism with kernel $\mathsf{c}_G(\sigma)$. $\square$[**A.**]

Assume (c). Then $1_G^{H_1}(1) = 1_G^{H_2}(1)$ implies $|H_1| = |H_2|$, and from **A** we obtain $|\Gamma_\sigma \cap H_1| = |\Gamma_\sigma \cap H_2|$ for all $\sigma \in G$.

Now assume (d) and set $\mathcal{C} = \{\Gamma_\sigma \mid \sigma \in G\}$. Then

$$|H_1| = \sum_{C \in \mathcal{C}} |C \cap H_1| = \sum_{C \in \mathcal{C}} |C \cap H_2| = |H_2|,$$

and **A** implies

$$1_G^{H_1}(\sigma) = \frac{|\Gamma_\sigma \cap H_1|\,|\mathsf{c}_G(\sigma)|}{|H_1|} = \frac{|\Gamma_\sigma \cap H_2|\,|\mathsf{c}_G(\sigma)|}{|H_2|} = 1_G^{H_2}(\sigma).$$

(c) $\Rightarrow$ (e) For $i \in \{1,2\}$ we consider the v-component $\mathcal{L}_{L/K,v}(s, 1_G^{H_i})$ of the Artin L function and obtain for $s \in \mathcal{H}_1$, using 7.8.3.2,

$$\begin{aligned} \mathcal{L}_{L/K,v}(s, 1_G^{H_i}) &= \prod_{\substack{v_i \in \mathsf{M}_{K_i} \\ v_i \mid v}} \mathcal{L}_{L/K_i,v_i}(s, 1_{H_i}) = \prod_{\substack{v_i \in \mathsf{M}_{K_i} \\ v_i \mid v}} \mathcal{L}_{K_i/K_i,v_i}(s, 1_1) \\ &= \prod_{\substack{v_i \in \mathsf{M}_{K_i} \\ v_i \mid v}} \frac{1}{1 - \mathfrak{N}(v_i)^{-s}} = \zeta_{K_i}^{(v)}(s) \end{aligned}$$

and thus $\zeta_{K_1}^{(v)} = \zeta_{K_2}^{(v)}$.

(e) $\Rightarrow$ (a) Let $i \in \{1,2\}$, $v \in \mathsf{M}_K^0$ and $\boldsymbol{f} = (f_1, \ldots, f_r)$ the decomposition type of v in K_i. It suffices to prove that $\boldsymbol{f}$ is uniquely determined by the semilocal zeta function $\zeta_{K_i}^{(v)}$. Let $v_1, \ldots, v_r \in \mathsf{M}_{K_i}$ be the places of K_i above v such that $f_j = f(v_j/v)$ and thus $\mathfrak{N}(v_i) = \mathfrak{N}(v)^{f_j}$ for all $j \in [1,r]$. For $s \in \mathcal{H}_1$ we obtain

$$\frac{1}{\zeta_{K_i}^{(v)}(s)} = \prod_{j=1}^{r}\Big(1 - \mathfrak{N}(v_j)^{-s}\Big) = \prod_{j=1}^{r}\Big(1 - \mathfrak{N}(v)^{-sf_j}\Big) = Q_i(\mathfrak{N}(v)^{-s})$$

where

$$Q_i = \prod_{j=1}^{r}(1 - X^{f_j}) = \prod_{j=1}^{r}\prod_{\nu=0}^{f_j-1}(1 - \zeta_{f_j}^{\nu}X) \in \mathbb{C}[X] \quad \text{with} \quad \zeta_{f_j} = \mathsf{e}^{2\pi\mathsf{i}/f_j}.$$

The polynomial Q is uniquely determined by the semilocal zeta functions $\zeta_{K_i}^{(v)}$, and $\boldsymbol{f}$ is uniquely determined by Q. $\square$

Arithmetically equivalent fields share many arithmetical invariants but need not be conjugate. We refer to [31] for details.

Bibliography

[1] E. Artin and J. Tate, *Class field theory*, Benjamin, 1967.

[2] N. Bourbaki, *General Topology I, II*, Addison-Wesley, 1966.

[3] H. Brückner, *Explizites Reziprozitätsgesetz und Anwendungen*, Vorlesungen aus dem Fachbereich Mathematik der Universität Essen (1979), 2.

[4] J.W.S. Cassels, *Global Fields*, Algebraic Number Theory (J. W. S. Cassels and A. Fröhlich, eds.), Academic Press, 1967.

[5] ———, *Local Fields*, London Math. Soc., 1986.

[6] N. Childress, *Class Field Theory*, Springer, 2007.

[7] C. W. Curtis and I. Reiner, *Methods of Representation Theory, Vol. I*, Wiley, 1990.

[8] P. Deligne, *Applications de la formule des traces aux sommes trigonometriques*, Lecture Notes in Math. **569** (1977), 168 – 232.

[9] M. Deuring, *Lectures on the Theory of Algebraic Functions of One Variable*, Springer, 1973.

[10] J. Dieudonné, *Treatise on Analysis*, Academic Press, 1976.

[11] B. Dwork, *Norm resiue symbol in local number fields*, Abh. Math. Sem. Univ. Hamburg **22** (1958), 180 – 190.

[12] I. B. Fesenko and S. V. Vostokov, *Local Fields and Their Extensions*, AMS, 2002.

[13] M. Fried and M. Jarden, *Field Arithmetic*, Springer, 2008.

[14] A. Frölich, *Local Fields*, Algebraic Number Theory (J. W. S. Cassels and A. Fröhlich, eds.), Academic Press, 1967.

[15] A. Geroldinger and F. Halter-Koch, *Non-Unique Factorizations. Algebraic, Combinatorial and Analytic Theory*, Pure and Applied Mathematics, vol. 278, Chapman & Hall/CRC, 2006.

[16] G. Gras, *Class Field Theory*, Springer, 2003.

[17] F. Halter-Koch, *A note on ray class fields of global fields*, Nagoya Math. J. **120** (1990), 61 – 66.

[18] ———, *An Invitation to Algebraic Numbers and Algebraic Functions*, Chapman & Hall/CRC, 2020.

[19] F. Halter-Koch, J. Klingen, and H. Stichtenoth, *Artin'sche L-Funktionen*, Seminarbericht Univ. Essen, 1976.

[20] H. Hasse, *Artin'sche Führer, Artin'sche L-Funktionen und Gauß'sche summen über endlich-algebraischen Zahlkörpern*, Acta Salamaticensia Ciencias Sec. Math. **4** (1954), 1 – 113.

[21] E. Hecke, *Eine neue Art von Zetafunktionen und ihre Beziehungen zur Verteilung von Primzahlen. Erste Mitteilung*, Math. Z. **1** (1918), 357 – 376.

[22] ———, *Eine neue Art von Zetafunktionen und ihre Beziehungen zur Verteilung von Primzahlen. Zweite Mitteilung*, Math. Z. **4** (1920), 11 – 51.

[23] ———, *Mathematische Werke*, Vandenhoeck & Ruprecht, 1970.

[24] E. Hewitt and K. A. Ross, *Abstract Harmonic Analysis*, Springer, 1963.

[25] K. Iwasawa, *On explicit formulas for the norm residue symbol*, J. Math. Soc. JPN **20** (1968), 151 – 165.

[26] ———, *Local Class Field Theory*, Oxford University Press, 1986.

[27] H. L. Montgomery J. C. Lagarias and A. M. Odlyzko, *A bound for the least prime ideal in the Chebotarev density theorem*, Invent. math. **54** (1979), 271 – 296.

[28] J. C. Jantzen and J. Schwermer, *Algebra*, Springer, 2006.

[29] G. J. Janusz, *Algebraic Number Fields*, Academic Press, 1996.

[30] J. L. Kelley, *General Topology*, Springer, 1975.

[31] N. Klingen, *Arithmetic Similarities*, Oxford Science Publ., 1998.

[32] H. Koch, *Galoissche Theorie der p-Erweiterungen*, VEB Deutscher Verlag der Wissenschaften, 1970.

[33] ———, *Algebraic Number Fields*, Number Theory II (A. N. Parshin and I. R. Shafarevich, eds.), Springer, 1992.

[34] J. C. Lagarias and A. M. Odlyzko, *Effective Versions of the Chebotarev Density Theorem*, Algebraic Number Fields (A. Fröhlich, ed.), Academic Press, 1977.

[35] E. Landau, *Über Ideale und Primideale in Idealklassen*, Math. Z. **2** (1918), 52 – 154.

[36] S. Lang, *Introduction to Algeabraic and Abelian Functions*, Springer, 1982.

[37] ———, *Algebra*, Addison-Wesley, 1993.

[38] ———, *Algebraic Number Theory*, Springer, 1994.

[39] F. Lorenz, *Einführung in die Algebra, II*, B. I., 1990.

[40] ———, *Algebraische Zahlentheorie*, B.I., 1993.

[41] J. Martinet, *Character theory and Artin L-functions*, Algebraic Number Fields (A. Fröhlich, ed.), Academic Press, 1977.

[42] C. Moreno, *Algebraic curves over finite fields*, Cambridge University Press, 1991.

[43] Y. Motohashi, *On Hecke's theta-formula*, Math. Japon. **13** (1968), 43 – 45.

[44] W. Narkiewicz, *Elementary and Analytic Theory of Algebraic Numbers*, Springer, 2004.

[45] ———, *The Story of Algebraic Numbers in the First Half of the 20th Century*, Springer, 2018.

[46] J. Neukirch, *Klassenkörpertheorie*, B.I., 1969.

[47] ———, *Neubegründung der Klassenkörpertheorie*, Math. Z. **186** (1969), 557 – 574.

[48] ———, *Algebraic Number Theory*, Springer, 1999.

[49] R. S. Pierce, *Associative Algebras*, Springer, 1982.

[50] G. Poitou, *Cohomologie galoisienne des modules finis*, Dunot, 1967.

[51] D. Ramakrishnan and R. J. Valenza, *Fourier Analysis on Number Fields*, Springer, 1999.

[52] R. Remmert, *Classical Topics in Complex Function Theory*, Springer, 1998.

[53] L. Ribes, *Introduction of Profinite Groups and Galois Cohomology*, Queen's University, 1978.

[54] L. Ribes and P. Zalesskii, *Profinite Groups*, Springer, 1991.

[55] J.-P. Serre, *Local Class Field Theory*, Algebraic Number Theory (J. W. S. Cassels and A. Fröhlich, eds.), Academic Press, 1967.

[56] ———, *Représentations linéaires des groupes finis*, Hermann, 1967.

[57] ———, *Local Fields*, Springer, 1995.

[58] H. Stichtenoth, *Algebraic Function Fields and Codes*, Springer, 2009.

[59] T. Tao, *Analysis II*, Hindustan Book Agency, 2006.

[60] J. Tate, *Global Class Field Theory*, Algebraic Number Theory (J. W. S. Cassels and A. Fröhlich, eds.), Academic Press, 1967.

[61] T. Tatuzawa, *On the Hecke-Landau L-series*, Nagoya Math. J. **16** (1960), 11 – 20.

[62] S. V. Vostokov, *Explicit form of the law of reciprocity*, Izv. Akad. Nauk SSSR Ser. Math. **42** (1978), 1288 – 1321.

[63] A. Weil, *Variétés Abéliennes et Courbes Algébriques*, Hermann, 1948.

[64] ———, *Basic Number Theory*, Springer, 1967.

Subject Index

List of Symbols

Printed in the United States
by Baker & Taylor Publisher Services